AF725189

THE GEOLOGICAL SOCIETY OF AMERICA

Geological Society of America
3300 Penrose Pl
P.O. Box 9140
Boulder, CO 80301
(303) 357-1000 • fax (303) 357-1073

February 2005

Re: "Evaporative differentiation of impact-produced melts: Laser-simulation experiments and comparison with impact glasses from the Logoisk crater," by M.V. Gerasimov, O.I. Yakovlev, Yu.P. Dikov, and F. Wlotzka (Chapter 20), *in* GSA Special Paper 384, *Large meteorite impacts III*

To Whom It May Concern:

Enclosed are errata sheets for pages 358, 359, and 364 showing corrected Figures 5 and 6. Our apologies for any inconvenience.

GSA Publications

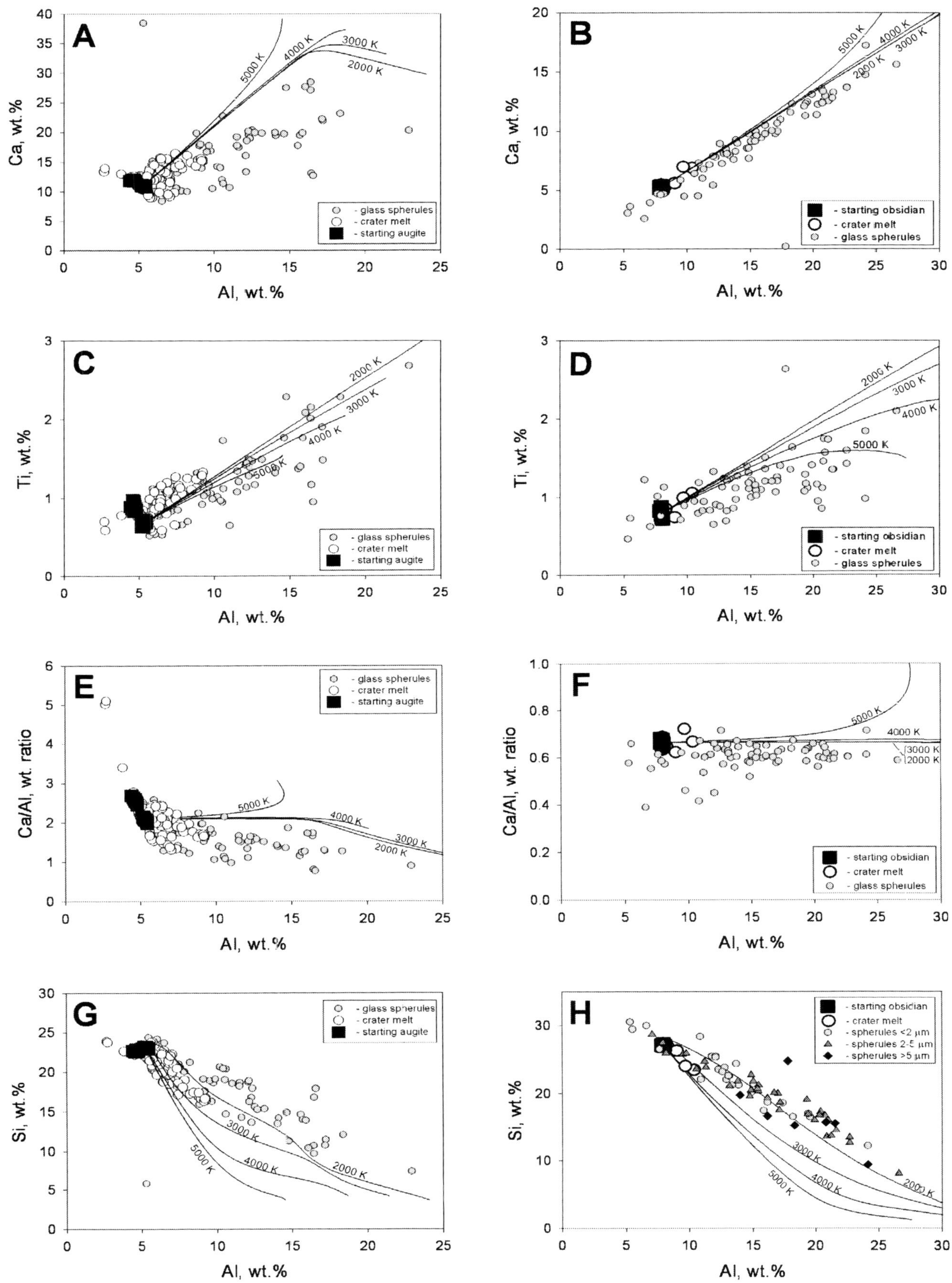

A
Ca, wt.%
Al, wt.%
5000 K
4000 K
3000 K
2000 K
- glass spherules
- crater melt
- starting augite
B
Ca, wt.%
Al, wt.%
5000 K
4000 K
3000 K
2000 K
- starting obsidian
- crater melt
- glass spherules
C
Ti, wt.%
Al, wt.%
2000 K
3000 K
4000 K
5000 K
- glass spherules
- crater melt
- starting augite
D
Ti, wt.%
Al, wt.%
2000 K
3000 K
4000 K
5000 K
- starting obsidian
- crater melt
- glass spherules
E
Ca/Al, wt. ratio
Al, wt.%
5000 K
4000 K
3000 K
2000 K
- glass spherules
- crater melt
- starting augite
F
Ca/Al, wt. ratio
Al, wt.%
5000 K
4000 K
3000 K
2000 K
- starting obsidian
- crater melt
- glass spherules
G
Si, wt.%
Al, wt.%
3000 K
5000 K
4000 K
2000 K
- glass spherules
- crater melt
- starting augite
H
Si, wt.%
Al, wt.%
3000 K
5000 K
4000 K
2000 K
- starting obsidian
- crater melt
- spherules <2 μm
- spherules 2-5 μm
- spherules >5 μm

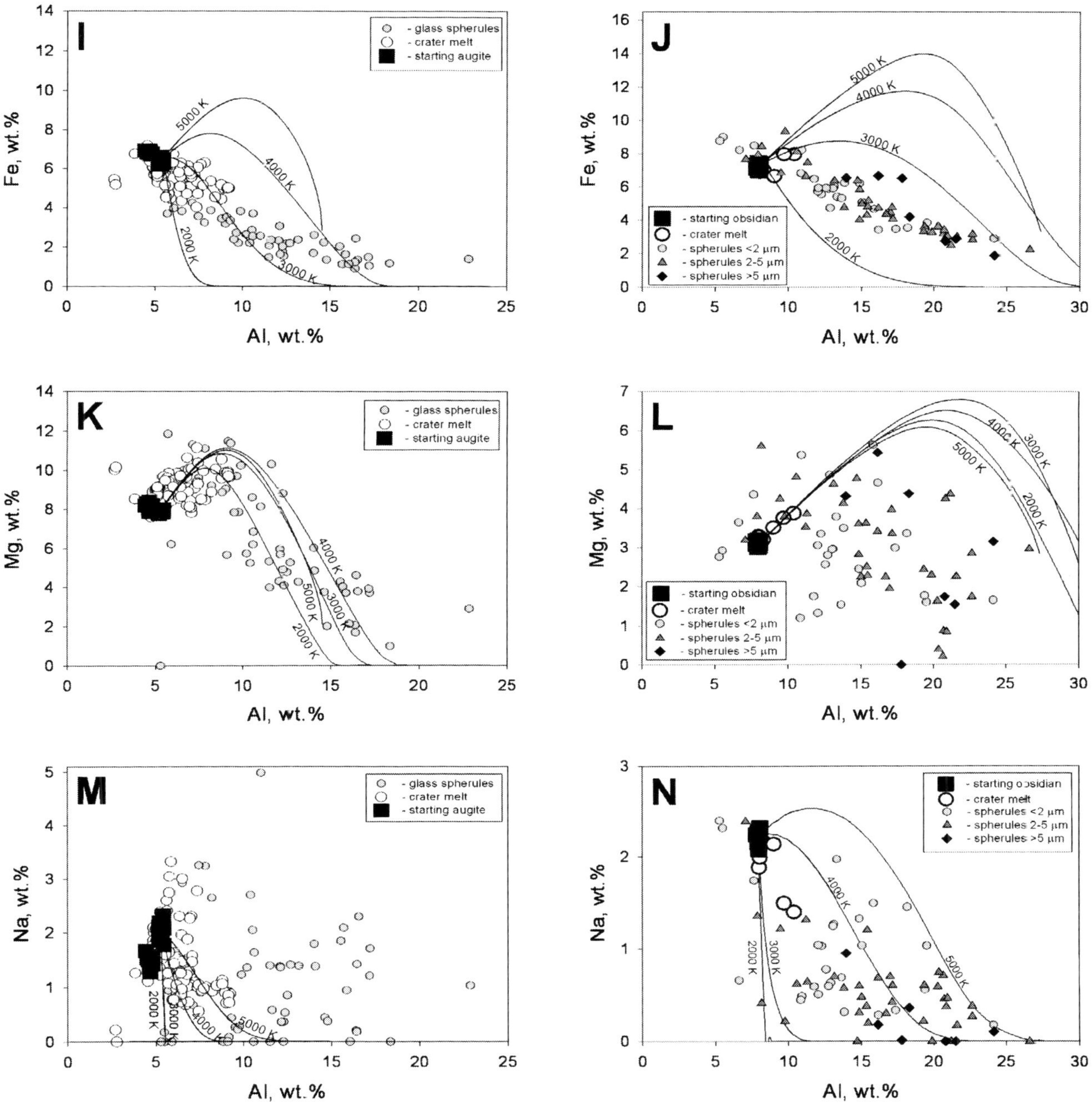

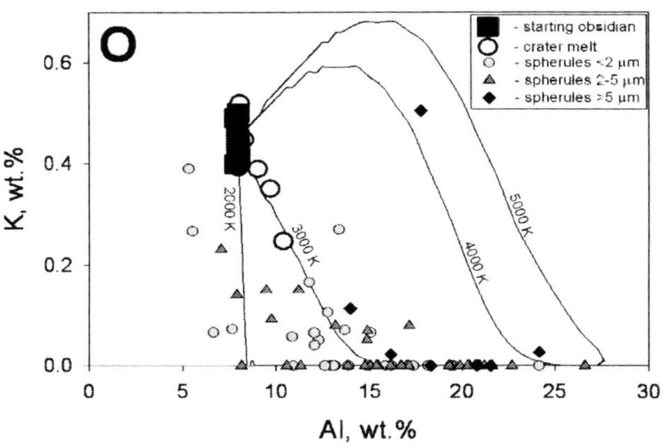

Figure 5 (*on this and previous page*). Chemical compositions of starting samples, crater melt, and glass spherules from laser pulse experiments. Figures A, C, E, G, I, K, and M show the Ca, Ti, Ca/Al, Si, Fe, Mg, and Na versus Al graphs respectively for experiments with augite. Figures B, D, F, H, J, L, N, and O show the Ca, Ti, Ca/Al, Si, Fe, Mg, Na, and K versus Al graphs, respectively, for experiments with obsidian. All 18 analyses for starting augite and all 10 analyses for starting obsidian are plotted in the graphs. Solid lines show the isothermal behavior of the average starting compositions at the temperatures indicated using the MAGMA code.

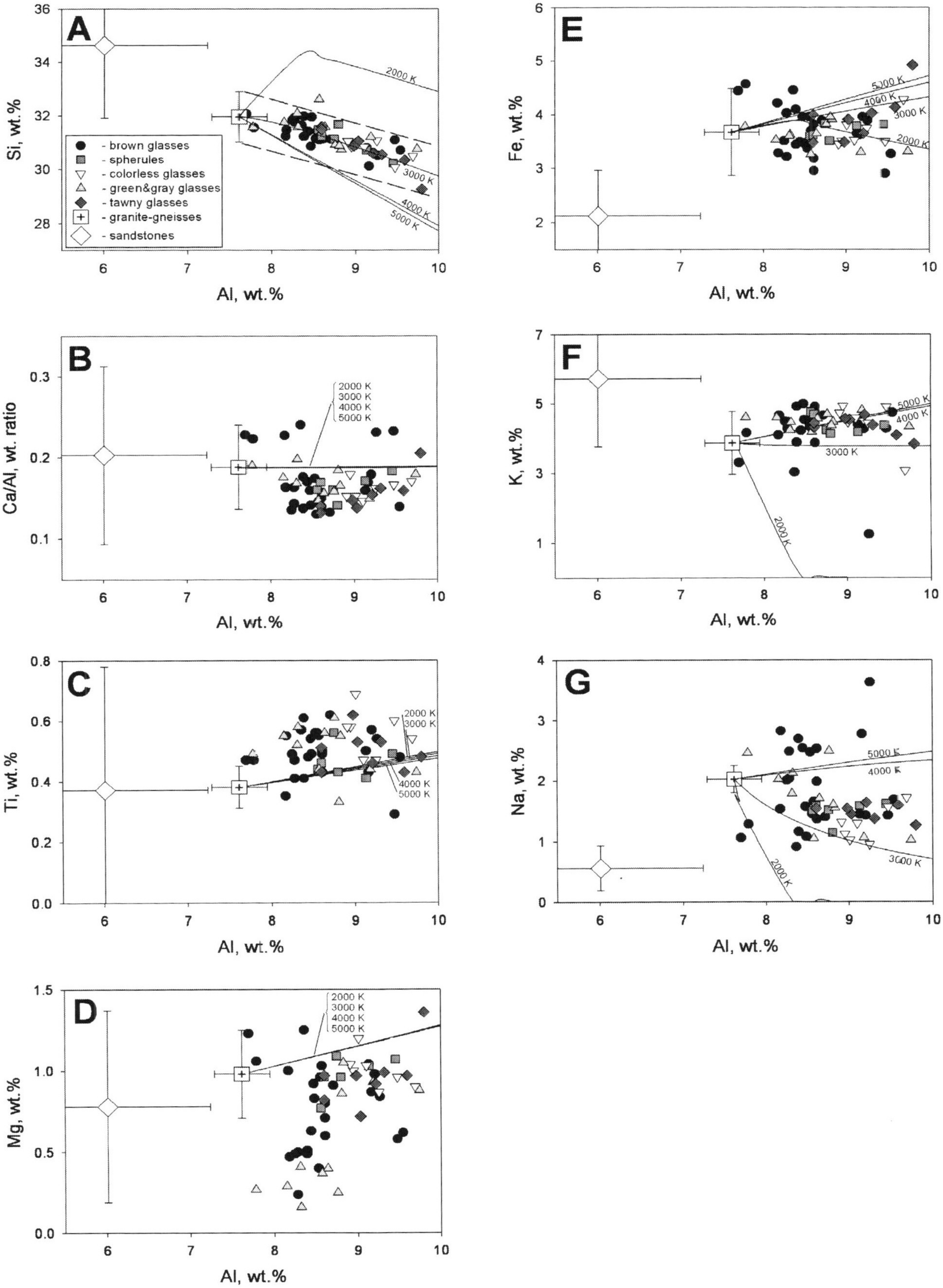

Figure 6. Chemical compositions of selected groups of impact glasses from suevites and of target granite-gneisses and sandstones from the Logoisk impact structure. Figures A, B, C, D, E, F, and G show the Si, Ca/Al, Ti, Mg, Fe, K, and Na vs. Al graphs, respectively. Solid lines show the isothermal behavior of the average granite-gneisses composition at indicated temperatures calculated using the MAGMA code.

Large meteorite impacts III

Edited by

Thomas Kenkmann
Institut für Mineralogie
Museum für Naturkunde
Humboldt-Universität, Berlin
Invalidenstrasse 43
D-10115 Berlin
Germany

Friedrich Hörz
National Aeronautics and Space Administration
Johnson Space Center, SR
Houston, Texas 77058
USA

Alex Deutsch
Institut für Planetologie
Universität Münster
D-48149 Münster
Germany

THE
GEOLOGICAL
SOCIETY
OF AMERICA

Special Paper 384

3300 Penrose Place, P.O. Box 9140 ▪ Boulder, Colorado 80301-9140 USA

2005

Published by The Geological Society of America, Inc.
3300 Penrose Place, P.O. Box 9140, Boulder, Colorado 80301-9140, USA
www.geosociety.org

Printed in U.S.A.

GSA Books Science Editor: Abhijit Basu

Library of Congress Cataloging-in-Publication Data

Large meteorite impacts III / edited by Thomas Kenkmann, Friedrich Hörz, Alex Deutsch.
p. cm. — (Special paper ; 384)
Includes bibliographic references and index.
ISBN 0-8137-2384-1 (pbk.)
1. Meteorite craters--Congresses. 2. Cratering--Congresses. 3. Planets--Origin—Congresses. 4. Impact--Congresses. I. Title: Large meteorite impacts 3. II. Kenkmann, Thomas, 1966- III. Hörz, Friedrich, 1940- IV. Deutsch, Alexander, 1950- V. Special papers (Geological Society of America) ; 384.

QB754.8.L38 2005
551.3'97—dc22

2004056412

Cover: Geometry of the central uplift of the Upheaval Dome impact crater, Utah. The three-dimensional visualization of a structure map shows the spatial construction of the strata boundary Lower–Upper Alibaba Member, Moenkopi Formation for the central part of the dome. The chosen boundary represents a good marker horizon in the field. The image is constructed making use of 3D-Move by Midland Valley Co.; viewing direction is from south-southwest.

10 9 8 7 6 5 4 3 2 1

Contents

The Planetary Perspective

Preface

The consequences of collisional processes are central to our understanding of planetary accretion, differentiation, and surface evolution, a lesson well taught by the first images and samples acquired during the Apollo area of lunar exploration. Subsequent missions to other solar system objects and an improved understanding of meteorites, including those from Mars, leave little doubt that hypervelocity impact was a dominant geologic process throughout the solar system in both space and time. Impact processes also play a significant role in the formation of atmospheres and hydrospheres. Its potential effects on the biosphere are amply illuminated by the impact-triggered mass extinctions at the Cretaceous-Tertiary (K-T) boundary. Lively debate focuses on the bombardment history of early Earth and its consequences for crustal evolution and early life. The impact process continues into the present as exemplified by the encounter of comet Shoemaker-Levy with Jupiter and ongoing astronomical observations of Earth-crossing objects. It follows that the study of impact processes has become a highly multidisciplinary undertaking involving and affecting most any aspect of geosciences.

While the analysis of planetary surface images provides insights into the history of the collisional record and into the gravity and size-dependent morphology of impact craters, it is the investigation of terrestrial craters that provides most of the ground truth needed to understand the actual impact process through field studies, geophysical investigations, petrographic characterizations, geochemical analyses, and experimental simulations. Terrestrial studies can provide only critical depth information for the complete description of, in part, rather complex structures which result from the sudden release of enormous amounts of energy contained in shock waves. In turn, these insights feed into theoretical considerations and associated numerical models, all combining into our current understanding of the cratering process and its geological and environmental effects, some yet to be discovered.

Current focus in terrestrial cratering research is on the detailed structural and geophysical characterization of select structures, amenable to improving our understanding of the subsurface mass movement during the excavation process, as well as during the gravity-induced collapse of the transient excavation cavity. This development was in part spurred by increasingly more detailed observations from large terrestrial craters, such as Sudbury, Vredefort, Chicxulub, Popigai, and others, as well as by vastly increased computer capabilities and associated fidelity of numerical models. This allows for meaningful, iterative feedback between field observations, theory, and numerical models, the latter becoming increasingly capable of calculating the shock propagation and deformation behavior of complex, layered target media in three dimensions and particularly over long integration times, commensurate with crater-growth times and gravitational collapse.

Obviously, the discovery and description of new impact craters on Earth continues to be an important activity, and increased sophistication of analytical instruments allowed for substantial progress in the identification and characterization of shock-modified rocks during the past decade. For instance, the search for impactor residues in impact melts was extended to the abundance and isotopic compositions of Os and Cr, allowing for the confident recognition of an extraterrestrial component at the trace level. Also, the initial discovery of shock-produced diamonds at the Popigai structure was extended to other craters and formation of diamond from graphite precursors seems common. Another central theme in the analysis of shocked rocks relates to the shock behavior of carbonates and sulfates; the liberation of CO, CO_2 and SO_x seems crucial to understanding the deterioration of the atmosphere and associated mass extinctions. Also, currently poorly understood, widespread breccias were identified that attest to disturbed carbonate platforms without apparent evidence for ballistic emplacement. Lastly, the hydrothermal alteration of terrestrial impact melts is being studied as a possible analogue to Mars, where impact-triggered, hydrothermal environments are speculated to be favorable sites for the evolution of life.

Starting with a detailed assessment of the K-T boundary during the Snowbird I Conference in 1981 (GSA Special Paper 190), the potential role of impact cratering on the environment and biosphere was periodically updated and synthesized via additional meetings and their proceedings (GSA Special Paper 247, 1988; GSA Special Paper 307, 1996; and GSA Special Paper 356, 2002). Additional workshops on large-scale cratering processes and planetary implications were held in 1994 (GSA Special Paper 293) and 1999 (GSA Special Paper 339). It was timely, therefore, to once again take stock of the latest developments in cratering research and to identify areas of fruitful, if not pressing, new inquiries. This volume constitutes the Proceedings of the resulting conference, held Aug. 5–7, 2003, in Nördlingen, Germany.

The conference was attended by 123 individuals from 21 different countries; approximately one-third of the attendants were either postdoctoral or graduate students, attesting to the broad appeal of the subject and assuring its continued health. Participation was lively, and the discussions were spirited, occasionally well beyond the formal presentations. The latter were complemented by two poster sessions. The medieval town of Nördlingen provided for sufficient charm and relaxation; local sponsors assured a generous dose of Bavarian hospitality. Being located in the center of the Ries crater, this venue also allowed for well-attended field trips to the Ries and Steinheim craters prior to and following the conference.

The Scientific Organizing Committee was composed of Co-Chairs Burkhard Dressler (Lunar and Planetary Institute, Houston) and Thomas Kenkmann (Humboldt University, Berlin), and included Alex Deutsch (University of Münster), Richard Grieve (Earth Science Sector, Natural Resources Canada), Robby Herrick (Lunar and Planetary Institute, Houston), Friedrich Hörz, (NASA, Johnson Space Center, Houston), Falko Langenhorst (University of Bayreuth), Jean Pohl (University of Munich), Buck Sharpton (University of Alaska, Fairbanks), and Dieter Stöffler (Humboldt University, Berlin). Members of the Local Organizing Committee were: Gisela Pösges and Michael Schieber (Rieskratermuseum, Nördlingen), Thomas Kenkmann (Humboldt University, Berlin), Jean Pohl (University of Munich), Peter Schiele (City of Nördlingen), and Falko Langenhorst (University of Bayreuth). All of these individuals worked diligently to assure a highly successful meeting.

The conference was sponsored by the German Science Foundation (DFG), the Lunar and Planetary Institute, Houston, and Humboldt University, Berlin. The City of Nördlingen provided generous support as well, as did the Barringer Co., Bavarian Ministry of Culture, Sparkasse Nördlingen, Raiffeisen-Volksbank Donau-Ries, Märker-Zementwerke AG, Münchener Rückversicherung and Bayerisches Geoinstitut, Bayreuth. Their contributions are gratefully acknowledged, as they provided travel stipends, crucial local logistics, refreshments, and other support.

The production of this Special Paper was organized by the undersigned, assisted by S. Biersack, Humboldt University, Berlin. We thank the contributing authors and reviewers, as well as the Geological Society of America and its publication staff, including Book Editor A. Basu, University of Indiana.

Special thanks go to the Lunar and Planetary Institute, S. Mackwell, Director, who encourages and supports topical conferences on collisional processes since Apollo. It is hoped that these proceedings continue the tradition of the earlier meetings and that they spawn new ideas and research that will be the topic of future conferences.

Thomas Kenkmann
Friedrich Hörz
Alex Deutsch

Geological Society of America
Special Paper 384
2005

Impact structures: What does crater diameter mean?

E.P. Turtle
Lunar and Planetary Laboratory, Department of Planetary Sciences, University of Arizona, Tucson, Arizona 85721-0092, USA, and *Planetary Science Institute, 1700 E. Fort Lowell, Suite 106, Tucson, Arizona 85719-2395, USA*

E. Pierazzo
Planetary Science Institute, 1700 E. Fort Lowell, Suite 106, Tucson, Arizona 85719-2395, USA

G.S. Collins*
Lunar and Planetary Laboratory, Department of Planetary Sciences, University of Arizona, Tucson, Arizona 85721-0092, USA

G.R. Osinski*
Lunar and Planetary Laboratory, Department of Planetary Sciences, University of Arizona, Tucson, Arizona 85721-0092, USA, and *Planetary Science Institute, 1700 E. Fort Lowell, Suite 106, Tucson, Arizona 85719-2395, USA*

H.J. Melosh
Lunar and Planetary Laboratory, Department of Planetary Sciences, University of Arizona, Tucson, Arizona 85721-0092, USA

J.V. Morgan
Department of Earth Science and Engineering, Imperial College London, London SW7 2AZ, UK

W.U. Reimold
Impact Cratering Research Group, School of Geosciences, University of the Witwatersrand, Private Bag 3, P.O. Wits 2050, Johannesburg, South Africa

ABSTRACT

The diameter of an impact crater is one of the most basic and important parameters used in energy scaling and numerical modeling of the cratering process. However, within the impact and geological communities and literature, there is considerable confusion about crater sizes due to the occurrence of a variety of concentric features, any of which might be interpreted as defining a crater's diameter. The disparate types of data available for different craters make the use of consistent metrics difficult, especially when comparing terrestrial to extraterrestrial craters. Furthermore, assessment of the diameters of terrestrial craters can be greatly complicated due to post-impact modification by erosion and tectonic activity. We analyze the terminology used to describe crater geometry and size and attempt to clarify the confusion over what exactly the term "crater diameter" means, proposing a consistent terminology to help avert future ambiguities. We discuss several issues of crater-size in the context of four large terrestrial examples

Present addresses: Collins—Department of Earth Science and Engineering, Imperial College London, South Kensington Campus, London SW7 2AZ, UK; Osinski—Canadian Space Agency, 6767 Route de l'Aeroport, Saint-Hubert, Quebec, J3Y 8Y9, Canada.

Turtle, E.P., Pierazzo, E., Collins, G.S., Osinski, G.R., Melosh, H.J., Morgan, J.V., and Reimold, W.U., 2005, Impact structures: What does crater diameter mean?, *in* Kenkmann, T., Hörz, F., and Deutsch, A., eds., Large meteorite impacts III: Geological Society of America Special Paper 384, p. 1–24. For permission to copy, contact editing@geosociety.org.

for which crater diameters have been disputed (Chicxulub, Sudbury, Vredefort, and Chesapeake Bay) with the aim of moving toward consistent application of terminology.

Keywords: impact craters, crater diameter, Vredefort, Sudbury, Chicxulub, Chesapeake Bay.

INTRODUCTION

The presence of impact craters on almost every solid surface in the solar system testifies to the importance of cratering as a geologic process. Nonetheless, the study of impact craters is still a relatively young field, and our understanding of the processes involved in crater formation remains far from complete. The current extent of our knowledge is derived from investigations spanning many disciplines, including geologic and geophysical exploration (both in situ and based on remote-sensing for Earth, whereas almost exclusively the latter for other planets); petrographic and geochemical analyses; small-scale laboratory and larger-scale nuclear explosion experiments; and numerical simulations. An unfortunate side effect of the cross-disciplinary nature of cratering investigations is that inconsistencies in terminology commonly arise because the data sets are so disparate. That the field is still young exacerbates the problem, as available data and interpretations thereof, as well as the terminology itself, are continuing to evolve.

Observations of terrestrial and extraterrestrial craters provide complementary data sets with which to study craters and the cratering process; however, the differences between these data sets complicate comparisons significantly. On one hand, features discernable in the types of data that can be acquired for terrestrial craters, where observations are made in situ and to a limited extent in three dimensions, are often impossible to detect or measure at comparable craters on other planets, where the only available data are acquired from orbit. On the other hand, terrestrial craters are subject to rapid erosion, tectonic modification, and burial, leaving few fresh examples, whereas craters on other planets are typically less modified (if not unmodified) and provide much information on crater morphology and structure, especially regarding relationships between geometric aspects of craters as a function of crater size and target parameters. It is therefore difficult to accurately compare a pristine extraterrestrial crater with the eroded remnant of a terrestrial crater, particularly if the original size or nature of the eroded crater is uncertain. This problem can be exacerbated by the use of generic terms like "crater size" and "crater diameter" in reference to impact structures that have been studied using different methods or that have complicated morphologies. For example, many impact structures, especially large or eroded ones, exhibit multiple concentric features, leading to ambiguity in establishing which feature is the best measure of crater size. The inconsistent use of terminology has led to considerable, and, in some cases, avoidable, controversy within the impact community.

Why does it matter if different sizes are quoted for a given crater? The precise definition of crater size may seem to be a relatively unimportant, semantic issue; however, misinterpretation of an impact structure's size can have significant implications. At our current level of understanding, the dimensions of many aspects of craters and the amount of energy required to create them are still determined by scaling based on information obtained from other craters. Although such scaling relationships are often not well constrained, they are currently some of the best tools available. For example, the diameter of a pristine crater is related to the kinetic energy of the projectile (e.g., Schmidt and Housen, 1987), but the nature of this dependence is not fully understood. Therefore, scaling relations were developed to allow derivation of impact energy based on measurements of the "final crater" diameter. However, if a different metric for crater diameter than the one for which the scaling relation was developed were to be used, the energy estimate could be incorrect by a significant factor. For example, the maximum and minimum diameters proposed for the Chicxulub impact structure in Mexico are ~295 km and ~150 km (Sharpton et al. [1993] and Morgan and Warner [1999], respectively; Table 1). Using scaling laws relating rim (or final crater) diameter to transient crater diameter (e.g., McKinnon and Schenk, 1985) and transient crater diameter to impact energy (e.g., Schmidt and Housen, 1987), these values imply impact energies that differ by an order of magnitude (~8.0×10^{24} J and ~8.0×10^{23} J, respectively). Accurate knowledge of the impact energy is paramount to assessing the environmental effects of an impact, and it is also vital as an input parameter for numerical simulations of crater formation. Our understanding of the consequences of impact cratering, as well as of the process itself, therefore depends on the use of consistent measurements of crater dimensions.

The complexity of the impact process, combined with our current limited knowledge, means that interpretation of the preserved structure of an impact crater (i.e., observations) and understanding the dynamic process by which it formed (i.e., theory and modeling) are intimately connected and cannot be considered separately. For example, until we fully understand how the outer, concentric faults observed in multi-ring impact craters form, identifying such features, unequivocally, in the field or in geophysical data can be problematical. However, observations of potential outer rings provide some of the only constraints we have for models of their genesis. Therefore, there is a need for any given metric of crater morphology, whether it is a field measurement, a distance derived from geophysical or remote-sensing observations, or the output of a computer simulation, to be described as accurately as possible, so that crater morphologies may be compared and contrasted meaningfully and with due caution.

It is not our intent to provide a method for determining crater size; indeed, the literature on the subject of crater size determi-

TABLE 1. DIAMETERS REPORTED FOR CHICXULUB

Diameter (km)	Feature (if specified)	Observation or method of determination	Reference(s)
180–210	–	Gravity and magnetic anomalies	Penfield and Camargo-Zanoguera, 1981, 1991; Hildebrand et al., 1991
295	–	Additional gravity data	Sharpton et al.,1993
170	Transient crater		
170	Internal crater	Gravity, magnetic, seismic co-analysis	Camargo-Zanoguera and Suarez-Reynoso, 1994
180	Structural rim	Physical properties of ejecta and melt sheet in combination with geophysics	Kring, 1995
200	–	Gravity and magnetic data	Espindola et al., 1995
180	–	New gravity data	Hildebrand et al., 1995
260	–	Surface features	Pope et al., 1997
80	Peak ring	Seismic survey	Morgan et al., 1997; Morgan and Warner, 1999; Morgan et al., 2000
150	Structural rim		
200	Outer ring		
250	Outer limit of deformation		

nation is quite extensive (in addition to many manuscripts that discuss this issue in the context of specific craters, general discussions include Croft [1985]; Schmidt and Housen [1987]; Chapter 7 of Melosh [1989], and references therein; and Grieve [1991]; to list only a few). Rather, we discuss the various metrics of impact structures that are relevant under different circumstances, and we attempt to codify these metrics to improve the clarity of future crater diameter discussions and the use of scaling laws. We also highlight four terrestrial cases in which ambiguities associated with crater size have caused controversy: Chicxulub, Sudbury, Vredefort, and Chesapeake Bay. Crater diameters can be derived based on many aspects of craters, including geophysical, petrologic, morphologic, and structural observations as well as scaling from experiments. Here, we have limited our discussion to a structural perspective with the intent of resolving discrepancies that have arisen primarily between geologic interpretations and numerical simulations (e.g., Herrick and Pierazzo, 2003). Our central tenet is that, to avoid the kinds of semantic misunderstanding that have plagued the field of impact cratering, it is necessary to be as explicit as possible. We hope to demonstrate the need for establishing a consistent terminology and, eventually, for developing methods to quantitatively relate observable parameters to impact parameters.

THE IMPACT CRATERING PROCESS AND BASIC TERMINOLOGY

Before discussing the important structural dimensions of an impact crater, it is helpful to review our current understanding of how the final crater structure develops and to define explicitly the terms for processes and morphologies that we use throughout this manuscript. This brief overview is intended simply to provide information necessary for the later discussion; a detailed review of the processes involved is outside the scope of this manuscript (see Melosh, 1989).

Impact cratering is a rapid, highly dynamic, continuous sequence of events that is essentially over in a matter of minutes. Nevertheless, the cratering process is conventionally divided into three stages, during which different physical processes dominate: (1) contact and compression (shock wave initiation and propagation), (2) crater excavation, and (3) crater collapse and modification.

Contact and Compression Stage

This initial stage begins at the moment the projectile contacts the target surface. It includes the generation of shock waves in both the projectile and the target, the propagation of the shock wave through the projectile, and the passage of the rarefaction wave back through the projectile. Contact and compression ends when the projectile has been unloaded from the shock state (e.g., Melosh, 1989). This stage of the cratering process is exceedingly brief, lasting only on the order of one second in the case of a 100-km-diameter transient crater (see discussion of excavation stage below) on Earth. The principal outcome of this stage is the transfer of most of the projectile's kinetic energy into kinetic and internal energy in both the projectile and the target. The internal energy heats both the projectile and the target, causing, for sufficiently strong shocks, melting and/or vaporization.

Excavation Stage

The passage of a shock wave first compresses and then releases the target rock from high pressure, leaving the target material with a residual velocity that is ultimately the result of the thermodynamic irreversibility of the shock (Melosh, 1989). The residual velocity is typically about one-fifth of the peak velocity in the expanding shock wave and is responsible for initiating the excavation flow, the movement of target material away from the point of contact that opens up the crater. The excavation flow is initially radial; however, interaction with rarefaction waves reflected from the free surface adds an upward component to the velocities of the target material, so that the excavation flow ejects some target material out of the crater on ballistic trajectories, while deeper material is driven radially downward and outward. The combination of these motions opens up what is often termed the *transient cavity* (Dence, 1968): the form of the opening, or collapsing, crater at any given instant during the impact event (Fig. 1). (We distinguish between the terms *transient cavity* and *transient crater*, using the former to refer to the evolving crater cavity and the latter to refer to an idealized shape defined by the maximum extent to which excavation proceeds in every direction; see further discussion below.) The excavation flow is retarded by the inertia of the surrounding, initially stationary, target material; dry friction (resistance to shear of granular materials); gravity; and any cohesive strength that the target material may retain. Gravity is the dominant parameter in large impacts (crater size depends on the target's surface gravity); excavation stops when there is insufficient energy left to drive material out of the crater.

Excavation does not cease simultaneously in all directions. Instead, the opening cavity continues to grow horizontally after vertical excavation has stopped (e.g., Fig. 1C and 1D). The term *transient crater* is generally used to refer to the idealized, theoretical construct defined by the maximum extent to which excavation proceeds in every direction. The transient crater is often represented as a paraboloid of revolution with a depth to diameter ratio of ~1:3. The transient crater diameter, which is measured with respect to the pre-impact surface, also corresponds to the diameter of the *excavated cavity*, the portion of the transient cavity from which material is expelled from the crater

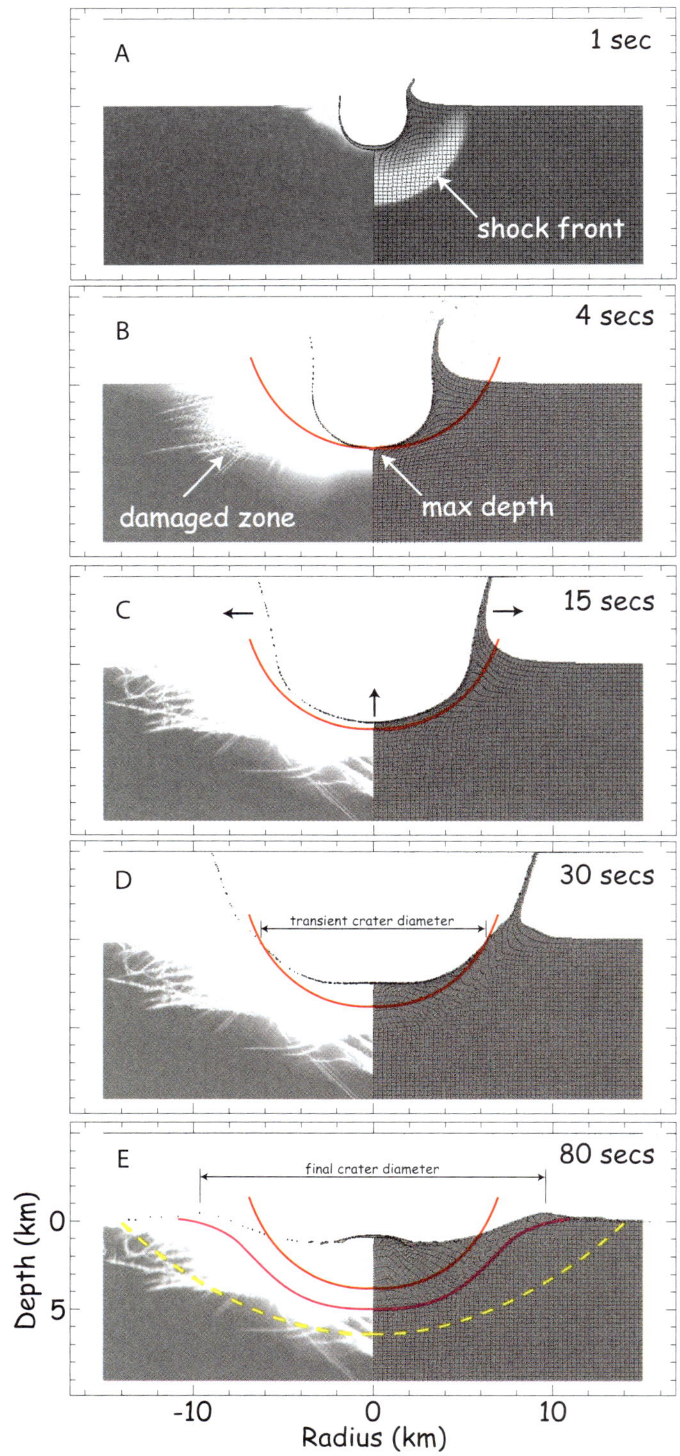

Figure 1. Illustration of the growth and collapse of the transient cavity from a simulation of an impact of a dunite projectile into a granite half-space, forming a ~12 km diameter transient crater. (A) The early growth is hemispherical—a result of the passage of the shockwaves and rarefaction-waves through the target. On the right side of the plot, light shading illustrates pressures above 10 GPa. The shading on the left side of the plot represents the extent of damage–the lighter, the shade the more extensive the damage. (B) The opening cavity reaches its maximum depth at 4 s. This point defines the depth of the transient crater, whose outline is shown in red, but at this point, the diameter of the opening cavity is still less than that of the idealized transient crater. (C) After ~15 s, the form of the opening cavity is more parabolic in profile; the crater floor has begun to rebound, while the diameter continues to grow—excavation and collapse are occurring simultaneously. By this stage, damage has reached its maximum extent. (D) By 30 s, the outward excavation flow has ceased; all motion, other than that of the ejecta curtain, is part of the inward collapse. The majority of deformation occurs within a zone well inside the damaged region. (E) The rim diameter is ~1.6 times the transient crater diameter. Note that two other modified regions can be delimited from the numerical simulation results: a highly disturbed region, characterized by total strains well in excess of 0.1 (pink curve), and a damaged zone (dashed, yellow curve), which marks the predicted boundary between fractured and unfractured target material.

on ballistic trajectories (as opposed to the portion from which material is displaced downward and outward). For small craters, the concept of the transient crater is quite useful because the final crater is only modified from the transient crater by debris sliding down the steepest part of the crater wall. However, in the case of large impact events where structural uplift of subcrater material and much more extensive slumping of the crater rim occur (see discussion of crater collapse and modification below), results of numerical models indicate that the transition from transient cavity growth to collapse cannot be defined explicitly; indeed, the processes of excavation and collapse can overlap (e.g., Fig. 1C and 1D), and the idealized transient crater may never actually exist. Nevertheless, the transient crater is an important concept because its size can be related to the impact energy (e.g., Schmidt and Housen, 1987). In addition, the idea that the early stages of the impact process, up to the onset of gravitational collapse, are fundamentally the same for all craters, regardless of size, is widely accepted. The excavation stage is much longer than the contact and compression stage, but is, nonetheless, quite a rapid process. Following Melosh (1989, p. 77), for a 100-km-diameter transient crater on Earth, the maximum extent of the opening cavity is achieved in roughly 80–100 s.

Modification Stage

The final stage of cratering, usually referred to as the modification stage or the collapse stage, involves processes that modify the transient cavity and shape the final crater. The degree to which collapse occurs is controlled by the post-shock material strength of the target and gravity, which acts to close the cavity. The balance between these competing influences determines the morphology of the final crater. Impact craters are generally classified as *simple* or *complex* (Dence, 1965). In the case of a simple crater, the strength of the rock surrounding the newly formed cavity prevents wholesale collapse, leaving a final crater form only little-modified from the transient crater. However, for craters that exceed a critical diameter, the target's intrinsic strength is overcome by gravitational forces. Quantitative study of many craters has shown that target strength is greatly diminished immediately after crater excavation, which has provoked much discussion regarding the mechanism by which strength reduction occurs (e.g., Melosh, 1977, 1982; McKinnon, 1978; Melosh and Gaffney, 1983; O'Keefe and Ahrens, 1993, 1999; Ivanov and Kostuchenko, 1997; Melosh and Ivanov, 1999), and the transient cavity collapses resulting in more complex morphologies.

Crater Morphologies

Simple Craters

Simple craters are circular, bowl-shaped depressions with raised rims (e.g., Meteor [Barringer] Crater, Arizona; Fig. 2). Part of the rim height is due to uplift of target material surrounding the crater and part is due to the ejecta blanket, which consists of an overturned flap of target material at the rim. The material that lines the transient cavity during excavation slides down the crater walls and accumulates in the center of the crater. Small-scale

Figure 2. Aerial photo of the simple crater, Meteor (Barringer) Crater, Arizona, United States, illustrating the morphology of a simple crater. The rim diameter is 1.2 km.

collapse also occurs where the rim is steep. Therefore, the final crater is slightly larger in diameter and shallower than the transient crater. Pristine examples of simple craters on the Moon exhibit an approximately parabolic profile, with a depth-to-diameter ratio of ~1:5 (Pike, 1977).

Complex Craters

Complex craters are significantly modified from the transient crater (e.g., Dence, 1968; Grieve et al., 1977; Grieve, 1991). The transition from simple to complex crater morphology depends primarily on the target's surface gravity, but also on its structure and material properties. The gravity dependence causes the transition to occur at different sizes on different planets. For craters that are sufficiently large, when the maximum depth of the transient crater is achieved, uplift of the crater floor begins before the opening cavity reaches its maximum radial extent (Fig. 1). As a result, complex craters have flat floors whose depths are not strongly dependent on crater diameter (e.g., Melosh, 1989). Above the floor rise *central peaks* (e.g., Theophilus, Moon; Fig. 3) that consist of bedrock that has been fractured into blocks, which are chaotically mixed in places, and elevated relative to its original stratigraphic position (Grieve et al., 1981; Pieters, 1982). The uplift occurs rapidly enough (several tens of seconds for a 100-km-diameter transient crater on Earth; Melosh, 1989, p. 142) that the breccia and melt that slide down the crater walls accumulate in an annulus around the central rise, rather than forming a central breccia lens as in simple craters, although some impact breccia can still be found on central peak complexes (e.g., Puchezh-Katunki, Russia). The uplift of material extends below the crater floor, and deep *structural uplift* is apparent in eroded complex craters on Earth (e.g., Vredefort; Grieve and Therriault, 2000; Gibson and Reimold, 2001). Rim terraces develop as wide blocks slump from the rim, enlarging the transient crater, typically by a factor of 1.5–2 (Grieve et al., 1981), to form the final crater.

For even larger craters, rather than a central peak, an internal ring of mountains (e.g., Greenaway, Venus; Fig. 4), often referred to as a *peak ring* or *inner ring* (Hartmann and Wood, 1971) can be formed (e.g., Grieve and Cintala, 1992; Ivanov and Kostuchenko, 1997; Collins et al., 2002). Peak rings are typically concentric with the crater center and rim, and occasionally surround a central peak. They are commonly discontinuous. Topographic profiles of peak rings are generally symmetric; i.e., the inward and outward slopes are approximately the same.

Some of the largest impact structures are surrounded by one or more concentric scarps or fractures (Hartmann and Kuiper, 1962; Fig. 5). These structures are called *multiple-* or *multi-ring craters* (or *multi-ring basins*, or sometimes simply *basins*). There are two basic morphological types, believed to be determined by the conditions under which the rings formed (e.g., Melosh and McKinnon, 1978). The first type exhibits a few to several inward-facing scarps with gentle outward slopes (e.g., Orientale, Moon; Fig. 5A). Based on seismic data (e.g., Morgan et al., 1997), the Chicxulub structure in Mexico has been interpreted to have this morphology. Craters of the second type exhibit tens to hundreds of closely spaced rings consisting of graben or outward-facing scarps surrounding a central, flat basin (e.g., Valhalla, Callisto; Fig. 5B). There is evidence that these rings also form quite rapidly (i.e., within minutes; Head, 1974; Prockter et al., 1999).

As discussed by Hartmann and Wood (1971; also Wood and Head, 1976, and Pike, 1985) the rings that define multi-ring craters are distinct from the peak rings described previously. Morphologically, peak rings generally appear mountainous and have symmetric topographic profiles (e.g., Fig. 4), whereas outer rings more strongly resemble scarps, having asymmetric topographic

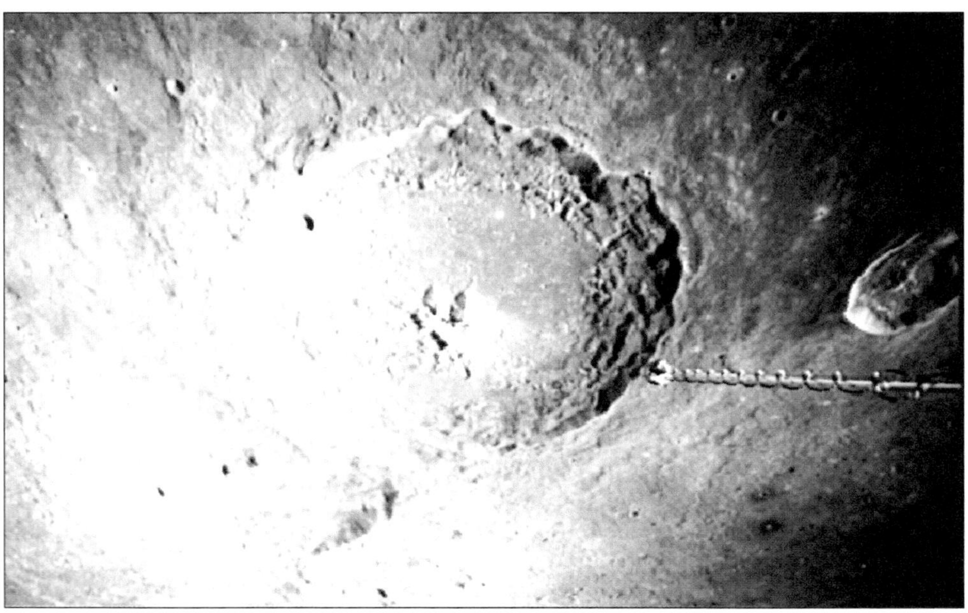

Figure 3. Apollo 16 metric mapping camera image (AS16-0692) of the lunar central peak crater Theophilus (11°S, 26°E), illustrating common features of complex craters such as terracing along the rim, a flat crater floor, and a central peak. The crater rim is 110 km in diameter. North is to the top of image, and the illumination is from the right.

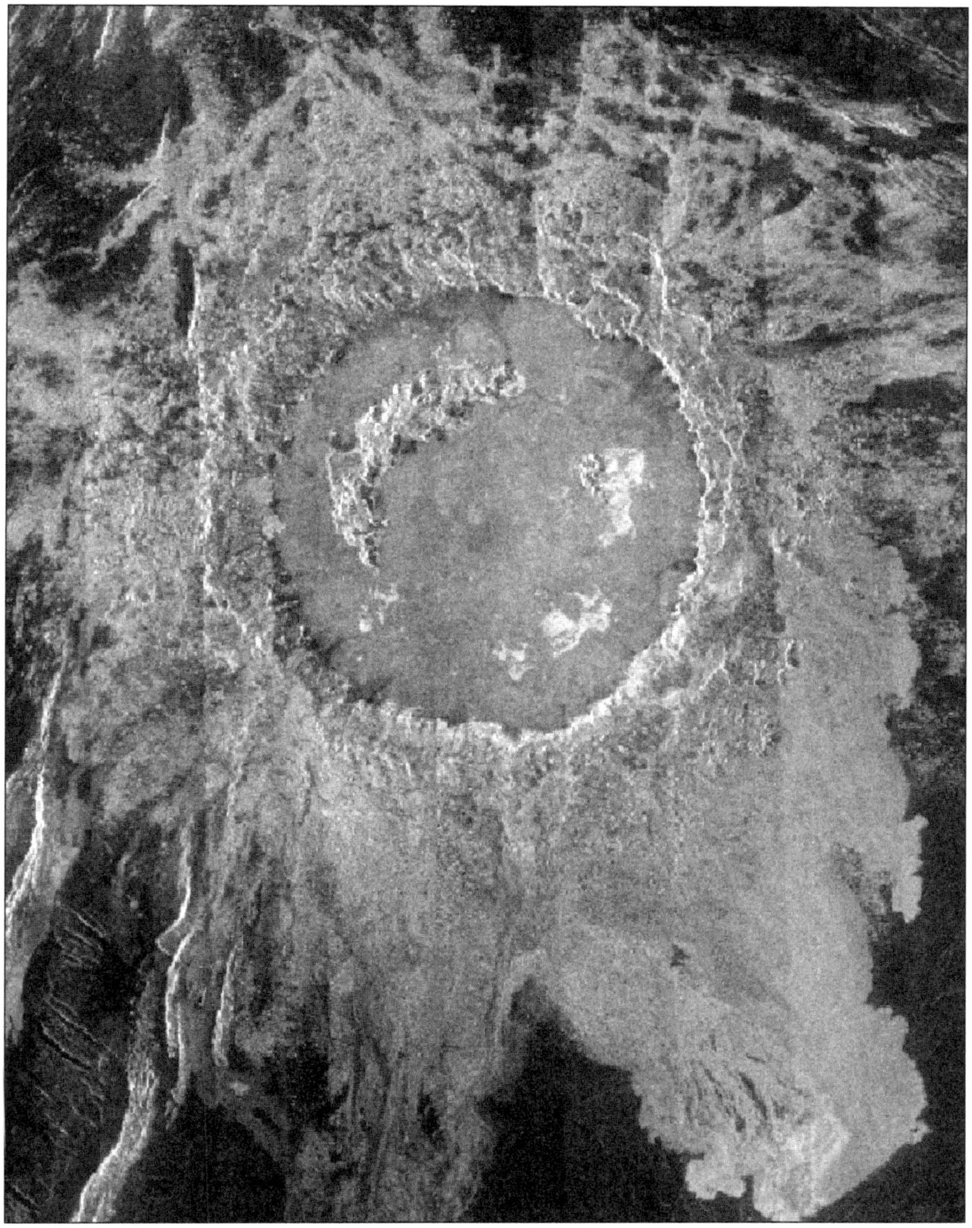

Figure 4. *Magellan* radar image (Full-Resolution Mosaicked Image Data Records [F-MIDRs] f25n143 and f20n145) of the Venusian peak ring crater Greenaway (23°N, 145°E), illustrating a discontinuous peak-ring. The crater rim diameter is 92 km and the peak ring diameter is 46 km. North is to the top of image.

profiles with a steep face on one side and a gentle slope on the other, or in some cases, graben (e.g., Fig. 5). Moreover, these two types of ring structures form under different conditions: peak rings form *within* the rim of the final, collapsed crater (perhaps due to collapse of an overheightened central peak; e.g., Collins et al., 2002) and consist of significantly disrupted material, whereas the rings that are characteristic of multi-ring craters form *outside* the final crater (as defined by the terraced crater rim formed by crater collapse, see recommended terminology below) and only locally disrupt preexisting features. For example, ridges that pre-date Europa's Callanish impact structure (Fig. 5C) are cut by the ring graben, but remain essentially undisturbed between the rings, whereas, within the crater center, any pre-impact structures have been completely erased. These observations provide evidence that different mechanisms are responsible for these two types of ring structures. Furthermore, the existence of external rings around much smaller impact structures suggests that their formation is strongly influenced by target structure and material properties. For example, Silverpit, a potential impact structure in the North Sea, is ~6 km in diameter, but is nonetheless multi-ringed, which has been attributed to an impact into a layer of brittle chalk overlying weak shales (Stewart and Allen, 2002; Collins et al., 2003). External, low relief, ring features that dip inward at shallow angles are also observed at the Libyan BP and

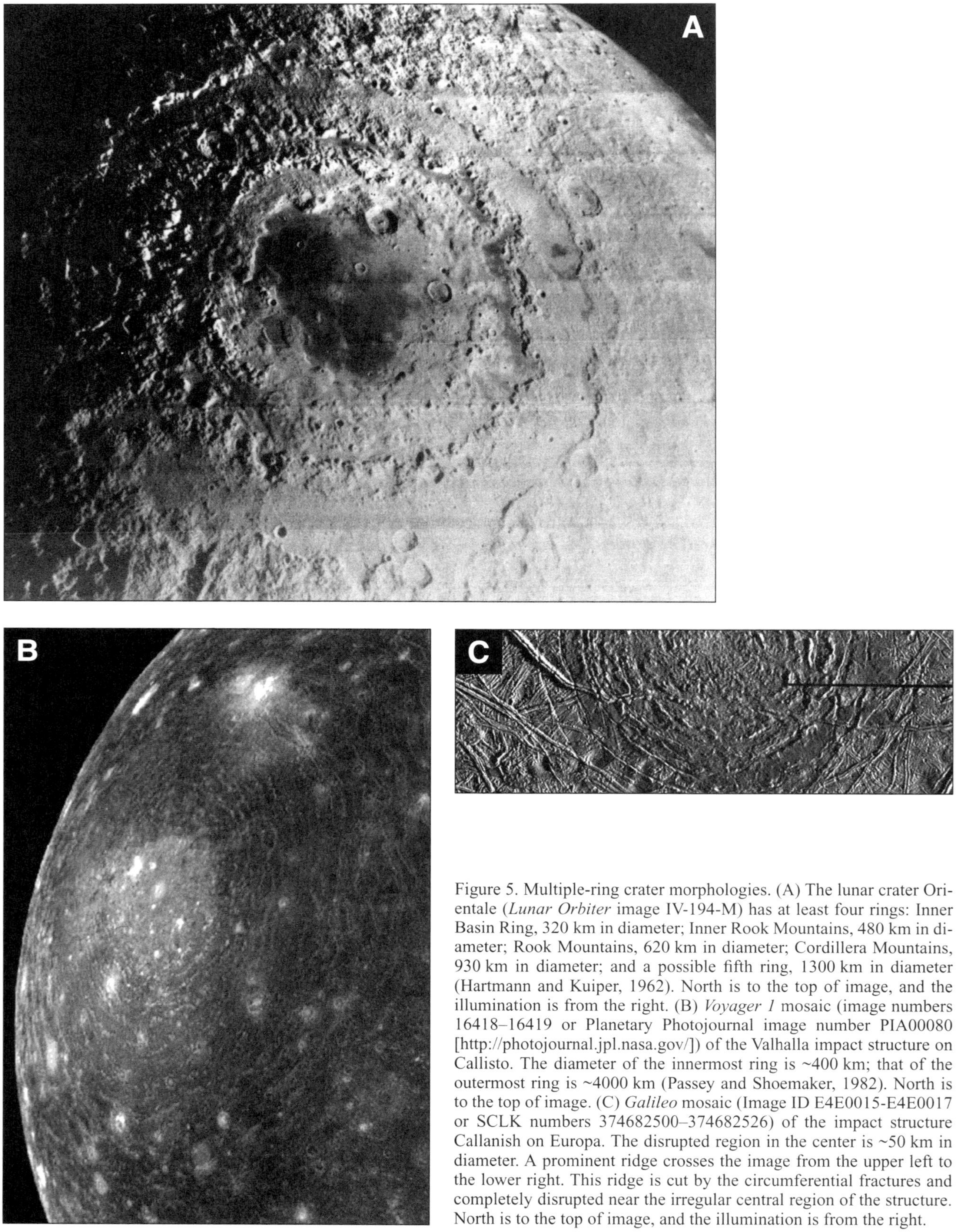

Figure 5. Multiple-ring crater morphologies. (A) The lunar crater Orientale (*Lunar Orbiter* image IV-194-M) has at least four rings: Inner Basin Ring, 320 km in diameter; Inner Rook Mountains, 480 km in diameter; Rook Mountains, 620 km in diameter; Cordillera Mountains, 930 km in diameter; and a possible fifth ring, 1300 km in diameter (Hartmann and Kuiper, 1962). North is to the top of image, and the illumination is from the right. (B) *Voyager 1* mosaic (image numbers 16418–16419 or Planetary Photojournal image number PIA00080 [http://photojournal.jpl.nasa.gov/]) of the Valhalla impact structure on Callisto. The diameter of the innermost ring is ~400 km; that of the outermost ring is ~4000 km (Passey and Shoemaker, 1982). North is to the top of image. (C) *Galileo* mosaic (Image ID E4E0015-E4E0017 or SCLK numbers 374682500–374682526) of the impact structure Callanish on Europa. The disrupted region in the center is ~50 km in diameter. A prominent ridge crosses the image from the upper left to the lower right. This ridge is cut by the circumferential fractures and completely disrupted near the irregular central region of the structure. North is to the top of image, and the illumination is from the right.

Oasis structures, which are 2–3.2 km and 5.1–18 km in diameter, respectively (Koeberl et al., 2005; McHone et al., 2002). Wagner et al. (2002) debated whether such outer concentric features represent the actual crater rim, ring faults outside of the actual rim, or expressions of the outer limit of the original ejecta blankets. These features could also represent faults resulting from inward slumping of rim terrace blocks during the modification stage.

THE MANY DIAMETERS OF AN IMPACT CRATER

Astronomical Observations

Galileo first observed lunar impact craters in 1610, at which time he described them as circular rimmed pits (Galilei, 1610). In the subsequent 300 years of astronomical observation, many measurements of the dimensions of these features were performed. Early workers were constrained to measure the diameter of the raised rim on opposite sides of the crater, i.e., the *rim-to-rim diameter*, or simply *rim diameter* (Fig. 6). It was also possible to determine the height of the rim above the crater floor and its height above the surrounding plains. Using such data, Baldwin (1949, 1963) showed that lunar craters were consistent with explosion crater statistics, convincing most workers that craters were impact, not volcanic, in origin. As more complex crater forms were discovered (such as multi-ring impact basins; Hartmann and Kuiper, 1962) and peak rings (Hartmann and Wood, 1971), and as terrestrial impact craters in various states of degradation were recognized (e.g., Shoemaker, 1960; Beals et al., 1963; Dence, 1965), it became more difficult to relate diameters measured for various features to the rim diameters of smaller or more pristine craters. The diameters of the central peaks protruding through the relatively flat floors of small complex craters were measured from where the peak meets the crater floor, and the diameters of the floors themselves were measured from where the floor meets the base of the rim terrace(s). In larger craters, the diameters of peak rings were measured from topographic high to topographic high (Hartmann and Kuiper, 1962). Crater dimensions were defined so that they could be easily measured by the technique at hand. They are not necessarily the most significant for understanding the mechanics of crater formation. Furthermore, some of the definitions themselves may cease to be appropriate for the largest impact structures: although circular arcs can be fit to the scarps concentric to such structures, the relation of these arcs to the final crater rim or to the transient crater from which we believe it formed is still obscure for the largest structures.

Experimental Cratering

A new set of crater dimension measurements came into use in the latter half of the twentieth century for the description of craters produced by large, high-explosive and nuclear detonations (Nordyke, 1961). Many of these artificial crater structures were carefully surveyed and excavated, providing investigators with detailed knowledge unavailable to astronomers studying the Moon. Explosion experts distinguished the "true crater," defined by the boundary between fractured and unfractured bedrock, from the "apparent crater," which is the surface manifestation of the crater. In addition, crater diameter is defined as the *apparent crater diameter* at the level of the pre-explosion ground surface, not at the rim, and so is smaller than the rim-to-rim diameter of the same crater. Similarly, the *crater depth* is the depth of the apparent crater's center below the pre-explosion surface. *Rim height* is also measured with respect to the pre-explosion surface.

Beginning ca. 1965 with the initial operation of the Ames Vertical Gun Range under the direction of Don Gault, impact cratering became a laboratory science. Extensive early work by Gault and his many collaborators clearly established the processes of impact crater excavation (Gault et al., 1968). Most of these laboratory-scale craters (dimensions 10 cm to perhaps 1 m) were formed in loose material, such as sand or pumice, in which elaborate target construction permitted the motions during crater excavation to be inferred. The craters thus formed underwent only very minimal amounts of collapse—one must consider them essentially frozen transient craters, as even the breccia lenses observed in simple craters are generally absent at this scale. For these craters, careful profiling of the final crater and sectioning of the target permitted the assignment of both morphologic dimensions (rim diameter, apparent depth and diameter, rim height) and sometimes structural diameter (true crater in cases of cohesive targets).

In these early sand-target experiments, no structural collapse was observed. Gault made one (unpublished) series of centrifuge collapse experiments on static crater forms (illustrated for the first time in Melosh, 1989), but no dynamic collapse experiments were performed. Robert Schmidt at Boeing was the first to use a geotechnical centrifuge to study cratering under elevated acceleration of gravity (Schmidt and Housen, 1987). By increasing the acceleration of gravity, the process of crater collapse could be observed during actual crater excavation using both explosive and impact energy sources. Schmidt made the important observation that, even with large gravitational accelerations that leave a nearly flat, collapsed final crater, the transient crater undergoes a transient deep excavation phase, as was suggested by the earlier sand target experiments. This observation is now generally recognized under the name "proportional growth" and agrees well with computer modeling results. While no new measures of crater diameter came out of this work, the experiments did show very clearly the stratigraphic uplift observed in terrestrial complex craters and underlined the large-scale smoothness of the collapse flow, which restored rock layers deeply underlying the crater to nearly their pre-impact locations, despite substantial initial displacement by the excavation flow.

Geologic Investigation

Detailed structural geologic mapping of terrestrial impact craters did not begin in earnest until the 1960s and 1970s. As a result, much of the initial interpretation drew analogies with the explosive detonation studies. References to true and apparent

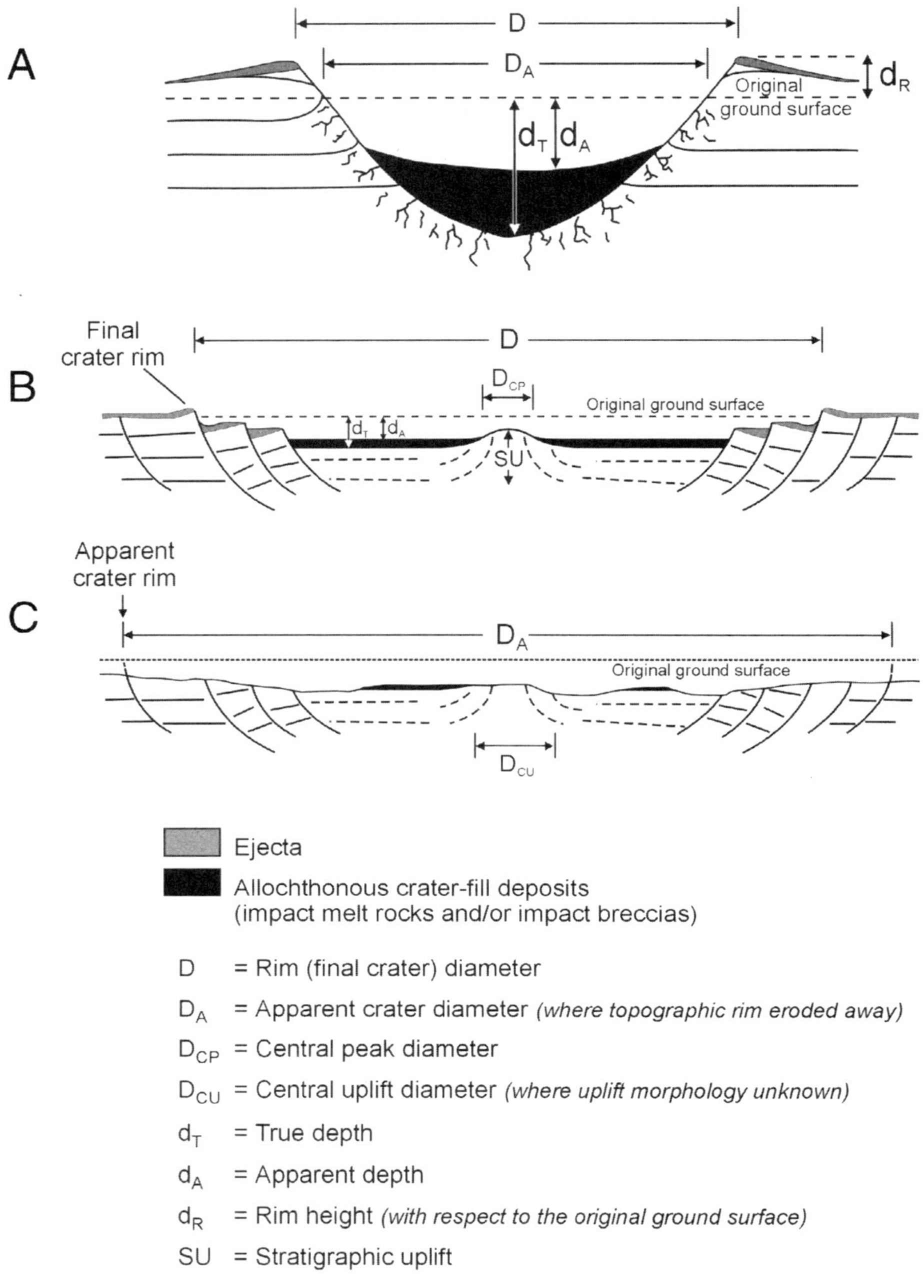

Figure 6. Series of schematic cross sections illustrating the various diameters and depths associated with hypervelocity impact craters. (A) For a fresh simple crater, the rim-to-rim diameter, D, is obvious; however, with increasing amounts of erosion, it will only be possible to measure the apparent diameter, D_A. Note that the apparent diameter should be determined with respect to the original pre-impact surface, which in deeply eroded simple craters may be difficult. Rim-to-floor depth, which is often quoted for craters, is the sum of d_R and d_A. (B and C) Schematic cross sections of a fresh (B) and eroded (C) complex impact structure. Due to erosion, it is often difficult to identify which of the inward facing faults in the eroded crater (C) corresponds to the final crater rim, D, in the original crater. Scaling from the apparent diameter, D_A, should be done cautiously. For complex structures eroded below the crater-fill deposits, it will not typically be possible to measure the central peak diameter, D_{CP}, but rather the diameter of the central uplift, D_{CU}. In this case, an estimate of the depth of erosion is important because the central uplift diameter will change considerably with depth (cf. the amount of measured stratigraphic uplift, SU).

crater diameters appear frequently in the geologic cratering literature from this time. Many impact craters studied, however, were significantly larger than any explosion craters and so several new metrics for impact craters were born.

The problem of defining crater diameter is particularly difficult for terrestrial craters because there is often significant post-impact erosion or tectonic overprinting, which may hinder morphological interpretation (especially for the oldest and largest structures). Structures in the basement rock, such as small fault offsets that would be concealed by thick ejecta deposits at fresh craters, may be exposed by erosion, further complicating interpretation. Thus, not only is there the problem of extrapolating the positions of observed features back to the pre-impact surface (which may not be well constrained), but there is also the difficulty of interpreting which structural feature corresponds to the final crater rim as defined by astronomical measurements of fresh craters (Figs. 6B and 6C). Few fresh terrestrial craters exist that can provide a basis for the correlation of structural features with surface morphology. We are still ignorant, for example, of how much structural deformation is hidden beneath the thick ejecta blanket immediately outside of the final crater rim of a complex crater. A geologist in the field is able to map out faults with only a few meters displacement. Would such faults be perceptible through the ejecta blanket around a fresh impact crater?

One of the most widely cited studies of terrestrial impact structures (Grieve et al., 1981) used the absence of preexisting target material within the crater to define the diameter of the *excavation cavity* (and, hence, the transient crater). This diameter was found to be related to the "apparent diameter" of the impact structure. Most of the apparent crater diameters from this study and subsequent geologic mapping of terrestrial craters are inferred from some significant structural feature that forms a semi-continuous ring around the crater center (Grieve et al., 1981). This feature is variously the outermost major fault with substantial offset along it, the most distal concentric occurrence of pseudotachylite, as at Sudbury (Thompson and Spray, 1994), or some other circular feature that would have been disturbed in connection with the impact. The problem is that these disparate features may not correspond to the same morphologic aspect of the fresh crater. For example, the features described above may correspond to the rim-to-rim diameter, the diameter of an external ring, or some other limit of deformation that has no direct analogue in astronomical studies. Furthermore, these diameters often need to be adjusted to account for the amount of erosion that is inferred to have taken place since the impact or for tectonic overprints that may have distorted the shape of the original crater (as observed, for example, at both Sudbury and Vredefort), thus compounding the uncertainty in associating structures documented in situ with those observed by remote sensing (i.e., between terrestrial and extraterrestrial craters).

In addition to using structural features to reconstruct the rim diameter, field studies of terrestrial craters have shown that the original impact melt volume scales with transient crater diameter (Grieve and Cintala, 1992); therefore, it may be possible to estimate the transient cavity diameter based on the amount of melt preserved. However, as Grieve and Cintala (1992) noted, the terrestrial impact record is still poorly documented, and estimating the original melt volume is difficult for the majority of craters that have been eroded to some degree.

Geophysical Investigation

Impact craters commonly exhibit strong geophysical signatures (for an overview, see Pilkington and Grieve, 1992). Most craters are characterized by a gravity low whose diameter increases with increasing crater size. Larger craters may also have a central gravity high (e.g., Manicouagan, Chicxulub, and Vredefort) that is most likely caused by structural uplift of denser rocks (e.g., Henkel and Reimold, 1998). Magnetic signatures at impact craters can originate from deposits of suevitic impact breccias (as in the Ries crater), impact melt rocks (as at Sudbury), and/or impact-induced thermal overprint of stratigraphically uplifted material (as at Vredefort; Henkel and Reimold, 2002). Magnetotelluric data from Chicxulub and the Ries crater indicate that there are zones of high conductivity beneath some impact structures that are probably related to the presence of fluids within the porous impact breccias and fractured crater floor. Pilkington and Hildebrand (2003) have recently shown that magnetic data can be used to place constraints on the diameter of the transient crater, which can be related to apparent crater diameter. However, there is considerable scatter in these data and the results should, therefore, be treated with caution until they can be confirmed using other parameters.

Seismic reflection and refraction data sets across impact craters provide structural and lithological information, which is particularly useful when the crater is buried. However, the acquisition of seismic data is expensive, and the data sets that are currently available for impact structures are variable in quality. The best seismic images of a crater to date have been obtained from three-dimensional tomographic refraction data across the Chicxulub crater (Morgan et al., 2002a). Seismic reflection data are particularly good at locating subvertical faults within the target rocks but, as in geological mapping of faults at surface, there is always some ambiguity in relating these to original morphological expression.

Although estimates of crater diameter that are obtained from geophysical data alone may well be accurate, they should always be interpreted with a degree of caution. Individual anomalies are strongly dependent on the geophysical properties of the target rocks, how these properties change with depth, and the extent of erosion, burial and post-impact deformation. Furthermore, there is an inherent non-uniqueness in modeling geophysical data; in many instances, multiple models can reproduce the observations equally well.

Numerical Modeling

With the advent of computers, numerical simulations have been employed to study the dynamic evolution of impact craters (e.g., Bjork, 1961; Kreyenhagen and Schuster, 1977; Anderson, 1987; O'Keefe and Ahrens, 1993; Melosh and Ivanov, 1999; Pierazzo and Melosh, 2000; Pierazzo and Collins, 2003; Collins et al., 2004). In principle, such models can record any desired measurement of a crater's shape; however, even today, resolution limitations make the final crater morphometry somewhat ambiguous. For example, no simulation to date has accurately reproduced terrace slumping of a crater wall, because the resolution required to simulate strain localization on this scale (meters to tens of meters) is prohibitively high. Instead, the numerical simulations model more distributed flow of the collapsing transient crater rim (e.g., Fig. 1). In such simulations it can be difficult to distinguish the exact location of the rim diameter, as no clear fault scarp is discernable. Estimates of rim diameter from numerical modeling, therefore, should be treated as approximations. On the other hand, numerical simulations can accurately predict crater dimensions at intermediate steps, so they can be used to predict the form, depth, and diameter of the opening and collapsing cavity.

Results from numerical simulations of impact events suggest two further metrics of a crater's size that may relate to observable parameters (see Fig. 1E): the diameter of the heavily-disturbed zone and the diameter of the damaged zone. The heavily-disturbed region is the portion of the target that deforms during collapse; it has the form of a shallow bowl with a broad lip (pink line in Fig. 1E), and is characterized by very high strain (>0.1). This zone may correspond to the source of the broad gravity, magnetic, and seismic-velocity anomalies associated with terrestrial impact structures (e.g., Pilkington and Hildebrand, 2003; Collins et al., 2004). The damaged zone corresponds to the spatial extent of fracturing predicted by the numerical simulation. This region may correspond to the "true crater" defined from nuclear explosion experiments. The limit of this region (dashed yellow line in Fig. 1E) is hard to define because of strain localization and is very sensitive to the constitutive equations used in the numerical model. Delimiting this zone from observation may be impossible due to the difficulty of distinguishing impact-related fracturing from non-impact fracturing outside of the final crater rim.

RECOMMENDED TERMINOLOGY

The somewhat different measures of crater dimensions described above are, generally, unambiguous within their respective areas of use. However, directly comparing measurements of crater size based on disparate data sets can lead to misinterpretation. For example, it is misleading and hazardous to apply the astronomically derived definition of rim-to-rim diameter to dimensions of eroded or only partially exposed terrestrial craters without great care in making sure that the same measurements are being applied to similar features. To avoid the kinds of semantic misunderstanding that have plagued this field, it is necessary to be as explicit as possible. Here, we make recommendations for specific definitions that correspond to several common measurements of crater-feature diameters (see also Fig. 6). If a feature is of unknown origin or category, we recommend that this be clearly stated. In the cases of heavily-modified craters, the degree of degradation also needs to be documented as well as possible. For example, the measured diameter of a depression that corresponds to an eroded impact crater, or the measured diameter of an albedo or surface-roughness feature that corresponds to a viscously relaxed crater, must be carefully explained and should not be over interpreted, especially in comparison with parameters that are better constrained.

Transient crater diameter: Defined as the diameter of the rim of the idealized transient crater measured at the pre-impact surface; that is, the maximum radial extent achieved by the opening cavity before rim collapse begins. The diameter of the transient crater is equivalent to the diameter of the excavated zone of target material (Fig. 1).

Rim (or final crater) diameter: Defined as the diameter of the *topographic* rim that rises above the surface for simple craters or above the outermost slump block not concealed by ejecta for complex craters (Fig. 6). The physical meaning of the rim-to-rim diameter is different in simple and complex craters as it refers to the outermost steep scarp that is recognizable, which is not necessarily the same position with respect to the transient crater diameter. The rim diameter is relatively easy to measure on most planetary bodies where the topographic rim is usually preserved due to low rates of erosion. On Earth, however, such pristine craters are rare. It should be noted that this rim-to-rim diameter is also the value normally quoted in computer modeling studies (e.g., Turtle and Pierazzo, 1998).

Apparent crater diameter: Defined as the diameter of the outermost ring of (semi-) continuous concentric normal faults, measured with respect to the pre-impact surface (i.e., accounting for the amount of erosion that has occurred). For the majority of terrestrial impact structures this will be the only measurable diameter. As discussed previously, it is not clear in all cases how this apparent diameter is related to the rim-to-rim diameter. Furthermore, some measurements of apparent diameter (e.g., Grieve et al., 1981) have been made with respect to the present surface, rather than accounting for erosion (although extrapolation to the location of the surface immediately after the impact surface can be fraught with uncertainty).

Central peak diameter: The diameter of a central peak measured at the contact with the surrounding crater floor (i.e., measured at the top of the allochthonous crater-fill breccias and melt rocks). For eroded craters, the central uplift diameter is less easily defined because distinguishing uplifted material depends on the availability of stratigraphic markers and their positions relative to the amount of vertical uplift and the amount of erosion.

Peak-ring diameter: Defined as the diameter of the peak of the often discontinuous, internal, mountainous ring that rises *above* the crater floor. To unequivocally call a ring-like feature in an eroded terrestrial structure a peak-ring, it must be proved that this ring originally protruded through the crater-fill deposits in the fresh crater. Where this is not possible, it is suggested that the term "inner" ring be used.

External ring diameters: Defined as the diameters of the crests of semi-continuous to continuous, concentric fault scarps external to the final crater rim. Structures outside the rim are not always apparent (e.g., they can be easily concealed by ejecta blankets or post-impact sedimentary or volcanic deposition). Furthermore, in some cases (e.g., Chicxulub; see subsequent discussion), external rings can stand topographically higher than the structural rim. However, for the purposes of crater scaling, the origins of features are more relevant than simply their elevations, so topography alone should not influence the identification of the final rim.

Outer limit of deformation: Defined as the maximum distance at which *impact-related* deformation can be identified. This metric clearly depends heavily on the types of data that can be acquired; only large features can be identified from orbit, whereas even faults with little offset can be identified if they are preserved in the terrestrial geologic record.

TERRESTRIAL EXAMPLES

To illustrate the difficulty in defining morphological features at a given terrestrial impact structure (in particular the rim diameter) we discuss four large terrestrial craters for which a certain degree of controversy has surrounded the issue of crater "size."

Chicxulub

The Chicxulub structure (Hildebrand et al., 1991), on the Yucatán Peninsula, México, was formed 65 Ma (Swisher et al., 1992). It is well preserved because the impact occurred on a stable carbonate platform and the crater was subsequently rapidly buried by ≤1 km of Tertiary sediments. Thus, the structure experienced little erosion; however, its burial makes access difficult. The geological structure was initially identified on the basis of circular geophysical anomalies during oil surveys in the 1950s (Cornejo-Toledo and Hernandez-Osuna, 1950). Its impact origin was only confirmed much later, through a combination of geophysical (Penfield and Camargo-Zanoguera, 1991), petrologic (including the identification of shocked quartz; Bohor et al., 1984), and geochemical analyses (Hildebrand et al., 1991; Kring and Boynton, 1992). Studies of the crater's morphology have been hampered by its burial (e.g., Sharpton et al., 1993; Pilkington et al., 1994; Morgan et al., 1997; Grieve and Therriault, 2000), and its size has been the source of heated discussions until recently.

The size of the Chicxulub structure (Table 1) was initially inferred to be 180–210 km. This diameter corresponds to a 30 mgal negative Bouguer gravity anomaly and a small (5–20 nT), short-wavelength magnetic anomaly (Penfield and Camargo-Zanoguera, 1981, 1991; Hildebrand et al., 1991), which lies outside a ~90-km-diameter, high-amplitude (100–1000 nT) magnetic anomaly. Sharpton et al. (1993) identified three to four rings in the gravity data and interpreted Chicxulub to be a ~295-km-diameter multi-ring crater formed from a 170-km-diameter transient crater (which would place the transient crater rim between the second and third rings). Camargo-Zanoguera and Suarez-Reynoso (1994) noted that the "internal crater" had a diameter of ~170 km from a combined analysis of gravity, magnetic, and seismic data, but did not come to a conclusion on a rim diameter. Kring (1995) used the physical properties of the K-T boundary impact ejecta and Chicxulub impact melt sheet to test the geophysical estimates of the crater's size, concluding that ~180 km is the best estimate for the diameter of the final crater rim. Independently, Espindola et al. (1995) inferred the presence of a single central peak in the Chicxulub structure and estimated the structure's diameter to be 200 km, based on an analysis of both gravity and magnetic data. Hildebrand et al. (1995) reanalyzed the gravity data set with five new, closely spaced, onshore gravity profiles to the southwest of the crater center. They examined the horizontal derivative of the Bouguer gravity anomaly (which emphasizes the effects of lateral density changes and suppresses regional gradients and broad features) over the structure, revealing several circular features, the most prominent of which is a steep gradient at a diameter of ~170 km that correlates with a ring of cenotes (sinkholes in the carbonate bedrock). No circular gradient features were evident beyond a radius of 90 km, so Hildebrand et al. (1995) concluded that the crater was ~180 km in diameter. The initial diameter estimate of Pope et al. (1993), 240 km, was later revised to 260 km based on surface topography across the crater (Fig. 7A; Pope et al., 1996). On a profile to the southeast of the crater, they noted a prominent depression that coincided with the ring of cenotes (~166 km diameter).

In 1996, the British Institutions Reflection Profiling Syndicate (BIRPS) conducted a marine seismic reflection survey, which provided a direct structural image of faulting and deformation within the target sediments along four offshore profiles (Morgan et al., 1997). From these data, Chicxulub was interpreted to be a large multi-ring crater, with three distinct types of rings: a peak ring, a crater rim, and at least one, possibly two, outer rings (Morgan and Warner, 1999). A schematic representation of the subsurface structure of Chicxulub is shown in Figure 7B. The peak ring is an almost symmetric topographic feature, rising from the crater floor with a diameter of ~80 km. The crater rim is the head scarp of the outermost slump-block (the second identified ring) with a diameter of ~150 km, which on some seismic profiles coincides with the edge of the basin in the Tertiary units. The outer ring has a diameter of ~195 km, which is also the diameter of the gravity low. The possible second outer ring has a diameter of ~250 km. The seismic data set was used to reconstruct the transient crater, which is interpreted to have been ~100 km in diameter, roughly coincident with the outer edge of the peak ring and the limit of the central magnetic anomaly (Morgan et al., 1997). This estimate of the transient crater diameter is particularly accurate because the seismic reflection profile clearly indicates the sharp, inward termination of flat-lying, pre-impact strata that define the excavation cavity and thus, after palinspastic reconstruction of the inward-slumped blocks, the transient crater diameter. The locations of rings in the offshore seismic data show a reasonable correlation with the onshore topography (Fig. 7), as well as with the gravity and magnetic data.

The combination of geophysical data sets has produced some consensus regarding the size of Chicxulub. There is now general agreement that Chicxulub is a multi-ring structure (e.g., Sharpton et al., 1993; Morgan et al., 1997) and that the rim diameter was 180–200 km and the transient crater diameter was 80–110 km (Hildebrand et al., 1998; Morgan et al., 2002b; Dressler et al., 2003). Strict adherence to the definition we propose above gives a rim diameter of 150 km. The controversy that raged over Chicxulub's size is a powerful illustration of the problems we highlight in this paper. For example, it is not clear that the rings observed in the seismic data would have been evident at the surface when the structure was fresh, making direct comparison between Chicxulub and large multi-ring basins on the Moon and Venus problematic. Furthermore, although the external rings in multi-ring craters appear to be formed by a different process than that which produces the final crater rim, we do not understand the formation of external rings well enough to provide unequivocal guidelines for identifying this type of ring. It is not surprising, therefore, that the "outer" and "exterior" rings in Figure 7B have both been interpreted to be the final crater rim. Unfortunately, until an accepted model for external ring formation exists, there will

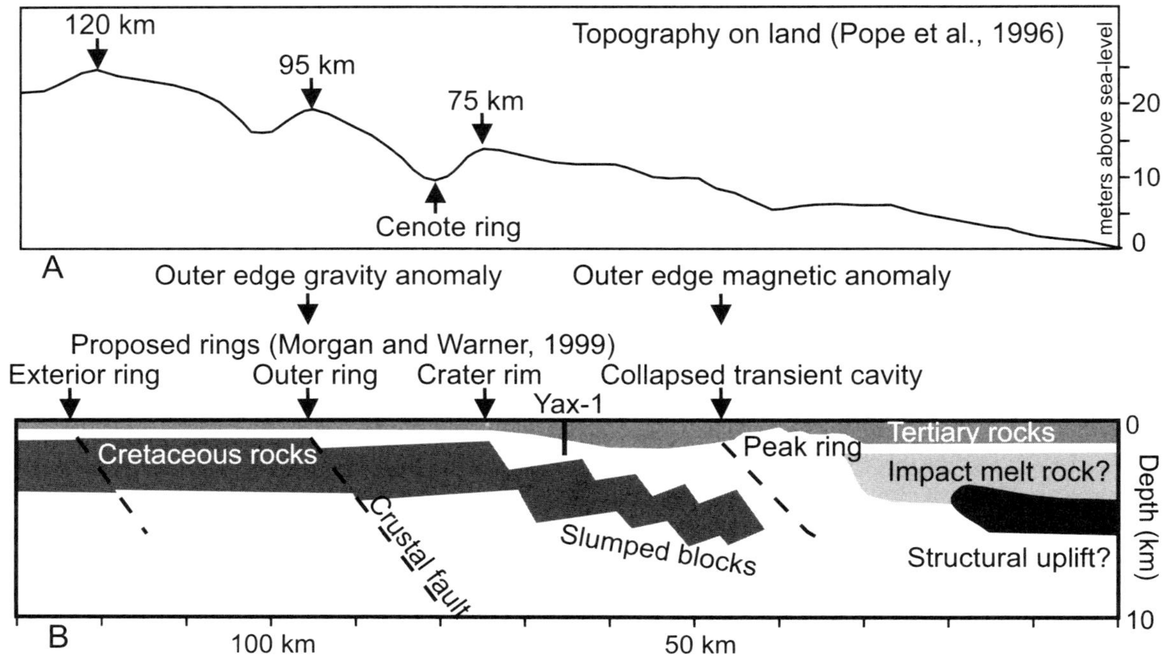

Figure 7. (A) Onshore topographic data along a bearing of 120° E from the center of the Chicxulub structure, redrawn from Pope et al. (1996). (B) A cartoon of the Chicxulub crater, after Morgan et al. (2002a). Mapping of the Tertiary and Cretaceous rocks is from four offshore seismic reflection profiles (e.g., Morgan and Warner, 1999). Structures within the central crater region are based on velocity data obtained from offshore/onshore three-dimensional tomographic seismic refraction data (Morgan et al., 2002a). Yax-1 indicates the location of a CSDP borehole. The edge of the 30 mgal gravity low correlates well with the outer ring. On two seismic profiles, the outer edge of the peak ring is coincident with the edge of the high-amplitude, 90-km-diameter, magnetic anomaly.

always be some ambiguity in distinguishing the final crater rim from external rings. Our recommendation is that this ambiguity be recognized and reported wherever interpretations are made.

Disagreement over the size of the Chicxulub crater was often the result of different interpretations of the geophysical and topographic data, as well as different definitions of the diameter of a complex crater. Such discrepancies have been responsible for significant uncertainties in estimating the transient crater size, and thus the impact energy, which is essential to understanding the environmental perturbations produced by this event, to which the Cretaceous-Tertiary mass extinction has been attributed (Sepkoski, 1994).

Sudbury

Of the three largest recognized impact structures on Earth, the 1.85-Ga Sudbury impact structure in Ontario, Canada, is particularly important because it is the best exposed and, although it has undergone erosion (e.g., Thompson et al., 1998; Grieve and Therriault, 2000), it is mostly intact. Sudbury has undergone a complicated tectonometamorphic evolution since its formation (e.g., Milkereit et al., 1992; Wu et al., 1995; Fueten and Redmond, 1997; Thompson et al., 1998; Spray et al., 2004). Nevertheless, it possesses a complete sequence from impact-damaged basement rocks, through footwall breccias and a ~2.5-km-thick impact melt sheet (the Sudbury Igneous Complex), to fallback/flowback breccias (the ~1.5-km-thick Onaping Formation) and overlying post-impact sedimentary cover (the Onwatin and Chelmsford Formations; Fig. 8) (Grieve and Therriault, 2000; Spray et al., 2004). The post-impact folding and thrusting deformed the Sudbury Igneous Complex into an elliptical shape (e.g., Wu et al., 1995). Four subconcentric, pseudotachylitic-breccia–rich zones have been identified (e.g., Dressler, 1984; Müller-Mohr, 1992; Butler, 1994; Spray et al., 2004).

An impact origin for Sudbury was first proposed by Dietz (1964) based on the presence of shatter cones and subsequently confirmed by French (1968) on the basis of planar deformation features (PDFs) in quartz. These early workers suggested a rim diameter of ~55 km based on the geometry of the Sudbury Igneous Complex. Subsequent studies revealed that the actual impact structure extended significantly beyond the Sudbury Igneous Complex, with estimates of ~140–200 km proposed (Table 2) (Peredery and Morrison, 1984; Lakomy, 1990; Grieve et al., 1991). It should be noted that these estimates are based on empirical relations derived at other large impact structures between various structural and shock metamorphic features (Table 2); i.e., the various diameters quoted were not correlated with features observed in the field. Higher estimates for Sudbury's rim

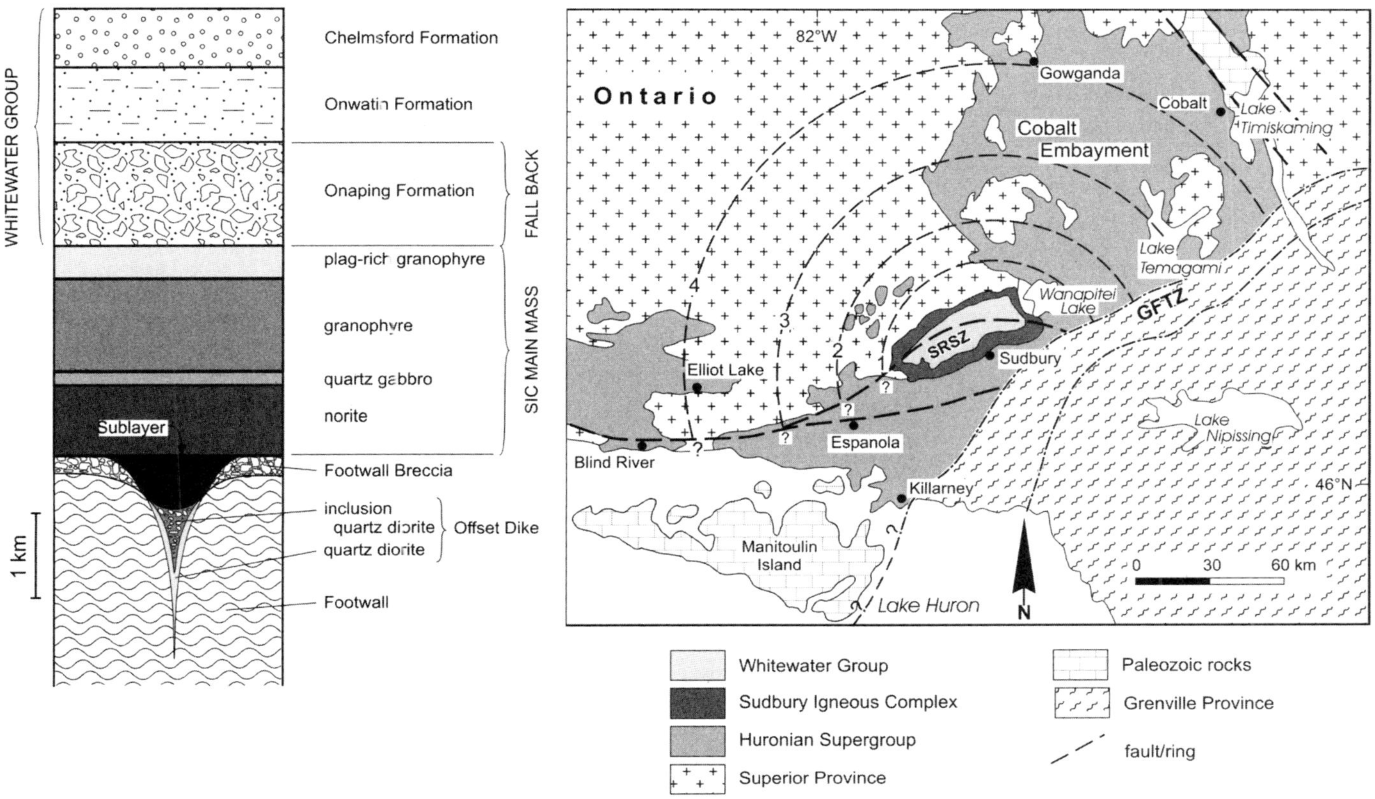

Figure 8. Simplified stratigraphic section of the Sudbury impact structure, from basement to overlying sedimentary cover. SIC—Sudbury Igneous Complex. The map shows concentric rings superimposed on the regional geology. SRSZ—South Range Shear Zone (Penokean Front); GFTZ—Grenville Front Tectonic Zone. The ring systems are obscured by post-impact tectonics to the south of SRSZ-related faults and the GFTZ. Modified after Spray et al. (2004).

diameter, ranging from ~220–280 km (e.g., Stöffler et al., 1989; Grieve, 1994; Deutsch et al., 1995), have also been suggested. The wide range of proposed values reflects ambiguities surrounding some of the original dimensions of the Sudbury structure, in particular the Sudbury Igneous Complex.

Over a century of geological and economic exploration, as well as impact-specific investigations carried out over the last few decades, have made it possible to crudely reconstruct the morphology of the Sudbury impact structure. Spray et al. (2004), using lineament analysis based on Landsat Multispectral Scanner data and regional geophysical data, document the presence of four concentric rings associated with the Sudbury structure (Fig. 8) (Butler, 1994; Spray et al., 2004). These rings occur at diameters of ~90 km (Ring 1), ~130 km (Ring 2), ~180 km (Ring 3), and ~260 km (Ring 4). Rings 1, 2, and 3 broadly correspond to zones of enhanced endogenic (E)-type pseudotachylite development (Dressler, 1984; Müller-Mohr, 1992; Thompson and Spray, 1994; Spray and Thompson, 1995; Wood and Spray, 1998; Spray et al., 2004) and thus have been hypothesized to be related to concentric fault systems that experienced very large displacements (Spray, 1998). The region between Rings 1 and 2 consists of an annular trough ~20 km wide. Within this trough, rocks have been down-faulted relative to those within Ring 1 (Tuchscherer and Spray, 2002). Rocks within Ring 1, which have experienced shock pressures between 5 and 10 GPa (Gibson and Spray, 1998; French, 1998), are interpreted to constitute the central uplift. Reconstruction of this deformed core, based on extrapolation from the more intact northern margin of the structure, indicates a central uplift diameter, D_{cu}, of ~90 km. The central uplift and annular trough are believed to roughly coincide with the transient crater, although some collapse may have occurred along the inward-facing walls of Ring 2 during the modification stage. It is suggested that the impact melt sheet would have originally occupied much of this area (Tuchscherer and Spray, 2002). Rings 3 and 4 lie outside the collapsed transient cavity, but it is not clear which, if either, coincides with the original final crater rim. Given that Ring 4 is the most distant recognizable feature found to date associated with the Sudbury impact structure, it represents the outer limit of known impact-related deformation. However, due to the lack of ground-truth information from the field (no attempt has yet been made to search for pseudotachylite in association with Ring 4 [Spray et al., 2004]), the significant amount

TABLE 2. DIAMETERS REPORTED FOR SUDBURY

Diameter (km)	Feature (if specified)	Observation or method of determination	Reference(s)
57	Apparent diameter*	Outer margin of SIC	Dietz, 1964, 1972; French, 1968
100		Extent of central uplift	French, 1970; Dence, 1972; Guy-Bray; 1972
100–120 155–240	Transient crater Apparent crater	Outer extent of breccias, SIC, shatter cones, and location of the ~7.5 GPa isobar	Stöffler et al., 1989
180–200	Final crater rim	Outer limit of shatter cones and PDFs in quartz as compared to other terrestrial impact structures	Lakomy, 1990
100–115 150–200†	Transient crater Final crater rim	Outer limit of shatter cones and PDFs in quartz	Grieve et al., 1991
100–130 175–228†	Transient crater Final crater rim	Outer limit of shatter cones, PDFs in quartz, and the SIC	Grieve, 1994
100–110 204–280†	Transient crater Final crater rim	Outer limit of shatter cones, PDFs in quartz, and the SIC	Deutsch et al. 1995
~90 ~130 ~260	Central uplift Transient crater External ring	Field-based mapping, Landsat MSS analysis and regional geophysics	Spray et al., 2004

Note: SIC—Sudbury Igneous Complex.

*Dietz (1964) placed the "apparent crater rim" at ~57 km diameter; however, from his Figure 10, using our definitions, this metric actually corresponds to the topographic crater rim and is, therefore, the rim diameter.

†These estimates for crater diameter are based on empirical scaling relationships (i.e., the diameters chosen were not correlated by features seen in the field).

of erosion, and post-impact tectonic deformation, it is not clear whether this outer ring structure corresponds to the rim diameter, the apparent crater diameter, or an external ring.

Vredefort

The Vredefort structure (see Gibson and Reimold, 2001, and references therein) is centered near the geographic midpoint of the Witwatersrand Basin in South Africa and consists of two major structural features: a central region of strongly uplifted Archean basement rocks, known as the Vredefort Dome, and a series of synclines and anticlines that encompass the Witwatersrand Basin. The Vredefort Dome comprises an Archean basement core, 40–50 km in diameter, which is surrounded by a ~20-km-wide ring of upturned and, in places, overturned supracrustal strata (Fig. 9). The structure's origin was debated for decades after it was described in detail by Hall and Molengraaff (1925) and Nel (1927). However, following the discoveries of coesite and stishovite (Martini, 1978) and a meteoritic component (Koeberl et al., 1996b) in the Vredefort Granophyre, which is believed to represent the remnant of the impact melt, and the confirmation of true PDFs in quartz (Leroux et al., 1994) and planar shock features in zircon (Kamo et al., 1996), Vredefort has become generally accepted to be an impact structure. The asymmetry of the Witwatersrand Basin could be the result of post-impact tectonic deformation coupled with regional tilting (Friese et al., 1995; Henkel and Reimold, 1998) or, alternatively, the regional stratigraphy may not have been horizontal at the time of impact (Lana et al., 2003). In the 2 b.y. since the impact (Spray et al., 1995; Kamo et al., 1996), between 5 and 11 km of material has been eroded from the central part of the structure (McCarthy et al., 1990; Gibson and Wallmach, 1995; Stevens et al., 1997; Therriault et al., 1997; Gibson and Reimold, 1999, 2001; Gibson, 2002), providing an opportunity to observe the deep structure of the central uplift of a very large impact crater.

The substantial erosion at Vredefort has removed all of the original surface features and, indeed, much of the crater structure, making it difficult to determine the diameter of the final crater rim and whether such features as a peak ring or external rings were present. The earliest estimate for the diameter, >50 km (Table 3), was made by Daly (1947) in the landmark paper in which he supported the idea of an impact origin for the structure. Clearly, this figure is directly related to the diameter then determined for the Vredefort Dome. Bishopp (1941) referred to a "granite plug" ~42 km in diameter. Grieve and Pesonen (1992) give a diameter of 140 km; however, how this value was derived is not clear. Based on a comparison of Corner et al.'s (1990) gravity profile of the Witwatersrand Basin, which indicates a 100-km-diameter central uplift, to central uplifts at other impact structures, Reimold and Wallmach (1991) reported that the Vredefort crater could have originally had an apparent diameter of 300–400 km. Rondot (1994) correlated the 102 km diameter of the annular syncline around Vredefort with the width of the ring graben seen around Charlevoix and estimated an original crater diameter of 160 km. Scaling from the 36 km of vertical uplift estimated by Hart et al. (1991), Grieve and Masaitis (1994) derived a crater diameter of 335 km. This value is consistent with their calculation of a diameter of 300 km based on scaling from a 75-km-diameter region of central uplift. However,

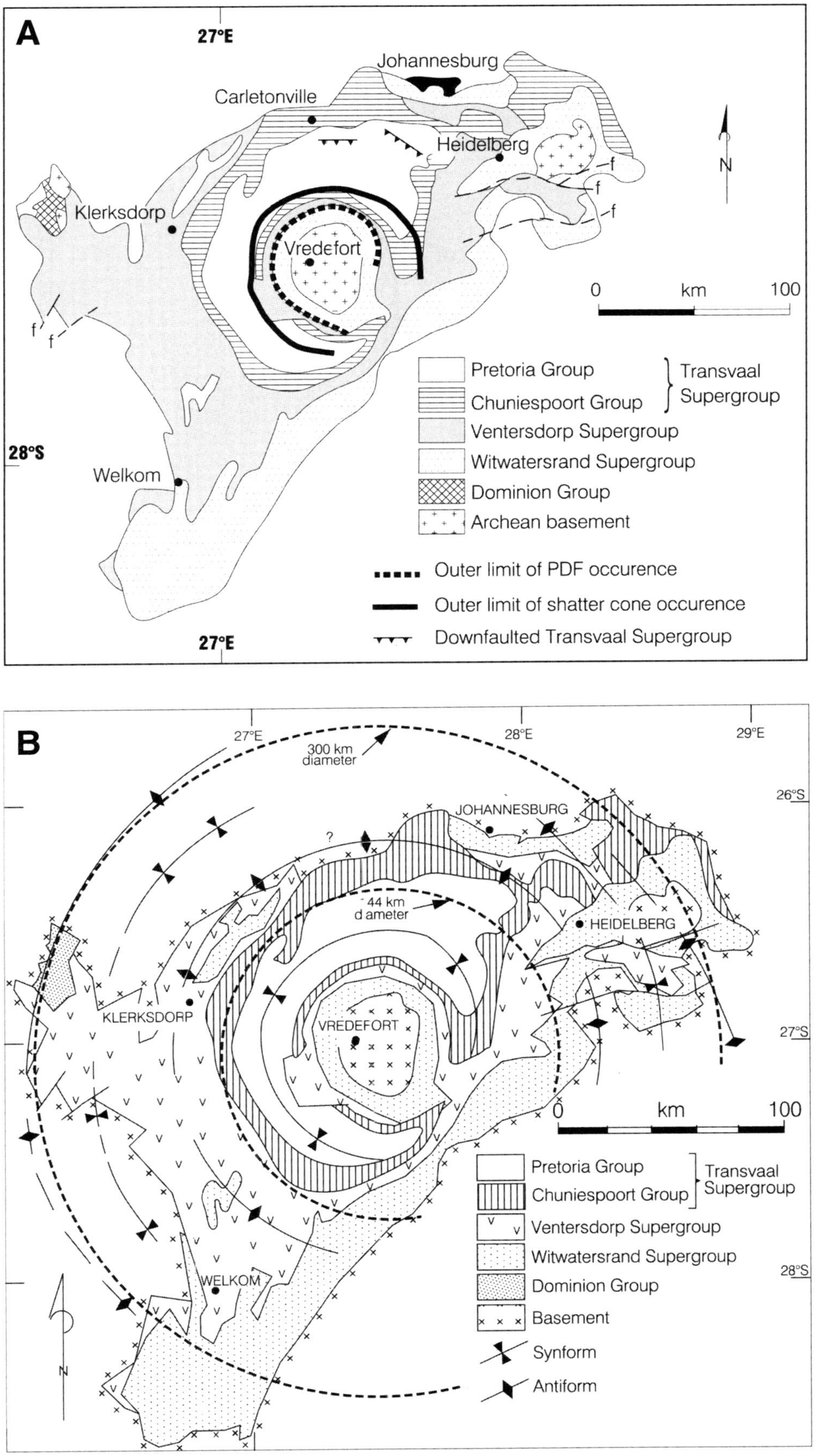

Figure 9 (*on this and following page*). (A) Geology of the Witwatersrand Basin around the Vredefort Dome, indicating the outer limits of shatter cone and planar deformation feature (PDF) occurrences (after Therriault et al., 1997), as well as downfaulted Transvaal Supergroup strata. Recent field work by F. Wieland and W.U. Reimold has revealed shatter cones significantly further north, up to ~10 km south of Carletonville. (B) Antiformal and synformal structures in the region of the Witwatersrand Basin. The northeast-southwest extension of the basin would match a 300 km diameter for the impact structure, but post-impact tectonic deformation would be necessary to explain the much shorter northwest-southeast dimension (e.g., Friese et al., 1995; Henkel and Reimold, 1998).

C

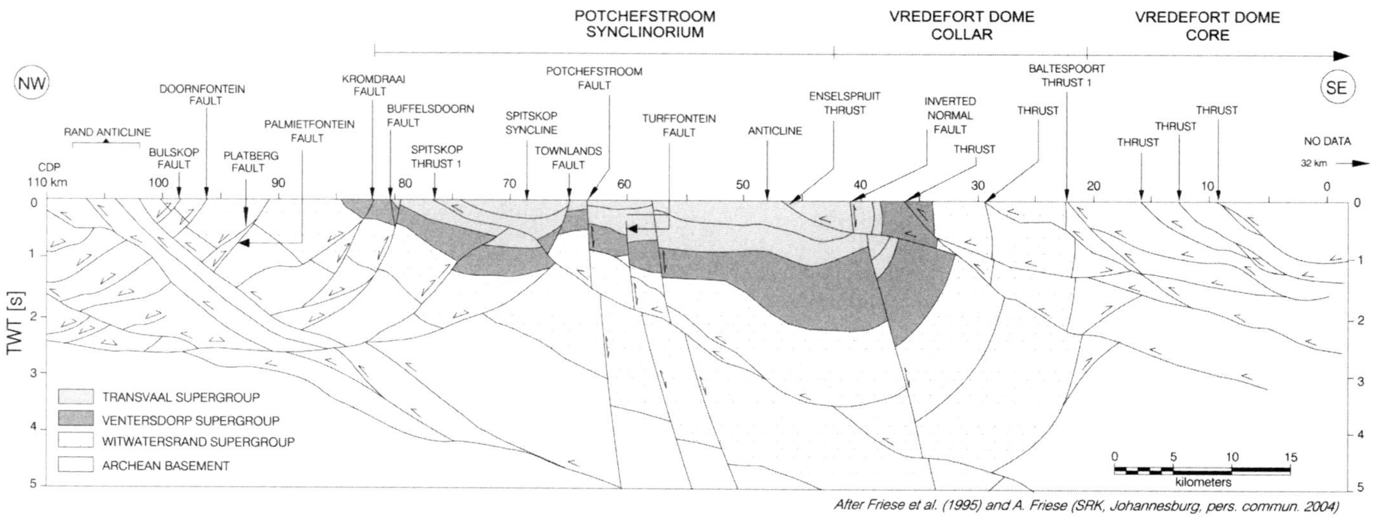

Figure 9 (*continued*). (C) Detailed interpretation of the Trans-Witwatersrand Basin seismic reflection profile (Durrheim et al., 1991) across the northwest Witwatersrand Basin and northwestern collar of the Vredefort Dome (modified after a profile courtesy of A. Friese, SRK Consulting, Johannesburg). Reflection seismic profiles HB-02W-89 and HB-02E/-02 EXT were also utilized in the production of this section (see also Friese et al., 1995). TWT—two-way traveltime, in seconds.

TABLE 3. DIAMETERS REPORTED FOR VREDEFORT

Diameter (km)	Feature (if specified)	Observation or method of determination	Reference(s)
>50	–	Exposed surface structures	Daly, 1947
300–400	–	Scaling from 100 km diameter central uplift	Reimold and Wallmach, 1991
160	–	Scaling from annular syncline	Rondot, 1994
335	–	Scaling from amount of vertical uplift	Grieve and Masaitis, 1994
300	–	Scaling from 75 km diameter central uplift	
121–152 200–300	Transient crater Final crater rim	Scaling from locations of shock features and downfaulted rocks	Therriault et al., 1997
285	Final crater rim	Scaling from vertical uplift	
284	Final crater rim	Scaling from 95 km diameter central uplift	
225–280	Final crater rim	Scaling from pseudotachylite dikes	
125 250–280	Transient crater Final crater rim	Geophysical models constrained by gravity, magnetic, and seismic data	Henkel and Reimold, 1998
200–300	–	Central uplift temperatures	Gibson, 2002
80–100	Transient crater	Shock pressure models constrained by observations	Turtle and Pierazzo, 1998
120–200	Structural rim		
~80	Transient crater	Impact heating models constrained by observations	Turtle et al., 2003
120–160	Structural rim		

Henkel and Reimold (1998) determined that the maximum amount of vertical uplift that could have occurred at Vredefort was 25 km. Therriault et al. (1997) compared the observed locations of shatter cones, PDFs in quartz, downfaulted original cover rocks, the outer limit of the central uplift, and concentric synclines and anticlines in the region of the Witwatersrand Basin to similar features at other complex impact structures and estimated that the (presumably final crater rim) diameter was between 270 and 300 km. Henkel and Reimold (1998) performed integrated geophysical modeling constrained by surface geology to demonstrate that cross sections of the Witwatersrand Basin could be consistent, stratigraphically and structurally, with a 250 to 300-km-diameter impact structure. In this model, the Vredefort Dome corresponds to the deeply eroded central uplift feature, and the Rand Anticline (the northern and northwestern margin of the Witwatersrand Basin) would represent the outer margin of the impact structure. Turtle and Pierazzo (1998) and Turtle et al. (2003) constrained predictions from numerical simulations using the observed distributions of shock features and impact heating around Vredefort and derived a transient crater diameter between 80 and 100 km, corresponding to a rim diameter between 120 and 200 km. Phillips et al. (1999) interpreted Landsat Thematic Mapper observations of arcuate ridges to indicate a diameter of 250 km, in agreement with the preferred value of Henkel and Reimold (1998). Most recently, W.U. Reimold discovered massive cataclasite on the southern slope of the Rand Anticline, ~90 km north of the center of the Vredefort Dome. This breccia has not been dated (indeed, the extensive alteration of the groundmass may preclude dating), but if linked to other massive breccias of Vredefort age in this region, its distance from the center of the dome may provide direct evidence for a minimum apparent crater diameter of 180 km, at the current depth of erosion.

Disagreement among the various diameter estimates has been attributed to a variety of factors, such as the assumption of different amounts of erosion and structural uplift, scaling from smaller craters, and assumptions of the shapes of low-pressure shock isobars. More recently, it has come to light that some of this discrepancy may lie instead in purely semantic differences in the use of the term "crater diameter"; in many cases, the term diameter is used with no specific definition of what exactly is meant (such ambiguities are unfortunately necessarily reproduced in the discussion above). In the modeling work of Turtle and Pierazzo (1998) and Turtle et al. (2003), the transient and rim diameters are explicitly used. However, in some other cases, the term diameter may actually refer to the apparent diameter or even to the maximum extent at which impact-related deformation is observed, an inconsistency that may resolve at least some of the disagreements.

Chesapeake Bay

The Chesapeake Bay impact structure was discovered as a result of an anomalous breccia layer in boreholes from the eastern seaboard of the United States (Poag et al., 1992, 1994; Powars et al., 1993). The breccia was correlated with the North American tektite strewn field (Poag et al., 1994; Poag and Aubry, 1995), and an impact origin was confirmed by the existence of shocked quartz and feldspar crystals (Koeberl et al., 1996a) and shocked zircon and reidite (Glass and Liu, 2001). At the time of impact (35.2–36.0 Ma), the crystalline basement was overlain by 600–1000 m of unconsolidated sediments, and the area was under 200–600 m of water. The structure was subsequently buried by ~500 m of sediments, but aside from erosion by the tsunami and backwash generated by the impact, it is well preserved.

Seismic exploration (Poag, 1996, 1997; Poag et al., 1994, 1999, 2002, 2004; Powars and Bruce, 1999) has revealed the various elements of the impact structure (Table 4): an inner basin in the crystalline basement that is ~38 km in diameter and more than 1 km deep; a possible central uplift ~5 km in diameter (Poag et al., 1999); a rather flat-floored, annular trough in the overlying sediments that is 20–30 km wide and as much as 600 m deep; and a series of faults and terraces in the sedimentary sequence with 300–1200 m of vertical offset, rising to a rim 40–45 km from the center of the structure (e.g., Figure 4.2A in Poag et al., 2004). Thus, the impact structure has the morphology of an inverted sombrero with a deeper basin in the crystalline basement surrounded by a shallower brim in the overlying sediments (Poag et al., 1994, 2004; Poag, 1996). Basement uplift of ~50–200 m at the margin of the inner crater was interpreted by Poag et al. (1994) to be a peak ring. Isolated faults with up to 200 m of offset are observed beneath the annular trough as well as outside of the outer rim to a diameter of ~160 km.

The rim diameter of the Chesapeake Bay crater is generally taken to be 85–90 km, the diameter of the outer rim. However, it is likely that the size of the outer rim has been enhanced relative to that of the inner rim due to the considerable contrast in material properties between the crystalline basement and the weaker overlying sediments (Poag et al., 2004). Furthermore, the basement uplift at the inner rim does not protrude above the crater floor. Thus, according to our strict definition, this feature is not a peak ring. Instead, the crater's morphology may be characteristic of marine impacts or impacts into targets consisting of rheologically weak material overlying more competent lithologies. If this is the case, the inner rim, with a diameter of ~38 km, may more closely represent the location of the final crater rim, were the crater to have formed entirely in a crystalline target. This idea is supported by recent numerical simulations of the dynamic development of the Chesapeake Bay impact crater (Collins and

TABLE 4. DIAMETERS REPORTED FOR CHESAPEAKE

Diameter (km)	Feature (if specified)	Observation or method of determination	Reference(s)
~38	Peak ring	Seismic survey	Poag et al., 1994
40	Transient crater	Seismic survey	Poag, 1996
~5	Central uplift	Seismic survey	Poag et al., 1999
30–40	Transient crater		Kenkmann, 2004
20–30	Transient crater		Hole et al., 2003
~85–90	Final crater rim	Seismic survey	Poag et al., 2004
~160	Outer faults		

Melosh, 2004; Crawford and Barnouin-Jha, 2004), which show that for a large enough contrast in material strength between the sedimentary and crystalline units, the sombrero morphology can be reproduced. Their results illustrate that the collapse of a ~25-km-diameter transient crater results in a deep inner basin, containing inwardly collapsed sediments and bounded in the basement by the remnant transient-cavity-rim uplift, and a shallower outer basin where deformation is confined to the sedimentary layer. The uncertainty surrounding the morphology of the Chesapeake Bay structure illustrates that target rheology, as well as erosion and multiple concentric rings, can complicate the interpretation of the final rim diameter.

SUMMARY

There are many ways to measure an impact structure; however, our current understanding of the cratering process makes it difficult to relate these observables to one another. In particular, the rim diameter of a crater is extremely difficult to determine if the crater has been eroded and/or is not well exposed, if the target has distinct rheologically variable layers, or if it has multiple concentric rings.

Terrestrial craters are never totally pristine: they are all eroded to some extent, and some have subsequently been buried. For the geologist in the field, identifying the original rim diameter of an impact structure is a complex task. If the topographic rim has been eroded away, one has to look to other structural features present in the target rocks. Outer, (semi-) continuous, ring structures can often be identified with detailed mapping (which is, unfortunately, lacking for many impact craters!). However, this measurement should not be interpreted as rim diameter unless strong evidence exists that the structural feature mapped corresponds to the fault scarp of the outermost slump block that was not concealed by ejecta. Detailed geologic mapping and high-resolution seismic reflection surveys can reveal crater-related fault systems that lie (or originally lay) beneath the ejecta blanket and that would not be identifiable in images of extraterrestrial craters. It is, therefore, difficult to directly compare an eroded terrestrial crater remnant with a pristine crater on another planet.

In situations where a crater is buried or geological outcrops are limited, geophysical surveys can provide significant information about the subsurface of an impact structure. The dimensions of anomalies derived from some of these geophysical techniques, such as gravity or magnetic surveys, are undoubtedly measures of a crater's scale. However, until a definitive link between structural elements of a crater and geophysical anomaly dimensions is found, it is not possible to determine an impact structure's rim-to-rim diameter from these techniques alone.

We still have much to learn about the impact process, particularly for the largest craters. Although gravitational collapse is the driving force behind both final crater rim and external ring formation, the processes that create these structures differ, and we do not understand the mechanism responsible for forming external rings well enough to provide unequivocal guidelines for identifying this type of ring. Furthermore, in situations where strong rheologic variations are present in the target, additional concentric features may form; in such situations, it becomes especially difficult to define and measure the rim diameter.

The central tenet of this paper is that interpreting the rim diameter of a terrestrial crater is more complicated than it might appear. Therefore, we stress that it is important to be as consistent and clear as possible when documenting crater measurements. Unqualified terms such as "crater diameter" or simply "diameter" are too ambiguous. The exact nature of the features being measured, or the interpretations thereof, should be specified and uncertainties in their values should be provided. Moreover, when scaling between different features, care must be taken to ensure that the starting values used are appropriate, so that the resulting values are reasonable. Until our understanding of impact cratering improves, there may be many cases in which discrepancies in crater "diameters" determined using different methods simply cannot be resolved. However, as we have shown here, some longstanding disagreements about crater size disappear when measurements are assessed consistently. In Table 5, we list our assessments of appropriate values for the different diameters we defined above for each of the craters we have discussed.

ACKNOWLEDGMENTS

The authors are sincerely grateful to John Spray for his substantial contributions to this manuscript and to William Hartmann, David Crown, David Kring, and Oleg Abramov for valuable discussions. They also thank Boris Ivanov and an anonymous reviewer for helpful suggestions from which the manuscript benefited significantly.

This manuscript is Planetary Science Institute contribution number 374 and Impact Cratering Research Group, University of the Witwatersrand, contribution number 69. This material is based upon work supported by the National Aeronautics and Space Administration under grant numbers NAG5-13429 and NAG5-11493 issued through the Planetary Geology and Geophysics Program.

REFERENCES CITED

Anderson, C.E., 1987, An overview of the theory of hydrocodes: International Journal of Impact Engineering, v. 5, p. 33–59.

Baldwin, R.B., 1949, The face of the Moon: Chicago, The University of Chicago Press, 239 p.

Baldwin, R.B., 1963, The measure of the Moon: Chicago, The University of Chicago Press, 488 p.

Beals, C.S., Innes, M.J.S., and Rottenberg, J.A., 1963, Fossil Meteorite Craters, *in* Middlehurst, B.M., and Kuiper, G.P., eds., The Moon meteorites and comets: Chicago, The University of Chicago Press, 810 p.

Bishopp, D.W., 1941, The geodynamics of the Vredefort Dome: Transactions of the Geological Society of South Africa, v. 44, p. 1–18.

Bjork, R.L., 1961, Analysis of the formation of Meteor Crater, Arizona. A preliminary report: Journal of Geophysical Research, v. 66, p. 3379–3387.

Bohor, B.F., Modreski, P.J., and Foord, E.E., 1984, Shocked quartz in the Cretaceous-Tertiary boundary clays: Evidence for global distribution: Science, v. 236, p. 705–709.

Butler, H.R., 1994, Lineament analysis of the Sudbury multiring impact structure, *in* Dressler, B.O., Grieve, R.A.F., and Sharpton, V.L., eds.,

TABLE 5. CRATER DIAMETERS

Crater	Diameters (km)						Erosion (km)
	Transient crater	Final rim	Central peak or uplift	Peak ring	External rings	Outer limit of deformation	
Barringer, Earth		1.2*	–	–	–	–	~0
Theophilus, Moon		~100		–	–	–	~0
Greenaway, Venus		92	–	46	–	–	~0
Orientale, Moon		320†	–	–	480, 620, 1300†	–	~0
Valhalla, Callisto		–	–	–	400–4000§	–	~0
Callanish, Europa	(15–33#)	(29–47#)	–	–	52–100**	–	~0
Sudbury, Earth	~90–130	~130–180	<~90	?	180?, 260?	~260	<6††
Chicxulub, Earth	~100	150	–	80	170–180, 195–200	250	0.3–0.4
Vredefort, Earth	~80–100	~120–200	~90–100§§	?	180##, 300***	180/300	5–11§§
Chesapeake, Earth	~25	~38	?	–	?	~160	~0

Note: Actual measured values are listed where possible. Numbers derived from scaling are given in parentheses. Features for which diameters are not relevant are indicated with –. Features for which diameter (or even existence) is not clear are indicated with ?.
*Shoemaker, 1987.
†Hartmann and Kuiper, 1962.
§Passey and Shoemaker, 1982.
#Moore et al., 2001.
**Turtle, 1998.
††Thompson et al., 1998; Spray et al., 2004.
§§McCarthy et al., 1990; Gibson and Wallmach, 1995; Stevens et al., 1997; Therriault et al., 1997; Henkel and Reimold, 1998; Gibson and Reimold, 1999; Gibson, 2002.
##Rand Anticline.
***Phillips et al., 1999.

Large Meteorite Impacts and Planetary Evolution: Geological Society of America Special Paper 293, p. 319–330.

Camargo-Zanoguera, A., and Suarez-Reynoso, G., 1994, Evidencia sísmica del cráter de impacto de Chicxulub: Boletín de la Asociación Mexicana de Geofisicos de Exploración, v. 34, p. 1–28.

Collins, G.S., and Melosh, H.J., 2004, Hydrocode Simulations of the Chesapeake Bay Impact: 35th Lunar and Planetary Science Conference, March 15–19, 2004, League City, Texas, abstract no. 1005.

Collins, G.S., Melosh, H.J., Morgan, J.V., and Warner, M.R., 2002, Hydrocode simulations of Chicxulub crater collapse and peak-ring formation: Icarus, v. 157, p. 24–33, doi: 10.1006/ICAR.2002.6822.

Collins, G.S., Ivanov, B.A., Turtle, E.P., and Melosh, H.J., 2003, Numerical Simulations of Silverpit Crater Collapse: Third International Conference on Large Meteorite Impacts, August 5–7, 2003, Nördlingen Germany, abstract no. 4126.

Collins, G.S., Melosh, H.J., and Ivanov, B.A., 2004, Modeling damage and deformation in impact simulations: Meteoritics & Planetary Science, v. 39, p. 217–231.

Cornejo-Toledo, A., and Hernandez-Osuna, A., 1950, Las anomalies gravimétricas en la cuenca salina del Istmo, planicie costera de Tabasco, Campeche y Península de Yucatán: Boletín de la Asociación Mexicana de Geologia y Petrología, v. 2, p. 453–460.

Corner, B., Durrheim, R.J., and Nicolaysen, L.O., 1990, Relationships between the Vredefort structure and the Witwatersrand Basin within the tectonic framework of the Kaapvaal craton as interpreted from regional gravity and aeromagnetic data: Tectonophysics, v. 171, p. 49–61, doi: 10.1016/0040-1951(90)90089-Q.

Crawford, D.A., and Barnouin-Jha, O.S., 2004, Computational investigations of the Chesapeake Bay impact structure: 35th Lunar and Planetary Science Conference, March 15–19, 2004, League City, Texas, Abstract no. 1757.

Croft, S.K., 1985, The scaling of complex craters, *in* Ryder, G., and Schubert, G., eds., Proceedings of the Fifteenth Lunar and Planetary Science Conference, March 12–16, 1984, Houston, Texas, Part 2: Journal of Geophysical Research, v. 90, Suppl., p. C828–C842.

Daly, R.A., 1947, The Vredefort ring-structure of South Africa: Journal of Geology, v. 55, p. 125–145.

Dence, M.R., 1965, The extraterrestrial origin of Canadian craters: Annals of the New York Academy of Sciences, v. 123, p. 941–969.

Dence, M.R., 1968, Shock zoning at Canadian craters: Petrography and structural implications, *in* French, B.M., and Short, N.M., eds., Shock metamorphism of natural materials: Baltimore, Maryland, Mono Book Corporation, p. 169–184.

Dence, M.R., 1972, Meteorite impact craters and the structure of the Sudbury Basin, *in* New developments in Sudbury geology: Toronto, Ontario, Geological Association of Canada Special Paper 10, Geological Association of Canada, p. 7–18.

Deutsch, A., Grieve, R.A.F., Avermann, M., Bischoff, L., Brockmeyer, P., Buhl, D., Lakomy, R., Möller-Mohr, R., Ostermann, M., and Stöffler, D., 1995, The Sudbury structure (Ontario, Canada): A tectonically deformed multi-ring impact basin: Geologische Rundschau, v. 84, p. 697–709, doi: 10.1007/S005310050034.

Dietz, R.S., 1964, Sudbury structure as an astrobleme: Journal of Geology, v. 72, p. 412–434.

Dietz, R.S., 1972, Sudbury astrobleme, splash emplaced sub-layer and possible cosmogenic ores: Geological Association of Canada Special Paper 10, p. 29–40.

Dressler, B.O., 1984, The effects of the Sudbury event and the intrusion of the Sudbury Igneous Complex on the footwall rocks of the Sudbury structure, *in* Pye, E., Naldrett, A., and Giblin, P., eds., The geology and ore deposits of Sudbury: Ontario, Ontario Geological Survey Special Volume 1, Ontario Ministry of Natural Resources, p. 97–136.

Dressler, B.O., Sharpton, V.L., Morgan, J., Buffler, R., Moran, D., Smit, J., Stöffler, D., and Urrutia, J., 2003, Investigating a 65-Ma-old smoking gun: Deep drilling of the Chicxulub impact structure: Eos (Transactions, American Geophysical Union), v. 84, p. 125, 130.

Durrheim, R.J., Nicolaysen, L.O., and Corner, B., 1991, A deep seismic reflection profile across the Archean-Proterozoic Witwatersrand Basin, South Africa: American Geophysical Union, Geodynamics Series, v. 22, p. 213–224.

Espindola, J.M., Mena, M., de La Fuente, M., and Campos-Enriquez, J.O., 1995, A model of the Chicxulub impact structure (Yucatan, Mexico) based on its gravity and magnetic signatures: Physics of the Earth and Planetary Interiors, v. 92, p. 271–278, doi: 10.1016/0031-9201(95)03022-6.

French, B.M., 1968, Sudbury structure, Ontario: Some petrographic evidence for an origin by meteorite impact, *in* French, B.M., and Short, N.M., eds., Shock metamorphism of natural materials: Baltimore, Maryland, Mono Book, p. 383–412.

French, B.M., 1970, Possible relations between meteorite impact and igneous petrogenesis, as indicated by the Sudbury structure, Ontario, Canada: Bulletin Volcanologique, v. 34, p. 466–517.

French, B.M., 1998, Traces of Catastrophe: Houston, Lunar and Planetary Institute Contribution 954, 120 p.

Friese, A.E.W., Charlesworth, E.G., and McCarthy, T., 1995, Tectonic processes within the Kaapvaal Craton during the Kibaran (Grenville) orogeny. Structural, geophysical and isotopic constraints from the Witwatersrand Basin and environs: Johannesburg, University of the Witwatersrand, Economic Geology Research Unit, Information Circular 292, 67 p.

Fueten, R., and Redmond, D.J., 1997, Documentation of a 1450 Ma contractional orogeny preserved between the 1950 Ma Sudbury impact structure and the 1 Ga old Grenville Orogenic front, Ontario: Geological Society of America Bulletin, v. 109, p. 268–279, doi: 10.1130/0016-7606(1997)1092.3.CO;2.

Galilei, G., 1610, Sidereus Nuncius: Venice, Italy, Thomas Baglionus, 58 p.

Gault, D.E., Quaide, W.L., and Oberbeck, V.R., 1968, Impact cratering mechanics and structures, *in* French, B.M., and Short, N.M., eds., Shock Metamorphism in Natural Materials: Baltimore, Mono Book Corp., p. 87–99.

Gibson, R.L., 2002, Impact-induced melting of Archean granulites in the Vredefort Dome, South Africa, I: Anatexis of metapelitic granulites: Journal of Metamorphic Geology, v. 20, p. 57–70, doi: 10.1046/J.0263-4929.2001.00358.X.

Gibson, R.L., and Reimold, W.U., 1999, The metamorphic fingerprint of large impact events: The example of the Vredefort Dome, South Africa: Meteoritics & Planetary Science, v. 34, p. A42–A43.

Gibson, R.L., and Reimold, W.U., 2001, The Vredefort impact structure, South Africa: the scientific evidence and a two-day excursion guide: Pretoria, Council of Geoscience, Memoir 92, 111 p.

Gibson, H.M., and Spray, J.G., 1998, Shock-induced melting and vaporization of shatter cone surfaces: evidence from the Sudbury impact structure: Meteoritics & Planetary Science, v. 33, p. 329–336.

Gibson, R.L., and Wallmach, T., 1995, Low pressure–high temperature metamorphism in the Vredefort Dome, South Africa: Anticlockwise pressure-temperature path followed by rapid decompression: Geological Journal, v. 30, p. 319–331.

Glass, B.P., and Liu, S., 2001, Discovery of high-pressure $ZrSiO_4$ polymorph in naturally occurring shock-metamorphosed zircon: Geology, v. 29, p. 371–373, doi: 10.1130/0091-7613(2001)0292.0.CO;2.

Grieve, R.A.F., 1991, Terrestrial impact—The record in the rocks: Meteoritics, v. 26, p. 175–194.

Grieve, R.A.F., 1994, An impact model of the Sudbury Structure, *in* Lightfoot, P.C., and Naldrett, A.J., eds., Proceedings of the Sudbury-Norilsk Symposium: Ottawa, Ontario Geological Survey Special Volume 5, p. 119–132.

Grieve, R.A.F., and Cintala, M.J., 1992, An analysis of differential impact crater scaling and implications for the terrestrial record: Meteoritics & Planetary Science, v. 27, p. 526–538.

Grieve, R.A.F., and Masaitis, V.L., 1994, The economic potential of terrestrial impact craters: International Geology Review, v. 36, p. 105–151.

Grieve, R.A.F., and Pesonen, L.J., 1992, The terrestrial impact cratering record: Tectonophysics, v. 216, p. 1–30, doi: 10.1016/0040-1951(92)90152-V.

Grieve, R.A.F., and Therriault, A., 2000, Vredefort, Sudbury, Chicxulub: Three of a kind?: Annual Reviews of Earth and Planetary Science, v. 28, p. 305–338.

Grieve, R.A.F., Dence, M.R., and Robertson, P.B., 1977, Cratering processes: As interpreted from the occurrence of impact melts, *in* Roddy, D.J., Pepin, R.O., and Merrill, R.B., eds., Impact and explosion cratering: New York, Pergamon Press, p. 791–814.

Grieve, R.A.F., Robertson, P.B., and Dence, M.R., 1981, Constraints on the formation of ring impact structures, based on terrestrial data, *in* Schultz, P.H., and Merrill, R.B., eds., Multi-Ring Basins: New York, Pergamon Press, p. 37–57.

Grieve, R.A.F., Stöffler, D., and Deutsch, A., 1991, The Sudbury structure: Controversial or misunderstood?: Journal of Geophysical Research, v. 96, p. 22,753–22,764.

Guy-Bray, J.V., 1972, New developments in Sudbury geology: Geological Association of Canada Special Publication 10, 124 p.

Hall, A.L., and Molengraaff, G.A.F., 1925, The Vredefort mountain land in the southern Transvaal and the northern Orange Free State: Verhandelingen der Koninklijke Akademie can Wetenschappen te Amsterdam, v. 24, p. 1–183.

Hart, R.J., Andreoli, M.A.G., Reimold, W.U., and Tredoux, M., 1991, Aspects of the dynamic and thermal history of the Vredefort cryptoexplosion structure: Implications for its origin: Tectonophysics, v. 192, p. 313–331, doi: 10.1016/0040-1951(91)90106-3.

Hartmann, W.K., and Kuiper, G.P., 1962, Concentric structures surrounding lunar basins: Communications of the Lunar and Planetary Laboratory, v. 1, p. 131–137.

Hartmann, W.K., and Wood, C.A., 1971, Moon: Origin and evolution of multi-ring basins: The Moon, v. 3, p. 3–78.

Head, J.W., 1974, Orientale multi-ringed basin interior and implications for the petrogenesis of lunar highland samples: The Moon, v. 11, p. 327–356.

Henkel, H., and Reimold, W.U., 1998, Integrated geophysical modelling of a giant complex impact structure: Anatomy of the Vredefort structure, South Africa: Tectonophysics, v. 287, p. 1–20, doi: 10.1016/S0040-1951(98)80058-9.

Henkel, H., and Reimold, W.U., 2002, Magnetic model of the central uplift of the Vredefort impact structure, South Africa: Journal of Applied Geophysics, v. 51, p. 43–62, doi: 10.1016/S0926-9851(02)00214-8.

Herrick, R., and Pierazzo, E., 2003, Results of the Workshop on Impact Cratering: Bridging the gap between modeling and observations: Houston, Texas, 7–9 February 2003, Lunar and Planetary Institute Contribution No. 1162, 156 p.

Hildebrand, A.R., Penfield, G.T., Kring, D.A., Pilkington, M., Jacobsen, S., and Boynton, W.V., 1991, Chicxulub crater: A possible Cretaceous-Tertiary boundary impact crater on the Yucatan Peninsula, Mexico: Geology, v. 19, p. 867–871, doi: 10.1130/0091-7613(1991)0192.3.CO;2.

Hildebrand, A.R., Pilkington, M., Connors, M., Ortiz-Aleman, C., and Chavez, R.E., 1995, Size and structure of the Chicxulub crater revealed by horizontal gravity gradients and cenotes: Nature, v. 376, p. 415–417, doi: 10.1038/376415A0.

Hildebrand, A.R., Pilkington, M., Ortiz-Aleman, C., Chavez, R.E., Urrutia-Fucugauchi, J., Connors, M., Graniel-Castro, E., and Camargo-Zanoguera, A., 1998, Mapping Chicxulub crater structure with gravity and seismic reflection data, *in* Grady, M.M., Hutchinson, R., McCall, G.J.H., and Rothery, D.A., eds., Meteorites: Flux with time and impact effects: Geological Society of London Special Publication 140, p. 177–193.

Hole, J.A., McGeary, S., Melosh, H.J., and Eriksson, S.C., 2003, Proposed seismic imaging and numerical modeling of the impact event at Chesapeake Bay: International Continental Scientific Drilling Program/U.S. Geological Survey Workshop on Deep Drilling of the Chesapeake Bay Impact Crater.

Ivanov, B.A., and Kostuchenko, V.N., 1998, Block oscillation model for impact craters collapse: 29th Lunar and Planetary Science Conference, March 17–21, 1997, Houston, Texas, Abstract no. 1654.

Kamo, S.L., Reimold, W.U., Krogh, T.E., and Colliston, W.P., 1996, A 2.023 Ga age for the Vredefort impact event and a first report of shock metamorphosed zircons in pseudotachylitic breccias and granophyre: Earth and Planetary Science Letters, v. 144, p. 369–387, doi: 10.1016/S0012-821X(96)00180-X.

Kenkmann, T., 2004, Structural comparison between the Chesapeake Bay impact crater, USA, and the Ries Crater, Germany; how did the central crater basin form? *in* Edwards, L.E., Horton, J.W., Jr., and Gohn, G.S., eds., Proceedings of ICDP-USGS workshop on Deep drilling in the central crater of the Chesapeake Bay impact structure, Virginia, USA, Herndon, Virginia, September 22–24, 2003, U.S. Geological Survey, Reston, Virginia, p. 30–31.

Koeberl, C., Poag, C.W., Reimold, W.U., and Brandt, D., 1996a, Impact origin of the Chesapeake Bay structure, and source of the North American tektites: Science, v. 271, p. 1263–1266.

Koeberl, C., Reimold, W.U., and Shirley, S.B., 1996b, Re-Os isotope and geochemical study of the Vredefort granophyre: Clues to the origin of the Vredefort structure, South Africa: Geology, v. 24, p. 913–916, doi: 10.1130/0091-7613(1996)0242.3.CO;2.

Koeberl, C., Reimold, W.U., and Plescia, J., 2005, BP and Oasis impact structures, Libya: Remote sensing and field studies, *in* Henkel, H., and Koeberl, C., eds., Impact Tectonics, Impact Studies Series, v. 6: Heidelberg, Springer-Verlag (in press).

Kreyenhagen, K.N., and Schuster, S.H., 1977, Review and comparison of hypervelocity impact and explosion cratering calculations, *in* Roddy, D.J., Pepin, R.O., and Merrill, R.B., eds., Impact and explosion cratering: New York, Pergamon Press, p. 983–1002.

Kring, D.A., 1995, The dimension of the Chicxulub impact crater and impact melt sheet: Journal of Geophysical Research, v. 100, p. 16,979–16,986, doi: 10.1029/95JE01768.

Kring, D.A., and Boynton, W., 1992, Petrogenesis of an augite-bearing melt rock in the Chicxulub structure and its relationship to K/T impact spherules in Haiti: Nature, v. 358, p. 141–144, doi: 10.1038/358141A0.

Lakomy, R., 1990, Distribution of impact induced phenomena in complex terrestrial impact structures; implications for transient cavity dimensions:

21st Lunar and Planetary Science Conference, March 12–16, 1990, Houston, Texas, p. 676–677.

Lana, C., Gibson, R.L., and Reimold, W.U., 2003, Impact tectonics in the core of the Vredefort dome, South Africa: Implications for central uplift formation in very large impact structures: Meteoritics & Planetary Science, v. 38, p. 1093–1107.

Leroux, H., Reimold, W.U., and Doukhan, J., 1994, A TEM investigation of shock metamorphism in quartz from the Vredefort dome, South Africa: Tectonophysics, v. 230, p. 223–239, doi: 10.1016/0040-1951(94)90137-6.

Martini, J.E.J., 1978, Coesite and stishovite in the Vredefort Dome, South Africa: Nature, v. 272, p. 715–717.

McCarthy, T.S., Stanistreet, I.G., and Robb, L.J., 1990, Geological studies related to the origin of the Witwatersrand Basin and its mineralization—An introduction and a strategy for research and exploration: South African Journal of Geology, v. 93, p. 1–4.

McHone, J.F., Greeley, R., Williams, K.K., Blumberg, D.G., and Kuzmin, R.O., 2002, Space shuttle observations of terrestrial impact structures using SIR-C and X-SAR radars: Meteoritics & Planetary Science, v. 37, p. 407–420.

McKinnon, W.B., 1978, An investigation into the role of plastic failure in crater modification, *in* Merrill, R.B., ed. The Moon and the inner solar system, Proceedings of the Ninth Lunar and Planetary Science Conference, March 13–17, 1978, Houston, Texas: New York, Pergamon, p. 3965–3973.

McKinnon, W.B., and Schenk, P.M., 1985, Ejecta blanket scaling on the Moon and inferences for projectile populations: 16th Lunar and Planetary Science Conference, March 1985, Houston, Texas, p. 544–545.

Melosh, H.J., 1977, Crater modification by gravity: A mechanical analysis of slumping, *in* Roddy, D.J., and Merrill, R.B., eds., Impact and explosion cratering: New York, Pergamon Press, p. 1245–1260.

Melosh, H.J., 1982, A schematic model of crater modification by gravity: Journal of Geophysical Research, v. 87, p. 371–380.

Melosh, H.J., 1989, Impact cratering: A geologic process: New York, Oxford University Press, 245 p.

Melosh, H.J., and Gaffney, E.S., 1983, Acoustic fluidization and the scale dependence of impact crater morphology: Journal of Geophysical Research, v. 88, supplement, p. A830–A834.

Melosh, H.J., and Ivanov, B.A., 1999, Impact crater collapse: Annual Review of Earth and Planetary Sciences, v. 27, p. 385–415, doi: 10.1146/ANNUREV.EARTH.27.1.385.

Melosh, H.J., and McKinnon, W.B., 1978, The mechanics of ringed basin formation: Geophysical Research Letters, v. 5, p. 985–988.

Milkereit, B., and Green, A., and The Sudbury Working Group., 1992, The geometry of the Sudbury structure from high-resolution seismic reflection profiling: Geology, v. 20, p. 807–811, doi: 10.1130/0091-7613(1992)0202.3.CO;2.

Morgan, J.V., and Warner, M.R., 1999, Chicxulub: The third dimension of a multi-ring impact basin: Geology, v. 27, p. 407–410, doi: 10.1130/0091-7613(1999)0272.3.CO;2.

Morgan, J.V., Warner, M.R., Brittan, J., Buffler, R., Camargo, A., Christeson, G.L., Denton, P., Hildebrand, A., Hobbs, R., Macintyre, H., Mackenzie, G., Maguire, P., Marín, L., Nakamura, Y., Pilkington, M., Sharpton, V.L., Snyder, D., Suarez, G., and Trejo, A., 1997, Size and morphology of the Chicxulub impact crater: Nature, v. 390, p. 472–476, doi: 10.1038/37291.

Morgan, J.V., Warner, M.R., Collins, G.S., Melosh, H.J., and Christeson, G.L., 2000, Peak ring formation in large impact craters: Earth and Planetary Science Letters, v. 183, p. 347–354, doi: 10.1016/S0012-821X(00)00307-1.

Morgan, J.V., Christeson, G., and Zelt, C., 2002a, Testing the resolution of a 3D velocity tomogram across the Chicxulub crater: Tectonophysics, v. 355, p. 215–226, doi: 10.1016/S0040-1951(02)00143-9.

Morgan, J., Warner, M., and Grieve, R.A.F., 2002b, Geophysical constraints on the size and structure of the Chicxulub impact crater, *in* Koeberl, C., and MacLeod, K.G., eds., Catastrophic events and mass extinctions: Impact and beyond: Geological Society of America Special Paper 356, p. 39–46.

Moore, J.M., Asphaug, E., Belton, M.J.S., Bierhaus, B., Breneman, H.H., Brooks, S.M., Chapman, C.R., Chuang, F.C., Collins, G.C., Giese, B., Greeley, R., Head, J.W., Kadel, S., Klaasen, K.P., Klemaszewski, J.E., Magee, K.P., Moreau, J., Morrison, D., Neukum, G., Pappalardo, R.T., Phillips, C.B., Schenk, P.M., Senske, D.A., Sullivan, R.J., Turtle, E.P., and Williams, K.K., 2001, Impact Features on Europa: Results of the Galileo Europa Mission (GEM): Icarus, v. 151, p. 93–111, doi: 10.1006/ICAR.2000.6558.

Müller-Mohr, V., 1992, Breccias in the basement of a deeply eroded impact structure, Sudbury, Canada: Tectonophysics, v. 216, p. 219–226, doi: 10.1016/0040-1951(92)90168-6.

Nel, L.T., 1927, The geology of the country around Vredefort: An explanation of the geological map: Pretoria, Geological Survey of South Africa, 134 p.

Nordyke, M.D., 1961, Nuclear craters and preliminary theory of the mechanics of explosive crater formation: Journal of Geophysical Research, v. 66, p. 3439–3459.

O'Keefe, J.D., and Ahrens, T.J., 1993, Planetary cratering mechanics: Journal of Geophysical Research, v. 98, p. 17,011–17,028.

O'Keefe, J.D., and Ahrens, T.J., 1999, Complex craters: Relationship of stratigraphy and rings to impact conditions: Journal of Geophysical Research, v. 104, p. 27,091–27,105, doi: 10.1029/1998JE000596.

Passey, Q.R., and Shoemaker, E.M., 1982, Craters and basins on Ganymede and Callisto: Morphological indicators of crustal evolution, *in* Morrison, D., and Matthews, M.S., eds., Satellites of Jupiter: Tucson, University of Arizona Press, p. 379–434.

Penfield, G.T., and Camargo-Zanoguera, A., 1981, Definition of a major igneous zone in the central Yucatán platform with aeromagnetics and gravity, *in* Society of Exploration Geophysicists, 51st Annual Meeting (abstracts): Tulsa, Oklahoma, SEG, p. 37.

Penfield, G.T., and Camargo-Zanoguera, A., 1991, Interpretation of geophysical cross sections on the north flank of the Chicxulub impact structure: 22nd Lunar and Planetary Science Conference, March 18–22, 1991, Houston, Texas, p. 1051–1052.

Peredery, W.V., and Morrison, G.G., 1984, Discussion of the origin of the Sudbury Structure, *in* Pye, E.G., Naldrett, A.J., and Giblin, P.E., eds., The geology and ore deposits of the Sudbury Structure: Toronto, Ontario Geological Survey Special Volume 1, p. 491–511.

Phillips, M.E., Bussell, M.A., Hart, R.J., and Andreoli, M.A.G., 1999, A remote-sensing and geological investigation of the Vredefort impact structure (South Africa) using Landsat Thematic Mapper imagery (Abstract): Meteoritics & Planetary Science, v. 34, p. A92–A93.

Pierazzo, E., and Melosh, H.J., 2000, Understanding oblique impacts from experiments, observations, and modeling: Annual Review of Earth and Planetary Sciences, v. 28, p. 141–167, doi: 10.1146/ANNUREV.EARTH.28.1.141.

Pierazzo, E., and Collins, G.S., 2003, A brief introduction to hydrocode modeling of impact cratering, *in* Henning, D., Burchell, M., and Claeys, P., eds., Impact studies: Cratering in marine environments and on ice: Berlin, Springer-Verlag, p. 323–340.

Pieters, C.M., 1982, Copernicus crater central peak—Lunar mountain of unique composition: Science, v. 215, p. 59–61.

Pike, R.J., 1977, Size-dependence in the shape of fresh impact craters on the moon, *in* Roddy, D. J., and Merrill, R.B., eds., Impact and explosion cratering: New York, Pergamon Press, p. 489–509.

Pike, R.J., 1985, Some morphologic systematics of complex impact structures: Meteoritics & Planetary Science, v. 20, p. 49–68.

Pilkington, M., and Grieve, R.A.F., 1992, The geophysical signature of terrestrial impact craters: Reviews of Geophysics, v. 30, p. 161–181.

Pilkington, M., and Hildebrand, A.R., 2003, Transient and disruption cavity dimensions of complex terrestrial impact structures derived from magnetic data: Geophysical Research Letters, v. 30, 2087, doi: 10.1029/2003GL018294.

Pilkington, M., Hildebrand, A.R., and Ortiz-Aleman, C., 1994, Gravity and magnetic field modeling and structure of the Chicxulub crater, Mexico: Journal of Geophysical Research, v. 99, p. 13,147–13,162, doi: 10.1029/94JE01089.

Poag, C.W., 1996, Structural outer rim of Chesapeake Bay impact crater: Seismic and borehole evidence: Meteoritics and Planetary Sciences, v. 31, p. 218–226.

Poag, C.W., 1997, The Chesapeake Bay bolide impact: A convulsive event in Atlantic Coastal Plain evolution: Sedimentary Geology, v. 108, p. 45–90, doi: 10.1016/S0037-0738(96)00048-6.

Poag, C.W., and Aubry, M.P., 1995, Upper Eocene impactites of the U.S. east coast: Depositional origins, biostratigraphic framework, and correlation: Palaios, v. 10, p. 16–43.

Poag, C.W., Powars, D.S., Poppe, L.J., Mixon, R.B., Edwards, L.E., Folger, D.W., and Bruce, S., 1992, Deep Sea Drilling Project site 612 bolide event: New evidence of a late Eocene impact-wave deposit and a possible impact site, U.S. east coast: Geology, v. 20, p. 771–774, doi: 10.1130/0091-7613(1992)020<0771:DSDPSB>2.3.CO;2.

Poag, C.W., Powars, D.S., Poppe, L.J., and Mixon, R.B., 1994, Meteoroid mayhem in Ole Virginny: Source of the North American tektite strewn field: Geology, v. 22, p. 691–694, doi: 10.1130/0091-7613(1994)022<0691:MMIOVS>2.3.CO;2.

Poag, C.W., Hutchinson, D.R., Colman, S.M., and Lee, M.W., 1999, Seismic expression of the Chesapeake Bay impact crater: Structural and morphologic refinements based on new seismic data, *in* Dressler, B.O., and Sharpton, V.L., eds., Large meteorite impacts and planetary evolution II: Geological Society of America Special Paper 339, p. 149–164.

Poag, C.W., Plescia, J.B., and Molzer, P.C., 2002, Ancient impact structures on modern continental shelves: The Chesapeake Bay, Montagnais, and Toms Canyon craters, Atlantic margin of North America: Deep-Sea Research II, v. 49, p. 1081–1102, doi: 10.1016/S0967-0645(01)00144-8.

Poag, C.W., Koeberl, C., and Reimold, W.U., 2004, The Chesapeake Bay Crater—Geology and geophysics of a Late Eocene submarine impact structure: Impact Studies Series: Heidelberg, Springer, 522 p.

Pope, K.O., Ocampo, A.C., and Duller, C.E., 1993, Surficial geology of the Chicxulub impact crater, Yucatán, Mexico: Earth, Moon, and Planets, v. 63, p. 93–104.

Pope, K.O., Ocampo, A.C., Kinsland, G.L., and Smith, R., 1996, Surface expression of the Chicxulub crater: Geology, v. 24, p. 527–530, doi: 10.1130/0091-7613(1996)024<0527:SEOTCC>2.3.CO;2.

Pope, K.O., Baines, K.H., Ocampo, A.C., and Ivanov, B.A., 1997, Energy, volatile production, and climatic effects of the Chicxulub Cretaceous/Tertiary impact: Journal of Geophysical Research, v. 102, p. 21,645–21,664, doi: 10.1029/97JE01743.

Powars, D.S., and Bruce, T.S., 1999, The effects of the Chesapeake Bay impact crater on the geologic frame-work and the correlation of hydrogeologic units of southeastern Virginia, south of the James river: U.S. Geological Survey Professional Paper 1612, 82 p.

Powars, D.S., Poag, C.V., and Mixon, R.B., 1993, The Chesapeake Bay "impact crater": Seismic and stratigraphic evidence (Abstract): Geological Society of America Abstracts with Programs, v. 25, no. 7, p. A-378.

Prockter, L.M., Antman, A.M., Pappalardo, R.T., Head, J.W., and Collins, G.C., 1999, Europa: Stratigraphy and geological history of the anti-Jovian region from Galileo E14 solid-state imaging data: Journal of Geophysical Research, v. 104, p. 16,531–16,540, doi: 10.1029/1998JE001015.

Reimold, W.U., and Wallmach, T., 1991, The Vredefort structure under discussion: South African Journal of Science, v. 87, p. 412–417.

Rondot, J., 1994, Recognition of eroded astroblemes: Earth Science Reviews, v. 35, p. 331–365, doi: 10.1016/0012-8252(94)90001-9.

Sepkoski, J., 1994, Extinction and the fossil record: Geotimes, v. 39, p. 15–17.

Sharpton, V.L., Burke, K., Camargo-Zanoguera, A., Hall, S.A., Lee, D.S., Marín, L.E., Suarez-Reynoso, G., Quezada-Muñeton, J.M., Spudis, P.D., and Urrutia-Fucugauchi, J., 1993, Chicxulub multiring impact basin: Size and other characteristics derived from gravity analysis: Science, v. 261, p. 1564–1567.

Schmidt, R.M., and Housen, K.R., 1987, Some recent advances in the scaling of impact and explosion cratering: International Journal of Impact Engineering, v. 5, p. 543–560.

Shoemaker, E.M., 1960, Penetration mechanics of high velocity meteorites, illustrated by Meteor Crater, Arizona: Report of the International Geological Congress, Session 21, Copenhagen, Norden, p. 418–434.

Shoemaker, E.M., 1987, Meteor Crater, Arizona, *in* Beus, S.S., ed., Centennial Field Guide volume 2: Rocky Mountain Section of the Geological Society of America, p. 399–404, doi: 10.1130/0-8137-5402-X(1987)002<0399:MCAXXX>2.0.CO;2.

Spray, J.G., 1998, Localized shock- and friction-induced melting in response to hypervelocity impact, *in* Grady, M.M., Hutchison, R., McCall, G.J.H., and Rothery, D.A., eds., Meteorites: Flux with time and impact effects: Geological Society [London] Special Publication 140, p. 171–180.

Spray, J.G., and Thompson, L.M., 1995, Friction melt distribution in a multi-ring impact basin: Nature, v. 373, p. 130–132, doi: 10.1038/373130A0.

Spray, J.G., Kelley, S.P., and Reimold, W.U., 1995, Laser probe ^{40}Ar/^{39}Ar dating of coesite- and stishovite-bearing pseudotachylytes and the age of the Vredefort impact event: Meteoritics, v. 30, p. 335–343.

Spray, J.G., Butler, H.R., and Thompson, L.M., 2004, Tectonic influences on the morphometry of the Sudbury impact structure: Implications for terrestrial cratering and modeling: Meteoritics & Planetary Science, v. 39, p. 287–301.

Stevens, G., Gibson, R.L., and Droop, G.T.R., 1997, Mid-crustal granulite facies metamorphism in the Central Kaapvaal craton: The Bushveld Complex connection: Precambrian Research, v. 82, p. 113–132, doi: 10.1016/S0301-9268(96)00043-5.

Stewart, S.A., and Allen, P.J., 2002, A 20-km-diameter multi-ringed impact structure in the North Sea: Nature, v. 418, p. 520–523, doi: 10.1038/NATURE00914.

Stöffler, D., Avermann, M., Bischoff, L., Brockmeyer, P., Deutsch, A., Dressler, B.O., Lakomy, R., and Müller-Mohr, V., 1989, Sudbury, Canada: Remnant of the only multi-ring (?) impact basin on Earth, *in* Wasson, J.T., ed., 52nd Meteoritical Society Meeting (abstracts): Meteoritics, v. 24, p. 328.

Swisher, C.C., Grajales-Nishimura, J.M., Montanari, A., Margolis, S.V., Claeys, P., Alvarez, W., Renne, P., Cedillo-Pardo, E., Maurrasse, F.J.-M.R., Curtis, G.H., Smit, J., and McWilliams, M.O., 1992, Coeval ^{40}Ar/^{39}Ar ages of 65.0 Million years ago from Chicxulub crater melt rock and Cretaceous-Tertiary boundary tektites: Science, v. 257, p. 954–958.

Therriault, A.M., Grieve, R.A.F., and Reimold, W.U., 1997, Original size of the Vredefort structure: Implications for the geological evolution of the Witwatersrand Basin: Meteoritics & Planetary Science, v. 32, p. 71–77.

Thompson, L.M., and Spray, J.G., 1994, Pseudotachylytic rock distribution and genesis within the Sudbury Structure, *in* Dressler, B.O., Grieve, R.A.F., and Sharpton, V.L., eds., Large Meteorite Impacts and Planetary Evolution: Geological Society of America Special Paper 293, p. 275–288.

Thompson, L.M., Spray, J.G., and Kelley, S.P., 1998, Laser-probe ^{40}Ar/^{39}Ar dating of pseudotachylyte from the Sudbury Structure: Evidence for post-impact thermal overprinting in the North Range: Meteoritics & Planetary Science, v. 33, p. 1259–1270.

Tuchscherer, M.G., and Spray, J.G., 2002, Geology, mineralization and emplacement of the Foy offset dike, Sudbury impact structure: Economic Geology and the Bulletin of the Society of Economic Geologists, v. 97, p. 1347–1368.

Turtle, E.P., 1998, Finite-element modeling of large impact craters: Implications for the size of the Vredefort structure and the formation of multiple ring craters [Ph.D. Thesis]: Tucson, University of Arizona, 176 p.

Turtle, E.P., and Pierazzo, E., 1998, Constraints on the size of the Vredefort impact crater from numerical modeling: Meteoritics & Planetary Science, v. 33, p. 483–490.

Turtle, E.P., Pierazzo, E., and O'Brien, D.P., 2003, Numerical modeling of impact heating and cooling of the Vredefort impact structure: Meteoritics & Planetary Science, v. 38, p. 293–303.

Wagner, R., Reimold, W.U., and Brandt, D., 2002, Bosumtwi impact crater, Ghana: A remote sensing investigation, *in* Plado, J., and Pesonen, L.J., eds., Impacts in Precambrian Shields. Impact Studies Series: Heidelberg, Springer, p. 189–210.

Wood, C.A., and Head, J.W., 1976, Comparison of impact basins on Mercury, Mars and the Moon, *in* Merrill, R.B., Morris, R.V., Rhodes, J.M., and Usselman, T.M., eds., Proceedings of the 7th Lunar and Planetary Science Conference, vol. 3, The Moon and other bodies: Houston, Texas, March 15–19, p. 3629–3651.

Wood, C.R., and Spray, J.G., 1998, Origin and emplacement of Offset Dykes in the Sudbury impact structure: constraints from Hess: Meteoritics & Planetary Science, v. 33, p. 337–347.

Wu, J., Milkereit, B., and Boerner, D., 1995, Seismic imaging of the enigmatic Sudbury structure: Journal of Geophysical Research, v. 100, p. 4117–4130, doi: 10.1029/94JB02647.

Manuscript Accepted by the Society 9 August 2004

Printed in the USA

Geological Society of America
Special Paper 384
2005

Magnetization on impact structures—Constraints from numerical modeling and petrophysics

H.A. Ugalde*
Department of Physics, University of Toronto, 60 St. George Street, Toronto, Ontario M5S 1A7, Canada

N. Artemieva*
Institute for Dynamics of Geospheres, Leninsky pr., 38, Building 6, 119334 Moscow, Russia

B. Milkereit*
Department of Physics, University of Toronto, 60 St. George Street, Toronto, Ontario M5S 1A7, Canada

ABSTRACT

Prominent magnetic anomalies over large impact craters are attributed to remanent magnetization as thermal effects induce extremely high Koenigsberger values (remanent to induced magnetization ratio, Q). Magnetization of impact melt rocks, breccias, and the rocks underneath the crater floor is related to the thermal evolution of large impact craters, from a single heat pulse to long-lived hydrothermal processes and associated alteration and mineral deposits. The magnetic signature observed on large impact structures can be primarily the aggregate of three effects: (1) composition and properties of target rocks, (2) modification of magnetic carriers due to high pressure-temperature (*P-T*) conditions, and (3) natural remanent magnetization (NRM). Numerical modeling is used to predict the pressure and temperature distribution for varying size craters ranging from a 1.5-km-diameter simple crater to a 90-km-diameter complex crater. We first compare the results of two hydrocodes: the Simplified Arbitrary Lagrangian Eulerian code, version B (SALEB), to produce the final crater shape after a vertical impact; and the Solid, Vapor, Air (SOVA) code to model the initial stage of an oblique impact and to evaluate differences in the volumes of highly shocked materials within the crater created by a vertical impact versus an oblique one. After defining the accuracy of the numerical modeling and reliability of the resulting *P-T* conditions predicted, numerical modeling is applied for 1–90 km diameter impact structures. The *P-T* distributions of the target material at its final position obtained with the SALEB code are used for geophysical predictions. The calculated maximum *P-T* values and their initial and final spatial distributions can be linked with geological and physical and/or chemical processes that lead to different geophysical signatures. For a 5 km diameter crater, brecciation ($P > 1$ GPa) affects all material in a 2.5 km hemisphere. Brecciation and fluid circulation trigger a hydrothermal system, which introduces chemical remanent magnetization (CRM) and modify magnetic carriers due to alteration. Pressures larger than 30 GPa induce a secondary component of remanent magnetization

*E-mails: Ugalde—ugalde@physics.utoronto.ca; Artemieva—nata_art@mtu-net.ru; Milkereit—bm@physics.utoronto.ca

Ugalde, H.A., Artemieva, N., and Milkereit, B., 2005, Magnetization on impact structures—Constraints from numerical modeling and petrophysics, *in* Kenkmann, T., Hörz, F., and Deutsch, A., eds., Large meteorite impacts III: Geological Society of America Special Paper 384, p. 25–42. For permission to copy, contact editing@geosociety.org.

(shock remanent magnetization, SRM). These *P-T* conditions are generally restricted to a radius of <0.5 km and depth <0.8 km. Therefore, SRM and melting are confined to small regions, occasionally not even discernible by a regional airborne magnetic survey, and easily removed by erosive processes.

Keywords: magnetic anomaly, numerical modeling, impact structure, petrophysics, forward modeling, remanent magnetization.

MOTIVATION

All terrestrial impact craters exhibit geophysical signatures (Pilkington and Grieve, 1992). Recent integration of results from seismic surveys, potential field studies, remote sensing, exploration drilling, and numerical modeling of impact processes can provide powerful constraints on the size and shape of the transient crater and give images of impact basin morphology. The purpose of this paper is to discuss common results on the magnetic signature of impact craters from recent geophysical studies across large to midsize impact structures like the 2006 Ma Vredefort (South Africa—300 km [Henkel and Reimold, 1998, 2002]), the 1850 Ma Sudbury (Canada—250 km [Wu et al., 1995]), the 65 Ma Chicxulub (Mexico—180 km [Morgan et al., 1997, 2002; Pilkington and Hildebrand, 2000; Rebolledo-Vieyra, 2001]), the 15 Ma Ries (Germany—24 km [Pohl et al., 1977]), and the 1 Ma Bosumtwi structures (Ghana—10 km [Plado et al., 2000]).

Given that 35% of the known impact structures on Earth are buried (Earth Impact Database, 2004), geophysics plays an important role in their reconnaissance and mapping. Of all the geophysical methods used for regional mapping, magnetics is the most useful due to its low cost/benefit ratio compared to gravity and seismic surveys. Although in most of the cases seismic data is able to map and define post-impact sedimentation, it has to be analyzed on a case-by-case basis, and its results cannot be extrapolated to a general case. The gravity signature can be described in most cases as a gravity low due to the density reduction in fractured rock volumes, plus the negative effect of low-density post-impact sediments as cover material. However, the use of gravity surveys is limited due to the lack of a finer sampling and resolution. In this paper, numerical modeling results are used to predict the size, shape, and source of magnetic anomalies. The initial conditions for post-impact thermal evolution of impact structures can be addressed by numerical modeling (Ivanov and Deutsch, 1999). With that, the geological processes responsible for geophysical signatures of impact structures can be addressed.

REVIEW OF THE MAGNETIC SIGNATURE OF IMPACT STRUCTURES

Two main reviews of the geophysical signature of terrestrial impact structures were done by Pilkington and Grieve (1992) and Grieve and Pilkington (1996). During the past 10 years, there have been many advances in geophysical techniques, such as data collection with higher resolution, data processing with better techniques and algorithms, and better hardware. Furthermore, over the last decade, new seismic, potential field, and borehole-logging data has been collected on many impact sites, such as Lake Bosumtwi (Karp et al., 2002; Scholz et al., 2002), Mjølnir (Barents Sea, Norway [Tsikalas et al., 1998, 1999, 2002]), Sudbury (Hearst and Morris, 2001; Morris, 2002; Mueller and Morris, 1995; Wu et al., 1995), Vredefort (Henkel and Reimold, 1998, 2002), Chicxulub (Morgan et al., 1997, 2002; Pilkington and Hildebrand, 2000; Rebolledo-Vieyra, 2001) and Popigai, Russia (Pilkington et al., 2002).

The current knowledge regarding the magnetic signature of impact craters will be addressed below.

The General Magnetic Signature of Impact Craters

The magnetic signature of terrestrial impact structures is the combined effect of (1) a disruption of the main magnetic trends due to shock and/or thermal demagnetization; and (2) high-amplitude and short wavelength magnetic anomalies in the center of midsize to large impact structures (Pilkington and Grieve, 1992). Figure 1 shows the gravity and magnetic signature of Chicxulub (Rebolledo-Vieyra, 2001). There is a clear pattern of high-amplitude and short wavelength magnetic anomalies in the center, which can be attributed to natural remanent magnetization (NRM) and a magnetic susceptibility increase caused by hydrothermal alteration (Rebolledo-Vieyra, 2001).

Although the focus of previous studies has been mainly on understanding the demagnetization processes observed in many impact structures (Pilkington and Grieve, 1992; Scott et al., 1997), borehole and petrophysical data demonstrate that NRM is the main cause for the high-amplitude and short wavelength anomalies that characterize midsize to large impact craters. The magnetic anomalies are related to the following geological units of an impact structure:

Crater floor: There are cases where the crater floor exhibits an enhanced magnetic signature. In the Sudbury case, the broad regional magnetic anomaly is ascribed to a more magnetic layer of the Levack Gneiss Complex (Hearst et al., 1994) subjacent to the Sudbury Igneous Complex, which is the differentiated impact melt layer. At the Manicouagan impact structure (Quebec), there is a magnetic anomaly larger than 3000 nT, which cannot be explained solely by induced magnetization and the measured magnetic susceptibilities. A localized, impact-generated magnetization produced in an uplifted mafic rock unit is suggested as the cause of this feature (Coles and Clark, 1978).

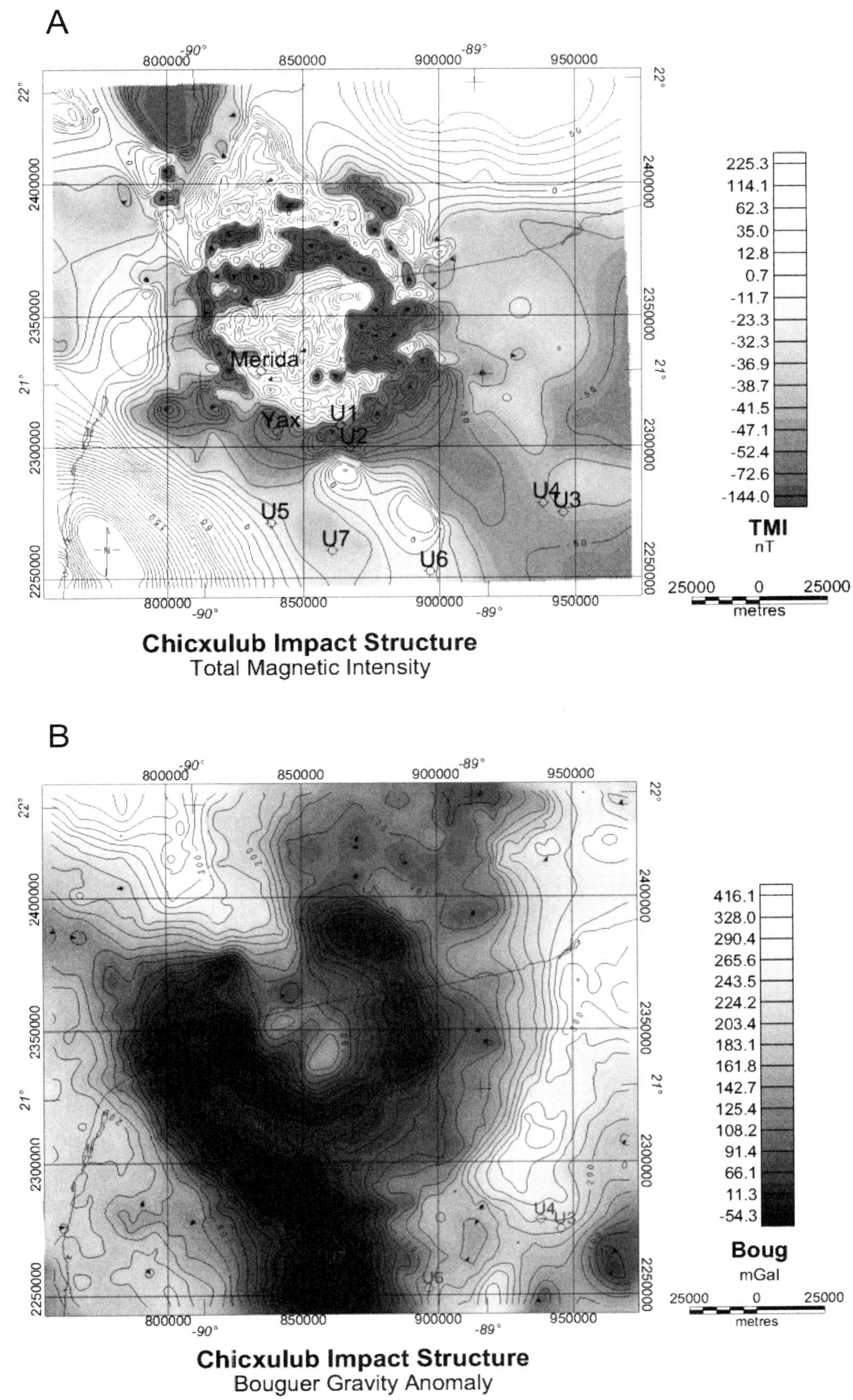

Figure 1. Chicxulub impact structure. (A) Total magnetic field. Contour interval, 10 nT. (B) Bouguer gravity anomaly. Contour interval, 5 mGal. The location of the shallow boreholes drilled by Universidad Nacional Autónoma de México (U) is shown (Rebolledo-Vieyra, 2001), as well as the Yaxcopoil-1 (Yax) borehole drilled by the International Continental Scientific Drilling Program. Location of the city of Merida is also shown. Magnetic and gravity data from Rebolledo-Vieyra (2001).

Impact melt rocks: At the Sudbury impact structure, magnetic properties of the Sudbury Igneous Complex are much higher than those of target rocks (Hearst et al., 1994). At Chicxulub, magnetic anomalies are assumed to be associated to impact melt rocks and suevite layers, which are presumed to be there although their thickness is not well constrained (Rebolledo-Vieyra, 2001).

Impact melt rocks and breccia: The Bosumtwi crater shows high-amplitude and high-frequency anomalies on the NE and SW of the structure. Plado et al. (2000) suggests that the impact structure, with its Fe-rich target (up to 9.2 wt% Fe_2O_3), contains melt rocks with high magnetization. The presence of a biotite-rich granite intrusion in the northern part of the lake is linked to ferromagnetic iron oxides decomposed from biotite by the shock and presented as an explanation of the large amplitude anomalies observed in the same area (Plado et al., 2000). The suevites in that part of the structure may have acquired their NRM by postshock thermochemical processes. In the Ries structure, the central part of the crater is characterized by negative anomalies ~−300 nT, which are related to the thick layer of reversely magnetized fall-back suevite below the lake sediments (Pohl et al., 1977). Some negative magnetic anomalies of smaller extent are associated with local fallout suevite packages below shallow sediments (Pohl et al., 1977).

Post-impact sediments: Hydrothermal alteration can dissociate Fe-Mg silicates to more magnetic mineral phases that will contribute to the magnetic signature of the usually non-magnetic sediments. This is the case at Chicxulub, where borehole logging measurements carried out in one of the Universidad Nacional Autónoma de México (UNAM) boreholes, UNAM 7, indicate increasing susceptibility with depth within the Tertiary sediments overlaying the impact breccias (Fig. 2; Rebolledo-Vieyra, 2001).

Magnetic Signature over Lunar and Extra-Terrestrial Craters

Vector magnetic field observations acquired by the *Mars Global Surveyor* show a direct correlation between the location of the magnetic sources and the ancient cratered terrain of the Martian highlands (Acuña et al., 1999). Observation of the Martian magnetic field shows that the three largest impact basins (>1000 km in diameter) Hellas, Argyre, and Isidis do not exhibit magnetic anomalies, suggesting that the internal dynamo must have ceased to operate prior to the impacts more than 4 Ga (Acuña et al., 1999). The Martian iron-rich core probably formed during the hot accretion of Mars, ca. 4.5 Ga, and subsequently cooled at a rate dictated by the overlying mantle. Mars' core dynamo operated much like that of the current Earth's, but it was probably limited in its duration to several hundred m.y. The dynamo might have stopped because of a change in the cooling rate of Mars' mantle or a switch in convective style that might have lead to mantle heating (Stevenson, 2001).

Azimuthally averaged magnetic intensity shows that the interior of the big Martian impact basins is completely demagnetized up to 80% of the basin radius, and partially demagnetized to ~1.2 times the radius (Mohit and Arkani-Hamed, 2004). As in terrestrial impact craters, the two mechanisms that could demagnetize such large areas are thermal and shock demagnetization. Thermal demagnetization operates by heating the rocks above their Curie temperature. If the rocks are exposed to a magnetic field during their cooling, they get remagnetized in the direction of that field (NRM). On the other hand, shock pressures around 1 GPa are enough to demagnetize any existent NRM in the rocks (shock demagnetization; Cisowski and Fuller, 1978). However, if the shock is applied under the presence of a magnetic field, an additional component of remanence parallel to the ambient field is recorded (shock remanent magnetization, SRM; Cisowski and Fuller, 1978; Halls, 1979). Numerical modeling estimations of the *P-T* conditions of impact events show that temperatures above the Curie isotherm for common ferromagnetic and ferrimagnetic minerals (580 °C for magnetite, 675 °C for hematite) cannot extend that far from the center of the impact. On the other hand, the shock pressure distribution calculated by Mohit and Arkani-Hamed (2004) and correlated with the magnetic intensity maps shows that pressures ~1–2 GPa are sufficient for total demagnetization, and partial demagnetization occurs at pressures down to 0.5 GPa. As magnetite undergoes substantial demagnetization at pressures >0.5 GPa (Cisowski and Fuller, 1978), while hematite-rich rocks show little shock demagnetization under similar pressures (Hargraves and Perkins, 1969), the mechanism involved should be a combination of both shock and thermal demagnetization, depending on the magnetic carrier mineral present (Mohit and Arkani-Hamed, 2004; Rochette et al., 2003).

Surface and satellite observations of lunar crustal magnetization and the remanent magnetization of lunar samples returned by the Apollo missions of 1969–1972 provide evidence for past magnetic fields on the Moon. More recently, the *Lunar Prospector* spacecraft provided a low altitude (15–30 km), improved mapping of the lunar crustal magnetic and gravity fields (Hood et al., 2001). Dynamo generation of the magnetic field in a molten, electrically-conducting core during an interval ca. 3.6–3.8 Ga is consistent with most of the observations but encounters theoretical difficulties associated with the deduced magnitude of the ancient field (Collinson, 1993). An alternative process is the generation of a transient magnetic field during meteorite impacts followed by thermoremanent or shock magnetization of debris and adjacent crust (Collinson, 1993; Hood et al., 2001).

Lunar magnetic fields measured by the *Lunar Prospector* over multi-ring impact basins and smaller craters (50 km to a few hundred km diameter) show that in both cases the primary signature is a decrease in magnetization up to 1.5–2 in basin radii, suggesting shock demagnetization (Halekas et al., 2002, 2003). Some Late Imbrian (older than ca. 3.85 Ga) large impact basins exhibit central magnetic anomalies as a secondary signature, which may be related to thermal remanence magnetization (TRM) of impact melt rocks and/or SRM in the central uplift (Halekas et al., 2003).

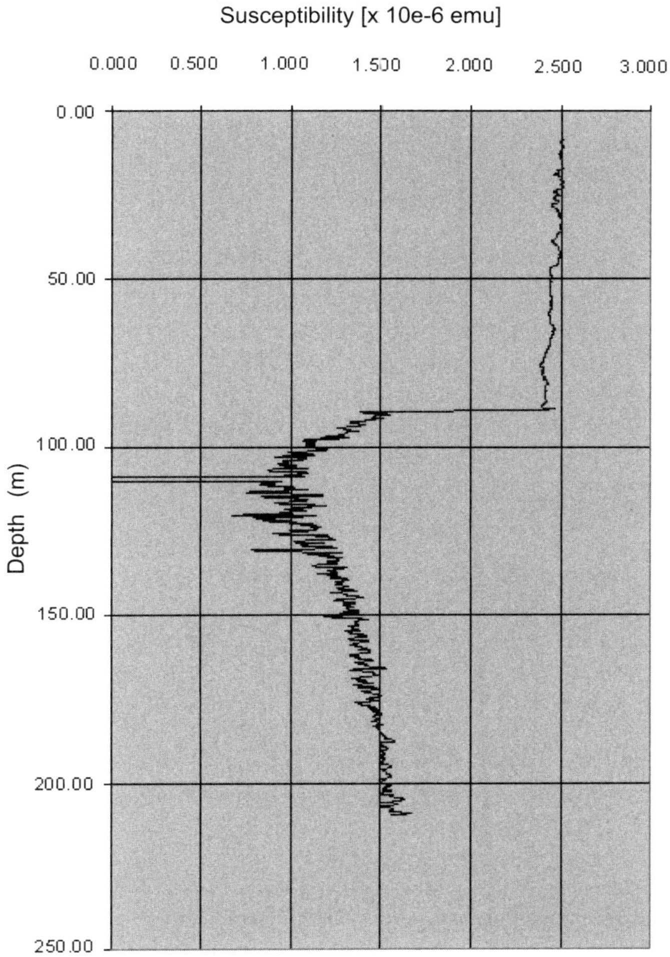

Figure 2. Borehole susceptibility log at the Universidad Nacional Autónoma de México (UNAM) UNAM 7 borehole in Chicxulub. Susceptibility is in 10^{-6} emu. At that depth, the stratigraphy consists of Tertiary sediments overlaying the impact breccias; the data indicate an increase in susceptibility with depth. Data collected by the University of Toronto, McMaster University, and UNAM in May 2002.

Processes that Cause and Modify the Magnetic Signature of Impact Structures

The compilation of literature data shows that the magnetic signature of an impact structure is not as simple as a reduction of magnetic properties of the target rocks. Evidence shows that NRM is the main effect responsible for high-amplitude and short wavelength anomalies on both terrestrial and lunar craters (Halekas et al., 2003).

The magnetic signature observed on large impact structures is thought to be a combination of three effects:

1. Composition and properties of target rocks. At identical impact conditions, the target controls the volume of impact melt, and the magnetic contrast of the melt rocks and breccias, and therefore, the associated magnetic anomaly (Plado et al., 1999).

2. Modification of magnetic carriers due to high impact–related *P-T* conditions (Chao, 1968; Henkel and Reimold, 2002; Pilkington and Grieve, 1992). This could either increase or decrease the magnetization of rocks. The passage of the shock wave may cause a dissociation of Mg-Fe silicates, for example into fine grained magnetite by shock thermal decomposition (STD; Fel'dman, 1994, 2001). Biotite decomposes into magnetite at pressures higher than 40 GPa (Chao, 1968) or when heated in air (Pohl, 1971). On the other hand, impact processes can destroy or modify the original magnetic minerals during melting and the subsequent cooling of the rocks. The distinction between both scenarios (increase or decrease of rock magnetization) is attributed to primary differences in the occurrence of magnetic minerals and the oxygen fugacity during rock cooling (Kukkonen et al., 1992).

3. NRM. As shown in Table 1, the Koenigsberger ratio of remanent over induced magnetization (Q) is generally much larger than 1. For impact structures, the NRM can be the aggregate of three components: TRM, SRM, and chemical remanent magnetization (CRM). In the case of complex craters, TRM is acquired on both impact and basement rocks due to cooling after of the heating and uplift of material (Henkel and Reimold, 2002) to below the Curie temperature in presence of a magnetic field. The resetting of any existent NRM due to temperatures above the Curie temperature of the magnetic carriers is known as thermal demagnetization. If the heated rocks cool down under the influence of a magnetic field, the demagnetization is temporary and a new TRM is acquired; if there is no ambient magnetic field during the cooling of the rocks, demagnetization will be permanent, as it is on Mars (Rochette et al., 2003; Mohit and Arkani-Hamed, 2004). Rapid cooling of the melt layer can result in the formation of paramagnetic impact glasses (Pohl, 1971), whereas TRM may be acquired in the case of a slowly cooled melt (Pilkington and Grieve, 1992). In the central area of a structure (Ugalde et al., 2003), SRM contributes as a secondary remanence component. It is proportional to the ambient field strength and decreases with distance from the impact point (Henkel and Reimold, 2002; Pilkington and Grieve, 1992). Although SRM is restricted to the low coercivity fraction of the minerals capable of carrying remanence (Cisowski and Fuller, 1978; Halls, 1979; Henkel et al., 2002), longer and more intense shock pulses may affect the higher coercivity fraction (Cisowski and Fuller, 1978). A SRM component has been recognized in Manicouagan (Coles and Clark, 1978) and at Slate Islands (Halls, 1979).

Pressures in the range of 1–5 GPa fracture and brecciate the rocks. Brecciation and fluid circulation trigger a hydrothermal system. This has two effects:

(*i*) Acquisition of CRM, which is controlled primarily by the oxygen fugacity (fO_2) and the cooling speed and only secondarily by bulk chemistry (Grant, 1985). High fO_2 generates higher magnetization intensities (Grant, 1985; Pilkington and Grieve, 1992). Rapid cooling allows most of the volatiles to escape, while Fe and Ti oxides stay in solid solution in equilibrium, facilitating low magnetic susceptibilities. On the other hand, high magnetic susceptibilities can be reached with the same chemical composition, but with

TABLE 1. MEASURED MAGNETIC PROPERTIES AT SELECTED IMPACT SITES

Structure	Rock type	k [× 10^{-6} SI]	NRM [mAm^{-1}]	Q
Lake Karikkoselkä, Finland (simple crater,	Impact breccia	225	2.3	0.22
Ø ≤ 2.4 km)	Fractured gray porphyry granite (740 m)	249	0.95	0.30
(Pesonen and Elo, 1999)	Less fractured gray porphyry granite (960 m)	297	0.76	0.08
	Unfractured gray porphyry granite (2300 m)	366	10.2	0.66
Kärdla, Estonia	Impact breccia (center)	2860	120	2
(Complex crater, Ø = 4 km)	Fractured granitic (center)	4580	380	2.3
(Plado et al., 1996)	Fractured granitic (CU)*	8290	290	0.8
	Fractured granitic (rim wall)	650	150	15.3
	Unfractured granitic	15750	770	2.9
Lake Bosumtwi, Ghana (Complex crater,	Target, granites and graywacke (mean value)	150	0.6	0.13
Ø = 10.5 km) (Plado et al., 2000)	Melt-rich suevite (mean value)	330	36.8	4.43
Kara, Russia	Argillite (target)	701	3.1	0.1
(Complex crater, Ø = 65–120 km)	Sandstones (target)	520	11	0.3
(Öhman et al., 2003)	Melt-matrix breccia	522	199	4.3
	Suevite	532	229	3.1
	Tagamite (melt)	586	134	3.7
Chicxulub, Mexico	Melt (Y6)†	12000	80–600	
(Complex crater, Ø ~ 180 km)	Melt breccia (Y6)	3000	120–350	0.12–14.99
(Pilkington and Hildebrand, 2000;	Ejecta breccia (U7)§	5.5		
Rebolledo-Vieyra, personal commun., 2003)	Tertiary carbonates	0–1		
	Bulk melt sheet		<600	
	Shallow melt sheet (center)		4000–5000	
Sudbury, Ontario	South Range			10–100
(Complex crater, Ø ~ 250 km)	Norite	4500	75	
(Hearst et al., 1994; Mueller and Morris, 1995)	Granophyre	8750	100	
	Altered granophyre	15000	75	
	Onaping	5000	560	
	Quartz-rich gabbro	35000	75	
	Chelmsford	10000		
	North Range			
	Norite	35000	750	
	Granophyre	25000	750	
	Onaping	25000	750	
Vredefort, South Africa	Granitic basement (16 sites)	1.59–249	180–131790	1.6–500
(Complex crater, Ø ~ 300 km, ca. 2.0 Ga)				
(Hart et al., 1995)				

Note: k—magnetic susceptibility; NRM—natural remanent magnetization; Q—Koenigsberger ratio of remanent to induced magnetization.
*CU—central uplift.
†Y6—borehole Yucatan-6.
§U7—borehole UNAM7.

a slower cooling rate (Grant, 1985). Biotite and amphibole can decompose to an assemblage of phases that often contain magnetite when they are exposed to high post-impact temperatures (Chao, 1968).

(*ii*) Dissociation of Fe-Mg silicates contributes with more magnetic mineral phases, whereas oxidation of preexisting magnetic carrier phases, such as magnetite, or titanohematite, toward hematite reduces magnetic susceptibilities (Grant, 1985; Henkel and Reimold, 2002). Impact-related hydrothermal systems have been recognized in Kärdla (Kirsimäe et al., 2002; Versh et al., 2003), Kara (Öhman et al., 2003), Chicxulub (Pilkington and Hildebrand, 2000), Lake Lappajärvi (Elo et al., 1992), and Sudbury (Ames et al., 1998). This is a characteristic feature of all craters due to the temperature gradient between lithologies, and since water is normally present in pores or hydrous phases.

Case Studies and Examples of Magnetization on Terrestrial Impact Structures

Table 1 shows magnetic susceptibility (k), NRM, and Q values for various size impact structures in different geological targets. Table 2 summarizes the main processes involved in the acquisition and modification of magnetization for the same impact structures.

Lake Karikkoselkä and Kärdla both show a decrease in magnetic susceptibility in impact formations; however, Lake Karikkoselkä shows CRM (Pesonen and Elo, 1999; Plado et al., 1996). In Kärdla, a secondary component of remanence was isolated that may correspond to SRM; unfortunately, the cores were not oriented to determine the origin of this component (Plado et al., 1996).

On Kara, tagamites (impact melt rocks) and impact melt breccias show large variations in their magnetic susceptibilities: both rock types have lower and higher values than the majority of target rocks (T. Öhman, 2003, personal commun.). Fresh glasses are a more coherent group with a higher magnetic susceptibility than target rocks. However, bulk suevites and altered glasses usually have lower susceptibilities than the target rocks and less variation than the melt lithologies (T. Öhman, 2003, personal commun.). For NRM, the amount of data is not as extensive as for susceptibilities, but certain features seem to be evident: melt rocks and fresh glasses have significantly higher NRM than target rocks; suevites also have slightly higher NRM. Altered glasses, which are much more common than fresh ones, have more or less the same low NRM as the target material. To summarize generally: the higher the degree of alteration, the lower the susceptibility and NRM (T. Öhman, 2003, personal commun.). Since there are no samples from the crater center where SRM might have an impact, its presence is unknown.

At the Chicxulub crater, magnetic susceptibilities increase toward the base of the suevite packages (Rebolledo-Vieyra, 2001), suggesting a hydrothermal origin. Paleomagnetic analysis of the sedimentary sequence between the impact breccias and the post-impact carbonate sequence on Yaxcopoil-1 shows a weak NRM, which is attributed to detrital remanent magnetization (DRM; Rebolledo-Vieyra and Urrutia-Fucugauchi, 2004). As for the impact breccias units on the same borehole, although their redeposition suggests DRM, the stability of demagnetization and the bigger NRM amplitude suggest that the acquisition mechanism is TRM (Rebolledo-Vieyra and Urrutia-Fucugauchi, 2004).

Vredefort is one of the largest known impact structures, with a diameter ≥250 km and a central uplift ~80 km wide. Thus, the area of magnetic resetting due to high temperatures was at least

TABLE 2. MAGNETIC PROCESSES INVOLVED IN THE GENERATION OF THE CURRENT MAGNETIC SIGNATURE OF THE IMPACT STRUCTURES OF TABLE 1

Structure	Modification of magnetic carriers	TRM	SRM	CRM (Hydrothermal)
Lake Karikkoselkä, Finland (Pesonen and Elo, 1999)	Yes, decreasing susceptibility in impact formations.	Melting constrained to small area.	No (high pressures constrained to small area). Secondary remanence is not on low coercivity and shows scatter.	Yes
Kärdla, Estonia (Plado et al., 1996)	Yes, decreasing susceptibility in impact formations.		Possibly (lack of orientation of cores).	Hydrothermal system detected, no CRM found.
Lake Bosumtwi, Ghana (Plado et al., 2000)	Yes, decomposed biotite from granite into ferromagnetic iron oxides.	Yes, both normal (N) and reverse (R) components observed in suevites.	Unknown	
Kara, Russia (Öhman et al., 2003)	Yes, due to alteration.			
Chicxulub, Mexico (Pilkington and Hildebrand, 2000; Rebolledo-Vieyra, 2001)		Yes. Big high-frequency anomalies on center.	Unknown	Yes, hydrothermal system might have caused increase on susceptibility toward base of package of suevites.
Sudbury, Ontario (Hearst et al., 1994)	Yes	Yes (Hearst et al., 1994).		Yes (Ames et al., 1998).
Vredefort, South Africa (Hart et al., 1995)		Yes (Q* >> 1) (Basement rocks demagnetized after uplift).		

Note: TRM—thermal remanent magnetization; SRM—shock remanent magnetization; CRM—chemical remanent magnetization.
*Q = Koenigsberger ratio of remanent to induced magnetization.

as big as the central uplift, a result that was seen in the NRM measurements recorded by Henkel and Reimold (2002).

Magnetic Source Locations and Mechanisms that Cause Magnetic Anomalies

The location of magnetic sources can be related to the mechanisms of their generation as follows:

Crater floor: Fracturing and brecciation occurs at pressures in the range 1 GPa < P < 5 GPa. Impact induced fractures, plus the presence of fluids, hot melt rocks, and the central uplift as a heat source are sufficient to trigger a hydrothermal system. Hydrothermal alteration can modify magnetic carriers toward hematite type minerals that are more remanent-proactive.

Impact melt rocks: Pressures >50 GPa melt wet granitic rocks. Before reaching the melting point, pressures >40 GPa are able to dissociate Mg-Fe silicates, for example into fine grained magnetite (Fel'dman, 1994, 2001). Therefore, it is highly probable that impact melt rocks have a greater susceptibility than the target rocks. Once the impact melt rocks cool down below the Curie temperature of the rock, they save the current direction of Earth's magnetic field, resetting the NRM component.

Impact melt breccias: The impact melt particles within the breccia have the same pressure and temperature conditions as impact melt rocks. The mechanisms of generation of magnetic signatures are hydrothermal alteration, NRM, and Mg-Fe dissociation due to high pressures. In the case of fragmental or lithic breccia, hydrothermal alteration and NRM are the main mechanisms that control magnetization. The amplitude of NRM is controlled largely by the composition of the target rocks.

Post-impact sediments: DRM and hydrothermal alteration are the main causes of magnetic anomalies in this zone of the structure.

NUMERICAL MODELING

A useful instrument for the analysis of geophysical anomalies at impact structures is numerical modeling of a high-velocity impact. The calculated values of maximum pressures and temperatures and their initial and final distributions in space can be linked to geological and physical and/or chemical processes that lead to different geophysical signatures. The source for the observed magnetic anomalies can then be referred to different units or processes: impact melt rocks, breccia, and post-impact alteration.

For this study, numerical modeling of a crater has been performed (1) with the Simplified Arbitrary Lagrangian Eulerian code, version B (SALEB), to produce the final crater shape after a vertical impact; and (2) with the Solid, Vapor, Air (SOVA) code to model the early stage of an oblique impact to estimate the amount of shock modified materials under various impact conditions. We first compare the results of the two hydrocodes and try to define the accuracy of the numerical modeling and reliability of the produced pressure-temperature distributions. Then, the values of maximum temperature and compression with respect to the final position of the target material within impact structures of 1–90 km in diameter, obtained with the SALEB code, are used for geophysical predictions.

Modeling

The two-dimensional SALE code, developed by Amsden et al. (1980) and recently substantially modified into SALEB by Ivanov (2003), includes a strength routine that employs different failure mechanisms to treat solid material, including gradual shear damage accumulation, thermal softening, and acoustic fluidization (Melosh and Ivanov, 1999). The code has been widely used to reproduce craters of various sizes on Earth and other planets (Collins et al., 2002; Ivanov and Artemieva, 2002; Ivanov and Deutsch, 1999; Melosh and Ivanov, 1999; Wünnemann and Ivanov, 2003).

The three-dimensional simulations of oblique impacts are carried out using the SOVA hydrocode (Shuvalov, 1999). The three-dimensional code models multimaterial, large deformation, and strong shock wave flow. The present version of the code works in hydrodynamic approximation (i.e., solids are treated as liquids without strength and viscosity) and cannot reproduce the final crater shape. However, because of its higher accuracy, it works better than SALEB at the initial stage of impact, when pressures are much higher than the material plastic limit and the strength is not as important.

The analytical equation of state (ANEOS) package (Thompson and Lauson, 1972) is used to prepare equation of state tables, describing pressure and temperature dependencies on the density and internal energy of the projectile and target materials over a wide range of both parameters (from cold solid state to hot plasma). These tables are then used by the SALEB or SOVA codes.

The modeled rock mechanics include a strength model for intact and damaged rocks with gradual shear failure, an instant tension failure, dry friction for damaged rocks (Byerlee et al., 1978), the decrease of strength and internal friction close to the melting temperature (Ohnaka, 1995), and an increase in shear strength with pressure increase (Lundborg, 1968). More details about the strength model in the SALEB code can be found in Wünnemann and Ivanov (2003).

To simulate the temporal decrease of friction in damaged rocks around a growing crater, the "block model" (a version of the general acoustic fluidization model) is used (Ivanov and Kostuchenko, 1997; Melosh and Ivanov, 1999). The scaling of the "block model" parameters has been published in Ivanov and Artemieva (2002).

The Eulerian kinematic description employed describes the instantaneous state of the material in each zone, but is unable to follow a point in the material through the mesh. Tracer particles (Pierazzo et al., 1997) are used for this purpose. The tracers are mass-less particles that move through the mesh using the local velocity at the current tracer position. They have no effect on the flow field but allow us to follow the thermodynamic history (density, pressure, temperature) of materials in time. Determination of shock melt and vapor volumes does not demand long term calculations until the final crater is formed. Instead, the total amounts of shock melt and vapor are determined by the value of entropy (or pressure), reached in a shock compression at the earliest stage of impact, as decompression is an isentropic process. Usually, the tracers mark each computational cell, and then the total amount of melt (or vapor) may be defined by summarizing the volumes of tracers with entropy corresponding to melting (vaporization) under normal conditions (for details, see Pierazzo et al., 1997). Tracers also allow definition of the final distribution of the materials within the crater, which initially were at different distances and, hence, were subjected to different compression. Later, *shock* melt may be "bulked up" by digestion of strongly, moderately, and weakly shocked material to become *impact* melt (for discussion of shock melt versus impact melt, see Pope et al., 2004).

The target consists of a homogeneous crystalline layer, modeled with the ANEOS equation of state for granite (Pierazzo et al., 1997). The target is in lithostatic equilibrium and has a low temperature gradient of 13 K/km. The projectile, striking vertically at 15 km/s, is modeled with the same equation of state (EOS), but is treated in the hydrocode as another material, with accurate definitions of the target-projectile boundaries. We vary the projectile size to produce the final crater, with sizes ranging from a 1.5 km diameter simple crater to a 90 km diameter complex crater (Fig. 6). In this paper, we will focus on the results of the 260 m diameter projectile, which produces a 5 km diameter crater (probable size of the Wanapitei structure, Canada; see L'Heureux et al., this volume). As the crater of this size is on the boundary between simple and complex craters on Earth, both complex and simple craters are simulated to predict the pressure and temperature distributions.

Analysis of Results from Two Hydrocodes

Both codes in use, SALEB and SOVA, work in Eulerian mode and include the same EOS, but are rather different in their internal structure, the accuracy of the differential equation approximations, the way in which they resolve shocks, etc. Thus, it is not a surprise that the results of the modeling are different. Figure 3 shows the pressure and temperature distribution within the target, referring to the initial position of the material. It is obvious that the SALEB code slightly underestimates the pressure values (i.e., in the SALEB code, the pressure decays more quickly than in the SOVA code) and substantially overestimates the temperature values (up to 200 K). Additional artificial overheating in the SALEB takes place during the ejection of the target material into the atmosphere; this overheating is visible in Figure 3 as the kinks on the 500 K and 373 K isotherms. The temperature values in the SOVA depend on the artificial viscosity. The usage of artificial viscosity allows the production of smoother numerical solutions but leads to overheating

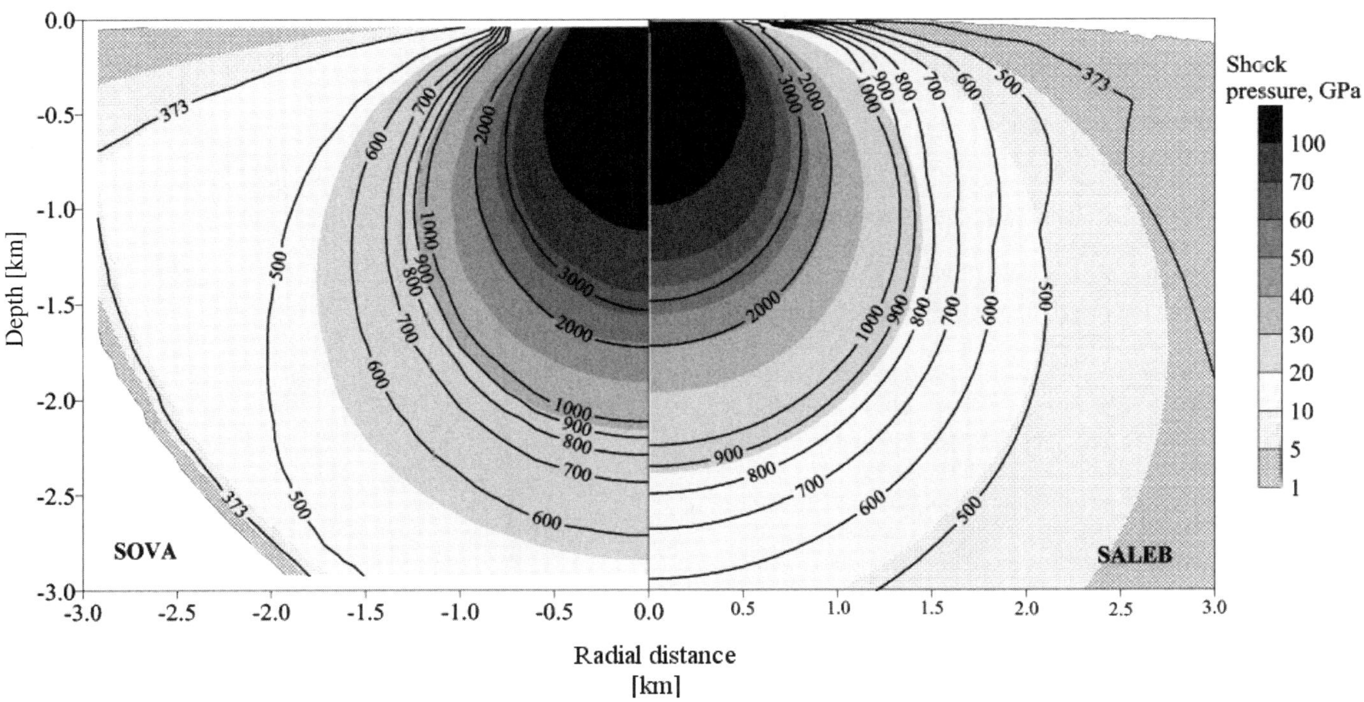

Figure 3. Pressure (gray scale) and temperature (contour lines) distribution with respect to initial material position after the 15 km/s vertical impact of a 760 m diameter projectile. The Solid, Vapor, Air (SOVA) code results are on the left; the Simplified Arbitrary Lagrangian Eulerian code, version B (SALEB) is on the right. The difference in the depth of the 800 K isotherms is ~250 m (0.3 times the projectile diameter, or 2% the final crater diameter).

in comparison with the ideal solution of Hugoniot equations (Amsden et al., 1980). Moreover, the temperature estimates are rather crude in both codes, as the ANEOS equation of state is not perfect and our knowledge of real temperatures in strong shock waves is restricted by the thermodynamical parameters that may be achieved in laboratory experiments (50–80 GPa). However, even these substantial discrepancies in the results take place in the background of high *P-T* gradients, which means that the thickness of the layers within a given temperature interval (e.g., 800–1000 K) is small with respect to the initial material position (as in Fig. 3) and will be even smaller in the final crater because of material motion outward during the excavation-collapse stages. We may consider these uncertainties as the error margin of the numerical models, similar to the accuracy in any geophysical device.

Shock Melt Production and Shock Modified Materials

The best fit for the ratio of the shock melt volume (V_m) to the projectile volume (V_{pr}) from the set of calculations of vertical impacts is $log(V_m/V_{pr}) = a + 3/2\mu\ log(U^2/E_m)$ (Pierazzo et al., 1997), with the coefficients $a = -0.595 \pm 0.064$, $\mu = 0.667 \pm 0.017$, and impact velocity U and melting energy $E_m = 5.2 \cdot 10^6$ J/kg for granite to granite impacts. Assuming an average impact velocity of the asteroid of 20 km/s, $V_m = (15–25)\ V_{pr}$. Our calculations with the SOVA code give a higher value of 26.9; however, it is impossible to compare these numerical estimates directly with the melt volume in real impact craters, because the projectile volume itself is an unknown (or poorly defined) value. Thus, we should establish another ratio, i.e., the ratio of the melt volume to the final crater volume, or find some dimensional connection between the melt volume and the final crater diameter, as the value of final crater volume is poorly defined itself due to erosion, infill by post-impact sediments, and other processes. For this purpose, we use the scaling law by Schmidt and Housen (1987), connecting projectile parameters and target properties. The transient cavity diameter is then associated with the observed crater rim diameter *D* through the Croft (1985) relation. Finally, from the known relationship between the projectile volume and melt volume $V_m = C(U)V_{pr}$, where the coefficient *C* is defined by numerical modeling, we have $V_m = 1.3 \cdot 10^{-2} CD^{3.26} U^{-1.7}$ (final crater diameter *D* is in km, impact velocity *U* in km/s, and melt volume V_m in km³).

It is a well-known fact that the melt volume grows more quickly ($V_m \propto D^3_{pr}$) than the crater volume ($D \propto D^3_{pr}$). The *C*-coefficient for a given projectile-target pair strongly depends on the impact velocity: it is equal to 26.9, 13.8, and 4.8 for 20, 15, and 11.2 km/s, respectively. However, this coefficient, combined with $U^{-1.7}$, leads to a rather weak dependence of the melt volume on the impact velocity: roughly two times the melt volume for the most probable velocity of 20 km/s in comparison to the lowest impact velocity of 11.2 km/s.

Comparisons of our estimates to the database from Grieve and Cintala (1992) and some new data are shown in Figure 4. Even for the lowest impact velocity of 11.2 km/s, numerical estimates are above available data. There are a few explanations for this fact. First of all, these numerical estimates are oversimplified, and a few additional factors may decrease or increase melt production: (1) complex target structure (e.g., a thick sedimentary layer) decreases the melt volume substantially (e.g., at the Ries and Logoisk craters); (2) a substantial portion (up to 50%) of the melt may be ejected from the crater and deposited later as tektites or (in mixture with solid ejecta) as suevitic breccias (the effect may be larger for large craters, in which a substantial part of the melt may be distributed); (3) a higher thermal gradient in the target increases the melt volume and may be substantial for large (~100 km in diameter) craters; (4) some additional amount of melt may be produced by mixing of the hot impact melt with hot rocks (digested rocks); (5) the scaling of the melt in oblique impacts is well established numerically for the known projectile size (Grieve and Cintala, 1992; Pierazzo and Melosh, 2000; Ivanov and Artemieva, 2002); however, the crater scaling for natural oblique impacts is not known yet and may be different from laboratory experiments (Chapman and McKinnon, 1986), depending, for example, on the impact velocity and the projectile and target types. On the other hand, geological estimates are probably underestimations and all later field studies suggest larger melt volumes (for Sudbury: 8000 km³ in Grieve et al. [1991]; 11,500–13,300 km³ in Stöffler et al. [1994]; and 21,800 km³ in Pope et al. [2004]).

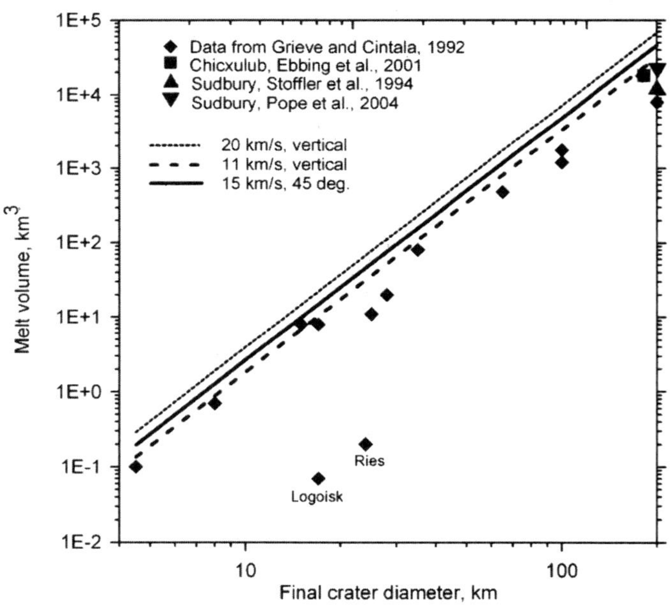

Figure 4. Melt volume versus final crater diameter. Diamonds show the geological estimates of *impact* melt volume (impact melt rocks, impact melt breccias, suevite, but not ejected melt) in Earth's impact craters. Database from Grieve and Cintala (1992) with additional points for Sudbury (Stöffler et al., 1994) and Chicxulub (Ebbing et al., 2001). Solid, dashed, and dotted lines show numerical estimates (Solid, Vapor, Air [SOVA] code) of the *shock* melt, which include ejected melt but do not include digestion of rocks in impact melt rocks. The Logoisk and Ries craters have substantially less melt because of the presence of a thick sedimentary layer.

The total amount of target material compressed above a given pressure level is shown in Figure 5. Impact obliquity decreases the amount of shock-modified material in all pressure intervals. However, this decrease is comparable with the accuracy of the hydrocodes applied, and the amount of shock-modified material in the SALEB modeling of a vertical impact is practically the same as in the SOVA modeling of a 45° impact. Thus, numerical models allow us to define the volume of the shock melt (as well as shock modified materials) for any impact conditions. However, the accuracy of the models is not high enough to look for the differences in P-T distributions in vertical or oblique impacts.

NUMERICAL MODELING RESULTS AND LINKS TO GEOPHYSICS

As shown in the previous section, both qualitative and quantitative connections can be established between modeled results and geophysical data. In this section, we restrict ourselves to a vertical impact model (the SALEB code), producing craters of different sizes (Fig. 6)—from 1 km (simple crater) through 5–10 km (complex crater), and up to 90–100 km (peak-ring crater).

Numerical Modeling Results

Figure 6 shows the pressure and temperature distributions for the different size craters modeled. For the 1.5 km crater, mineral fracturing, shatter cones, and rock fracturing (1 GPa < P < 8 GPa) occurs in a radius of up to 600 m from the impact center and at depths from 400 m to 600 m at the center of the structure. Pressures >30 GPa, related to transformation of individual minerals to diaplectic glasses and quartz conversion to coesite (French, 1998), are confined to small patches within a zone 150 m from the center and from the surface of the crater (at 270 m from the original surface) until 350 m depth. Rock melting (P > 50 GPa) is limited to little occurrences within the same area. Morphologically, the main remark is the formation of a of 270-m-deep depression that subsequently can be filled with sediments.

The 12 km crater exhibits the common morphology of a complex crater: development of a central uplift and a much flatter cavity. In this case, mineral and rock fracturing (1 GPa < P < 8 GPa) goes as far as 4 km from the center and as deep as 5 km. Pressures in the range of 20–30 GPa (minerals transformation) extend to a radius of 2 km from the center and within depths of 0.5–1 km. Rock melting (P > 50 GPa) is still minor.

The 90 km large complex crater shows a vast area where fracturing and brecciation of rocks and minerals could happen (P > 1 GPa): more than 50 km depth beneath the crater center and down to 18 km depth near the crater rim, located 40 km from the crater center. However, this is just the instantaneous fracturing after the passage of the shock wave. Later on, lithostatic pressure is enough at 8 km depth to seal the fractures and reduce the porosity (Pilkington and Grieve, 1992). Pressures >20 GPa are observed in an area from 15 km to 25 km depth at the center to 10 km depth 12 km away from the center. Rock melting goes on in the center of the structure from 15 km deep to the surface and up to 25 km away from the impact point.

The pressure and temperature distributions for the 5 km final-diameter complex crater and the ~4 km diameter simple crater (impact conditions are identical with the exception of the acoustic fluidization model applied for the complex crater) are shown in Figures 7 and 8, based on both initial (Fig. 7) and final (Fig. 8) target material geometry. Pressures in the range of 2–8 GPa (mineral fracturing, shatter cones, and rock fracturing) are found in a radius of up to 1.3 km from the impact center and at depths from 1 to 1.8 km, whereas brecciation (P > 1 GPa) will occur to a depth of 2.5 km. Pressures >30 GPa (transformation to diaplectic glasses, quartz conversion to coesite) are constrained to the ~200-m-thin central part (<500 m) of the final crater, but also incorporate material that originally was at depths of 0.6 km and as far as 0.3 km away from the center of the impact. Rock melting (P > 50 GPa) is confined to a radius smaller than 0.25 km and depths less than 0.4 km from the initial target geometry. Partial vaporization (P > 150 GPa) will involve material originally at a distance <100 m from the impact point. For the simple crater (Fig. 8, left) the range of pressures and temperatures is the same, although the final geometry of the structure is different: the simple bowl-shaped crater is ~300 m deeper compared to the complex crater with the central uplift (Fig. 8, right).

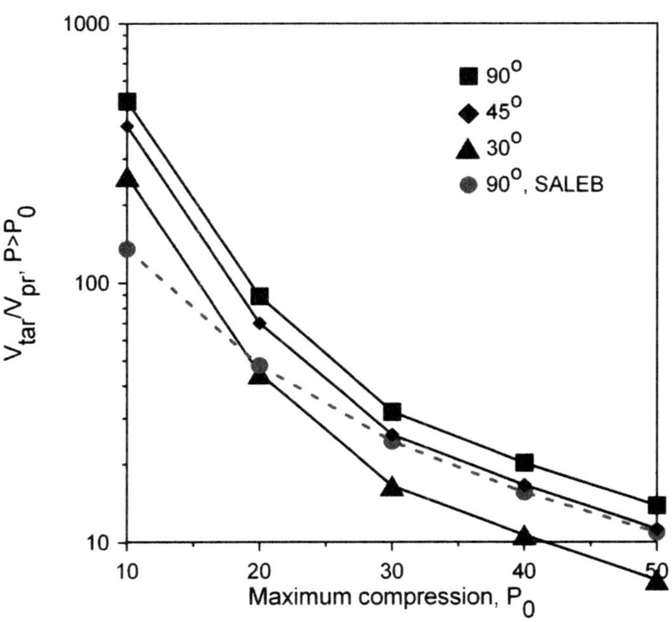

Figure 5. Ratio of target material volume, compressed above a certain pressure level, to projectile volume for various impact angles. For the vertical impact, we compare the Simplified Arbitrary Lagrangian Eulerian code, version B (SALEB), results (gray line) and the Solid, Vapor, Air (SOVA) code results.

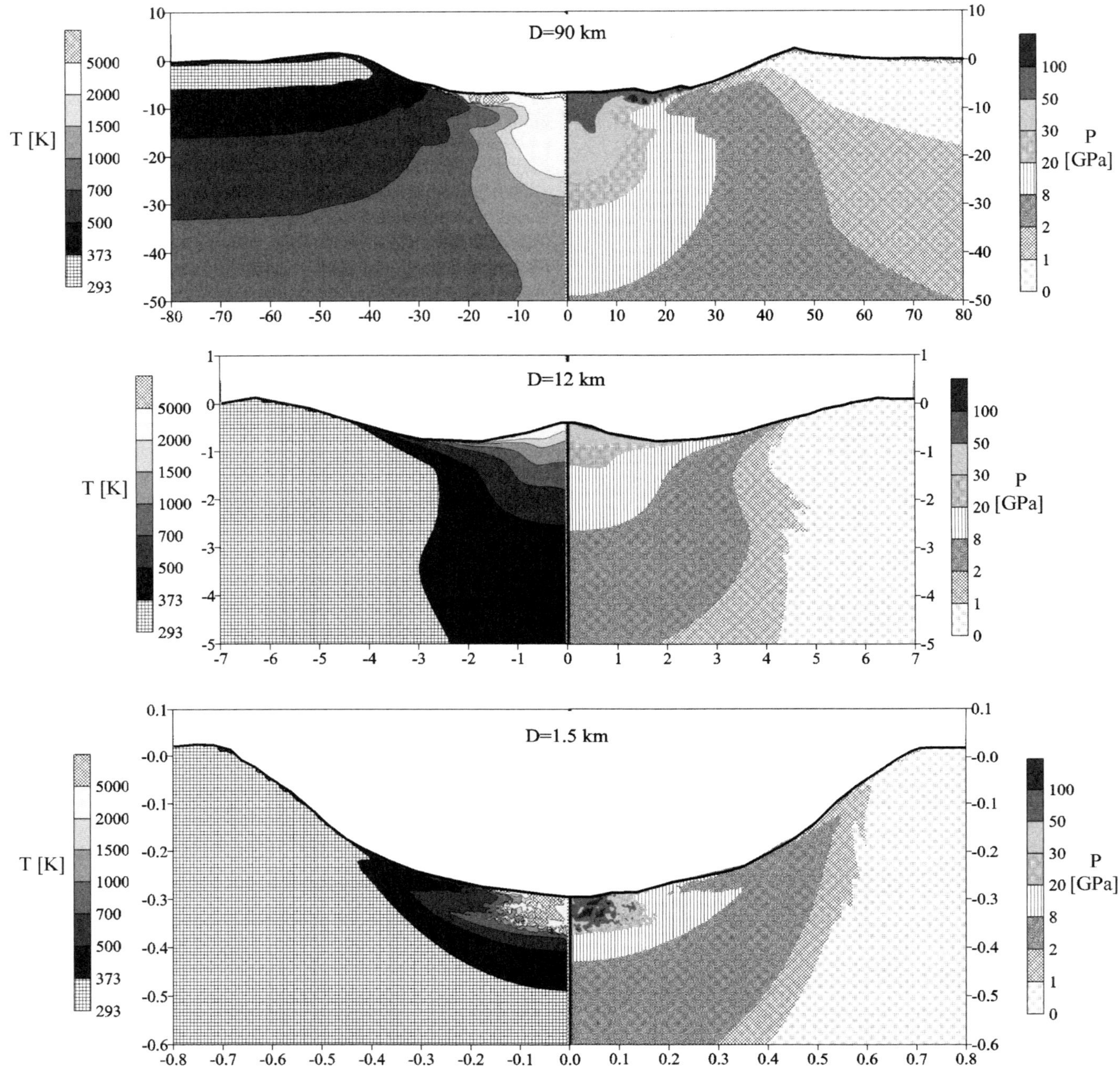

Figure 6. Maximum pressure (right, in GPa) and temperature (left, in K) for three different sized craters. Top: 90 km diameter crater; middle: 12 km diameter crater; bottom: 1.5 km diameter crater. Computation with the Simplified Arbitrary Lagrangian Eulerian code, version B (SALEB), code.

Links to Geophysics

The *P-T* distribution shown in Figure 6 can be related to the processes that modify the magnetic properties of rocks and minerals.

The amount of impact melt rocks ($P > 50$ GPa) remaining in the 1.5 km crater after the modification stage is negligible, as is dissociation of Mg-Fe silicates. SRM ($P > 30$ GPa) is constrained to little patches in the center of the structure, up to 150 m from the center and from 270 to 350 m depth. However, the 373 K and higher temperature isotherms extend from 500 m depth at the center up to a 400 m radius near the surface. Therefore, for this crater size, the three most important elements that will cause and/or modify the magnetic signature are (1) NRM removal and

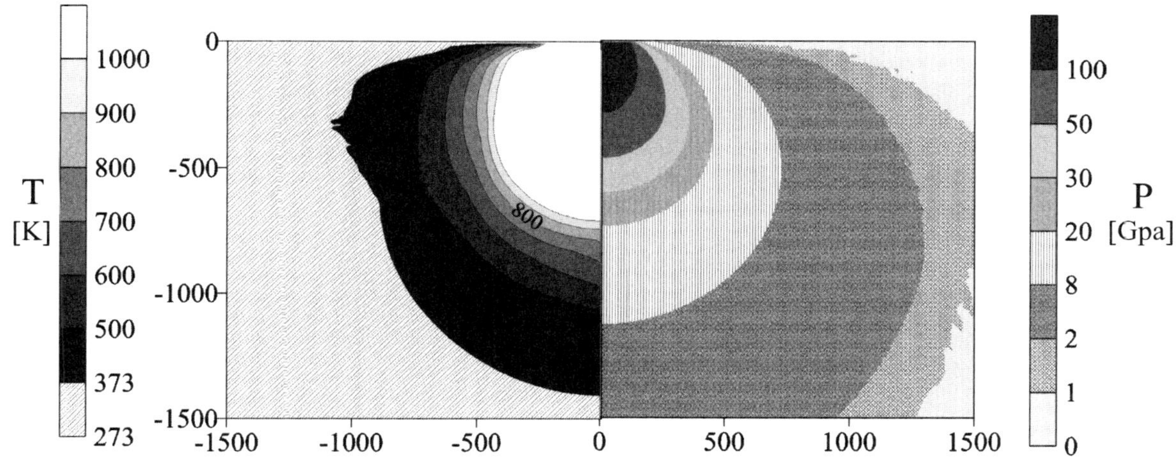

Figure 7. Maximum pressure (right, in GPa) and temperature (left, in K) for a 5 km diameter crater with respect to the initial position of the target material. Computation with the Simplified Arbitrary Lagrangian Eulerian code, version B (SALEB), code.

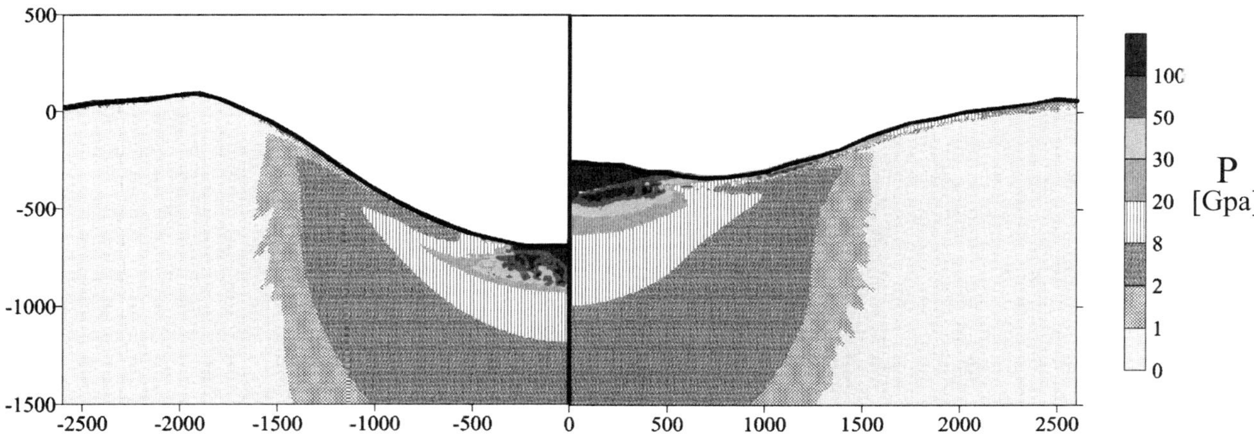

Figure 8. Contours of the maximum pressure in the target with respect to the final position of the rocks in the modified crater. The final crater diameter is 5 km. A complex crater is on the right; a simple crater on the left. The thick black line shows the "visible" crater shape.

resetting because of shock ($P > 1$ GPa); (2) sedimentary infill of the crater, usually with lower magnetic susceptibilities than the target rocks and with possibilities of DRM depending on the sediments deposited and the conditions of the deposition; (3) the fracture and brecciation zone is big enough and there is an adequate amount of heat to allow fluid circulation and start a hydrothermal system that could yield to CRM and/or alter the original minerals toward a more or less magnetic facies.

In the 12 km crater, the three factors mentioned above are equally or even more important: rock fracturing and NRM resetting occurs in this case in a zone with a 4 km radius and 5 km depth. The transformation of minerals due to shock and SRM assumes a more important role. Moreover, as the amount of impact melt rocks increases, this translates into new lithological units unobserved in smaller craters. Cooling of the central uplift to temperatures <100 °C can take less than 500–1000 yr after the impact (e.g., for the 4 km diameter Kärdla crater [Versh et al., 2003]).

The 90 km crater is by far the more complex, taking into account the scale of the processes. Rock melting involves rocks from 5 km to 12 km depth at the center; SRM takes place up to a radius of 15 km and in the zone of 12 km to 25 km depth at the center; the heat source for any subsequent hydrothermal system (>373 K) extends all over the crater, from 5 km depth at a distance of 80 km from the center to near the surface at a radius of 30 km, as well as all the effects mentioned before for the smaller craters (NRM resetting; sedimentary infill with eventual DRM; fracturing and brecciation that creates space for CRM; SRM in the center of the structure). Cooling of the melt sheet is proportional to its thickness, $t_c \sim h_m^2$. For a first order estimation, we can consider this melt sheet as a cylinder with diameter proportional to the final crater diameter, then $V_m \sim D^2 h_m$. As we have shown earlier, the melt volume V_m itself is proportional to $D^{3.26}$. Combining these estimates, we have $t_c \sim D^{2.52}$. Thus, if cooling time for Sudbury with a final diameter of ~250 km is estimated as 0.3–2 m.y. (Ivanov and Deutsch, 1999), then the melt sheet in a 90 km diameter crater reaches the solidus in 20–150 k.y.

From the above statements on the magnetic signature of impact structures, it is now possible to separate their incidence in time and space. As an aid to understand impact and post-impact processes that yield geophysical anomalies, a synthetic model was created and its gravity and magnetic response was computed through two-dimensional forward modeling.

Modeling was done for a crater with a final diameter of 5 km. The target consists of two layers, air and hard rock, with equally spaced vertical diabase dikes every 500 m (Fig. 9A). Density and magnetization parameters used for the modeling are listed in Table 3.

The numerical model shown in Figure 8 was used to create the post-impact geometry of the resultant rocks, while *P-T* predictions allowed estimation of the post-impact petrophysical parameters (Table 3). As the synthetic model was applied to estimate the true size of the Wanapitei Lake impact crater (L'Heureux et al., this volume), starting parameters for host rock and diabase dikes were chosen to be representative of Sudbury Huronian metasediments and Sudbury dikes, respectively. Magnetization parameters were taken from Hearst et al. (1994) and density from McGrath and Broome (1994). There are no petrophysical measurements for these lithologies in its pre- and post-impact states in the area, so the post-impact parameters were estimated from different sources. A density decrease of 150 kg/m^3 was assumed due to fracturing (Pilkington and Grieve, 1992) on a region delimited by 1 GPa < P < 5 GPa, which is predicted by numerical modeling. Plado et al. (1996) showed a 50% decrease in magnetic susceptibility between impact-metamorphosed lithologies inside and outside the Kärdla crater; the same 50% decrease was applied to the pre-impact values on this model. Cisowski and Fuller (1978) demonstrated that a shock pressure of 1 GPa is sufficient to reset any existent remanent magnetization. Next, the remanence magnetization intensity (M_r) was set to zero for the region delimited by P > 1 GPa. Impact breccia density was taken from Dence and Popelar (1972); again, Kärdla was used to estimate the magnetic susceptibility and remanence intensity of the impact breccia (Plado et al., 1996); remanence inclination and declination were taken from remanence magnetization measurements on dike swarms near Sudbury by Strangway (1960). Density and magnetic susceptibility of sediments were estimated from petrophysical tables.

The highly magnetic vertical diabase dikes exhibit two effects: (1) they are melted and vaporized, and therefore removed, up to 300 m depth in the center of the crater; (2) the shock pressure reduces its magnetic susceptibility, magnetic remanence, and density through fracturing up to 1800 m in the center and to a radius of 1500 m. The fractured basement with a reduced density extends from 600 m to 1800 m depth in the center and up to a radius of 1900 m. A unit of breccia of low density but higher susceptibility than the target lies on top of the fractured basement, from 200 m to 550 m depth in the center and up to 1100 m radius. Low density and susceptibility sediments occupy a cavity up to 200 m deep in the center.

The post-impact model gives rise to a 6 mGal negative gravity anomaly, assuming that the stations are on the surface. The magnetic model shows no signature of the diabase dikes (or "magnetic low") on a 1700 m radius. This argument has been used by L'Heureux et al. (this volume) to constrain the size of the Wanapitei crater.

Figure 10 shows a sketch of an impact structure based on numerical modeling results, including the different elements that build geophysical anomalies and their spatial occurrence.

CONCLUSIONS

The geophysical signature of terrestrial impact structures has been reviewed. The magnetic signature and high amplitude and short wavelength anomalies in the center of large structures are primarily caused by NRM. Numerical modeling helps to establish a connection between variables (pressure, temperature, melt volume estimate) that characterize impact processes and the magnetic signatures of large impact craters. In addition, numerical modeling defines favorable settings for post-impact hydrothermal processes and alteration above and below the floor of the craters. Since an important component of the magnetic signature is attributed to melt, magnetic data can be a potential indicator for melt volume estimations. This can have significant consequences on resource exploration, since ~25% of the known terrestrial impact structures are related to some form of economic deposits. Exploration of epigenetic deposits, related to hydrothermal alteration and/or formation of an enclosed topographic basin, or the long-term flow of hydrocarbons into the structural trap formed by the impact structure (Grieve and Masaitis, 1994), can be aided by this model. Until now, not much importance has been given to NRM as a main contributor for magnetization of impact structures. The main reason for this is the lack of petrophysical data and the usual assumption of potential field modeling and interpretation: that all magnetization is induced. This assumption usually leads to erroneous results of modeling and inversion of magnetic data.

A simple synthetic model that resembles the magnetic and gravimetric signature of a 5 km diameter impact crater was presented. The variations from pre- to post-impact rock properties were estimated spatially from numerical modeling, and the computed anomalies appear to be similar in shape and amplitude to what is observed on Wanapitei, a crater of similar size. However, the results have to be interpreted carefully, since the model does not incorporate later processes like hydrothermal alteration that may modify the magnetic properties of rocks significantly. Moreover, because it is a two-dimensional model, it does not take into account any lateral variation in petrophysical properties.

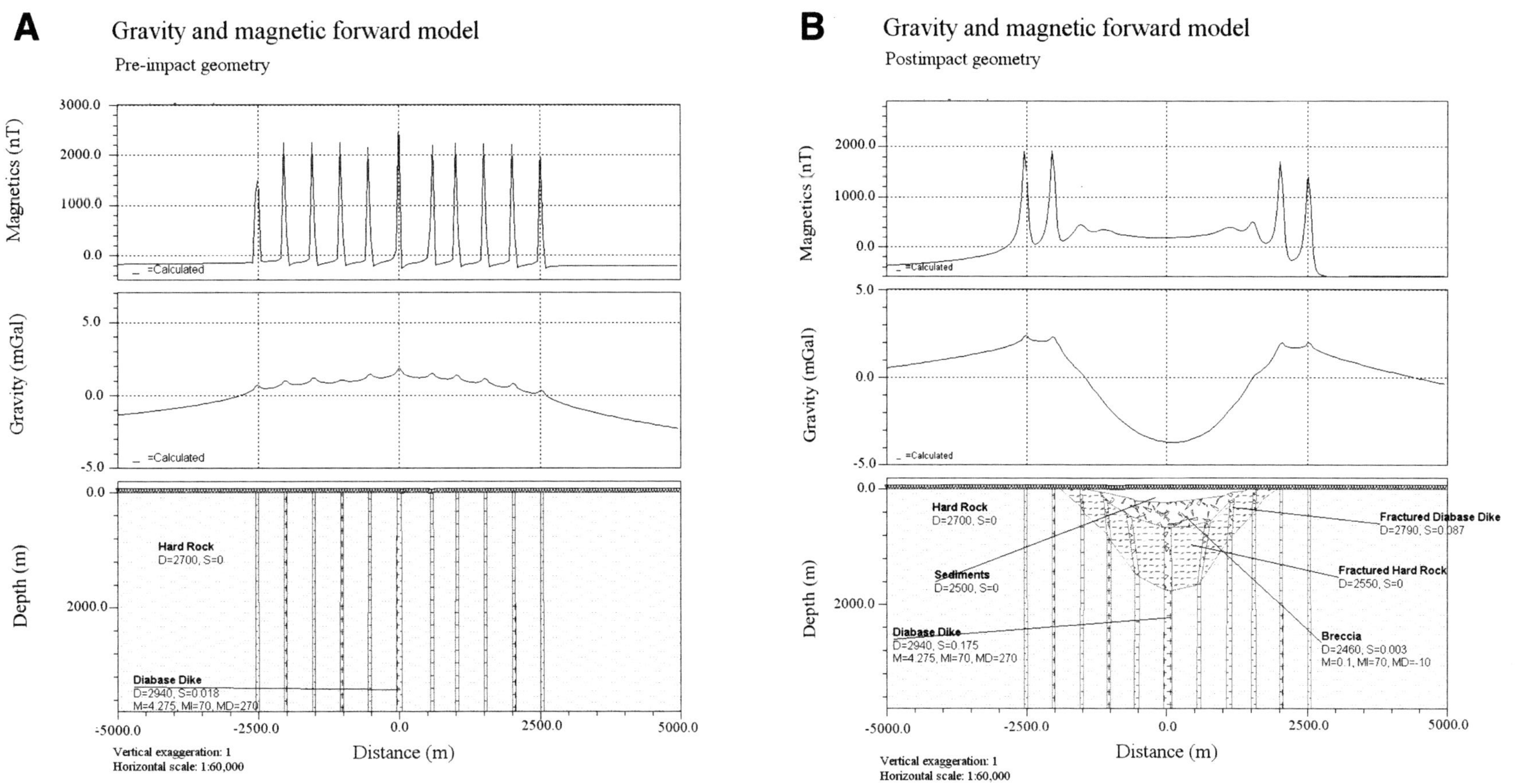

Figure 9. Synthetic magnetic and gravity model for a 5-km-sized crater. (A) Pre-impact geometry: two layers (air and hard-rock) with vertical diabase dikes as markers, (B) Post-impact geometry. Numerical modeling results are used to predict impact-related changes in the physical rock properties; they are listed in Table 3. See text for details.

TABLE 3: PETROPHYSICAL PARAMETERS USED FOR THE 5 KM CRATER FORWARD MODELING

Unit	Pre-impact			Post-impact		
	k (S.I.)	NRM	ρ (kg/m³)	k (S.I.)	NRM	ρ (kg/m³)
Hard rock	0.0003		2700			
Diabase dike	0.175	M_r = 4.275 I = 70° D = 270°	2940			
Fractured hard rock				0.0003		2550
Fractured diabase dike				0.0875	M_r = 0	2790
Impact breccia				0.002	M_r = 0.1 I = 70° D = −10°	2460
Sediments				0.0002	M_r = 0	2500

Note: k—magnetic susceptibility; NRM—natural remanent magnetization; described in the model as intensity (M_r, in A/m), inclination (I, in degrees), and declination (D, in degrees); ρ—rock density.

The estimates described above should be treated mainly qualitatively, not quantitatively. Comparison of numerical results with geological and geophysical data is not simple and straightforward. The reconstruction of impact scenarios should include the analysis of the final crater morphology, the melt/shock-modified material production and the gravity/magnetic anomalies altogether. Even in the case of such a multidisciplinary approach, the results may not be conclusive. This is especially true in the case of multi-ring basins where it is still not well understood what the final crater diameter should be (see Turtle et al., this volume).

ACKNOWLEDGMENTS

We thank T. Öhman and W.A. Morris for the petrophysical data they provided for Kara and Sudbury, respectively. We also appreciate the thoughtful comments of K. Wünneman, G. Collins,

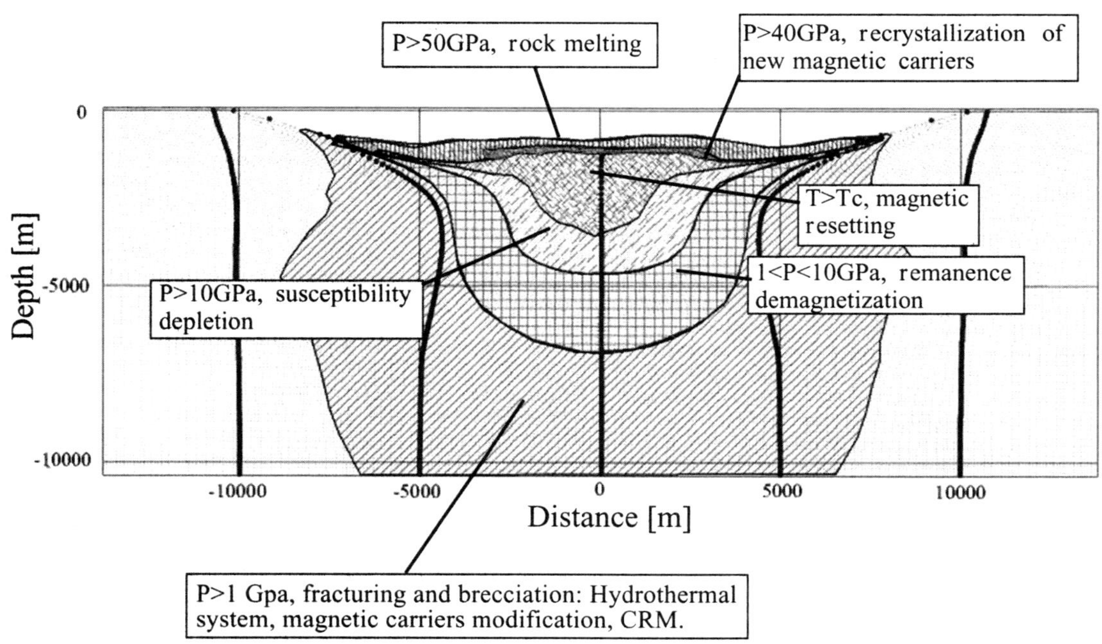

Figure 10. Sketch showing the source of magnetic anomalies. The vertical lines represent the diabase dikes used as markers. The geological processes responsible for changes in the magnetic properties of the target rocks are located spatially based on the numerical model *P-T* predictions. 1 GPa < *P* < 5 GPa: fracturing and brecciation. *P* < 10 GPa: existing remanence demagnetized. *P* > 10 GPa: susceptibility depletion. *P* > 30 GPa: Magnetic resetting. CRM—chemical remanence magnetization.

and A. Deutsch that significantly improved the paper. B. Ivanov kindly helped with the SALE modeling. Funding for N. Artemieva was provided by a University of Toronto grant. The Chicxulub borehole magnetic survey was funded by DFG Grant MI558/5-1 to Milkereit and Janle. H. Ugalde is supported by the Natural Sciences and Engineering Research Council (NSERC) of Canada.

REFERENCES CITED

Acuña, M.H., Cannemey, J.E.P., Ness, N.F., Lin, R.P., Mitchell, D., Carlson, C.W., McFadden, J., Anderson, K.A., Reme, H., Mazelle, C., Vignes, D., Wasilewski, P., and Cloutier, P., 1999, Global distribution of crustal magnetization discovered by the Mars Global Surveyor MAG/ER experiment: Science, v. 284, p. 790–793, doi: 10.1126/SCIENCE.284.5415.790.

Ames, D.E., Watkinson, D.H., and Parrish, R.R., 1998, Dating of a regional hydrothermal system induced by the 1850 Ma Sudbury impact event: Geology, v. 26, p. 447–450, doi: 10.1130/0091-7613(1998)0262.3.CO;2.

Amsden, A.A., Rupel, H.M., and Hirt, C.W., 1980, SALE: A simplified ALE computer program for fluid flow at all speeds: Los Alamos National Laboratories, LA-8095.

Byerlee, J., Mjachkin, V., Summers, R., and Voevoda, O., 1978, Structures developed in fault gouge during stable sliding and stick-slip: Tectonophysics, v. 44, p. 161–171, doi: 10.1016/0040-1951(78)90068-9.

Chao, E.C.T., 1968, Pressure and temperature histories of impact metamorphosed rocks—based on petrographic observations, *in* French, B.M., and Short, N.M., eds., Shock Metamorphism of Natural Materials: Baltimore, Mono Book Corp., p. 135–158.

Chapman, C.R., and McKinnon, W.B., 1986, Cratering of planetary satellites, *in* Burns, J., and Matthews, M.S., eds., Satellites: Tucson, University of Arizona Press, p. 492–580.

Cisowski, S.M., and Fuller, M., 1978, The effect of shock on the magnetism of terrestrial rocks: Journal of Geophysical Research, v. 83, no. B7, p. 3441–3458.

Coles, R.L., and Clark, J.F., 1978, The central magnetic anomaly, Manicouagan Structure, Quebec: Journal of Geophysical Research, v. 83, no. B6, p. 2805–2808.

Collins, G.S., Melosh, H.J., and Morgan, J.V., 2002, Hydrocode simulations of Chicxulub crater collapse and peak-ring formation: Icarus, v. 157, p. 24–33, doi: 10.1006/ICAR.2002.6822.

Collinson, D.W., 1993, Magnetism on the Moon—A Lunar core dynamo or impact magnetization?: Surveys in Geophysics, v. 14, p. 89–118.

Croft, S.K., 1985, The modification stage of basin formation: Conditions of ring formation, *in* Schultz, P.H., and Merrill, R.B., eds., Multi-ring basins: Proceedings of Lunar and Planetary Science: New York, Pergamon Press, p. 227–257.

Dence, M.R., and Popelar, J., 1972, Evidence for an impact origin for Lake Wanapitei, Ontario, *in* Guy-Bray, J.V., ed., New developments in Sudbury geology: Geological Association of Canada Special Paper 10, p. 117–124.

Earth Impact Database, 2004, Planetary and Space Science Centre, University of New Brunswick, http://www.unb.ca/passc/ImpactDatabase (accessed 7 June 2004).

Ebbing, J., Janle, P., Kouloris, J., and Milkereit, B., 2001, 3D gravity modeling of the Chicxulub impact structure: Planetary and Space Science, v. 49, p. 599–609, doi: 10.1016/S0032-0633(01)00005-8.

Elo, S., Jokinen, T., and Soininen, H., 1992, Geophysical investigations of the Lake Lappajärvi impact structure, western Finland: Tectonophysics, v. 216, p. 99–109, doi: 10.1016/0040-1951(92)90158-3.

Fel'dman, V.I., 1994, The conditions of shock metamorphism, *in* Dressler, B.O., Grieve, R.A.F., and Sharpton, V.L. eds., Large Meteorite Impacts and Planetary Evolution: Geological Society of America Special Paper 293, p. 121–132.

Fel'dman, V.I., 2001, Opacitization or shock thermal decomposition?: Lunar and Planetary Science, v. 32, Abstract 1164.

French, B.M., 1998, Traces of a catastrophe: a handbook of shock metamorphic effects in Terrestrial Meteorite Impact Structures, LPI Contribution no 120: Houston, Lunar and Planetary Institute, 120 p.

Grant, F.S., 1985, Aeromagnetics, geology and ore environments, 1, Magnetite in igneous, sedimentary and metamorphic rocks: An overview: Geoexploration, v. 23, p. 303–333, doi: 10.1016/0016-7142(85)90001-8.

Grieve, R.A.F., and Cintala, M.J., 1992, An analysis of differential impact melt-crater scaling and implications for the terrestrial crater record: Meteoritics, v. 27, p. 526–538.

Grieve, R.A.F., and Masaitis, V.L., 1994, The economic potential of impact craters: International Geology Review, v. 36, no. 2, p. 105–151.

Grieve, R.A.F., and Pilkington, M., 1996, The signature of terrestrial impact craters: AGSO Journal of Australian Geology & Geophysics, v. 16, no. 4, p. 399–420.

Grieve, R.A.F., Stöffler, D., and Deutsch, A., 1991, The Sudbury structure: Controversial or misunderstood?: Journal of Geophysical Research, v. 96, no. E5, p. 22,753–22,764.

Halekas, J.S., Mitchell, D.L., Lin, R.P., Hood, L.L., Acuña, M.H., and Binder, A.B., 2002, Demagnetization signatures of lunar impact craters: Geophysical Research Letters, v. 29, n. 13, 1645, doi: 10.1029/2001GL013924.

Halekas, J.S., Lin, R.P., and Mitchell, D.L., 2003, Magnetic fields of lunar multi-ring impact basins: Meteoritics & Planetary Science, v. 38, no. 4, p. 565–578.

Halls, H.C., 1979, The Slate Islands meteorite impact site: A study of shock remanent magnetization: Geophysical Journal of the Royal Astronomical Society, v. 59, p. 553–591.

Hargraves, J.W., and Perkins, W.E., 1969, Investigations of the effects of shock on natural remanent magnetization: Journal of Geophysical Research, v. 74, p. 2576–2589.

Hart, R.J., Hargraves, R., Andreoli, M., Tredoux, M., and Doucoré, C.M., 1995, Magnetic anomaly near the center of the Vredefort structure: Implications for impact-related magnetic signatures: Geology, v. 23, no. 3, p. 277–280, doi: 10.1130/0091-7613(1995)0232.3.CO;2.

Hearst, R., and Morris, W.A., 2001, Regional gravity setting of the Sudbury Structure: Geophysics, v. 66, no. 6, p. 1680–1690, doi: 10.1190/1.1487110.

Hearst, R., Morris, W.A., and Thomas, M., 1994, Magnetic interpretation along the Sudbury structure lithoprobe transect: Geophysical Research Letters, v. 21, no. 10, p. 951–954, doi: 10.1029/93GL02610.

Henkel, H., and Reimold, W.U., 1998, Integrated geophysical modeling of a giant, complex impact structure: anatomy of the Vredefort Structure, South Africa: Tectonophysics, v. 287, p. 1–20, doi: 10.1016/S0040-1951(98)80058-9.

Henkel, H., and Reimold, W.U., 2002, Magnetic model of the central uplift of the Vredefort impact structure, South Africa: Journal of Applied Geophysics, v. 51, p. 43–62, doi: 10.1016/S0926-9851(02)00214-8.

Henkel, H., Reimold, W.U., and Koeberl, C., 2002, Magnetic and gravity model of the Morokweng impact structure: Journal of Applied Geophysics, v. 49, p. 129–147, doi: 10.1016/S0926-9851(01)00104-5.

Hood, L.L., Zakharian, A., Halekas, J., Mitchell, D.L., Lin, R.P., Acuña, M.H., and Binder, A.B., 2001, Initial mapping and interpretation of lunar crustal magnetic anomalies using Lunar Prospector magnetometer data: Journal of Geophysical Research, v. 106, p. 27,825–27,840, doi: 10.1029/2000JE001366.

Ivanov, B.A., 2003, Complex crater formation: verification of numerical models, *in* Impact cratering: Bridging the gap between modeling and observations: Houston, Lunar and Planetary Institute Workshop, February 7–9, 2003, LPI contribution no. 1155, paper no. 8026.

Ivanov, B.A., and Artemieva, N., 2002, Numerical modeling of the formation of large impact craters, *in* Koeberl, C., and McLeod, K.G., eds., Catastrophic Events and Mass Extinctions: Impacts and Beyond: Geological Society of America Special Paper 356, p. 619–630.

Ivanov, B.A., and Deutsch, A., 1999, Sudbury impact event: Cratering mechanics and thermal history, *in* Dressler, B.O. and Sharpton, V.L., eds., Large meteorite impacts and planetary evolution II: Geological Society of America Special Paper 339, p. 389–398.

Ivanov, B.A., and Kostuchenko, V.N., 1997, Block oscillation model for impact crater collapse: Lunar and Planetary Science, v. 28, Abstract 1655, CD-ROM.

Karp, T., Milkereit, B., Janle, J., Danuor, S.K., Pohl, J., Berckhemer, H., and Scholz, C.A., 2002, Seismic investigation of the Lake Bosumtwi impact crater: Preliminary results: Planetary and Space Science, v. 50, p. 735–743, doi: 10.1016/S0032-0633(02)00049-1.

Kirsimäe, K., Suuroja, S., Kirs, J., Kärki, A., Polikarpus, M., Puura, V., and Suuroja, K., 2002, Hornblende alteration and fluid inclusions in Kärdla impact crater, Estonia: Evidence for impact-induced hydrothermal activity: Meteoritics & Planetary Science, v. 37, p. 449–457.

Kukkonen, I.T., Kivekäs, L., and Paananen, M., 1992, Physical properties of kärnäite (impact melt), suevite and impact breccia in the Lappajärvi meteorite crater, Finland: Tectonophysics, v. 216, p. 111–122, doi: 10.1016/0040-1951(92)90159-4.

Lundborg, N., 1968, Strength of rock-like materials: International Journal of Rock Mechanics and Mining Science, v. 5, p. 427–454, doi: 10.1016/0148-9062(68)90046-6.

McGrath, P.H., and Broome, H.J., 1994, Gravity model for the Sudbury structure along the lithoprobe seismic line: Geophysical Research Letters, v. 21, no. 10, p. 955–958, doi: 10.1029/93GL02247.

Melosh, H.J., and Ivanov, B.A., 1999, Impact crater collapse: Annual Review of Earth and Planetary Science, v. 27, p. 385–415, doi: 10.1146/ANNUREV.EARTH.27.1.385.

Mohit, P.S., and Arkani-Hamed, J., 2004, Impact demagnetization on Martian crust: Icarus, v. 168, no. 2, p. 305–317, doi: 10.1016/J.ICARUS.2003.12.005.

Morgan, J., Warner, M., and The Chicxulub Working Group, 1997, Size and morphology of the Chicxulub impact crater: Nature, v. 390, p. 472–476, doi: 10.1038/37291.

Morgan, J.V., Christeson, G.L., and Zelt, C.A., 2002, testing the resolution of a 3D velocity tomogram across the Chicxulub crater: Tectonophysics, v. 355, p. 215–226, doi: 10.1016/S0040-1951(02)00143-9.

Morris, W.A., 2002, The Sudbury Structure: A circular impact crater?: Geophysical Research Letters, v. 29, no. 20, doi: 10.1029/2002GL015255.

Mueller, E., and Morris, W.A., 1995, A 3-D model of the Sudbury igneous complex: Society of Exploration Geophysicists Annual Meeting, October 8–13, 1995, p. 777–780.

Öhman, T., Lorenz, K., Pesonen, L.J., Badjukov, D., Raitala, J., Elo, S., and Ojala, K., 2003, Kara impact structure, Russia: Recent developments in petrophysical and geochemical studies: 3rd International Conference on Large Meteorite Impact Craters, August 5–7, 2003: Nördlingen, Germany, abstract no. 4071.

Ohnaka, M., 1995, A shear failure strength law of rock in the brittle-plastic transition regime: Geophysical Research Letters, v. 22, p. 25–28, doi: 10.1029/94GL02791.

Pesonen, L.J., and Elo, S., 1999, Lake Karikkoselkä impact structure, central Finland: New geophysical and petrographic results, *in* Dressler, B.O. and Sharpton, V.L., eds., Large meteorite impacts and planetary evolution II: Geological Society of America Special Paper 339, p. 131–147.

Pierazzo, E., and Melosh, H.J., 2000, Melt production in oblique impacts: Icarus, v. 145, p. 252–261, doi: 10.1006/ICAR.1999.6332.

Pierazzo, E., Vickery, A.M., and Melosh, H.J., 1997, A reevaluation of impact melt production: Icarus, v. 127, p. 408–423, doi: 10.1006/ICAR.1997.5713.

Pilkington, M., and Grieve, R.A.F., 1992, The geophysical signature of terrestrial impact craters: Reviews of Geophysics, v. 30, no. 2, p. 161–181.

Pilkington, M., and Hildebrand, A.R., 2000, Three-dimensional magnetic imaging of the Chicxulub Crater: Journal of Geophysical Research, v. 105, no. B10, p. 23,479–23,491, doi: 10.1029/2000JB900222.

Pilkington, M., Pesonen, L.J., Grieve, R.A.F., and Masaitis, V.L., 2002, Geophysics and petrophysics of the Popigai Impact Structure, Siberia, *in* Plado, J., and Pesonen, L.J., eds., Impacts in Precambrian Shields: Berlin, Springer-Verlag, p. 87–108.

Plado, J., Pesonen, L.J., Elo, S., Puura, V., and Suuroja, K., 1996, Geophysical research on the Kärdla impact structure, Hiiumaa Island, Estonia: Meteoritics & Planetary Science, v. 31, p. 289–298.

Plado, J., Pesonen, L.J., Koeberl, C., and Elo, S., 2000, The Bosumtwi meteorite impact structure, Ghana: A magnetic model: Meteoritics & Planetary Science, v. 35, p. 723–732.

Plado, J., Pesonen, L.J., and Puura, V., 1999, Effect of erosion on gravity and magnetic signatures of complex impact structures: Geophysical modeling and applications, *in* Dressler, B.O. and Sharpton, V.L., eds., Large meteorite impacts and planetary evolution II: Geological Society of America Special Paper 339, p. 229–240.

Pohl, J., 1971, On the origin of the magnetization of impact breccias on Earth: Zeitschrift für Geophysik, v. 37, p. 549–555.

Pohl, J., Stöffler, D., Gall, H., and Ernston, K., 1977, The Ries Impact Crater, *in* Roddy, D.J., Pepin, R.O. and Merrill, R.B., eds., Impact and Explosion Cratering: New York, Pergamon Press, p. 343–404.

Pope, K.O., Wieffer, S.W., and Ames, D.E., 2004, Empirical and theoretical comparisons of the Chicxulub and Sudbury impact structures: Meteoritics & Planetary Science, v. 39, no. 1, p. 97–116.

Rebolledo-Vieyra, M., 2001, Magnetoestratigrafia y Paleomagnetismo del crater de impacto de Chicxulub [Ph.D. thesis]: Mexico City, Universidad Nacional Autónoma de México, 120 p.

Rebolledo-Vieyra, M., and Urrutia-Fucugauchi, J., 2004, Magnetostratigraphy of the impact breccias and post-impact carbonates from borehole Yaxcopoil-1, Chicxulub impact crater, Yucatan, Mexico: Meteoritics & Planetary Science, v. 39, p. 6–7.

Rochette, P., Hood, L., Fillion, G., Ballou, R., and Ouladdiaf, B., 2003, Impact demagnetization by phase transition on Mars: Eos (Transactions, American Geophysical Union), v. 84, p. 561–568.

Schmidt, R.M., and Housen, K.R., 1987, Some recent advances in the scaling of the impact and explosion cratering: International Journal of Impact Engineering, v. 5, p. 543–560.

Scholz, C.A., Karp, T., Brooks, K.M., Milkereit, B., Amoako, P., and Arko, J., 2002, Pronounced central uplift identified in the Bosumtwi impact structure, Ghana, using multichannel seismic reflection data: Geology, v. 30, p. 939–942, doi: 10.1130/0091-7613(2002)0302.0.CO;2.

Scott, R., Pilkington, M., and Tanczyk, E., 1997, Magnetic investigations of the West Hawk, Deep Bay and Clearwater impact structures, Canada: Meteoritics & Planetary Science, v. 32, p. 293–308.

Shuvalov, V., 1999, 3D Hydrodynamic code SOVA for interfacial flows, application to thermal layer effect: Shock Waves, v. 9, p. 381–390, doi: 10.1007/S001930050168.

Stevenson, D.J., 2001, Mars' core and magnetism: Nature, v. 412, no. 6843, p. 214–219, doi: 10.1038/35084155.

Stöffler, D., Deutsch, A., Avermann, M., Bischoff, L., Brockmeyer, P., Buhl, D., Lakomy, R., and Mueller-Mohr, V., 1994, The formation of the Sudbury Structure, Canada: Toward a unified impact model, *in* Dressler, B.O., Grieve, R.A.F., and Sharpton, V.L., eds., Large Impact Craters and Planetary Evolution: Geological Society of America Special Paper 293, p. 303–318.

Strangway, D.W., 1960, Magnetic properties of some Canadian diabase dike swarms [Ph.D. thesis]: Toronto, Ontario, University of Toronto, 134 p.

Thompson, S.L., and Lauson, H.S., 1972, Improvements in the Chart D radiation-hydrodynamic CODE III: Revised analytic equations of state: Albuquerque, New Mexico, USA, Sandia National Laboratory Report SC-RR-71 0714, 119 p.

Tsikalas, F., 2002, The Mjølnir marine impact crater porosity anomaly: Deep-Sea Research, v. II, no. 49, p. 1103–1120.

Tsikalas, F., Gudlaugsson, S., and Faleide, J., 1998, The anatomy of a buried complex impact structure: The Mjølnir Structure, Barents Sea: Journal of Geophysical Research, v. 103, no. B12, p. 30,469–30,483, doi: 10.1029/97JB03389.

Tsikalas, F., Gudlaugsson, S., Faleide, J., and Eldholm, O., 1999, Mjølnir Structure, Barents Sea: A marine impact crater laboratory, *in* Dressler, B.O. and Sharpton, V.L., eds., Large meteorite impacts and planetary evolution II: Geological Society of America Special Paper 339, p. 193–204.

Ugalde, H., Artemieva, N., and Milkereit, B., 2003, Numerical modeling and petrophysical constraints on the magnetic signature of impact structures: 3rd International Conference on Large Meteorite Impact Craters, August 5–7, 2003: Nördlingen, Germany, Abstract no. 4017.

Versh, E., Kirsimäe, K., Jõeleth, A., and Plado, J., 2003, Impact induced hydrothermal system at Kärdla crater: development and biological consequences: 3rd International Conference on Large Meteorite Impact Craters, August 5–7, 2003: Nördlingen, Germany, Abstract no. 4120.

Wu, J., Milkereit, B., and Boerner, D.E., 1995, Seismic imaging of the enigmatic Sudbury structure: Journal of Geophysical Research, v. 100, no. B3, p. 4117–4130, doi: 10.1029/94JB02647.

Wünnemann, K., and Ivanov, B.A., 2003, Numerical modeling of the impact crater depth-diameter dependence in an acoustically fluidized target: Planetary and Space Science, v. 51, p. 831–845, doi: 10.1016/J.PSS.2003.08.001.

MANUSCRIPT ACCEPTED BY THE SOCIETY 9 AUGUST 2004

Printed in the USA

Geological Society of America
Special Paper 384
2005

Using vertical dikes as a new approach to constraining the size of buried craters: An example from Lake Wanapitei, Canada

E. L'Heureux*
H. Ugalde
B. Milkereit
Department of Physics, University of Toronto, 60 St. George, Toronto, Ontario M5S 1A7, Canada

J. Boyce
W. Morris
School of Geography and Geology, McMaster University, 1280 Main St. West, Hamilton, Ontario L8S 4K1, Canada

N. Eyles
Department of Physical & Environmental Sciences, University of Toronto at Scarborough, 1265 Military Trail, Toronto, Ontario M1C 1A4, Canada

N. Artemieva
Institute for Dynamics of Geospheres, Moscow, Russian Academy of Sciences, Leninskii pr. 38-6, Moscow 117334, Russia

ABSTRACT

Lake Wanapitei, located within the Southern Province of Ontario, Canada, provides the setting for a unique study of an impact crater situated within a shield environment. Evidence for the 7.5-km-diameter Wanapitei impact includes a circular Bouguer gravity low centered over the central area of the lake and features of shock metamorphism in samples of glacial drift found on the southern shores. Geophysical studies of craters in hard-rock environments are often limited by the lack of markers used for exploration; this may be overcome with the use of the large igneous dike swarms that characterize Archean terrains. The 1.2 Ga Sudbury dike swarm predates the impact that is suggested to have generated Lake Wanapitei and provides the setting for a study to constrain the size and location of the impact crater. The swarm is clearly visible on aeromagnetic maps as high amplitude, linear features, suggesting they could be used as vertical markers indicative of structural changes having an effect on target rock susceptibilities.

To fully establish the size of the crater, a total field magnetic map was produced to trace the Sudbury dikes through the proposed crater center. A gap in their signature, expressed as a 100 nT low, 2–3 km in width, constrains the size of the crater to <5 km. Numerical modeling suggests that a crater of this size will demagnetize target rocks, producing a low in the total magnetic field, up to a maximum diameter of 3 km. Dikes

*elizabeth.lheureux@utoronto.ca

L'Heureux, E., Ugalde, H., Milkereit, B., Boyce, J., Morris, W., Eyles, N., and Artemieva, N., 2005, Using vertical dikes as a new approach to constraining the size of buried craters: An example from Lake Wanapitei, Canada, *in* Kenkmann, T., Hörz, F., and Deutsch, A., eds., Large meteorite impacts III: Geological Society of America Special Paper 384, p. 43–50.

within the central crater structure will be excavated, vaporized, and melted down to a depth of 1.3 km.

Keywords: Wanapitei, dike swarm, Sudbury, impact structure, magnetic anomaly, Superior Province.

INTRODUCTION

Lake Wanapitei is located within the Huronian Supergroup of the Southern Province in northern Ontario, Canada, and is bounded on its west side by the eastern rim of the Sudbury impact structure (Fig. 1A). The lake was suggested to have been generated by a meteorite impact, with a resultant crater 7.5 km in diameter (Dence and Popelar, 1972). Identification of an impact crater is commonly based on crater morphology, macroscopic and microscopic shock metamorphic features, and the presence of specific lithologies such as suevitic breccias and geophysical anomalies. In stratified targets, marker beds help to deduce the degree of impact-induced tilting and faulting of the rocks. Identification of craters in crystalline rock environments such as the Canadian Shield is made more difficult by the lack of prominent marker horizons in geophysical data sets; identifying buried structures in a nonsedimentary setting often reduces diagnostic approaches to analysis of potential field data.

The Wanapitei region is transected by several Precambrian dike swarms, namely the Matachewan and Sudbury swarms. Archean provinces such as the Canadian Shield are generally characterized by these large swarms of igneous dikes that are thought to emanate from mantle plumes (Ernst et al., 1995). The geometrical and geophysical properties of these dikes can provide key constraints on any structural geological model. Evidence suggests that upon intrusion most dikes are near vertical and have a preferred orientation that corresponds to the regional stress field applicable at the time of intrusion (Morris and Tanczyk, 1978). By defining variations in the current geometry and orientation of a portion of a larger dike swarm it is possible to unravel complex tectonic histories (see Halls et al., 1994; or Siddorn and Halls, 2002). The vertical dikes extend several kilometers below the surface; hence, simple uplift is marked by a change in the width of the thermal contact aureole (Schwarz and Buchan, 1989). Traditionally, these large collections of vertical dikes have been used in studies of supercontinent cyclicity (Yale and Carpenter, 1998; Mertanen et al., 1999) or structural deformation (West and Ernst, 1991; Buchan and Ernst, 1994).

Igneous dike swarms are commonly associated with prominent magnetic features that can be traced over several kilometers and are easily imaged on aeromagnetic maps (see Fig. 1B). In the Wanapitei region, dikes of the Sudbury swarm were intruded ca. 1.2 Ga (Krogh et al., 1987), prior to the suggested impact that lead to the formation of Lake Wanapitei. In this study, we took advantage of the geological setting offered by the Sudbury dike swarm to investigate a buried meteorite impact crater, using the dikes as vertical markers for constraining crater dimensions.

Geologic Setting and Evidence for Impact

The impact crater studied lies entirely within the central, circular portion of the 12 km diameter Wanapitei Lake, and is believed to be of medium size (with a diameter of ~7.5 km; see Fig. 1) (Dence and Popelar, 1972). The lake is situated on the border between the Southern and Superior Provinces, with target rocks of early Proterozoic age (Dressler and Reimold, 2001). It is surrounded by Huronian sedimentary rocks that are intruded by the northwest-trending Sudbury dikes, aged 1238 Ma (Krogh et al., 1987). Dence and Popelar (1972) were the first to present topographic, petrographic, and geophysical evidence in support of the impact theory. The circular shape of Lake Wanapitei was one of the first and most obvious clues indicating its origin. The authors used this observation as well as the drainage pattern of the lake to support their approach, citing a concentric pattern of streams and smaller lakes within 5 km of Wanapitei as further proof. This pattern is most obvious on the western side of the lake and coincides with the indented eastern rim of the Sudbury basin (Fig. 1). Dressler (1982) observed an apparently concentric pattern of joints and fractures peripheral to the lake as well as shatter cones on the islands and up to 5 km northwest of the lake. However, it is not clear whether the shatter cones should be attributed to the Wanapitei or the Sudbury impact events as the lake is located within the larger Sudbury structure.

The most conclusive evidence for an impact origin of the Wanapitei structure comes from petrographic studies of cobbles and boulders of suevitic breccias found in glacial drift south of Lake Wanapitei. These breccias contain impact melt fragments and clasts that exhibit shock metamorphic features. The first samples were found by M.R. Dence in the area south of the lake, north of Skead, followed by more discoveries around the Massey Bay area (see Fig. 1A). Dence and Popelar (1972) argued that the breccias originated from the bottom of the lake, stating that the boulders are weakly compacted and disintegrate upon handling. The boulders are also concentrated around the southern shores of the lake: Pleistocene glaciation is suggested to have scoured the bottom of the lake and transported the boulders of breccia southward.

Petrographic studies reveal target rock clasts in the suevitic breccias that exhibit planar deformation features in quartz and feldspar, and glass fragments with lechatelierite (Dence and Popelar, 1972; Dence et al., 1974; Grieve and Ber, 1994; Dressler et al., 1997). Coesite, a high pressure modification of silica, was identified by Dence and Popelar (1972) in a strongly shocked quartzite. Dressler (1982) observed planar deformation features in quartz grains in bedrock at only three locations southwest of the lake (Fig. 1A). Based on glass samples and the

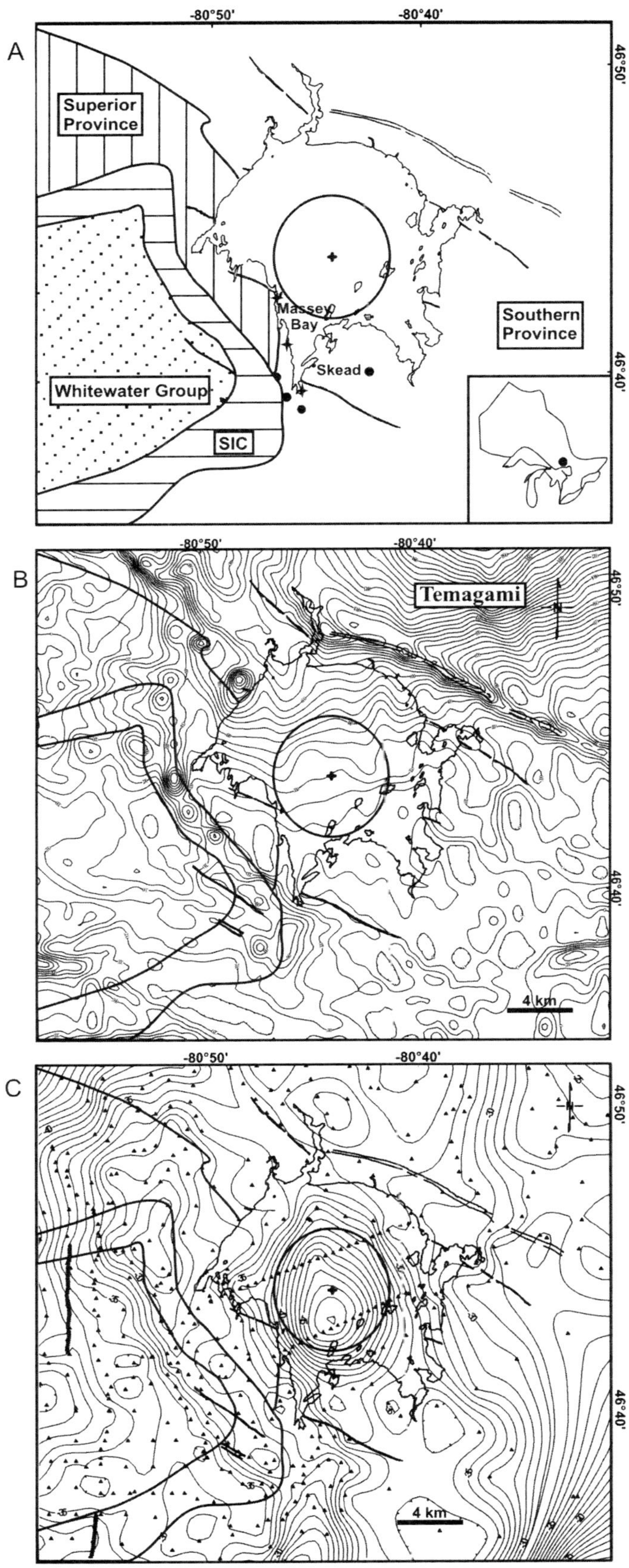

necessary pressure for the formation of coesite, Dence et al. (1974) estimated that minimum shock pressures were between 40 and 50 GPa. Dressler et al. (1997) estimated shock pressures to be 50–60 GPa and temperatures to be >1700 °C, based on the absence of feldspar and the presence of glass in samples. A study was performed by Winzer et al. (1976) to determine the age of three samples consisting of breccia with glass veins, showing no visible signs of recrystallization or alteration. Two K/Ar dates were obtained from one glass vein and one whole rock glassy boulder: these revealed ages of 37.8 ± 1.6 Ma and 36.0 ± 1.6 Ma, respectively. Bottomley (1982) dated the samples at 36.1 ± 0.5 Ma based on ^{40}Ar-^{39}Ar dating of what appeared to be "melt textured rock." Evan et al. (1993) studied samples from the collection of Dence and Popelar and found evidence suggesting that the impacting body was chondritic (LL, L, or C1). Wolf et al. (1980) analyzed samples from the collection of the Gravity and Geodynamics Division, Earth Physics Branch, Department of Energy, Mines and Resources, Ottawa, Canada, for Ir and Os abundances. Interelement ratios suggest a C1-C2 or an LL-chondrite as an impacting body.

As the crater is submerged, there are few constraints on its actual size: until recently, its suggested diameter has been based on one gravity survey performed on the ice covered lake in 1969 (Dence and Popelar, 1972) (Fig. 1C). The data were corrected for water mass using 1968 bathymetry measurements from the Ontario Department of Lands and Forests; however, no terrain corrections were applied. From the final Bouguer anomaly map, a circular low of ~15 mGal was observed covering almost the entire area of the lake. The 11-km-diameter bowl-shaped anomaly is of significant amplitude, yet constrained by only a few observations (see Fig. 1C). The relatively uniform magnetic field, as seen by aeromagnetic maps flown at a line spacing of 800 m (Fig. 1B), did not seem to indicate the presence of an intrusive body (Dence and Popelar, 1972).

Vertical Markers

The study of impact craters by geophysical methods often focuses on the recognition of offsets of local features such as seismic reflectors; such offsets can be used to trace faults. Seismic surveys make use of contrasting lithologies to map the tilting of slumped blocks or overturned strata at the rim of the crater, to

Figure 1. Lake Wanapitei regional setting. Northwest-trending lines represent Sudbury dikes and the 7.5-km-diameter circle indicates the impact crater proposed by Dence and Popelar (1972). (A) Geology. Stars indicate locations of planar deformation features in quartz; dots indicate samples of suevite and shock metamorphosed rocks (extracted from Dressler et al., 1997). SIC—Sudbury Igneous Complex. (B) Regional magnetic field; contour interval is 50 nT. The large Temagami magnetic anomaly extends southward into Lake Wanapitei. (C) Regional Bouguer gravity; contour interval is 1 mGal. Triangles indicate station locations. The circular anomaly has an amplitude of −15 mGal. Lake outline and dikes extracted from Dressler (1982).

trace a central uplift, and to identify zones of breccia infill (Mazur et al., 2000). Potential field anomalies reveal impact structures by their typically circular geometry, accompanied by disruptions in local trends. To study the Wanapitei impact, we used dikes of the Sudbury swarm as well-defined vertical markers indicative of such disruptions. The Precambrian olivine diabase dikes have a maximum width of 120 m in the lake area and exhibit a strong, linear total field magnetic anomaly (Figs. 1B and 2B). They appear to be more resistant to erosion than surrounding rocks in the immediate vicinity of the lake and are found in several outcrops on both the east and west shores.

A total field magnetic survey was conducted on Wanapitei in an attempt to track the diabase dikes through the proposed crater location. According to Ugalde et al. (this volume), the magnetic signature observed over large impact structures is primarily the combination of three effects: (1) composition and properties of target rocks; (2) modification of magnetic carriers due to high *P-T* conditions; and (3) natural remanent magnetization (NRM). It was our goal to constrain the outline of the crater based on a comparison of the measured magnetic field to models of pressure effects on the magnetic signature of the dikes (Fig. 3). To map the submerged dikes as well as to constrain gravity and magnetic anomalies, the thickness of young sediments was monitored with a high frequency single channel seismic survey (see Fig. 2C).

The results of the magnetic survey are shown in Figure 4A. 48 north-south lines were surveyed with a spacing of 500 m over

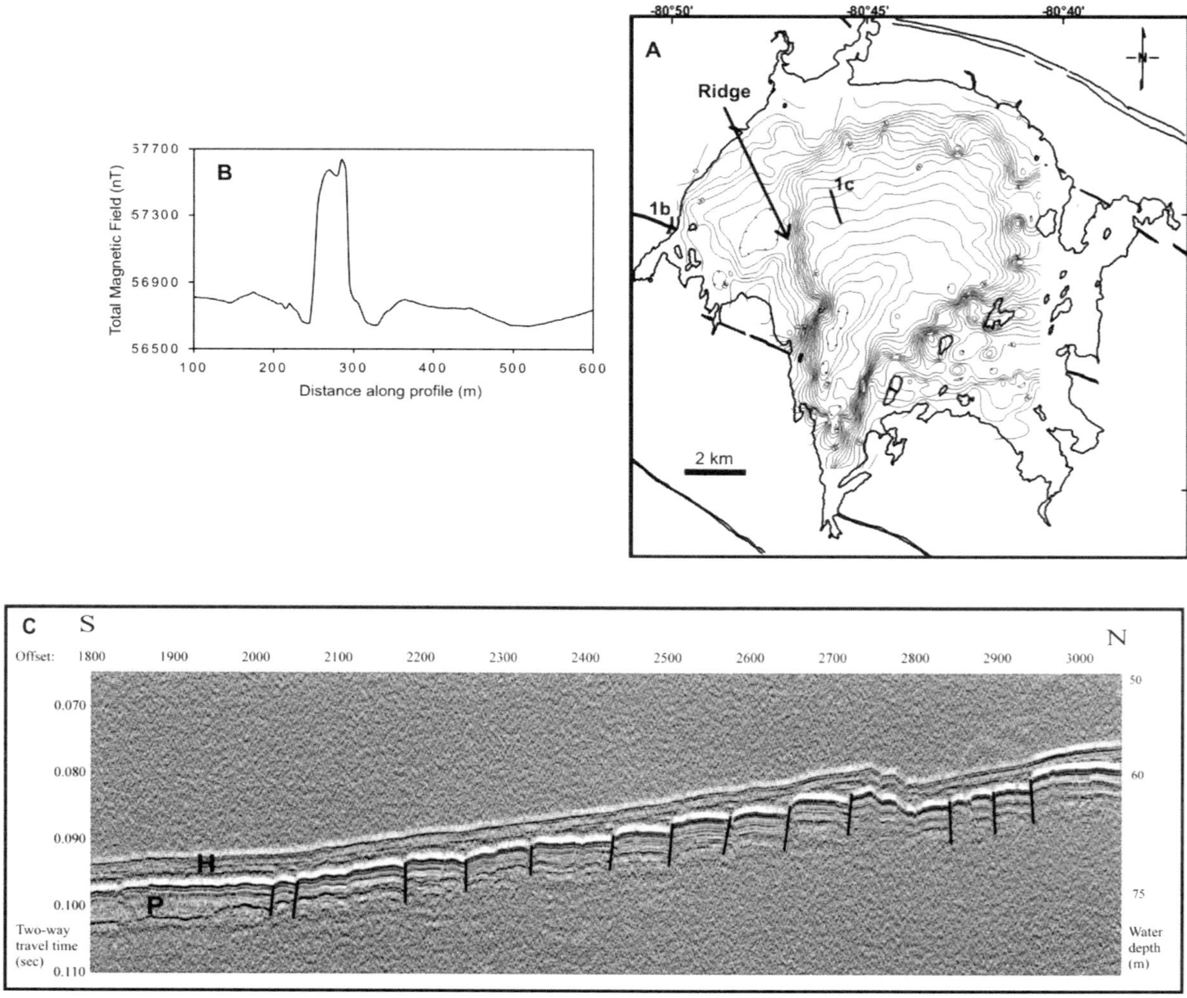

Figure 2. (A) Bathymetry over Lake Wanapitei (5 m contour lines) acquired with a 200 kHz echo sounder. The deepest part of the lake runs through the southern bay and reaches depths of ~100 m. Black lines labeled 1b and 1c indicate the locations of the magnetic and seismic profiles shown in B and C, respectively. (B) Magnetic signature of the Sudbury dike located on the west side of the lake. The dike exhibits a maximum amplitude of 1000 nT as compared to its granitic host rocks. (C) High frequency, single channel seismic profile through the central basin of Lake Wanapitei. Profile is ~1200 m in length, reaching up to 70 m water depth. Faulting occurs in Pleistocene sediments. H—Holocene sediments; P—Pleistocene sediments.

the entire area of the lake, and 12 east-west lines with uneven spacing were collected for leveling purposes. The data were acquired with a proton precession magnetometer with a sampling rate of 4 Hz, corresponding to a 1.4–2.4 m inline sample spacing. Bathymetry (Fig. 2A) was monitored concurrently with a 200 kHz echo sounder and global positioning system (GPS) positioning. The seismic data were collected with a high resolution digital sub-bottom profiler, with a depth penetration reaching up to 60 m after processing. Results show that the thickness of Pleistocene and Holocene sediments varies greatly throughout the lake, with maximum thicknesses of ~40–50 m in the western section of the lake (west of the ridge indicated in Fig. 2B). Furthermore, an area of ~5–6 km in diameter in the center of the lake is characterized by a thin, faulted layer of Pleistocene sedimentation (Fig. 2C).

DISCUSSION OF RESULTS

The total magnetic field over the lake correlates with the aeromagnetic map; it is dominated by the large Temagami magnetic anomaly to the north, which extends southward to cover approximately half of the lake area (see Figs. 1B and 4). The anomaly, situated in the Precambrian aged rocks of the Southern Province (Milkereit and Wu, 1996), is not affected by the Wanapitei impact. Toward the southern end of the lake, the magnetic field continues to decrease in amplitude. The Sudbury dikes are imaged as discontinuous, linear 75–100 nT anomalies trending northwest (Fig. 4). The gap in their signature coincides with an ~100 nT low in the magnetic field of 2–3 km in diameter. There is no evidence for a central uplift or coherent melt body in the magnetic field. Although direct evidence of the impact is not seen in the shallow penetrating seismic data of this study, profiles show that the area described by the gap in dike signature is covered by a thin, faulted and disturbed sediment layer of Pleistocene age (Fig. 2C). This may be the result of differential compaction and slumping of post-impact sediments above brecciated target rocks and fallback material. The same effect is observed in seismic profiles over the 10 km diameter impact crater in Lake Bosumtwi, Ghana. Lacustrine sediments overlying the crater are vertically fractured in the area of the faulted central uplift (Scholz et al., 2002), indicating that structural damage due to an impact has an effect on deposition and compaction of post-impact sediments.

Numerical modeling was used to simulate pressure conditions in a crater with a final diameter of 5 km (Ugalde et al., this volume) (Fig. 3). The model crater was generated with a stony asteroid projectile of 260 m in diameter striking vertically into a crystalline target at a velocity of 15 km/s. For a more complete description of the numerical model, please see Ugalde et al. (this volume). The results show that the originally vertical markers representing the Sudbury dikes are affected by the impact in the complex crater as follows:

1. Strain: the vertical markers in the center of the complex structure are removed by excavation and vaporization down to a depth of 500 m below the original target surface (0 level in Fig. 3). Markers located at 1000 m from the center of the structure are displaced outward down to a depth of 800 m.

2. A decrease in susceptibility and natural remanence: elevated pressures and temperatures are assumed to correlate with a reduction in susceptibility. Thus, we can infer that a decrease in susceptibility occurs as far down as 1000 m (where pressures are

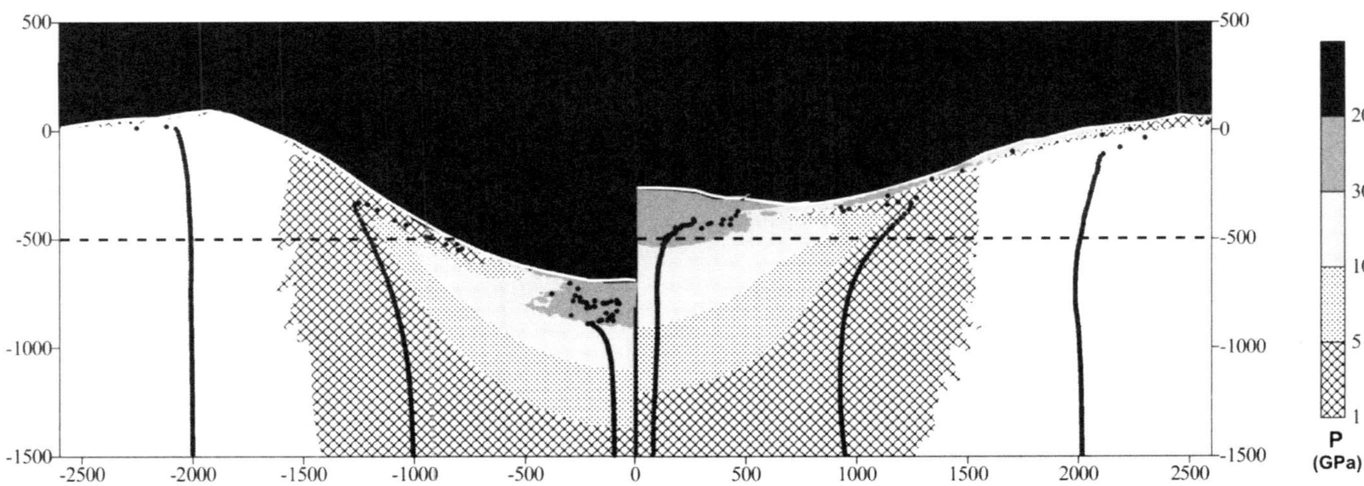

Figure 3. Results of a numerical model displaying pressure conditions in a 5-km-diameter impact crater (based on Ugalde et al., this volume). Pressures are in GPa, distances in m. The left half of the figure represents a simple crater; the right half is a complex crater. The dashed line indicates the assumed present erosion level (Dence and Popelar, 1972); thick vertical lines represent the Sudbury dikes. 1 GPa < P < 5 GPa = fracturing and brecciation (French, 1998). P < 10 GPa = existing remanence demagnetized (Cisowski and Fuller, 1978). P > 10 GPa = susceptibility depletion (Pilkington and Grieve, 1992). P > 30 GPa = magnetic resetting. Material above 200 GPa is vaporized. Dikes are demagnetized or vaporized in a central area with a diameter of ~3 km.

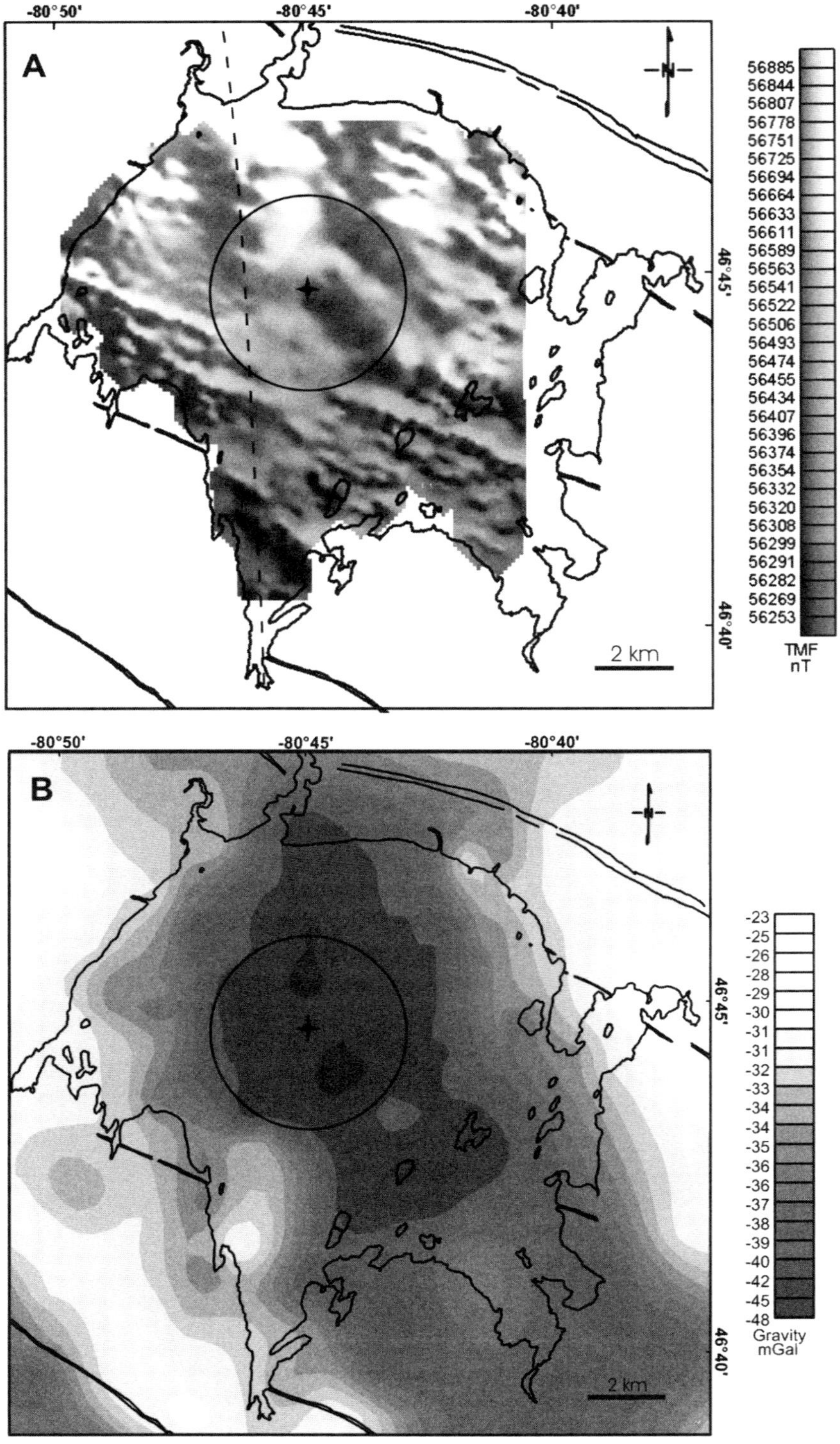

Figure 4. (A) Total magnetic field map over Lake Wanapitei. Sudbury dikes are imaged as linear, 75–100 nT anomalies striking northwest. Map is shaded from 45° declination, 0° inclination. A 5 km crater where the dike signature is disrupted and the location of a possible fault zone running north-south through the western part of the lake are indicated. (B) New gravity data collected during the winters of 2003 and 2004 (Ugalde et al., 2004). Increased sampling shows that the shape of the anomaly is concordant to the regional geologic framework rather than simply circular.

greater than 10 GPa [Pilkington and Grieve, 1992]). Remanence demagnetization (at pressures less than 10 GPa [Cisowski and Fuller, 1978]) affects the target structure down to ~1300 m below the target surface and as far as 1200 m from the center. Subsequently, these rocks may reacquire remanence through NRM as they cool.

3. Low magnetization sedimentary infill: post-impact sediments may fill the crater basin up to a depth of 300 m and out to a radius of 2000 m. Unless they build up significant detrital remanent magnetization (DRM), they will have a lower magnetic susceptibility and NRM than the crystalline basement.

4. Hydrothermal alteration: fracturing and brecciation intensively occur in a pressure interval of 1–5 GPa (French, 1998), out to a radius of 1500 m and depth >1500 m. The fractured rocks provide space for fluid circulation, and a hydrothermal system driven by small inclusions of melt could either increase or decrease the original magnetization (Ugalde et al., this volume).

The results of the model indicate that the processes generating an impact crater with a final diameter of 5 km have enough energy to decrease the magnetization of the vertical dikes out to a diameter of 3 km. The low in the magnetic field is the combined result of susceptibility depletion as described, excavation of the upper portion of the dikes (with subsequent infill of suevitic breccia and post-impact sediments), and brecciation of target rocks producing a randomization of magnetic vector orientations. Magnetic forward modeling of target rocks after the impact demonstrates that although the dikes may still be present at a depth of 500 m below the present-day surface, their magnetic signature is greatly reduced due to the petrophysical changes induced by the impact (Ugalde et al., this volume). Although the observed discontinuity in the magnetic anomaly of the dikes does not cross the center of the structure (Fig. 4), the possible location and size of the Wanapitei impact crater can be roughly constrained by projecting the gap in the magnetic field toward the center of impact. Based on the previously described model and the absence of a dike signature in the central part of the lake, we propose that the Wanapitei impact crater is less than 5 km in diameter. The magnetic discontinuity is located within the central section of the lake (Fig. 4), coinciding with the greatest depths (see bathymetry in Fig. 2A). The unconsolidated and faulted material below and above the impact crater will produce a negative gravity anomaly in the center of the lake.

Recently, new gravity data was collected that provides a finer sampling of the anomaly over Lake Wanapitei (Ugalde et al., 2004). The final Bouguer map, including terrain corrections, shows a smaller, less circular low in the central area of the lake (Fig. 4B). The anomaly no longer has the large characteristic bowl shape that distinguished it from the regional field.

Structural Deformation Unrelated to the Impact

Further interpretation with respect to the structure of Lake Wanapitei and a possible relation to the eastern rim of the Sudbury basin (Fig. 1A) can be derived from the results of these surveys. The most apparent feature is seen in the bathymetry of the lake (Fig. 2): a ridge separating the central basin from the western "lobe." This separation has a distinct north-south trend and coincides with the southeastern shore of the lake, as well as with an inferred fault running through Massey Bay (Dressler, 1982). It may be the extension of the regional Onaping-Matagami fault zone that runs more than 450 km northward (Buchan and Ernst, 1994) and produces 7–8 km of left-lateral offset. This fault is between 2.2 and 1.9 Ga in the Sudbury-Wanapitei region (Buchan and Ernst, 1994), thereby predating the formation of the dikes, but it may have gone through subsequent periods of reactivation. It is therefore suggested that a large fault zone runs north-south through Wanapitei Lake (Fig. 4B), exerting control over the emplacement of the Sudbury dikes as well as acting as a zone of weakness influencing Pleistocene glacial activity. Valleys of glacial till extend more than 25 km to the south of the lake along the same path (Burwasser, 1979). It is estimated that up to 500 m of erosion have affected the area since the emplacement of the Wanapitei crater 37 Ma (Dence and Popelar, 1972). The trend of the whole lake structure is comparable to the eastern rim of the Sudbury basin (see Fig. 1), suggesting that the present-day shape and size of Wanapitei Lake is due rather to regional deformation and glacial action than to the original meteorite impact.

CONCLUSIONS

By modeling the effects of impact on the magnetic signature of a crystalline target, it is possible to estimate the extent of the crater making use of markers in the terrain. In the case of the Wanapitei structure, vertical dikes with a characteristically high, linear total field magnetic anomaly were used as indicators to outline the crater. The method proved useful in reducing the estimated diameter of the crater based on attenuation of the dikes' magnetic signature. Large swarms such as the Sudbury dikes exist in most Archean craton environments and have typically been used for large scale deformation or continental studies. It has been demonstrated here that through detailed images of the dikes, small disruptions in their signature can be identified and related to impact events.

The detailed magnetic field map over Lake Wanapitei serves to constrain the possible location of the impact crater by the gap in the linear signature of the Sudbury dikes, which are used as vertical indicators of the structural damage produced by meteorite impact. The area delineated by the gap is characterized by a faulted layer of sediment and is located in the central part of the lake's main basin. The ridge that separates the lake into eastern and western sections follows a north-south trend that is comparable to that of the East Range in the Sudbury Igneous Complex. It corresponds to the extension of a documented fault zone that runs for several kilometers north of the Wanapitei area (Buchan and Ernst, 1994). Glacial scouring has produced valleys of till that reveal the zone of weakness associated with the fault and the excavation by glaciers during the Pleistocene (see Burwasser, 1979). It is our conclusion that the Wanapitei impact crater is confined to an area less than 5 km in diameter and that the character of the magnetic and gravity fields over the lake suggests a simple geometry (there is no evidence for a central uplift). The lake itself takes its semi-circular shape from regional deformation associated with faulting and glacial scouring.

ACKNOWLEDGMENTS

We thank Mike Doughty for conducting the seismic survey, Wei Qian for the magnetometer surveys, and Erick Adam for his help with seismic data processing. Comments and suggestions from T. Kenkmann, B. Dressler, and an anonymous reviewer were greatly appreciated. Data collection on Lake Wanapitei was funded by the National Sciences and Engineering Research Council (Canada).

REFERENCES CITED

Bottomley, R.J., 1982, ^{40}Ar-^{39}Ar dating of melt rock from impact craters [Ph.D. thesis]: Toronto, University of Toronto, 103 p.

Buchan, K.L., and Ernst, R.E., 1994, Onaping fault system: Age constraints on deformation of the Kapuskasing structural zone and units underlying the Sudbury structure: Canadian Journal of Earth Sciences, v. 31, p. 1197–1205.

Burwasser, G.J., 1979, Quaternary geology of the Sudbury basin area, district of Sudbury: Ontario Geological Survey Report 181, 103 p.

Cisowski, S.M., and Fuller, M., 1978, The effect of shock on the magnetism of terrestrial rocks: Journal of Geophysical Research, v. 83, no. B7, p. 3441–3458.

Dence, M.R., and Popelar, J., 1972, Evidence for an impact origin for Lake Wanapitei, Ontario, *in* Guy-Bray, J.V., ed., New developments in Sudbury geology: Geological Association of Canada Special Paper 10, p. 117–124.

Dence, M.R., Robertson, P.B., and Wirthlin, R.L., 1974, Coesite from the Lake Wanapitei crater, Ontario: Earth and Planetary Science Letters, v. 22, p. 118–122, doi: 10.1016/0012-821X(74)90071-5.

Dressler, B.O., 1982, Geology of the Wanapitei Lake area, district of Sudbury: Ontario Geological Survey Report 213, 131 p.

Dressler, B.O., and Reimold, W.U., 2001, Terrestrial impact melt rocks and glasses: Earth Science Reviews, v. 56, p. 205–284, doi: 10.1016/S0012-8252(01)00064-2.

Dressler, B.O., Crabtree, D., and Schuraytz, B.C., 1997, Incipient melt formation and devitrification at the Wanapitei impact structure, Ontario, Canada: Meteoritics & Planetary Science, v. 32, p. 249–258.

Ernst, R.E., Head, J.W., Parfitt, E., Grosfils, E., and Wilson, L., 1995, Giant radiating dyke swarms on Earth and Venus: Earth Science Reviews, v. 39, p. 1–58, doi: 10.1016/0012-8252(95)00017-5.

Evan, N.J., Gregoire, D.C., Grieve, R.A.F., Goodfellow, W.D., and Veizer, J., 1993, Use of platinum-group elements for impactor identification: Terrestrial impact craters and Cretaceous-Tertiary boundary: Geochimica et Cosmochimica Acta, v. 57, p. 3737–3748, doi: 10.1016/0016-7037(93)90152-M.

French, B.M., 1998, Traces of catastrophe: A handbook of shock-metamorphic effects in terrestrial meteorite impact structures, LPI Contribution no 120: Houston, Lunar and Planetary Institute, 120 p.

Grieve, R.A.F., and Ber, T.J., 1994, Shocked lithologies at the Wanapitei impact structure, Ontario, Canada: Meteoritics, v. 29, p. 621–631.

Halls, H.C., Palmer, H.C., Bates, M.P., and Phinney, W.C., 1994, Constraints on the nature of the Kapuskasing structural zone from the study of Proterozoic dyke swarms: Canadian Journal of Earth Sciences, v. 31, p. 1182–1196.

Krogh, T.E., Corfu, F., Davis, D.W., Dunning, G.R., Heaman, L.M., Kamo, S.L., Machado, N., Greenhough, J.D., and Nakamura, E., 1987, Precise U-Pb isotopic ages of diabase dykes and mafic to ultramafic rocks using trace amounts of baddeleyite and zircon, *in* Halls, H.C. and Fahrig, W.F., eds., Mafic dyke swarms: Geological Association of Canada Special Paper 34, p. 147–152.

Mazur, M.J., Stewart, R.R., and Hildebrand, A.R., 2000, The seismic signature of meteorite impact craters: CSEG Recorder, v. 25, p. 10–16.

Mertanen, S., Halls, H.C., Vuollo, J.I., Pesonen, L.J., and Stepanov, V.S., 1999, Paleomagnetism of 2.44 Ga mafic dykes in Russian Karelia, eastern Fennoscandian shield—implications for continental reconstructions: Precambrian Research, v. 98, p. 197–221, doi: 10.1016/S0301-9268(99)00050-9.

Milkereit, B., and Wu, J., 1996, Seismic image of an early Proterozoic rift basin: Tectonophysics, v. 264, p. 89–100, doi: 10.1016/S0040-1951(96)00119-9.

Morris, W.A., and Tanczyk, E.I., 1978, Precambrian dykes: A preferred orientation of intrusion: Nature, v. 275, p. 120–121.

Pilkington, M., and Grieve, R.A.F., 1992, The geophysical signature of terrestrial impact craters: Reviews of Geophysics, v. 30, p. 161–181.

Scholz, C.A., Karp, T., Brooks, K.M., Milkereit, B., Amoako, P.Y.O., and Arko, J.A., 2002, Pronounced central uplift identified in the Bosumtwi impact structure, Ghana, using multichannel seismic reflection data: Geology, v. 30, p. 939–942, doi: 10.1130/0091-7613(2002)0302.0.CO;2.

Siddorn, J.P., and Halls, H.C., 2002, Variation in plagioclase clouding intensity in Matachewan dykes: Evidence for the exhumation history of the northern margin of the Sudbury Igneous Complex: Canadian Journal of Earth Sciences, v. 39, p. 933–942, doi: 10.1139/E02-003.

Schwarz, E.J., and Buchan, K.L., 1989, Identifying types of remanent magnetization in igneous contact zones: Physics of the Earth and Planetary Interiors, v. 58, p. 155–162, doi: 10.1016/0031-9201(89)90051-4.

Ugalde, H., L'Heureux, E., Milkereit, B., and Lachapelle, R., 2004, Measuring gravity on ice: an example from Wanapitei Lake, Ontario, Canada [expanded abstract]: SEG International Exposition and 74th Annual Meeting.

West, G.F., and Ernst, R.E., 1991, Evidence from aeromagnetics on the configuration of Matachewan dykes and the tectonic evolution of the Kapuskasing structural zone, Ontario, Canada: Canadian Journal of Earth Sciences, v. 28, p. 1797–1811.

Winzer, S.R., Lum, R.K.L., and Schuhmann, S., 1976, Rb, Sr and strontium isotopic composition, K/Ar age and large ion lithophile element abundances in rocks and glasses from the Wanapitei Lake impact structure: Geochimica et Cosmochimica Acta, v. 40, p. 51–57, doi: 10.1016/0016-7037(76)90193-9.

Wolf, R., Woodrow, A.B., and Grieve, R.A.F., 1980, Meteoritic material at four Canadian impact craters: Geochimica et Cosmochimica Acta, v. 44, p. 1015–1022, doi: 10.1016/0016-7037(80)90289-6.

Yale, L.B., and Carpenter, S.J., 1998, Large igneous provinces and giant dike swarms: Proxies for supercontinent cyclicity and mantle convection: Earth and Planetary Science Letters, v. 163, p. 109–122, doi: 10.1016/S0012-821X(98)00179-4.

Manuscript Accepted by the Society 9 August 2004

Geological Society of America
Special Paper 384
2005

Enhancement of magnetic signatures of impact structures

Duncan R. Cowan*
Cowan Geodata Services, 12 Edna Road, Dalkeith, Western Australia 6009, Australia

Gordon R.J. Cooper*
School of Geosciences, University of the Witwatersrand, Johannesburg 2050, South Africa

ABSTRACT

Aeromagnetic surveys are a useful tool in the detection and analysis of terrestrial impact structures. Although gravity anomalies provide clearer and simpler signatures of impact craters, large regional-scale aeromagnetic surveys are more widely available.

A combination of many factors, such as the nature of the target rocks, the impact-related magnetization, and effects of crater fill and post-impact cover results in a great variation of magnetic signatures in the terrestrial impact craters. In crystalline basement targets, the most common signature of a complex impact structure is a magnetic low with a central peak or ring uplift magnetic anomaly. Contributions to the magnetic signature include demagnetization, shock remagnetization, and thermal and chemical remanent magnetization effects. Impact craters in sedimentary targets usually are of small magnetic amplitude, depending on the lithology. The origin of the magnetic signatures in sedimentary targets is not well understood.

Enhancement of magnetic signatures of terrestrial impact structures using filtering techniques is an important part of detection and analysis. Derivative and derivative-based (such as sunshading) techniques, along with separation filtering, are probably the most used methods. Here we present our new developments of algorithms for fractional order derivatives and circular shaded relief that have dramatically improved filter results. The fractional derivative order can be varied to optimize the separation of the impact magnetic signature. Given a chosen center location, the circular shaded relief algorithm treats all directions equally, thus preventing fade-out of features subparallel to the shading direction evident in conventional shaded relief. Unlike Hough transform based algorithms, the circular sunshading method is not sensitive to the radius of the circular feature being searched for, and no radius parameter is specified during the data processing

We illustrate the new fractional derivative and circular shaded relief algorithms using selected Australian and Canadian impact crater data sets involving both crystalline basement and sedimentary targets.

Keywords: Aeromagnetic surveys, circular shaded relief, fractional order derivatives, impact structures, magnetization.

*E-mails: cowangeo@bigpond.net.au; cooperg@geosciences.wits.ac.za.

Cowan, D.R., and Cooper, G.R.J., 2005, Enhancement of magnetic signatures of impact structures, *in* Kenkmann, T., Hörz, F., and Deutsch, A., eds., Large meteorite impacts III: Geological Society of America Special Paper 384, p. 51–65. For permission to copy, contact editing@geosociety.org.

INTRODUCTION

Aeromagnetic surveys play an important role in the search for terrestrial impact structures. Interpretation of remote sensing and digital elevation images is often very effective for outcropping structures. However, potential field methods can provide information on structures that may be completely covered and have no geomorphic expression. Although gravity anomalies provide clearer and simpler signatures of impact craters, large, regional scale aeromagnetic surveys are more widely available and offer great potential for recognition of additional impacts, especially those concealed by post-impact sediments. Terrestrial craters are constantly being affected by erosion, sedimentation, and tectonic activity, and only ~175 terrestrial impact craters have been recognized, the majority in geologically stable cratons of North America, Europe, Africa, and Australia (Grieve et al. 1995, Pilkington and Grieve, 1992).

Impact craters can be divided into two groups based on morphostructure: simple and complex. Simple craters are relatively small, bowl-shaped depressions with an upraised and fractured rim and with depth-to-diameter ratios of ~1:5. Complex craters are larger, with a central uplift formed as the cavity floor rebounds upward. The transition diameter between simple and complex craters on Earth depends on rock type and amounts to 2 km in sedimentary rocks and 4 km in crystalline lithologies (Pilkington and Grieve, 1992).

Magnetic signatures of terrestrial impact craters vary greatly, reflecting the composition of the target rocks, the impact-related magnetization, and effects of crater fill and post-impact cover (Pilkington and Grieve, 1992). The most common crater signature is a magnetic low, ranging in amplitude from a few nT up to a few hundred nT. In a complex impact crater, the central peak or ring uplift of basement may produce strong magnetic anomalies with amplitudes of several 100 nT. The presence of melt sheets can produce high amplitude magnetic anomalies often with significant remanent magnetization. The magnetic signature may be due to shock demagnetization, shock remagnetization, and thermal and chemical remanent magnetization effects and is often multicomponent. Impact craters in sedimentary targets usually have very subdued magnetic relief, depending on the lithology. Amplitudes of a few nT up to 10 nT are common.

Enhancement of magnetic signatures of terrestrial impact structures using filtering techniques is an important part of detection and analysis. In crystalline basement, it can be difficult to recognize impact magnetic signatures in an active magnetic background such as a greenstone belt or where other sources, such as small granitoids, also produce magnetic lows. In a sedimentary target area, it is necessary to recognize subtle low-amplitude crater anomalies that may be masked by high-amplitude crystalline basement anomalies, as well as by other intrasedimentary anomalies. Derivatives and shaded relief techniques, along with separation filtering, are probably the most commonly used methods. Recent development of algorithms for fractional horizontal and vertical derivatives and circular shaded relief has dramatically improved filter results (Cooper, 2003; Cooper and Cowan, 2003a; Cooper and Cowan, 2003b). The fractional derivative order can be varied to optimize separation of the impact magnetic signature. The circular shaded relief algorithm treats all directions equally, preventing the fade-out for features subparallel to the shading direction, which are evident in conventional shaded relief.

In this paper, the new fractional derivative, circular shaded relief, and separation filtering algorithms are illustrated using very different impact crater data sets. The Neoproterozoic (estimated 590 Ma) Acraman impact structure in South Australia is located in Mesoproterozoic Gawler Range volcanics and coeval weakly magnetic Hiltaba suite granitoids. The size of the Acraman structure is still controversial (e.g., Williams et al., 1996) and the problem is to distinguish impact-related magnetic anomalies in an area of magnetically variable basement.

The Triassic (212 Ma) Manicouagan impact structure in Quebec, Canada, is estimated to have had an original diameter of 100 km and is a deeply eroded structure in crystalline basement rocks. The problem there is to separate the impact-related magnetic anomalies from those due to adjacent granite-greenstone structures (Roest and Pilkington, 1993). Both Landsat imagery and digital elevation data from the Shuttle Radar Topographic Mission (Cowan and Cooper, 2003) provide complementary information to aid magnetic interpretation.

The Mesozoic (estimated 90 Ma) 12-km-diameter Yallalie impact structure (Dentith et al., 1999; Hawke et al., 2003) is located ~200 km north of Perth, Western Australia, in Mesozoic sedimentary rocks of the Perth Basin. The problem at Yallalie is to enhance the very low-amplitude anomalies.

Finally, we applied the new algorithms to a circular magnetic anomaly at Camooweal in Queensland, Australia. This anomaly represents a possible late Neoproterozoic (600 Ma) impact structure around 30 km in diameter showing multiple rings, but the feature is covered by a thick sequence of Palaeozoic sediments (Glikson, 1996).

Enhancement Techniques

Filtered images play an important role in aeromagnetic interpretation, enabling the interpreter to highlight features of interest and suppress unwanted information. Image processing of aeromagnetic data is most effective when dealing with shallow source depths and rock types with relatively large contrasts in magnetization. However, as depth to source increases, or where the rocks are only weakly magnetic, magnetic expression may be very subtle, requiring specialized filters to highlight features of interest. Both amplitude (tone) and textural information are essential in recognizing patterns that characterize impact structures. Filters such as conventional derivatives and reduction to the pole are widely available and most geophysicists are familiar with them (Gunn, 1975). However, we describe below some new algorithms, which offer improved recognition of magnetic anomalies due to terrestrial impacts. The new filters are illustrated using aeromagnetic data from the Archaean Yilgarn Shield in Western Australia, covering a major granite-greenstone dome structure (Fig. 1). The data are from the Leonora SH51-1 1:250,000 map sheet, Geological Survey of Western Australia, 1975, 1st edition.

Fractional Gradients

Both horizontal and vertical gradients of magnetic and gravity data are used routinely to enhance the edges of anomalies or as input to interpretation techniques such as Euler deconvolution. The most commonly used gradients are of first and second order. Higher orders are used less frequently, due to noise problems. A generalized approach using fractional gradients provides maximum flexibility, as the order of the fractional derivative can be selected to match the data (Cooper and Cowan, 2003a). Different order fractional derivatives can be combined to produce red-green-blue images (Fig. 1F), or the order of the derivative varied across a data set, and this can be a significant aid to the interpretation of the data (Cooper and Cowan, 2003c, 2004). Fractional gradients also allow the use of an improved sunshading algorithm that is less affected by noise than the standard method (Cooper and Cowan, 2003c).

Sunshading

Shaded relief, or sunshading, is a popular tool commonly used to enhance aeromagnetic data (Dods et al., 1985). Sunshading considers the magnetic intensity values as if they represent a topographic surface that is illuminated by a light source (or "sun"), which has a specified azimuth and elevation. The reflectance of the light by the surface is calculated and imaged. Features with strike directions orthogonal to the sun's azimuth reflect the light strongly and are consequently enhanced, whereas those lying parallel to the direction of the sun do not reflect the light and become less apparent as a result. Sunshading is a variety of high-pass filtering and so enhances both detail and noise in the data. If the result of sunshading an image is too noisy for easy interpretation, a possible solution is to use horizontal gradients of lower order (i.e., less than one). Conversely, if the data are smooth enough, then higher order gradients may be used. Fractional order gradients may be easily calculated in the frequency domain and then the reflectance can be calculated (Cooper and Cowan, 2003a).

The standard sunshading algorithm, which uses a constant sun azimuth, can be modified to enhance features that lie either on or orthogonal to radius vectors that pass through a chosen origin position. This enhancement can be very effective for accentuating circular anomalies (Cooper, 2003; Cooper and Cowan, 2003b). Given a user-selected circle center, the sun azimuth is adjusted across the image, allowing the detection of a circular feature of any radius. The algorithm, referred to as circular sunshading, can use fractional gradients. Circular gradients can also be calculated and used to enhance circular features. The contrast between images of conventional shaded relief and the new circular shaded relief algorithm (Fig. 1B and 1C) is striking, with the circular shading providing much clearer resolution. In addition, it is possible to image radial features passing through the center of the dome structure (Fig. 1D)

Separation Filtering

Separation (or layer) filtering of magnetic data allows deconvolution of the effects of causative sources that occur around a particular level (Cowan and Cowan, 1993). The measured magnetic field is a superposition of effects from sources at different levels as well as noise, so one must separate out the component of interest. In the simple case, this is the familiar regional-residual separation problem in which the regional field represents deep, thick sources and the residual field represents shallow, thin sources. We assume that the observed field is simply the sum of the regional field, the residual field, and a noise component. However, it is possible to extend the model to a general multilayer case and deconvolve the effects of sources at a number of distinct levels by separation filtering (Cowan and Cowan, 1993). Separation filtering depends fundamentally on the concept of random distributions of sources within discrete layers and assumes that there is no statistical difference in response along each ideal layer and no correlation between the distributions in each layer. In practice, it is usually impossible to achieve a complete separation, but, qualitatively, separation filtering is often very effective in highlighting magnetic anomalies of interest. The results of separating near-surface magnetic sources can be seen by comparing the total magnetic intensity data (Fig. 1A) with a separation filter grayscale image (Fig. 1E). Fractional order derivatives provide objective, flexible approaches to shallow-layer separation filtering as an alternative to spectral-matched filtering and differential field continuation (Cowan and Cooper, 2004). The order of the fractional derivative can be selected to match the data and optimize enhancement of the shallow-field component, avoiding the uncertainties in selecting spectral-matched filter parameters (Fig. 1F). The Matlab code for circular sunshading is available at http://www.mathworks.com/matlabcentral/fileexchange/loadFile.do?objectId=3092&objectType=file.

MAGNETIC SIGNATURES OF SELECTED IMPACT STRUCTURES

Acraman, South Australia

The Acraman impact structure, located at 32°01′S, 135°27′E on the Gawler Craton of South Australia, was first identified from circular features in satellite imagery. Its impact origin was later verified from the presence of shock metamorphic features. Acraman is located in Mesoproterozoic Gawler Range Volcanics, a suite of acid lavas and pyroclastic ash flows. The Yardea Dacite, near the top of the volcanic sequence, is mapped at the surface over much of the structure (Williams et al., 1996), although the magnetic low is covered by Lake Acraman (inset to Fig. 2). The

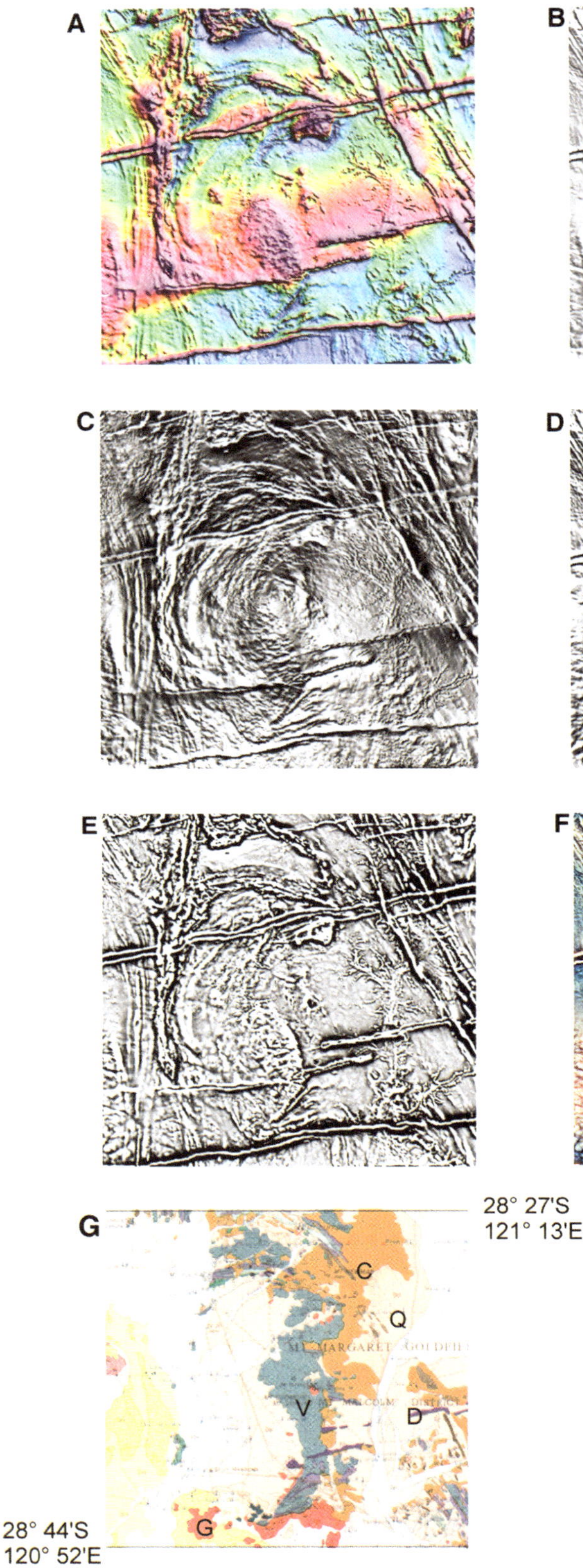

Figure 1. Images of granite-gneiss-greenstone terrain, Yilgarn Shield, Western Australia, to illustrate new filters. Image is 34 × 33 km. (A) Total magnetic intensity. (B) Conventional shaded relief illuminated from the NE. (C) Circular shaded relief illuminated from the center of the image. (D) Radial image illuminated from the center of the image. (E) Separation filter image. (F) Red-green-blue image of derivatives of order 0.5, 1.0, and 1.5. (G) Geology: C—Tertiary laterite; G—granitoids; D—Proterozoic dykes; Q—alluvium; V—mafic volcanics.

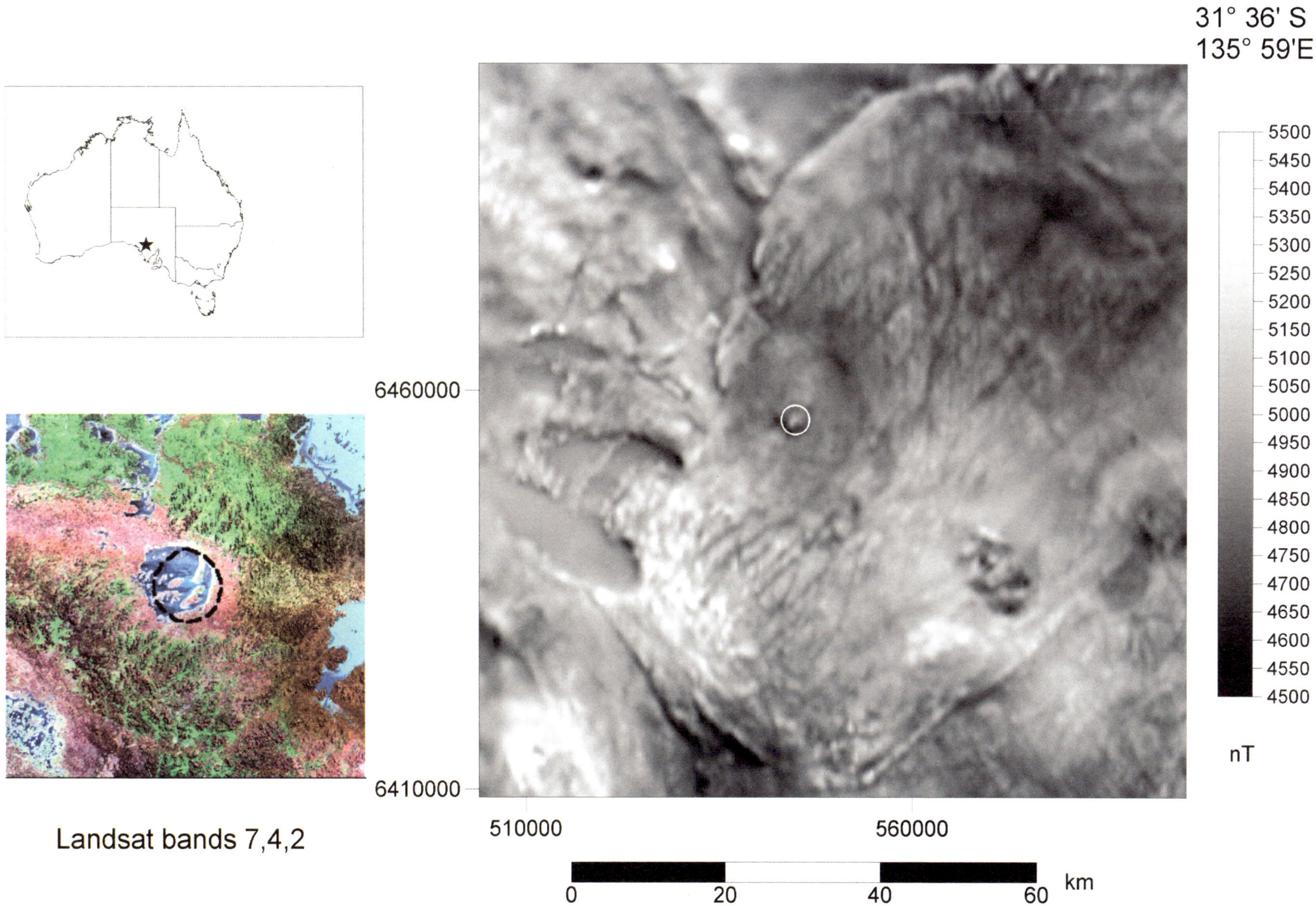

Figure 2. Residual magnetic intensity grayscale image of the Acraman impact structure and surrounding area. The image covers an area of ~91 × 91 km. Light tones indicate higher values of the magnetic field; dark tones indicate lower values. The center of the structure is marked by a white circle. The coordinates are UTM zone 53S. The inset shows a Landsat band 7,4,2 scene.

structure is located in a topographic depression (digital elevation three-dimensional image; inset to Fig. 3). Palaeomagnetic studies of the ejecta-bearing late Neoproterozoic Bunyeroo Formation in the Adelaide fold belt, South Australia (Schmidt and Williams, 1996) indicate that the Acraman impact occurred in the late Neoproterozoic, ca. 590 Ma.

We applied a range of filters to a grid of the total magnetic intensity data from a regional aeromagnetic survey acquired for the South Australian Department of Mines and Energy (now PIRSA) as part of the South Australian Exploration Initiative (Drexel et al. 1993). The data were acquired along north-south flight lines, 400 m apart, at a nominal elevation of 80 m. Our objective was to enhance the central circular magnetic low and highlight any outer circular features. We produced a grayscale image of residual magnetic intensity (Fig. 2) and a grayscale image of fractional vertical gradients of orders 0.5, 0.7, and 0.9 (Fig. 3). Acraman is in the center of the images and appears as a circular structure 18 km in diameter, with dark tones showing reduced magnetic amplitude. The circular reduced-amplitude anomaly clearly disrupts the northeast magnetic lineament pattern in the surrounding dacites. Note the small high-amplitude dipolar magnetic anomaly in the center of the low. The fractional vertical-gradient image enhances the circular magnetic low and the central small magnetic high associated with the impact structure, but there is no clear evidence of any outer rings or other impact-related anomalies. The magnetic data suggest that Acraman is a small impact structure with an 18 km circular magnetic low zone and a small central magnetic dipolar anomaly ~4 km in diameter. The circular magnetic low is consistent with shock demagnetization of the Yardea dacite, producing a decrease in magnetic susceptibility relative to undisturbed dacite (Williams et al., 1996). The central dipolar magnetic anomaly shows significant remanent magnetization (Williams et al., 1996) and may be due to a small melt pool or remagnetized uplifted basement. The magnetic data do not show any evidence of a much larger impact structure, as suggested by some authors (Williams et al. 1996). Assuming the actual crater size is the extent of the topographic depression, it appears that only the central part of the structure has a magnetic expression.

Manicouagan, Quebec, Canada

The Manicouagan impact structure located at 51°23′N, 68°42′W represents the remnants of one of the largest impact craters still preserved on the surface of Earth. Lake Manicouagan surrounds the central uplift, which is ~70 km in diameter, and is composed of impact-brecciated rock surrounded by erosion-resistant metamorphic and igneous rocks. A range of filters was applied to a grid of the total magnetic intensity data from an old Geological Survey of Canada regional aeromagnetic survey. The data were collected along north-south lines, with 800-m line spacing, 60 m sampling along line, and show leveling problems. Our objective was to enhance the central circular magnetic anomaly and highlight any outer circular features. We produced a grayscale image of residual magnetic intensity (Fig. 4) and an image of the fractional order 0.5 vertical gradient (Fig. 5). We used a fractional order of 0.5 vertical gradient because a conventional order 1 derivative amplified leveling problems in the data and produced marked north-south stripes. The residual magnetic-intensity data show a relatively subdued positive circular anomaly over the central uplift with amplitudes ~100 nT and a rugose or mottled texture. A small, higher amplitude magnetic anomaly seen in the center of the central uplift corresponds to a topographic high. The small, high amplitude anomaly has amplitudes up to 1600 nT and is clearly remanent (Roest and Pilkington, 1993). The vertical gradient image has sharpened up the response of the central uplift, and highlights the rugose magnetic texture with no clear trend directions (Fig. 5). The central uplift anomaly at Manicouagan is a weak positive magnetic anomaly rather than a magnetic low, like that associated with the Acraman structure. Images of Landsat bands 7,4,2 and Shuttle Radar Topographic Mission digital elevation (insets to Figs. 4 and 5) provide complementary information that help to highlight the impact magnetic structure and contrast the differences between the impact and the elliptical, magnetically quiet anorthosite bodies.

Simplified geology of the area (inset to Fig. 5) shows the Triassic impact breccia/melt pool structure within folded and faulted Grenville Province target rocks. Glaciation and other erosional processes have reduced the visible morphological visible extent of the crater. The original diameter is estimated to have been ~100 km. The impact that formed Manicouagan occurred at 212 Ma, toward the end of the Triassic (Hodych and Dunning, 1992).

Yallalie, Western Australia

The Yallalie structure, located at 30°27′S, 115°46′E in Mesozoic sediments of the Perth Basin, Western Australia, was first identified from seismic reflection data (Dentith et al., 1999). Although recent aeolian sands cover the structure, it does coincide with a circular topographic depression (inset to Fig. 6). Dentith et al. (1999) describe how image processing of digital elevation data has allowed the recognition of concentric and radial structures extending as far as 40 km from the center of the depression. No shock metamorphic features have so far been found, but aeromagnetic data provide a convincing argument for its impact origin (Hawke et al., 2003). The seismic reflection data show continuity of the stratigraphy below the Yallalie structure, supporting a cause for its creation from above (Dentith et al., 1999). We applied a range of filters to a grid of the total magnetic intensity aeromagnetic data. The aeromagnetic data, acquired as part of a University of Western Australia research project, were collected along north-south flight lines at 200 m line spacing using a horizontal gradiometer equipped aircraft. We produced a residual magnetic intensity grayscale image (Fig. 6) and a fractional-gradient circular shaded relief image (Fig. 7) that show a small central peak and five concentric ring magnetic anomalies. A three-dimensional image (Fig. 8) shows the fractional gradient circular shaded relief draped on the residual magnetic intensity.

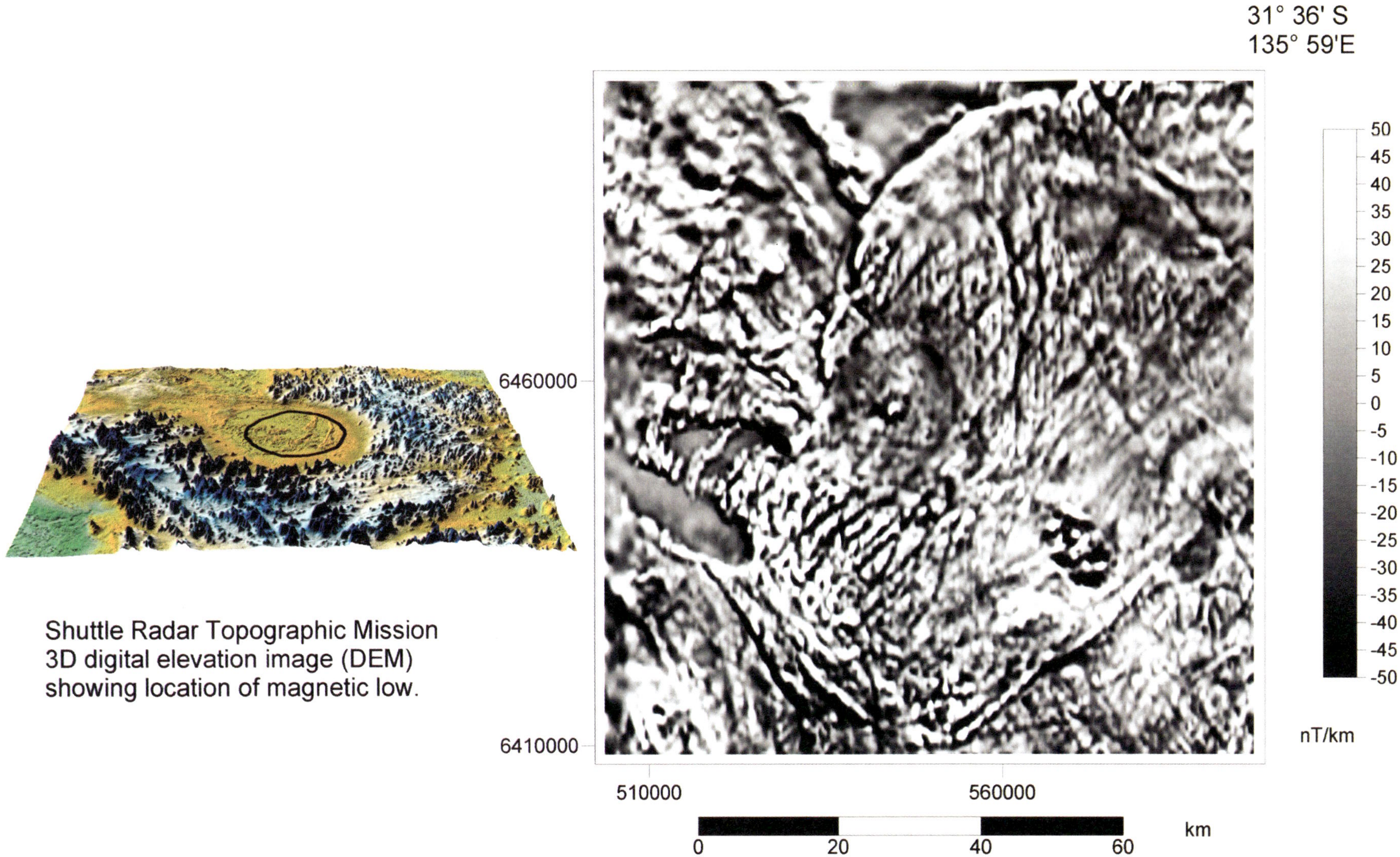

Figure 3. Fractional order vertical derivative grayscale image of the Acraman impact structure and surrounding area. The image covers an area of ~91 × 91 km. The image was created by blending fractional derivatives of order 0.5, 0.7, and 0.9 to highlight the impact structure. Light tones indicate higher values of the magnetic field gradient; dark tones indicate lower values. The inset shows Shuttle Radar Topography Mission (SRTM) digital elevation data viewed from the south with the outer limit of the magnetic low marked. The SRTM data show a much larger diameter topographic depression than the magnetic anomaly data.

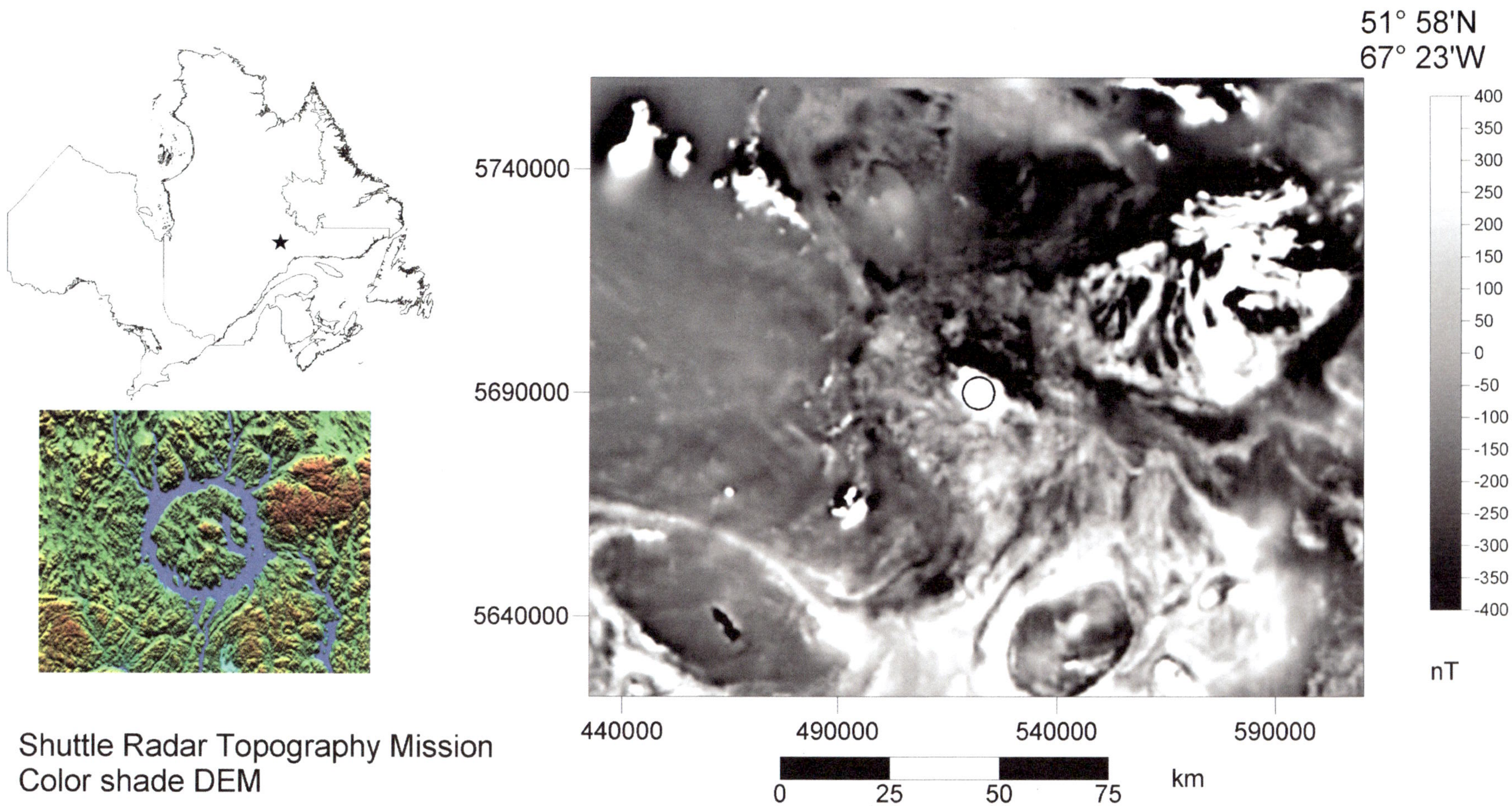

Figure 4. Residual magnetic intensity grayscale image of the Manicouagan impact structure and surrounding area. The image covers an area of ~178 × 138 km. Light tones indicate higher values of the magnetic field; dark tones indicate lower values. The center of the impact is marked by a black circle. The coordinates are UTM zone 19N. The inset shows Shuttle Radar Topography Mission digital elevation.

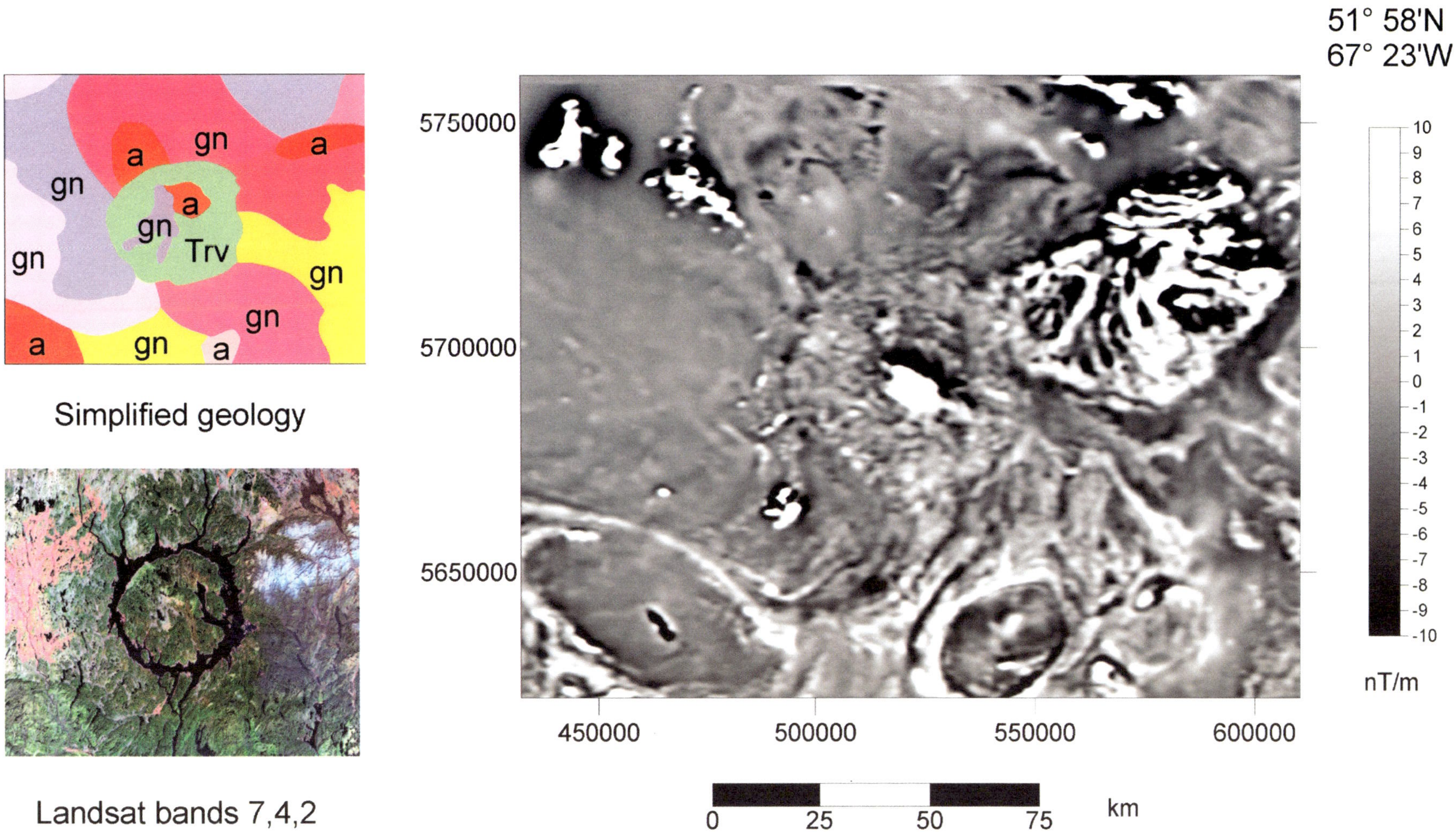

Figure 5. Image of the fractional order vertical derivative of the Manicouagan impact structure and surrounding area. The image covers an area of ~178 × 138 km. Light tones indicate higher values of the magnetic field gradient; dark tones indicate lower values. We used a fractional vertical derivative order of 0.5 to reduce the effects of leveling errors. The insets show the simplified geology and a Landsat band 7,4,2 scene. Geology: a—anorthosite; gn—undifferentiated gneisses; Trv—Triassic impact breccia/melt sheet.

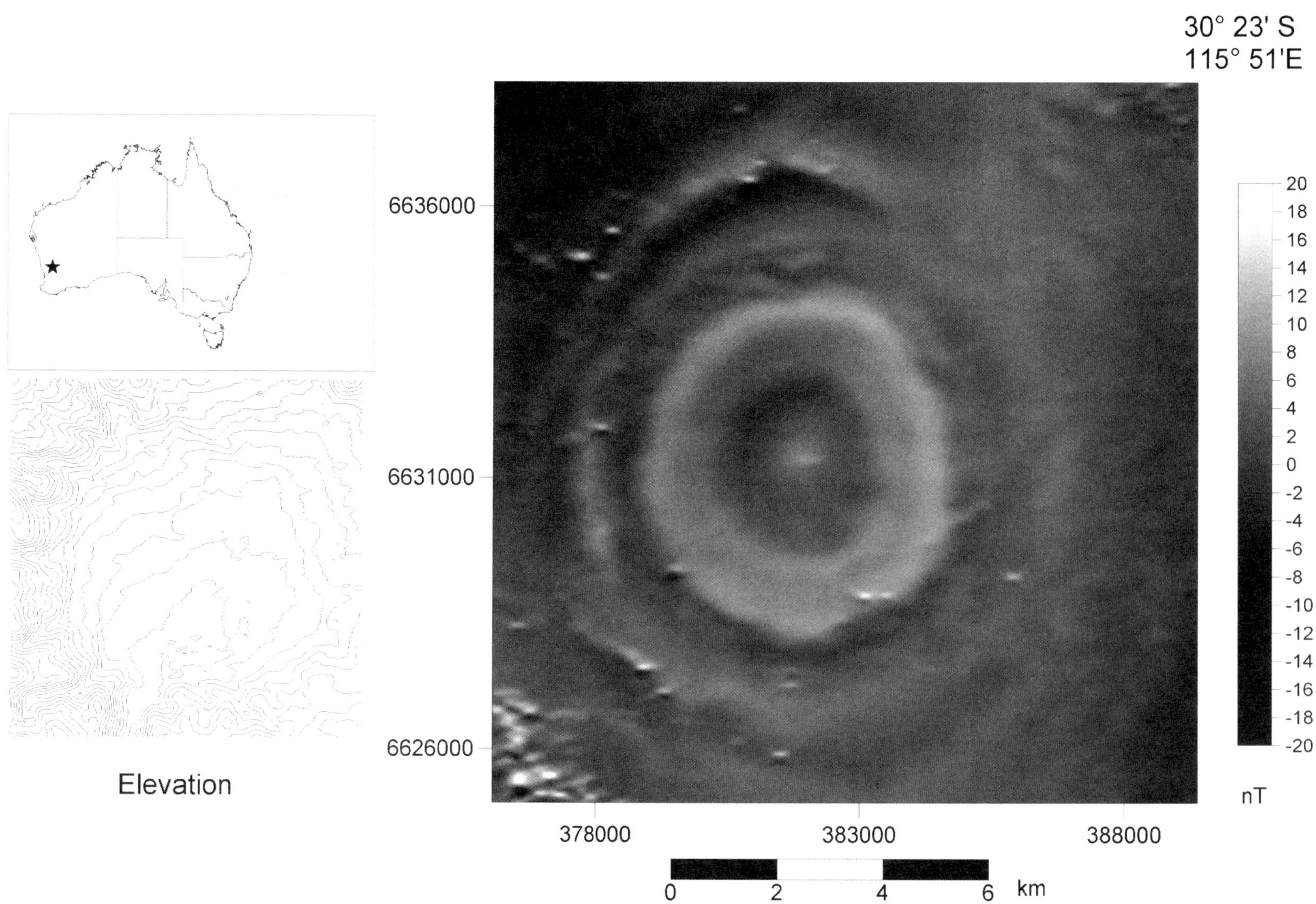

Figure 6. Residual magnetic intensity grayscale image of the Yallalie structure and surrounding area. The image covers an area of ~13 × 13 km. Light tones indicate higher values of the magnetic field; dark tones indicate lower values. The image shows five concentric magnetic anomalies and a central magnetic high. The inner magnetic ring, with a diameter of 3.4 km, is interpreted as the structure's central uplift. The outer rings are interpreted as fault terraces, based on correlation with seismic data. The diameter of the outer ring is close to the size of the structure, based on seismic interpretation. Coordinates are in UTM zone 50S. The inset shows contours of the altimetric digital elevation produced from the airborne survey global positioning system height and radar altimeter data. The digital elevation data show a clear topographic depression corresponding to the structure.

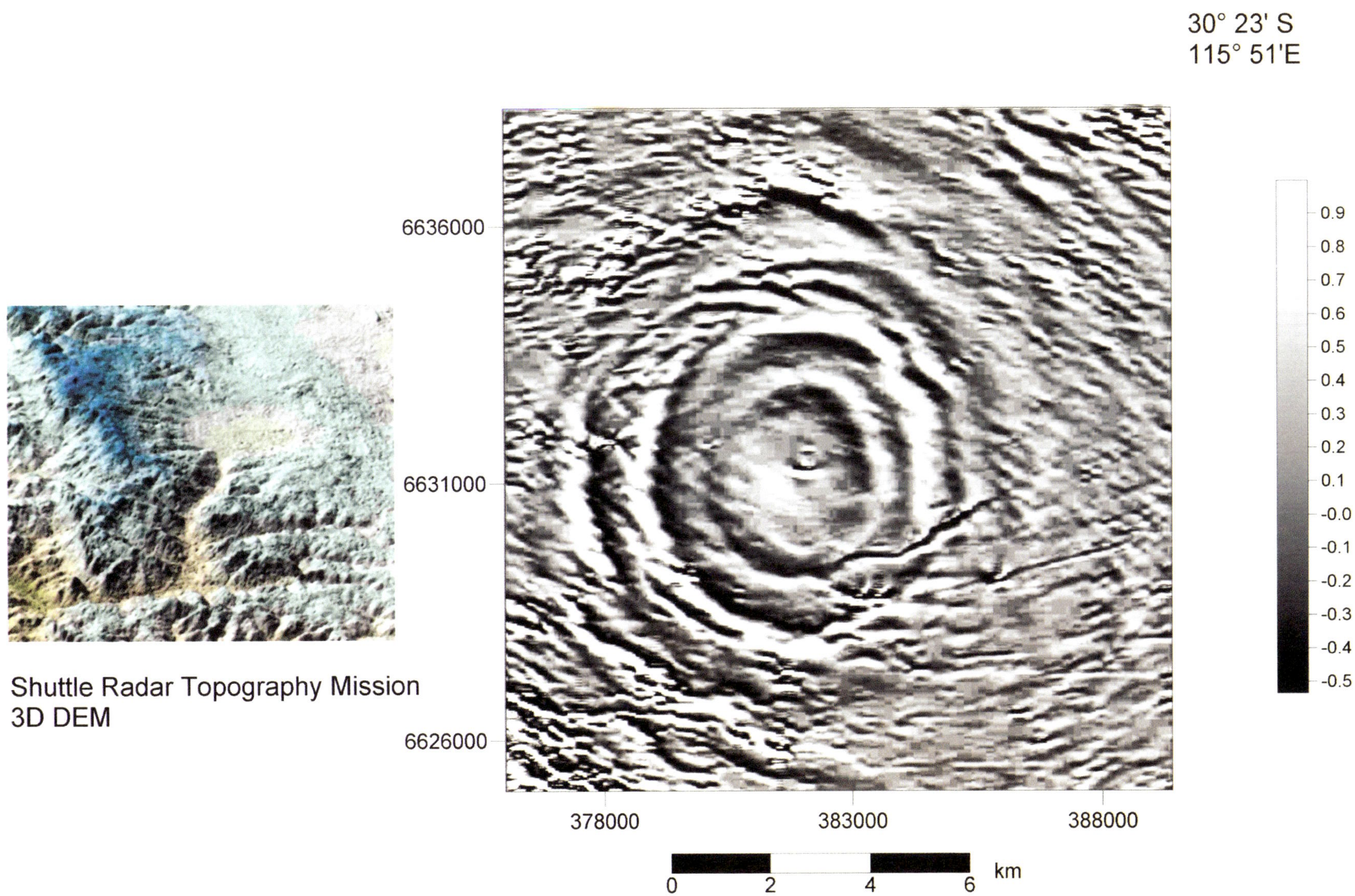

Figure 7. Fractional gradient circular shaded relief grayscale image of the Yallalie impact structure and surrounding area. The image covers an area of ~13 × 13 km. Light tones indicate higher values of the shaded magnetic field; dark tones indicate lower values. The image highlights the concentric rings seen in Figure 6 and also enhances the small central magnetic high, which is a possible melt pool anomaly. The inset Shuttle Radar Topography Mission digital elevation data shows a topographic depression with greater extent than the ring structure.

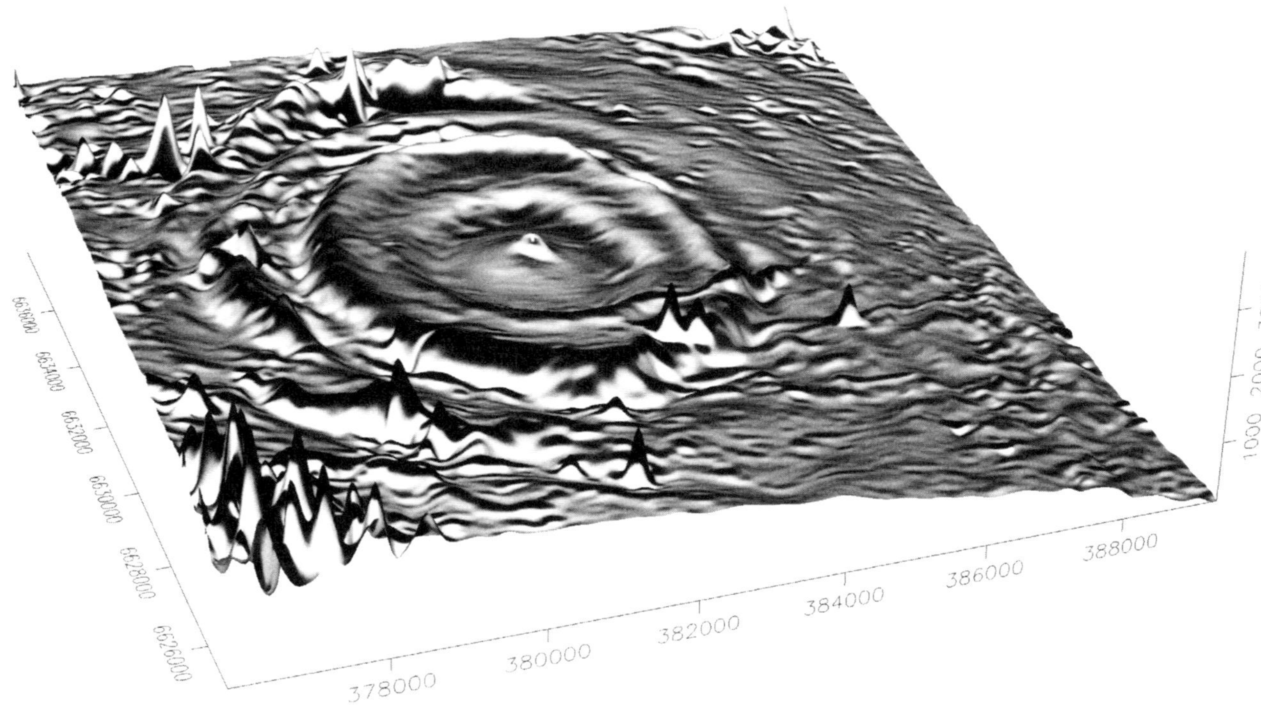

Figure 8. Three-dimensional image of the circular shaded fractional order vertical derivative draped on residual magnetic intensity of the Yallalie impact structure and surrounding area. The image covers an area of ~13 × 13 km. Light tones indicate higher values of the magnetic field gradient; dark tones indicate lower values. We used a fractional vertical derivative order of 1.5.

We also contoured the altimetric digital elevation data to illustrate the coincidence between the aeromagnetic ring anomalies and the topographic depression (inset to Fig. 6). The topographic depression is at an average level of 180 m, with the surrounding ridges up to 280 m.

The circular shade images have dramatically improved resolution of the ring anomalies. Although anomaly amplitudes are quite low (4–12 nT), the multiple subcircular ring anomalies are compelling, indicating a crater diameter of 12 km. These ring anomalies closely correlate with faults interpreted from seismic data, which indicate a central uplift and a series of terraces, similar to those of the Silverpit structure in the North Sea (Stewart and Allen, 2002). The ring anomalies are slightly elliptical in plan, which may reflect anisotropy in the target rocks or an oblique impact. There is evidence of post-impact faulting in the east of the images. Lower amplitudes in the eastern part of the structure are due to increased thickness of cover. The source of the ring magnetic anomalies is unknown, but the very low amplitudes would be consistent with a hematite source or very low volume percent of magnetite or maghemite.

Camooweal, Queensland, Australia

Camooweal is a circular magnetic anomaly feature located at 19°15′S, 138°07′E in Palaeozoic sediments in NW Queensland. The structure has no geomorphic expression and is completely covered by Cambrian limestones; its impact origin has not been confirmed. We applied a range of filters to a grid of the residual magnetic intensity data, collected along north-south flight lines at 300 m line spacing. The total magnetic intensity data are dominated by long wavelength basement anomalies (Fig. 9), so we used separation filtering to remove the basement signal. An order 0.8 fractional vertical gradient of the separation filter data (Fig. 10) shows a central area of low magnetic relief and several concentric ring magnetic anomalies. The combination of separation filtering (to deconvolve the effects of magnetic sources around the circular magnetic anomaly) and a fractional derivative of order 0.8 provides optimum source resolution. Magnetic source depth estimation suggests a depth >800 m, so the ability of filter techniques to highlight low-amplitude magnetic signatures of a possible impact structure is impressive. The west-northwest linear anomalies in the northeast of the image are within the basement.

CONCLUSIONS

Aeromagnetic surveys play an important role in the detection and analysis of terrestrial impact structures. Magnetic signatures of terrestrial impact craters are quite variable, reflecting the composition of the target rocks, the impact-related effects on

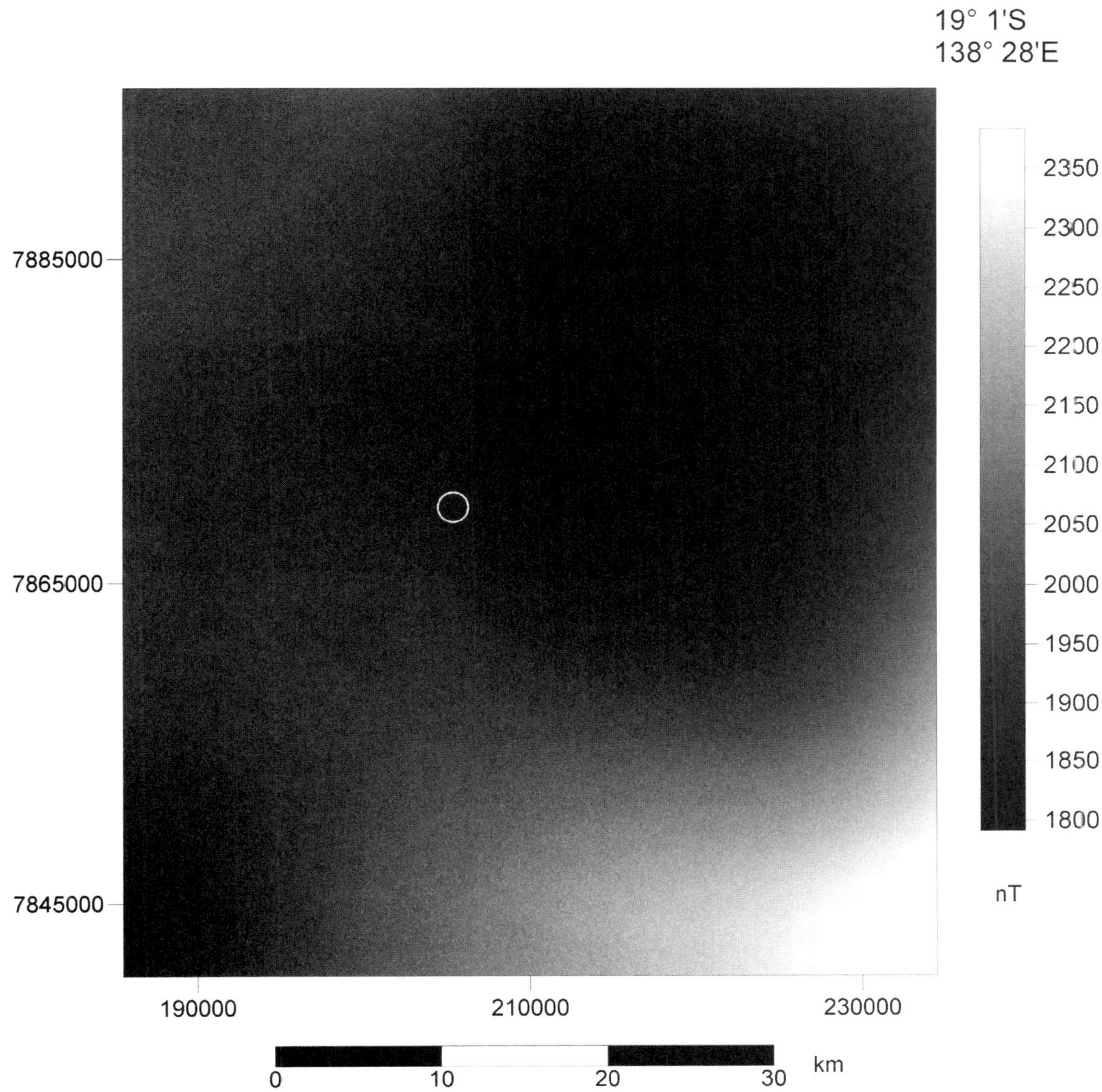

Figure 9. Total magnetic intensity grayscale image of the Camooweal magnetic anomaly and surrounding area. The image covers an area of ~49 × 55 km. Light tones indicate higher values of the magnetic field; dark tones indicate lower values. The center of the magnetic anomaly is marked by a white circle. Coordinates are in UTM zone 54S.

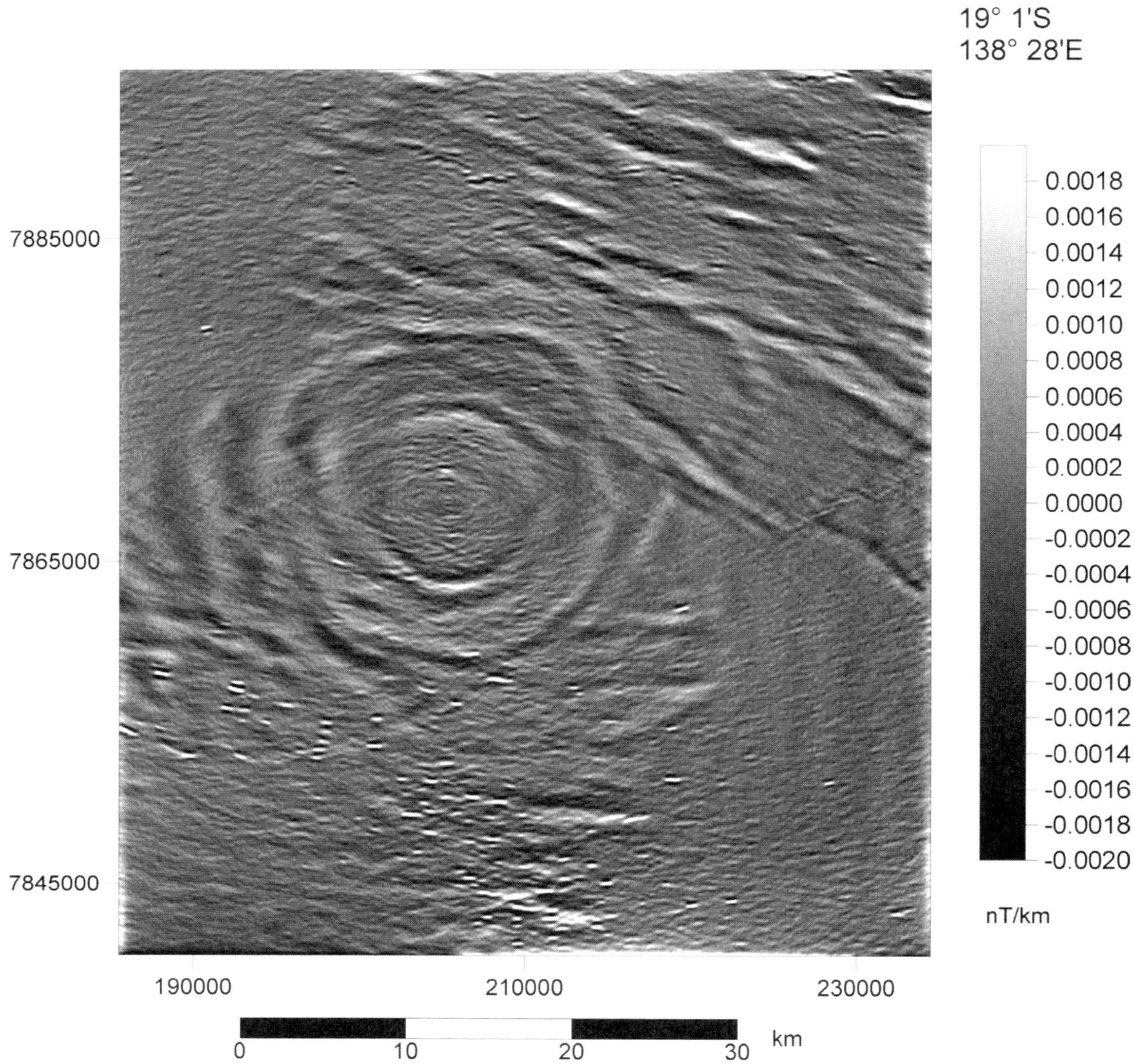

Figure 10. Fractional-order vertical derivative separation filter grayscale image of the Camooweal magnetic anomaly and surrounding area. The image covers an area of ~49 × 55 km. Light tones indicate higher values of the magnetic field gradient; dark tones indicate lower values.

magnetization, effects of crater fill, and post-impact cover. In crystalline basement rocks, the signature is usually a magnetic low with a central magnetic anomaly or ring anomalies. Impact craters in sedimentary targets usually have subdued magnetic relief depending on the lithology, and anomaly amplitudes of a few nT up to 10 nT are typical. Our filter results show that enhancement of magnetic signatures of terrestrial impact structures is an important part of detection and analysis. A range of filters is required to handle the highly variable magnetic signatures of terrestrial impact structures and also to cope with variable-quality data. Recent development of algorithms for fractional horizontal and vertical derivatives and circular shaded relief has dramatically improved filter results. The fractional derivative order can be varied to optimize separation of the magnetic signature of impact structures. Because the circular sunshading algorithm was specifically designed to enhance circular features, it consequently gives much better results with impact structures than does the standard sunshading algorithm, which was designed to enhance linear features.

Despite low anomaly amplitudes, it appears that magnetic anomalies due to impacts in sedimentary targets have more distinctive signatures than impact anomalies in crystalline basement rocks.

ACKNOWLEDGMENTS

We thank the Department of Earth and Geographical Sciences, the University of Western Australia, for permission to use the

Yallalie aeromagnetic data. Helpful reviews from referees Lauri Pesonen and C. Wylie Poag and associate editor, Alex Deutsch, have improved both the style and content of the paper.

REFERENCES CITED

Cooper, G.R.J., 2003, Feature detection using sun shading: Computers & Geosciences, v. 29, p. 941–948, doi: 10.1016/S0098-3004(03)00091-8.

Cooper, G.R.J., and Cowan, D.R., 2003a, The application of fractional calculus to potential field data: Exploration Geophysics, v. 34, p. 51–56.

Cooper, G.R.J., and Cowan, D.R., 2003b, Feature detection using sunshading: Extended abstracts: Adelaide, Australian Society of Exploration Geophysicists 16th Geophysical Conference and Exhibition, February 2003, CD-ROM.

Cooper, G.R.J., and Cowan, D.R., 2003c, Sunshading geophysical data using fractional order horizontal gradients: The Leading Edge, v. 22, p. 204–205, doi: 10.1190/1.1564523.

Cooper, G.R.J., and Cowan, D.R., 2004, Filtering using variable order vertical derivatives: Computers & Geosciences, v. 30, p. 455–459, doi: 10.1016/J.CAGEO.2004.03.001.

Cowan, D.R., and Cooper, G.R.J., 2003, The Shuttle Radar Topography Mission—a new source of near-global digital elevation data: Australian Society of Exploration Geophysicists Preview, v. 107, p. 27–29.

Cowan, D.R., and Cooper, G.R.J., 2004, Separation filtering using fractional order gradients: Extended abstracts, Australian Society of Exploration Geophysicists 17th Geophysical Conference and Exhibition, August 2004, Sydney (CD-ROM).

Cowan, D.R., and Cowan, S., 1993, Separation filtering applied to aeromagnetic data: Exploration Geophysics, v. 24, p. 429–436.

Dentith, M.C., Bevan, A.W.R., Backhouse, J., Featherstone, W.E., and Koeberl, C., 1999, Yallalie: A buried structure of possible impact origin in the Perth Basin, Western Australia: Geological Magazine, v. 136, p. 619–632, doi: 10.1017/S0016756899003386.

Dods, S.D., Teskey, D.J., and Hood, P.J., 1985, The new series of 1:1,000,000-scale magnetic-anomaly maps of the Geological Survey of Canada: Compilation techniques and interpretation, *in* Hinze, W.J., ed., The utility of regional gravity and magnetic anomaly maps: Tulsa, Oklahoma, Society of Exploration Geophysicists, p. 69–87.

Drexel, J.F., Preiss, W.V., and Parker, A.J., eds., 1993, The geology of South Australia, Volume 1, The Precambrian: Geological Survey of South Australia, Bulletin 54, p. 15.

Glikson, A.Y., 1996, A compendium of Australian impact structures, possible impact structures and ejecta occurrences: Australian Geological Survey Organization Journal of Australian Geology & Geophysics, v. 16, p. 373–375.

Grieve, R., Rupert, J., Smith, J., and Therriault, A., 1995, The record of terrestrial impact cratering: GSA Today, v. 5, no. 10, p. 189–196.

Gunn, P.J., 1975, Linear transformations of gravity and magnetic fields: Geophysical Prospecting, v. 23, p. 300–312.

Hawke, P.J., Buckingham, A.J., and Dentith, M.C., 2003, Origin of magnetic anomalies associated with the Yallalie impact structure, Perth Basin, Western Australia: Extended abstracts: Adelaide, Australian Society of Exploration Geophysicists 16th Geophysical Conference and Exhibition, February 2003 (CD-ROM).

Hodych, J.P., and Dunning, G.R., 1992, Did the Manicouagan impact trigger end-of-Triassic mass extinction?: Geology, v. 20, p. 51–54, doi: 10.1130/0091-7613(1992)0202.3.CO;2.

Pilkington, M., and Grieve, R.A.F., 1992, The geophysical signature of terrestrial impact craters: Reviews of Geophysics, v. 30, p. 161–181.

Roest, W.R., and Pilkington, M., 1993, Identifying remanent magnetization effects in magnetic data: Geophysics, v. 58, p. 653–659, doi: 10.1190/1.1443449.

Schmidt, P.W., and Williams, G.E., 1996, Palaeomagnetism of the ejecta-bearing Bunyeroo Formation, late Neoproterozoic, Adelaide fold belt, and the age of the Acraman impact: Earth and Planetary Science Letters, v. 144, p. 347–357, doi: 10.1016/S0012-821X(96)00169-0.

Stewart, S.A., and Allen, P., 2002, A 20-km-diameter multi-ringed impact structure in the North Sea: Nature, v. 418, p. 520–523, doi: 10.1038/NATURE00914.

Williams, G.E., Schmidt, P.W., and Boyd, D.M., 1996, Magnetic signature and morphology of the Acraman impact structure, South Australia: Australian Geological Survey Organization Journal of Australian Geology & Geophysics, v. 16, p. 431–442.

Manuscript Accepted by the Society 9 August 2004

Printed in the USA

Geological Society of America
Special Paper 384
2005

Is Ries crater typical for its size? An analysis based upon old and new geophysical data and numerical modeling

K. Wünnemann*
J.V. Morgan
Department of Earth Science and Engineering, Imperial College London, London SW7 2AZ, UK

H. Jödicke
Institut für Geophysik, Westfälische Wilhelms–Universität Münster, Correnstraße 24, 48149 Münster, Germany

ABSTRACT

Considering the Ries crater as an example for middle-sized complex impact structures on Earth, we use geophysical data to examine the structure underneath the crater and simulate the formation process by numerical modeling. In contrast to previous investigations, we show by reanalyzing seismic refraction profiles that some clues for structural uplift exist beneath the crater in a range of at least 1 km. We propose that the average P-wave velocity inside the crater is lower than outside the crater from the surface to ~2.2 km depth, but that below this level, the velocity increases beneath the center, presumably due to uplifted basement rocks. In addition, we utilized magnetotelluric depth sounding to investigate the deep electrical structure beneath the crater. Two-dimensional inversion models of the data show anomalously high conductivity beneath the crater. Our best model features a zone of presumably brine-filled fractures in open pore space to a depth of ~2 km. Furthermore, our numerical modeling results for the crater formation are consistent with surface and subsurface observations in the vicinity of the crater. In order to explain the structural differences between similarly sized craters and Ries, we investigate the sensitivity of crater shape and subsurface structure to varying target compositions. We show that for a reasonable range of constitutive material and acoustic fluidization parameters the model calculations produce a large variety of different crater shapes, even for the same amount of impact energy. In contrast to the conventional estimate of crater diameters, our results suggest that Ries crater is comparable in size with the Bosumtwi and Zhamanshin crater. Despite their apparent lack of similarity at first look, Ries and Bosumtwi are closely matched in terms of transient crater size (inner ring), aspect ratio, and structural elements, and we conclude that they both represent typical complex crater structures of the terrestrial impact record for their size range.

Keywords: impact cratering, numerical modeling, Ries, magnetotelluric, Bosumtwi, crater morphology.

*Present address: Lunar and Planetary Science Lab, University of Arizona, 1629 E. University Blvd., Tucson AZ 85721, USA; wunnema@lpl.arizona.edu.

Wünnemann, K., Morgan, J.V., and Jödicke, H., 2005, Is Ries crater typical for its size? An analysis based upon old and new geophysical data and numerical modeling, *in* Kenkmann, T., Hörz, F., and Deutsch, A., eds., Large meteorite impacts III: Geological Society of America Special Paper 384, p. 67–83. For permission to copy, contact editing@geosociety.org.

INTRODUCTION

As a basic principle, impact crater morphology is a function of crater size. A major change from simple to complex morphology occurs at a well defined threshold diameter depending on target properties and the given gravity field (e.g., Grieve and Pesonen, 1992). The shape of simple craters is relatively consistent and does not change much with increasing size. In contrast, complex craters show greater variety in morphology and can be roughly subdivided with increasing diameter into (1) central peak craters, (2) peak ring craters, and (3) multi-ring basins (e.g., Melosh, 1989). On Earth it seems to be more difficult to assign a typical crater form to a given crater size than on most other planetary bodies. The lack of consistency between similarly sized terrestrial craters makes it more difficult to understand the dynamics of complex crater formation. A further aggravating circumstance is the fact that most crater structures on Earth are not pristine. Post-impact modifications like erosion and tectonic overprint can prevent, in many cases, an accurate reconstruction of the crater structure and determination of the crater diameter. Thus, it is difficult to define a typical morphology for a given crater size on Earth. In addition, the inhomogeneous composition of the target, which is presumably more complex than on any other planetary body, probably has a fundamental effect on the formation process and might be at least partially responsible for the lack of structural similarity between similarly sized craters.

The investigation of the subsurface of terrestrial impact structures provides important clues to understanding the crater formation process. Thus, it is crucial to know whether a particular crater is a relatively typical example for its size or whether it is a structure where the formation was heavily influenced by local target conditions.

The objective of the present paper is to discuss the structure and morphology of the well preserved and extensively investigated Ries crater in Germany in comparison to other terrestrial craters. Present structural models of the Ries are based on drilling and a variety of geophysical explorations that were mainly carried out in the late 1960s and early 1970s (Angenheister and Pohl, 1969). Here we present a reanalysis of a subset of the seismic refraction data, aimed at investigating the apparent absence of structural uplift beneath the crater center. We also show new magnetotelluric depth soundings that complement the existing data set and provide a model of the electrical resistivity structure deep beneath the crater. Furthermore, we conducted a suite of numerical models of the formation process. Although the field of impact crater modeling has become an important part of modern impact research (Ivanov and Artemieva, 2002), there has been no recent attempt to model the formation of the Ries crater (Stöffler et al., 2001). Computer simulations of the formation process provide important clues to evaluate the influence of varying target compositions on the cratering process and the final crater morphology. For this reason, in our dynamic models of the formation of Ries crater, we tested different assumptions regarding structural composition and material properties of rocks in the vicinity of the crater.

Our analysis is then exemplified by a comparison of Ries with Bosumtwi and Zhamanshin craters, revealing important structural and morphological aspects of middle-sized impact craters.

THE RIES CRATER

The Ries crater is a middle-sized complex crater located ~120 km northwest of Munich (Fig. 1). It was formed ca. 15 Ma (Bolten and Müller, 1969; Gentner et al., 1963) and due to relatively low erosion rates in this area the crater is well preserved. Since 1961, when Shoemaker and Chao (1961) proved its impact origin, the Ries has been intensively investigated, making it one of the best-studied craters on Earth. An excellent review was published by Pohl et al. (1977), but since then no major progress, in terms of the investigation of the deep subsurface structure and understanding the formation of the crater, has been achieved.

The crater is characterized by an almost circular, relatively flat inner basin, 12 km in diameter, which is covered by post-impact lake sediments surrounded by an inner ring. Topographically, this inner ring is only visible as a circular chain of isolated crystalline hills standing ~50 m above the basin. Adjacent to the plains of post-impact lake sediments surrounded by the inner ring, hummocky relief extends to the outer morphological crater rim with an approximate diameter of 24–26 km. This area mainly consists of displaced megablocks up to 1 km in size and is believed to represent a system of concentric normal faults, with the outermost fault referred to as the tectonic or crater rim.

The near-surface structure, to about ~1 km depth, is well investigated. Under the post-impact sediments, with a maximum depth of ~350 m in the central basin (Ernstson, 1974), a 300–400-m-thick suevite layer (a polymict breccia primarily derived from differently shocked basement clasts with cogenetic impact melt clasts and bombs) covers the crystalline basement (Pohl et al., 1977). Suevite deposits up to 200 m in thickness are also found outside the inner ring and are located on the top of the Bunte breccia (derived from near-surface target rocks; cf. Pohl et al., 1977).

The pre-impact stratigraphy of the target is made up of sedimentary deposits, 620–750 m thick, overlying a crystalline basement composed mainly of granites (Pohl et al., 1977). Immediately after the projectile hit this target, a transient crater was formed, and its diameter is believed to have been coincident with the inner ring. The transient cavity would have been roughly one-third of the diameter; e.g., ~4 km deep (Melosh, 1989). Since the crater floor now lies less than ~1 km below the surface in the crater center, some kind of modification process of the transient crater must have taken place. According to the size of the transient cavity, the collapse is most likely associated with structural uplift. However, to date there are no clues in the geophysical data as to the shape and extent of such a structural uplift. In this section, we provide reanalyzed seismic refraction data and new magnetotelluric data to investigate the subsurface beneath the central basin.

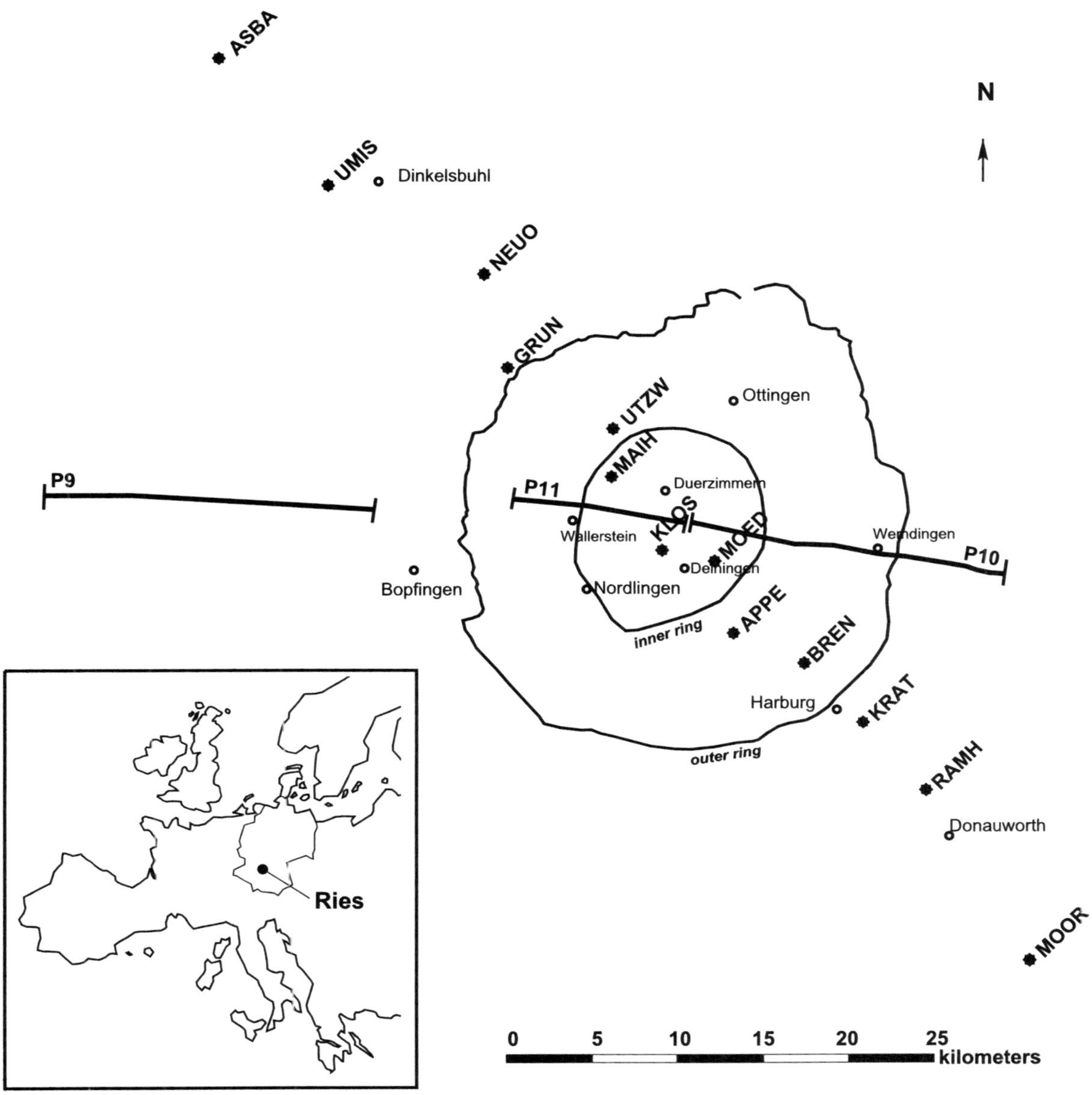

Figure 1. Map of the new magnetotelluric (MT) and the seismic refraction profiles acquired in 1968 (Angenheister and Pohl, 1969) in the Ries crater area. Asterisks mark the positions of MT stations. Bold solid lines indicate the location of the seismic profiles (P9, P10, P11). The circular solid lines display the approximate positions of the inner (crystalline) ring and the outer (tectonic) ridge (Bayerisches Geologisches Landesamt, 1999).

Seismic Refraction Analysis

Several seismic refraction profiles have been acquired across the crater (Angenheister and Pohl, 1969; Pohl and Will, 1974). The authors have concluded that the average velocities inside the crater are lower than those outside, and this low-velocity–zone extends to 3–6 km below the surface (Pohl and Will, 1974; Ernstson and Pohl, 1977). At most complex craters, structural uplift is observed in the crater center and, when not observed in outcrop, can often be detected by its geophysical signature. Structural uplift is typically associated with an increase in seismic velocity, increase in density, and possibly a change in the magnetic or electrical signature (e.g., Pilkington and Grieve, 1992; Karp et al., 2002; Morgan et al., 2002). Thus, a low-velocity–zone in the crater center at Ries apparently indicates an absence of structural uplift at this crater.

All the seismic refraction data are along two-dimensional profiles that are a few tens of kilometers long (Fig. 1), and most consist of a single shot at one end that is recorded at receivers spaced a few kilometers apart. To be able to constrain lateral changes in velocity, multiple shots are required into multiple receivers, and velocity resolution is directly related to the shot or receiver spacing, whichever is the largest (Zelt, 1998). Hence, single shot profiles provide no lateral resolution. Profile 11 (P11 in Fig. 1) is reversed—a shot was fired at each end. This is the only profile that has crossing ray paths, and thus the only profile with any lateral resolution—albeit quite limited. Profiles 9 and 10 were acquired in an expanding spread configuration, in which a number of single shots were recorded on single receivers, with the shot and receiver being moved to maintain a constant midpoint. In expanding spread profiles, there are no crossing ray paths and, because of this, they are conventionally modeled with a one-dimensional interpretation (e.g., Fowler et al., 1989). From the experimental geometries of the current suite of seismic refraction data, it is clear that lateral changes in velocity cannot be well constrained. Given the limitations of these data, we have adopted the most objective approach—that is to find the best-fit one-dimensional velocity model that fits the data. Here we reanalyze the best of these data: refraction profiles 9, 10, and 11. Profile 9 is entirely outside the crater rim, profile 10 mostly inside the crater rim, and profile 11 entirely inside the crater rim (see Fig. 1).

P-wave velocity usually increases with depth, and upper crustal continental rocks typically have velocities of between 6.0 and 6.2 km s^{-1} (e.g., Christensen and Mooney, 1995). The basement rocks at Ries are granitic; fully compacted granites also have velocities of between 6.0 and 6.2 km s^{-1} (Birch, 1961). With this in mind, we have replotted the refraction data with a reduction velocity of 6 km s^{-1} (Fig. 2). In such plots, arrivals with an apparent velocity of 6 km s^{-1} appear as horizontal lines and are easy to identify, and any difference in the appearance of refractions from layers with this velocity are easy to spot. The vertical lines in these plots are picked first-arrival travel times. The picks for profiles 9 and 10 are taken directly from Pohl and Will (1974). Profile 11 data is directly from Angenheister and Pohl (1969); these data show a low signal to noise ratio, and first-arrivals were not easy to pick.

As mentioned above, we have obtained the best-fit one dimensional velocity structure for each profile (Figs. 2 and 3). The program RAYINVR has been used to ray-trace the data (Zelt and Smith, 1992), and the straight lines through the arrivals in Figure 2 are the calculated travel times. Profiles 9 and 10 are both very simple and easy to interpret as the picked arrivals are clearly aligned along three straight lines with different apparent velocities (V_a in Fig. 2). These plots suggest a subsurface consisting of three layers, with the velocity of the top two layers being higher for profile 9 (3.3 and 5.35 km s^{-1}) than profile 10 (2.0 and 4.2 km s^{-1}; Fig. 3). Profile 11 shows more lateral variation, and the best-fit straight lines have a larger misfit. This suggests there are either lateral changes in velocity across this profile or some of the boundaries between individual layers are dipping. However, the most striking difference between the profiles inside and outside the crater is the point at which the phase with an apparent velocity of 6 km s^{-1} becomes a first arrival. Outside the crater (Fig. 2A), this first occurs at ~23 km, whereas inside the crater, it occurs between 6 and 8 km (Fig. 2B–D). This observation strongly suggests that the depth at which velocities of 6 km s^{-1}

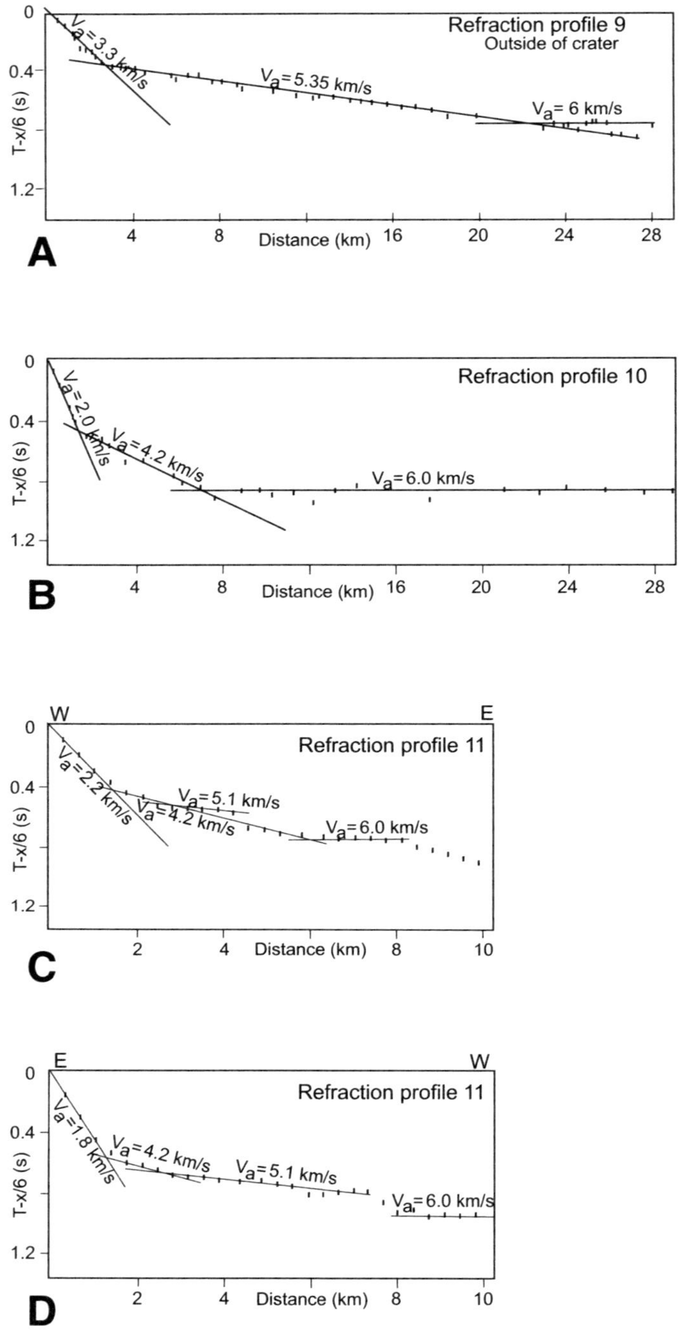

Figure 2. Traveltime plots of seismic refraction data along (A) profile 9, (B) profile 10, and (C) profile 11, with a shot at the western end of profile, and (D) profile 11, with a shot at the eastern end of profile. The horizontal scale is the distance from shot to receiver—note the different scaling for the two upper and lower profiles. The vertical scale is the time taken to travel from shot to receiver, plotted in reduced time, which equals the true arrival time (TAT) minus distance (X) divided by 6 ([TAT – X]/6). A subset (~30%) of the picked first-arrival traveltimes is plotted as tick marks. When consecutive arrival times align along a straight line, the P-waves are traveling at a constant apparent velocity (V_a). Straight lines are calculated first-arrival traveltimes for the one-dimensional velocity models shown in Figure 3. The V_a values are apparent P-wave velocities of picked arrivals.

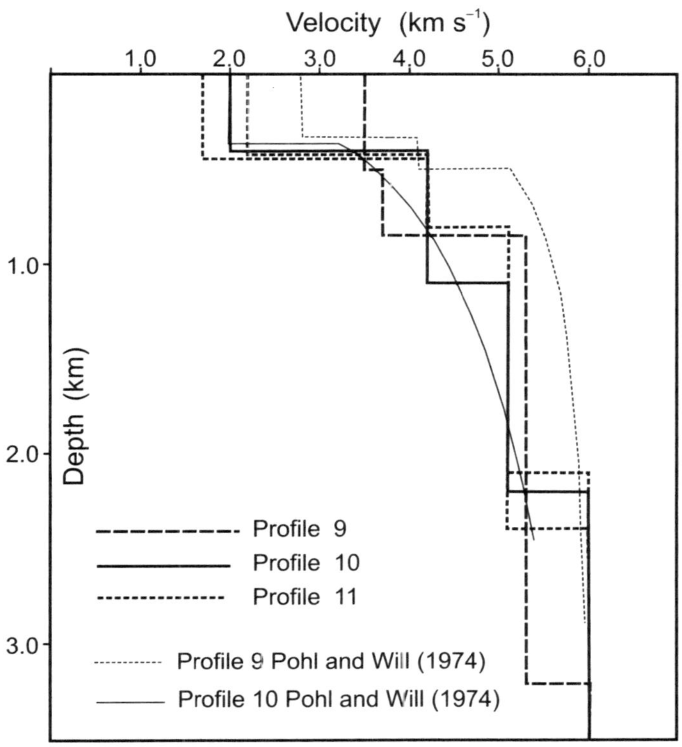

Figure 3. P-wave velocity/depth profiles determined from modeling of arrivals shown in Figure 2 for profiles 9, 10, and 11. For comparison, velocity models for profiles 9 and 10 from Pohl and Will (1974) are also displayed.

first occur is deeper outside the crater than inside the crater. In our one-dimensional interpretation of the data, velocities of 6 km s^{-1} are reached between 2.1 and 2.4 km depth for the three profiles inside the crater, whereas outside the crater this velocity boundary is at ~3.2 km (Fig. 3). This depth is in agreement with a seismic refraction model presented by Pohl et al. (1977), in which the 6 km s^{-1} velocity contour is at ~3.1 km depth to the southeast, northeast, and northwest of the crater.

From our analyses of the refraction data, we suggest that the average velocity inside the crater is lower than outside the crater from the surface to ~2.2 km depth. Moreover, the apparent elevation of the 6 km s^{-1} velocity boundary by ~1 km within the crater center supports the proposal that there might be structural uplift here. We acknowledge that it is possible to construct other velocity models with dipping interfaces and/or with lateral velocity changes that show no structural uplift (e.g., Ernstson and Pohl, 1977) and that these might fit the data equally as well. However, such lateral changes are not well constrained by the current data set. The linearity of the arrivals in refraction profiles 9 and 10 (Fig. 2A and 2B) do not suggest strong lateral changes in velocity, although we recognize that they could easily be generated from a velocity model with one or more dipping interfaces. Taking all these factors into consideration, we conclude that the seismic refraction data are consistent with a structural uplift of ~1 km at the Ries crater.

Magnetotelluric Investigation of the Deep Subsurface of the Crater

In order to further investigate the deep structure beneath the Ries crater, the distribution of electrical conductivity was studied by means of magnetotelluric (MT) measurements during two short field campaigns in 1999 and 2000. Utilizing the induction effect of natural variations in Earth's magnetic field for electromagnetic deep sounding, the MT method is particularly capable of detecting highly conductive structures in the deep subsurface. Within the generally highly resistive upper crustal environment of the Moldanubian belt (ERCEUGT-Group, 1992) such structures would be readily visible, allowing an estimate of the extension of porous or fractured zones to depth. An early MT study by Haak et al. (1977) confirmed the existence of highly conductive lake deposits inside the crater, but the deep electrical conductivity structure beneath the crater was not analyzed. Other MT studies of impact structures in Sweden (Siljan; Zhang et al., 1988) and Canada (Charlevoix; Mareschal and Chouteau, 1990) revealed zones of increased electrical conductivity below the craters, indicating the existence of fractures that might enable fluid circulation at depth. At the Ries crater, we attempted to find constraints from electrical parameters that could support the presence of basement uplift or deep reaching fractures, as favored by Pohl et al. (1977).

The natural time-varying electric and magnetic fields were measured at 13 MT sites along a NW-SE–trending profile (Fig. 1); i.e., perpendicular to the geological strike of the Moldanubian gross structure (cf. Haak et al., 1977). The entire length of the profile extends over 73 km, with an average spacing between the measurement points of 10 km outside the crater and 2–5 km between the outer and the inner ring. The MT data, covering a period range of 0.25–4096 s, were recorded in three bands at sampling rates of 16 Hz, 2 s, and 32 s. Due to the high level of cultural noise, typical of populated areas, the data quality of the short period band was too low at some sites to be used for modeling, especially inside the crater structure.

To obtain magnetotelluric transfer functions, as displayed in Figure 4 in the form of resistivity and phase curves, standard MT data processing was applied (cf. Jording et al., 2000). It includes statistical frequency analysis methods as well as decomposition procedures in order to test for static shift effects. Considering the circular crater structure, a three-dimensional MT study would have been appropriate; however, the dimensionality analysis of the data gave skewness values <0.3 for all sites and periods, which means that two-dimensional modeling may be applied as an approximation (e.g., Bahr, 1991). Furthermore, due to the lack of noticeable static shift effects, it was not necessary to correct for them. The coordinate systems of all individual data sets were rotated by the same −45° angle into the system parallel to the large-scale geologic strike approximating the profile direction. Hereby, the E- and the

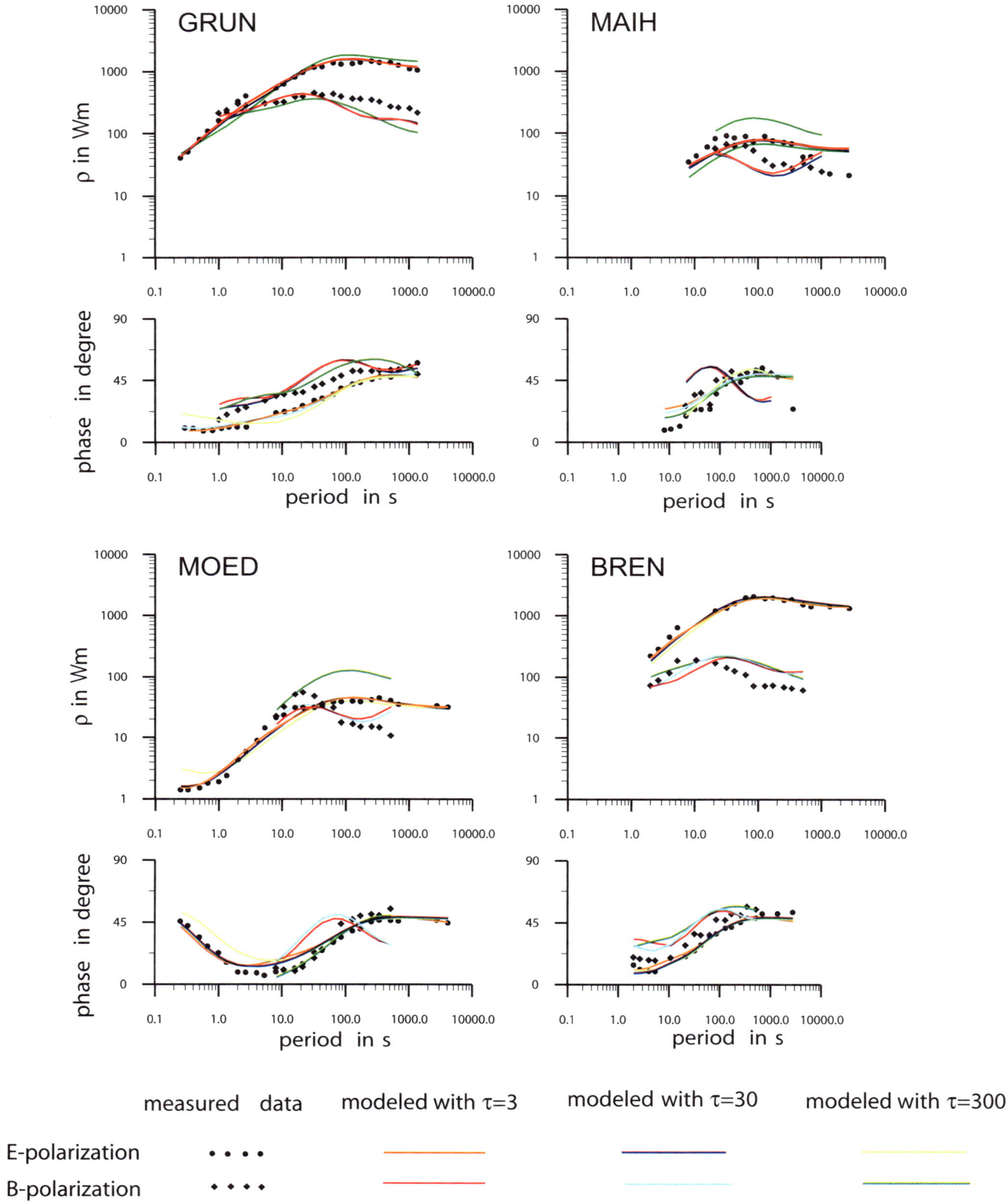

Figure 4. Examples of 4 magnetotelluric apparent resisitivity (ρ) and phase curves. Each chart shows a comparison of measured data (tick marks) and curves obtained from MT-models I (τ = 3), II (τ = 30), and III (τ = 300) (see Fig. 5). For the location of the stations (GRUN, MAIH, MOED and BREN) see Figure 1.

B-polarization modes were determined, at least roughly, as needed for two-dimensional modeling (cf., Jording et al., 2000).

Using the inversion algorithm by Mackie et al. (1997), resistivity and phase data of the E- and the B-polarization modes were inverted simultaneously. Special emphasis was placed on the analysis of the influence of the regularization parameter, τ, which enables us to find models varying from best fit of the local sites (small τ) to lateral smoothing of the resistivity structure as a whole (large τ). As a result of our calculations, three models (I, II, III) with different values of τ (3, 30, 300) are presented in Figure 5. The corresponding model fits of four selected sites from the inner crater range are displayed in Figure 4. Taking the limitations of the two-dimensional approximation into account, the overall misfit of the models (root mean square [rms]-error) is reasonable (Fig. 5). The overall fit is small for the models with small or medium τ (rms = 7.49, 7.23) and somewhat larger for the model with large τ (rms = 11.27). Influenced by the outer two-dimensional effect of the gross geological structure, the data fit at the sites GRUN and BREN near the outer crater ring is remarkably better than for the central, possibly three-dimensionally influenced sites MAIH and MOED. The main effect of the conductivity anomaly is clearly evidenced by relatively low apparent resistivity values of 10–100 Ωm in the inner ring compared to 100–1000 Ωm near or outside the outer rim (Fig. 4).

Independently of τ, all models show a near surface (< ~1 km) high-conductivity zone (3–10 Ωm) within the crater, centered at the position of its inner ring (Fig. 5). These resistivity values are in good agreement with the resistivity values of the lake sediments (~1–4 Ωm) and the suevite layer (~10 Ωm), known from borehole measurements in the deep drill hole Forschungsbohrung Nördlingen (Ernstson and Pohl, 1974) and by geoelectric deep soundings inside the inner ring (Ernstson, 1974; Blohm et al., 1977). At larger depths, another conductive structure appears in models I and II, right below the crater center, reaching to depths of 4 and 6 km, respectively (Fig. 5). This structure is clearly separated from the surface layer, whereas the equivalent structure in model III is connected with the surface layer, reaching a maximum depth of ~2 km only.

Strikingly, in all three models, the resulting resistivity minimum (or maximum of the electrical conductivity) at larger depths is shifted from the morphological crater center toward the northwest by 2–3 km (note the dashed lines indicating the inner ring and outer rim in Fig. 5). This shifted position was also found for the minimum Bouguer anomaly related to the Ries crater (Jung and Schaaf, 1967; Kahle, 1969) as shown by the values of the uncorrected Bouguer anomaly along the MT profile extracted from the Bouguer gravity map of Jung and Schaaf (1967) (top of Fig. 5). Any convincing explanation of this discrepancy is missing so far. The first idea, namely that it might be caused by an oblique northwest directed impact, can be ruled out due to the distribution of ejecta (tektites). The moldavite (tektite) strewn field extends from the crater center east–northeastward, and according to earlier modeling (Artemieva et al., 2002) it is most likely that the impact was oblique coming from a west-southwest direction.

Various attempts have been made to model the gravity anomaly of the Ries (Jung and Schaaf, 1967; Kahle, 1969, 1970), the most satisfactory of which is the refined model of Kahle (1969). This model explains the crater-related gravity minimum by means of a series of four cylindrical layers (I–IV) with decreasing density contrasts (≥−0.49, −0.25, −0.20, −0.15 g/cm^3) and radii (11.0, 5.0, 3.5, 2.0 km) and thicknesses of 0.5 km each (Fig. 6). The geometric dimensions of this model are in good agreement with the conductivity distribution of model III.

A direct comparison in model III (Fig. 6) between density data and conductivity, both depending mainly on porosity, is suitable at least for larger depths, where a simple power law (Archie's formula) may be applied to roughly estimate the porosity needed to explain the observed conductivity. The resulting values, between 10% and 20% for the suevite layer and the uppermost part of the basement, assuming high saline water (0.3–1 Ωm) as supposed by Blohm et al. (1977), are also generally in good agreement with the results of the density model. Further support for model III comes from purely magnetotelluric arguments. In model III, the phase fits of the inner crater sites are superior to those of models I and II in spite of their smaller rms errors (Fig. 4). The phase fit is more valuable because phases are less distorted by static shift or biased by correlated noise effects than are the apparent resistivity values. Based on these arguments, we may favor model III, although we have to be aware that all models (I–III) represent more or less equivalent solutions. The choice of model III would exclude deep-reaching fractures below the crater, in good agreement with the seismic model, if the resistivity increase at 2–3 km depth is interpreted as a transition to dense and resistive basement rocks. On the contrary, models I and II would indicate the existence of a fractured zone underneath the crater expanding to depths of 4 or 6 km, respectively.

NUMERICAL MODELING OF CRATER FORMATION

We utilized numerical modeling to simulate the formation of Ries crater and to investigate the influence of varying target composition on the formation process for similarly sized impact structures in general. We used the well-known SALE-3MAT (Simplified Arbitrary Lagrangian-Eulerian 3 Material) hydrocode (Amsden et al., 1980), including some major changes to the original version (e.g., stress- and multi-material-extension; Ivanov et al. 1997; Wünnemann and Lange, 2002) to compute the entire formation process. All calculations begin with the first contact of the impactor and last until the main dynamic motions have ceased. The impact velocity is 18 km s^{-1}, consistent with the mean impact velocity on Earth (O'Keefe and Ahrens, 1994). The target is composed of granite basement covered by an 800-m-thick sediment layer. The thermodynamic behavior of the different materials is calculated by the Tillotson equation of state (EOS; Tillotson, 1962) using material properties of a wet tuff for the sediments and by the analytical equation of state (ANEOS; Thompson and Lauson, 1972) for the granitic basement. Apart from the thermodynamics, the elasto-plastic response of the

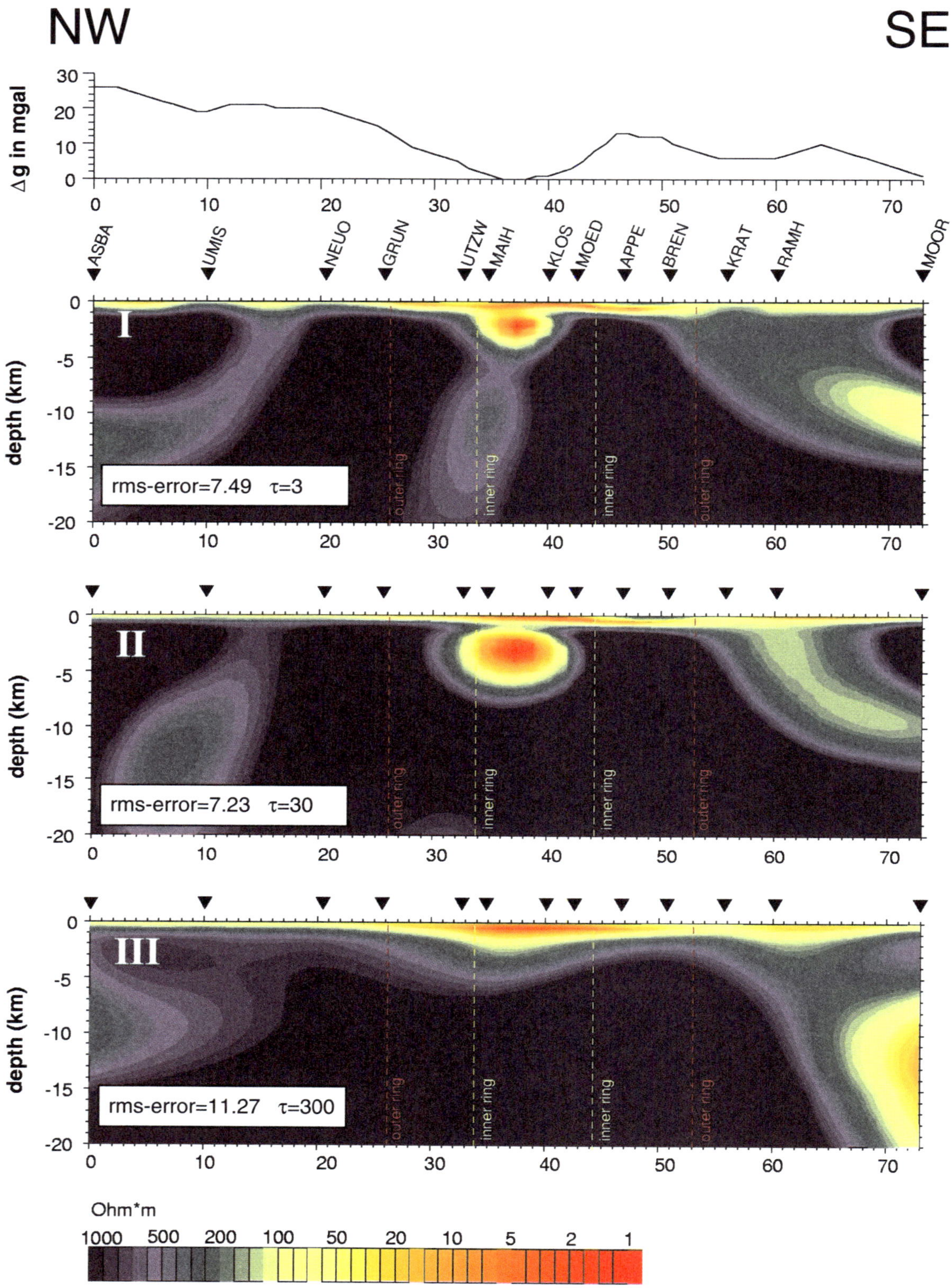

Figure 5. Two-dimensional MT-models (I–III) after Mackie et al. (1997) for different regularization parameters τ, with overall root mean square error (rms-error), compared with the Bouguer anomaly along the MT profile (top; cf. Fig. 1), extracted from the Bouguer gravity map (Jung and Schaaf, 1967). For the location of stations (ASBA, UMIS, NEUO, GRUN, UTZW, MAIH, KLOS, MOED, APPE, BREN, KRAT, RAMH, MOOR) see map in Figure 1.

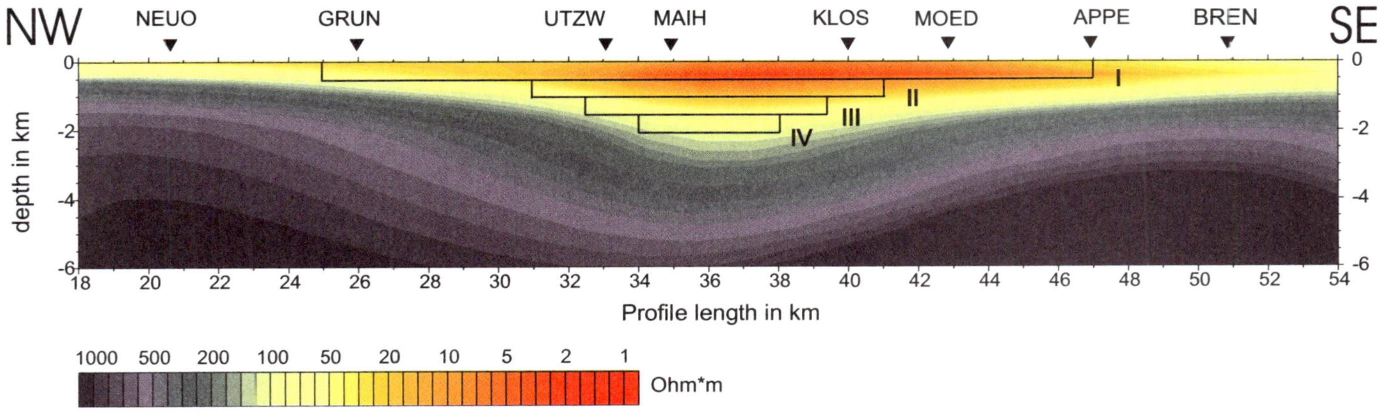

Figure 6. Comparison between MT model III (Fig. 5) and the gravity model consisting of cylindrical layers (I–IV), redrawn after Kahle (1969) along the MT profile (cf. Fig. 1). For the location of stations (NEUO, GRUN, UTZW, MAIH, KLOS, MOED, APPE, BREN) see map in Figure 1.

material to a given stress field is crucial for the collapse of the transient cavity and the formation of the final crater shape. We utilized standard constitutive models to describe the strength properties of rocks that are exposed to strong deformations (e.g., Ivanov et al., 1997). In these models, the yield strength Y is a function of pressure P and temperature T, and is mainly controlled by the coefficient of internal friction ϕ and the cohesion C. In addition, an acoustic fluidization model was used to explain the temporary fluid-like behavior of rocks in the vicinity of the crater (Melosh, 1979). The behavior of acoustically fluidized matter is mainly determined by the viscosity η and the decay time T_{dec}, which describes how long the viscous state of the material lasts until the vibration (acoustic wave field) is damped down. Both parameters, η and T_{dec}, are strongly linked with the fragmentation state (fragment size) of the rocks beneath the structure.

For a full description of the rheology and especially of the acoustic fluidization model utilized in the present study, we refer to Wünnemann and Ivanov (2003), Melosh and Ivanov (1999), and Melosh (1979). The covered ranges for all material parameters as well as all important model specifications are summarized in Tables 1 and 2.

Dynamic Model of Ries Crater Formation

The numerical model that best matches the present crater morphometry is shown in Figure 7. Snapshots at different stages

TABLE 1. PARAMETERS USED IN THE NUMERICAL MODELS OF THE RIES

Parameters	Values
Impactor:	
Diameter	1100 m
Impact velocity	18 km s^{-1}
Material/EOS	Granite, ANEOS
Top layer:	
Material/EOS	Wet tuff properties/Tillotson (1962)
Cohesion C_S	0, 10 MPa
Dry friction coefficient ϕ_S	0.5, 0.9
Thickness l	800 m, 2000 m
Basement:	
Material/EOS	Granite, ANEOS
Cohesion C_B	10 MPa
Dry friction coefficient ϕ_B	0.9
Model set-up:	
Number of cells nx × ny (radial, vertical)	350 × 350 cells
Number of extended cells top, right, bottom	40, 70, 70
Spatial increment (high resolution area)	50 m

Note: EOS—equation of state; ANEOS—analytical equation of state.

TABLE 2. ACOUSTIC FLUIDIZATION PARAMETERS USED IN MODELS A, B, C, AND D (FIGS. 9 AND 10)

Model description	Acoustic fluidization parameters	
	Viscosity η	Decay time T_{dec}
(A) Best-fit model (Fig. 9)	1.50 GPa s	33.0 s
(B) Wünnemann and Ivanov (2003) (Fig. 10B)	0.70 GPa s	16.5 s
(C) Ivanov and Artemieva (2002), Bosumtwi model (slightly modified), (Fig. 10C)	0.07 GPa s	11.0 s
(D) As (3), but longer lasting vibration (compare to Chicxulub morphology, Collins et al. [2002]) (Fig. 10D)	0.07 GPa s	22.0 s

of the cratering process illustrate the crater formation. In each frame the left-hand side shows tracers arranged in horizontal layers that depict the stratigraphy and the uplift of the target rocks due to the collapse of the transient cavity. On the right-hand side, the density distribution of the rocks is indicated. The change in density between the overlying rocks and the crystalline basement is pronounced. In the model (Fig. 7), the inner ring has a diameter of ~12 km and is formed from uplifted and outwardly displaced basements rocks. The megablock zone between the inner ring and outer rim, which is basically characterized by concentrically faulted separate blocks, is represented in the model by a continuously inclined overlying layer, which extends to a radial distance of ca. 12 km. The shape of the modeled crater is generally consistent with the observed crater morphometry. In addition, the thermodynamic history and the displacement of rocks in the vicinity of the impact structure are illustrated in Figure 8. On the left-hand side of this figure, the maximum pressure each tracer particle has experienced throughout the impact process is shown with respect to the final position of the tracer. Peak pressures above 46 and 56 GPa usually cause incipient and complete melting, respectively, in granite after release from shock pressure (Pierazzo et al., 1997). Thus, the top layer inside the inner crater is composed of melt-rich material that is roughly coincident with the suevite layer in the Ries crater. The right side illustrates the distance traveled by each tracer relative to its initial location, where all distances are scaled between 0 and 1. The total displacement is accumulated along the path a tracer follows throughout the entire cratering process (not to be mistaken with the distance between the initial and the final position of a tracer). A hemispherical zone of strong tracer displacement (>0.25 of maximum displacement) can be identified to a depth of ~6 km (Fig. 8). It is likely that rocks inside this area are exposed to strong deformations causing brittle fracturing. The upper 2 km are intensively fractured (completely damaged) according to the large magnitude of displacement (>0.75 of maximum displacement). Below this level, down to a depth of 6 km, fracturing is less pronounced (displacement: 0.25–0.75). The rocks in this region have experienced shock pressures <10 GPa (Fig. 7), which is a typical range where fracturing by release from shock pressure occurs (e.g., Grieve et al., 1996). Our best-supported conductivity model, based on magnetotelluric measurements (III, Figure 5), is consistent with this assumption concerning the extent of the fragmentation zone. The tracer layers on the left-hand side in Figure 7 indicate that rocks are uplifted by 1.5–2 km within the crater center, which is roughly in agreement with the stratigraphic uplift of at least 1 km derived from the analyses of the refraction data (Fig. 2 and 3). Therefore, the model presented in Figure 7 agrees in both crater morphometry and interior crater structure with the observations at the Ries crater.

Influence of Different Target Properties on Crater Morphology

The large-scale rheology of rocks is complex and cannot be reproduced in the laboratory with measurements on small samples. Even the same kind of rock can exhibit different rheological properties, especially in terms of cohesion and internal friction, depending on its exact mineralogical composition and also on its thermodynamic and mechanical history. Different constitutive properties as well as varying compositions of different kinds of target rocks may have a strong influence on the formation process and could well be responsible for the lack of structural similarities of similarly sized craters on Earth.

Although we have included a sedimentary layer (wet tuff EOS parameters in Table 1) covering the crystalline basement (granite EOS parameters) in the calculations yielding the best-fit results, shown in A-1 in Figure 9, the strength properties for both materials are the same. We used a cohesion of 10 MPa and a coefficient of internal friction of 0.9 in both layers. Although these parameters suggest a higher strength than expected for sediments, they lead to the best agreement with the observed morphology, in particular the rise of the inner ring above the pre-impact surface level. Usually, sediments are assumed to exhibit much weaker strength properties (smaller internal friction coefficient and smaller cohesion) than basement rocks. Furthermore, assuming cohesive strength in both layers (sediments and basement) contradicts the usual assumption that rocks in the vicinity of the crater are heavily fractured, possessing no or very little cohesive strength. Much weaker strength properties for the sediment layer ($C_S = 0$, $\phi_S = 0.5$) result in the crater morphology shown in A-2 in Figure 9. While the upper layer slumps further inward, the crystalline inner ring collapses further outward above

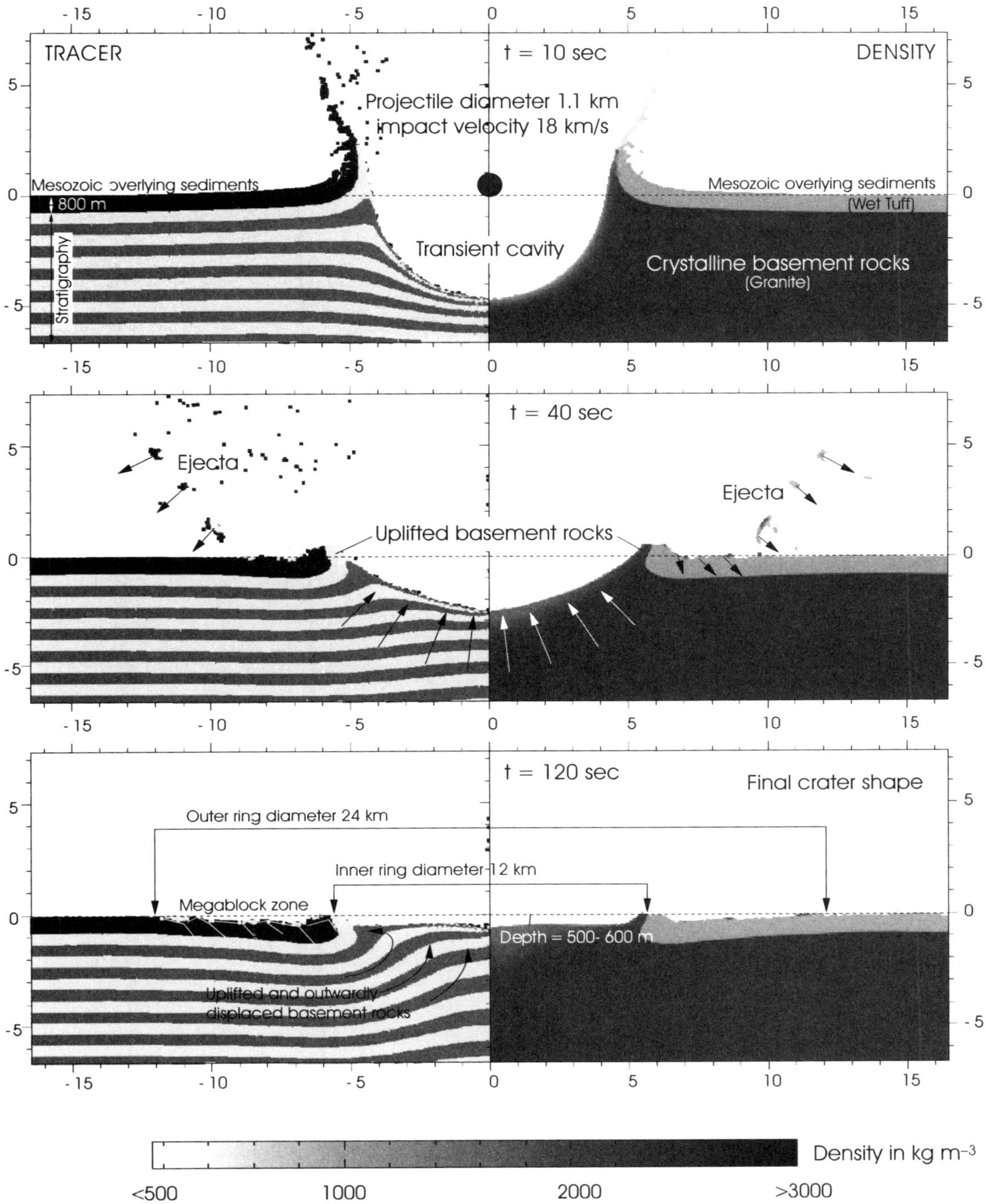

Figure 7. Series of snapshots of our best-fit model A illustrating the formation process of the Ries crater (see Tables 1 and 2 for specific model parameters). All distances are given in km. The left frame shows tracer particles arranged in horizontal layers to illustrate the stratigraphic uplift. The uppermost layer (black) represents the sediment layer. On the right frame, the density distribution is shown.

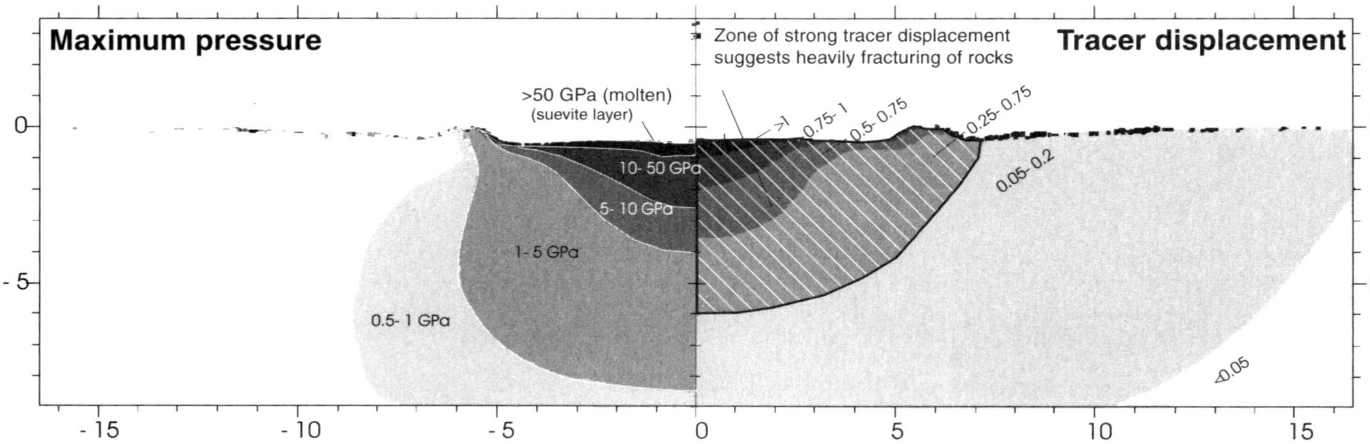

Figure 8. Final crater shape of model A (compare to Fig. 7). The left frame indicates the maximum pressure each tracer has experienced during the crater formation. The right frame shows the accumulated distance traveled by the tracers from their initial positions (not to be mistaken with distance between initial and final position; see text for further explanation). All displacement distances are scaled between 0 and 1.

the sediments (bidirectional material movements), giving rise to an inversion of the stratigraphy. The overall morphology appears to be flatter, featuring less distinct topographic elements than in the upper model (A-1). For instance, the distinct topographic elevation of the inner ring in model A-1 does not appear in model A-2. Finally the influence of increasing thickness (l = 2 km) of the soft ($C_S = 0$, $\phi_S = 0.5$) sediment layer on the top of a crystalline basement is demonstrated in A-3 in Figure 9. No bedrock is uplifted to the surface, although a kind of a dome structure is formed beneath the upper layer. Sediments slump into the crater cavity and over the central dome, covering any topographic evidence of a peak or inner ring structure. The area where the final surface level is below the pre-impact level extends much further outward than in A-1 or A-2, so the final crater structure appears somewhat larger.

Our numerical simulations (Fig. 9) demonstrate, in agreement with other studies (e.g., Shuvalov et al., 2002), that varying mechanical properties in different material layers can dramatically affect the morphology of the final crater structure. The weaker the material in the upper layers, the smoother the resulting crater morphology, and topographic elements are less distinct. However, the structure beneath the crater does not change significantly with variations in the near-surface material strength.

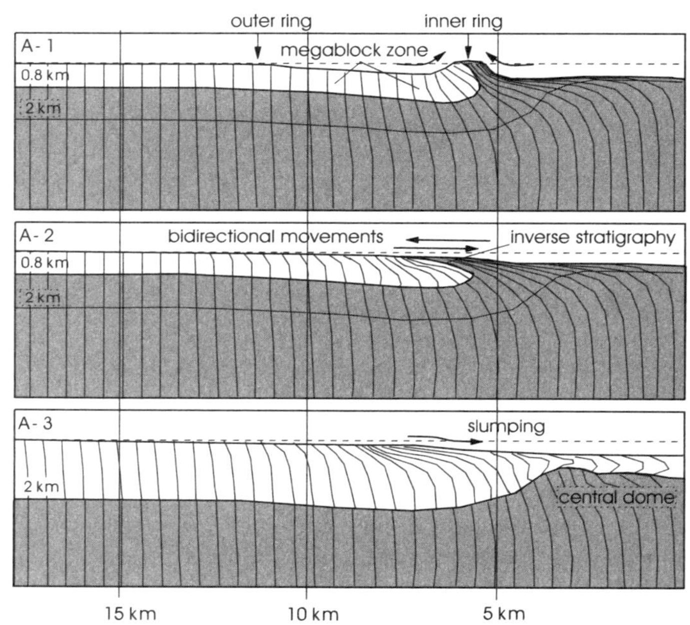

Figure 9. Comparison of the final crater shape of three models utilizing different strength properties and spatial extensions of the upper (sedimentary) layer. Model A-1: l = 0.8 km, C_S = 10 MPa, ϕ_S = 0.9; Model A-2: l = 0.8 km, C_S = 0 MPa, ϕ_S = 0.5; Model A-3: l = 2 km, C_S = 0 MPa, ϕ_S = 0.5 (see text for an explanation of the variables; see also Tables 1 and 2). All models were carried out using the same impact energy.

Influence of Acoustic Fluidization Parameters on Final Crater Form

Although investigations of the subsurface of crater structures confirm the existence of a fragmentation zone deep beneath the surface, the actual state of the rocks in terms of fragment size is unknown and can only be estimated. Unfortunately, the parameters utilized in the acoustic fluidization model, such as the viscosity η and the decay time T_{dec} of the vibration, are particularly dependent on these rock properties in the vicinity of the crater. Different estimates of the mean fragment size have been proposed, mainly by adjusting the parameter choice so the models fit best with the observations (Wünnemann and Ivanov, 2003; Ivanov and Artemieva, 2002). Apart from modeling studies, deep drilling of the 40 km Puchezh-Katunki structure reveals

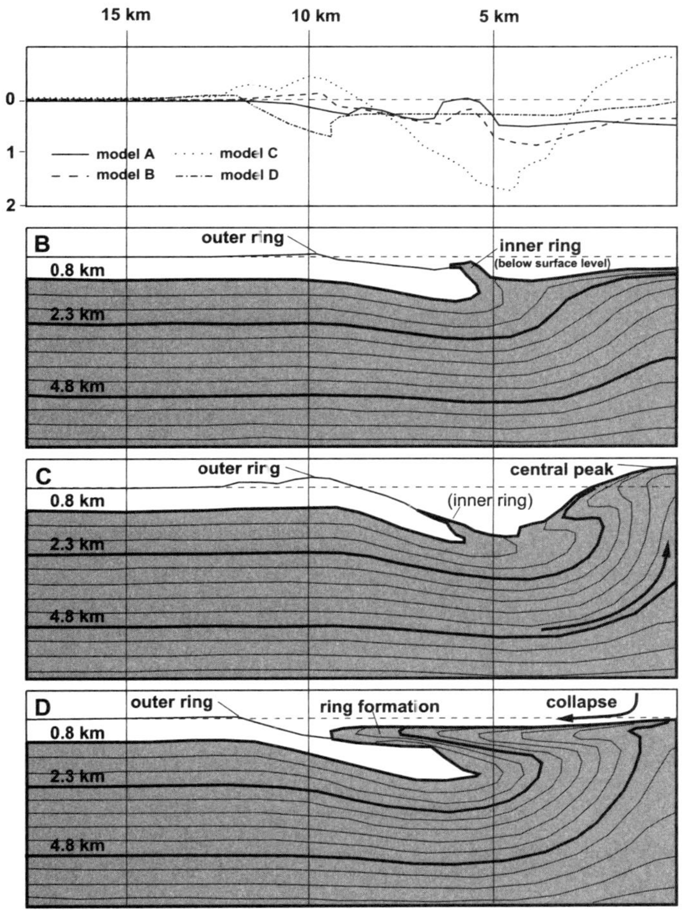

Figure 10. Comparison of the final crater shape of three models (B, C, D) utilizing different acoustic fluidization parameters (η, T_{dec}; see Table. 2). The strength properties correspond to the values in model A-1 ($C_S = 10$ MPa, $\phi_S = 0.9$). The uppermost frame shows a diagram comparing the final crater topography of all models in this figure (model B, C, D) and from Figure 7 (model A-1). All models were carried out using the same impact energy.

a system of blocks with an average size of ~100 m (Ivanov et al., 1996). The investigation of the size of ejected fragments (Kocharyan et al., 1996) and the variety of parameter choices in numerical models suggests that fragment size is a function of crater diameter. In larger crater structures, the block size is assumed to be much larger than in smaller craters and, therefore, the acoustically fluidized state lasts longer and the rock fragments behave more viscously. As Wünnemann and Ivanov (2003) show, it is possible to scale η and T_{dec} as a linear function of projectile size to reproduce the observed depth-diameter dependence of impact craters. However, the parameter choice for similarly sized craters is not unique. More or less similar results arise from a low viscosity and short decay time as from a high viscosity with a longer lasting vibration. In order to investigate the sensitivity of the crater shape to the choice of η and T_{dec}, we have modeled the final crater morphology for a broad range of plausible parameters for Ries-sized craters. The applied values are displayed in Table 2 and compared with those proposed in other numerical modeling studies. Nearly identical results to our best-fit model (model A: η = 1.5 GPa s, T_{dec} = 33 s) arise if the values proposed by Wünnemann and Ivanov (2003) are applied (η = 0.7 GPa s and T_{dec} = 16.5 s; Fig. 10B) using the suggested scaling law for acoustic fluidization parameters as a function of projectile size. Only the crater floor inside the inner ring is slightly more uplifted and is convex in shape. Moreover, the inner ring no longer rises above the pre-impact surface level. Utilizing parameters close to the range used in the Bosumtwi crater model (Ivanov and Artemieva, 2002; Table 2 herein) results in a final crater shape that differs clearly from the previous models, featuring a well defined central peak (Fig. 10C). In addition, the topographic evidence for the crystalline rim of the transient cavity (inner ring) is missing in this model, and the size of the crater structure is determined clearly by the outer (tectonic) crater rim at ~10 km radius. Because the Bosumtwi crater is smaller than the Ries, η was chosen to be twice as large and T_{dec} less than half, following the model by Ivanov and Artemieva (2002). Assuming an even longer vibration time (twice as long) with the same η yields a final crater shape in which the central peak is collapsed (Fig. 10D). The collapse gives rise to the formation of a peak ring at ~9 km, and the outer edge of the crater extends to a diameter of ~12 km. This model is similar to that proposed for the Chicxulub crater by Collins et al. (2002).

The presented results clarify that the final crater morphology is very sensitive to the assumed parameter choice of the acoustic fluidization model. The values proposed by Wünnemann and Ivanov (2003) might be the most reliable ones, because they have been verified over a broad range of crater sizes to be consistent with observed depth/diameter ratios. However, the general simplicity of the model (acoustic fluidization parameters are presumed to be a function of distance to the point of impact) and the lack of information about fragment sizes does not allow more precise statements. Furthermore, the rim of the final crater structure appears at different radii in all four models (Figs. 9 and 10). While models A and B show clear evidence for an inner ring structure coincident with the size of the transient cavity, models C and D are characterized by a more distinct tectonic crater rim and are lacking any topographic clues of the former transient cavity.

DIFFERENCES IN SUBSURFACE STRUCTURE AND SHAPE OF SIMILARLY SIZED CRATERS ON EARTH

When we examine the crater morphology and subsurface structure of the Ries and similarly sized terrestrial craters, we see surprising dissimilarities in terms of crater morphometry and characteristics of structural elements (e.g., inner rings and outer rims, central peaks, stratigraphic uplift). In some cases this can be explained by post-impact modifications like erosion and tectonic adjustment (e.g., Gosses Bluff, Australia; Milton et al., 1972). In order to focus upon differences caused by varying target conditions, we compare Ries crater with the two youngest and most pristine complex crater structures of Earth's impact record: the 0.87 Ma Zhamanshin crater, Kazakhstan (Masaitis et al., 1985), and the 1.07 Ma Bosumtwi crater in Ghana (Koeberl et al., 1997). Examples of topographic profiles are given in Figure 11. Both structures are believed to be well preserved and it is unlikely that any morphological differences can be explained by variations in the extent of erosion (Garvin and Schnetzler, 1994). The apparent diameter of Zhamanshin is ~14 km. The crater is located in a semi-desert, featuring a sedimentary platform environment with a typical erosion and infill amount of 0.02 mm/yr (Masaitis et al., 1985). For the 10.5 km (diameter of most pronounced morphological annular ring) Bosumtwi crater, which was formed on an equatorial crystalline shield region, the erosion rate is estimated to be slightly higher, between 0.06 and 0.14 mm/yr (Garvin and Schnetzler, 1994). In both cases, the quoted size of the structure refers to the most noticeable annular elevation, which is not necessarily equivalent to the real size of the crater. Therefore, we use the terminology "apparent crater rim" or "diameter" since it is not clear whether we compare similar structural elements. We believe that in the case of Bosumtwi and Zhamanshin, the apparent crater rim represents a similar structural element as the inner ring at Ries, which is ~12 km in diameter and almost the same size. As the inner ring structure of Ries is assigned to the approximate size of the transient cavity, we assume that the crater rims of Bosumtwi and Zhamanshin are roughly coincident with the size of the former transient cavity as well. Therefore, we affirm that all three craters are comparable in size. This idea is generally supported by the fact that the diameter of the transient cavity is an appropriate measure for the amount of impact energy, and in previous numerical models of the formation of Bosumtwi crater, the assumed amount of impact energy (750 m projectile) is in the same range as used in this study for the Ries models (Karp et al., 2003, 2004).

We note that the definition of crater size, especially for the size range of craters addressed in the present work, appears sometimes confusing. The Ries crater exhibits a well defined zone of normal faulting marking the maximum extent of the structure (classical definition of diameter D = 24 km) and a less morphologically pronounced inner ring that is assigned to the size of the transient cavity (D_t = 12 km). In contrast, Bosumtwi does not show clear evidence for an outer tectonic rim; hence, the diameter of the structure is indicated by the apparent crater rim. However, some clues exist for the presence of an outer ridge at a radius of 8–10.5 km defined by subtle topography (Wagner et al., 2002), which may be the equivalent of the tectonic rim at Ries. Note that in the case of Ries, the topographic elevation of the crystalline ring is widely obliterated by erosion and does not represent the dominating morphological element of the crater structure. Nevertheless, for the purpose of crater comparison, we use the diameter D_t of the inner ring structure because of the similar size of equivalent structural elements (assumed approximate size of transient cavity) in all three examples.

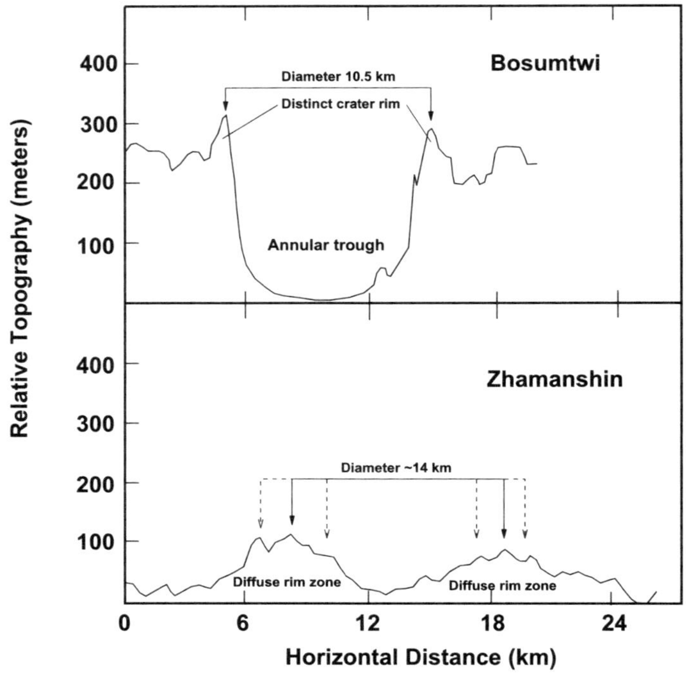

Figure 11. Topographic profiles through the centers of Bosumtwi and Zhamanshin, out to a distance of two radii from the crater rim crest. Morphometric values used in the text were derived from a large number of profiles through the craters (after Garvin and Schnetzler, 1994).

The most distinguishing feature between the three crater structures is the fact that the morphology of Zhamanshin appears to be much smoother, lacking clear, superficial crater-like topographic expressions. The aspect ratio (depth/diameter, where depth d as measured from the rim crest to the apparent crater floor) is d/D_t = 0.013, given by Garvin and Schnetzler (1994), and is based on the remote sensing signature. In contrast, Bosumtwi exhibits a well defined U-shape morphology with a steep-sided rim. Recent seismic reflection data identify a small peak structure in the center of the crater (Karp et al., 2002; Scholz et al., 2002). The aspect ratio is 0.048 (d/D_t = 0.5/10.5), where the depth is measured to the top of the breccia and melt layer. Based on remote sensing data, Garvin and Schnetzler (1994) specify a value of d/D_t = 0.029 for Bosumtwi, without, however, taking into account the thickness of the lake sediments (post-impact infill amount) in the crater. The higher erosional level of Ries crater allows only an approximation of the original crater depth by considering the thickness of the lake sediments (max 350 m; Ernstson, 1974) and an estimation of the former rim crest height. In order to determine a more sophisticated value for the depth of the Ries crater, we utilize empirical equations (scaling laws) that are based on the morphometry of selected (low erosional level) terrestrial complex craters. Grieve and Pesonen (1992) suggest two different relationships for crater depth as a function of diameter in sedimentary and crystalline targets:

$$d = 0.12\, D^{0.30} = 0.311 \text{ km (sedimentary)} \tag{1}$$

and

$$d = 0.15\, D^{0.43} = 0.588 \text{ km (crystalline)}, \tag{2}$$

where D is the diameter of the outer tectonic rim of Ries ($D = 24$ km). Although the sedimentary layer may have had an influence on the formation of the crater structure, the Ries crater was formed in a mainly crystalline target. Therefore, the aspect ratio for Ries is 0.049 (d/D_t; d = 0.588 km, D_t = 12 km). This value is in agreement with the observed depth/diameter ratio for Bosumtwi ($d/D_t = 0.048$). The differences in crater morphometry between Ries and Bosumtwi craters are mainly the existence of a more pronounced megablock zone adjacent to the inner basin and a less distinct topographic expression of the inner ring at Ries, which can be explained. Whereas the latter is most likely due to the higher amount of erosion at Ries, we believe that normal faulting at Ries (megablock zone) might have been caused by the weaker material properties of the overlying sediments, which support slumping into the crater cavity as the models in the previous section ("Numerical Modeling of Crater Formation") show (cf. Fig. 9). In comparison, Zhamanshin is unusually flat (aspect ratio $d/D_t = 0.013$), and Garvin and Schnetzler (1994) suggest this might be a consequence of weak or water-saturated target materials. Our model in Figure 9 confirms that a weak target causes craters having a smaller aspect ratio.

Ries and Bosumtwi formed under more or less comparable conditions with only slight differences in impact energy. As their aspect ratios are quite similar, one would expect their subsurface structure to be similar, too. In both cases, the crater basin is covered by a breccia or suevite layer. The thickness of this layer has a maximum extent of 300–400 m at Ries (Pohl et al., 1977) and 400–600 m at Bosumtwi (Karp et al., 2002). At Bosumtwi, it is uncertain whether the central peak structure is formed from breccia or whether it is actually structurally uplifted crystalline basement. Ries shows some clues for structural uplift underneath the crater but no topographic evidence for a central peak structure. In both craters, seismic investigations reveal similar P-wave velocities of 3–3.4 km s^{-1} in the brecciated layer (Angenheister and Pohl, 1976; Karp et al., 2002). Velocity models of greater depths are not yet available for Bosumtwi crater.

At first view, all three examples of complex craters discussed in this section look dissimilar in terms of crater morphology. But despite their differences in shape, Ries and Bosumtwi show a strong degree of resemblance due to their comparable (apparent) diameter (transient cavity), estimated aspect ratios, and target compositions. We believe that both craters are typical examples of small to middle-sized complex crater structures, and despite the lack of any obvious central elevation both craters are most similar to the type of central peak craters. For this reason, they represent appropriate natural laboratories on Earth, allowing the exploration of the crater formation processes.

SUMMARY AND CONCLUSIONS

The shape and structure of middle-sized complex impact craters on Earth vary considerably, and it is difficult to define characteristic topographic and structural elements. The numerical models of the formation of Ries crater show that the final crater shape is very sensitive to the constitutive properties of the target rocks. Sedimentary material is assumed to be less resistant against shear failure than crystalline basement rocks. For this reason, craters formed in sedimentary target compositions are relatively flat, whereas craters in mainly crystalline rocks exhibit more pronounced structural elements. This is demonstrated by the numerical models as well as by the comparison of the present morphology of Zhamanshin, Bosumtwi, and Ries craters, which were all formed by similarly energetic impacts. In the present study, we have used the diameter of the transient crater, which can be approximated by the extent of the inner most topographic elevation (inner ring or rim) in the three given examples, as a measure for crater size (or impact energy). The outermost ridge is usually used to determine the diameter of an impact structure, but as exemplified for Bosumtwi and Ries, this definition can cause a wrong assessment of the crater size.

The numerical models show that the crater morphology is highly sensitive to the strength profile of the near-surface rocks and the fluidization state of the rocks (acoustic fluidization parameters) during crater formation. In this respect, a similar amount of impact energy can cause different degrees of rim collapse and structural uplift. Accordingly, we propose that Ries and Bosumtwi are similarly sized craters and the inner ring structure of Ries is equivalent to the (interpreted) apparent crater rim at Bosumtwi. The presence of a sedimentary layer overlying crystalline basement at Ries might have supported the formation of a more pronounced megablock zone, which is more or less lacking at Bosumtwi.

Apart from matching the surface structure, our model for the formation of the Ries crater is also consistent with the subsurface observations. The structural uplift in the numerical model is ~1.5–2 km under the crater center and thus is consistent with our interpretation of the seismic refraction experiments. An estimate of the fragmentation zone beneath the crater based on the intensity of material movements in the dynamic model correlates with the anomalously high conductivity to a depth of 2 km due to interconnected brine-filled pores and fractures revealed by our magnetotelluric study and reduced seismic velocities to approximately the same depth. Gravity modeling in previous studies also showed that the observed mass deficit could be located entirely in the upper 2 km. Furthermore, the dynamic model gives evidence for molten material in the upper 300–500 m beneath the crater floor, which can be related to the suevite layer underneath the lake sediments detected by drilling and geoelectrical depth soundings. We conclude that the presented model agrees well with the observed crater morphometry and subsurface structure.

We should note, however, that our interpretation of the geophysical signature is not unique; in particular, it is not consistent with previous studies. For example, Pohl et al. (1977) proposed that the observed gravity low is caused by a hemispherical fragmentation zone beneath the crater up to a depth of ~5 km, which is the reason for reduced seismic P-wave velocities to similar depth. This interpretation would be consistent with one of our models of the electrical conductivity distribution (model II, Fig. 5). However, our new interpretation is mainly based on our well-supported MT model of the electrical conductivity beneath the crater, which is in sound agreement with our reinterpreted velocity data and the early gravity models by Kahle (1969). Therefore, we support the assumption of a high degree of fracturing induced by release from shock pressure and plastic deformation remaining in the upper ~2 km, causing the observed reduced seismic velocities, the gravity low, and likewise the increased electrical conductivity. Below this level, basement rocks are uplifted by ~1–1.5 km, giving rise to increased seismic velocities in this area.

The numerical models show that both fragmentation and structural uplift of dense basement material accompanied the formation of the Ries crater. Hence, the geophysical signatures of both processes are superimposed. Different models of the subsurface, however, especially of the extent of the fragmentation zone, may be possible, and the present geophysical observations cannot constrain one unique model clearly.

ACKNOWLEDGMENTS

We are very grateful to the Deutsche Forschungsgemeinschaft (DFG) providing Kai Wünnemann with a scholarship to stay at Imperial College London, making the collaboration with J. Morgan possible (DFG gran Wu 355/3-1). The authors thank J. Pohl for providing the seismic reflection data and supporting us with valuable information on the Ries crater. Furthermore we appreciate the support by A. Jording during the magnetotelluric field campaigns and his assistance in processing the acquired data. The electromagnetic measurements were supported by the DFG-Graduiertenkolleg "Entstehung und Entwicklung des Sonnensystems" of the Deutsche Forschungsgemeinschaft. We are also very thankful to E. Turtle and F. Tsikalas for the very valuable and helpful review of the manuscript. Finally, we want to thank A. Deutsch for his great support editing the paper.

REFERENCES CITED

Amsden, A.A., Ruppel, H.M., and Hirt, C.W., 1980, SALE: A simplified ALE computer program for fluid flows at all speeds: LA-8095 Report: Los Alamos, NM, Los Alamos National Laboratories, 101 p.

Angenheister, G., and Pohl, J., 1969, Die seismischen Messungen im Ries von 1948–1969: Geologica Bavarica, v. 61, p. 304–326.

Angenheister, G., and Pohl, J., 1976, Results of seismic investigation in the Ries crater area (Southern Germany), *in* Giese, P., Prodehl, C., and Stein, A., eds., Explosion Seismology in Central Europe; Data and Results: Berlin, Springer-Verlag, p. 290–302.

Artemieva, N., Pierazzo, E., and Stöffler, D., 2002, Numerical modeling of tektite origin in oblique impacts: Implication to Ries-Moldavites strewn field: Bulletin of the Czech Geological Survey, v. 77, p. 31–39.

Bahr, K., 1991, Geological noise in magnetotelluric data: A classification of distortion types: Physics of the Earth and Planetary Interiors, v. 66, p. 24–38, doi: 10.1016/0031-9201(91)90101-M.

Bayerisches Geologisches Landesamt, 1999, Geologie des Rieses: München, Roland Eichborn Verlag, CD-ROM.

Birch, F., 1961, The velocity of compressional waves in rocks to 10 kilobars (Part II): Journal of Geophysical Research, v. 66, p. 2199–2224.

Blohm, E.-K., Friedrich, H., and Homilius, J., 1977, Ein Ries-Profil nach geoelektrischen Tiefensondierungen: Geologica Bavarica, v. 75, p. 381–393.

Bolten, R., and Müller, D., 1969, Das Tertiär im Nördlinger Ries und in seiner: Umgebung: Geologica Bavarica, v. 61, p. 87–130.

Christensen, N.I., and Mooney, W.D., 1995, Seismic velocity structure and composition of the continental crust: A global view: Journal of Geophysical Research, v. 100, B7, p. 9761–9788.

Collins, G.S., Melosh, H.J., Morgan, J.V., and Warner, M.R., 2002, Hydrocode simulations of Chicxulub Crater collapse and peak-ring formation: Icarus, v. 157, p. 24–33, doi: 10.1006/ICAR.2002.6822.

Ernstson, K., 1974, The structure of the Ries crater from geoelectric depth soundings: Journal of Geophysics, v. 40, p. 639–659.

Ernstson, K., and Pohl, J., 1974, Einige Kommentare zu den bohrlochgeophysikalischen Messungen in der Forschungsbohrung Nördlingen 1973: Geologica Bavarica, v. 72, p. 81–90.

Ernstson, K., and Pohl, J., 1977, Neue Modelle zur Verteilung der Dichte und Geschwindigkeit im Ries-Krater: Geologica Bavarica, v. 75, p. 355–371.

ERCEUGT-Group, 1992, An electrical resistivity crustal section from the Alps to the Baltic Sea (central segment of the EGT): Tectonophysics, v. 207, p. 123–139, doi: 10.1016/0040-1951(92)90474-K.

Fowler, S.R., White, R.S., Spence, G.D., and Westbrook, G.K., 1989, The Hatton Bank continental margin: Deep structure from two-ship expanding spread seismic profiles: Geophysical Journal, v. 96, p. 295–309.

Garvin, J.B., and Schnetzler, C.C., 1994, The Zhamanshin impact feature: A new class of complex craters? *in* Dressler, B.O., Grieve, R.A.F., and Sharpton, V.L., eds., Large Meteorite Impacts and Planetary Evolution: Geological Society of America Special Paper 293, p. 249–257.

Gentner, W., Lippolt, H.J., and Schaeffer, O.A., 1963, Argonbestimmung an Kaliummineralien-XI. Die Kalium-Argon-Alter der Gläser des Nördlinger Rieses und der böhmischmährischen Tektite: Geochimica et Cosmochimica Acta, v. 27, p. 191–200, doi: 10.1016/0016-7037(63)90058-9.

Grieve, R.A.F., and Pesonen, L.J., 1992, The terrestrial impact cratering record: Tectonophysics, v. 216, p. 1–30, doi: 10.1016/0040-1951(92)90152-V.

Grieve, R.A.F., Langenhorst, F., Stöffler, D., 1996, Shock metamorphism of quartz in nature and experiment: II Significance in geoscience: Meteoritics and Planetary Science, v. 31, p. 6–35.

Haak, V., Berktold, A., Wiesner, H., Knödel, K., and Rodemann, H., 1977, Zur Verteilung der elektrischen Leitfähigkeit im Ries, ermittelt aus Magneto-Tellurik-Messungen: Geologica Bavarica, v. 75, p. 395–400.

Ivanov, B.A., and Artemieva, N.A., 2002, Numerical modeling of the formation of large impact craters, *in* Koeberl, C., and MacLeod, K., eds., Catastrophic events and mass extinctions: Impact and beyond: Geological Society of America Special Paper 356, p. 619–629.

Ivanov, B.A., Kocharyan, G.G., Kostuchenko, V.N., and Kirjakov, A.F., and Pevzner, L., 1996, Puchezh Katunki impact crater: Preliminary data on recovered core block structure: Lunar and Planetary Science Conference XXVI, Houston, Texas, Abstract 1655, p. 589.

Ivanov, B.A., Deniem, D., and Neukum, G., 1997, Implementation of dynamic strength models into 2D hydrocodes: Application for atmospheric breakup and impact cratering: International Journal of Impact Engineering, v. 17, p. 375–386.

Jording, A., Ferrari, L., Arzate, J., and Jödicke, H., 2000, Crustal variations and terrane boundaries in southern Mexico as imaged by magnetotelluric transfer functions: Tectonophysics, v. 327, p. 1–13, doi: 10.1016/S0040-1951(00)00166-9.

Jung, K., and Schaaf, H., 1967, Gravimetermessungen im Nördlinger Ries und seiner Umgebung, Abschätzung der gesamten Defizitmasse: Zeitschrift für Geophysik, v. 33, p. 319–345.

Kahle, H.-G., 1969, Abschätzung der Störmasse im Nördlinger Ries: Zeitschrift für Geophysik, v. 35, p. 317–345.

Kahle, H.-G., 1970, Deutung der Schwereanomalie im Nördlinger Ries: Zeitschrift für Geophysik, v. 36, p. 601–606.

Karp, T., Milkereit, B., Janle, P., Danuor, K.S., Pohl, J., Berckhemer, H., and Scholz, C.A., 2002, Seismic investigation of the Lake Bosumtwi impact crater: preliminary results: Planetary and Space Science, v. 50, p. 735–743, doi: 10.1016/S0032-0633(02)00049-1.

Karp, T., Artemieva, N.A., and Milkereit, B., 2003, Seismic investigation and numerical modeling of the lake Bosumtwi impact crater: Conference on

"Impact Cratering: Bridging the Gap between Modeling and Observations," Houston, Texas, February 2003: Lunar and Planetary Institute, Abstract 1155, p. 42.

Karp, T., Artemieva, N.A., and Milkereit, B., 2004, Pre-drilling investigation of the lake Bosumtwi impact crater: Constrains from geophysical and numerical modeling: Lunar and Planetary Science Conference XXXV, Houston, Texas, March 2003, Abstract no. 1282.

Kocharyan, G.G., Kostuchenko, V.N., and Ivanov, B.A., 1996, Mechanics of rock massive disruption: Implementation to planetary cratering process: Lunar and Planetary Science Conference XXVI, Houston, Texas, March 1996, Abstracts, p. 677–678.

Koeberl, C., Bottomley, R., Glass, B.P., and Storzer, D., 1997, Geochemistry and age of Ivory Coast tektites and microtektites: Geochimica et Cosmochimica Acta, v. 61, p. 1745–1772, doi: 10.1016/S0016-7037(97)00026-4.

Mackie, R., Rieven, S., and Rodi, W., 1997, Users manual and software documentation for two-dimensional inversion of magnetotelluric data: Cambridge, Massachusetts, Massachusetts Institute of Technology, Earth Resources Laboratory, 14 p.

Mareschal, M., and Chouteau, M., 1990, A magnetotelluric investigation of the structural geology beneath Charlevoix crater, Quebec: Physics of the Earth and Planetary Interiors, v. 60, p. 120–131, doi: 10.1016/0031-9201(90)90254-U.

Masaitis, V.L., Mashchak, M.S., and Yezerskiy, V., 1985, Lifetime of impact features under different geologic conditions: Izvestia Academii Nauk SSSR, Seriya Geology, v. 2, p. 109–114.

Melosh, H.J., 1979, Acoustic fluidization: A new geologic process?: Journal of Geophysical Research, v. 84, p. 7513–7520.

Melosh, H.J., 1989, Impact cratering: A geological process: New York, Oxford University Press, 245 p.

Melosh, H.J., and Ivanov, B.A., 1999, Impact crater collapse: Annual Review of Earth and Planetary Science, v. 27, p. 385–415, doi: 10.1146/ANNUREV.EARTH.27.1.385.

Milton, D.J., Barlow, B.C., Brett, R., Brown, A.R., Glikson, A.Y., Manwaring, E.A., Moss, F.J., Sedmik, E.C.E., Van-Son, J., and Young, G.A., 1972, Gosses Bluff structure, Australia: Science, v. 175, p. 1199–1207.

Morgan, J.V., Warner, M.R., and Grieve, R.A.F., 2002, Geophysical constraints on the size and structure of the Chicxulub impact crater, *in* Koeberl, C., and MacLeod, K., eds., Catastrophic events and mass extinctions: Impact and beyond: Geological Society of America Special Paper 356, p. 39–46.

O'Keefe, J.D., and Ahrens, T.J., 1994, Impact induced melting on planetary surfaces, *in* Dressler, B.O., Grieve, R.A.F., Sharpton, V.L., eds., Large Meteorite Impacts and Planetary Evolution: Geological Society of America Special Paper 293, p. 103–109.

Pierazzo, E., Vickery, A.M., and Melosh, H.J., 1997, A reevaluation of impact melt production: Icarus, v. 127, p. 408–423, doi: 10.1006/ICAR.1997.5713.

Pilkington, M., and Grieve, R.A.F., 1992, The geophysical signature of terrestrial impact craters: Reviews of Geophysics, v. 30, p. 161–181.

Pohl, J., and Will, M., 1974, Vergleich der Geschwindigkeitsmessungen im Bohrloch der Forschungsbohrung Nördlingen 1973 mit seismischen Tiefensondierungen innerhalb und außerhalb des Ries: Geologica Bavarica, v. 72, p. 75–80.

Pohl, J., Stöffler, D., Gall, H., and Ernstson K., 1977, The Ries impact crater, *in* Roddy, D.J., Pepin, R.O., and Merrill, R.B., eds., Impact and explosion cratering: New York, Pergamon Press, p. 343–404.

Scholz, C.A., Karp, T., Milkereit, B., Amoako, P.Y.O., and Arko, J.A., 2002, Pronounced central uplift identified in the Bosumtwi impact structure, Ghana, using multichannel seismic reflection data: Geology, v. 30, p. 939–942, doi: 10.1130/0091-7613(2002)0302.0.CO;2.

Shoemaker, E.M., and Chao, E.C.T., 1961, New evidence for the impact origin of the Ries Basin, Bavaria, Germany: Journal of Geophysical Research, v. 66, p. 3371–3378.

Shuvalov, V., Dypvik, H., and Tsikalas, F., 2002, Numerical simulation of the Mjolnir marine impact crater: Journal of Geophysical Research, v. 107, doi: 10.1029/2001JE001698.

Stöffler, D., Artemieva, N.A., Pierazzo, E., and Ivanov, B.A., 2001, Ries crater Germany: Geology and numerical modeling of impact cratering [abs.]: Meteoritics and Planetary Science, v. 36, Supplement, p. A199.

Thompson, S.L., and Lauson, H.S., 1972, Improvements in the Chart D radiation-hydrodynamic code 3: Revised analytic equation of state: Report SC-RR-71 0714: Albuquerque, New Mexico, Sandia Laboratories, 119 p.

Tillotson, J.M., 1962, Metallic equation of state for hypervelocity impact: General Atomic Report GA-3216: San Diego, Advanced Research Project Agency, 141 p.

Wagner, R., Reimold, W.U., and Brandt, D., 2002, Bosumtwi impact crater Ghana: A remote sensing investigation, *in* Plado, J., and Pesonen, L.J., eds., Meteorite impacts in Precambrian shields: Berlin, Springer-Verlag, 336 p.

Wünnemann, K., and Ivanov, B.A., 2003, Numerical modeling of the impact crater depth-diameter dependence in an acoustically fluidized target: Planetary and Space Science, v. 51, p. 831–845, doi: 10.1016/J.PSS.2003.08.001.

Wünnemann, K., and Lange, M.A., 2002, Numerical modeling of impact induced modifications of the deep-sea floor: Deep-Sea Research II, v. 49, p. 969–981.

Zelt, C.A., 1998, Lateral velocity resolution from three-dimensional seismic refraction data: Geophysical Journal International, v. 135, p. 1101–1112, doi: 10.1046/J.1365-246X.1998.00695.X.

Zelt, C.A., and Smith, R.B., 1992, Seismic travel-time inversion for 2D crustal velocity structures: Geophysical Journal International, v. 108, p. 16–34.

Zhang, P., Rasmussen, T.M., and Pederson, L.B., 1988, Electric resistivity structure of the Siljan impact region: Journal of Geophysical Research, v. 93, p. 6485–6501.

Manuscript Accepted by the Society 9 August 2004

Printed in the USA

Geological Society of America
Special Paper 384
2005

Structure and formation of a central uplift: A case study at the Upheaval Dome impact crater, Utah

Thomas Kenkmann*
Andreas Jahn
Dirk Scherler
Institut für Mineralogie, Museum für Naturkunde, Humboldt-Universität Berlin, Invalidenstrasse 43, D-10115 Berlin, Germany

Boris A. Ivanov
Institute for Geodynamics and Geospheres, Russian Academy of Science, Leninsky Prospect 38-1, 119334 Moscow, Russia

ABSTRACT

The internal structures of central uplifts of impact craters are among the most complex geologic features within Earth's crust. Upheaval Dome, Utah, is used as a reference and case study to display the internal geometry of a central uplift and to deduce mechanisms of uplift formation in impact craters within a sedimentary, siliciclastic target. Geological and structural data gained from our high-resolution mapping of the central part of the structure were combined with topographic data in an ArcGIS database. A three-dimensional visualization of the geometry of faults and strata within the central uplift is presented and interpreted with respect to their deformation history. Central uplift formation is induced by an inward and upward directed convergent flow of the crater floor during gravity-driven collapse of the transient crater cavity. Radial folds and a concentric stacking of imbricated thrust slices are prominent deformation features and result from a constrictive strain pattern. The arrangement of structural elements in the inner part of the Upheaval Dome roughly displays some bilateral symmetry, trending northwest. The dominance of northwest-dipping reverse faults indicates a material transport of top to the southeast, which may be caused by an oblique impact. Fault planes commonly dip steeply and are bent due to a passive distortion after activation. The macroscopic coherence of large target units and blocks and the anisotropy of the layered target cause remarkable deviations from an ideal convergent flow field. Stratified siliciclastic rocks are commonly deformed by localized brittle faulting, and massive sandstones are deformed by a distributed cataclastic flow. During crater collapse, pervasively crushed sandstones will flow locally as a granular medium, resulting in the formation of dikes. Acting as lubricants, they accommodate the complex mesoscale folding and faulting of the neighboring strata. A standard numerical model of impact cratering was designed for comparison with the observed structures and to estimate impact parameters like initial crater size, amount of erosion, and the time of impact. The best fit between model and field data is found when the White Rim Sandstone is buried ~2000 m beneath the

*thomas.kenkmann@museum.hu-berlin.de

Kenkmann, T., Jahn, A., Scherler, D., and Ivanov, B.A., 2005, Structure and formation of a central uplift: A case study at the Upheaval Dome impact crater, Utah, *in* Kenkmann, T., Hörz, F., and Deutsch, A., eds., Large meteorite impacts III: Geological Society of America Special Paper 384, p. 85–115. For permission to copy, contact editing@geosociety.org.

target surface. This most likely corresponds to an Upper Cretaceous age of Upheaval Dome during deposition of the Mancos shales. The initial diameter of the Upheaval Dome impact crater would have been ~7–8 km.

Keywords: impact crater, central uplift, Upheaval Dome, crater modification, numerical modeling, faulting.

INTRODUCTION

Terrestrial impact structures with diameters >3–4 km are called "complex" impact craters because their final shapes markedly deviate from the crater cavity that is initially formed by the impact. These craters display a central uplift, collapsed crater walls, and a flattened crater floor, leading to a decreased depth to diameter ratio with respect to "simple" bowl-shaped craters (e.g., Melosh, 1989; Wünnemann and Ivanov, 2003). In terrestrial craters of >25 km (e.g., the Ries crater, Germany), the central uplift is often present as a peak ring that surrounds a central basin. The formation of a central uplift has sometimes been attributed to a rebound effect. This description, however, implies that the material responds elastically. Melosh and Ivanov (1999) have shown that an elastic rebound is of subordinate importance during central uplift formation, in particular for larger craters with a relatively large volume of shock damaged rocks. A comparison of the simple-to-complex transition size of craters on planets and their satellites clearly indicates that the ultimate driving force for crater modification and central uplift formation is gravity. To explain the magnitude of inward and upward motion of rocks, a temporary strength degradation must occur (e.g., Melosh, 1977; McKinnon, 1978). Phenomenologically, the formation of a central uplift is similar to the central jet that develops when a raindrop hits a water surface. This similarity suggests a fluid-like, strengthless behavior of the rocks during crater formation. More precisely, the rheology matches a Bingham fluid (viscous response above some critical yield stress). Mechanisms known as acoustic fluidization (Melosh, 1979, 1989), block oscillation (Ivanov and Kostuchenko, 1997), and thermal weakening (O'Keefe and Ahrens, 1993) were proposed to explain this temporary soft state of the rock. Ivanov and Deutsch (1999) showed that the latter process alone is not sufficient to explain crater collapse and that it cannot account for the collapse of small and medium-sized complex craters. Acoustic fluidization provides an approach in which the temporary decrease of rock friction facilitates transient cavity collapse. The efficiency of this effect decreases toward the end of the collapse, and the fluid-like material regains the strength properties of a fragmented solid, thereby defining the final crater shape (see the review by Melosh and Ivanov, 1999).

The formation of a central uplift is always accompanied by the subsidence of the crater rim and the formation of major terraces (e.g., Melosh, 1989). These terraces are thought to be the product of slumps that occur along listric normal faults. The simultaneous interplay of the inward slumping cavity walls on one side and the fluid-like uplifting of the crater floor on the other produces a complex deformation pattern that has not been investigated in detail to date.

Compiling data from 14 terrestrial impact craters, Grieve et al. (1981) found that the stratigraphic uplift in a complex impact crater obeys the phenomenological relationship

$$SU = 0.06\, D_c^{\,1.1},$$

where SU is the stratigraphic uplift and D_c is the final diameter in km. This relationship was further refined by Cintala and Grieve (1998) to

$$SU = 0.086\, D_c^{\,1.03}\,.$$

They took into account that the minimum depth of origin of central peak material should coincide with the maximum depth of melting (Ivanov et al., 1982; Basilevsky et al., 1983). The transition from central peak craters to peak ring craters that occurs on Earth at a critical diameter ranging from ~25 km (Ries) to ~60–80 km (Siljan, Chesapeake Bay) is possibly caused by the disproportionately large amount of melt and a prolonged period of the fluidized state that leads to the oversteepening and subsequent collapse of the unstable flanks of the central mount (Morgan et al., 2000). During this collapse, material moves outward and overrides the inward moving parts of the transient crater walls.

In the past decade, much progress in understanding central uplift formation has been achieved by the use of numerical models (e.g., Melosh and Ivanov, 1999; Collins et al. 2002). They provide an important background for geologists who try to unravel the kinematic and dynamic history of a particular structure based on detailed observations and measurements in the field. Recent numerical simulations of crater collapse implement important algorithms for mechanical behavior (e.g., for acoustic fluidization, thermal softening, and brittle damage). They are capable of providing an idealized particle flow field and the state of stress, strain, and temperature for each time increment (e.g., Wünnemann and Ivanov, 2003). However, these models are commonly two-dimensional (assuming radial symmetry) and of limited spatial resolution. A direct correlation between model and field observation (meso and macroscale deformation features) is difficult because deformation localization is a widespread process in nature. An adequate implementation of strain localization in the cratering flow is still problematical. Despite all these limitations, it is necessary to combine model and observation-based approaches to achieve the best results. The present study is guided by this philosophy.

This study will demonstrate that lithology and rheological anisotropies play a dominant role for the formation and geometry of faults. Larger faults often bound rigid or internally deformed and bent blocks or slices. Their centric motion and mutual interaction lead to significant deviations from the ideal convergent flow field and, thus, are major causes for deviation from a simple, onion-skin geometry of a central dome with the oldest rocks exposed in the center, surrounded by successively younger strata, which is also expected from numerical models. Additionally, Schultz and Anderson (1996) emphasized that the impact angle has a pronounced effect on the final geometry of a central uplift. They found that the central uplift offsets uprange from the geometric center of a crater and may be breached into two ridges that trend parallel to the trajectory. Ekholm (1999), however, showed in a statistical analysis of Venusian craters that there is no empirical evidence for a systematic uprange offset of a central peak in craters formed by oblique impacts.

Among the ~170 terrestrial impact structures that are presently known, there are only ~40 where a central uplift is exposed and accessible. The number of craters further decreases to ~20 when considering only those craters where the central uplift is mapped in detail and on the basis of appropriate marker beds. The requirement of marker beds restricts a detailed structural analysis to craters in sedimentary targets. Structural studies, of different resolution, of central uplifts or some of their parts exist for the following medium-sized impact craters: Flynn Creek, United States, 3.6 km (Roddy, 1979); Decaturville, United States, 6 km (Offield and Pohn, 1979); Upheaval Dome, United States, 6 km (Kriens et al., 1999); Kelly West, Australia, ~10 km (Shoemaker and Shoemaker, 1996); Spider, Australia, 13 km (Shoemaker and Shoemaker, 1996); Sierra Madera, United States, 13 km (Wilshire et al., 1972); Kentland, United States, 13 km (Laney and Van Schmus, 1978); Gosses Bluff, Australia, 22 km (Milton et al., 1972, 1996); and Strangways, Australia, 26 km (Shoemaker and Shoemaker, 1996). In addition to size, target composition, stratification, other impact conditions differ from one crater to another along with the degree of preservation and the level of exhumation. Without denying the uniqueness of each impact, a number of structural features are common to all of these craters, as compiled by Kenkmann (2002). Generally, the intensity of faulting and folding increases from the periphery to the center of a central uplift. Radial transpression ridges (Kenkmann and von Dalwigk, 2000), in which material is uplifted or squeezed out in different modes, accommodate the converging mass flow. Folds may become tight and subvertical or reclined, and their development is often linked to the propagation of faults. Faulting is more effective in accommodating the constrictive strain (two equally large axes of principal shortening are oriented horizontally, the principal axis of extension is oriented vertically) during formation of a central dome with respect to folding; this particularly holds for the inner part of a central uplift. A symmetric stack of steeply dipping and curved thrust slices is recorded from the Spider structure (Shoemaker and Shoemaker, 1996). Upheaval Dome provides the unique possibility for studying these common structural features of complex impact craters in detail. Summarizing these observations, it becomes obvious that geologists are confronted with a unique type of deformation, which neither occurs in any orogeny, diapirism, nor any other endogenic process. This applies not only to its rapid single-slip formation event, but also to the upward directed, constrictive particle displacement paths.

The deeply eroded Upheaval Dome impact crater, with its strong relief, provides the opportunity to view different depths of the crater floor. The purpose of this paper is to present the internal structure of the central uplift of a small to medium-sized complex impact crater in a sedimentary target and to deduce the principles of its formation. The present study is devoted to inferring the architecture of the central uplift of Upheaval Dome with the greatest possible accuracy. To achieve this goal, more detailed mapping than presented by Jackson et al. (1998) and Kriens et al. (1999) was required. The flat-lying and undeformed rocks of the Colorado Plateau provide marker horizons that are well suited for high-resolution mapping. A major goal was to analyze and visualize the complex pattern of faults, folds, and strata at a high resolution by the use of modern processing techniques and, in doing so, to better understand central uplift formation. Of particular interest is the development of the style of deformation from a target's surface down to the roots of a collapsing crater. Since a terraced crater rim indicates the movement of slump blocks on discrete fault surfaces, their extension toward greater depths is not without problems. We also addressed the question of how strata thickening and thinning as well as faulting and folding is accommodated on different scales during high strain deformation, and we provide a rationale to infer displacements on the basis of a volume balancing technique. The microscopic aspects, however, will be discussed in a separate, later paper. Furthermore, we also focused on specific deformation features observed at Upheaval Dome, such as the "tongue-shaped" rock masses of Wingate Sandstone, seen at the flanks of the central depression (Kriens et al., 1999) and the abundant clastic dikes in all stratigraphic levels, which have long been recognized (e.g., Shoemaker and Herkenhoff, 1984). In order to reconstruct important aspects of the crater, such as the original crater size, the time of impact, and the amount of erosion since the impact, we tried to correlate the presently exposed remnants of the crater with a numerical model. The results might explain the absence of definitive shock features such as planar deformation features (PDFs) at Upheaval Dome, but indicate elevated pressures much higher than possible lithostatic pressures. The latter were inferred from a microstructural analysis of the White Rim Sandstone (Kenkmann, 2003). The presented reconstruction of the crater is in conflict with previous assumptions by Kriens et al. (1999).

GEOLOGICAL AND TECTONIC SETTING

Stratigraphy and Regional Geological Context

The Upheaval Dome structure (Fig. 1) in southeastern Utah, United States, is located on the Colorado Plateau near

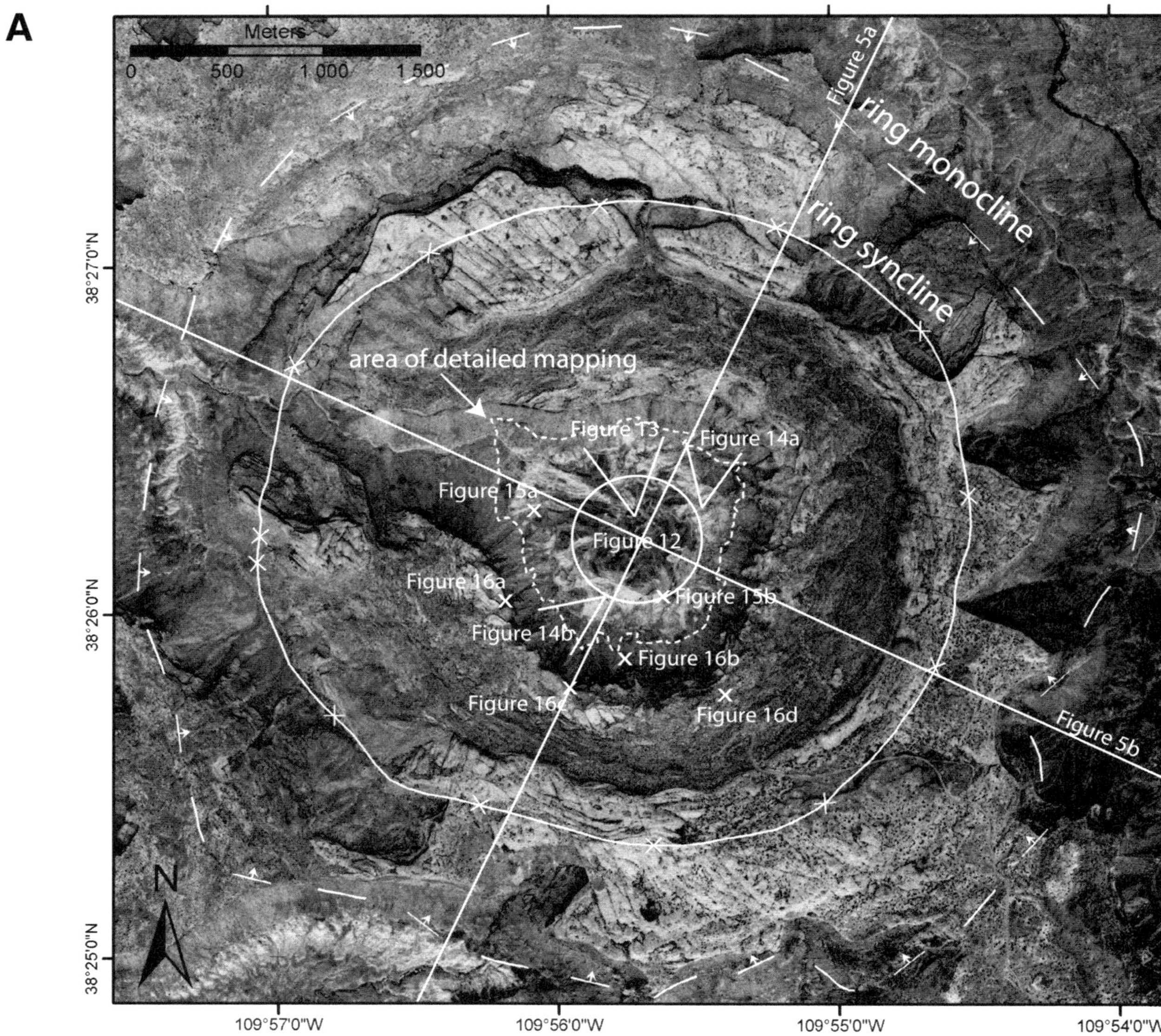

Figure 1 (*on this and following page*). (A) Aerial photograph of Upheaval Dome. Dashed line surrounds the area that was mapped at high resolution. Locations and viewing directions of specific field photos are given. The circle marks the location of the circular cross section in Figure 12. Two straight lines indicate the traces of the linear cross sections in Figure 5. (B) Simplified geological map of the Upheaval Dome structure, Utah, modified after Kriens et al. (1999). The inner part of the dome inside the Chinle Formation–Wingate Sandstone contact is the focus of this study. Oldest rocks of uppermost Permian age are exposed in the center of the dome; youngest strata (Navajo Sandstone) belong to the Jurassic and are located along the ring syncline of the structure.

the western boundary of the Paradox basin. This basin formed during the Pennsylvanian Ancestral Rocky Mountains orogeny, which also led to the formation of the nearby Uncompahgre uplift (Kluth and Coney, 1981; Marshak et al., 2000). Several hundred meters of evaporites (Paradox Formation) accumulated from middle to late Pennsylvanian in this basin, followed by the deposition of clastic sediments in a continental environment (Cutler Group) during the Permian (Stokes, 1988) (Fig. 2). Subsequent salt tectonism mainly led to the formation of gentle, basin-parallel anticlines that decrease in amplitude away from the Uncompahgre highlands (Jackson et al., 1998). Triassic sedimentation in fluvial to shallow marine environments resulted in the fine-grained Moenkopi and Chinle Formation, predominantly sandstones that are exposed in the innermost part of Upheaval Dome (Fig. 3). During the Jurassic, thick units of porous eolian sandstones (Wingate and Navajo Sandstones) as well as layered fluvial sediments (Kayenta Formation) were deposited in continental environments and make up the youngest units outcropping in the structure; all younger strata were removed by post-Laramide erosion.

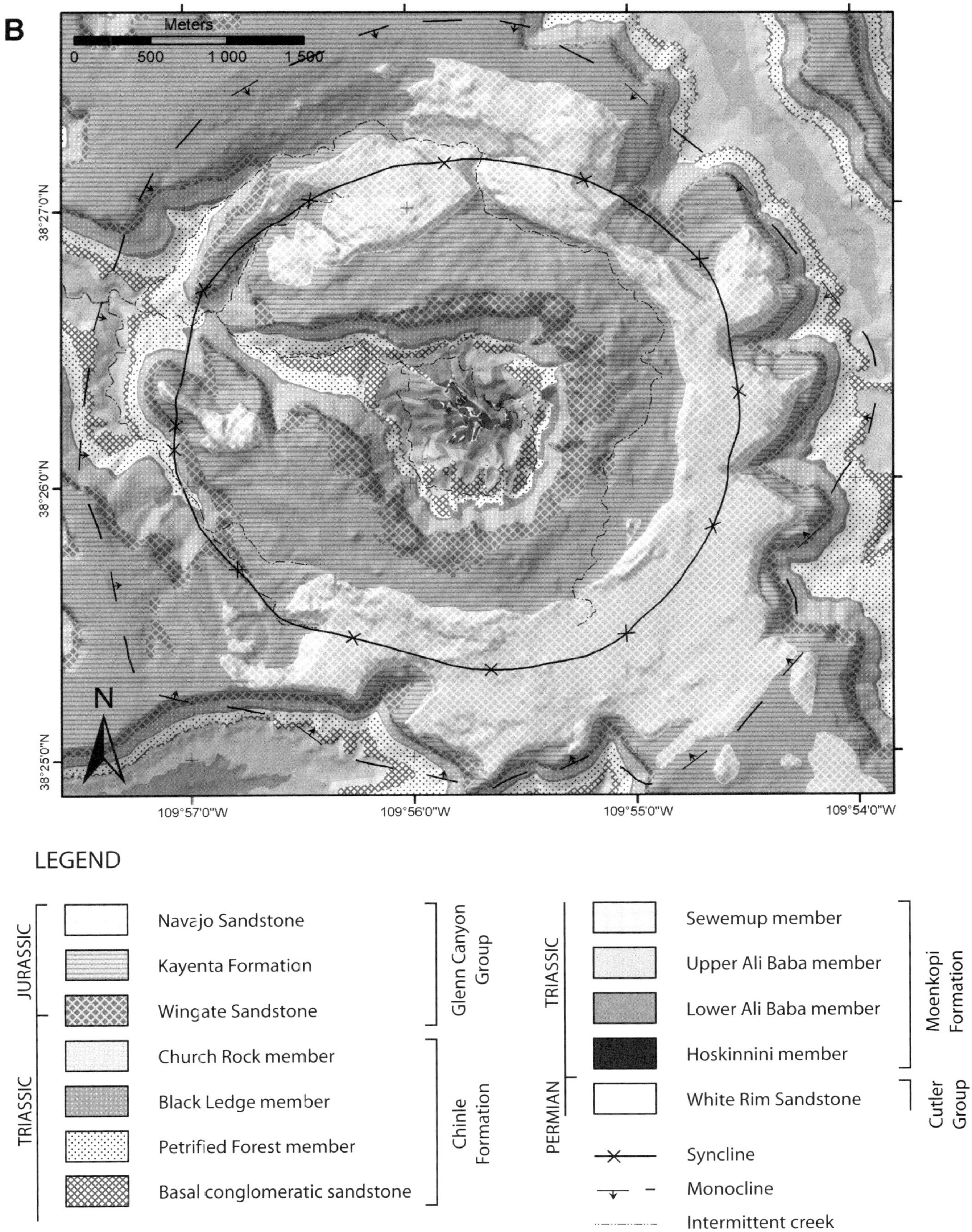
B
Meters
0
500
1 000
1 500
38°27'0"N
38°26'0"N
38°25'0"N
109°57'0"W
109°56'0"W
109°55'0"W
109°54'0"W
N
LEGEND
JURASSIC
TRIASSIC
Navajo Sandstone
Kayenta Formation
Wingate Sandstone
Glenn Canyon Group
Church Rock member
Black Ledge member
Petrified Forest member
Basal conglomeratic sandstone
Chinle Formation
TRIASSIC
PERMIAN
Sewemup member
Upper Ali Baba member
Lower Ali Baba member
Hoskinnini member
Moenkopi Formation
White Rim Sandstone
Cutler Group
Syncline
Monocline
Intermittent creek

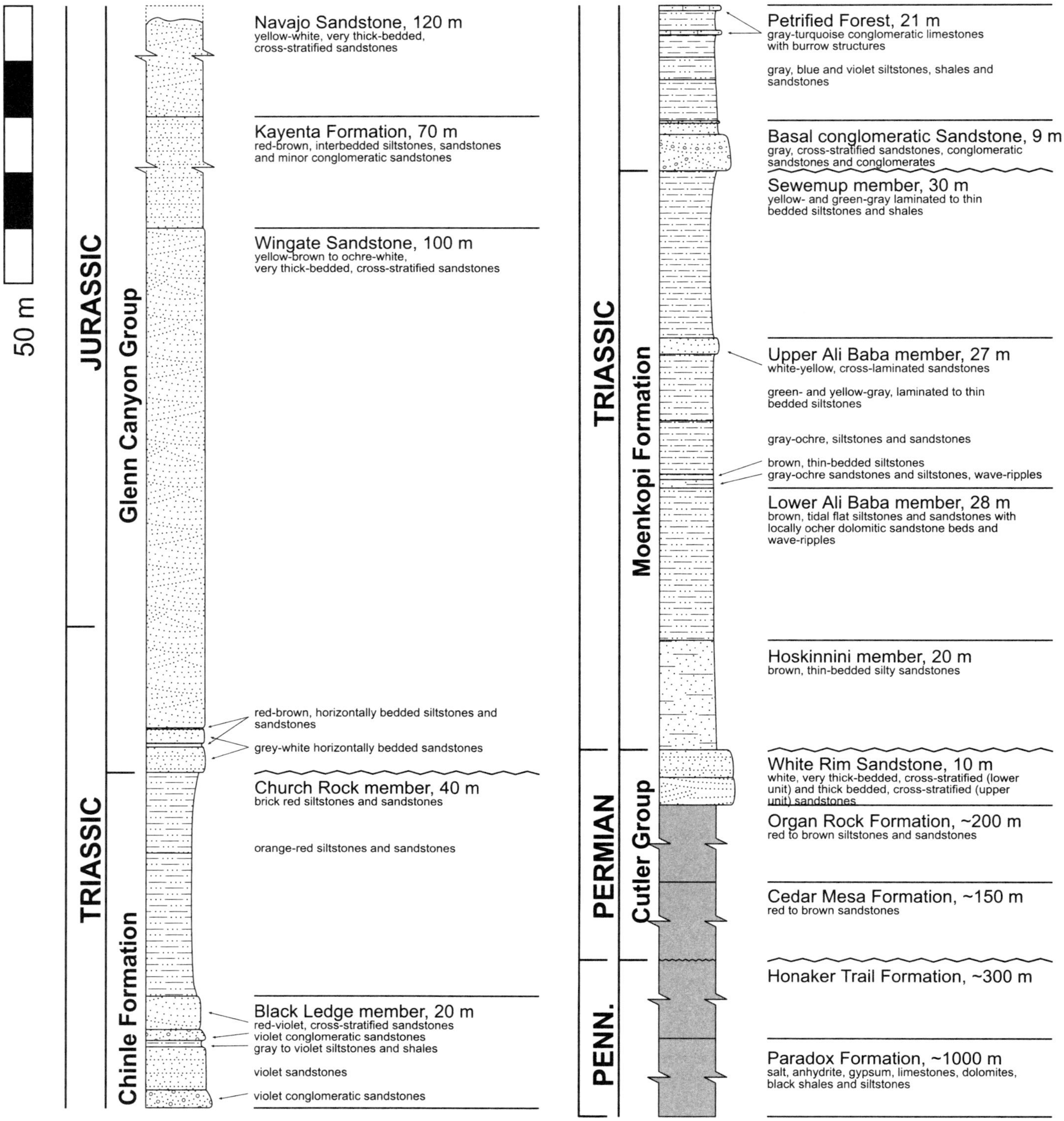

Figure 2. Stratigraphic column of the formations present at Upheaval Dome. Gray shaded formations do not crop out. Data were compiled from (1) stratigraphic field study of Wingate Sandstone to Moenkopi Formation in the Holeman Spring basin, south of Upheaval Dome; (2) stratigraphic field study of lower Chinle to Moenkopi Formations at Upheaval Bottom near the Green River; (3) stratigraphic field study of middle to lower Chinle Formation in the northern Syncline Valley at the northeastern edge of Upheaval Dome; (4) acoustic well log data at Buck Mesa #1, located at the east-southeastern edge of Upheaval Dome for Permian Formations (Kanbur et al. 2000), (5) data from Jackson et al. (1998) for thicknesses of Kayenta Formation, Navajo Sandstone and hiatus within Triassic to Pennsylvanian strata.

Figure 3. Wide-angle panoramic view across the inner part of the dome showing the area of investigation. The center of the photographs is to the northwest. The circular escarpment is formed by the Wingate Sandstone. Prominent cliffs in the center are formed by the White Rim Sandstone.

With the onset of the late Cretaceous Laramide orogeny, several broad anticlines developed in the sedimentary cover above reactivated basement faults (Dickinson et al., 1988; Marshak et al., 2000; Bump and Davis, 2003) and ended a long period of regional sedimentation. During that time, the entire Colorado Plateau experienced a strong uplift. Based on a geographic information system (GIS) approach, Pederson et al. (2002) estimated an uplift of ~3000 m and an exhumation of 1500–2000 m for the area of investigation. Adjacent Laramide uplifts (San Rafael Swell, Uncompahgre Uplift) as well as Paleogene intrusions are more than 50 km away from Upheaval Dome and do not affect the local structural geology. Minor bending (~1°) of the country rocks some 2 km northeast of the structure within the southwest-trending Grays Pasture Syncline is attributed to salt tectonics and adds to a gentle dip of ~1° to the northwest (Huntoon et al., 1982).

The circumstance that the sedimentary and tectonic history in the area of investigation has not been affected by much deformation since Precambrian times makes it possible to attribute the majority of deformation features at Upheaval Dome to the impact event. Pervasive subvertical northwest and north-striking joint systems postdate the deformation at Upheaval Dome and are an expression of the regional stress field. A large fracture and fissure zone, called Roberts Rift, exists at 22–32 km from Upheaval Dome and shows a subradial trend to the crater. Huntoon and Shoemaker (1995) supposed that this fissure was formed by impact-induced overpressured fluid compartments in the Pennsylvanian Paradox section and attendant hydraulic fracturing of the overlying strata. Although the temporal linkage is not proven yet, it could indicate the presence of subsurface fluids during the impact event.

Geology of Upheaval Dome

Shoemaker and Herkenhoff (1983, 1984) identified Upheaval Dome as a deeply eroded, complex impact crater. The present topography at Upheaval Dome is entirely erosional in nature and does not reflect the original crater form. The exposures lie well below the original crater surface, as will be shown later. In previous concurrent hypotheses, Upheaval Dome was considered to be of cryptovolcanic origin (Bucher, 1936) or the result of salt diapirism (McKnight, 1940; Mattox, 1968; Huntoon et al., 1982). One of the greatest problems with the salt diapir interpretation for Upheaval Dome was the absence of any diapiric structure remotely like Upheaval Dome anywhere within the 40,000 km^2 part of the Paradox Basin that is underlain by the Pennsylvanian Paradox salt section. There are merely a few small-scale uplifts exposed at the base of the Colorado valley that are formed within the Paradox formation itself. An explanation of Upheaval Dome as a remnant of a pinched-off salt doming scenario was presented by Jackson et al. (1998). To explain Upheaval Dome in this way is particularly difficult because there is neither evidence that salt moved through the core of the structure nor are there any relict fragments of the Paradox and hanging wall formations (i.e., Pennsylvanian Honaker

Trail and Permian Cedar Mesa and Organ Rock Formations), which should be found within the throat of salt passage (Huntoon, 2000). Kriens et al. (1997, 1999) and Huntoon (2000) extended the structural work by Shoemaker and Herkenhoff (1983, 1984) and described deformation features and strain indicators that are all consistent with an impact origin.

Upheaval Dome, located within Canyonlands National Park, consists of a central uplift of ~2.5 km diameter, surrounded by a ring-syncline with a diameter of 3.5 km and a crater rim monocline of ~5 km diameter. The latter defines the extent of the structure (Fig. 1). The inner part of the central uplift forms a morphological depression of ~1.4 km diameter (Fig. 3). The exposed strata comprise rocks of uppermost Permian age in the very center (Cutler group, White Rim Sandstone) to strata of Triassic (Moenkopi, Chinle Formation) and Jurassic age (Wingate Sandstone, Kayenta Formation, Navajo Sandstone) (Fig. 2), which are located at progressively increasing distance from the center (Fig. 1). The Navajo Sandstone is preserved along the ring-syncline axis. Shoemaker and Herkenhoff (1983, 1984), Kriens et al. (1999), and Huntoon (2000) recognized a series of centric-dipping listric faults at the perimeter of the structure that flatten within the ring-syncline and then rise toward the central uplift. Displacement on these faults is found to evolve from normal to bedding-parallel, and eventually reverses in the central uplift and is thought to be the reason for thinning of the Wingate Sandstone in several parts near the perimeter. While the layered Kayenta Formation displays complex internal thrusting (see Fig. 4) on branches of these listric faults, the massive Wingate and Navajo Sandstone show radial folds, plunging away from the center (Kriens et al., 1999). These structures are interpreted as signs of convergent material flow. Further evidence for the impact theory was given by Kanbur et al. (2000) from a seismic study. They identified (1) <100 m relief for the top of the Paradox Formation, contrary to a salt diapirism scenario; (2) listric normal faults with displacements toward the center of the dome; and (3) deformation of the central uplift ceasing with increasing depth. The seismic refraction data suggested that all deformation was limited to structural levels above the Paradox. Based on acoustic well logging (Buck Mesa #1) close to the ring-syncline, they identified the Paradox salt at a depth of 1160–2200 m. Unfortunately, the resolution of the seismic reflection studies is too low to give further details of the structural geometry of the central uplift at depth.

The debate on Upheaval Dome's origin lasted so long because unequivocal evidence, like shock features, was lacking. A fracture pattern resembling a shatter cone was described by Kriens et al. (1999). Vesicle-rich nodules on top of the Navajo Sandstone, interpreted as ballistic, once molten ejecta (Kriens et al., 1999), were proven to be unrelated to the impact (Koeberl et al., 1999). Planar microstructures in quartz of the White Rim Sandstone were presented by Kriens et al. (1999) and Kenkmann and Scherler (2002). However, TEM investigations (Kenkmann, 2003) showed that they do not represent amorphous lamellae, and thus are not PDFs. Kenkmann (2003), however, demonstrated by comparison with pressure-calibrated deformation microstructures of a similar, experimentally deformed rock, that the White Rim Sandstone was deformed at elevated effective confining pressures in excess of 250 MPa. The inferred pressure was shown to be inconsistent with the possible lithostatic pressures for the region.

APPLIED TECHNIQUES

Mapping and Digital Processing

The central part of Upheaval Dome, defined by the steep escarpments of the Wingate Sandstone, was newly mapped at a scale of 1:2500 (the outline of the mapped area is shown in Fig. 1A). To enable comparisons with earlier maps, we discriminated the same stratigraphic units as Kriens et al. (1999), but slight deviations regarding the definition of the lithological boundaries occur for some units. Numerous marker horizons enable the detailed mapping of the dome. At sites where appropriate outcrops were lacking, a rather sensitive way to map different strata was found to be the soil color. We took only those faults into account that have a displacement of more than 2 m. The degree of detail during the mapping is strongly dependent on the accessibility and visibility of the differently sized structures. The elevated parts are especially difficult to access, with debris-covered slopes of more than 45° inclination.

Digital elevation data with a resolution of 10 m was taken from the U.S. Geological Survey (Upheaval Dome Quadrangle Utah, 7.5 min topographic series) map published in 1988, compiled from aerial photographs taken in 1981 (U.S. Geological Survey, 1988). We combined this topographic data with the geological and structural data from the field in an ArcGIS database (version 8). The geological and structural data consist of polygons (e.g., outcropping units), lines (e.g., faults) and points (e.g., dips). Using the software 3D-Analyst by ESRI (380 New York Street, Redlands, CA 92373–8100, USA), we constructed a three-dimensional digital terrain model (3D-DTM) covered with geological signatures. Fault planes and the structure map are constructed using the strike trajectories. Their visualization is performed by exporting the data into the software 3D-Move (Midland Valley Exploration Ltd., 14 Park Circus, Glasgow G3 6AX, UK). The dip map was interpolated via the inverse-distance-weighted method from point data. Besides our own mapping of the inner part of the dome, much structural information outside the central depression (Wingate Sandstone, Kayenta Formation, and Navajo Sandstone) was taken from the map by Kriens et al. (1999). The cross sections and folds through the crater are constructed using the geometric Faill-rules (Woodward et al., 1989). However, due to the non-plane strain deformation, they do not represent balanced cross sections.

Our sampling included all lithological units and reference samples from outside the crater (Scherler, 2004). About 90 standard polished thin sections were prepared to investigate the petrography and fabric with an optical microscope and a scanning electron microscope (JEOL JSM 6300) in backscattered-electron mode.

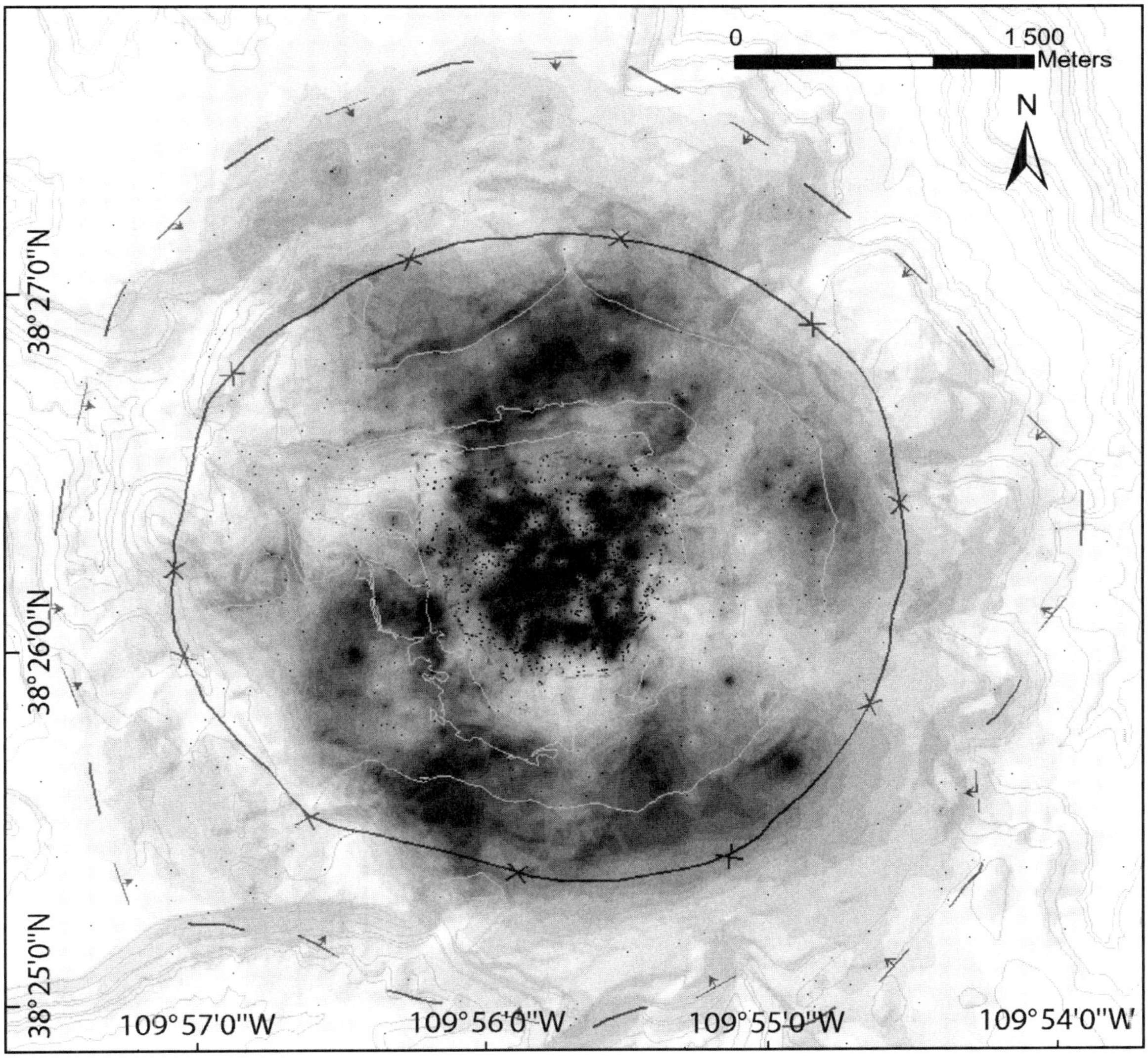

Figure 4. Dip map of the Upheaval Dome structure. The gray shading code corresponds to different dip values, white means flat-lying, and black represents vertical dip. Dots are measurement points. Note that the average dip in the west-northwestern and east-southeastern sectors deviates from the rest of the structure.

Numerical Modeling

We used a numerical model to compare structural data gained in the field with those predicted by modeling and to reconstruct the original crater from its present relics. The aim of modeling was to estimate the initial size of the crater and to determine the amount of erosion. Both of these problems have no unique solution due to model simplifications and the lack of field data above the erosion level. However, the comparison of target strata deformation (e.g., dipping of target layers, position of the syncline axis) with the modeling results allows us to provide a consistent model that is in agreement with both observation in the field and the dynamics and physics of the cratering process.

Hydrocode

We use the Simplified Arbitrary Lagrangian Eulerian code, version B (SALEB) hydrocode which is based on the Simplified Arbitrary Lagrangian Eulerian (SALE) code developed by the Los Alamos Laboratory (see the original SALE version by Amsden et al., 1980). The original SALE code was heavily modified to enable the modeling of two different materials. The reconstruction of the material interfaces in each cell at any time step is constructed in the same way as in the Solid, Vapor, Air (SOVA) hydrocode (Shuvalov, 1999; Shuvalov et al., 1999). For large scale events, such as Upheaval Dome, the presence of the atmosphere plays a subordinate role, affecting only the motion of small fragments in the ejecta curtain, which are of no interest here. We therefore did not include an atmosphere and modeled the target rocks and projectile with identical materials.

Equation of State and Material Strength

The thermodynamic relations between pressure, temperature, and density of a material are defined in the equation of state (EOS). We used the relatively simple analytic Tillotson's EOS (Tillotson, 1962; Allen, 1967; Melosh, 1989) for describing a projectile and a target; our model runs lasted sufficiently long to trace the crater growth and modification. The original EOS version (Tillotson, 1962; Melosh, 1989) defines pressure for a given specific internal energy and density of the material. To estimate temperature, we used direct calculations of "cold" and "thermal" energy. The explicit value of the thermal energy allows us to estimate the temperature assuming a proper value of the heat capacity, averaged in the temperature range of interest (see details in Ivanov et al., 2001).

For porous sandstones, comprising major parts of the Upheaval Dome's target, both the EOS and the mechanical properties are very complicated. In addition to natural heterogeneities presented, for example, by diverse bedding, the water saturation of the pore space is unknown. However, the high pressure of the pore fluid could compensate for the lithostatic pressure and may result in very low effective confining pressure and the reduced strength of sandstones (e.g., Ricard and Bercovici, 2003).

The initial model run has been conducted using Tillotson's EOS with coefficients for a "limestone"—some generic competent rock with low pressure bulk modulus of 40 GPa and density 2700 kg m^{-3} (see Table 1; Rock-1). The reason for this choice was that at the late stage of crater growth and collapse, the compressibility of a target is not an important parameter, as the fragmented target material flows like an incompressible medium with some friction. However, the material compressibility (or, more exactly, the material EOS in compression) may significantly control the shock pressure decay and, finally, the velocity field in the excavation flow around a growing crater. To investigate how the model results depend on the initial material compressibility, we constructed a set of model equations of state. It is still not possible to consider initial porosity and strength of porous materials in the SALEB code. For this reason, we constructed the modified Tillotson EOS for a porous sandstone as if it were a homogeneous material with experimentally measured initial density and bulk modulus. To reach a reasonable fit to experimental data for initially porous rocks, the cubic term for the density related pressure was added to the original Tillotson formulation (see Appendix). Hugoniot adiabats are known for Moenkopi and Coconino sandstones (initial density of 2250 and 2030 kg m^{-3}, initial bulk modulus 6 GPa, resulted from the approximation of experimental data by Roddy et al., 1980) and for an unspecified sandstone with the initial density of 2460 kg m^{-3} (a few Hugoniot data points measured by Trunin et al., 1988). Using these experimental data as a basis, we constructed three Tillotson parameter sets that mimic the Hugoniots of sandstone up to particle velocities of ~6 km s^{-1}. Hence, for vertical impact, we used these approximated Tillotson EOSs for an impact velocity ~12 km/s. The appendix summarizes our Tillotson EOS modification, and Table 1 gives the parameters used in the present modeling.

TABLE 1. MATERIAL EOS AND STRENGTH PARAMETERS USED IN THE MODELING

Model rock designation	Rock-1 (limestone)	SS-1 (sandstone)	SS-2a (sandstone)	SS-2b (sandstone)
Tillitson equation of state parameters (see notations in Melosh, 1989)				
ρ_0, kg m^{-3}	2700	2460	2020	2020
A, GPa	40	30	6	6
B, GPa	67	10	5	30
C, GPa	–	40	9	24
E_0, MJ kg^{-1}	10	10	30	–
a	0.5	.5	0.5	0.1
b	0.6	.6	0.3	0
α	5	5	5	5
β	5	5	5	5
E_{iv}, MJ kg^{-1}	2.5	3.5	3.5	3.5
E_{cv}, MJ kg^{-1}	14	20	20	20
Strength and elastic parameters (see notations in Collins et al., 2004)				
Poisson ratio	0.25	0.3	0.25	0.25
Cohesion, Y_0, MPa	50.	20.	5.	5.
High-pressure plastic limit, Y_M, MPa	800	1000	1000	1000
Non-damaged "friction," μ_i	1.5	0.95	0.95	0.95
Damaged rock cohesion	0.	0.	0.	0.
Damaged rock friction, μ_d	0.6	0.5	0.43	0.43
Melting temperature, K	1500	1750	1750	1750
Thermal softening parameter, a_m	1.2	1.2	1.2	1.2

The model assumes that intact rocks have some finite shear strength at zero ambient pressure (cohesion). At enhanced pressures, the shear strength increases before it reaches a constant value in the pressure region of several GPa, where plastic deformation is assumed. We assume that the intact rock strength grows with pressure according to the Lundborg approximation (Lundborg, 1968). The detailed representation of shear damage accumulation and transition from the intact to the damaged rock with some friction strength employing SALEB is published by Collins et al. (2004).

Acoustic Fluidization

In addition to thermal softening of rocks (e.g., O'Keefe and Ahrens, 1993), we used acoustic fluidization (Melosh and Ivanov, 1999) with appropriate scaling for variable crater size. The latter feature of the model is the most significant for the crater formation analysis. The "block model" used here is a specific version of the general acoustic fluidization model proposed by Melosh (1979, 1989). The formulation of the "block model" version (Ivanov and Kostuchenko, 1997), and comparison with Melosh's original model, is presented by Melosh and Ivanov (1999). According to the "block model," the low frequency oscillations of rock fragments ("blocks") periodically reduce contact forces between adjacent blocks; hence, the friction is decreased and slip occurs if oscillations are strong enough. When shear stress, τ, is larger than the so-called "Bingham limit," the motion of a system of oscillating blocks may be presented as viscous motion with an effective limiting kinematic viscosity,

$$\nu_{lim} = c_{af}\, h^2/T,$$

where h is the characteristic block size, T is the period of oscillations, and c_{af} is a numerical coefficient in the range from 4 to 8 (depending on internal model assumptions). Another important parameter of the "block" model is the decay time of the block oscillations, T_{dec}. This parameter is closely connected to the quality factor, Q, which is, by definition, the ratio of the energy stored per cycle to the energy lost over the same period:

$$Q = T/T_{dec}.$$

In the numerical modeling for each "Lagrangian" particle, the initial vibrational amplitude was recorded as a fraction of the maximum particle velocity in the stress wave. This value decays exponentially with the characteristic time T_{dec}. Hence, in the initial stages of transient cavity growth and modification, fragmented rocks around the crater behave as a viscous fluid (with a viscosity ν_{lim}). At later times, the assumed oscillations decay and fragmented rocks behave as a cohesionless medium with a dry friction coefficient. For more details of the model, refer to Melosh and Ivanov (1999), Ivanov and Artemieva (2002), and Wünnemann and Ivanov (2003).

The parametric study of the acoustic fluidization (AF) model and the scaling of model parameters for a wide range of crater diameters have been discussed by Wünnemann and Ivanov (2003). From this study (using Rock 1), we derived the values of $\nu_{lim} = 1.6 \cdot 10^4$ m²/s and $T_{dec} = 16$ s, which are suitable for the case of Upheaval Dome with a diameter of $D_c \sim 7$–8 km. This set of model parameters reproduces the desired crater morphology of a complex crater with a central uplift. For model runs with a simulated porous sandstone (SS-1, SS-2a, and SS-2b), we repeated the parameter fit for the AF parameters to reproduce approximately the same crater rim diameter: for sandstones with a friction lower than Rock-1, a larger crater results (at constant projectile conditions). After ~20 model runs for porous sandstones, the following parameters have been found: a projectile of Rock-1 properties has a diameter of 400 m and $\nu_{lim} = 1.0 \cdot 10^4$ m²/s and $T_{dec} = 10$ s. This compares with a 500 m impactor (of Rock 1) encountering a target entirely composed of Rock 1. These parameters and projectiles result in approximately the same rim diameter (~7 km) and general crater morphology. However, the central uplift is larger in the sandstone compared to the Rock-1 target.

RESULTS OF THE FIELD STUDY

Large-Scale Structure

Strata Layering and Thickness Variations

The geological map (Fig. 1), adopted from Kriens et al. (1999) for the outer regions, displays a circular crater rim monocline and crater ring syncline axis. Between both lines, all strata gently dip inward into the crater (Fig. 5). A thinning of the Kayenta Formation and the Wingate Sandstone (Shoemaker and Herkenhoff, 1984) can be observed in the cliffs of the Upheaval and Trail Canyon and at Holeman Spring in the South of the structure. This thinning is attributed to low-angle normal faults that truncate these formations and along which the hanging wall units have been displaced toward the center (Fig. 6). Their occurrence roughly coincides with the ring-monocline. Figure 7 shows elevation data of several strata boundaries as a function of distance to the center. These data were collected from the digital topographic and geological map in the western sector of the structure. This sector was chosen because the Upheaval Canyon transects the majority of the stratigraphic units. The scattering of the data is caused by (1) varying topographic heights at different localities; (2) mismatches between the DTM and the digitized geological signatures (i.e., unit boundaries); and (3) increasing difficulties in assessing a precise elevation value when approaching steep slopes. Therefore, a primary statistical data reduction was needed, by averaging the values in 200 m bins of radial distance from the impact center and calculating the standard deviations for each bin. To smooth the curve, the so-called "sliding average" procedure was applied. The layer thickness was calculated for those bins where an upper and lower strata boundary is recorded. The accuracy of the thickness values is estimated as the square root of the sum of the standard deviations for the elevation data.

The graph (Fig. 7) displays the stratigraphic uplift or subsidence at any distance from the crater center. The stratigraphic

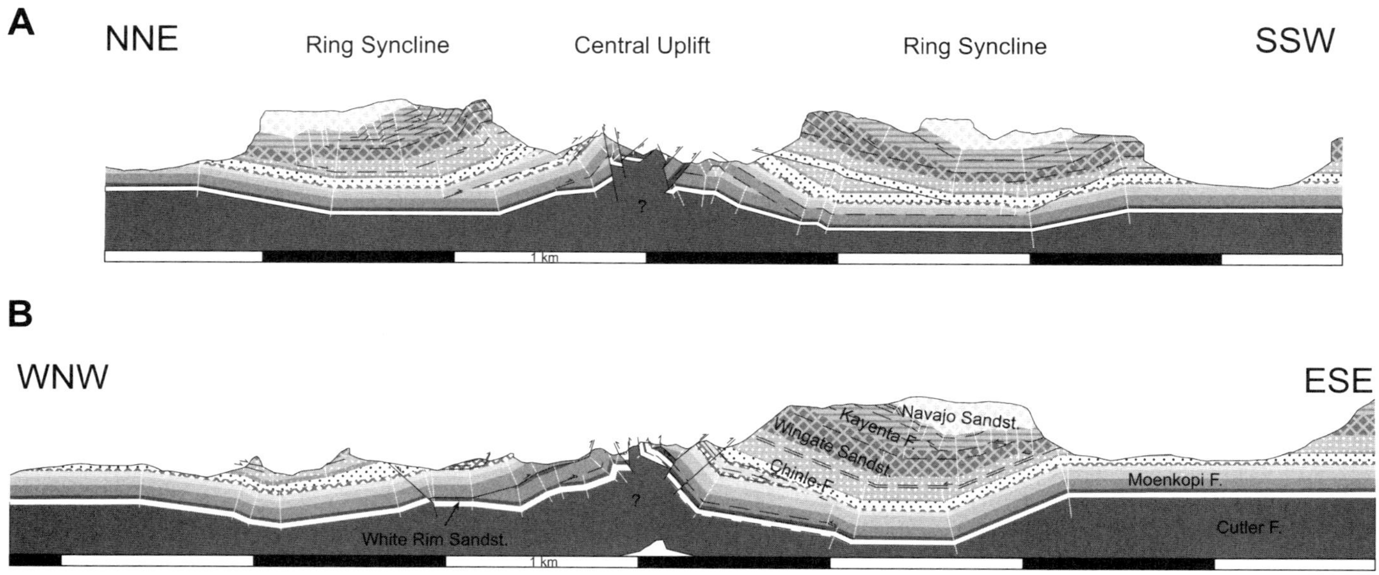

Figure 5. Two cross sections across Upheaval Dome, trending (A) north-northeast–south-southwest and (B) west-northwest–east-southeast. The profiles are entirely based on the geological map by Kriens et al. (1999) and were constructed before the field campaign. Note the modest inconsistencies with the new high-resolution map of Figure 8. Both sections show the prominent crater features: crater rim monocline, ring syncline, and a central uplift.

Figure 6. Photograph of a low-angle normal fault at the southern rim of the crater. The fault is exposed along its strike direction and cuts through the Kayenta Formation and the Wingate Sandstone. The former is downfaulted; the latter appears thinned. The uppermost unit is the Navajo Sandstone residing along the ring syncline. The view is toward the north (Holeman spring).

subsidence within the syncline is on the order of 50 m with respect to the elevation outside the crater for different layers of the Chinle Formation but increases to ~150 m for the Wingate-Kayenta stratigraphic boundary layer. This means that the syncline becomes an increasingly prominent feature from footwall to hanging wall units. Further inward, past the ring-syncline, the layers dip outward away from the center. Within the Kayenta Formation, there is large scatter in the dip of individual strata, with locally steep dip values (Fig. 4). This scatter can be attributed to intraformational faults and folds that account for a thickening of the entire sequence. Shearing occurred along ramps and bedding-parallel planes (flats) and led to associated folding (Fig. 5). The style of thickening within Wingate Sandstone is less clear, due to its massive nature, the lack of easily recognized sedimentary structures, and its difficult accessibility. The stratigraphic uplift at 900 m distance from the center is about ~100 m for the Chinle Formation and slightly higher for the Wingate Sandstone.

The stratigraphic uplift within the center of the dome is measured for the White Rim Sandstone to be ~250 m.

At a distance of 800–1500 m from the crater center along the exposures of the Kayenta Formation, the dip map shows low dip-angle sectors in the west-northwest and east-southeast (Fig. 4). Even though a canyon cuts through the structure and exposes a deeper level in the west-northwestern sector, which

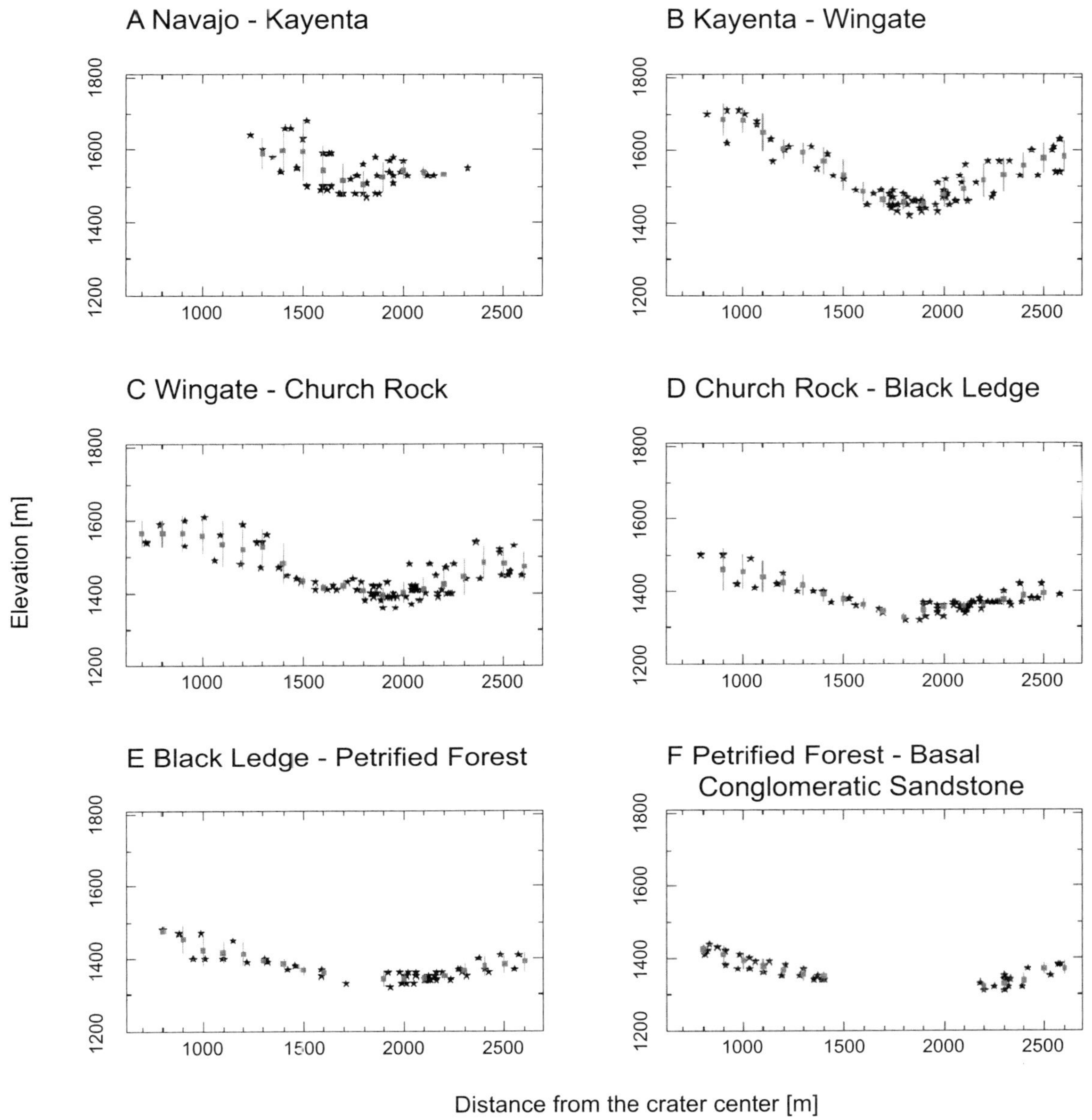

Figure 7. (A–F) Elevation data of different stratigraphic boundaries of the western sector of the structure as a function of radial distance from the crater's center. Mean elevations and standard deviations were calculated for 200 m horizontal bins.

makes a comparison difficult, this effect might not be attributed to topographic height because it is also seen in the east-southeastern sector.

While the layers in the external areas of the central uplift generally dip outward and strike concentrically, this does not hold for the core of the central uplift (within the central morphological depression) (Figs. 8 and 9). Dipping of the strata often changes on meter scales due to mesoscale folding and faulting, which outline and define tilted blocks and folded areas (Fig. 10). The structure maps of Figure 11 display the orientation and dip of a single marker bed (the lithological boundary of the Upper and Lower Alibaba member, Moenkopi Formation) in the inner part of the central uplift. For the layer construction and extrapolation to heights above and beneath the present topography (Fig. 11C), several geometrical constraints were considered, such as the estimated thicknesses of hanging wall and footwall strata, the fault geometry, etc. Also, the larger the distance between the constructed marker bed and the present topography, the larger any possible error. The maps show that strata dip and strike is controlled by large faults that separate different thrust slices. Despite

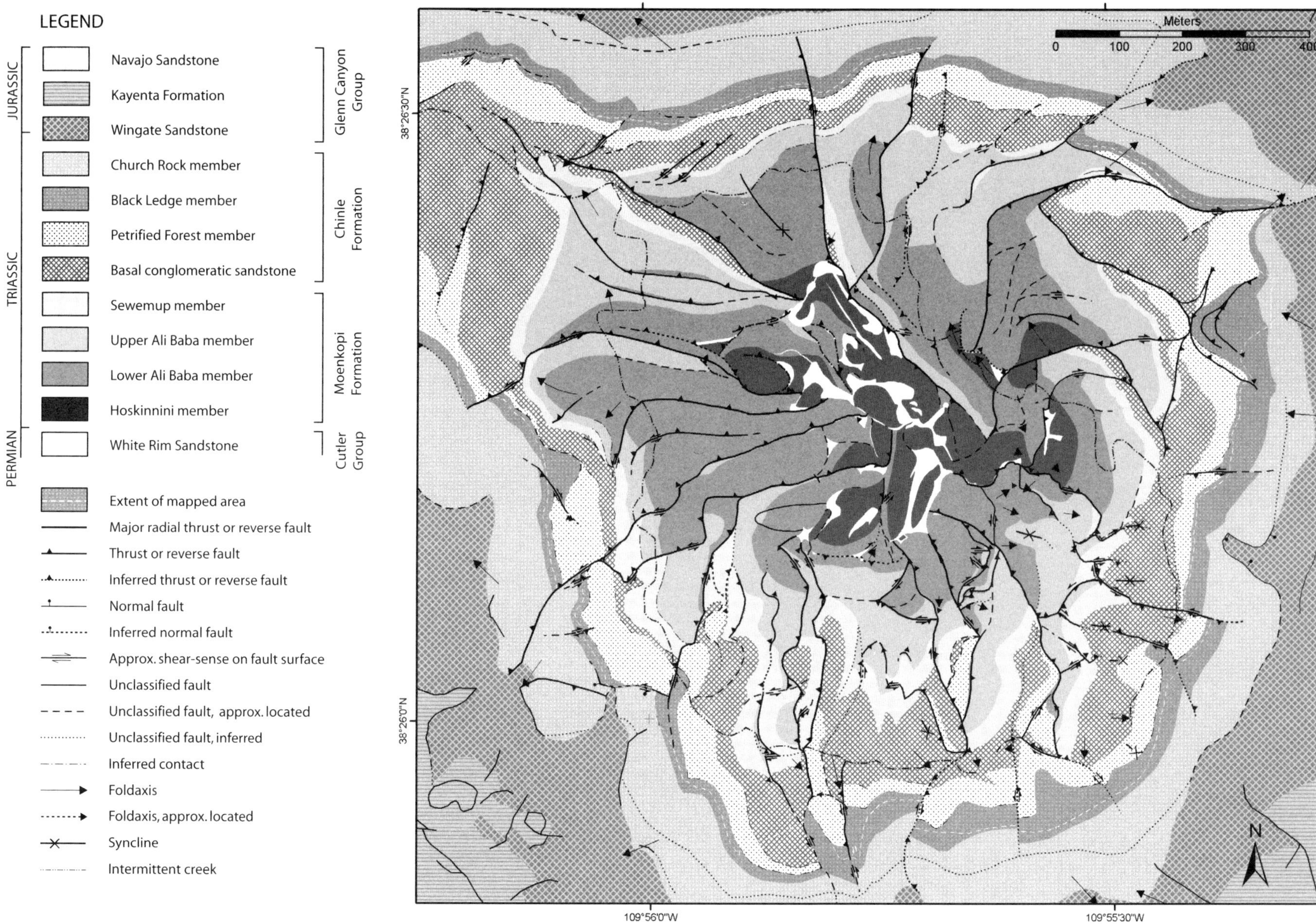

Figure 8. New, high resolution geological map of the central part of the Upheaval Dome impact structure, Canyonlands National Park, Utah, United States (this work). Structural and geological data outside the mapped area are taken from the map by Kriens et al. (1999).

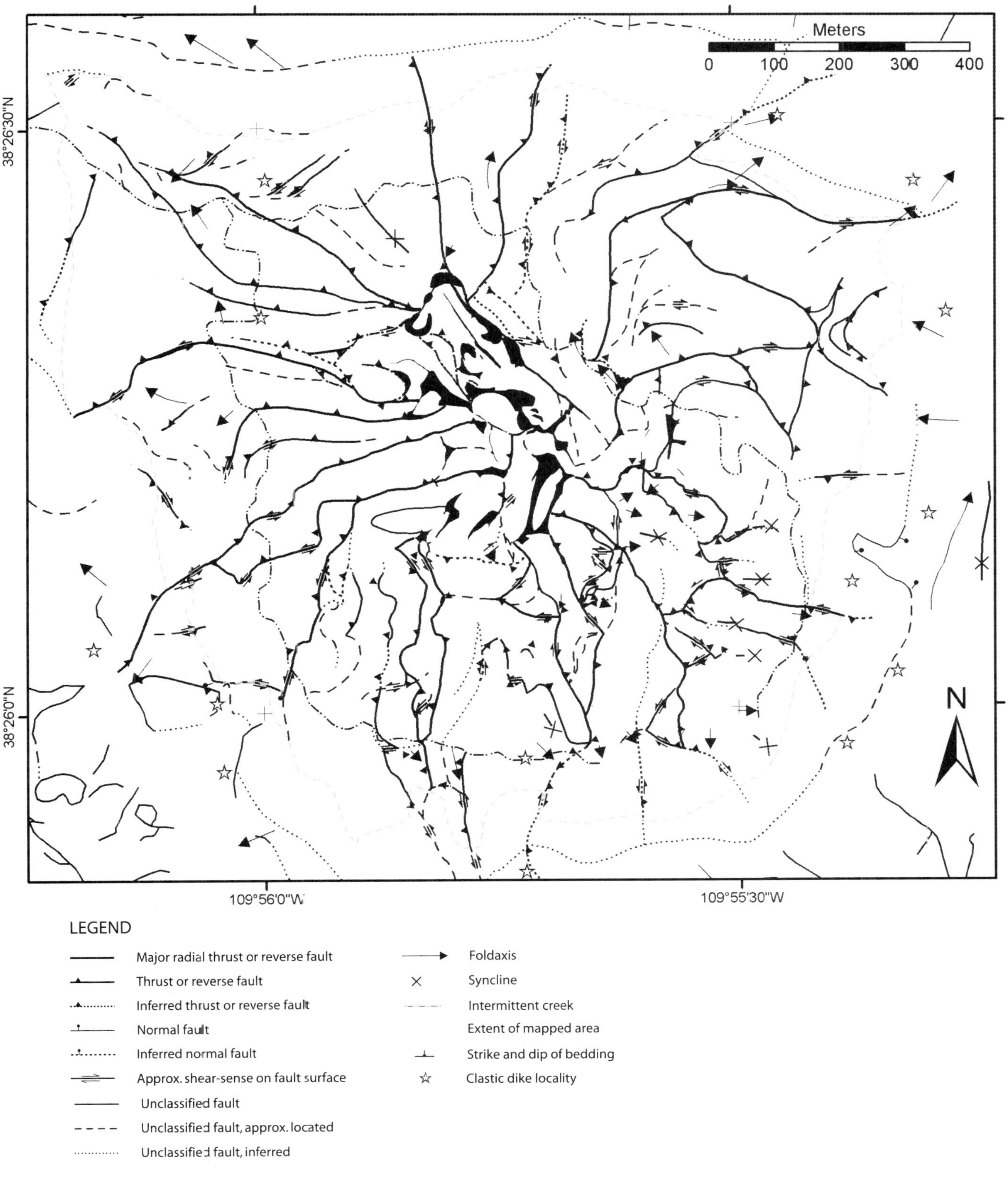

Figure 9. Tectonic map of the central part of the Upheaval Dome impact structure, Canyonlands National Park, Utah.

Figure 10. Three-dimensional visualization of the major fault zones of Figure 9 using 3D-Move (see text).

considerable scatter, a transition from substantially concentric strikes to radial strikes can be observed with decreasing distance from the center within several slices of the inner dome (e.g., Fig. 11A). This rotation is best exposed in rocks of the Moenkopi Formation (Sewemup to Hoskinnini members) (Fig. 9). These radially oriented strata show a tendency to trend parallel to the faults and an increasing dip toward the center (Fig. 4).

Faulting

The readily visible, and thus mappable, faults in the Kayenta Formation tend to show a concentric strike pattern (Kriens et al., 1999), while those of the inner part of the dome commonly trend in a radial fashion (Figs. 1, 8–10). This major change in fault orientation is not a function of horizontal distance from the center alone. One must also take into account that the structure is eroded to a deeper level in the center. While concentric shortening, occurring on reverse faults or radial folds, is localized at the perimeter, the fault pattern gets more complex and diffuse toward the center. Restricting our view to the central dome (Figs. 8–10), where a large number of faults were mapped, the major ones are traceable from the margin of the mapped area toward the center, and they allow us to distinguish different blocks or slices that are bound by these faults. The term "block" is probably a misnomer in this respect because it suggests an internally rigid behavior and a complete isolation by faults. This is not the case in any of the "blocks." They are all internally deformed and may remain in contact with adjacent blocks by bridges of strongly deformed rocks. The three-dimensional visualization of the structure map (Fig. 11D) clearly demonstrates the distorted character of the different slices. The major faults with relative large offsets were also constructed and extrapolated three-dimensionally (Fig. 10). Although a radial arrangement of the faults is dominant, deviations from this orientation are frequent. Many faults show a change in strike, some displaying strong curvature, in particular in the northeastern sector and, to a lesser degree, in the west-southwestern sector. This strong bending seems to have its cause in the formation of downward displaced masses of Wingate Sandstone (see next section). Faults dip steeply to vertically, with changing dip azimuths along a single fault (Fig. 10). There is a tendency in both the southwestern and northeastern sector for the faults to dip

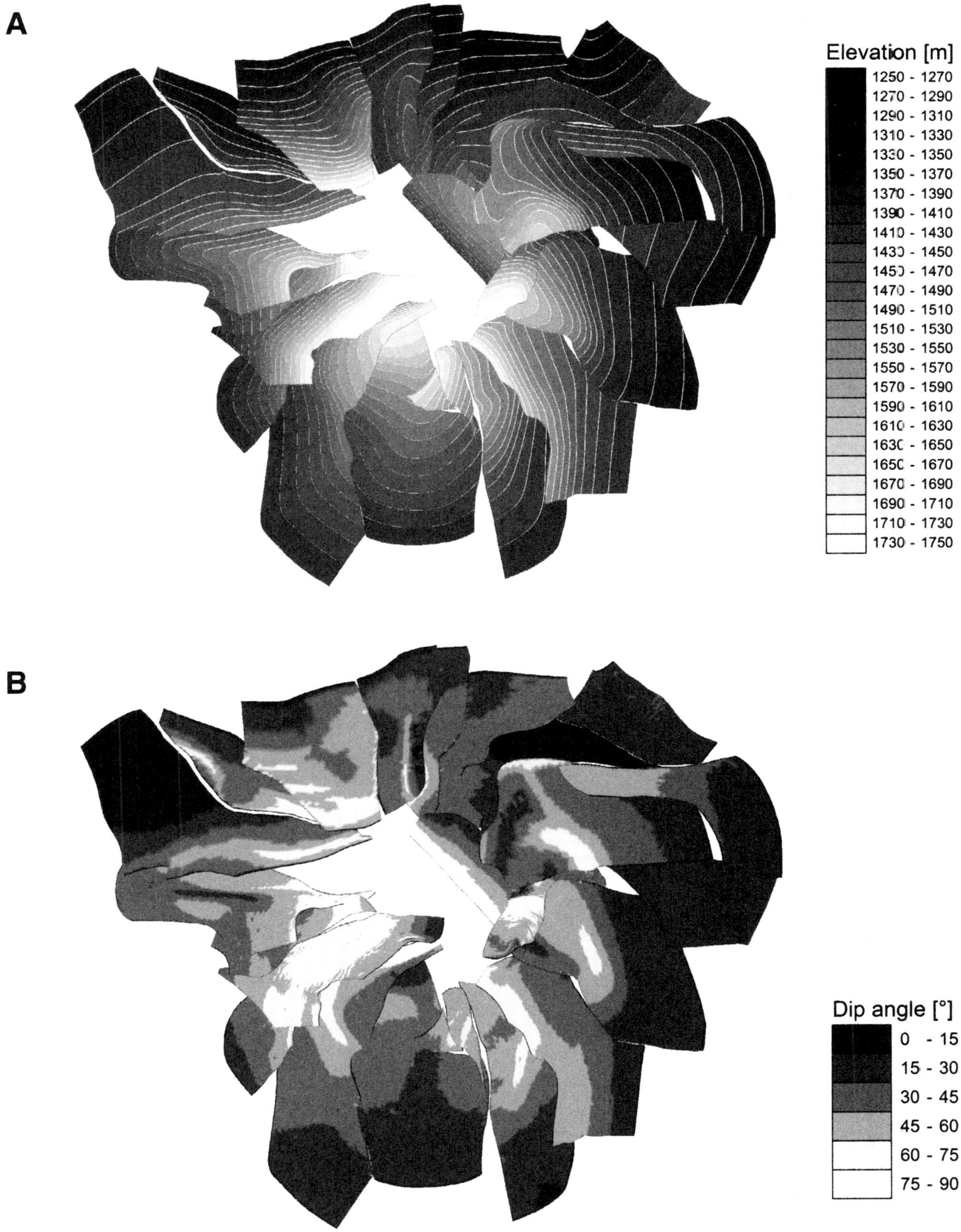

Figure 11 (*on this and following page*). Structure map of the strata boundary Lower/Upper Alibaba Member, Moenkopi Formation, for the central part of the Upheaval Dome impact crater. The chosen boundary represents a good marker horizon in the field. (A) Contour intervals and isolines indicate different elevations of the layer. (B) Contour intervals indicating the dip angle of the above contact.

Figure 11 (*continued*). (C) Three-dimensional visualization of the structure map using 3D-Move. Viewing direction is from the east-northeast. (D) Three-dimensional visualization of the structure map using 3D-Move with the viewing direction from the south-southwest.

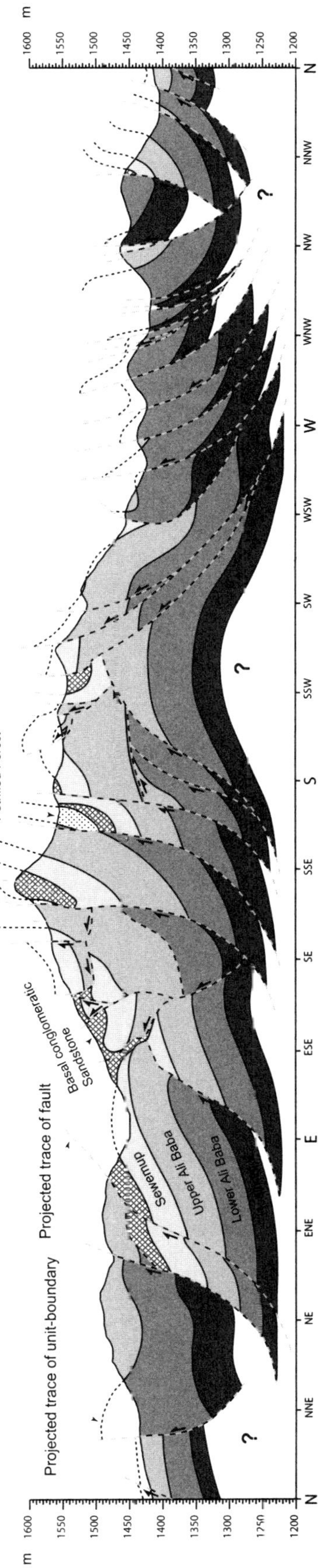

Figure 12. Circular cross section constructed along a perimeter of 350 m from the geometrical crater center at R593540/H4254640. The cross section shows the uniform stacking mode of reverse faults in the western and eastern sector, a synform at northwest, a complex faulted antiform at south-southwest, and a positive flower structure in the southeast. Due to dip of strata away from the center, layers appear thickened. For further explanation, refer to the text.

and have a convex shape, roughly toward the northwest (Figs. 9 and 10). The circular cross section (Fig. 12), qualitatively constructed along a perimeter of 350 m from the geometrical crater center (38°26′13.8″N; 109°55′42.4″W) (Fig. 1B), shows that these steep reverse faults form stacks of imbricated slices that dominantly indicate a transport from top-to-the-south to east, while north to west-dipping faults are less common (Figs. 8–10). The displacements along the faults increase toward the center and reach vertical offsets of 100–200 m (Fig. 10). The fault arrangement appears to show some symmetry on both sides of a large northwest-trending reverse fault that restricts the majority of the White Rim Sandstone occurrence to the southwest of it. This fault divides the center of the dome in a dominant and higher uplifted southwestern ridge that was thrusted onto a smaller northeastern ridge (Figs. 8 and 9). The offsets along the fault planes increase toward the center. Sometimes faults are seen to develop from radial striking folds in the periphery, eventually indicating a thrust direction by their convergence. A direct measurement of the slip vectors, using, for instance, striations on fault planes, was rarely possible. A rough determination of the shear sense and the magnitude of displacement was possible where lithological boundaries are offset. In some cases, an oblique slip vector with a dominant reverse and a subordinate strike-slip component can be assumed.

The tectonic situation in the southern and southeastern sector, which gives insights into a topographically higher level, is more complex (Scherler, 2004). Faults are interconnected and branch into sub-faults. The circular cross section (Fig. 12) suggests that branched faults merge into detachment faults at depth. Approaching the innermost part of the dome, many faults terminate at the contact to the White Rim Sandstone. The complex deformation pattern in the center of the dome, where many radial faults meet each other, required high strain deformation to accommodate incompatibilities of space. It was shown earlier that the White Rim Sandstone was capable of accumulating this strain by flowing like a granular medium during crater collapse and acting as a lubricant (Kenkmann, 2003).

Folding

Folds are frequent in the central part of Upheaval Dome. The most striking features are outward plunging open and symmetrical anticlines of radial strike that are best exposed along the base of the northern escarpment of the inner depression. They are formed within the Wingate Sandstone (Fig. 13) and sometimes extend to layered rocks of the Chinle Formation. These folds are described in detail by Kriens et al. (1999), who attributed their formation to concentric buckling induced by the convergent flow field during crater collapse. Synclines between the anticlines are wider and less pronounced.

In sharp contrast to the regular, concentric buckle folds are the irregular, "tongue-shaped" rock masses of Wingate Sandstone in the Church Rock and Black Ledge members of the Chinle Formation (Fig. 14). Their occurrence coincides with inclined slopes formed in Wingate Sandstone, which usually is a cliff-forming

Figure 13. Radial anticlines observed along the circular escarpment of the Wingate Sandstone. They are described by Kriens et al. (1999) and are believed to be formed by concentric buckling during convergent inward flow of rock.

unit, as seen at the northern flank of the inner depression. Along three prominent occurrences (northeast, east, southwest), several other folds are inferred. Since no marker beds exist in the Wingate Sandstone, it is difficult to assess if disharmonic folding is accommodating this downward displacement, or if other processes of material transfer occurred. These "tongues" are sites where several sandstone dikes emanate into the footwall units and they also act as nuclei for faulting in the footwall sequences (Scherler et al., 2003; Scherler, 2004). At the southwestern locality, normal faulting, seemingly induced by the downward displacement of Wingate rock masses in the hanging wall, postdates and overprints reverse faulting (Fig. 14B). The strong bending of the major fault in the northeastern sector and the splitting into two segments (Figs. 9 and 10), as well as a rotational movement of the large block just south of it, may also be caused by the nearby downward displacement in the Wingate Sandstone. These structures are absent along the northern Wingate cliffs. Above the eastern "tongue," a subhorizontal pair of a large syncline and anticline is trending north-south (concentric). The frequent folds in the Kayenta Formation (Kriens et al., 1999) have a generally smaller size and are often linked to nearby faults.

Folds in the inner part of the dome exhibit various morphological shapes (Fig. 15). Fold axes often plunge steeply, thus representing vertical and reclined folds. Folds become tight to isoclinal (Fig. 15A). They are typically meters to tens of meters in size, thus representing mesoscale deformation features associated with faulting. Fold axes are sometimes bent. While a tendency to a radial fold strike and outward plunge exists in the outer parts of the dome, this does not hold for the inner part. As previously mentioned, layers often turn from a concentric strike into a radial strike direction (Fig. 11). The radial distance at which this bending occurs varies and causes the formation of gently plunging, open synclines and anticlines.

There are several examples where a fold is exposed in the hanging wall unit of a radial fault with its fold axis parallel to the fault strike. These folds are interpreted to be formed during the propagation of the associated faults and successive slip on them. With increasing constriction and decreasing distance from the crater center, bent layers might evolve to folds which eventually develop into reverse faults. The proposed symmetric arrangement of structural elements is bound in the northwest by a likewise plunging syncline (Fig. 12). The large, northwest-trending ridge in the northern to eastern part of the dome (see section on faulting), is in fact a large anticline, with its shape modified by some radial faults.

Dikes

Clastic dikes of different lithologies occur throughout the structure but are concentrated in the center where the White Rim Sandstone, as the lowermost outcropping unit, forms a complex dike network (Kriens et al., 1999; Kenkmann, 2003). Seemingly, only those formations that consist of relatively porous lithologies are capable of forming dikes. The geometry and distribution of the White Rim Sandstone dike network within the central part of the Dome is extremely complex and was mapped in detail by Kenkmann (2003). Striking features of the White Rim Sandstone occurrences are extreme thickness variations, blind terminations, frequent embranchments at nodule-like points, and isolated occurrences. The White Rim Sandstone has almost completely lost its internal coherence during the emplacement process. The proximity of the White Rim Sandstone dike network to the overlying Hoskinnini member suggests relatively short transport and injection distances during dike formation.

Dikes unrelated to White Rim Sandstone are generally smaller and the determination of their protoliths is mainly based on their color. Protoliths are most often sandstones (Fig. 16). As

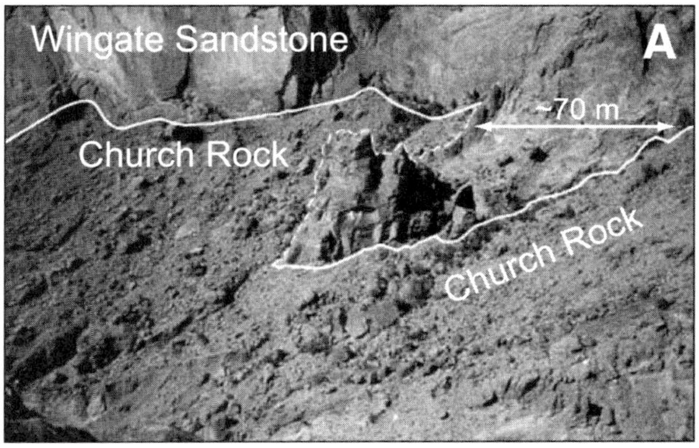

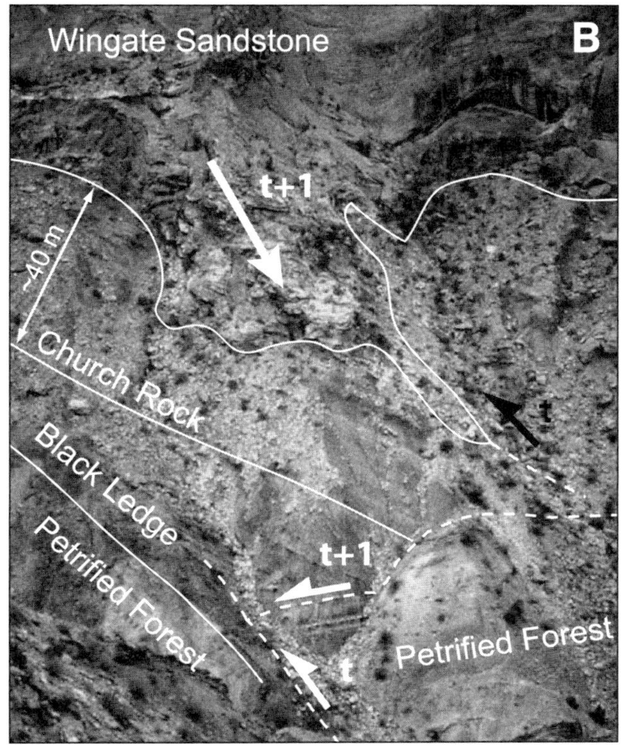

Figure 14. Irregular "tongue-shaped" rock masses of Wingate Sandstone in the Church Rock and Black Ledge members of the Chinle Formation. The downward displaced masses induced reverse faulting in the footwall units. (A) Northeastern sector. (B) Southwestern sector.

those, the Wingate Sandstone and sandy units of the Chinle and Moenkopi Formation are inferred source rocks. Kriens et al. (1999) also report several clastic dikes in the Kayenta Formation and Navajo Sandstone. Meter-sized dikes of Wingate Sandstone can be found in the footwall strata, in particular at those sites where the "tongue-shaped" rock masses are displaced into the Chinle Formation. Sandstone units in the Chinle Formation (Black Ledge Member, Basal conglomeratic sandstone member) and Moenkopi Formation (individual layers of the Upper and Lower Ali Baba member) were also, at least in part, mobilized. The occurrence of sandstone-dikes within fold hinges is observed in different localities and suggests a genetic link to the folding process (Figs. 15A, 16A). In many localities, where different blocks or slices meet each other, sandstones may display pronounced thickening, resulting in a massive appearance and the loss of sedimentary structures. This indicates a loss of coherence. In this context it should be mentioned that the "tongue-shaped" rock masses of the Wingate Sandstone in the underlying Chinle

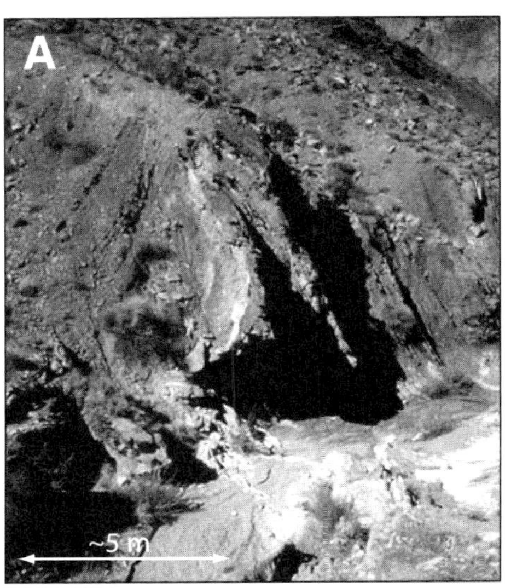

Figure 15. (A) Photograph of a tight anticline within the lower Alibaba member, Moenkopi Formation. The outcrop is exposed at the western creek and exhibits a partly fluidized sandstone in the hinge of the fold. Sandstone dikes emanate from the hinge. (B) Small-scale fault propagation fold formed within the basal conglomeratic sandstone, Chinle Formation.

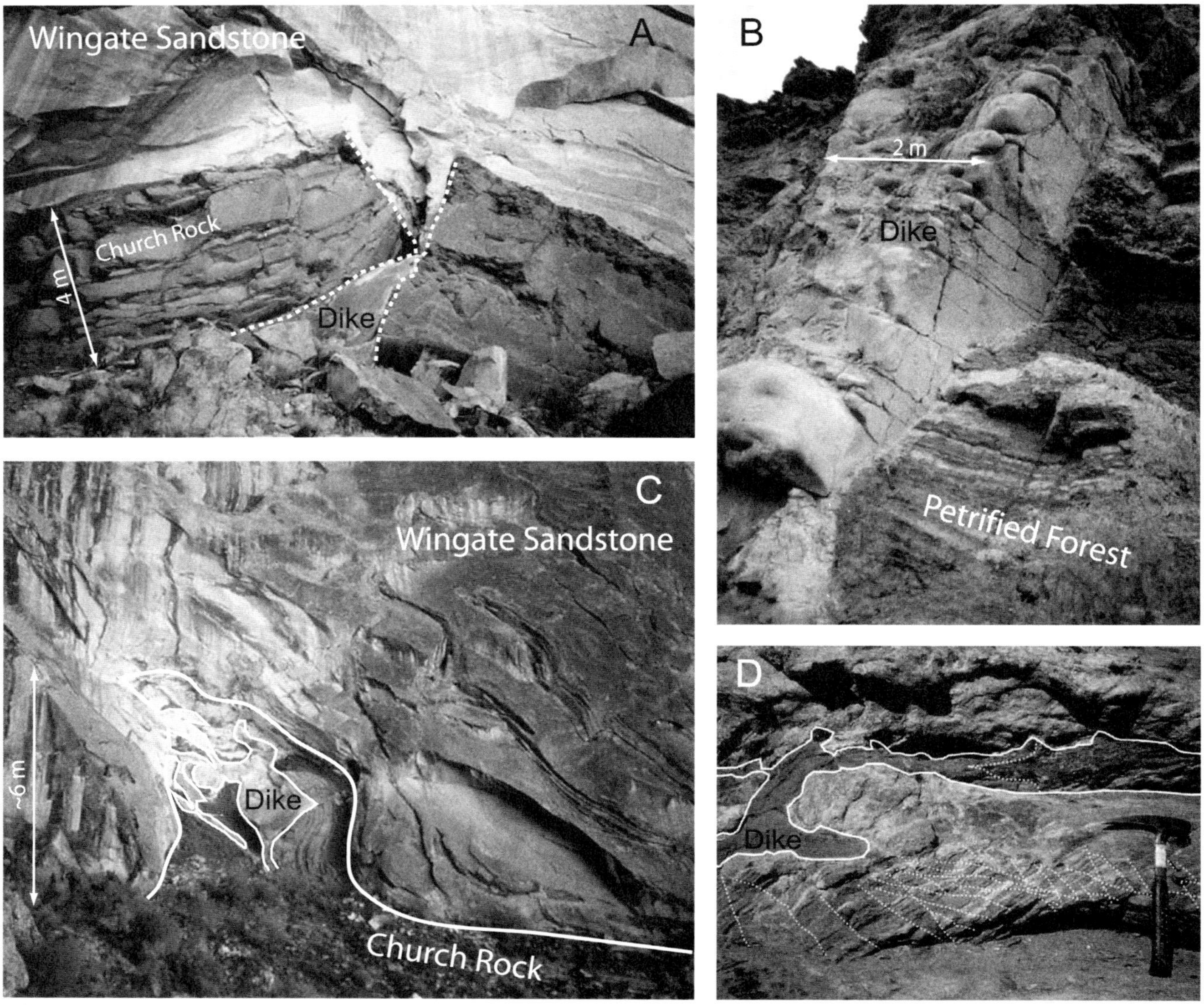

Figure 16. (A) Dike-like extrusion of fluidized sandstone in the hinge of an open fold. The photograph shows the lithological boundary between the massive Wingate Sandstone and the underlying stratified Church Rock member of the Chinle Formation. (B) A meter-thick sandstone dike that crosscuts the stratified sequence of Moenkopi Formation. (C) Deformed contact of massive Wingate Sandstone with the underlying Church Rock member of the Chinle Formation in the southwestern corner of the inner depression. A clastic Wingate dike resides in an isolated, pocket-like occurrence of Church Rock, the latter being incorporated into the hanging wall. To the right of the dike are steeply dipping, bent layers from the basal parts of the Wingate unit. A smaller fault to the right, as deduced from an offset of a silty layer (arrow), is not further traceable and exemplifies a localized, apparent ductile deformation of Wingate Sandstone. (D) A sandstone dike (which probably stems from the underlying Wingate Sandstone) intrudes into brittlely deformed Kayenta Formation.

Formation at the perimeter of the inner depression appear to have undergone some ductile deformation similar to the dike rocks.

Mesoscale and Microscale Structure

In principle, shear zones and fractures at Upheaval Dome developed on all scales of observation and may obey self-similarity rules, much like tectonic shear zones. Closely spaced fracturing is abundant in many parts of the structure and even though only small individual offsets (mm-cm) can be observed, their accumulation might result in significant strain on a macroscale. On a dm- to m-scale, intraformational thrusting in a ramp-flat-geometry (Fig. 15B) with associated folding is observed in stratified units such as the Kayenta Formation or the Chinle and Moenkopi formations.

A microstructural investigation of the White Rim Sandstone is given by Kenkmann (2003). Further microstructural analyses of other lithologies will be the subject of a separate paper and are

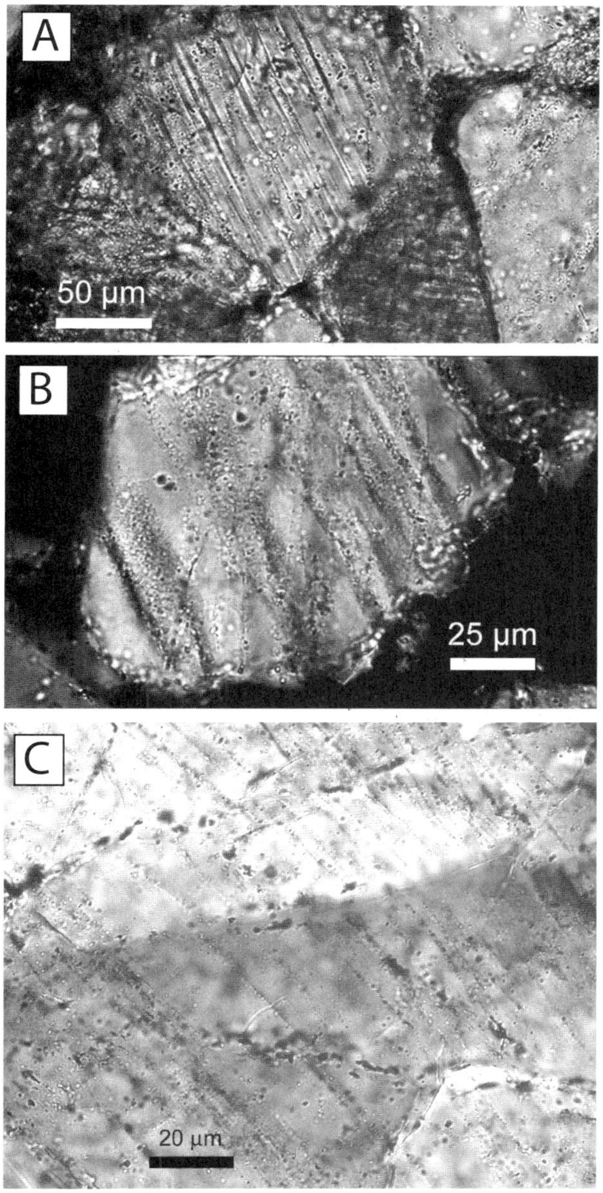

Figure 17. (A) Deformed quartz grain of the White Rim Sandstone displaying one set of parallel planar microstructures. (B) Two sets of sub-planar features: The vertical trending set is defined by trails of fluid inclusions; the second set, trending from the upper left to the lower right, represents deformation bands and is associated with undulose extinction. (C) Planar microstructures most commonly display microfractures lined up with fluid inclusions. Note that they are not perfectly parallel. A low-angle grain boundary is crossing the photograph from left to right.

beyond the scope of this contribution. However, to understand the macroscopic deformational behavior, some general notes regarding the micromechanical response of different lithologies are necessary. The exposed target lithologies are dominantly siliciclastic rocks. (An exception are carbonate-dominated layers of the Petrified Forest Member, Chinle Formation). Differences in the style of deformation occur most obviously between the massive, porous sandstones (Wingate and Navajo Sandstone) and the layered sequences of the Chinle, Moenkopi and Kayenta Formations.

Deformation within Sandstone Dikes and Highly Deformed Sandstones

The White Rim Sandstone and sandstone dikes of other lithologies display a massive appearance. Thin section observations reveal that deformation within these sandstone-dikes is pervasive and occurs on grain-scales. Deformation microstructures include reduced grain sizes, reduced porosities, the dominance of angular fragments, and an increased ratio of crushed and comminuted material relative to primary grains. All of this indicates grain crushing, breakup of grain boundaries, and collapse of pore space. Quartz grains exhibit abundant micro-fractures and planar microstructures (Fig. 17), as shown by Kenkmann (2003). Trails of fluid inclusions along planar microstructures and micro-fractures may suggest the presence of a fluid phase during the deformation event.

Another unit that exhibits strong internal deformation, which leads to an apparent ductile behavior, is the basal conglomeratic sandstone member of the Chinle Formation. As a prominent morphologic lithology, it forms large parts of outward-dipping slopes in the east and south of the inner depression. Abundant intraformational folds and faults (Fig. 15B) at unresolvable scales result in a sheet of rock that accommodated much higher strains than the neighboring thin to medium bedded strata.

Transitional stages between intact and fluidized sandstones also occur (Fig. 16C). Weakly to moderately deformed sandstones remain grain supported; the crushed matrix is of subordinate importance or forms isolated pockets between the grains. Macroscopic fluidization correlates with larger, interconnected volumes of comminuted matrix, progressively leading to an isolation of relic grains and a loss of internal coherence. The comminuted matrix is often cemented by calcite that seems unrelated to the impact.

Stratified Siliciclastic Rocks

Stratified siliciclastic rocks of the Kayenta, Chinle, and Moenkopi Formation can often display narrowly-spaced networks of microfaults. These are associated with localities of high strain that become more dominant toward the center. Micro-faults have variable length and widths and form networks of different geometries. The gouge zones of the micro-faults are characterized by reduced grain sizes and the dominance of angular grain fragments. The Church Rock sample in Figure 18 was collected from near the contact of the "tongue-shaped" Wingate rock mass in the E. The photographs show a tight fracture pattern of two sections oriented perpendicular to each other. A regular stair-stepping geometry of parallel microfaults with similar shear displacements and uniform shear sense is contrasted by a more lenticular and anastomosing, interconnected network (Fig. 18). The stair-stepping arrangement of micro-faults can potentially

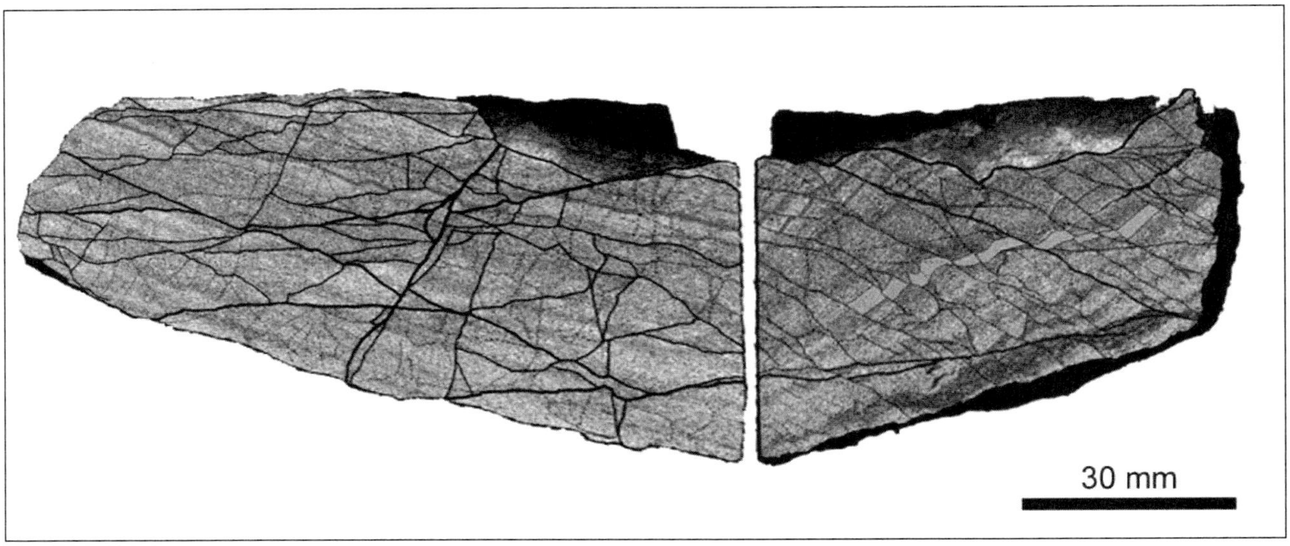

Figure 18. Microphotographs illustrating tight fracture patterns perpendicular to each other. A regular stair-stepping geometry of parallel micro-faults of similar shear displacements and uniform shear sense is contrasted with a more lensoid and anastomosing interconnected network.

accommodate the macroscopic dipping of individual strata without rotation of individual layers, which is often observed in craters. Small-scale folding of granular rocks is generally accommodated by brittle deformation and localized slip on complex shear zone networks. The collective deformation on the fracture networks causes the apparent flexural flow with limb tightening and fold-axis thickening. Peak strains within fold hinges can cause the dike-like extrusion of material (Fig. 16A).

RECONSTRUCTION OF THE IMPACT CRATER

The part of Upheaval Dome that is presently exposed is merely the relic of a larger impact structure (Shoemaker and Herkenhoff, 1984) that was heavily affected by erosion as a consequence of regional uplift of the Colorado Plateau (Pederson et al., 2002). To deduce the original size of the crater, the amount of erosion, and thus its most likely stratigraphic age, we have tried to superpose a standard numerical model of a typical impact crater onto the preserved part of the structure. Details of the model were given earlier. The model allows recording of the magnitude of the initial shock compression and the ensuing rock displacement. The final position of initially horizontal strata can be compared and fitted to the above field observations.

The starting model for a generic competent rock target (Rock-1) is that of a vertical impact of a 0.5 km diameter stony (Rock-1) projectile with a velocity of 12 km s^{-1} that results in the formation of a transient crater with a maximum depth of ~1.4 km. The transient crater collapse leads to a final crater with a rim-crest diameter of 7.5–8.0 km and a diameter at the present surface level of ~5.8 km. Modeling for simulated porous sandstones produces similar craters with a smaller projectile (diameter of 0.4 km for the same projectile material and impact velocity). Lagrangian tracers embedded in the initially horizontal target layers allow us to visualize the layer distortion beneath the crater and to compare the model with data from the DTM (Fig. 19). Three lines of fit between the numerical model and DTM were made: (1) Pressure isobars of the numerical model were compared with pressure estimates inferred from the observed deformation microstructures and their experimental equivalents. Pressure isobars are then used to estimate a minimum depth from the initial target surface of the exposed stratum during the impact. (2) The measured elevation and thickness of strata at certain distances from the crater center were compared with the position and thickness of artificial layers within the model to provide the best match with regard to depth. (3) Other structural features like the general dipping of strata or the position of the syncline axis were compared with the model to find the depth with the best match.

1. Definitive shock features are lacking at Upheaval Dome. Kenkmann (2003) found that the presently exposed White Rim Sandstone has suffered an effective confining pressure of 250 MPa $\ll \sigma_n - p_f < 3–5$ GPa during deformation, where σ_n is the applied stress and p_f is the pore fluid pressure. This estimate is based on a comparison of the deformation microstructure of the White Rim Sandstone with a very similar, experimentally deformed sandstone. Thus, the shock wave attenuated to magnitudes below the shock limit necessary to form PDFs in quartz (Stöffler and Langenhorst, 1994), but was still well above any possible lithostatic pressure of this region. The model results shown in Figure 19 display remarkable differences regarding the position of the shock isobars. The lack of shock features constrains the minimum amount of erosion d to

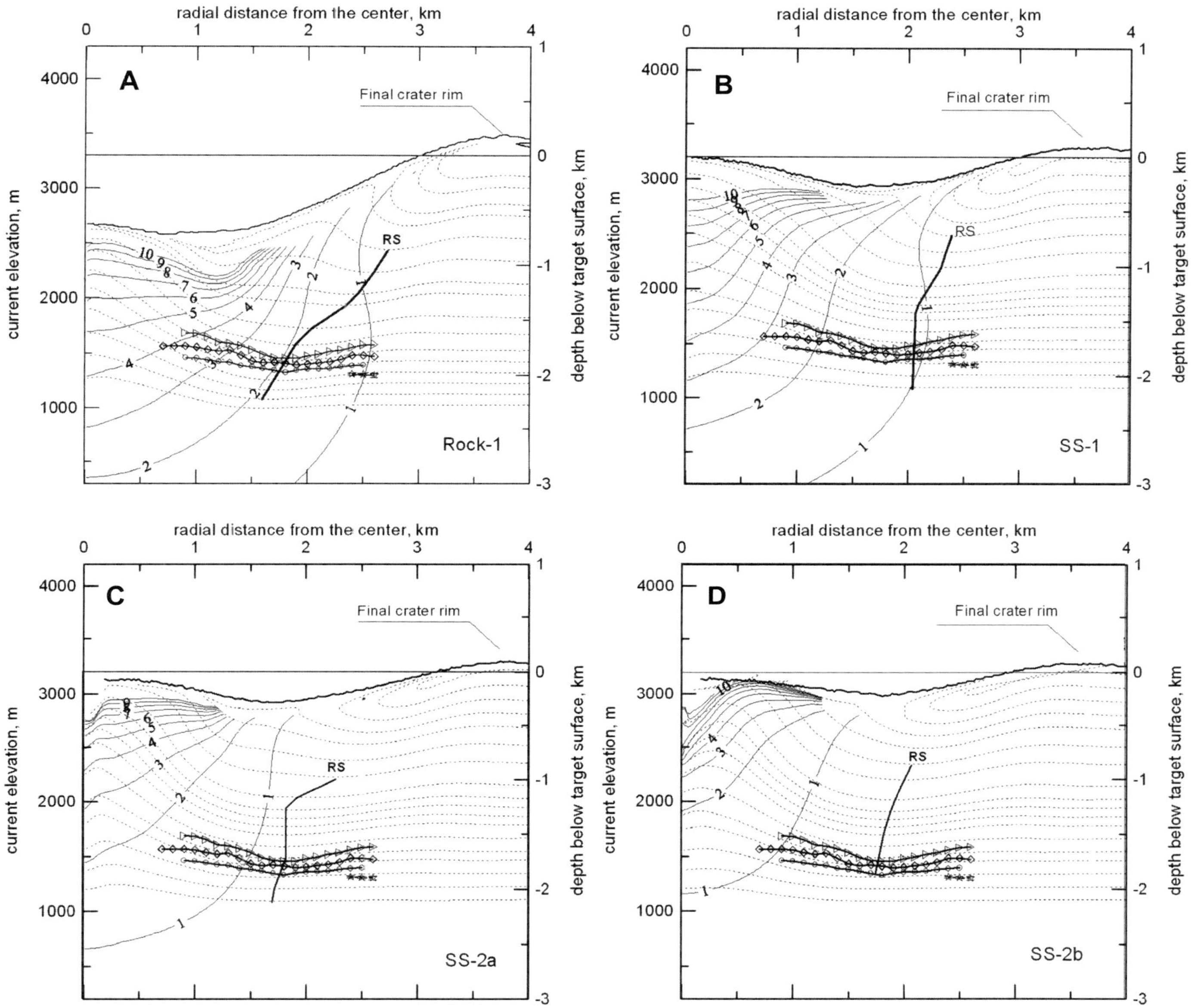

Figure 19. Superposition of the numerical model with digital terrain model (DTM) data for several strata. A similarity between model (dashed lines) and DTM data (triangles—transition Wingate Sandstone–Kayenta Formation; diamonds—transition Church Rock Member–Wingate Sandstone; circles—transition Black Ledge Member–Church Rock Member; stars: transition Black Ledge Member–Petrified Forest Member) for dipping, evolution of thickness, and specifically the position of the syncline axis is found at the shown depth. Thin curves with numerical labels 1–10 show final position of isobars (in GPa). The thick curve labeled RS shows the position of the ring syncline in the model. All variants are for vertical impacts at 12 km s^{-1} of a meteoroid presented as Rock-1, with Tillotson equation of state (EOS) parameters listed in Table 1 (density of 2700 kg m^{-3}). The target materials are described by the Tillotson EOS for the following materials: (A) Rock-1 (same as the projectile); (B) sandstone with density 2460 kg m^{-3} (SS-1); (C) sandstone with density 2020 kg m^{-3} and an initial Gruneisen coefficient of 0.8 (SS-2a); (D) the same as SS-2a but with a low initial Gruneisen coefficient of 0.1 (SS-2b). For all cases, crater size and the final layer geometry are similar. The main difference is the decay of shock pressure with distance from the point of impact. The range of maximum shock pressure in exposed rocks varies from 1 to 4 GPa in (A) to less than 1 GPa in (D) and is the main effect of various equations of state.

~1 km according to model SS-2 (Fig. 19C and 19D) and to 1.5 and 2.0 km according to models SS-1 and Rock-1, respectively (Fig. 19A and 19B). Assuming an oblique impact such as 45°, specific pressure isobars would shift to higher stratigraphic levels (Pierazzo and Melosh, 1999).

2. A change in thickness of Triassic and Jurassic strata was documented by several investigators (Shoemaker and Herkenhoff, 1984; Jackson et al., 1998; Kriens et al., 1999; Kenkmann and Scherler, 2002) (Figs. 5 and 7). The numerical models also show a thinning and thickening of marker beds. The best fit regarding changes in layer thicknesses is found to occur when the Wingate-Kayenta boundary (W-K) is placed at a depth of ~1.7 km. This corresponds to a burial of the White Rim Sandstone of 2.0 km below the impact surface (Fig. 19).

3. In both model and observed cross sections (Figs. 5, 7, and 19), the distance of the syncline axis to the crater center decreases with increasing depth. A best fit between the syncline axis of the numerical model and the actual cross section is found by assuming a depth of ~2 km for the White Rim Sandstone below the initial target surface. The same depth can be derived by comparing the observed dip of strata with those produced by the model. The different models presented in Figure 19 show similar geometries of the marker beds. However, model SS-2b (Fig. 19D) shows the best match with the field data.

Combining the different fitting approaches between model and observation (Figs. 19), the best fit is found when the White Rim Sandstone is buried at a depth of ~2000 m from the target surface. Considering the stratigraphy of the Colorado Plateau in the Canyonlands region, this would place the impact into upper Cretaceous times, during deposition of the Mancos shales. The initial diameter of the Upheaval Dome impact was ~7–8 km.

The match between numerical model and structural data shows a reasonably good correlation, yet it is not sufficiently quantitative to determine precisely and uniquely the amount of erosion. Some precautionary notes regarding the model should be mentioned. The model is two-dimensional and considers an axially symmetric crater. Deviations from this axial symmetry are mentioned above. The mechanical properties within the model are uniform and no predefined zones of weakness were implemented. This means the model does not consider the important rheological differences of the layered sequences and the massive sandstone units. This distinction plays an important role, as not only the style of deformation but also the deformation mechanisms may change and lead to deviations from rather simple geometries. Involving a thin veneer of weak or fluidized material can seriously affect the final geometry of the entire model crater, as test runs have shown. Further restrictions come from the partly constructed Tillotson EOS as described earlier and in the appendix. However, the different fitting approaches and estimates of the amount of erosion clearly show that a large amount of erosion must have occurred since the impact, which prohibits that Upheaval Dome formed during the deposition of the Jurassic Carmel Formation that rests on top of the Navajo Sandstone as proposed by Alvarez et al. (1998).

DISCUSSION

Accommodation Mechanisms of Convergent Flow

A crater collapse causes the involved rocks to move on ideally convergent trajectories. Upheaval Dome provides an opportunity to study different volume-conserving modes of deformation as a function of lithology and distance from the crater center. Bulk thinning and thickening is an important accommodation mechanism for most strata and appears to be most effective within the ring-syncline. In fact, the ring-syncline might be the result of thinning further outward and a thickening further inward. The formation of the monocline could be associated with (1) the sagging and subsidence of rocks induced by the thinning of strata and accumulation of rocks within the central uplift, and (2) drag folding of the strata during shear displacement along the normal faults. Concentric buckling as a consequence of inward motion is particularly effective along the periphery of the central uplift where constrictive strain is still moderate. The fold width and amplitude should change as a function of the thickness of the involved strata and of the contrast in competency to the neighboring strata. When concentric buckling cannot fully accommodate the inward motions, radial reverse faults start to develop and will extend in area and displacement toward the center.

Kenkmann and Scherler (2002) described the thickening and thinning of strata with a power law function of the form:

$$h_f/h_i = A\,R_f^{-b},$$

with h_f being the thickness of a layer at a radius R_f from the impact center, h_i being the initial layer thickness, and A and b being empirical parameters that can be determined in radial geologic cross sections. The volume V of each stratigraphic unit in a sector α is given by

$$V = \alpha\,\pi/360° \int A^{2/b}/h^{2/b}\,dh.$$

Solving the integral in the boundaries of h_f and h_i, one obtains:

$$V = \alpha\,\pi/360°\,A^{2/b}\,[\{h_f^{-2/b+1}/(-2/b+1)\} - \{h_i^{-2/b+1}/(-2/b+1)\}].$$

Calculating volumetrically equivalent cylinder segments with the initial thickness h_i and radius R_i, the amount of inward displacement $R_i - R_f$ can be estimated, which is necessary to balance the volume. Using this procedure, the amount of inward motion for the Kayenta, Wingate, and Chinle Formations was estimated by Kenkmann and Scherler (2002), which ranges from a few decameters to several hundreds of meters from the footwall to the hanging wall units, respectively. This inward motion is partly compensated by inward shear along low-angle normal faults of the periphery that are connected with inward thrust and reverse faults surrounding the central uplift. These shear zones are particularly frequent in the layered Kayenta Formation, but seem to be absent in the massive Wingate Sandstone. For the latter, a

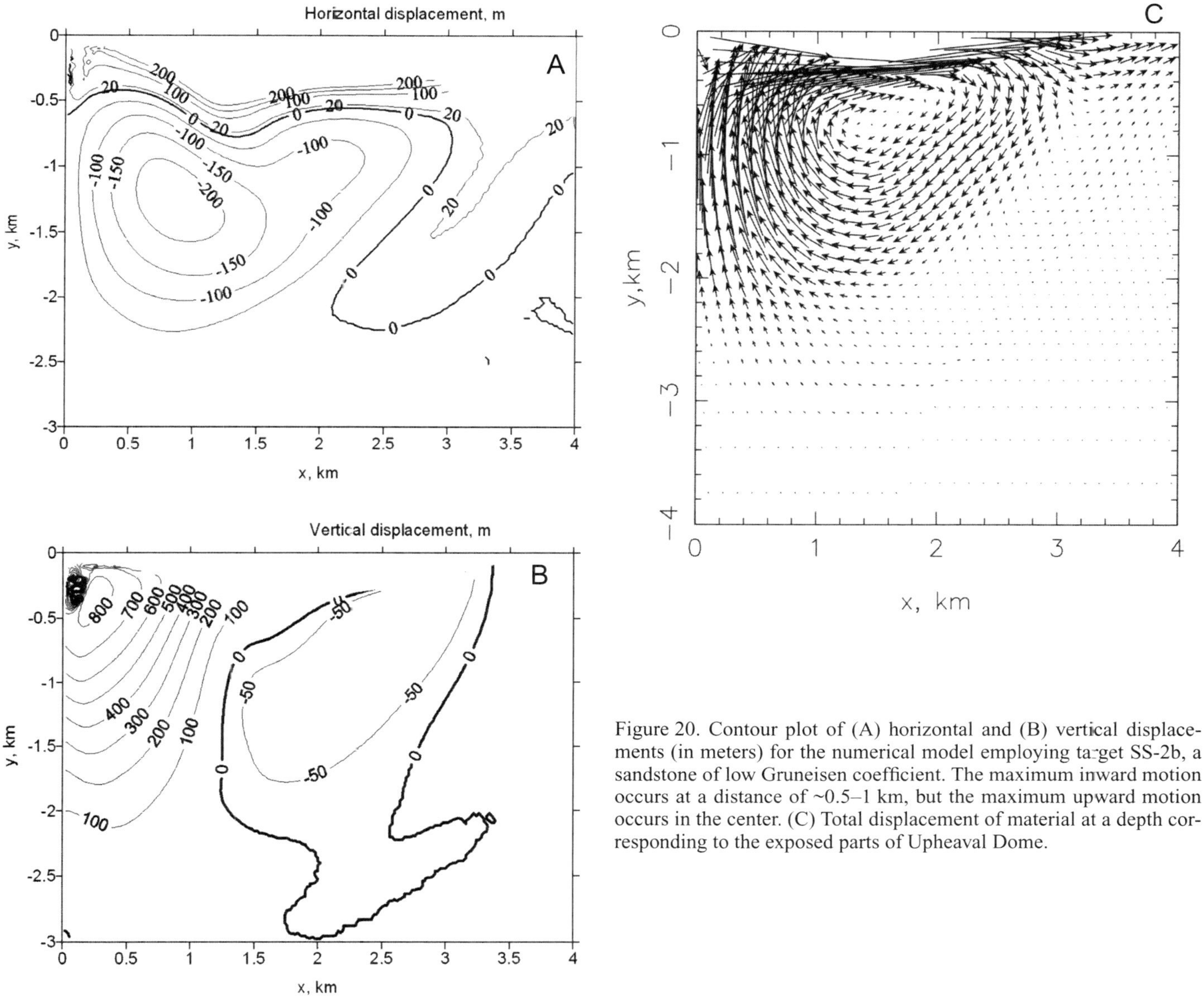

Figure 20. Contour plot of (A) horizontal and (B) vertical displacements (in meters) for the numerical model employing target SS-2b, a sandstone of low Gruneisen coefficient. The maximum inward motion occurs at a distance of ~0.5–1 km, but the maximum upward motion occurs in the center. (C) Total displacement of material at a depth corresponding to the exposed parts of Upheaval Dome.

more pervasive thinning and thickening must be assumed. However, the simple geometric model neither takes into account that individual strata are thinned during the growth of the transient cavity, nor considers volume changes due to dilatancy. A more thorough treatment of the volumetrically consistent kinematics during crater collapse is provided by the numerical model. Figure 20 shows the amount of (A) horizontal, (B) vertical, and (C) total displacement. The horizontal displacement reaches its maximum of ~200 m at a distance of 0.5–1 km from the crater center while the vertical displacement is largest in the center. At a depth corresponding to the present level of erosion at Upheaval Dome, the calculated amount of uplift is ~200 m (Fig. 20B), which is in good agreement to the measured stratigraphic uplift of the White Rim Sandstone.

Deformation Styles and Mechanisms

The attenuating shock wave and the subsequent deformations during the crater modification stage damaged the lithologies in different modes: rocks were either deformed by localized brittle faulting on widely different scales or by pervasive flow at scales of individual grains. The latter occurs within the frequent clastic dikes. Within the hinges of folds, where local excursions of strain and pressure were dynamically built up (Figs. 15A and 16A), material may be extruded, to give rise to dikes. This is achieved by the transition from shear localization to distributed cataclastic flow. The mechanism and degree of deformation found in the rocks varies considerably and appears to show a strong dependence on the setting within the collapsing crater and

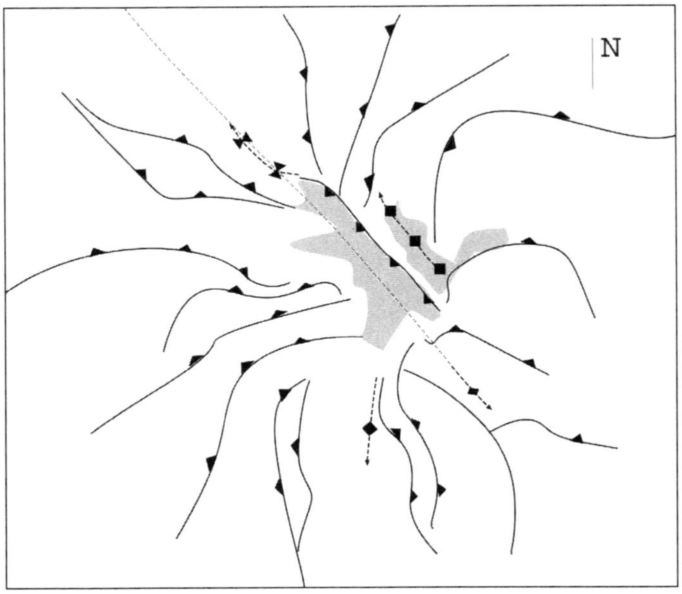

Figure 21. Sketch showing some important structural features of the inner part of the central uplift that suggest some bilateral symmetry.

the radial distance to the center. A rock unit in close proximity to a high strain locality (i.e., fault zone, dike, or fold hinge) will suffer strong deformation, which is represented as a closely-spaced fracture network or even as a clastic dike. These localities grow in number and distribution when approaching the center with its many interconnected faults that could not all be mapped quantitatively. A view at the schematic circular cross section (Fig. 12) shows, for the topographically lower parts in the west, that the offsets along the numerous clockwise dipping faults are quite small. Some of these faults are not present further away from the center, indicating activation by increasing constriction and a more homogeneous strain distribution toward the center. Figure 11 suggests that 100–200 m of vertical offset is common between the thrust slices near the center.

A primary cause of the distribution or the localization of deformation in different lithologies may be the presence or absence of preexisting pore space: when the attenuated shock wave approaches porous rocks, pore space may readily collapse due to extreme impedance contrasts between voids and competent grains, favoring pervasive grain crushing. In contrast, nonporous materials might be shattered to variable degrees and form discrete shear zones. The subsequent material flow induced by the shock wave causes a continuation of deformation at local ambient pressures along the fractured domains. Pervasively crushed rocks (e.g., initially porous sandstones) would most likely behave as an unconsolidated rock without cohesion and may flow like a granular medium during crater collapse. A prerequisite for granular flow is the development of transient dilatancies in the material. This is enabled during crater collapse by the circumstance that seismic energy is stored in the target and causes vibrations and fluctuating pressures. Nonporous target rocks, which did not gain a pervasive deformation overprint during shock propagation, reactivated discrete fractures and transformed them into disseminated networks that eventually coalesced into larger shear zones. Hence, large parts of the rock may remain unaffected by deformation.

Modes of Strength Degradation

Upheaval Dome is a structure that is highly suited to the investigation of the cause of the strong reduction of rock strength that must occur during crater collapse and central peak formation. Possible modes of strength degradation include (1) acoustic fluidization, (2) effects of fluids, (3) shock damage at microscales, (4) thermal softening, and (5) friction properties.

1. The recognition of opposite shear senses on parallel faults, which might be indicative of oscillating blocks, can be seen occasionally. However, anomalously thick fault gouges, which might develop by multidirectional fault activation, were not unambiguously identified. The presence of granular flow features within the dikes supports the theory of acoustic fluidization.

2. Transient fluid overpressures may be generated during shock propagation and are likely to favor the tensile failure of rocks. A large hydraulic fracture 20 km outside Upheaval Dome, called Roberts Rift (Huntoon and Shoemaker, 1995), is possibly linked to the impact event. Abundant fluid inclusions, decorating micro cracks in quartz of the White Rim Sandstone, also indicate that fluids were present during the impact event. Enhanced fluid pressures may also have favored the dike emplacement.

3. As mentioned before, the propagation of shock and rarefaction waves will damage target rocks and can influence their strength properties. Cracking results from tensile stress concentration around flaws during the compressive pulse and the tensile portion of the stress wave (Ahrens and Rubin, 1993), and is therefore particularly effective in porous lithologies. The intensity of fracturing depends on the loading rate and loading duration. According to Grady and Kipp (1987), the radial dependence of crack damage obeys $D \propto r^{-1.5}$. The resulting failure strength of the shock-damaged rock is only a fraction of the unshocked rock strength (He and Ahrens, 1993).

4. Shock heating as a prominent cause for strength reduction (thermal softening; O'Keefe and Ahrens, 1999) at Upheaval Dome can be ruled out. The investigated rocks do not show any microstructures that indicate elevated temperatures and thermally activated deformation processes. The apparent absence of pseudotachylites also excludes the possibility of softening due to localized frictional heating.

5. The onset of rock deformation during crater collapse is controlled by the frictional strength of sliding, rather than by the fracture strength of the shock damaged rock. Being a dynamic property, friction is state and rate dependent. Thus, friction can decrease and increase when fluctuations in normal stress occur during acoustic fluidization. Velocity weakening and slip weakening (Scholz, 1990) are further effects that can potentially reduce

the coefficient of friction and, hence, cause a degradation of the bulk strength of the target rocks. However, it is very difficult to prove this in the field. Scaling properties of faults (thickness versus length versus displacement of a fault) may provide the key to unravel this, but, unfortunately, this has not yet been done.

Asymmetry of the Upheaval Dome

The geometric arrangement of some structural features in the inner part of the Upheaval Dome suggests some bilateral rather than centro-symmetric relationships. They are given in the form of (Fig. 21): (1) stacks of imbricated slices with a dominant south-directed thrust component that are bound by an uplifted syncline in the northwest; and (2) the major exposures of the White Rim Sandstone and the Hoskinini Member of the Chinle Formation (the two lowermost units in the center of the dome) occur along a northwest-trending ridge that was thrusted onto a northwest-trending anticline (exposing less of these units) (gray shaded areas in Fig. 21). Asymmetry is further observed regarding the lack of downward displaced "tongue-shaped" rock masses of Wingate Sandstone at the northern escarpment and low dip-angle sectors in the west-northwest and east-southeast of the outer part of the dome.

In the southeastern parts of the structure, the situation is more complex, as it exposes a higher topographic level that shows no related symmetry. Rather, in the south one might find the other end of a symmetrical arrangement of imbricated slices that are bound by an anticline. It is not clear, though, how the faults evolve toward greater depth and whether a regular pattern of imbricated slices can be found or not. A similar, even more pronounced arrangement was found to occur at the Spider structure, Australia (Shoemaker and Shoemaker, 1996). This similarity indicates that the structural features and fault arrangement are not a specific but a general characteristic of complex impact craters of this size. The stack of slices with identical thrust directions was interpreted in terms of a remnant momentum that carried material downrange, thus leading to a preferred thrusting in one direction. Although this "remnant momentum" cannot be quantified, it provides a reasonable explanation for the preferred thrust direction. With respect to Upheaval Dome, this would mean that the projectile came from approximately the northwest. Based on numerical models on oblique impacts (Ivanov and Artemieva, 2002), it is known that a crater, while keeping a circular outline, shifts downrange as it grows and collapses. This, in turn, means that the collapse starts uprange and propagates downrange with time. Geologically speaking, this time sequence of collapse means that slices and slumps of collapsing material from the uprange sector predominantly fed the rising central uplift and lead to the formation of imbricated slices.

Alternatively, a crater asymmetry can be caused by heterogeneities of the target rocks (Melosh, 1989). Preexisting fractures can play a dominant role, as can be seen by the square plan-view of Meteor Crater, Arizona. There are two sets of fractures clearly visible in aerial photographs outside the Upheaval Dome. An east-southeast/west-northwest and a south-southeast/north-northwest trending fracture set bisects at an angle of ~45° and crosscuts the entire crater structure without showing any structural effect. Thus, these fractures clearly postdate the impact and are most likely related to the uplift of the Colorado Plateau.

CONCLUSIONS

Upheaval Dome is a deeply eroded impact crater with a presently exposed diameter of 5 km. A comparison and fitting of the exposed structure to a numerical cratering model suggests that the White Rim Sandstone, a distinct marker bed on the Colorado Plateau and in the center of the dome, was buried ~2000 m deep when the impact occurred. The structure was most likely formed during the deposition of the Jurassic Mancos Shales and had an original size of 7–8 km.

Upheaval Dome provides insights into different levels below the initial floor of a small to medium-sized complex impact crater within a sedimentary target and it can be used as a case study to understand the geometry and formation of central uplifts. Based on detailed mapping of the inner part of the dome, it was possible to reconstruct the internal structure of the central uplift in a certain depth range. Radial folds and a stacking of imbricated thrust slices are prominent deformation features of the central uplift and result from a constrictive strain pattern. Fault planes commonly dip steeply and are curved due to passive distortion after activation. The arrangement of structural elements in the inner part of the dome displays some crude bilateral symmetry, trending northwest. The dominance of northwest dipping reverse faults indicates a material transport top-to-the-southeast, which may be caused by an oblique impact. Depending on lithology, rocks were either deformed by localized brittle faulting on different scales or by pervasive flow at the scale of grains. The latter occurs as common clastic dikes and within fold hinges, where local excursions of strain and pressure were built up dynamically.

ACKNOWLEDGMENTS

This work was funded by the Deutsche Forschungsgemeinschaft (DFG) grant KE-732/6-1. The Alexander von Humboldt award for B.A. Ivanov enabled his collaboration with the German team. We are grateful to the United States Department of the Interior National Park Service for providing scientific research and collection permits. Many thanks to H.R. Knöfler for preparing thin sections. U. Riller provided access to 3D-Move. We are grateful to J. Plescia and T. Ahrens, who provided very constructive reviews, and to F. Hörz for editorial help.

APPENDIX

Tillotson EOS to Model the Sandstone Hugoniot at Low Shock Pressures

According to the Tillotson EOS in the compression region where density, ρ, is greater than the initial density, ρ_0, pressure, P, is a function of the density and specific internal energy, E:

$$P = A\mu + B\mu^2 + \rho E\left(a + \frac{b}{\frac{E}{E_o\eta^2}+1}\right) \quad \text{(A1)}$$

where $\eta = \rho/\rho_0 - 1$ and $\mu = \eta - 1$. To fit the Hugoniot to porous media, we add the cubic term to Equation (A1):

$$P = A\mu + B\mu^2 + C\mu^3 + \rho E\left(a + \frac{b}{\frac{E}{E_o\eta^2}+1}\right) \quad \text{(A2)}$$

Equation A2 provides a better fit to experimental Hugoniot curves than (A1). However, the fit of the Hugoniot does not guarantee a proper description of the thermal properties. We use equations (A1) and (A2) for porous media with the cautionary caveat that at low pressures (~50 MPa), which exists during the modification stage, the main role is material flow with friction forces. Additional parameters used in the model calculations are listed in Table 1.

We note the inherent problems of the preliminary modeling employed in this paper. First, the single equation of state for a porous medium is unable to quantitatively reproduce pore collapse in and phase transitions of silica. Second, the Tillotson EOS [equations (A1) and (A2)] cannot properly reproduce details of the increase in sound speed due to hydrostatic compression. Experiments show that for sandstone with an initial porosity of 18% (bulk density 2140 kg m^{-3}), the longitudinal seismic velocity increases from ~3.2 km s^{-1} at ambient pressure to ~4.3 km s^{-1} at 200 MPa (Johnston and Toksöz, 1980). At lower pressures of ~50 MPa, the Poisson ratio is increasing dramatically in many sandstones (e.g., Dvorkin et al., 1999). All of these effects in porous target rocks deserve much more detailed future study.

REFERENCES CITED

Ahrens, T.J., and Rubin, A.M., 1993, Impact-induced tensional failure in rock: Journal of Geophysical Research, v. 98, p. 1185–1203.

Allen, R.T., 1967, Equation of state of rocks and minerals: General Dynamics, Special Nuclear Effects Lab, Defense Atomic Support Agency, DA49-146-XZ-462.

Alvarez, W., Staley, E., O'Connor, D., and Chan, M.A., 1998, Synsedimentary deformation in the Jurassic of southeastern Utah: A case of impact shaking?: Geology, v. 26, p. 579–582, doi: 10.1130/0091-7613(1998)0262.3.CO;2.

Amsden, A.A., Ruppel, H.M., and Hirt, C.W., 1980, SALE: A Simplified ALE Computer Program for Fluid Flow at All Speeds: LA-8095 Report: Los Alamos, New Mexico, Los Alamos National Laboratories, 101 p.

Basilevsky, A.T., Ivanov, B.A., Florensky, K.P., Yakolev, O.I., Fel'dman, V.I., et al., 1983, Impact craters on the Moon and other planets: Moscow, Nauka Press, 200 p (in Russian, English technical translation, 1985, National Aeronautics and Space Administration, TM 77667).

Bucher, W.H., 1936, Cryptovolcanic structures in the United States: 16th International Geological Congress, Report Session 2, p. 1055–1084.

Bump, A.P., and Davis, G.H., 2003, Late Cretaceous–early Tertiary Laramide deformation of the northern Colorado Plateau, Utah and Colorado: Journal of Structural Geology, v. 25, p. 421–440, doi: 10.1016/S0191-8141(02)00033-0.

Cintala, M.J., and Grieve, R.A.F., 1998, Scaling impact melting and crater dimensions: Implications for the lunar cratering record: Meteoritics & Planetary Science, v. 33, p. 889–912.

Collins, G.S., Melosh, H.J., Morgan, J.V., and Warner, M.R., 2002, Hydrocode simulations of Chicxulub Crater collapse and peak-ring formation: Icarus, v. 157, p. 24–33, doi: 10.1006/ICAR.2002.6822.

Collins, G.S., Melosh, H.J., and Ivanov, B.A., 2004, Modeling damage and deformation in impact simulations: Meteoritics & Planetary Science, v. 39, p. 217–231.

Dickinson, W.R., Klute, M.A., Hayes, M.J., Janecke, S.U., Lundin, E.R., McKittrick, M.A., and Olivares, M.D., 1988, Paleogeographic and paleotectonic setting of Laramide sedimentary basins in the central Rocky Mountain region: Geological Society of America Bulletin, v. 100, p. 1023–1039, doi: 10.1130/0016-7606(1988)1002.3.CO;2.

Dvorkin, J., Mavko, G., and Nur, A., 1999, Overpressure detection from compressional- and shear-wave data: Geophysical Research Letters, v. 26, p. 3417–3420, doi: 10.1029/1999GL008382.

Ekholm, A.G., 1999, Crater features diagnostic of oblique impacts: the central peak offset: Houston, Texas, Lunar and Planetary Science Conference 30, Abstract 1706.

Grady, D.E., and Kipp, M.E., 1987, Dynamic rock fragmentation, *in* Atkinson, B.K., ed., Fracture mechanics of rock: San Diego, Academic Press, p. 429–475.

Grieve, R.A.F., Robertson, P.B., and Dence, M.R., 1981, Constraints on the formation of ring impact structures, *in* Schultz, P.H., and Merill, R.B., eds., Multiring basins: New York, Pergamon, p. 37–57.

He, H., and Ahrens, T.J., 1993, Mechanical properties of shock damaged rocks: International Journal of Rock Mechanics and Mining Science & Geomechanics Abstracts, October 1994, v. 31, no. 5, p. 525–533.

Huntoon, P.W., 2000, Upheaval Dome, Canyonlands, Utah: Strain indicators that reveal an impact origin, *in* Sprinkel, D.A., Chidsey Jr., T.C., and Anderson, P.B., eds., Geology of Utah's parks and monuments: Utah Geological Association Publication 28, p. 619–628.

Huntoon, P.W., and Shoemaker, E.M., 1995, Roberts Rift, Canyonlands, Utah: A natural hydraulic fracture caused by comet or asteroid impact: Ground Water, v. 33, p. 561–569.

Huntoon, P.W., Billingsley Jr., G.H., and Breed, W.J., 1982, Geologic map of Canyonlands National Park and vicinity, Utah: Canyonlands National and History Association, Moab, Utah, scale 1:62,500.

Johnston, D.H., and Toksöz, M.N., 1980, Ultrasonic P and S wave attenuation in dry and saturated rocks under pressure: Journal of Geophysical Research, v. 85, p. 925–936.

Ivanov, B.A., and Artemieva, N.A., 2002, Numerical modeling of the formation of large impact craters, *in* Koeberl, C., and MacLeod, K.G., eds., Catastrophic events and mass extinctions: Impacts and beyond: Geological Society of America Special Paper 356, p. 619–630.

Ivanov, B.A., and Deutsch, A., 1999, Sudbury impact event: Cratering mechanics and thermal history, *in* Dressler, B.O., and Sharpton, V.L., eds., Large meteorite impacts and planetary evolution: Geological Society of America Special Paper 339, p. 389–397.

Ivanov, B.A., and Kostuchenko, V.N., 1997, Block oscillation model for impact crater collapse [abs.]: Houston, Texas, Lunar and Planetary Science Conference 28, p. 631–632.

Ivanov, B.A., Basilevsky, A.T., and Sazonoya, L.V., 1982, Formation of the central uplift in meteoritic craters: Meteoritika, v. 40, p. 60–81 (in Russian, English technical translation, 1986, National Aeronautics and Space Administration TM 88427).

Ivanov, B.A., Langenhorst, F., Deutsch, A., and Hornemann, U., 2001, How strong was impact-induced CO_2 degassing in the Cretaceous-Tertiary event? Numerical modeling of shock recovery experiments, *in* Koeberl, C., and MacLeod, K.G., eds., Catastrophic events and mass extinctions: Impact and beyond: Geological Society of America Special Paper 356, p. 587–594.

Jackson, M.P.A., Schulz-Ela, D.D., Hudec, M.R., Watson, I.A., and Porter, M.L., 1998, Structure and evolution of Upheaval Dome: a pinched-off salt diapir: Geological Society of America Bulletin, v. 110, p. 1547–1573, doi: 10.1130/0016-7606(1998)1102.3.CO;2.

Kanbur, Z., Louie, J.N., Chávez-Pérez, S., Plank, G., and Morey, D., 2000, Seismic reflection study of Upheaval Dome, Canyonlands National Park, Utah: Journal of Geophysical Research, v. 105, p. 9489–9505, doi: 10.1029/1999JE001131.

Kenkmann, T., 2002, Folding within seconds: Geology, v. 30, p. 231–234.

Kenkmann, T., 2003, Dike formation, cataclastic flow, and rock fluidization during impact cratering: An example from the Upheaval Dome structure, Utah: Earth and Planetary Science Letters, v. 214, p. 43–58, doi: 10.1016/S0012-821X(03)00359-5.

Kenkmann, T., and von Dalwigk, I., 2000, Radial transpression ridges: A new structural feature of complex impact craters: Meteoritics & Planetary Science, v. 35, p. 1189–1202.

Kenkmann, T., and Scherler, D., 2002, New structural constraints on the Upheaval Dome impact crater: Houston, Texas, Lunar and Planetary Science Conference 33, Abstract 1037.

Kluth, C.F., and Coney, P.J., 1981, Plate Tectonics of the Ancestral Rocky Mountains: Geology, v. 9, p. 10–15.

Koeberl, C., Plescia, C.L., Hayward, C.L., and Reimold, W.U., 1999, A petrographical and geochemical study of quartzose nodules, country rocks, and dike rocks from the Upheaval Dome structure, Utah: Meteoritics & Planetary Science, v. 34, p. 861–868.

Kriens, B.J., Shoemaker, E.M., and Herkenhoff, K.E., 1997, Structure and kinematics of a complex impact crater, Upheaval Dome, southeast Utah: Brigham Young University Geology Studies, v. 42, p. 19–31.

Kriens, B.J., Shoemaker, E.M., and Herkenhoff, K.E., 1999, Geology of the Upheaval Dome impact structure, southeast Utah: Journal of Geophysical Research, v. 104, p. 18,867–18,887, doi: 10.1029/1998JE000587.

Laney, J.E., and Van Schmus, W.R., 1978, A structural study of the Kentland, Indiana, impact site: Houston, Texas, Proceedings of the Lunar and Planetary Science Conference 9, p. 2609–2639.

Lundborg, N., 1968, Strength of rock-like materials: International Journal of Rock Mechanics and Mining Science, v. 5, p. 427–454.

Marshak, S., Karlstrom, K., and Timmons, J M., 2000, Inversion of proterozoic extensional faults: An explanation for the pattern of Laramide and ancestral Rockies intracratonic deformation, United States: Geology, v. 28, p. 735–738, doi: 10.1130/0091-7613(2000)0282.3.CO;2.

Mattox, R.B., 1968, Upheaval Dome, a possible salt dome in the Paradox basin, Utah: Geological Society of America Special Paper 88, p. 331–347.

McKinnon, W.B., 1978, An investigation into the role of plastic failure in crater modification: Houston, Texas, Proceedings of the Lunar and Planetary Science Conference 9, p. 3965–3973.

McKnight, E.T., 1940, Geology of the area between Green and Colorado Rivers Grand and San Juan Counties, Utah, United States: Geological Survey Bulletin 908, 147 p.

Melosh, H.J., 1977, Crater modification by gravity: A mechanical analysis of slumping, *in* Roddy, J., Pepin, R.O., and Merill, R.B., eds., Impact and explosion cratering: New York, Pergamon Press, p. 1245–1260.

Melosh, H.J., 1979, Acoustic fluidization: A new geologic process?: Journal of Geophysical Research, v. 84, p. 7513–7520.

Melosh, H.J., 1989, Impact cratering: A geological process: New York, Oxford University Press, 245 p.

Melosh, H.J., and Ivanov, B.A., 1999, Impact crater collapse: Annual Review of Earth and Planetary Sciences, v. 27, p. 385–415, doi: 10.1146/ANNUREV.EARTH.27.1.385.

Milton, D.J., Barlow, B.C., Brett, R., Brown, A.R., Glikson, A.Y., Manwaring, E.A., Moss, F.J., Sedmik, E.C.E., van Son, J., and Young, G.A., 1972, Gosses Bluff impact structure, Australia: Science, v. 175, p. 1199–1207.

Milton, D.J., Glikson, A.Y., and Brett, R., 1996, Gosses Bluff—a latest Jurassic impact structure, central Australia. Part I: Geological structure, stratigraphy, and origin: AGSO Journal of Australian Geology & Geophysics, v. 16, p. 453–486.

Morgan, J.V., Warner, M.R., Collins, G.S., Melosh, H.J., and Christeson, G.L., 2000, Peak-ring formation in large impact craters: Geophysical constraints from Chicxulub: Earth and Planetary Science Letters, v. 183, p. 347–354, doi: 10.1016/S0012-821X(00)00307-1.

Offield, T.W., and Pohn, H.A., 1979, Geology of the Decaturville impact structure: U.S. Geological Survey Professional Paper 1042, 48 p.

O'Keefe, J.D., and Ahrens, T.J., 1993, Planetary cratering mechanics: Journal of Geophysical Research, v. 98, p. 17,011–17,028.

O'Keefe, J.D., and Ahrens, T.J., 1999, Complex craters: Relationship of stratigraphy and rings to impact conditions: Journal of Geophysical Research, v. 104, p. 27,091–27,104, doi: 10.1029/1998JE000596.

Pederson, J.L., Mackley, R.D., and Eddleman, J.L., 2002, Colorado Plateau uplift and erosion evaluated using GIS: GSA Today, v. 12, no. 8, p. 4–10, doi: 10.1130/1052-5173(2002)012<0004:CPUAEE>2.0.CO;2.

Pierazzo, E., and Melosh, H.J., 1999, Hydrocode modeling of Chicxulub as an oblique impact event: Earth and Planetary Science Letters, v. 165, p. 163–176, doi: 10.1016/S0012-821X(98)00263-5.

Ricard, Y., and Bercovici, D., 2003, Two-phase damage theory and crustal rock failure: The theoretical 'void' limit, and the prediction of experimental data: Geophysical Journal International, v. 155, p. 1057–1064, doi: 10.1111/J.1365-246X.2003.02112.X.

Roddy, D.J., 1979, Structural deformation at the Flynn Creek impact crater, Tennessee: Houston, Texas, Lunar and Planetary Science Conference 10, p. 2519–2534.

Roddy, D.J., Schuster, S.H., Kreyenhagen, K.N., and Orphal, D.L., 1980, Computer code simulations of the formation of Meteor Crater, Arizona—Calculations MC-1 and MC-2: New York, Pergamon Press, Proceedings of Lunar and Planetary Science Conference, v. 3, p. 2275–2308.

Scherler, D., 2004, Upheaval Dome Struktur, Utah, USA: Detailkartierung, 3D-Modellierung, Strukturanalyse [M.S. thesis]: Berlin, Humboldt-Universität, 108 p.

Scherler, D., Jahn, A., and Kenkmann, T., 2003, Structural investigations in the central uplift of the Upheaval Dome impact crater, Utah: 3rd International Conference on Large Meteorite Impacts, Abstract 4072, CD-ROM.

Scholz, C.H., 1990, The mechanics of earthquakes and faulting: Cambridge, UK, Cambridge University Press, 439 p.

Schultz, P.H., and Anderson, R.R., 1996, Asymmetry of the Manson impact structure: Evidence for impact angle and direction: Geological Society of America Special Paper 302, p. 397–417.

Shoemaker, E.M., and Herkenhoff, K.E., 1983, Impact origin of Upheaval Dome, Utah: Eos (Transactions, American Geophysical Union), v. 44, p. 747.

Shoemaker, E.M., and Herkenhoff, K.E., 1984, Upheaval Dome impact structure: Houston, Texas, Lunar and Planetary Science Conference 15, p. 778–779.

Shoemaker, E.M., and Shoemaker, C.S., 1996, The Proterozoic impact record of Australia: AGSO Journal of Australian Geology & Geophysics, v. 16, p. 379–398.

Shuvalov, V.V., 1999, 3D hydrodynamic code SOVA for interfacial flows, application to thermal layer effect: Shock Waves, v. 9, p. 381–390, doi: 10.1007/S001930050168.

Shuvalov, V.V., Artemieva, N.A., and Kosarev, I.B., 1999, 3D hydrodynamic code SOVA for multi-material flows, application to Shoemaker-Levy 9 comet impact problem: International Journal of Impact Engineering, v. 23, p. 847–858, doi: 10.1016/S0734-743X(99)00129-3.

Stöffler, D., and Langenhorst, F., 1994, Shock metamorphism of quartz in nature and experiment: I. Basic observation and theory: Meteoritics, v. 29, p. 155–181.

Stokes, W.L., 1988, Geology of Utah: Occasional Paper of the Utah Museum of Natural History, v. 6, 308 p.

Tillotson, J.H., 1962, Metallic equations of state for hypervelocity impact: General Atomic Report GA-3216, 141 p.

Trunin, R.F., Simakov, G.V., Dudoladov, I.P., Telegin, G.S., and Trusov, I.P., 1988, Compressibility of rocks in shock waves: Fizika Zemli, v. 1, p. 52–58 (in Russian).

U.S. Geological Survey, 1988, Upheaval Dome Quadrangle Utah 7.5 Minute Series (Topographic): United States Department of the Interior Geological Survey Provisional Edition 1988.

Wilshire, H.G., Offield, T.W., Howard, K.A., and Cummings, D., 1972, Geology of the Sierra Madera cryptoexplosion structure, Pecos County, Texas: U.S. Geological Survey Professional Paper 599-H, 42 p.

Woodward, N.B., Boyer, S.E., and Suppe, J., 1989, Balanced geological cross-sections: An essential technique in geological research and exploration: American Geophysical Union, Short Course in Geology, Washington, v. 6, 132 p.

Wünnemann, K., and Ivanov, B.A., 2003, Numerical modeling of the impact crater depth-diameter dependence in an acoustically fluidized target: Planetary and Space Science, v. 51, p. 831–845, doi: 10.1016/J.PSS.2003.08.001.

Manuscript Accepted by the Society 9 August 2004

Printed in the USA

Geological Society of America
Special Paper 384
2005

Eastern rim of the Chesapeake Bay impact crater: Morphology, stratigraphy, and structure

C. Wylie Poag
U.S. Geological Survey, 384 Woods Hole Road, Woods Hole, Massachusetts 02543-1598, USA

ABSTRACT

This study reexamines seven reprocessed (increased vertical exaggeration) seismic reflection profiles that cross the eastern rim of the Chesapeake Bay impact crater. The eastern rim is expressed as an arcuate ridge that borders the crater in a fashion typical of the "raised" rim documented in many well preserved complex impact craters. The inner boundary of the eastern rim (rim wall) is formed by a series of crater-facing, steep scarps, 15–60 m high. In combination, these rim-wall scarps represent the footwalls of a system of crater-encircling normal faults, which are downthrown toward the crater. Outboard of the rim wall are several additional normal-fault blocks, whose bounding faults trend approximately parallel to the rim wall. The tops of the outboard fault blocks form two distinct, parallel, flat or gently sloping, terraces. The innermost terrace (Terrace 1) can be identified on each profile, but Terrace 2 is only sporadically present. The terraced fault blocks are composed mainly of nonmarine, poorly to moderately consolidated, siliciclastic sediments, belonging to the Lower Cretaceous Potomac Formation. Though the ridge-forming geometry of the eastern rim gives the appearance of a raised compressional feature, no compelling evidence of compressive forces is evident in the profiles studied. The structural mode, instead, is that of extension, with the clear dominance of normal faulting as the extensional mechanism.

Keywords: Chesapeake Bay, impact crater, stratigraphy, structure, morphology.

INTRODUCTION

The Chesapeake Bay impact crater has been documented primarily on the basis of >2000 km of seismic reflection profiles and eight continuously cored boreholes (Poag et al., 1994, 2004; Poag, 1996, 1997; Powars and Bruce, 1999). Multi-channel, two-channel, and single-channel seismic reflection data provide three different levels of stratigraphic resolution and depth of penetration in the western sector of the crater. No multi-channel profiles are available, however, for the eastern sector—that part located under the inner continental shelf, east of the Delmarva Peninsula (Figs. 1 and 2). Poag (1996) and Poag et al. (1999, 2004) interpreted the broad stratigraphic and structural aspects of the crater outer rim and found a significant difference between the eastern and western sectors. The rim in the eastern sector displays a distinctive "raised" ridge, whereas only scattered possible remnants of a raised ridge are present in the western sector.

The steep, inward-facing wall of the Chesapeake Bay crater is formed by stratified beds of the Potomac Formation (Fig. 2), a pre-impact deposit of Lower Cretaceous nonmarine sediments, characterized seismically by discontinuous, moderate to high-amplitude, horizontal or gently tilted, parallel reflections (Fig. 3). The crater-filling Exmore breccia, in contrast, typically lacks stratal reflections, but instead, characteristically

Poag, C.W., 2005, Eastern rim of the Chesapeake Bay impact crater: Morphology, stratigraphy, and structure, *in* Kenkmann, T., Hörz, F., and Deutsch, A., eds., Large meteorite impacts III: Geological Society of America Special Paper 384, p. 117–130. For permission to copy, contact editing@geosociety.org.

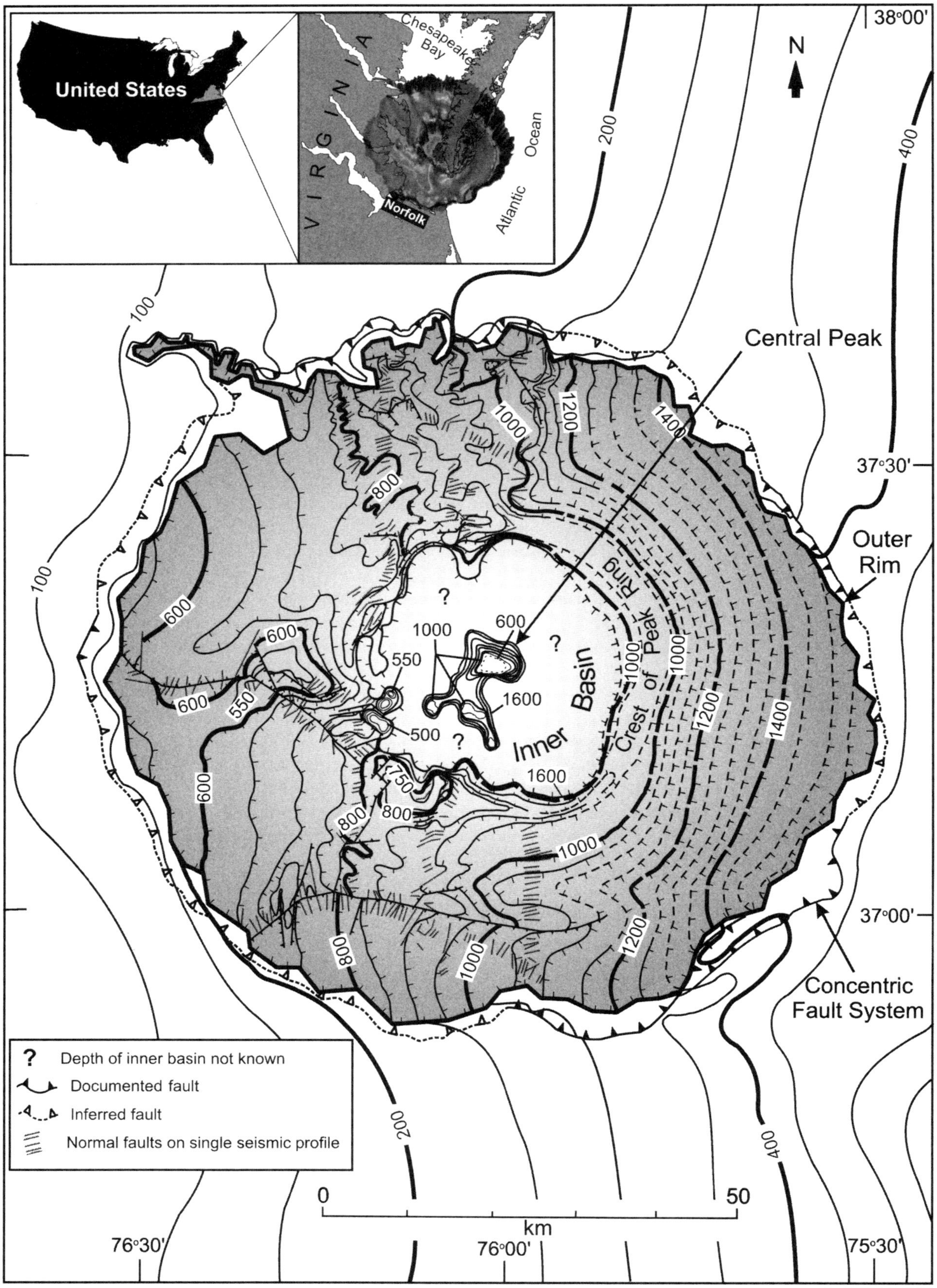

Figure 1. Location and structure map of Chesapeake Bay impact crater. Shaded area represents surface of crystalline basement. All crater-fill deposits (Exmore breccia and displaced megablocks) have been removed. Contour interval 50 m. Structure from Figure 4.1 of Poag et al. (2004).

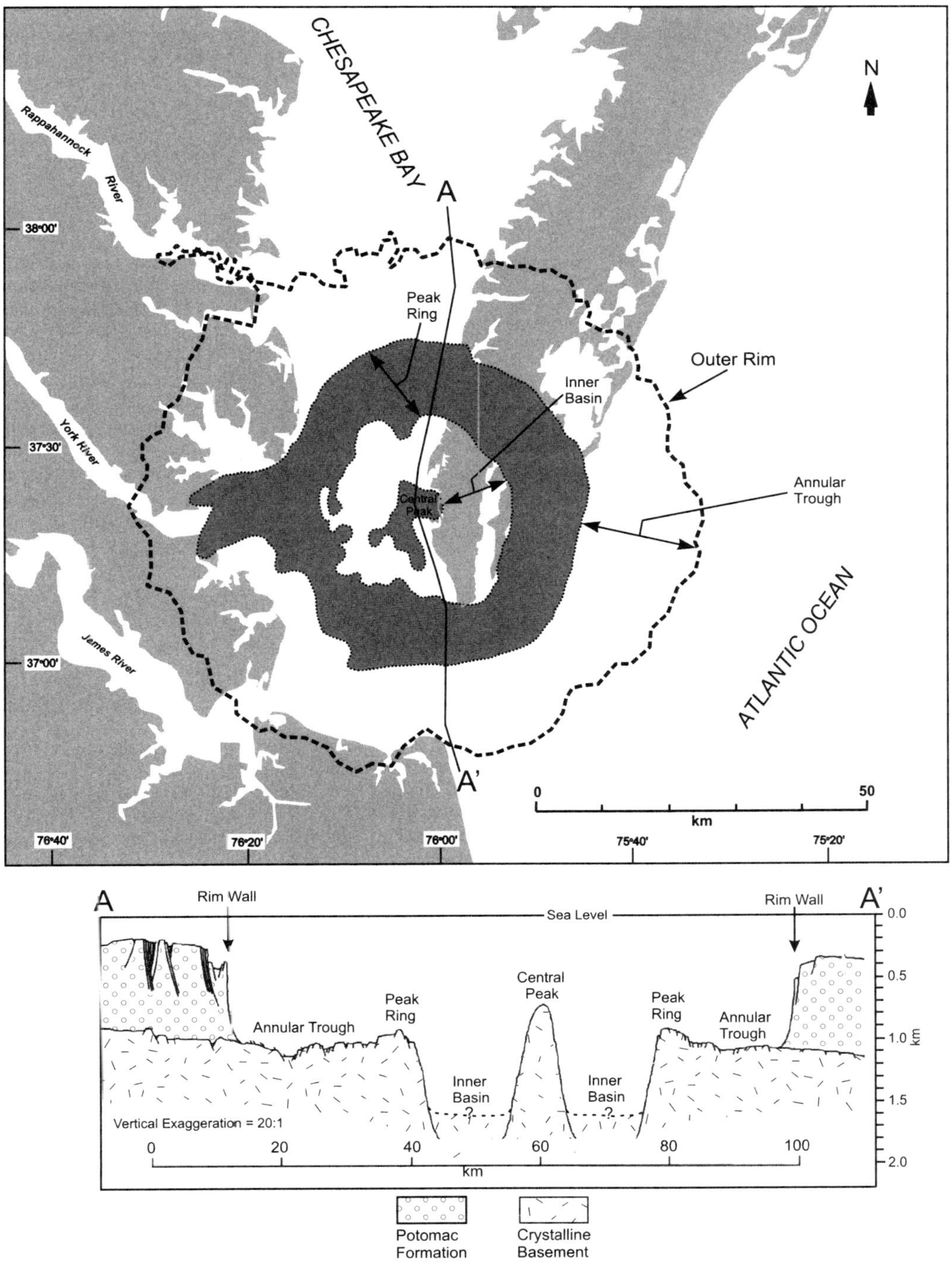

Figure 2. Map and cross section, showing morphological and structural features of Chesapeake Bay impact crater. Exmore breccia, displaced megablocks, and postimpact deposits have been removed from depth-scaled cross section (modified from Figure 4.2A of Poag et al., 2004).

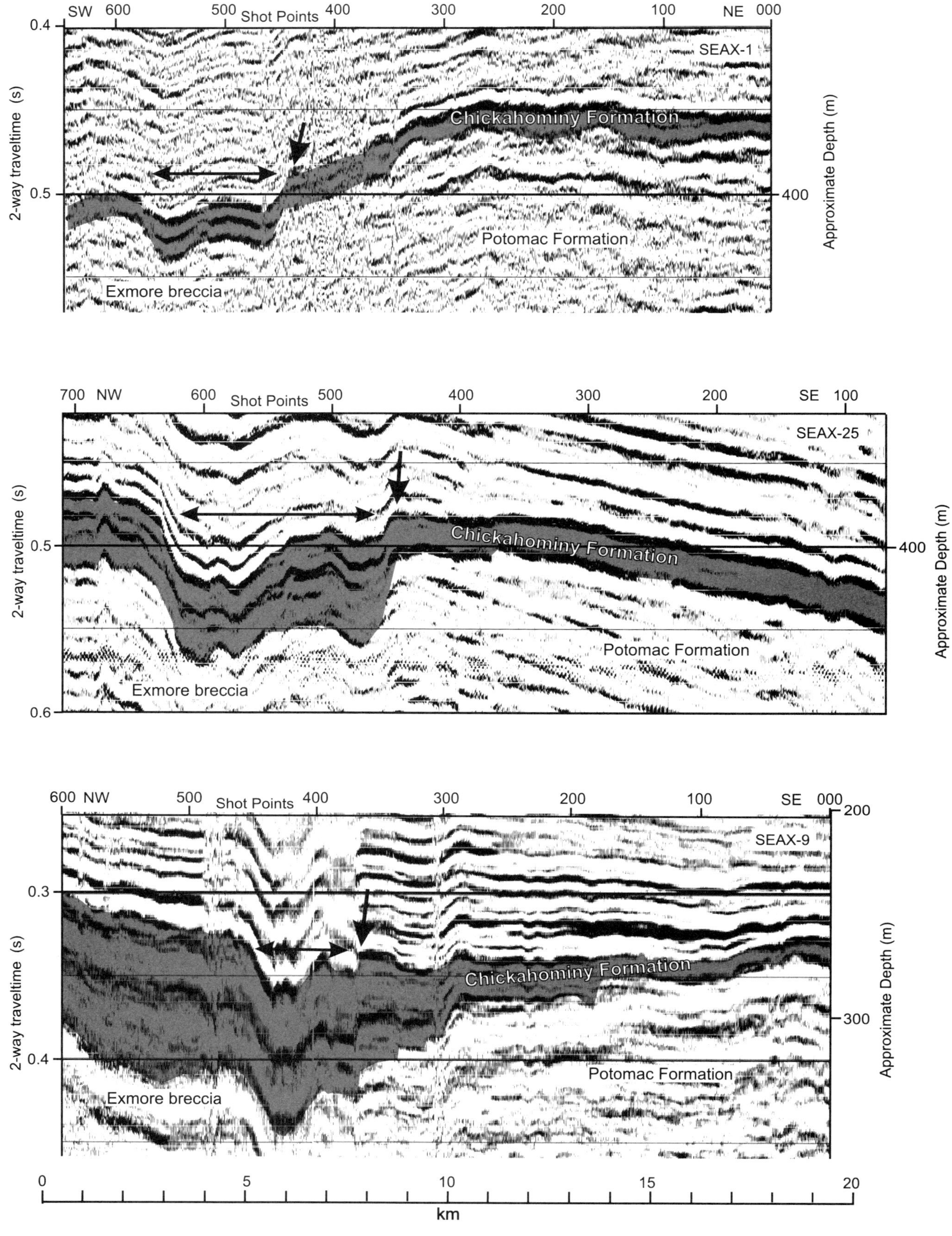

Figure 3. Three examples of single-channel seismic reflection profiles that show how Chickahominy Formation (initial postimpact formation) sags and thickens into the crater as it crosses the rim wall (heavy arrow) in the eastern sector of Chesapeake Bay impact crater. All profiles are at the same scale; vertical exaggeration is ~30:1. Note that the internal multiples in Exmore breccia obscure the typical chaotic seismic signature that distinguishes breccia from stratified Potomac sediments in the western sector of crater. Double-headed arrow indicates trough (ring graben) in Exmore breccia, which encircles the entire crater margin at the base of the rim-wall scarp. See Figure 5 for location of profiles.

displays chaotic hyperbolic reflections. However, this chaotic breccia signature does not appear on single and two-channel profiles. On these profiles, seismic ringing and internal multiple reflections, which originate in the stratified post-impact sediments, give the Exmore breccia a pseudostratified signature (Fig. 3; see also the CD-ROM published by Poag et al., 2004, for many good examples of these features). Therefore, it is especially difficult to distinguish the chaotic breccia (Exmore breccia) from the stratified rim wall (Potomac Formation) in the eastern sector of the crater. To overcome this difficulty, Poag et al. (2004, chapters 4 and 7) showed that a proxy geometric feature could be used to identify the location of the rim wall in those areas lacking multi-channel seismic data. The proxy consists of abrupt structural sagging and thickening of the Chickahominy Formation (the initial post-impact formation), which take place where the Chickahominy Formation crosses the rim wall (Fig. 3). The sagging results from differential subsidence—relatively rapid post-impact subsidence of the Chickahominy Formation due to compaction of the underlying unconsolidated breccia—versus relatively slow subsidence over the more consolidated Potomac Formation. The Chickahominy Formation thickens concomitantly, because the crater originally was a closed seafloor depression, which collected sediment. Differential subsidence of the underlying breccia created additional sediment accommodation space, which further thickened the Chickahominy.

Poag et al. (2004) used structure maps of the Exmore breccia (Fig. 4A) and Chickahominy Formation (Fig. 4B) to show that the eastern sector of the crater exhibits a raised rim. These maps, however, have a 50 m contour interval, which provides minimal resolution of the rim morphology and stratigraphy. My objective here is to improve morphologic, stratigraphic, and structural definition of the raised rim by (1) increasing the vertical exaggeration of the seismic reflection profiles, and (2) constructing a new structure map with a 10 m contour interval. These improvements, in turn, allow a better assessment of the origin of the raised rim.

METHODS

The U.S. Geological Survey (USGS) and the National Geographic Society collected the single-channel profiles used herein in 1996, using the research vessel *Seaward Explorer* (SEAX). The primary seismic source for the SEAX survey was a Generator-Injector airgun (GI gun; Seismic Systems Inc.). This system did not image the crystalline basement surface, however, because the basement is deeper (>1.5 km) than the maximum penetration of the seismic signal in the eastern sector of the crater (Fig. 1). The *Seaward Explorer* data were processed by Myung Lee (USGS, Denver, Colorado).

The USGS and the Lamont-Doherty Earth Observatory collected a two-channel profile in 1998, using the research vessel *Maurice Ewing*. The seismic source for the *Ewing* survey was a tuned array of six airguns with chamber capacities of 80, 120, 145, 200, 305, and 500 in^3, at 2000 psi. This profile successfully imaged the crystalline basement surface. The two-channel data were processed by the Lamont-Doherty Earth Observatory.

For this study, I increased the vertical exaggeration of the seismic profiles from 10:1 (used by Poag et al., 2004) to ~30:1, a process that also exaggerates the steepness of nonvertical structural and morphological features, such as fault scarps and terraces (Stone, 1991). References to these features herein are based on their appearance on the exaggerated profiles and should be regarded, for the most part, as *relative* comparisons. Many of the features I discuss are quite subtle at true scale.

I used the same method employed by Poag et al. (2004, chapter 3) to convert seismic two-way traveltime to depth. Root mean square velocities calibrated with borehole stratigraphy indicate that the seismic velocity at the depth of the eastern crater rim is ~1600 m/s; therefore, the two-way traveltime (measured in milliseconds), multiplied by 0.8, yields the approximate depth in meters.

LOCATING THE OUTER RIM

Following Poag et al. (2004), I use the sagging and thickening of the Chickahominy Formation (Figs. 3 and 4) to assist in locating the rim wall on six single-channel profiles and one two-channel profile (Figs. 5, 6, and 7). The track lines for six of these seven profiles run approximately perpendicular to the crater rim (profile SEAX-1 is diagonal to the rim), and thereby, when measured at true scale, would yield nearly true slopes for the topographic surface of the rim. In order to illustrate more clearly the changes in Chickahominy geometry at the crater rim, however, I use a ~30:1 vertical exaggeration (Figs. 6 and 7). The increased exaggeration also clarifies stratal relationships between the pre-impact Potomac Formation that forms the crater rim, the syn-impact Exmore breccia inside the crater, and the post-impact Chickahominy Formation and younger deposits overlying the crater. In Figures 6 and 7, I have stripped the Chickahominy Formation and all post-impact deposits from the seven vertically exaggerated profiles to emphasize the morphological, stratigraphic, and structural features of the crater rim. A few additional single-channel profiles (described by Poag et al., 2004, chapter 4) from the eastern sector of the crater also provide data for the structure map of this area (Fig. 8).

Figure 3 illustrates sections of three of the seismic profiles (representing a northern, central, and southern crossing of the rim) to show the geometric variability of the Chickahominy Formation as it sags and thickens into the crater. These profiles also demonstrate the difficulty in distinguishing (seismically) the late Eocene Exmore breccia from the Lower Cretaceous Potomac Formation, because extensive ringing and interbed multiples produced in the overlying strata typically mask the chaotic signature of the breccia. The profiles in Figure 3 also illustrate the cross sectional geometry of the ring graben (double headed arrows) that encircles the crater along the base of the outer rim escarpment (Poag et al., 2004; chapter 4).

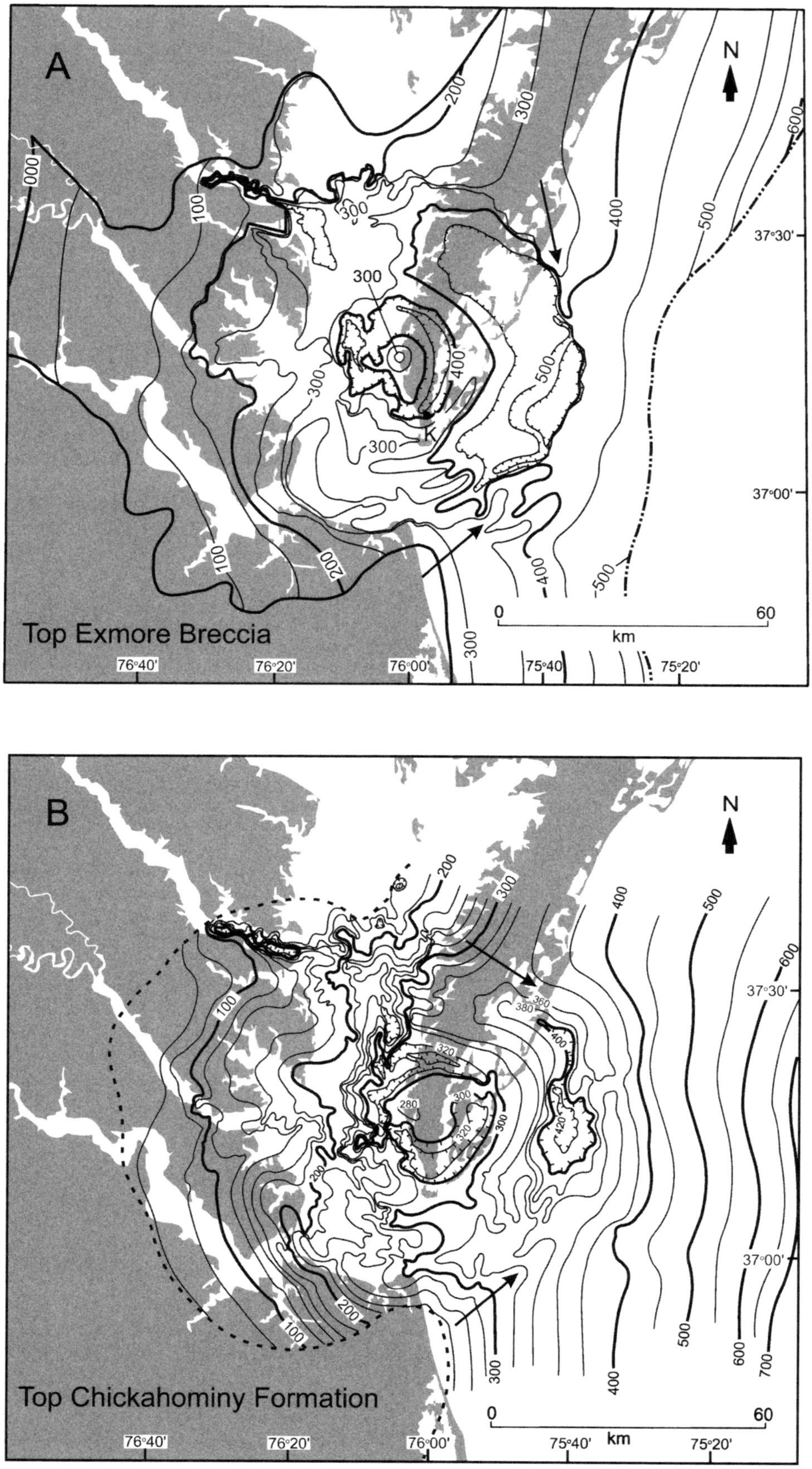

Figure 4. (A) Structure map of the upper surface of the Exmore breccia in the Chesapeake Bay impact crater. Contour interval: 50 m. Arrows indicate the location of the raised rim in the eastern sector of the crater. The dashed line indicates the approximate seaward extent of Exmore breccia, as determined from seismic reflection profiles. From Figure 6.11 of Poag et al. (2004). (B) Structure map of the upper surface of the Chickahominy Formation, which is the initial postimpact formation to cover Chesapeake Bay impact crater. Contour interval: 50 m. Arrows indicate location of the raised rim in the eastern sector of the crater. The dashed line indicates the approximate landward limit of Chickahominy Formation. From Figure 7.8 of Poag et al. (2004).

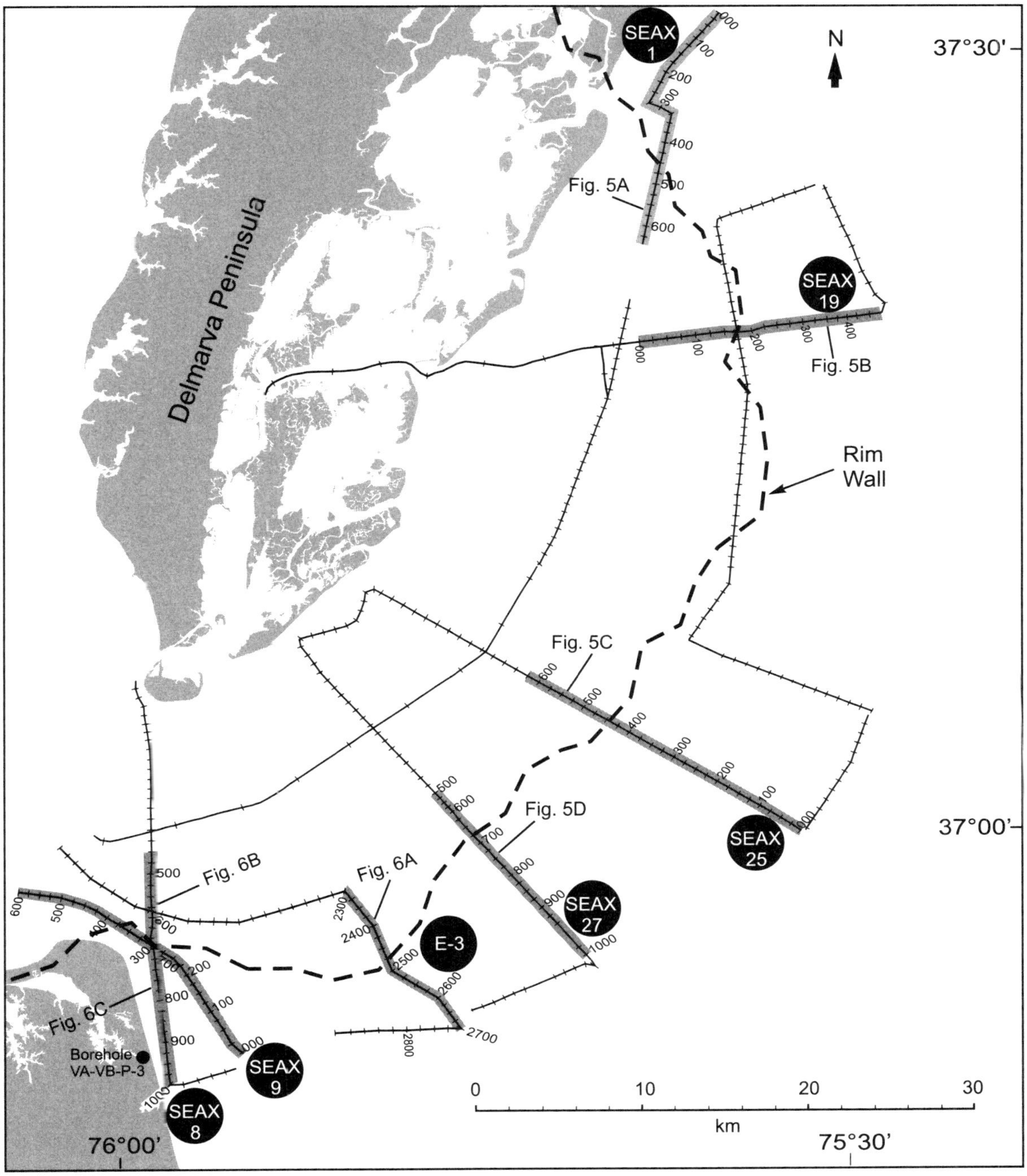

Figure 5. Location map showing tracklines (labeled); shaded lines and shot points indicate seven seismic reflection profiles analyzed in this study (Figs. 3, 6, and 7). Borehole VA-VB-P-3 is the closest of the numerous boreholes used to calibrate profile SEAX-8 and, by correlation, the other six profiles. Additional unshaded segments of these and other tracklines are included to show where seismostratigraphic units can be traced throughout the study area.

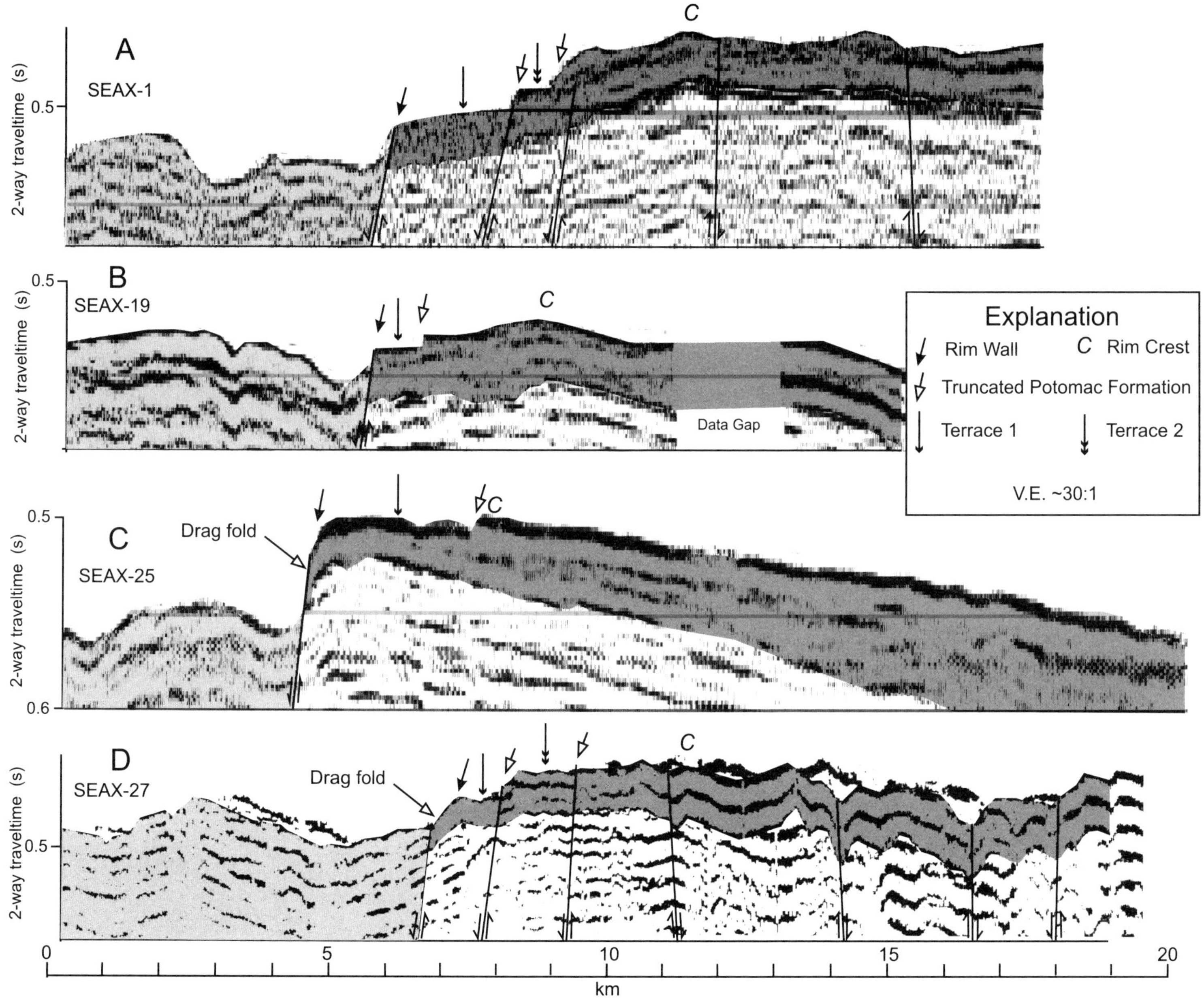

Figure 6. Geologic interpretation of four seismic reflection profiles that cross the eastern rim of the Chesapeake Bay impact crater. Profiles are arranged from north (A) to south (D). All postimpact deposits have been removed. The lighter shaded interval at left is the Exmore breccia. The darker shaded interval at right corresponds to the dominantly coarse-sand section at the top of the Potomac Formation (drilled in corehole VA-VB-P-3). Diagonal distance from rim wall to rim crest as measured on profile SEAX-1 (A) is 5.25 km, but perpendicular distance (from Fig. 8) is 3.5 km (given in Table 3). All profiles are at the same scale; vertical exaggeration is ~30:1. Note the drag folds on the footwall block of the rim-wall scarp (C and D). See Figure 5 for location of profiles.

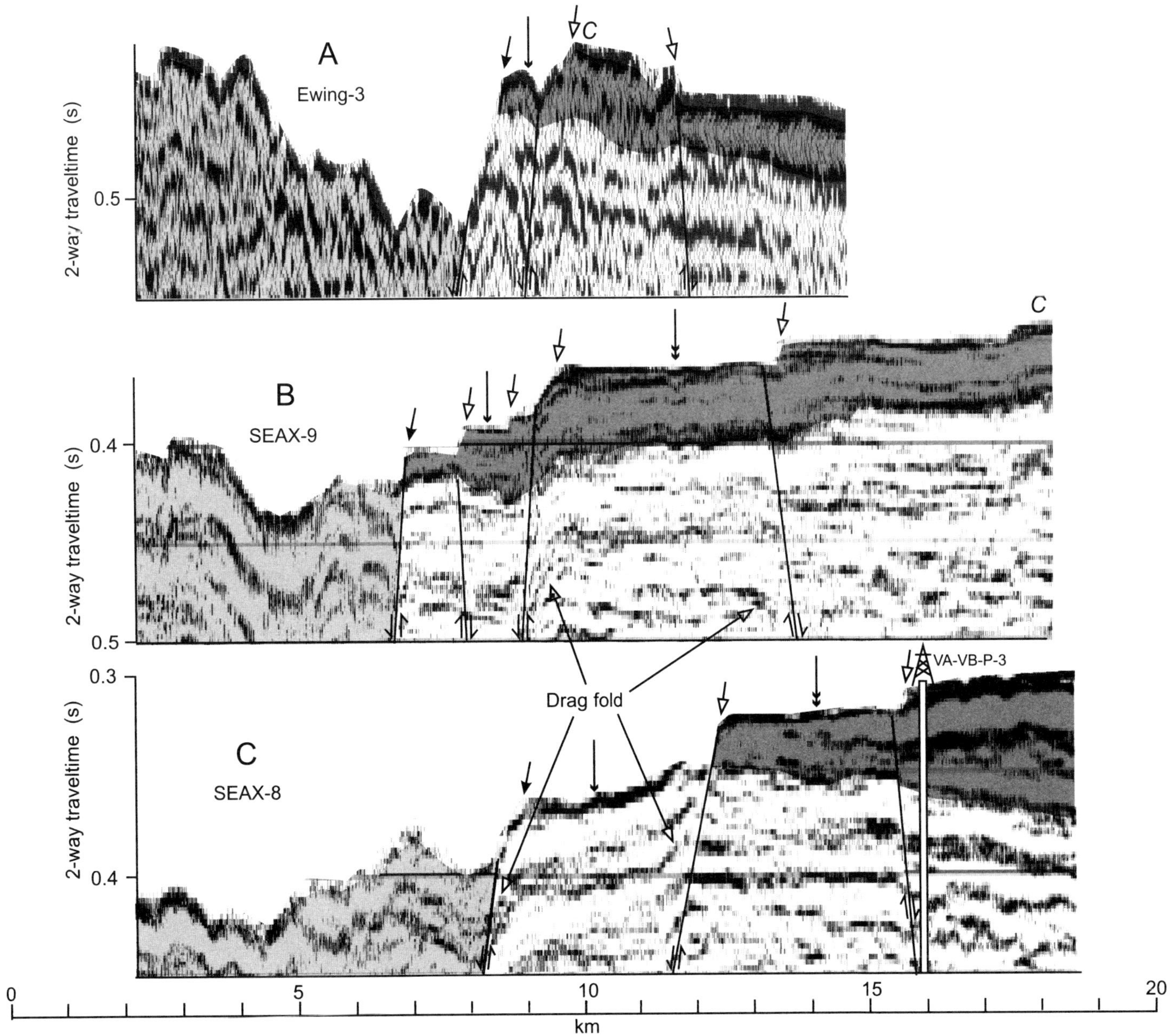

Figure 7. Geologic interpretation of three seismic reflection profiles that cross the eastern rim of the Chesapeake Bay impact crater. Profiles are arranged from north (A) to south (C). All postimpact deposits have been removed. The lighter shaded interval at left is the Exmore breccia. The darker shaded interval at right corresponds to dominantly the coarse-sand section at the top the of Potomac Formation (drilled in corehole VA-VB-P-3, which is located ~1 km west of SEAX-8; see Fig. 5). All profiles are at the same scale; vertical exaggeration is ~30:1. Symbols as in Figure 6. Note the drag folds that indicate relative motion along fault planes (B and C). See Figure 5 for location of profiles.

PHYSICAL FEATURES OF THE OUTER RIM

Rim Morphology

In the eastern sector of the Chesapeake Bay crater, the outer rim displays considerable morphological complexity and variability. Two main morphostructural components of the rim can be recognized here. The first component of the outer rim is the *rim wall*, which can be defined as the innermost, crater-facing, fault scarp that separates the Potomac Formation from the main body of crater-fill breccia (Exmore breccia; Figs. 2, 3, 6, and 7). The second component of the eastern outer rim is the *rim crest*, which can be defined as the maximum elevation of the sedimentary ridge that borders the eastern sector of the crater (Figs. 3, 6, 7, and 8).

Several broad-scale features are characteristic of most seismic profiles used in this study. First of all, the vertical relief on top of the Exmore breccia is generally greater than that on top of the Potomac Formation, except where the latter is faulted at the rim wall. Also, a notable trough of variable width (2–8.5 km) is present on the breccia surface, inboard of the rim wall. Where it contacts the breccia, the rim-wall scarp ranges from 15 m to 60 m in height above the breccia surface (average height 28 m; Table 1). The scarp is highest in the southern part of the study area (profiles SEAX-25 and Ewing-3), and lowest to the north (profiles SEAX-1 and SEAX-19; Figs. 6, 7, and 8).

All profiles studied exhibit a flat or gently sloping terrace (Terrace 1), formed in the upper part of the Potomac Formation at the top of the rim-wall scarp (Figs. 6 and 7). Terrace 1 is 0.75–6.5 km wide (average width 3.54 km; Table 2). The surface of Terrace 1 is cut by several smaller terraces on profile SEAX-9.

Four of the profiles (SEAX-1, SEAX-27, SEAX-9, SEAX-8) exhibit a second terrace (Terrace 2) outboard of Terrace 1 (Figs. 6 and 7). Terrace 2 varies from 1 km to 8 km in width (average width 2.5 km; Table 2). Terrace 1 is separated from Terrace 2 by a relatively low scarp, 5–20 m high (average height 11 m). Terrace 2 is separated from the outboard, upward, slope of the rim by another low scarp, 5–10 m high (average height 5 m; Table 2).

Six of the seven profiles (not SEAX-8) cross the rim crest (maximum elevation of the rim), but at varying distances from the rim wall (range 1.25–11.5 km; average distance 4.3 km; Table 3). The rim crest is nearest the rim wall on SEAX-19 and SEAX-27 (3 km; Figs. 6 and 7; Table 3), and farthest from the rim wall on SEAX-9 (11.5 km; Figs. 7 and 8; Table 3). The depth (below sea level) of the rim crest is >420 m between profiles SEAX-19 and SEAX-25, and decreases to <370 m to the north (profile SEAX-1), and to <250 to the south (profiles SEAX-8, SEAX-9; Figures 6, 7, and 8; Table 3). This pattern of variable rim-crest elevation is due, in part, to the eastward regional dip of the sediments and crystalline basement surface into the offshore Baltimore Canyon trough, which, thereby, tilts the crater to the east (Poag et al., 2004, chapter 2; Fig. 1 herein).

TABLE 1. HEIGHT OF RIM-WALL SCARP IN EASTERN SECTOR OF CHESAPEAKE BAY IMPACT CRATER

Seismic reflection profile	Height of rim-wall scarp (m)
SEAX-1	15
SEAX-19	15
SEAX-25	45
SEAX-27	15
E-3	60
SEAX-9	20
SEAX-8	25
Average	28

Note: SEAX—*Seaward Explorer* data; E—*Ewing* survey data.

TABLE 2. WIDTH OF RIM TERRACES AND HEIGHT OF BOUNDING SCARPS IN EASTERN SECTOR OF CHESAPEAKE BAY IMPACT CRATER

Seismic profile	Rim terrace 1		Rim terrace 2	
	Scarp height (m)	Width (km)	Scarp height (m)	Width (km)
SEAX-1	10	4.5	10	1
SEAX-19	5	1.75	none	none
SEAX-25	5	5.5	none	none
SEAX-27	10	0.75	5	2
E-3	15	1.5	none	none
SEAX-9	15	4.25	10	8
SEAX-8	20	6.5	10	6.5
Average	11	3.54	5	2.5

Note: SEAX—*Seaward Explorer* data; E—*Ewing* survey data.

TABLE 3. HEIGHT OF RIM CREST AND DISTANCE FROM RIM-WALL SCARP IN EASTERN SECTOR OF CHESAPEAKE BAY IMPACT CRATER

Seismic reflection profile	Height of rim crest (m)	Distance from rim-wall scarp (km)
SEAX-1	60	3.5*
SEAX-19	30	3.0
SEAX-25	45	3.0
SEAX-27	30	3.75
E-3	75	1.25
SEAX-9	60	11.5
Average	50	4.3

Note: SEAX—*Seaward Explorer* data; E—*Ewing* survey data.
*This is perpendicular distance measured from Fig. 8; diagonal distance on Fig. 6A is 5.25 km.

Vertical relief, measured from the foot of the rim-wall scarp (surface of Exmore breccia) to the apex of the rim crest, ranges from 30 m to 75 m (average relief 54 m) and is consistently greatest in the southern part of the study area (profiles E-3, SEAX-9, SEAX-8; Fig. 7; Table 3).

Rim Stratigraphy

The lithostratigraphic and chronostratigraphic relationships interpreted from the seismic profiles are based on calibration with numerous boreholes drilled inside and outside the Chesapeake Bay crater (Powars and Bruce, 1999; Poag et al., 2004, chapters 1–6). The nearest borehole to my study area is VA-VB-P-3 (482 m TD; Figs. 8 and 9), located 1 km west of profile SEAX-8 (Figs. 5 and 7; Poag et al., 2004, chapter 1). The lower 232 m of this borehole, located near the rim crest, penetrated a succession of nonmarine, Lower Cretaceous, siliciclastic sediments, mainly hematitic, sideritic, and cherty, coarse to medium sands and micaceous clays of the Potomac Formation (Fig. 9). The general seismostratigraphic expression of the Potomac Formation (in the western sector of the crater, as well) is a succession of low, moderate, and high-amplitude, subcontinuous to discontinuous, horizontal to gently tilted, parallel to convergent (onlapping) reflections (Figs. 3, 6, and 7). This expression is typical of stratified, nonmarine (fluvial, lacustrine), sedimentary deposits, as borne out by the strata encountered in borehole VA-VB-P-3 and many other boreholes in the southeastern Virginia Coastal Plain (Powars and Bruce, 1999; Poag et al., 2004, chapter 1). A dominantly coarse sand deposit constitutes the upper 40 m of the Potomac Formation in borehole VA-VB-P-3 (Fig. 9) and produces distinctive bounding reflections on the seismic profiles (Figs. 3, 6, and 7). Internal stratal reflections in this coarse sand unit are truncated at the outboard edges of Terraces 1 and 2, and the entire sand unit is missing from Terrace 1 on profile SEAX-8 (Fig. 7).

At 250 m in borehole VA-VB-P-3, an erosional unconformity separates the Potomac Formation from marine, fine to medium sands and pyritic, diatomaceous clays of middle Miocene age (Calvert Formation; Fig. 9). The Chickahominy and other Paleogene formations have been removed at this unconformity, which represents a hiatus of ~96 m.y.

Rim Structure

On each seismic profile of this study, the prominent rim-wall scarp is formed by the footwall of a normal fault that is downthrown toward the annular trough of the Chesapeake Bay crater (Figs. 2, 6, and 7). At true scale, the slope of the rim-wall scarp on the illustrated profiles ranges approximately from 15° to 20°. Previous work demonstrated that these faults are segments of a single, giant (85-km-diameter), ring-fault system, whose intersecting footwalls form the irregularly circular rim wall of the crater (Poag, 1996; Poag et al., 1999, 2004). On most profiles, Terraces 1 and 2 are seen to be the tops of incipient megablocks, which have not completely detached from the rim wall. Multichannel seismic profiles in the western sector of the crater show that many completely detached megablocks are present inboard of the rim wall and are buried by the Exmore breccia. These detached blocks cannot be distinguished, however, on most single and two-channel profiles, including all profiles that cross the eastern rim.

Likewise, most (but not all) of the terrace-bounding scarps are formed by the footwalls of normal faults. A few faults are downthrown away from the crater (e.g., Fig. 7), so in these cases, the scarp outboard from Terrace 2 is formed by a hanging wall rather than a footwall (e.g., SEAX-8, SEAX-9). These outer rim terraces in the eastern sector of the Chesapeake Bay crater are similar in dimensions, stratigraphy, and structure to those described by Poag et al. (2004, chapter 4) from the northern rim of the crater (their Figures 4.3A and 4.3B). The terrace-bounding faults and most other normal faults that cut through the Potomac Formation (Figs. 6 and 7) have true-scale dips of ~30°–60°.

The intersections of the rim-wall scarp and the crystalline basement are not imaged on most of the profiles in the eastern sector of the crater. The deeper parts of profiles E-3, SEAX-9, and SEAX-8, however, as well as most profiles in the western sector of the crater, show clearly that rim-wall scarps terminate against the surface of crystalline basement (not illustrated here, but shown by Poag et al., 2004, chapter 4). The basement surface acts as a glide plane, along which the detached sedimentary megablocks slide and slump into the annular trough. Differential fault motion among the detached and partially detached megablocks is indicated by drag folds. These distinctive drag features can be seen on several of the eastern sector profiles where the stratal reflections of the Potomac Formation intersect the fault planes (e.g., Figs. 6C, 6D, 7B and 7C).

DISCUSSION

In most previous studies of submarine impact craters, including Mjølnir crater: Tsikalas et al. (1998); Montagnais crater: Jansa et al. (1989); Lockne crater: Lindström et al. (1996); Ormö and Lindström (2000); Manson crater: Koeberl and Anderson (1996); and Chesapeake Bay crater: Poag et al. (2004), little or no trace of compressional uplift or stratigraphically overturned beds has been identified at the outer rim. On the other hand, the extensive record of subaerial terrestrial and extraterrestrial craters indicates that raised rims are a normal product of compressional forces during hypervelocity impact (Melosh, 1989). To explain this anomalous contrast, most investigators have concluded that surgeback effects during implosion of the transient water cavity would have eroded the original raised rim (and overturned beds) while enlarging the crater.

The profiles analyzed in this study show no compelling evidence of structural uplift, inverted stratigraphy, or other features to indicate that compressional forces have modified the outer rim in the eastern sector of the Chesapeake Bay crater. Though some profiles (e.g., SEAX-25 and SEAX-8) clearly show that stratification within the Potomac Formation dips eastward outboard from the crater and reverses to westward dip close to the rim-wall scarp, this reversal is not a compressional feature. It is produced by a combination of westward downfaulting along the rim wall, superimposed on eastward depositional and structural dip. Eastward dip is further accentuated on some profiles by normal faults that are downthrown to the east (e.g., Figs. 7A and 7C). Thus,

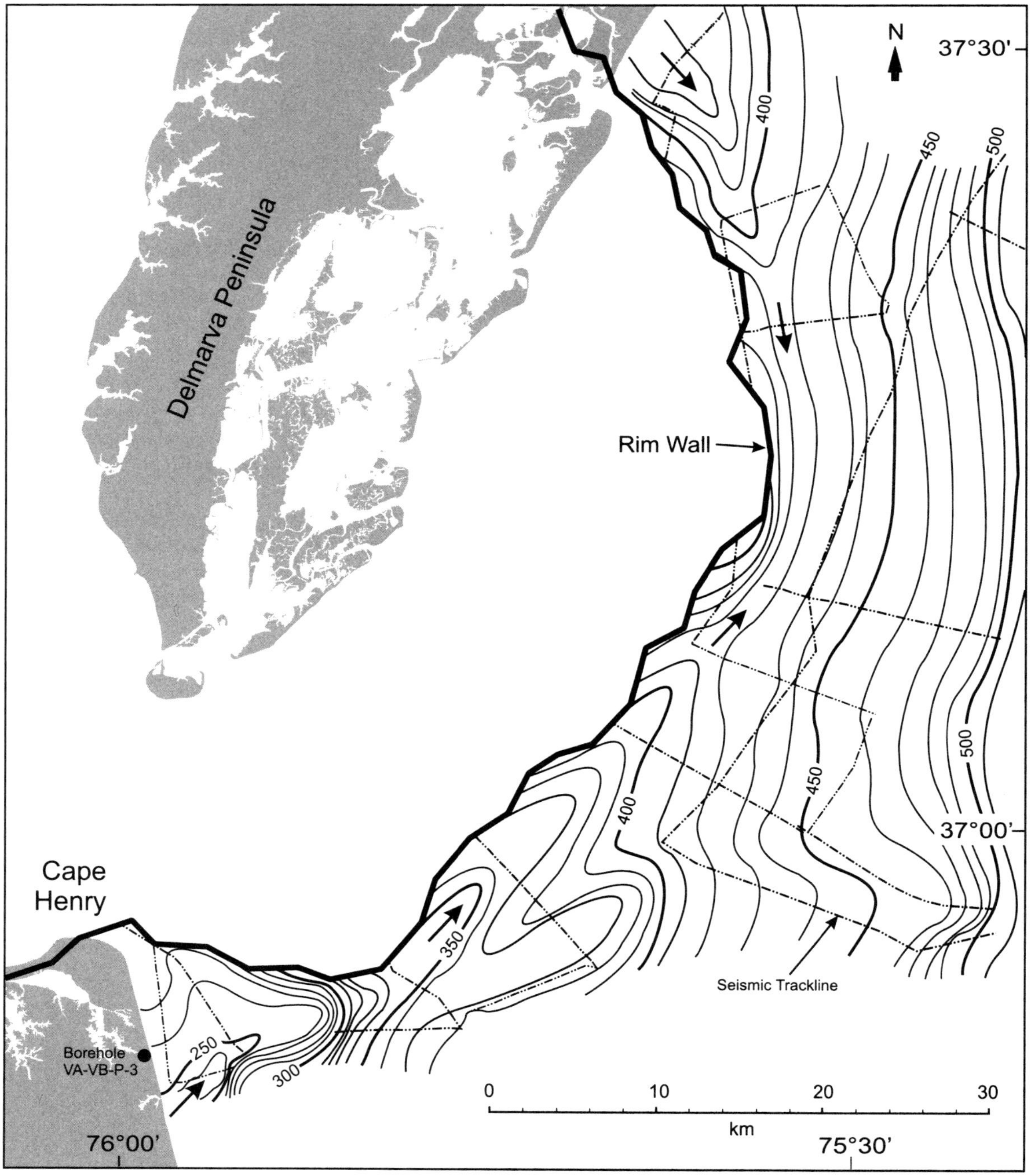

Figure 8. Structure map of the eastern rim of Chesapeake Bay impact crater, based on interpretation of seismic reflection profiles. Arrows indicate the location of the rim crest. Contour interval: 10 m. Figure includes additional profile tracklines, which allowed mapping of the rim structure between and beyond profiles illustrated in Figures 6 and 7. Compare with Figure 1.

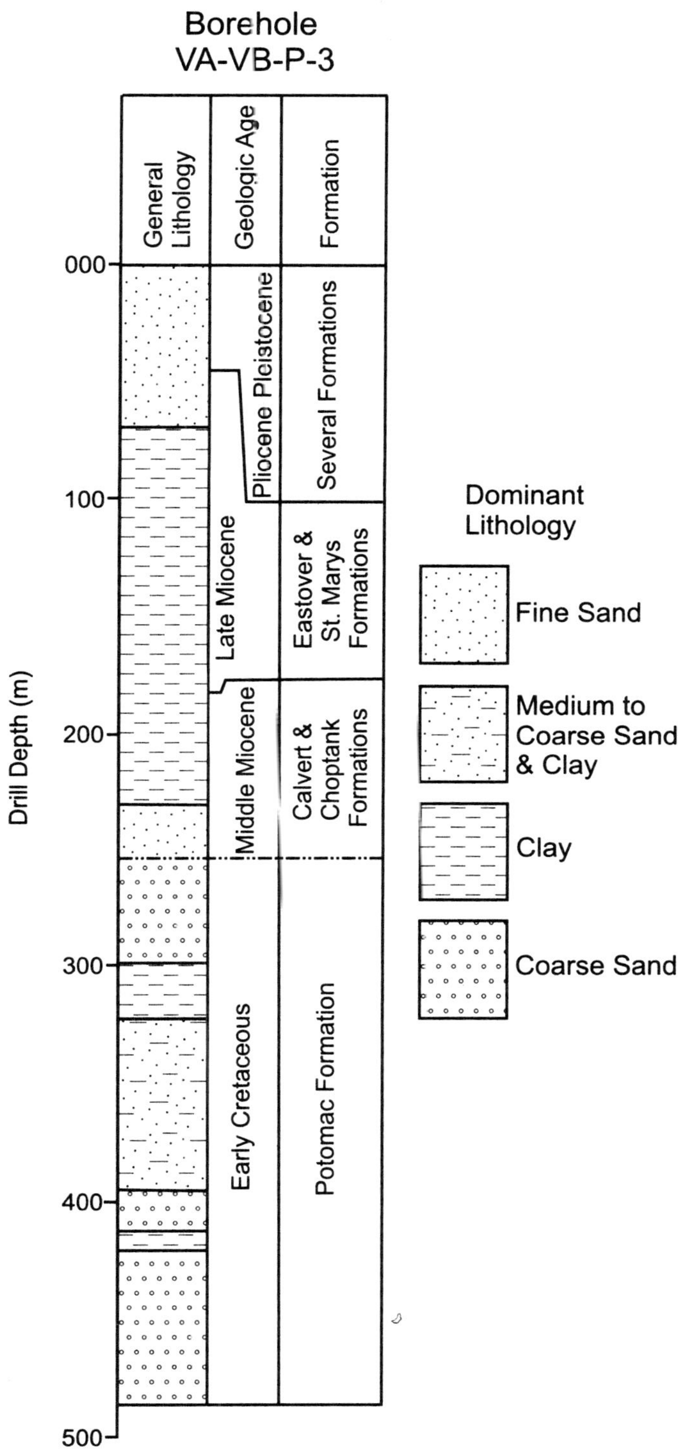

Figure 9. Stratigraphic column for borehole VA-VB-P-3, which calibrates the seismic profiles studied here (see Figs. 5 and 8 for location of borehole). Stratigraphy after Brown et al. (1972).

morphologically, the eastern rim of Chesapeake Bay crater is expressed as a raised ridge, but its internal stratigraphic and structural architecture give no hint of a compressional origin.

Instead, the pervading structural mode at the crater rim is extensional and dominated by normal faulting. Most prominent faults (those identifiable on seismic profiles) are parts of a crater-encircling concentric fault system (Poag et al., 2004, chapter 4; Figs. 1 and 2) formed during late-stage collapse of the sedimentary crater margin. This collapse produced two groups of fault-bounded megablocks. Group one contains detached megablocks, which have dropped fully (often rotating) into the annular trough of the crater (inboard of the rim wall; Poag et al., 2004, chapters 4 and 6). Group two contains incipient megablocks (outboard of the rim wall), which are fault-bounded, but not yet detached from the outer rim (Figs. 6 and 7).

Radial gullies attributed to surgeback flow are prominent erosional features at several submarine crater sites (Ormö and Lindström, 2000; von Dalwigk and Ormö, 2001; Suuroja et al., 2002), but only sparse evidence of such gullies has been inferred in the Chesapeake Bay crater (Poag et al., 2004). The seismic profiles studied here imaged no surgeback gullies either. However, the profiles are spaced relatively far apart (10–25 km), so large gullies actually could be present without having been crossed by any profile.

CONCLUSIONS

Seven vertically stretched seismic reflection profiles (vertical exaggeration ~30:1) across the eastern rim of the Chesapeake Bay impact crater provide new insights for interpretation of rim morphology, stratigraphy, and structure. In the study area, the crater rim is expressed morphologically as a curved ridge (concave toward the crater), typical of raised rims on most terrestrial and planetary impact craters. The inner (toward the crater) wall of this ridge is a steep footwall fault scarp, whose height averages 28 m above the surface of the crater-fill breccia (Exmore breccia). The rim-wall scarp is the principal boundary between two groups of megablocks: (1) an outboard group, which has not fully detached from the crater rim, and (2) an inboard group, which has fully detached, having slid and slumped into the annular trough and been buried by the Exmore breccia. Only the outboard megablocks are imaged on the profiles of this study. The tops of these outboard megablocks typically display two parallel, relatively flat terraces.

The rim wall in the eastern sector of the crater, as in the western sector, is constructed of nonmarine siliciclastic sediments assigned to the Lower Cretaceous Potomac Formation. Seismostratigraphic expression of the Potomac Formation reveals characteristic drag folds that reflect relative motion along the fault planes and also clearly indicates that the Potomac strata have been truncated at the rim wall and along the outboard terraces.

The profiles examined in this study reveal no compelling evidence that compressional forces contributed to the construction of the eastern "raised" rim of the crater. Instead, the rim exhibits dominantly extensional features, mainly normal faults, which are downthrown mainly toward the crater.

ACKNOWLEDGMENTS

I am grateful to Debbie Hutchinson, Steve Colman, Dave Foster, Tommy O'Brien, Barry Irwin, Dave Nichols, Jeremy Loss, John Evans, Nancy Soderberg, Tom Aldrich, Joe Newell, and John Diebold for collecting the seismic reflection data. The Committee for Research and Exploration of the National Geographic Society provided funding for the seismic surveys. I thank Steve Colman, Dave Foster, Jeff Plescia, Greg Gohn, and an anonymous reviewer for providing constructive criticisms of an early version of the manuscript.

REFERENCES CITED

Brown, P.M., Miller, J.A., and Swain, F.M., 1972, Structural and stratigraphic framework and spatial distribution of permeability of the Atlantic Coastal Plain, North Carolina to New York: U.S. Geological Survey Professional Paper 796, p. 1–79.

Jansa, L.F., Pe-Piper, G., Robertson, B.P., and Friedenreich, O., 1989, Montagnais: A submarine impact structure on the Scotian shelf, eastern Canada: Geological Society of America Bulletin, v. 101, p. 450–463, doi: 10.1130/0016-7606(1989)1012.3.CO;2.

Koeberl, C., and Anderson, R.R., eds., 1996, The Manson impact structure, Iowa: Anatomy of an impact crater: Geological Society of America Special Paper 302, 468 p.

Lindström, M., Sturkell, E.F.F., Törnberg, R., and Ormö, J., 1996, The marine impact crater at Lockne, central Sweden: Geologiska F'reningens i Stockholm F'rhandlingar, v. 118, p. 193–206.

Melosh, H.J., 1989, Impact cratering—A geologic process: Oxford University Press, New York, 245 p.

Ormö, J., and Lindström, M., 2000, When a cosmic impact strikes the seabed: Geological Magazine, v. 137. p. 67–80.

Poag, C.W., 1996, Structural outer rim of the Chesapeake Bay impact crater: Seismic and bore hole evidence: Meteoritics & Planetary Science, v. 31, p. 218–226.

Poag, C.W., 1997, The Chesapeake Bay bolide impact: A convulsive event in Atlantic Coastal Plain evolution, *in* Segall, M.P., Colquhoun, D.J., and Siron, D., eds., Evolution of the Atlantic Coastal Plain—Sedimentology, stratigraphy, and hydrogeology: Sedimentary Geology, v. 108, p. 45–95.

Poag, C.W., Powars, D.S., Poppe, L.J., and Mixon, R.B., 1994, Meteoroid mayhem in Ole Virginny: Source of the North American tektite strewn field: Geology, v. 22, p. 691–694, doi: 10.1130/0091-7613(1994)0222.3.CO;2.

Poag, C.W., Hutchinson, D.R., Colman, S.M., and Lee, M.W., 1999, Seismic expression of the Chesapeake Bay impact crater: Structural and morphologic refinements based on new seismic data, *in* Dressler, B.O., and Sharpton, V.L., eds., Impact cratering and planetary evolution II: Geological Society of America Special Paper 339, p. 149–164.

Poag, C.W., and Koeberl, C., and Reimold, W.U., 2004, The Chesapeake Bay Crater—Geology and geophysics of a late Eocene submarine impact structure: Springer-Verlag, Heidelberg, 522 p., with CD-ROM.

Powars, D.S., and Bruce, T.S., 1999, The effects of the Chesapeake Bay impact crater on the geologic framework and the correlation of hydrogeologic units of the lower York–James Peninsula, Virginia: U.S. Geological Survey Professional Paper 1612, 82 p.

Stone, D.S., 1991, Analysis of scale exaggeration on seismic profiles: American Association of Petroleum Geologists Bulletin, v. 75, p. 1161–1177.

Suuroja, K., Suuroja, S., All, T., and Floden, T., 2002, Kärdla (Hiiumaa Island, Estonia)—the buried and well-preserved Ordovician marine impact structure: Deep-Sea Research Part II, v. 49, p. 1121–1144.

Tsikalas, F., Gudlaugsson, S.T., and Faleide, J.I., 1998, Collapse, infilling, and postimpact deformation at the Mjølnir impact structure, Barents Sea: Geological Society of America Bulletin, v. 110, p. 537–552, doi: 10.1130/0016-7606(1998)1102.3.CO;2.

von Dalwigk, I., and Ormö, J., 2001, Formation of resurge gullies at impacts at sea: The Lockne crater, Sweden: Meteoritics & Planetary Science, v. 36, p. 359–369.

Manuscript Accepted by the Society 9 August 2004

Geological Society of America
Special Paper 384
2005

Original diameter and depth of erosion of the Popigai impact crater, Russia

V.L. Masaitis*
M.V. Naumov
M.S. Mashchak
Karpinsky Geological Institute, 199106 Sredny prospekt 74, St. Petersburg, Russia

ABSTRACT

Previous and recent geological, topographic, drilling, and geophysical data are summarized to acknowledge the 100 km original size of the relatively young and well-preserved multi-ring Popigai impact structure and to estimate the depth of its erosion. The relics of the bedrock rim are traced in the northern and western sectors of the crater, allowing the correct evaluation of the crater diameter. The depth of erosion varies in different crater sectors and concentric zones from 50 to 100 m to 300 m. The most significant erosion occurred on the crater rim. The crater rim has a complex inner structure and is composed of outwardly dragged and uplifted blocks of target rocks overlain by ejecta. During the 35 m.y. that have passed after the crater formation, ~1000–1200 km^3 of unconsolidated allogenic breccias were eroded from the ejecta blanket on the crater rim and from the upper layers of breccias and suevites of the central depression.

Keywords: Popigai impact crater, depth of erosion, impact craters, morphology, Russia, structural features.

INTRODUCTION

Various estimates of the diameters of large terrestrial impact structures remain questionable because most of the known craters are considerably modified by erosion or remain buried underneath younger sediments. Therefore, it is impossible to establish their primary rim-to-rim diameter (the diameter of the final crater, or the original diameter) exactly. As a result, diameters of most of the ancient terrestrial craters are usually measured using remaining structural features. In contrast, the Popigai impact structure that formed at the end of the Eocene in the northern part of the Siberian platform represents a well-preserved large impact structure with some relics of the elevated crater rim topography around the crater depression and remnants of the ejecta blanket.

Since the beginning of the 1970s, the Popigai crater has been well studied by means of geological mapping, geophysical surveys, and drilling (Masaitis et al., 1998, and references therein). The degree of preservation of the original Popigai crater structure and the numerous investigations carried out in the area have enabled researchers to estimate its original diameter and the depth of erosion with a high degree of precision. During the earlier period of investigations, the diameter of the Popigai crater was estimated to be around 100 km, but other crater diameter assessments were also presented (Dence, 1972; Pike, 1985). In this paper, we summarize geological, topographic, and geophysical data used to positively substantiate the size of the original Popigai crater.

*victor_masaitis@vsegei.ru

Masaitis, V.L., Naumov, M.V., and Mashchak, M.S., 2005, Original diameter and depth of erosion of the Popigai impact crater, Russia, *in* Kenkmann, T., Hörz, F., and Deutsch, A., eds., Large meteorite impacts III: Geological Society of America Special Paper 384, p. 131–140. For permission to copy, contact editing@geosociety.org.

There are some uncertainties regarding evaluation of morphometric parameters such as the diameter of large complex lunar craters and the explanation of some of their formation mechanisms (Melosh, 1989). In addition, many questions remain unanswered regarding formation of certain structural features in large terrestrial craters. The data from the Popigai crater presented below can be used to further constrain models of impact cratering.

GENERAL GEOLOGY OF THE POPIGAI CRATER

The Popigai crater (Fig. 1) is situated on the northern edge of the Siberian platform in a two-layered target that is composed of Archean and Lower Proterozoic crystalline basement overlain by a 1.5-km-thick cover of flat-lying Upper Proterozoic, Lower and Upper Paleozoic, and Mesozoic sediments (Masaitis et al., 1998, 1999; Masaitis, 2002b). Crystalline rocks are represented mainly by various gneisses and schists having general NW-SE strike and by minor granite intrusions. The sedimentary sequence is composed of siliciclastic and carbonate deposits and intruded by dolerite sills and dikes of Early Triassic age.

The Popigai crater is a multi-ring impact structure. Its main morphostructural characteristics are (1) a circular center depression (D = 40 km); (2) a peak ring (D = 45 km); (3) an annular trough (outer diameter D = 72 km); and (4) an outer flat terrace, where disturbed target rocks and relics of allogenic breccia may be observed (outer diameter D = 100 km). Small remnants of the ejected allogenic breccia were found as far as 75 km away from the crater's center; most of them are located north of the rim zone.

The brecciated and shock-metamorphosed crystalline rocks of the uplifted crater bottom form the peak ring. They are exposed in the northwestern sector of the crater, while in the other sectors they are buried beneath impactites and breccias. The inner slope of the peak ring rises more gently than the outer slope. Impactites and breccias fill the annular trough and the central depression; they are also found on the terrace, where partly disturbed sedimentary rocks occur. These disturbed rocks become less frequent with increasing distance from the center and disappear close to the outer edge of the ring terrace.

The ~2-km-thick crater fill consists of allogenic lithic breccias and impactites; the latter include tagamites (impact melt rocks) and suevites, both of which are diamond bearing. Impactites and breccias also occur in the vicinity of the crater as remnants of the ejecta blanket. The general sequence formed by allogenic breccias and impactites from the bottom upward is as follows: polymict allogenic breccia (predominantly blocky breccia, or megabreccia) in the lower part; alternating lens-like and sheet-like bodies of tagamites and suevites in the midsection; suevites and fine-grained polymict allogenic breccia (microbreccia) in the upper part. Chilled or partly crystallized impact melt that forms the matrix of tagamites (70–98 vol% of the rock) and chilled glass fragments in suevites (10–90 vol% of the rock) were produced by impact melting of gneisses basement with minor admixture of material from the platform cover (Masaitis et al., 1998).

The upper part of the crater fill shows features of sorting and relatively regular deposition (obscure-bedding, graded bedding, etc.). Buried dune-like features were observed within these deposits close to the crater's edge (Masaitis, 2003). Most likely, these features were formed on the fresh surface of the ejecta blanket outside the crater rim.

Neogene and Quaternary, mostly lacustrine, alluvial, and fluvio-glacial gravels and sands overlie the central part of the crater depression; they also occur in other places within the crater. Their total thickness locally reaches 200 m.

The age of the Popigai crater was established by Ar-Ar dating of impact melt rocks (tagamites and impact glasses) as end-Eocene (35.7 Ma; Bottomley et al., 1997). The impact event occurred in a continental setting, probably in either a semi-arid or humid climate conditions (Plotnikova, 1990).

TOPOGRAPHIC FEATURES OF THE PRESERVED CRATER

The circular topographic features of the crater area were observed long before the impact origin of the structure was recognized (Kiryushina, 1959). During the initial stage of prospecting at the beginning of the 1970s, possible remnants of the initial impact crater topography were revealed and the structural outer rim was outlined in some places. This allowed an estimate of the original crater diameter (diameter of final crater) to be ~100 km (Masaitis et al., 1976), although some authors citing these data suggested a significantly smaller diameter because the circular depression is only ~70 km across (Dence, 1972).

This circular topographic low (30–80 m absolute height) that represents the partly modified apparent crater bottom is the characteristic feature of the Popigai impact structure (Fig. 2). This area is flat, but is marked in the western part by arcuate ridges of ~150–200 m in height (200–250 m in absolute height) that are parallel to the crater's edge. The drainage system of the crater area shows radial and circumferential features, having inherited principal structural patterns of the impact structure (peak ring, ring trough, etc.). Some river valleys probably follow faults that may have formed during the late modification stage or more recent tectonic movements (Mashchak and Naumov, 2005).

The remnants of the original structural rim are observed in the western, northern, and partly in the northeastern sectors. They are located at a distance of ~10–15 km from the outer edge of the central topographic depression, which they encircle. They occur as discontinuous hills of ~50–150 m in height with small outcrops and rock debris on their slopes and summits. The original crater rim had a complex structure and is composed of displaced and uplifted target rocks (bedrock structural rim) that were buried under the allogenic breccia (embankment rim), which is now strongly eroded.

In the northern and northeastern sectors, the remnants of the lower part of this complex rim form hills consisting of thrust-faulted Cambrian carbonate rocks. The absolute height of these

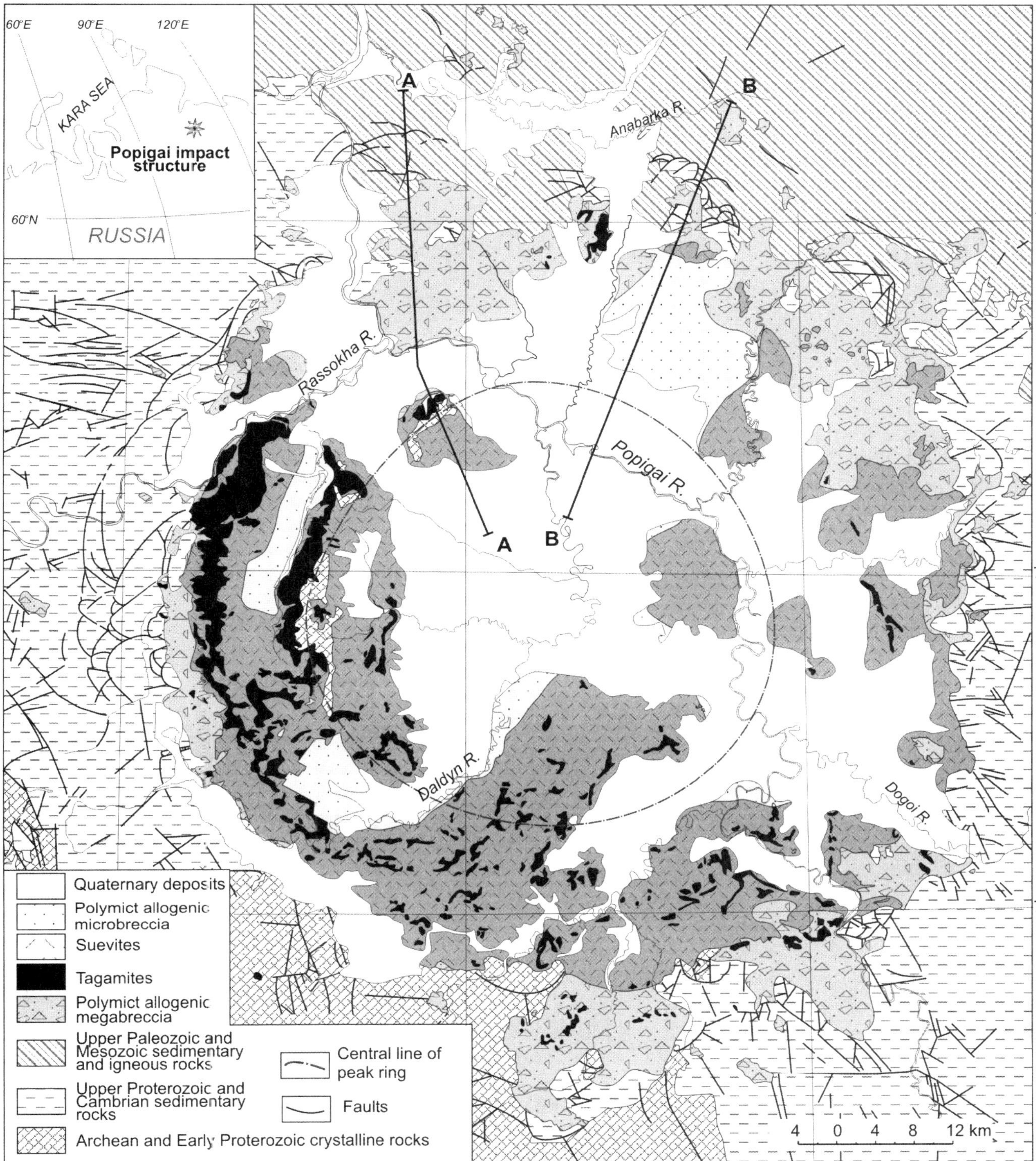

Figure 1. Geological map of the Popigai impact crater (partially modified after Masaitis et al., 1998). A–A and B–B are geological cross sections; see Figure 3.

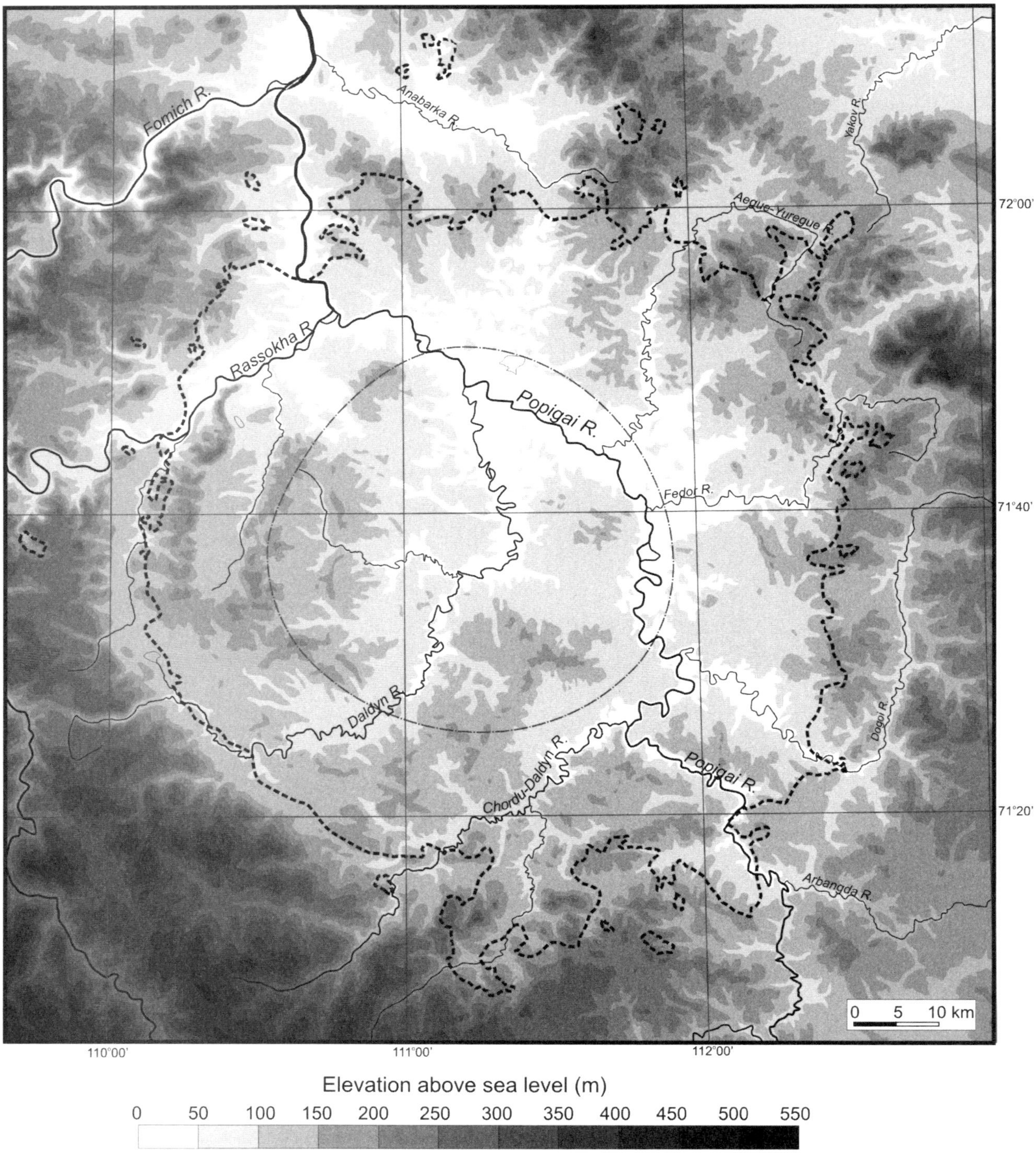

Figure 2. Topographic map of the Popigai impact crater. Dotted line: extent of the impactites and allogenic breccias; dashed line: central line of peak ring.

hills reaches 250–350 m in some places, while the maximum local relief in the vicinity is ~100–150 m. Thus, the topographic uplift of the bedrock in the rim zone is ~100–200 m. However, it could have been as high as 200–250 m before erosion took place. Small fragments of disintegrated ejected allogenic material (shocked Upper Proterozoic quartzites, Archean gneisses, particles of impact melt rock) are sometimes found scattered over the ground at the base of the hills and on their slopes.

This arcuate chain of hills in the northern and northeastern sectors can be traced up to 60–70 km from northwest to northeast, and their summits are located at a distance of ~50 km from the crater's center. Small patches of ejected allogenic breccia were mapped in the area northeast and east of these hills (Fig. 1). This breccia belongs to the relics of the ejecta blanket connected with the former embankment rim and probably represents the remnants of its outer gentle slope.

Similar features are observed in the northwestern sector. The isolated hills, composed of uplifted blocks of Upper Proterozoic sandstones, are ~250 m above sea level. The summits of the hills are located at a distance of 46–48 km from the crater's center, but the erosion level is deeper compared to the northeastern sector. The remnants of the ejecta blanket of blocky allogenic breccia are dispersed on the hill flanks. Taking into account that the eroded embankment rim crest was originally located somewhat farther from the crater's center relative to the position of the structural rim crest, it is evident that the distance from the crater's center to the embankment rim crest was ~50 km.

In the eastern and southern sectors of the crater, reliable relics of the former structural rim are not observed. Though numerous faults and irregular folds were mapped here, imbricated overlapping of disturbed sediments is rare. Thrusts of the Upper Proterozoic quartzites over the Cambrian dolomites were observed on the eastern slope of the Popigai river valley in the southeastern sector. These thrusts are locally overlain with allogenic breccia.

Two cross sections in the northern sector illustrate the interrelations between the topographic features and the inner geological structure of the target rocks and the crater fill (Fig. 3). They are discussed in detail below.

STRUCTURAL PATTERNS OF THE OUTER CRATER ZONE

The reconstruction of the outer crater zone is based on data from geological mapping, drilling, and partly from geophysical surveying. The relief of the base of the allogenic breccia and impactites (Fig. 4) was traced close to the edge of the annular trough and in the area of the outer terrace, where this base intersects the present surface. This reconstruction shows that the +150 m contour line of the surface of the target rocks may be considered as the outer border of the terrace outlined by discontinuous remnants of the bedrock of the crater rim. In the northeastern sector, the distance of this contour line from the crater's center is ~50 km, but does not exceed 48 km in the northern and northwestern sectors. Several radial troughs filled with ejecta material cause the festoon-like shape of the +150 m contour line.

This study also shows that the discontinuous distribution of the bedrock forming the structural rim occurred before the deposition of the ejected allogenic material. The surface irregularities are characteristic of the crater's outer zone that embraces the terrace, the structural rim on its outer border, and areas outside the rim. Some of these irregularities possibly inherited the pre-impact relief of the target rocks, although most of them were formed during cratering. This is proved by the presence of isolated, round depressions of ~2–4 km width filled with allogenic megabreccia (e.g., small remnants of this breccia outside of the structural rim in the western sector).

The inner part of the structural rim and parts of the terrace zones are characterized by numerous arcuate thrusts, dipping toward the crater's center (Masaitis et al., 1976, 1998). These thrusts display multiple imbrications with lateral displacements from some hundreds of meters to several kilometers (Fig. 3). Well-documented collapses of the crater wall and terrace formation are rare. Nevertheless, such terrace-like subsidence can be observed in outcrops on the eastern bank of the Popigai River, downstream of the Rassokha River mouth (northern edge of the ring trough). These features reflect the complex outward and, in places, inward motion that occurred during the excavation and early modification stages of cratering.

Similar outward and inward motion of bedrock material in the rim zone was revealed by drilling in the Kaluga impact crater, Central Russia. In this crater, subsidence of the terraces occurred after the uplift of the edge of the transient crater and the associated outward displacement and imbrication of the crystalline basement (Masaitis, 2002a).

GRAVITY FIELD AND ITS IRREGULARITIES

Gravity field data over the Popigai crater and the surrounding region have been published (Masaitis et al., 1998; Pilkington et al., 2002). The gravity anomaly of more than −30 mGal, associated with the crater area, is one of the highest compared to any other terrestrial impact structure. This anomaly consists of a central low surrounded by a gravity high, which, in turn, is encircled by a gravity low. The anomalies correspond with the central depression, the peak ring, and the annular trough of the impact crater.

More detailed gravity mapping (Fig. 5) shows the presence of a semicircular discontinuous high in the northern part of the crater area surrounding the circular low that reflects the annular trough. This arcuate positive anomaly coincides in general with the arcuate discontinuous structural rim described above. The axis of this semicircular high is located at a distance of ~49–51 km from the crater center.

The study of the inner irregularities of this discontinuous semicircular gravity high reveals some bilateral symmetry with an axis striking at 340° NW. Its western and eastern branches

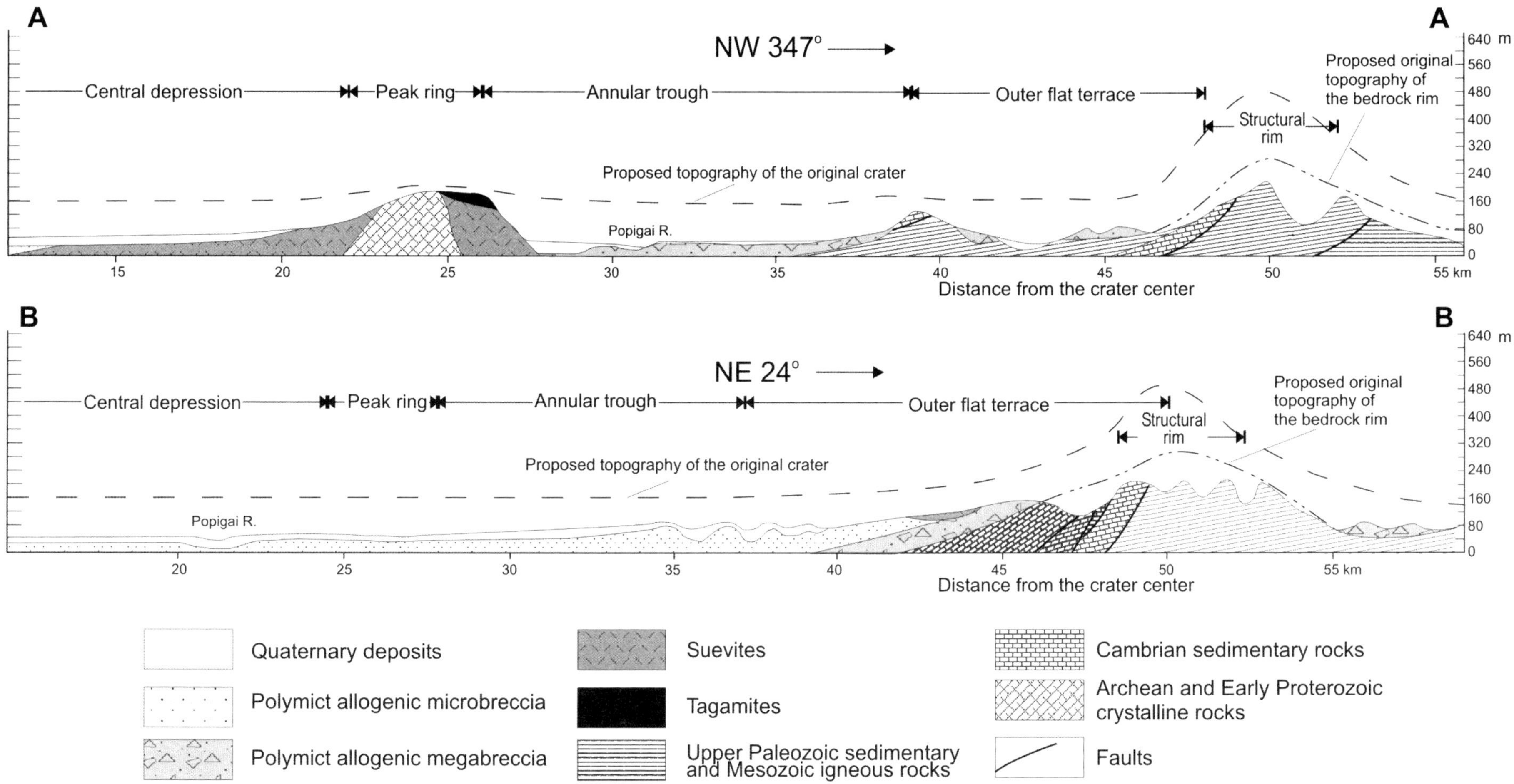

Figure 3. Topographic and geologic cross sections in the northern sector of the Popigai crater (profile locations noted on Fig. 1).

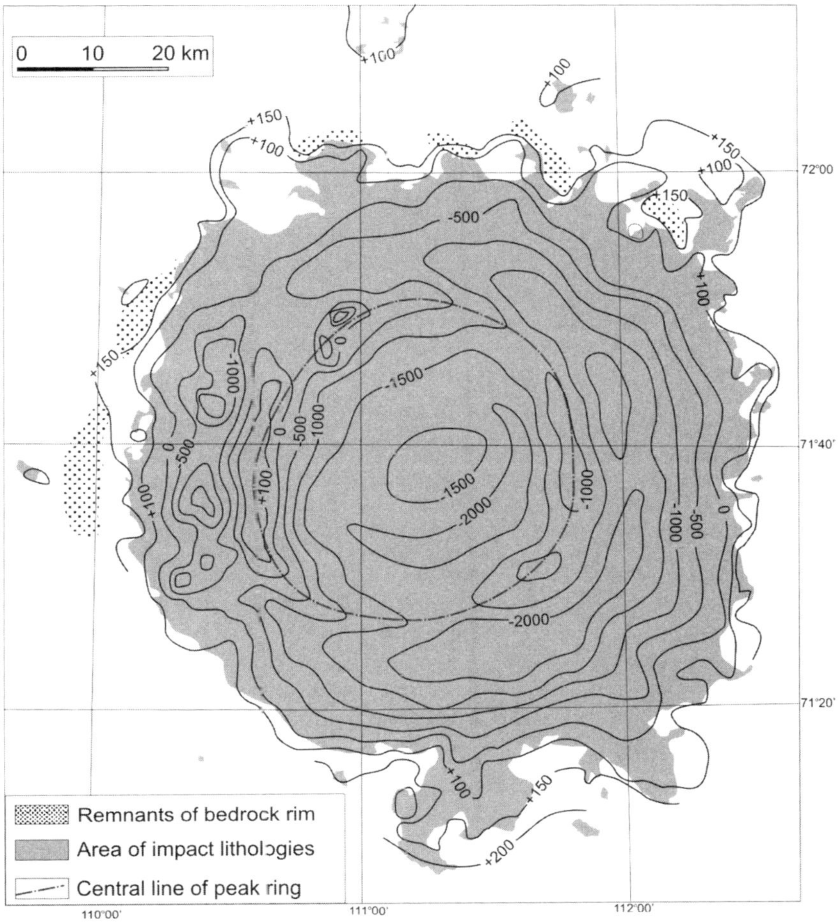

Figure 4. Map of isohypses of the base of impactites and allogenic breccias inside the Popigai crater.

consist of several smaller elongated positive anomalies, each extending to 6–12 km. The axes of these "subanomalies" (or second order anomalies) are aligned at angles of about 15°–30° relative to the axis of the principal semicircular high; thus, these "subanomalies" look like two en echelon–shaped patterns reflecting multiple imbrications of uplifted blocks of crystalline basement and the overlying sediments. The multiple series of arcuate thrust faults dividing relics of the structural rim in the outer zone of the crater's northern part reflect the displacements of underground crystalline basement. The average size of displaced blocks of Cambrian and Upper Proterozoic sediments is ~2–4 km; they are lens-like in three dimensions. The thrust surfaces are listric, and the whole structural feature reflects the outward motion of imbricated slab-like masses of rocks that form the bedrock uplift in the rim zone.

The discontinuous gravity high is not traced in the southern part of the crater's outer zone. This difference between the patterns of the gravity field in the northern and southern parts of the outer zone may partly reflect regional differences in density of the crystalline rocks in the areas north-northwest and south-southeast of the crater (Fig 5).

Nevertheless, in the southern part of the crater, the negative low that coincides with the annular trough changes to a positive anomaly outwards. Small, more intense highs are observed only in some places and most likely coincide with small uplifted and overthrusted crystalline blocks. These small patterns are located at a distance of ~50 km from the crater's center.

DEPTH OF EROSION

During the last 35 m.y., erosion and accumulation of post-impact deposits accompanied by some tectonic movements have modified the initial morphology of the Popigai impact crater (Plotnikova, 1990). The piled up crater rim, the main part of the ejecta blanket outside the crater, and the upper horizons of the crater fill were destroyed. The discovery of impact diamonds dispersed in some riverbeds far from the crater indicates that the suevite-bearing ejecta blanket originally extended as far as two crater radii or more from the crater's center, being later almost completely eroded.

The estimation of the heights of the final crater rim (H) of large terrestrial impact craters is problematic. There are

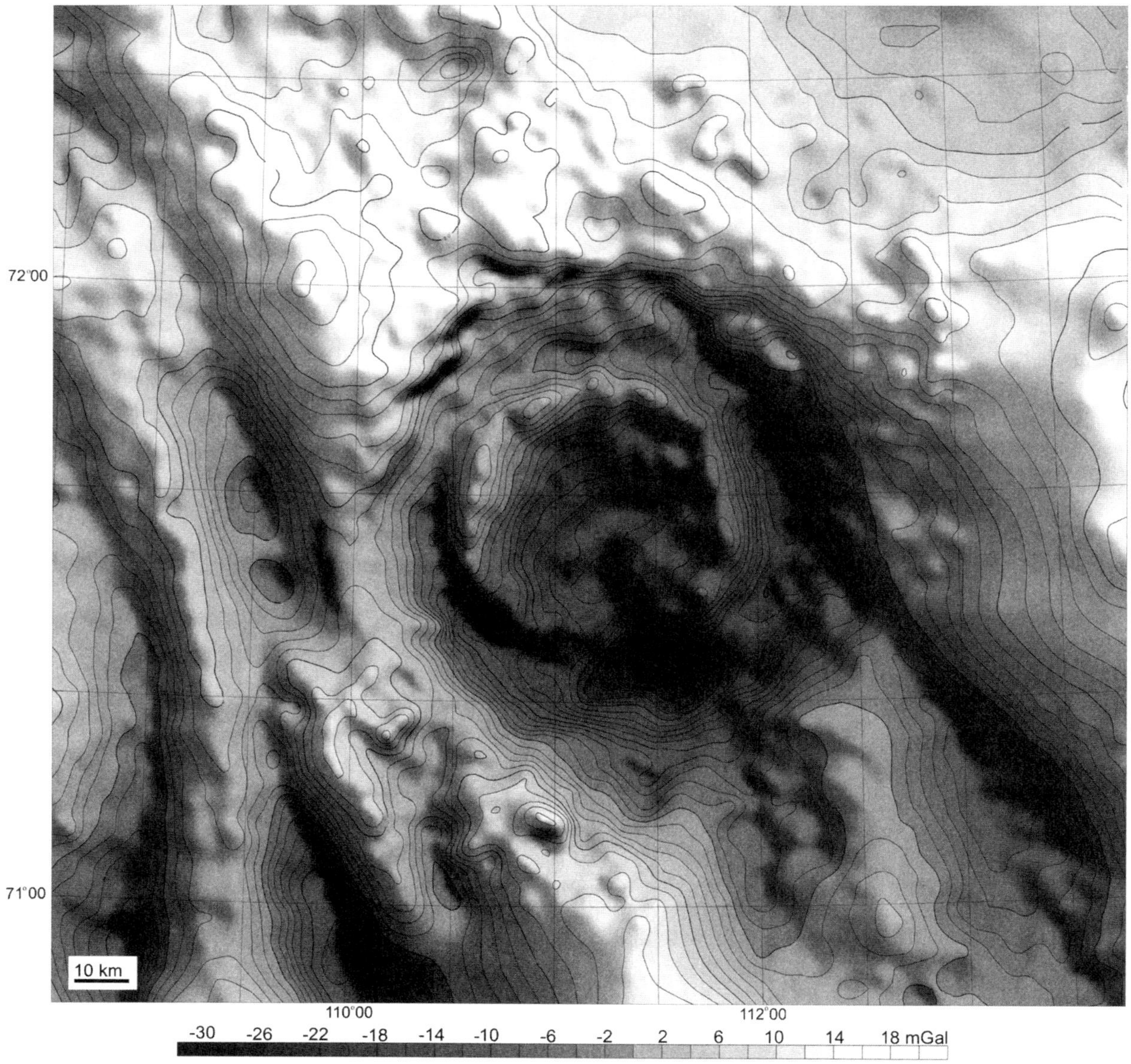

Figure 5. Gravity map (Bouguer anomaly) of the Popigai impact crater and adjacent area, extracted from the gravity map of the Taymyr region (courtesy of V.V. Koshevoi). Compiled by Institute VNII (Vserossiisky Naucno-Issledovatel'sky Institut Geofisiki) Geophysika, based on the data of the Geophysical Expedition number 3, "Krasnoyarskgeologia."

estimations related to the large lunar craters (Pike, 1977; Melosh, 1989), but their application to terrestrial craters is questionable. The morphology of large complex craters on the Moon and on Earth is different, and this prevents applying the ratios between diameter and rim height compiled for lunar craters directly to terrestrial ones. A correction is needed, taking into account the differences in gravity fields and the presence of atmosphere. Moreover, there is no exact evaluation of the ratio between heights of the structural (h_s) and piled-up material (or thickness of ejecta, T) of the final rim, although some data (Pike, 1976; Melosh, 1989) show that it can vary from 1:1 to 2:3.

Some suggestions can be made based on data obtained from examination of a few relatively fresh craters that have preserved remnants of their original rims partially overlain with ejected material. Data from the Flynn Creek crater (Roddy, 1977), which has a diameter of ~4 km, show that structural rim uplift is ~12 m and the thickness of ejecta on the rim varies between 50 and 100 m. Another example comes from the study of the Kaluga impact crater, which has a diameter of ~15 km (Masaitis, 2002a). The buried remnants of the structural bedrock rim in this crater have a height of ~170–250 m and in some places the allogenic breccia accumulates on the rim to ~30–100 m. After crater collapse took place at the modification stage, the structural rim was apparently destroyed in these parts by an inflow of resurgent wave that washed out most of the piled-up allogenic breccia (Masaitis, 2002a).

According to the geological map of the Ries crater (D = 26 km) and its relief model (Schmidt-Kaler et al., 1970), the remnants of slightly uplifted bedrock in the southern and western crater sectors (areas of highest topography) lie ~50–100 m above the pre-crater level. These topographic highs coincide with the so-called "tectonic rim" (Stöffler, 1977; Pohl et al., 1977). The thickness of adjacent remnants of ejected allogenic polymict breccia does not exceed a few tens of meters, but originally it could have been up to 100–150 m or more in the rim crest, since the preserved thickness of allogenic breccia (Bunte Breccia), based on drilling data at a distance of 20–30 km from the crater's center, was measured to reach 50–80 m locally (Hörz et al., 1977). This means that erosion in the remnants of the bedrock rim was insignificant, and the final bedrock rim was relatively flat before erosion.

These examples show that the height of the final structural rim, h_s, of complex terrestrial craters with diameters, D, between 4 and 26 km varies slightly and is given by

$$h_s = kD/100,$$

where k is a coefficient that varies between 0.4 and 1.3. The coefficient seems to be lower in larger craters. The estimated total heights of the crater rim can exceed the obtained values by a factor of two. All these data enable us to make a first approximation of the original height of the Popigai crater rim and depth of its erosion.

According to the data presented above, the estimated average of the original height of the Popigai crater's complex rim was ~350 m above the pre-crater level. Taking into account that the height of the structural rim was ~50–150 m, the accumulated material (allogenic breccia, including microbreccia and suevites) composing the upper part of the rim probably had a total thickness of up to 150–250 m. It is possible that this thickness was not the same all around the crater due to azimuthal irregularities of ejection. Comparing the positions of the structural rim crest and the piled up rim crests, it could be suggested that the latter occurred farther from the crater's center and that the rim-to-rim primary crater diameter was slightly larger than 100 km. Additional proof is based on the supposed width of this rim. If we compare the width (W) of the rim with the widths of the rims of the complex lunar craters with similar diameter ($W = 0.5D$), then all remnants of allogenic breccia that are now distributed around the Popigai crater should be regarded as material from the outer slope of its piled-on rim.

According to Pike (1977), the final diameter/rim height ratio (D/H) shows that the height of a complex lunar crater rim equals $H = 0.236\ D^{0.399}$. However, this does not match the geological and morphological observations made on the terrestrial craters cited above. The deviation is possibly caused by differences in the mode of ejecta transport. In case of a fluidized flow of ejected material, the structural rim can be partially cut off, resulting in a lowered total height of the rim. However, this does not occur in the case of ballistic transport.

It seems more realistic to apply the formula for the estimation of the crater's rim breccia thickness from McGetchin et al. (1973) to the terrestrial craters. This formula is given by:

$$T = 0.14R^{0.74},$$

where T is the thickness of the ejecta on the crater's rim and R is the crater's radius. The obtained values for T are close to the geological observations made in craters mentioned above and in the Popigai crater as well, but are probably slightly overestimated.

A thorough study of the recent evolution of the Popigai crater relief (Plotnikova, 1990) showed that it was subjected to several cycles of denudation and accumulation during Neogene and Quaternary times. It is possible that the notable variations in the amount of denudation inside the impact crater are caused by uneven block displacements during a late modification stage (Mashchak and Naumov, 2005). The maximum depth of erosion occurred in the rim area, where most of the ejected material and parts of the structural rim itself were removed. This depth was estimated to be as much as 300 m or more in the southwestern sector (where the structural rim was completely destroyed and the target rocks were also partly eroded) and ~200–250 m in the northwestern sector. This difference was caused by the irregular tilting of the crater area. The denudation was less significant in the inner depression of the crater. Originally, the semicircular ridges in the western part of the central depression were buried under not less than 50–100 m of loose ejecta (microbreccia and suevites). However, in some places their summits were exposed on the surface.

The thickness of these loose rocks in the central part of the depression before erosion took place could have been up to 100–150 m. A similar amount of erosion for the Popigai crater during the Oligocene-Quaternary time interval was established based on the study of the evolution of the original relief (Plotnikova, 1990).

The total volume of breccias and suevites removed from crater area (taking into account the rate of erosion during the recent geological periods) is ~1000–1200 km^3. The main portion of this lost volume was composed of unconsolidated allogenic breccia (microbreccia or coptoclastite) that formed the uppermost layer of the crater fill. In general, ~20% of all the impact rocks were eroded.

CONCLUSIONS

The 35 Ma Popigai crater is one of the better preserved large impact structures on Earth. The preservation of much of the outer rim allows an accurate interpretation of its original size and the depth of its erosion. All data gained from the analysis of topography, geological record, and gravity measurements show that the original rim-to-rim diameter of the Popigai crater (i.e., diameter of the final crater formed at the end of the early modification stage) may be estimated at 100 km, but probably was slightly (2–3 km) larger.

The original complex rim (final crater rim) was composed of the structural uplift of disturbed target rocks covered with ejected allogenic material (lithic breccias and suevites). The average total height of the rim is interpreted to have been ~350 m, but probably was not uniform all around the crater. The thickness ratio between the structural uplift of bedrocks and the accumulated allogenic material varies between 0.25 and 1. The well-preserved remnants of the structural rim in the northern part of the outer crater zone have a complex inner structure characterized by uplifts combined with thrusts. These dislocations affect both the crystalline basement and sedimentary cover and were produced by lateral drag of the uppermost layers of target rocks.

The crater area was subjected to several cycles of erosion and accumulation of sand and gravel derived from disintegration of impact rocks (lithic breccias and impactites) as well as other rocks from surroundings areas. Loose breccia and unconsolidated suevites were eroded first, at a time when solid impactites formed by chilling and crystallization of impact melt (tagamites) were more stable. The depth of erosion was irregular and may be estimated to vary in different crater areas, between 300 m (southern outer zone), 100–250 m (area of structural rim in the northern portion), and 50–150 m (central depression). Some of the irregularities in erosion and denudation were caused by post-impact and more recent tectonic movements.

The defined principal data of the original Popigai impact structure morphology and depth of its erosion provide a base for a more precise reconstruction of the cratering parameters and the cratering event. Moreover, these data may be helpful for a morphological reconstruction of other impact craters on Earth having a size similar to Popigai.

ACKNOWLEDGMENTS

We are grateful to Prof. Roger L. Gibson, Dr. David S. Powars, and Dr. Thomas Kenkmann, whose patient reviews led to considerable improvements of the paper.

REFERENCES CITED

Bottomley, R., Grieve, R., York, D., and Masaitis, V.L., 1997, The age of the Popigai impact event and its relation to events on the Eocene/Oligocene boundary: Nature, v. 388, p. 365–368, doi: 10.1038/41073.

Dence, M.R., 1972, The nature and significance of terrestrial impact structures: International Geological Congress, XXIV session, Section 15: Montreal, Planetology, p. 77–89.

Hörz, F., Gall, H., Hüttner, R., and Oberbeck, V.R., 1977, Shallow drilling in the "Bunte Breccia" impact deposits, Ries Crater, Germany, *in* Roddy, D.J., Pepin, R.O., and Merrill, R.B., eds., Impact and explosion cratering: New York, Pergamon Press, p. 425–448.

Kiryushina, M.T., 1959, The manifestation of Mesocenozoic volcanism in the northern region of the Siberian platform (in Russian): Izvestiya AN SSSR, serya geol., no.1, p. 15–22.

Masaitis, V.L., 2002a, The middle Devonian Kaluga impact crater (Russia): New interpretation of marine setting: Deep-Sea Research, Part II, Topical Studies in Oceanography, v. 49, no. 6, p. 1157–1169, doi: 10.1016/S0967-0645(01)00142-4.

Masaitis, V.L., 2002b, Popigai crater: General geology, *in* Plado, J., and Pesonen, L.J., eds., Impacts in Precambrian shields: Berlin, Springer-Verlag, p. 81–86.

Masaitis, V.L., 2003, Obscure-bedded ejecta facies from the Popigai impact structure, Siberia: Lithological features and mode of origin, *in* Koeberl, C., and Martinez-Ruiz, F., eds., Impact markers in the stratigraphic record: Berlin, Springer-Verlag, p. 137–162.

Masaitis, V.L., Mikhaylov, M.V., and Selivanovskaya, T.V., 1976, The Popigai meteorite crater: Washington, D.C., English technical translation, National Aeronautics and Space Administration F-16,900, 167 p.

Masaitis, V.L., Mashchak, M.S., Raikhlin, A.I., Selivanovskaya, T.V., and Shafranovsky, G.I., 1998, Diamond-bearing impactites of the Popigai astrobleme: St. Petersburg, All-Russia Geological Research Institute (VSEGEI) Press, 178 p. (in Russian).

Masaitis, V.L., Naumov, M.V., and Mashchak, M.S., 1999, Anatomy of the Popigai impact crater, Russia, *in* Dressler, B.O., and Sharpton, V.L., eds., Large meteorite impacts and planetary evolution II: Geological Society of America Special Paper 339, p. 1–18.

Mashchak, M.S., and Naumov, M.V., 2005, Late modification-stage tectonic deformation of the Popigai impact structure, Russia, *in* Koeberl C., and Henkel, H., eds., Impact tectonism: Berlin, Springer-Verlag (in press).

McGetchin, T.R., Settle, M., and Head, J.W., 1973, Radial thickness variation in impact crater ejecta: Implications for lunar basin deposits: Earth and Planetary Science Letters, v. 20, p. 226–236, doi: 10.1016/0012-821X(73)90162-3.

Melosh, H.J., 1989, Impact cratering: A geological process: Oxford, Oxford University Press, 245 p.

Pike, R.J., 1976, Ejecta thickness, rim uplift, energy type, and depth of energy release: Symposium on planetary cratering mechanics, LPI Contribution 259: Houston, Texas, Lunar and Planetary Institute, p. 105–107.

Pike, R.J., 1977, Size-dependence in the shape of fresh impact craters on the moon, *in* Roddy D.J., Pepin R.O., and Merrill R.B., eds., Impact and explosion cratering: New York, Pergamon Press, p. 489–510.

Pike, R.J., 1985, Some morphologic systematics of complex impact structures: Meteoritics, v. 20, p. 49–68.

Pilkington, M., Pesonen, L.J., Grieve, R.A.F., and Masaitis, V.L., 2002, Geophysics and petrophysics of the Popigai impact structure, Siberia, *in* Plado, J., and Pesonen, L.J., eds., Impacts in Precambrian shields: Berlin, Springer-Verlag, p. 87–107.

Plotnikova, M.I., 1990, Essay of the post-Oligocene history of the Popigai impact morphostructure: Meteoritika, v. 49, p. 154–163 (in Russian).

Pohl, J., Stöffler, D., Gall, H., and Ernston, K., 1977, The Ries impact crater, *in* Roddy D.J., Pepin R.O. and Merrill R.B., eds., Impact and explosion cratering: New York, Pergamon Press, p. 343–404.

Roddy, D.J., 1977, Tabular comparison of the Flynn Creek impact crater, United States, Steinheim impact crater, Germany, and Snowball explosion crater, Canada, *in* Roddy D.J., Pepin R.O., and Merrill R.B., eds., Impact and explosion cratering: New York, Pergamon Press, p. 125–162.

Schmidt-Kaler, H., Treibs, W., and Hüttner, R., 1970, Geologische übersichtskarte des Rieses und seiner umgebung 1:100 000; exkursionführer zur geologischen übersichtskarte des Rieses, 1:100 000: München, Bayerisches Geologisches Landesamt, 68 p.

Stöffler, D., 1977, Research drilling Nördlingen 1973: Polymict breccias, crater basement, and cratering model of the Ries impact structure: Geologica Bavarica, v. 75, p. 443–458.

Manuscript Accepted by the Society 9 August 2004

Geological Society of America
Special Paper 384
2005

Topography over the Chicxulub impact crater from Shuttle Radar Topography Mission data

Gary L. Kinsland*
Energy Institute, University of Louisiana at Lafayette, Lafayette, Louisiana 70504, USA

Kevin O. Pope
Geo Eco Arc Research, 16305 St. Mary's Church Road, Aquasco, Maryland 20608, USA

Manuel Hurtado Cardador
Instituto Mexicano del Petróleo, Eje Central Lazaro Cardenas 152, Mexico D.F. 07730, México

Gordon R.J. Cooper
Department of Geophysics, University of Witwatersrand, Johannesburg 2050, South Africa

Duncan R. Cowan
Cowan Geodata Services—Consulting Geophysicists, 12 Edna Road, Dalkeith, Western Australia 6009, Australia

Michael Kobrick
SRTM Project Scientist, Jet Propulsion Laboratory, Pasadena, California 91109, USA

Gary Sanchez
Regional Application Center, University of Louisiana at Lafayette, Lafayette, Louisiana 70504, USA

ABSTRACT

Shuttle Radar Topography Mission (SRTM) data over the Chicxulub impact crater are imaged and compared to previously available topography data. While the two data sets contain different biases related to variations in terrain and vegetation cover, the correspondence of the two sets supports earlier interpretations that the complex structure of the buried crater is expressed in the topography of the northwestern Yucatán Peninsula, México.

Keywords: Chicxulub, SRTM, topography, impact crater.

INTRODUCTION

The Chicxulub structure was first suggested as an impact crater from magnetic and gravity data (Penfield and Camargo, 1981); however, topography and the distribution of sinkholes (the Cenote Ring) were important in the early recognition that Chicxulub is the site of the end-Cretaceous impact (Pope et al., 1991, 1993). Hildebrand et al. (1995) noted that over at least part of the extent of the Cenote Ring, the cenotes occur in a valley. Pope et al. (1996) used elevation data from a gravity database developed by Petróleos Mexicanos (PEMEX) in the 1950s and 1960s to produce the first detailed topographic map of the region. This topographic analysis revealed a series of subtle concentric troughs both inside and outside of the main trough that contains

*glkinsland@louisiana.edu

Kinsland, G.L., Pope, K.O., Hurtado Cardador, M., Cooper, G.R.J., Cowan, D.R., Kobrick, M., and Sanchez, G., 2005, Topography over the Chicxulub impact crater from Shuttle Radar Topography Mission data, *in* Kenkmann, T., Hörz, F., and Deutsch, A., eds., Large meteorite impacts III: Geological Society of America Special Paper 384, p. 141–146. For permission to copy, contact editing@geosociety.org.

the Cenote Ring (Pope et al., 1996), confirming that the buried Chicxulub impact crater has a complex surface expression. Correlations between the topography and detailed gravity data led Kinsland et al. (2000) to propose that many of the surface features and their associated gravity anomalies are the result of solution along fractures propagated upward into the Tertiary carbonates from buried impact structures.

Some of the topographic features we interpreted in our analyses of the PEMEX data are subtle (Pope et al., 1996; Kinsland et al., 2000) (Fig. 1). Also, the PEMEX data contain gaps and are based on an uneven sample distribution (Fig. 2). Therefore, uncertainties remained and several of our interpretations of buried crater structures have been questioned (Connors et al., 1996). In this paper, we compare the previous PEMEX topographic data with the recently acquired Shuttle Radar Topography Mission (SRTM) data over northwestern Yucatán and offer some new interpretations of the surface expression of the buried Chicxulub impact crater.

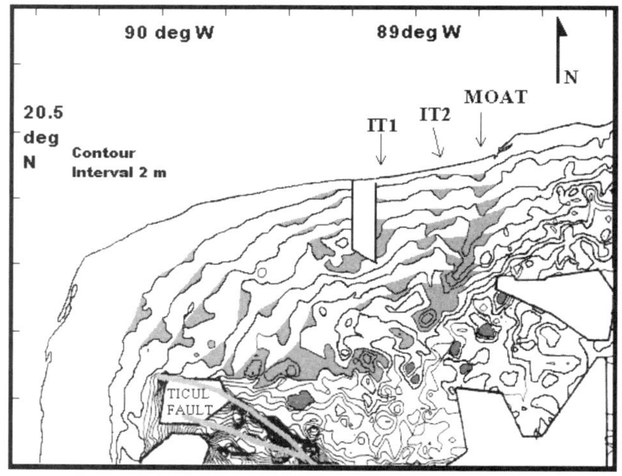

Figure 1. Topography of the Chicxulub impact crater region as derived from the location and elevation data in the Petróleos Mexicanos gravity data files. The continuous, semi-concentric valleys in the data are highlighted by gray infilling. This figure is modified after Kinsland et al. (2000), who adapted the figure from Pope et al. (1996).

SRTM DATA

In February 2000 the United States of America National Aeronautic and Space Administration (NASA) flew a space shuttle *Endeavor* mission, which collected C-band radar interferometric topographic data over Earth from 60° N latitude to 56° S latitude (Farr and Kobrick, 2001). Portions of these data, Shuttle Radar Topography Mission (SRTM) data, were processed with 90 m pixels and made available to selected principal investigators ahead of the scheduled batch processing of the global data set. We received data over the Yucatán Peninsula in an area covering six one-degree squares (corners of the composite area are at 20° N, 91° W and 22° N, 88° W), which show the topography over the crater for first time with uniform data coverage. We merged the six one-degree square data sets supplied to us by the Jet Propulsion Laboratory and color-coded the elevations (Fig. 3) to best display the topographic features of the buried crater (Kinsland et al., 2002, 2003). In Figure 4, we superimpose the interpretations of Kinsland et al. (2000) over the topographic data of Figure 3. Figure 5 is an image produced by subjecting the SRTM data to fractional vertical gradient processing as described in Cooper and Cowan (2003). In this image, the fractional vertical derivatives 0.3, 0.5, and 0.7 of the SRTM elevation data are, respectively, represented by the colors red, green, and blue. This processing has emphasized the shorter wavelength features in the topography data (as well as some obvious linear artifacts of the data collection and processing system). Figure 6 compares the resolution of the SRTM and PEMEX data sets.

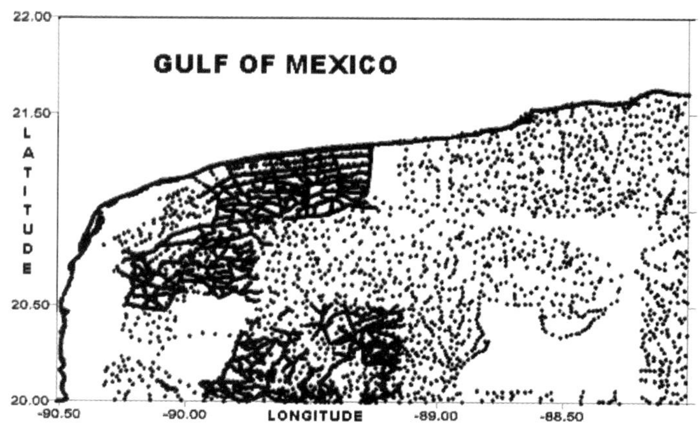

Figure 2. Locations of the elevation points utilized to create Figure 1. The data are from gravity data files collected mostly by Petróleos Mexicanos in the 1950s and 1960s. The uneven distribution of locations is the result of some areas being covered only by decimated data sets available from the United States Defense Mapping Agency as the original, much denser, data were lost to fire in México. The shoreline points in this figure were added from separate digital data and were not included in the creation of the other contours.

The SRTM topography data are produced by C-band radar (~6 cm wavelength) signals bounced off Earth's surface. Studies of C-band radar data over the Yucatán Peninsula indicate that at this wavelength vegetation penetration is good for most land cover types except the closed canopy forests (Pope et al., 2001). Therefore, the SRTM data represent the ground surface if there is little vegetation, but represent the upper canopy in dense forests. This situation can be readily seen along the Yucatán's northwest coast (Fig. 3), where islands of tall trees (known as petenes) in the marshes and mudflats appear as small elevated areas up to 20 m high. Elsewhere in the SRTM data, patches of forest distort the elevation information and complicate interpretations. For example, agricultural plots appear in the SRTM data as low rectangles surrounded by vegetation a few meters high. (This is most

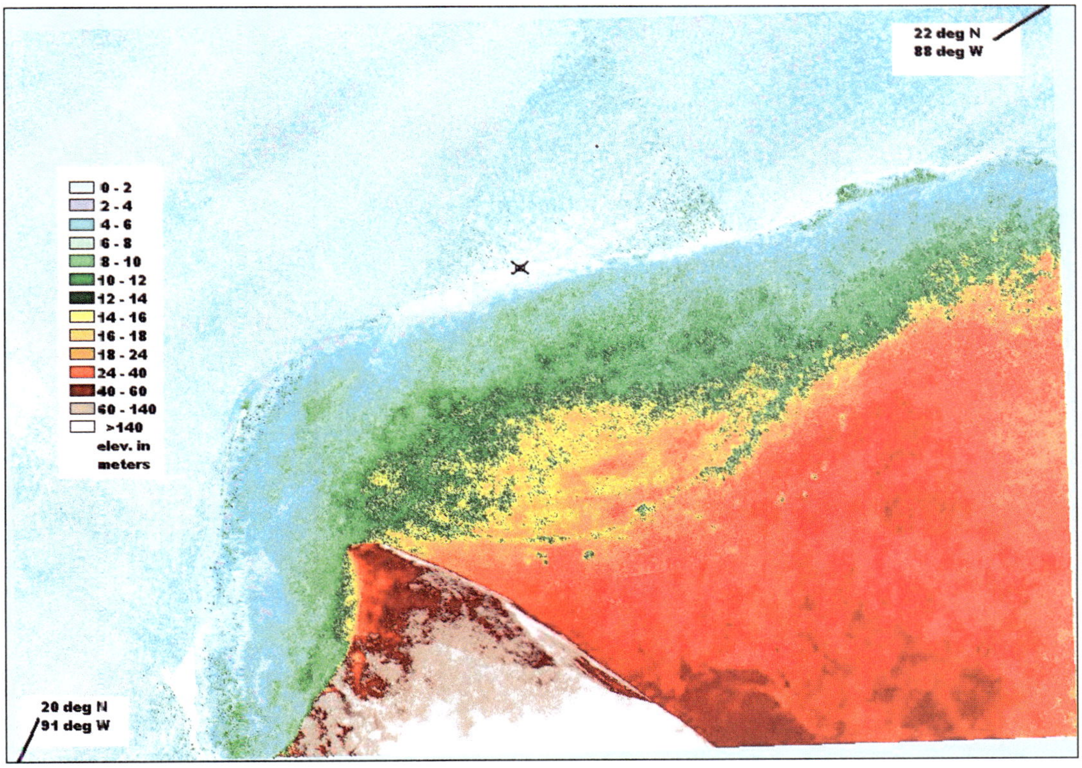

Figure 3. Shuttle Radar Topography Mission (SRTM) data. These are 90 m resolution data processed for our research effort by the Principal Investigator processing flow. The edges of the data within the figure are true north-south and east-west with north at the top. The corners are labeled with latitude and longitude, indicating that the figure is 3° long × 2° lat in extent. The scale varies in this projection; however, the figure is ~311 km in the east-west direction and ~223 km in the north-south direction. The center of the Cenote Ring as inferred by Pope et al. (1993) is indicated at 21.32 °N and 89.52 °E.

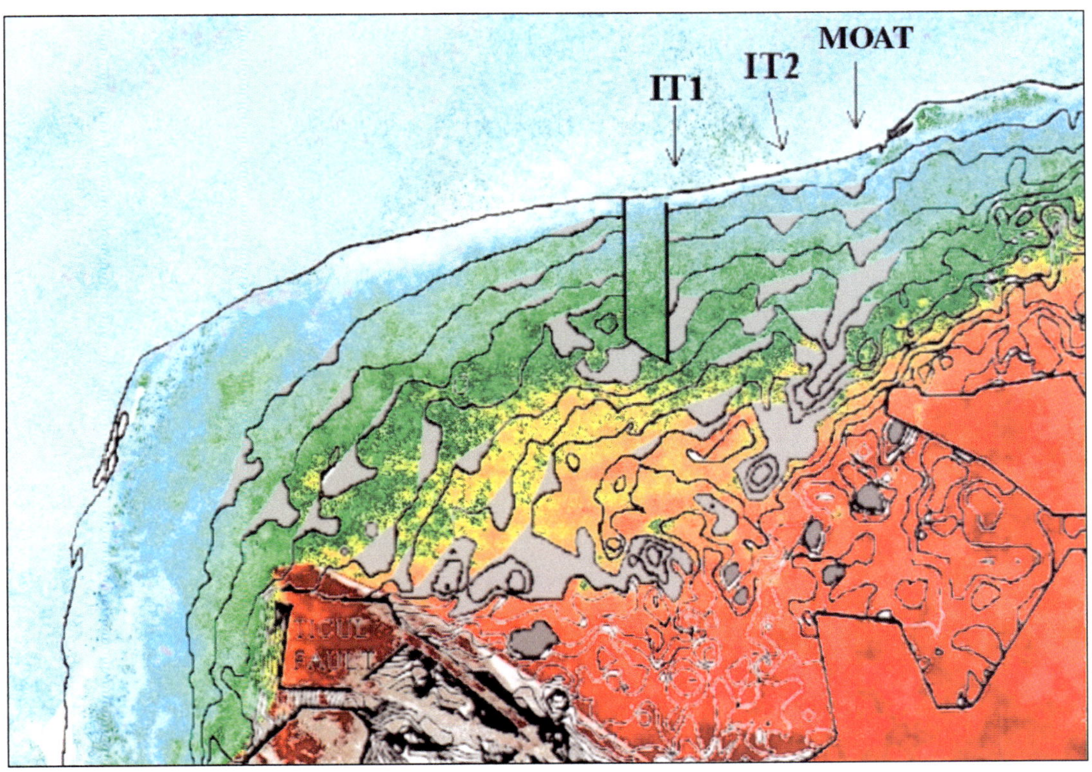

Figure 4. Figure 1 superimposed on Figure 3. Figure 1 was manually adjusted to fit Figure 3 and superimposed so as to illustrate the correlation of the two different data sets.

Figure 5. The Shuttle Radar Topography Mission (SRTM) elevation data have been subjected to fractional vertical derivative processing (Cooper and Cowan, 2003). The fractional vertical derivatives 0.3, 0.5, and 0.7 are represented as red, green, and blue in this image, which emphasizes the shorter wavelength features in the SRTM topography data. The area covered is the same as in Figure 3.

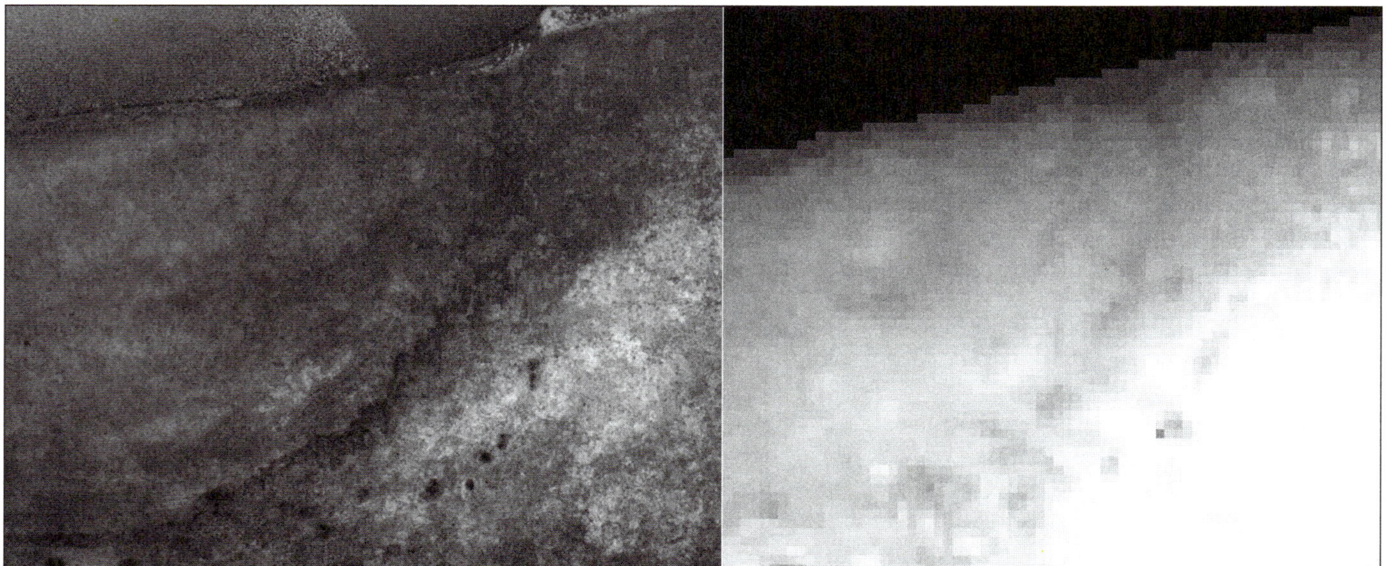

Figure 6. Comparison of the Shuttle Radar Topography Mission (SRTM; left) and Petróleos Mexicanos (PEMEX; right) topographic data. The SRTM data are circular sunshade data (Cooper, 2003). The PEMEX data have been gridded and the elevation represented in grayscale. Note that the gridded resolution of the PEMEX data is much less than the 90 m resolution of the SRTM data. For scale, compare to Figure 3.

clearly evidenced in Fig. 5.) In the dissected karstic topography, the SRTM data tend to underrepresent the relief and overrepresent the average elevation because the vegetation tends to be thicker in the sediment filled valleys, thereby producing anomalously high elevations, while the remaining limestone uplands are less vegetated, resulting in more accurate characterization of those elevations. Additional discrepancies exist between the PEMEX and SRTM data because the PEMEX data were mostly acquired along roads and trails that avoided swampy (low) and rugged (high) terrain, resulting in a data set with less relief.

Given these biases in both the PEMEX and SRTM data, one should not expect, and one does not find, exact correspondence between elevation values; for example, the 10 m contour in the PEMEX data does not correspond exactly with the upper limit of the 8–10 m color band in the SRTM data. The useful correlation that does exist between the two data sets is that of relative relief. A few additional notes on the SRTM data in Figures 3 and 5 are: (1) much of the white along the shoreline is the effect of shallow water waves, (2) the patch of white along the western shore is a data gap, and (3) the linear features off the north coast that form a faint triangle are artifacts of the orbital geometry of the shuttle.

TOPOGRAPHIC EXPRESSION OF THE CHICXULUB CRATER

The crater's overall topographic expression, seen in both Figures 3 and 5, is that of a shallow bowl, truncated by the coast, with a low relief interior, a deep concentric trough, and an elevated rim with deep solution pits (dolines) and low arcuate ridges. As noted by Pope et al. (1996), the most prominent feature is the conspicuous "moat" or valley, in places more than 8 m deep, which contains the major Cenote Ring and has a radius of ~83 km (Figs. 3 and 5). The two inner troughs (radii of ~41 and ~62 km) reported by Pope et al. (1996) appear as irregular depressions ~2 m deep and, as suggested by Kinsland et al. (2000), are not clearly concentric features. These putative troughs are expressed in Figure 3 as indentations of lower elevation colors into the higher elevation colors. These features are not distinct in part because the color scheme in Figure 3 was chosen to emphasize the higher relief crater-related topographic features. While the nature of these inner troughs remains speculative, we note that they correlate well with fault scarps (Morgan et al., 1997) of the buried crater (Pope et al., 2004). The outer trough proposed by Pope et al. (1996), with a radius of ~103 km, shows up more clearly in the SRTM data than in the PEMEX data due to complete coverage by the SRTM in the southeastern portion of the crater. This trough also correlates well with a major fault scarp of the Chicxulub crater (Morgan et al., 2002; Pope et al., 2004), which extends through the entire crust (e.g., Morgan and Warner, 1999). The ~129 km radius topographic rim of the crater inferred from the PEMEX topography data by Pope et al. (1996) also correlates well with the outermost (thrust) fault of the Chicxulub impact crater (Morgan et al., 2002; Pope et al., 2004).

Just inside the topographic rim proposed by Pope et al. (1996) are numerous large arcuate ridges in the SRTM data that have not been previously noted (Figs. 3 and 5). These ridges are similar in width, spacing, and relief to beach ridges of Pleistocene and Holocene age seen in Landsat TM and SIR-C/X-SAR imagery (Pope et al., 1996). Similar arcuate ridges are seen in the SRTM data in the vicinity of the Ticul fault and inside the Cenote Ring, but the ones adjacent to and outside of the outer trough are the most extensive. We propose that these arcuate ridges are indeed fossil Tertiary beach ridges that formed as sea level dropped in the late Eocene and exposed the buried crater rim. The fact that these ridges are not perfectly concentric with the crater may be due to tilting of the peninsula in the Paleocene-Eocene or perhaps differential erosion of the rim by longshore currents as it was first submerged and then emerged.

CONCLUSIONS

We have presented SRTM data, both color-coded elevation data and fractional vertical derivative data. The surface topography over the buried Chicxulub impact crater is clearly expressed in these data sets. Comparison of these data with the topographic data and interpretations of Pope et al. (1996) and Kinsland et al. (2000) supports their earlier conclusions that the complex structure of the buried crater is expressed in the present day topography and extends well beyond a diameter of 200 km.

ACKNOWLEDGMENTS

We wish to acknowledge Utsav Agarwal for his help with some of the images of this paper and Cathy Bishop for her computer and logistical support.

REFERENCES CITED

Connors, M., Hildebrand, A.R., Pilkington, M., Ortiz-Aleman, C., Chavez, R.E., Urrutia-Fucugauchi, J., Graniel-Castro, E., Camara-Zi, A., Vasquez, J., and Halpenny, J.F., 1996, Yucatan karst features and the size of the Chicxulub crater: Geophysical Journal International, v. 127, p. F11–F14.

Cooper, G.R.J., 2003, Feature detection using sunshading: Computers & Geosciences, v. 29, no. 8, p. 941–948, doi: 10.1016/S0098-3004(03)00091-8.

Cooper, G.R.J., and Cowan, D.R., 2003, The application of fractional calculus to potential field data: Exploration Geophysics, v. 34, no. 1, p. 51–56.

Farr, T., and Kobrick, M., 2001, The Shuttle Radar Topography Mission: Eos (Transactions, American Geophysical Union), v. 82, p. 47.

Hildebrand, A.R., Pilkington, M., Connors, M., Ortiz-Aleman, C., and Chavez, R.E., 1995, Size and structure of the Chicxulub crater revealed by horizontal gravity gradients and cenotes: Nature, v. 376, p. 415–417, doi: 10.1038/376415a0.

Kinsland, G.L., Hurtado, M., and Pope, K.O., 2000, Detection of groundwater conduits in limestones with gravity surveys: Data from the area of the Chicxulub impact crater, Yucatan Peninsula, Mexico: Geophysical Research Letters, v. 27, no. 8, April, p. 1223–1226.

Kinsland, G.L., Hurtado Cardador, M., Urrutia Fucugauchi, J., Cruz Cisneros, M., and Long, Z.R., 2002, New Gravity and Topographic Data over the Chicxulub Impact Structure: Geological Society of America Abstracts with Programs, v. 34, no. 6, p. 264.

Kinsland, G.L., Sanchez, G., Kobrick, M., and Hurtado Cardador, M., 2003, Elements of the Chicxulub Impact structure as revealed in SRTM and surface

GPS topographic data: Abstracts, 34th Lunar and Planetary Science Conference: Houston, Lunar and Planetary Institute, CD-ROM, Abstract 1156.

Morgan, J., and Warner, M., 1999, Chicxulub: The third dimension of a multi-ring impact basin: Geology, v. 27, p. 407–410, doi: 10.1130/0091-7613(1999)0272.3.CO;2.

Morgan, J., Warner, M., and Chicxulub Working Group, 1997, Size and morphology of the Chicxulub impact crater: Nature, v. 390, p. 472–476.

Morgan, J., Warner, M., and Grieve, R., 2002, Geophysical constraints on the size and structure of the Chicxulub impact crater, *in* Koeberl, C. and MacLeod, K.G., eds., Catastrophic events and mass extinctions: Impacts and beyond: Geological Society of America Special Paper 356, p. 39–46.

Penfield, G.T., and Camargo, Z.A., 1981, Definition of a major igneous zone in the central Yucatan platform with aeromagnetics and gravity: Los Angeles, Society of Exploration Geophysics 51st Annual International Meeting, Technical Program, Abstracts and Biographies, p. 37.

Pope, K.O., Ocampo, A.C., and Duller, C.E., 1991, Mexican site for K/T impact crater?: Nature, v. 351, p. 105, doi: 10.1038/351105a0.

Pope, K.O., Ocampo, A.C., and Duller, C.E., 1993, Surficial geology of the Chicxulub impact crater, Yucatan, Mexico: Earth, Moon, and Planets, v. 63, p. 93–104.

Pope, K.O., Ocampo, A.C., Kinsland, G.L., and Smith, R., 1996, Surface expression of the Chicxulub crater: Geology, v. 24, p. 527–530, doi: 10.1130/0091-7613(1996)0242.3.CO;2.

Pope, K.O., Rejmankova, E., and Paris, J.F., 2001, Spaceborne imaging radar-C (SIR-C) observations of groundwater discharge and wetlands associated with the Chicxulub impact crater, northwestern Yucatan Peninsula, Mexico: Geological Society of America Bulletin, v. 113, no. 3, p. 403–416, doi: 10.1130/0016-7606(2001)113<0403:SIRCSC>2.0.CO;2.

Pope, K.O., Kieffer, S.W., and Ames, D.E., 2004, Empirical and theoretical comparisons of the Chicxulub and Sudbury impact structures: Meteoritics & Planetary Science, v. 39, p. 97–116.

Manuscript Accepted by the Society 9 August 2004

Printed in the USA

Geological Society of America
Special Paper 384
2005

Recent research on the Chesapeake Bay impact structure, USA—Impact debris and reworked ejecta

J. Wright Horton Jr.*
U.S. Geological Survey, MS 926A, 12201 Sunrise Valley Drive, Reston, Virginia 20192, USA

John N. Aleinikoff
Michael J. Kunk
U.S. Geological Survey, MS 963, Denver Federal Center, Denver, Colorado 80225, USA

Gregory S. Gohn
Lucy E. Edwards
Jean M. Self-Trail
David S. Powars
U.S. Geological Survey, MS 926A, 12201 Sunrise Valley Drive, Reston, Virginia 20192, USA

Glen A. Izett
Department of Geology, College of William and Mary, 3012 East Whittaker Close, Williamsburg, Virginia 23285, USA, and
U.S. Geological Survey Emeritus

ABSTRACT

Four new coreholes in the western annular trough of the buried, late Eocene Chesapeake Bay impact structure provide samples of shocked minerals, cataclastic rocks, possible impact melt, mixed sediments, and damaged microfossils. Parautochthonous Cretaceous sediments show an upward increase in collapse, sand fluidization, and mixed sediment injections. These impact-modified sediments are scoured and covered by the upper Eocene Exmore beds, which consist of highly mixed Cretaceous to Eocene sediment clasts and minor crystalline-rock clasts in a muddy quartz-glauconite sand matrix. The Exmore beds are interpreted as seawater-resurge debris flows. Shocked quartz is found as sparse grains and in rock fragments at all four sites in the Exmore, where these fallback remnants are mixed into the resurge deposit. Crystalline-rock clasts that exhibit shocked quartz or cataclastic fabrics include felsites, granitoids, and other plutonic rocks. Felsite from a monomict cataclasite boulder has a sensitive high-resolution ion microprobe U-Pb zircon age of 613 ± 4 Ma. Leucogranite from a polymict cataclasite boulder has a similar Neoproterozoic age based on muscovite $^{40}Ar/^{39}Ar$ data. Potassium-feldspar $^{40}Ar/^{39}Ar$ ages from this leucogranite show cooling through closure (~150 °C) at ca. 261 Ma without discernible impact heating. Spherulitic felsite is under investigation as a possible impact melt. Types of crystalline clasts, and exotic sediment clasts and grains, in the Exmore

*whorton@usgs.gov

Horton, J.W., Jr., Aleinikoff, J.N., Kunk, M.J., Gohn, G.S., Edwards, L.E., Self-Trail, J.M., Powars, D.S., and Izett, G.A., 2005, Recent research on the Chesapeake Bay impact structure, USA—Impact debris and reworked ejecta, *in* Kenkmann, T., Hörz, F., and Deutsch, A., eds., Large meteorite impacts III: Geological Society of America Special Paper 384, p. 147–170. For permission to copy, contact editing@geosociety.org.

vary according to location, which suggests different provenances across the structure. Fractured calcareous nannofossils and fused, bubbled, and curled dinoflagellate cysts coexist with shocked quartz in the Exmore, and this damage may record conditions of heat, pressure, and abrasion due to impact in a shallow-marine environment.

Keywords: Chesapeake, impact, crater, ejecta, cataclastic, shock, resurge, SHRIMP, argon, geochronology, dinoflagellate, nannofossil.

INTRODUCTION

The Chesapeake Bay impact structure is a late Eocene complex crater formed in a continental shelf environment on the Atlantic margin of Virginia, United States. This large, buried impact structure underlies the southern part of Chesapeake Bay, its surrounding peninsulas, and a small part of the western Atlantic Ocean (Fig. 1). The Chesapeake Bay structure is the largest known impact crater in the United States and the seventh largest known on Earth (Earth Impact Database, 2004). It may be Earth's best preserved example of a large impact structure formed in a predominantly siliciclastic continental shelf environment. The Chesapeake Bay impact structure coincides with an unusual region of saline groundwater and has a profound influence on groundwater flow and quality, including salinity, across one of the fastest growing urban centers on the east coast of North America (Hampton Roads Planning District Commission, 1999; Sanford, 2003; McFarland and Bruce, 2005). A review of current knowledge is given by Poag (1999), Powars and Bruce (1999), Powars (2000), Poag et al. (2004), and Horton et al. (2005b).

The Chesapeake Bay impact structure has an average diameter of ~85 km, ranging from ~80–95 km. It has the shape of an "inverted sombrero" and consists of a central crater, also termed the inner basin, surrounded by a flat-floored annular trough (Poag et al., 1999a; Powars and Bruce, 1999; Powars, 2000). The central crater likely approximates the former location and diameter of the impact's transient crater. The multilayered impact target consisted of seawater (maximum depth <340 m) underlain by poorly consolidated, water-saturated, lower Tertiary and Cretaceous sediments (400 m to more than 750 m thick) and crystalline basement rocks (Horton et al., 2005b).

The central crater is variously interpreted on the basis of geophysical data to be ~30–38 km in diameter (Powars and Bruce, 1999; Powars et al., 2003) or 35–40 km in diameter (Poag et al., 1999a). Interpretations of seismic reflection data suggest that the floor of the transient crater penetrated crystalline basement to a depth ~1.3 km below the rim of the existing central crater and ~1.6 km below sea level. The seismic data also suggest that a 15 to 20-km-diameter mass of crystalline rock rose as a central uplift to ~900 m above the central crater floor (Poag et al., 1999b, 2002).

The annular trough surrounds the central crater and adds ~24 km to the crater radius (Poag et al., 1994). The inner margin of the annular trough is characterized by uplifted basement rocks and has been interpreted by Poag et al. (2002) as an irregular peak ring. The outer margin of the annular trough is roughly circular and is characterized by a terraced zone of inwardly slumped fault blocks (Poag, 1996; Powars and Bruce, 1999). The outer margin of the annular trough is delineated by seismic reflection profiles, which cross the margin at 61 locations (Poag et al., 2002), and is generally considered to be the edge of the crater. Outside the annular trough, an outer fracture zone is ~35 km in width and contains irregularly spaced concentric and radial faults (Powars, 2000; Powars et al., 2002). All features of the impact structure are well preserved beneath a 150–400-m-thick blanket of post-impact Tertiary and Quaternary sediments (Poag et al., 1994; Powars and Bruce, 1999).

The Chesapeake Bay impact structure is a likely source of the glassy material in tektites (bediasites and georgiaites) and microtektites of the North American strewn field (Poag et al., 1994; Koeberl et al., 1996; Glass, 2002). Distal ejecta in this strewn field also include shocked quartz and zircon, high pressure silica polymorphs coesite and stishovite, and reidite (a high pressure polymorph of zircon) (Bohor et al., 1988; Koeberl and Glass, 1988; Glass, 1989, 2002; Glass and Liu, 2001; Glass et al., 1986, 1995, 2002; Kamo et al., 2002).

Recently drilled cores from the western annular trough at Bayside in Mathews County (728.5 m deep), at the National Aeronautics and Space Administration (NASA) Langley Research Center in Hampton (635.1 m deep), and at North in Mathews County (435.1 m deep), and a core from the outer rim at Watkins School in Newport News (300.3 m deep), are located 8, 19, 24, and 27 km, respectively, outside the central crater (Fig. 1). The Bayside, Langley, and North cores penetrated impact-disrupted, seawater-resurge, and post-impact sediments; however, only the Bayside and Langley cores sampled complete crater sections that include crystalline basement. The Watkins School core penetrated a thin interval of resurge sediments below post-impact sediments and above undisrupted pre-impact sediments. These cores provided the samples used for this study of impact debris and reworked ejecta in the Chesapeake Bay impact structure.

STRATIGRAPHY OF THE WESTERN ANNULAR TROUGH

Figure 2 summarizes the stratigraphy and structure of the western margin of the annular trough, as interpreted from the U.S. Geological Survey (USGS)–NASA Langley corehole, the Watkins School corehole, and a connecting seismic reflection profile. Pre-impact Cretaceous unconsolidated sediments overlie

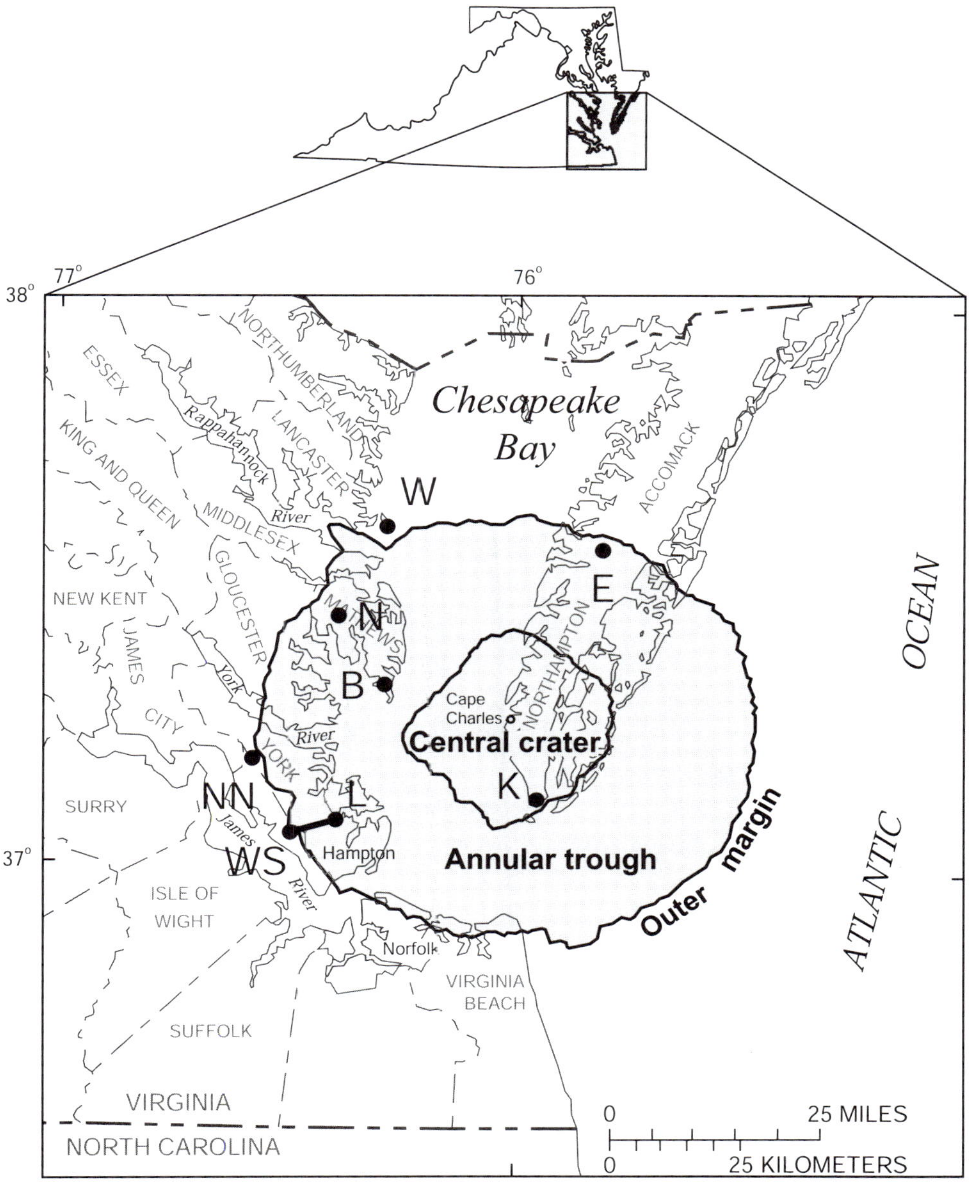

Figure 1. Map locating major features of the Chesapeake Bay impact structure, four recent coreholes in southeastern Virginia that recovered samples used in this study (L—USGS-NASA Langley corehole at Hampton; B—Bayside corehole in Mathews County; N—North corehole in Mathews County; WS—Watkins School corehole at Newport News), and earlier coreholes in southeastern Virginia that recovered impact debris at Exmore (E), Kiptopeke (K), Newport News (NN), and Windmill Point (W). Line connecting WS and L marks location of cross section in Figure 2. Location of the central crater and outer margin are from Powars and Bruce (1999). Illustration modified from Powars et al. (2002) and Edwards and Powars (2003).

granite in the USGS-NASA Langley core. Extensive impact disruption (fluidization, downward sediment injection, and fracturing) increases upward in the pre-impact sedimentary section. The disrupted Cretaceous sediments are overlain by seawater-resurge sediments of the Exmore beds that are capped by silt interpreted to have settled from the water column following the cessation of impact processes. Post-impact Tertiary and Quaternary sediments overlie the Exmore beds (Gohn et al., 2002).

Autochthonous Crystalline Basement

Autochthonous crystalline basement rocks were recovered below the sedimentary sections in the USGS-NASA Langley core (Fig. 2) from 626.3 m (top) to 635.1 m (total depth) and in the Bayside core from 708.9 m (top) to 728.5 m (total depth) (Horton et al., 2002a, 2005a). Basement rocks in both cores are pale red, medium-grained monzogranites that are nonfoliated and pervasively chloritized (Horton et al., 2002a, 2005a). Chlorite is the principal mafic mineral in the granites; traces of relict biotite remain in the granite at Bayside. The top of the granite in each core is weathered but not saprolitized. Features diagnostic of shock metamorphism were not found in these autochthonous granites. SHRIMP U-Pb zircon data for the basement granites indicate Neoproterozoic crystallization ages of 625 ± 11 Ma (2σ) at Bayside and 612 ± 10 Ma (2σ) at Langley (Horton et al., 2002a, 2002b, 2005a).

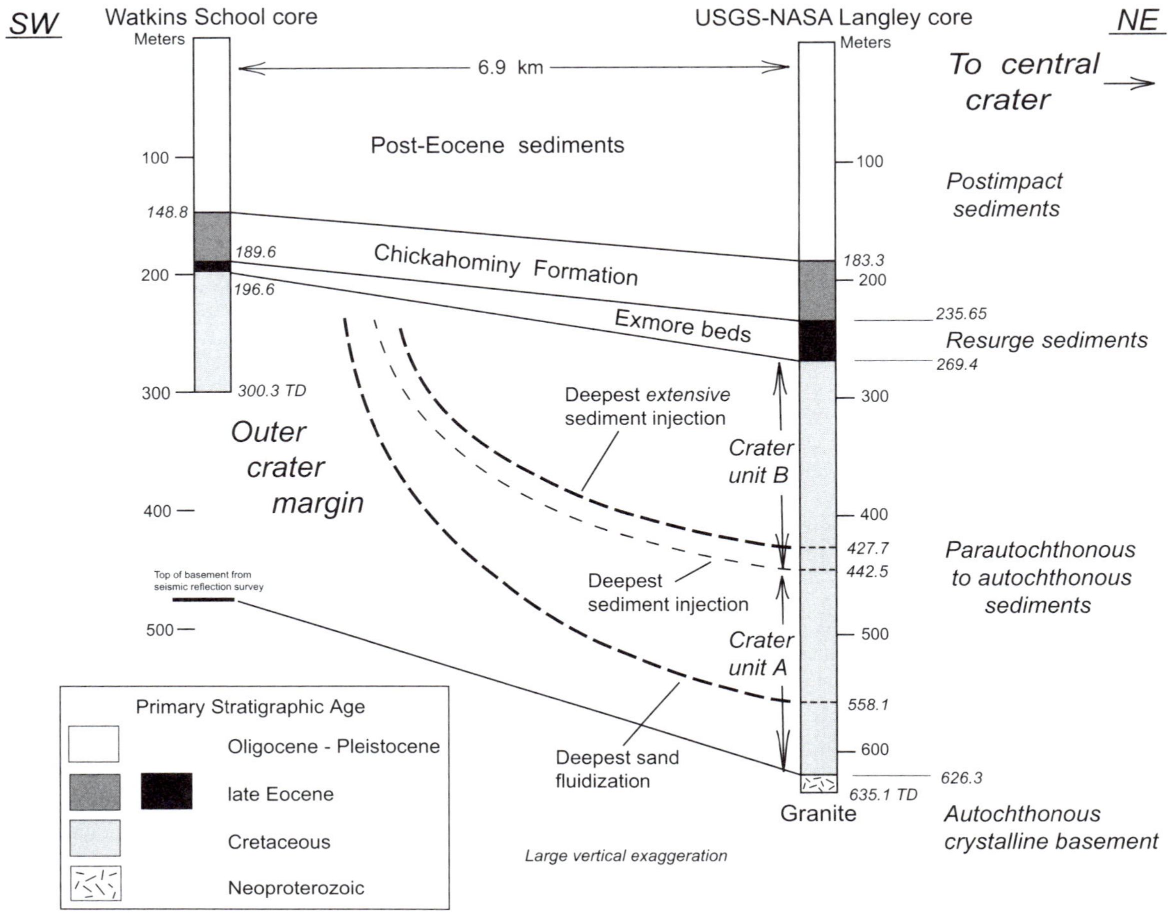

Figure 2. Schematic cross section illustrating the stratigraphy and structure of the western margin of the annular trough. The line of section follows a seismic-reflection survey (Catchings et al., 2005) and connects the USGS-NASA Langley corehole (L in Fig. 1) and the Watkins School corehole (WS in Fig. 1). Granite in the Langley core is overlain by pre-impact Cretaceous sediments that show upward-increasing disruption by sand fluidization, slumping, and injection of exotic sediments. The Cretaceous sediments in the Watkins School core do not show impact disruption. The Cretaceous sediments are overlain by mixed resurge sediments of the Exmore beds, which are in turn covered by post-impact sediments.

Fission-track ages ($\pm 2\sigma$) of zircons, dating cooling below ~235 °C (362 ± 49 Ma and 332 ± 44 Ma at Bayside; 375 ± 44 Ma at Langley), and apatites, dating cooling below ~90–100 °C (239 ± 79 Ma and 198 ± 40 Ma at Bayside; 184 ± 32 Ma at Langley), show no impact-related thermal disturbance in the basement granites (Horton et al., 2002b, 2005a). The absence of shocked minerals and discernible impact heating in the autochthonous basement granites from the Bayside and Langley cores provides boundary constraints for computational models of the impact process.

Many veins, fractures, and faults in both granites contain epidote, albite, and chlorite. These minerals probably formed at temperatures higher than those recorded by pre-impact fission-track ages of apatite from the granites (Horton et al., 2002a, 2002b, 2005a), and their formation is interpreted to predate the impact event. The number of fractures in the granite cores is not unusual in comparison to cores from the nearby Appalachian Piedmont (Horton et al., 2005a). Impact-related faults, evident in overlying sediment cores and in the seismic profiles (Catchings et al., 2001, 2005; Gohn et al., 2005), are unconfirmed in the basement granites from these cores. Younger granites of late Paleozoic age and rift basins containing sedimentary rocks of Triassic to Jurassic age are concealed beneath the Coastal Plain near the crater (Russell et al., 1985; Horton et al., 1991, 2005b).

Autochthonous and Parautochthonous Sediments

Within the annular trough, the pre-impact Lower to Upper Cretaceous fluvial sediments include a basal "crater unit A" (informal name) (Fig. 2) consisting of autochthonous to parautochthonous block-faulted material, which Gohn et al. (2005) subdivided into a non-fluidized lower part and a variably fluidized upper part. An overlying, parautochthonous "crater unit B" (informal name) consists of blocks of Lower Cretaceous sediment disrupted by zones of extensive fluidization and injection of sediment that includes exotic, mixed-aged particles (Gohn et al., 2002, 2005; Frederiksen et al., 2005). The base of crater unit B, defined as the deepest occurrence of glauconite, is interpreted to represent the lowest zone of infiltration by Tertiary marine sediment from above. Where injection and mixing pervade the entire thickness of Cretaceous sediments, as in the Bayside core 9 km outside the central crater (Fig. 1), the lower autochthonous to parautochthonous crater unit A is not recognized, and crater unit B lies directly on granite basement.

In the Langley core, Cretaceous fluvial sediments that occur directly above the autochthonous granite are nearly pristine. However, because overburden confining pressure decreased upward in the Cretaceous section, water-saturated sand beds became increasingly fluidized (Fig. 3A), and clay beds became more intensely fractured upsection (vertically upward) during the impact process (Gohn et al., 2005). At the top of the pre-impact section, Upper Cretaceous and lower Tertiary marine sediments are missing from their normal stratigraphic positions in cores within the annular trough, but their disaggregated remnants are mixed into the upper part of the parautochthonous Cretaceous sediments, as discussed in a later section.

Two trends in the style and magnitude of impact deformation are apparent within the annular trough (Gohn et al., 2002). First, the overall magnitude of deformation decreases both downward and laterally outward (Fig. 2). Second, there is a differential response of the sand layers versus the more competent clay layers at shallow depths within the parautochthonous sediments (upper part of crater unit A and crater unit B; Gohn et al., 2005). Acoustic fluidization (Melosh, 1989; Melosh and Ivanov, 1999) of fractured rocks and sediments was probably important in the formation of the central crater. However, the liquefaction structures that formed in water-saturated, poorly consolidated sands of the annular trough more closely resemble those associated with paleoseismic shaking (e.g., Obermeier et al., 2001; Olson et al., 2003).

Sand fluidization occurred at deeper levels and was more pervasive close to the central crater in the Bayside core than it is farther from the central crater in the Langley core. No sand fluidization occurred in the pre-impact section located just beyond the outer crater margin in the Watkins School core (Fig. 2). This progressive upward and outward decrease in fluidization is consistent with a decay of impact energy with distance from the transient crater. At the outer crater margin, the overburden stress may have been too large to permit this strength reduction and resulting viscous flow except near the surface (Gohn et al., 2002).

Hence, the base of the dissociated and mixed parautochthonous sediments of upper crater unit A and crater unit B is a yield horizon or zone defined by the overburden load. This horizon rises stratigraphically away from the crater center and defines the boundary of the strength crater in the sense of Croft (1981), which marks the transition from thoroughly impact-dissociated materials to materials that were only ruptured or elastically deformed by the impact (Croft, 1981; his Figure 4). This horizon does not correspond in the outer annular trough to the subjacent incremental increase in density that occurs at the contact between pre-impact crystalline rocks and overlying Cretaceous sediments. Nor is the resulting outer margin of the strength crater necessarily at the outer limit of faulting produced by late-stage gravity-driven crater collapse.

Resurge Sediments of the Exmore Beds

The Exmore beds (Fig. 2) consist mainly of granule to cobble-sized lithoclasts suspended in an unsorted and unstratified matrix of muddy quartz-glauconite sand (Gohn et al., 2005). The clasts represent all known pre-impact basement and sedimentary units in the impact target area (Figs. 3B and 3C). Most clasts consist of Cretaceous, Paleocene, and Eocene sediments, but sparse crystalline-rock clasts also are present (Frederiksen et al., 2005; Gohn et al., 2005; Horton and Izett, 2005). The only observed sedimentary structure in the Exmore beds is normal "coarse-tail" grading of the larger lithoclasts. Some large sediment clasts have irregular shapes and centimeter-thick rinds, suggesting a fallback ejecta origin.

Clasts in the Exmore beds are supported by a matrix of fossiliferous, muddy, quartz-glauconite sand. The matrix is derived from pre-impact Lower Cretaceous fluvial sediments and glauconitic Upper Cretaceous and lower Tertiary marine sediments of the target area. Reworked Late Cretaceous marine microfossils are present only in the upper part of the Exmore in the Langley core, whereas early Tertiary microfossils are present throughout this section.

The Exmore beds are interpreted to be subaqueous debris flows generated by catastrophic resurge of ocean water into the transient cavity created by the impact (Gohn et al., 2005; Horton, et al., 2005c). Some clasts and matrix components in the upper part of crater unit B at Langley, and in most of crater unit B at Bayside, are indistinguishable from Exmore components, and these materials are interpreted to be syn-impact injection products.

In a proposed international classification of impactites (Stöffler and Grieve, 1994, 2003), the Exmore beds consist of "allochthonous (allogenic) lithic impact breccia" (although this classification is not useful for distinguishing the characteristics of a marine crater resurge deposit), and crater unit B consists of "autochthonous (authigenic) impact breccia" (although parautochthonous would be more accurate).

Post-Impact Sediments

The transition from syn-impact to post-impact deposition is gradational; impact-related debris was still settling out of the water column and washing into the partially filled crater while normal marine sedimentation resumed. The post-impact stratigraphic record (Edwards et al., 2005; Poag and Norris, 2005; Poag et al., 2004; Powars et al., 2005) began with rapid sedimentation throughout deposition of the Chickahominy Formation (Fig. 2), reflecting the filling or partial filling of the remnant crater. The informally named lower Oligocene Delmarva beds (Powars et al., 1992) and Drummonds Corner beds (Powars et al., 2005) are known only from the vicinity of the impact

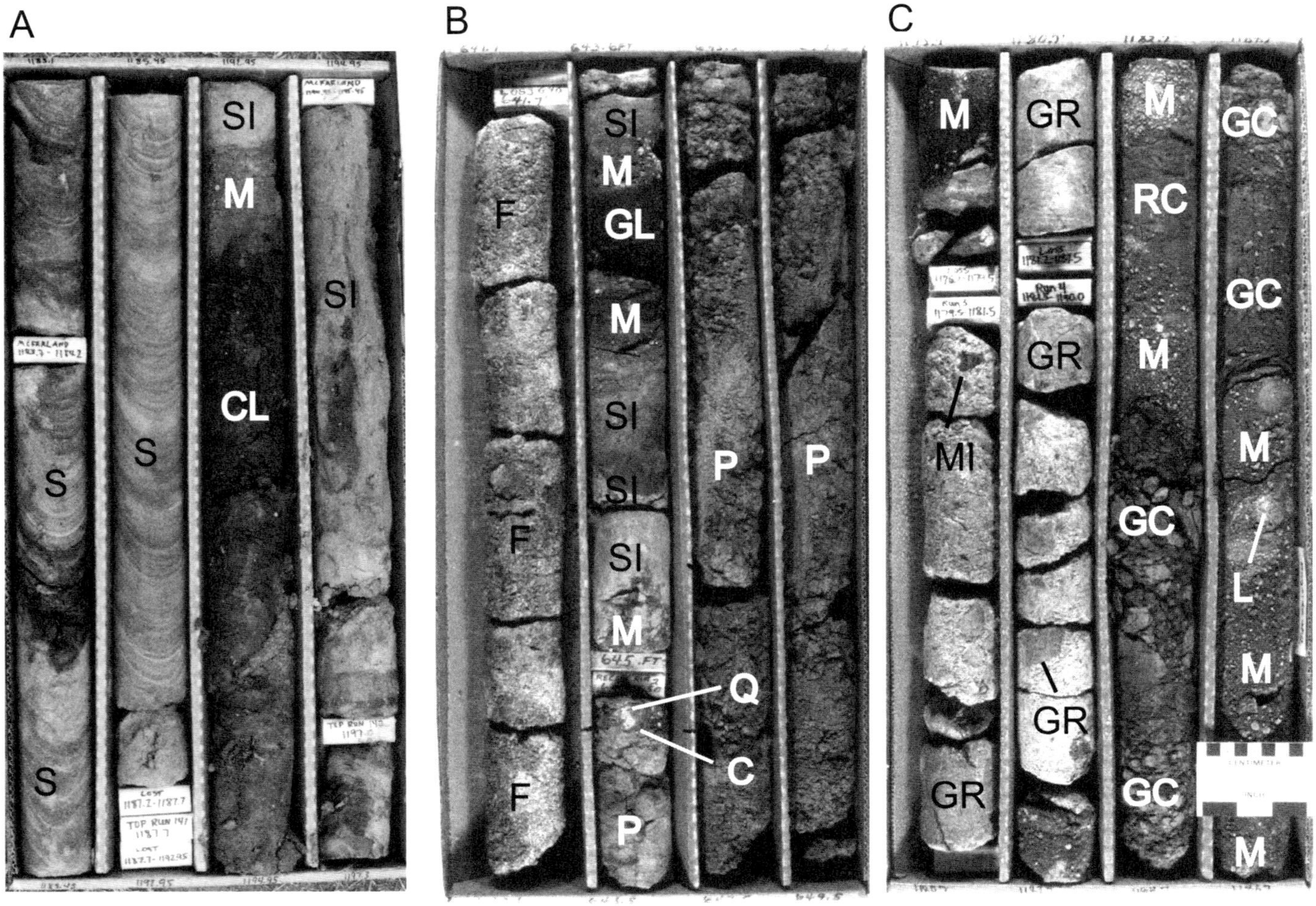

Figure 3. Photographs of selected drill cores. Boxes (inner trays) are 2.0 ft (0.61 m) long. Depth increases from the upper left to the lower right in each box; depths marked on boxes are in feet. (A) Disrupted parautochthonous sediment from the annular trough, showing blocks of moderately dipping, laminated sand (S), disrupted clay (CL), and contorted, steeply dipping clayey silt (SI) of the Cretaceous Potomac Formation. Muddy quartz-glauconite matrix (M) occurs locally between blocks. Unit described by Gohn et al. (2005) as crater unit B. USGS-NASA Langley core, box 126 from 364.9 to 360.6 m (logged 1197.3 ft to 1183.1 ft). (B) Resurge sediment (Exmore beds) from just outside the outer margin of the crater, showing a boulder of monomict felsite cataclasite (F) or cohesive breccia derived from the Coastal Plain basement, two broken cobbles (locally injected by sandy quartz-glauconite matrix) of clayey silt (SI) from the Cretaceous Potomac Formation, a dark cobble of glauconitic clayey silt (GL), and a rounded detrital quartz pebble (Q) in sandy quartz-glauconite matrix (M). The Exmore beds have a sharp, irregular basal contact (C) with underlying autochthonous, undisturbed clayey silt of the Potomac Formation (P). Watkins School core, Recore-1 box from 198.0 to 195.6 m depth (logged 649.5–641.7 ft). The felsite cataclasite boulder (F) spans the section of core from 196.2 to 195.6 m depth (logged 643.6 ft to 641.7 ft) and includes U-Pb geochronology sample WS642 (sawed half of core from 195.97 to 195.6 m; logged 642.95–641.7 ft). (C) Resurge sediment (Exmore beds) from the inner part of the annular trough ~9 km from the edge of the central crater, showing a boulder of polymict cataclasite (cohesive breccia) composed of abundant leucogranite fragments (GR) and fewer fragments of mafic igneous rock (MI), irregular cobble-sized clasts of red clay (RC) and gray clay (GC) derived from the Cretaceous Potomac Formation, a limestone pebble (L), and glauconitic matrix (M). The cataclasite boulder has monomict central and lower parts composed of leucogranite (GR) fragments, and a polymict upper part that contains the fragments of mafic igneous rock (MI) in addition to leucogranite. Bayside core 2, box 2 from 361.7 to 358.4 m (logged 1186.7–1175.7 ft). The polymict cataclasite boulder is the largest crystalline clast interpreted to be reworked fallback ejecta in any of the drill cores. It spans the section of core from the exposed matrix contact at 360.4 m (logged 1182.5 ft) to 359.5 m (logged 1179.5 ft) depth, where the top of the boulder was lost in drilling (a section within the boulder was also lost in drilling). The largest fragment of homogeneous leucogranite within the cataclasite boulder at 360.2–359.8 m (logged 1181.7–1180.5 ft) includes $^{40}Ar/^{39}Ar$ geochronology sample BA1181 (sawed half of core from 360.0 to 359.8 m; logged 1181.2–1180.5 ft).

structure in Virginia, which suggests continuing post-impact subsidence of the crater and the resulting increased preservation of these sediments. Similarly, the upper Oligocene Old Church Formation contains a more complete stratigraphic record within the crater than is known from outcrops or shallow cores outside the limits of the crater. During Oligocene and early and middle Miocene time, sea level changes produced unconformities that punctuate the stratigraphic record. Deposition of the late Miocene St. Mary's Formation and Eastover Formation records a second episode of rapid sedimentation. These units show nearly continuous sediment accumulation in sections above the impact structure and thus reveal a more complete section than those found in classic localities outside the structure. The Pliocene Yorktown Formation and Pleistocene Tabb Formation indicate discontinuous sedimentation during the late Cenozoic.

Reworking of microfossils, especially dinoflagellates, is a prominent feature in many samples from the post-impact sediments. Some reworked specimens show impact-related damage, as described by Edwards and Powars (2003), Self-Trail (2003), and Frederiksen et al. (2005). Other reworked specimens clearly postdate the impact. Continued exhumation of sediment beds along faults may have facilitated the reworking of older material into the younger units filling the Chesapeake Bay impact structure (Edwards et al., 2005).

IMPACT DEBRIS AND REWORKED EJECTA

Impact debris observed in drill cores from the western part of the annular trough consists of the autochthonous and parautochthonous sections of variably fluidized, fractured, and mixed pre-impact sediments in crater units A and B, and the seawater-resurge sediments of the Exmore beds. We use the term "ejecta" in the general sense of Stöffler and Grieve (1994, 2003) for rocks that have been ejected ballistically from an impact crater. However, in the western annular trough of the Chesapeake Bay impact structure, materials that are interpreted to be fallback ejecta occur only as reworked clasts in sedimentary deposits. These materials are found only in the Exmore beds, with rare exceptions, and they include shocked minerals (quartz and feldspar), cataclastically deformed crystalline-rock fragments, possible impact melt, and damaged microfossils.

Shocked Minerals (Quartz and Feldspar)

Gram-size samples of sandy matrix were taken from the Exmore beds of the Langley core between depths of 250.1 m and 236.9 m. These samples were sieved, and the grains finer than 0.5 mm were prepared for microscopic examination as described by Horton and Izett (2005). Immersion-oil mounts of the dried grains were prepared by placing several hundred grains in 1.544 index oil without a cover glass. Individual quartz grains suspected to be shock metamorphosed, because of their appearance under a binocular microscope, were removed from the oil, cleaned, and dried. Selected grains were glue-mounted on the tip of a steel wire spindle and inserted into the immersion cell of a spindle stage (Wilcox, 1959). This device is ideal for the study of shock-metamorphic features in separate mineral grains because planar deformation features can be rotated to the vertical, and their orientations in relation to the optic axis (coincident with the *c*-axis in uniaxial crystals) can be measured and plotted on a stereonet with a precision of about ±2° (Wilcox, 1959). Techniques for preparing and manipulating quartz grains on the spindle stage and for estimating the relative abundance (percentage) of shocked grains per sample are described in Wilcox (1959), Izett (1990), and Horton and Izett (2005).

The Langley, Bayside, North, and Watkins School cores were searched for lithic clasts composed of crystalline rock. Most crystalline-rock clasts found in and just below the Exmore beds were rounded detrital pebbles. Thin sections of these rounded pebbles revealed no unusual high strain-rate fabrics or shock-induced features (Horton and Izett, 2005). A much smaller population of crystalline-rock clasts had angular shapes or cataclastic fabrics, and these clasts were examined in thin sections. Quartz and feldspar grains were separated from several cataclastic clasts, processed as described above, and rotated on a spindle stage to further verify the presence or absence of shock-induced planar deformation features (PDFs).

Quartz and potassium feldspar are the chief mineral components of the clasts found in the Exmore beds. Most of these grains appear structureless and show normal, sharp to slightly undulatory extinction under a petrographic microscope in cross-polarized light. Some grains show anomalous extinction, and others exhibit Böhm lamellae. Significantly, rare quartz grains contain multiple intersecting sets of PDFs diagnostic of shock metamorphism (Fig. 4). As many as five sets occur in some quartz grains. The grains are subangular, commonly 0.2–0.3 mm in diameter, and generally colorless. Spindle stage measurements of PDFs in these quartz grains most commonly have poles inclined at ~23° to the *c*-axis, which is the rhombohedral omega (ω) direction $\{10\bar{1}3\}$.

Quartz and feldspar are the only minerals identified in these sediment and clast samples that have convincing shock-induced PDFs (Fig. 4). Figures 4A and 4B show examples of shock-metamorphosed quartz grains from the USGS-NASA Langley core.

Shock-metamorphosed quartz is scarce in the Exmore beds. Horton and Izett (2005) estimated the relative abundance to be no more than one shocked grain in several thousand quartz grains in samples from the Langley core, based on the number of immersion-oil slides per sample and the numbers of total grains and shocked grains per slide. The limited number of shocked grains from the Langley core was insufficient for a statistically meaningful assessment of crystallographic orientations, although the orientations of planar deformation features measured on a spindle stage are consistent with shock metamorphism (Horton and Izett, 2005). Some individual rock fragments in the other cores, however, contain multiple shocked grains and may be amenable to statistical analysis and shock-pressure estimates using the methods described and critiqued by Grieve et al. (1996).

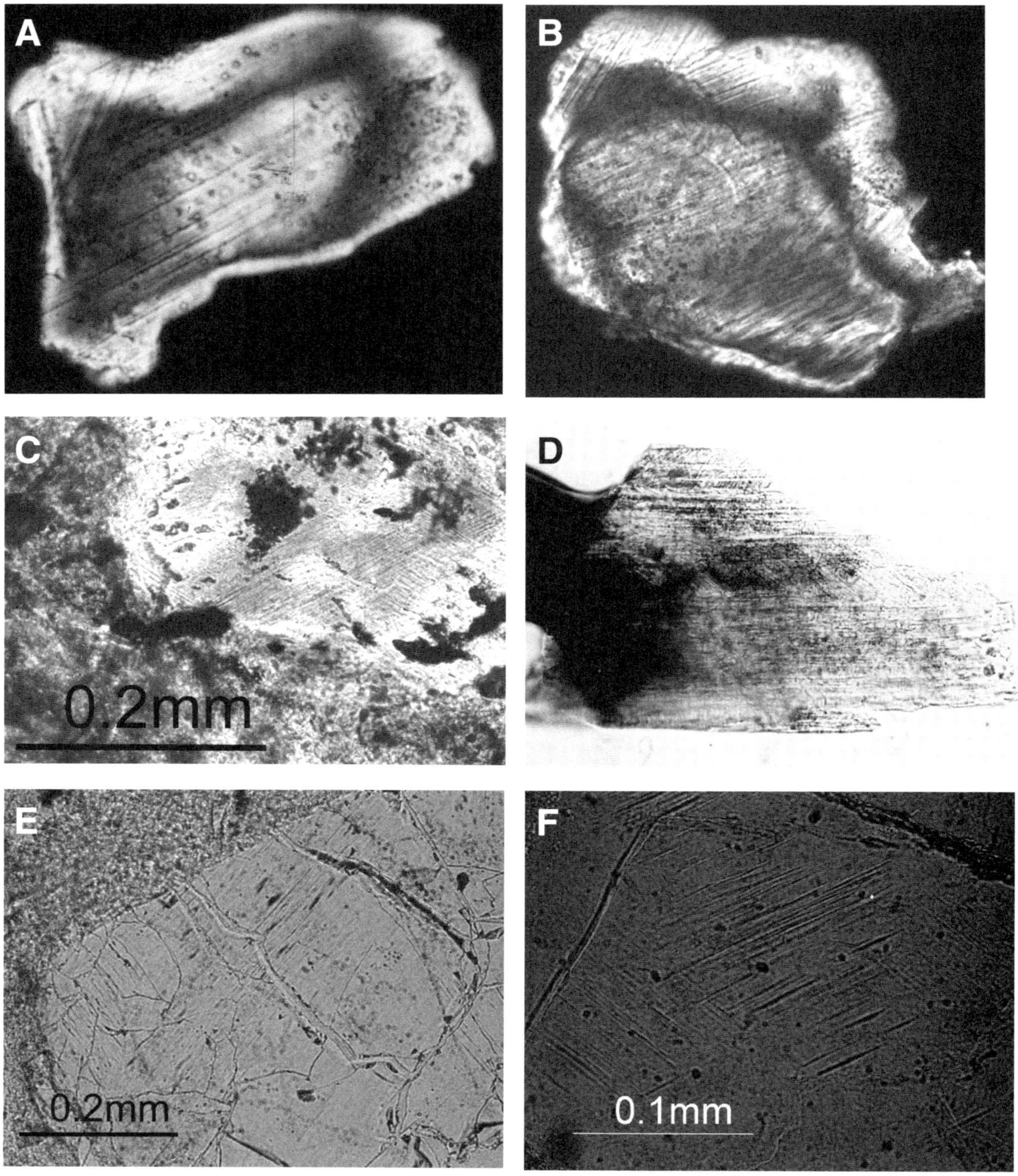

Figure 4. Examples of shock-metamorphic planar deformation features (PDFs) in quartz and feldspar from the Exmore beds in four recent drill cores. (A) Matrix quartz grain (0.13 mm length, 2 sets of PDFs) on spindle stage in cross-polarized light; USGS-NASA Langley core, 236.9 m depth, sample NL777.3. (B) Matrix quartz grain (0.23 mm length, 2 sets of PDFs) on spindle stage in cross-polarized light; USGS-NASA Langley core, 250.1 m depth, sample NL820.6, grain #1. (C) Quartz (2 sets of PDFs) from altered igneous clast, thin section in cross-polarized light; Bayside core, 320.4 m depth, sample BA1051.2. (D) Plagioclase from same altered igneous clast as C, showing chevron pattern of planar deformation features across albite twins (pictured horizontally) on spindle stage in cross-polarized light (short dimension of grain ~0.10 mm); Bayside core, 320.4 m depth, sample BA1051.2. (E) Quartz (2 sets of PDFs) in fractured leucogranite clast, thin section in plane-polarized light; North core, 233.9 m depth, sample UN767.3, thin section 2. (F) Quartz (2 sets of PDFs) in fine-grained granitoid, thin section in plane-polarized light; Watkins School corehole, washed interval from 194.5 to 192.6 m depth, sample WSR632-638B1-2.

The individual quartz grains that have shock-induced PDFs are mostly subangular and lack the rounded shapes that would indicate derivation from a detrital sedimentary target. This suggests that they are mainly particles of crystalline basement, or possibly fragments of smashed detrital grains, that were excavated by the impact.

Cataclastically Deformed Crystalline-Rock Fragments

Cataclastic rocks consist of fragments bounded by fractures, and they are produced at high strain rates by mechanical crushing, faulting, and fracturing of existing rocks. The cataclastic rocks found in impact structures are identical in many respects to those found in fault zones (e.g., Snoke and Tullis, 1998). They are not diagnostic criteria for an impact origin; however, where they are associated with features such as shock-metamorphosed minerals or a recognizable crater, they are likely to be impact related.

Clasts of crystalline rock that have cataclastic fabrics and/or shocked minerals are sparsely distributed throughout the Exmore beds. They are interpreted to be rock fragments that were excavated and ejected from the transient (central) crater and subsequently mixed into sediments of the resurge debris flows.

Cataclastic Fabrics

This study follows the descriptive terminology of Sibson (1977) and Snoke and Tullis (1998) in which cataclastic rocks that show evidence of primary cohesion (cataclasite series) are subdivided according to the volume percent of matrix between fragments as protocataclasite (<50%), cataclasite (50%–90%), and ultracataclasite (>90%), and those lacking primary cohesion are subdivided into breccia (visible fragments >30% of rock) and gouge (visible fragments <30%).

Lithic clasts of crystalline rock that have angular shapes or cataclastic fabrics are distributed throughout the Exmore beds, and one (shown in Fig. 5A) was found 6.4 m below the Exmore beds in crater unit B of the Langley core. Horton and Izett (2005) found clasts larger than the nominal 6.4 cm core diameter only in the lower half of the Exmore beds and below.

Figure 5A shows a cataclasite having pervasive sets of through-going microfaults and fractures. Figure 5B shows dark veinlets of microbreccia or pseudotachylyte(?) containing broken crystals and fragments of the felsite host rock.

Shock-metamorphosed quartz is an integral part of the cataclastic fabric in some of the studied crystalline-rock clasts. This relationship indicates that the cataclastic fabric was produced by the impact. With a few notable exceptions, only a few quartz grains in any thin section have multiple sets of intersecting planar deformation features. Shocked quartz and feldspar crystals in lithic clasts of crystalline rock are shown in Figure 4C–4F. All of these crystals have intersecting sets of parallel planar deformation features that are interpreted to be shock-induced. Planar deformation features viewed in thin sections on a flat stage, as in Figure 4C, 4E, and 4F, tend to be oblique to the microscope axis and thus not as clear in photographs as those oriented on a spindle stage (Fig. 4A, 4B, and 4D).

Features of lithic clasts considered to be likely reworked fallback ejecta include macroscopic to microscopic cataclastic fabrics such as internal faults and fractures, veins of microbreccia or pseudotachylyte(?), and shock-induced planar deformation features in quartz.

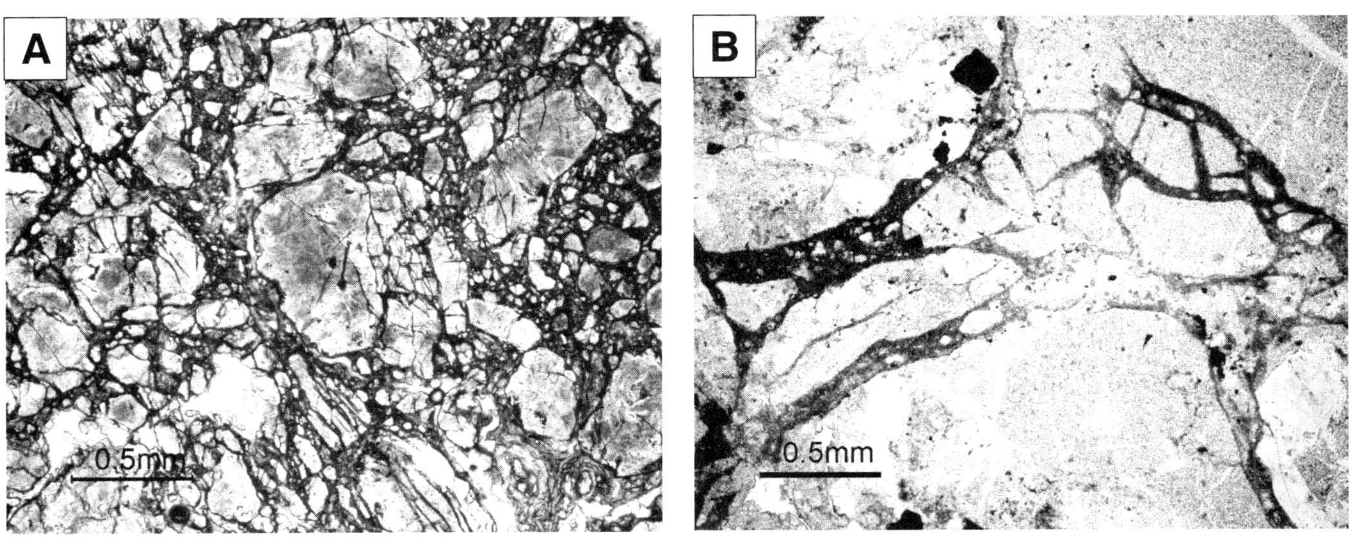

Figure 5. Cataclastic microfabrics in lithic clasts of crystalline rock (photomicrographs). (A) Cataclasite showing pervasive, throughgoing microfaults and fractures in a clast that also contains shocked quartz as part of the cataclastic fabric; clast from injected, mixed sediment 6.4 m below the base of Exmore beds in crater unit B (USGS-NASA Langley core, 275.8 m depth, sample NL905.0, thin section 4). (B) Crosscutting veins of microbreccia or devitrified pseudotachylyte containing broken crystals of feldspar and quartz, and fragments of the felsite host rock, in a clast from the Exmore beds (North core, 241.1 m depth, sample UN790.9, thin section 2). Thin sections viewed in plane-polarized light.

Types of Crystalline-Rock Ejecta

Crystalline-rock fragments interpreted to be fallback ejecta include felsite and a variety of felsic to mafic plutonic rocks (Horton et al., 2002b, 2003). The cored boulders from the Exmore beds that are shown in Figure 3B and 3C are the largest masses of crystalline-rock ejecta recovered in any of the drill cores.

Figure 3B shows a boulder of monomict felsite cataclasite (cohesive breccia) from the Watkins School core. The boulder is composed of angular fragments bounded by fractures and internal faults (Fig. 6A), and the fragments consist entirely of felsite having an aphanitic-porphyritic texture (Fig. 6B). The brittle deformation is pervasive at macroscopic to microscopic scales and so intense that relatively few fragments can be matched across the fractures.

Figure 3C shows a boulder of polymict cataclasite (cohesive breccia) from the Bayside core. The boulder contains abundant fragments of leucogranite, a few fragments of mafic igneous rock, minor amounts of highly strained calcite, and clear, unstrained, fracture-filling calcite (Fig. 7A and 7B). The leucogranite fragments have granophyric texture that is cut by internal microfaults (Fig. 7C). The character of brittle deformation in this rock is similar to that of the felsite cataclasite in Figure 6 in that it is pervasive at a range of scales and sufficiently intense that few fragments can be matched across fractures. The fabric indicates that intense brittle deformation is responsible for the polymict character, although details of the mixing mechanisms are not yet understood. The largest fragment of homogeneous leucogranite within the polymict cataclasite boulder at 360.2–359.8 m includes geochronology sample BA1181 (sawed half of core from 360.0 to 359.8 m).

Some crystalline-rock clasts are interpreted as fallback ejecta because they contain shocked quartz and/or cataclastic fabrics. In the Langley core, all lithic clasts found to contain quartz with shock-induced planar deformation features consist of cataclastically deformed felsite having aphanitic to aphanitic-porphyritic textures. Moreover, except for one granodiorite clast (NL802.07), all lithic clasts in the Langley core having cataclastic fabrics without confirmed shocked minerals also consist of felsite with similar textures. Thus, nearly all of the crystalline-rock fragments confirmed or interpreted to be fallback ejecta are composed of variably porphyritic felsite. The contrast between relatively uniform crystalline-ejecta compositions at the Langley site and more varied compositions at the Bayside, North, and Watkins School sites suggests that the fallback ejecta were distributed unevenly (Horton and Izett, 2005). The impact event provided a remarkable sampling tool by excavating an enormous volume of target rock, including little known basement terranes (Horton et al., 1991; Rankin, 1994; Sheridan et al., 1999) and scattering fragments where they can be retrieved at shallower levels.

Some of the felsite clasts have spherulitic texture (Horton and Izett, 2005). Each spherulite consists of a spherical cluster of acicular plagioclase crystals radiating from a central point, and

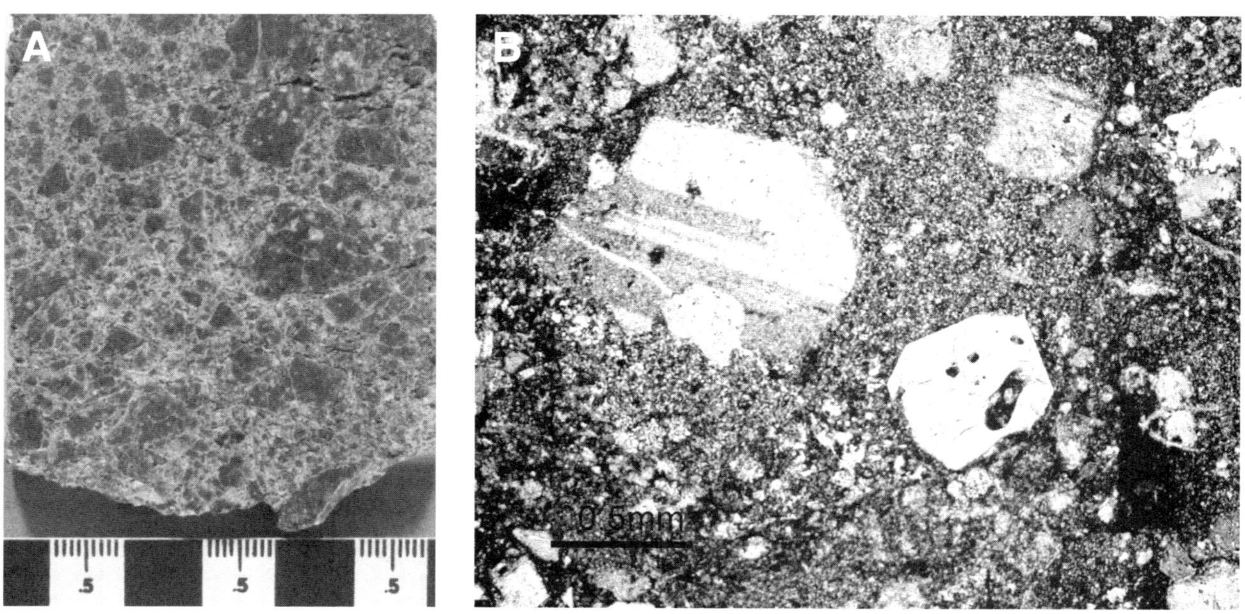

Figure 6. Part of monomict felsite cataclasite boulder from the Exmore beds; Watkins School corehole, Recore-1, 196.02 m depth, sample WSR643.1. (A) Cataclastic fabric showing angular fragments bounded by crosscutting fractures and faults as viewed on a cut surface; scale in cm with mm subdivisions. (B) Felsite fragments within the cataclasite showing aphanitic groundmass and phenocrysts of feldspar and quartz. The phenocrysts are euhedral to subhedral and locally have holes and embayments filled with aphanitic igneous groundmass. Thin section view in cross-polarized light. The boulder, which is shown in Figure 3B, includes U-Pb geochronology sample WS642.

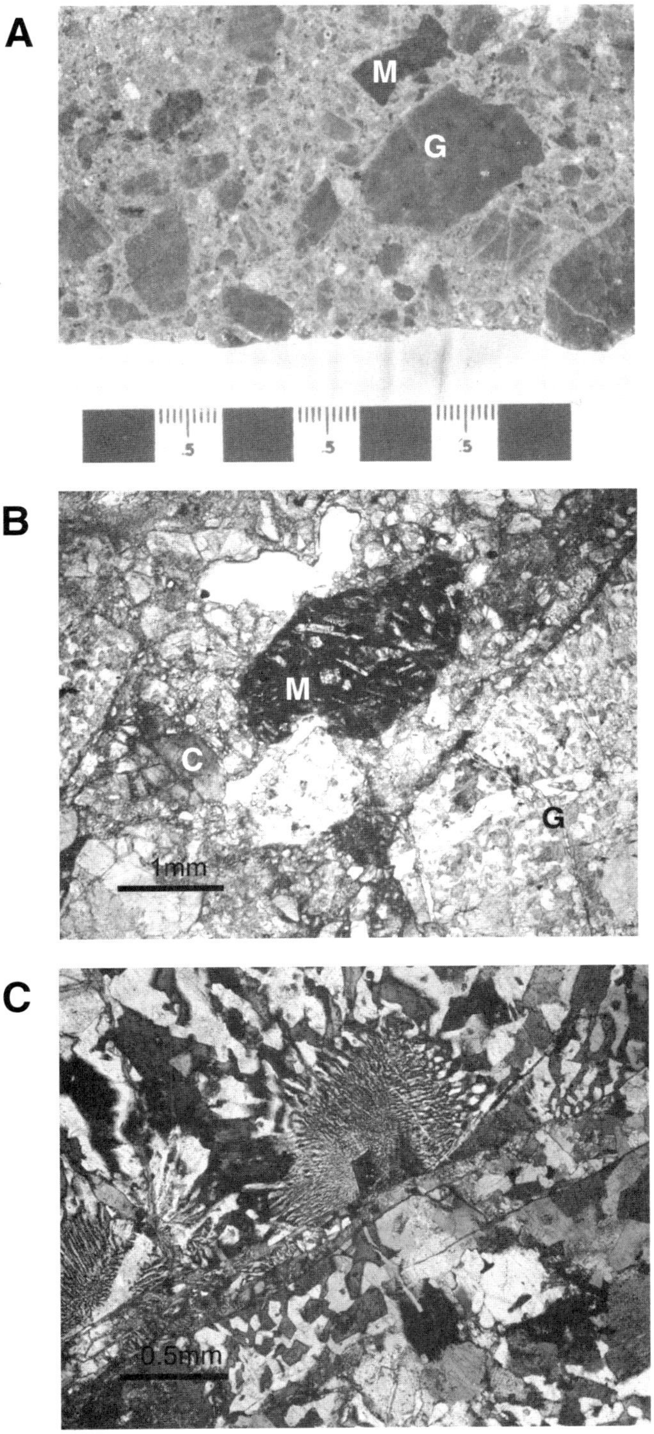

Figure 7. Features in a polymict cataclasite boulder from the Exmore beds, Bayside core. (A) Cataclastic fabric showing angular fragments of leucogranite (G) and mafic igneous rock (M), viewed on cut surface; scale in cm with mm subdivisions. (B) Angular fragments of granophyric leucogranite (G), dark, mafic igneous rock (M), and highly strained calcite (C); thin section viewed in plane-polarized light. (C) The granophyric texture within leucogranite fragments is cut by through-going microfaults; thin section viewed in cross-polarized light. The boulder, which is shown in Figure 3C, includes $^{40}Ar/^{39}Ar$ geochronology sample BA1181.

some have potassium-feldspar–rich outer rims. The spherulites vary in size, even within a thin section, and the largest observed diameter is ~0.4 mm. Spherulitic textures commonly form in volcanic glasses by devitrification (Ross and Smith, 1961) and are locally preserved in metamorphosed volcanic rocks of the southeastern United States (e.g., Allen and Wilson, 1968). However, one of the four felsite clasts with spherulitic devitrification texture encountered in the Langley core contains shocked quartz (Horton and Izett, 2005). Dating of feldspars by the $^{40}Ar/^{39}Ar$ method (in progress) may help to determine if the spherulitic texture formed in an Eocene impact melt or a pre-impact target rock.

Horton and Izett (2005) present chemical analyses for a porphyritic felsite clast from the Exmore beds and for the granite basement at Langley. In the LeBas et al. (1986) chemical classification of volcanic rocks, the felsite is classified as rhyolite. Both the rhyolite clast and the basement granite at Langley are slightly peraluminous. They have essentially identical alumina saturation indices [A/CNK = $Al_2O_3/(CaO+Na_2O+K_2O)$, mol proportion] of 1.1, as well as corundum in the CIPW norms (Horton et al., 2005a; Horton and Izett, 2005). Trace element concentrations in the rhyolite fragment and in the granite are similar in other respects, and trace element discrimination diagrams (not shown) for identifying the tectonic settings of granite emplacement (Pearce et al., 1984; Harris et al., 1986) consistently indicate a volcanic arc setting for both rocks.

SHRIMP U-Pb Zircon Dating of Felsite in Monomict Cataclasite

Sample WS642 is part of a single boulder of monomict felsite cataclasite or cohesive breccia found within the Exmore beds of the Watkins School corehole at Newport News, Virginia (Figs. 3B and 6). This sample is a sawed half of the core (HQ nominal diameter 6.4 cm) from 195.97 to 195.6 m (logged 642.95–641.7 ft) depth within the boulder, which spans the section of core from 196.2 to 195.6 m (logged 643.6 ft to 641.7 ft) depth. Figure 6A shows the cataclastic fabric of the rock as viewed on the cut surface, and Figure 6B shows the igneous aphanitic-porphyritic texture within the felsite fragments as viewed in thin section. About 2 kg of this sample were processed for zircon by standard mineral separation procedures, including crushing, pulverizing, and concentration of heavy minerals using methylene iodide and a magnetic separator. A Wilfley table was not used in order to conserve the limited amount of material. Zircons to be dated were hand picked, mounted in epoxy, ground to about half-thickness using 1,500 grit wet-dry sandpaper, and polished with 6 μm and 1 μm diamond suspension. Zircon standard R33 (419 Ma monzodiorite from the Braintree complex, Vermont; R. Mundil, 1999, personal commun.; S. Kamo, 2001, personal commun.) was used to correct Pb/U ratios for instrumental fractionation. All grains were imaged in cathodoluminescence and photographed in both transmitted and reflected light prior to SHRIMP measurements to identify pristine areas for analysis and to determine if multiple age components (i.e., cores and overgrowths) were present (Fig. 8). These zircons were dated using the USGS–Stanford sensitive high-resolution ion microprobe–reverse geometry (SHRIMP-RG). Analytical procedures followed the methods described in Williams (1998). Raw data were reduced using the SQUID program of Ludwig (2001). Concentrations of U and Th (Table 1) are believed to be accurate to ~20%. A Tera-Wasserburg plot of ^{204}Pb-corrected isotopic data (plotted as 2σ error ellipses, Fig. 9) was used only for visual assessment of the data array to determine which points to include in the age calculation. The age of sample WS642 was determined by (1) calculating the weighted average of the ^{206}Pb/^{238}U ages (using the ^{207}Pb–common Pb correction method; Compston et al., 1984), shown in the Figure 9 inset (2σ error bars), and (2) calculating a concordia age (Ludwig, 2003), which takes into account the uncertainties in both ^{206}Pb/^{238}U and ^{207}Pb/^{206}Pb ages of concordant data.

Zircons from the felsite are light tan and euhedral, have length to width ratios of ~2–5, and exhibit fine, concentric, oscillatory zoning in cathodoluminescence (Fig. 8). They have morphologic characteristics of volcanic origin, including fluid inclusion tubes that are mostly parallel to the *c*-axis (e.g., Chesner, 1998) (Fig. 8). A total of 16 spots on 16 grains (one spot per grain) were analyzed. Uranium concentrations are relatively low and have a limited range (29–84 ppm) (Table 1). The ^{206}Pb/^{238}U ages range from 603.4 ± 6.1 Ma to 633.8 ± 6.4 Ma (1σ). The weighted average age of data from all 16 analyses is 615 ± 6 Ma (mean square of the weighted deviates, MSWD = 2.8). Exclusion of two slightly older analyses (grains 9 and 13) results in a more precise age of 613 ± 4 Ma (MSWD = 1.6) as shown in Figure 9. Omission of these data based on age alone is subjective and may be inappropriate because these grains are morphologically similar to other dated grains from this sample. However, the age of 613 ± 4 Ma is interpreted to indicate a Neoproterozoic crystallization age for this felsic volcanic rock, which we conclude was derived from the impact target.

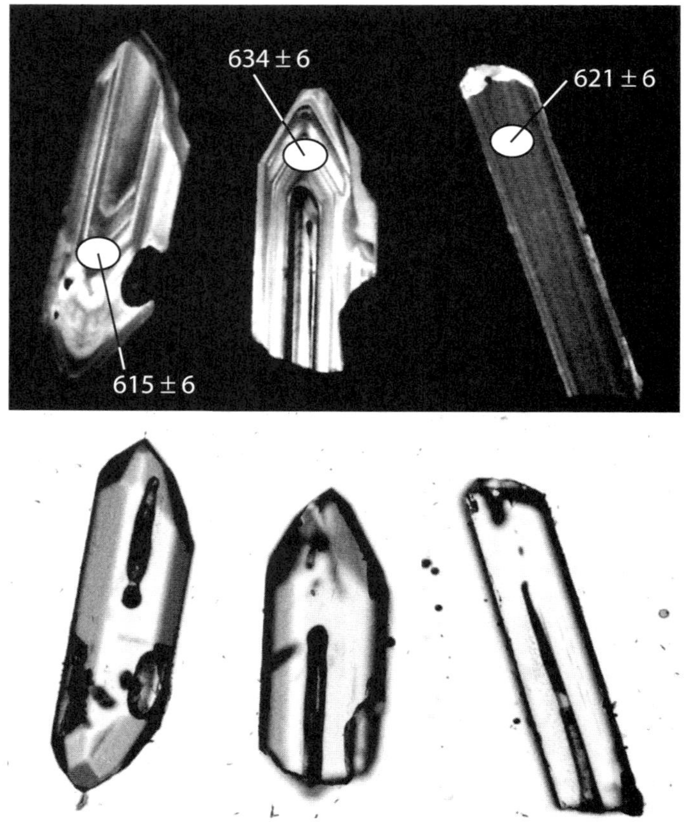

Figure 8. Examples of zircon crystals analyzed to determine sensitive high-resolution ion microprobe (SHRIMP) U-Pb zircon age of felsite in monomict cataclasite (sample WS642; Figs. 3B and 6). Upper image: Cathodoluminescence images show oscillatory internal zones of relatively high (lighter) and low (darker) uranium content; SHRIMP spot diameter ~25 µm. Lower image: Matching transmitted-light photomicrographs show external morphology; fluid inclusion tubes parallel to c-axis are typical, although not diagnostic, of volcanic origin (e.g., Chesner, 1998).

^{40}Ar/^{39}Ar Age Spectrum of Leucogranite in Polymict Cataclasite

The largest mass of crystalline fallback ejecta to be recovered from the Exmore beds in any of the drill cores is a boulder of polymict cataclasite or cohesive breccia from the Bayside core (Fig. 3C). The boulder spans the section of core from the exposed matrix contact at 360.4 m (logged 1182.5 ft) to 359.5 m (logged 1179.5 ft), where the top was lost in drilling. The polymict cataclasite contains abundant fragments of granophyric leucogranite and fewer fragments of other rocks (Fig. 7). Part of the largest homogeneous leucogranite fragment in the polymict cataclasite (sample BA1181 = sawed half of core from 360.0 to 359.8 m; logged 1181.2–1180.5 ft) was processed for dating by ^{40}Ar/^{39}Ar age spectrum techniques.

Muscovite and potassium feldspar fractions from the leucogranite sample (BA1181) were prepared using standard techniques, including crushing and grinding, sieving, magnetic separation, density separation, ultrasonic cleaning, and hand picking, to an apparent purity of >99%. The samples were then re-sieved to remove fines and washed in acetone, alcohol, and de-ionized water in an ultrasonic cleaner. The potassium feldspar is altered and has a cloudy appearance in thin section. The alteration is probably to clay or sericite, and could be a result of lower greenschist facies metamorphism and/or chemical weathering evident in the core samples.

The samples were analyzed at the U.S. Geological Survey's thermochronology laboratory in Denver, Colorado. The samples were packaged in Cu foil capsules and sealed under vacuum into fused silica vials before irradiation in the U.S. Geological Survey's TRIGA reactor (Dalrymple et al., 1981) to convert a portion of their ^{39}K to ^{39}Ar. To monitor this conversion, packets of the standard MMhb-1 hornblende were intercalated with the samples. MMhb-1 has an age at 519.4 ± 2.4 Ma (Alexander et al., 1978; Dalrymple et al., 1981).

The samples dated for this study were analyzed by a MAP 216 mass spectrometer using the ^{40}Ar/^{39}Ar age spectrum dating method. Procedural blanks were subtracted from the analyses

TABLE 1. SHRIMP U-Th-Pb DATA FOR ZIRCON FROM FELSITE IN MONOMICT CATACLASITE BOULDER, SAMPLE WS642

Sample	Measured $^{204}Pb/^{206}Pb$	Measured $^{207}Pb/^{206}Pb$	% common ^{206}Pb	U (ppm)	Th/U	$^{206}Pb/^{238}U$ (Ma)	Error* (Ma)	$^{238}U/^{206}Pb$†	Error* %	$^{207}Pb/^{206}Pb$†	Error* %
WS642-1.1	0.000136	0.0605	0.01	40	1.13	618.0	8.4	9.96	1.4	0.0585	3.2
WS642-2.1	0.000071	0.0593	–	58	0.85	629.2	7.2	9.78	1.2	0.0583	2.1
WS642-3.1	0.000104	0.0628	0.34	54	1.05	603.4	6.1	10.18	1.0	0.0613	2.6
WS642-4.1	–	0.0605	–	61	1.75	604.1	7.4	10.15	1.2	0.0625	2.3
WS642-5.1	0.000000	0.0617	0.19	68	1.66	609.8	5.4	10.06	0.9	0.0617	1.6
WS642-6.1	0.000187	0.0617	0.18	50	0.99	612.4	5.7	10.05	0.9	0.0590	2.7
WS642-7.1	–	0.0623	0.25	47	1.62	614.5	6.0	9.97	1.0	0.0630	2.0
WS642-8.1	0.000034	0.0621	0.23	72	1.05	613.9	6.8	9.99	1.1	0.0616	1.8
WS642-9.1	–	0.0616	0.09	30	1.32	633.8	6.4	9.64	1.1	0.0645	3.6
WS642-10.1	0.000094	0.0597	–	63	0.83	606.1	5.4	10.16	0.9	0.0584	1.9
WS642-11.1	0.000074	0.0618	0.19	46	1.15	614.1	5.8	10.00	1.0	0.0607	2.2
WS642-12.1	–	0.0603	0.00	84	0.84	614.0	5.8	10.01	1.0	0.0603	1.5
WS642-13.1	–	0.0610	0.02	82	1.22	633.1	5.4	9.69	0.9	0.0610	1.6
WS642-14.1	0.000193	0.0605	–	29	1.15	627.4	6.6	9.82	1.1	0.0577	3.7
WS642-15.1	–	0.0610	0.11	56	0.99	605.0	5.5	10.15	0.9	0.0610	1.8
WS642-16.1	0.000075	0.0604	–	59	2.03	620.9	5.6	9.90	0.9	0.0593	2.5

*1σ error.
†Radiogenic ratios, corrected for common Pb, based on the Stacey and Kramers (1975) model.

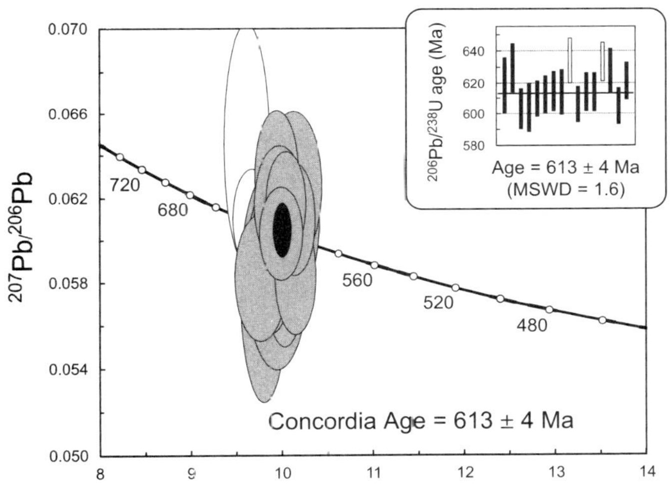

Figure 9. Tera-Wasserburg concordia plot and weighted averages plot of sensitive high-resolution ion microprobe U-Pb isotopic data from zircons in felsite (sample WS642, Watkins School core) from the monomict cataclasite boulder shown in Figures 3B and 6. Concordia age, shown by black-filled error ellipse, calculated by Isoplot/Ex (Ludwig, 2003) from data plotted as gray-shaded error ellipses. The $^{206}Pb/^{238}U$ ages are shown in inset. Data excluded from age calculations shown as open bars. All symbols plotted with 2σ errors.

prior to data reduction. All data reduction for these samples was accomplished using a version of the computer program Mass Spec (Deino, 2001). All estimates of analytical precision are at the 1σ level. Decay constants used are those recommended by Steiger and Jäger (1977). Corrections for the production of interfering reactor-produced argon isotopes from Ca, K, and Cl are those given in Dalrymple et al. (1981) and Roddick (1983). Details of the argon analytical technique can be found in Haugerud and Kunk (1988) and McDougall and Harrison (1999).

Muscovite from sample BA1181 (Fig. 10A; Table 2) has a fairly simple age spectrum, climbing rapidly from 86 Ma in the 550 °C step to 606 Ma in the 1025 °C step, and climbing only slightly in apparent age throughout the rest of the spectrum to a maximum of 616 Ma in the 1215 °C step. The age spectrum does not develop an age plateau as defined by Fleck et al. (1977) and modified by Haugerud and Kunk (1988). Apparent K/Ca and K/Cl ratios also climb rapidly in the lower temperature interval, suggesting the inclusion of minor amounts of another mineral phase that is richer in calcium and chlorine than the muscovite. The 606–616 Ma range of the 1025–1225 °C steps is interpreted to represent the cooling of this sample through ~350 °C (the argon closure temperature for muscovite). This age range is consistent with simple cooling following emplacement of the leucogranite in the central region of the impact target that was generally contemporaneous with the emplacement of granites in the outer region of the impact target, as determined by SHRIMP U-Pb zircon dating of granites in the Bayside (625 ± 11 Ma) and Langley (612 ± 10 Ma) cores (Horton et al., 2001, 2002a, 2005a). Although the rapid climb in age in the lower temperature portion of the spectrum could represent a partial thermal resetting of the sample, it is more likely due to the inclusion of another phase that degassed at lower temperatures.

The age spectrum of potassium feldspar from sample BA1181 is more complex (Fig. 10B; Table 3). The apparent age decreases from >2 Ga in the 500 °C step to a minimum of ca. 261 Ma in the 1125 °C step and then climbs to a maximum of

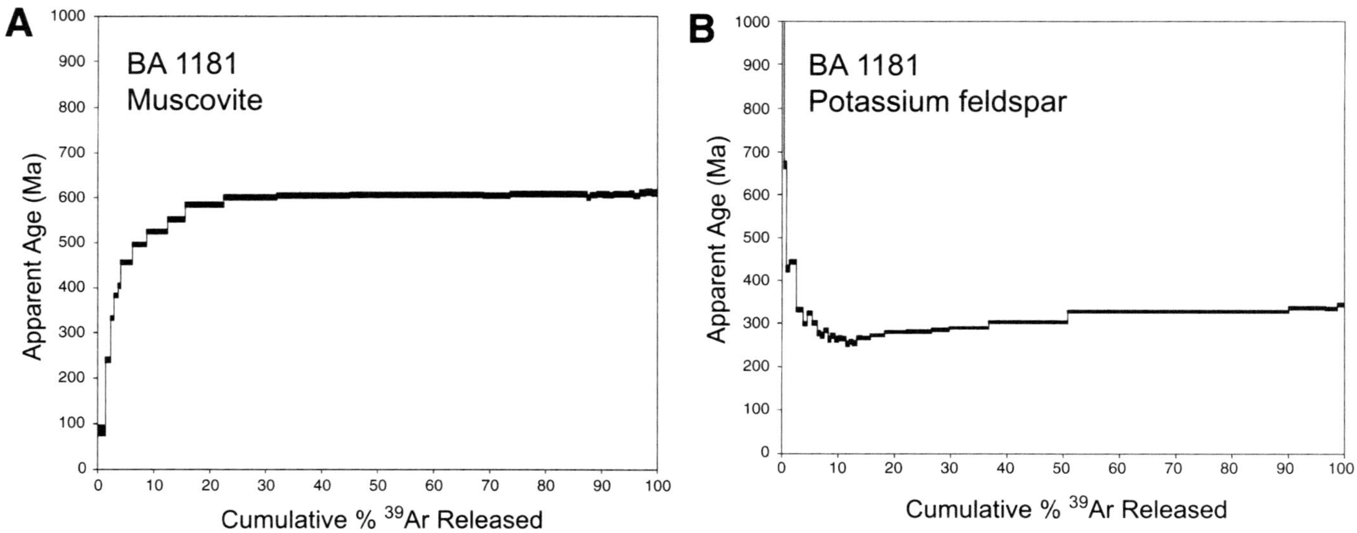

Figure 10. The $^{40}Ar/^{39}Ar$ age spectra of muscovite (A) and potassium feldspar (B) from a granophyric leucogranite fragment (sample BA1181, Bayside core) from polymict cataclasite shown in Figures 3C and 7. Each horizontal box represents an individual age measurement. The height of the box is the 2σ analytical precision in the apparent ages.

TABLE 2. RESULTS OF $^{40}Ar/^{39}Ar$ ANALYSES FOR LEUCOGRANITE SAMPLE BA1181, MUSCOVITE

Heating step	Temp (°C)	^{39}Ar (% Total)	$^{40}Ar^{*\dagger}$ (%)	$^{39}Ar_K$[§] ($\times 10^{-14}$ mol)	$^{40}Ar^*/^{39}Ar_K$	Apparent K/Ca	Apparent K/Cl[#]	Apparent age (Ma)	Error ($\pm 1\sigma$)
A	550	1.4	7.5	0.178171	12.534	5.45	***	86.00	5.44
B	600	0.9	55.1	0.116084	36.618	5.51	***	240.55	2.49
C	650	0.6	84.3	0.072930	51.815	7.20	***	331.67	3.04
D	700	0.7	92.0	0.089123	60.536	8.55	1	381.96	2.48
E	750	0.5	93.5	0.067222	65.353	9.37	3	409.14	3.25
F	800	2.0	96.1	0.250412	74.099	16.61	7	457.48	1.31
G	850	2.6	97.7	0.323510	81.565	39.22	14	497.74	1.28
H	900	3.8	98.1	0.473491	86.972	41.81	15	526.35	1.13
I	950	3.1	96.5	0.385998	92.131	43.84	19	553.23	1.14
J	975	6.8	95.9	0.859645	98.333	70.57	28	585.02	0.82
K	1000	9.5	98.1	1.198561	101.657	73.53	34	601.83	0.80
L	1025	13.1	99.0	1.642576	102.539	86.43	39	606.26	0.78
M	1050	9.6	99.3	1.207556	102.893	79.43	37	608.04	0.86
N	1070	14.3	97.1	1.793528	102.829	72.83	39	607.72	0.72
O	1085	4.8	99.2	0.604361	102.541	53.97	30	606.27	0.94
P	1095	13.6	99.1	4.516205	103.285	76.57	40	610.00	0.72
Q	1105	0.7	99.7	0.217814	101.506	62.50	25	601.07	1.31
R	1115	1.2	99.3	0.386555	102.831	52.25	30	607.73	1.21
S	1125	1.1	99.5	0.369079	103.299	53.82	27	610.07	1.02
T	1135	1.1	99.0	0.353729	102.972	61.35	29	608.43	1.13
U	1145	0.8	99.0	0.257307	102.870	42.00	29	607.92	1.39
V	1155	1.0	99.3	0.336018	103.180	58.04	26	609.48	1.15
W	1165	1.0	99.2	0.323362	103.342	58.07	25	610.29	1.08
X	1175	1.0	99.1	0.322065	103.153	51.47	22	609.34	1.00
Y	1185	0.8	99.7	0.250628	103.808	36.43	18	612.62	1.32
Z	1195	1.0	99.2	0.342254	102.767	53.79	20	607.41	1.26
AA	1205	1.1	99.3	0.356375	104.062	56.05	19	613.89	1.16
BB	1215	1.1	99.3	0.379916	104.438	39.12	19	615.77	1.12
CC	1225	0.9	98.9	0.313163	103.934	35.12	20	613.25	1.39
Total Gas								582.10	
No Plateau									

Note: J = 0.003895 ± 0.35%; sample weight = 0.0050 g.
[†]Radiogenic ^{40}Ar.
[§]Potassium derived ^{39}Ar.
[#]Asterisks (***) indicate that the ratio was below the detection limit of the mass spectrometer.

TABLE 3. RESULTS OF $^{40}Ar/^{39}Ar$ ANALYSES FOR LEUCOGRANITE SAMPLE BA1181, K-FELDSPAR

Heating step	Temp (°C)	^{39}Ar (% Total)	$^{40}Ar^{*†}$ (%)	$^{39}Ar_K$§ (×10^{-14} mol)	$^{40}Ar^{*}/^{39}Ar_K$	Apparent K/Ca	Apparent K/Cl#	Apparent age (Ma)	Error (±1σ)
A	500	0.0	6.3	0.001771	530.860	0.18	***	2030.62	514.43
B	550	0.1	56.7	0.068055	194.203	16.89	2	1021.55	19.80
C	600	0.1	85.4	0.056027	235.859	10.77	2	1181.55	21.45
D	650	0.3	93.1	0.150350	194.081	23.59	2	1021.07	8.69
E	700	0.5	93.6	0.229075	116.314	27.73	3	677.99	6.72
F	725	0.5	96.5	0.255207	69.665	79.18	5	435.77	6.72
G	800	1.3	97.9	0.627241	71.922	187.97	10	448.27	2.78
H	825	1.1	97.3	0.549893	52.366	95.24	21	336.99	3.30
I	850	0.8	98.4	0.388776	47.208	69.40	35	306.45	4.65
J	850	0.9	97.6	0.465076	51.162	47.26	23	329.90	3.85
K	900	0.9	95.3	0.441854	47.471	27.25	21	308.02	4.09
L	940	0.6	98.6	0.282910	43.662	115.07	63	285.15	6.44
M	960	0.5	98.7	0.268632	43.063	34.34	28	281.53	6.88
N	980	0.8	97.4	0.379231	44.687	21.48	25	291.34	4.86
O	1000	0.5	100.0	0.260329	41.567	52.77	33	272.45	7.14
P	1025	0.7	100.0	0.332400	42.691	30.88	26	279.28	5.63
Q	1050	0.6	100.0	0.301282	41.401	52.69	355	271.44	6.21
R	1075	0.6	100.0	0.312573	41.749	27.71	32	273.56	5.89
S	1100	0.7	100.0	0.333020	41.625	28.51	32	272.81	5.57
T	1125	0.6	99.1	0.308208	39.704	23.43	22	261.09	5.98
U	1150	0.7	99.0	0.336183	40.472	29.04	27	265.78	5.46
V	1175	0.7	96.8	0.359133	39.867	32.74	29	262.08	5.30
W	1200	0.9	99.9	0.440011	41.702	125.79	23	273.27	4.19
X	1225	1.5	96.5	0.725793	41.198	38.39	17	270.21	2.60
Y	1250	2.5	97.3	1.237196	42.164	69.06	21	276.08	1.52
Z	1275	3.9	98.3	1.935693	43.477	80.58	31	284.04	1.02
AA	1300	4.5	97.6	2.248906	43.569	102.04	7	284.59	0.92
BB	1325	3.1	97.3	1.524685	44.198	82.71	36	288.39	1.27
CC	1350	7.1	97.6	3.539414	44.815	80.26	5	292.11	0.63
DD	1375	14.0	97.9	6.965706	47.175	86.06	32	306.25	0.45
EE	1450	39.2	98.3	19.449198	51.211	154.32	36	330.19	0.38
FF	1500	6.6	98.3	3.259957	52.859	130.04	18	339.87	0.68
GG	1550	2.1	97.4	1.050820	52.495	138.70	11	337.74	1.73
HH	1650	1.1	94.4	0.536488	54.471	59.52	8	349.30	3.32
Total Gas								322.10	
No Plateau									

Note: J = 0.003922 ± 0.35%; sample weight = 0.0045 g.
†Radiogenic ^{40}Ar.
§Potassium derived ^{39}Ar.
#Asterisks (***) indicate that the ratio was below the detection limit of the mass spectrometer.

349 Ma in the 1650 °C step. The ~1.8 b.y. decrease in apparent age in the first 12% of the age spectrum represents the inclusion of extraneous argon in the potassium feldspar, and the apparent ages of these steps have no geologic meaning. The ca. 349 Ma maximum age of the 1650 °C step is interpreted to represent cooling of the sample through ~250 °C. While this could represent very slow cooling from the time of emplacement of the leucogranite, it is more likely that this age represents cooling from a low-grade thermal event during the Late Devonian (Acadian?) or Early Carboniferous periods. Preservation of old cooling ages in the muscovite spectrum indicates that the temperature of this thermal event could not have exceeded the closure temperature of muscovite (350 °C). The ca. 261 Ma minimum age of the 1125 °C step is interpreted to represent the maximum age of cooling of the sample through potassium feldspar closure at ~150 °C.

In conclusion, muscovite in a leucogranite fragment from the polymict cataclasite boulder in the Exmore beds (Bayside core) has $^{40}Ar/^{39}Ar$ age spectrum data (Fig. 10A) consistent with Neoproterozoic U-Pb ages of autochthonous basement granites beneath the Coastal Plain sediments at Langley and Bayside (Horton et al., 2002b, 2005a). The age spectrum shows no evidence for Paleozoic metamorphism greater than ~350 °C. Potassium feldspar from that clast (Fig. 10B) has an $^{40}Ar/^{39}Ar$ age spectrum indicating cooling through ~250 °C at ca. 349 Ma and a maximum age of cooling through potassium feldspar closure (~150 °C) at ca. 261 Ma, respectively, without discernible impact heating. The cooling history of this leucogranite from the polymict cataclasite boulder resembles that reported for the nearby basement granite at Langley (Horton et al., 2005a).

Distribution of Crystalline-Rock Ejecta

The crystalline-rock fragments interpreted to be fallback ejecta include felsite and a variety of felsic to mafic plutonic rocks, as described above. In the Langley core, these fragments consist mainly of felsite in contrast to more diverse assemblages in other cores, indicating that fallback material in the Exmore beds is distributed unevenly (Horton et al., 2003). If so, the geographic distribution of impact-derived rock fragments may provide clues to the pre-impact distribution of rock types and their relations in the target area.

Detailed studies of the USGS-NASA Langley core (Horton and Izett, 2005) indicate that the largest impact-derived rock fragments are in the lower part of the Exmore beds, suggesting that, in some areas, resurge sediments of the Exmore beds are crudely size graded (Horton et al., 2003; Horton and Izett, 2005). The largest boulders of crystalline rock were found near the base of the Exmore beds in the Watkins School core (Fig. 3B), but higher in the Exmore in the Bayside core (Fig. 3C).

The presence of a cataclastic rock fragment containing shock-metamorphosed quartz 6.4 m below the Exmore beds (sample NL905.0, shown in Fig. 5A) indicates injection of Exmore resurge sediments into underlying mixed sediments in the upper part of crater unit B (Horton and Izett, 2005). The shocked and/or cataclastic crystalline-rock ejecta in the Exmore beds likely were derived by resurge-current erosion of a proximal ejecta blanket or by direct ejecta fallback into the active submarine debris flows.

Tectonic Setting of the Target Crystalline Rocks

Crystalline rocks beneath the Atlantic Coastal Plain in the vicinity of the Chesapeake Bay impact structure are poorly known (Horton et al., 1991). Hence, consideration of the Neoproterozoic crystalline rocks in the impact crater coreholes may help to constrain geologic correlations and tectonic interpretations. The radiometrically dated felsite ejecta and the cored basement granites differ in age from older (e.g., Mount Rogers Formation) and younger (e.g., Catoctin Formation) Neoproterozoic igneous rocks associated with Laurentia (Table 4; Su et al., 1994; Aleinikoff et al., 1995; Fetter and Goldberg, 1995). However, igneous rocks of similar age (Table 4) are found in magmatic-arc terranes of the Appalachian Piedmont and beneath the Atlantic Coastal Plain of the eastern United States. The dated felsite and granites of the target region are similar in age to some of the older igneous rocks in the Carolina, Spring Hope, and Roanoke Rapids terranes of the Appalachian Piedmont and in the Hatteras terrane beneath the Atlantic Coastal Plain (Table 4).

Possible Impact Melt

Samples of possible impact melt from the drill cores are under investigation. In addition, target materials in the Chesapeake Bay impact structure are a likely source for tektites of the North American strewn field (Koeberl et al., 1996; Glass, 2002).

Drill-Core Samples

Poag (2002, his Figure 4) and Poag et al. (2004, their Figures 6.24A and 6.24B) describe and illustrate millimeter-scale pyrite microstructures with hemispherical depressions found in disaggregated sediment samples from the top of the Exmore beds in the Langley core (sample 3 from 235.87 to 235.85 m depth and sample 4 from 235.85 to 235.82 m depth). The hemispherical depressions have the form of partial molds around spherical objects that were ~0.5–1.0 mm in diameter but are now missing from the samples. Poag (2002) interpreted the inferred spherical objects as impact glass microspherules in a fallout layer at the top of the Exmore beds. In hope of isolating the microspherules for chemical and isotopic analysis, we resampled pieces from the same interval in the Langley core (no longer intact because of previous sampling), impregnated them with epoxy, and prepared polished thin sections to examine the pyrite microstructures in situ. Glass microspherules were not observed in any of the samples.

Pyrite aggregates observed in the thin sections have the same morphology (hemispherical depressions) as the pyrite microstructures reported from the earlier disaggregated samples. In the thin sections, however, the pyrite surrounds and forms molds around glauconite pellets as shown in Figure 11. The aggregates of pyrite crystals occupy former pore spaces around the glauconite pellets and fractures within glauconite pellets. Glauconite-pyrite associations are known elsewhere in the Atlantic Coastal Plain, in association with localized reducing chemical environments (Bohlke and Denver, 1995).

At this time, there is no unequivocal evidence for a distinct fallout layer of microspherules at the top of the Exmore beds in the Langley core. An alternative interpretation, presented here, is that the silty, pyrite-bearing layer that immediately overlies coarser resurge debris flows of the Exmore beds is a late-stage deposit formed by settling of suspended particles from the water column. Such a deposit would be similar to part of the Loftarstone that overlies coarser resurge deposits of the Lockne Breccia in the Lockne crater of central Sweden (Sturkell, 1998).

Poag et al. (1992) reported vesicular tektite glass in samples from the corehole at Exmore, Virginia (Fig. 1). More recently, Poag (2002, his Figure 5) and Poag et al. (2004, their Figure 6.32) described and illustrated spherical objects, about one millimeter or smaller in diameter, in thin sections of poorly consolidated sediment of the Exmore beds from the Exmore and Newport News coreholes (Fig. 1). They interpreted these objects as glass microspherules (Poag, 2002) and as "the first known occurrence of proximal, glassy, impact-melt microspherules in a submarine impact crater" (Poag et al., 2004). Further tests, such as measurements of index of refraction, chemical analysis, and $^{40}Ar/^{39}Ar$ isotopic analysis are needed to confirm and characterize the suspected glass.

Microspherulitic matrix in some felsite clasts is evidence of high-temperature devitrification of either impact melt or older volcanic rock (Horton et al., 2001, 2003; Horton and Izett, 2005).

TABLE 4. ISOTOPIC AGES OF SELECTED NEOPROTEROZOIC IGNEOUS ROCKS FOR COMPARISON

Unit dated	Age (Ma)	Isotopic system and method	Reference
Chesapeake block			
felsite ejecta, Watkins School core, Va.	613 ± 4	$^{206}Pb/^{238}U$ SHRIMP zircon	1
granite basement, USGS-NASA Langley core, Va.	612 ± 10	$^{206}Pb/^{238}U$ SHRIMP zircon	5, 9
granite basement, Bayside core, Va.	625 ± 11	$^{206}Pb/^{238}U$ SHRIMP zircon	5
Carolina terrane			
granite of Flat River complex, N.C.	613.4 +2.8/–2	U-Pb zircon, upper intercept	7
diorite of Flat River complex, N.C.	613.9 +1.6/–1.5	U-Pb zircon, upper intercept	7
Osmond granite gneiss, N.C.	612.4 +5.2/–1.7	U-Pb zircon, upper intercept	7
granodiorite near Clarkesville, Va.	602 ± 9	U-Pb zircon, upper intercept	4
granite of Chapel Hill pluton, N.C.	633 +2/–1.5	U-Pb zircon, upper intercept	7
felsic gneiss of Hyco Formation, N.C.	619.9 +4.5/–3	U-Pb zircon, upper intercept	7
metarhyolite of Hyco Formation, N.C.	615.7 +3.7/–1.9	U-Pb zircon, upper intercept	7
felsic metatuff of Hyco Formation, Va.	621 ± 8	U-Pb zircon, upper intercept	4
felsic crystal metatuff of Hyco Formation, Va.	616 ± 4	U-Pb zircon, upper intercept	4
Spring Hope terrane			
gneiss at Mill Creek, N.C.	620 ± 9	U-Pb zircon, upper intercept	2
felsic crystal tuff near Spring Hope, N.C.	590 ± 3	U-Pb zircon, upper intercept	2
Roanoke Rapids terrane			
metatonalite of Roanoke Rapids complex, N.C.	668	$^{207}Pb/^{206}Pb$ zircon, discordant	3
metatonalite intruding(?) Easonburg formation, N.C.	607	$^{207}Pb/^{206}Pb$ zircon, discordant	3
Hatteras terrane			
amphibole quartz monzonite at Camp LeJeune, N.C.	630 ± 39	Rb/Sr whole rock	6
garnet-cordierite-biotite granite at Stumpy Point, N.C.	583 ± 46	Rb/Sr whole rock	6
Goochland terrane			
Fine Creek Mills granite, Va. (A-type)	629 +4/–5	U-Pb zircon, lower intercept	8
granite (SF98-2), Va. (A-type)	630 +9/–10	U-Pb zircon, lower intercept	8
granite (SF99-11), Va. (A-type)	600 +7/–9	U-Pb zircon, lower intercept	8
granite (SF99-20), Va. (A-type)	588 +9/–12	U-Pb zircon, lower intercept	8
Laurentia			
metarhyolite, Catoctin Formation, Pa.	564 ± 9	U-Pb zircon, upper intercept	10
metarhyolite, Mount Rogers Formation, Va.	758 ± 12	U-Pb zircon, upper intercept	10

Note: SHRIMP—sensitive high-resolution ion microprobe. References: 1—this paper; 2—Goldberg, 1994; 3—Horton and Stern, 1994; 4—Horton et al., 1999; 5—Horton et al., 2002b; 6—Russell et al., 1981; 7—Wortman et al., 2000; 8—Owens and Tucker, 2003; 9—Horton et al., 2005a; 10—Aleinikoff et al., 1995.

$^{40}Ar/^{39}Ar$ isotopic analyses of these rocks, including one that appears to be vesicular, are in progress.

Tektite Source

Pre-impact target rocks of the Chesapeake Bay impact structure are considered to be a likely source for impact melt glass in tektites of the North American strewn field (Poag et al., 1994, 2004; Koeberl et al., 1996, 2001; Glass, 2002). Furthermore, late Eocene tektites and microtektites from several sites in this field have Nd model ages (620–670 Ma, Shaw and Wasserburg, 1982; 630 Ma, Ngo et al., 1985), indicating that these distal ejecta were derived from a Neoproterozoic source similar in age or slightly older than the radiometrically dated felsite ejecta and basement granites discussed in this report, or from later sediments that originated from such a source. Trace element geochemistry of a felsite (rhyolite) clast from the Exmore beds, and of the granite basement in the Langley core, neither confirm nor rule out crystalline basement in the Chesapeake Bay impact structure as a partial source component for tektites of the North American strewn field (Horton and Izett, 2005). Since the uppermost target material is a more likely tektite source, Koeberl et al. (2001) and Poag et al. (2004) compared the chemistry of North American tektites to some Cenozoic sedimentary formations in the area. At this stage, the lack of a good chemical match suggests the need to sample and compare the chemistry of additional near-surface target sediments (Poag et al., 2004).

Mixed Sediment Clasts and Matrix

The normal stratigraphic succession of the Virginia Coastal Plain outside the crater to the west consists of Cretaceous non-

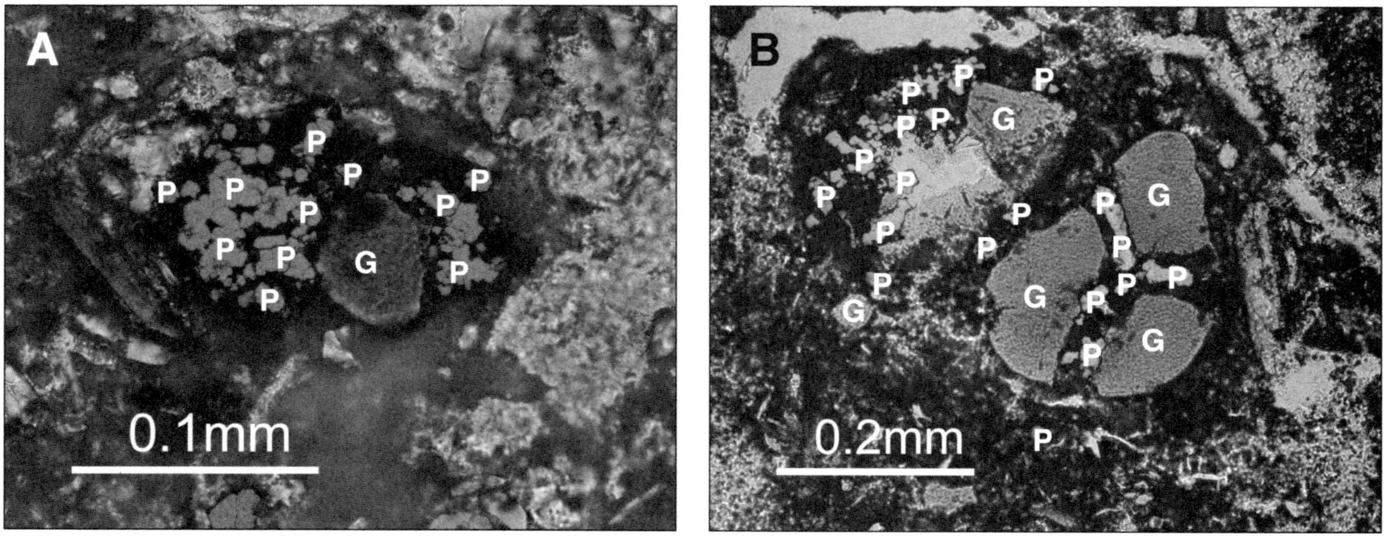

Figure 11. Aggregates of partly oxidized pyrite (P) in former pore spaces that surround and partly enclose spheroidal glauconite pellets (G). The pyrite microstructures form molds around the glauconite in the shape of hemispherical depressions. Samples are from the uppermost Exmore beds, USGS-NASA Langley core. Both thin sections are shown as simultaneously illuminated by transmitted light (for non-opaque minerals) and reflected light (for pyrite). (A) Partly oxidized pyrite aggregate (P, length ~0.2 mm) enclosing round glauconite pellet (G, diameter 0.05 mm); 235.88 m depth, sample NL773.9-B. (B) Partly oxidized pyrite aggregate (P, length ~0.5 mm) enclosing and filling fractures in round glauconite pellet (G, diameter 0.2–0.3 mm); 235.92 m depth, sample NL774.0-1A. In these examples, the aggregates of pyrite crystals occupy former pore spaces around the glauconite pellets, which could have contributed to a highly localized, reducing chemical environment.

marine sands and clays of the Potomac Formation, overlain by mostly unconsolidated, lower Paleocene, upper Paleocene, and lower Eocene siliciclastic sediments, which are overlain by relatively cohesive middle Eocene limestone. Major unconformities punctuate the Paleocene and Eocene record. Cretaceous marine sediments are known from cores to the north and south of the crater. Within the crater, the Potomac Formation occurs as autochthonous and parautochthonous blocks grading upward into, and injected by, a jumble of mixed-age material. The mixed-age material consists of muddy, gravelly, quartz-glauconite sand matrix and recognizable blocks (Gohn et al., 2005).

Fossil dinoflagellate floras from the glauconitic matrix of the impact deposits (Exmore beds) are unusual, and the mixture of stratigraphic ages (Fig. 12A and 12B) is notable. Late Cretaceous, early and late Paleocene, and early, middle, and late Eocene fossil specimens are found mixed together, and the proportions of the variously aged specimens are highly variable among samples. The mixture of ages gives some samples apparent high diversities. The variability among samples highlights the differing mixtures of local and exotic components.

In the Exmore beds, both the sedimentary clasts and the matrix contain microfossils indicating ages that are not found locally in Virginia Coastal Plain cores. Frederiksen et al. (2005) report middle Eocene fossils in the Exmore matrix that are both older and younger than recognized in middle Eocene sediments of the Piney Point Formation. The Piney Point sediments are widespread north, west, and south of the crater, but are not intact within it (Powars and Bruce, 1999). In the inferred shelf setting, known unconformities should become conformable seaward, and the inner or eastern parts of the target area should contain stratigraphic intervals missing in the Virginia Coastal Plain cores outside the crater to the west. Disaggregated remnants of these sediments, now removed, are evidently mixed into the injected matrix between blocks of Lower Cretaceous sediment in crater unit B as well as resurge deposits of the Exmore beds. A discrete interval bearing marine Cretaceous microfossils that could only have come from the north, south, or east is known from the Exmore beds in the Langley core (Frederiksen et al., 2005). As is the case with crystalline basement rocks, the impact event served as a remarkable sampling tool by excavating an enormous volume of sediment and redistributing it both vertically and laterally.

Resurge sediment from the Exmore beds apparently was injected and mixed into the variably liquefied and slumped underlying sediments in the upper part of crater unit B (Gohn et al., 2005), where Horton and Izett (2005) found shocked quartz in a single clast of felsic impact breccia (at 275.8 m depth), and where Frederiksen et al. (2005) reported the only matrix samples from crater unit B found to contain Tertiary microfossils. However, zones containing glauconite of marine origin, presumably of Tertiary age, indicate deeper injections of exotic marine sediment in the matrix between parautochthonous blocks of fluvial-deltaic sediment from the Cretaceous Potomac Formation. This interpretation has been tested in the Bayside core, where glauconite-bearing matrix as deep as 654 m contains mixed-aged Eocene dinocyst fragments.

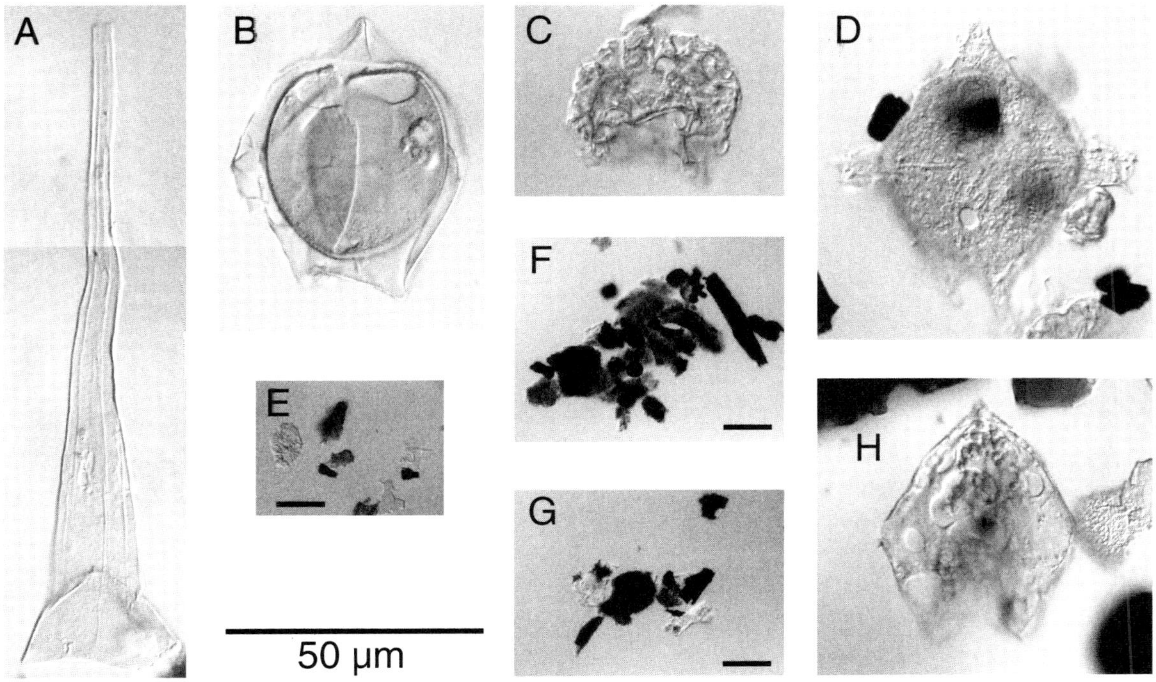

Figure 12. Damaged dinoflagellate cysts from the Chesapeake Bay impact structure. (A) *Odontochitina*, operculum, a genus restricted to the Cretaceous (Kiptopeke core). (B) *Deflandrea phosphoritica*, a genus and species restricted to the Cenozoic, broken specimen from the same sample as A (Kiptopeke core). (C) Miscellaneous chorate form, not identifiable to genus, tightly curled (Exmore core). (D) *Dracodinium varielongitudum*, species highly degraded and pitted, specimen from the same sample as C (Exmore core). (E) *Cordosphaeridium funiculatum* and miscellaneous fragments (Bayside core). (F, G) Fused organic debris (USGS-NASA Langley core). (H) *Deflandrea phosphoritica*, partially melted specimen; compare with B (Watkins School core). Large scale bar is for A–D and H; scale bar shown is 50 μm for E and 100 μm for F and G. Sample locations: (A, B) Kiptopeke core, 405.3 m midpoint, logged 1329.8–1329.6 ft, sample R4371 A. (C, D) Exmore core, 393.5 m midpoint, logged 1291–1290.8 ft, sample R3948. (E) Bayside core, 653.6 m midpoint, logged 2144.8–2144.1 ft, sample R6190 CD. (F) USGS-NASA Langley core, 236.9 m midpoint, logged 773.85–773.75 ft, sample R6110 DC, Exmore beds. (G) USGS-NASA Langley core, 278.4 m midpoint, logged 913.4–913.3 ft, sample R6110 DH, crater unit B. (H) Watkins School core, 136.0 m midpoint, logged 446.4–446.0 ft, sample R6230 E, reworked into the Oligocene Old Church Formation.

Impact-Damaged Microfossils

In addition to shocked minerals, marine microfossils that are interpreted to be damaged by the impact are present in the Exmore beds of the Langley core (Edwards and Powars, 2003; Self-Trail, 2003; Frederiksen et al., 2005). Fractured calcareous nannofossils and fused, bubbled, and curled dinoflagellate cysts occur in the Exmore matrix, where the coexistence of these highly unusual, degraded microfossils and shocked quartz particles indicates an impact origin (Edwards and Powars, 2003; Self-Trail, 2003). The unusual damage preserved in these marine microfossils may record conditions of heating, pressure, and/or abrasion during the oceanic impact.

Fossil dinoflagellates from the Exmore matrix of the Langley core exhibit anomalous preservation (Edwards and Powars, 2003). In the Langley and other cores, impact damage to individual specimens includes beading, breaking, bubbling, clumping, curling, folding, fusing, melting, and/or pitting in a variety of combinations (Fig. 12A–12H). Within a single sample, a wide range of preservation types may be observed. From core to core, or within a single core, variations in preservation types can be used to place constraints on local impact conditions. In the Bayside core, the specimens in the lowest matrix sample at 653.6 m are of mixed ages and are broken into small fragments but not clumped, bubbled, or melted (Fig. 12E), contrasting sharply with the Langley core samples. Within the Langley core, specimens are more highly clumped near the top of the Exmore beds (Fig. 12F, 236.9 m) than in the matrix of mixed sediments beneath the Exmore beds in the upper part of crater unit B (Fig. 12G, 278.4 m). Damaged specimens may also be reworked, yet again, into higher units (Fig. 12H).

Samples of matrix sediment from the Exmore beds in the USGS-NASA Langley core also contain unusual specimens of the calcareous nannofossil *Discoaster multiradiatus*, which Self-Trail (2003) interprets to be shock fractured. These specimens typically have pentagonal shapes and fracture angles of 110° to 130°, although the mechanism is not understood.

DISCUSSION AND CONCLUSIONS

The Chesapeake Bay impact event excavated previously unknown rocks and sediments from the central crater and scattered fragments where they could be sampled for study by drilling in other areas.

Implications for the Atlantic Coastal Plain Basement

Studies of crystalline basement and basement-derived clasts sampled by drilling in the impact structure can help in understanding the regional geology of poorly known terranes that underlie the Atlantic Coastal Plain and continental margin.

Neoproterozoic Volcanic Arc

Felsite from a monomict cataclasite boulder, interpreted as fallback ejecta that originated in the central crater, has a Neoproterozoic crystallization age of 613 ± 4 Ma based on SHRIMP U-Pb zircon data. Despite similarities in age, the volcanic origin of this rock obviously differs from the intrusive origin of granites that underlie the annular trough at Bayside and Langley. Chemical analyses of a small clast of similar aphanitic-porphyritic felsite from the Exmore beds indicate a slightly peraluminous, rhyolitic composition and emplacement in a volcanic-arc setting (Horton and Izett, 2005).

Leucogranite from a polymict cataclastic boulder in the Exmore beds has a muscovite $^{40}Ar/^{39}Ar$ age spectrum that exhibits high-temperature steps of 606–616 Ma, and these steps are interpreted to represent cooling through ~350 °C. We interpret this leucogranite, excavated and ejected from the central crater, to be penecontemporaneous with the Neoproterozoic autochthonous basement granites beneath the annular trough at Bayside and Langley (625 ± 11 Ma and 612 ± 10 Ma, respectively, by SHRIMP U-Pb zircon dating; Horton et al., 2001, 2002a, 2005a). The granite at Langley is a peraluminous monzogranite and geochemical data indicate that it was emplaced in a volcanic-arc setting (Horton et al., 2005a; Horton and Izett, 2005), unlike granitoids of similar age in the Goochland terrane (Table 4), which are chemically distinct (Owens and Tucker, 2003).

Regional Tectonic Models

Recent tectonic models of eastern North America have interpreted little-known basement rocks in the Chesapeake Bay target region alternatively as a northern extension of the Roanoke Rapids volcanic-arc terrane (Horton et al., 1991); as a remnant of the Gondwanan Archean crust of northwest Africa, left behind when the Atlantic Ocean opened (Lefort and Max, 1991); or as Mesoproterozoic (Grenvillian) basement of Laurentia (Sheridan et al., 1999).

The dated felsite and leucogranite boulders from the Exmore beds, and the autochthonous basement granites at Bayside and Langley, resemble Neoproterozoic igneous rocks of peri-Gondwanan volcanic-arc terranes to the south (Horton et al., 2002a, 2005a). This similarity raises doubts about the Chesapeake Bay suture shown on earlier tectonic maps (Horton et al., 1991, and references therein) and suggests an unmapped suture between the terrane sampled in these cores and Mesoproterozoic Laurentian basement beneath the New Jersey Coastal Plain (Horton et al., 2002a, 2005a). Additional chemical and isotopic analyses will help to test possible relations of allochthonous crystalline clasts in the Exmore beds and autochthonous basement rocks to other Neoproterozoic volcanic-intrusive suites and terranes in southeastern North America (Table 4) and generate implications for tectonic models such as the proposed Chesapeake Bay indenter of Archean crust proposed by Lefort and Max (1991).

Implications for the Target Sediments

Major impact effects on the target sediments include in situ deformation of Cretaceous fluvial sediments, as discussed in an earlier section, and redistribution of pre-impact marine sediments, some of which represent newly recognized stratigraphic units.

Redistribution of Pre-Impact Marine Sediments

Pre-impact Upper Cretaceous and lower Tertiary marine sediments are missing from their normal stratigraphic position above Lower Cretaceous to basal Upper Cretaceous fluvial sediments in all cores recovered in the annular trough (Powars and Bruce, 1999; Gohn et al., 2005). However, these compositionally and paleontologically distinct sediments can be recognized even where they have been disaggregated and mixed into other impact-modified or impact-generated sedimentary sections. Hence, the glauconitic sediments with faunas and floras of mixed ages that constitute the Exmore matrix are interpreted to represent scour of the Upper Cretaceous and lower Tertiary sections within and outside the annular trough. Similarly, exotic glauconitic sediments found in sediment dikes (matrix zones) within crater unit B in the Langley and North cores are interpreted as injected exotic sediments from the now-missing Upper Cretaceous and lower Tertiary marine section and the Exmore resurge sediments (Gohn et al., 2005). The presence of shock-metamorphosed quartz in a felsite fragment from crater unit B below the Exmore beds in the Langley core is additional evidence for at least limited injection of Exmore resurge sediments into the parautochthonous Cretaceous sediments of crater unit B (Gohn et al., 2005; Horton and Izett, 2005). The extreme disruption and transportation of the Upper Cretaceous and lower Tertiary marine sediments from the top of the pre-impact sedimentary section are consistent with an impact origin (Gohn et al., 2005), as are distinctive deformation features seen in marine microfossils of the Exmore beds (Edwards and Powars, 2003; Self-Trail, 2003).

Newly Recognized Pre-Impact Deposits

The resurge deposits of the Exmore beds consist of a chaotic mixture of sediment clasts and matrix that contain fossils representative of the known pre-impact Early Cretaceous to late Eocene stratigraphic units in the target area (Edwards and

Powars, 2003; Frederiksen et al., 2005). However, a few Exmore clasts and matrix contain fossils that indicate the presence of lower middle Eocene, uppermost middle Eocene, and lowermost upper Eocene strata that were previously unrecognized in the southeastern Virginia Coastal Plain (Frederiksen et al., 2005). Strata of these ages apparently were present in the impact target area. In addition, Late Cretaceous marine microfossils in the Exmore matrix in cores from the western part of the annular trough (Frederiksen et al., 2005) indicate westward resurge transport of Upper Cretaceous sediments that are present north, south, and east, but not west, of the impact structure (Powars and Bruce, 1999; Powars, 2000).

Implications for the Impact Event

Shocked Quartz

Results of this study are consistent with previous studies of shocked quartz from the Exmore beds of the Chesapeake Bay impact structure. Measured crystallographic orientations of planar deformation features in shocked quartz from earlier coreholes at Exmore, Kiptopeke, Newport News, and Windmill Point (Fig. 1) indicate that the $\{10\bar{1}3\}$ and $\{10\bar{1}2\}$ orientations are dominant (Koeberl et al., 1996; Reimold et al., 2002; Horton et al., 2004; Poag et al., 2004). Reimold et al. (2002) concluded that most of the grains were shocked at 10–20 GPa, consistent with the observed 1 or 2 sets of planar deformation features per grain, and that the Exmore beds contain a mixture of material from different levels in the evolving crater.

Planar deformation features diagnostic of shock metamorphism are documented here in quartz from the Exmore beds at four additional localities, including the Langley corehole (Horton and Izett, 2005) and the Bayside, North, and Watkins School coreholes (Fig. 4). These results extend the number of sites in the impact structure where shocked quartz is confirmed to eight, providing a strong body of evidence that the Exmore beds in and near the Chesapeake Bay impact structure are of impact origin. The shocked quartz grains in the Exmore matrix are inferred to be mainly particles of excavated crystalline basement, because they lack the rounded shapes that would indicate derivation from a clastic sedimentary target.

The relative proportion of shocked to unshocked quartz grains in Exmore matrix sediments is very low in comparison to some other impact-related deposits (e.g., Izett, 1990, his Table 9), indicating that the shock-metamorphosed grains are mixed into and diluted by an enormous volume of other sediment. This observation is consistent with the character of the Exmore beds as mixed sediments produced from seawater-resurge currents (Gohn et al., 2005; Horton et al., 2005b). Although materials interpreted as fallback ejecta are present in sediments of the Exmore beds, a distinct ejecta blanket has not been found intact in the coreholes within or just outside the annular trough. These observations suggest the scouring and removal of near-field ejecta by resurge erosion as documented in other marine impact craters (e.g., Von Dalwigk and Ormö, 2001).

Possible Impact Melt Rocks

Some felsite clasts, including one that contains shocked quartz, contain spherulites that are interpreted to be devitrification products, either of an Eocene impact melt or of older pre-impact volcanic rock in the target area (Horton and Izett, 2005). Koeberl et al. (1996) interpreted similar spherulitic texture as impact melt because the spherulitic matrix contains "clasts" of shocked quartz and feldspar. Alternatively, this texture could be a pre-impact feature of volcanic target rocks similar to those described by Allen and Wilson (1968) in the North Carolina Piedmont. $^{40}Ar/^{39}Ar$ dating of spherulitic felsite is in progress to test these hypotheses.

Distribution of Ejecta Types

The monotonous population of impact-derived felsite clasts in the Exmore beds at the Langley site, in contrast to other core sites in the western annular trough, suggests that the original fallback ejecta were distributed unevenly. With more sampling, the geographic distribution of impact-derived clast populations may provide clues to the original distribution of rock types excavated from the central crater.

Impact Thermal History

The $^{40}Ar/^{39}Ar$ age spectra of potassium feldspar from leucogranite in a polymict cataclasite boulder of the Exmore beds (Fig. 10B) indicate a maximum age of final cooling through feldspar closure (~150 °C) at ca. 261 Ma. Thus, despite the intense brittle deformation in this rock, which is interpreted as fallback ejecta that originated from the central crater, the argon system shows no evidence for sustained impact heating of the rock.

ACKNOWLEDGMENTS

U.S. Geological Survey (USGS) investigations of the Chesapeake Bay impact structure are conducted in cooperation with the Hampton Roads Planning District Commission, the Virginia Department of Environmental Quality, the Department of Geology at the College of William and Mary, and the National Aeronautics and Space Administration (NASA) Langley Research Center. The Hampton Roads Planning District Commission and the USGS provided funds for the drilling of the USGS-NASA Langley, North, Bayside, and Watkins School coreholes. The Virginia Department of Environmental Quality provided funds for the drilling of the North and Bayside coreholes and provided extensive operational support at these sites and the USGS-NASA Langley site. The NASA Langley Research Center provided extensive operational and logistical support for the drilling operations at the Langley Center. We thank Wylie Poag (USGS) for accompanying Horton in the collection of samples shown in Figure 11, and Daniel Milton (USGS) and Bevan French (Smithsonian Institution) for insightful discussions of shock metamorphic features. Reviews by David King (Auburn University), Christian Koeberl (University of Vienna, Austria), Nancy Naeser (USGS), and Jens Ormö (Centro de Astrobiológica, Spain) significantly improved the manuscript.

REFERENCES CITED

Aleinikoff, J.N., Zartman, R.E., Walter, M., Rankin, D.W., Lyttle, P.T., and Burton, W.C., 1995, U-Pb ages of metarhyolites of the Catoctin and Mount Rogers formations, Central and Southern Appalachians—Evidence for two pulses of Iapetan rifting: American Journal of Science, v. 295, no. 4, p. 428–454.

Alexander, E.C., Jr., Mickelson, G.M., and Lanphere, M.A., 1978, Mmhb-1: A new $^{40}Ar/^{39}Ar$ dating standard, *in* Zartman, R.E., ed., Short papers of the Fourth International Conference, Geochronology, Cosmochronology, Isotope Geology 1978: U.S. Geological Survey Open-File Report 78-701, p. 6–8.

Allen, E.P., and Wilson, W.F., 1968, Geology and mineral resources of Orange County, Virginia: Raleigh, North Carolina Geological Survey Bulletin 81, 58 p.

Bohlke, J.K., and Denver, J.M., 1995, Combined use of groundwater dating, chemical, and isotopic analyzes to resolve the history and fate of nitrate contamination in two agricultural watersheds, Atlantic Coastal Plain, Maryland: Water Resources Research, v. 31, no. 9, p. 2319–2339, doi: 10.1029/95WR01584.

Bohor, B.F., Betterton, W.J., and Foord, E.E., 1988, Coesite, glass, and shocked quartz at DSDP Site 612: Evidence for nearby impact in the late Eocene [abs.]: Houston, Texas, Lunar and Planetary Institute, Proceedings of the 19th Lunar and Planetary Science Conference, p. 114–115.

Catchings, R.D., Saulter, D.E., Powars, D.S., Goldman, M.R., Dingler, J.A., Gohn, G.S., Schindler, J.S., and Johnson, G.H., 2001, High-resolution seismic reflection/refraction images near the outer margin of the Chesapeake Bay impact crater, York-James Peninsula, southeastern Virginia: U.S. Geological Survey Open-File Report 01-407, 18 p.

Catchings, R.D., Powars, D.S., Gohn, G.S., and Goldman, M.R., 2005, High-resolution seismic reflection image of the Chesapeake Bay impact structure, NASA-Langley Research Center, Hampton, Virginia, chap. I, *in* Horton, J.W., Jr., Powars, D.S., and Gohn, G.S., eds., Studies of the Chesapeake Bay impact structure—The USGS-NASA Langley corehole, Hampton, Virginia, and related coreholes and geophysical surveys: U.S. Geological Survey Professional Paper 1688 (in press).

Chesner, C.A., 1998, Petrogenesis of the Toba Tuffs, Sumatra, Indonesia: Journal of Petrology, v. 39, p. 397–438, doi: 10.1093/petrology/39.3.397.

Compston, W., Williams, I.S., and Meyer, C., 1984, U-Pb geochronology of zircons from lunar breccia 73217 using a sensitive high-resolution ion-microprobe: Proceedings of the 14th Lunar Science Conference: Journal of Geophysical Research, v. 98B, p. 525–534.

Croft, S.K., 1981, The excavation stage of basin formation: A qualitative model, *in* Proceedings of the conference on multi-ring basins: Formation and evolution, Lunar and Planetary Institute, Houston, 1980: Geochimica et Cosmochimica Acta, Supplement 15, p. 207–225.

Dalrymple, G.B., Alexander, E.C., Lanphere, M.A., and Kraker, G.P., 1981, Irradiation of samples for $^{40}Ar/^{39}Ar$ dating using the Geological Survey TRIGA reactor: U.S. Geological Survey Professional Paper 1176, 55 p.

Deino, A.L., 2001, Users manual for Mass Spec v. 5.02: Berkeley, California, Berkeley Geochronology Center Special Publication 1a, 119 p.

Earth Impact Database, 2004, http://www.unb.ca/passc/ImpactDatabase/, (accessed: 28 January 2004).

Edwards, L.E., and Powars, D.S., 2003, Impact damage to dinocysts from the late Eocene Chesapeake Bay event: Palaios, v. 18, no. 3, p. 275–285.

Edwards, L.E., Barron, J.A., Brewster-Wingard, G.L., Bukry, D., Bybell, L.M., Cronin, T.M., Poag, C.W., and Weems, R.E., 2005, Paleontology of the upper Eocene to Quaternary stratigraphic section in the USGS-NASA Langley core, Hampton, Virginia, chap. H, *in* Horton, J.W., Jr., Powars, D.S., and Gohn, G.S., eds., Studies of the Chesapeake Bay impact structure—The USGS-NASA Langley corehole, Hampton, Virginia, and related coreholes and geophysical surveys: U.S. Geological Survey Professional Paper 1688 (in press).

Fetter, A.H., and Goldberg, S.A., 1995, The age and geochemical characteristics of bimodal magmatism in the Neoproterozoic Grandfather Mountain rift basin: Journal of Geology, v. 103, p. 313–326.

Fleck, R.J., Sutter, J.F., and Elliot, D.H., 1977, Interpretation of discordant $^{40}Ar/^{39}Ar$ age spectra of Mesozoic tholeiites from Antarctica: Geochimica et Cosmochimica Acta, v. 41, p. 15–32, doi: 10.1016/0016-7037(77)90184-3.

Frederiksen, N.O., Edwards, L.E., Self-Trail, J.M., Bybell, L.M., and Cronin, T.M., 2005, Paleontology of the impact-modified and impact-generated sediments in the USGS-NASA Langley core, Hampton, Virginia, chap. D, *in* Horton, J.W., Jr., Powars, D.S., and Gohn, G.S., eds., Studies of the Chesapeake Bay impact structure—The USGS-NASA Langley corehole, Hampton, Virginia, and related coreholes and geophysical surveys: U.S. Geological Survey Professional Paper 1688 (in press).

Glass, B.P., 1989, North American tektite debris and impact ejecta from DSDP Site 612: Meteoritics, v. 24, p. 209–218.

Glass, B.P., 2002, Distal impact ejecta from the Chesapeake Bay impact structure: Geological Society of America Abstracts with Programs, v. 34, no. 6, p. 465.

Glass, B.P., and Liu, S., 2001, Discovery of high-pressure $ZrSiO_4$ polymorph in naturally occurring shock-metamorphosed zircons: Geology, v. 29, no. 4, p. 371–373, doi: 10.1130/0091-7613(2001)0292.0.CO;2.

Glass, B.P., Hall, C.M., and York, D., 1986, $^{40}Ar/^{39}Ar$ laser-probe dating of North American tektite fragments from Barbados and the age of the Eocene-Oligocene boundary: Chemical Geology, v. 59, p. 181–186.

Glass, B.P., Koeberl, C., Blum, J., Senftle, F., Izett, G.A., Evans, B.J., Thorpe, A.N., Povenmire, H., and Strange, R.L., 1995, A Muong Nong-type Georgia tektite: Geochimica et Cosmochimica Acta, v. 59, p. 4071–4082, doi: 10.1016/0016-7037(95)00290-G.

Glass, B.P., Liu, S., and Leavens, P.B., 2002, Reidite: An impact-produced high-pressure polymorph of zircon found in marine sediments: American Mineralogist, v. 87, p. 562–565.

Gohn, G.S., Powars, D.S., Quick, J.E., Horton, J.W., Jr., and Catchings, R.D., 2002, Variation of impact response with depth and lithology, outer annular trough of the Chesapeake Bay impact structure, Virginia Coastal Plain: Geological Society of America Abstracts with Programs, v. 34, no. 6, p. 465.

Gohn, G.S., Powars, D.S., Bruce, T.S., and Self-Trail, J.M., 2005, Physical geology of the impact-modified and impact-generated sediments in the USGS-NASA Langley core, Hampton, Virginia, chap. C, *in* Horton, J.W., Jr., Powars, D.S., and Gohn, G.S., eds., Studies of the Chesapeake Bay impact structure—The USGS-NASA Langley corehole, Hampton, Virginia, and related coreholes and geophysical surveys: U.S. Geological Survey Professional Paper 1688 (in press).

Goldberg, S.A., 1994, U-Pb geochronology of volcanogenic terranes of the eastern North Carolina Piedmont—Preliminary results, *in* Stoddard, E.F., and Blake, D.E., eds., Geology and field trip guide—Western flank of the Raleigh metamorphic belt, North Carolina, Carolina Geological Society Field Trip Guidebook 1994, Raleigh, North Carolina, November 5–6, 1994: Raleigh, North Carolina Geological Survey, p. 13–17.

Grieve, R.A.F., Langenhorst, F., and Stöffler, D., 1996, Shock metamorphism of quartz in nature and experiment—II, Significance in geoscience: Meteoritics, v. 31, p. 6–35.

Hampton Roads Planning District Commission, 1999, 35 million year old meteor crater affects the region's ground water: Chesapeake, Virginia: Hampton Roads Planning District Commission Special Report, Issue, v. 2, p. 2–5.

Harris, N.B.W., Pearce, J.A., and Tindle, A.G., 1986, Geochemical characteristics of collision zone magmatism, *in* Coward, M.P., and Ries, A.C., eds., Collision tectonics: London, Geological Society Special Publication 19, p. 67–81.

Haugerud, R.A., and Kunk, M.J., 1988, ArAr*, a computer program for reduction of ^{40}Ar-^{39}Ar data: U.S. Geological Survey Open-File Report 88-261, 68 p.

Horton, J.W., Jr., and Izett, G.A., 2005, Crystalline-rock ejecta and shocked minerals of the Chesapeake Bay impact structure, USGS-NASA Langley core, Hampton, Virginia, with supplemental constraints on the age of impact, chap. E, *in* Horton, J.W., Jr., Powars, D.S., and Gohn, G.S., eds., Studies of the Chesapeake Bay impact structure—The USGS-NASA Langley corehole, Hampton, Virginia, and related coreholes and geophysical surveys: U.S. Geological Survey Professional Paper 1688 (in press).

Horton, J.W., Jr., and Stern, T.W., 1994, Tectonic significance of preliminary uranium-lead ages from the eastern Piedmont of North Carolina: Geological Society of America Abstracts with Programs, v. 26, no. 4, p. 21.

Horton, J.W., Jr., Drake, A.A., Jr., Rankin, D.W., and Dallmeyer, R.D., 1991, Preliminary tectonostratigraphic terrane map of the central and southern Appalachians: U.S. Geological Survey Miscellaneous Investigations Series Map I-2163, scale 1:2,000,000.

Horton, J.W., Jr., Aleinikoff, J.N., Burton, W.C., Peper, J.D., and Hackley, P.C., 1999, Geologic framework of the Carolina slate belt in southern Virginia—Insights from geologic mapping and U-Pb geochronology: Geological Society of America Abstracts with Programs, v. 31, no. 7, p. A-476.

Horton, J.W., Jr., Aleinikoff, J.N., Izett, G.A., Naeser, C.W., and Naeser, N.D., 2001, Crystalline rocks from the first corehole to basement in the Chesapeake Bay impact structure, Hampton, Virginia: Geological Society of America Abstracts with Programs, v. 33, no. 6, p. A-448.

Horton, J.W., Jr., Aleinikoff, J.N., Izett, G.A., Naeser, N.D., Naeser, C.W., and Kunk, M.J., 2002a, Crystalline basement and impact-derived clasts from three coreholes in the Chesapeake Bay impact structure, southeastern Virginia: Eos (Transactions, American Geophysical Union), v. 83, no. 19, Spring Meeting Supplement, Abstract T21A-03, p. S351.

Horton, J.W., Jr., Kunk, M.J., Naeser, C.W., Naeser, N.D., Aleinikoff, J.N., and Izett, G.A., 2002b, Petrography, geochronology, and significance of crystalline basement rocks and impact-derived clasts in the Chesapeake Bay impact structure, southeastern Virginia: Geological Society of America Abstracts with Programs, v. 34, no. 6, p. 466.

Horton, J.W., Jr., Gohn, G.S., Edwards, L.E., Self-Trail, J.M., Powars, D.S., Kunk, M.J., and Izett, G.A., 2003, Recent research in the Chesapeake Bay impact crater, USA—Part 2. Reworked impact ejecta and impact debris: Third International Conference on Large Meteorite Impacts, August 5–7, 2003, Nördlingen, Germany, LPI Contribution 1167, Abstract 4051.

Horton, J.W., Jr., Koeberl, C., Reimold, W.U., Aleinikoff, J.N., Gohn, G.S., Izett, G.A., Kunk, M.J., Naeser, C.W., Naeser, N.D., Poag, C.W., and Powars, D.S., 2004, Review of petrography, geochemistry, and geochronology—Chesapeake Bay crater, *in* Edwards, L.E., Horton, J.W., Jr., and Gohn, G.S., compilers, ICDP-USGS Workshop on Deep Drilling in the Central Crater of the Chesapeake Bay Impact Structure, Virginia, USA: Proceedings Volume: U.S. Geological Survey Open-File Report 2004-1016, CD-ROM version, 33 slides with explanatory notes (PowerPoint, 3 MB).

Horton, J.W., Jr., Aleinikoff, J.N., Kunk, M.J., Naeser, C.W., and Naeser, N.D., 2005a, Petrography, structure, age, and thermal history of granitic Coastal Plain basement in the Chesapeake Bay impact structure, USGS-NASA Langley core, Hampton, Virginia, chap. B, *in* Horton, J.W., Jr., Powars, D.S., and Gohn, G.S., eds., Studies of the Chesapeake Bay impact structure—The USGS-NASA Langley corehole, Hampton, Virginia, and related coreholes and geophysical surveys: U.S. Geological Survey Professional Paper 1688 (in press).

Horton, J.W., Jr., Powars, D.S., and Gohn, G.S., 2005b, Studies of the Chesapeake Bay impact structure—Introduction and discussion, chapter A, *in* Horton, J.W., Jr., Powars, D.S., and Gohn, G.S., eds., Studies of the Chesapeake Bay impact structure—The USGS-NASA Langley corehole, Hampton, Virginia, and related coreholes and geophysical surveys: U.S. Geological Survey Professional Paper 1688 (in press).

Izett, G.A., 1990, The Cretaceous/Tertiary boundary interval, Raton Basin, Colorado and New Mexico, and its content of shock-metamorphosed minerals—Evidence relevant to the K-T boundary impact-extinction theory: Geological Society of America Special Paper 249, 100 p.

Kamo, S.L., Krough, T.E., Glass, B.P., and Liu, S., 2002, U-Pb study of shocked zircons from the North American microtektite layer: Houston, Texas, Lunar and Planetary Institute, 33rd Lunar and Planetary Science Conference Abstracts, LPI Contribution 1109, abstract 1643, 2 p.

Koeberl, C., and Glass, B.P., 1988, Chemical composition of North American microtektites and tektite fragments from Barbados and DSDP Site 612 on the continental slope off New Jersey: Earth and Planetary Science Letters, v. 87, p. 286–292, doi: 10.1016/0012-821X(88)90016-7.

Koeberl, C., Poag, W.C., Reimold, W.U., and Brandt, D., 1996, Impact origin of the Chesapeake Bay structure and the source of the North American tektites: Science, v. 271, p. 1263–1266.

Koeberl, C., Kruger, F.J., Poag, C.W., and Allsopp, H., 2001, Geochemistry of surficial sediments near the Chesapeake Bay impact structure and the search for source rocks of the North American tektites: Houston, Texas, Lunar and Planetary Institute, 32nd Lunar and Planetary Science Conference Abstracts, LPI Contribution 1080, abstract 1333, 2 p.

LeBas, M.J., Le Maitre, R.W., Streckeisen, A., and Zanettin, B., 1986, A chemical classification of volcanic rocks based on the total alkali-silica diagram: Journal of Petrology, v. 27, p. 745–750.

Lefort, J.P., and Max, M.D., 1991, Is there Archean crust beneath Chesapeake Bay?: Tectonics, v. 10, p. 213–226.

Ludwig, K.R., 2001, SQUID, 1.02, A user's manual: Berkeley, California, Berkeley Geochronology Center Special Publication 2, 21 p.

Ludwig, K.R., 2003, User's manual for Isoplot 3.00: A geochronological toolkit for Microsoft Excel: Berkeley, California, Berkeley Geochronology Center Special Publication 4, 73 p.

McDougall, I., and Harrison, T.M., 1999, Geochronology and thermochronology by the $^{40}Ar/^{39}Ar$ method: New York, Oxford University Press, 269 p.

McFarland, E.R., and Bruce, T.S., 2005, Distribution, origin, and resource-management implications of ground-water salinity along the western margin of the Chesapeake Bay impact structure in eastern Virginia, chap. K, *in* Horton, J.W., Jr., Powars, D.S., and Gohn, G.S., eds., Studies of the Chesapeake Bay impact structure—The USGS-NASA Langley corehole, Hampton, Virginia, and related coreholes and geophysical surveys: U.S. Geological Survey Professional Paper 1688 (in press).

Melosh, H.J., 1989, Impact cratering—A geological process: New York, Oxford University Press, 245 p.

Melosh, H.J., and Ivanov, B.A., 1999, Impact crater collapse: Annual Review of Earth and Planetary Sciences, v. 27, p. 385–415, doi: 10.1146/annurev.earth.27.1.385.

Ngo, H.H., Wasserburg, G.J., and Glass, B.P., 1985, Nd and Sr isotopic compositions of tektite material from Barbados and their relationship to North American tektites: Geochimica et Cosmochimica Acta, v. 49, p. 1479–1485, doi: 10.1016/0016-7037(85)90297-2.

Obermeier, S.F., Pond, E.C., and Olson, S.M., 2001, Paleoliquefaction studies in continental settings—Geologic and geotechnical factors in interpretations and backanalysis: U.S. Geological Survey Open-File Report 01-029, 75 p.

Olson, S.M., Green, R.A., and Obermeier, S.F., 2003, Geotechnical analysis of paleoseismic shaking using liquefaction features: Part I. Major updating of analysis techniques: U.S. Geological Survey Open-File Report 03-307, 33 p.

Owens, B.E., and Tucker, R.D., 2003, Geochronology of the Mesoproterozoic State Farm gneiss and associated Neoproterozoic granitoids, Goochland terrane, Virginia: Geological Society of America Bulletin, v. 115, p. 972–982, doi: 10.1130/B25258.1.

Pearce, T.H., Harris, N.B.W., and Tindale, A.G., 1984, Trace element discrimination diagrams for the tectonic interpretation of granitic rocks: Journal of Petrology, v. 25, p. 956–983.

Poag, C.W., 1996, Structural outer rim of Chesapeake Bay impact crater—Seismic and borehole evidence: Meteoritics & Planetary Science, v. 31, p. 218–226.

Poag, C.W., 1999, Chesapeake invader, Discovering America's giant meteorite crater: Princeton, New Jersey, Princeton University Press, 183 p.

Poag, C.W., 2002, Synimpact-postimpact transition inside Chesapeake Bay crater: Geology, v. 30, p. 995–998, doi: 10.1130/0091-7613(2002)0302.0.CO;2.

Poag, C.W., and Norris, R.D., 2005, Stratigraphy and paleoenvironments of early postimpact deposits—USGS-NASA Langley core, Chesapeake Bay impact crater, chap. F, *in* Horton, J.W., Jr., Powars, D.S., and Gohn, G.S., eds., Studies of the Chesapeake Bay impact structure—The USGS-NASA Langley corehole, Hampton, Virginia, and related coreholes and geophysical surveys: U.S. Geological Survey Professional Paper 1688 (in press).

Poag, C.W., Powars, D.S., Poppe, L.J., Mixon, R.B., Edwards, L.E., Folger, D.W., and Bruce, S., 1992, Deep Sea Drilling Project Site 612 bolide event—New evidence of a late Eocene impact-wave deposit and a possible impact site, U.S. East Coast: Geology, v. 20, p. 771–774.

Poag, W.C., Powars, D.S., Poppe, L.J., and Mixon, R.B., 1994, Meteoroid mayhem in Ole Virginny—Source of the North American tektite strewn field: Geology, v. 22, p. 691–694, doi: 10.1130/0091-7613(1994)0222.3.CO;2.

Poag, C.W., Hutchinson, D.R., Colman, S.M., and Lee, M.W., 1999a, Seismic expression of the Chesapeake Bay impact crater: Structural and morphologic refinements based on new seismic data, *in* Dressler, B.O., and Sharpton, V.L., eds., Large meteorite impacts and planetary evolution II: Geological Society of America Special Paper 339, p. 149–164.

Poag, C.W., Plescia, J.B., and Molzer, P.C., 1999b, Chesapeake Bay impact structure: Geology and geophysics [abs.], *in* Gersonde, R., and Deutsch, A., eds., ESF-IMPACT Workshop, Oceanic Impacts: Mechanisms and Environmental Perturbations, April 15–17, 1999, Abstracts: Berichte zur Polarforschung (Reports on Polar Research), v. 343, p. 79–83.

Poag, C.W., Plescia, J.B., and Molzer, P.C., 2002, Ancient impact structures on modern continental shelves: The Chesapeake Bay, Montagnais, and Toms Canyon craters, Atlantic margin of North America: Deep-Sea Research, Part II, Topical Studies in Oceanography, v. 49, p. 1081–1102, doi: 10.1016/S0967-0645(01)00144-8.

Poag, C.W., Koeberl, C., and Reimold, W.U., 2004, The Chesapeake Bay impact crater—Geology and geophysics of a late Eocene submarine impact structure: New York, Springer-Verlag, 522 p. plus CD-ROM.

Powars, D.S., 2000, The effects of the Chesapeake Bay impact crater on the geologic framework and the correlation of hydrogeologic units of southeastern Virginia, south of the James River: U.S. Geological Survey Professional Paper 1622, 53 p.

Powars, D.S., and Bruce, T.S., 1999, The effects of the Chesapeake Bay impact crater on the geological framework and correlation of hydrogeologic units of the lower York-James Peninsula, Virginia: U.S. Geological Survey Professional Paper 1612, 82 p.

Powars, D.S., Mixon, R.B., and Bruce, T.S., 1992, Uppermost Mesozoic and Cenozoic geologic cross section, outer Coastal Plain of Virginia, *in* Gohn, G.S., ed., Proceedings of the 1988 U.S. Geological Survey Workshop on the Geology and Geohydrology of the Atlantic Coastal Plain: U.S. Geological Survey Circular 1059, p. 85–101.

Powars, D.S., Johnson, G.H., Edwards, L.E., Horton, J.W., Jr., Gohn, G.S., Catchings, R.D., McFarland, E.R., Izett, G.A., Bruce, T.S., Levine, J.S., and Pierce, H.A., 2002, An expanded Chesapeake Bay impact structure, eastern Virginia: New corehole and geophysical data: Houston, Texas, Lunar and Planetary Institute, 33rd Lunar and Planetary Science Conference Abstracts, LPI Contribution 1109, abstract 1034, 2 p.

Powars, D.S., Gohn, G.S., Catchings, R.D., Horton, J.W., Jr., and Edwards, L.E., 2003, Recent research in the Chesapeake Bay impact crater, USA—Part 1. Structure of the western annular trough and interpretation of multiple collapse structures: Third International Conference on Large Meteorite Impacts, August 5–7, 2003, Nördlingen, Germany, LPI Contribution 1167, Abstract 4053.

Powars, D.S., Bruce, T.S., Edwards, L.E., Gohn, G.S., Self-Trail, J.M., Weems, R.E., Johnson, G.H., Smith, M.J., and McCartan, C.T., 2005, Physical stratigraphy of the upper Eocene to Quaternary postimpact section in the USGS-NASA Langley core, Hampton, Virginia, chap. G, *in* Horton, J.W., Jr., Powars, D.S., and Gohn, G.S., eds., Studies of the Chesapeake Bay impact structure—The USGS-NASA Langley corehole, Hampton, Virginia, and related coreholes and geophysical surveys: U.S. Geological Survey Professional Paper 1688 (in press).

Rankin, D.W., 1994, Continental margin of the eastern United States; past and present, *in* Speed, R.C., ed., Phanerozoic evolution of North American continent-ocean transitions: Boulder, Colorado, Geological Society of America, DNAG Continent-Ocean Transect Volume, p. 129–218.

Reimold, W.U., Koeberl, C., and Poag, C.W., 2002, Chesapeake Bay impact crater: Petrographic and geochemical investigations of the impact breccia fill: Geological Society of America Abstracts with Programs, v. 34, no. 6, p. 466.

Roddick, J.C., 1983, High precision calibration of ^{40}Ar-^{39}Ar standards: Earth and Planetary Science Letters, v. 47, p. 887–898.

Ross, C.S., and Smith, R.L., 1961, Ash-flow tuffs: Their origin, geologic relations, and identification: U.S. Geological Survey Professional Paper 336, 81 p.

Russell, G.S., Russell, C.W., Speer, J.A., and Glover, L., III, 1981, Rb-Sr evidence of latest Precambrian to Cambrian and Alleghanian plutonism along the eastern margin of the sub-Coastal Plain Appalachians, North Carolina and Virginia: Geological Society of America Abstracts with Programs, v. 13, no. 7, p. 543.

Russell, G.S., Speer, J.A., and Russell, C.W., 1985, The Portsmouth granite, a 263 Ma postmetamorphic biotite granite beneath the Atlantic Coastal Plain, Suffolk, Virginia: Southeastern Geology, v. 26, no. 2, p. 81–93.

Sanford, W.E., 2003, Heat flow and brine generation following the Chesapeake Bay bolide impact: Journal of Geochemical Exploration, v. 78-79, p. 243–247, doi: 10.1016/S0375-6742(03)00108-0.

Self-Trail, J.M., 2003, Shock-wave-induced fracturing of calcareous nannofossils from the Chesapeake Bay impact crater: Geology, v. 31, p. 697–700, doi: 10.1130/G19678.1.

Shaw, H.F., and Wasserburg, G.J., 1982, Age and provenance of the target materials for tektites and possible impactites as inferred from Sm-Nd and Rb-Sr systematics: Earth and Planetary Science Letters, v. 60, p. 155–177, doi: 10.1016/0012-821X(82)90001-2.

Sheridan, R.E., Maguire, T.J., Feigenson, M.D., Patino, L.C., and Volkert, R.A., 1999, Grenville age of basement rocks in Cape May, N.J. well—New evidence for Laurentian crust in U.S. Atlantic Coastal Plain basement Chesapeake terrane: Journal of Geodynamics, v. 27, p. 623–633, doi: 10.1016/S0264-3707(98)00015-5.

Sibson, R.H., 1977, Fault rocks and fault mechanisms: Journal of the Geological Society (London), v. 133, p. 191–231.

Snoke, A.W., and Tullis, J., 1998, An overview of fault rocks, *in* Snoke, A.W., Tullis, J., and Todd, V.R., eds., Fault-related rocks, A photographic atlas: Princeton, New Jersey, Princeton University Press, p. 3–18.

Stacey, J.S., and Kramers, J.D., 1975, Approximation of terrestrial lead isotope evolution by a two-stage model: Earth and Planetary Science Letters, v. 26, p. 207–226, doi: 10.1016/0012-821X(75)90088-6.

Steiger, R.H., and Jäger, E., 1977, Subcommission on geochronology—Convention on the use of decay constants in geo- and cosmochronology: Earth and Planetary Science Letters, v. 36, p. 359–362, doi: 10.1016/0012-821X(77)90060-7.

Stöffler, D., and Grieve, R.A.F., 1994, Classification and nomenclature of impact metamorphic rocks: A proposal to the IUGS Subcommission on the Systematic of Metamorphic Rocks, *in* Montanari, A. and Smit, J., eds., Post-Östersund Newsletter: European Science Foundation Scientific Network on Impact Cratering and Evolution of Planet Earth, Strasbourg, p. 9–15.

Stöffler, D., and Grieve, R.A.F., 2003, Towards a unified nomenclature of metamorphic petrology: 11. Impactites. A proposal on behalf of the IUGS Subcommission on the Systematics of Metamorphic Rocks, Web version of 30.04.2003, 9 p.

Sturkell, E.F.F., 1998, The marine Lockne impact structure, Jamtland, Sweden: A review: Geologische Rundschau, v. 87, p. 253–267, doi: 10.1007/s005310050208.

Su, Q., Goldberg, S.A., and Fullagar, P.D., 1994, Precise U-Pb ages of Neoproterozoic plutons in the southern Appalachian Blue Ridge and their implications for the rifting of Laurentia: Precambrian Research, v. 68, p. 81–95, doi: 10.1016/0301-9268(94)90066-3.

Von Dalwigk, I., and Ormö, J., 2001, Formation of resurge gullies at impacts at sea: The Lockne crater, Sweden: Meteoritics & Planetary Science, v. 36, p. 359–369.

Wilcox, R.E., 1959, Use of the spindle stage for determination of principle indices of refraction: American Mineralogist, v. 44, p. 1272–1293.

Williams, I.S., 1998, U-Th-Pb geochronology by ion microprobe, *in* McKibben, M.A., Shanks, W.C., III, and Ridley, W.I., eds., Applications of microanalytical techniques to understanding mineralizing processes: Reviews in Economic Geology, v. 7, p. 1–35.

Wortman, G.L., Samson, S.D., and Hibbard, J.P., 2000, Precise U-Pb zircon constraints on the earliest magmatic history of the Carolina terrane: Journal of Geology, v. 108, p. 321–338, doi: 10.1086/314401.

Manuscript Accepted by the Society 9 August 2004

Printed in the USA

Geological Society of America
Special Paper 384
2005

Chicxulub impact ejecta deposits in southern Quintana Roo, México, and central Belize

Kevin O. Pope*
Geo Eco Arc Research, 16305 St. Mary's Church Road, Aquasco, Maryland 20608, USA

Adriana C. Ocampo*
Jet Propulsion Laboratory, California Institute of Technology, 4800 Oak Grove Drive, Pasadena, California 91109, USA

Alfred G. Fischer
Department of Earth Sciences, University of Southern California, University Park, Los Angeles, California 90089, USA

Francisco J. Vega
Instituto de Geología, Universidad Nacional Autónoma de México, Ciudad Universitária, México D.F. 04510, México

Doreen E. Ames
Geological Survey of Canada, Natural Resources Canada, 601 Booth Street, Ottawa, Ontario K1A 0E8, Canada

David T. King Jr.
Department of Geology, Auburn University, Auburn, Alabama 36849, USA

Bruce W. Fouke
Richard J. Wachtman
Department of Geology, University of Illinois at Urbana–Champaign, Urbana, Illinois 61801, USA

Gunther Kletetschka
Department of Physics, Catholic University of America, Washington D.C., USA; Institute of Geology, Academy of Sciences, Prague, Czech Republic, and *NASA Goddard Space Flight Center, Greenbelt, Maryland 20771, USA*

ABSTRACT

Discoveries of Chicxulub impact ejecta of the Albion Formation in road cuts and quarries in southern Quintana Roo, México and Belize, broaden our understanding of ejecta depositional processes in large impacts. There are numerous new exposures of ejecta near the Río Hondo in Quintana Roo México, located at distances of 330–350 km from the center of the Chicxulub crater. A single ejecta exposure was discovered near Armenia in central Belize, 470 km from Chicxulub. The Albion Formation is composed of two lithostratigraphic units: the spheroid bed and diamictite bed, originally identified at Albion Island, Belize. The new spheroid bed exposures range from

*E-mail, Pope: kpope@starband.net. *Note*: Pope and Ocampo contributed jointly to this paper.

Pope, K.O., Ocampo, A.C., Fischer. A.G., Vega, F.J., Ames, D.E., King, D.T., Jr., Fouke, B.W., Wachtman, R.J., and Kletetschka, G., 2005, Chicxulub impact ejecta deposits in southern Quintana Roo, México, and central Belize, *in* Kenkmann, T., Hörz, F., and Deutsch, A., eds., Large meteorite impacts III: Geological Society of America Special Paper 384, p. 171–190. For permission to copy, contact editing@geosociety.org.

2 to 5 m thick and are composed of altered glass fragments, accretionary lapilli, and pebble-sized carbonate clasts in a fine-grained calcite matrix. The base of the spheroid bed is exposed at Ramonal South in México and at Albion Island and Armenia in Belize, and at all three locations, the spheroid bed was deposited on a weathered karst land surface that had emerged in the Late Cretaceous. The new diamictite bed exposures are composed of altered glass fragments and carbonate clasts up to 9.0 × 3.2 m in size. In all but one of the new exposures, the diamictite bed extends to the surface with observed thicknesses up to 8 m. At Agua Dulce in México, the weathered top of the diamictite bed is overlain by thin-bedded Tertiary carbonates. No diamictite bed is found in Armenia, where the spheroid bed is overlain with a limestone conglomerate containing altered glass shards and shocked quartz. These discoveries indicate that ejecta are emplaced in large terrestrial impacts by at least two distinct flows: (1) an initial flow involving a volatile-rich cloud of fine debris similar to a volcanic pyroclastic flow, which extends >4.7 crater radii (the spheroid bed), and (2) a later flow of coarse debris that may not extend much beyond 3.6 crater radii (the diamictite bed). The former deposit we attribute to material entrained in the impact vapor plume, and the latter to the turbulent collapse of the ejecta curtain.

Keywords: Chicxulub crater, impact ejecta, Cretaceous-Tertiary boundary, Mexico; Belize.

INTRODUCTION

Out-of-crater ejecta deposits from the Chicxulub impact are the best preserved of any large crater on Earth, and they provide an ideal opportunity to study impact processes. The distal, global ejecta deposits from Chicxulub are well known through decades of research related to the Cretaceous-Tertiary (K-T) boundary and mass extinction of life apparently caused by the impact. Studies in northern Belize reveal that a portion of the ejecta blanket is exposed south of the crater (Ocampo et al., 1996; Pope et al., 1999; Smit, 1999; Fouke et al., 2002; King and Petruny, 2003).

In this paper we report on our ongoing study of Chicxulub impact ejecta surface exposures, including new outcrops in southern Quintana Roo, México and central Belize (Fig. 1). These outcrops provide the only known exposures of the outer portion (3.3–4.7 crater radii) of the ejecta blanket from a large impact crater. Impact ejecta deposits in this region comprise the Albion Formation, which is subdivided into two lithostratigraphic sub-units following previous work at Albion Island, Belize (Ocampo et al., 1996; Pope et al., 1999). These two sub-units are named the spheroid bed and diamictite bed, following the common use of the term "bed" to designate small formation sub-units (e.g., Salvador, 1994).

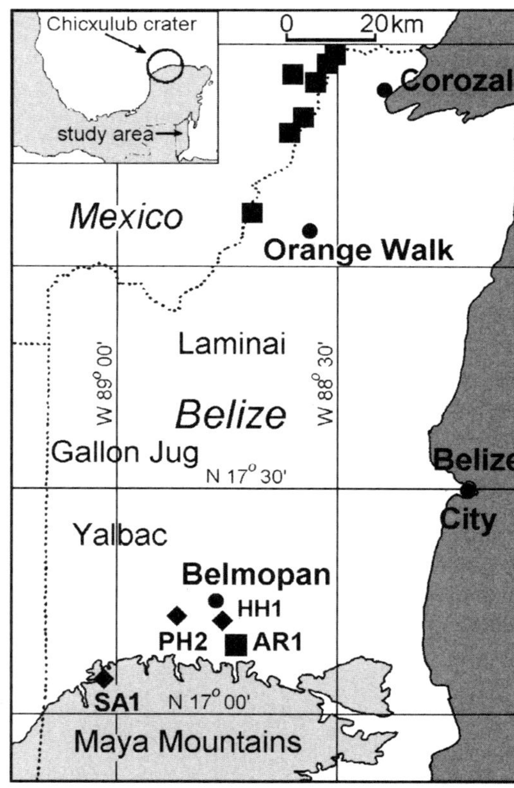

Figure 1. Map of northern Belize and southern Quintana Roo, México, showing Chicxulub impact ejecta outcrops. Square symbols mark location of Albion Formation ejecta outcrops (see Figure 2 for detailed locations of ejecta outcrops along the México-Belize border; not all can be shown at this scale). Diamonds mark other locations mentioned in text. AR1—Armenia 1, PH2—Pook's Hill 2, SA1—San Antonio 1, and HH1—Hummingbird Highway 1. Lightly shaded region at bottom of map is area above 200 m elevation.

SOUTHERN QUINTANA ROO

Geological maps of México indicate that the Cretaceous is not exposed in the Mexican portion of the Yucatán Peninsula (INEGI, 1987). Nevertheless, our 1997 and 2001 geological reconnaissance along the highway from Ucum to La Unión in southern Quintana Roo, México, discovered several outcrops of Chicxulub impact ejecta and exposures of the underlying Upper Cretaceous Barton Creek Formation (Fig. 2). These new

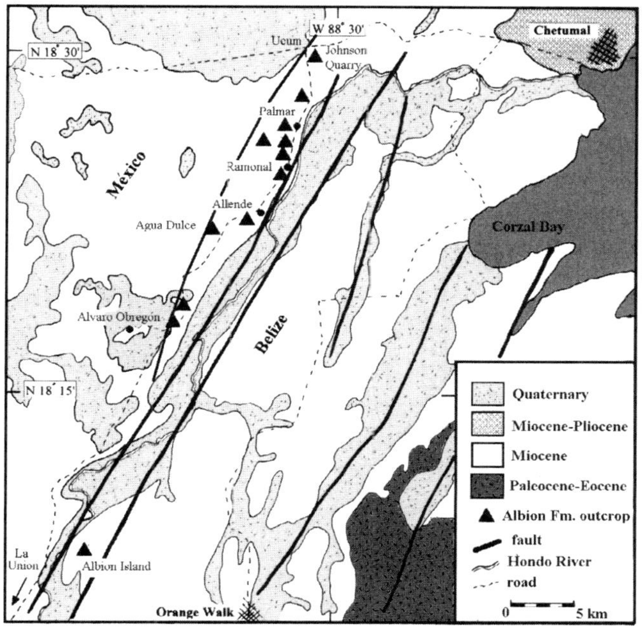

Figure 2. Geological map of southern Quintana Roo, México and northern Belize showing locations of new Albion Formation ejecta outcrops (triangles), as well as the original discovery site at Albion Island (some triangles mark multiple sites, see Appendix for list of outcrop coordinates). Recent research demonstrates that the Upper Cretaceous carbonates are exposed in several areas formerly mapped as Miocene in both México and Belize. Mexican portion of map from Instituto Nacional de Estadistica, Geografia, e Informatica (INEGI, 1987). Belize portion from Cornec (1986).

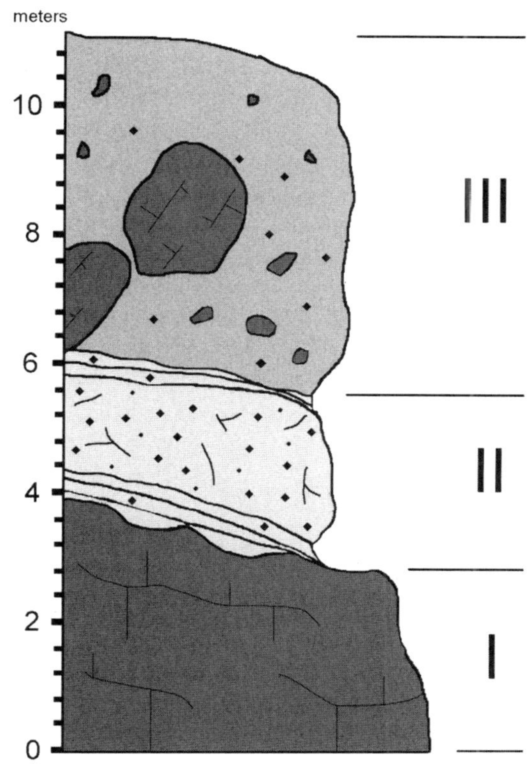

Figure 3. Ramonal South stratigraphic section. (I) Deeply weathered Barton Creek Formation dolomite with iron oxide staining, calcite veins, and abundant solution cavities. (II) 5%–20% green clay clasts 1–20 mm diameter; 5% micritic carbonate clasts including white chalky limestone clasts 5–20 mm diameter with rare clasts up to 4 cm in diameter; rare (~1%) carbonate accretionary lapilli 1–20 mm in diameter; common arcuate fracture planes with thin clay linings. At the base of stratum II are ~1% thin flakes of micritic limestone 1–5 cm in diameter (rip-up clasts?) within multiple layers of orange, purple, and red clay and calcite silts (shear plane?). Similar orange, purple, and red clay and calcite silts in thin layers (shear plane?) at top of unit. (III) 30% micritic limestone and dolomite cobbles and boulders (up to 3 m in diameter) supported in a matrix of calcite and dolomite silt; 10% green clay clasts 1–20 mm in diameter, commonly in lenses.

ejecta outcrops lie ~330–350 km southeast from the center of the Chicxulub crater. The ejecta deposits belong to the Albion Formation previously identified on Albion Island in nearby northern Belize (Ocampo et al., 1996; Pope et al., 1999). The Albion Formation exposures in Quintana Roo are found in a series of road cuts along the highway that parallels the Río Hondo and in nearby quarries between the towns of Ucum and Alvaro Obregón (Fig. 2). Reconnaissance of road cuts and quarries further south from Alvaro Obregón to La Unión produced no Albion Formation outcrops. Locations of 15 sites with confirmed impact ejecta in this region are given in Appendix 1. The closest site to Chicxulub is the Johnson quarry site (Fig. 2), which contains Albion Formation diamictite deposits 330 km from the center of Chicxulub. The combined Albion Formation exposures in Belize and Quintana Roo indicate that a surface or near-surface discontinuous sheet of impact debris is preserved for ~40 km along the Río Hondo, covering at least a 70 km^2.

We conducted detailed studies of four locations: Ramonal North, Ramonal South, Agua Dulce, and Alvaro Obregón (Figs. 3–6), which all lie 335–350 km from Chicxulub. The basal contact of the Albion Formation with the underlying Barton Creek Formation is well exposed at Ramonal South (Fig. 3). The only site with an exposed upper contact of the Albion Formation is Agua Dulce, where the weathered top of the ejecta deposit is overlain by Tertiary limestones (Fig. 6). The other two sites, Ramonal North and Alvaro Obregón, have good exposures of ejecta, but no upper or lower contact.

Ramonal South

Field Observations

The Barton Creek Formation at Ramonal South exhibits extensive karst weathering with local relief of 3–10 m in the surface of a heavily recrystallized limestone with iron-oxide staining, caliche deposits, abundant vugs and travertine deposits. Well-preserved fossils are rare due to the recrystallization, but several Late Cretaceous nerineid gastropods (Pope et al., 1999)

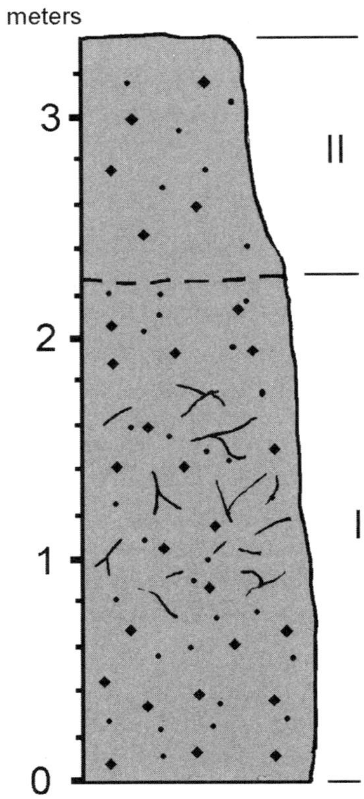

Figure 4. Ramonal North stratigraphic section. (I) 30%–35% green clay clasts 2–40 mm in diameter (mean ~4–6 mm), many clasts have elongated vesicles; 20% carbonate accretionary lapilli 1–25 mm in diameter, most lapilli are coated with red iron oxide and many have cores of limestone and green clay fragments; 1%–3% angular micritic limestone clasts 0.5–5 cm in diameter with rare clasts up to 20 cm; 40% calcite silt matrix; common arcuate fracture planes with thin clay linings. (II) Weathered top of section; 1% green clay clasts 2–20 mm in diameter; 10% carbonate accretionary lapilli 1–25 mm in diameter; 1%–3% angular micritic limestone clasts 0.5–5 cm in diameter; 85% calcite silt matrix; abundant roots.

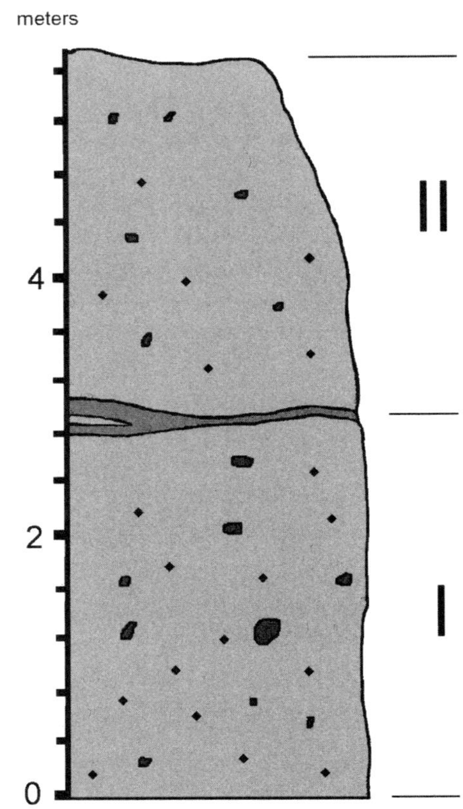

Figure 5. Alvaro Obregón stratigraphic section. (I) 30% sub-angular to sub-rounded micritic dolomite and limestone pebbles and cobbles up to 25 cm in diameter; 10% green clay clasts 3–20 mm in diameter. (II) 20% sub-angular to sub-rounded micritic dolomite and limestone pebbles and cobbles up to 30 cm in diameter; 5% green clay clasts. These two units are separated by a horizontal layer of orange calcitic clay with slickensides (shear plane?).

were found in Barton Creek outcrops between Ramonal South and the town of Allende. This confirms that the Cretaceous is exposed in southern Quintana Roo. Ongoing research in Quintana Roo has recovered a rather diverse Upper Cretaceous fauna in the Barton Creek Formation. Several species of bivalves and gastropods are being studied. A new species of the gastropod *Aporrhais* represents the first record for that genus in the Cretaceous of the Gulf Coast and Caribbean Provinces (Vega et al., 2001). Although the stratigraphic range for this genus extends from Lower Cretaceous to Recent, specimens from Quintana Roo show similarities to species from the Maastrichtian of Montana and Germany.

Overlying the Barton Creek Formation at Ramonal South is a 2-m-thick section of the Albion Formation spheroid bed, which is overlain by a 5-m-thick section of the diamictite bed that extends to the surface (Figs. 3 and 7). The upper and lower contacts of the spheroid bed are marked by 2–20-cm-thick layers of orange, purple, and green calcitc clay with common slickensides. These clay layers appear to be shear planes. There is an abrupt appearance of large micritic dolomite clasts above the upper shear plane and one large dolomite boulder rests immediately on top (Fig. 7). Field tests with HCl indicate that the base of the spheroid bed is calcitic, and becomes progressively more dolomitic near the top. The diamictite bed matrix is dolomitic.

Laboratory Analyses

We examined 12 thin sections from the top of the Barton Creek Formation and from the spheroid and diamictite beds at Ramonal South. Here we describe the textures and key aspects of the mineralogy. Estimates of the percentages (area) of major constituents were made using photomicrographs and percentage charts. The top of the Barton Creek exhibits abundant iron oxide staining, secondary coarse calcite filling fractures and vugs,

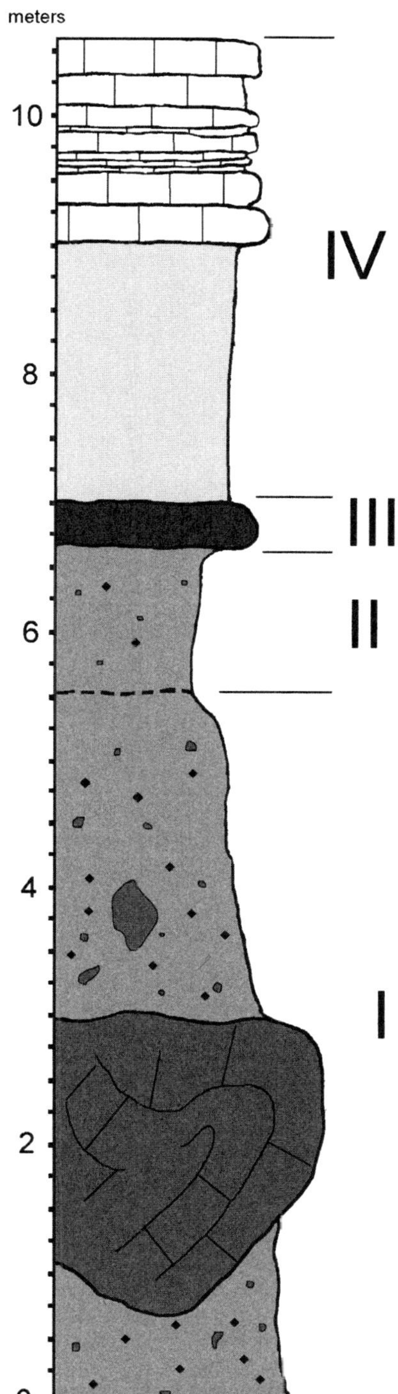

Figure 6. Agua Dulce stratigraphic section. (I) 25% sub-angular to sub-rounded micritic dolomite and limestone pebbles, cobbles, and boulders up to 3.2 × 9.0 m in size; 5%–10% green clay clasts; fewer (5%) cobbles and boulders in the upper part. (II) 10% micritic dolomite clasts 1–15 cm in diameter; 1%–3% green and orange clay clasts 0.5–2 cm in diameter; 85% matrix of coarse calcite silt with orange and purple iron oxide straining (weathered version of stratum I); (III) calcite cemented caliche with large 2–3 mm calcite crystals; (IV) weathered white micritic dolomite overlying pink and tan, thin-bedded, micritic limestone.

and polycrystalline calcite replacement of dolomite rhombs (de-dolomitization) (Fig. 8), all consistent with a subaerially weathered surface. The main stratum of the spheroid bed (stratum II, Fig. 3) contains 20% yellowish-green clay with spherulitic devitrification features and relict vesicles typical of altered glass (Fig. 9). The spheroid bed contains 20% carbonate clasts of various types, including foraminifera fragments (Fig. 9), and 3% accretionary lapilli (core-type, Schumacher and Schmincke, 1991) composed of aggregates of detrital calcite grains 10–125 µm in size (Figs. 9). The matrix of the spheroid bed is composed of detrital calcite grains similar in size and character to that of the accretionary lapilli aggregates. Some portions of the matrix are altered to interlocking grains of dolomite, especially near the top of the bed. Thin section analysis of the diamictite bed reveals that it is pervasively dolomitized, but contains 10% clay and 20% dolomite clasts (>0.5 mm) of various types (smaller clasts are completely recrystallized) (Fig. 10).

Ramonal North

Field Observations

The Ramonal North road cut exposes a 3.6 m section of the spheroid bed in a dip in Ucum–La Unión highway ~200 m north of Ramonal South (Figs. 4 and 11). Neither the top nor the base of the spheroid bed is exposed in the road cut, but the Albion Formation diamictite bed crops out ~30 m behind the road cut, suggesting the top of this exposure may be near the upper contact with the diamictite bed. The base of the Ramonal section lies ~13 m below the base of the spheroid bed at Ramonal South. We interpret this elevation difference as a dip in the paleo-karst landscape, as the exposed surface of the Barton Creek Formation at Ramonal South is sloping toward Ramonal North and this surface is observed to undulate up and down in numerous road cuts along the Ucum–La Unión highway.

Laboratory Analyses

We analyzed 17 thin sections from the Ramonal North spheroid bed. Yellowish-green clay fragments comprise 20% of the bed and contain common relict vesicles and spherulitic devitrification features typical of altered glass (Figs. 12). No preserved glass was found. Calcite accretionary lapilli of diverse types comprise 10% of the bed, and range in size from 0.2 to 5.0 mm, and average ~1.0 mm (Figs. 13–15). There are many examples of what are sometimes classified as mantled lapilli—lithic cores coated with fine ash (Fig. 15), and both rim-type and core-type accretionary lapilli—aggregates of fine detrital grains, with or without a fine-grained rim (Schumacher and Schmincke, 1991). We refer to all types as accretionary lapilli, in part for simplicity, but also because it is not always apparent whether the core is an aggregate or lithic fragment when the latter is rounded and has a similar lithology to the aggregate. Rim-type accretionary lapilli are the most abundant type (Fig. 13), but core-type are also common (Fig. 14), as are ones with limestone (Fig. 15) and clay (altered glass) lithic cores (Fig. 14). The detrital calcite grains in the lapilli aggregates range

Figure 7. Ramonal South outcrop photograph showing the spheroid bed overlaying the Barton Creek Formation and the diamictite bed overlaying the spheroid bed. Note large boulder resting on contact between spheroid and diamictite beds.

in size from 150 μm to <10 μm, and the outermost rims are typically composed of very fine particles (<10 μm). Detrital grains are nearly all calcite, but clay fragments (altered glass?) make up 1%–3% of the aggregate. Detrital carbonate clasts (>100 μm) of various lithologies, including foraminifera tests (Fig. 14), comprise 5% of the spheroid bed in thin section and occur in approximately the same size range as the accretionary lapilli, although there is no discernable lower size limit (there is a continuum between these clasts and the matrix). The matrix, arbitrarily defined here as that fraction <100 μm, comprises ~60%–70% of the thin sections and is composed mostly of detrital carbonate grains and a few percent clay. The matrix and the lapilli aggregate are essentially the same composition.

Two samples of the clay (altered glass fragments) from within the spheroid bed were analyzed by X-ray diffraction after air drying, glycolation, and heating to 550 °C, using a Philips PW1710 X-ray diffractometer at the Geological Survey of Canada. Both were found to be smectite with traces of mixed layer smectite-illite.

Alvaro Obregón

Field Observations

The road cut at Alvaro Obregón exposes 7.5 m of the diamictite bed (Fig. 5) separated by a sub-horizontal (dips slightly south) orange calcitic clay layer a few centimeters thick with slickensides on the upper and lower surfaces. This clay layer is positioned ~2.6 m up from the base of the exposure and extends the length (~100 m) of the road cut (Fig. 16). As at Ramonal South, we interpret this clay layer as a shear plane. Rare cobbles in the diamictite bed have striated surfaces, polish, and accretionary rinds. There is no upper or lower diamictite bed contact

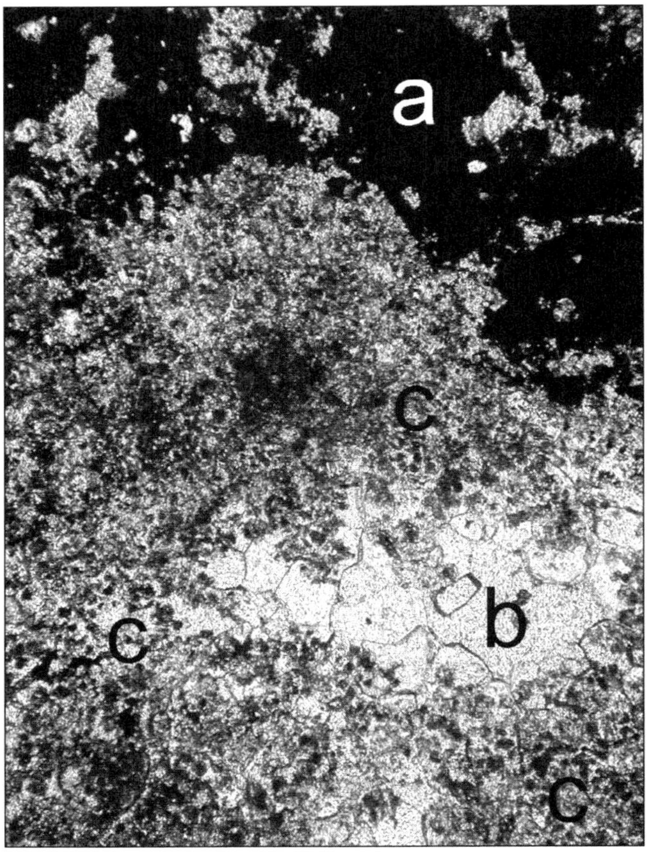

Figure 8. Photomicrograph of thin section from the upper contact of the Barton Creek Formation at Ramonal South showing (a) iron oxide staining; (b) secondary coarse calcite filling cavities; and (c) polycrystalline calcite (light) replacing dolomite (dark cores). Plane polarized transmitted light. Field of view: 2.8 mm wide.

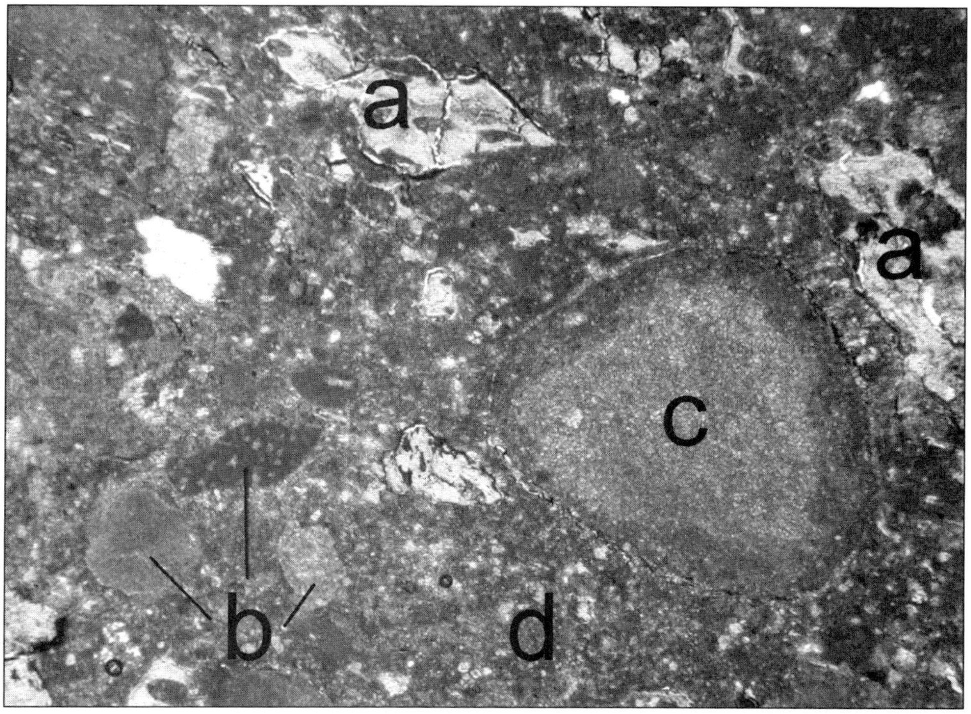

Figure 9. Photomicrograph of thin section from the Ramonal South spheroid bed showing (a) clay fragments with elongated vesicles filled with carbonate matrix; (b) micritic limestone clasts (center arrow foraminifera); (c) accretionary lapilli; (d) matrix of detrital carbonate grains. Plane polarized transmitted light. Field of view: 4.4 mm wide.

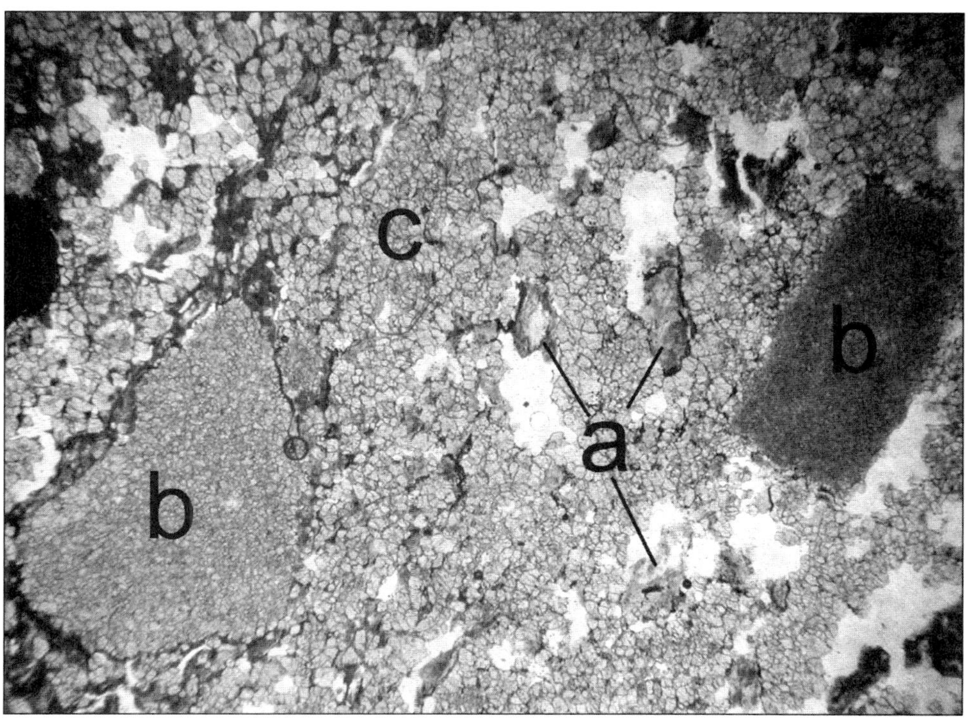

Figure 10. Photomicrograph of thin section from the base of the diamictite bed Ramonal South showing (a) clay fragments; (b) recrystallized coarse grain dolomite (left) and fine grained limestone (right) clasts; and (c) coarse grained dolomite matrix (recrystallized). Plane polarized transmitted light. Field of view: 6.0 mm wide.

Figure 11. Hand sample of Ramonal North spheroid bed showing abundant accretionary lapilli and clay fragments (altered glass) and common limestone clasts.

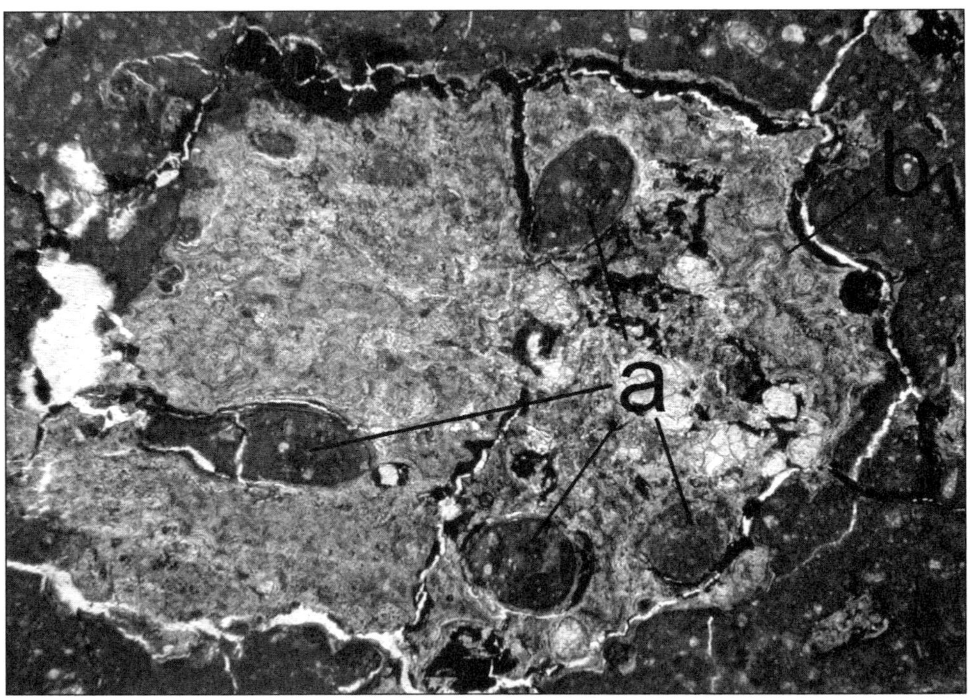

Figure 12. Photomicrograph of thin section from the Ramonal North spheroid bed showing a large clay fragments with (a) elongated vesicles filled with carbonate matrix; and (b) spherulitic devitrification features. Plane polarized transmitted light. Field of view: 4.2 mm wide.

exposed at Alvaro Obregón. Tests with dilute HCl indicate that the upper stratum is mostly dolomitized, whereas the base is only partially dolomitized and contains several micritic limestone clasts 1–20 cm in diameter and a slightly calcitic matrix.

Laboratory Analyses

We analyzed seven thin sections from the Alvaro Obregón diamictite bed. The bed contains 10% yellowish-green clay fragments, which like those in the spheroid bed, possess relict vesicles and spherulitic devitrification features typical of altered glass (Fig. 17). Angular to sub-rounded carbonate clasts >100 μm in size comprise 25% of the thin sections, some of which retain poorly preserved foraminifera (Fig. 18). Most carbonates are recrystallized to dolomite, which makes it difficult to discern clast boundaries for the smaller clasts as they merge with the 10–40 μm recrystallized dolomite matrix. The predominant clast lithology is a light brown dolomite with ~20–40-μm-sized interlocking dolomite crystals and indistinct clast-matrix boundaries (Fig. 18).

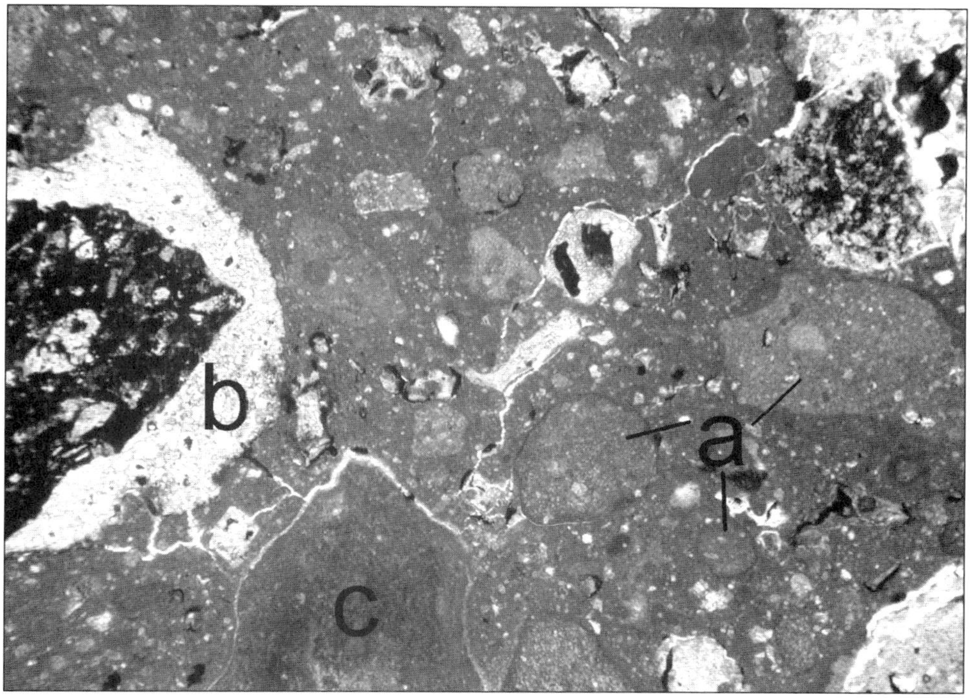

Figure 13. Photomicrograph of thin section from the Ramonal North spheroid bed showing (a) accretionary lapilli composed of coarse carbonate detrital grains with a very thin, fine (<10 μm) accretionary rim (rim-type); (b) lithic clast with a spherulitic altered glass rim; and (c) angular limestone clast. Plane polarized transmitted light. Field of view: 5.9 mm wide.

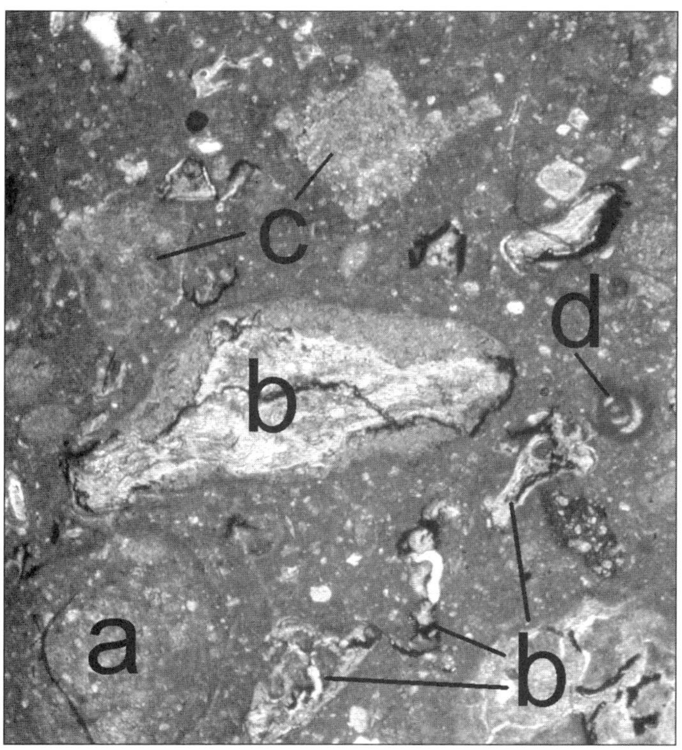

Figure 14. Photomicrograph of thin section from the Ramonal North spheroid bed showing (a) accretionary lapilli composed of coarse carbonate detrital grains (core type); (b) clay clasts with relict vesicles and flow banding (center), note altered glass fragment with accretionary rind (center); (c) carbonate clasts; and (d) foraminifera test. Plane polarized transmitted light. Field of view: 3.0 mm wide.

Agua Dulce

Field Observations

The road cut at Agua Dulce extends for ~500 m and exposes an 8-m-thick sequence of the diamictite bed (Fig. 6) with many large micritic dolomite boulders (Fig. 19). The diamictite bed here is similar to that found at Alvaro Obregón, except for the abundance of large boulders (compare the photos in Figs. 16 and 19). Tests with dilute HCl indicate that the diamictite bed clasts and matrix at Agua Dulce are only partially dolomitized. The Agua Dulce section contains one very large dolomite boulder (9.0 × 3.2 m) that has highly deformed bedding (Fig. 19). The upper (~2 m) part of the diamictite bed has fewer large cobbles and clay clasts, but the change is gradational. The top ~1 m of the diamictite bed is a weathered horizon capped with well-indurated caliche. Overlying this caliche are over 10 m of thin-bedded dolomite and limestone.

Laboratory Analyses

We analyzed seven thin sections from the Agua Dulce diamictite bed. The bed contains 15% yellowish-green clay fragments with the same relict vesicles and spherulitic devitrification features indicative of altered glass found at Alvaro Obregón (Fig. 20). Angular to sub-rounded carbonate clasts >100 μm in size comprise 25% of the thin sections, some with preserved foraminifera (Fig. 20). Some carbonate clasts are recrystallized to dolomite, and the previously noted clasts of light brown dolomite with 20–40 μm-sized interlocking dolomite crystals and indistinct clast boundaries are present (Fig. 20). Dolomite clasts

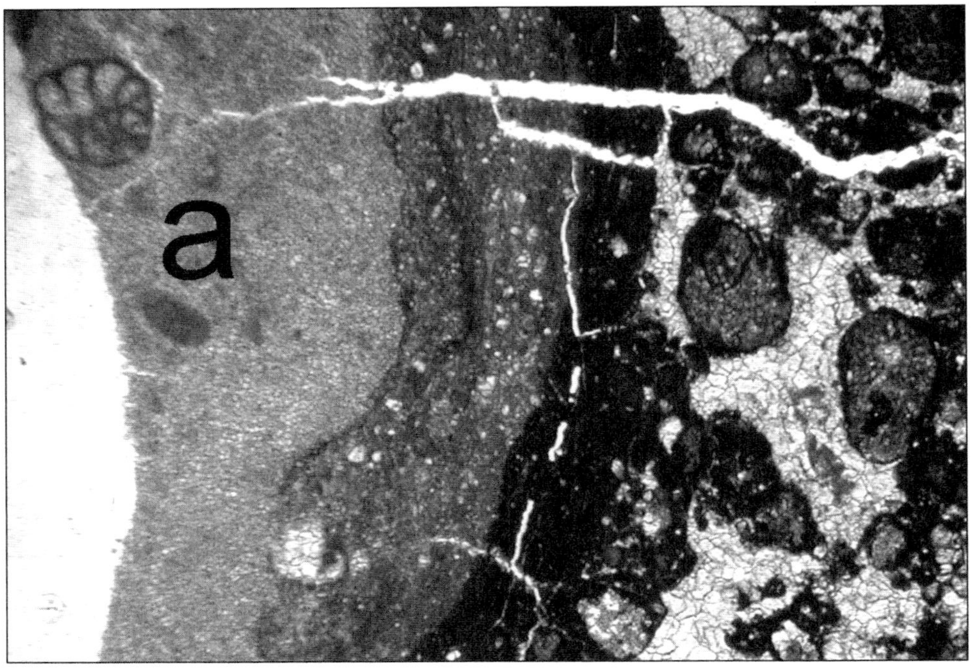

Figure 15. Photomicrograph of thin section from the Ramonal North spheroid bed showing a large accretionary lapilli with (a) a micritic limestone lithic core with a foraminifera test that is encased by multiple layers of accreted detrital calcite grains. Outer dark color is due to iron oxide staining of outer layers. Plane polarized transmitted light. Field of view: 4.3 mm wide.

Figure 16. Alvaro Obregón road cut exposing Albion Formation diamictite bed. Note horizontal calcitic clay layer (shear plane?) along the center of the outcrop. Exposure is 7.5 m high.

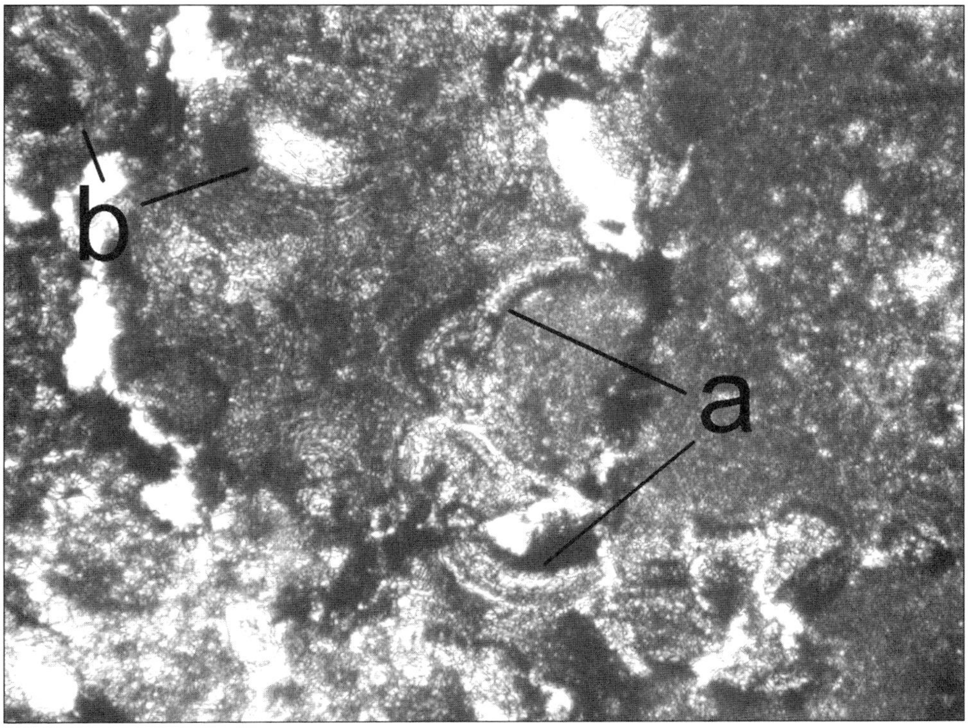

Figure 17. Photomicrograph of thin section from near the base of the Alvaro Obregón diamictite bed showing a clay fragment (altered glass) with (a) relict vesicles filled with matrix and (b) spherulitic devitrification features. Plane polarized transmitted light. Field of view: 1.9 mm wide.

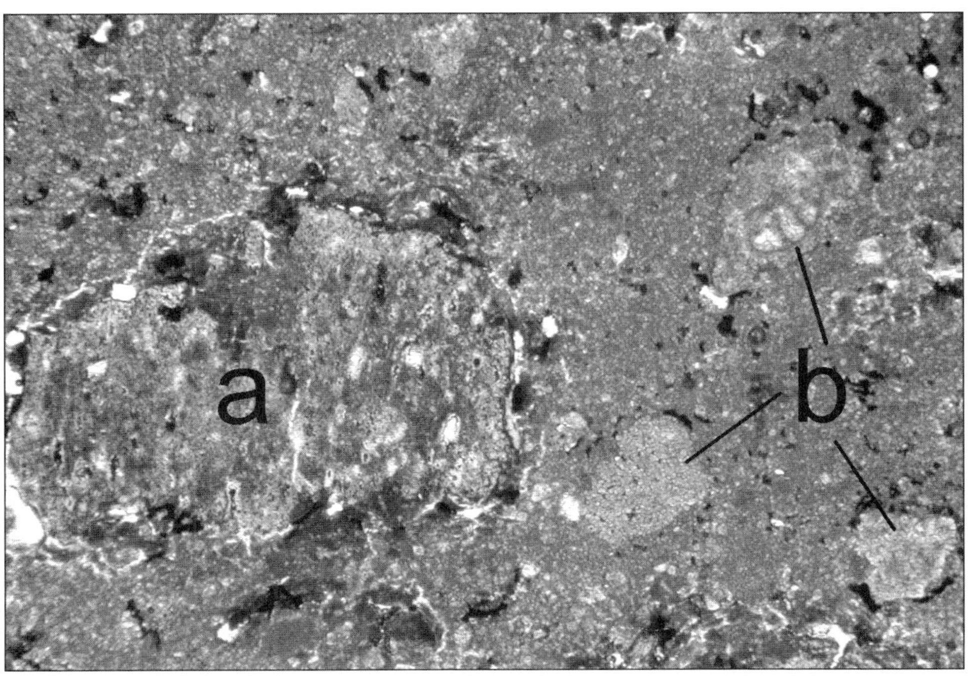

Figure 18. Photomicrograph of thin section from near the base of the Alvaro Obregón diamictite bed showing (a) a clay fragment (altered glass) with elongated vesicles; and (b) lithic clasts of coarse dolomite, one with a foraminifera test, in a mostly fine grained calcite and dolomite matrix. Field of view: 4.2 mm wide.

are not as abundant as at Alvaro Obregón, and micritic limestone clasts are common. The carbonate matrix is finer-grained at Agua Dulce (less altered) than at Alvaro Obregón, and small clast boundaries are more distinct.

Four clay separates (altered glass fragments) were analyzed by X-ray diffraction and are dominated by smectite with traces of mixed layered smectite-illite.

NORTH-CENTRAL BELIZE

The discovery of Chicxulub impact ejecta along the Belize-México border prompted our search for more distal ejecta deposits further south, in Belize. The preservation of Chicxulub ejecta on Albion Island, an uplifted block of Cretaceous carbonate platform sediments associated with the Río Hondo fault zone (Ocampo et al., 1996; Pope et al., 1999), suggests that other juxtaposed fault-blocks of Cretaceous and Tertiary platform carbonates in north-central Belize may preserve remnants of Chicxulub ejecta. Nevertheless, our reconnaissance of Cretaceous blocks near Gallon Jug, Laminai, and Yalbac (Fig. 1) produced no evidence of impact ejecta. Thus, if Chicxulub ejecta deposits existed in this region, they have apparently been stripped by erosion.

Late Cretaceous and early Tertiary platform carbonates are well known in central Belize (Ower, 1928; Dixon, 1956; Flores 1952a, 1952b; Cornec, 1985; King et al., 2004). Marine sedimentation is thought to be discontinuous across the K-T boundary (Flores, 1952b), but the stratigraphic relationship between Cretaceous and Tertiary strata in Belize has not been studied in much detail. Upper Cretaceous Barton Creek Formation outcrops with rudist fragments were identified by Flores (1952a, 1952b) along the base of the Maya Mountains (Fig. 1). Paleocene sediments have not been previously identified in this region, but outcrops of shallow marine Paleocene carbonates are found 35 km northeast of Belmopan (Flores 1952a, 1952b). Lower Eocene limestones outcrops of the El Cayo Group are common along the northern flanks of the Maya Mountains (Flores 1952a, 1952b). A major uplift of the Maya Mountains began in the early Tertiary involving the reactivation of the northern boundary fault of the Maya Mountains (Fig. 1) (Bateson and Hall, 1977).

We examined stratigraphic sections along the northern flanks of the Maya Mountains in the vicinity of juxtaposed Cretaceous and Tertiary blocks shown on the geologic map of Belize (Cornec, 1986). The complex tectonism and the thick tropical vegetation made if difficult to trace contacts between sections. The upper contact of the Barton Creek Formation in this region is a highly irregular karst surface with relief of 10–30 m and exposures of caves and rubble-filled solution cavities. The Barton Creek carbonates near the upper contact are a coarse, recrystallized limestone with abundant calcite-filled fractures and vugs, but grade down-section into dolomitized limestones. Good exposures of the weathered top of the Barton Creek are found at Pook's Hill 2 (PH2), Hummingbird Highway 1 (HH1), and San Antonio 1 (SA1) (Fig. 1). At 15 locations we found the Barton Creek Formation overlain by 10–30 m of pebbles, cobbles, and

Figure 19. Agua Dulce road cut with large dolomite boulders in the Albion Formation diamictite bed. All have major axis roughly parallel to the ground surface. Upper two boulders both ~6–7 m long, lower one is 9 m (outline in white, black lines mark extreme folding in relict beds).

boulders supported in a matrix of red clay and coarse calcite silt. The coarse size fraction (>1 cm) of this unit is a mixture of angular to rounded clasts, but sub-rounded clasts dominate. Clast sizes range from sub-millimeter grains to 2 m diameter boulders, but most clasts are <30 cm in diameter. The clast lithology is a diverse mixture of carbonates, including micritic limestone, coarse crystalline limestone, laminated dolomite, breccia, and calcareous mudstone. The coarse crystalline limestone is the most common lithology, and is probably derived from the weathered Barton Creek Formation. We interpret this unit as a regolith or saprolite resulting from the karst weathering of the Barton Creek Formation, but we recognize that its genesis may well be complex and not solely the product of in situ weathering given the diverse clast lithologies.

Tertiary sediments overly the Barton Creek regolith at Pook's Hill 2 (PH2, Fig. 1), but the contact is not well exposed and is marked by an abrupt change in the lithology of clasts eroding out of the bulldozed upper slopes of a hill. These Tertiary sediments are a cream-colored fossiliferous dolomite with abundant rocky-shore bivalves and gastropods, including *Mytilus*

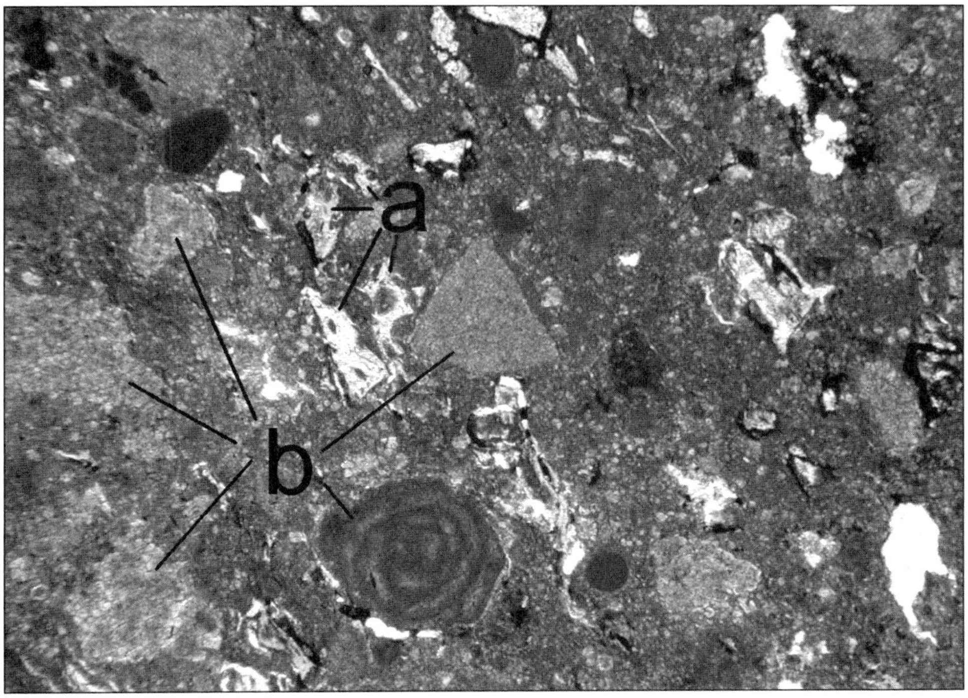

Figure 20. Photomicrograph of thin section from middle of the Agua Dulce diamictite bed showing (a) clay fragments (altered glass) with matrix filled vesicles; and (b) lithic clasts of coarse dolomite with indistinct boundaries (grains on left) and micritic limestone (grains on right), one with a foraminifera test, in a mostly fine grained calcite and dolomite matrix. Plane polarized transmitted light. Field of view: 5.1 mm wide.

sp., *Septifer* sp., and *Puncturella* sp. This fauna marks the beginning of the Tertiary marine transgression on the edge of the Maya Mountains, and a similar assemblage is typical of Paleocene transgression sequences in the southern United States (e.g., Palmer and Palmer, 1937).

We conclude that, as at Albion Island and Ramonal South, the northern flanks of the Maya Mountains were uplifted and subjected to karst weathering at the end of the Cretaceous. This finding suggests that remnants of Chicxulub ejecta may be found associated with the regolith that caps this weathered surface. Thus far our research has confirmed Chicxulub ejecta associated with this regolith cap at only one location, near the village of Armenia, 470 km from Chicxulub.

Armenia

Field Observations

Near the village of Armenia we found a 5-m-thick exposure of the Albion Formation spheroid bed overlying the terminal Cretaceous regolith in a road cut along the Hummingbird Highway south of Belmopan (Figs. 1, 21, and 22). The spheroid bed contains abundant, angular green and red altered glass fragments and accretionary lapilli in a calcite silt matrix (Fig. 23). Underlying the spheroid bed is 50–150-cm-thick stratum of calcitic red clay with weathered limestone cobbles overlying an ~5-m-thick indurated regolith deposit. The highly oxidized state (deep red color), clay texture, and prismatic structure of the calcitic red clay stratum suggest that it is a paleosol. Deeply weathered Barton Creek Formation exposures occurs in road cuts just north of Armenia, but at Armenia the base of the regolith is a shear zone, perhaps formed by the collapse of a solution cavity or by a fault. Overlying the spheroid bed is an ~5 m-thick matrix supported limestone pebble and cobble conglomerate (Fig. 22).

Laboratory Analyses

We analyzed 10 thin sections from the Armenia spheroid bed and underlying clay stratum. Analyses of the calcitic red clay of the paleosol shows that it contains 10% small rounded limestone clasts 0.2–3 mm in diameter and lenticular clay clasts in sharp contact with the overlying spheroid bed (Fig. 24). Analyses of samples from the Armenia spheroid bed reveal well-preserved primary textures of the accretionary lapilli and clay clasts similar to those found at Ramonal North. Greenish and reddish brown clay fragments comprise ~20% of the bed and contain common relict vesicles and spherulitic devitrification features typical of altered glass (Figs. 25). Portions of the larger altered glass fragments are translucent, lack devitrification features, have patches that are nearly isotropic, and thus are probably palagonite, an early alteration product of glass.

Calcite accretionary lapilli of diverse types comprise 20% of the spheroid bed in thin section, range in size from 0.3 to 5.0 mm, and average 1.0 mm (Fig. 25). Rim-type, core-type, and ones with lithic cores are all common, but as at Ramonal North, rim-type accretionary lapilli predominate. The detrital calcite grains that comprise the accretionary lapilli range in size from ~140 μm to <10 μm. Clay fragments comprise <1% of the lapilli aggregate, but angular clay lithic cores, some with relict vesicles, are common (Fig. 23). Detrital micritic limestone clasts (>100

Figure 21. Armenia road cut showing Albion Formation spheroid bed resting on deeply weathered Upper Cretaceous Barton Creek Formation. Just below the spheroid bed is a dark red calcitic clay interpreted as a terminal Cretaceous paleosol. Overlying the spheroid bed are 5 m of limestone conglomerate.

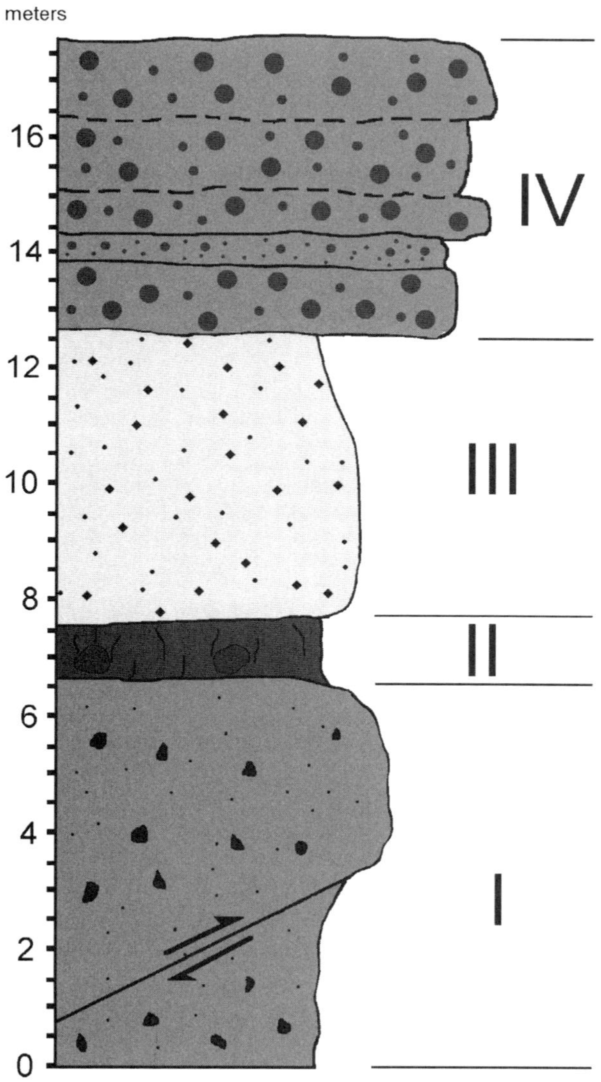

Figure 22. Armenia stratigraphic section. (I) 20% sub-rounded to angular micritic limestone and dolomite pebbles and cobbles supported in a red calcitic clay matrix; abundant secondary calcite in fractures and cavities. This stratum is in fault contact with a breccia unit with similar clasts and matrix, but includes large blocks of the overlying strata. (II) dark red calcitic red clay with 10% small rounded and angular pebbles and grains 1–5 mm. There are several large (10–30 cm) deeply weathered cobbles and a faint prismatic soil structure. (III) 20%–30% green clay clasts, many with vesicles; 10%–20% calcite accretionary lapilli 1–60 mm in diameter, (mean 5 mm), most larger (>0.5 cm) lapilli have cores of limestone or clay clasts; 1%–3% micritic limestone clasts 10–80 mm in diameter (mean 12 mm). (IV) 60%–75% sub-rounded micritic limestone, rare chert and dolomite, pebbles and cobbles 1–50 cm in diameter (mean ~5–10 cm); many cobbles exhibit a fine polish, striations, gouges, penetrating grains, and hinge fractures; 25%–40% matrix of sand and silt-sized limestone and clay clasts rarely occurring as sandy lenses; the stratum is weakly consolidated, but there are several layers that are cemented with calcite cement.

Figure 23. Hand sample of Armenia spheroid bed showing abundant accretionary lapilli and clay fragments (altered glass)—dark fragments in light colored detrital calcite matrix. Note that the large accretionary lapilli has a core of clay (altered glass, dark in photo).

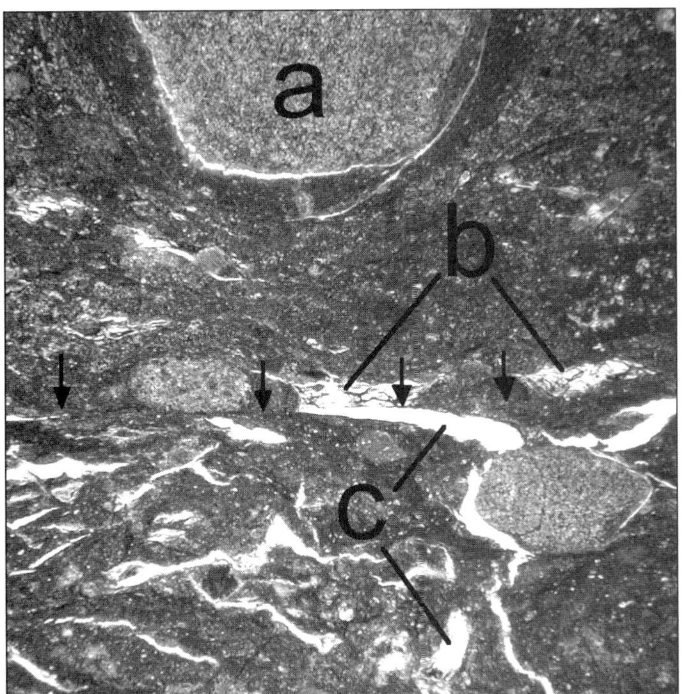

Figure 24. Photomicrograph of thin section of the basal contact of the Armenia spheroid bed showing a sharp contact (arrows) and (a) abrupt appearance of accretionary lapilli; (b) translucent clay clasts (altered glass; note minor mixing [one rounded limestone grain resting above contact]); (c) root casts, some with intact roots, probably modern. Plane polarized transmitted light. Field of view: 4.4 mm wide.

μm) comprise 5% of the spheroid bed in thin section. About 50%–60% of the spheroid bed is a fine-grained calcite matrix with essentially the same composition and size distribution as the accretionary lapilli aggregate.

One clay sample (altered glass fragment) from the Armenia spheroid bed was analyzed with X-ray diffraction and found to be predominately an interstratified mixed layer illite-smectite with a large portion composed of illite.

We examined 15 slides with grain residues prepared from acid leached samples of the Armenia spheroid bed matrix. Residues were mounted in oil, which allows the grains to be rotated to view all axes, but not to measure the angle. We found no shocked quartz in the spheroid bed (detrital quartz was quite rare in these samples). We also examined 10 slides with grain residues prepared in the same fashion from samples of the matrix of the conglomerate bed that immediately overlies the Armenia spheroid bed. We found nine detrital quartz grains with multiple sets (8 with 2 sets, 1 with 3 sets) of planar deformation features (PDFs) indicative of impact shock (Figs. 26A–26C), and several more with a single set of PDFs. We calculate that there are ~4 grains with multiple PDFs per gram of matrix. Vesicular glass shards (50–500 μm), now altered to clay, are also a common conglomerate matrix constituent at Armenia (Fig. 26D). The relatively abundant shocked quartz in the conglomerate bed is unlikely reworked from the spheroid bed, as the latter lacks shocked quartz. Presumably, the conglomerate contains ejecta reworked form another source, or is itself an impact deposit. Further research is needed to test these two possibilities.

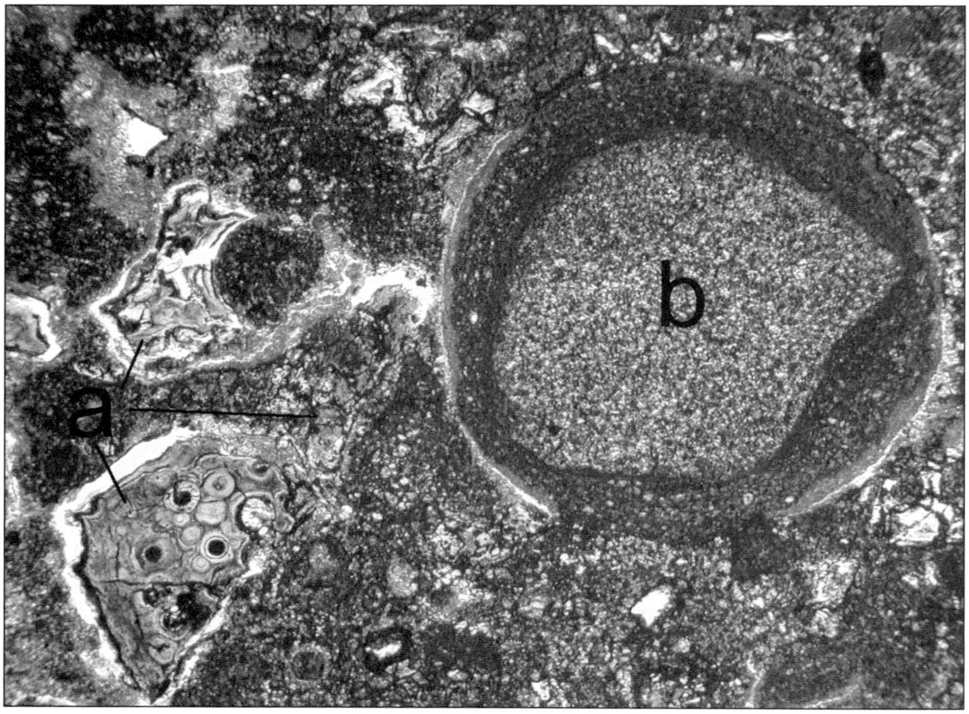

Figure 25. Photomicrograph of thin section from the lower portion of the Armenia spheroid bed showing (a) clay clasts with matrix filled vesicles and spherulitic devitrification features and (b) rim-type accretionary lapilli. Plane polarized transmitted light. Field of view: 2.7 mm wide.

COMPARISONS WITH OTHER EJECTA DEPOSITS

The new Albion Formation outcrops in Quintana Roo and central Belize show close similarities with those from Albion Island and help confirm that Chicxulub ejecta are exposed over a broad area at the base of the Yucatán Peninsula. Our interpretation that these deposits are impact ejecta is based primarily upon the occurrence of abundant altered vesicular glass fragments and carbonate accretionary lapilli, which cannot be explained by any known volcanic activity in the region. Their association with the Chicxulub impact is established by the large thickness and dominant carbonate lithology, which are appropriate for the nearby Chicxulub crater, and by their stratigraphic position resting on a terminal Cretaceous land surface. These criteria are critical, since carbonate breccias are common near the K-T boundary in this region. These new sites also help confirm a regional subdivision of the Albion Formation, originally proposed from Albion Island (Ocampo et al., 1996), into a basal spheroid bed and overlying diamictite bed. The thickest deposit remains the one at Albion Island, where recent (2001) quarrying has exposed a 17-m-thick section.

An important aspect of the new outcrops is that they are much less altered than those on Albion Island and thus better preserve primary textures and mineralogies. The Albion Formation matrix, clasts, and accretionary lapilli on Albion Island are all pervasively dolomitized (Ocampo et al., 1996; Pope et al., 1999; Fouke et al., 2002). The better preserved spheroid bed outcrops at Ramonal North and Armenia confirm that the original carbonate mineralogy of this unit was calcite, and that the carbonate "spheroids" are, without a doubt, accretionary lapilli analogous to ones found in volcanic pyroclastic deposits. Similarly, the altered glass fragments with vesicular textures are much better preserved at Ramonal North and Armenia than in the spheroid bed at Albion Island, although it appears that little to no true glass remains. The diamictite bed at Alvaro Obregón and Agua Dulce is also slightly less dolomitized than its equivalent at Albion Island, and we speculate that the matrix of this unit was also originally calcite. Likewise, it is clear that the diamictite bed contains both limestone and dolomite clasts, but their original abundances are difficult to determine given the pervasive postdepositional dolomitization.

The only other reported Chicxulub ejecta deposit with a discrete bed of accretionary lapilli is at Guayal, in Tabasco, México, located ~550 km southwest from the center of Chicxulub (Grajales-Nishimura et al., 2000; Salge et al., 2000). While the texture of the Guayal lapilli is similar to those of the Albion Formation spheroid bed at Ramonal North and Armenia, the composition is very different. The Guayal lapilli contain shocked quartz grains (Grajales-Nishimura et al., 2000) and are reported to be 64% quartz, 19% clay, 9% calcite, and 8% dolomite (Salge et al., 2000), but much of this quartz may be secondary (Griscom et al., 1999). This abundance of quartz and paucity of calcite, if primary, is in marked contrast to the Albion Formation spheroid bed. It should be noted that that the Guayal lapilli bed was deposited in deep water, contains many fragmented lapilli, and is associated with a massive slump of debris triggered by the impact, and thus it may not be a primary deposit. No accretionary lapilli beds have been reported

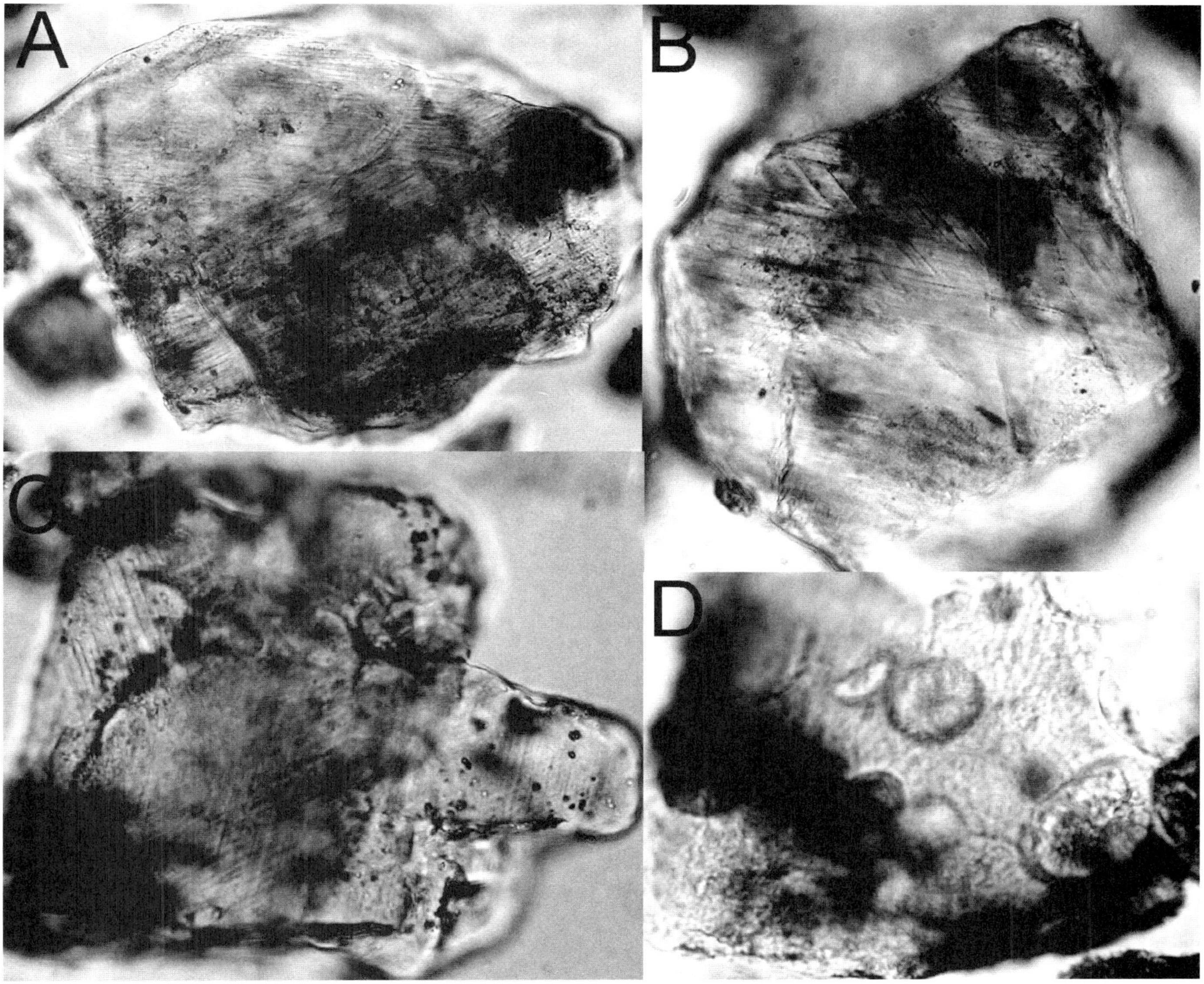

Figure 26. (A–C) Detrital shocked quartz grains with multiple sets of planar deformation features (PDFs). (D) Altered glass shard with vesicles and spherulitic devitrification features, all from the matrix of the limestone conglomerate bed overlying the Armenia spheroid bed. All grains mounted in oil and shown in plane polarized transmitted light. (A) 240 μm, 2 sets PDFs; (B) 215 μm, 2 sets PDFs; (C) 140 μm, 3 sets PDFs; (D) 160 μm.

from deposits closer to the Chicxulub crater, in ejecta deposits drilled by Universidad Nacional Autónoma de México (UNAM) and Petróleos Mexicanos (Pemex) (Sharpton et al., 1996; Corrigan, 1998), or by the Intercontinental Drilling Program (ICDP) (Dressler et al., 2003). Our own examination of the UNAM cores in 2003 likewise found no accretionary lapilli. Calcite accretionary lapilli are found in Chicxulub ejecta deposits at sites more distal than Guayal in northeastern México (Smit et al., 1996), but these lapilli are apparently rare and do not form discrete beds.

Accretionary lapilli have been reported from the crater-fill suevite of several other impact craters, for example Ries (Graup, 1981), Sudbury (Muir and Peredery, 1984), Manson (Witzke and Anderson, 1996), and Popigai (Masaitis, 2003). No accretionary lapilli occur in the out-of-crater ejecta at Ries. Carbonate lapilli similar to those in the Albion Formation occur in discrete beds in the Alamo breccia impact deposits in Nevada, which, like Chicxulub, is interpreted as a carbonate platform impact (Warme et al., 2002). Nevertheless, these lapilli beds are more like the one at Guayal, as they have been redeposited by high-energy submarine flows, and their exact location with respect to the impact site is not known.

None of these other impact accretionary lapilli deposits provide a good analogue for the Albion Formation spheroid bed, and the best comparison is with volcanic pyroclastic deposits.

Massive deposits of mixed mantled, core, and rim-type accretionary lapilli, which characterize the Albion Formation spheroid bed, have their closest analogues with ash-rich pyroclastic flows or surges produced by large phreatomagmatic eruptions (e.g., Schumacher and Schmincke, 1991). The main contrast is that large phreatomagmatic volcanic eruptions do not typically produce both abundant silicate glass and calcite ash, yet an impact into a mixed silicate-carbonate target would.

The Albion Formation diamictite bed is similar to, yet distinct from, the more proximal ejecta deposits cored near the Chicxulub rim by UNAM and Pemex (e.g., Sharpton et al., 1996, 1999; Corrigan, 1998). These more proximal deposits are divided into two units: (1) a basal polymict breccia called Bunte-type breccias (after the Bunte Breccia of the Ries crater, e.g., Hörz et al., 1983), which contains ~40%–60% sulfate, ~30%–40% carbonate, and <1%–4% silicate basement; and (2) an upper suevitic breccia (a name also derived from Ries) containing 40%–60% silicate basement (Sharpton et al., 1996, 1999). The Chicxulub suevitic breccia extends ~140 km from the crater's center (Sharpton et al., 1999), while the Chicxulub Bunte–type breccia extends more than 200 km (Sharpton et al., 1996). The Albion Formation diamictite bed, while technically classified as a suevitic breccia (mixture of melted and unmelted clasts), contains far less silicate material than is found in the Chicxulub suevitic breccias near the rim, and considerably more melt than is found in the Chicxulub Bunte–type breccia. Furthermore, the 40%–60% sulfate component of the Chicxulub Bunte–type breccia is mostly missing in the diamictite bed, although it is possible that the sulfate once present has been leached.

There are no ejecta blanket deposits from other terrestrial impact craters that are directly comparable to the Albion Formation. The ejecta from the Ries crater, which possesses one of the best studied examples of a terrestrial ejecta blanket, are not preserved beyond 3 crater radii, whereas the Albion Formation outcrops described in this paper extend from 3.3 to 4.7 crater radii. It should also be added that Chicxulub is much larger than Ries (200 km versus 26 km diameter).

The diamictite bed contains <<1% unmelted deep target material (only traces have been found; Ocampo et al., 1996; Pope et al., 1999), and some unknown percentage of shallow target carbonate primary ejecta. This aspect matches well with the studies of Ries, where the Bunte Breccia contains <1% deep target and ~34% shallow target material (Hörz et al., 1983). The main contrast is that the diamictite bed contains 10% shock-melted deep target material (the altered glass), which is absent in the Ries Bunte Breccia. Another contrast between the diamictite bed and the Ries Bunte Breccia is that there is little evidence of erosion at the base the Albion Formation diamictite bed, given the intact spheroid bed at all three basal exposures. The emplacement of the Bunte Breccia apparently eroded over 50 m of basal Tertiary unconsolidated sands (Hörz et al., 1977). The Albion Formation diamictite bed is best described as a mixture of the Bunte-type and suevitic breccias, this fact, the lack of erosion at the base, and the deposition of pyroclastic-like flows (the spheroid bed) all indicate that ejecta from large craters are emplaced at distances greater than 3 radii by processes somewhat different from those proposed for Ries crater (e.g., Hörz et al., 1983). We suggest that at these greater distances there is more mixing of ejecta, less influence of secondary cratering, and emplacement is by less erosive surface flows.

CONCLUSIONS

We have presented evidence that Chicxulub impact ejecta are preserved in numerous locations in Quintana Roo, México and Belize, at distances of 330–470 km from the center of Chicxulub. It can be debated whether or not thick deposits of the Albion Formation ejecta once ringed the Chicxulub crater and thus formed part of the "continuous ejecta" as seen in images of other terrestrial planets and defined at Ries crater (e.g., Hörz et al., 1983). Thick ejecta deposits have only been found close to the rim and south of Chicxulub (UNAM and Pemex cores, Albion Formation), and this is the only area where typical ejecta blanket deposits are expected, as this region had the only dry land at the time of impact. All other proximal ejecta were emplaced in deep water (e.g., Smit, 1999). We argue that the Albion Formation diamictite bed is part of the Chicxulub continuous ejecta, given its extent (>70 km^2) and thickness (>17 m). It is less clear that the Albion Formation spheroid bed should be considered part of the continuous ejecta where it occurs without the overlying diamictite bed, since there is only one site, Armenia, with this stratigraphy. No Albion Formation diamictite bed deposits have been found at distances greater than 3.6 crater radii (Albion Island), and we propose that this may be the approximate limit for the Chicxulub continuous ejecta. We speculate that the abundance of debris along the Río Hondo may represent the terrestrial equivalent of a terminal thickened rampart, as seen along the outer edge of fluidized ejecta blankets on Mars (e.g., Carr et al., 1977).

The discovery of Albion Formation spheroid bed deposits at a distance of 4.7 crater radii (Armenia) demonstrates that this unit was significantly more extensive than the diamictite bed. The limestone cobble bed with shocked quartz and altered impact glass overlying the spheroid bed at Armenia may be impact related, or contain reworked ejecta; research is ongoing to determine which scenario applies. The limited number of spheroid bed outcrops—four, with the most distal one, Armenia, >100 km from the others—precludes any definitive discussion of the distribution of this unit, but some conclusions are warranted. First, we observed no significant differences in thickness, texture, or composition between Ramonal North and Armenia. This apparent homogeneity across >100 km suggests that the ash cloud that produced this accretionary lapilli bed was immense and that the flow evolved little over this distance. Applying a simple energy line model developed for large pyroclastic flows (Sheridan, 1979), suggests that the ash cloud that produced pyroclastic flows at distances of 500 km reached an elevation of ~100 km, not unreasonable for a large impact. The abundance of shock-melted basement (20%–30% altered glass) in deposits

470 km from the impact center is perhaps best explained by the explosive dispersal of shock melt by volatile expansion (Kieffer and Simonds, 1980). The occurrence of this glass with accretionary lapilli, which form in water-rich, ash-laden clouds, indicates that the first and fastest ejecta to leave the crater may have been from massive steam explosions that produced gas-rich, pyroclastic-like flows (Wohletz and Sheridan, 1983).

In summary, we propose that the Albion Formation spheroid bed is a product of explosive vapor plume expansion that dispersed a large volume of shock melt. The vapor was water-rich and upon expansion, condensed to help form the large mass of accretionary lapilli, which were deposited in a mode similar to pyroclastic flows. We interpret the diamictite bed to be derived from the turbulent mixing and collapse of the main ejecta curtain, which moved as a surface flow over the vapor plume deposits until halting somewhere near the current México and Belize border.

ACKNOWLEDGMENTS

This work was supported by the National Aeronautics and Space Administration Exobiology Program, California Institute of Technology, Jet Propulsion Laboratory, The Planetary Society and Belize expedition participants, Natural History Museum of Los Angeles County, and the European Space Agency. We also acknowledge the kind support of the Belizean Geology and Petroleum Office and Pook's Hill Lodge.

APPENDIX

Locations of Albion Formation Outcrops

(1) Albion Island N18° 07.84′ W88° 41.13′
Belize
(2) Armenia N17° 08.93′ W88° 44.07′
Belize
(3) Johnson Quarry N18° 29.73′ W88° 30.89′
México
(4) Sacxán N18° 27.96′ W88° 31.23′
México
(5) Palmar 1 N18° 27.15′ W88° 31.67′
México
(6) Palmar 2 N18° 26.93′ W88° 32.17′
México
(7) Palmar 3 N18° 26.85′ W88° 33.49′
México
(8) Palmar 4 N18° 26.68′ W88° 32.47′
México
(9) Palmar 5 N18° 26.38′ W88° 31.85′
México
(10) Ramonal North N18° 26.01′ W88° 31.90′
México
(11) Ramonal South N18° 25.71′ W88° 31.96′
México
(12) Allende 1 N18° 22.94′ W88° 33.95′
México
(13) Allende 2 N18° 22.83′ W88° 34.27′
México
(14) Allende 3 N18° 22.52′ W88° 34.82′
México
(15) Sabidos N18° 20.47′ W88° 36.33′
México
(16) Agua Dulce N18° 20.16′ W88° 36.66′
México
(17) Alvaro Obregón N18° 17.88′ W88° 37.185′
México

REFERENCES CITED

Bateson, J.H., and Hall, I.H.S., 1977, The geology of the Maya Mountains: Institute of Geological Studies, Natural Environmental Research Council, Overseas Memoir, v. 3, p. 1–38.

Carr, M.H., Crumpler, L.S., Cutts, J.A., Greeley, J.E., Guest, J.E., and Masursky, H., 1977, Martian impact craters and emplacement of ejecta by surface flows: Journal of Geophysical Research, v. 82, p. 4055–4065.

Cornec, J.H., 1985, Note on the Provisional Geological Map of Belize at the scale of 1:250,000: Petroleum Office, Ministry of Natural Resources, Belmopan, Belize, p. 22.

Cornec, J.H., 1986, Geological Map of Belize: Geology and Petroleum Office, Ministery of Natural Resources, Belmopan, Belize, scale 1:250,000.

Corrigan, C.M., 1998, The Composition of Impact Breccias from the Chicxulub Impact Crater, Yucatan Peninsula, Yucatan, Mexico [M.S. thesis]: East Lansing, Michigan State University, p. 125.

Dixon, C.G., 1956, Geology of Southern British Honduras: Belize City, Belize Government Printer, p. 92.

Dressler, B.O., Sharpton, V.L., and Marin, L.E., 2003, Chicxulub YAX-1 impact breccias: Whence they come? [abs.]: Lunar and Planetary Institute, Houston, Texas, USA, Lunar and Planetary Science, v. 34, abstract no. 1259 (CD-ROM).

Flores, G., 1952a, Geology of northern British Honduras: American Association of Petroleum Geologists Bulletin, v. 36, p. 404–409.

Flores, G., 1952b, Summary Report of the Preliminary Geological Studies of the area N of 17° N Latitude, British Honduras: Freeport, Bahamas Exploration Company, LTD, 35 p.

Fouke, B.W., Zerkle, A.L., Alvarez, W., Pope, K.O., Ocampo, A.C., Wachtman, R.J., Grajales-Nishimura, J.M., Claeys, P., and Fischer, A.G., 2002, Cathodoluminescence petrography and isotope geochemistry of KT impact ejecta deposited 360 km from the Chicxulub crater, at Albion Island, Belize: Sedimentology, v. 49, p. 117–138, doi: 10.1046/j.1365-3091.2002.00435.x.

Grajales-Nishimura, J.M., Cedillo-Pardo, E., Rosales-Domínguez, C., Padilla-Avila, P., Sánchez-Ríos, A., Morán-Zenteno, D.J., Alvarez, W., Claeys, P., Ruíz-Morales, J., and García-Hernández, J., 2000, Chicxulub impact: The origin of reservoir and seal facies in the southeastern Mexico oil fields: Geology, v. 28, p. 307–310, doi: 10.1130/0091-7613(2000)0282.3.CO;2.

Graup, G., 1981, Terrestrial chondrules, glass spherules and accretionary lapilli from the suevite, Ries crater, Germany: Earth and Planetary Science Letters, v. 55, p. 407–418, doi: 10.1016/0012-821X(81)90168-0.

Griscom, D.L., Beltran-Lopez, V., Merzbacher, C.I., and Bolden, E., 1999, Electron spin resonance of 65-million-year-old glasses and rocks from the Cretaceous-Tertiary boundary: Journal of Non-Crystalline Solids, v. 253, p. 1–22, doi: 10.1016/S0022-3093(99)00340-3.

Hörz, F., Gall, H., Hüttner, R., and Oberbeck, V.R., 1977, Shallow drilling in the Bunte Breccia deposits, Ries Crater, Germany, *in* Roddy, D.J., Pepin, R.O., and Merrill, R.B., eds., Impact and Explosion Cratering: New York, Pergamon, p. 425–488.

Hörz, F., Ostertag, R., and Rainey, D.A., 1983, Bunte Breccia of the Ries: Continuous deposits of large impact craters: Reviews of Geophysics and Space Physics, v. 21, p. 1667–1725.

INEGI (Instituto Nacional de Estadistica, Geografia, e Informatica), 1987, Carta Geologica, Chetumal Sheet, Mexico 1:250,000, 1 sheet.

Kieffer, S.W., and Simonds, C.H., 1980, The role of volatiles and lithology in the impact cratering process: Review of Geophysics and Space Physics, v. 18, p. 143–181.

King, D.T., Jr., and Petruny, L.W., 2003, Stratigraphy and sedimentology of coarse diamictite breccia beds within the Cretaceous-Tertiary boundary section, Albion Island, Belize, *in* Koeberl, C., and Martinez-Ruiz, F., eds., Impact Markers in the Stratigraphic Record: Berlin, Springer-Verlag, p. 203–227.

King, D.T., Jr., Pope, K.O., and Petruny, L.W., 2004, Stratigraphy of Belize, north of the 17th parallel: Transactions of the Gulf Coast Association of Geological Societies, v. 54, p. 289–304.

Masaitis, V.L., 2003, Obscure-bedded ejecta facies from the Popigai impact structure, Siberia: Lithological features and mode of origin, *in* Koeberl, C., and Martinez-Ruiz, F., eds., Impact Markers in the Stratigraphic Record: Berlin, Springer-Verlag, p. 137–162.

Muir, T.L., and Peredery, W.V., 1984, The Onaping Formation, *in* Pye, E.G., Naldrett, A.J., and Giblin, P.E., eds., The Geology and Ore Deposits of the Sudbury Structure: Geological Survey of Ontario, p. 139–204.

Ocampo, A.C., Pope, K.O., and Fischer, A.G., 1996, Ejecta blanket deposits of the Chicxulub crater from Albion Island, Belize, *in* Ryder, G., Fastovsky, D., and Gartner, S., eds., The Cretaceous-Tertiary Event and Other Catastrophes in Earth History: Geological Society of America Special Paper 307, p. 75–88.

Ower, L., 1928, Geology of British Honduras: Journal of Geology, v. 36, p. 494–509.

Palmer, K. van W., 1937, The Claibornian Scaphopoda, Gastropoda and Dibrachiate Cephalopoda of the Southern United States: Bulletin of American Paleontology, v. 7, p. 1–730.

Pope, K.O., Ocampo, A.C., Fischer, A.G., Alvarez, W., Fouke, B.W., Webster, C.L., Jr., Vega, F.J., Smit, J., Fritsche, A.E., and Claeys, Ph., 1999, Chicxulub impact ejecta from Albion Island, Belize: Earth and Planetary Science Letters, v. 170, p. 351–364, doi: 10.1016/S0012-821X(99)00123-5.

Salge, T., Tagle, R., and Claeys, P., 2000, Accretionary lapilli from the KT boundary site of Guayal, Mexico: Preliminary insights of expansion plume formation: Abstracts of the 63rd Annual Meteoritical Society Meeting, Chicago, Illinois, Abstract 5124.

Salvador, A., ed., 1994, International Stratigraphic Guide, 2nd ed: International Union of Geological Sciences and Geological Society of America, 214 p.

Schumacher, R., and Schmincke, H.-U., 1991, Internal structure and occurrence of accretionary lapilli—A case study at Laacher See Volcano: Bulletin of Volcanology, v. 53, p. 612–634.

Sharpton, V.L., Marin, L.E., Carney, J.L., Lee, S., Ryder, G., Schuraytz, B.C., Sikora, P., and Spudis, P.D., 1996, A model of the Chicxulub impact basin based on evaluation of geophysical data, well logs, and drill core samples, *in* Ryder, G., Fastovsky, D., and Gartner, S., eds., The Cretaceous-Tertiary Event and Other Catastrophes in Earth History: Geological Society of America Special Paper 307, p. 55–74.

Sharpton, V.L., Corrigan, C.M., Marin, L.E., Urrutia-Fucugauchi, J., and Vogel, T.A., 1999, Characterization of impact breccias from the Chicxulub impact basin: Implications for excavation and ejecta emplacement [abs.]: Houston, Texas, USA, Lunar and Planetary Institute, Lunar and Planetary Science, v. 30, abstract no. 1515 (CD-ROM).

Sheridan, M.F., 1979, Emplacement of pyroclastic flows: A review, *in* Chapin, C.E., and Elston, W.E., eds., Ash-Flow Tuffs: Geological Society of America Special Paper 180, p. 125–136.

Smit, J., 1999, The global stratigraphy of the Cretaceous-Tertiary boundary impact ejecta: Annual Reviews of Earth and Planetary Science, v. 27, p. 75–113, doi: 10.1146/annurev.earth.27.1.75.

Smit, J., Roep, T.B., Alvarez, W., Montanari, A., Claeys, P., Grajalez-Nishimura, J.M., and Bermudez, J., 1996, Coarse-grained, clastic sandstone complex at the K/T boundary around the Gulf of Mexico: Deposition by tsunami waves induced by the Chicxulub impact? *in* Ryder, G., Fastovsky, D., and Gartner, S., eds., The Cretaceous-Tertiary Event and Other Catastrophes in Earth History: Geological Society of America Special Paper 307, p. 151–182.

Vega, F.J., Perrilliat, M.C., Ocampo, A., and Pope, K., 2001, Aporrhaid gastropod of latest Cretaceous age from the upper part of the Barton Creek Formation below Chicxulub ejecta blanket in southern Mexico: VII North American Paleontological Convention, Program and Abstracts, p. 129.

Warme, J.E., Morgan, M., and Kuehner, H.-C., 2002, Impact-generated carbonate accretionary lapilli in Late Devonian Alamo Breccia, *in* Koeberl, C., and MacLeod, K.G., eds., Catastrophic Events and Mass Extinctions: Impacts and Beyond: Geological Society of America Special Paper 356, p. 489–504.

Witzke, B.J., and Anderson, R.R., 1996, Sedimentary-clast breccias of the Manson impact structure, *in* Koeberl, C., and Anderson, R.R., eds., The Manson Impact Structure: Geological Society of America Special Paper 302, p. 115–144.

Wohletz, K.H., and Sheridan, M.F., 1983, Martian rampart crater ejecta: Experiments and analysis of melt-water interaction: Icarus, v. 56, p. 15–37, doi: 10.1016/0019-1035(83)90125-2.

Manuscript Accepted by the Society 9 August 2004

Printed in the USA

Geological Society of America
Special Paper 384
2005

Chicxulub impact ejecta from the Cretaceous-Paleogene (K-P) boundary in northeastern México

Peter Schulte*
Geologisches Institut, Universität Karlsruhe (TH), Kaiserstrasse 12, D-76128 Karlsruhe, Germany

Agnes Kontny
Geologisch-Paläontologisches Institut, Universität Heidelberg, Im Neuenheimer Feld 234, D-69120 Heidelberg, Germany

ABSTRACT

The combined petrological and rock magnetic study of the Cretaceous-Paleogene (K-P) boundary in northeastern México revealed compositionally and texturally complex Chicxulub ejecta deposits. The predominant silicic ejecta components are Fe-Mg–rich chlorite and Si-Al-K–rich glass spherules with carbonate inclusions and schlieren. Besides these silica phases, the most prominent ejecta component is carbonate. Carbonate occurs as lithic clasts, accretionary lapilli, melt globules (often with quench textures), and as microspar. The composition of the spherules provides evidence for a range of target rocks of mafic to intermediate composition, presumably situated in the northwestern sector of the Chicxulub impact structure. The abundance of carbonate ejecta suggests that this area received ejecta mainly from shallow, carbonate-rich lithologies. Rare µm-sized metallic and sulfidic Ni-Co–rich inclusions in the spherules indicate a possible contamination by meteoritic material. This complex composition underlines the similarities of ejecta in NE México to Chicxulub ejecta from K-P sections worldwide.

Although the ejecta display a great variability, the magnetic susceptibility, remanence, and hysteresis properties of the ejecta deposits are fairly homogeneous, with dominantly paramagnetic susceptibilities and a weak ferromagnetic contribution from hematite and goethite. The absence of spinels and the ubiquitous presence of hematite and goethite points to high oxygen fugacity during the impact process. The microfacies and internal texture of the ejecta deposits show welding and fusing of components, as well as evidence for liquid immiscibility between silicic and carbonate melts. No evidence for binary mixing of ejecta phases was found. Therefore, Chicxulub ejecta in NE México probably derived from less energetic parts of the ejecta curtain. However, welding features of ejecta particles and enclosed marl clasts and/or benthic foraminifera from a siliciclastic environment suggest interaction of the—still hot—ejecta curtain with northern Mexican shelf sediments. In addition, an initial ground surge–like ejecta-dispersion mode seems possible.

Keywords: northeastern México, Chicxulub impact, Cretaceous-Paleogene boundary, K-P, ejecta, spherules, petrography, mineral chemistry, rock magnetic properties.

*Present address, Schulte: Institut für Geologie, Mineralogie und Geophysik, Ruhr-Universität Bochum, Universitätsstrasse 150, D-44801 Bochum, Germany, peter.schulte@ruhr-uni-bochum.de.

Schulte, P., and Kontny, A., 2005, Chicxulub impact ejecta from the Cretaceous-Paleogene (K-P) boundary in northeastern México, *in* Kenkmann, T., Hörz, F., and Deutsch, A., eds., Large meteorite impacts III: Geological Society of America Special Paper 384, p. 191–221. For permission to copy, contact editing@geosociety.org.

INTRODUCTION

The accurate knowledge of target lithologies and the physical processes involved in crater formation is not only crucial for estimating the environmental consequences of large impacts (e.g., Toon et al., 1997; Pierazzo et al., 1998; Gupta et al., 2001; Luder et al., 2003), it also allows for the correlation of distal ejecta to specific cratering events (e.g., French, 1998; Kettrup et al., 2003). In this context, the study of the Chicxulub impact event on the Yucatán peninsula, southern México, is important for understanding the Cretaceous-Paleogene (K-P) mass extinctions (see papers in Ryder et al., 1996; Koeberl and MacLeod, 2002). Critical issues include (1) the amount of dust and climatically active gases released by this impact (e.g., Maruoka and Koeberl, 2003; Pierazzo et al., 2003), and (2) the severity of its immediate disastrous effects (e.g., Kring and Durda, 2002; Robertson et al., 2004). Resolving both issues requires a detailed assessment of the pre-impact geology of Yucatán and ejecta formation and distribution processes. Access to the Chicxulub crater is, however, hindered by a several hundred meter–thick sequence of Paleogene sediments, and to date, only a few drill cores have provided access to the Yucatán subsurface (e.g., Morgan et al., 1997; Claeys et al., 2003; Kettrup and Deutsch, 2003; Stöffler et al., 2004). These drill cores revealed a range of melt phases and lithic clasts within the impact breccias and suevite, pointing to a geologically complex basement beneath the Yucatán, which is covered by carbonaceous sediments. However, the compositional variability of the basement with depth and in radial directions, as well as ejecta mixing, volatilization (fractionation), and homogenization processes and the subsequent dispersal are still poorly constrained (e.g., Hörz et al., 2002; Griscom et al., 2003; Kettrup and Deutsch, 2003).

Further insights into the Yucatán subsurface and into cratering processes may derive from the study of Chicxulub ejecta in K-P boundary sites (see Figs. 1 and 2; e.g., Koeberl and Martínez-Ruiz, 2003; Kettrup and Deutsch, 2003). Specifically, the up to 1-m-thick ejecta deposits at the K-P boundary in NE México provide an excellent opportunity to study the properties of Chicxulub ejecta across an area of several hundred kilometers (Figs. 2 and 3). These K-P sections are situated only ~700–1000 km northwest of the crater center and, hence, their study may improve the data for the northern sector of Chicxulub, which has not yet been drilled. Previous investigations of the K-P sections in this area have been primarily focused on a siliciclastic depositional sequence associated with the ejecta deposits. This depositional sequence is considered to be either impact-generated tsunamite (e.g., Smit et al., 1996; Smit, 1999; Alegret et al., 2001), a sea-level lowstand deposit (e.g., Stinnesbeck et al., 1996; Keller et al., 1997), or a sequence of impact-induced debris-flows

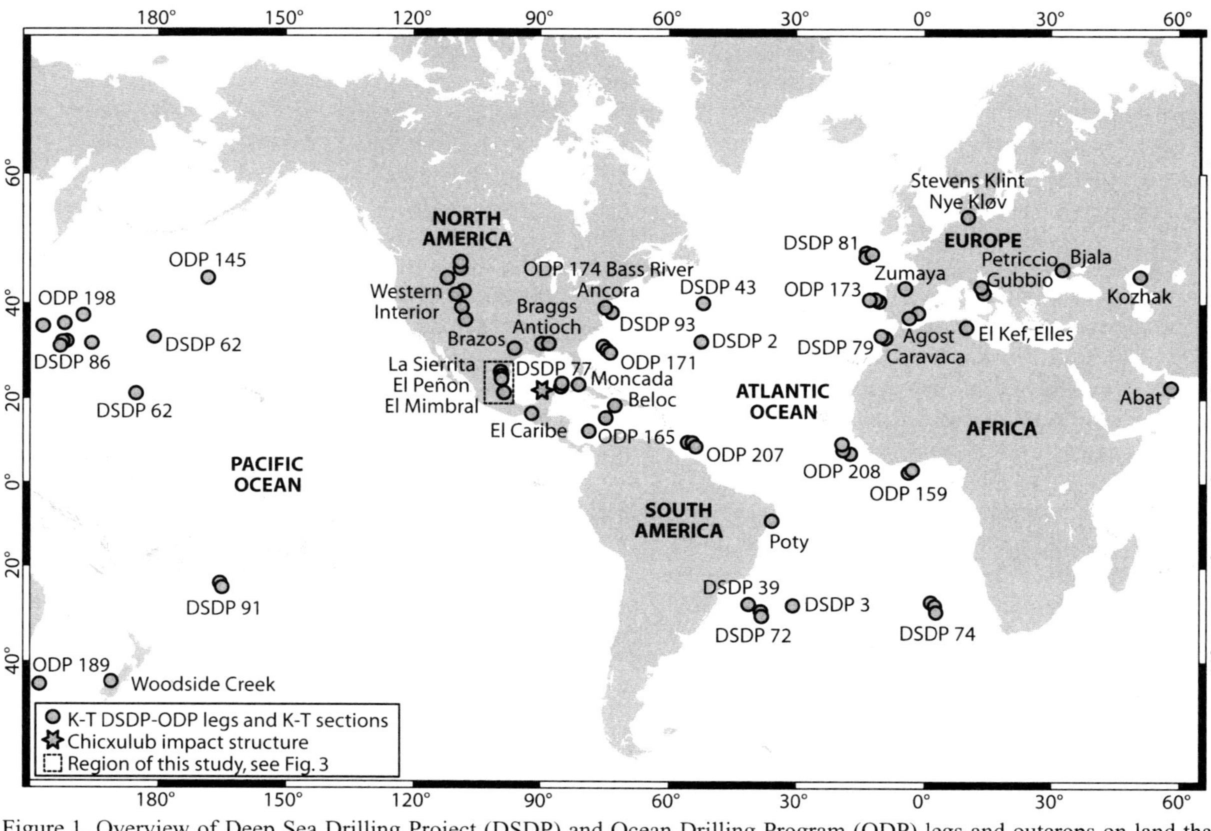

Figure 1. Overview of Deep Sea Drilling Project (DSDP) and Ocean Drilling Program (ODP) legs and outcrops on land that include Chicxulub impact ejecta at the K-P boundary.

(Bohor, 1996; Schulte, 2003). Smit et al. (1996) provided electron microprobe data, focusing on glassy ejecta particles with a composition similar to andesite with CaO-rich schlieren. Additional studies by Sharpton et al. (1996) and Kettrup and Deutsch (2003) extended the compositional range of the ejecta phase into the fields of mafic rocks.

None of these studies, however, combined the textural and compositional investigation of ejecta phases and rock magnetic methods (magnetic susceptibility, remanence, hysteresis properties), both of which are the major aims of the present study.

LOCALITIES AND GEOLOGICAL SETTING

The Mexican K-P boundary outcrop areas are situated in the Gulf coastal plain of NE México in the States of Nuevo León and Tamaulipas. For this study, the K-P outcrops at La Sierrita, El Peñon, and El Mimbral were geologically mapped and sampled (Figs. 3 and 4).

The La Sierrita outcrop area is found adjacent to the small hamlet of La Sierrita near the road that connects General Terán and Vaquería (Figs. 3A and 3B). The exposures are situated along the flanks of low-lying hills to the south and the north of the Mesa Juan Perez (25°13′45″N, 99°31′05″W).

The El Peñon outcrop area (25°58′05″N, 99°12′50″W) is located close to the unpaved road from Linares to the hamlet of Burgos (Figs. 3A and 3C). The outcrops are located ~2 km northwest of the Porvenir Lake at a series of low-lying hills west of the main road close to the Rancho El Peñon.

The El Mimbral outcrop (23°13′10″N; 98°40′05″W) is found ~10 km southeast of Llera de Canales on the southwestern bank of the El Mimbral Creek and can be reached via an unpaved road from Mexican Highway 81 to the east at the radio antenna (Fig. 3A).

The lithology of the K-P succession in NE México (see also Smit et al., 1996; Stinnesbeck et al., 1996; Schulte et al., 2003; Schulte, 2003) is generally characterized by the intermittent presence of a channelized or sheet-like spherule and sand-siltstone

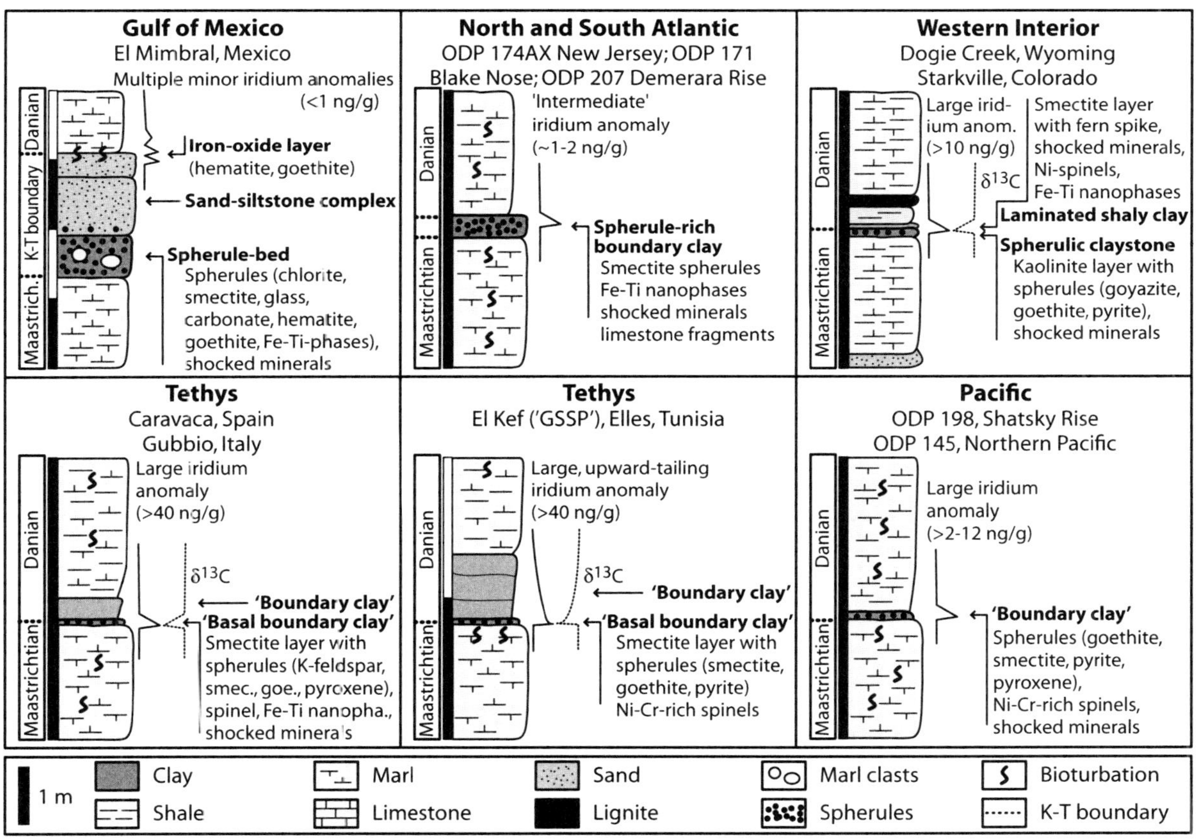

Figure 2. Generalized lithology of the K-P boundary in different ocean basin and terrestrial settings, including mineralogical and geochemical markers. Note the absence of a distinct K-P boundary clay and dispersion of the "classical" boundary markers over an interval of several meters in the Gulf of México as an example of areas proximal to the Chicxulub impact structure. This "dispersion" is in contrast with the El Kef K-P section, Tunisia, the "Global Stratotype Section and Point (GSSP)" for the K-P boundary, which is typical for distal settings. The negative $\delta^{13}C$ excursion and iridium anomaly at the K-P boundary are drawn schematically; the peak of the magnetic susceptibility (not shown) almost parallels the excursion of the Ir anomaly. Additional details and references are given in the text.

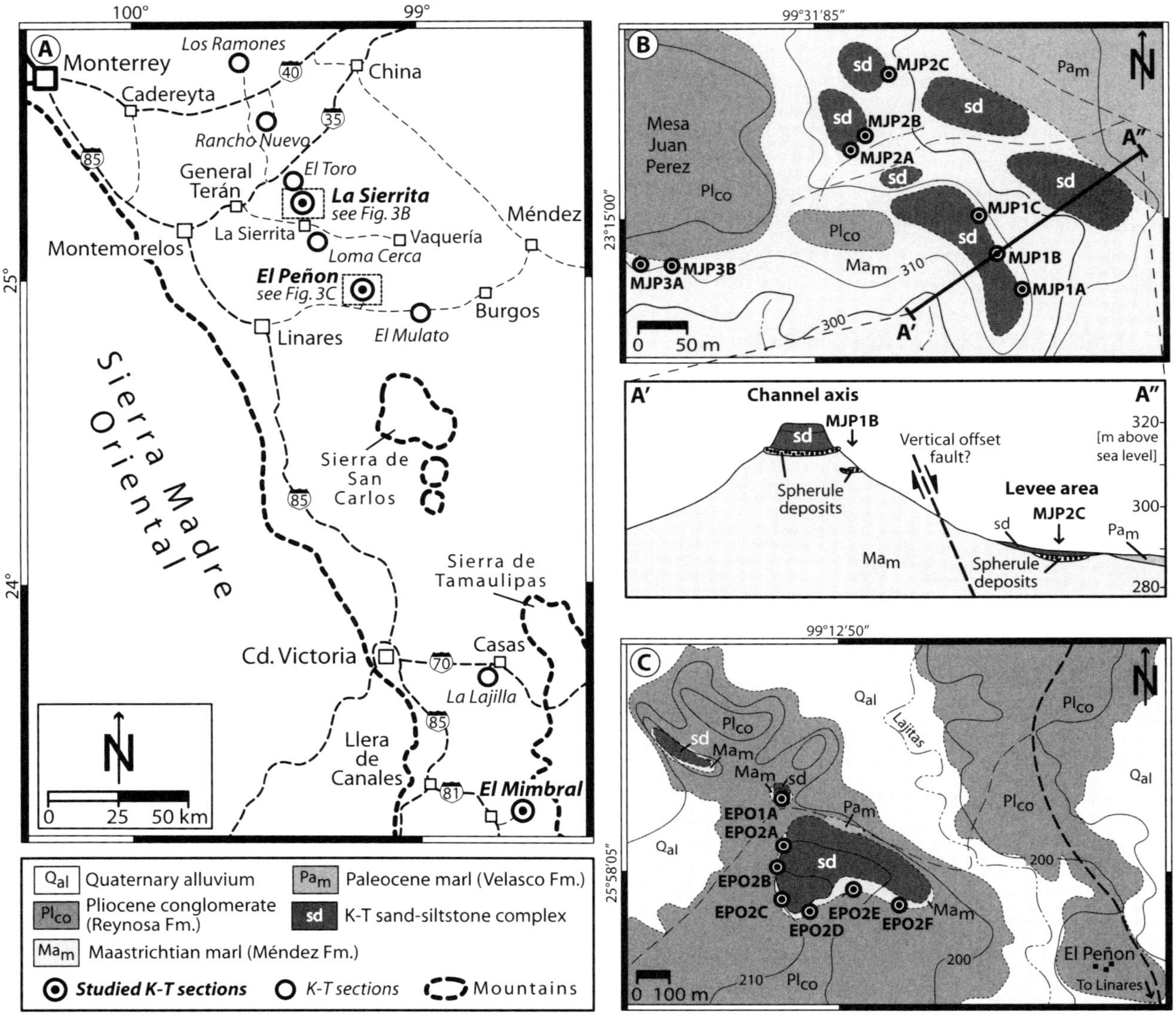

Figure 3. Topographic and geologic maps with locations of K-P sections referred to in this study. Legend provided in the lower right. (A) Topographic overview of NE México. (B) Geological map of the Mesa Juan Perez area close to La Sierrita with a cross-cut A′–A″ through the Mesa Juan Perez, showing the slumping-sliding and tectonization of the K-P boundary units. (C) Geological map of the El Peñon area.

deposit (named K-P sand-siltstone complex), which overlies uniform Maastrichtian Méndez marls of uppermost Maastrichtian age (biozone CF1, *Plummerita hantkeninoides*). Recent studies by Schulte (2003) revealed that this K-P sand-siltstone complex was rapidly emplaced by multiple debris flows. Three distinct units characterize the K-P sand-siltstone complex in NE México (Fig. 4):

1. A centimeter to meter-thick basal spherule deposit (unit 1), with intercalated thin layers of quartzose calcareous or foraminiferal sandstone.

2. The overlying unit 2 is 0.2–10 m thick, structureless-massive, or slightly laminated sandstone consisting of quartz, limestone clasts, feldspar, and occasionally foraminifera tests. Unit 2 frequently contains large marl clasts, plant debris, and rare spherules at its base.

3. The topmost unit 3 comprises 0.1–1.5-m–thick alternating silt-sandstones with a similar composition as in unit 2 but with abundant benthic and mainly planktic foraminiferal tests. The top of the unit is orange-stained, current-rippled, and bioturbated by *Chondrites*, *Planolites*, and *Ophiomorpha*.

The sheet-like channel-levee part of the sand-siltstone complex is commonly dislocated relative to the central channel ("thalweg"), with a vertical offset of several meters (see Fig. 3B). Also, individual segments of the central channel may be displaced vertically against each other by up to 20 m, and in these locations, slumped, "brecciated," or "lens-like" spherule deposits are locally mixed with Méndez marls (Figs. 3B and 4; Soria et al., 2001; Schulte et al., 2003). Detailed sedimentological, petrological, mineralogical, and geochemical analysis, however,

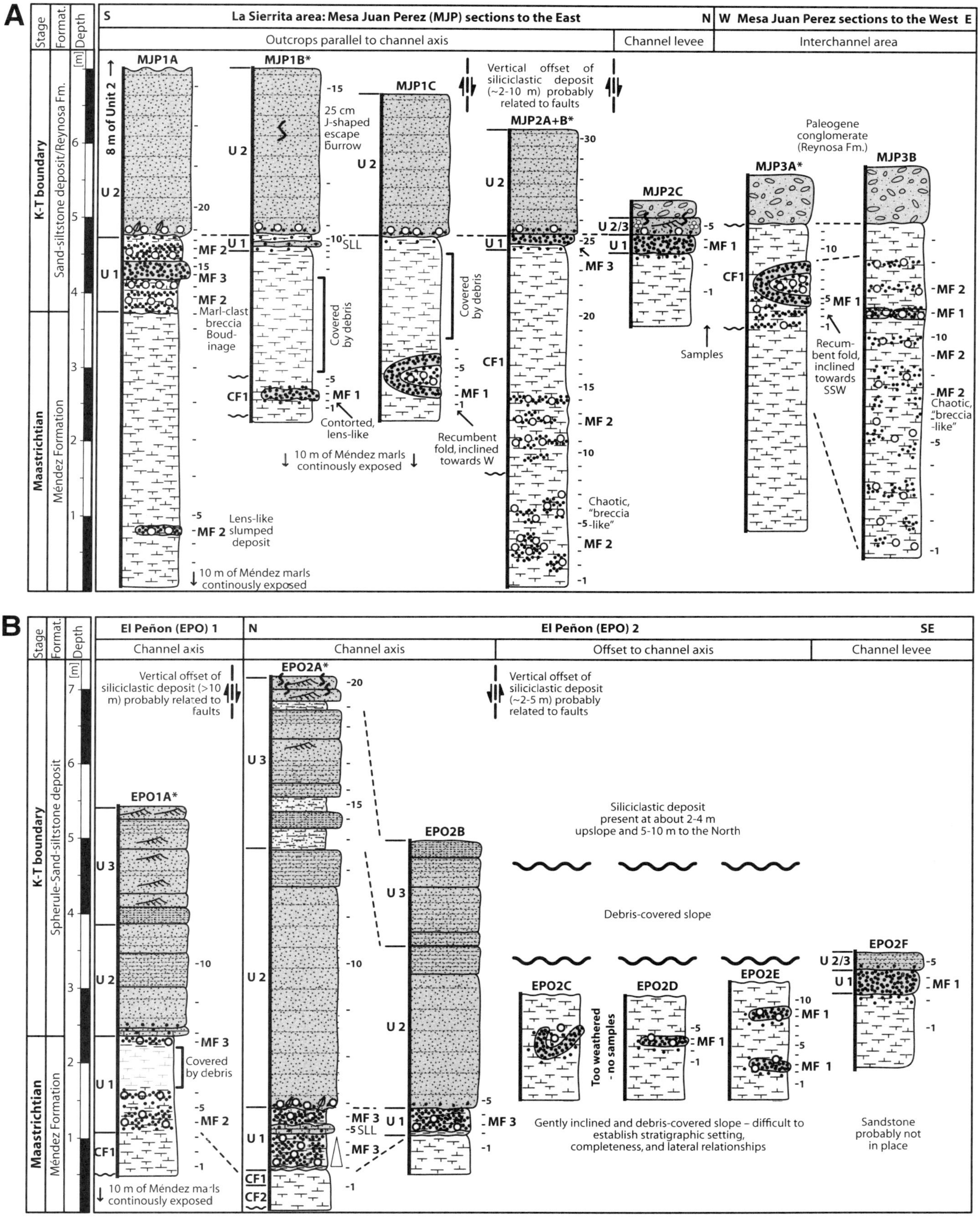

Figure 4 (*on this and following page*). Lithology and stratigraphy of sections with spherule deposits (A) in the Mesa Juan Perez, La Sierrita area, (B) in the El Peñon area, and (C) at El Mimbral (with a key to the lithologic symbols; modified after Schulte 2003). Biostratigraphic assignation of sections marked with an asterisk adopted from López-Oliva and Keller (1996), Keller et al. (1997), and Schulte et al. (2003). Note the lithologic variability and lateral discontinuity of slumped and slided spherule deposits mixed with Méndez marls, probably due to slope instability generated by the Chicxulub event.

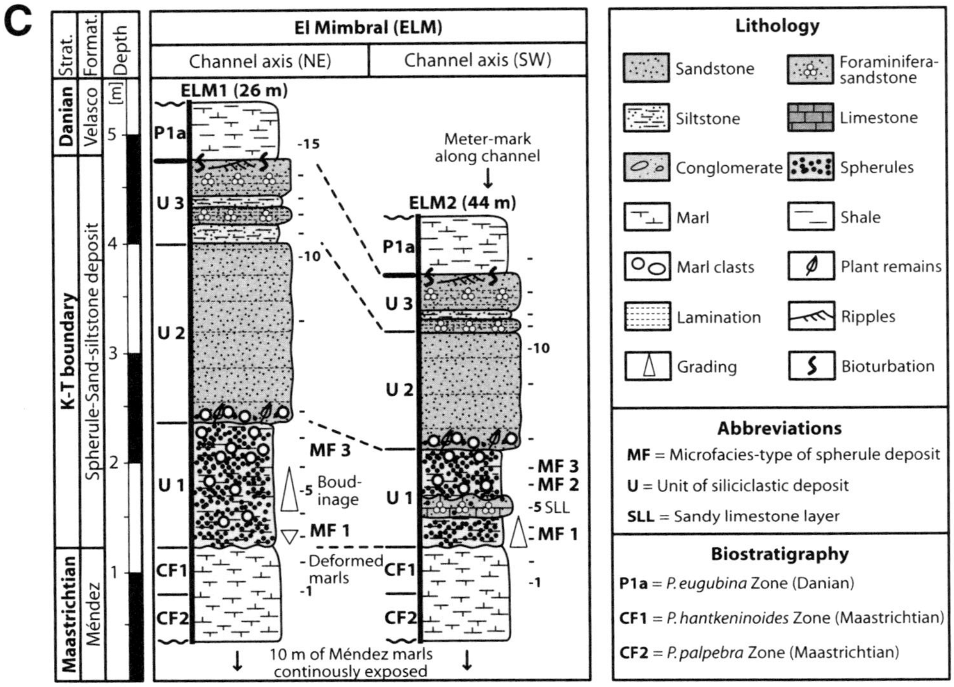

Figure 4 (*continued*).

revealed that the disturbed spherule deposits all originated from the Chicxulub event at the K-P boundary via immediate and/or later reworking (Soria et al., 2001; Schulte, 2003; and below).

In most sections, there is an erosional surface at the top of the K-P sand-siltstone complex with no sediments preserved above, and, only in few sections, marls of the Paleocene Velasco Formation (early Danian biozone P1a, *Parvularugoglobigerina eugubina*) disconformably overlie unit 3 (Fig. 4). Locally, the base of the Velasco shale is made of a 2–10-cm-thick orange-stained and indurated limestone devoid of microfossils (earliest Danian biozone P0). Elevated platinum group element (PGE) concentrations are present in this layer (0.3–0.9 ng/g iridium), as well as in the topmost part of unit 3 (0.3–0.4 ng/g Ir; cf., Smit et al., 1996).

Besides the locations investigated for this study, comparable K-P sequences have been reported from other sites in NE México, including Rancho Nuevo, Los Ramones, Loma Cerca, El Mulato, and La Lajilla, shown in Figure 3A, though with variable thickness and petrological composition. However, they usually show less slumping and consequently have only one spherule deposit at the base of the sand-siltstone complex (Smit et al., 1992b, 1996; Stinnesbeck et al., 1993, 1996; Keller et al., 1994, 1997; Soria et al., 2001).

STATE OF KNOWLEDGE—CHICXULUB MELT ROCKS AND EJECTA AT THE K-P BOUNDARY

In proximal settings (e.g., at the Gulf of México and adjacent areas), the centimeter to meter thick Chicxulub ejecta layer consists mostly of spherules, accompanied by shocked and unshocked minerals, and fine-grained carbonate and clayey phases, overlain by a layer containing the characteristic Ir-anomaly (Fig. 2 and Izett, 1991; Bohor and Glass, 1995; Smit et al., 1996; Ortega-Huertas et al., 2002; Schulte et al., 2003). In more distal settings to Chicxulub, the millimeter to centimeter thick K-P clay layer is also accompanied by a strong iridium anomaly as well as by peak values of the magnetic susceptibility, though even there, the ejecta and the Ir-anomaly are occasionally separated (Fig. 2; Worm and Banerjee, 1987; Montanari, 1991; Martínez-Ruiz et al., 1997; Rocchia and Robin, 1998; Wdowiak et al., 2001; Bauluz et al., 2004).

Three types of ejecta phases are present in K-P boundary layers: (1) "microtektite"-like mm-cm–sized, vesiculated, glassy, clayey and carbonate spherules, indicating silicic and carbonate-bearing progenitor phases; (2) ~0.1-mm–sized microcrystalline spherules ("microkrystites") of pyroxene, K-feldspar, or iron-oxide crystals interpreted as alteration products of unknown (mafic?) precursor lithologies (e.g., Montanari, 1991; Smit et al., 1992a; Smit, 1999; Bohor and Glass, 1995; Martínez-Ruiz et al., 1997; Wdowiak et al., 2001); and (3) a fine-grained ferruginous and iridium-enriched clayey phase or carbonate phase that also bears abundant shocked minerals (see Fig. 2).

Smit et al. (1992a) and Pollastro and Bohor (1993) proposed that microtektite spherules formed exclusively from molten target rocks and were dispersed ballistically by the impact ejecta curtain in regions close to the Chicxulub impact site (e.g., Montanari and Koeberl, 2000). In contrast, microkrystites may have formed by condensation from the vaporized target rock and meteoritic materials in the expanding vapor plume cloud ("fireball": Pollastro and Bohor, 1993). The proposed elevation of the microkrystites well above the atmosphere presumably led to their almost global distribution concomitant with the iridium-rich dust (e.g., Alvarez et al., 1995; Croskell et al., 2002; Kring and Durda, 2002). These different distribution paths may also account for the observation that the iridium anomaly frequently is found atop of the ejecta spherules (e.g., in North America; see Fig. 2) or is even separated from the spherule deposits by slumped and channelized shelf sediments (e.g., in the Gulf of México area; this study).

The composition of glass and clayey spherules suggests that they derived from silicate rocks of gneiss-granite–like composition with an admixture of variable amounts of calcium and sulfur-rich lithologies (e.g., Izett, 1991; Koeberl and Sigurdsson, 1992; Bohor and Glass, 1995; Stüben et al., 2002). This composition is congruent with the basement (gneiss, schist, granites) and carbonate-rich (limestone, anhydrite) sedimentary cover of the Yucatán peninsula (e.g., Schuraytz et al., 1994), as shown by isotope and geochemical studies of Chicxulub melt rocks and ejecta (e.g., Blum et al., 1993; Kettrup and Deutsch, 2003). However, studies of Sm-Nd isotope ratios in Chicxulub melt rocks also indicate the admixture of mafic precursor phases to account for the observed mixing trend (Kettrup et al., 2000). Findings of amphibolite clasts in suevite drilled at the Chicxulub structure further corroborate these results (Kring and Boynton, 1992; Kring et al., 2003; Kettrup and Deutsch, 2003; Pilkington et al., 2004).

STATE OF KNOWLEDGE—MAGNETIC PROPERTIES OF THE K-P BOUNDARY

In settings distal to the Chicxulub impact, the mm-thick K-P boundary layer is characterized by a distinct anomaly of the magnetic susceptibility that is associated with peak values of iridium and siderophile elements, with Fe-rich and Fe-Ti–rich nanophases, as well as with microkrystites containing magnetite and Ni-rich spinels in a goethite-hematite host phase (Fig. 2; Worm and Banerjee, 1987; Morden, 1993; Griscom et al., 1999, 2003; Wdowiak et al., 2001; Bralower et al., 2002). The magnetic susceptibility of this basal layer is approximately proportional to the contents of the microkrystites and may be additionally affected by goethite and hematite in the boundary clay, which both increase its coercivity (Worm and Banerjee, 1987). The microkrystites have been interpreted to form under low oxygen fugacity, probably in the ejecta plume rising well above the atmosphere, whereas goethite and hematite are characteristic of a more oxidizing environment (Wdowiak et al., 2001; Bhandari et al., 2002; Verma et al., 2002). The concentration of hematite and goethite nanophases is also approximately correlated with the iridium content of the boundary clay, indicating that their formation is not exclusively linked to alteration, but rather is related to the impact process (Fig. 2; Verma et al., 2002).

The Chicxulub proximal ejecta layer in southern México and throughout Belize consists of various spherules and glass shards, each showing characteristic magnetic properties (Kletetschka et al., 2001) from highly paramagnetic spherules with no ferromagnetic components to slightly paramagnetic spherules with soft ferromagnetic components. Mössbauer studies on black Beloc glass spherules from Haiti by Thorpe et al. (1994) revealed that magnetic properties and the Fe^{3+}/Fe^{2+} ratio for these glass spherules and common tektites are essentially the same. According to Werner and Borradaile (1998), tektites are dominated by paramagnetism of the silicate glass. Ferromagnetic contributions are negligible because formation of oxides was suppressed during rapid quenching, similar to what is observed in submarine hyaloclastites (e.g., Kontny et al., 2003).

Magnetic studies on basement breccias from Chicxulub drill cores indicate the occurrence of Fe-rich titanomagnetite of single or pseudo-single domain states as the dominant magnetic carrier in addition to hemo-ilmenite and secondary magnetite (Urrutia-Fucugauchi et al., 1994; Steiner and Walker, 1996; Pilkington et al., 2004). The hemo-ilmenite may have formed from the oxidation of magnetite, probably related to the hydrothermal processes following the impact.

In conclusion, these data suggest different sources and formation mechanisms of magnetic minerals in suevite and ejecta from the Chicxulub impact: magnetic components (1) from crustal rocks, (2) of extraterrestrial origin, and (3) as secondary alteration products.

METHODS

Optical microscopy of resin-stained thin sections was performed for ~80 ejecta samples to assess the textural and mineralogical properties.

Wavelength-dispersive spectroscopy (WDS) and energy-dispersive spectroscopy (EDS) electron microprobe analyses (EPMA), as well as backscattered electron (BSE) images were performed with a CAMECA SX50 microprobe on polished, carbon-coated thin sections at the Laboratory for Electron Microscopy, University of Karlsruhe. The SX50 microprobe is equipped with four crystal spectrometers: Quantitative (WDS) microprobe analyses were carried out using the crystals TAP (Si, Al), PET (Ti, Ca, K), RAP (Mg, Na) and LiF (Fe, Mn). All quantitative major element analyses were calibrated with the following standards: Fe—Fe_2O_3; Si—wollastonite; Mg—MgO; K—orthoclase; Ca, Al—anorthite; Ti—$MnTiO_3$; Na—albite; Ni—NiO. Accelerating voltage was set to 15–20 kV with a primary beam of 15 nA, and counting times of 20–40 s were used per element. Detection limits are in the range between 0.5 and 1 wt%. Oxide percentages were calculated using the ZAF correction program and natural silicate and oxide standards.

Polished thin sections were examined on a CITL CCL 8200mk3 cold cathodoluminoscope, which operates at a voltage of 20 kV and with a beam current of 380 μA, at the Geological Institute of the University of Heidelberg.

The volume magnetic susceptibility (κ) from ejecta, marl, and sandstone samples was determined in low fields (300 A/m, 1200 Hz) with a Kappabridge KLY-3 (AGICO). Three aliquots with 7 g bulk powder of every sample were measured and a mean value was calculated.

The temperature-dependent magnetic susceptibility (κ-T curve) was measured for the temperature range −192–700 °C in an argon atmosphere (110 ml/min) to avoid oxidation during heating using the KLY-2 Kappabridge (AGICO), fitted with the CS-2/CS-L furnace apparatus (Hrouda, 1994). The sensitivity of this Kappabridge is given with 4×10^{-8} (in SI units) using a standard pickup unit for 10 cm^3 nominal volume (KLY-2.1). During the cooling run (−192–0 °C) and the subsequent heating/cooling cycles from room temperature to 700 °C (heating rate 10 °C/min), temperatures and susceptibilities were recorded. The raw data were corrected for the empty furnace and normalized to the susceptibility magnitude at 0 °C (low temperature) and room temperature (high temperature).

The remanent magnetization was measured on samples, which were prepared by coring cylinders with a diameter of 2.5 cm and length of 2.1 cm from well-indurated specimens. The measurements were made with a JR5A spinner magnetometer (AGICO). The stability of magnetic phases has been analyzed by detailed stepwise alternating field (AF) demagnetization at room temperature up to fields of 160 mT at the rock magnetic laboratory at Heidelberg University and by acquisition of isothermal remanent magnetization (IRM) performed in fields up to 1000 mT at the ETH Zürich. Hysteresis measurements were obtained with a Princeton Instruments MicroMag in fields up to 1.2 T at Bremen University, and hysteresis loops were obtained by measuring the induced magnetization of the sample during the application of a large field.

RESULTS

The ejecta deposits in NE México, which are the subject of this study, are considerably more complex than previously reported (Smit et al., 1992b; Smit et al., 1996). They show a range of characteristic sedimentary microfacies textures and bear a variety of different ejecta components.

Microfacies of the Ejecta Deposits

The ejecta deposits in NE México consist of three different microfacies types (Table 1; Fig. 5), though gradual transitions between these microfacies exist locally.

Microfacies Type 1

Microfacies type 1 has a dense, chaotic, grain-supported texture, resembling a sedimentary polymict microbreccia, though occasionally constituents are loosely "floating" in gray microspar with a distinct matrix-supported texture (Figs. 5A, 5B, and 5D). Constituents include all types of ejecta particles (spherules, fragments, carbonate clasts, and accretionary lapilli-like clasts), as well as marl clasts and occasional foraminifera. Microfacies type 1 is generally devoid of terrigenous detritus. In this microfacies, many spherules and fragments appear welded together and large clusters are present with concave-convex contacts and concomitant deformation of internal structures. In addition, spherules and fragments, as well as marl clasts, appear to have been reshaped plastically along their contacts and large cm-sized filaments envelop other constituents (e.g., marl clasts and benthic foraminifera), giving an "amalgamated" appearance to this microfacies. Benthic foraminifera in the ejecta mainly comprise miliolids and rotaliids; the genera *Anomalina*, *Cibicidoides*, *Lenticulina*, *Neoflabelina*, *Tritaxia*, and *Palmula* have been identified (R. Speijer, 2000, personal commun.). Most of these benthic foraminifera genera, however, are not typical carbonate platforms inhabitants (e.g., Yucatán), but are common on Cretaceous-Paleogene siliciclastic shelves, including the northeastern Mexican shelf (Murray, 1991).

Microfacies Type 2

Microfacies type 2 is distinguished from type 1 by its micritic (marly) matrix and a mud-supported texture, revealing a variable mixture with Méndez marls. In contrast to microfacies 1, welding and amalgamation of constituents are rarely observed. The main constituents of microfacies 2 and their relative abundances are similar to microfacies type 1 (Table 1). Terrestrial detritus is present only in the fine fraction (<63 μm) of the marl matrix. In this microfacies type, mainly Late Cretaceous planktic foraminifera are present (globotruncanids, heterohelicids), whereas benthic foraminifera are rare and no identifiable individuals have been detected.

Microfacies Type 3

Microfacies type 3 shows a grain-supported to matrix-supported texture with microspar (Fig. 5C). Spherule deposits of this microfacies type have spherules, fragments, carbonate clasts, and accretionary lapilli, besides a variable amount of detritus (5–25%) such as quartz and feldspar (Table 1). Rounded marl clasts are present as minor components. Like microfacies type 2, welding and amalgamation of constituents are rarely present. Planktic or benthic foraminifera are rare in this microfacies and comprise similar genera as found in microfacies type 1.

The classification of the ejecta deposits in the individual outcrops into the microfacies types 1–3 is included in Figure 4A–4C. At the La Sierrita area, all types of spherule deposits are present. However, the occurrence of microfacies type 1 is not restricted to the slumped and slided spherule deposits within Méndez marls, but has also been observed at the base of the sand-siltstone deposits (unit 1). An analogous situation occurs in the El Peñon outcrops, though microfacies type 1 in this area is restricted to the slumped spherule deposits within Méndez marls. The meter-thick unit 1 at the base of the siliciclastic deposits in the El Mimbral section shows an excellently preserved microfacies type 1 in its lower part with an increasing sand content toward its upper part (hence classified as microfacies type 3).

Petrography of the Ejecta Components

The main components of the ejecta deposits in NE México are carbonate and silicic spherules and glass shards (Figs. 5 and 6). Additional, though minor, ejecta components are fine-grained lapilli-like particles with concentric accretionary features (Figs. 6E

TABLE 1. MODAL COMPOSITION OF SPHERULE DEPOSITS FROM THE LA SIERRITA, EL PEÑON, AND EL MIMBRAL AREA IN NE MÉXICO DETERMINED BY THIN-SECTION ANALYSIS

	Microfacies type	Fragments (vol%)	Spherules (vol%)	Carbonate clasts (vol%)	Detritus (Qtz/Fdsp) (vol%)	Marl clasts	Ø of spherules (mm)	
							mean	max
La Sierrita	MF 1	30–70	15–40	15–30	<5	rare to common	~0.8	8
	MF 2	25–35	20–40	10–20	<5	frequent*	~1	8
	MF 3	30–60	5–20	15–30	5–20	rare*	~0.8	8
El Peñon	MF 1	10–20	15–30	30–60	<5	rare to common	~0.6	15
	MF 3	15–20	5–20	30–60	510	rare*	~0.9	14
El Mimbral	MF 1	20–30	15–40	10–30	<5	rare to common	~0.8	16
	MF 2	10–15	10–30	10–30	<5	frequent*	~0.6	10
	MF 3	20–30	5–20	10–30	5–10	rare*	~0.7	9

Note: Qtz—quartz; Fdsp—fe dspar. Note the large variability of composition but the similarity in ranges of all three microfacies types (MF).
*Indicates the presence of large marl clasts (1–25 cm).

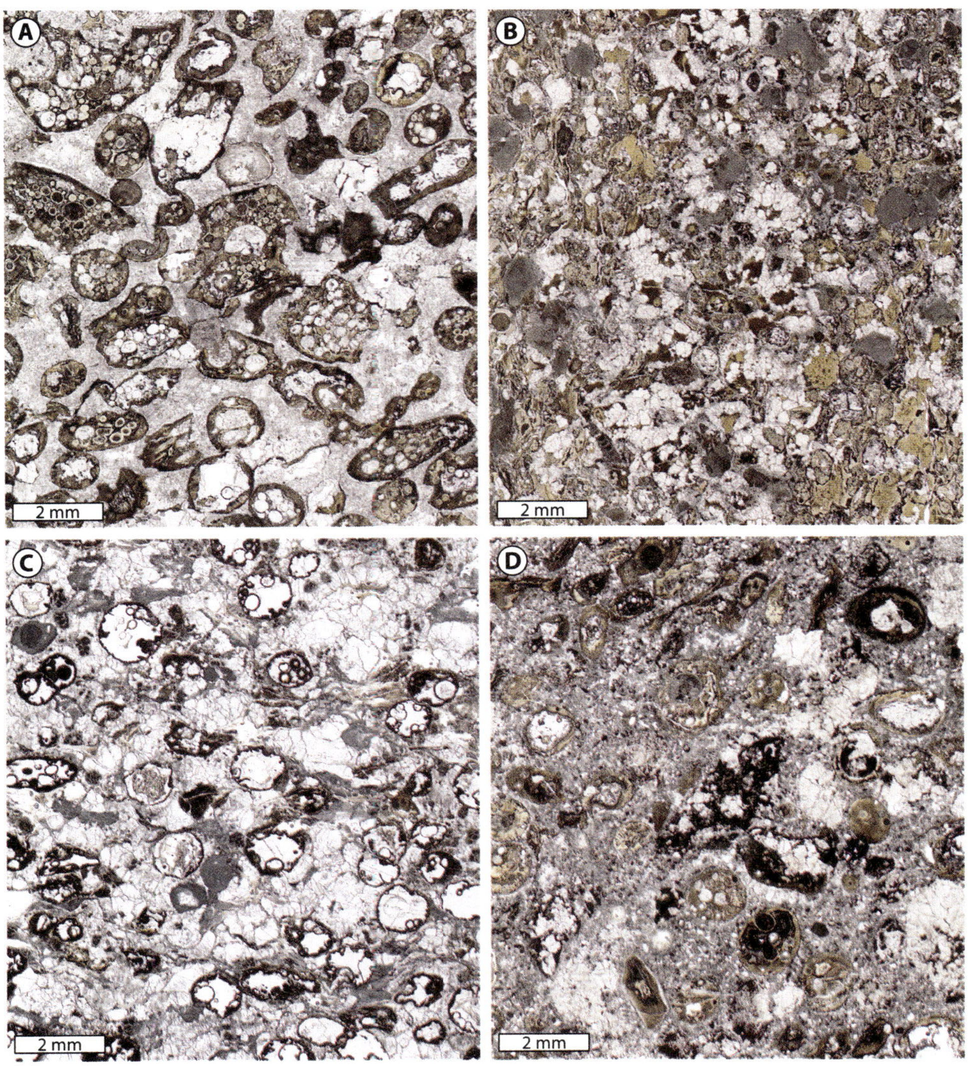

Figure 5. Thin section microphotographs of spherule deposits in NE México (plane polarized light). (A) "Grainstone"-like microfacies type 1 from La Sierrita with spherules, (ejecta-) fragments, and carbonate clasts within microspar. The matrix-supported texture suggests that some carbonates had to be part of the original deposit (sample MJP1B-3). (B) "Dense" variety of microfacies type 1 with less spherules than in (A), but with more (ejecta-) fragments and enclosed limestone and accretionary lapilli grains (sample MAS1-2). (C) Microfacies type 1 from El Mimbral with abundant spherules, limestone clasts, and blocky carbonate (sample MIM1-5). (D) Microfacies type 3 with spherules, fragments, carbonate, clasts, and minor fine-grained terrigeneous debris (e.g., quartz, feldspar) (sample MJP1A-14).

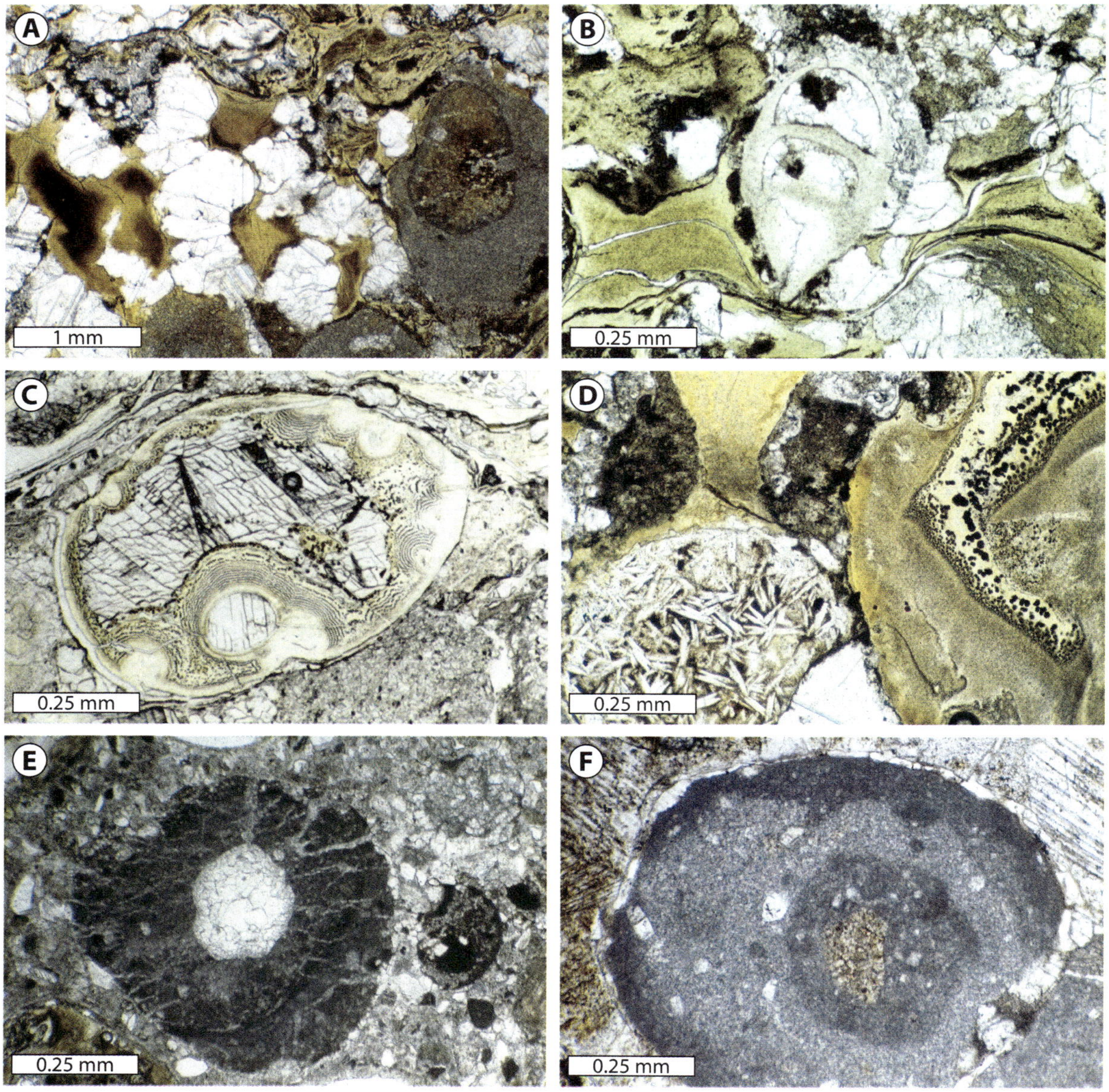

Figure 6. Thin section microphotographs of spherule deposits in NE México (plane polarized light). (A) Microbreccia-like texture of ejecta beds in the La Sierrita area with large cauliflower-shaped calcite crystals enclosing ejecta fragments to the left and an accretionary lapilli-like grain to the right (sample MAS1-2). (B) Ribbon-like chlorite ejecta fragment from La Sierrita (sample MAS1-2) that encloses a benthic foraminifera. (C) Spherule from El Mimbral (sample ELM1-3) showing distinct, garland-shaped opaque lamellae. These lamellae are made of submicrometer-sized, microcrystalline Fe-Ti–rich phases and rutile, and show a spherule-inward increase in grain size. (D) Lath-shaped altered K-feldspar crystals with characteristic slanted end-phases in a spherule from La Sierrita (sample MJP3A-4). (E) and (F) Lithoclasts from La Sierrita with concentric internal structure interpreted as accretionary lapilli-like ejecta (samples MJP1A-14 and MAS2-2).

and 6F). Along with these ejecta phases, limestone clasts, intrabasinal marl clasts, benthic and planktic foraminifera, and extrabasinal siliciclastic detritus, such as quartz and feldspar, are important constituents.

Spherules and Glass Shards

The morphologies and surface features of spherules and glass shards are highly variable in all ejecta deposits and many features are gradational, making a rigorous classification difficult (Figs. 5 and 6). Therefore, they may be considered as end-members of a "morphological mixing line" between *spherules*, which are generally round, elongate, or teardrop to dumbbell-shaped, and *fragments*, which are irregularly shaped, blocky, or ribbon-like forms. Occasionally, spherules and fragments are broken, with fractures that cut through internal structures and cavities, confirming a primary origin of these internal features. The

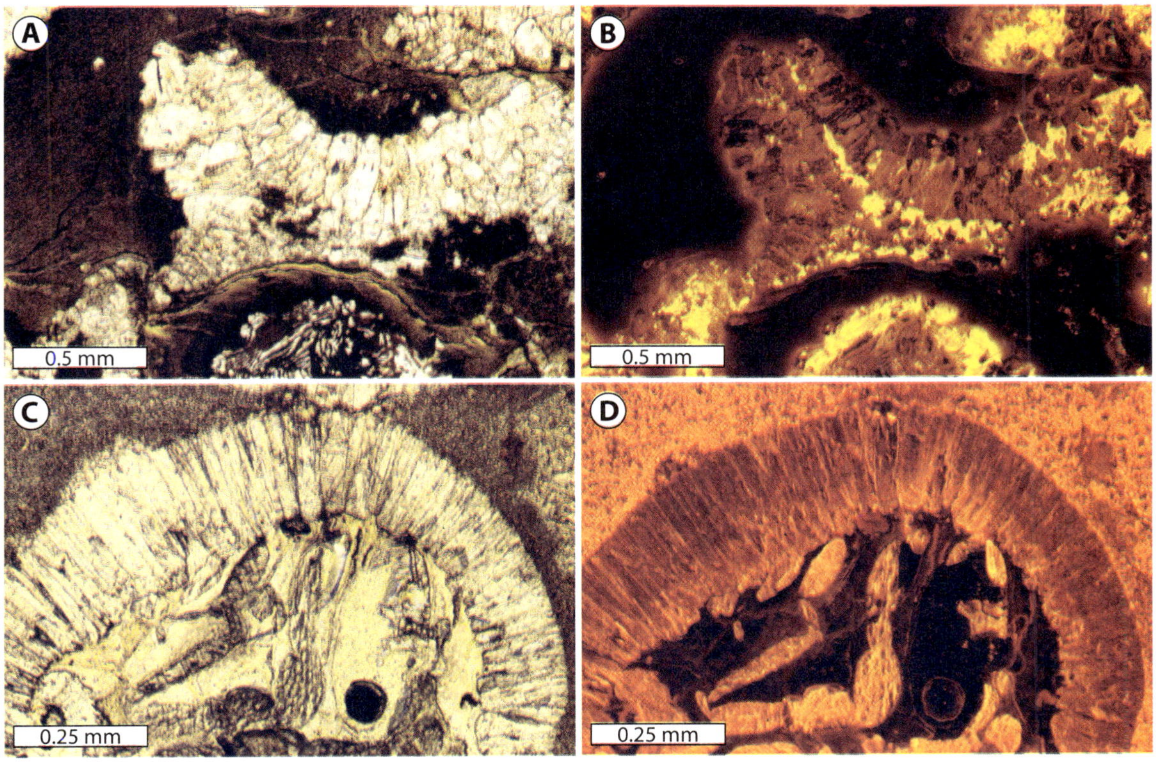

Figure 7. Representative transmitted light (left) and cathodoluminescence (right) photomicrographs of spherule deposits. (A) and (B) Radially arranged blocky calcite crystals with typical "cauliflower-texture" enclosing ejecta fragments (sample MJP3A-6). The blocky calcite shows typical dark red to brown luminescence colors; the matrix is of bright yellow luminescence. (C) and (D) Part of a carbonate-silicate spherule (sample EPO2A-3). This spherule shows a prominent radially arranged calcite overgrowth ("feathery" calcite).

relative amount of spherules to fragments varies between 1:1 and 3:1, and these particles are generally between under one millimeter to about 1 cm, up to a maximum of 2 cm in size.

The characteristic morphological features of spherules and their internal textures, such as vesicles and schlieren as outlined below, indicate an origin as droplets from highly fluid melt with subsequent cooling (Shelley, 1993; Engelhardt et al., 1995). The abundance of spherical particles suggests surface tension control on their shape, implying low viscosity, probably at temperatures well above the solidus (Freundt and Schmincke, 1998). The same applies to the angular fragments with similar flow textures, though their irregular shapes and less abundant vesicles point to a different thermal history and indicate either fragmentation in a ductile-rigid state by turbulence in the air or by thermal stress due to quenching with water, which are well compatible with an origin by impact (Glass et al., 1997).

The ejecta particles generally consist of one or more distinct compositional phases (Figs. 5 and 6), including (1) chlorite with intense green and brown-orange colors and high birefringence; (2) an translucent isotropic glass phase, partly altered to a sheet silicate with high birefringence indicative for illite or mica; and (3) a carbonate phase that may also fill cavities and that forms thick rims having outward-radiating fine or coarse crystals. The amount of these phases in each ejecta particle is different, and the particles may consist, for instance, entirely of chlorite, glass, or carbonate (homogeneous particles). Conversely, all three phases may be present within an individual particle (so-called "tektite in tektite" configuration).

Cathodoluminescence investigations show that the carbonate within ejecta particles as well as the carbonate globules have a very low, dark ("brick") red to dull brown luminescence or show even total absence of luminescence (Fig. 7). The brick red or dull brown luminescence color of the radially arranged carbonate crystals stands out clearly against the yellow to orange colored luminescence of the microspar matrix; no concentric zoning has been observed that could give proof of multiple stages of diagenetic precipitation of carbonate. The dull luminescence may be related to "luminescence quenching" due to the enhanced amount of Fe and Mg (0.5–1 wt%) in the calcite compared to the enclosing matrix (e.g., Machel, 2000).

Within the ejecta particles, the phases are arranged in the following modes: (1) schlieren with a distinct flow pattern or globular structures of round, elongate, or twisted shape (Fig. 8A), (2) "foam" or "pumice-like" with stretched-out and attenuated films between carbonate and silicate or silicate and silicate (Fig. 8D), (3) circular or brain-like structures with alternating

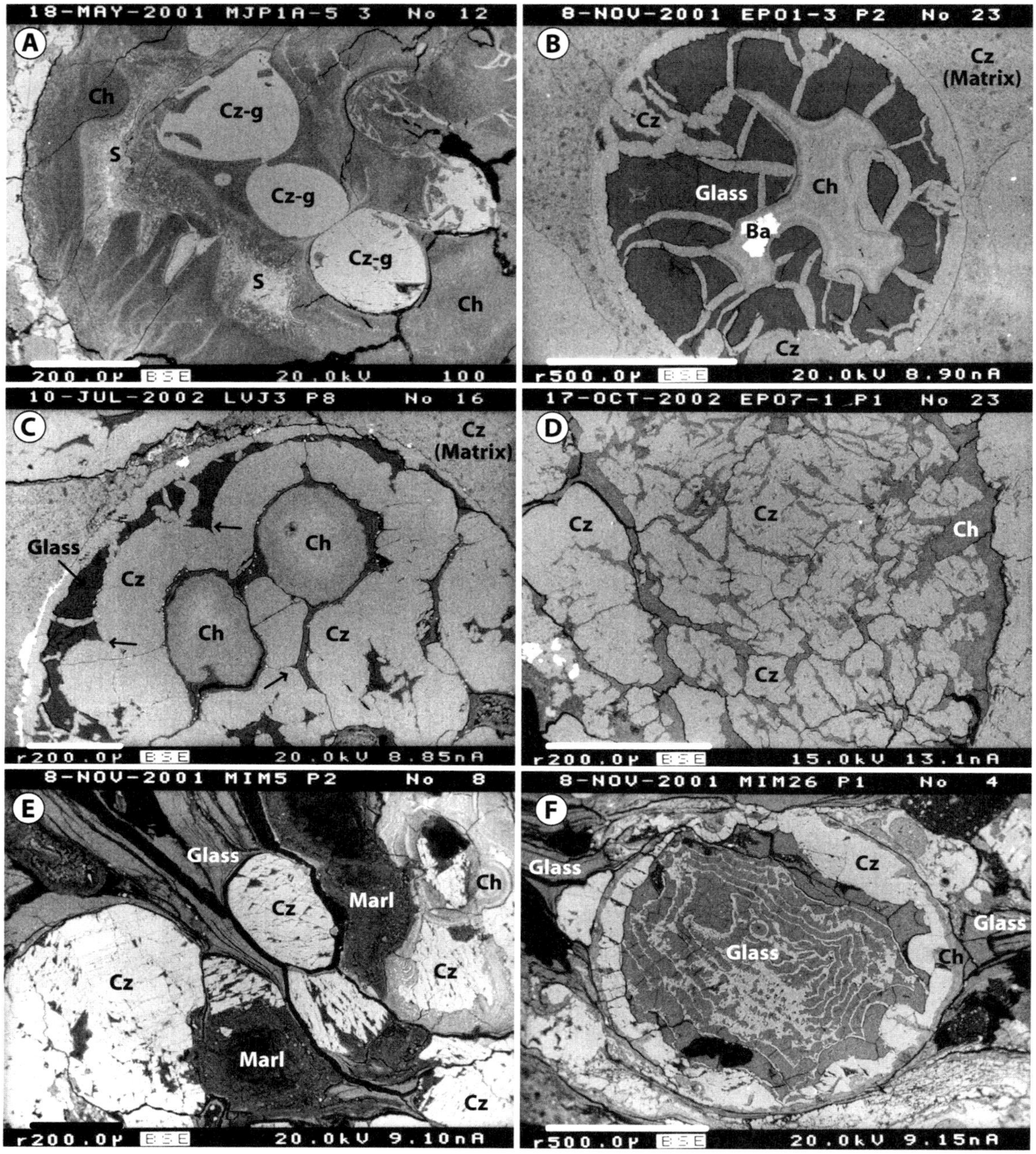

Figure 8. Backscattered electron images of spherule deposits from NE México. S—Fe, K, and Ti-rich schlieren; Cz—calcite; Ch—chlorite; Cz-g—calcite globules; Ba—baryte. (A) Chlorite-spherule from La Sierrita with internal Ti and K-rich schlieren and collapsed cavities (vesicles?). (B) Si-Al-K–rich glass spherule from El Peñon with rim of radially arranged calcite and irregular, internal calcite veins. (C) Si-Al-K–rich glass-carbonate spherule from El Peñon with two rounded inclusions of chlorite (bright areas) and characteristic "marble-like" texture of the carbonate and the curved menisci (arrows) that delimit the glass-carbonate boundary. (D) Irregular "foamy" chlorite-carbonate spherule from El Peñon with complex intermingled carbonate and silicic phases. (E) Large ribbon-like Si-Al-K–rich glass fragments from El Mimbral that enclose carbonate globules and limestone fragments. (F) Spherule from El Mimbral with typical "tektite-in-tektite" pattern consisting of three different phases: Si-Al-K–rich glass in the center with abundant garland-shaped cracks filled with calcite; this glass is surrounded by blocky calcite, which in turn is enclosed by chlorite.

carbonate-silicic phases (Fig. 8F), and (4) distinct carbonate veins (Figs. 8B and 8F). Schlieren, which are characteristic of impact glasses, show a flow pattern (Fig. 8A) and document the violent mixing of melt from different origins and rapid quenching (Engelhardt et al., 1995; See et al., 1998). Such fast melting and cooling prevents homogenization and crystal growth (Glass, 1990; Izett, 1991). The broken and irregular outlines of silicic cores within carbonate ejecta clasts suggest that cooling of the silicates was finished before they were enveloped by carbonate (Fig. 8F).

Additional internal structures in ejecta particles include vesicles and microliths as well as inclusions of opaque minerals, micrite, marl, and foraminifera. Vesicles reach up to 75% of the spherule volume and have smooth internal surfaces, exhibiting peeled "onion ring–like" opaque coatings that have tiny hematite or rutile crystals (Figs. 5 and 6C). They are filled with calcite, micrite, and zeolites. Most calcite infillings are coarse sparite, and no concentrically zoned drusy spar or geopetal mineralization was observed. Larger vesicles are embedded within smaller ones, resembling composite spherules.

The microliths in ejecta particles are pseudomorphically replaced by calcite or chlorite (Fig. 6D). We distinguish between (1) thin skeletal and branching crystals that are analogous to pyroxene and olivine textures observed in impact glasses of mafic composition (e.g., Jones et al., 2000; Hörz and See, 2000); (2) characteristic crisscross patterns with twinned individuals, similar to quenched K-feldspar (Shelley, 1993); and (3) dark, radial, hair-like crystals, analogous to pyroxene whiskers observed in impact glass from the Ries impact structure (Engelhardt et al., 1995).

Opaque minerals in Fe- and Mg-rich chlorite spherules and fragments are goethite, hematite, rutile, and Fe-Ti oxides (Fig. 9). They occur as tiny µm-sized grains that constitute either single, isolated crystals, occasionally with dendritic or skeletal shape, or they are arranged as garland-shaped lamellae. Goethite may also occur as agglomerated rounded masses, preferentially inside cavities. These opaque phases are present in the carbonate phase in spherules, as well as in carbonate globules (Figs. 9B and 9C). In contrast, the sparitic or micritic matrix to the ejecta particles contains significantly less abundant opaque phases. The goethite and hematite crystals are ~5–20 µm in size and frequently zoned. Goethite and hematite are either authigenic crystals or have replaced magnetite or pyrite during alteration, which both would explain the rather unusual cubic and hexagonal crystal shapes (Figs. 9A and 9B).

In chlorite spherules, the prominent garland-shaped Ti- and Fe-rich lamellae often have a distinct onion ring–like appearance (Figs. 6C, 9D, and 9E). The grain sizes in these lamellae range from submicrometer in the interior up to 5–10 µm in the outer part of the chlorite spherules. The presence of lamellae in tiny broken spherule fragments "swimming" in the carbonate blebs inside the spherules indicates a primary origin of these features and consequently suggests prolonged cooling inside the grain, probably related to the enlarged thermal diffusivity of the carbonate melt when compared to siliceous melt (e.g., Treiman, 1989).

Quite outstanding petrological textures of the Fe-rich chlorite spherules are 10–40-µm-sized cruciform, lath-shaped, and dendritic-skeletal crystals of Fe-Ti–rich silicates (Fig. 9F) with a composition close to chlorite (see below). Their crystal shape and arrangement are similar to spinifex textures in the rims of oceanic pillow basalts and, by analogy, indicate rapid quenching from a melt (e.g., Kontny et al., 2003).

Lapilli-like Accretionary Spherules

The concentric lapilli-like spherules are similar to accretionary lapilli from pyroclastic deposits. They have rounded shapes as well as more irregular outlines with grain sizes from 0.5 mm to >2 mm (Figs. 6E and 6F). The lapilli are made of fine-grained carbonate and clayey particles and show a pronounced concentric structure with several continuous layers. Subsequent layers may show slight variations in grain size and color and document multiple episodes of accretion. Some lapilli grains have distinct cores of either irregular limestone fragments or calcite aggregates, or even spherule fragments.

Accretionary lapilli have been reported from impact ejecta deposits of the Ries impact structure, Germany (Graup, 1981); the Alamo impact structure, Texas (Warme et al., 2002); several Russian impact craters (Masaitis, 1994); and also from Chicxulub ejecta in the Gulf of México area (e.g., Pope et al., 1999; Salge et al., 2000).

Mineral Chemistry of Ejecta Components

Thin-sections of spherule deposits were examined using EPMA (BSE images and EDS) to get an overview and to determine representative ejecta particles for quantitative analysis (Figs. 8 and 9). In a second step, some 15–20 individual points were analyzed by WDS for each phase on selected ejecta particles to assure good average values and to get a measure for the compositional variability in a thin section (Tables 2–4). In a third step, linescans with a point-to-point distance of 2 µm and 50–400 points each were made to assess possible zonation or inhomogeneities (Fig. 10). Based on methods outlined in Newman and Brown (1987), molar abundances of cations were calculated (Tables 2–4; Fig. 11). EPMA linescans reveal that individual melt phases in ejecta particles are mostly well confined and generally occur as distinct irregular inclusions, globules, or schlieren in the respective host phase (Fig. 10). Boundaries between phases are limited to a few µm; gradational transitions are absent.

Silicic Phases

The silicic phases within ejecta particles show a wide range for the SiO_2 (10–60 wt%), MgO (1–18 wt%), Fe_2O_3 (1–45 wt%), and K_2O (0.1–9 wt%) contents. They are chlorites, chlorite mixed-layers, potassium-rich glass, and mica.

Chlorite and chlorite mixed-layers. Spherules and fragments now consisting of chlorite contain 24–42 wt% SiO_2, 16–24 wt% Al_2O_3, 16–32 wt% FeO, and 8–14 wt% MgO. The alkali content is generally below 1 wt%; some ejecta clasts have CaO contents

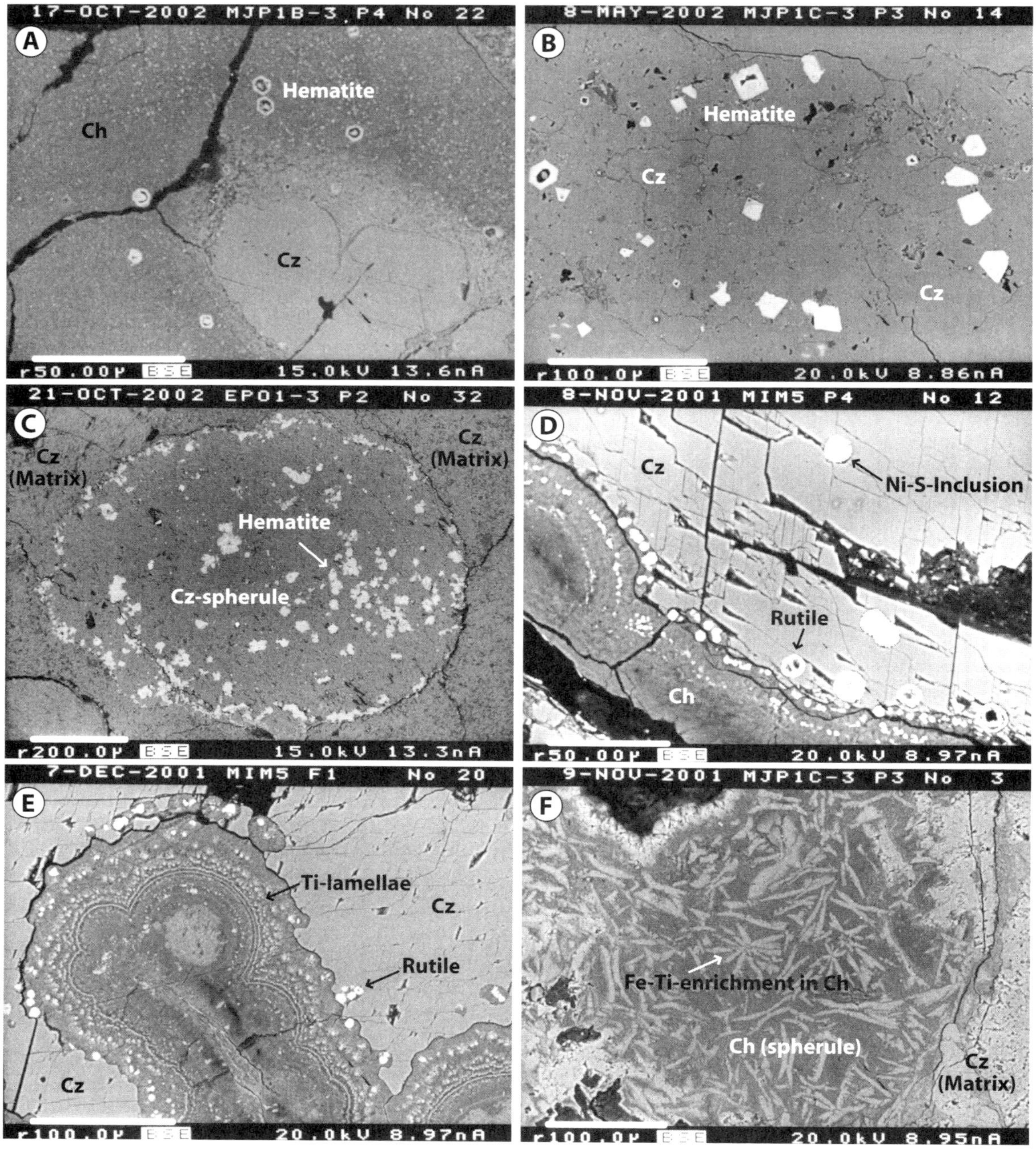

Figure 9. Backscattered electron images showing textures of opaque minerals from spherule deposits in NE México. Cz—calcite; Ch—chlorite. (A) and (B) 5–15 µm-sized hematite crystals in ejecta particles (A in chlorite; B in calcite). Some crystals have a zoned composition with a distinct Si (5–10 wt%) and Ni (~0.4 wt%) enrichment in the core. (C) Carbonate globule with Fe-oxides (goethite and hematite). (D) Ti-rich (rutile) and Ni-S–rich inclusions in a spherule from El Mimbral. (E) Ti and Fe-rich lamellae in a chlorite spherules with rhythmic garland-shaped textures. Crystal sizes of Fe-Ti–rich phases, as well as the Ti content, increase toward the carbonate-rich interior of the ejecta particles with rutile as an end member; the brightest points have the highest Ti-contents. (F) Skeletal growth of Fe-Ti–rich chlorite (up to 50 µm in size) within a chlorite spherule, resembling a spinifex texture. Note the hourglass-shaped and cruciform Fe-Ti–rich phases.

TABLE 2. CHLORITE COMPOSITION* AND FORMULAE CALCULATED STOICHIOMETRICALLY FROM REPRESENTATIVE ELECTRON MICROPROBE ANALYSES OF CHICXULUB EJECTA PARTICLES FROM SPHERULE DEPOSITS IN NE MÉXICO

Locality	La Sierrita					El Peñon					El Mimbral			
Number	(1)	(2)	(3)	(4)	(5)	(6)	(7)	(8)	(9)	(10)	(11)	(12)	(13)	(14)
SiO_2	26.39	26.36	27.09	25.84	38.94	29.49	24.99	30.56	28.45	26.95	32.38	34.91	28.04	41.44
TiO_2	0.04	0.02	0.10	0.07	0.05	0.01	0.03	0.01	0.03	0.03	0.13	0.07	0.05	0.18
Al_2O_3	21.51	22.29	19.26	23.17	21.35	21.81	20.03	21.75	23.45	22.33	17.67	19.45	17.18	18.28
FeO	27.59	29.21	33.83	28.27	18.85	23.47	32.29	23.85	24.67	25.75	22.12	17.69	30.07	15.48
MgO	12.35	10.43	5.58	11.17	7.79	11.01	10.07	11.44	10.93	12.35	14.28	14.60	12.11	13.04
CaO	0.22	0.10	2.32	0.16	2.99	0.42	0.54	0.35	0.39	0.18	0.86	1.06	0.97	1.23
Na_2O	0.01	0.01	0.09	0.01	0.14	0.02	0.03	0.09	0.04	0.10	0.23	0.11	0.08	0.17
K_2O	0.05	0.17	0.35	0.25	0.76	0.21	0.12	0.22	0.24	0.08	0.27	0.22	0.20	1.28
Total	88.16	88.59	88.63	88.94	90.86	86.56	88.19	88.30	88.34	87.86	88.11	88.24	88.90	91.18
Numbers of cations per half structural unit cell normalized to $O_{10}(OH)_8$														
Si	2.79	2.79	2.99	2.71	3.74	3.08	2.58	3.12	2.93	2.82	3.32	3.45	3.01	3.90
Al(IV)	1.21	1.21	1.01	1.29	0.26	0.92	1.42	0.88	1.07	1.18	0.68	0.55	0.99	0.10
Σ (Tetra)	4.00	4.00	4.00	4.00	4.00	4.00	4.00	4.00	4.00	4.00	4.00	4.00	4.00	4.00
Al(VI)	1.47	1.57	1.50	1.58	2.16	1.76	1.54	1.74	1.78	1.58	1.45	1.72	1.19	1.93
Fe	2.44	2.58	3.13	2.48	1.52	2.05	2.96	2.04	2.13	2.26	1.90	1.46	2.70	1.22
Mg	1.94	1.64	0.92	1.75	1.12	1.71	1.64	1.74	1.68	1.93	2.18	2.15	1.94	1.83
Ti	0.01	0.01	0.02	0.01	0.01	0.01	0.01	0.01	0.01	0.01	0.02	0.01	0.01	0.03
Σ (Octa)	5.85	5.81	5.56	5.82	4.80	5.54	6.15	5.53	5.60	5.79	5.56	5.36	5.85	5.02
Ca	0.01	0.01	0.14	0.01	0.15	0.02	0.03	0.02	0.02	0.01	0.05	0.06	0.06	0.06
Na	0.01	0.01	0.01	0.01	0.01	0.01	0.01	0.01	0.01	0.01	0.02	0.01	0.01	0.02
K	0.01	0.02	0.04	0.03	0.08	0.02	0.01	0.02	0.03	0.01	0.02	0.01	0.01	0.02
Σ (Inter)	0.02	0.02	0.19	0.04	0.25	0.05	0.05	0.05	0.05	0.03	0.09	0.08	0.07	0.09
Fe/(Fe+Mg)	0.56	0.61	0.77	0.59	0.58	0.54	0.64	0.54	0.56	0.54	0.47	0.40	0.58	0.40
T (°C)	328	328	262	353	–	234	394	221	282	316	158	114	256	–

Note: Analyses (5) and (14) correspond to di,trioctahedral chlorites, whereas all other analysis correspond to trioctahedral chlorites. T (°C) gives the formation temperature according to the method of Cathelineau (1988) by using the empirical equation: T (°C) = –61.9229 + (321.9772 × Al(IV)).
Σ (Tetra)—sum of tetrahedral sites; Σ (Octa)—sum of octahedral sites; Σ (Inter)—sum of interlayered sites. Total iron is given as FeO.
*Composition noted as wt%.

of up to 4 wt%, correlated with MgO contents of ~5–7 wt%. Calculated formulas normalized to $O_{10}(OH)_8$ (Table 2) yield reasonable cation sums allowing identification of this phase as chlorite and chlorite-smectite, in keeping with results of bulk rock X-ray diffraction (XRD) by Schulte et al. (2003). The oxide totals are ~86–90 wt%, which is typical for chlorites (Newman and Brown, 1987). Based on the distribution of Mg^{2+} and Fe^{2+} in the octahedral sites and the tetrahedral occupancy by Al, the following chlorite varieties are distinguished (cf. Bailey, 1980; Fig. 11 herein):

1. Fe/(Fe+Mg) = 0.3–0.49, corresponding to Mg-rich chlorite (Fe-chlinochlore). This chlorite shows very low Al on the tetrahedral site ($Al^{IV} \leq 0.8$), and has been detected in the ejecta particles from El Mimbral and the La Sierrita area.

2. Fe/(Fe+Mg) = 0.5–0.59, corresponding to intermediate chlorite (Mg-chamosite). It has high Al occupancy on the tetrahedral site (Al^{IV} = 0.8–1.6), and occurs in ejecta deposits from all studied regions.

3. Fe/(Fe+Mg) = 0.6–0.8, corresponding to Fe-rich chlorite (Fe-rich Mg-chamosite). It shows a wide range of Al occupancies on the tetrahedral site. This chlorite variety has only been found in the La Sierrita area.

The cation sums for the octahedral sites Σ(octa) per formula unit vary between 4.6 and 5.9 (Table 2). According to Bailey (1980) and Newman and Brown (1987), a Σ(octa) per formula unit between 5.6 and 5.9 corresponds to a trioctahedral chlorite, whereas 4.6–5.2 corresponds to di,trioctahedral chlorites. The low Σ(octa) per formula unit of the di,trioctahedral chlorites, their high silica content (Si = 3.3–3.9 per formula unit), and their high Al^{VI} to Al^{IV} ratio indicate that illite or smectite interlayers are present (Shau et al., 1991). Illite or smectite interlayers may also account for the presence of Ca, Na, and K compared to the more "pure" chlorites shown above (Newman and Brown, 1987).

Potassium-rich glass and mica. The glass and the mica fraction of spherules and fragments consist of 42–55 wt% SiO_2, 25–32 wt% Al_2O_3, and 5–9 wt% K_2O. This phase has low amounts of FeO and MgO (<2–5 wt% each), and the oxides sum up to 88–96 wt% (Table 3). These oxide total values are relatively low when compared to common impact or volcanogenic glass analysis, implying the presence of volatile phases due to hydration (Heider et al., 1993; Koeberl, 1994; Osinski and Spray, 2003). Remarkably, the stoichiometric formulas obtained from the oxide-weight percentage yield reasonable cation sums pointing to a "white" mica-like mineralogy (Table 3; Newman and Brown, 1987), though only the most hydrous Si-Al-K–rich phases (<90 wt% total) show high-order interference colors indicative of mica. Particles consisting of this phase have been found in all spherule deposits, though they are most abundant in the ejecta deposits from the El Peñon and El Mimbral K-P sections, and rarely present at the K-P boundary of La Sierrita.

TABLE 3. "WHITE MICA" FORMULAE CALCULATED STOICHIOMETRICALLY FROM REPRESENTATIVE ELECTRON MICROPROBE ANALYSES* OF CHICXULUB EJECTA PARTICLES FROM SPHERULE DEPOSITS IN NE MÉXICO

Locality	La Sierrita		El Peñon		El Mimbral	
Number	(1)	(2)	(3)	(4)	(5)	(6)
SiO_2	50.79	50.17	53.02	53.69	53.34	52.66
TiO_2	0.31	0.17	0.58	0.6	0.34	0.19
Al_2O_3	30.37	30.14	23.21	21.79	25.45	28.7
FeO	1.34	1.34	4.12	4.62	2.75	2.32
MgO	2.06	2.2	4.22	4.56	2.92	3.29
CaO	0.63	0.6	0.35	0.36	0.53	0.64
Na_2O	0.13	0.13	0.11	0.09	0.06	0.03
K_2O	7.37	7.12	8.14	8.18	6.81	7.39
Total	93.02	91.92	93.83	93.99	92.21	95.23
Numbers of cations per half structural unit cell normalized to $O_{10}(OH)_2$						
Si	3.32	3.32	3.50	3.55	3.52	3.38
Al(IV)	0.68	0.68	0.50	0.45	0.48	0.62
Σ (Tetra)	4	4.00	4	4.00	4	4.00
Al(VI)	1.66	1.67	1.30	1.24	1.51	1.55
Fe	0.07	0.07	0.23	0.26	0.15	0.12
Mg	0.20	0.22	0.42	0.45	0.29	0.31
Ti	0.03	0.02	0.06	0.06	0.03	0.02
Σ (Octa)	1.96	1.97	2.00	2.01	1.98	2.01
Ca	0.02	0.02	0.01	0.01	0.02	0.02
Na	0.01	0.01	0.01	0.01	0.01	0.01
K	0.52	0.50	0.58	0.58	0.48	0.51
Σ (Inter)	0.55	0.53	0.59	0.60	0.51	0.53
Fe/(Fe+Mg)	0.27	0.25	0.35	0.36	0.35	0.28

Note: Σ (Tetra)—sum of tetrahedral sites; Σ (Octa)—sum of octahedral sites; Σ (Inter)—sum of interlayered sites.
*Composition noted as wt%.

Carbonate Ejecta Phases

The EPMA linescans show that carbonate globules, inclusions, and feathery or cauliflower-like overgrowths often contain a distinct amount of FeO (0.4–1 wt%) and MgO (0.2–0.6 wt%) (Fig. 10A). The Fe-Mg contents vary within a spherule and usually are close to zero at the outermost rim. Notably, this Fe-Mg enrichment is not related to the composition of the enclosing silicic phase(s). The MnO content of the carbonate is small (<0.1–0.2 wt%), but may occasionally reach up to 0.4 wt%. In addition, a distinct SiO_2 content, up to a few wt%, may be present (Fig. 10).

Textural domains in ejecta particles. Several textural domains in the ejecta show a markedly different composition, compared to the mineralogically and geochemically defined phases in the spherules and fragments outlined above.

The schlieren in spherules and fragments are enriched either in Fe, Ti, or K (Fig. 8A), and the globules are slightly enriched in Fe and Mg, compared to the hosting chlorite or glass phase. The schlieren and globules document small-scale heterogeneities of the melt composition and suggest that quenching suppressed completion of the mixing process (Delano and Hanson, 1996). Schlieren are also a prominent feature of spherules from the K-P boundary in Beloc, Haiti, though these schlieren are characterized by high Ca and S contents (e.g., Izett, 1991; Bohor and Glass, 1995; Stüben et al., 2002). The globules may be interpreted as exsolution or segregation of distinct phases in the melt, according to similar observations in igneous rocks (e.g., Philpotts, 1990) and impact-generated glasses (Delano and Hanson, 1996). Such globules are unique to spherules from NE México and have not been found in other K-P localities to date.

The hematite crystals have SiO_2 contents of ~5–10 wt% and occasionally a NiO content of ~0.4 wt% (Table 4). The tiny, opaque crystals in the garland-shaped lamellae are rich in Fe and Ti and become essentially pure rutile (TiO_2) at the edge of the carbonate interior of a spherule (Table 4). This characteristic pattern of inward-directed increasing crystal sizes and Ti contents indicates progressive expelling of incompatible elements (e.g., Ti) from the melt during rapid cooling. The cruciform, lath-shaped and dendritic crystals shown in Figure 9F have compositions close to chlorite and a TiO_2 content of up to ~8 wt%.

The composition of the globular or irregular µm-sized Fe-Ni-Co–rich metallic or Ni-rich sulfides is given in Figure 12. The µm-sized metallic or sulfidic Ni-Co–rich inclusions may indicate contamination by projectile material as condensed droplets (see Koeberl, 1998; Hart et al., 2002). An analogous explanation may apply to the µm-sized P-Ni–rich inclusions found in the El Mimbral spherule bed. They may be indicative of the phosphide minerals schreibersite $(Fe,Ni)_3P$, barringerite $(Fe,Ni)_2P$, or rhabdite (Fe,Ni)P, common in extraterrestrial material (Japel et al., 2002). Ni-enriched iron oxides have also been reported by Soria et al. (2001) for spherules from NE México and are common in microkrystites from the K-P boundary clay (e.g., Montanari, 1991; Martínez-Ruiz et al., 1997; Bauluz et al., 2000).

Rock Magnetic Properties of Ejecta Deposits

Magnetic Susceptibility

The magnetic susceptibility of a rock reflects the contents of diamagnetic (e.g., calcite), paramagnetic (e.g., chlorite), and ferromagnetic (e.g., magnetite) phases per volume or mass unit (Dunlop and Özdemir, 1997). Diamagnetic minerals have low negative susceptibilities (below about -10^{-6} SI), paramagnetic

TABLE 4. REPRESENTATIVE CHEMICAL COMPOSITIONS OF HEMATITE AND Fe-Ti PHASES*
FROM CHICXULUB EJECTA PARTICLES IN NE MÉXICO

Locality	La Sierrita		El Mimbral						
Number	(1)	(2)	(3)	(4)	(5)	(6)	(7)	(8)	(9)
Type	Hematite	Hematite	Hematite	Fe-Ti	Fe-Ti	Fe-Ti	Fe-Ti	Rutile	Rutile
SiO_2	5.15	4.67	4.03	7.63	3.72	2.60	2.25	1.64	1.63
TiO_2	0.04	0.07	0.03	56.16	65.41	72.14	76.98	91.46	88.80
Al_2O_3	0.06	0.63	0.64	6.45	3.82	3.05	2.15	0.25	0.46
Cr_2O_3	0.02	0.05	0.05	<0.01	<0.01	<0.01	<0.01	0.03	0.04
Fe_2O_3	82.35	81.72	83.81	0.00	0.00	0.00	<0.01	2.02	2.68
FeO	4.52	3.72	1.80	21.94	20.42	15.77	11.23	<0.01	<0.01
NiO	0.31	<0.01	0.35	<0.01	0.02	0.12	0.04	0.01	0.04
MgO	0.35	0.21	0.13	2.43	0.31	0.30	0.18	0.00	0.06
CaO	0.47	0.92	1.75	1.15	1.66	2.53	2.80	3.78	3.75
Na_2O	0.02	0.06	0.05	0.06	0.11	0.06	0.11	0.07	0.05
K_2O	0.02	0.03	<0.01	0.09	0.04	0.03	<0.01	0.03	0.01
Totals	93.31	92.09	92.65	95.91	95.51	96.60	95.74	99.29	97.52
Si	0.141	0.131	0.112	0.284	0.141	0.096	0.082	0.022	0.022
Ti	0.001	0.001	0.001	1.571	1.863	2.000	2.115	0.931	0.922
Al	0.002	0.021	0.021	0.283	0.170	0.133	0.093	0.004	0.007
Cr	<0.001	0.001	0.001	<0.001	<0.001	<0.001	<0.001	<0.001	<0.001
Fe_3	01.73	1.719	1.755	0.000	0.000	0.000	0.000	0.021	0.028
Fe_2	0.1.4	0.087	0.042	0.682	0.647	0.486	0.343	<0.001	<0.001
Ni	0.007	<0.001	0.008	<0.001	0.001	0.004	0.001	<0.001	<0.001
Mg	0.014	0.009	0.005	0.135	0.018	0.016	0.010	<0.001	0.001
Ca	0.014	0.028	0.052	0.046	0.067	0.100	0.110	0.055	0.055
Na	0.001	0.003	0.003	0.004	0.008	0.004	0.008	0.002	0.001
K	0.001	0.001	<0.001	0.004	0.002	0.001	<0.001	0.001	<0.001
Σ (cations)	2	2	2	3.009	2.916	2.841	2.761	1.035	1.039

Note: The mineral formula is calculated assuming stoichiometry. Hematite: 2 cations and 3 oxides; Fe-Ti: 3 cations and 5 oxides (pseudobrookite); rutile: 1 cation and 2 oxides.
*in wt%.

minerals have low to intermediate susceptibilities (10^{-6}–10^{-3} SI), and ferromagnetic minerals show very high susceptibilities (>10^{-3} SI) (Hunt et al., 1995). Therefore, the volume magnetic susceptibility is a sensitive tool either to monitor various lithologies or to detect even low amounts of ferromagnetic phases.

The ejecta deposits from NE México show uniform low, paramagnetic values of the mass-related magnetic susceptibility (Fig. 13). Samples from the La Sierrita area display magnetic susceptibility values that range between 6 and 30 × 10^{-6} SI/g. The microfacies type 1 ejecta deposits cover the broadest range from ~8–30 × 10^{-6} SI/g, whereas the microfacies type 2 show the lowest values (7–14 × 10^{-6} SI/g). The microfacies type 3 show similar high susceptibility values as microfacies type 1, but a more limited range (14–24 × 10^{-6} SI/g). The samples from the El Peñon sections (microfacies type 1 and 3) show a smaller range (from 6 to 16 × 10^{-6} SI/g), whereas the samples from the El Mimbral locality (microfacies type 1 and 3) generally have heterogeneous susceptibilities (between 5 and 28 × 10^{-6} SI/g). The magnetic susceptibility values of the Méndez marls and the sand-siltstones are significantly lower and show a more limited range from 0 to 10 × 10^{-6} SI/g, compared to the susceptibilities of the ejecta deposits. Therefore, compared to the host lithologies, the elevated magnetic susceptibility of ejecta deposits in NE México is caused by the occurrence of Fe-rich spherules and fragments. The susceptibility is, in good approximation, proportional to the iron content (Fig. 13D, inset), suggesting dominance of paramagnetic contributions, and since the bulk of iron is found in chlorite spherules (25–35 wt% FeO), the susceptibility is directly proportional to the abundance of these spherules.

In addition to chlorite, opaque inclusions in the spherules and fragments, including Fe-Ti–rich phases (Fe-bearing rutile and probably members of the ferropseudobrookite-pseudobrookite solid solution) and, more importantly, Fe-oxides (hematite, goethite), further enhance the magnetic susceptibility. In combination, these paramagnetic and ferromagnetic phases account for the susceptibility values of ~15–30 × 10^{-6} SI/g of the spherule deposits. Evidence for ferrimagnetic phases other than goethite and hematite was not found by optical methods and therefore temperature-dependent magnetic susceptibility and remanence measurements were performed.

Temperature-Dependent Magnetic Susceptibility

Temperature-dependent magnetic susceptibility (κ-T) allows identification of ferromagnetic phases by monitoring their

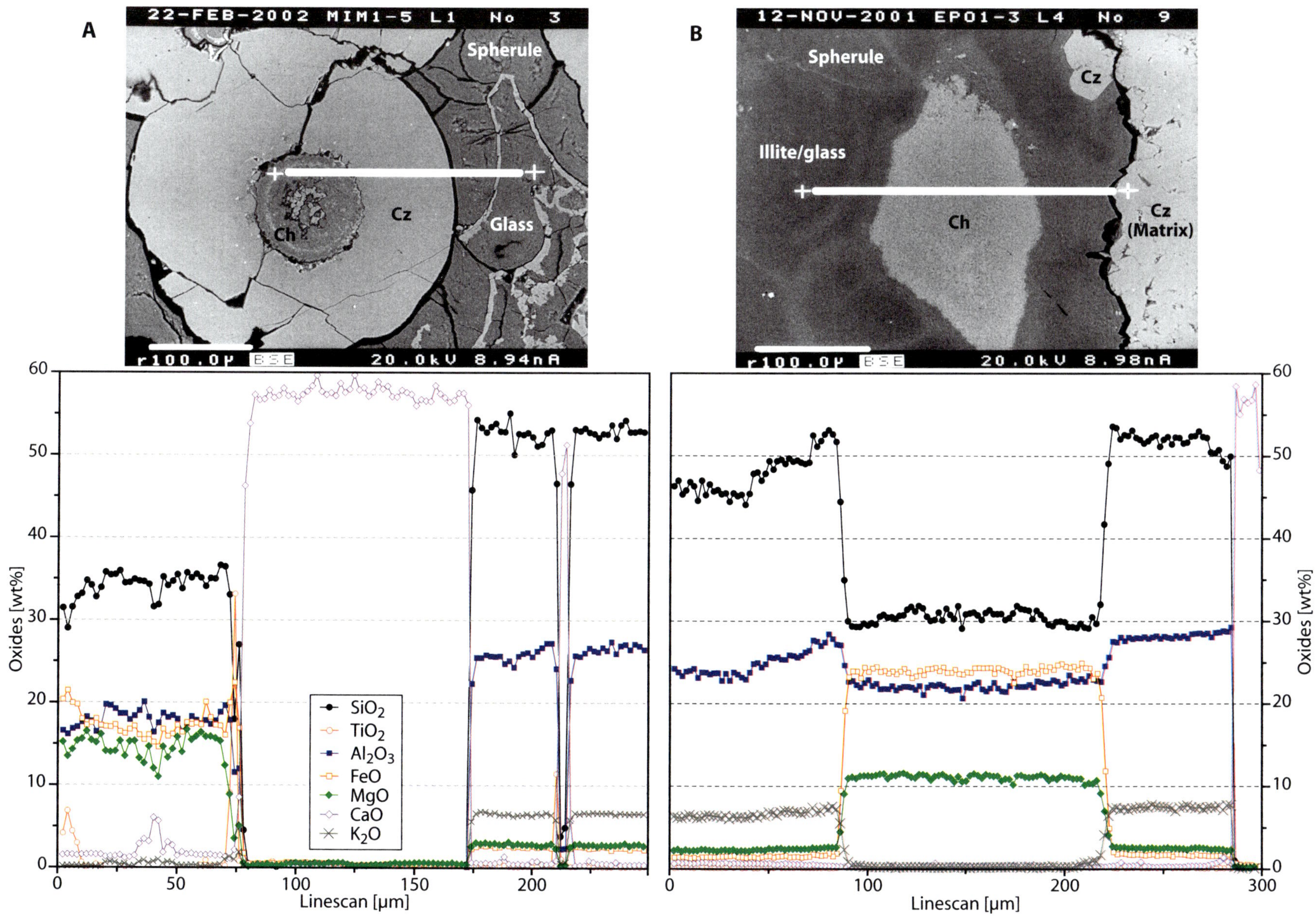

Figure 10. Backscattered electron images and representative quantitative electron microprobe linescans of ejecta spherules from spherule deposits at (A) El Mimbral, showing chlorite, calcite, and Si-Al-K glass, and (B) El Peñon with chlorite inclusion within Si-Al-K–rich glass. Cz—calcite; Ch—chlorite.

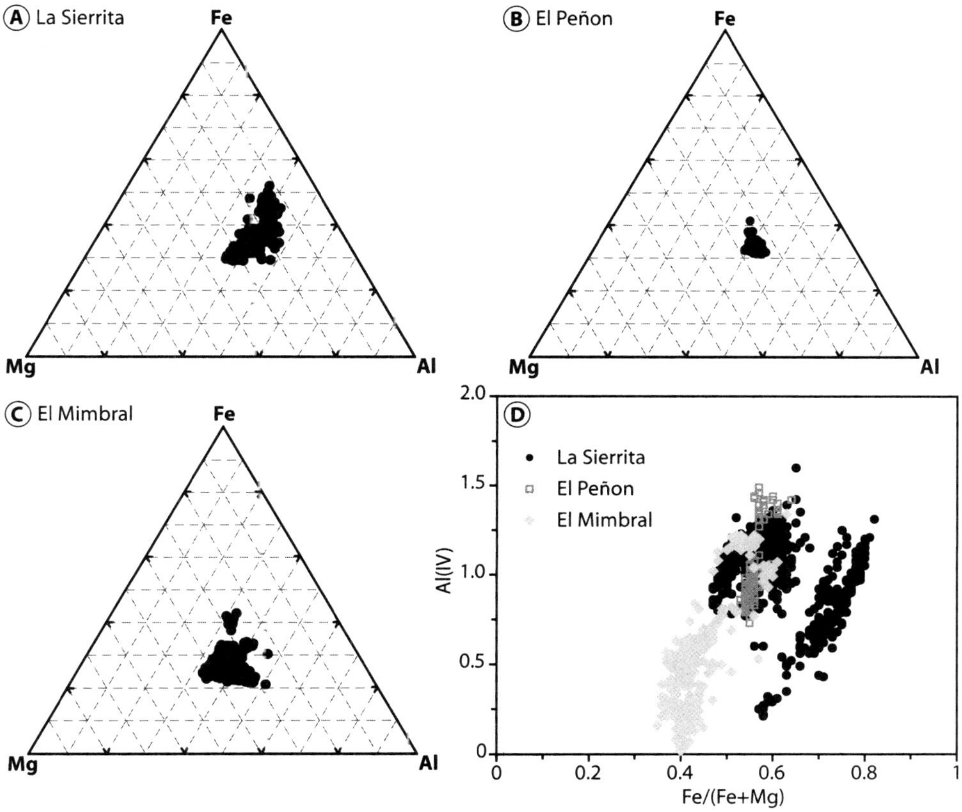

Figure 11. Ternary Fe-Mg-Al diagrams of chlorites from spherule deposits of NE México. (A) Mesa Juan Perez (La Sierrita) area. (B) El Peñon area. (C) El Mimbral. (D) Plot of Fe/(Fe+Mg) versus Al on tetrahedral site (Al^{IV}).

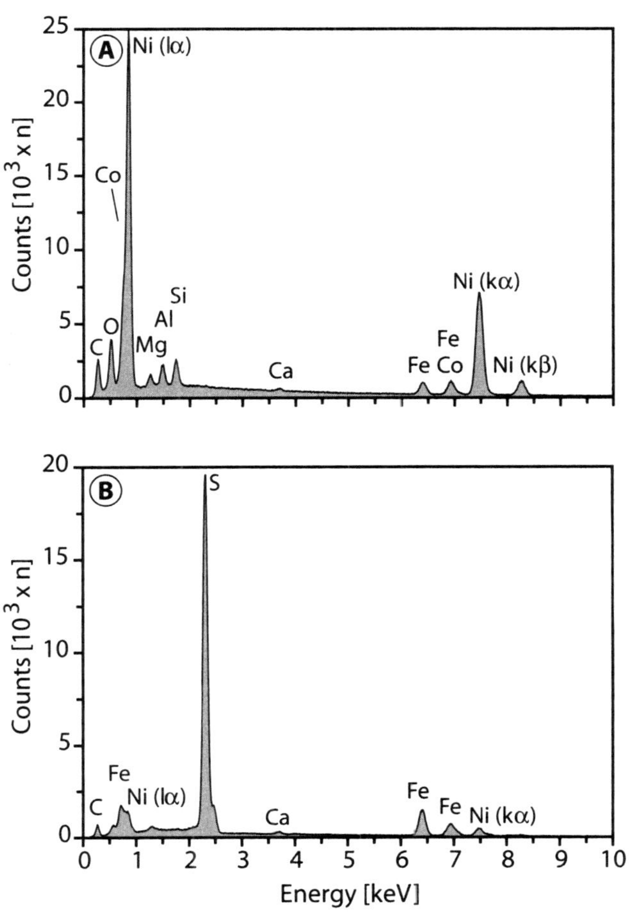

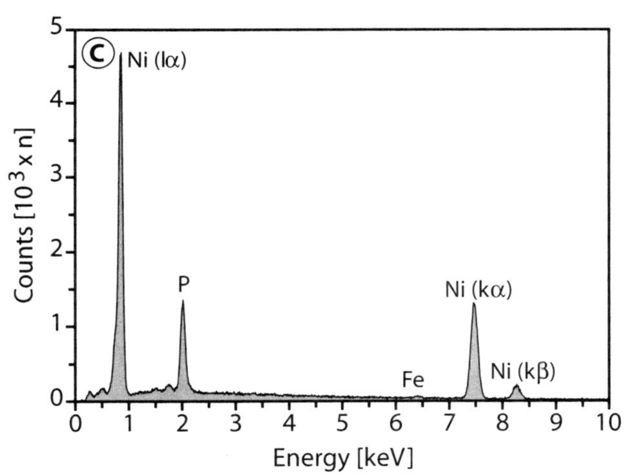

Figure 12. Energy-dispersive X-ray fluorescence spectra of metallic and sulfidic Ni-rich and Co-rich inclusions in spherules and carbonates from La Sierrita and El Mimbral (with 15 keV acceleration voltage). (A) Sample MAS2-2. (B) Sample MAS1-3. (C) Sample MIM1-5. The weak Mg, Al, Si, and Fe peaks in (A) and Ca peaks in (B) are derived from the enclosing spherule and carbonate, respectively. lα, mα, kα, kβ, and mβ are the characteristic X-ray lines of the respective element.

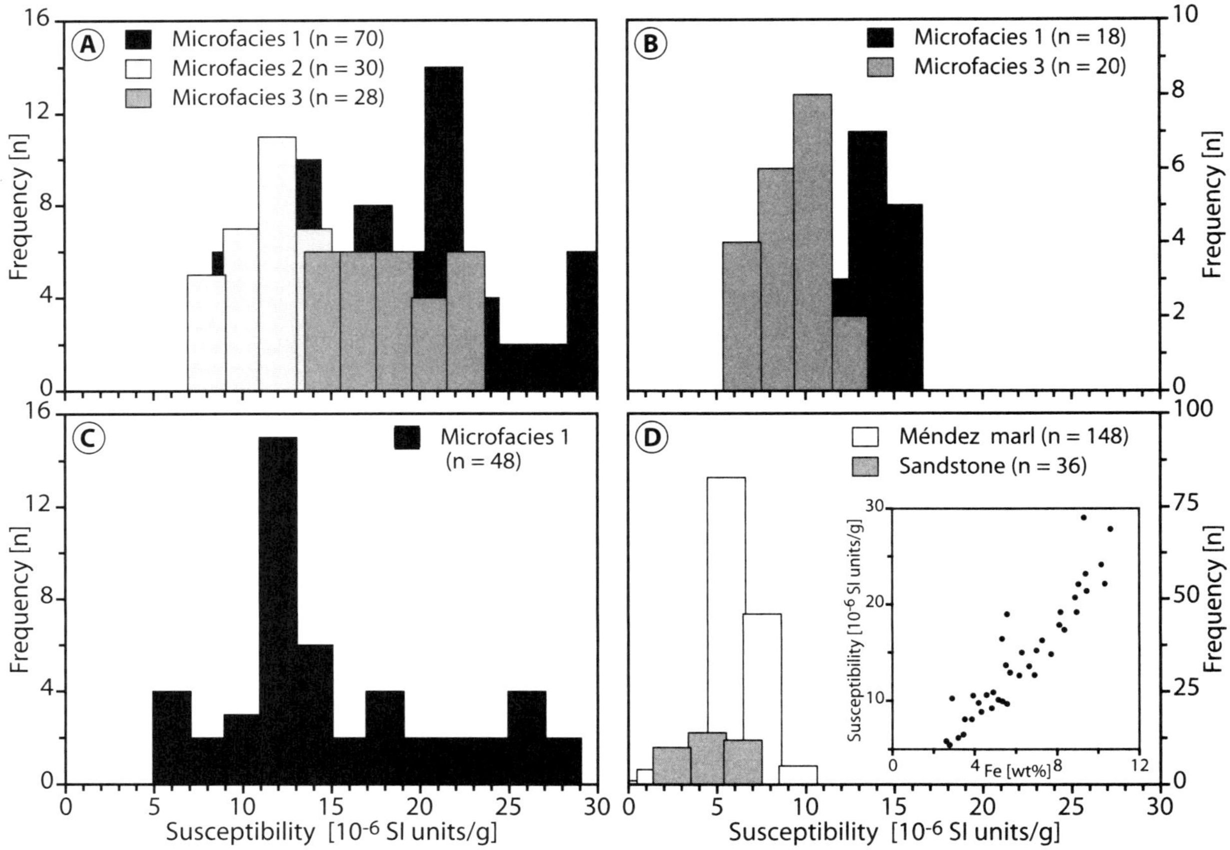

Figure 13. Frequency distribution of magnetic susceptibility for spherule deposits according to their microfacies and outcrop localities. (A) La Sierrita area. (B) El Peñon area. (C) El Mimbral. Compare to (D), the enclosing marls and sandstones. The inset in (D) shows a scatterplot with the bulk Fe content (from wavelength-dispersive spectroscopy analysis; Schulte, 2003) versus the magnetic susceptibility, illustrating the good correlation between both.

characteristic transition temperatures (e.g., Curie or Néel temperature; Dunlop and Özdemir, 1997). The κ-*T* curves in Figure 14 show that the samples from spherule deposits in NE México behave in very similar ways. The κ-*T* from −192–700 °C shows low values and an exponential decrease, indicating a mainly paramagnetic behavior. However, a small peak is superimposed at −80 °C on the paramagnetic curve, either suggesting a Curie or Néel temperature (of a phase similar to pseudobrookite?) or reflecting grain-size effects of a not yet identified ferromagnetic phase (Dunlop and Özdemir, 1997).

The heating leg of several samples shows a small peak at 560–580 °C that may be diagnostic for the presence of fine-grained (single domain) magnetite (Morden, 1993). The heating is associated with a pronounced change from the original grayish-brown color to brick red. The heating and cooling runs are irreversible, with much higher susceptibilities displayed in the cooling run. The cooling leg displays two Curie temperatures at 570 °C and 480 °C. Such behavior is typical for mineral transformations during heating of paramagnetic iron-bearing minerals like Fe-rich carbonates or pyrite, or transformations of hematite-goethite into maghemite or magnetite-near spinel (e.g., Just et al., 2005). No characteristic Curie temperature related either to goethite or to hematite has been detected; this absence could be linked to the observed impurities (Si, Al, Ni), to nonstoichiometry (Table 4), or the rather small amount of these phases in the samples of the ejecta deposits (Fig. 9).

Isothermal Remanent Magnetization

IRM acquisition and alternating field (AF) demagnetization are methods employed for inferring magnetic domain behavior, for instance to characterize magnetically "soft" (e.g., magnetite) and "hard" (e.g., hematite) phases (e.g., Lowrie, 1990; Dunlop and Özdemir, 1997).

The initial intensity of the natural remanent magnetization (NRM) of the samples MJP3A-1 (microfacies type 1) and MJP1A-14 (microfacies type 3) is very low and shows maximum values of ~0.3–0.7 mA/m (Fig. 15). In fields below 10 mT, the alternating field demagnetization of NRM displays an initial rapid decrease of ~50% of the original intensity, whereas in fields up to 160 mT, no further demagnetization occurs. The initial decrease

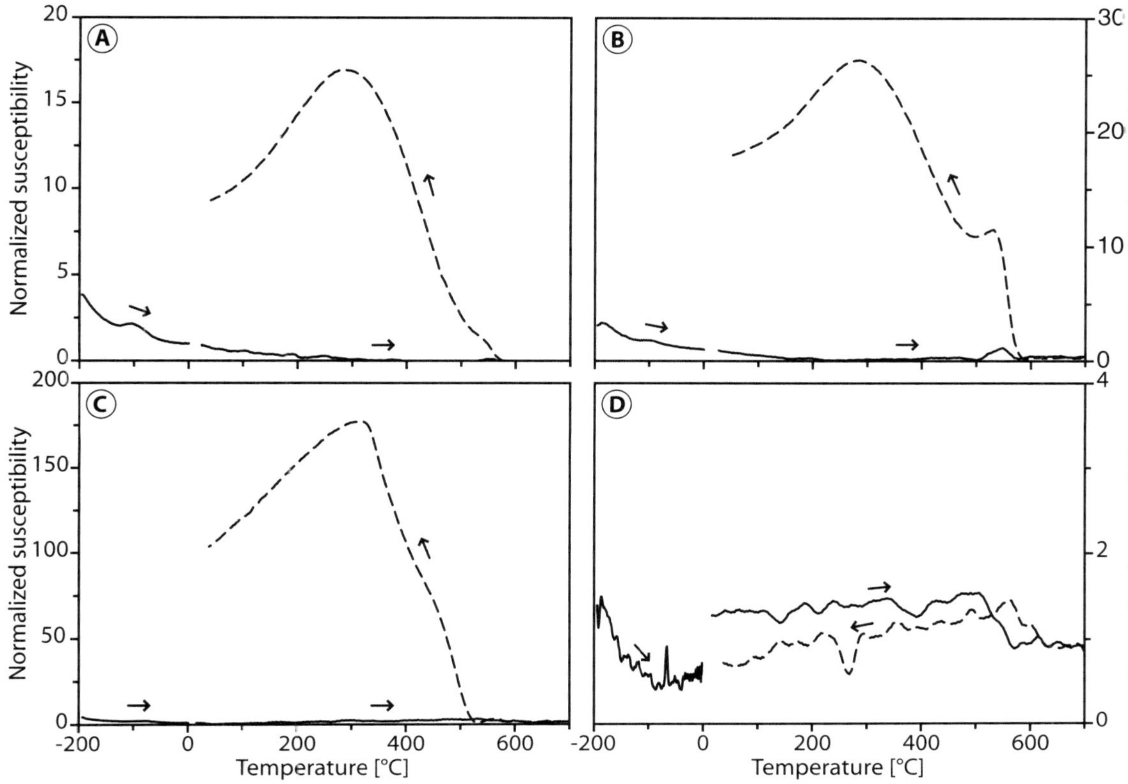

Figure 14. Typical curves showing the temperature dependence of the magnetic susceptibility of spherule deposits in NE México. (A) Microfacies type 1 (sample MJP1B-3). (B) Microfacies type 1 (sample LDP3-14). (C) Microfacies type 3 (sample MJP1A-14). (D) Sandstone (sample MJP1B-1).

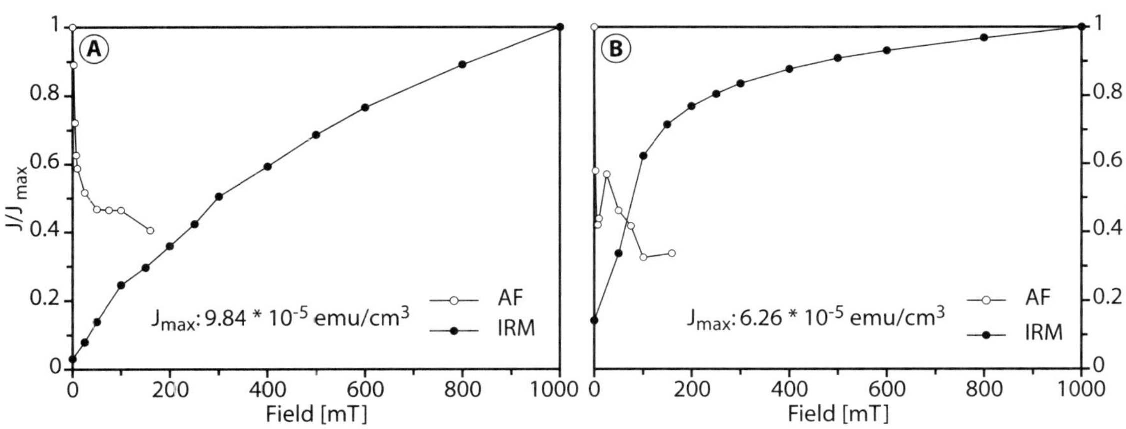

Figure 15. Normalized alternating field demagnetization and isothermal remanent magnetization for typical ejecta samples from the La Sierrita area comprising (A) microfacies type 1 (sample MJP3A-1; J_{max} = 9.84*10^{-5} emu/cm^3) and (B) microfacies type 3 (sample MJP1A-14; J_{max} = 6.26*10^{-5} emu/cm^3).

of NRM is probably related to a weak viscous remanent magnetization. The induced remanent magnetization (IRM) acquisition curves in Figure 15 show that some samples are nearly saturated in fields of 1000 mT while others are not. The IRM data confirm the dominance of high coercive minerals; for instance, hematite and goethite (Dunlop and Özdemir, 1997). The determination of coercivity from hysteresis loop measurements was only successful in a few samples from microfacies type 1 and 3 while most other samples showed a purely paramagnetic behavior (Fig. 16). For these samples, coercivity ranges between 6 and 7.3 mT, and the ratio of saturation remanence and saturation magnetization, M_r/M_s, is between 0.02 and 0.07, while the saturation magnetization (0.01–0.03 Am^2kg^{-1} at 300 mT) is very low. However, the M_r/M_s versus coercivity ratio (remanent coercivity/saturation coercivity, H_{cr}/H_c, = 4.8–6.9) indicates pseudo–single domain to multidomain behavior. Microscopic observations have shown grain sizes up to 20 µm for hematite, which is suggested to be the main carrier of the minor ferromagnetic behavior. According to Dunlop and Özdemir (1997), the critical single domain size is 15 µm, while Kletetschka and Wasilewski (2002) suggest a metastable single domain to multidomain transition in hematite at 100 µm. The remanence data are in accordance with hematite as the main carrier of remanence.

In conclusion, the rock magnetic investigations indicate that the magnetic mineral assemblage seems to be relatively homogeneous between the different microfacies and localities. The paramagnetic behavior clearly dominates the ferromagnetic contribution.

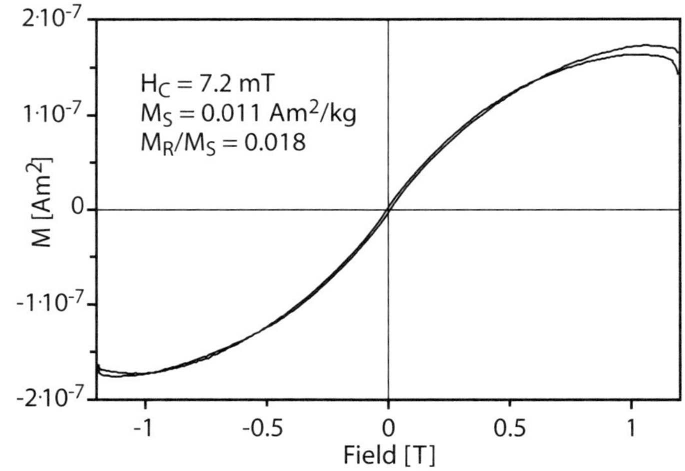

Figure 16. Hysteresis loop measurements for a representative microfacies type 1 ejecta sample from the La Sierrita area (sample MJP1B-3) indicating hematite as ferromagnetic phase. H_c—coercivity; M_s—saturation magnetization; M_R—remanence.

DISCUSSION

The proximal Chicxulub ejecta deposits in NE México are complex and exhibit various petrologic and geochemical features that allow the evaluation of (1) potential target lithologies at the impact site, (2) ejecta formation and deposition processes, and (3) the relationship to distal Chicxulub ejecta in the K-P boundary clay.

Chicxulub Target Lithologies

The reconstruction of precursor phases to the ejecta from NE México is essential to define the primary high-temperature, impact-related textures and mineralogy, and distinguish then post-impact diagenetic effects.

Alteration of Ejecta Components

The data presented here indicate that spherules and glass fragments from NE México were quenched from impact melts of variable but predominantly mafic to intermediate composition. The ubiquitous presence of secondary chlorite, chlorite-mixed layers, Fe-oxides and Fe-hydroxides, and the low oxide totals of the Si-Al-K–rich glass all result from alteration of the originally glassy phases (see Fisher and Schmincke, 1984; Berger et al., 1994; Mørk et al., 2001; Utzmann et al., 2002). However, the following observations suggest that alteration alone does not explain the observed compositional trends:

1. Elements such as Ti, Al, and Fe are relatively insoluble in natural waters of almost neutral pH, and their amounts are therefore assumed to remain largely unchanged during rock alteration, though some solution and precipitation may occur (e.g., Daux et al., 1994; Stroncik and Schmincke, 2002; Utzmann et al., 2002).

2. The excellent preservation of ejecta components with delicate microstructures, including geochemically distinct internal domains such as schlieren or globules (see Figs. 6–10), is a strong argument against a destructive pervasive alteration (e.g., Mørk et al., 2001). Particularly, the "microtektite in microtektite" configuration, with sharply contrasting neighboring compositional phases found in single ejecta particles is indicative of a "pristine" character. The good preservation of petrological features of the ejecta components may be explained by their rapid embedding in the clay-rich Méndez marls via slumping or by rapid covering by the sandstone units. Also, the large amounts of carbonate from the ejecta may have further promoted early cementation that predated compaction by sediment overload (Mørk et al., 2001), leading to the enhanced preservation of textures.

3. Hydrothermal influx that could explain the chloritization of spherules and fragments was neither recognized in this study nor has it been reported from elsewhere in the Méndez Formation (Perez Cruz, 1993). According to Keller et al. (1997, p. 419), the 500–1000 m of sediments, which are assumed to have overlain the clastic K-P sequence in this region, are "...insufficient to cause significant transformation of clay minerals (e.g., from smectite to illite-smectite mixed layers and to illite)." This interpretation is in agreement with the results of Hower et al. (1976), who suggested that a significant clay mineral transformation in

buried sediments of the Gulf of México usually occurs only at depths in excess of 3 km.

Consequently, we interpret the various silicate phases to indeed represent ejecta phases of originally different composition. This conclusion is in keeping with observations by Claeys et al. (2003), who found large compositional variability in the melt fragments of the suevite of drill hole Yucatán 6 site in the Chicxulub crater. Details of the alteration, however, are poorly understood. Studies on hydrothermal and metamorphic rocks by Cathelineau (1988) and de Caritat et al. (1993) showed that the composition of chlorites is mostly dependent on the pressure and temperature during their genesis. Specifically, the increased substitution of Al for Si in the tetrahedral site is clearly correlated with formation temperature. Elevated formation temperature results in higher substitution, almost independent of the source lithology (e.g., de Caritat et al., 1993). This geothermometer gives a distinct range of formation temperatures for the chlorites in the Chicxulub ejecta. The Al^{IV} occupation of the trioctahedral chlorites corresponds to a formation temperature of ~200–400 °C (Table 2), indicating that chlorite formation did not occur under post-depositional, low-grade, diagenetic conditions. The maximum burial depth of the ejecta in the Burgos Basin of NE México was well below 1000 m, corresponding to temperatures of only ~50–80 °C when assuming a normal temperature gradient. The Al^{IV} occupation of the di,trioctahedral chlorites with interlayers of smectite or illite, however, indicates significantly lower temperatures (~50 °C) or even unrealistic temperatures below 0 °C. Hence, this geothermometer is not applicable for chlorite-mixed layer minerals, and they may have formed during weathering.

In the absence of further evidence of in situ hydrothermal overprint, as outlined above, we conclude that the high formation temperatures of the chlorite point to an auto-hydrothermal alteration during the impact event rather than formation during alteration. This proposed origin of the chlorites is in agreement with the extremely fine state of crystallization of the chlorite as indicated by the very low crystallinity index of 0.36 ± 0.05 °2Θ (XRD analysis in Schulte, 2003); such low crystallinity indexes are typical for chlorites that formed under high temperatures (e.g., Arkai and Sadek Ghabrial, 1997; Shata and Hesse, 1998). An auto-hydrothermal origin is also in keeping with the related formation of rutile and hematite that requires hydrothermal conditions with high oxygen fugacities to transform the glass into these mineral phases. These results are sustained by EMP data for smectites formed from impact glass in fallout suevites from the Ries provided by Osinski (2003). Several petrographic characteristics of the Ries smectite ejecta particles, including evidence for liquid immiscibility, a porphyroclastic structure with intermingled glass-smectite, and fresh glass adjacent to smectite particles, appear to be incompatible with a secondary in situ hydrothermal origin. In agreement with Osinski (2003), we conclude that the chlorite did not form directly from the melt, but instead originated during immediate devitrification of the hot glass, presumably upon contact with the sea surface. However, in contrast to the setting of the Ries on land, for NE México, the contact of hot glass with the sea surface in the Gulf of México has probably enhanced devitrification and transformation of glass into chlorite, analogous to observations on the behavior of pumice (Manville et al., 1998).

Constraints on Potential Target Lithologies

The mineralogical and geochemical composition of the ejecta particles in NE México indicates carbonates and a suite of mafic to intermediate lithologies as precursor rocks (Tables 2 and 3). Although the basement of the Yucatán peninsula is rather poorly known, granite to gneiss seem to dominate, as reflected by the abundance of these lithologies in the Chicxulub suevite and impact breccia (e.g., Sharpton et al., 1996). However, according to $^{143/144}Nd/^{87/86}Sr$ isotope systematics of the Chicxulub melt, provided by Kettrup et al. (2000, p. 1229), "...at least one additional intermediate to mafic precursor component is required to explain the data." Mafic components are quite rare in the Chicxulub suevite and impact breccia and only few amphibolite, pyroxenite, and diabase clasts have been found (Kettrup and Deutsch, 2003; Kring et al., 2003). This range of mafic lithologies fits the mafic rocks known from the southeastern part of the Gulf of México (e.g., DSDP Leg 77; Schlager et al., 1984), the Yucatán peninsula (Bateson and Hall, 1977; Rebolledo-Vieyra et al., 2000), and the Maya Mountains in Belize (Bateson and Hall, 1977). They mostly occur as dikes or intrusions in the attenuated (transitional) continental crust that comprises large parts of the Yucatán peninsula and the southernmost Gulf of México. For instance, at DSDP Hole 538, several hundred kilometers to the north of the Yucatán peninsula, post-tectonic diabase dikes intruded an association of variably mylonitic felsic gneisses and interlayered amphibolites (Dallmeyer, 1984). Our findings in NE México suggest that comparable mafic precursor rocks are also present in the northwestern sector of Chicxulub.

In addition, Kettrup and Deutsch (2003) have shown a close similarity of the Rb-Sr and Sm-Nd systematics between the amphibolite in the Chicxulub suevite and the chlorite spherules from NE México (La Lajilla). The range of element abundances and diverse element ratios (Si/Al, Fe/Mg) found in the Fe-Mg–rich chlorite spherules from NE México does not allow an unequivocal correlation to amphibolite precursors (Table 2). Yet, complex mixing and fractionation processes within the impact melt, as observed at various impact structures (Mittlefehldt et al., 1992; Hörz et al., 1998; Koeberl, 1998; See et al., 1999), may lead to melt compositions that cannot directly be correlated to specific target rock melts. For instance, the presence of Fe-Ti-K–rich schlieren in spherules and glassy fragments may reflect partially digested mineral melts (e.g., hornblende, biotite).

The precursor phase to the Si-Al-K–rich glass is more difficult to constrain, since there are few rock types with similar low Si/Al ratios (Table 3). Therefore, either melt-fractionation processes during the impact have to be considered (see above) and/or K-feldspar–rich to mica-rich source rocks (e.g., schist) provided high amounts of potassium and aluminum to the melt. Such Si-Al-K–rich rocks have so far only rarely been reported from the Yucatán peninsula and from the Chicxulub impact breccia, though Claeys et al. (2003) found Al-K–rich glass in the upper suevite of drill hole Yucatán 6. In addition, K-feldspar–rich biotite leucogranites occur in the Maya Mountains, central Belize (Steiner and Walker, 1996). A contribution of similar Silurian leucogranitic rock types from the Yucatán basement to the Chicxulub ejecta seems possible, particularly since López-Ramos (1975) reported a Silurian Rb-Sr age of ca. 410 Ma for a schist fragment in the core Yucatán No. 1. This age is in accordance with the U-Pb age of 418 ± 6 Ma published for shocked zircons from the K-P boundary in the Western Interior, from the ejecta deposits at Beloc, as well as from suevites from inside the Chicxulub structure (Krogh et al., 1993a; Krogh et al., 1993b; Kamo and Krogh, 1995).

The high amount of allochthonous carbonate (40–70 wt%) in bulk rock ejecta that occurs as (1) microcrystalline sparite, (2) distinct former melt phase (e.g., globules with "feathery" calcite or in spherules), and (3) limestone clasts implies the substantial dispersion of carbonate by the Chicxulub event. In addition, the high amount of carbonate indicates that NE México ejecta was derived mainly from the sedimentary cover of the Yucatán peninsula, and hence, shallower target areas. Evidence for a significant contribution by sulfur-rich lithologies, however, was not found.

The origin of the microcrystalline carbonate in the spherule deposits (microspar in the matrix-supported microfacies type 1) is not yet clear. An exclusive diagenetic origin seems unlikely because of the partly matrix-supported textures of these spherule deposits (cf. Fig. 5). Therefore, the microcrystalline carbonate may have originated from dust dispersed by the impact event and/or as secondary product derived from back-reaction of CO_2 and solid CaO, as suggested by Deutsch et al. (1998) and Agrinier et al. (2001).

Ejecta Formation and Deposition Processes

The petrological features of the spherules provide not only compelling evidence for an impact-origin of the spherule deposits, but also constrain the physical conditions during and immediately following the impact. Additional information on impact processes derives from the microfacies of the spherule beds and their rock magnetic properties.

Liquid Immiscibility in Ejecta Particles

The ejecta particles are largely made of multiple silicate phases and carbonates, which all show a range of particular petrological features, including (1) irregular and broken silica particles with abrasion textures in the cores of spherules, which are mutually enclosed by layers of carbonate and silicate phases (Fig. 8); (2) the elongated and twisted forms of the Fe-Mg globules in spherules, suggesting a liquid precursor state (Fig. 5A); and (3) "knife-sharp" transitions between compositionally distinct silicic phases (Fig. 10).

In addition, the carbonate phase reveals further characteristic textures: (1) individual calcite globules are confined to the spherule deposits and show no traces of later infilling by alteration (marble-like carbonate texture; Figs. 8C, 8D, 9D, and 9E); (2) emulsion-like textures of carbonate intermingling with silicate (Fig. 8D); (3) coalesced calcite globules in spherules and ejecta-fragments (Fig. 8A); and (4) curved concave-convex menisci with sharp boundaries between spherules, fragments, and carbonate overgrowths or inclusions (Fig. 8C and 8F). These textures are interrupted along fractures of broken spherules, fragments, and carbonates and therefore predate alteration.

According to data provided by Graup (1999), Jones et al. (2000), and Osinski and Spray (2001), these microtextures are indicative of liquid immiscibility between silicate and carbonate melts. Consequently, the presence of multiple geochemical phases in single ejecta particles provides evidence for the existence of compositionally different melts (cf. Fig. 8F). The overgrowths on spherules and fragments by "feathery" calcite (Fig. 7) are in keeping with this interpretation, pointing to rapid cooling of a carbonate melt (Graup, 1999). Ejecta particles with analogous textures are present in suevite from the Ries (Graup, 1999), the Haughton (Martinez et al., 1994; Osinski and Spray, 2001), and the Chicxulub impact structures (Jones et al., 2000; Claeys et al., 2003).

Reasonable explanations for the ubiquitous presence of immiscibility features observed in the spherule deposits at NE México derive from the physical properties of the melt phases. For polymerized silicic melts, physical properties depend on the amount of polymer modifiers (e.g., Mg) versus polymer formers (e.g., Si), which result in specific liquidus temperatures, for instance ~750 °C for a granitic and ~1100 °C for a basaltic melt at atmospheric pressure, as well as in a lower viscosity for basaltic melts (e.g., Philpotts, 1990). In contrast, carbonate melts are unpolymerized ionic liquids and do not associate with the silicate anions in the melt, resulting in the effect known as "carbonate-silicic liquid immiscibility" (e.g., Graup, 1999). The liquidus temperature of carbonate melts is ~500–600 °C, several hundred degrees lower than for silicate melts (e.g., Osinski and Spray, 2001). Therefore, carbonate melts remain in a liquid state longer, whereas the silicic melt fraction has already quenched to glasses or is in a more viscous state (Barker, 1989). This behavior may explain the thick carbonate overgrowths on spherules and fragments, indicating that the carbonate melt was a hot liquid, whereas the silicate melt was at least partly solid.

This "delayed" crystallization of the carbonate phase may be enhanced by the low thermal diffusivity of carbonate melt, which is more than a magnitude smaller than for basaltic melt, so that carbonatite melts will cool more slowly than siliceous melts (Dawson et al., 1990). Moreover, the viscosity of carbonate melts is extremely low (10^{-2} poise) and is comparable to the viscosity of water (Treiman, 1989), resulting in an enormous wetting power with the potential to infiltrate thin cracks or porous media. These physical properties may explain the complex carbonate veins observed in the spherules (Graup, 1999; Osinski and Spray, 2001). The rims of molten carbonate may have induced thermal stress inside the cooler host grains, leading to their partial melting and disruption and the subsequent intrusion of carbonate into the cracks (Figs. 8B and 8F). Hence, the particular multiple coated ejecta particles ("tektite in tektite" textures) record the time-temperature path from ejection of the melt until cooling at the water surface, well comparable to "coated gneiss" bombs observed at the Popigai impact structure (Masaitis and Deutsch, 1999; Kettrup et al., 2003). Such impact glass shells may have originated when the ejected spherules and fragments passed through different parts of the turbulent vapor cloud above the growing crater (Kerschhofer et al., 2000).

In conclusion, these observations provide unequivocal evidence not only for the silicic glasses, but also for large parts of the calcite as initially being in a liquid state. The existence of compositionally distinct silicic and carbonaceous melt phases, however, suggests that in this part of the ejecta curtain, the thermal energy, turbulences, and pressure may have been already low enough to preclude further (complete) melt homogenization. The occurrence of chemically different melt fragments in the upper part of the Chicxulub suevite has also been interpreted in terms of the absence of widespread melt homogenization (Claeys et al., 2003), analogous to data from the fallout-suevite of the Ries (Osinski, 2003).

It is remarkable that, with the exception of marl and carbonate clasts, no lithic clasts (e.g., basement rocks) or diaplectic glass clasts have been observed in the spherule deposits. Therefore, the spherule deposits appear to be comprised entirely of former liquid-state glasses, indicative of "shock state IV," comprising shock pressures of 60–80 GPa and postshock temperatures of 1300–3000 °C (Stöffler, 1971; Engelhardt et al., 1997; French, 1998).

Scenario for Ejecta Dispersal

Welding and amalgamation of ejecta particles as well as the enveloping of carbonate and marl clasts and benthic foraminifera could have resulted from compaction by sediment overload (Tucker and Wright, 1992; Collinson, 1994; Bohor, 1996). However, the textural features revealed by our study (e.g., random grain orientation, accretionary lapilli, three-dimensional mesh-like fabric) are not compatible with an exclusively diagenetic origin of these features by sediment overload. Consequently, the amalgamated and welded ejecta particles may have formed by accretion and mingling of distinct melt droplets during ejection in the expanding ejecta curtain or in the rising/collapsing plume, as commonly observed in ejecta deposits from other craters (See et al., 1998; Hörz et al., 2002). Moreover, the "enveloping" of marl clasts and benthic foraminifera—which both presumably derived from the NE Mexican siliciclastic shelf—by ejecta particles is remarkable: it implies that ejecta was still in a hot and viscous state when it arrived at the shelf area and that there was direct contact of hot ejecta with marine shelf sediments. Consequently, we assume that some areas of the NE Mexican shelf were blanketed from seawater during passage of the ejecta curtain or the collapsing plume.

In addition, and equivalent to observations on pyroclastic flows from explosive volcanoes, the hot ejecta curtain may have caused large steam explosions upon entering shallower marine settings (e.g., Cas and Wright, 1991; Freundt and Schmincke, 1998; Allen and Cas, 2001). Such steam explosions may also have induced the secondary local redistribution of ejecta phases and marine sediments, as shown schematically in Figure 17A. However, little experience exists with ejecta curtain behavior in large-scale marine impact events. It is not clear if the Chicxulub ejecta curtain behaved similarly to a pyroclastic flow, traveling large distances on the sea surface by "gliding" on a thin layer of steam over the water. Such a nuée-ardente-like or glowing-avalanche-like ejecta-distribution mechanism has been proposed for the uppermost units of the Chicxulub suevite (Claeys et al., 2003; Kring et al., 2003). Likewise, the ejecta curtain may also have entered the water column up to a certain depth, shoving the water forward and enhancing the interaction of ejecta with seawater and marine sediments on the temporarily exposed Mexican shelf (Fig. 17B). For Chicxulub, the marine conditions at and around the impact site, as well as the turbulent mixing and heating of seawater during crater-outward movement of the ejecta curtain, may have additionally favored the generation of accretionary lapilli by providing large amounts of vapor in the ejecta curtain or plume.

Compared to the Chicxulub ejecta deposits in Beloc that have an FeO content below 4 wt% (bulk rock; Stüben et al., 2002) and very low paramagnetic susceptibility values (Senftle et al., 1993; Thorpe et al., 1994), magnetic data for ejecta deposits in NE México are more variable: They reflect the compositional differences of the ejecta phases, with input of ferruginous and, hence, potential magnetic phases. In addition, the susceptibility values suggest a different ejecta dispersal mechanism for the Haitian glass spherules. In the spherules from Haiti, formation of iron oxides was suppressed by rapid quenching (see also Werner and Borradaile, 1998), whereas the ubiquitous presence of iron and the probably slower cooling of NE Mexican ejecta yielded recrystallization of paramagnetic glass into paramagnetic chlorite, clay minerals, pseudobrookite-like phases, rutile, hematite, and goethite.

The low bulk rock susceptibility and near absence of magnetite in the spherule deposits of NE México both indicate that quenching followed by autohydrothermal alteration under strongly oxidizing conditions might have prevented formation of magnetic phases or destroyed them, if they were already present. This hypothesis agrees with the inferred fluid-vapor–rich, highly oxidizing environment during the ejecta-distribution outlined above. However, the occurrence of goethite indicates that diagenetic effects might have further enhanced the oxidization of ferrimagnetic phases.

Proximal versus Distal Chicxulub Ejecta

The variable mineralogy and geochemistry of the suevite from within Chicxulub (e.g., Claeys et al., 2003; Stöffler et al., 2004) and of ejecta in NE México (this study) both reflect the complex geology of the Yucatán peninsula. However, a variable mineralogy and geochemistry is also an essential feature of ejecta found in the K-P boundary clay in settings that are more distal

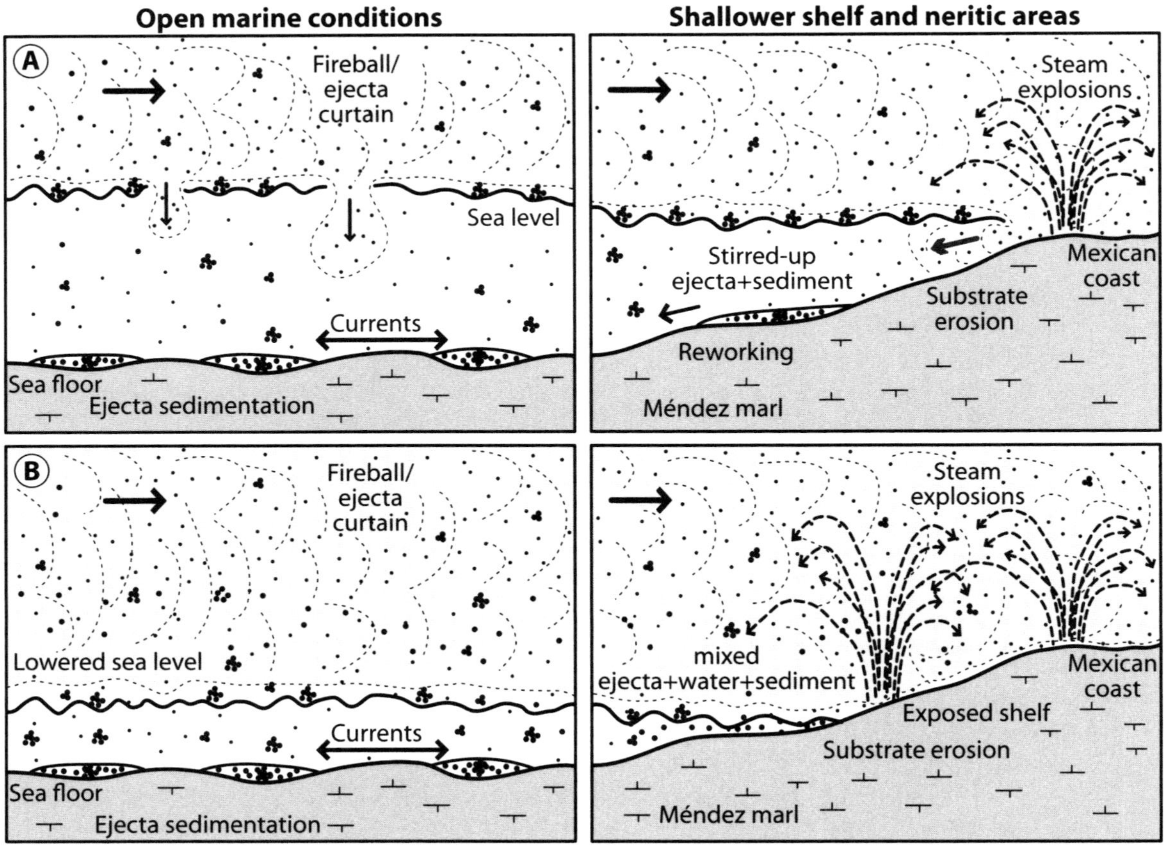

Figure 17. Schematic models for the ejecta dispersal by the Chicxulub impact event. They are based on the presumption that the ejecta curtain of the Chicxulub impact event (A) behaved similar to a pyroclastic flow, gliding over the water for a considerable distance, and/or (B) entered the water column to a certain depth, shoving the water away. Modified after Cas and Wright (1991) and Freundt and Schmincke (1998).

(Figs 1 and 2). Besides relict Ca-rich glass fragments, mostly altered to smectite with Ca as the dominant interlayer cation reflecting the composition of the parent glass (e.g., Bauluz et al., 2000, 2004), the K-P boundary clay generally bears a variety of impact-derived components: Si-K–rich (K-feldspar) spherules, glauconite spherules, iron and iron-titanium oxides, and fibro-radial pyroxene spherules (Fig. 2; Izett, 1990; Montanari, 1991; Schmitz, 1992; Bohor and Glass, 1995; Martínez-Ruiz et al., 1997; Ortega-Huertas et al., 1998, 2002; Bauluz et al., 2000, 2004; Wdowiak et al., 2001). This wide compositional spectrum of Chicxulub ejecta has long been a matter of intense debate in terms of the origin of these ejecta phases from either mafic (e.g., DePaolo et al., 1983) or felsic continental crust (e.g., Sharpton et al., 1990). The mineralogical and geochemical data from the Chicxulub suevite and the ejecta in NE México provide reasonable explanations for this variability in distal settings to Chicxulub, though the link of these ejecta phases to possible generation and distribution modes, for instance condensation from the vapor plume, has yet to be established (Smit et al., 1992a; Evans et al., 1994; Griscom et al., 1999, 2003).

CONCLUSIONS

1. In NE México, Chicxulub ejecta deposits at the K-P boundary are either locally mixed with late Maastrichtian marls by slump-slides and liquefaction, or, more commonly, occur at the base of a channelized sand-siltstone complex, resulting from rapidly deposited multiple debris flows.

2. The 0.2–1-m-thick ejecta deposits consist of mm-cm sized spherical to drop-shaped, vesiculated spherules, angular to ribbon-like "glassy" fragments, and abundant carbonate. The microfacies and internal texture of the ejecta deposits show a variety of features, including welding and fusing of components, and evidence for liquid immiscibility between silicic and carbonate melts. Welded and amalgamated ejecta constituents, as well as the enveloping of marl and benthic foraminifera by spherules, suggest interaction of hot ejecta with marine sediments on the NE Mexican shelf.

3. Ejecta spherules and fragments are characterized by a Fe-Mg–rich chlorite and Si-Al-K–rich glass composition. They contain Fe-Ti-K–rich schlieren, Fe-Mg–rich globules,

and rare µm-sized metallic and sulfidic Ni-Co–rich inclusions. Therefore, ejecta phases originated mostly from mafic lithologies, with minor contributions from intermediate to felsic rocks and carbonates; a meteoritic component may be present as well. No evidence for binary mixing of ejecta phases was found. We derive a considerable compositional variability of the target rocks in the northwestern sector of the Chicxulub structure. However, correlation of the ejecta phases in NE México to distinct target lithologies from the impact melt breccia and suevite has yet to be established. Specifically, the Si-Al-K–rich glasses are difficult to relate to target lithologies as they show a low Si/Al ratio of ~1.5 and high amounts of potassium (6–9 wt%), pointing to either a mica/illite–rich or a K-feldspar–rich precursor phase.

4. The high amount of CaO (25–35 wt%) in the spherule deposits indicates that mainly ejecta from shallow carbonate-rich target lithologies is present. Carbonate-rich ejecta phases were dispersed as unshocked clasts or as melt. The microspar in the spherule deposits and their matrix-supported texture indicate that microcrystalline carbonate and spherules were ejected and deposited simultaneously. Evidence for a significant contribution by anhydrite or gypsum was not found.

5. Differences in ejecta composition exist between locations to the north (La Sierrita, El Peñon), with a chlorite-dominated mineralogy, and the southernmost outcrop at El Mimbral, where Si-K–rich glasses and illite-ejecta particles are the main ejecta phases. Major element data for bulk samples support this result. The compositional trends indicate localized patterns of melt dispersion from specific sectors in the target area.

6. The variable mineralogy and geochemistry of ejecta found in NE México are congruent with the wide range of ejecta phases found at the K-P boundary in proximal to distal settings. The range of target rocks at the Chicxulub impact site and the alteration of ejecta phases by various impact and post-impact processes provide reasonable explanations for this variety.

8. The spherule deposits exhibit elevated magnetic susceptibility compared to the host lithologies. This elevated susceptibility reflects the occurrence of Fe-rich chlorite spherules.

9. The ubiquitous occurrence of hematite and goethite in the spherule deposits points to highly oxidizing conditions during or after ejecta formation, or later, in keeping with the inferred ground surge–like ejecta dispersion mode. Such oxidizing conditions may have prevented the formation of a stronger ferromagnetic phase (magnetite) or destroyed it. Hematite in the ejecta particles formed probably with chlorite from the glass. Remanence measurements indicate the presence of only hematite or goethite as high coercitive phases at room temperature. Evidence for ferrimagnetic phases other than goethite and hematite is rare and restricted to small peaks in the κ-T curves at 560–580 °C. This magnetite probably formed during the heating procedure in the laboratory and was not a primary phase.

10. The small grain sizes of Fe-Ti–rich phases and dendritic-skeletal crystals ("spinifex texture") are indicative of rapid quenching from a melt, although these minerals may have formed by hydrothermal alteration after primary phases. The very fine-grained Fe-Ti phase (pseudobrookite?) that could not be further identified in this study may explain the characteristic small peak at −80 °C in the low-temperature susceptibility.

ACKNOWLEDGMENTS

This study is part of the first author's Ph.D. thesis and was supported by the Deutsche Forschungsgemeinschaft via the International Continental Drilling Program (ICDP) (STI 128/7-1,2 and KO 1514/1). Thierry Adatte, Doris Stüben, and particularly Wolfgang Stinnesbeck are acknowledged for support and comments on early parts of this paper. We are also indebted to Günther Kletetschka (National Aeronautics and Space Administration [NASA] Goddard Space Flight Center, Greenbelt, Maryland, USA) for constructive comments. Jana Just kindly provided the isothermal remanent magnetization measurements. Carsten Vahle and Gesine Lorenz are thanked for their support in the magnetic laboratory and with the cathodoluminescence microscope at the Geologisch-Paläontologisches Institut, Heidelberg. Thomas Frederichs is acknowledged for his support and discussions during the stay of A.K. at the Institut für Marine Geophysik, Universität Bremen. Special thanks go to Volker Zibat from the Laboratory for Electron Microscopy and to Stefan Unrein from the sedimentary laboratory, both Universität Karlsruhe. In addition, we thank Armin Schafhauser from the Universität Karlsruhe as well as Lutz Hecht, Tobias Salge, and Axel Wittmann from the Humboldt University Berlin for encouraging discussions and support. We are also very grateful to the editor, Alexander Deutsch, and the reviewers Henning Dypvik and David T. King for their constructive comments.

REFERENCES CITED

Agrinier, P., Deutsch, A., Schärer, U., and Martinez, I., 2001, Fast back-reactions of shock-released CO_2 from carbonates: An experimental approach: Geochimica et Cosmochimica Acta, v. 65, no. 15, p. 2615–2632, doi: 10.1016/S0016-7037(01)00617-2.

Alegret, L., Molina, E., and Thomas, E., 2001, Benthic foraminifera at the Cretaceous-Tertiary boundary around the Gulf of Mexico: Geology, v. 29, no. 10, p. 891–894, doi: 10.1130/0091-7613(2001)0292.0.CO;2.

Allen, S.R., and Cas, R.A.F., 2001, Transport of pyroclastic flows across the sea during the explosive, rhyolitic eruption of the Kos Plateau, Tuff, Greece: Bulletin of Volcanology, v. 62, no. 6-7, p. 441–456, doi: 10.1007/s004450000107.

Alvarez, W., Claeys, P., and Kieffer, S.W., 1995, Emplacement of Cretaceous-Tertiary boundary shocked quartz from Chicxulub crater: Science, v. 269, p. 930–935.

Arkai, P., and Sadek Ghabrial, D., 1997, Chlorite crystallinity as an indicator of metamorphic grade of low-temperature meta-igneous rocks: A case study from the Bukk Mountains, Northeast Hungary: Clay Minerals, v. 32, no. 2, p. 205–222.

Bailey, S.W., 1980, Summary of recommendations of AIPEA nomenclature committee: Clay Minerals, v. 15, p. 85–93.

Barker, D.S., 1989, Field relations of carbonatites, *in* Bell, K., ed., Carbonatites: Genesis and evolution: London, Unwin Hyman, p. 38–69.

Bateson, J.H., and Hall, I.H.S., 1977, The geology of the Maya Mountains, Belize, Overseas Memoir: London, Institute of Geological Sciences, 43 p.

Bauluz, B., Peacor, D.R., and Elliott, W.C., 2000, Coexisting altered glass and Fe/Ni oxides at the Cretaceous/Tertiary boundary, Stevns Klint (Denmark): Direct evidence of meteorite impact: Earth and Planetary Science Letters, v. 182, p. 127–136, doi: 10.1016/S0012-821X(00)00245-4.

Bauluz, B., Peacor, D.R., and Hollis, C.J., 2004, TEM study of meteorite impact glass at New Zealand Cretaceous-Tertiary sites: Evidence for multiple impacts or differentiation during global circulation?: Earth and Planetary Science Letters, v. 219, no. 3-4, p. 209–219, doi: 10.1016/S0012-821X(04)00011-1.

Berger, G., Claparols, C., Guy, C., and Daux, V., 1994, Dissolution rate of a basalt glass in silica-rich solutions: Implications for long-term alteration: Geochimica et Cosmochimica Acta, v. 58, no. 22, p. 4875–4886, doi: 10.1016/0016-7037(94)90218-6.

Bhandari, N., Verma, H.C., Upadhyay, C., Tripathi, A., and Tripathi, R.P., 2002, Global occurrence of magnetic and superparamagnetic iron phases in Cretaceous-Tertiary boundary clays, *in* Koeberl, C., and MacLeod, K.G., eds., Catastrophic events and mass extinctions: Impacts and beyond: Geological Society of America Special Paper 356, p. 201–212.

Blum, J.D., Chamberlain, C.P., Hingston, M.P., Koeberl, C., Marín, L.E., Schuraytz, B.C., and Sharpton, V.L., 1993, Isotopic comparison of K/T boundary impact glasses with melt rock from the Chicxulub and Manson impact structures: Nature, v. 364, p. 325–327, doi: 10.1038/364325a0.

Bohor, B.F., 1996, A sediment gravity flow hypothesis for siliciclastic units at the K/T boundary, northeastern Mexico, *in* Ryder, G., Fastovsky, D., and Gartner, S., eds., The Cretaceous-Tertiary boundary event and other catastrophes in Earth history: Geological Society of America Special Paper 307, p. 183–195.

Bohor, B.F., and Glass, B.P., 1995, Origin and diagenesis of K/T impact spherules—From Haiti to Wyoming and beyond: Meteoritics, v. 30, p. 182–198.

Bralower, T.J., Premoli Silva, I., and Malone, M.J., 2002, Shipboard Scientific Party: Leg 198 summary: College Station, Texas, Proceedings of the Ocean Drilling Program, Initial Results, v. 198, 148 p.

Cas, R.A.F., and Wright, J.V., 1991, Subaqueous pyroclastic flows and ignimbrites: An assessment: Bulletin of Volcanology, v. 53, p. 357–380, doi: 10.1007/BF00280227.

Cathelineau, M., 1988, Cation site occupancy in chlorites and illites as a function of temperature: Clay Minerals, v. 23, p. 471–485.

Claeys, P., Heuschkel, S., Lounejeva-Baturina, E., Sanchez-Rubio, G., and Stöffler, D., 2003, The suevite of drill hole Yucatán 6 in the Chicxulub impact crater: Meteoritics & Planetary Science, v. 38, p. 1299–1317.

Collinson, J.D., 1994, Sedimentary deformation structures, *in* Maltman, A.J., ed., The geological deformation of sediments: London, Chapman & Hall, p. 95–125.

Croskell, M.S., Warner, M., and Morgan, J., 2002, Annealing of shocked quartz during atmospheric re-entry: Geophysical Research Letters, v. 29, no. 20, p. 1–4, doi: 10.1029/2001GL014382.

Dallmeyer, R.D., 1984, $^{40}Ar/^{39}Ar$ ages from a pre-Mesozoic crystalline basement penetrated at Holes 537 and 538a of the Deep Sea Drilling Project Leg 77, southeastern Gulf of Mexico: Tectonic implications: College Station, Texas, Deep Sea Drilling Program, Initial Reports, v. 77, p. 497–504.

Daux, V., Crovisier, J.L., Hemond, C., and Petit, J.C., 1994, Geochemical evolution of basaltic rocks subjected to weathering: Fate of the major elements, rare earth elements, and thorium: Geochimica et Cosmochimica Acta, v. 58, no. 22, p. 4941–4954, doi: 10.1016/0016-7037(94)90223-2.

Dawson, J.B., Pinkerton, H., Norton, G.E., and Pyle, D.M., 1990, Physical properties of alkali carbonatite lavas: Data from the 1988 eruption of Oldoinyo Lengai, Tanzania: Geology, v. 18, p. 260–263, doi: 10.1130/0091-7613(1990)018<0260:PPOACL>2.3.CO;2.

de Caritat, P., Hutcheon, I., and Walshe, J.L., 1993, Chlorite geothermometry: A review: Clays and Clay Minerals, v. 41, p. 219–239.

Delano, J.W., and Hanson, B., 1996, Liquid immiscibility: Cause of compositional heterogeneity in tektites: Lunar and Planetary Science, v. 27, p. 305–306.

DePaolo, D.J., Kyte, F.T., Marshall, B.D., O'Neil, J.R., and Smit, J., 1983, Rb-Sr, Sm-Nd, K-Ca, O, and H isotopic study of Cretaceous-Tertiary boundary sediments, Caravaca, Spain: Evidence for an oceanic impact site: Earth and Planetary Science Letters, v. 64, no. 3, p. 356–373, doi: 10.1016/0012-821X(83)90096-1.

Deutsch, A., Schärer, U., and Agrinier, P., 1998, Evidence for backreacted carbonates in distant ejecta of the Chicxulub impact event? An experimental approach [abs]: Lunar and Planetary Science, v. 29, Abstract 1363 (CD-ROM).

Dunlop, D.J., and Özdemir, Ö., 1997, Rock magnetism: Fundamentals and frontiers: Cambridge, Cambridge University Press, 573 p.

Engelhardt, W., Arndt, J., Fecker, B., and Pankau, H.G., 1995, Suevite breccia from the Ries crater, Germany: Origin, cooling history and devitrification of impact glasses: Meteoritics, v. 30, p. 279–293.

Engelhardt, W., Arndt, J., Fecker, B., and Pankau, H.G., 1997, Suevite breccia from the Ries impact crater, Germany: Petrography, chemistry and shock metamorphism of crystalline rock clasts: Meteoritics & Planetary Science, v. 32, p. 545–554.

Evans, N.J., Gregoire, D.C., Goodfellow, W.D., Miles, N., and Veizer, J., 1994, The Cretaceous-Tertiary fireball layer, ejecta layer and coal seam: Platinum-group element content and mineralogy of size fractions: Meteoritics, v. 29, p. 223–235.

Fisher, R.V., and Schmincke, H.-U., 1984, Pyroclastic rocks: Heidelberg, Springer, 472 p.

French, B.M., 1998, Traces of catastrophe: A handbook of shock-metamorphic effects in terrestrial meteorite impact structures, LPI Contribution No. 954: Houston, Texas, Lunar and Planetary Institute, 130 p.

Freundt, A., and Schmincke, H.-U., 1998, Emplacement of ash layers related to high-grade ignimbrite P1 in the sea around Gran Canaria: Scientific Results: College Station, Texas, Proceedings of the Ocean Drilling Program, v. 174, p. 201–218.

Glass, B.P., 1990, Tektites and microtektites: Key facts and inferences: Tectonophysics, v. 171, p. 393–404, doi: 10.1016/0040-1951(90)90112-L.

Glass, B.P., Muenow, D.W., Bohor, B.F., and Meeker, G.P., 1997, Fragmentation and hydration of tektites and microtektites: Meteoritics & Planetary Science, v. 32, no. 3, p. 333–341.

Graup, G., 1981, Terrestrial chondrules, glass spheres and accretionary lapilli from the suevite, Ries Crater, Germany: Earth and Planetary Science Letters, v. 55, p. 407–418, doi: 10.1016/0012-821X(81)90168-0.

Graup, G., 1999, Carbonate-silicate liquid immiscibility upon impact melting: Ries Crater, Germany: Meteoritics & Planetary Science, v. 34, p. 425–438.

Griscom, D.L., Beltrán-López, V., Merzbacher, C.I., and Bolden, E., 1999, Electron spin resonance of 65-million-year-old glasses and rocks from the Cretaceous/Tertiary boundary: Journal of Non-Crystalline Solids, v. 253, no. 1-3, p. 1–22, doi: 10.1016/S0022-3093(99)00340-3.

Griscom, D.L., Beltrán-López, V., Pope, K.O., and Ocampo, A.C., 2003, New geochemical insights from electron-spin-resonance studies of Mn^{2+} and SO_3^- in calcites: Quantitative analyses of Chicxulub crater ejecta from Belize and southern Mexico with comparison to limestones from distal Cretaceous-Tertiary boundary sites, *in* Koeberl, C., and Martínez-Ruiz, F., eds., Impact markers in the stratigraphic record: Impact studies: Heidelberg, Springer, p. 229–270.

Gupta, S.C., Ahrens, T.J., and Yang, W., 2001, Shock-induced vaporization of anhydrite and global cooling from the K/T impact: Earth and Planetary Science Letters, v. 188, no. 3-4, p. 399–412, doi: 10.1016/S0012-821X(01)00327-2.

Hart, R.J., Cloete, M., McDonald, I., Carlson, R.W., and Andreoli, M.A.G., 2002, Siderophile-rich inclusions from the Morokweng impact melt sheet, South Africa: Possible fragments of a chondritic meteorite: Earth and Planetary Science Letters, v. 198, no. 1-2, p. 49–62, doi: 10.1016/S0012-821X(02)00497-1.

Heider, F., Koerner, U., and Bitschene, P.R., 1993, Volcanic ash particles as carriers of remanent magnetization in deep-sea sediments from the Kerguelen Plateau: Earth and Planetary Science Letters, v. 118, no. 1-4, p. 121–134, doi: 10.1016/0012-821X(93)90163-4.

Hörz, F., and See, T. H., 2000, Quenched olivines and pyroxenes in impact melts from Meteor Crater, AZ [abs]: Lunar and Planetary Science, v. 31, Abstract 1737 (CD-ROM).

Hörz, F., See, T. H., Yang, V., and Mittlefehldt, D. W., 1998, Major element composition of ballistically dispersed melt particles from Meteor Crater, AZ [abs]: Lunar and Planetary Science, v. 29, Abstract 1777 (CD-ROM).

Hörz, F., Mittlefehldt, D.W., See, T.H., and Galindo, C., 2002, Petrographic studies of the impact melts from Meteor Crater: AZ: Meteoritics & Planetary Science, v. 37, no. 4, p. 501–532.

Hower, J.C., Eslinger, E., Hower, M.E., and Perry, E.A., 1976, Mechanism of burial metamorphism of argillaceous sediment: 1. Mineralogical and chemical evidence: Geological Society of America Bulletin, v. 87, p. 725–737.

Hrouda, F., 1994, A technique for the measurement of thermal changes of MS of weakly magnetic rocks by the CS-2 apparatus and KLY-2 kappabridge: Geophysical Journal International, v. 118, p. 604–612.

Hunt, C.P., Moskowitz, B.M., and Banerjee, S.K., 1995, Magnetic properties of rocks and minerals, *in* Ahrens, T.J., ed., A handbook of physical constants: Rock physics and phase relations: AGU Reference Shelf: San Francisco, California, American Geophysical Union, p. 189–204.

Izett, G.A., 1990, The Cretaceous/Tertiary boundary interval, Ration basin, Colorado and New Mexico, and its content of shock-metamorphosed minerals: Evidence relevant to the K/T boundary impact-extinction event: Geological Society of America Special Paper 249, 100 p.

Izett, G.A., 1991, Tektites in Cretaceous-Tertiary boundary rocks on Haiti and their bearing on the Alvarez impact extinction hypothesis: Journal of Geophysical Research, v. 96, no. E4, p. 20,879–20,905.

Japel, S., Prewitt, C., Boctor, N., and Veblen, D., 2002, Iron-nickel phosphides at high pressures and temperatures [abs]: Eos (Transactions, American Geophysical Union), v. 83(47), Fall Meeting Supplement, Abstract MR61A-1025.

Jones, A.P., Claeys, P., and Heuschkel, S., 2000, Impact melting of carbonates from the Chicxulub impact crater, *in* Gilmour, I., and Koeberl, C., eds., Impacts and the early Earth: Lecture notes in earth sciences: Heidelberg, Springer, p. 343–361.

Just, J., Kontny, A., de Wall, H., and Hirt, A., 2005, Development of magnetic fabrics during hydrothermal alteration in the Soultz-sous-Forêts granite from the EPS-1 drilling, Upper Rhine Graben, *in* Martín-Hernández, F., Lüneburg, C.M., Aubourg, C., and Jackson, M., eds., Magnetic Fabric: Methods and Applications: Geological Society of London Special Publication 238, p. 509–526.

Kamo, S.L., and Krogh, T.E., 1995, Chicxulub crater: Source for shocked zircon crystals from the Cretaceous-Tertiary boundary layer, Saskatchewan: Evidence from new U-Pb data: Geology, v. 23, p. 281–284, doi: 10.1130/0091-7613(1995)023<0281:CCSFSZ>2.3.CO;2.

Keller, G., Stinnesbeck, W., and López-Oliva, J.-G., 1994, Age, deposition and biotic effects of the Cretaceous/Tertiary boundary event at Mimbral, NE Mexico: Palaios, v. 9, p. 144–157.

Keller, G., López-Oliva, J.-G., Stinnesbeck, W., and Adatte, T., 1997, Age, stratigraphy and deposition of near-K/T siliciclastic deposits in Mexico: Relation to bolide impact?: Geological Society of America Bulletin, v. 109, p. 410–428, doi: 10.1130/0016-7606(1997)109<0410:ASADON>2.3.CO;2.

Kerschhofer, L., Kettrup, B., Deutsch, A., and Masaitis, V., 2000, Al-rich orthopyroxenes in impact melt coatings of gneiss bombs from Popigai, Russia [abs]: Lunar and Planetary Science, v. 31, Abstract 1360 (CD-ROM).

Kettrup, B., and Deutsch, A., 2003, Geochemical variability of the Yucatán basement: Constraints from crystalline clasts in Chicxulub impactites: Meteoritics & Planetary Science, v. 38, p. 1079–1092.

Kettrup, B., Deutsch, A., Ostermann, M., and Agrinier, P., 2000, Chicxulub impactites: Geochemical clues to the precursor rocks: Meteoritics & Planetary Science, v. 35, p. 1229–1238.

Kettrup, B., Deutsch, A., and Masaitis, V.L., 2003, Homogeneous impact melts produced by a heterogeneous target? Sr-Nd isotopic evidence from the Popigai crater, Russia: Geochimica et Cosmochimica Acta, v. 67, no. 4, p. 733–750, doi: 10.1016/S0016-7037(02)01143-2.

Kletetschka, G., and Wasilewski, P.J., 2002, Grain size limit for SD hematite: Physics of the Earth and Planetary Interiors, v. 129, no. 1-2, p. 173–179, doi: 10.1016/S0031-9201(01)00271-0.

Kletetschka, G., Wasilewski, P.J., Ocampo, A.C., and Pope, K.O., 2001, Magnetic properties of the ejecta blanket from the Chicxulub impact crater: Analog for robotic exploration of similar deposits on Mars [abs.]: American Geophysical Union, Spring Meeting, Boston, Massachusetts, #P32A-07.

Koeberl, C., 1994, Tektite origin by hypervelocity asteroidal or cometary impact: Target rocks, source craters, and mechanisms, *in* Dressler, B.O., Grieve, R.A.F., and Sharpton, V.L., eds., Large meteorite impacts and planetary evolution: Geological Society of America Special Paper 293, p. 133–151.

Koeberl, C., 1998, Identification of meteoritic components in impactites, *in* Grady, M.M., Hutchison, R., McCall, G.J.H., and Rothery, D.A., eds., Meteorites: Flux with time and impact effects: London, Geological Society Special Publication 140, p. 133–153.

Koeberl, C., and MacLeod, K.G., editors, 2002, Catastrophic events and mass extinctions: Impacts and beyond: Geological Society of America Special Paper 356, 746 p.

Koeberl, C., and Martínez-Ruiz, F.C., editors, 2003, Impact markers in the stratigraphic record: Impact Studies: Heidelberg, Springer, 347 p.

Koeberl, C., and Sigurdsson, H., 1992, Geochemistry of impact glasses from the K/T boundary in Haiti: Relation to smectites and a new type of glass: Geochimica et Cosmochimica Acta, v. 56, p. 2113–2129, doi: 10.1016/0016-7037(92)90333-E.

Kontny, A., Vahle, C., and de Wall, H., 2003, Characteristic magnetic behavior of subaerial and submarine lava units from the Hawaiian Scientific Drilling Project (HSDP-2): Geochemistry, Geophysics, Geosystems, v. 4, no. 2, p. 1–31, doi: 10.1029/2002GC000304.

Kring, D.A., and Boynton, W.V., 1992, Petrogenesis of an augite-bearing melt rock in the Chicxulub structure and its relationship to K/T impact spherules in Haiti: Nature, v. 358, p. 141–143, doi: 10.1038/358141a0.

Kring, D.A., and Durda, D.D., 2002, Trajectories and distribution of material ejected from the Chicxulub impact crater: Implications for postimpact wildfires: Journal of Geophysical Research, v. 107, no. E8, p. 1–22.

Kring, D.A., Hörz, F., and Zürcher, L., 2003, Initial assessment of the excavation and deposition of impact lithologies exposed by the Chicxulub Scientific Drilling Project, Yaxcopoil, Mexico [abs]: Lunar and Planetary Science, v. 34, Abstract 1641 (CD-ROM).

Krogh, T.E., Kamo, S.L., and Bohor, B.F., 1993a, Fingerprinting the K/T impact site and determining the time of impact by U-Pb dating of single shocked zircons from distal ejecta: Earth and Planetary Science Letters, v. 119, no. 3, p. 425–429, doi: 10.1016/0012-821X(93)90150-8.

Krogh, T.E., Kamo, S.L., Sharpton, V.L., Marin, L.E., and Hildebrand, A.R., 1993b, U-Pb ages of single shocked zircons linking distal K/T ejecta to the Chicxulub crater: Nature, v. 366, p. 731–734, doi: 10.1038/366731a0.

López-Oliva, J.-G., and Keller, G., 1996, Age and stratigraphy of near-K/T boundary clastic deposits in northeastern Mexico, *in* Ryder, G., Fastovsky, D., and Gartner, S., eds., The Cretaceous-Tertiary boundary event and other catastrophes in Earth history: Geological Society of America Special Paper 307, p. 227–242.

López-Ramos, E., 1975, Geological summary of the Yucatán Peninsula, *in* Nairn, A.E.M., and Stehli, F.G., eds., Ocean basins and margins: The Gulf of Mexico and the Caribbean: New York, Plenum Press, p. 257–282.

Lowrie, W., 1990, Identification of ferromagnetic minerals in a rock by coercivity and unblocking temperature properties: Geophysical Research Letters, v. 17, no. 2, p. 159–162.

Luder, T., Benz, W., and Stocker, T.F., 2003, A model for long-term climatic effects of impacts: Journal of Geophysical Research, v. 108, no. E7, p. 1–12, doi: 10.1029/2002JE001894.

Machel, H.G., 2000, Application of cathodoluminescence to carbonate diagenesis, *in* Pagel, M., Barbin, V., Blanc, P., and Ohnenstetter, D., eds., Cathodoluminescence in geosciences: Heidelberg, Springer, p. 271–301.

Manville, V., White, J.D.L., Houghton, B.F., and Wilson, C.J.N., 1998, The saturation behaviour of pumice and some sedimentological implications: Sedimentary Geology, v. 119, no. 1-2, p. 5–16, doi: 10.1016/S0037-0738(98)00057-8.

Martinez, I., Agrinier, P., Schärer, U., and Javoy, M., 1994, A SEM-ATEM and stable isotope study of carbonates from the Haughton impact crater, Canada: Earth and Planetary Science Letters, v. 121, p. 559–574, doi: 10.1016/0012-821X(94)90091-4.

Martínez-Ruiz, F., Ortega-Huertas, M., Palomo-Delgado, I., and Acquafredda, P., 1997, Quench textures in altered spherules from the Cretaceous-Tertiary boundary layer at Agost and Caravaca, SE Spain: Sedimentary Geology, v. 113, no. 1-2, p. 137–147, doi: 10.1016/S0037-0738(97)00057-2.

Maruoka, T., and Koeberl, C., 2003, Acid-neutralizing scenario after the Cretaceous-Tertiary impact event: Geology, v. 31, p. 489–492, doi: 10.1130/0091-7613(2003)031<0489:ASATCI>2.0.CO;2.

Masaitis, V.L., 1994, Impactites from Popigai Crater, *in* Dressler, B.O., Grieve, R.A.F., and Sharpton, V.L., eds., Large meteorite impacts and planetary evolution: Geological Society of America Special Paper 293, p. 153–162.

Masaitis, V. L., and Deutsch, A., 1999, Popigai: Gneiss bombs coated with impact melt—heating in the fireball? [abs]: Lunar and Planetary Science, v. 30, Abstract 1237 (CD-ROM).

Mittlefehldt, D.W., See, T.H., and Hörz, F., 1992, Dissemination and fractionation of projectile materials in the impact melts from the Wabar Crater, Saudi Arabia: Meteoritics, v. 27, p. 361–370.

Montanari, A., 1991, Authigenesis of impact spheroids in the K/T boundary clay from Italy: New constraints for high-resolution stratigraphy of terminal Cretaceous events: Journal of Sedimentary Petrology, v. 61, p. 315–339.

Montanari, A., and Koeberl, C., 2000, Impact stratigraphy, Lecture notes in earth sciences: Heidelberg, Springer, 364 p.

Morden, S.J., 1993, Magnetic analysis of K/T boundary layer clay from Stevns Klint, Denmark: Meteoritics, v. 28, p. 595–599.

Morgan, J.V., Warner, M.R., Brittan, J., Buffler, R.T., Camargo, A.Z., Christeson, G.L., Denton, P., Hildebrand, A.R., Hobbs, R., Macintyre, H., Mackenzie, G.D., Maguire, P.K.H., Marín, L.E., Nakamura, T., Pilkington, M.,

Sharpton, V.L., Snyder, D., Suarez, G., and Trejo, A., 1997, Size and morphology of the Chicxulub impact crater: Nature, v. 390, p. 472–476, doi: 10.1038/37291.

Mørk, M.B.E., Leith, D.A., and Fanavoll, S., 2001, Origin of carbonate-cemented beds on the Naglfar Dome, Vøring Basin, Norwegian Sea: Marine and Petroleum Geology, v. 18, p. 223–234, doi: 10.1016/S0264-8172(00)00062-3.

Murray, J.W., 1991, Ecology and palaeoecology of benthic foraminifera: London, Longman Scientific, 397 p.

Newman, A.C.D., and Brown, G., 1987, The chemical constitution of clays, *in* Newman, A.C.D., ed., Chemistry of clays and cay minerals: Mineralogical Society Monograph: London, Longman Scientific, p. 1–128.

Ortega-Huertas, M., Palomo-Delgado, I., Martínez-Ruiz, F., and Gonzalez, I., 1998, Geological factors controlling clay mineral patterns across the Cretaceous-Tertiary boundary in Mediterranean and Atlantic sections: Clay Minerals, v. 33, p. 483–500, doi: 10.1180/000985598545651.

Ortega-Huertas, M., Martínez-Ruiz, F., Palomo-Delgado, I., and Chamley, H., 2002, Review of the mineralogy of the Cretaceous-Tertiary boundary clay: Evidence supporting a major extraterrestrial catastrophic event: Clay Minerals, v. 37, p. 395–411, doi: 10.1180/0009855023730054.

Osinski, G.R., 2003, Impact glasses in fallout suevites from the Ries impact structure, Germany: An analytical SEM study: Meteoritics & Planetary Science, v. 38, no. 11, p. 1641–1667.

Osinski, G.R., and Spray, J.G., 2001, Impact-generated carbonate melts: Evidence from the Haughton structure, Canada: Earth and Planetary Science Letters, v. 194, no. 1-2, p. 17–29, doi: 10.1016/S0012-821X(01)00558-1.

Osinski, G.R., and Spray, J.G., 2003, Impact melt rocks from the Ries impact structure, Germany: Preliminary results of an analytical SEM study [abs]: Meteoritics & Planetary Science, v. 4, Abstract 1719 (CD-ROM).

Perez Cruz, G.A., 1993, Geologic evolution of the Burgos basin, Northeastern Mexico [PhD thesis]: Houston, Texas, Rice University, 155 p.

Philpotts, A.R., 1990, Principles of igneous and metamorphic rocks: New Jersey, Prentice Hall, 498 p.

Pierazzo, E., Kring, D.A., and Melosh, H.J., 1998, Hydrocode simulations of the Chicxulub impact event and the production of climatically active gases: Journal of Geophysical Research, v. 103, no. E12, p. 28,607–28,625, doi: 10.1029/98JE02496.

Pierazzo, E., Hahmann, A.N., and Sloan, L.C., 2003, Chicxulub and climate: Radiative perturbations of impact-produced S-bearing gases: Astrobiology, v. 3, no. 1, p. 99–118, doi: 10.1089/153110703321632453.

Pilkington, M., Ames, D.E., and Hildebrand, A.R., 2004, Magnetic mineralogy of the Yaxcopoil-1 core, Chicxulub crater: Meteoritics & Planetary Science, v. 39, no. 7, p. 831–842.

Pollastro, R.M., and Bohor, B.F., 1993, Origin and clay-mineral genesis of the Cretaceous/Tertiary boundary unit, Western Interior of North America: Clays and Clay Minerals, v. 41, no. 1, p. 7–25.

Pope, K.O., Ocampo, A.C., Fischer, A.G., Alvarez, W., Fouke, B.W., Webster, C.L., Vega, F.J., Smit, J., Fritsche, A.E., and Claeys, P., 1999, Chicxulub impact ejecta from Albion Island, Belize: Earth and Planetary Science Letters, v. 170, no. 4, p. 351–364, doi: 10.1016/S0012-821X(99)00123-5.

Rebolledo-Vieyra, M., Vera-Sánchez, P., Urrutia-Fucugauchi, J., and Marin, L.E., 2000, Physical characteristics of deposition of impact breccias Pan-African basement affinities of Chicxulub crater [abs.], Catastrophic events and mass extinctions: Impacts and beyond: LPI Contribution 1053: Houston, Texas, Lunar and Planetary Institute, p. 180–181.

Robertson, D.S., McKenna, M.C., Toon, O.B., Hope, S., and Lillegraven, J.A., 2004, Survival in the first hours of the Cenozoic: Geological Society of America Bulletin, v. 116, p. 760–768, doi: 10.1130/B25402.1.

Rocchia, R., and Robin, E., 1998, L'iridium à la limite Crétaceé-Tertiaire du site d'El Kef, Tunisie: Bulletin de la Société Géologique de France, v. 169, p. 515–526.

Ryder, G., Fastovsky, D., and Gartner, S., eds., 1996, The Cretaceous-Tertiary event and other catastrophes in Earth history: Geological Society of America Special Paper 307, 556 p.

Salge, T., Tagle, R., and Claeys, P., 2000, Accretionary lapilli from the K-T boundary site of Guayal, Mexico: Preliminary insights of expansion plume formation [abs.]: 63rd Annual Meeting, Abstract 5124 (CD-ROM).

Schlager, W., Buffler, R.T., Angstadt, D., and Phair, R., 1984, Geologic history of the southeastern Gulf of Mexico: College Station, Texas, Initial Reports, Deep Sea Drilling Project, v. 77, p. 715–738.

Schmitz, B., 1992, Chalcophile elements and Ir in continental Cretaceous-Tertiary boundary clays from the Western Interior of the USA: Geochimica et Cosmochimica Acta, v. 56, p. 1695–1703, doi: 10.1016/0016-7037(92)90235-B.

Schulte, P., 2003, The Cretaceous-Paleogene transition and Chicxulub impact ejecta in the northwestern Gulf of Mexico: Paleoenvironments, sequence stratigraphic setting, and target lithologies [Ph.D. thesis]: University of Karlsruhe, 204 p., www.ubka.uni-karlsruhe.de/cgi-bin/psview?document=2003/bau-geo/10.

Schulte, P., Stinnesbeck, W., Stüben, D., Kramar, U., Berner, Z., Keller, G., and Adatte, T., 2003, Fe-rich and K-rich mafic spherules from slumped and channelized Chicxulub ejecta deposits at the northern La Sierrita area, NE Mexico: International Journal of Earth Sciences, v. 92, p. 114–142.

Schuraytz, B.C., Sharpton, V.L., and Marín, L.E., 1994, Petrology of impact-melt rocks at the Chicxulub multiring basin, Yucatán, Mexico: Geology, v. 22, p. 868–872, doi: 10.1130/0091-7613(1994)022<0868:POIMRA>2.3.CO;2.

See, T.H., Wagstaff, J., Yang, V., Hörz, F., and McKay, G.A., 1998, Compositional variation and mixing of impact melts on microscopic scales: Meteoritics & Planetary Science, v. 33, p. 937–948.

See, T.H., Galindo, C., Golden, D.C., Yang, V., Mittlefehldt, D.W., and Hörz, F.P., 1999, Major-element composition of ballistically dispersed melts from Meteor Crater, AZ [abs.]: Lunar and Planetary Science, v. 30, Abstract 1633 (CD-ROM).

Senftle, F.E., Thorpe, A.N., May, L., Barkatt, A., Adel Hadadi, M.A., Marbury, G.S., Izett, G.A., Sigurdsson, H., and Maurrasse, F.J., 1993, Magnetic properties and Mössbauer analyses of glass from the K-T boundary, Beloc, Haiti: Lunar and Planetary Science, v. 24, p. 1275–1276.

Sharpton, V.L., Schuraytz, B.C., Burke, K., Murali, A.V., and Ryder, G., 1990, Detritus in K/T boundary clays of western North America: Evidence against a single oceanic impact, *in* Sharpton, V.L., and Ward, P.D., eds., Global catastrophes in Earth history: An interdisciplinary conference on impacts, volcanism, and mass mortality: Geological Society of America Special Paper 247, p. 349–357.

Sharpton, V.L., Marín, L.E., Carney, J.L., Lee, S., Ryder, G., Schuraytz, B.C., Ikora, P., and Spudis, P.D., 1996, A model of the Chicxulub impact basin based on evaluation of geophysical data, well-logs, and drill core samples, *in* Ryder, G., Fastovsky, D., and Gartner, S., eds., The Cretaceous-Tertiary boundary event and other catastrophes in Earth history: Geological Society of America Special Paper 307, p. 55–74.

Shata, S., and Hesse, R., 1998, A refined XRD method for the determination of chlorite composition and application to the McGerrigle Mountains anchizone in the Quebec Appalachians: Canadian Mineralogist, v. 36, p. 1525–1546.

Shau, Y.H., Feather, M.E., Essene, E.J., and Peacor, D.R., 1991, Genesis and solvus relations of submicroscopically intergrown paragonite and phengite in a blueschist from Northern California: Contributions to Mineralogy and Petrology, v. 106, p. 367–378.

Shelley, D., 1993, Igneous and metamorphic rocks under the microscope: London, Chapman & Hall, 445 p.

Smit, J., 1999, The global stratigraphy of the Cretaceous-Tertiary boundary impact ejecta: Annual Reviews of Earth and Planetary Sciences, v. 27, p. 75–113, doi: 10.1146/annurev.earth.27.1.75.

Smit, J., Alvarez, W., Montanari, A., Swinburn, N.H.M., Van Kempen, T.M., Klaver, G.T., and Lustenhouwer, W.J., 1992a, "Tektites" and microkrystites at the Cretaceous Tertiary boundary: Two strewn fields, one crater?: Lunar and Planetary Science, v. 22, p. 87–100.

Smit, J., Montanari, A., Swinburn, N.H.M., Alvarez, W., Hildebrand, A.R., Margolis, S.V., Claeys, P., Lowrie, W., and Asaro, F., 1992b, Tektite-bearing, deep-water clastic unit at the Cretaceous-Tertiary boundary in northeastern Mexico: Geology, v. 20, p. 99–103, doi: 10.1130/0091-7613(1992)020<0099:TBDWCU>2.3.CO;2.

Smit, J., Alvarez, W., Montanari, A., Claeys, P., and Grajales-Nishimura, J.M., 1996, Coarse-grained, clastic sandstone complex at the K/T boundary around the Gulf of Mexico: Deposition by tsunami waves induced by the Chicxulub impact?, *in* Ryder, G., Fastovsky, D., and Gartner, S., eds., The Cretaceous-Tertiary boundary event and other catastrophes in Earth history: Geological Society of America Special Paper 307, p. 151–182.

Soria, A.R., Liesa, C.L., Mata, M.P., Arz, J.A., Alegret, L., Arenillas, I., and Meléndez, A., 2001, Slumping and a sandbar deposit at the Cretaceous-Tertiary boundary in the El Tecolote section (northeastern Mexico): An impact-induced sediment gravity flow: Geology, v. 29, p. 231–234, doi: 10.1130/0091-7613(2001)029<0231:SAASDA>2.0.CO;2.

Steiner, M.B., and Walker, J.D., 1996, Late Silurian plutons in Yucatán: Journal of Geophysical Research, v. 101, no. B8, p. 17,727–17,735, doi: 10.1029/96JB00174.

Stinnesbeck, W., Barbarin, J.M., Keller, G., López-Oliva, J.-G., Pivnik, D.A., Lyons, J.B., Officer, C.B., Adatte, T., Graup, G., Rocchia, R., and Robin, E., 1993, Deposition of channel deposits near the Cretaceous-Tertiary boundary in northeastern Mexico: Catastrophic or "normal" sedimentary deposits?: Geology, v. 21, p. 797–800, doi: 10.1130/0091-7613(1993)021<0797:DOCDNT>2.3.CO;2.

Stinnesbeck, W., Keller, G., Adatte, T., López-Oliva, J.-G., and MacLeod, N., 1996, Cretaceous-Tertiary boundary clastic deposits in NE Mexico: Bolide impact or sea-level lowstand?, *in* MacLeod, N., and Keller, G., eds., The Cretaceous-Tertiary boundary mass extinction: Biotic and environmental events: New York, Norton Press, p. 471–517.

Stöffler, D., 1971, Progressive metamorphism and classification of shocked and brecciated crystalline rocks at impact craters: Journal of Geophysical Research, v. 76, no. E23, p. 5541–5551.

Stöffler, D., Hecht, L., Kenkmann, T., Schmitt, R.T., Tagle, R.A., and Wittmann, A., 2004, Petrography, composition, shock metamorphism, and geology of the allochthonous impact formations at the ICDP drilling site Yaxcopoil-1, Chicxulub impact crater, Mexico: Meteoritics & Planetary Science, v. 39, no. 7, p. 1035–1068.

Stroncik, N.A., and Schmincke, H.-U., 2002, Palagonite—A review: International Journal of Earth Sciences, v. 91, no. 4, p. 680–697, doi: 10.1007/s00531-001-0238-7.

Stüben, D., Kramar, U., Berner, Z., Eckhardt, J.-D., Stinnesbeck, W., Keller, G., Adatte, T., and Heide, K., 2002, Two PGE anomalies above the Cretaceous/Tertiary boundary at Beloc/Haiti: Geochemical context and consequences for the impact scenario, *in* Koeberl, C., and MacLeod, K.G., eds., Catastrophic events and mass extinctions: Impacts and beyond: Geological Society of America Special Paper 356, p. 163–188.

Thorpe, A.N., Senftle, F.E., May, L., Barkatt, A., Adel-Hadadi, M.A., Marbury, G.S., Izett, G.A., and Maurrasse, F.R., 1994, Comparison of the magnetic properties and Mössbauer analysis of glass from the Cretaceous-Tertiary boundary, Beloc, Haiti, with tektites: Journal of Geophysical Research, v. 99, no. E5, p. 10,881–10,886, doi: 10.1029/94JE00638.

Toon, O.B., Zahnle, K., Morrison, D., Turco, R.P., and Covey, C., 1997, Environmental perturbations caused by the impacts of asteroids and comets: Reviews of Geophysics, v. 35, no. 1, p. 41–78, doi: 10.1029/96RG03038.

Treiman, A.H., 1989, Carbonatite magma: Properties and processes, *in* Bell, K., ed., Carbonatites: Genesis and evolution: London, Unwin Hyman, p. 89–104.

Tucker, M.E., and Wright, V.P., 1992, Carbonate sedimentology: Oxford, Blackwell Science, 482 p.

Urrutia-Fucugauchi, J., Marín, L.E., and Sharpton, V.L., 1994, Reverse polarity magnetized melt rocks from the Cretaceous/Tertiary Chicxulub Structure, Yucatán Peninsula, Mexico: Tectonophysics, v. 237, no. 1-2, p. 105–112, doi: 10.1016/0040-1951(94)90161-9.

Utzmann, A., Hansteen, T.H., and Schmincke, H.-U., 2002, Trace element mobility during sub-seafloor alteration of basaltic glass from Ocean Drilling Program site 953 (off Gran Canaria): International Journal of Earth Sciences, v. 91, no. 4, p. 661–679, doi: 10.1007/s00531-001-0247-6.

Verma, H.C., Upadhyay, C., Tripathi, A., Tripathi, R.P., and Bhandari, N., 2002, Thermal decomposition pattern and particle size estimation of iron minerals associated with the Cretaceous-Tertiary boundary at Gubbio: Meteoritics & Planetary Science, v. 37, p. 901–909.

Warme, J.E., Morgan, M., and Kuehner, H.-C., 2002, Impact-generated carbonate accretionary lapilli in the Late Devonian Alamo breccia, *in* Koeberl, C., and MacLeod, K.G., eds., Catastrophic events and mass extinctions: Impacts and beyond: Geological Society of America Special Paper 356, p. 489–504.

Wdowiak, T.J., Armendarez, L.P., Agresti, D.G., Wade, M.L., Wdowiak, Y.S., Claeys, P., and Izett, G.A., 2001, Presence of an iron-rich nanophase material in the upper layer of the Cretaceous-Tertiary boundary clay: Meteoritics & Planetary Science, v. 36, p. 123–133.

Werner, T., and Borradaile, G.J., 1998, Homogeneous magnetic susceptibilities of tektites: Implications for extreme homogenization of source material: Physics of the Earth and Planetary Interiors, v. 108, p. 235–243, doi: 10.1016/S0031-9201(98)00098-3.

Worm, H.U., and Banerjee, S.K., 1987, Rock magnetic signature of the Cretaceous-Tertiary boundary: Geophysical Research Letters, v. 14, p. 1083–1086.

Manuscript Accepted by the Society 9 August 2004

Printed in the USA

Geological Society of America
Special Paper 384
2005

Stable isotope record of post-impact fluid activity in the core of the Yaxcopoil-1 borehole, Chicxulub impact structure, Mexico

Lukas Zurcher*
David A. Kring
Lunar and Planetary Laboratory, University of Arizona, 1629 E. University Boulevard, Tucson, Arizona 85721, USA

Mark. D. Barton
David Dettman
Mark Rollog
Department of Geosciences, University of Arizona, 1040 E. Fourth Street, Tucson, Arizona 85721, USA

ABSTRACT

Carbon, oxygen, and hydrogen isotope results from carbonate and silicate fractions of altered core samples from the Yaxcopoil-1 borehole drilled into the 65 Ma Chicxulub impact crater provide constraints on the physico-chemical parameters of the hydrothermal solutions, and their likely origin.

Yaxcopoil-1 impactites were initially permeated with calcite and halite at ambient temperature. This was followed by thermal metamorphism (diopside after igneous augite) and widespread Na-K metasomatism (feldspar after igneous plagioclase), which were overprinted by abundant lower-temperature clay and calcite.

Silicate fraction isotopic values have $\delta^{18}O_{SMOW}$ values between 10 and 23‰ indicating important isotopic exchange between impact melt (~8‰) and Cretaceous limestone (~26‰). Heavier $\delta^{18}O$ values occur over depth intervals with intense feldspar alteration (813–833 m and 864–872 m). The δD_{SMOW} values (–34 to –54‰) are chiefly influenced by smectite abundance and roughly mirror $\delta^{18}O$ values. Carbonate fraction $\delta^{18}O_{SMOW}$ values (22–30‰) are controlled by calcite contents, and several exceed the limestone signature. Most $\delta^{13}C_{PDB}$ (–1 to +2‰) values also cluster around that of local limestone, but a number are significantly lighter (down to –7‰).

Isotopic and fluid inclusion results indicate hydrothermal fluid temperatures between 270 and 100 °C, high salinities (~20%), and minor kerogen contents. These data are compatible with mineralogical constraints, which further support an increase in oxidation state with decreasing temperature. Isotopic data point to a saline CO_2-bearing fluid mixed with small amounts of reduced carbon, and decarbonation and infiltration processes. Combined results are most consistent with a basinal oil-field saline brine that was driven by impact-induced heat.

Keywords: Chicxulub impact crater, Yaxcopoil-1 impactite sequence, hydrothermal paragenesis, mineral compositions, stable isotopes, fluid sources, saline oil-field brine.

*lzurcher@geo.arizona.edu

Zurcher, L., Kring, D.A., Barton, M.D., Dettman, D., and Rollog, M., 2005, Stable isotope record of post-impact fluid activity in the core of the Yaxcopoil-1 borehole, Chicxulub impact structure, Mexico, *in* Kenkmann, T., Hörz, F., and Deutsch, A., eds., Large meteorite impacts III: Geological Society of America Special Paper 384, p. 223–238. For permission to copy, contact editing@geosociety.org.

INTRODUCTION

Continuous core recovered from the Yaxcopoil-1 borehole by the Chicxulub Scientific Drilling Program (CSDP; http://www.icdp-online.de/sites/chicxulub/index/) provided a new opportunity to evaluate the impact-related hydrothermal activity at the 65 Ma Chicxulub crater (see Fig. 1). The structure was first recognized during geophysical surveys of the petrochemical potential of the Yucatán Peninsula (Cornejo-Toledo and Hernandez-Osuna, 1950). The presence of associated igneous rocks was reported later by Penfield and Camargo-Zanoguera (1981), but it was not until the discovery of diagnostic shock-metamorphic features that the structure was identified as an impact crater (Hildebrand et al., 1991; Kring et al., 1991; Kring and Boynton, 1992) of K-T boundary age. The age was then confirmed with radiometric techniques by Swisher et al. (1992) and Sharpton et al. (1992).

Impact-generated fluid regimes can encompass large volumes of Earth's crust, and the hydrothermal system at Chicxulub (~180 km diameter) may be one of the most extensive and best preserved for study. Impact-related hydrothermal systems are important because fluid processes influence the thermal and chemical evolution of an impact crater. Furthermore, these hydrothermal fluids can lead to formation of economic mineral deposits (i.e., Sudbury; Ames et al., 1998), and may have played an important role in the origin and evolution of early life (Kring, 2000a, 2000b).

This study presents the results of a C, O, and H stable isotope survey carried out on carbonate and silicate fractions of samples collected from the altered impactite sequence at Yaxcopoil-1. Combined with geological, mineralogical, and geochemical observations (Zurcher and Kring, 2004), these data allowed us to place further constraints on the physico-chemical parameters of the hydrothermal fluid (temperature, redox state, alkalinity, and salinity). We used this information to evaluate possible mixing, likely sources, as well as extent of post-hydrothermal isotopic exchange.

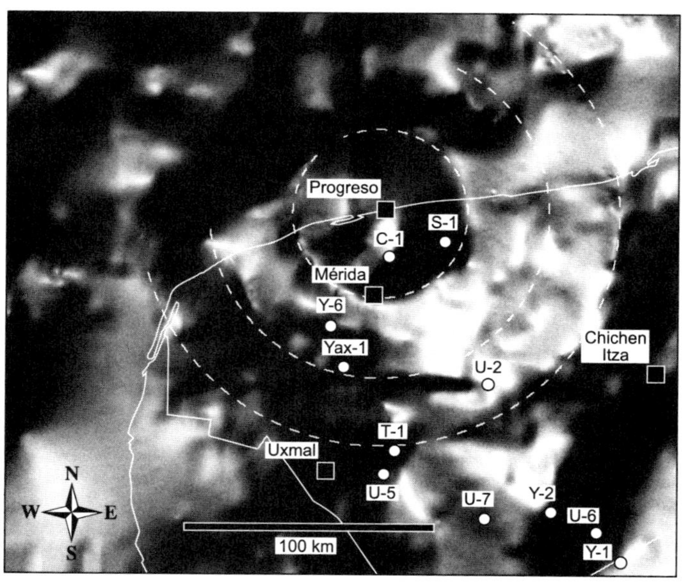

Figure 1. Bouguer gravity shaded relief map of the Chicxulub impact crater (image generated from Figure 2 in Hildebrand et al., 1991). Dashed circles mark concentric gravity anomalies. The locations of the Yaxcopoil-1 and other boreholes are shown in white circles, and selected localities are depicted by black squares.

Figure 1 shows the position of the Yaxcopoil-1 borehole relative to other boreholes in the Chicxulub impact crater. The Yaxcopoil-1 site is located ~63 km from the center of the crater. The Bouguer gravity shaded relief map (after Hildebrand et al., 1991) illustrates that the Yaxcopoil-1 site lies beyond the peak ring in the annular trough, some 20 km outside the limit of the estimated transient crater rim (Kring, 1995; Morgan et al.,1997). The nearby Yucatán-6 borehole is located ~17 km north-northwest of Yaxcopoil-1 on or just beyond the peak ring, and is also some 5 km outside the limit of the transient crater rim. Comparisons between the Yaxcopoil-1 and the Yucatán-6 sites are drawn throughout this study, because the latter was affected by similar hydrothermal alteration products (Kring and Boynton, 1992; Schuraytz et al., 1994), and because stable isotopic data have also been collected there (Kettrup et al., 2000).

The Yaxcopoil-1 borehole was drilled to a depth of 1511 m, penetrating Tertiary cover, ~100 m of impactites, and underlying megablocks of Cretaceous target carbonates and evaporites. The ~100 m impactite sequence consists of six units (Dressler et al., 2003). From top to bottom they were described as redeposited suevite (Unit 1: 13.39 m thick), suevite (Unit 2: 14.86 m thick), chocolate-brown melt breccia (Unit 3: 22.94 m thick), suevitic breccia (Unit 4: 15.26 m thick), green monomict autogene melt breccia (Unit 5: 23.86 m thick), and variegated polymict allogenic-clast melt breccia (Unit 6: 10.02 m thick). In this paper, we report isotopic results obtained from 16 samples collected from units 1 through 5.

Previous stable isotope studies on Chicxulub ejecta include Fouke et al. (2002), Kettrup et al. (2000), Dissmann et al. (1998), Smit (1997), Pope et al. (1996), Chaussidon et al. (1996), Blum and Chamberlain (1992), Blum et al. (1993), Koeberl (1993), Sharpton and Marin (1993), and Perry et al. (1992). However, none of these papers directly addressed the isotopic results in combination with the effects of hydrothermal alteration. Thus, in this paper we contribute new stable isotopic data, and examine the results in concert with the controls exerted by the hydrothermal alteration processes that affected the impactite section at Yaxcopoil-1 (Zurcher and Kring, 2004; see also Ames et al., 2004; Hecht et al., 2004; Lüders and Rickers, 2004). Previously identified igneous and hydrothermal mineralogy, paragenesis, and textural relations (Zurcher and Kring, 2004) enabled us to make an assessment of the relative volumes of alteration associations, which we utilized to better understand their relative contributions and influences on stable isotope results.

HYDROTHERMAL ALTERATION AT YAXCOPOIL-1

The understanding of hydrothermal alteration presented in Zurcher and Kring (2004) allowed an adequate evaluation of the characteristics of the fluids involved in the post-impact hydrothermal event. Without the temperature and mineralogical constraints provided by that investigation, interpretations of any isotopic data would be speculative. Thus, we first summarize the findings of that study here.

Figure 2 shows the impactite stratigraphy at Yaxcopoil-1, the relative location of the studied samples, and the estimated fracture (i.e., veins) abundances in the core. Samples are labeled by YAX-1 followed by a number, which identifies the depth of the sample in meters. Figure 2 also summarizes relative abundances, distribution in space, and compositional aspects of the main alteration phases determined from petrographic and electron microprobe analyses. The early to late time-sequence of introduction of selected alteration minerals is generally represented from left to right in the diagram.

At the Yaxcopoil-1 site, an impact melt (Unit 5) and overlying impact melt-rich breccias exhibit primary silicate minerals with comparable compositions (Kring et al., 2004; see also Dressler et al., 2004; Tuchscherer et al., 2004; Wittmann et al., 2004). Melt samples in Unit 5 and melt fragments in the breccias above are composed of compositionally equivalent microcrystalline augite ($Aeg_2Fs_{11}En_{40}Wo_{46}$), plagioclase ($Or_3Ab_{43}An_{54}$), subordinate K-feldspar, and other minor silicate phases in a groundmass of probable original glass. Primary accessory phases, including apatite, magnetite, ilmenite, and zircon, are also preserved in the impact melt unit as well as melt fragments in the overlying breccias.

Cross-cutting relations showed that the impactites were initially permeated with secondary mineral phases that originated at ambient-temperature. The timing of the introduction of calcite, halite clots in breccia matrix, gaylussite ($Na_2Ca(CO_3)_2 \cdot 5H_2O$) stringers, as well as traces of oil is consistent with modification of the impactite section shortly after its deposition, and before the onset of the hydrothermal regime. As observed under the microscope, however, this earliest secondary event encompasses a very minor fraction of the rock volume.

Calc-silicate (diopside-hedenbergite after igneous augite) and feldspar metasomatism (widespread Na and K for Ca exchange after igneous plagioclase) developed as part of the early hydrothermal stage. Vein-controlled K-feldspar with biotite, quartz, and calcite envelopes with minor magnetite ± (bornite [Cu_5FeS_4]) ± hematite, titanite, and Cl-bearing apatite were introduced with and after metasomatic feldspar. Metasomatic feldspar ± (magnetite) also occurs in diffuse replacement fronts that commonly occupy much of the rock volume. The latter style of alkali exchange could have occurred at lower temperatures. Although highly variable and locally widespread from sample to sample, these early silicate alteration associations correspond, on average, to not more than ~15% of the rock volume.

Extensive late alteration associations overprinted the early hydrothermal stage. Abundant smectite (saponite and montmorillonite, as well as minor illite) replaced probable primary glass (and probable chlorite after glass) and most mafic minerals, whereas widespread calcite was deposited as veins and open-space fillings. Several generations of calcite are apparent. In general, earlier calcite contains lower magnesium fractions that later calcite, with later calcite commonly being more abundant in Unit 1 and the upper half of Unit 2 of the impactite section. Together, these late clay and carbonate alteration associations represent up to 80% of the rock volume. However, calcite and clay contents are indirectly proportional.

The most intensely altered zone in the impactite section occurs in Unit 3 and the lower part of Unit 2. Unit 3 exhibits a low density of fracturing (see Fig. 2), implying that the permeability in this breccia was largely porosity-related. A more isolated zone of intense alteration is located in Unit 5. Here, permeability was more fracture-controlled. These intense alteration zones exhibit the best-preserved calc-silicate replacement and the best-developed Na and K metasomatism. These zones also contain apatites that are relatively depleted in chlorine and sulfur (implying that chlorine partitioned into an aqueous phase), and where smectite, conversely, contains the highest Fe/(Mg+Fe*) and Cl/F ratios. Relative to the alteration zone in Unit 5, moreover, smectite in the more intensely altered zone of Unit 3 and Unit 2 shows, on average, higher Fe/(Mg+Fe*) ratios (see Fig. 2). Metasomatic feldspar and clay abundances are indirectly proportional, with feldspar and magnetite being more abundant over the most intensely altered zones, consistent with the results of Pilkington et al. (2004) and Urrutia-Fucugauchi et al. (2004).

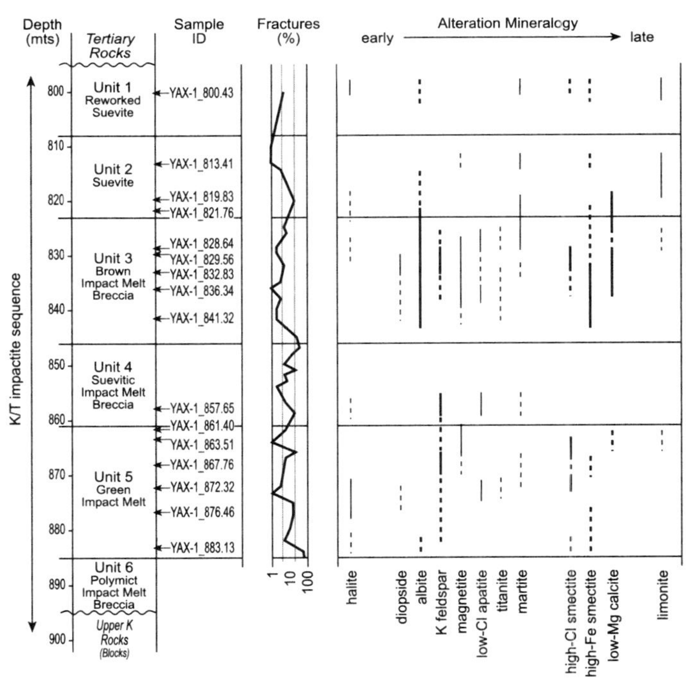

Figure 2. Impactite stratigraphy, relative location of studied samples, estimated fracture (i.e., veins) abundances, and depth-distribution of alteration minerals in the Yaxcopoil-1 core (modified from Zurcher and Kring, 2004). Albite, K-feldspar, smectite, and calcite (wide solid and dashed lines) are most abundant.

Post-hydrothermal ambient-temperature alteration included clay formation and partial oxidation of iron oxides to limonite (particularly toward the top of the impactite section). Recrystallized halite, rare zeolite, barite, and chalcopyrite in vugs and open spaces, as well as carbonate-apatite after hydrothermal apatite are also present. Francolite and limonite suggested weathering after the hydrothermal event and before deposition of the Tertiary marine sediments. The presence of sparse vermiculite, barite, and chalcopyrite further supported an influence of diagenetic processes. From petrographic observations, however, the extent of these processes appears to have been limited.

The Zurcher and Kring (2004) study confirmed that primary minerals of the impact melt and associated melt-rich breccias at Yaxcopoil-1 have a composition consistent with that of a slightly alkalic andesite. For a melt of andesitic composition, however, elevated sodium contents in augite, somewhat higher-than-expected anorthite fractions in igneous plagioclase, as well as important sulfur concentrations (~2 wt% SO_3) in primary apatite suggested that the siliceous impact melt assimilated significant fractions of carbonate and evaporite. As will be seen later, this exchange is also expressed in the $\delta^{18}O$ isotope values of silicate fractions at Yaxcopoil-1.

Early secondary calc-silicate, feldspar metasomatism, and extensive late phyllosilicate and calcite formation strongly suggested that the hydrothermal event evolved from a high-temperature (>300 °C) to a low temperature (<300 °C) regime. The composition of alteration minerals indicated that the hydrothermal fluids progressively became more iron- and chlorine-rich (Cl/F > 10), that sulfur was flushed out of the section, and that the oxidation state increased substantially (from pyroxene and magnetite to titanite and hematite stable assemblages) as temperature decreased through time.

From petrographic observations it was not possible to evaluate how widespread post-hydrothermal near-ambient-temperature clay formation may have been. However, stable isotope results from silicate fractions did provide useful constraints on its extent at Yaxcopoil-1.

ISOTOPIC METHODS

The microcrystalline nature and complex replacement history observed in Yaxcopoil-1 samples made isotopic analyses of individual mineral separates unfeasible. Thus, a less adequate but simpler sample preparation technique was adopted and tested.

Samples were dry-cut with an ISOMET saw and pulverized to ~200 mesh with agate mortar and pestle. In between samples, we ground high-purity silica and cleaned the equipment with acetone, methanol, and distilled water. Sample splits for $\delta^{18}O_{SMOW}$ and δD_{SMOW} analyses of silicate fractions were acidified in 8% HCl to free the sample of the carbonate fraction. To test for the influence of carbonate contents on the analyses, representative carbonate-rich and carbonate-poor doubles were also analyzed without acidification. Only carbonate-rich samples showed a difference of >1‰, indicating that in most instances, the isotopic contribution of carbonate to the silicate fraction was satisfactorily removed. We attribute the remaining discrepancies in the results from doubles to the contribution of other heterogeneities in the sample. For hydrogen isotope analyses of silicate fractions, we also made both ~200 mesh and several silt-size splits to test for phyllosilicate interlayer water loss during preparation. Here, the results showed a variation of <5‰, only somewhat higher than the analytical precision.

Carbonate $\delta^{18}O_{SMOW}$ and $\delta^{13}C_{PDB}$ values were determined on whole rock powders. Analysis of a relatively carbonate-poor double showed no important change in the $\delta^{18}O$ result (<1‰), indicating that the analytical procedure for carbonates did not liberate significant oxygen contained in silicate minerals. However, an important change of close to 2‰ in the $\delta^{13}C$ value was observed. Consistent with mineralogical observations, the relatively large change in this $\delta^{13}C$ value was attributed to the presence of small amounts of hydrocarbon in the sample.

Silicate $\delta^{18}O_{SMOW}$ values were obtained by conventional techniques using bromine pentafluoride (BrF_5) as the fluorinating agent. Three to six mg of sample powder was heated to 650 °C in sealed nickel vessels for 6–10 h in the presence of excess BrF_5 to generate O_2 (after Clayton and Mayeda, 1963). The extracted O_2 was then reacted with hot platinized graphite to produce CO_2. Calculated yields were typically between 95 and 100%. The isotopic composition of the CO_2 was then measured on a Finnigan MAT Delta S mass spectrometer. Samples were calibrated against in-house and public standards and cross-calibrated with NBS-28 ($\delta^{18}O_{SMOW}$ = 9.6‰). Analyses were repeated two to three times on separate fractions of the same powder. Results were reproducible to better than ±0.2‰.

Silicate δD_{SMOW} values were determined on whole rock samples combusted at 1400 °C using a Finnigan TCEA (thermal combustion elemental analyzer) coupled to a Finnigan Delta PlusXL continuous-flow gas-ratio mass spectrometer. Values were calibrated against internal standards and NIST SRM 8540 (SMOW). Analyses were repeated two to three times on separate fractions of the same powder. The precision of the results was better than ±2.5‰.

The ~200 mesh whole rock powders for $\delta^{18}O$ and $\delta^{13}C$ analyses of carbonate fractions were reacted with dehydrated phosphoric acid under vacuum at 70 °C. Isotopic ratios were measured using a KIEL-III automated carbonate preparation device coupled to a Finnigan MAT 252 gas-ratio mass spectrometer. The isotope ratio measurements were calibrated against NBS-19 and NBS-18 standards. Repeat analyses on separate fractions of the same powder indicated a precision of ±0.1‰ for $\delta^{18}O$ and ±0.06‰ for $\delta^{13}C$.

STABLE ISOTOPE RESULTS

Oxygen and Hydrogen Isotopes in Silicate Fractions

Overall, the relative abundance of metasomatic feldspar and smectite over other silicate minerals heavily influenced the oxygen and hydrogen isotopic values observed. Oxygen and hydrogen isotope results of silicate fractions are listed in Table 1. The location of samples in the impactite sequence and their oxygen and hydrogen isotopic compositions are depicted in Figure 3A and 3B.

Silicate fractions have $\delta^{18}O_{SMOW}$ values between 10 and 23‰. The values are above the oxygen signature determined for the impact melt rocks at the Chicxulub-1 (C-1) borehole in the central part of the Chicxulub crater (~8‰; Blum et al., 1993), and observed in impact glass ejected from the crater and recovered in the Haitian Cretaceous-Tertiary boundary (~6‰; Blum and Chamberlain, 1992). Moreover, some of the $\delta^{18}O$ values approximate the oxygen signature of local carbonate (~26‰), consistent with extensive isotopic exchange between the impact melt and the target lithology. A similar but less extensive $\delta^{18}O$ shift was observed in silicate-dominated samples from the Yucatán-6 borehole (~10–13‰; Kettrup et al., 2000).

Figure 3A shows that relative $\delta^{18}O$ lows occur in samples that contain the best developed calc-silicate replacement (see Fig. 3). In contrast, relative $\delta^{18}O$ highs occur between samples YAX-1_832.83 and YAX-1_813.41 in Unit 3 and Unit 2, and to some extent also in sample YAX-1_867.76 in Unit 5. These samples occur in zones where feldspar alteration is most abundant in the impactite section (see Fig. 3). These lows and highs are consistent with the early rock-buffered thermal-metamorphic, and the late fluid-dominated hydrothermal stages, respectively.

This study provides the first hydrogen isotope data from impactites in the Chicxulub crater. The δD_{SMOW} values of silicate fractions (Fig. 3B) exhibit a range between −34 and −54‰ (see Table 1). Comparison of Figure 3A and Figure 3B shows that the depth distribution of the δD values roughly mirrors that of the $\delta^{18}O$ values over the most intense alteration zone in the upper part of the impactite section, where smectites are more iron-rich. This is consistent with the fact that, for a given temperature, Fe-rich hydrated phases commonly concentrate less deuterium relative to Mg-rich phases (Suzuoki and Epstein, 1976). An equivalent δD departure to lighter values is not evident in the narrower alteration zone of Unit 5, where magnesium-rich smectite is more abundant (see Fig. 3). The lack of parallelism between the oxygen and hydrogen isotope values over these depth intervals is also consistent with the temperature-dependent effects on the formation of

TABLE 1. OXYGEN AND HYDROGEN ISOTOPE ANALYSES OF SILICATE FRACTIONS

Stratigraphy	Sample ID	$\delta^{18}O^{silicate}_{SMOW}$ (‰)	$\delta^{18}O^{silicate}$ σ*	$\delta D^{silicate}_{SMOW}$ (‰)	$\delta D^{silicate}$ σ*
Unit 1	YAX-1_800.43	17.41	0.02	−51.59	1.06
Unit 2	YAX-1_813.41	18.96	0.02	−46.61	3.10
	YAX-1_816.86	13.97	0.02	−36.74	1.50
	YAX-1_819.83	20.99§	0.02	N.D.†	N.D.†
	YAX-1_819.83	16.48	0.02	−41.69	1.34
	YAX-1_821.76	16.99	0.01	N.D.†	N.D.†
	YAX-1_821.76	N.D.†	N.D.†	−41.46	1.50
	YAX-1_823.05	15.24	0.02	−51.31	1.50
Unit 3	YAX-1_828.64	20.19	0.02	−47.83	1.50
	YAX-1_829.56	16.80	0.01	−54.06	3.10
	YAX-1_832.83	22.94	0.01	−41.41	3.10
	YAX-1_836.34	17.11	0.02	−37.96	1.50
	YAX-1_841.32	14.04	0.02	−37.11	3.10
	YAX-1_841.32	14.97§	0.02	−34.99§	1.50
Unit 4	YAX-1_857.65	14.30§	0.03	−40.75	2.44
	YAX-1_857.65	13.08	0.02	−34.22§	1.34
Unit 5	YAX-1_861.40	13.34	0.02	−34.12§	3.10
	YAX-1_861.40	14.66§	0.02	−36.00	1.50
	YAX-1_863.51	14.49§	0.02	−41.20	3.10
	YAX-1_863.51	11.80	0.02	−39.40§	1.50
	YAX-1_867.76	17.34	0.02	−39.78	1.50
	YAX-1_872.32	9.93	0.02	−40.82	1.50
	YAX-1_876.46	12.97§	0.03	−45.05§	3.10
	YAX-1_876.46	10.82	0.02	−50.25	1.34
	YAX-1_883.13	18.59	0.02	N.D.†	N.D.†

*The analytical precision is ±0.2‰ for $\delta^{18}O$ and ±2.5‰ for δD (1 sigma).
†N.D—no data.
§Not acidified and slit-size double for oxygen and hydrogen, respectively.

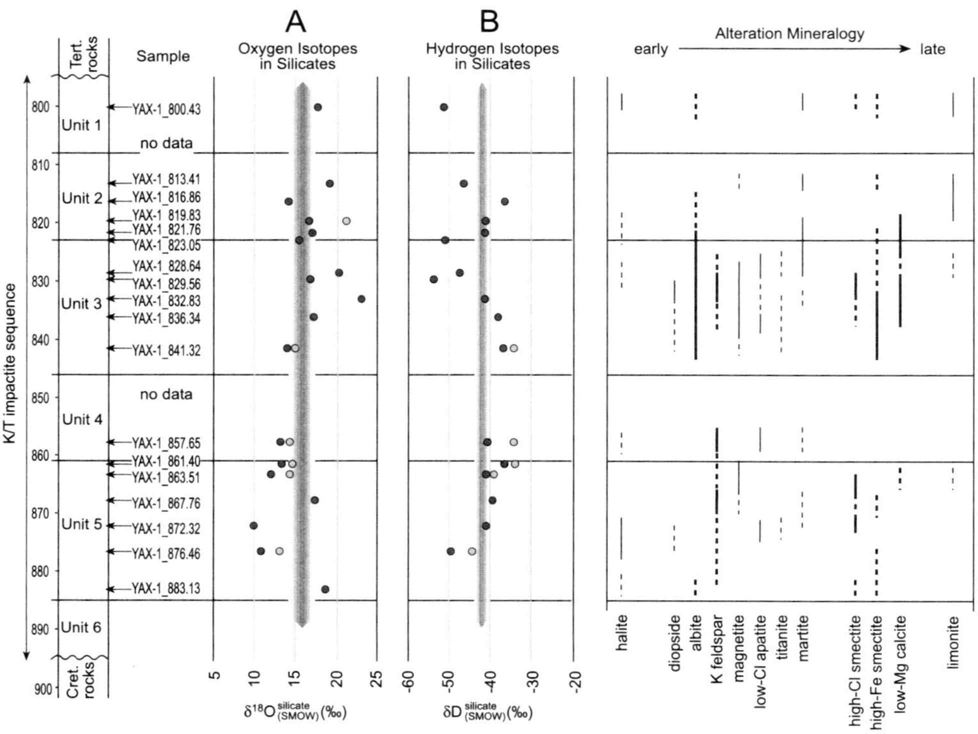

Figure 3. (A) Depth-distribution of oxygen isotope results on silicate fractions. Light-gray circles depict analyses of doubles prepared without acidification. Shaded range represents the mean of the isotopic data and the 5% statistical significance level (excluding non-acidified doubles). (B) Depth-distribution of hydrogen isotope results on silicate fractions. Light-gray circles depict analyses of silt-sized doubles. Shaded range represents the mean of the isotopic data and the 5% statistical significance level (excluding silt-sized doubles).

late smectite relative to those of early calc-silicate and feldspar. Furthermore, smaller shifts to heavier δD values appear to occur about zones where fracture densities are higher toward the top and bottom of the intense alteration zone in Unit 3 and Unit 2 and about the contact between Unit 5 and Unit 4 (see Fig. 2).

The δD and $\delta^{18}O$ values of the raw rock data (see points plotted in Fig. 5; this figure will be addressed in detail in a subsequent section) show that most of the oxygen and hydrogen isotopic values plot well to the right of the kaolinite line at 25 °C. This indicates that most values have not been significantly affected by later lower-temperature supergene processes, consistent with the very slow rates of isotopic exchange that occur at low temperatures. Partial modification of these signatures is only apparent in the upper parts of the impactite sequence, where limonite after iron oxide is more manifest.

Oxygen and Carbon Isotopes in Carbonate Fractions

The contents of calcite largely controlled the isotopic values obtained from carbonate fractions. Oxygen and carbon isotope results are given in Table 2, and isotopic compositions as a function of depth are illustrated in Figure 4A and 4B.

Carbonate fractions exhibit heavier $\delta^{18}O_{SMOW}$ values (22–30‰) relative to silicate fractions. This is consistent with extensive isotopic exchange between the hydrothermal fluid and target carbonate rocks. In general, $\delta^{18}O$ values that are heavier than the local carbonate host occur toward the top of the impactite section, where late relatively Mg-rich calcite is more abundant. It is noteworthy that all $\delta^{18}O$ values obtained at Yaxcopoil-1, however, are higher than the Yucatán-6 data (10–20‰; Kettrup et al., 2000).

Figure 4A shows that within the intense alteration zone of Unit 3 and Unit 2, the lowest $\delta^{18}O$ values occur in samples YAX-1_836.34, YAX-1_832.83, YAX-1_821.76, and YAX-1_819.83. Samples YAX-1_829.56 and YAX-1_828.64 are also located within this intense alteration zone, but exhibit heavier $\delta^{18}O$ values. These samples contain early low-Mg calcite. A minor $\delta^{18}O$ low also occurs in and above the more calcite-poor alteration zone in Unit 5 (samples YAX-1_872.32 and YAX-1_867.76). Here, however, the data are difficult to evaluate given that the contrast is not statistically significant. Additional low $\delta^{18}O$ values are contained in sample YAX-1_857.65 that is K-feldspar–rich and very calcite-poor, and in sample YAX-1_883.13 that contains a tectonic (i.e., fault-related) breccia with late (unrelated to impact-induced hydrothermal event?) translucent calcite rather than the typical turbid calcite observed in all other samples. Conversely, the highest $\delta^{18}O$ values occur in a higher-T albite ± magnetite and calcite-rich sample (YAX-1_841.32), and an albite ± (apatite) vein with a calcite-filled center (YAX-1_800.43). Calcite in these two samples is late and Mg-rich. Thus, the changes in $\delta^{18}O$ values of carbonate fractions appear to conform to compositional changes, which are in turn dependent on temperature.

Figure 4B shows that with the exception of samples YAX-1_867.76 (−1.2‰), YAX-1_863.51 (−7.2‰; see Table 2), and YAX-1_861.40 (−0.95‰), which contain traces of kerogen, $\delta^{13}C_{PDB}$ values from samples at Yaxcopoil-1 occupy a narrow range between 0 and +2‰. The $\delta^{13}C$ value of the local target

TABLE 2. OXYGEN AND CARBON ISOTOPE ANALYSES OF CARBONATE FRACTIONS

Stratigraphy	Sample ID	$\delta^{18}O^{carbonate}_{SMOW}$ (‰)	$\delta^{18}O^{carbonate}$ σ*	$\delta^{13}C^{carbonate}_{PDB}$ (‰)	$\delta^{13}C^{carbonate}$ σ*
Unit 1	YAX-1_800.43	28.76 (vn)†	0.09	1.78 (vn)†	0.04
	YAX-1_800.43	26.09	0.05	1.82	0.02
Unit 2	YAX-1_813.41	27.51	0.08	2.07	0.02
	YAX-1_816.86	26.57	0.06	1.86	0.06
	YAX-1_819.83	23.63	0.06	1.90	0.01
	YAX-1_821.76	23.05	0.06	1.64	0.07
	YAX-1_823.05	24.66	0.04	1.56	0.03
Unit 3	YAX-1_828.64	27.19	0.08	1.36	0.04
	YAX-1_829.56	25.90	0.05	1.44	0.04
	YAX-1_832.83	21.75	0.09	1.18	0.04
	YAX-1_836.34	22.71	0.03	1.34	0.04
	YAX-1_841.32	29.85	0.09	1.77	0.02
Unit 4	YAX-1_857.65	23.61	0.07	0.39	0.08
Unit 5	YAX-1_861.40	24.46§	0.06	0.61§	0.07
	YAX-1_861.40	25.01	0.05	–0.95	0.03
	YAX-1_863.51	26.12	0.04	–7.15	0.02
	YAX-1_867.76	25.18	0.09	–1.20	0.03
	YAX-1_872.32	25.46	0.07	0.78	0.02
	YAX-1_876.46	26.27	0.07	0.50	0.08
	YAX-1_883.13	23.36	0.04	1.95	0.03

*The analytical precision is ±0.1‰ for $\delta^{18}O$ and ±0.06‰ for $\delta^{13}C$ (1 sigma).
†Calcite vein.
§Carbonate-poor double.

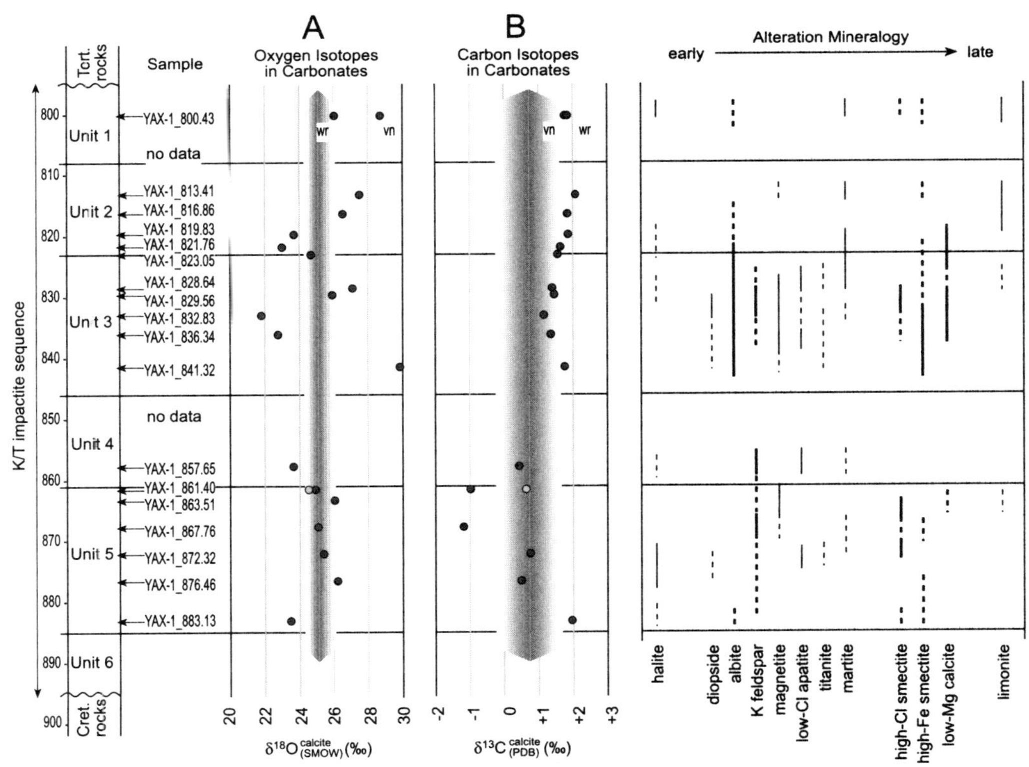

Figure 4. (A) Depth-distribution of oxygen isotope results on carbonate fractions. (B) Depth-distribution of carbon isotope results on carbonate fractions. Light-gray circles depict analysis of double. Shaded range represents the mean of the isotopic data and the 5% statistical significance level. wr—wall rock calcite; vn—calcite vein.

carbonate is +2‰. As with oxygen isotopes from carbonate fractions, $^{13}C/^{12}C$ ratios in Yaxcopoil-1 samples are heavier than the values between −1 and −5‰ determined in samples at Yucatán-6 by Kettrup et al. (2000). However, like at Yaxcopoil-1, the lighter $\delta^{13}C$ (and heavier $\delta^{18}O$) values in Yucatán-6 samples are reported to contain carbonaceous material (data plotted in Fig. 6), consistent with partial oxidation of organic carbon derived from the hydrocarbon-bearing target rocks below. At Yaxcopoil-1, too, oil-bearing fractures and oil-filled pores were observed in Unit 6 (samples YAX-1_900.44, YAX-1_910.60, and YAX-1_916.39), and in the target carbonates below (i.e., sample YAX-1_1398.53).

Relative to carbonate fraction $\delta^{18}O$ values, a modest $\delta^{13}C$ low at Yaxcopoil-1 also appears to occur over the most altered zone in Unit 3. Nevertheless, again, the contrast of this anomaly is not sufficiently significant to be evaluated with confidence. The higher $\delta^{13}C$ values apparent toward the top of the impactite section (and also in the translucent calcite-bearing breccia in sample YAX-1_883.13 at the bottom of Unit 5) correspond to samples that contain late Mg-rich calcite.

Overall, both silicate and carbonate fraction isotopic data exhibit the largest variations about the intense alteration zones in Unit 3 and Unit 2, and in Unit 5. These results not only reflect the influence of mineralogy but also of temperature-dependent fractionation.

DISCUSSION

In combination with mineralogical studies, carbon, oxygen, and hydrogen isotope results provided a wealth of information on the composition of hydrothermal solutions, processes, and fluid sources (Valley, 1986). In the following sections, we use isotopic results in conjunction with petrographic and mineral compositional constraints to infer fluid parameters (temperature, redox state, pH, salinity), influence of mixing, decarbonation, infiltration, and water-rock interaction, as well as possible post-hydrothermal isotopic modifications. This information is then used to deduce the likely origin of the fluids responsible for the hydrothermal alteration in the impactite section at Yaxcopoil-1.

Post-Hydrothermal Modification of Isotopic Signatures

In order to assess the significance of the isotopic results, we evaluated first the possible effects of post-hydrothermal processes on the isotopic signatures. The silicate fraction isotopic results are shown in the δD versus $\delta^{18}O$ plot of Figure 5. Filled squares represent the raw rock data. As mentioned previously, only one sample (YAX-1_832.83) plots near the projected kaolinite line at 25°C of Savin and Epstein (1970). Five additional samples (YAX-1_829.56, YAX-1_828.64, YAX-1_813.41, and YAX-1_800.43) cluster about $\delta^{18}O$ and δD values of 17 and

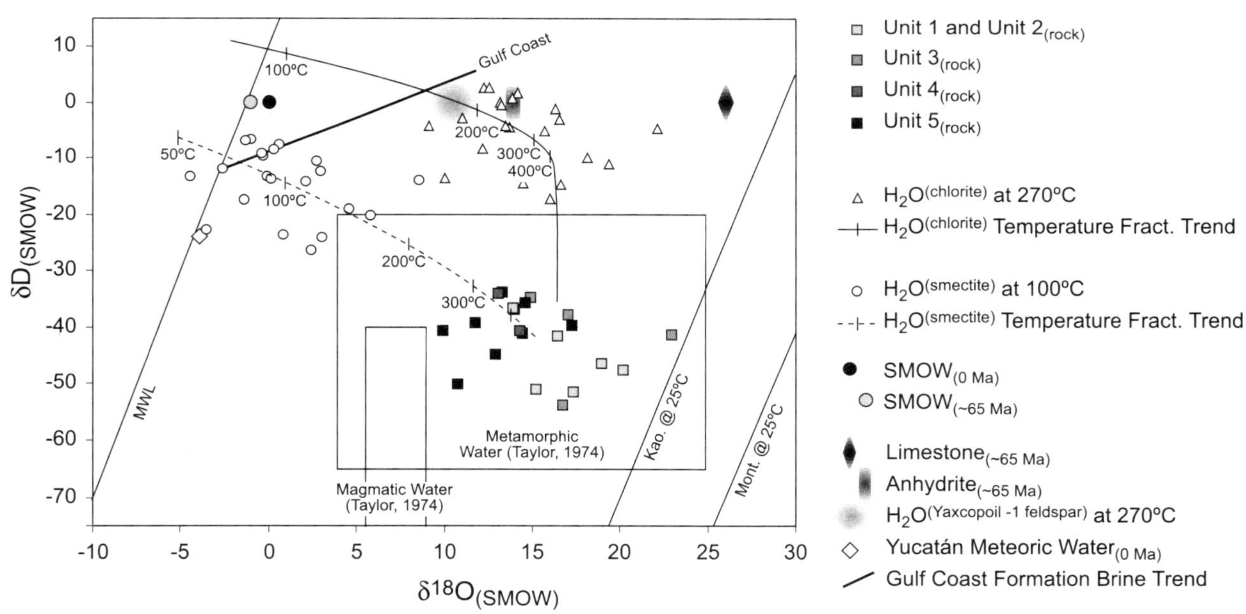

Figure 5. δD_{SMOW} vs. $\delta^{18}O_{SMOW}$ plot of isotope results on silicate fractions. Filled squares depict the distribution of the raw rock data. Open circles show waters in equilibrium with smectite at 100 °C computed with the fractionation factors of Yeh (1980) and Yeh and Savin (1977). Open triangles represent waters in equilibrium with chlorite at 270 °C calculated with the fractionation factors of Cole and Ripley (1999) and Graham et al. (1984) for oxygen and hydrogen, respectively. The Meteoric Water Line (MWL) of Craig (1961), the kaolinite and montmorillonite lines at 25 °C of Savin and Epstein (1970), limestone (filled diamond), anhydrite (filled rectangle), present-day northern Yucatán meteoric water (open diamond), present-day SMOW (black-filled circle), and Upper Cretaceous SMOW (gray-filled circle) are included for reference. The black heavy line represents the Gulf Coast oil brine isotopic trend extracted from Barnes (1979).

−50‰, respectively. A common feature of these six samples is that they are located toward the top of the impactite sequence, where post-hydrothermal weathering is most evident. Thus, the isotopic composition of these samples was likely affected by variable post-hydrothermal isotopic exchange related to near ambient-temperature processes. The other 13 samples, however, plot well to the left of the kaolinite line (and certainly the montmorillonite line), strongly suggesting that these values are the product of higher-temperature processes.

Weathering before the onset of Tertiary marine sediment deposition as well as diagenetic processes that must have been active at Yaxcopoil-1 throughout the Cenozoic appear to have had a limited influence on the preexisting isotopic compositions. Preservation of isotopic signatures is also reported in studies of porphyry copper deposits (Sheppard et al., 1969), where the isotopic values of hydrothermal clays that formed at higher temperatures are not significantly affected by later ambient-temperature supergene processes. Similarly, research on fine-grained Paleocene weathering-related clays has shown that these clays have not equilibrated with present-day ground waters (Taylor, 1974). Consequently, we believe that the original hypogene feldspar- and smectite-dominated isotopic signatures produced by the impact-related hydrothermal event are largely preserved, particularly in the bottom half of the impactite section.

Temperature and Depth of Formation of the Hydrothermal Event

Temperature is the single most important factor in isotopic fractionation. Independent temperature estimates at Yaxcopoil-1 from fluid inclusion work by Lüders et al. (2003) and Lüders and Rickers (2004) place the best constraints available to date. We summarize their results in the following paragraph.

Fluid inclusions in late calcite from the impactite sequence at Yaxcopoil-1 suggest temperatures slightly in excess of 100 °C. These inclusions commonly contain one fluid phase, but vapor bubbles occupying up to 3% of the volume are also present. Fluid inclusions in hydrothermal quartz veins (with anhydrite selvages) just 100 m below the base of the impactite section, in contrast, typically exhibit two phases with or without halite. A number of these fluid inclusions contain hydrocarbons, show high salinities (~20 wt% eq. NaCl), and exhibit homogenization temperatures of up to 270 °C. The presence of halite in parts of the impactite rock column supports these data and implies that fluid salinities may have locally exceeded 26 wt% (halite saturation). It is worth noting that halite is widespread in the target carbonates below the impactite section.

The fluid inclusion data from quartz are consistent with the presence of a higher temperature hydrothermal stage, in part represented by early K-feldspar ± (quartz) ± magnetite veins in the impactite section (the mineralogy of these early veins changes downward to K-feldspar ± pyrite ± [quartz] veins about the contact with the target carbonate, and to anhydrite ± pyrite ± [quartz] veins in the carbonate rocks below). The lower temperatures suggested by fluid inclusions in calcite are also in agreement with the late lower-*T* hydrothermal smectite and calcite overprint observed at Yaxcopoil-1. Fluid inclusion salinities and temperatures further suggest that the pressure of emplacement of the hydrothermal products must have been low (on the order of one hundred bars). Even though a particularly vapor-rich phase is not present in fluid inclusions, it appears that temperatures in excess of 250 °C would have been hard to achieve without boiling in this shallow environment.

These independent temperature data provided the necessary framework to test the probable isotopic compositions of waters in equilibrium with given alteration minerals, and attempt deduction of the likely sources of the hydrothermal fluids at Yaxcopoil-1. Constraints on fluid composition, the possible influence of these parameters on the isotopic results, and other processes are examined first.

Composition of Hydrothermal Fluids

The chemical state of hydrothermal fluids is controlled by the initial composition and by subsequent changes caused by reactions with wall rocks as well as with previously deposited minerals. Mixing of fluids from different sources or boiling also change the chemical state. Therefore, inference of fluid composition requires consideration of several interacting variables. In the following sections, we present the most relevant arguments used to constrain the complex nature of the impact-related hydrothermal solutions at Yaxcopoil-1.

Redox State

The presence of sulfides as well as hydrocarbons, and the absence of iron oxides in the target carbonate rocks below the impactite section suggest relatively reduced initial conditions. However, the presence of anhydrite ± pyrite ± (quartz) veins points to a sulfate-stable fluid that was already quite oxidized before reaching the more mafic (and probably relatively reducing) impactite host.

The near-absence of sulfides and increasing hematite to magnetite ratios from bottom to top in the impactite section indicate that the oxidation state increased as the hydrothermal fluid evolved (Zurcher and Kring, 2004) from higher to lower temperatures. Furthermore, decreasing Mg/(Mg+Fe*) ratios in phyllosilicates also support more oxidizing conditions over the most intensely altered zones.

The light $\delta^{13}C$ values from the upper part of Unit 5, and in particular the exceptional $\delta^{13}C$ value of −7.2‰, however, strongly suggest a significant isotopic contribution from reduced hydrocarbons sourced in the carbonate section below. Because the isotopic composition of this calcite would differ substantially from that of the CH_4-bearing solution from which it precipitated (Ohmoto and Rye, 1979), a mixture of only a small amount of reduced carbon with the otherwise rather oxidized hydrothermal fluid could account for the large $\delta^{13}C$ shift observed.

Salinity

The presence of halite clots and recrystallized halite in a number of samples from the impactite section (Zurcher and Kring, 2004), the widespread occurrence of halite in the carbonate rocks below, as well as the fluid inclusion data reviewed above (Lüders et al., 2003), indicate that the fluids involved in the alteration process at Yaxcopoil-1 were saline or even hypersaline (halite-saturated). Calculated $\delta^{18}O$ and δD values of hydrothermal waters are likely influenced by the salt concentration in the fluid, because salt affects the activity coefficients of the aqueous species (Taylor, 1974). Even though salinity effects are negligible when compared with the uncertainty inherent in the equilibrium fractionation constants, it has been shown, for example, that $\delta^{18}O$ and δD values of waters in the Red Sea increase as much as 1 and 10‰, respectively, for every 5‰ increase in salinity (Craig, 1966). This positive correlation results from evaporative processes. Thus, part of the shift to heavier isotopic values at Yaxcopoil-1 can be attributed to this effect.

Alkalinity

The carbonate fraction isotopic results obtained from samples of the impactite section are shown in the $\delta^{13}C$ versus $\delta^{18}O$ plot of Figure 6. The abundance of an aqueous species (and therefore the $\delta^{13}C$ value of the fluid) is largely a function of temperature, pH, and redox state (Rye and Ohmoto, 1974). The figure shows that, with exception of the kerogen bearing samples, carbon isotope data at Yaxcopoil-1 are uniform, and suggests that there was substantial dissolution and/or carbon isotopic exchange between the fluid and the target limestone. Given the uniform values and the oxidized hydrothermal fluids at Yaxcopoil-1, the data further suggest that the fluids predominantly contained an oxidized carbon species (i.e., CO_2 or $H_2CO_{3(ap)}$; Ohmoto, 1972; Ohmoto and Rye, 1979), with total carbon $\delta^{13}C$ values similar to the ones obtained for calcite (0 to +2‰). Under these conditions, the hydrothermal solution in the impactite section must have had a near neutral to slightly alkaline pH.

Hydrothermal Processes

Besides mixing, evaporation and/or boiling, combined examination of the $\delta^{13}C$ and $\delta^{18}O$ values from this study, the Yucatán-6 data of Kettrup et al. (2000), as well as petrographic observations strongly suggest that both decarbonation and infiltration reactions took place at these two similarly altered sites.

Decarbonation

Thermal decarbonation is a commonly occurring natural process that is known to produce ^{18}O and ^{13}C depletions (Valley, 1986). This process results from loss of CO_2 from carbonate-bearing rocks during high-temperature replacement reactions. A decarbonation isotope exchange model between heated water with a $\delta^{18}O$ value around 7‰ in the impact melt, and local limestone (continuous curve in Fig. 6) illustrates that the spread of $\delta^{18}O$ values between Yucatán-6 and Yaxcopoil-1 could have developed in part from decarbonation processes. This observation has important implications because it reconciles the apparent compositional contrast between the Yucatán-6 and the Yaxcopoil-1 data. However, several $\delta^{13}C$ values plot significantly below the expected decarbonation trend, indicating that the carbon isotopic signatures in these samples were also influenced by the composition of the fluid. As mentioned above, small amounts of organic carbon mixed with the hydrothermal fluid likely accounted for the large shift to lighter $\delta^{13}C$ values. This is represented by the vertical dotted lines in Figure 6.

Under this scenario it is worth noting that, using the calcite-water $^{18}O/^{16}O$ fractionation factor of Friedman and O'Neil (1977), early calcite formation could have taken place around ~320 °C at Yucatán-6, but only around 90 °C at Yaxcopoil-1. Consequently, the spread in heavier-than-local limestone $\delta^{18}O$ values (26–30‰; see Fig. 6) in Yaxcopoil-1 samples probably occurred by dissolution and reprecipitation of calcite from progressively ^{13}C- and ^{18}O-enriched CO_2 (Shieh and Taylor, 1969) at lower temperatures. This process probably removed much of the isotopic evidence for previous decarbonation processes at Yaxcopoil-1. Even though it is possible that these heavy $\delta^{18}O$ values could have also been produced by infiltration of Tertiary fluids (see epoch averages in Fig. 6) much after the impact event, it is also true that none of these $\delta^{18}O$ values is heavier than those of Maastrichtian (74–65 Ma) carbonates. In any case, the temperatures derived from carbon isotopic data reflect a range that corresponds to that estimated from fluid inclusion data, even though these temperatures are the result of mixing of several stages of calcite precipitation.

Infiltration

Infiltration of fluids with carbonate-dominated isotopic compositions into the impactite sequence at Yaxcopoil-1 is evident from hydrothermal calcite, which filled vugs, occupied open spaces in breccia matrix, and precipitated as veins in the impact melt unit and overlying suevitic breccias. This appears to be well reflected by the carbonate fraction isotopic data at Yaxcopoil-1 that lie along the narrow subhorizontal trend ranging between $\delta^{18}O$ and $\delta^{13}C$ values of 23–26‰ and 1–2‰, respectively. An infiltration isotope exchange model between Cretaceous seawater and limestone (dashed curve in Fig. 6) illustrates that $\delta^{18}O$ values could have been representative of a range from ~85% water (10‰) at Yucatán-6, to close to 0% water (26‰) at Yaxcopoil-1. The amount of water suggested by the infiltration model at Yucatán-6 appears unrealistic, and $\delta^{13}C$ values, particularly those of Kettrup et al. (2000), plot significantly below the expected infiltration trend. Thus, isotopic signatures, in part, must have also been the product of additional processes, including the one associated with the extent of fluid flux and associated isotopic dilution over different depth intervals in the impactite section.

Fluid-Rock Interaction

Oxygen-isotope data on silicate fractions allowed us to calculate a possible scenario for the minimum amount of water that migrated through the rock during the hydrothermal event at

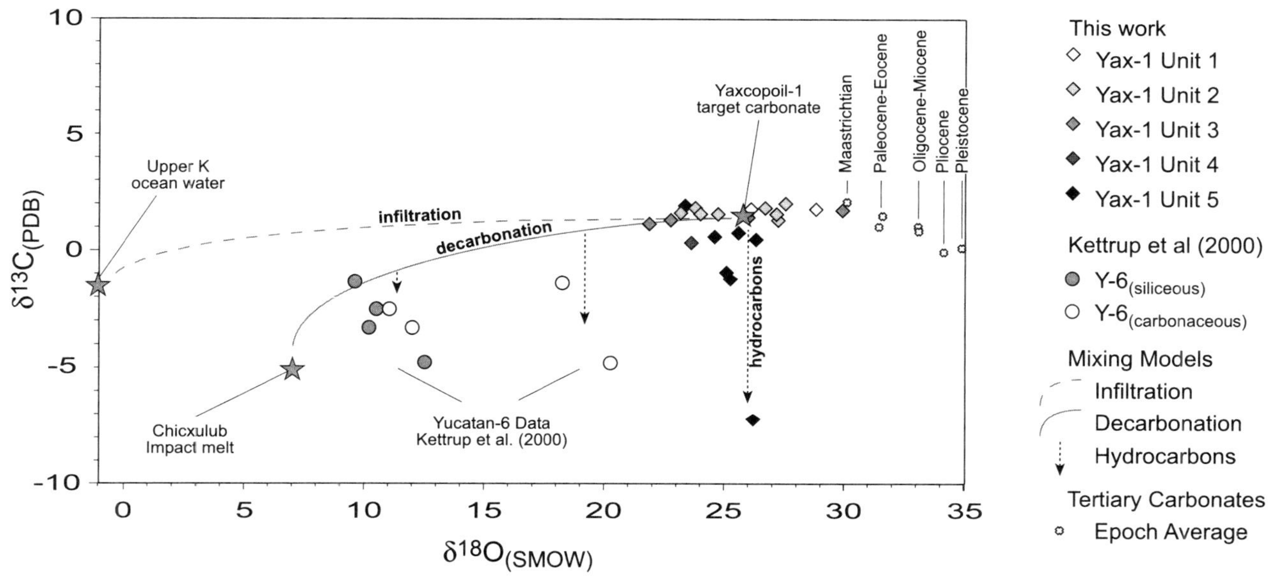

Figure 6. $\delta^{13}C_{PDB}$ versus $\delta^{18}O_{SMOW}$ plot of isotope results on carbonate fractions. The diamonds show Yaxcopoil-1 samples (this study). The Yucatán-6 data (circles) of Kettrup et al. (2000) are plotted for comparison. The Chicxulub impact melt isotopic signature is the average of that of Blum and Chamberlain (1992) and Blum et al. (1993) for Haiti black glass and Chicxulub-1 impact melt, respectively. The Yaxcopoil-1 target carbonate is from this study, and the ocean water is from estimated upper Cretaceous SMOW. Tertiary carbonate average isotopic signatures were extracted from Figure 2 in Zachos et al. (2001).

Yaxcopoil-1 (after Taylor, 1979). Calculations assume that the final fluid composition depends on isotopic equilibration with the host rock at a given temperature. Because equilibrium alteration mineral assemblages were not positively identified in the impactite section, we consider water-rock estimates only as rough proxies to actual fluid flows.

The scenario assumes that the initial fluid was a formation brine that largely equilibrated with the Cretaceous carbonate host at a $\delta^{18}O$ value of 26‰, and that the initial rock was the impactite with a $\delta^{18}O$ value of 6‰. Water-rock ratios were obtained at 270 °C using calculated waters in equilibrium with albite (albite-water and K-feldspar–water fractionation factors are very similar) after Zheng (1993), and at 100 °C using waters in equilibrium with smectite after Yeh (1980). The calculated ratios are given in Table 3, and their depth distribution is depicted in Figure 7.

Under this scenario, calculations showed that the highest water-rock ratios (w/r_{closed} ~2 and w/r_{open} ~1.0) occurred about the intense alteration zone in Unit 3 and Unit 2, with the highest fluid fluxes occurring about the middle and lower parts. In contrast, water-rock ratios in the relatively less altered and more fracture-controlled zone in Unit 5 were lower (<1), and channeled through the middle part. A decrease in temperature to 100 °C somewhat decreases the average water-rock ratio to ~1 and 0.7 for the closed and open systems, respectively. An additional high fluid flux zone is located at the bottom of Unit 5, where the tectonic (fault-related) breccia with translucent calcite occurs. Overall, the model presented above illustrates the extent to which isotopic signatures in the fluid must have also been affected by dilution effects related to relative fluid fluxes.

Origin of Hydrothermal Fluid

Knowledge of the parameters presented above enabled us to place better constraints on the likely sources of the hydrothermal fluids at Yaxcopoil-1. For this purpose, we tested fluid isotopic signature scenarios from fluid inclusion-derived temperatures and $\delta^{18}O$ and δD data from silicate fractions. These involved calculations of waters in equilibrium with two hydrous phases. Smectite is an obvious low-temperature (i.e., 100 °C) hydrous candidate, because it is compatible with the mineralogical data. Smectite-dominated clays typically form at temperatures below 120 °C (i.e., Harvey and Brown, 2000). A good higher-temperature (i.e., 270 °C; as indicated by fluid inclusions) hydrous candidate is chlorite, because it was identified in Yaxcopoil-1 samples by RAMAN and electron microprobe analyses, and deduced by compositional and petrographic relations to be one of the likely precursors to smectite (Zurcher and Kring, 2004). Even though chlorite is very scarce in the core at present, it was probably abundant previous to being replaced by clay. Chlorite microanalyses are also reported from the similarly-altered Yucatán-6 core by Lounejeva-Baturina (2001). In addition, chlorite it is also known to be widespread at other impact-related hydrothermal systems (Naumov, 2002). We discuss the proposed high-temperature scenario first.

Table 4 lists and Figure 5 illustrates the calculated waters in equilibrium with chlorite (open white triangles) at 270 °C using the fractionation factors of Cole and Ripley (1999) and Graham et al. (1984) for oxygen and hydrogen, respectively. It is noteworthy that the waters in equilibrium with chlorite also coincide acceptably with the average composition Yaxcopoil-1 metasomatic feldspar-

TABLE 3. WATER-ROCK RATIOS (W/R) CALCULATED FROM OXYGEN ISOTOPE DATA

Stratigraphy	Sample ID	$W/R_{closed\ system}$* at 270°C	$W/R_{open\ system}$* at 270°C	$W/R_{closed\ system}$* at 100°C	$W/R_{open\ system}$* at 100°C
Unit 1	YAX-1_800.43	0.82	0.60	0.50	0.40
Unit 2	YAX-1_813.41	1.05	0.72	0.61	0.47
	YAX-1_816.86	0.46	0.38	0.30	0.26
	YAX-1_819.83	0.71	0.53	0.44	0.36
	YAX-1_821.76	0.77	0.57	0.47	0.39
	YAX-1_823.05	0.57	0.45	0.37	0.31
Unit 3	YAX-1_828.64	1.27	0.82	0.70	0.53
	YAX-1_829.56	0.74	0.56	0.46	0.38
	YAX-1_832.83	2.02	1.10	0.97	0.68
	YAX-1_836.34	0.78	0.58	0.48	0.39
	YAX-1_841.32	0.46	0.38	0.31	0.27
Unit 4	YAX-1_857.65	0.39	0.33	0.26	0.23
Unit 5	YAX-1_861.40	0.41	0.34	0.27	0.24
	YAX-1_863.51	0.30	0.26	0.20	0.19
	YAX-1_867.76	0.81	0.59	0.49	0.40
	YAX-1_872.32	0.18	0.17	0.13	0.12
	YAX-1_876.46	0.23	0.21	0.16	0.15
	YAX-1_883.13	0.99	0.69	0.58	0.46
	Average:	0.72	0.52	0.43	0.35
	Maximum:	2.02	1.10	0.97	0.68
	Minimum:	0.18	0.17	0.13	0.12

*Calculated after Taylor (1979).

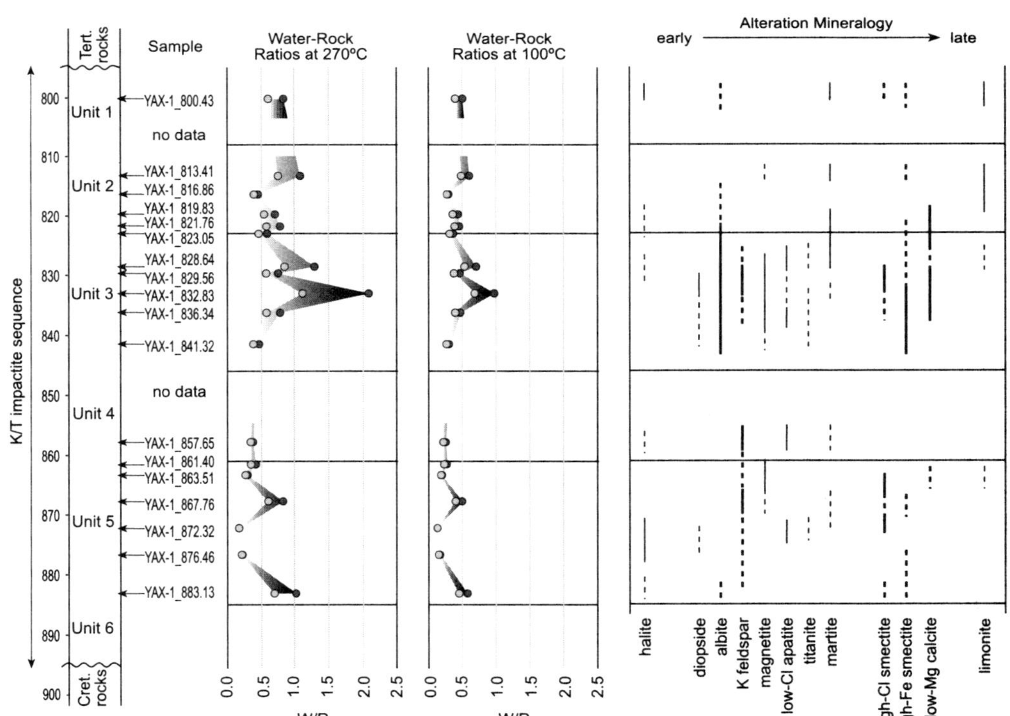

Figure 7. (A) Depth-distribution of calculated water-rock ratios at 270 °C. (B) Depth-distribution of calculated water-rock ratios at 100 °C. Albite-water and smectite-water fractionation factors of Zheng (1993) and Yeh (1980), respectively. Light-gray and dark-gray circles represent the open and closed systems of Taylor (1979).

TABLE 4. ISOTOPIC COMPOSITION OF HYDROTHERMAL FLUID CALCULATED FROM OXYGEN AND HYDROGEN ISOTOPE DATA

Stratigraphy	Sample ID	Fluid @ 270°C		Fluid @ 100°C	
		$\delta^{18}O$*	δD†	$\delta^{18}O$§	δD#
Unit 1	YAX-1_800.43	16.64	−15.05	3.07	−24.06
Unit 2	YAX-1_813.41	18.18	−10.07	4.62	−19.08
	YAX-1_816.86	13.20	−0.20	−0.37	−9.21
	YAX-1_819.83	20.21	N.D.**	6.65	N.D.**
	YAX-1_819.83	15.70	−5.15	2.14	−14.16
	YAX-1_821.76	16.21	N.D.**	2.65	N.D.**
	YAX-1_821.76	N.D.**	−4.92	N.D.**	−13.93
	YAX-1_823.05	14.47	−14.77	0.90	−23.78
Unit 3	YAX-1_828.64	19.41	−11.29	5.85	−20.30
	YAX-1_829.56	16.02	−17.52	2.46	−26.53
	YAX-1_832.83	22.16	−4.87	8.60	−13.88
	YAX-1_836.34	16.34	−1.42	2.77	−10.43
	YAX-1_841.32	13.26	−0.57	−0.30	−9.58
	YAX-1_841.32	14.19	1.55	0.63	−7.46
Unit 4	YAX-1_857.65	13.52	−4.21	−0.04	−13.22
	YAX-1_857.65	12.31	2.32	−1.26	−6.69
Unit 5	YAX-1_861.40	12.56	2.42	−1.00	−6.59
	YAX-1_861.40	13.88	0.54	0.32	−8.47
	YAX-1_863.51	13.71	−4.66	0.15	−13.67
	YAX-1_863.51	11.02	−2.86	−2.54	−11.87
	YAX-1_867.76	16.56	−3.24	3.00	−12.25
	YAX-1_872.32	9.15	−4.28	−4.41	−13.29
	YAX-1_876.46	12.19	−8.51	−1.37	−17.52
	YAX-1_876.46	10.04	−13.71	−3.52	−22.72
	YAX-1_883.13	17.81	N.D.**	4.25	N.D.**

*Chlorite-water fractionation factor of Cole and Ripley (1998).
†Chlorite-water fractionation factor Graham et al (1984).
§Smectite-water fractionation factor of Yeh (1980).
#Smectite-water fractionation factor of Yeh and Savin (1977).
**N.D.—no data.

derived water. The plot shows that the calculated fluid could be ocean or connate (seawater trapped in pores) water that experienced a $\delta^{18}O$ shift to heavier values, compatible with substantial exchange with limestone (dark-shaded rhomboid), and consistent with isotopic results. However, the fluid composition indicates that seawater alone could not have been a precursor source, because relative to ocean water the salinity of the hydrothermal fluid was close to one order of magnitude higher than that of seawater. In contrast, upper Cretaceous anhydrite (gray-shaded rectangle) with a typical oxygen isotope signature of around 14‰ overlaps well with the calculated data. Anhydrite most likely contributed to the observed isotopic signature, because it constitutes ~25% of the rock below the impactite sequence (Dressler et al., 2003). Additional estimates based on boreholes near the crater suggest that evaporites could amount to as much as 50% of the Cretaceous target lithology (see discussion in Pierazzo et al., 1998). Consequently, an evaporitic source would be consistent with both the high salinity and the isotopic values of the calculated fluids. This δD and $\delta^{18}O$ scenario is also compatible with fluids found in closed evaporite-bearing basins (Clayton et al., 1966).

Temperature is, of course, a key factor to be considered in further evaluating the sources examined above. The temperature fractionation trend depicted in Figure 5 (thin solid curve) was calculated from the average isotopic values of the fluid in equilibrium with chlorite. The graph shows that an increase in temperature of 150 °C would only modestly alter the isotopic composition of the fluids. However, a similar decrease in temperature would elevate the deuterium and lower the ^{18}O contents of the fluids substantially, to a point close to the meteoric water line. Thus, the 270 °C temperature selected appears reasonable.

The scenario whereby the isotopic signatures are related to smectite is explored in the following paragraphs. Figure 5 and Table 4 show the calculated waters in equilibrium with smectite (open white circles) at 100 °C using the fractionation factors of Yeh (1980) and Yeh and Savin (1977) for oxygen and hydrogen, respectively. The graph shows that the calculated fluid exhibits a δD and $\delta^{18}O$ shift that plots on and to the right of the meteoric water line, and suggests that surface-derived waters were probably involved. Present-day meteoric water oxygen and hydrogen isotope ratios measured in the northern part of the Yucatán

Peninsula are around –4 and –24 (open diamond), respectively (IAEA/WMO, 2001). This isotopic signature probably has not changed much since the Chicxulub impact event, because the position of the meteoric water line is believed to have remained unaffected for the past 150 Ma (Taylor, 1974). The isotopic shift to the right and up of present-day meteoric water, however, implies that if surface-derived groundwater formed part of the fluid at Yaxcopoil-1, then it must have also been modified. Once more, the effects of evaporation and high salinity likely explain the isotopic shift to heavier values, because the only surface-derived waters that do not plot on the meteoric water line are, again, those of closed basins that have undergone appreciable evaporation (Hitchon and Friedman, 1969).

The effects of temperature under this scenario are shown by the thin stippled line in Figure 5. As opposed to the water in equilibrium with chlorite, an increase of just 100 °C for water in equilibrium with smectite would substantially change the isotopic signature into the metamorphic water field. This possibility is unlikely because smectite (as opposed to smectite-illite) rarely forms above 150 °C, and because waters of metamorphic origin hardly ever make it to shallow environments such as the one that prevailed at Yaxcopoil-1 during the hydrothermal event. A decrease in temperature of just 50 °C is also unrealistic because it would shift much of the calculated data to the left of the meteoric water line, where very few fluids plot and where their origin is not well understood (i.e., Frape et al., 1984). Consequently, the 100 °C temperature selected also appears reasonable.

The models above seem to place adequate limits on the fluid source of the hydrothermal solutions at Yaxcopoil-1. Stable isotope results in conjunction with mineralogical and compositional data are most consistent with an oil-bearing saline formation brine as a source for the hydrothermal fluid at Yaxcopoil-1. Different oil-field brines exhibit a very wide range in $\delta^{18}O$, δD, and salinity values, but the waters within a sedimentary basin of a given field follow a distinct isotopic trend, such as the one exhibited by the brine in the Gulf Coast, just a few hundred kilometers north of Chicxulub (see heavy black line in Figure 5; extracted from Figure 6.4 in Barnes, 1979). Oil-field basinal brines are the combined result of mixing of connate waters of buried unmetamorphosed sedimentary rocks and meteoric waters, where fractionation effects are affected by evaporation, membrane filtration, and/or reactions involving hydrocarbons (i.e., Clayton et al., 1966).

CONCLUSIONS

We have used isotopic methods in conjunction with an evaluation of hydrothermal mineralogy to constrain the sources that were involved in the impact-generated alteration event at Yaxcopoil-1. This has yielded an appreciation of the complex evolution of fluids about an impact crater. We expect that similar processes were likely in other large impact sites and that they may have been important in the early evolution of Earth. The main conclusions from this study are presented as follows:

1. Most $\delta^{18}O$ and δD results of silicate fractions of Yaxcopoil-1 drill core samples plot well to the left of the montmorillonite or kaolinite lines at 25 °C, strongly suggesting that these values are the product of higher-temperature processes. Consequently, the original silicate isotopic signatures produced by the impact-related hydrothermal event appear to be largely preserved.

2. $\delta^{18}O$ and δD results of silicate fractions of Yaxcopoil-1 samples are compatible with mineralogical and fluid inclusion constraints, which show that the hydrothermal fluid was saline to hypersaline, and that it evolved from a higher-temperature (~300 °C) to a lower-temperature (~100 °C) regime. Isotopic results exhibit relative heavier $\delta^{18}O$ and lighter δD values, respectively, over zones of intense hydrothermal alteration, reflecting the dominant alteration mineralogy as well as formation temperature.

3. $\delta^{18}O$ and $\delta^{13}C$ results of carbonate fractions are also compatible with mineralogical and fluid inclusion constraints, which indicate that there was substantial carbon isotopic exchange between the fluid and the target limestone, and that the solutions were oxidized and CO_2-bearing, but that they also mixed with small amounts of organic carbon.

4. In combination with isotopic results from the Yucatán-6 hole, $\delta^{18}O$ and $\delta^{13}C$ results from Yaxcopoil-1 further suggest that higher-temperature decarbonation reactions took place. At Yaxcopoil-1, however, dissolution and reprecipitation of calcite from progressively ^{13}C- and ^{18}O- enriched CO_2 likely occurred during lower-temperature infiltration processes.

5. Calculation of waters in equilibrium with chlorite at 270 °C show that the fluid source could have been seawater that experienced a $\delta^{18}O$ shift to heavier values, compatible with exchange with limestone or equilibration with an evaporitic source (i.e., connate water). The involvement of the latter source appears evident, given the high fluid salinities observed. Calculation of waters in equilibrium with smectite at 100 °C further show that the fluid could have also evolved from surface-derived water. The $\delta^{18}O$ and δD shift to heavier values with respect to meteoric water, however, requires the input of evaporation and, again, high salinity.

6. Stable isotope results from this study are most consistent with a fluid derived from a mixture of sources, much like the basinal brines that develop in oil fields such as the Gulf Coast just a few hundred kilometers north of Chicxulub. These relatively deep oil-field brines typically exhibit temperatures between 70 and 180 °C. In the otherwise near-surface impactite sequence at Yaxcopoil-1, however, the higher temperatures indicated by mineralogy and fluid inclusions strongly support the addition of impact-induced heat.

ACKNOWLEDGMENTS

We thank the International Continental Drilling Program, Universidad Nacional Autónoma de México, and the Chicxulub Scientific Drilling Project team for producing the Yaxcopoil-1 core. We express our gratitude to the anonymous referees for the thorough reviews of the manuscript, and A. Deutsch and

T. Kenkmann for the editorial support. This work was sponsored by National Science Foundation grant EAR-0126055.

REFERENCES CITED

Ames, D.E., Watkinson, D.H., and Parrish, R.R., 1998, Dating of a regional hydrothermal system induced by the 1850 Ma Sudbury impact event: Geology, v. 26, p. 447–450, doi: 10.1130/0091-7613(1998)0262.3.CO;2.

Ames, D.E., Kjarsgaard, I.M., Pope, K.O., Dressler, B.O., and Pilkington, M., 2004, Secondary alteration of the impactite and mineralization in the basal Tertiary sequence, Yaxcopoil-1, Chicxulub impact crater, Mexico: Meteoritics and Planetary Science, v. 39, no. 7, p. 1145–1167.

Barnes, H.L., 1979, Geochemistry of hydrothermal ore deposits: New York, USA, John Wiley and Sons, Inc., second edition, 798 p.

Blum, J.D., and Chamberlain, C.P., 1992, Oxygen isotope constraints on the origin of impact glasses from the Cretaceous-Tertiary Boundary: Science, v. 257, no. 5073, p. 1104–1107.

Blum, J.D., Chamberlain, C.P., Hingston, M.P., Koeberl, C., Marin, L.E., Schuraytz, B.C., and Sharpton, V.L., 1993, Isotopic comparison of K/T boundary impact glass with melt rock from the Chicxulub and Manson impact structures: Nature, v. 364, no. 6435, p. 325–327, doi: 10.1038/364325a0.

Chaussidon, M., Sigurdsson, H., and Metrich, N., 1996, Sulfur and boron isotope study of high-Ca impact glasses from the K-T boundary; constraints on source rocks, *in* Ryder, G., Fastovsky, D., and Gartner, S., eds., The Cretaceous-Tertiary event and other catastrophes in Earth history: Geological Society of America Special Paper 307, p. 253–262.

Clayton, R.N., and Mayeda, T.K., 1963, The use of bromine pentafluoride in the extraction of oxygen from oxides and silicates for isotopic analysis: Geochimica et Cosmochimica Acta, v. 27, no. 1, p. 43–52, doi: 10.1016/0016-7037(63)90071-1.

Clayton, R.N., Friedman, I., Graf, D.L., Mayeda, T.K., Meents, W.F., and Shimp, N.F., 1966, The origin of saline formation waters: I. isotopic composition: Journal of Geophysical Research, v. 71, no. 16, p. 3869–3882.

Cole, D.R., and Ripley, E.M., 1999, Oxygen isotope fractionation between chlorite and water from 170–350C: A preliminary assessment based on partial exchange and fluid/rock experiments: Geochimica et Cosmochimica Acta, v. 63, p. 449–457, doi: 10.1016/S0016-7037(99)00071-X.

Cornejo-Toledo, A., and Hernandez-Osuna, A., 1950, Las anomalias gravimétricas en la cuenca salina del Istmo, planicie costera de Tabasco, Campeche y península de Yucatán: Boletín de la Asociación Mexicana de Geológos Petroleros, v. 2, no. 7, p. 453–460.

Craig, H., 1961, Isotopic variations in meteoric waters: Science, v. 133, p. 1702–1703.

Craig, H., 1966, Isotopic composition and origin of the Red Sea and Salton Sea geothermal brines: Science, v. 154, no. 3756, p. 1544–1548.

Dissmann, B., Agrinier, P., Ostermann, M., and Deutsch, A., 1998, Chicxulub; PEMEX Well Y-6; geochemistry (REE, εNd, εSr, $\delta^{18}O$, $\delta^{13}C$) of impact melt breccias and suevites: 29th Lunar and Planetary Science Conference, Houston, Texas, USA, Abstract no. 1610.

Dressler, B.O., Sharpton, V.L., Morgan, J., Buffler, R., Moran, D., Smit, J., Stoeffler, D., and Urrutia, J., 2003, Investigating a 65-Ma-Old smoking gun; deep drilling of the Chicxulub impact structure: Eos (Transactions, American Geophysical Union), v. 84, no. 14, p. 125, 130.

Dressler, B.O., Sharpton, V.L., Schwandt, C.S., and Ames, D.E. 2004, Impactites of the Yaxcopoil-1 drilling site, Chicxulub impact structure: Petrography, geochemistry, and depositional environment: Meteoritics and Planetary Science, v. 39, no. 6, p. 857–878.

Fouke, B.W., Zerkle, A.L., Alvarez, W., Pope, K.O., Ocampo, A.C., Wachtman, R.J., Nishimura, J.M.G., Claeys, P., and Fischer, A.G., 2002, Cathodoluminescence petrography and isotope geochemistry of KT impact ejecta deposited 360 km from the Chicxulub Crater, at Albion Island, Belize: Sedimentology, v. 49, no. 1, p. 117–138, doi: 10.1046/j.1365-3091.2002.00435.x.

Frape, S.K., Fritz, P., and McNutt, R.H., 1984, Water-rock interaction and chemistry of groundwaters from the Canadian shield: Geochimica et Cosmochimica Acta, v. 48, p. 1617–1627, doi: 10 1016/0016-7037(84)90331-4.

Friedman, I., and O'Neil, J.R., 1977, Compilation of stable isotope fractionation factors of geochemical interest, *in* Fleisher, M., ed., Data of geochemistry: U.S. Geological Survey, Professional Paper 440-KK, 12 p., 49 figs.

Graham, C.M., Atkinson, J., and Harmon, R.S., 1984, Hydrogen isotope fractionation in the system chlorite-water: NERC 6th Progress Report of Research 1981–1984, NERC Publication Series D, no. 25, p. 139.

Harvey, C., and Brown, P., 2000, Mixed layer clays in geothermal systems and their effectiveness as mineral geothermometers: Proceedings, World Geothermal Congress, Kyushu-Tohoku, Japan, p. 1201–1205.

Hecht, L., Wittmann, A., Schmitt, R.T., and Stöffler, D., 2004, Composition of impact melt particles and the effects of post-impact alteration in suevitic rocks at the Yaxcopoil-1 drill core, Chicxulub Crater, Mexico: Meteoritics and Planetary Science, v. 39, no. 7, p. 1169–1186.

Hildebrand, A.R., Penfield, G.T., Kring, D.A., Pilkington, M., Camargo-Zanoguera, A., Jacobsen, S.B., and Boynton, W.V., 1991, Chicxulub Crater; a possible Cretaceous-Tertiary boundary impact crater on the Yucatan Peninsula, Mexico: Geology, v. 19, p. 867–871, doi: 10.1130/0091-7613(1991)0192.3.CO;2.

Hitchon, B., and Friedman, I., 1969, Geochemistry and origin of formation waters in the western Canada sedimentary basin-I. Stable isotopes of oxygen, carbon, and sulfur: Geochimica et Cosmochimica Acta, v. 33, p. 1321–1358, doi: 10.1016/0016-7037(69)90178-1.

IAEA/WMO, 2001, Global network for isotopes in precipitation, the GNIP database: http://es.ucsc.edu/%7Egbowen/OIPC_Main.html.

Kettrup, B., Deutsch, A., Ostermann, M., and Agrinier, P., 2000, Chicxulub impactites: Geochemical clues to the precursor rocks: Meteoritics & Planetary Science, v. 35, p. 1229–1238.

Koeberl, C., 1993, Chicxulub Crater, Yucatán; tektites, impact glasses, and the geochemistry of target rocks and breccias: Geology, v. 21, p. 211–214, doi: 10.1130/0091-7613(1993)0212.3.CO;2.

Kring, D.A., 1995, The dimensions of the Chicxulub impact crater and impact melt sheet: Journal of Geophysical Research, v. 100, no. E8, p. 19979–19986, doi: 10.1029/95JE01768.

Kring, D.A., 2000a, Impact events and their effect on the origin, evolution, and distribution of life: GSA Today, v. 10, no. 8, p. 1–7.

Kring, D.A., 2000b, Impact-induced hydrothermal activity and potential habitats for thermophilic and hyperthermophilic life, *in* Koeberl, C., ed., Catastrophic events and mass extinctions; impacts and beyond: Houston, Texas, USA, Lunar and Planetary Institute Contribution, p. 106–107.

Kring, D.A., and Boynton, W.V., 1992, The petrogenesis of an augite-bearing melt rock in the Chicxulub structure and its relationship to K/T impact spherules in Haiti: Nature, v. 358, no. 6382, p. 141–144.

Kring, D.A., Hildebrand, A.R., and Boynton, W.V., 1991, The petrology of an andesitic melt rock and a polymict breccia from the interior of the Chicxulub structure, Yucatan, Mexico: Houston, Texas, USA, 22nd Lunar and Planetary Science Conference, p. 755–776.

Kring, D.A., Hörz, F., Zurcher, L., and Urrutia-Fucugauchi, J., 2004, Impact lithologies and their emplacement in the Chicxulub impact crater: initial results from the Chicxulub Scientific Drilling Project, Yaxcopoil, Mexico: Meteoritics & Planetary Science, v. 39, no. 6, 879–897.

Lounejeva-Baturina, E., 2001, Mineralogía y petrología de las rocas fundidas del crater Chicxulub, México [Master's Thesis]: México, D.F., México, Universidad Nacional Autónoma de México, 112 p.

Lüders, V., and Rickers, K., 2004, Fluid inclusion evidence for impact-related hydrothermal fluid and hydrocarbon migration in Cretaceous sediments of the ICDP-Chicxulub drill core Yaxcopoil-1: Meteoritics and Planetary Science, v. 39, no. 7, p. 1187–1197.

Lüders, V., Horsfield, B., Kenkmann, T., Mingram, B., and Wittmann, A., 2003, Hydrocarbons and aqueous fluids in Cretaceous sediments of the IDDP-Chicxulub drill core YAX-1: Houston, Texas, 34th Lunar and Planetary Science Conference, Abstract no. 1378.

Morgan, J., Warner, M., Brittan, J., Buffler, R., Camargo, A., Christeson, G., Denton, P., Hildebrand, A., Hobbs, R., Macintyre, H., Mackenzie, G., Maguire, P., Marin, L.E., Nakamura, Y., Pilkington, M., Sharpton, V.L., Snyder, D., Suarez, G., and Trejo, A., 1997, Size and morphology of the Chicxulub impact crater: Nature, v. 390, no. 6659, p. 472–476, doi: 10.1038/37291.

Naumov, M.V., 2002, Impact-generated hydrothermal systems; data from Popigai, Kara, and Puchezh-Katunki impact structures, *in* Plado, J., and Pesonen Lauri, J., eds., Impacts in Precambrian shields: Berlin, Federal Republic of Germany, Springer, p. 117–171.

Ohmoto, H., 1972, Systematics of Sulfur and Carbon Isotopes in Hydrothermal Ore Deposits: Economic Geology and the Bulletin of the Society of Economic Geologists, v. 67, no. 5, p. 551–578.

Ohmoto, H., and Rye, R.O., 1979, Isotopes of sulfur and carbon, *in* Barnes, H.L., ed., Geochemistry of hydrothermal deposits: John Wiley & Sons, p. 509–567.

Penfield, G.T., and Camargo-Zanoguera, A., 1981, Definition of a major igneous zone in the central Yucatán platform with aeromagnetics and gravity: Society of Exploration Geophysicists Technical Program, Abstracts and Biographies, v. 51, p. 37.

Perry, E.C., Winter, D.J., Sagar, B., and Wu, B., 1992, The Chicxulub Structure; surface manifestation and possible sulfur isotope signature: Abstracts of papers submitted to the 23rd Lunar and Planetary Science Conference, Houston, Texas, USA, p. 1057–1058.

Pierazzo, E., Kring, D.A., and Melosh, H.J., 1998, Hydrocode simulation of the Chicxulub impact event and the production of chemically active gases: Journal of Geophysical Research, v. 103, no. E12, p. 28607–28625, doi: 10.1029/98JE02496.

Pilkington, M., Ames, D.E., and Hildebrand, A.R., 2004, Magnetic mineralogy of the Yaxcopoil-1 core, Chicxulub crater: Meteoritics and Planetary Science, v. 39, no. 6, p. 831–841.

Pope, K.O., Ocampo, A.C., Fischer, A.G., Morrison, J., and Sharp, Z., 1996, Carbonate condensates in the Chicxulub ejecta deposits from Belize: Houston, Texas, USA, Abstracts of papers submitted to the 27th Lunar and Planetary Science Conference, p. 1045–1046.

Rye, R.O., and Ohmoto, H., 1974, Sulfur and carbon isotopes and ore Genesis: a review: Economic Geology and the Bulletin of the Society of Economic Geologists, v. 69, no. 6, p. 826–842.

Savin, S.M., and Epstein, S., 1970, The oxygen and hydrogen isotope geochemistry of clay minerals: Geochimica et Cosmochimica Acta, v. 34, p. 43–64, doi: 10.1016/0016-7037(70)90150-X.

Schuraytz, B.C., Sharpton, V.L., and Marin, L.E., 1994, Petrology of impact-melt rocks at the Chicxulub multiring basin, Yucatan, Mexico: Geology, v. 22, p. 868–872, doi: 10.1130/0091-7613(1994)0222.3.CO;2.

Sharpton, V.L., and Marin, L.E., 1993, Recent studies of the Chicxulub multiring impact basin, Yucatán, Mexico: Washington, D.C., USA, American Geophysical Union 1993 fall meeting, p. 386.

Sharpton, V.L., Dalrymple, G.B., Marin, L.E., Ryder, G., Schuraytz, B.C., and Urrutia-Fucugauchi, J., 1992, New links between the Chicxulub impact structure and the Cretaceous/Tertiary boundary: Nature, v. 359, no. 6398, p. 819–821, doi: 10.1038/359819a0.

Sheppard, S.M.F., Nielsen, R.L., and Taylor, H.P., 1969, Oxygen and hydrogen isotope ratios of clay minerals from porphyry copper deposits: Economic Geology and the Bulletin of the Society of Economic Geologists, v. 64, p. 755–777.

Shieh, Y.N., and Taylor, H.P., 1969, Oxygen and carbon isotope studies of contact metamorphism of carbonate rocks: Journal of Petrology, v. 10, p. 307–331.

Smit, J., 1997, Predictive elements of large-body impacts in geologic history, *in* Schlager, W., and Kroonenberg, S., eds., Predictions in geology: Berlin-Heidelberg-New York, Springer International, p. 464–470.

Suzuoki, T., and Epstein, S., 1976, Hydrogen isotope fractionation between OH-bearing minerals and water: Geochimica et Cosmochimica Acta, v. 40, no. 10, p. 1229–1240, doi: 10.1016/0016-7037(76)90158-7.

Swisher, C.C., Grajales-Nishimura, J.M., Montanari, A., Margolis, S.V., Claeys, P., Alvarez, W., Renne, P., Cedillo-Pardo, E., Maurrasse, F.J.-M.R., Curtis, G.H., Smit, J., and McWilliams, M.O., 1992, Coeval $^{40}Ar/^{39}Ar$ ages of 65.0 million years ago from Chicxulub Crater melt rock and Cretaceous-Tertiary boundary tektites: Science, v. 257, p. 954–958.

Taylor, H.P., 1974, The application of oxygen and hydrogen isotope studies to problems of hydrothermal alteration and ore deposition: Economic Geology and the Bulletin of the Society of Economic Geologists, v. 69, no. 6, p. 843–883.

Taylor, H.P., 1979, Oxygen and hydrogen isotope relationships in hydrothermal mineral deposits, *in* Barnes, H.L., ed., Geochemistry of hydrothermal deposits: John Wiley & Sons, p. 236–277.

Tuchscherer, M.G. Reimold, W.U., Koeberl, C., Gibson, R.L., and de Bruin, D., 2004, First petrographic results on impactites from the Yaxcopoil-1 borehole, Chicxulub structure, Mexico: Meteoritics and Planetary Science, v. 39, no. 6, p. 899–930.

Valley, J.W., 1986, Stable isotope geochemistry of metamorphic rocks, *in* Valley, J.W., Taylor, H.P., and O'Neil, J.R., eds., Stable isotopes in high temperature geological processes: Washington, D.C., USA, Mineralogical Society of America Reviews in Mineralogy, v. 16, p. 445–489.

Yeh, H., 1980, D/H ratios and late-stage dehydration of shales during burial: Geochimica et Cosmochimica Acta, v. 44, p. 341–352, doi: 10.1016/0016-7037(80)90142-8.

Yeh, H., and Savin, S.M., 1977, Mechanism of burial metamorphism of argillaceous sediments: 3. O-isotope evidence: Geological Society of America Bulletin, v. 88, p. 1321–1330.

Urrutia-Fucugauchi, J., Soler-Arechalde, A.M., Rebolledo-Vieyra, M., and Vera-Sanchez, P., 2004, Paleomagnetic and rock magnetic study of the Yaxcopoil-1 impact breccia sequence, Chicxulub impact crater, Mexico: Meteoritics and Planetary Science, v. 39, no. 6, p. 843–856.

Wittmann, A., Kenkmann, T., Schmitt, R.T., Hecht, L., and Stöffler, D., 2004, Impact-related dike breccia lithologies in the ICDP drill core Yaxcopoil-1, Chicxulub impact structure, Mexico: Meteoritics and Planetary Science, v. 39, no. 6, p. 931–954.

Zachos, J.C., Pagani, M., Sloan, L., Thomas, E., and Billups, K., 2001, Trends, rhythms and aberrations in global climate 65 Ma to present: Science, v. 292, p. 686–693, doi: 10.1126/science.1059412.

Zheng, Y.F., 1993, Calculation of oxygen isotope fractionation in anhydrous silicate minerals: Geochimica et Cosmochimica Acta, v. 57, p. 1079–1091, doi: 10.1016/0016-7037(93)90042-U.

Zurcher, L., and Kring, D.A., 2004, Hydrothermal alteration in the core of the Yaxcopoil-1 borehole, Chicxulub impact structure, Mexico: Meteoritics and Planetary Science, v. 39, no. 7, p. 1199–1221.

Manuscript Accepted by the Society 9 August 2004

Printed in the USA

Geological Society of America
Special Paper 384
2005

Linking large impacts, gas hydrates, and carbon isotope excursions through widespread sediment liquefaction and continental slope failure: The example of the K-T boundary event

Simon Day*
Mark Maslin
Environmental Change Research Centre and Benfield Hazard Research Centre, University College London, 26, Bedford Way, London, WC1H 0AP, UK

ABSTRACT

We present a mechanism linking large impacts, such as Chicxulub, to significant continental sedimentary slope failures and gas hydrate releases, and hence to carbon isotope excursions. Extensive continental margin failures and seabed sediment liquefaction at the Cretaceous-Tertiary boundary up to thousands of kilometers from Chicxulub have been linked to this impact in previous studies: here we analyze the implied seismic shaking and explore its effects in terms of gas hydrate release from failed continental margin sediments. Paleoseismic analysis of published studies of liquefaction and slope failure at the Cretaceous-Tertiary boundary in North America and adjacent regions suggests that, due to low seismic attenuation in plate interior rocks, there was a sufficient seismic forcing to cause the observed widespread sediment liquefaction and failure along tens of thousands of kilometers of continental slope. The implied magnitude of the impact-related seismicity (equivalent to an earthquake with moment magnitude ≈11) is shown to be broadly consistent with the characteristics of the Chicxulub impact structure. An extended period of post-impact liquefaction and slope failure may account for the observed complexity of Cretaceous-Tertiary boundary sequences in Mexico and North America. We favor a seismic shaking model for the triggering of slope failure over previous models implicating impact generated tsunamis, because the shallow-water Chicxulub impact itself is now recognized as an inefficient tsunami source.

We have calculated the potential storage of gas hydrates based on known environmental conditions during the Cretaceous. This suggests that slope failures caused by the Chicxulub impact could have released between 300 and 1300 GtC (best estimate ~700 GtC) of methane from the destabilization of gas hydrates. This would produce a global carbon isotopic excursion of between –0.5‰ and –2‰ (best estimate ~–1‰). This compares well with the observed carbon isotopic excursion of –1‰ in the planktonic foraminifera records across the K-T boundary. This large release of methane may also account for the recently reconstructed very high atmospheric pCO_2 levels

*Present address: Institute of Geophysics and Planetary Physics, Department of Earth Sciences, University of California, 1156 High Street, Santa Cruz, California 95064, USA.

Day, S., and Maslin, M., 2005, Linking large impacts, gas hydrates, and carbon isotope excursions through widespread sediment liquefaction and continental slope failure: The example of the K-T boundary event, *in* Kenkmann, T., Hörz, F., and Deutsch, A., eds., Large meteorite impacts III: Geological Society of America Special Paper 384, p. 239–258. For permission to copy, contact editing@geosociety.org.

after the Cretaceous-Tertiary boundary as our estimated gas hydrate releases could have increased atmospheric carbon dioxide by a maximum of 600–2300 ppm (best estimate ~1200 ppm). This mechanism linking impacts to carbon isotope excursions may apply to other significant excursions, such as that at the Paleocene-Eocene Thermal Maximum. The difficulty in identifying impact craters means that many of the other abrupt carbon isotope excursions found in the geological record could be related to impacts and not to climatic changes.

Keywords: K-T boundary, carbon isotope excursion, gas hydrates, continental slope failure, impact paleoseismicity.

INTRODUCTION

Carbon Isotope Excursions at the Cretaceous-Tertiary Boundary and Other Extinction Events

One of the characteristic features of the Cretaceous-Tertiary (K-T) boundary extinction event is a marked but short-lived change or boundary excursion in the stable isotopic composition of carbon in carbonate sediments deposited in the earliest Tertiary (e.g., Boersma and Shackleton, 1981; Shackleton and Hall, 1984; Zachos and Arthur, 1986; Zachos et al., 1989; 2001; Stott and Kennett, 1989). Planktonic foraminifera $\delta^{13}C$ records indicate a significant shift to light or depleted values, meaning that the sediment is enriched in the light isotope ^{12}C. Similar, short-lived ^{12}C enrichments have been identified at several other turnover events in the history of life on Earth, such as the Paleocene-Eocene Thermal Maximum, or PETM (e.g., Zachos et al., 1993, 2001; Thomas and Shackleton, 1996; Dickens, 2003). In this paper we focus on the K-T boundary event but also consider these other events below. Under normal circumstances, ^{12}C is preferentially incorporated in organic material or biomass, particularly through the process of photosynthesis, depleting carbonate sediments in this isotope. The K-T boundary excursion is therefore interpreted as indicating a sudden and massive input of organic ^{12}C-enriched carbon into the atmosphere, which is then oxidized, dissolved into seawater, and finally precipitated out as carbonate sediment. This interpretation is supported by the observation that the excursion is accompanied by disturbance of the normal depth gradient in $\delta^{13}C$, indicated by differences between carbon isotope ratios of planktonic and benthonic microfossils (Ivany and Salawitch, 1993). At the K-T boundary, the planktonic foraminifera indicate the surface waters become enriched in ^{12}C, while the benthic foraminifera show an enrichment of the bottom water in ^{13}C (see Fig. 1). This is interpreted as indicating input of isotopically light organic carbon from the atmosphere into the surface layers of the ocean on the one hand and a reduction in the rate of oxidation of isotopically light organic carbon to dissolved carbonate in deep water on the other. We consider the actual amount of organic carbon that must have been involved in the K-T boundary excursion, to produce the observed isotopic changes, in more detail below. A key difference between the K-T boundary carbon isotope excursions and other effects such as the PETM should also be noted. In contrast to the K-T excursion, at the PETM, both the surface and deep water become enriched in ^{12}C (e.g., Zachos et al., 1993; 2001; Thomas and Shackleton, 1996; Dickens, 2003), which has implications for the speed of the release of gas hydrates and other influences of the carbon releasing event on the oceans.

Explanations for the inferred association of these negative carbon isotope boundary excursions with impact events generally fall into one of two categories, both of which have problems:

1. The impact destroys terrestrial and other biomass on the surface of the Earth (e.g., through global forest fires triggered by the heat emitted from impact ejecta as it reenters the atmosphere; Wolbach et al., 1990; Melosh et al., 1990; Robertson et al., 2004) that is converted to relative light $\delta^{13}C$ atmospheric carbon dioxide and then to carbonate. The maximum possible isotopic and climatic effect in these models is limited by the amount of biomass available at the surface of Earth to be consumed in the event; the actual effects reflect the proportion of biomass burned and the fate of the burned material, particularly the relative proportions of carbon dioxide and charcoal produced. Estimates of these parameters vary widely, contributing to wide uncertainty in the estimate of proportion of biomass burned (Ivany and Salawitch, 1993), from <25% to 100%, and probable inconsistency between the isotopically based estimates and estimates based on the occurrence of soot in K-T boundary deposits (Wolbach et al., 1990; Belcher et al., 2003).

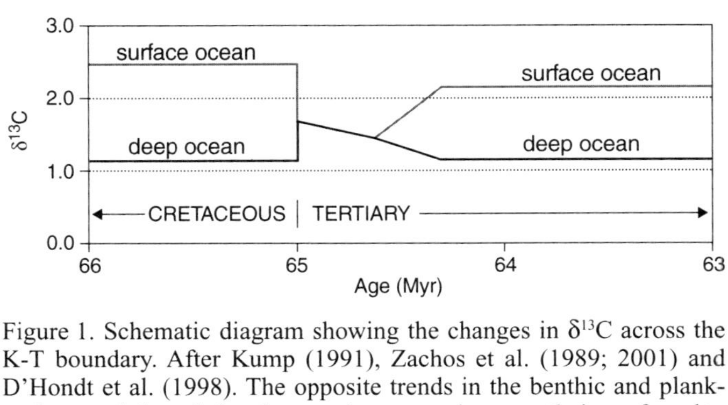

Figure 1. Schematic diagram showing the changes in $\delta^{13}C$ across the K-T boundary. After Kump (1991), Zachos et al. (1989; 2001) and D'Hondt et al. (1998). The opposite trends in the benthic and planktonic records have led to the post-impact environment being referred to as the Strangelove Ocean (Kump, 1991).

2. A strong climatic positive feedback caused by the impact, acting on the terrestrial environment, results in a catastrophic release either of biomass or of fossil organic carbon. In recent years, the most widely proposed process of this kind has been the thermal destabilization of solid methane gas hydrate trapped in sub-seafloor sediments, leading to massive hydrocarbon gas releases. In these models, the positive feedback is provided by the dependence of gas hydrate stability upon temperature: an initial release of gas hydrate leads to warming of the atmosphere (methane and its oxidation product, carbon dioxide, being powerful greenhouse gases), hence warming of the ocean and sub-seafloor sediments, hence more gas hydrate release, and hence more warming (e.g., Thomas et al., 2002). The principal problem with this hypothesis as applied to the K-T boundary is the accumulating evidence that the initial effect of the impact was a cooling, possibly lasting for as much as 2000 years (Galeotti et al., 2004), due initially to blocking of solar radiation by atmospheric soot from fires and/or sulfate aerosols from evaporites in the target sequence (Pope et al., 1997) and sustained by changes in deep ocean circulation (Galeotti et al., 2004), which would tend in particular to keep suboceanic gas hydrates cool.

Here we examine an alternative mechanism, in which a catastrophic external event—a large impact such as the Chicxulub impact at the K-T boundary—releases fossil organic carbon through its effect on the solid Earth. Such a mechanism was first suggested by Max et al. (1999), who proposed that impact-generated tsunamis would produce cyclic pressure variations in the sediments on the continental slope, and in particular that the passage of tsunami wave troughs would lead to explosive blowouts of the temporarily overpressured free gas reservoirs beneath the hydrate layers. However, the efficiency of the impact itself as a primary tsunami source is questionable, owing to the location of the impact site in very shallow subtidal or lagoonal waters (Ocampo et al., 1996; Grajales-Nishimura et al., 2000; Dressler et al., 2003); we consider the issue of secondary tsunamis produced indirectly by the impact further below. Furthermore, estimates of sub-hydrate gas masses cannot be constrained by stability arguments as is the case with the hydrate layer itself, and so it is difficult to quantify the Max et al. (1999) hypothesis. Here we propose a mechanism that centers upon seismic waves generated by the impact itself and by impact crater modification processes, and their effect on the mechanical stability of sequences of hydrate-bearing sediments on the adjacent continental slopes.

We proceed by first reviewing the geological evidence for seismic disturbance, sediment liquefaction and slope instability at the K-T boundary over a wide region centered upon Chicxulub, and using this to define the zonation of seismic effects around the impact site; then by analyzing structural geological and theoretical aspects of impact-related deformation and seismicity to show how this disturbance may have been be related to the processes of formation and modification of the Chicxulub impact structure. The processes responsible for the regional seismic shaking are important for our analysis because of implications for the inferred duration and intensity of seismic shaking, that in turn have consequences for the nature, location, and extent of continental slope instability associated with the impact event. The final step of the analysis involves quantification of the isotopic effects of the gas hydrate release that we infer to have been associated with this continental slope instability.

Distribution of Sediment Liquefaction and Slope Failure Phenomena Associated with the Chicxulub Impact

Figure 2 shows the distribution of K-T boundary sections onshore in central and north America (modified after Smit, 1999; with additional data from Ocampo et al., 1996; Grajales-Nishimura et al., 2000; Busby et al., 2002; Kiyokawa et al., 2002; Stoffer, 2002; Schulte et al., 2002), as well as Ocean Drilling Program (ODP) sites in adjacent ocean sections where the K-T boundary section has been identified and characterized, and mapped units (horizon A* from Norris et al., 2000, and Norris and Firth, 2002; Campeche breccia units from Alvarez et al., 1992, Bralower et al., 1998, and Kiyokawa et al., 2002) interpreted as products of various types of large-scale slope failures. In addition to these sites, limited slope failure has been inferred along the eastern margin of the Atlantic Ocean from the occurrence of smaller-scale slump features in chalk beds at the K-T boundary in ODP hole 398D off the coast of Portugal (Norris and Firth, 2002).

Taken together, these define three different types of paleoseismic phenomena whose stratigraphic ages, as interpreted by the authors of the studies of these sections, imply association with the Chicxulub impact and the K-T boundary. (We consider paleoseismic aspects of the hypothesis that the Chicxulub impact does not coincide with the K-T boundary and that multiple impacts occurred in a short time interval around that time [Keller et al., 2004] further below.) These are: in situ liquefaction and faulting; slope instability involving deep-seated collapse of indurated platform margins; and slope failure involving poorly lithified submarine sediments.

1. ***In situ liquefaction.*** Intense in-situ liquefaction of water-saturated, especially submarine sediments is common in K-T boundary sections up to some 2500 km from Chicxulub at the time of the impact. (Note that the effects of subsequent plate motions have been restored in Figure 2 and Figure 3, below.) Multiple liquefaction structures and layers, interbedded with mass flow units in some cases (Smit et al., 1996), suggest multiple liquefaction and mass flow episodes (Adatte et al., 1996), not necessarily separated by large intervals, but in some cases this may reflect liquefaction of particular strata within a sedimentologically varied sequence. At Millers Ferry in Alabama (Olsson et al., 1996), the topmost Maastrichtian chalk layer exhibits abundant small-scale normal faulting, suggesting liquefaction and spreading in an underlying unlithified unit. At a number of these sites, most notably at El Mimbral, El Tecolote, and La Sierrita in northeastern Mexico (Smit et al., 1992, 1996; Soria et al., 2001; Schulte et al., 2002), at Millers Ferry in Alabama (Olsson et al., 1996), and Sage Creek Arch in South Dakota (Stoffer, 2002), intense liquefaction affects not only the pre-impact shallow-marine sediments

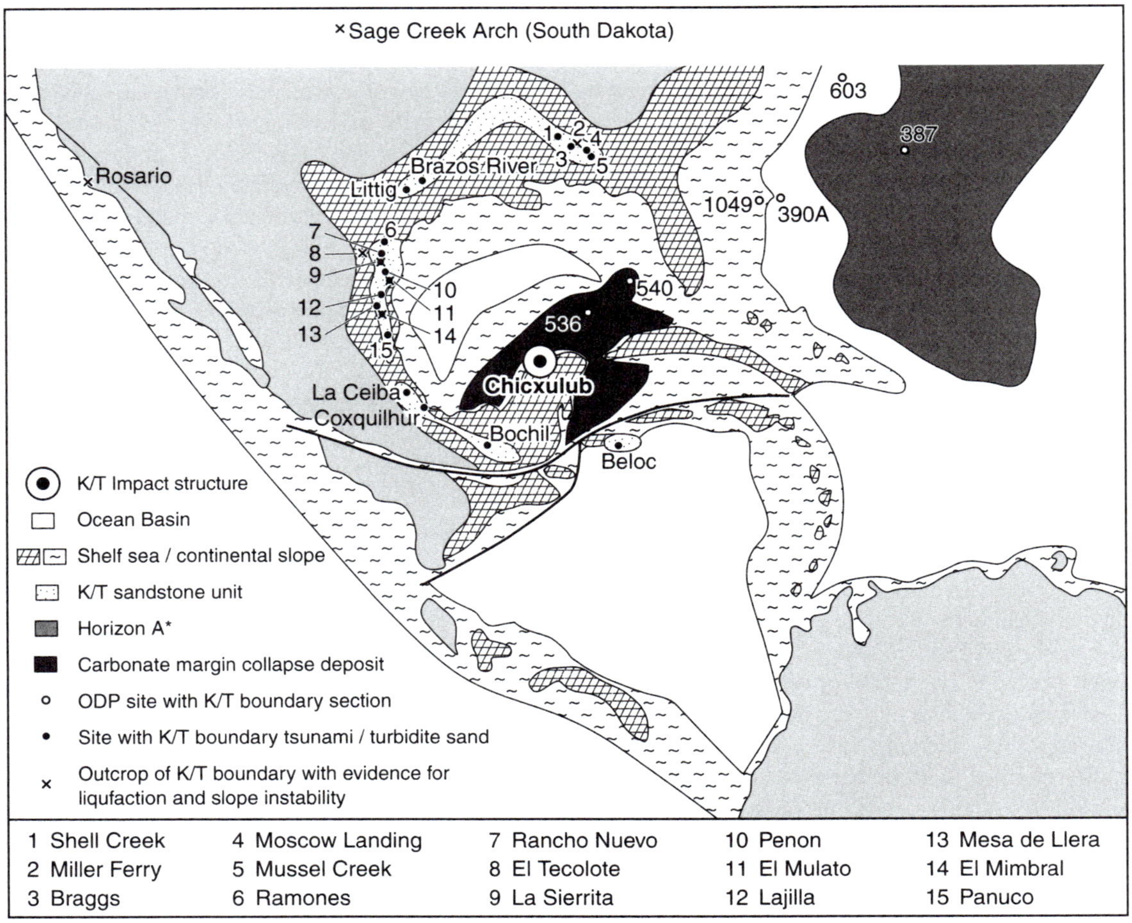

Figure 2. Distribution of evidence for syn- and post-impact seismic liquefaction and slope instability, associated with the Chicxulub impact structure, in North and Central America. Adapted from Smit (1999) with additional data from references discussed in text. Distribution of carbonate platform margin collapse deposits is approximate.

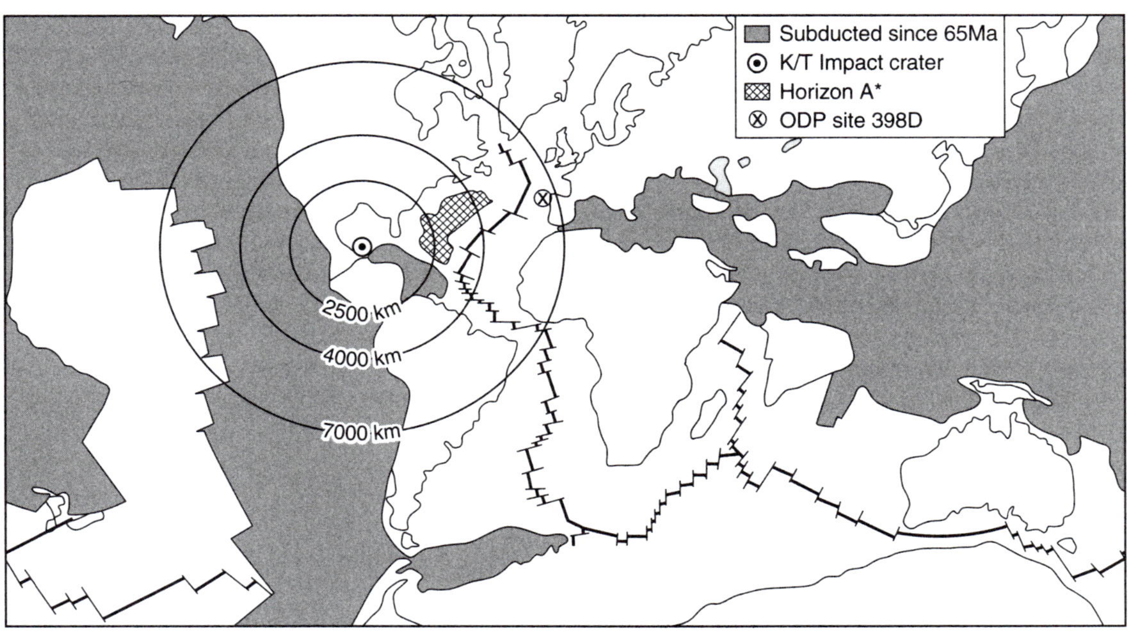

Figure 3. Distant seismic effects of the Chicxulub impact: 2500 km and 7000 km radius rings represent outer limits of evidence for liquefaction and slope instability associated with the impact. Adapted from Smit (1999) with additional data from Norris et al. (2000) and Norris and Firth (2002).

but also at least the lower units of impact ejecta and tsunami (or tsunamigenic turbidite [Bohor, 1996]) deposits. Perhaps the closest marine K-T boundary section to Chicxulub without major liquefaction is that at Bass River, New Jersey (Olsson et al., 1997), some 2500 km distant, although this is only a borehole section and so would not necessarily detect evidence of liquefaction if it was not developed pervasively. The liquefaction of ejecta layers requires that the liquefaction persisted for a period at least equal to the ballistic travel time of the ejecta (tens of minutes; Melosh, 1989) to the top of the atmosphere above the deposition sites, plus the time required for settling of the ejecta through the atmosphere and the water column. The latter depends critically upon the dynamics of settling of dense particle suspensions. Such settling may be accelerated by density-driven convective plumes if the concentration of accumulating sediment in the water column is sufficiently rapid (Carey, 1997). Similarly, the liquefaction of tsunami deposits requires the persistence of liquefaction over a period comparable to the travel times of the tsunami waves, but since the source of these is not defined for individual sections and may be closer to the site of sedimentation than the impact site itself (see below) this does not necessarily provide a good time constraint beyond indicating liquefaction developing over a period of tens of minutes if not hours. It should be noted that, while in general, liquefaction of preexisting sediments may persist for hours after an earthquake as pore fluid overpressures gradually dissipate (as discussed below) and so the duration of liquefaction of these sediments does not constrain the duration of shaking, in some of these K-T boundary sections (El Mimbral, La Sierrita, and Sage Creek in particular) the liquefaction is seen to have occurred in sediments that had not yet been deposited at the time of the impact and the directly impact-related seismic shock (Smit et al., 1992, 1996; Schulte et al., 2002; Stoffer, 2002). This suggests persistent seismic shaking of the sequences during the period of ejecta sedimentation.

2. ***Deep-seated collapse of indurated platform margins.*** Exceptionally thick (up to several hundreds of meters) and extensive (up to several hundreds of kilometers long at outcrop or subcrop) breccia and calcareous sandstone deposits including abundant lithified clasts of platform-forming carbonate occur in the Gulf of Mexico, east of the Yucatan peninsula, and in uplifted sections in Cuba (Alvarez et al., 1992, Bralower et al., 1998, Grajales-Nishimura et al., 2000, Kiyokawa et al., 2002). These are associated with deep-seated failure of the margins of the Campeche marine platform, for example, cutting down as far as Turonian rocks of the continental slope in the area of the east Yucatan margin now exposed in western Cuba (Tada et al., 2002), where exposed still in situ the platform rocks are also deeply fractured (Ocampo et al., 1996). In many locations and boreholes the collapse breccia is directly overlain by coarse impact ejecta (Grajales-Nishimura et al., 2000), as are the fractured rocks of the platform in Belize (Ocampo et al., 1996), implying that the collapse of the platform margin followed immediately after the impact and before arrival of the coarse, poorly sorted ejecta sheet some tens to a few hundreds of seconds later (Alvarez et al., 1992). There is little evidence for further failures of the platform margin after this initial period, in the form of multiple breccia layers or breccia layers interbedded with the impact ejecta sheet. Tsunami wave action reworked the finer-grained units at the top of the mass failure sequence in Cuba (Kiyokawa et al., 2002), at the top of the overlying ejecta sheet in the Campeche area (Grajales-Nishimura et al., 2000), and deposited sequences within the mass failure scar (Tada et al., 2002), again indicating that bedrock instability was confined to the period immediately after the impact.

3. ***Extensive slope failure involving unlithified continental margin sediments.*** Large-scale failures of unlithified sediments on the continental slopes, at the K-T boundary, have been documented not only from close to the impact site in the Caribbean and Gulf of Mexico (e.g., Alvarez et al., 1992; Bralower et al., 1998; Soria et al., 2001) but also from the Pacific margin of North America (Busby et al., 2002) and especially from the margins of the Atlantic Ocean (Klaus et al., 2000; Norris et al., 2000; Norris and Firth, 2002). In some cases (Busby et al., 2002, note that the Rosario site was relatively close to Chicxulub at the time of impact) these failures are complex and involve sheets of cohesive sediment as well as disaggregated sediment (perhaps from unlithified sediments in layered sequences that failed and took the interlayered lithified units with them). In contrast, features interpreted as deep-seated K-T boundary slumps at Blake Nose (Norris et al., 2001; Norris and Firth, 2002) appear to be earlier Maastrichtian slump structures (MacLeod et al., 2003). In most cases, these more distant failures appear to have been relatively shallow-seated, affecting the top few hundreds of meters of the Late Cretaceous sequences at most. Nonetheless, the Atlantic margin failures in particular were extremely large, resulting in sedimentation from turbidity currents of a layer of disaggregated chalk debris (reflector A*, Norris et al., 2000) over the whole end-Cretaceous western Atlantic ocean floor between North America and the mid-Atlantic ridge, an area of some 3,900,000 km^2 (Figs. 2 and 3). The volume of this deposit, based on its thickness in Deep Sea Drilling Project (DSDP) boreholes (Norris et al., 2000), may be as much as several tens of thousands of cubic kilometers, making it the largest landslide deposit known. It consists of very fine-grained, generally less than silt grade (apart from sand grade impact spherules), fully disaggregated chalk-rich sediment, implying that the slope failures affected unlithified sediments only and that the failed sediment masses disaggregated and developed into turbidity currents. A resedimented chalk horizon in the K-T boundary sequence on the Demerara Rise (Erbacher et al., 2004) suggests that diluted sediment from the turbidity currents, lofted into suspension, may have clouded the Atlantic Ocean as far south and east as northeast South America. K-T boundary deposits produced by slumping of the European continental margin have been identified off Spain at DSDP site 398D (Norris and Firth, 2002). These structures are interpreted as representing relatively small-scale, local slump features by Norris and Firth (2002), in contrast to the large-scale slope failures seen on the western margin of the North Atlantic, suggesting that they developed close to the outer

limit of the zone of impact-related slope failure associated with Chicxulub. This is consistent with the lack of disturbance seen in well-studied K-T boundary sections further east along the margins of the Tethyan ocean (for example, at El Kef in Tunisia; Smit et al., 1997; Galeotti et al., 2004), and more generally in the distal region of impact related sedimentation defined by Smit (1999), that covers the rest of Earth. Nevertheless, the indications are that some impact-related slope failure occurred up to 7000 km from the impact site (Fig. 3). In further contrast to the proximal breccias, Norris et al., (2000) show that the A* deposit was emplaced as two or more turbidity current flows over a period long enough for the start of crater ejecta accumulation on the ocean floor (hours to perhaps as much as several days, although this upper limit is critically dependent upon assumptions regarding rates of settling of dense ejecta-laden sediment plumes in the ocean [Carey, 1997]). In all, at least 10,000 km and perhaps as much as 20,000 km length of continental margin, including large sections of the North American (both Atlantic and Pacific), northern South American, and West African continental margins, are likely to have been affected by large-scale slope failure in multiple events occurring immediately after the Chicxulub impact. Klaus et al., (2000) document a transect across the Blake Spur that indicates that a down-slope section some 50 km wide was affected by slope failure. Even if the width of margin affected at more distant sites and along the narrower Pacific continental margin was smaller, an area of a few hundred thousand to as much as a million km^2 of continental margin sediments is likely to have been involved in slope failure to a depth of a few hundred meters below the sea bed (from the seismic lines of Klaus et al. [2000] and volume balance with the reflector A* deposit discussed above). The bulk of these slope failures occurred in water depths of some hundreds of meters to 2–3 km, and thus broadly in the region of gas hydrate occurrence.

In summary, therefore, a wealth of evidence exists for the occurrence of far-reaching effects of seismic waves generated by the effects of the impact at the K-T boundary. This evidence defines zones more or less centered on the Chicxulub impact crater, consistent with the usual inference that this crater is the product of the K-T boundary impact event and inconsistent with suggestions (for example, Stinnesbeck and Keller, 1996; Keller et al., 2004) that multiple impacts, some even larger than that which formed the Chicxulub crater, occurred around the end of the Cretaceous. If this were the case, paleoseismic evidence of these events would be expected and in particular would greatly complicate the stratigraphy of the many K-T boundary sections distant from Chicxulub (see review in Smit, 1999), most of which are structurally undisturbed (see Hagstrum and Abbott [2002] for a possible exception). Conversely, evidence of comparable, very large regions of seismic shaking distant from Chicxulub in sequences immediately above or below the K-T boundary might provide evidence for such additional impacts even if the source craters have been lost by subduction.

Formal paleoseismological analysis of the Chicxulub event is complicated by the difficulties of scaling from mere terrestrial earthquakes to the much more energetic impact and the need for differences in seismic attenuation rates owing to the intraplate location of the Chicxulub crater, coupled with problems relating to the duration of the seismic shaking associated with the impact that we consider further below. Nevertheless, we here attempt a tentative paleoseismic analysis of Chicxulub by comparing the liquefaction and slope failure phenomena identified as being associated with the Chicxulub impact with comparable but more recent phenomena associated with terrestrial earthquakes.

Distribution of Sediment Liquefaction and Slope Failure Phenomena Associated with Large, Historical Terrestrial Earthquakes as a Basis for a Tentative Paleoseismic Analysis of the Chicxulub Impact–Related Seismicity

Sediment liquefaction and slope failure are important, frequently damaging secondary phenomena associated with moderate to large terrestrial earthquakes, and an extensive literature exists relating to controls on their occurrence and also their use in paleoseismology, or the reconstruction of prehistoric and pre-instrumental (before widespread use of seismometers) earthquakes (see Yeats et al., 1997; Obermeier, 1996; Jibson, 1996, for reviews). Unsurprisingly, the area affected by these phenomena in any given earthquake increases with its magnitude, but there are a number of complicating factors that need to be considered when attempting to transfer results from this literature to interpretation of the K-T boundary liquefaction and slope instability phenomena. In addition to the review papers noted above, Keefer (1984) provides a survey of data relating to liquefaction and slope failure occurrence on land as a function of seismic intensity and other parameters of more recent earthquakes, while Youd and Perkins (1987) and Johnston (1996a, 1996b) provide particularly important studies of factors affecting the relationship between earthquake magnitude and the distribution of liquefaction and slope failure phenomena. On the basis of these we emphasize the following points in particular.

1. ***Seismic wave attenuation is lower in the stable plate interiors than at plate margins.*** Most very large earthquakes of terrestrial origin, and almost all of the largest earthquakes that have been instrumentally recorded (Kanamori, 1977), occur on the boundaries of Earth's tectonic plates or, more precisely, in plate boundary zones composed of intensely deformed and fractured rocks. However, a number of the most widely felt and most widely damaging large earthquakes, such as the 1755 Lisbon Earthquake, the 1811–1812 New Madrid Earthquakes, and the 1886 Charleston Earthquake (see reviews in Johnston, 1996a, 1996b), have affected the continental interiors: for example, the Lisbon Earthquake produced notable, damaging landslides up to 900 km from the inferred epicenter. The greater geographical extent of the effects of these events as compared to the earthquakes of comparable or even greater magnitude located on plate margins reflects lower rates of seismic attenuation in the continental interiors, as quantified by Johnston (1996a) among others. Empirical equations relating earthquake magnitude to distance

from source and observed liquefaction or slope failure effects of varying severity (quantified, for example, by the liquefaction severity index, or LSI, of Youd and Perkins [1987]), developed using data from plate margin earthquakes tend to underestimate the distances over which the plate interior can result in damage. As noted by Youd and Perkins (1987), this effect may be the cause of a factor of ~2 to 3, in disagreement with an earlier and less detailed study (Youd and Perkins, 1978), relating maximum distance (from earthquake to liquefaction site) to earthquake magnitude (see especially Figure 8 of Youd and Perkins [1987]). Thus, the seismic effects of impacts such as Chicxulub may have been particularly widespread, not just because of their magnitude but also because, unlike earthquakes, most impacts occur in the plate interiors.

2. ***Occurrence and severity of sediment liquefaction and slope failure is highly sensitive to sediment mechanical strength and pore water saturation.*** Keefer (1984) showed that occurrence of landslides associated with earthquakes, as a function of felt seismic intensity, is highly variable. In large part, this reflects the wide variation in rock and sediment strength, slope conditions, and groundwater saturation. Layered sequences with variations in permeability between layers may be particularly susceptible to failure as a result of coseismic and post-seismic fluid movements leading to overpressurization beneath impermeable layers (see quantification of these effects by Biscontin et al., 2004). The finite rate of fluid migration and hence overpressurization of layered sediments in this situation means that such slope failure may occur well after the end of seismic shaking. Jibson et al. (1994) discuss examples of slope failure occurring up to 3–5 days after the earthquake to which they are attributed. Such delays make it difficult to determine the duration of seismic shaking associated with the Chicxulub impact from the stratigraphic relationships of slope failure and liquefaction to ejecta deposition, but also indicate the potential for producing complex sequences of repeated slope failure and ejecta deposition in K-T boundary sequences. The mixed slope failure deposits at Rosario, including sheets of indurated sediment in an unlithified sediment matrix (Busby et al., 2002) may fit this pattern. Turning to spatial variation in landslide development, in the subaerial environment the development of coseismic liquefaction, and the intensity of liquefaction and resulting sediment movement, is closely associated with the occurrence of water-saturated, unlithified sediments, most especially river and delta alluvium in terrestrial settings (Youd and Perkins, 1987). Relatively little data exists on the occurrence of coseismic landslides and sediment liquefaction in the submarine environment, although the existing data (Keefer, 1984; Hampton et al., 1996) are consistent with what might be expected, that unlithified submarine sediments (being water-saturated) are especially susceptible to liquefaction and slope failure. The observed trend associated with Chicxulub, that deep-seated failures affecting slopes in lithified rocks such as carbonate platform margins only occurred relatively close to the impact site, whereas liquefaction and failure in unlithified sediments close to the seabed was much more widespread, is consistent with this.

3. ***Slope failures in unlithified sediments are more widespread than liquefaction, especially when seismic shaking is long-lasting.*** Studies of large, historical earthquakes in stable continental regions indicate that landslides of various types occur at even greater distances from the earthquake epicenters than does liquefaction. For example, Johnston (1996b) reviews evidence from the 1811–1812 New Madrid, 1886 Charleston, and 1755 Lisbon earthquakes and concludes that landsliding occurs at epicentral distances three to five times greater than is the case for sediment liquefaction: up to 900 km away in the case of the Lisbon earthquake. This arises because liquefaction occurs below near-horizontal surfaces whereas landslides occur on slopes subject to a continuing gravitational load, entirely independent of the earthquake, that takes effect as the material making up the slope is weakened by repeated cycles of seismic acceleration (see Biscontin et al. [2004] for a quantitative examination of this effect in the context of submarine slope failures). Thus, the longer the seismic shaking and the greater the number of acceleration cycles, which increases markedly with earthquake magnitude (Youd and Perkins, 1987), the greater the maximum distance of landsliding relative to the maximum distance of liquefaction from the earthquake epicenter. Thus, the observation that K-T boundary slope failure is more widespread and extensively developed (at distances up to ~7000 km) around the Chicxulub impact site than liquefaction on subhorizontal seafloors (at distances up to ~2500 km) may point to an extended period of seismic shaking associated with the impact.

The paleoseismic literature provides a number of empirical relationships between terrestrial earthquake magnitude and maximum distances of liquefaction and slope failure phenomena. These can be tentatively extrapolated to estimate the equivalent magnitude of the seismic source, located at Chicxulub, that is implied by the observed distribution of paleoseismic effects. A major problem with applying these relationships is that we have little data on the engineering properties of the dominant sediment types involved at the time of the impact, in particular the unlithified, water-saturated calcareous oozes of the chalk horizons that failed on the Atlantic margins. It seems reasonable to suppose, however, that these were as susceptible to failure as, or even more susceptible than, the weakest water saturated sediments studied in the modern environment: certainly the extended period of failure of these sediments implied by the complex relationships of the resedimented chalk horizons to the impact ejecta layers (Norris et al., 2000; Norris and Firth, 2002) is similar to the failure mode of the weakest modern sediments (Jibson et al., 1994). The water-saturated, unlithified clastic sands that failed and liquefied around the Gulf of Mexico are also likely to have been highly susceptible to failure, but the engineering properties of the indurated margins of the carbonate banks are not constrained and we therefore do not attempt to analyze the failure of these.

Youd and Perkins (1987) define a Liquefaction Severity Index (LSI) based on the abundance and scale of sand boils, clastic dikes, and other evidence of liquefaction at a particular site (Youd and Perkins, 1987; Table 4). Comparison of this table with the abundant meter- to decameter-scale liquefaction effects observed at the K-T boundary sites described from North America by various authors as discussed above and shown in Figure 2 suggests that all of these sites are characterized by LSI values in the range 50–100, at distances of up to 2500 km from Chicxulub. Note, however, that the outer bound is defined in part by the lack of evidence for liquefaction in the Bass River borehole section (Olsson et al., 1997), so some uncertainty attaches to this limit because of the very limited sample supplied by the borehole. Youd and Perkins (1987), using the data for plate boundary earthquake-related liquefaction effects, suggest the relation:

$$M_w = (\log_{10}LSI + 1.86\log_{10}R + 3.49)/0.98 \qquad (1)$$

where M_w is the moment magnitude of the earthquake (see below for further discussion of this definition of the magnitude of earthquakes and its applicability to the Chicxulub event) and R is the distance from the seismic source in kilometers. Note that in view of the only semiquantitative nature of LSI it is not appropriate to assign precise error bars to an estimate obtained using this relationship. Assigning an LSI of 50 to the most distant sites and assuming as an upper bound that this extends up to 3000 km from Chicxulub gives an M_w estimate of 11.7 for the event, using this equation. It should however be noted that this is likely to be an overestimate because of the reduced rate of seismic attenuation in the plate interiors, as noted above; extrapolation of the different, less well constrained relationship (with M_w varying with ~1.5 $\log_{10}R$) developed by Youd and Perkins (1978) using a mixture of data from plate margin and plate interior earthquakes suggests an M_w of 11 or less for the K-T boundary seismic source. The reduced distance limit for pervasive liquefaction of ~2500 km suggested by the Bass River borehole section (Olsson et al., 1997) also implies a lower M_w but the effect of this uncertainty is small compared to the uncertainty in the seismic energy attenuation rate.

Turning to the case of the slope failures, the key parameter in the case of seismically triggered failure of the very weakest water saturated sediments is the Arias Intensity I_a, or time integral of the square of seismic acceleration at the failure site (Arias, 1970; Harp and Wilson, 1995); stronger sediments and rocks are only affected by accelerations over a certain limit related to their strength, but as noted above, the unlithified chalk oozes involved in the bulk of the distant slope failures at the K-T boundary are likely to have had a minimal strength. For such materials the lower limit of I_a required to trigger slope failure is ~0.2 ms^{-1} (Jibson and Keefer, 1993). Wilson and Keefer (1985) again use data from historical plate margin earthquakes to define an empirical relationship between Arias Intensity, distance from source R at which that intensity is developed, and earthquake magnitude:

$$M_w = \log_{10}I_a + 2\log_{10}R + 4.1 \qquad (2)$$

Assuming that the most distant slope failure at the K-T boundary that we consider here, the small-scale slumping in DSDP Hole 398D (Norris and Firth, 2002), some 7000 km from Chicxulub at the time, is associated with an Arias Intensity of 0.2 ms^{-1}, this yields an M_w estimate of 11.0 for the K-T boundary seismic source. Again, lower intraplate seismic attenuation rates will tend to cause this to be an overestimate of the seismic magnitude of the event.

With these estimates and the points made above on the likely sources and magnitudes of uncertainties associated with them in mind, we now turn to a consideration of the Chicxulub impact as a seismic source. This will provide a basis for considering whether it is realistic to relate the liquefaction and slope instability phenomena at the K-T boundary to the seismic effects of the impact, or whether another mechanism, such as the cyclic pressurization/depressurization tsunami wave mechanism proposed by Max et al. (1999), is required. In particular, it is necessary to consider whether the extrapolation from terrestrial earthquakes is valid, or whether the difference in magnitude between these and the Chicxulub impact, in terms of total energy released, is accompanied by differences in the nature of the source, in particular its duration and attenuation characteristics, that make such extrapolation invalid.

Impact-Related Seismicity: Syn-Impact and Post-Impact?

Impacts on Earth of large asteroids or comets, forming complex craters (Melosh and Ivanov, 1999, for a review) with an initial or transient diameter of the order of 3 km or more, are inferred to produce seismic energy in two ways.

First and most obvious is the impact itself. The fraction of impactor kinetic energy converted to seismic energy is likely to be small, but is poorly constrained owing to problems of scaling from much smaller impact-like events such as underground nuclear explosions (also likely to be more efficient seismic sources than impacts because they are confined) and laboratory scale impact experiments. Even within the range of magnitudes of large impact events, differing efficiencies of energy absorbing processes within the target rock mass are exemplified by the disproportionate increase in impact melt rock volume with impact size (Melosh, 1989; Melosh and Ivanov, 1999; Grieve and Therriault, 2000). Schultz and Gault (1975) suggest kinetic-to-seismic energy conversion efficiencies in the range 10^{-2} to 10^{-5}. Toon et al. (1997) consider kinetic-to-seismic energy conversion efficiencies in the range 10^{-2} to 10^{-4} and calculate that the initial impact at Chicxulub would have released seismic energy equivalent to an earthquake with moment magnitude M_w between 11.3 and 10.0. For comparison, the largest known terrestrial earthquake, the Chile earthquake of 1960, had a moment magnitude M_w = 9.5. Since

$$M_w = 2/3 \log_{10} M_0 - 6.0 \qquad (3)$$

(Kanamori, 1977; Hanks and Kanamori, 1979), this means that the difference in terms of seismic moment between Chicxulub and Chile is that the latter is of the order ten to one thousand times smaller in terms of the seismic moment M_0: a unit change in M_w corresponds to a factor of 30 change in seismic moment.

This peculiarity of the moment magnitude scale results from the need to maintain approximate consistency between moment magnitude and instrumental earthquake magnitude scales (Kanamori, 1977).

Comparison of the initial impact seismic signal with earthquake seismic signals is further complicated by differences in the signal generation process and in the efficiencies of non-seismic energy absorbing mechanisms. The initial impact seismic energy from expansion of the transient cavity in a Chicxulub-sized impact is released into the Earth over a few tens of seconds (Melosh, 1989; Melosh and Ivanov, 1999), rather than the few hundreds of seconds rupture and strong-motion shaking times typical of the largest known terrestrial earthquakes (Weichert, 1960; Hansen, 1966; Hansen et al., 1966; Barrientos and Ward, 1990), and from a much smaller region (transient cavity surface area of ~30,000 km^2) than would be the case with an earthquake of equivalent magnitude (rupture plane area of ~1,000,000 km^2, from the seismic moment–rupture size relationship discussed below). Furthermore, the growing transient cavity is embedded within a region of shattered rock produced as the leading shock wave exceeds the strength and elastic limit of deformation of the rock and is partly absorbed in fracturing, deforming, and even melting the rock mass. This permanent inelastic deformation produces, amongst other features, the shatter cones that are a characteristic feature of the rocks within and beneath large impact structures, in the region outside the transient cavity limit but within the final structure (Milton, 1977). This high level of seismic attenuation associated with passage of large-amplitude, intense seismic waves, much more intense than any associated with terrestrial earthquakes, makes application of the computer simulations of Boslough et al. (1996) to the real Earth problematic.

The second mechanism of seismic energy generation associated with impacts operates after the impact, as the initial transient crater is transformed into its broader and shallower final form. Large impact craters, with diameters greater than a lower limit around 3 km in the terrestrial gravity field, are unstable features and undergo post-impact modification to produce various forms of complex crater, such as central peak craters, peak-ring craters and (generally, the largest craters of all), multi-ring craters. Melosh (1989), Spray (1997), and Melosh and Ivanov (1999), review different mechanisms that may be involved in the development processes of these different crater morphologies. Of the different types of structure, multi-ring craters are the least well understood (Melosh and Ivanov, 1999, p. 390; Grieve and Therriault, 2000) in the context of impact crater dynamics, but the presence of discrete fault zones defining the outer ring structures means that the crater modification process is amenable to analysis in terms of more conventional seismic sources. Estimates of seismic moment released during the formation of these faults therefore provide a useful lower bound on the seismic magnitude of the overall impact event and also allow a more detailed comparison with the source characteristics of terrestrial earthquakes.

Here we are primarily concerned with multi-ring craters, as the buried Chicxulub crater shows multiple ring faults on seismic profiles (Morgan et al., 1997; Brittan et al., 1999; Snyder and Hobbs, 1999) and corresponding ring-shaped gravity structure (Pilkington et al., 1994; Hildebrand et al., 1998). The outermost faults are a ring ("outer ring" of Snyder and Hobbs, 1999) of inward-facing normal faults at a distance of 90–100 km from the center of the structure with near-surface displacements of around 500 m and a zone at around 125–130 km from the center containing both outward-directed thrust faults and normal faults ("exterior ring" of Snyder and Hobbs, 1999) with a net thrust displacement in the near surface of around 200 m. Both sets of faults curve downward and inward to the base of the crust and may offset the Moho and the uppermost mantle rocks toward their inner limits around 40 km and 60 km from the center of the structure, respectively (Morgan et al., 1997; Snyder and Hobbs, 1999). The thrust faults are most likely the outermost expressions of syn-impact deformation (consistent with theoretical modeling of syn-impact deformation incorporating damage-related weakening; O'Keefe et al., 2001; O'Keefe and Ahrens, 2002). They were subsequently reactivated as normal faults during crater collapse and multi-ring basin formation except for the low-angle tip section where the thrust faults intersected the surface, producing the characteristic isolated ridges at 125–130 km radial distance on the seismic profiles (see, for example, Fig. 3 of Snyder and Hobbs, 1999). Thus the total displacement that occurred on these faults during the development of the multi-ring basin may be substantially greater than the net displacement now observed. Furthermore, the conical, inward-dipping geometry of both the outer and exterior ring faults means that the displacement on them must have increased inward in order to accommodate the convergent component of inward movement. This must also have been associated with tangential shortening, perhaps on radial faults: these would not be well-imaged on the largely radial seismic profiles across Chicxulub but are a prominent feature (Scott and Benn, 2001) of the comparable (Grieve and Therriault, 2000) Sudbury impact structure. The high reflectivity of the Chicxulub fault zones in seismic profiles (Morgan et al., 1997) suggests the presence of thick sheets of frictional melt rock or pseudotachylite in the fault zones. While the nature of the ring faults at Chicxulub is poorly known owing to the lack of exposure, the much older but similarly sized Sudbury impact structure in Canada is deeply eroded and exposes much evidence for the mechanisms and timing of post-impact crater modification (Grieve et al., 1991; Spray and Thompson, 1995; Scott and Spray, 2000; Scott and Benn, 2001). Ring structures around Sudbury are characterized by superfaults (Spray, 1997) with a few to several kilometers of displacement across thick sheets, tens of meters to kilometers thick, of frictional melt rocks or pseudotachylites. Frictional melting on faults, particularly on the very large scales at which it occurred at Sudbury, is indicative of large, rapid seismogenic displacements. We therefore argue that it is appropriate to estimate seismic energy released during post-impact crater modification at Chicxulub assuming that all movement on seismically well-imaged faults (meaning those inferred to contain pseudotachylite) was seismogenic. Within the outer ring faults, a zone of terrace blocks between 70 and 40 km from the center of the structure and individually between 5 and 10 km wide represent the innermost coherent, fault-bounded bodies of rock at Chicxulub. Displacements on the faults bounding these total some 4000 m (Snyder and Hobbs, 1999), but the fault zones are not so well imaged in the seismic sections This may be because of their steeper dip angle but may also be because pseudotachylites are not so well developed on them despite the greater displacements. Melosh and Ivanov (1999) argue that the displacement of these blocks may be interpreted in terms of low-friction movement within an acoustically fluidized flow, although the blocks are individually substantially larger than can be accommodated in their model (~10 km as opposed to ~200 m), so it is less certain whether the motion of these blocks was as efficiently seismogenic as appears to have been the case with the outer faults.

Table 1 summarizes the geometrical features of these structures and their estimated contributions to the seismic moment released during the crater modification process. Fault areas are calculated assuming that the faults can be approximated as truncated inverted cone shapes, neglecting any minor curvature of the fault surfaces, in which case the area A of the inclined fault surface is defined by the standard geometrical formula relating this to the outer and inner radii of the truncated cone, r_o and r_i, and the axial separation h of the truncating surfaces (in this case taken as equivalent to the maximum depth to which the faults can be traced).

$$A = \pi (r_o + r_i)\sqrt{((r_o - r_i)^2 + h^2)} \quad (4)$$

Any contribution to the total seismogenic faulting from lower-angle faults or other structures linking the inclined faults to the central uplift of the structure is neglected, as these structures are not clearly imaged on the seismic profiles through Chicxulub (Snyder and Hobbs, 1999; Morgan et al., 1997) and so it is not known whether the deformation in this zone (which is geometrically necessary) involved seismogenic slip on pseudotachylite-bearing fault zones (as developed in the central uplift at Vredefort; Reimold and Colliston, 1994), or whether it involved more pervasive or even fluidized flow (Melosh, 1989; Melosh and Ivanov, 1999). Similarly, as noted above, we neglect any contributions from radial faults. Neglecting such contributions will cause the estimate of total crater modification-related seismicity below to be a moderate underestimate, as discussed below. Slip u on the faults increases inward with decreasing radial distance r from the center of the structure as $u = u_o(r_o/r)$, where subscript "o" denotes the values observed at the outer limit of each structure where the faults intersect near-surface layered rocks as discussed above; this needs to be allowed for when calculating the seismic moment M_0 released by the fault movement, is normally defined as:

$$M_0 = \mu u A \quad (5)$$

(Aki, 1966), where μ is the shear modulus of elasticity of the faulted rocks (generally taken as $3*10^{10}$ Nm^{-2} for the crust and $7*10^{10}$ Nm^{-2} for the mantle; although some fraction of the Chicxulub fault surfaces appear to cut uppermost mantle rocks (Snyder and Hobbs, 1999), we use the lower value throughout here in order to provide a conservative estimate of seismic moment released), and u is the average slip over a fault surface A. For a perfectly conical fault, this is most easily done by calculating seismic moment from the equivalent displacement on a planar surface:

$$M_0 = \mu u_o(2\pi r_o)\sqrt{((r_o - r_i)^2 + h^2)} \quad (6)$$

where the r and h terms are as defined for Equation (4).

The results presented in Table 1 imply that the movements on the outer and exterior ring faults, that we have argued definitely moved seismogenically, were equivalent on their own to an earthquake of $M_w = 10.0$; while if the movements on the terrace faults were also seismogenic, the equivalent single earthquake rises to $M_w = 10.3$. Note that these will tend to be lower estimates for a variety of reasons: radial faulting and faulting in the central uplift are neglected, as noted above, and reversing movements (first thrust, then normal) on the exterior faults and possibly other faults are neglected. All of these fault movements are inferred to have occurred during a geologically very brief period during the impact itself (in particular, early thrust movements on the exterior ring fault) and subsequent formation of the multi-ring basin. Modeling by O'Keefe and Ahrens (2002) of the formation and modification of the Chicxulub structure suggests that the entire development of the structure would have taken ~300 seconds, comparable with the duration of the largest terrestrial earthquakes as noted above. Depending on the efficiency of direct conversion of impact energy to seismic energy during the growth

TABLE1. CALCULATION OF SEISMIC MOMENT RELEASED BY DISPLACEMENTSON THE MAIN CHICXULUB RING STRUCTURES

Structure	Outer radius R_o (km)	Inner radius R_i (km)	Depth h (km)	Area A (km^2)	Equivalent area $2\pi R_o\sqrt{(R_o-R_i)^2+h^2)}$	Near-surface offset u (m)	Seismic moment M_0 (Nm)	Equivalent moment magnitude M_w
Exterior ring	125	60	30	41600	62000	200	$3.7*10^{23}$	9.71
Outer ring	100	40	30	29500	46000	500	$6.9*10^{23}$	9.89
Inner ring	70	40	30	14660	18600	4000	$2.23*10^{24}$	10.23
TOTAL	–	–	–	–	–	–	$3.29*10^{24}$	10.34

of the transient cavity, which, as discussed above, is poorly known, the total seismic moment generated by movement on the discrete fault systems could be as little as ~3% of the total seismic moment associated with the impact or as much as ~80%.

Although the seismic moment equations were developed for tectonic earthquakes, in which displacements are a small fraction of fault length (around 10^{-4} to 10^{-5}; Wells and Coppersmith, 1994) and the fault ruptures terminate at sides and base within the Earth, rather than in the shattered rocks beneath the axis of the crater, they have been successfully applied to seismicity associated with the formation of volcanic calderas such as Katmai in 1912 (Abe, 1992). These are also characterized by ring faults with large displacements and terminations in less-than-solid rocks, and may have many features in common with post-impact faulting leading to multi-ring basin formation, as also noted by Spray (1997). The successful application of the seismic moment equations to caldera formation suggests that they can also be appropriately applied to fault movements associated with multi-ring impact basin formation.

For the overall purpose of this paper, the importance of this analysis of the different contributions to the seismic moment released during the impact and subsequent crater modification is that it provides a check upon the assumption implicit in the tentative paleoseismic analysis, saying that the impact was broadly comparable to a very large earthquake, not only in the moment released but also in terms of the duration of seismic shaking and the number and intensity of acceleration cycles, since it is these that control the development of liquefaction and slope failure in sediments as discussed above (see especially Youd and Perkins, 1987; Biscontin et al., 2004). Theoretical modeling (Melosh, 1989; Melosh and Ivanov, 1999; also O'Keefe et al., 2001; O'Keefe and Ahrens, 2002) suggests that development of the transient cavity itself is broadly complete in a period of the order of a few tens of seconds, significantly shorter than the few hundreds of seconds typical of the very largest terrestrial earthquake ruptures. However, the theoretical modeling also indicates that formation of the inner ring fault-bounded terraces and the outer ring faults might be expected to take of the order of a few hundred seconds, while completion of movements on the exterior faults might be expected to take even longer, since the models consistently show outward propagation of the deformation from the central structure. Borehole data from the central part of Chicxulub indicate that the impact melt sheet and crater fill breccias (comprising fallback, backwash, and possibly collapsed central uplift material; Pilkington et al., 1994) are not cut by underlying faults. Spray (1997) and Melosh and Ivanov (1999) interpret this as implying that faulting in at least the central area was completed during the period of minutes before accumulation of fallback began, but this is problematic as the impact melt would not record fault movements until it is solidified, and similarly the poorly consolidated crater fill breccias may have deformed by intergranular slip rather than by development of discrete faults.

Whatever the precise duration of the seismic shaking associated with the impact crater modification phase, it does, however, seem comparable to the duration of seismic shaking associated with very large earthquakes, and so we argue that the application of the paleoseismic analysis methods outlined above is reasonable, if not definitively valid. Note, however, that the extended period of liquefaction and slope failure implied by the complex stratigraphy and structure of the K-T boundary sections within a few thousand kilometers of Chicxulub does not provide a strong independent estimate of the duration of seismic shaking since, as noted above, both phenomena may persist or be significantly delayed relative to the shaking as a result of progressive pore fluid migration and pressurization (Jibson et al., 1994; Biscontin et al., 2004).

The results of this section indicate that the Chicxulub impact was broadly comparable at the very least to a $M_w = 10$ earthquake, most probably to at least a $M_w = 10.3$ earthquake and, when the effects of the initial impact and the crater modification faulting are taken together, to an earthquake closer to $M_w = 11$ than $M_w = 10$. Although there are large and in some cases poorly constrained uncertainties in this impact structure based analysis, we note that the results are broadly consistent with the paleoseismic analysis in the previous section and conclude that seismic shaking associated with the Chicxulub impact and subsequent impact crater modification is a reasonable mechanism for producing the widespread slope failures that occurred at the K-T boundary in the North American and North Atlantic regions. Conversely, if there were numerous and even larger impacts associated with the K-T boundary (Keller et al., 2004) we would expect widespread and even more severe seismic effects to be evident in K-T boundary sections in other parts of Earth.

Slope Failures and Tsunamis: Cause and Effect?

Before moving on to consider the environmental effects of continental slope failures at the K-T boundary, it is necessary to briefly consider alternative mechanisms for triggering the widespread continental slope failures and liquefaction at the K-T boundary, in particular the causal relationship between the slope failures and the well-documented giant tsunamis associated with the Chicxulub impact. Originally identified in the northern Gulf of Mexico region on the basis of tsunami deposits (Bourgeois et al., 1988), K-T boundary tsunamis have now been inferred over a much wider area, including the entire eastern margin of North America (Norris and Firth, 2002) on the basis of slow recovery of shallow-water marine faunas after the impact (Jablonski, 1998). This is despite the fact that the impact occurred in water that was very shallow compared to the crater depth and will therefore have been an inefficient tsunami source compared to impacts into the deep ocean (Ward and Asphaug, 2000). The analysis of Matsui et al. (2002) predicts a large tsunami associated with rapid filling of the impact crater, but their model assumes an unrealistically large water depth on the Yucatan carbonate platform (200 m as compared to the subtidal depths implied by studies of the latest Cretaceous sequences [Ocampo et al., 1996; Grajales-Nishimura et al., 2000]). Furthermore, deposition of the proximal ejecta sheet would have further reduced the water depth and preservation of the ejecta sheet in many outcrops and boreholes is inconsistent with rapid sheet flows filling the crater and subsequent propagation of a giant tsunami back out over the platform top.

Even if it is assumed that the impact may have directly generated very large near-source tsunamis, around the Gulf of Mexico in particular (Matsui et al., 2002), the situation for the margins of the North Atlantic and the Rosario area of Baja California is more problematic, partly because of the greater distance from the impact site but also because of the shadowing effects of areas of land and shallow sea (where very large tsunamis, with amplitudes >100 m, would tend to break), the distribution of which is shown in Figures 2 and 3. The potential for directly impact-generated tsunamis to trigger the distant slope failures through cyclic variation in pore fluid pressure is therefore limited. At large distances from the impact point, we consider it more likely that (seismically triggered) submarine landslides generated tsunamis, rather than the other way around, although it is possible that the first generation of very large landslides along any given section of continental margin triggered a cascade of failures nearby through both local seismic shaking and tsunami effects, until slope failure was complete.

The inference that directly impact-generated tsunamis are likely to have been limited in scale and also geographically limited to the Gulf of Mexico and other seas immediately adjacent to the Chicxulub impact site poses severe problems for the model of Max et al. (1999), which, as noted above, proposes that pressure variations associated with large tsunamis caused blowouts of methane from free gas reservoirs beneath gas hydrate layers. We turn, therefore, to the alternative hypothesis that seismically triggered slope failures, involving unlithified sediments on the continental slopes, may have been responsible for gas hydrate releases at the K-T boundary.

Slope Failures and Gas Hydrate Release

Very large continental slope failures during the Quaternary have frequently been linked to releases of gas hydrates stored in continental slope sediments (either as cause or effect) and to negative carbon isotopic excursions in marine carbonates (for example, Haq, 1998; Maslin et al., 1998; Bryn et al., 2003; Kennett et al., 2000, 2003). The link between slope failure, gas hydrate release, and carbon isotopes has also been proposed for events prior to the Quaternary, most notably the Paleocene-Eocene Thermal Maximum (PETM) (Katz, et al., 2001; Rohl et al., 2000; Dickens, 2001, 2003).

Key features of gas hydrates from the perspective of potential links between large-scale continental slope failures and isotope excursions are their limited stability range, buoyancy, and extreme stable isotopic characteristics. Recent reviews of their general characteristics and environmental significance are provided by Haq (1998), Kvenvolden (1998), Dickens (2001), and Kennett et al. (2003). Here we emphasize aspects relevant to their likely occurrence and distribution in the latest Cretaceous, to their likely fate in the large-scale but shallow-seated slope failures that we have inferred to have been triggered by seismic shaking associated with the Chicxulub impact, and to the environmental and isotopic consequences of their release.

Gas hydrates are ice-like deposits containing a mixture of water and gas. They are non-stoichiometric compounds—that means water molecules (so-called structural molecules) form cage-like structures in which gas molecules are trapped as guest molecules. The most common gas is methane (CH_4) in naturally occurring gas hydrates, although H_2S, CO_2 and other hydrocarbons can occur. In the following analysis, we neglect these minor gases and focus on methane-bearing gas hydrate. Gas hydrates are stable under high pressures and relatively low temperatures: in the modern glaciated Earth, they are found underneath the oceans and in permafrost regions, but the latter reservoir is likely to have been negligible in the warmer end-Cretaceous world. The combination of pressure and temperature constraints limit subsea gas hydrate occurrence largely to a narrow zone within the shallower parts of sediment sequences on the continental slopes; the limits of this stability zone are indicated schematically in Figure 4, and we discuss below how these limits may have varied between the modern and end-Cretaceous oceans. Coring and the occurrence of distinctive bottom-simulating reflectors (BSRs), produced by the reflection of seismic waves off interfaces between hydrate-bearing sediment layers and underlying warmer sediments containing free methane gas trapped by the overlying gas hydrate, indicate that at the present day, gas hydrates are near ubiquitous in continental slope sediments that lie within the gas hydrate stability field: there could be as much as 10,000 gigatons of carbon stored naturally as gas hydrates (Kvenvolden, 1998), or over ten times the amount of carbon currently in the atmosphere. Apart from suitable temperature and pressure conditions, gas hydrate formation depends on how much gas, primarily CH_4, is available. Most of the methane found in the oceanic sediments is produced by decomposition of organic components or by bacterial reduction of CO_2 in sediments, although in some cases it may also be a product of thermo-catalytic alteration processes in more deeply buried sediments. CH_4 production is highest at the continental margins. Here, high plankton productivity and high rates of input of organic material from the continents yield large amounts of organic matter, which becomes the basis for a production of gas in the sediment. Gas hydrate is, therefore, found at almost all passive and active continental margins. Due to its production principally through bacterial processes, the methane released from gas hydrates is characteristically very depleted in ^{13}C with a $\delta^{13}C$ value of between −40‰ and −100‰ (see references cited in both Maslin and Thomas, 2003 and Dickens, 2001) and thus is an obvious candidate for large-scale changes in the carbon isotopic budget of the oceans.

A further consequence of the limited pressure and temperature range over which gas hydrates are stable is the potential for rapid decomposition of the hydrates, releasing the caged gases, if they are taken out of this stability field. This may occur by environmental change, particularly sea level falls (pressure

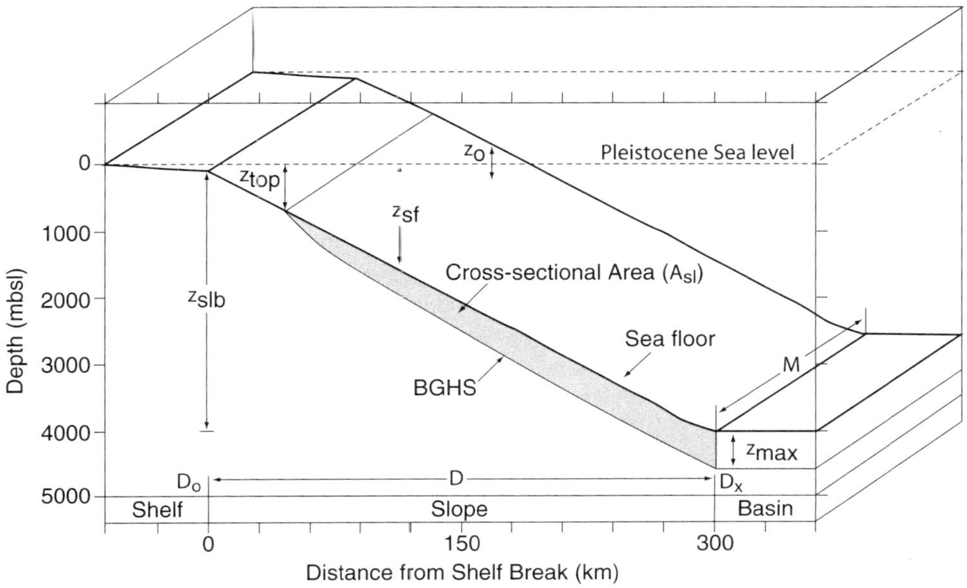

Figure 4. Calculation of gas hydrate potential are based on the dimensions of the lens shaped area (A_{sl}) between the sea floor and the base of the gas hydrate stability zone (BGHS) across any given continental margin (M). When Z_0 = depth if sea level relative to standard sea level, Z_{sf} = depth of sea floor below standard sea level, Z_{top} = depth of top of the hydrate stability zone below standard sea level, Z_{slb} = depth of the base of slope below standard sea level, Z_{max} = maximum depth of the gas hydrate stability zone, D = horizontal distance perpendicular to a margin (M), when D_0 = the initial distance and D_x = distance to the base of slope (adapted from Dickens, 2001).

reductions) or temperature increases—hence the "clathrate gun" hypothesis linking gas hydrate release to deglaciation (Kennett et al., 2003)—or to rapid, large-scale sediment mass movements. These sediment movements may release methane from hydrates in one or both of two mechanisms.

1. If the sediment failure initially occurs in the sequence above the hydrate stability zone, the removal of the weight of the displaced sediment (over the weight of an equivalent depth of water) will cause gas hydrates in sediments below the newly formed transient collapse scar to dissociate (Haq, 1998; Kvenvolden, 1998; Kvenvolden and Lorenson, 2001; Bryn et al., 2003). This can be explosive as pressure effects are transferred through the sediment rapidly and thus can enlarge the sediment collapse as large quantities of hydrate turn to gas, producing progressively deeper slope failures or discrete blow-out structures similar to those described by, amongst others, Max et al. (1999).

2. Sediment failures that are more deep-seated may cut directly into the hydrate stability zone and so directly displace the gas hydrate bearing sediments. While a number of authors have pointed out that the sediments involved in the slope failures will tend to move to higher pressures and thus into the hydrate stability field, in many cases these slope failures are disintegrative (Hampton et al., 1996) and develop into large-volume turbidity currents or grain flows. In such cases, while carbonate or other lithic grains settle out as ocean floor deposits, low-density, buoyant hydrate clasts will tend to rise in the water column and ultimately float upwards through the low-pressure limit of hydrate stability and there decompose, releasing methane to the surface layers of the ocean or even directly to the atmosphere. On a small scale, such floating may involve individual hydrate grains or nodules as demonstrated experimentally by Paull et al. (2003); we consider buoyant plume phenomena that may affect larger natural failures of hydrate-bearing sediment further below. Evidence for geologically recent sediment failures and associated gas hydrate release can be found on many continental margins (e.g., Bouriak, et al., 2000; Laberg et al., 2000; Bryn et al., 2003; Maslin et al., 1998, 2004).

A Calculation of the Amount of Gas Hydrate Released at the K-T Boundary and the Consequent Carbon Isotope Excursion

A key first step in analyzing the scale and isotopic consequences of gas hydrate releases that may have been associated with the continental slope sediment failures triggered by the Chicxulub impact is an estimate of the abundance of gas hydrates in continental slope sediment sequences in the latest Cretaceous. Dickens (2001) studied the possible changes in the carbon cycle during the PETM and has modeled the potential abundance of gas hydrates stored globally during and after the Paleocene. Evidence from benthic foraminifera oxygen isotopes suggests that deep-water oceanic temperatures were very similar between the Cretaceous and Paleocene (Zachos et al., 2001). Hence we suggest the range of estimates produced by Dickens (2001) can be used as a basis for our discussion of the potential hydrate release at the K-T boundary.

Table 2 shows that the modern oceanic sediments globally have a greater potential gas hydrate occupancy than their Paleocene equivalents. This is because the warmer ocean and high sea levels during the Paleocene reduced the cross-sectional area of the continental margin which gas hydrates can occupy (Fig. 4). Using the Paleocene reconstruction as a good approximation of the Cretaceous ocean, we present two possible scenarios, based on the discussion above: one with a disrupted margin of

TABLE 2. CALCULATION OF POTENTIAL RELEASE OF GAS HYDRATE AND INFLUENCE ON THE OCEANIC CARBON ISOTOPE RECORD FOR THE K-T BOUNDARY

Time Period	Cross-sectional area (km^2)	Porosity	Margin length (1000 km)	Potential global occupancy by volume of gas hydrate (10^6 km^3)	Mass of gas hydrates 5% occupancy (GtC)	Mass of gas hydrates 10% occupancy (GtC)	Global oceanic $\delta^{13}C$ 5% occupancy (‰)	Global oceanic $\delta^{13}C$ 10% occupancy (‰)
Modern (global)	35	50	200	2.63	12,000	23,000	n/a	n/a
Paleocene and Cretaceous (global)	15	50	150	1.13	5,000	10,000	n/a	n/a
K-T boundary Scenario 1	15	50	10	n/a	330	670	–0.5	–1
K-T boundary Scenario 2	15	50	20	n/a	670	1330	–1	–2

10,000 km and another of 20,000 km. However, there are a number of other assumptions that have to be made in order to estimate the amount of gas hydrate released post impact.

The first of these assumptions is the assumed occupancy of the pore space by gas hydrates, as this will control the amount of methane released per m^3 of failed sediment. Various authors have used figures as widely different as 5% to 20% for calculating the modern occupation of the pore space above the BSR by gas hydrates (see review in Dickens, 2001). The best estimate we have at the moment for modern occupancy lies between 5% and 10%, which we use as our range for the K-T calculations (see Table 2). It is also worth considering that for the past 2.5 Ma, sea level and oceanic temperatures have fluctuated between glacial and interglacial periods (up to 120 m and 10 °C) releasing the most vulnerable stores of gas hydrate (Kennett et al., 2003). These extreme fluctuations in the oceanic condition controlling the stability of gas hydrates may not have occurred prior to the K-T event. So we could envisage a long-term build-up of gas hydrates over hundred of thousands of years in the sediment producing at least 10% occupancy within the hydrate stability zone, in addition to large accumulations of free gas trapped beneath the hydrate layer.

The second assumption is that the gas hydrate is released efficiently from the disrupted sediment; as in the first instance the descent of the landslides from the continental slope to the deep ocean would carry the gas hydrate to high pressure conditions under which gas hydrate would be expected to be more stable. This objection has been raised widely in the context of models for gas hydrate release from continental slope landslides, but the validity of this argument is dependent upon the transport mechanism of the landslides and whether they travel as coherent slumps in which the hydrate remains trapped, or as disintegrative sediment landslides (Hampton et al., 1996) from which buoyant hydrate grains or clasts may escape. In the specific case of the K-T boundary landslides, Norris et al. (2000) show that the mechanism of sediment transport responsible for the laterally extensive chalk layers at the K-T boundary in the western Atlantic involved disintegrative landslides that developed into giant turbidity currents from which the dense but very small chalk particles settled out as individual grains. Buoyant gas hydrate grains, on the other hand, would tend to rise in the turbidity currents and perhaps even produce convective lift off of the turbidity current once a sufficiently high fraction of the chalk grains had settled out, in the reverse process to the rapid sinking of dense sediment laden plumes (Carey, 1997). Large-scale lift-off of buoyant plumes from turbidity currents, equivalent to the buoyant rise of hot ash clouds from pyroclastic flows (Sparks et al., 1993), has been inferred from the characteristics of graded hemipelagic mud layers associated with the giant freshwater turbidites produced by meltwater discharge events (Heinrich events) from the Laurentian ice sheets (Hesse et al., 2004). The resedimented chalk mud layer on the Demerara rise, associated with the K-T boundary (Erbacher et al., 2004), may represent the record of similar turbid but buoyant plumes associated with escape of hydrate particles en masse from the giant turbidity currents responsible for the deposition of the A* reflector resedimented chalk bed (Norris et al., 2000).

Alternatively, larger gas hydrate particles could rise individually (Paull et al., 2003) after the currents had come to a halt. In either case, the fully disaggregated nature of the landslides would leave the hydrate grains free to rise in the ocean and escape to the surface. The only fraction of gas hydrate that would remain on the seabed would be finely disseminated hydrate adhering to chalk grains (despite abrasion of the grains by impacts in the flow) in a proportion of chalk to hydrate such that the aggregates were not buoyant.

So if we assume the occupancy of gas hydrate in the potential space on the continental margin to be between 5% and 10%, and assume that the gas hydrate is released efficiently from the disrupted sediment, the gas hydrates could change the global oceanic $\delta^{13}C$ inventory by –0.5‰ to –2‰ (given an ocean carbon total of ~40,000 GtC and an isotopic composition of gas hydrates of –60‰; see Maslin and Thomas, 2003, and Dickens, 2001). This compares well with the possible change of –1‰ observed in the planktonic foraminifera records across the K-T boundary (see Fig. 1).

The result of our analysis linking these disparate lines of evidence is that we have an indication of broad consistency between, on the one hand, the estimated seismic effects of the Chicxulub impact based on the size and geometry of the impact structure and the estimated seismic effects of the impact based on the previously documented extent of sediment liquefaction and submarine slope failure, and on the other hand, consistency between the documented extent of sediment liquefaction and submarine slope failure and the carbon isotopic excursion associated with the impact. We argue, therefore, that it is possible to relate the K-T boundary carbon isotopic excursion quantitatively (at least at the order of magnitude level of accuracy) to the size, location, and seismic effects of the Chicxulub impact. This does not exclude the possibility that other effects of the impact such as the burning of terrestrial biomass may also have contributed to the isotope excursion, although it does suggest that such burning may not have been as extensive or as efficient in producing carbon gases as supposed (Wolbach et al., 1990; Melosh et al., 1990; Ivany and Salawitch, 1993; Robertson et al., 2004). However, our conclusion does suggest that there is little scope for positive feedbacks in the immediate post-impact environment to produce runaway releases of more carbon gases, consistent with the inference that the environment cooled rather than warmed in the immediate aftermath of the impact (Galeotti et al., 2004). Similarly, the isotopic data leaves little scope for multiple impacts at the end of the Cretaceous (Keller et al., 2004) to produce widespread continental slope sediment failures and gas hydrate releases from other parts of Earth's surface.

Implications for the Detailed Characteristics of the K-T Boundary and Other Isotope Excursions in Earth History

If the mechanism outlined above is realistic, it has a number of implications. As noted at the beginning of this paper, the K-T boundary carbon isotope excursion is different from others, such as the PETM; as the planktonic foraminifera indicates, the surface waters become enriched in ^{12}C, while the benthic foraminifera show a ^{13}C enrichment of the bottom water (Fig. 1). In contrast, at the PETM, both the surface and deep water become enriched in ^{12}C (e.g., Zachos et al., 1993, 2001; Thomas and Shackleton, 1996; Dickens, 2003). This has two implications for the K-T events. First, it implies that there could have been significant changes in surface water productivity and/or bottom water anoxia during the K-T event that did not occur during the PETM. The main way to get an enrichment of ^{13}C in the benthic foraminifera but not of the planktonic foraminifera is to reduce the amount of surface water organic matter enriched in ^{12}C being oxidizing in the deep to CO_2. This can occur either by reducing the amount of organic matter reaching the deep ocean or by reducing the oxidation of that organic matter due to anoxic conditions. There is evidence for both reduced surface water productivity and anoxic-hypoxic bottom water conditions after the K-T event (Coccioni et al., 1993; Kaiho et al., 1999; Norris et al., 2001; Culver, 2003; Alegret et al., 2001, 2002, 2003). The second implication is that at the K-T boundary, methane released from gas hydrates must have moved through the water column extremely rapidly, preventing consumption, oxidation, and dissolution within the sediments and the lower water column (e.g., Paull et al., 2003). This is consistent with the concept that the methane may have risen to the surface of the ocean while still in the hydrates, in buoyant, hydrate-laden plumes. If this occurred then most of the methane from the gas hydrates would have reached the atmosphere where it would have been oxidized to carbon dioxide (Dickens, 2001). Some of this depleted $\delta^{13}C$ atmospheric carbon dioxide would then have exchanged with the surface oceans dissolved CO_2, thus causing the negative excursion observed in the planktonic foraminifera. This may also explain the cause of the recently reconstructed very high atmospheric pCO_2 levels after the K-T boundary (Beerling, et al., 2002), as our estimated gas hydrate releases (330–1330 GtC) would have increased atmospheric carbon dioxide by 575–2275 ppm. It should be noted, however, that exact levels of pCO_2 and the timing of their fluctuations are still under debate (Nordt et al., 2002) and so we cannot yet model the fate of the hydrates once they reached the surface layer of the ocean in detail

The required rapid release of gas hydrate methane supports our proposed impact seismicity-triggered continental slope failure mechanism. It also strongly suggests that there were other major changes associated with the impact, such as reduced oceanic productivity and anoxic-hypoxic conditions which would account for the observed heavier carbon isotope benthic foraminifera record (Figs. 1 and 5). In this interpretation, the reduced oceanic productivity and anoxic condition in the intermediate and bottom waters will initially make the benthic foraminifera calcite isotopically heavy (Fig. 5). Calculations have shown that this reduction in surface water productivity is not sufficient to explain the isotopically light record in the planktonic foraminifera (e.g., Ivany and Salawitch, 1993). Hence we suggest the

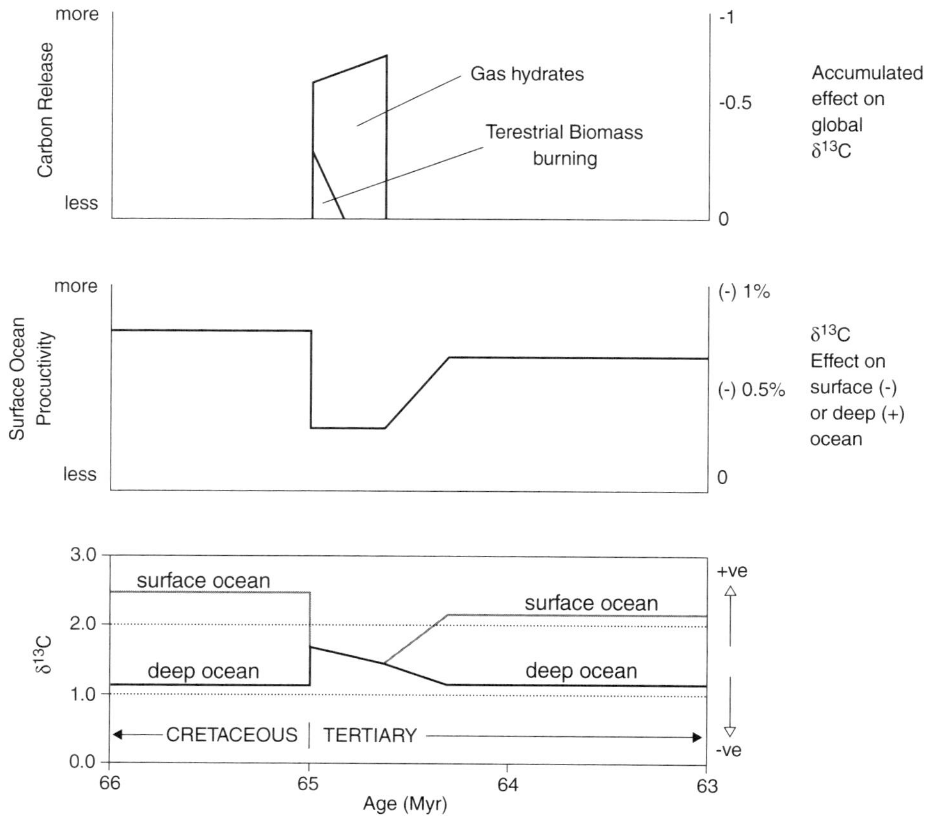

Figure 5. Schematic diagram showing the changes in $\delta^{13}C$ across the K-T boundary (after Kump, 1991; Zachos et al., 1989, 2001; and D'Hondt et al., 1998) and the possible combination of changes in surface water productivity (reinforced by bottom water anoxic conditions), gas hydrate release, and biomass burning which may account for the observed changes.

planktonic foraminifera record is initially a combination of both reduced surface water productivity and the release of very light carbon via gas hydrate oxidation and terrestrial biomass burning (Fig. 5). However, both the planktonic and benthic foraminifera carbon isotopic records get progressively lighter in the earliest Paleocene; this implies the whole ocean carbon reservoir getting lighter. This may be due to either (1) the progressive release of gas hydrates after the impact due to post-impact super global warming (Beerling et al., 2002) after the initial cooling (Galeotti et al., 2004); and/or (2) the return to the ocean of isotopically light atmospheric carbon dioxide which was produced by the oxidation of post-impact release of gas hydrate methane in to the atmosphere. The removal of a light carbon event can take hundreds of thousands of years (Norris and Rohl, 1999; Rohl et al., 2000), and this is entirely consistent with carbon isotopic record. After a few hundred thousand years, the surface water productivity recovers, and the normal separation between benthic and planktonic carbon isotopes is resumed (Fig. 5).

We therefore argue that the unusual stratification of the earliest Paleocene ocean is responsible for the reversal of the normal foraminiferal C isotopic gradient at this time and hence the unusual characteristics of the K-T boundary isotopic excursion compared to, for example, the PETM isotope excursion. Thus, some other isotope excursions in the geological record may also have been produced by the impact seismicity-triggered slope failure and gas hydrate release mechanism that we have analyzed in this paper. For example, there is now evidence for an impact coeval with the PETM (Schmitz, et al., 1997; Dolenec et al., 2001; Kent et al., 2003), and so impact-related seismicity may also be the cause for the initial release of gas hydrates at the PETM. Because of the difficulty in finding impact craters we cannot rule out the possibility that many of the marked carbon isotope excursions found in the geological record may be impact-related and not climatic as has been recently suggested (Dickens, 2003).

CONCLUSIONS

We have shown that there is broad consistency between, on the one hand, the estimated seismic effects of the Chicxulub impact based on the size and geometry of the impact structure and the estimated seismic effects of the impact based on the previously documented extent of sediment liquefaction and submarine slope failure at distances of up to, respectively, around 2500 km and 7000 km from the Chicxulub impact site; and on the other hand, consistency between the documented extent of sediment liquefaction and submarine slope failure, affecting up to 20,000 km length of continental margin, and the carbon isotopic excursion associated with the impact. This magnitude of slope failure could have released between 300–1300 Gt of gas hydrates: this can account for much of the carbon isotope excursion observed in planktonic foraminifera at the Cretaceous-Tertiary boundary. We argue, therefore, that it is possible to relate the K-T boundary carbon isotopic excursion quantitatively (at least at the order-of-magnitude level of accuracy) to the size, location, and seismic effects of the Chicxulub impact. It does not appear necessary to invoke destabilization of gas hydrates by tsunami waves from the impact itself, which does not seem to have been an efficient tsunami source, although tsunamis produced by the first slope failures may have assisted in producing a cascade of successive slope failures on adjacent margin segments. Our analysis does not exclude the possibility that other effects of the impact, such as the burning of terrestrial biomass, may also have contributed to the isotope excursion, although it does suggest that such burning may not have been as extensive or as efficient in producing carbon gases as supposed. However, our conclusion does suggest that there is little scope for positive feedbacks in the immediate post-impact environment to produce runaway releases of more carbon gases over those released in the immediate aftermath of the impact. Neither the paleoseismic record nor the isotopic data leave much scope for the idea that multiple impacts, some much larger than Chicxulub, may have occurred around the time of the K-T boundary. Finally, we suggest this mechanism linking impacts to carbon isotope excursion may be applicable to other significant carbon isotopic excursions.

ACKNOWLEDGMENTS

The authors would like to thank G. Dickens, T. Kenkmann, C. Koeberl, and J. Melosh for their thoughtful reviews, which greatly improved this manuscript, and Erik Asphaug for illuminating discussions of impact crater modification processes. We also thank the staff of the University College London Department of Geography Drawing Office for help in preparation of the diagrams.

REFERENCES CITED

Abe, K., 1992, Seismicity of the caldera-making eruption of Mount Katmai, Alaska in 1912: Bulletin of the Seismological Society of America, v. 82, p. 175–191.

Adatte, T., Stinnesbeck, W., and Keller, G., 1996, Lithostratigraphic and mineralogical correlations of near-K/T boundary clastic sediments in Northeastern Mexico: Implications for origin and nature of deposition, *in* Ryder, G., Fastovsky, D., and Gartner, S., eds., The Cretaceous-Tertiary Event and Other Catastrophes in Earth History, Geological Society of America Special Paper 307, p. 211–226.

Aki, K., 1966, Generation and propagation of G waves from the Niigata earthquake of June 16, 1964, Part 2; estimation of earthquake movement, released energy and stress-strain drop from the G wave spectrum: Bulletin of the Earthquake Research Institute, v. 44, p. 73–88.

Alegret, L., Molina, E., and Thomas, E., 2001, Benthic foraminifera at the Cretaceous/Tertiary boundary around the Gulf of Mexico: Geology, v. 29, p. 891–894, doi: 10.1130/0091-7613(2001)0292.0.CO;2.

Alegret, L., Arenillas, I., Arz, J.A., Liesa, C., Meléndez, A., Molina, E., Soria, A.R., and Thomas, E., 2002, The Cretaceous/Tertiary boundary: sedimentology and micropaleontology at El Mulato section, NE Mexico: Terra Nova, v. 14, p. 330–336, doi: 10.1046/j.1365-3121.2002.00425.x.

Alegret, L., Molina, E., and Thomas, E., 2003, Benthic foraminiferal turnover across the Cretaceous/Paleogene boundary at Agost (southeastern Spain): paleoenvironmental inferences: Marine Micropaleontology, v. 48, p. 251–279.

Alvarez, W., Smit, J., Lowrie, W., Asaro, F., Margolis, S.V., Claeys, P., Kastner, M., and Hildebrand, A.R., 1992, Proximal impact deposits at the Cretaceous-Tertiary boundary in the Gulf of Mexico: a restudy of DSDP Leg 77 Sites 536 and 540: Geology, v. 20, p. 697–700, doi: 10.1130/0091-7613(1992)0202.3.CO;2.

Arias, A., 1970, A measure of earthquake intensity, *in* Hansen, R.J., ed., Seismic design for Nuclear Power Plants: Cambridge, Massachusetts, Massachusetts Institute of Technology Press, p. 438–483.

Barrientos, S.E., and Ward, S.N., 1990, The 1960 Chile earthquake; inversion for slip distribution from surface deformation: Geophysical Journal International, v. 103, p. 589–598.

Beerling, D.J., Lomax, B.H., Royer, D.L., Upchurch, G.R., and Kump, L.R., 2002, An atmospheric pCO_2 reconstruction across the Cretaceous-Tertiary boundary from leaf megafossils: Proceedings of the National Academy of Sciences of the United States of America, v. 99, p. 7836–7840, doi: 10.1073/pnas.122573099.

Belcher, C.M., Collinson, M.E., Sweet, A.R., Hildebrand, A.R., and Scott, A.C., 2003, Fireball passes and nothing burns—the role of thermal radiation in the Cretaceous-Tertiary event: evidence from the charcoal record of North America: Geology, v. 31, p. 1061–1064, doi: 10.1130/G19989.1.

Biscontin, G., Pestana, J.M., and Nadim, F., 2004, Seismic triggering of submarine slides in soft cohesive soil deposits: Marine Geology, v. 203, p. 341–354, doi: 10.1016/S0025-3227(03)00314-1.

Boersma, A., and Shackleton, N., 1981, Oxygen and carbon isotope variations and planktonic foraminifer depth habitats, Late Cretaceous to Paleocene, central Pacific: Initial Reports of the Deep Sea Drilling Project, v. 62, p. 513–526.

Bohor, B.F., 1996, A sediment gravity flow hypothesis for siliciclastic units at the K/T boundary, northeastern Mexico, *in* Ryder, G., Fastovsky, D., and Gartner, S., eds., The Cretaceous-Tertiary Event and Other Catastrophes in Earth History: Geological Society of America Special Paper 307, p. 183–195.

Boslough, M.B., Chael, E.P., Trucano, T.G., Crawford, D.A., and Campbell, D.L., 1996, Axial focusing of impact energy in the Earth's interior: a possible link to flood basalts and hotspots, *in* Ryder, G., Fastovsky, D., and Gartner, S., eds., The Cretaceous-Tertiary Event and Other Catastrophes in Earth History: Geological Society of America Special Paper 307, p. 541–550.

Bouriak, S., Vanneste, M., and Saoutkine, A., 2000, Inferred gas hydrates and clay diapirs near the Storegga Slide on the southern edge of the Vøring plateau, offshore Norway: Marine Geology, v. 163, p. 125–148, doi: 10.1016/S0025-3227(99)00115-2.

Bourgeois, J., Hansen, T.A., Wiberg, P.L., and Kauffman, E.G., 1988, A tsunami deposit at the Cretaceous-Tertiary boundary in Texas: Science, v. 241, p. 567–570.

Bralower, T.J., Paull, C.K., and Leckie, R.M., 1998, The Cretaceous-Tertiary boundary cocktail: Chicxulub impact triggers margin collapse and extensive sediment gravity flows: Geology, v. 26, p. 331–334, doi: 10.1130/0091-7613(1998)0262.3.CO;2.

Brittan, J., Morgan, J., Warner, M., and Marin, L., 1999, Near-surface seismic expression of the Chicxulub impact crater, *in* Dressler, B.O., and Sharpton, V. L., eds., Large Meteorite Impacts and Planetary Evolution II: Geological Society of America Special Paper 339, p. 269–279.

Bryn, P., Solheim, A., Berg, K., Lien, R., Forsberg, C.F., Haflicason, H., Ottesen, D., and Rise, L., 2003, The Storegga complex; repeated large scale sliding in response to climatic cyclicity, *in* Locat, J., and Meinert, J., eds., Submarine mass movements and their consequences: Advances in Natural and Technological Hazards Research Series: Dordrecht, Kluwer Academic Publishers, p. 215–222.

Busby, C.J., Yip, G., Blikra, L., and Renne, P., 2002, Coastal landsliding and catastrophic sedimentation triggered by Cretaceous-Tertiary bolide impact: A Pacific margin example?: Geology, v. 30, p. 687–690, doi: 10.1130/0091-7613(2002)0302.0.CO;2.

Carey, S., 1997, Influence of convective sedimentation on the formation of widespread tephra falls in the deep sea: Geology, v. 25, p. 839–842, doi: 10.1130/0091-7613(1997)0252.3.CO;2.

Coccioni, R., Fabbrucci, L., and Galeotti, S., 1993, Terminal Cretaceous deep-water benthic foraminiferal decimation, survivorship and recovery at Caravaca (SE Spain): Paleopelagos, v. 3, p. 3–24.

Culver, S., 2003, Benthic foraminifera across the Cretaceous-Tertiary (K-T) boundary: a review: Marine Micropaleontology, v. 47, p. 177–226, doi: 10.1016/S0377-8398(02)00117-2.

D'Hondt, S., Donaghay, P., Zachos, J.C., Luttenberg, D., and Lindinger, M., 1998, Organic carbon fluxes and ecological recovery from the Cretaceous-Tertiary mass extinction: Science, v. 282, p. 276–279, doi: 10.1126/science.282.5387.276.

Dickens, G.R., 2001, The potential volume of oceanic methane hydrates with variable external conditions: Organic Geochemistry, v. 32, p. 1179–1193, doi: 10.1016/S0146-6380(01)00086-9.

Dickens, G.R., 2003, Rethinking the global carbon cycle with a large dynamic and microbially mediated gas hydrate capacitor: Earth and Planetary Science Letters, v. 213, p. 169–183, doi: 10.1016/S0012-821X(03)00325-X.

Dolenec, T., Pavsic, J., and Lojen, S., 2001, Ir anomalies and other elemental markers near the Paleocene-Eocene: Terra Nova, v. 2, p. 199–204.

Dressler, B.O., Sharpton, V.L., Morgan, J., Buffler, R., Moran, D., Smit, J., Stoeffler, D., and Urrutia, J., 2003, Investigating a 65-Ma-old smoking gun: deep drilling of the Chicxulub impact structure: Eos (Transactions, American Geophysical Union), v. 84, p. 125, 130.

Erbacher, J., Mosher, D., Malone, M., and ODP Leg 207 Scientific Party, 2004, Drilling probes past carbon cycle perturbations on the Demerara rise: Eos (Transactions, American Geophysical Union), v. 85, p. 57–63.

Galeotti, S., Brinkhuis, H., and Huber, M., 2004, Records of post–Cretaceous-Tertiary boundary millennial-scale cooling from the western Tethys: a smoking gun for the impact—winter hypothesis?: Geology, v. 32, p. 529–532, doi: 10.1130/G20439.1, doi: 10.1130/G20439.1.

Grajales-Nishimura, J.M., Cedillo-Pardo, E., Rosales-Dominguez, C., Moran-Zenteno, D.J., Alvarez, W., Claeys, P., Ruiz-Morales, J., Garcia-Hernandez, J., Padilla-Avila, P., and Sanchez-Rios, A., 2000, Chicxulub impact: The origin of reservoir and seal facies in the southeastern Mexico oil fields: Geology, v. 28, p. 307–310, doi: 10.1130/0091-7613(2000)0282.3.CO;2.

Grieve, R.A.F., and Therriault, A., 2000, Vredefort, Sudbury, Chicxulub: Three of a kind?: Annual Review of Earth and Planetary Sciences, v. 28, p. 305–328, doi: 10.1146/annurev.earth.28.1.305.

Grieve, R.A.F., Stöffler, D., and Deutsch, A., 1991, The Sudbury structure: controversial or misunderstood?: Journal of Geophysical Research, v. 96, p. 22753–22764.

Hagstrum, J.T., and Abbott, D., 2002, Evidence for a large bolide impact in the Proto-Pacific Ocean preceding the Chicxulub impact by about 2 million years (Abstract): Eos (Transactions, American Geophysical Union), Fall Meeting Supplement, v. 83, OS22C–0297.

Hampton, M.A., Lee, H.J., and Locat, J., 1996, Submarine landslides: Reviews of Geophysics, v. 34, p. 33–59, doi: 10.1029/95RG03287.

Hanks, T.C., and Kanamori, H., 1979, A moment magnitude scale: Journal of Geophysical Research, v. 84, p. 2348–2350.

Hansen, W.R., 1966, The Alaska earthquake of 1964: Nature, v. 215, p. 348–351.

Hansen, W.R., Eckel, E.B., Schaem, W.E., Lyle, R.E., George, W., and Chance, G., 1966, The Alaska earthquake, March 27, 1964: field investigations and reconstruction effort: U.S. Geological Survey Professional Paper 541, 111 p.

Haq, B., 1998, Natural gas hydrates: searching for the long-term climate and slope stability records, *in* Henriet, J-P., and Mienert, J., eds., 1998, Gas Hydrates: Relevance to world margin stability and climate change: Geological Society [London] Special Publication 137, p. 303–318.

Harp, E.L., and Wilson, R.C., 1995, Shaking intensity for rock falls and slides; evidence from 1987 Whittier Narrows and Superstition Hills earthquake strong-motion records: Bulletin of the Seismological Society of America, v. 85, p. 1739–1757.

Hildebrand, A.R., Pilkington, M., Ortiz-Aleman, C., Chavez, R., Urrutia-Fucugauchi, J., Connors, M., Graniel-Castro, E., Camara-Zi, A., Halpenny, J., and Niehaus, D., 1998, Mapping Chicxulub crater structure with gravity and seismic reflection data, *in* Grady, M. M., Hutchinson, R., McCall, G.J.H., and Rothery, D.A., eds., Meteorites: flux with time and impact effects: Geological Society of London Special Publication 140, p. 177–193.

Hesse, R., Rashid, H., and Khodabaksh, S., 2004, Fine-grained sediment lofting from meltwater-generated turbidity currents during Heinrich events: Geology, v. 32, p. 449–452, doi: 10.1130/G20136.1, doi: 10.1130/G20136.1.

Ivany, L.C., and Salawitch, R.J., 1993, Carbon isotopic evidence for biomass burning at the K-T boundary: Geology, v. 21, p. 487–490, doi: 10.1130/0091-7613(1993)0212.3.CO;2.

Jablonski, D.J., 1998, Geographic variation in the molluscan recovery from the end-Cretaceous extinction: Science, v. 279, p. 1327–1330, doi: 10.1126/science.279.5355.1327.

Jibson, R.W., 1996, Using landslides for paleoseismic analysis, *in* McCalpin, J.F., ed., Paleoseismicity: San Diego, Academic Press, p. 397–438.

Jibson, R.W., and Keefer, D.K., 1993, Analysis of the seismic origin of landslides—Examples from the New Madrid seismic zone: Geological Society of America Bulletin, v. 105, p. 521–536, doi: 10.1130/0016-7606(1993)1052.3.CO;2.

Jibson, R.W., Prentice, C.S., Borissoff, B.A., Rogozhin, E.A., and Langer, C.J., 1994, Some observations of landslides triggered by the 29 April 1991 Racha earthquake, Republic of Georgia: Bulletin of the Seismological Society of America, v. 84, p. 963–973.

Johnston, A.C., 1996a, Seismic moment assessment of earthquakes in stable continental regions—II, Historical Seismicity: Geophysical Journal International, v. 125, p. 639–678.

Johnston, A.C., 1996b, Seismic moment assessment of earthquakes in stable continental regions—III, New Madrid 1811–1812, Charleston 1886 and Lisbon 1755: Geophysical Journal International, v. 126, p. 314–344.

Kaiho, H., Kajiwara, Y., Tazaki, K., Ueshima, M., Takeda, N., Kawahata, H., Arinobu, T., Ishiwatari, R., Hirai, A., and Lamolda, M.A., 1999, Oceanic primary productivity and dissolved oxygen levels at the Cretaceous/Tertiary boundary: their decrease, subsequent warming, and recovery: Paleoceanography, v. 14, p. 511–524, doi: 10.1029/1999PA900022.

Kanamori, H., 1977, The energy release in great earthquakes: Journal of Geophysical Research, v. 82, p. 2981–2987.

Katz, M.E., Cramer, B.S., Mountain, G.S., Katz, S., and Miller, K.G., 2001, Uncorking the bottle: What triggered the PETM methane release: Paleoceanography, v. 16, p. 549–562, doi: 10.1029/2000PA000615.

Keefer, D.K., 1984, Landslides caused by earthquakes: Geological Society of America Bulletin, v. 95, p. 406–421.

Keller, G., Adatte, T., and Stinnesbeck, W., Rebolledo-Vieyra, M., Urrutia-Fucugauchi, J., Kramar, U., and Stuben, D., 2004, Chicxulub impact predates the K-T boundary mass extinction: Proceedings of the National Academy of Sciences, v. 101, p. 3753–3758.

Kennett, J., Cannariato, K.G., Hendy, I.L., and Behl, R.J., 2000, Carbon isotopic evidence for methane hydrate instability during Quaternary interstadials: Science, v. 288, p. 128–133, doi: 10.1126/science.288.5463.128.

Kennett, J., Cannariato, K.G., Hendy, I.L., and Behl, R.J., 2003, Methane hydrates in Quaternary climate change: The clathrate gun hypothesis: Washington, D.C., American Geophysical Union, 216 p.

Kent, D.V., Cramer, B.S., Lanci, L., Wang, D., Wright, J.D., and Van der Voo, R., 2003, A case for a comet impact trigger for the Paleocene/Eocene thermal maximum and carbon isotope excursion: Earth and Planetary Science Letters, v. 211, p. 13–26, doi: 10.1016/S0012-821X(03)00188-2.

Kiyokawa, S., Tada, R., Iturralde-Vinent, M., Tajika, E., Yamamoto, S., Oji, T., Nakano, Y., Goto, K., Takayama, H., Garcia Delgado, D., Diaz Otero, C., Rojas-Consuegra, R., and Matsui, T., 2002, Cretaceous-Tertiary boundary sequence in the Cacarajicara Formation, western Cuba: An impact-related, high-energy, gravity-flow deposit, *in* Koeberl, C., and MacLeod, K.G., eds., Catastrophic Events and Mass Extinctions: Impacts and Beyond. Geological Society of America Special Paper 356, p. 125–144.

Klaus, A., Norris, R.D., Kroon, R., and Smit, J., 2000, Impact-induced mass wasting at the K-T boundary: Blake Nose, western North Atlantic: Geology, v. 28, p. 319–322, doi: 10.1130/0091-7613(2000)0282.3.CO;2.

Kump, L., 1991, Interpreting carbon-isotope excursions: Strangelove Oceans: Geology, v. 19, p. 299–302, doi: 10.1130/0091-7613(1991)0192.3.CO;2.

Kvenvolden, K.A., 1998, A primer on the geological occurrence of gas hydrate, *in* Henriet, J-P., and Mienert, J., eds., Gas Hydrates: Relevance to world margin stability and climate change: Geological Society [London] Special Publication 137, p. 9–30.

Kvenvolden, K.A., and Lorenson, T.D., 2001, The Global Occurrence of Natural Gas Hydrates, *in* Paull, C.K., and Dillon, W.P., eds., Natural Gas Hydrates: Occurrence, Distribution and Detection: American Geophysical Union Geophysical Monograph, v. 124, p. 3–18.

Laberg, J.S., Vorren, T.O., Dowdeswell, J.A., Kenyon, N.H., and Taylor, J., 2000, The Andoya slide and the Andoya Canyon, northeastern Norwegian-Greenland Sea: Marine Geology, v. 162, p. 259–275, doi: 10.1016/S0025-3227(99)00087-0.

MacLeod, K.G., Fullagar, P.D., and Huber, B.T., 2003, $^{87}Sr/^{86}Sr$ test of the degree of impact-induced slope failure in the Maastrichtian of the western North Atlantic: Geology, v. 31, p. 311–314, doi: 10.1130/0091-7613(2003)0312.0.CO;2.

Maslin, M.A., and Thomas, E., 2003, Balancing the deglacial global carbon budget: The hydrate factor: Quaternary Science Reviews, v. 22, p. 1729–1736, doi: 10.1016/S0277-3791(03)00135-5.

Maslin, M.A., Mikkelsen, N., Vilela, C., and Haq, B., 1998, Sea-level– and gas-hydrate–controlled catastrophic sediment failures of the Amazon Fan: Geology, v. 26, p. 1107–1110, doi: 10.1130/0091-7613(1998)0262.3.CO;2.

Maslin, M.A., Owen, M., Day, S., and Long, D., 2004, Linking continental-slope failures and climate change: Testing the clathrate gun hypothesis: Geology, v. 32, p. 53–56, doi: 10.1130/G20114.1, doi: 10.1130/G20114.1.

Matsui, T., Imamura, F., Tajika, E., Nakano, Y., and Fujisawa, Y., 2002, Generation and propagation of a tsunami from the Cretaceous-Tertiary impact event, *in* Koeberl, C., and MacLeod, K. G., eds., Catastrophic Events and Mass Extinctions: Impacts and Beyond: Geological Society of America Special Paper 356, p. 69–77.

Max, M.D., Dillon, W.P., Nishimura, C., and Hurdle, B.G., 1999, Sea-floor methane blow-out and global firestorm at the K-T boundary: Geo-Marine Letters, v. 18, p. 285–291, doi: 10.1007/s003670050081.

Melosh, H.J., 1989, Impact Cratering: A Geologic Process: New York, Oxford University Press, 256 p.

Melosh, H.J., and Ivanov, B.A., 1999, Impact crater collapse: Annual Review of Earth and Planetary Sciences, v. 27, p. 385–415, doi: 10.1146/annurev.earth.27.1.385.

Melosh, H.J., Schneider, N.M., Zahnle, K., and Latham, D., 1990, Ignition of global wildfires at the Cretaceous/Tertiary boundary: Nature, v. 343, p. 251–254, doi: 10.1038/343251a0.

Milton, D.J., 1977, Shatter cones—an outstanding problem in shock mechanics, *in* Roddy, D.J., Pepin, R.O., and Merril, R.B., eds., Impact and Explosion Cratering: New York, Pergamon Press, p. 703–714.

Morgan, J., Warner, M., and the Chicxulub Working Group, 1997, Size and morphology of the Chicxulub impact crater: Nature, v. 390, p. 472–476.

Nordt, L., Atchley, S., and Dworkin, S.I., 2002, Paleosol barometer indicates extreme fluctuations in atmospheric CO_2 across the Cretaceous-Tertiary boundary: Geology, v. 30, p. 703–706, doi: 10.1130/0091-7613(2002)0302.0.CO;2.

Norris, R.D., and Firth, J.V., 2002, Mass wasting of the Atlantic continental margins following the Chicxulub impact event, *in* Koeberl, C., and MacLeod, K.G., eds., Catastrophic Events and Mass Extinctions: Impacts and Beyond: Geological Society of America Special Paper 356, p. 79–95.

Norris, R.D., and Rohl, U., 1999, Carbon cycling and chronology of climate warming during the Paleocene/Eocene transition: Nature, v. 401, p. 775–778, doi: 10.1038/44545.

Norris, R.D., Firth, J.V., Blusztajn, J.S., and Ravizza, G., 2000, Mass failure of the North Atlantic margin triggered by the Cretaceous-Paleogene bolide impact: Geology, v. 28, p. 1119–1122, doi: 10.1130/0091-7613(2000)0282.3.CO;2.

Norris, R.D., Klaus, A., and Kroon, D., 2001, Mid-Eocene deep water, the late Paleocene Thermal maximum and continental slope mass wasting during the Cretaceous-Paleogene impact, *in* Norris, R.D., Klaus, A., and Kroon, D., eds., Western North Atlantic Palaeogene and Cretaceous palaeoceanography: Geological Society [London] Special Publication 183, p. 23–48.

Obermeier, S.F., 1996, Using liquefaction-induced features for paleoseismic analysis, *in* McCalpin, J.F., ed., Paleoseismicity: San Diego, Academic Press, p. 331–396.

Ocampo, A.C., Pope, K.O., and Fischer, A.G., 1996, Ejecta blanket deposits of the Chicxulub crater from Albion Island, Belize, *in* Ryder, G., Fastovsky, D., and Gartner, S., eds., The Cretaceous-Tertiary Event and Other Catastrophes in Earth History: Geological Society of America Special Paper 307, p. 75–88.

O'Keefe, J.D., and Ahrens, T., 2002, Chicxulub impact simulation demonstrates virtually all observed crater features using shock damaged rock model: Eos (Transactions of the American Geophysical Union), Fall Meeting Supplement Abstract, v. 83, P22B–0406.

O'Keefe, J.D., Stewart, S.T., Lainhart, M.E., and Ahrens, T., 2001, Damage and rock—volatile mixture effects on impact crater formation: International Journal of Impact Engineering, v. 26, p. 543–553, doi: 10.1016/S0734-743X(01)00112-9.

Olsson, R.K., Liu, C., and van Fossen, M., 1996, The Cretaceous-Tertiary catastrophic event at Millers Ferry, Alabama, *in* Ryder, G., Fastovsky, D., and Gartner, S., eds., The Cretaceous-Tertiary Event and Other Catastrophes in Earth History: Geological Society of America Special Paper 307, p. 263–277.

Olsson, R.K., Miller, K.G., Browning, J.V., Habib, D., and Sugarman, P.J., 1997, Ejecta layer at the Cretaceous-Tertiary boundary, Bass River, New Jersey (Ocean Drilling Program Leg 174AX): Geology, v. 25, p. 759–762, doi: 10.1130/0091-7613(1997)0252.3.CO;2.

Paull, C.K., Brewer, P.G., Ussler, W., III, Peltzer, E.T., Rehder, G., and Clague, D., 2003, An experiment demonstrating that marine slumping is a mechanism to transfer methane from seafloor gas-hydrate deposits into the upper ocean and atmosphere: Geo-Marine Letters, v. 22, p. 198–203, doi: 10.1007/s00367–002–0113-y.

Pilkington, M., Hildebrand, A.R., and Ortiz-Aleman, C., 1994, Gravity and magnetic field modeling and structure of the Chicxulub Crater, Mexico: Journal of Geophysical Research, v. 99, p. 13147–13162, doi: 10.1029/94JE01089.

Pope, K.O., Baines, K.H., Ocampo, A.C., and Ivanov, B.A., 1997, Energy, volatile production, and climate effects of the Chicxulub Cretaceous/Tertiary impact: Journal of Geophysical Research, v. 102, p. 21645–21664, doi: 10.1029/97JE01743.

Reimold, W.U., and Colliston, W.P., 1994, Pseudotachylites of the Vredefort Dome and the surrounding Witwatersrand Basin, South Africa, *in* Dressler, B.O., Grieve, R.A.F., and Sharpton, V.L., eds., Large meteorite impacts and planetary evolution: Geological Society of America Special Paper 293, p. 177–196.

Robertson, D.S., McKenna, M.C., Toon, O.B., Hope, S., and Lillegraven, J.A., 2004, Survival in the first hours of the Cenozoic: Geological Society of America Bulletin, v. 116, p. 760–768, doi: 10.1130/B25402.1, doi: 10.1130/B25402.1.

Rohl, U., Bralower, T.J., Norris, R.D., and Wefer, G., 2000, New chronology for the late Paleocene thermal maximum and its environmental implications: Geology, v. 28, p. 927–930, doi: 10.1130/0091-7613(2000)0282.3.CO;2.

Schmitz, B., Asaro, F., Molina, E., Monecht, S., von Salis, K., and Speijer, R., 1997, High-resolution iridium profiles across the latest Paleocene benthic extinction event at Zumaya, Spain: Paleogeography, Paleoclimatology and Paleoecology, v. 133, p. 49–68, doi: 10.1016/S0031-0182(97)00024-2.

Schulte, P., Stinnesbeck, W., Kontny, A., Stuben, D., Kramar, U., and Harting, M., 2002, Multiple (immiscible) melt phases of mafic composition in Chicxulub impact ejecta from northeastern Mexico: New Constraints on target lithologies [abs.]: Eos (Transactions, American Geophysical Union), Fall Meeting Supplement, v. 83, OS22C–0293.

Schultz, P.H., and Gault, D.E., 1975, Seismic effects from major basin formation on the Moon and Mercury: Moon, v. 12, p. 159–177.

Scott, R.G., and Benn, K., 2001, Peak-ring collapse accommodated by impact melt filled transfer faults, Sudbury impact structure, Canada: Geology, v. 29, p. 747–750, doi: 10.1130/0091-7613(2001)0292.0.CO;2.

Scott, R.G., and Spray, J.G., 2000, The South Range breccia belt of the Sudbury impact structure: a possible terrace collapse feature: Meteoritics & Planetary Science, v. 35, p. 505–520.

Shackleton, N., and Hall, M.A., 1984, Carbon isotope data from Leg 74 sediments: Initial Reports of the Deep Sea Drilling Project, v. 74, p. 613–619.

Smit, J., 1999, The Global Stratigraphy of the Cretaceous-Tertiary Boundary Impact Ejecta: Annual Review of Earth and Planetary Sciences, v. 27, p. 75–113, doi: 10.1146/annurev.earth.27.1.75.

Smit, J., Montanari, A., Swinburne, N.H.M., Alvarez, W., Hildebrand, A.R., Margolis, S.V., Claeys, P., Lowrie, W., and Asaro, F., 1992, Tektite-bearing, deep-water clastic unit at the Cretaceous-Tertiary boundary in northeastern Mexico: Geology, v. 20, p. 99–103, doi: 10.1130/0091-7613(1992)0202.3.CO;2.

Smit, J., Roep, T.B., Alvarez, W., Montanari, A., Claeys, P., Grajales-Nishimura, J.M., and Bermudez, J., 1996, Coarse-grained clastic sandstone complex at the K-T boundary around the Gulf of Mexico: Deposition by tsunami waves induced by the Chicxulub impact? *in* Ryder, G., Fastovsky, D., and Gartner, S., eds., The Cretaceous-Tertiary Event and Other Catastrophes in Earth History: Geological Society of America Special Paper 307, p. 151–182.

Smit, J., Keller, G., Zargouni, F., Razgallah, S., Shimi, M., Ben Abdelkader, O., Ben Haj Ali, N., and Ben Salem, H., 1997, The El Kef sections and sampling procedures: Marine Micropalaeontology, v. 29, p. 69–72, doi: 10.1016/S0377-8398(96)00039-4.

Snyder, D.B., and Hobbs, R.W., 1999, Deep seismic reflection profiles across the Chicxulub crater, *in* Dressler, B.O., and Sharpton, V.L., eds., Large Meteorite Impacts and Planetary Evolution II: Geological Society of America Special Paper 339, p. 263–268.

Soria, A.R., Liesa, C.L., Mata, M.P., Arz, J.A., Alegret, L., Arenillas, I., and Melendez, A., 2001, Slumping and a sandbar deposit at the Cretaceous-Tertiary boundary in the El Tecolote section (northeastern Mexico): An impact-induced sediment gravity flow: Geology, v. 29, p. 231–234, doi: 10.1130/0091-7613(2001)0292.0.CO;2.

Sparks, R.S.J., Bonnecaze, R.T., Huppert, H.E., Lister, J.R., Hallworth, M.A., Mader, H., and Phillips, J., 1993, Sediment-laden gravity currents with reversing buoyancy: Earth and Planetary Science Letters, v. 114, p. 243–257, doi: 10.1016/0012-821X(93)90028-8.

Spray, J.G., 1997, Superfaults: Geology, v. 25, p. 579–582, doi: 10.1130/0091-7613(1997)0252.3.CO;2.

Spray, J.G., and Thompson, L.M., 1995, Friction melt distribution in a multi-ring impact basin: Nature, v. 373, p. 130–132, doi: 10.1038/373130a0.
Stinnesbeck, W., and Keller, G., 1996, K-T boundary coarse-grained siliciclastic deposits in northeastern Mexico and northeastern Brazil: Evidence for megatsunami or sea-level changes? *in* Ryder, G., Fastovsky, D., and Gartner, S., eds., The Cretaceous-Tertiary Event and Other Catastrophes in Earth History: Geological Society of America Special Paper 307, p. 197–209.
Stoffer, P.W., 2002, Evidence for a widespread disruption layer associated with the Cretaceous-Tertiary boundary in the Upper Fox Hills Formation throughout the Badland National Park region of South Dakota [abs.]: Eos (Transactions, American Geophysical Union), Fall Meeting Supplement, v. 83, PP11A–0300.
Stott, L.D., and Kennett, J.P., 1989, New constraints on early Tertiary Paleoproductivity from carbon isotopes in Foraminifera: Nature, v. 342, p. 526–529, doi: 10.1038/342526a0.
Tada, R., Nakano, Y., Iturralde-Vinent, M.A., Yamamoto, S., Kamata, T., Tajika, E., Toyoda, K., Kiyokawa, S., Garcia Delgado, D., Oji, T., Goto, K., Takayama, H., Rojas-Consuegra, R., and Matsui, T., 2002, Complex tsunami waves suggested by the Cretaceous-Tertiary boundary deposit at the Moncada section, western Cuba, *in* Koeberl, C., and MacLeod, K. G., eds., Catastrophic Events and Mass Extinctions: Impacts and Beyond: Geological Society of America Special Paper 356, p. 109–123.
Thomas, E., and Shackleton, N.J., 1996, The Paleocene-Eocene benthic foraminiferal extinction and stable isotope anomalies, *in* Knox, R.O'B., Corfield, R., and Dunay, R. E., eds., Correlations of the early Paleogene in Northwest Europe: Geological Society of London Special Publication 101, p. 401–411.
Thomas, D.J., Zachos, J.C., Bralower, T.J., Thomas, E., and Bohaty, S., 2002, Warming the Fuel for the Fire: Evidence for the thermal dissociation of methane hydrate during the Paleocene-Eocene thermal maximum: Geology, v. 30, p. 1067–1070, doi: 10.1130/0091-7613(2002)0302.0.CO;2.
Toon, O.B., Zahnle, K., Morrison, D., Turco, R.P., and Covey, C., 1997, Environmental perturbations caused by the impacts of asteroids and comets: Reviews of Geophysics, v. 35, p. 41–78, doi: 10.1029/96RG03038.
Ward, S.N., and Asphaug, E., 2000, Asteroid impact tsunami: a probabalistic hazard assessment: Icarus, v. 145, p. 64–78, doi: 10.1006/icar.1999.6336.
Weichert, W., 1960, Further observations of geologic and geomorphic changes resulting from the catastrophic earthquake of May 1960, in Chile: Bulletin of the Seismological Society of America, v. 53, p. 1253–1257.
Wells, D.L., and Coppersmith, K.J., 1994, New empirical relationships among magnitude, rupture length, rupture width, rupture area, and surface displacement: Bulletin of the Seismological Society of America, v. 84, p. 974–1002.
Wilson, R.C., and Keefer, D.K., 1985, Predicting areal limits of earthquake-induced landsliding, *in* Ziony, J.I., ed., Evaluating earthquake hazards in the Los Angeles region—an Earth Science perspective: U.S. Geological Survey Professional Paper 1360, p. 316–345.
Wolbach, W.S., Gilmour, I., and Anders, E., 1990, Major wildfires at the Cretaceous-Tertiary boundary, *in* Sharpton, V.L., and Ward, P., eds., Global Catastrophes in Earth History: Geological Society of America Special Paper 247, p. 391–400.
Yeats, R.S., Sieh, K., and Allen, C.R., 1997, The geology of earthquakes: New York, Oxford University Press, 534 p.
Youd, T.L., and Perkins, D.M., 1978, Mapping liquefaction-induced ground failure potential: Journal of Geotechnical Engineering of the American Society of Civil Engineers, v. 104, p. 433–446.
Youd, T.L., and Perkins, D.M., 1987, Mapping of liquefaction severity index: Journal of Geotechnical Engineering of the American Society of Civil Engineers, v. 113, p. 1374–1392.
Zachos, J.C., and Arthur, M.A., 1986, Paleoceanography of the Cretaceous-Tertiary boundary event: inferences from stable isotopic and other data: Paleoceanography, v. 1, p. 5–26.
Zachos, J.C., Arthur, M.A., and Dean, W.E., 1989, Geochemical evidence for suppression of pelagic marine productivity at the Cretaceous/Tertiary boundary: Nature, v. 337, p. 61–64, doi: 10.1038/337061a0.
Zachos, J.C., Lohmann, K.C., Walker, J.C.G., and Wise, S.W., 1993, Abrupt climate change and transient climates during the Paleogene: a marine perspective: Journal of Geology, v. 101, p. 191–213.
Zachos, J.C., Pagani, M.N., Sloan, L., Thomas, E., and Billups, K., 2001, Trends, rhythms and aberrations in global climate 65 Ma to present: Science, v. 292, p. 686–693, doi: 10.1126/science.1059412.

Manuscript Accepted by the Society 9 August 2004

Printed in the USA

Geological Society of America
Special Paper 384
2005

Late Devonian Alamo Impact, southern Nevada, USA: Evidence of size, marine site, and widespread effects

Jared R. Morrow*
Department of Earth Sciences, University of Northern Colorado, Greeley, Colorado 80639, USA

Charles A. Sandberg
U.S. Geological Survey, Box 25046, MS 939, Federal Center, Denver, Colorado 80225-0046, USA

Anita G. Harris
1523 East Hillsboro Boulevard #1023, Deerfield Beach, Florida 33441, USA

ABSTRACT

The early Late Devonian (early Frasnian) Alamo Impact targeted an oceanic, off-platform site in southern Nevada, excavating a crater with a final diameter of 44–65 km. The original crater is now dismembered and buried beneath younger rocks. Consequently, its size and site must be deduced through multiple converging lines of geological and paleontological evidence. Previous and new evidence includes the catastrophically emplaced Alamo Breccia, tsunamites, shock-metamorphosed quartz grains, carbonate accretionary lapilli, an iridium anomaly, sub-Breccia clastic injection, deep-water Breccia channels, and ejecta material. We now demonstrate, on the basis of conodont microfossils in carbonate ejecta clasts within lapillistone blocks and widely distributed shocked-quartz and lithic-clast ejecta within the upper part of the Breccia, that the Alamo Impact excavated down at least into Upper Cambrian strata, at a depth of 1.7 km, and possibly into the underlying Proterozoic–Lower Cambrian Prospect Mountain Quartzite, ~2.5 km beneath the Late Devonian seafloor. Distal tsunamites and probable ejecta are now documented as far north as Devils Gate, northern Nevada, and as far northeast as the Confusion Range, western Utah. A charcoal-bearing, early Frasnian estuarine deposit in the Bighorn Mountains, Wyoming, may provide the first evidence of an Alamo Impact fallout-generated forest fire. Our new data further document the widespread effects and size of the Alamo Impact, and constrain the likely present position of the tectonically transported crater to an area between the Timpahute and Hot Creek Ranges, southern Nevada.

Keywords: Late Devonian, Alamo Impact, Nevada, conodonts, impact ejecta.

*jared.morrow@unco.edu

Morrow, J.R., Sandberg, C.A., and Harris, A.G., 2005, Late Devonian Alamo Impact, southern Nevada, USA: Evidence of size, marine site, and widespread effects, *in* Kenkmann, T., Hörz, F., and Deutsch, A., eds., Large meteorite impacts III: Geological Society of America Special Paper 384, p. 259–280. For permission to copy, contact editing@geosociety.org.

INTRODUCTION

Ongoing research during the past decade, aided in part by important advances within the study of marine impact processes (e.g., Jansa, 1993; Oberbeck et al., 1993; Ormö and Lindström, 2000; Dypvik and Jansa, 2003), has greatly increased our understanding of the geologic history and genesis of the early Late Devonian (early Frasnian, ca. 367 Ma) Alamo Impact Event. As evidenced by the catastrophically deposited Alamo Breccia and related impact phenomena, the Alamo Event was produced by a relatively large impact into an oceanic setting west of an extensive, Late Devonian carbonate platform bordering western North America, including the present area of eastern Nevada and western Utah (Fig. 1). Following original discovery (Warme et al., 1991) and description of the onshore Alamo Breccia, numerous studies have documented its impact-related origin and the magnitude, paleogeographic setting, chronology, offshore effects, and distal evidence (e.g., Warme, 1991; Sandberg and Warme, 1993; Leroux et al., 1995; Warme and Sandberg, 1995, 1996; Kuehner, 1997; Sandberg et al., 1997; Morrow et al., 1998; Warme and Kuehner, 1998; Chamberlain, 1999; Morrow and Sandberg, 2001; Morrow et al., 2001; Sandberg et al., 2002, 2003; Warme et al., 2002; Koeberl et al., 2003; Warme, 2004). The Alamo Breccia was formally named as a member of the upper Middle

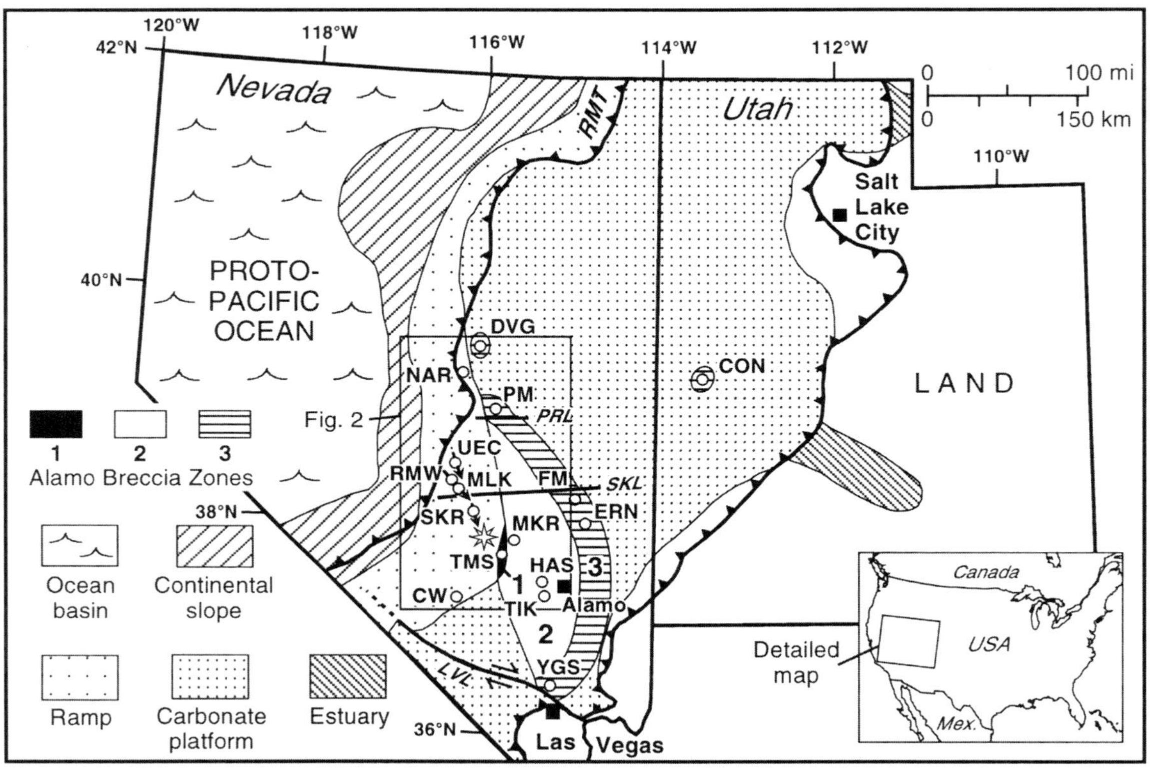

Figure 1. Paleotectonic map (partly restored) of Nevada and Utah during early Frasnian *punctata* Zone, showing possible site of tectonically eastward-transported, oceanic Alamo Impact crater (star symbol) and distribution of Alamo Impact Breccia and related event deposits. Alamo Breccia Zones 1, 2, and 3 form semicircular pattern on ramp (Zone 1), outer carbonate platform (Zone 2), and peritidal, inner platform (Zone 3). Also shows schematically the location and current direction of four deep-water channels of Alamo Breccia (see Fig. 2). Selected major post-breccia structural features: latest Devonian to earliest Mississippian Roberts Mountains thrust (RMT), which significantly displaced transitional- and oceanic-facies Devonian rocks, including the Alamo crater; and three Tertiary lineaments that affected distribution of Alamo Breccia: Las Vegas lineament (*LVL*); Silver King lineament (*SKL*); and Pancake Range lineament (*PRL*). Alamo Event localities, mentioned in text or used to constrain known extent of Zone 3, are shown by small circles: CW—Carbonate Wash; ERN—East Ridge North; FM—Fox Mountain; HAS—Hancock Summit (type locality of Alamo Breccia); MLK—Milk Spring; MKR—Meeker Peak; NAR—Northern Antelope Range; PM—Portuguese Mountain; RMW—Rawhide Mountain West; SKR—Streuben Knob; TIK—Tikaboo Peak; TMS—Tempiute Mountain South; UEC—Upper Empire Canyon; YGS—Yucca Gap South. Thin beds of sandy, carbonate-clast microbreccia at Devils Gate (DVG), Nevada, and at Little Mile-and-a-Half Ridge (CON), Confusion Range, western Utah, are now classified as Zone 3 Alamo Breccia. Because of the close proximity of many studied sections, only Zone 1 and 2 localities that are discussed in text can be shown. Area covered by Figure 2 is indicated. Modified from Morrow and Sandberg (2003, fig. 1).

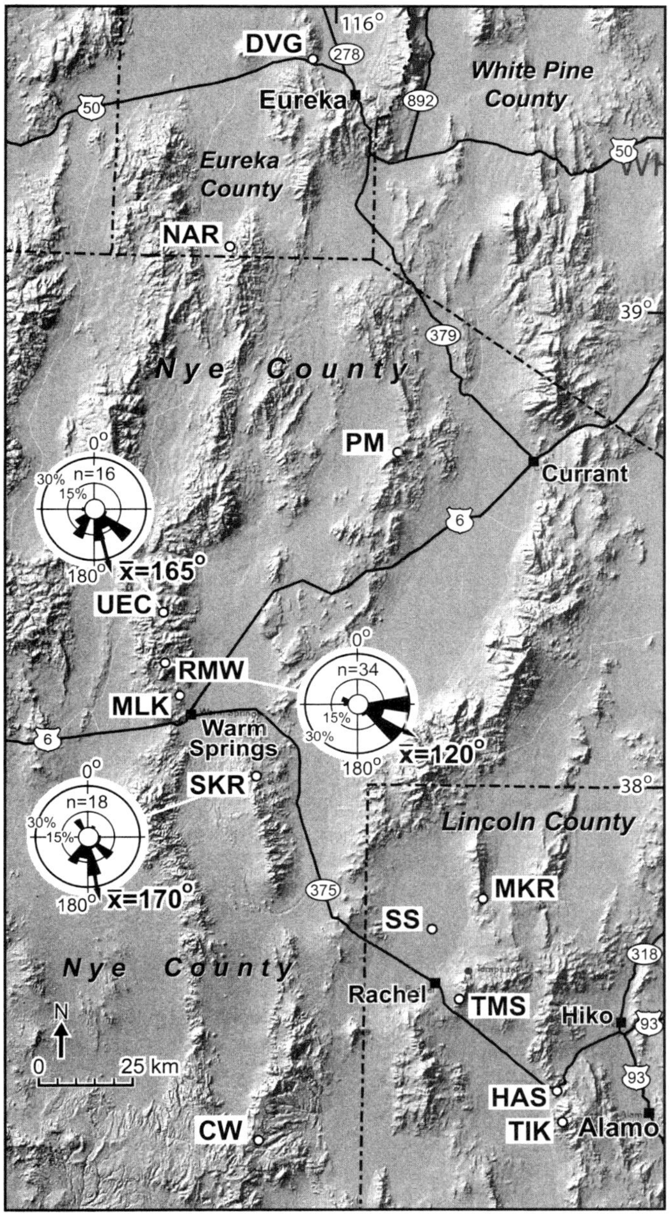

Figure 2. Detailed shaded relief digital elevation map of a portion of southern Nevada (Fig. 1), showing offshore and selected onshore Alamo Event localities discussed in text. Locality abbreviations are the same as in Figure 1. Shows location of possible breccia spire (SS) near Sand Spring, also discussed in text. Rose diagrams, showing paleocurrent directions based on tabular-clast imbrication (corrected for regional dip), are given for localities Upper Empire Canyon (UEC) and Rawhide Mountain West (RMW), southern Hot Creek Range, and Streuben Knob (SKR), Reveille Range; number of clast measurements (n) and mean paleocurrent direction (x-bar) are indicated at each locality. Base map from Johnson (1998).

to Upper Devonian Guilmette Formation and Devils Gate Limestone by Sandberg et al. (1997).

The Alamo Impact crater, which may have been located originally in the vicinity of the southern part of the Roberts Mountains thrust (Fig. 1, RMT), is obscured by post-Devonian geologic processes that include probable fault-dismemberment, eastward tectonic transport, post-Cretaceous extension, and burial beneath younger sedimentary and volcanic rocks. However, at least nine separate lines of geologic, paleontologic, petrologic, and geochemical evidence verify an impact origin for the Alamo Event, define its magnitude and extent, and constrain the depth and location of its original crater. These lines of evidence are reviewed and summarized herein, with emphasis placed on recent discoveries and results that bear directly both on documenting new distal phenomena related to the event and on reconstructing the depth and location of the Alamo crater.

Three new lines of evidence document the distal effects of the Alamo Event. (1) Occurrence of scattered, altered, hematite-studded quartz grains, which are probably Alamo Event–derived ejecta, in distal areas where the Breccia is not present, e.g., the Northern Antelope Range (Figs. 1 and 2, locality NAR). (2) Confirmation that thin, fine-grained, graded carbonate breccia beds at Devils Gate, Nevada, and in the Confusion Range, western Utah (Fig. 1, localities DVG and CON, respectively), represent distal Alamo Breccia deposits, documented by conodont biostratigraphic dating and the presence of altered, hematite-studded quartz grains. (3) Interpretation of a charcoal-bearing, lower Frasnian estuarine deposit as the first evidence of possible forest fires ignited by very distant Alamo Impact ejecta in the Bighorn Mountains, Wyoming.

Also, four recent finds shed new light on the depth and location of the original crater. (1) Biostratigraphically and paleogeographically diagnostic Late Cambrian to early Late Devonian conodont microfossils were recovered from impact-ejected carbonate lithic fragments within carbonate accretionary lapilli-bearing (lapillistone) blocks in the upper part of the Breccia at Tikaboo Peak (Figs. 1 and 2, locality TIK). Middle Ordovician conodonts were also found in two of the lower-slope Alamo Breccia channels (Figs. 1 and 2, localities MLK and UEC). (2) Abundant, altered lithic fragments of argillite, chert, and quartzite, probably ejected from lower Paleozoic or possibly upper Proterozoic target formations by the impact, were found within lapillistone beds and scattered throughout the upper parts of both the offshore and onshore Breccia. (3) Abundant polycrystalline grains of finely crystalline quartzite or chert, possibly derived from the upper Proterozoic (Precambrian Z) to Lower Cambrian Prospect Mountain Quartzite, were found within shock-metamorphosed, quartz-sand populations in the upper part of the Breccia. (4) Five deep-water, upper- and lower-slope Alamo Breccia channels, positioned radially around the impact site and filled by probable crater-directed resurge deposits, were located and current directions of some were determined (Figs. 1 and 2).

These four converging lines of evidence allow us to locate the likely present location of the crater, or at least of its upper slice, in the area west of the Late Devonian carbonate platform margin, between the west end of the present-day Timpahute Range and the southern Hot Creek Range (Figs. 1 and 2, localities TMS and RMW, respectively). Furthermore, this evidence permits us to document a minimum impact excavation depth of 1.7 km (Fig. 3), indicating that the final Alamo crater was at least 44 km in diameter, as detailed below.

EVIDENCE FOR LATE DEVONIAN ALAMO EVENT

Previously Documented Evidence

Megabreccia, Seismic Effects, and Tsunamites

Initial reports of an anomalous megabreccia in southern Nevada interpreted that it had a relatively limited areal distribution and perhaps was produced by an impact (Warme, 1991; Warme et al., 1991). As fieldwork progressed, however, it became apparent that the megabreccia, which comprises a lower interval of large, ~5–10-m-long slabs in a matrix of smaller clasts (Unit B) and an upper interval of normally graded, capping beds (Unit A), is only the upper part of a much thicker breccia sequence (Sandberg and Warme, 1993). Bedded stromatoporoid biostromes, which initially had been considered to form the substrate of the breccia, were subsequently recognized to be the top beds of huge carbonate megablocks (Unit C), as much as 500 m long and 40 m high, that are separated from the unbrecciated substrate by a 1- to >7-m-thick, fluidized detachment zone (Unit D), comprising a bed of carbonate diamictite (Warme and Sandberg, 1995, 1996).

The four units were described by Warme and Sandberg (1995, 1996) and Warme and Kuehner (1998), who showed that they had a much wider distribution, and thus were the products of a larger impact. They divided the Alamo Breccia areally into three zones forming partial rings around the impact site (Fig. 1). Zone 1, the distribution of which is still only partially documented, comprises breccia fragments, generally smaller than 2 m in diameter (Kuehner, 1997). These fragments spilled downslope and fanned off the edge of the carbonate platform as deposits at least 100 m thick. Zone 2, composed of megabreccia averaging 80 m in thickness, forms an arcuate pattern and is the only part of the Alamo Breccia that contains the huge megablocks. Zone 3, composed of tsunami uprush deposits, 10 m grading outward to <1 m in thickness, was deposited across the inner platform and peritidal areas. Warme (2004) recognized eight lithologically and genetically distinct types of Alamo Breccia within Zones 1–3.

Shock waves generated by the impact, in addition to probably creating a series of circular and radiating fractures, traveled horizontally through the carbonate platform and created a seaward-sloping detachment surface. The combination of the two sets of fractures with the detachment surface produced the three dimensions of huge megablocks. These megablocks then slid or were jostled by continued shock waves or a megatsunami down the slope. Based on stratigraphic thickness measurements relative to the top of the Fox Mountain Formation, which directly underlies the Guilmette Formation, this slope had a very low seaward gradient, probably <1° over the 70-km width of Zone 2 (Warme and Sandberg, 1995, 1996; Warme and Kuehner, 1998).

The megatsunami created by the Alamo Impact may have reached an initial wave height of 300 m or greater (Warme and Sandberg, 1995). Succeeding tsunami waves of decreasing intensity deposited a sequence of normally graded tsunamites (Unit A) at the top of the Alamo Breccia (Kuehner, 1997; Warme and Kuehner, 1998). The base of each succeeding tsunamite generally incised the top of, or locally cut completely through, the preceding tsunamite.

Shock-Metamorphosed Quartz Grains

As a unique and geologically durable indicator of the extreme pressure-temperature conditions resulting from hypervelocity impact (Stöffler and Langenhorst, 1994; Grieve et al., 1996; French, 1998), shock-metamorphosed quartz grains have been the most critical line of evidence establishing an impact origin for the Alamo Event (e.g., Leroux et al., 1995). Petrographic examination of isolated quartz grains within the Breccia matrix (Ackman, 1991; Warme and Kuehner, 1998; Warme, 2004) and of acid-insoluble quartz-grain concentrates from bulk Unit A Breccia conodont samples showed the common occurrence of multiple sets of inclusion-rich, planar microstructures within grains. Subsequent transmission electron microscopy analysis (Leroux et al., 1995), thin-section petrographic evaluation (Warme and Sandberg, 1995, 1996; Kuehner, 1997; Warme and Kuehner, 1998; Morrow and Sandberg, 2001; Warme et al., 2002), and U-stage microscope measurements (Morrow and Sandberg, 2001; Warme et al., 2002) of these planar features confirmed the presence of multiple sets of highly decorated, shock-produced planar deformation features (PDFs).

As previously reported (Warme and Sandberg, 1995, 1996; Kuehner, 1997; Morrow et al., 1998, 2002; Warme and Kuehner, 1998; Morrow and Sandberg, 2001, 2002; Sandberg et al., 2002; Warme et al., 2002), shocked-quartz grains have been found as clastic components within: (1) onshore and offshore Alamo Breccia, (2) sub-Breccia clastic injection dikes and sills, and (3) carbonate accretionary lapilli. They have also been found (4) in Zone 1, in situ within siliciclastic formations below the Breccia, possibly including the Middle Ordovician Eureka Quartzite (Morrow and Sandberg, 2001). As reported herein, possible shocked-quartz ejecta are now recorded (5) in northern Nevada and western Utah.

A characteristic property of Alamo quartz grains, both with and without PDFs, is the abundance of relatively large, subhedral to euhedral hematite crystals, pseudomorphous after pyrite and rarely after magnetite. These abundant and distinctive crystals stud surfaces of grains and are imbedded within them (Figs. 4 and 5; Warme and Sandberg, 1995, 1996; Warme and Kuehner, 1998; Morrow and Sandberg, 2001; Warme et al., 2002). Similar grains do not occur in units above and below the Breccia and in laterally equivalent, Upper Devonian formations outside the event area. Although their origin remains undocumented, the

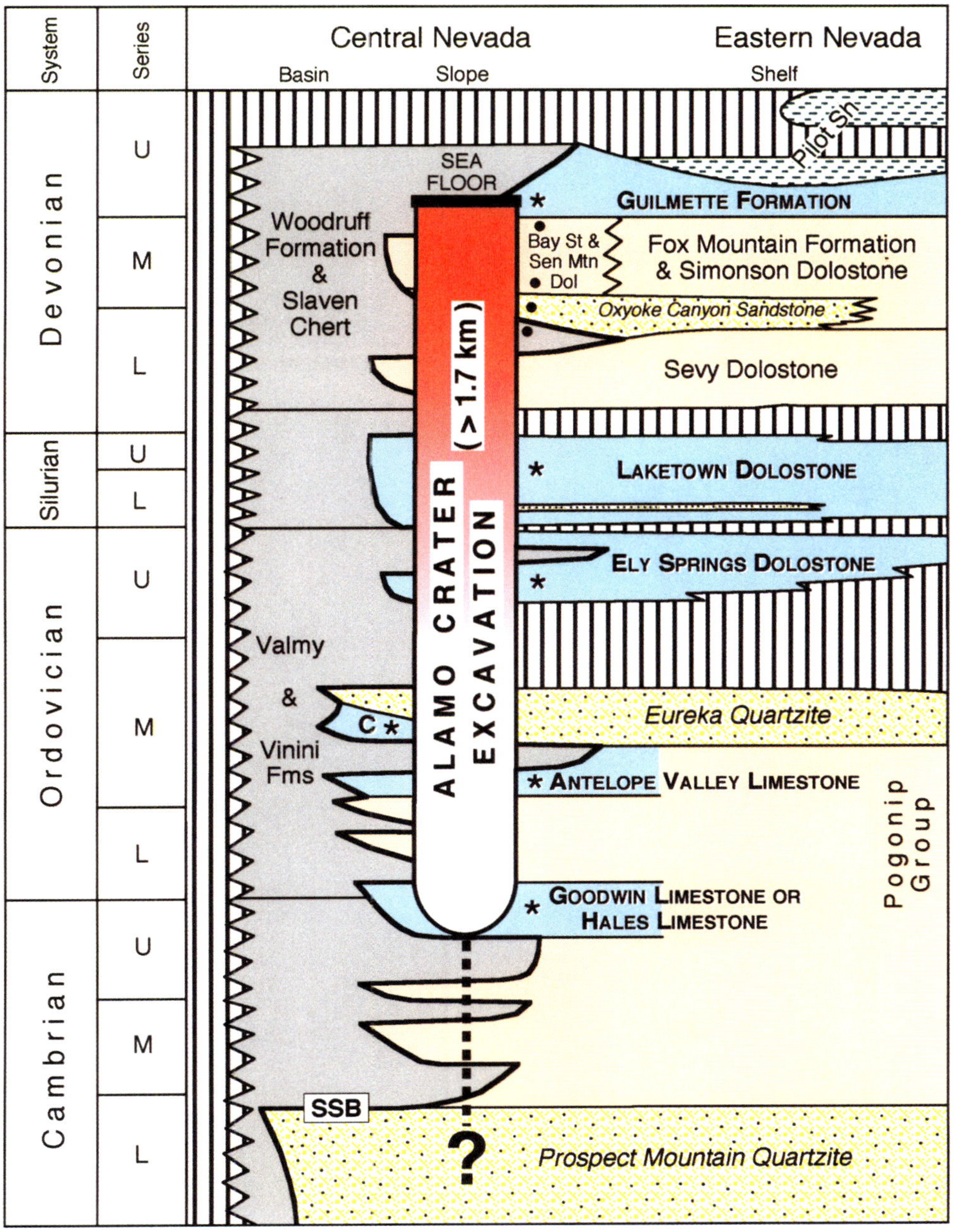

Figure 3. Generalized Alamo crater excavation diagram showing formations that contributed clasts, megabreccia, quartz grains, and conodonts to the Alamo Breccia and that document the minimum excavation depth. Area of central Nevada to the west of inferred crater site on Late Devonian (early Frasnian) seafloor (heavy black bar) is compressed laterally in comparison to shelf area in eastern Nevada. SSB—shelf-slope break. Bay St and Sen Mtn Dol—Bay State Dolostone and Sentinel Mountain Dolostone, respectively. Heavy vertical lines indicate unconformities within pre- and post-impact rock sequences. Six formations labeled in bold uppercase type and indicated by * contributed conodonts recovered from lapillistone blocks in upper part of Unit B of Alamo Breccia; C—Copenhagen Formation. Sandy units labeled in italics and shown by dotted pattern were probable or possible sources of shock-metamorphosed quartz grains found in lapillistone blocks and Units B and A of Alamo Breccia. Guilmette Formation was also source of megablocks that constitute Unit C in Zone 2 of Alamo Breccia and of bulbous stromatoporoids that form virtual conglomerates within Unit B and in deep-water localities, such as Streuben Knob, Milk Spring, and Upper Empire Canyon (Figs. 1 and 2). Tongues of Woodruff Formation and Slaven Chert in vicinity of crater contributed common clasts of argillite, black chert, and black quartzite within Unit B. Valmy and Vinini Formations, primarily west of crater, may have contributed small clasts of similar lithologies found in lapillistone blocks and Unit A. Column of black dots on east side of crater shows stratigraphic position of sandy, breccia-filled injection dikes and sills beneath the Alamo Breccia in Zone 1.

studded grains are a valuable tool for recognizing and correlating widespread Alamo Event ejecta (e.g., Morrow and Sandberg, 2002, 2003).

Carbonate Accretionary Lapilli

Impact-related carbonate accretionary lapilli (Warme et al., 2002) are a volumetrically minor but extremely important constituent of the Alamo Breccia, comprising another important line of evidence supporting its impact origin (Kuehner, 1997; Warme and Kuehner, 1998; Warme et al., 2002). The detailed physical characteristics of the lapilli, their stratigraphic and geographic occurrence within the Alamo Breccia, and their complex genesis and depositional history as part of the Alamo Impact are described by Warme et al. (2002).

Iridium Anomaly

Minor iridium enrichment has been detected at two depositional sites within the Alamo Breccia (Warme and Sandberg, 1995, 1996; Warme and Kuehner, 1998; Koeberl et al., 2003): (1) in the very fine-grained, upper part of the normally graded Unit A tsunamite beds capping the Breccia; and (2) within carbonate accretionary lapillistone blocks distributed in the upper

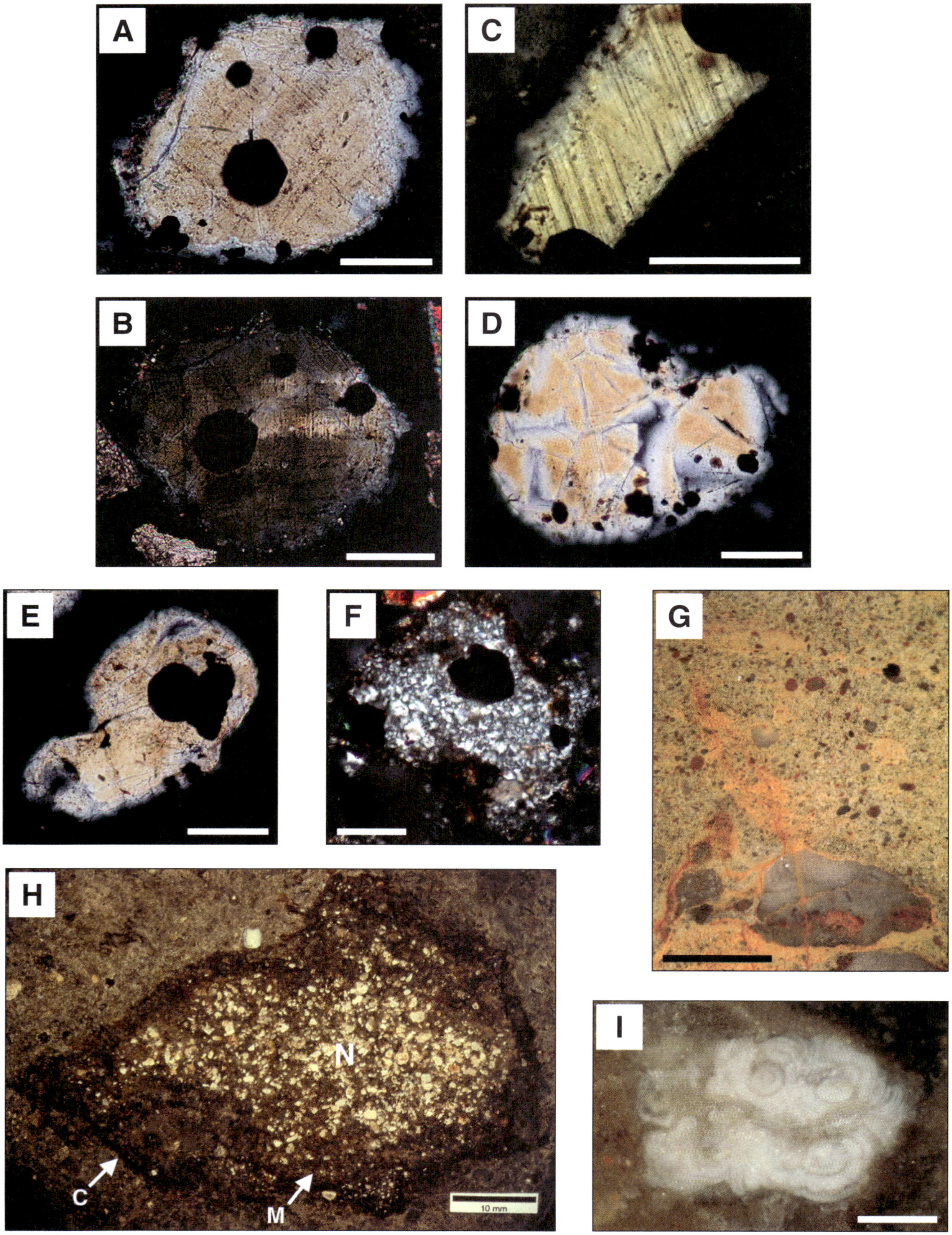
A
C
B
D
E
F
G
H
N
C
M
10 mm
I

part of Unit B and within Unit A. Warme and Sandberg (1995, 1996) and Warme and Kuehner (1998) reported maximum iridium concentrations of 133–139 parts per trillion (ppt) ~1.5 m beneath the top of Alamo Breccia Unit A in the Worthington Mountains (Figs. 1 and 2, locality MKR). This low level of iridium enrichment may be potentially significant when compared against the measured background iridium levels of ≤25 ppt.

Injected Dikes and Sills

At least 300 m beneath the Alamo Breccia in Zone 1 at Tempiute Mountain (Figs. 1 and 2, locality TMS), the underlying Middle Devonian Bay State and Sentinel Mountain Dolostones and Middle to Lower Devonian Oxyoke Canyon Sandstone are pervasively deformed, vug-filled, heavily veined, and cross-cut by numerous anastomosing clast- and sand-filled sedimentary dikes and sills (Warme and Sandberg, 1995, 1996; Kuehner, 1997; Sandberg et al., 1997; Warme and Kuehner, 1998). Passage of an impact-generated seismic wave (Kuehner, 1997; Sturkell and Ormö, 1997; Warme and Kuehner, 1998) may explain the observed country-rock deformation and upward injection of quasi-fluidized lithic fragments and quartz grains into the sedimentary dikes and sills.

New and Corroborative Lines of Evidence

Ejected and Redeposited Conodonts and Lithic Clasts

Conodonts, phosphatic marine microfossils, recovered from carbonate beds above and below, as well as within the Alamo Breccia, played the initial role in dating it biostratigraphically as within the early Late Devonian (early Frasnian) *punctata* Zone, demonstrating its almost instantaneous emplacement, and determining a change in water depth from shallow (~10 m) to deep (>100 m) before and after its emplacement (Sandberg and Warme, 1993; Sandberg and Morrow, 1998a). They provided strong supporting evidence that the Breccia was produced by the marine, off-platform impact of a bolide (Warme and Sandberg, 1995, 1996; Sandberg et al., 1997, 2002; Morrow and Sandberg, 2003). This evidence includes the chaotic mixing of coeval *punctata* Zone onshore and offshore faunas and older Paleozoic conodonts, enclosed in fragments of the target strata, within offshore-transported megabreccia. Conodonts also were used for preliminary estimation of the crater depth and diameter (Sandberg and Morrow, 1998b; Morrow et al., 1998). Herein, we present new data provided by conodonts ejected from the crater and redeposited within Alamo Breccia lapillistone beds and deep-water channels.

The lapillistone beds were fused as the ejecta quickly cooled after settling from the impact cloud, and within a few hours were broken into blocks by the megatsunami and redeposited within the Breccia (Warme et al., 2002). The blocks are composed of carbonate accretionary lapilli, shocked-quartz grains, and angular, pebble or smaller size carbonate, argillite, chert, and quartzite lithic fragments, which are often rimmed and shattered. A few of the larger carbonate lithic fragments fortuitously preserve highly microfractured, very fragile, Late Devonian to latest Cambrian conodonts. Nannofossils, which were microfractured, welded, melted, and deformed by shock, heat, and ballistic transport, have also been used to document impact excavation and to identify target formations in a recent study of the late Eocene Chesapeake Bay impact structure, southeastern Virginia (Edwards and Powars, 2003; Self-Trail, 2003).

Besides their utility for precise biochronologic dating, especially for the Late Devonian Epoch (Sandberg and Ziegler, 1996), determining marine paleogeographic settings, and providing data on the geochemistry of ancient oceans, conodonts are also valuable thermal indices (Epstein et al., 1977; Rejebian et al., 1987). The trace amount of organic matter sealed within these centrifugally laminated, transparent apatitic skeletal elements changes color during thermal progradation. The color alteration of conodonts is both time and temperature related and has been

Figure 4. (A–C) Thin-section photomicrographs of shocked, hematite-bearing quartz sand grains from onshore, Zone 2 localities of Alamo Breccia Member; crossed polars, scale bars = 0.1 mm. (A) Grain showing at least three sets of partially decorated planar deformation features (PDFs) (Figs. 1 and 2, locality TIK), sample TIK-1. (B) Same grain as in A, rotated to show mottled, irregular extinction pattern, which is characteristic of most Alamo Breccia shock-altered quartz grains. (C) Grain showing one very well-developed set (with basal, or (0001), crystallographic orientation) and one moderately developed set (with probable {1013} crystallographic orientation) of decorated PDFs (Figs. 1 and 2, locality Meeker Peak (MKR); also see Morrow and Sandberg, 2001, fig. 3), sample MKR-3A. (D) Thin-section photomicrograph of highly fractured, hematite-studded quartz sand grain from distal Zone 3 Alamo Event bed, lower part of Guilmette Formation, Little Mile-and-a-Half Ridge, Confusion Range, western Utah (Fig. 1, locality CON); sample CON-44, crossed polars, scale bar = 0.1 mm. (E) Thin-section photomicrograph of highly fractured, inclusion- and hematite-rich quartz sand grain reworked on migrating forebulge from Alamo Event bed into much younger Frasnian Late *rhenana* Zone upper tongue of Fenstermaker Wash Formation, Northern Antelope Range (Figs. 1 and 2, locality NAR; Sandberg et al., 2003, fig. 6); sample NAR-A-1, crossed polars, scale bar = 0.1 mm. (F) Thin-section photomicrograph of hematite-rich, finely crystalline quartzite or chert grain from distal Zone 3 Alamo Event bed, lower member of Devils Gate Limestone, Devils Gate (Figs. 1 and 2, locality DVG); sample DVG-21D-R, crossed polars, scale bar = 0.1 mm. (G) Oriented, polished slab photomicrograph showing normally graded microbreccia channel deposit containing abundant heterolithic, hematite-rimmed and altered clasts and quartz grains, lower member of Devils Gate Limestone, Devils Gate (Figs. 1 and 2, locality DVG); sample DVG-21D-R, unpolarized reflected light, scale bar = 5 mm. (H) Polished slab photograph of spindle-shaped, lithic ejecta bomb, which comprises a central nucleus (N), enveloping mantle (M), and outer crust (C), further described in text; Milk Spring, southern Hot Creek Range (Figs. 1 and 2, locality MLK); sample MLK-8A, polarized reflected light, scale bar = 10 mm. (I) Photomicrograph of lithologic chip sample containing light-colored, globular, concentrically microlaminated ejecta grain of possible melt origin; from lapillistone bed, upper unit B, type Alamo Breccia Member, Hancock Summit (Figs. 1 and 2, locality HAS); sample HAS-33, unpolarized reflected light, scale bar = 0.5 mm.

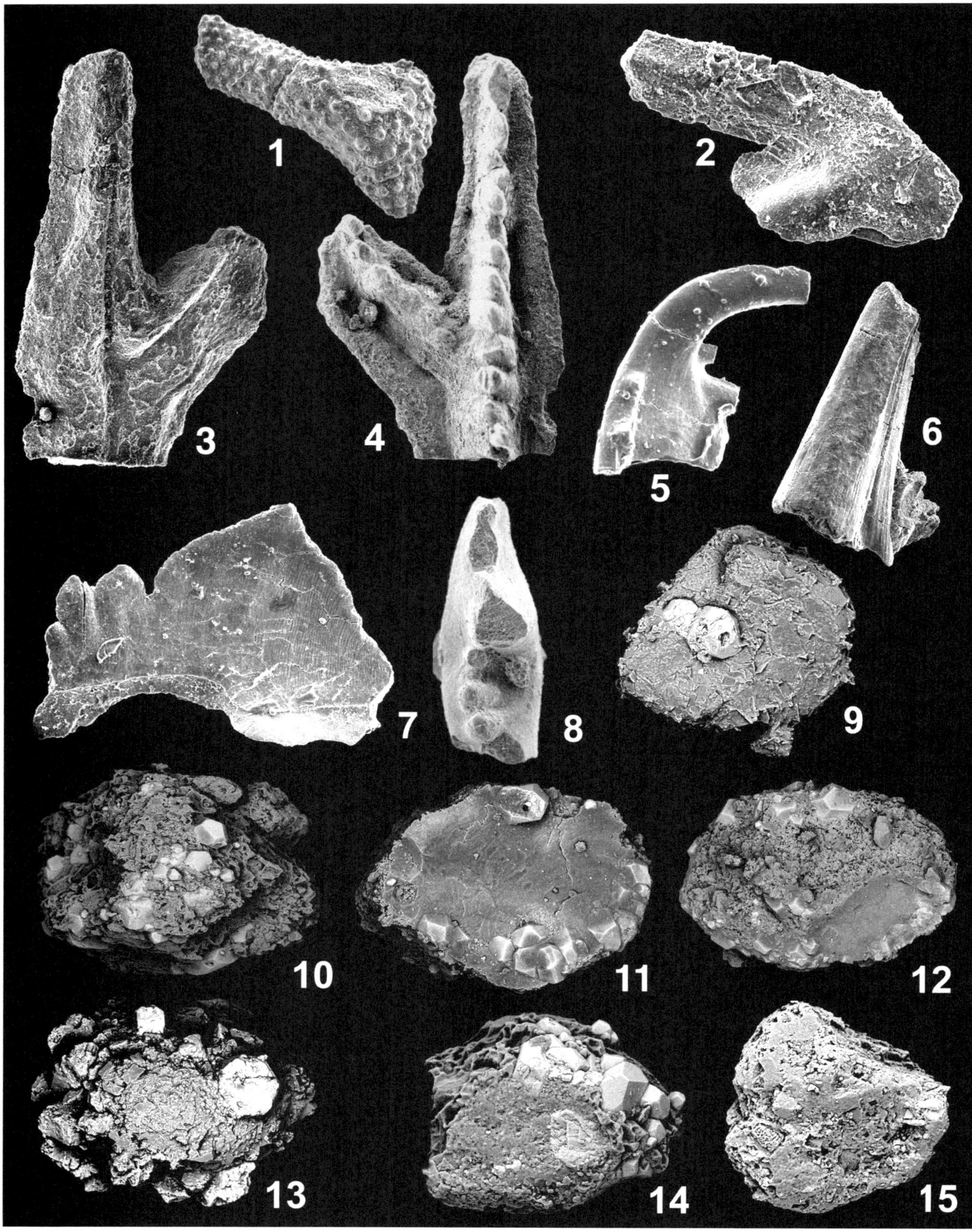
1
2
3
4
5
6
7
8
9
10
11
12
13
14
15

documented down deep drill holes, adjacent to igneous intrusions, and reproduced during temperature- and pressure-controlled laboratory experiments. Epstein et al. (1977) presented a conodont color alteration index (CAI) on a scale of 1 (least altered) to 8 (most altered). With increasing temperature, conodonts change from colorless to pale yellow to amber, light brown, dark brown, and finally black in the range of ~50–300 °C as a result of carbonization of the trace amount of organic matter trapped within their skeletal elements. In the range of ~350–600 °C, black conodonts change to grayish black, gray, opaque white, and finally crystal clear as a result of loss of organic matter (by oxidation and volatilization) and release of water of crystallization accompanied by increasing apatite crystallinity (Königshof, 1992).

Regionally, conodonts from the Alamo Breccia and enclosing lower Upper Devonian rocks in Nevada, east of the Roberts Mountains thrust, have CAI values in the range of 3–3.5. Overlying Devonian and Mississippian to Triassic rocks have progressively lower CAI values (most commonly 3–1.5) because of decreasing depths of burial, whereas underlying Devonian to Cambrian rocks have higher CAI values (most commonly 3.5–5) because of increasing depths of burial. Conodonts within fragments of shattered formations redeposited in lapillistone blocks differ, however, in displaying a reverse CAI gradient, which is probably due to brief, intense, impact-induced shock heating that decreased with depth. Thus, Devonian conodonts ejected from strata in the upper part of the crater have CAI values of 3.5–4.5, whereas those from Silurian to Cambrian strata in the underlying part have CAI values decreasing downward from 3.5 to 1. The anomalously low CAI values of the ejected Silurian to Cambrian conodonts probably resulted because they were protectively enclosed in rimmed or case-hardened micrite or dolomicrite ejecta fragments within the lapillistone blocks and hence were not affected by later burial and regional metamorphism, which increased the CAI of conodonts in enclosing rocks.

Approximately 10 m beneath the top of the Alamo Breccia at Tikaboo Peak (Figs. 1 and 2, locality TIK), two isolated lapillistone blocks within upper Unit B yielded conodonts from the lower Upper Devonian part of the Guilmette Formation, the Silurian Laketown Dolostone, the Upper Ordovician Ely Springs Dolostone, the lower Middle Ordovician Antelope Valley Limestone, and the lowermost Ordovician to uppermost Cambrian Goodwin Limestone or Upper Cambrian Hales Limestone (Fig. 3). Some of the better-preserved conodonts that could be subjected to scanning electron microscope (SEM) photomicrography are illustrated in Figure 5 together with an Ordovician microfossil, *Milaculum* sp., of unknown infinity. The two Tikaboo Peak lapillistone blocks contain a chaotic mixture of whole and broken lapilli; pebble or smaller size, rimmed, altered carbonate, argillite, chert, and quartzite lithic fragments, most of probable ejecta origin; scattered quartz silt and sand grains, some displaying PDFs; irregular dark, opaque mineral-rich flame-like structures; and poorly sorted, relatively fine-grained matrix (Figs. 6 and 7).

The Tikaboo Peak blocks represent the main occurrence where exotic, pre-Devonian conodonts have been recovered from within lapillistone. Other sampled lapillistone blocks from Breccia sections at nearby Hancock Summit (Figs. 1 and 2, locality HAS) and at Mount Irish, located 27 km north of Hancock Summit, yielded only Devonian conodonts. The Tikaboo blocks document a rare case where the lapillistone was not pervasively dolomitized and where relatively large, abundant, lithic ejecta fragments preserved conodonts. Other scattered occurrences of ejecta-derived Ordovician conodonts within Unit A were documented by Warme and Sandberg (1995, 1996) and Morrow et al. (1998).

Conodonts from the lapillistone blocks provide important new constraints on the depth of crater penetration and on crater location. The oldest recovered conodonts are long, slender speci-

Figure 5. (1–8) Scanning electron microscope (SEM) photomicrographs of pre–Alamo Impact *Milaculum* and conodonts that were ejected from crater and fell out with other ejecta to form lapillistone bed, two broken blocks of which were redeposited in close proximity in upper part of Unit B of Alamo Breccia at Tikaboo Peak, Pahranagat Range, Nevada (Figs. 1 and 2, locality TIK). 1—*Milaculum* sp., microfossil of unknown affinity, upper view, ×250; sample TIK-3; late Early to Late Ordovician. 2—oistodontiform M element, lateral view, ×275; sample TIK-3R; Ordovician; derived from Pogonip Group. 3–4—*Pterospathodus procerus*, lower and upper views, anterior part of Pa element, ×155; sample TIK-1; Early Silurian (latest Llandoverian) *Pterospathodus celloni* Zone into *Pt. amorphognathoideus* Zone; derived from Laketown Dolostone. 5—Sc element, lateral view, ×180; sample TIK-3R; Late Ordovician or Silurian; derived from Ely Springs or Laketown Dolostone. 6—*Panderodus* sp., lateral view, ×130; sample TIK-3R; Middle Ordovician to Middle Devonian. 7—*Histiodella sinuosa*, lateral view, shock-wave-fractured Pa element, ×215; sample TIK-3R; early Middle Ordovician *Histiodella sinuosa* Zone into *Hi. holodentata* Zone; derived from Antelope Valley Limestone. 8—*Pandorinellina insita* advanced form, upper view of broken Pa element, ×175; sample TIK-3R; early late Devonian (early Frasnian) *punctata* Zone; derived from shallow-water, pre–Alamo Breccia lower part of Guilmette Formation. (9–13) Backscatter SEM photomicrographs of probable shocked, hematite-studded (after subhedral to euhedral pyritohedron crystals of pyrite) quartz grains from Unit A of Alamo Breccia at Meeker Peak (MKR), Worthington Mountains; Alamo Breccia lapillistone block at Tikaboo Peak (TIK); and Unit B of Alamo Breccia at Streuben Knob (SKR), Reveille Range; all in Nevada (Figs. 1 and 2). Identical grains from these localities, containing well-developed planar deformation features, are shown in thin section in Figure 4. 9—Exterior of grain without overgrowths, ×325; sample MKR-3A. 10—Exterior of grain with quartz overgrowths, ×160; sample TIK-1. 11—Interior of grain broken with conchoidal fracture, ×160; sample TIK-1. 12—Partly broken grain exposing interior at lower right; ×95; sample TIK-1. 13—Exterior of grain, ×185; sample SKR-10. 14–15—Backscatter SEM photomicrographs of similar highly fractured, hematite-studded, very fine sand-sized grains, from distal Zone 3 Alamo Event deposit, lower part of Guilmette Formation, Little Mile-and-a-Half Ridge (CON), Confusion Range, western Utah (Fig. 1). 14—Exterior of agglomerated grain, ×460; sample CON-45R. 15—Exterior of agglomerated grain incorporating 20-μm-long fossil fragment with canals (calcareous alga?) at lower left, ×390; sample CON-45R.

mens of the coniform *Teridontus* sp. This form of *Teridontus* is restricted to the uppermost Cambrian to lowermost Ordovician Skullrockian Stage of the Ibexian Series. Thus, they could have been derived either from the Goodwin Limestone, which straddles the Cambrian-Ordovician boundary, or from the underlying Hales Limestone. According to James F. Miller (2003, personal commun.), however, the long slender form of *Teridontus* occurs abundantly in the Hales Limestone in the Tybo area, mapped by Quinlivan and Rogers (1974), in the southern Hot Creek Range. As this is the closest area to the probable original impact site, we infer that the Hales Limestone was a more likely source of *Teridontus* and that the Alamo Impact excavated strata at least into the Upper Cambrian sequence. On the basis of the regional stratigraphic thickness of these lower Paleozoic formations (Quinlivan and Rogers, 1974; Stewart, 1980; Poole et al., 1992), we calculate that the crater excavation depth was at least 1.7 km, exceeding the 1.5 km of an earlier estimate (Morrow et al., 1998).

We do not know how much lower into the Cambrian sequence the impact might have excavated because older Cambrian conodonts are tiny and less likely to be preserved, especially in a sequence that becomes increasingly argillaceous and sandy with fewer carbonate interbeds downward. Argillaceous lithic fragments found in the lapillistone blocks are more highly altered than carbonate fragments. Moreover, widely distributed polycrystalline quartz grains in the Breccia (e.g., Fig. 4F) and some unidentified quartzite fragments in the lapillistone blocks and in the offshore Breccia deposits might have been derived from the Prospect Mountain Quartzite (Fig. 3) or from underlying Precambrian rocks.

Middle Ordovician conodonts recovered from two deep-water Alamo Breccia channels provide compelling evidence that

Figure 6. Outcrop photograph of highly deformed carbonate accretionary lapillistone block within upper part of Unit B, ~10 m stratigraphically beneath top of Alamo Breccia, Tikaboo Peak (Figs. 1 and 2, locality TIK; sample TIK-3). Block contains chaotic mixture of abundant non-bedded, sub-spherical lapilli (L) rimmed by light-colored, very fine-grained, microlaminated crusts; larger, subangular, alteration-rimmed lithic fragments (F); irregular, dark-colored, flame-like patches of opaque mineral-rich carbonate (P); and Breccia matrix. Alteration-rimmed lithic debris consists of ejected carbonate, argillite, chert, and sandstone or quartzite target-rock fragments (e.g., Fig. 7). Pencil scale is 10 cm long.

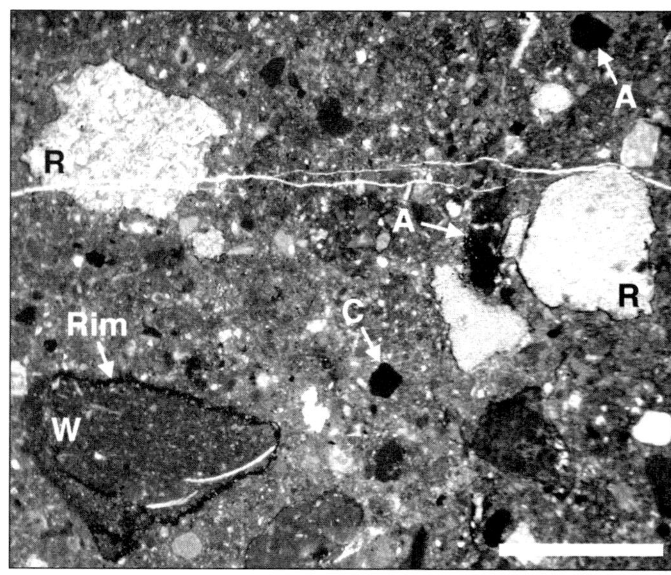

Figure 7. Thin-section photomicrograph of portion of Alamo Breccia lapillistone block, Tikaboo Peak (Figs. 1 and 2, locality TIK, sample TIK-3; Fig. 6), showing heterolithic mixture of clasts, including probable target-rock ejecta fragments such as those that yielded important pre-impact conodonts (Fig. 5). Clasts include abundant carbonate rock fragments, recrystallized carbonate (R), hematite-rich argillite (A), dark chert (C), and isolated quartz grains, some displaying planar deformation features (e.g., Fig. 4A). Brachiopod-bearing skeletal wackestone clast (W) is surrounded by sharp, dense, hematite-rich alteration rim (indicated), which is a common feature of lithic clasts associated with lapilli. Slide is cross-cut by later, thin calcite vein. Plane-polarized light, scale bar = 2 mm.

some lithic clasts in these channels were derived from the west side of the crater. The conodont *Appalachignathus delicatulus* was found near the top of the deep-water channel deposit at Upper Empire Canyon (Figs. 1 and 2, locality UEC). The range of this late Middle Ordovician, moderately deep-water species, which is not commonly reported in Nevada, would correlate with an interval from the base through at least the succeeding 70 m of Member B of the Copenhagen Formation. The Copenhagen occurs mainly in the eastern Monitor Range (range west of locality UEC) and westernmost Antelope Range.

The early Middle Ordovician conodont *Histiodella serrata* was recovered from the top of the Alamo Breccia deep-water channel at Streuben Knob (Figs. 1 and 2, locality SKR). This species ranges from the upper part of the *sinuosa* Zone through the lower part of the *holodentata* Zone, but it is slightly younger than *Histiodella sinuosa*, which is illustrated in Figure 5, no. 7. In central Nevada, *Histiodella serrata* occurs mainly in the Antelope Valley Formation exposed in the Monitor, Toquima, and Groom Ranges.

Petrography and Stratigraphic Sources of Alamo Quartz Grains

Under petrographic examination, many quartz grains from Alamo Event deposits display one or more distinctive petrographic properties, including (1) irregular optical extinction; (2) curvilinear microfractures; (3) planar fractures (PFs); (4) inclusion-rich, decorated PDFs; (5) polycrystalline microfabric; and (6) subhedral to euhedral hematite crystal inclusions (Table 1). Although the presence of PDFs is the only petrographic criterion that uniquely identifies a shock-metamorphic origin, numerous reports have documented the common development and association of these other petrographic properties in experimentally and naturally shocked quartz (e.g., French and Short, 1968; Stöffler and Langenhorst, 1994; Grieve et al., 1996; French, 1998; and references therein). This close association between PDFs and other unusual optical properties is further demonstrated for Alamo Event quartz grains (Table 1; Fig. 4).

PDFs are present in up to 31% of the grains examined (Table 1; Figs. 4A–4C). Although based on a limited number of samples, quartz grain populations with the highest percentages

TABLE 1. SELECTED QUARTZ GRAIN PROPERTIES, ALAMO BRECCIA*

Locality: (See Fig. 1)	ERN (n = 500)	MKR (n = 500)	SKR (n = 271)	TIK (n = 500)	TMS (n = 500)	TMS Breccia dike† (n = 150)
Property:	%	%	%	%	%	%
Unaltered	39	1	1	16	3	45
Irregular extinction§	13	20	15	4	18	2
Curved fractures	11	12	28	14	22	1
Planar fractures	27	64	82	74	64	34
PDFs#	17	29	20	5	31	6
PDF sets/grain:						
>3	1	1	1	<1	2	<1
2–3	10	20	14	3	15	5
1	6	8	5	2	14	1
PDF decoration**	L = 13 M = 42 H = 45	L = 15 M = 53 H = 32	L = 58 M = 33 H = 9	L = 58 M = 25 H = 17	L = 53 M = 35 H = 12	L = 66 M = 44 H = 0
Polycrystalline	21	55	33	2	15	9
Hematite inclusions††	22	85	10	46	15	35
Other inclusions§§	2	14	54	30	9	<1

*Determined petrographically from point-count analysis of grain-mount thin sections; n = number of grains analyzed.
†Grains recovered ~150 m beneath Alamo Breccia, within sub-Breccia clastic injection dike.
§Includes single-crystal grains showing undulose and mosaic extinction patterns.
#Planar deformation features (PDFs).
**A relative measure of the quantity of micro-inclusions defining PDF planes, normalized to total number of PDF-bearing grains for that locality: L—low; M—medium; H—high.
††Grains containing subhedral to euhedral hematite inclusions after pyrite (abundant) and magnetite (rare).
§§Distinct high-relief, high-birefringence, non-opaque mineral inclusions within grains.

of PDFs occur at localities closer to the inferred crater site (i.e., localities MKR, SKR, and TMS, Figs. 1 and 2). The Alamo quartz PDFs (Leroux et al., 1995; Morrow and Sandberg, 2001, fig. 3; Warme et al., 2002) record both basal and higher index crystallographic orientations and relative abundances identical to those from other proven sedimentary-target impacts (e.g., Bunch, 1968; French et al., 1974; Bohor et al., 1987; Koeberl et al., 1996, 1997).

Another characteristic property of Alamo quartz is the relative abundance of polycrystalline grains within the total quartz-grain population (Table 1). This quartz, which has the form of both rounded sand grains and angular fragments, appears optically as very finely crystalline quartzite or coarsely crystalline chert (e.g., Fig. 4F) and displays the other distinctive properties similar to those found in single-crystal Alamo quartz grains. Tiny, transparent, high-relief inclusions are a common feature in Alamo quartz grains also containing PDFs (Table 1). Based on their optical properties and on estimated minimum Alamo Impact shock pressures of 16–20 GPa (Leroux et al., 1995; Morrow and Sandberg, 2001; Warme et al., 2002), some of these inclusions may consist of the high-pressure quartz mineral polymorphs coesite or stishovite (Kieffer et al., 1976; Grieve et al., 1996; French, 1998).

Petrographic, spindle-stage, immersion-oil analysis of 40 Alamo quartz grains from sample localities ERN, MKR, SKR, and TMS (Figs. 1 and 2) documented slightly lowered mean refractive index and birefringence values, which are consistent with the expected changes in bulk optical properties due to shock metamorphism (Stöffler and Langenhorst, 1994). In these Alamo grains, which also commonly showed irregular extinction, PFs, PDFs, and hematite inclusions, the mean refractive indices (n) ranged from 1.544 to 1.547 and the corresponding birefringence (δ) varied from 0.005 to 0.008 (Morrow and Sandberg, 2001).

Regionally widespread siliciclastic units that were penetrated by the crater and contributed quartz grains to the Breccia include the Oxyoke Canyon Sandstone, the Eureka Quartzite, and possibly the Prospect Mountain Quartzite (Fig. 3). Dikes and sills injected through the Sentinel Mountain Dolostone beneath Zone 1 Breccia at Tempiute Mountain (Figs. 1 and 2, locality TMS) connect the Alamo Breccia with the underlying Oxyoke Canyon Sandstone, clearly documenting it as a major quartz sand source (Warme and Sandberg, 1995, 1996; Kuehner, 1997; Warme and Kuehner, 1998). Optically, quartz sand grains within the sub-Breccia injections at Tempiute Mountain contain a lower percentage of PDFs and appear less altered overall than grains from the overlying Breccia (Table 1). This suggests that many of the allochthonous quartz grains within the main Breccia body experienced higher shock levels than those preserved within the sub-Breccia dikes and sills, and thus were probably derived from sources closer to, or within, the crater.

Outside of areas where extensive diagenetic or hydrothermal alteration has occurred, the Oxyoke Canyon Sandstone and Eureka Quartzite are composed of moderately well-sorted, fine- to medium-grained quartz arenite dominated by relatively clear, single-crystal quartz grains displaying sharp to slightly undulose extinction. Quartz populations in Breccia samples, however, display a much higher percentage of polycrystalline grains than can be accounted for from the grain suites observed in the Oxyoke, the Eureka, or the sub-Breccia dikes and sills (Table 1). This observation suggests that there may have been another, presumably deeper, stratigraphic source for the polycrystalline Alamo quartz.

Petrographic examination of the Prospect Mountain Quartzite exposed at the northeastern end of the Groom Range (~12 km southwest of Tempiute Mountain, locality TMS, Figs. 1 and 2), at Ruby Hill (~12 km southeast of Devils Gate, locality DVG, Figs. 1 and 2), and at Prospect Mountain, type section of the Prospect Mountain Quartzite (~15 km southeast of Devils Gate, locality DVG, Figs. 1 and 2), shows the presence of up to 30% polycrystalline quartzite or meta-chert lithic grains, which are texturally similar to allochthonous quartz grains within the Breccia. This quartzite or other laterally equivalent siliciclastic units, which were widespread over much of central and eastern Nevada (Poole et al., 1992), could have been a viable source for polycrystalline quartz entrained into the Breccia.

Offshore Channel-Fill Deposits

Relatively deep-water, erosively emplaced Alamo Breccia channel or lens deposits are now documented at five off-platform localities west of the main onshore body of the Breccia: Upper Empire Canyon, Rawhide Mountain West, Milk Spring, Streuben Knob, and Carbonate Wash (Figs. 1 and 2, localities UEC, RMW, MLK, SKR, and CW, respectively; Morrow et al., 2001). The regional downcutting by the Alamo Breccia across the carbonate-platform and upper-slope settings continues and increases seaward. Middle Devonian lower-slope deposits beneath Breccia channels to the west are much older than Middle and Upper Devonian rocks beneath the Breccia on the carbonate platform. At UEC, RMW, and SKR, the middle *punctata* Zone Breccia downcuts into Middle Devonian, lower-slope carbonate facies as old as *hermanni-cristatus* Zone (Fig. 8; Morrow et al., 2001, fig. 2). This incision results in a time gap of ~3.5 m.y. (Sandberg et al., 2002, fig. 1). The offshore Breccia channels are unequivocally linked to the Alamo Event on conodont biostratigraphy, depositional fabric, heterolithic clast composition, shocked-quartz grains, and impact ejecta such as target-rock lithic fragments and an accretionary bomb.

The thickness of off-platform, deep-water Alamo Breccia deposits ranges from <10 m to >40 m. Laterally, the Breccia changes thickness abruptly, evidencing localized deposition within wide channels that probably were oriented radially to the impact site. Where exposed, the lower contacts of the offshore Breccia channels are either very sharp or consist of anastomosing, Breccia-filled dikes and sills that crosscut the underlying units. The upper contact of the Breccia channels is also sharp. Very fine-grained, normally graded dolosiltstone and dolomudstone of the uppermost Breccia, which is commonly altered to jasperoid, is disconformably overlain by either turbiditic, debris-flow, lower-slope rocks of the upper Frasnian part of the Woodruff Formation (Figs. 1 and 2, localities RMW, MLK, and SKR), or by coarse-grained siliciclastics of the Mississippian

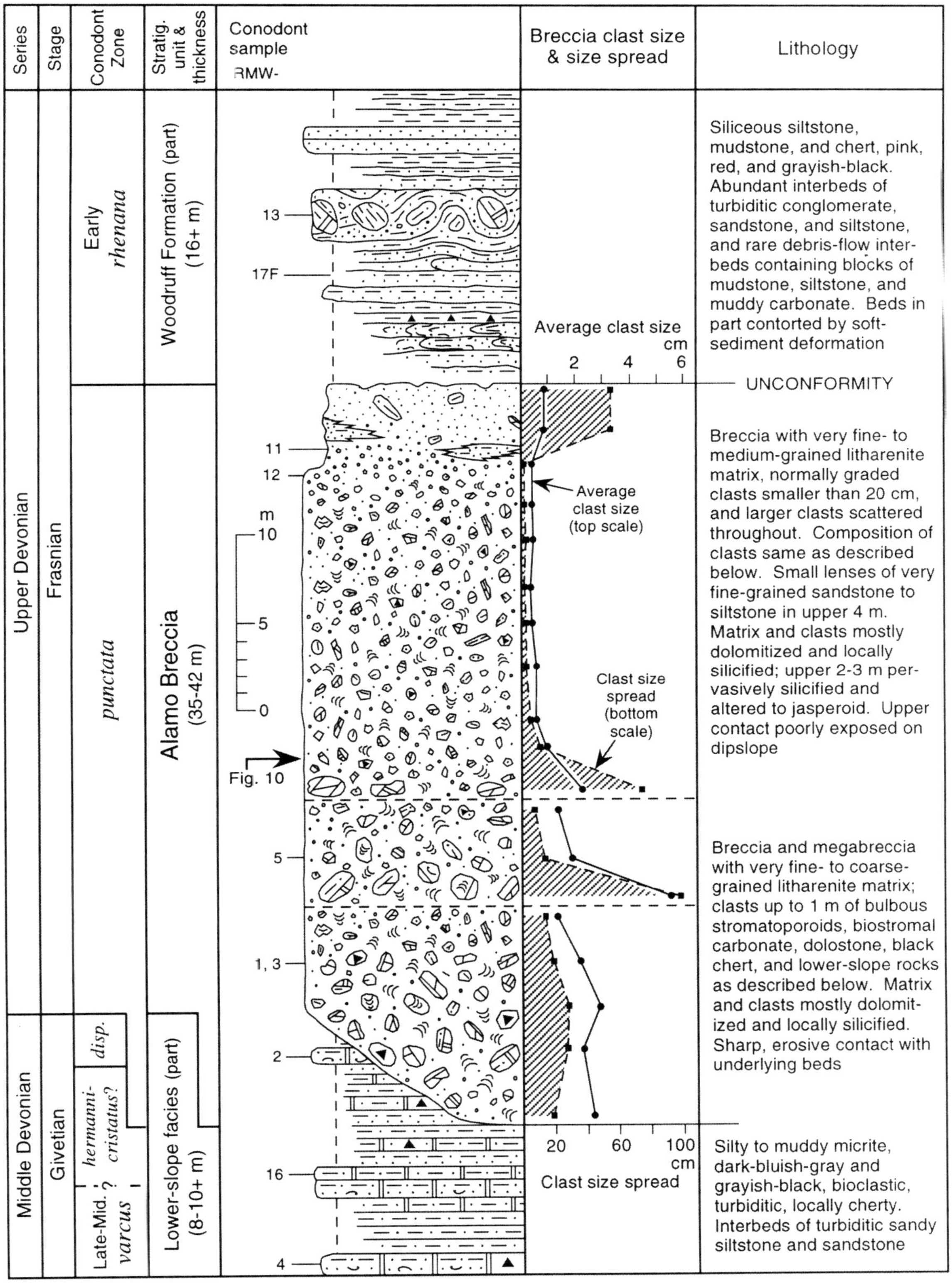

Figure 8. Composite stratigraphic section of deep-water Alamo Breccia exposed on northwest side of Rawhide Mountain (Figs. 1 and 2, locality RMW), southern Hot Creek Range, secs. 11 and 14, T. 5 N., R. 49-1/2 E. (unsurveyed), Tybo 1:24,000 quadrangle, Nye County, Nevada. Shows position of Alamo Breccia erosively emplaced into underlying Middle Devonian, lower-slope facies and unconformably overlain by lower-slope Woodruff Formation. Also shows positions of close-up view of Breccia outcrop (Fig. 10) and of nine bulk conodont samples and one conodont bedding-plane sample (RMW-17F) used to constrain biostratigraphic age of sequence. Average Breccia clast size and clast-size spread, which are based on 19 2-m-long stretchline measurements through the Breccia, are also shown (method after Kuehner, 1997).

Eleana Formation (Figs. 1 and 2, locality UEC) (Quinlivan and Rogers, 1974; Morrow et al., 2001, fig. 2). At RMW and MLK, the disconformity capping the Breccia represents a time gap of ~2 m.y., based on the recovery of late Frasnian Early *rhenana* Zone conodonts from carbonate nodules and debris-flow blocks in the lower Woodruff Formation directly overlying the Breccia.

At Rawhide Mountain West, a well-exposed Alamo Breccia channel has a maximum thickness of 42 m (Figs. 8 and 9). Internally, the Breccia here consists of a heterolithic mixture of platform- and off platform-derived clasts, including abundant bulbous stromatoporoid heads, up to 1 m across (Figs. 8 and 10). Although similar to the onshore Alamo Breccia in terms of clast composition, the offshore Breccia contains significantly more deep-water clasts and possible target-rock lithologies such as dark limestone, argillite, black chert, and quartzite. Stretchline clast-size measurements taken through the Breccia at RMW show that the sequence comprises at least three crudely graded, fining-upward units (Fig. 8; Morrow et al., 2002). The upper unit, which makes up more than half of the total thickness, is capped by an ~5-m-thick interval of fining-upward, pebble- to silt-size lithic clasts in a muddy to sandy, dolomitized matrix that is locally altered to jasperoid. Distinctive features in the upper part of this unit include scattered lenses or pockets of irregularly laminated, coarser grained sandstone and siltstone, and larger

Figure 9. Photograph taken toward south of 40-m-high, north-south striking cliff of deep-water Alamo Breccia exposed on northwest side of Rawhide Mountain (Figs. 1 and 2, locality RMW; Fig. 8). Regional dip of beds is ~45° to the west (right). Partially covered slope beneath cliff contains Middle Devonian, limestone turbidite-rich, lower-slope rocks. Tree-covered dipslope above Breccia contains scattered exposures of unconformably overlying turbiditic and debris-flow–rich beds in the Upper Devonian lower Woodruff Formation. Laterally, the Breccia thickness and localized depth of downcutting into underlying rocks decrease abruptly, suggesting deposition in localized, erosive channels. Interval of close-up view of Breccia outcrop (Fig. 10) is indicated by bracket.

Figure 10. Outcrop photograph of central part of deep-water Alamo Breccia cliff (Fig. 9) exposed on northwest side of Rawhide Mountain (Figs. 1 and 2, locality RMW; Fig. 8). The lower and central parts of the Breccia contain up to 20% whole and broken, shallow stromatoporoid heads deposited in a chaotic mixture of angular and tabular, platform- and off platform-derived lithic clasts up to 1 m across. Although roughly similar in fabric and composition to onshore-deposited Alamo Breccia, the deep-water Breccia lacks the megablocks characteristic of Zone 2 deposits, is much finer grained, and contains a significantly higher percentage of slope-derived lithic clasts. Dip-corrected flat clast-imbrication measurements from the Breccia at this locality give a mean paleocurrent direction of S. 60° E. (Fig. 2), which is interpreted to be toward the impact crater site. Photograph taken ~20 m stratigraphically above base of Breccia; scale is 10 cm long.

lithic clasts, up to ~60 cm long, which are concentrated in the uppermost 1–3 m. The laminated lenses of coarser sediment may represent isolated cut-and-fill structures or possibly dewatering structures formed during compaction and localized remobilization of saturated, fine-grained sediment.

Isolated, fault-bounded exposures of deep-water Alamo Breccia exposed near Milk Spring (Figs. 1 and 2, locality MLK) show similar internal fabric and composition to other offshore Breccia deposits, although the composite thickness of the unit is only ~8 m (Fig. 11). Here, at least four coarsely graded, fining-upward units are present internally within the Breccia. The lowest unit, as thick as 6 m, contains relatively coarse platform and off-platform clasts, up to 50 cm across, of bulbous stromatoporoids, carbonate rock, argillite, chert, and quartzite. Near the top of this interval, an unusual, 6-cm-long, zoned ejecta bomb (Fig. 4H) was recovered. The three overlying, normally graded units are much thinner and finer grained, attaining a combined thickness of 4 m or less. Within the upper units, rare, isolated carbonate accretionary lapilli are present. As at locality RMW, the upper graded unit of the Breccia at MLK contains isolated pockets of coarser grained sandstone and siltstone, together with scattered lithic pebbles, up to ~15 cm across, which are floating in a much finer grained, dolosiltstone and dolomudstone matrix. Large, scattered lithic clasts, which include bulbous stromatoporoid heads and stromatoporoid-rich breccia blocks, also occur in the upper part of an Alamo Breccia channel exposed at Upper Empire Canyon (Figs. 1 and 2, locality UEC).

The presence of multiple, normally graded, fining-upward beds capping the offshore Breccia indicates that deposition of these units was not a simple, single event. The stratigraphic position of fining-upward beds above ejecta material, e.g., the bomb at MLK, and the occurrence of relatively large, in part platform-derived clasts within the uppermost fining-upward bed of the sequences suggest that post-event, submarine mass-wasting processes may have played a significant role in modifying the original deep-water impact deposits. Post-event collapse of the

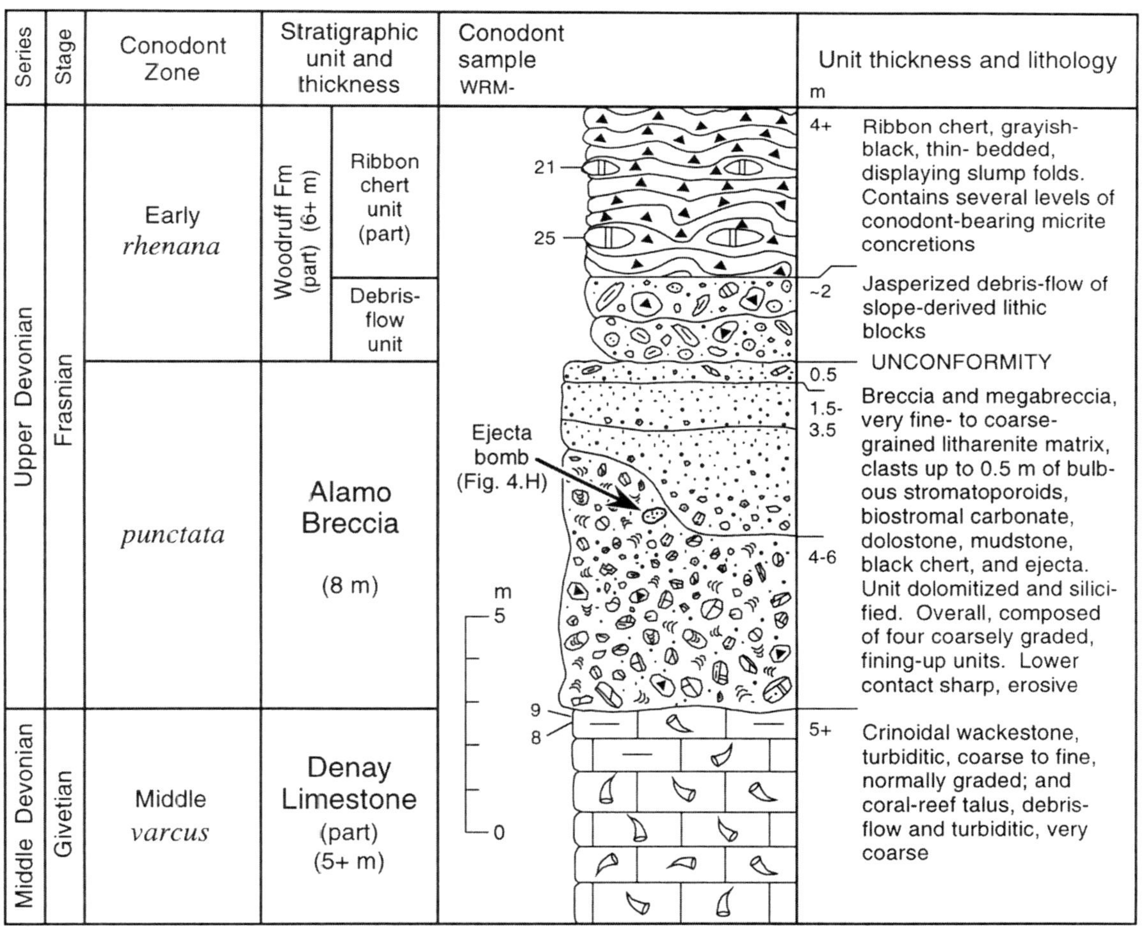

Figure 11. Composite stratigraphic section of deep-water Alamo Breccia exposed 2–2.4 km southeast of Milk Spring (Figs. 1 and 2, locality MLK), southern Hot Creek Range, S 1/2 s. 6 and N 1/2 s. 7, T. 4 N., R. 50 E. (unsurveyed), Warm Springs NW 1:24 000 quadrangle, Nye County, Nevada. Because of fault contacts above and below the Breccia, the stratigraphy of and conodont dating of underlying and overlying units is projected northward from the well-documented Warm Springs South section (Sandberg et al., 1997), located ~7 km south of Milk Spring.

unstable, impact-shattered carbonate-platform margin could have rapidly transported large quantities of debris to the deep-water setting, generating additional tsunami as a result. The importance and widespread effects of post-impact submarine mass wasting are well documented for other marine impact events (e.g., Bralower et al., 1998; Norris and Firth, 2002; Dypvik and Jansa, 2003).

Paleocurrent measurements, based on tabular, imbricated clasts within the lower half of the offshore Breccia at UEC, RMW, and SKR (Fig. 2), indicate that these units were deposited by south- to southeast-flowing currents, which were probably constrained within radially oriented, crater-directed resurge channels or gullies (cf., von Dalwigk and Ormö, 2001).

Sub-Breccia Dikes and Sills

Clastic dikes and sills beneath the Alamo Breccia have been recently documented at two additional localities. These localities, which presumably were more distant from the impact site than the sub-Breccia injection zones at Tempiute Mountain, occur beneath onshore, Zone 3 Breccia at Portuguese Mountain (Figs. 1 and 2, locality PM) and under offshore, deep-water Breccia at Upper Empire Canyon (Figs. 1 and 2, locality UEC; Morrow et al., 2001). At Portuguese Mountain, the clastic sediment-filled structures consist of branching, downward-tapering dikes that extend ~20 cm into underlying biostromal dolostone of the lower Guilmette Formation. Internally, these dikes are laterally zoned and filled with very fine- to fine-grained quartz sand and fine-grained, heterolithic matrix, which were sourced from the overlying Alamo Breccia. Below the base of the Breccia at Upper Empire Canyon, anastomosing, tapering dikes and sills, as wide as 30 cm, penetrate ~5 m into underlying Middle Devonian, lower-slope limestone. These dikes and sills are filled with a laterally zoned mixture of quartz silt, very fine-grained quartz sand, and very fine-grained sand- to pebble-size, dolomitized carbonate breccia that is identical in composition to the overlying Alamo Breccia. Quartz grains within these dikes and sills, which include both single-crystal and polycrystalline microtextures, display irregular optical extinction, abundant PFs, rare PDFs, and abundant hematite inclusions.

At Portuguese Mountain and Upper Empire Canyon, the genesis of clastic dikes and sills, filled from above by Alamo Breccia–derived debris, differs from that recorded by the clastic injection dikes and sills at Tempiute Mountain. Instead of being formed by fluidization and upward injection of underlying shocked formations, these dikes and sills may represent seismically induced fractures that were infilled from above, either passively or forcefully, by unconsolidated Alamo Breccia debris. Forcefully injected, downward-infilled clastic dikes and sills, which are possible analogs to those described herein, were documented for the Chicxulub impact at a distal site along the Tombigbee River, western Alabama (Stewart, 2003).

Spindle-Shaped Bomb

An exotic ejecta bomb was recovered ~3 m below the top of deep-water Alamo Breccia exposed near Milk Spring (Figs. 1 and 2, locality MLK; Fig. 11; Morrow and Sandberg, 2002). As documented by serial sections, the 6-cm-long, tapered, twisted, spindle-shaped bomb (Fig. 4H) is zoned internally with a relatively large central nucleus; a thin, enveloping mantle; and an outer, discontinuous crust.

Reflected-light and thin-section examination of cross sections through the bomb reveals important features of its three parts. (1) The central nucleus consists of poorly sorted, heterolithic carbonate and siliceous lithic fragments, <3 mm in diameter; rare quartz silt and very fine- to fine-grained quartz sand; phosphatic grains; light-colored, sub-rectangular clay pseudomorphs after possible altered feldspar, with long axes <3 mm in length; and light-colored, concentrically microlaminated aggregate clumps, <1.5 mm in diameter. (2) The enveloping, twisted mantle, which is ~3–5 mm across, is composed of lithic fragments, <1.5 mm in diameter; common to abundant, medium- to coarse-grained quartz sand grains, rarely displaying PFs and PDFs; and concentrically microlaminated aggregate clumps, <1.5 mm in diameter. (3) The thin outer, discontinuous crust, <2 mm across, consists of a plastically deformed, anastomosing rim of dense, cryptocrystalline hematite imbedded with quartz silt.

The zoned internal microstratigraphy of the bomb attests to a complex accretional history during its passage through the impact ejecta cloud, perhaps partly analogous to that described for the smaller, more regular, carbonate accretionary lapilli (Warme et al., 2002). The bomb nucleus may have accreted initially as an undifferentiated mass within the heated cloud, which contained a heterogeneous mixture of ejected lithic fragments, isolated individual mineral grains, pulverized and vaporized carbonate, water vapor, and possibly minor melt. During its travel through the ejecta cloud, the unconsolidated mass accreted additional material onto its outer surface, forming the mantle. Unlike the graded, fining-outward Alamo lapilli mantles, the bomb mantle coarsens outward, with an increase in medium and coarse quartz sand toward the crust. This suggests either that the outer portion of the bomb was centrifugally sorted and graded by rotation, or that the mantle was acquired in a different setting where coarser sand was more abundant. Rotation of the mass would have also promoted the formation of its tapered, spindle shape. During descent, the now-differentiated bomb accreted an incomplete and probably ductile outer crust that was entwined into the surrounding unlithified Breccia matrix upon deposition. The outer edge of the crust retains a branching, feathery texture, giving evidence that the bomb was not significantly reworked by currents after its deposition.

Distal Ejecta and Lateral Extent of Zone 3

Distal Alamo Event–related ejecta occurring beyond the onshore Breccia Zones 1, 2, and 3, and the offshore Breccia-filled channels are documented in the Northern Antelope Range (Figs. 1 and 2, locality NAR). At NAR, isolated, fractured, hematite-studded quartz grains, which are petrographically identical to grains recovered from the Breccia at all other localities, are preserved in a micritic turbidite bed within the basal part of the lower tongue of the Woodruff Formation (Sandberg et al., 2003). This turbidite is dated as *punctata* Zone, the same age as the Alamo Breccia elsewhere. During the event, Zone 3 Alamo Breccia or ejected quartz sand was also probably deposited onto the nearby proto-Antler forebulge, which bordered the eastern margin of the developing Woodruff basin. This is evidenced by the common co-occurrence of fractured, hematite-studded quartz grains (Fig. 4E), which are identical to Alamo grains, and reworked *punctata* Zone and other Frasnian conodonts within younger debris-flow beds of the Woodruff Formation and overlying Fenstermaker Wash Formation (Sandberg et al., 2003, fig. 6).

Recent conodont biostratigraphic, petrographic, and SEM analyses have confirmed that two previously unclassified Alamo Event localities at Devils Gate, Nevada, and in the Confusion Range, western Utah (Fig. 1, localities DVG and CON, respectively) represent distal equivalents of the Alamo Breccia. At Devils Gate, the Alamo Breccia is represented by a 29-cm-thick bed of slightly sandy, graded, calcareous diamictite with abundant iron oxide-stained grains and matrix (Fig. 4G), which is present ~120 m below the top of the lower member of the Devils Gate Limestone. Previously identified as a possible Alamo Breccia–equivalent bed based on its unusual lithology and preliminary conodont biostratigraphic age (Sandberg et al., 1997, 2003), the recent evaluation of additional conodont specimens and the discovery of typical fractured and hematite-studded Alamo quartz grains (Fig. 4F) within this bed confirm its link to the event.

In the Confusion Range, the distal Alamo Breccia consists of a 6-m-thick interval of thin-bedded, medium- to coarse-grained carbonate turbidite and breccia, occurring ~170 m above the base of the Guilmette Formation. Recently analyzed conodont samples taken through this interval confirm a middle *punctata* Zone age for the turbiditic breccia, which also yields abundant fractured, inclusion-rich, hematite-studded quartz silt to fine-grained quartz sand grains (Fig. 4D).

The quartz grains in the Confusion Range show two significant differences when compared to typical Alamo Breccia grain populations. First, the 150 µm and smaller fraction at locality CON contains a much larger proportion of hematite-rich and altered quartz than the larger (225–250 µm) fine sand fraction, which is a trend not observed at other localities. This observation suggests that there may have been current size sorting of the Alamo Impact–sourced quartz during deposition of this distal event bed, either by water or by aerial transport through the ejecta cloud before entrainment into the Breccia. The second major difference is that most of the smaller, fine sand grains, mainly 150 µm in diameter, are heavily studded by crystals of hematite, pseudomorphous after pyrite pyritohedrons, and are apparently agglomerates of tiny fragments. These tiny fragments, many only 1–5 µm in diameter (Fig. 5, no. 14), include probable silicified microfossils (Fig. 5, no. 15). The composite nature of these small quartz grains suggests an accretionary origin within the ejecta cloud, further supporting a component of aerial transport.

Based on the confirmation of these two new Alamo Event deposits, which probably represent distal, shallow-water, Zone 3 tsunamites, we are now able to extend the north-south extent of Zone 3 from Devils Gate to just south of the Las Vegas lineament (Warme, 2004), a distance of ~390 km (Fig. 1). With recognition of Alamo Event beds in the Confusion Range, the apparent width of the arcuate Zone 3 is now ~200 km, measured across the carbonate platform in a northeast-to-southwest direction from locality CON toward the inferred present crater location (Fig. 1).

Other Possible Alamo Impact–Related Features

Possible Melt Grains

Small, enigmatic, globular silicate grains have been documented at several places in the Alamo Breccia, including within the interior of the ejecta bomb (Figs. 1 and 2, locality MLK), within carbonate accretionary lapillistone blocks at several localities, and as isolated grains recovered from acid-insoluble residues of Breccia upper Unit B and Unit A bulk samples. The light-colored, globular to elongate grains, which are 1–3 mm across, comprise aggregate masses of tiny, concentrically laminated spheroids (Fig. 4I) that locally appear stretched or plastically deformed. The outer rims of the globular grains are sharp and dense, but irregular, locally penetrating into the enclosing fine-grained Breccia matrix. In thin section, these grains appear to be composed of very finely polycrystalline to cryptocrystalline quartz, which may have formed by secondary replacement or recrystallization of the original grain material. Optically, the outer rims are cryptocrystalline and opaque.

The direct association of these unusual grains with the bomb and the lapillistone suggests that they were formed and deposited together with other ejecta material into the Breccia matrix. There is no evidence that such grains originated within larger lithic clasts, which were subsequently broken up by the event, releasing the grains into the Breccia. Further, these spheroids do not resemble quartz-filled vugs or other common by-products of burial diagenesis, such as botryoidal quartz cement.

These enigmatic grains may evidence an as yet unproven melt phase within the Alamo Breccia. A wide range of possible melt products are produced during impact into saturated, volatile-rich, marine targets, including small melt particles widely dispersed within the resulting impact deposits (Kieffer and Simonds, 1980; Dypvik and Jansa, 2003). Burial alteration of ancient impact deposits would further mask the presence of any original melt material. Therefore, the globular grains within the Alamo Breccia require additional analytical study to further constrain their likely mode of origin.

Possible Spire

Near the center of Sand Spring Valley, located north of Rachel, Nevada, is an unusual, isolated outcrop (Fig. 2, locality SS) that was mapped previously as possible Lower Cambrian Zabriskie Quartzite (Stewart and Carlson, 1978). This outcrop, which is located at the top of a small hill surrounded by Cenozoic valley-fill sediments, consists of a resistant, 5-m-high spire of breccia comprised mainly of vertically oriented clasts. Wholly unlike typical Zabriskie Quartzite, this anomalous breccia spire is composed of silicified, heterolithic, fractured clasts, as long as 20 cm, which are infilled with clastic quartz grains and silicified matrix (Fig. 12A).

Under petrographic examination, the breccia fabric is characterized by pervasively fractured clasts that are penetratively filled by detrital quartz silt and quartz sand, which suggest formation under dynamic, potentially high-pressure conditions (Fig. 12B). The silicified clasts contain fossil ghosts of possible ostracods, brachiopods, and pelmatozoan fragments, indicating an original limestone composition for some of the clasts. The fracture-filling quartz sand is mostly angular to subrounded, contains scattered PFs, and occurs as both whole and comminuted grains. Although the spire has been locally altered to jasperoid, the highly fractured fabric is present throughout the entire exposure.

The anomalous occurrence of this breccia spire within the center of Sand Spring Valley and its pervasively fractured fabric suggest an unusual and possibly violent mode of formation. Its site is near the postulated present location of the dismembered Alamo crater (Fig. 1); therefore, further investigation is warranted to evaluate a possible link between this anomalous spire and the impact event.

Possible Fallout-Generated Forest Fire

An isolated estuarine deposit preserved directly below the Upper Devonian Jefferson Formation on the north wall of Cottonwood Canyon, Bighorn Mountains, Wyoming, may provide the first evidence of a forest fire generated by fallout from the Alamo Impact. This unique deposit of spirorbal limestone, which is located ~925 km northeast of the inferred present Alamo crater site, contains abundant *Spirorbis* tubes, fish remains, carbonized wood, plant impressions, spores, and megaspores (Sandberg, 1963, 1967). The deposit is most likely correlative with the Maywood Formation in southwestern Montana (Sandberg and McMannis, 1964), which is biostratigraphically dated as early Late Devonian (early Frasnian) *punctata*

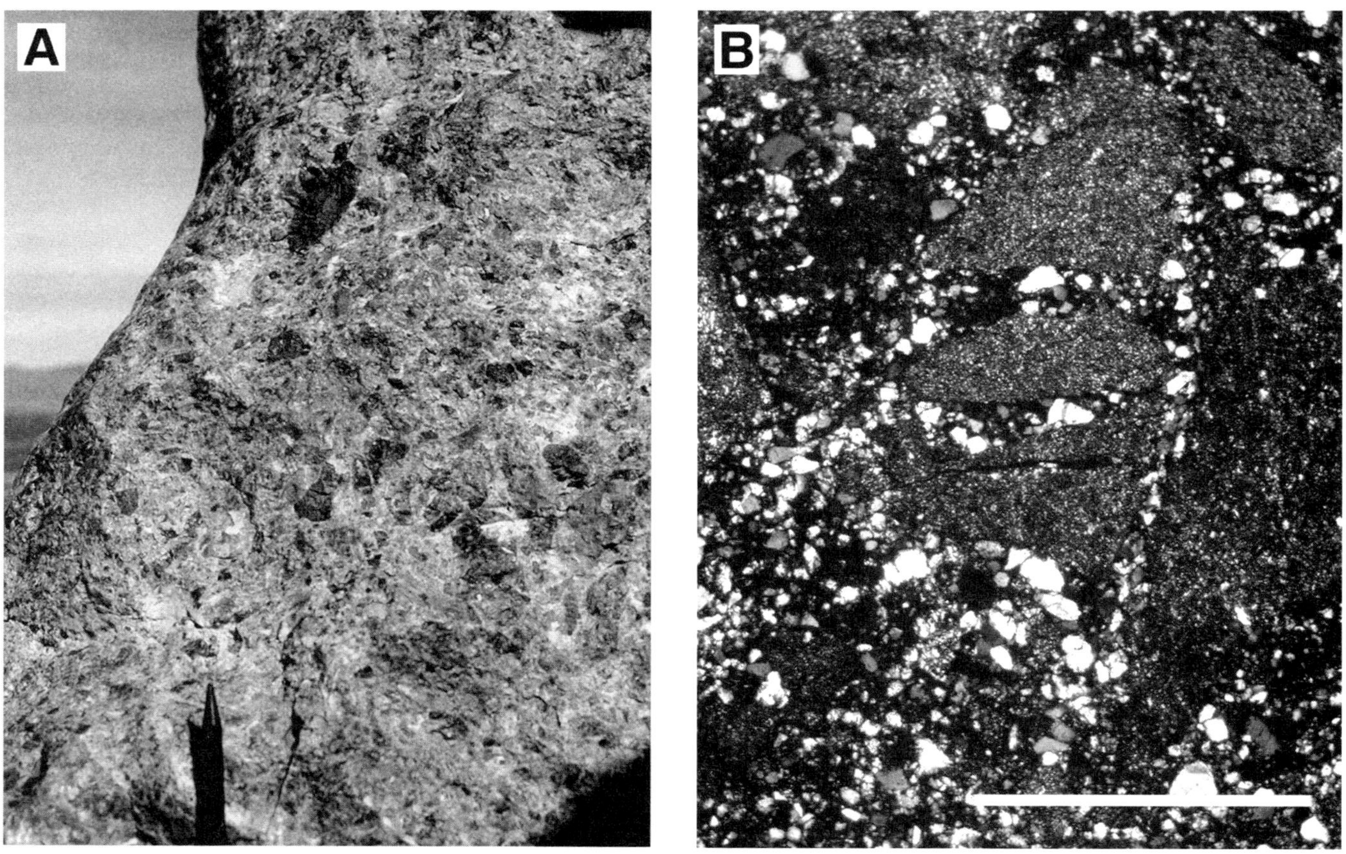

Figure 12. Images of isolated, 5-m-high, silicified breccia spire exposed near Sand Spring, center of Sand Spring Valley (Fig. 2, locality SS). (A) Outcrop photograph of breccia spire consisting of vertically oriented, highly fractured or shattered clasts up to 20 cm across, which are infilled with silicified matrix and fractured clastic quartz grains. Pencil scale is 10 cm long. (B) Thin-section photomicrograph of breccia showing highly fractured, very finely crystalline, pervasively silicified clasts and abundant detrital quartz matrix, which is composed of silt and very fine- to fine-grained, angular to subangular sand, in part fractured. The detrital quartz matrix appears to have been injected into the fractured clasts as they were broken. The broken clasts show a variable amount of movement after fracturing, with some clast fragments virtually in place and others moved, rotated, and incorporated into the comminuted matrix. Locally, fossil ghosts are preserved within clasts. Crossed polars, scale bar = 1 mm.

Zone and is therefore the same age as the Alamo Event. The carbonized wood in the deposit occurs as rounded charcoal balls, such as those carried by streams transporting woody debris from modern forest fires. Thus, although the coeval forest fire that produced the charcoal balls may be coincidental, it provides compelling evidence that this deposit and similar deposits in the western United States should be investigated more fully for other evidence of the Alamo Impact.

ALAMO CRATER SIZE AND LOCATION

Calculation of Crater Size

On the basis of a minimum Alamo Impact excavation depth of 1.7 km, inferred from the stratigraphic level of the deepest, Upper Cambrian conodonts ejected by the impact and preserved within lapillistone blocks, it is possible to estimate the minimum Alamo crater size by applying basic crater scaling calculations. A conservative, rule-of-thumb relationship between maximum depth of excavation (H_{exc}), transient crater depth (H_t), and transient crater diameter (D_t) is given by Melosh (1989, p. 78, equation 5.5.4):

$$H_{exc} \approx 1/3(H_t) \approx 1/10(D_t). \quad (1)$$

For the Alamo Impact, then, H_{exc} = 1.7 km, $H_t \approx$ 5.1 km, and $D_t \approx$ 17 km. The ratio between transient crater diameter (D_t) and final apparent crater diameter (D_A) for marine impact craters is tabulated by Dypvik and Jansa (2003, p. 313, table 2), on the basis of previously documented marine impact craters with dimensions similar to that expected for the Alamo crater. Using an average of their tabulated values, the ratio $D_A/D_t \approx 2.6$, which indicates a minimum final Alamo crater diameter ≈44 km. This value is consistent with calculations given by Morrow et al. (1998), which were based on crater scaling relationships derived primarily from subaerial impact structures.

If quartz-grain and lithic-clast ejecta within the Alamo Breccia are proven to have been derived from the upper Proterozoic to Lower Cambrian Prospect Mountain Quartzite, this would indicate a minimum $H_{exc} \approx$ 2.5 km, based on the regional stratigraphic thickness of lower Paleozoic rocks in central Nevada (Poole et al., 1992). In this case, the final Alamo crater would have reached a minimum diameter of ~65 km. Most of the currently documented marine impact events occurred in epicontinental settings, where they penetrated and excavated a target consisting of both sedimentary and crystalline-basement rocks (Ormö and Lindström, 2000; Dypvik and Jansa, 2003). To date, no crystalline-basement lithic clasts have been identified in the Alamo Breccia. Given the inferred location of the Alamo Impact over a thick geosynclinal sequence of continental-margin sedimentary and metasedimentary rocks, which have a cumulative thickness >6 km (Stewart, 1980; Poole et al., 1992), it is unlikely that crystalline-basement ejecta will ever be found in Alamo Event deposits, even if the impact is proven to be much larger than our current minimum estimates. Assuming an impact excavation depth of 6 km and applying the crater scaling calculations used above, this implies a maximum, outer size limit of ~150 km for the final apparent crater diameter.

As noted by French (1998), the final diameter of an impact crater is generally 20–30 times the diameter of the projectile that produced it. This is consistent with the impact crater scaling law calculations summarized by Melosh (1989, p. 121, table 7.2) for impact craters with a transient crater diameter greater than ~16 km. Using a minimum final Alamo crater diameter of 44 km, this implies an impactor with a diameter of 1.5–2.2 km. This range must represent a minimum value, however, because the impact energy absorbed by the overlying water column, which is estimated to have been at least 300 m deep at the Alamo Impact site (Warme and Sandberg, 1995, 1996), is not considered here. Simplified analyses of the water column effect, based on the discussion in Ormö and Lindström (2000), suggest that the Alamo impactor was ~10% larger in diameter than is indicated by its minimum crater size alone.

Sandberg et al. (1997), Warme and Kuehner (1998), and Koeberl et al. (2003) suggest a possible cometary composition for the Alamo impactor. This is based on the relatively low iridium concentrations measured within the Breccia and the lapillistone, and on the apparent lack of exotic, stony or metallic fragments recognized within Breccia outcrops, within Breccia petrographic samples, and within the acid-insoluble conodont residues concentrated from nearly 200 kg of bulk Breccia material. With an inferred final diameter of 44–65 km, the Alamo crater lies in a size range where the probability of a comet-impact origin is equal to, or greater than, that for an asteroid-impact origin (Shoemaker et al., 1990).

Crater Location

The paleogeographic settings of ejected conodonts recovered from the lapillistone blocks help to constrain the position of the Alamo crater. Shallow-water conodonts such as *Pandorinellina insita* (Fig. 5, no. 8), derived from the pre–Alamo Breccia part of the Guilmette Formation, show that the eastern edge of the crater must have penetrated the western edge of the early Late Devonian carbonate platform (Fig. 3). Conodonts from the Goodwin or Hales Limestone provide a western limit to the western edge of the crater. The Goodwin Limestone starts to change facies westward, downslope from the Late Cambrian platform, to the cherty and more clastic, upper-slope Hales Limestone in the Tybo area (Quinlivan and Rogers, 1974). These formations grade to deeper water, lower-slope, more argillaceous, cherty limestone, and shale farther westward (Cook and Taylor, 1977) and eventually to the chert- and siltite-dominated, oceanic Vinini and Valmy Formations. Thus, the present location of the tectonically transported crater, or at least of its thrust-faulted, sliced-off upper part, is most likely constrained to the area between the edge of the carbonate platform and the southern Hot Creek Range, e.g., locality RMW (Figs. 1 and 2). Supporting evidence for this location is provided by a conodont from the Copenhagen Formation, which occurs northwest of the southern Hot Creek Range.

We speculate that the original position of the crater lay in the vicinity of the leading edge of the Roberts Mountains thrust (Fig. 1, RMT), somewhere west or northwest of the inferred present location of the dismembered upper part of the crater, as depicted in Figure 1. Although the exact amount and foci of Late Devonian to Mesozoic crustal shortening across this area are very difficult to assess (Stewart, 1980), regional, eastward-directed crustal shortening probably totaled ~200 km west of the Roberts Mountains thrust and ~60–100 km east of the Roberts Mountains thrust, to the western limit of Laramide-age thrusting (Poole et al., 1992). In estimating the original crater position, we have also considered the possibility of additional dismemberment produced by >50 km of post-Cretaceous extension (Wernicke, 1992). The net effect of regional shortening and later extension would have been to dismember the crater across thrust faults and to transport its faulted segments eastward, to their present inferred position west of the carbonate platform margin (Fig. 1).

CONCLUSIONS

Our reassessment of older lines of evidence and interpretation of new data on ejected conodonts and lithic fragments, altered and shock-metamorphosed quartz grains, offshore channel deposits, distal onshore Breccia deposits, and probable distal ejecta lead us to these conclusions regarding the Alamo Impact Event:

(1) The Alamo impactor struck an offshore, oceanic slope-to-basin setting west of the carbonate platform margin, excavating down at least into Upper Cambrian rocks and possibly into the underlying Proterozoic to Lower Cambrian Prospect Mountain Quartzite.

(2) The partial, semicircular rings formed by Alamo Breccia deposits in Zones 1, 2, and 3 are restricted to the carbonate platform and supratidal platform areas; they cannot be extended farther west because Breccia deposits there consist of isolated, deep-water, radially directed channel-fill deposits, which have a depositional history distinctive from the onshore deposits.

(3) The occurrence of distal, Zone 3 Alamo Breccia tsunamites is now confirmed in northern Nevada and western Utah, indicating that the arcuate Zone 3 may be ~390 km long in a north-south direction and ~200 km wide in an east-west direction across the carbonate platform.

(4) Ejecta from the Alamo Impact fell at least as far north as northern Nevada, probably into western Utah, and possibly as far northeast as northern Wyoming, where they may have ignited forest fires.

(5) Crater scaling calculations based on a minimum impact excavation depth of 1.7 km indicate that the final Alamo crater was at least 44 km in diameter. If material from the Prospect Mountain Quartzite was excavated by the impact, the final crater would have reached at least 65 km in diameter.

(6) The original site of the Alamo Impact crater is inferred to be near the present position of the leading edge of the Roberts Mountains thrust. The tectonically sliced off and eastward-transported upper part of the Alamo crater now lies buried somewhere between the Timpahute Range and the southern Hot Creek Range.

(7) The bolide that produced the Alamo Impact Event had a minimum diameter range of 1.5–2.2 km; current evidence suggests that it was probably a comet. Although too small to have produced an immediate biotic mass extinction, the Alamo Impact may have been a major contributor to a series of closely spaced impact events that led to the end-Frasnian (F-F) mass extinction ~3 m.y. later.

ACKNOWLEDGMENTS

The U.S. Geological Survey is gratefully acknowledged for its continuing laboratory and field support and for assisting with publication costs. The University of Northern Colorado Sponsored Programs and Academic Research Center and the donors of The Petroleum Research Fund, American Chemical Society (ACS-PRF #34827), are acknowledged for support of Morrow's research. Isabelle K. Brownfield and William J. Betterton greatly assisted us with laboratory analyses and photography of Alamo Event-related samples. Bryan E. Musgrave, Kelly A. Jenkins, and Thomas M. Hillman provided valuable field assistance with measuring Alamo Breccia clast sizes and clast orientations. F.G. (Barney) Poole has contributed his considerable expertise on the geology of Nevada and Utah. He and the late Willi Ziegler are gratefully acknowledged for their assistance with field studies during the past 35 years. John E. Warme, Hans-Christian Kuehner, and Alan K. Chamberlain helped immeasurably with our interpretations of Alamo Breccia sedimentology and stratigraphy, and with our understanding of the genesis of the Alamo Event. Friedrich P. Hörz, John E. Warme, and Philippe Claeys are gratefully acknowledged for their thorough, detailed reviews of the manuscript.

REFERENCES CITED

Ackman, B.W., 1991, Stratigraphy of the Guilmette Formation, Worthington Mountains and Schell Creek Range, southeastern Nevada [M.S. thesis]: Golden, Colorado School of Mines, 207 p.

Bohor, B.F., Modreski, P.J., and Foord, E.E., 1987, Shocked quartz in the Cretaceous-Tertiary boundary clays: Evidence for a global distribution: Science, v. 236, p. 705–709.

Bralower, T.J., Paull, C.K., and Leckie, R.M., 1998, The Cretaceous-Tertiary boundary cocktail: Chicxulub impact triggers margin collapse and extensive sediment gravity flows: Geology, v. 26, p. 331–334, doi: 10.1130/0091-7613(1998)0262.3.CO;2.

Bunch, T.E., 1968, Some characteristics of selected minerals from impact craters, *in* French, B.E., and Short, N.M., eds., Shock metamorphism of natural materials: Baltimore, Mono Book Corporation, p. 413–432.

Chamberlain, A.K., 1999, Structure and stratigraphy of the Timpahute Range, Nevada [Ph.D. thesis]: Golden, Colorado School of Mines, 564 p.

Cook, H.E., and Taylor, M.E., 1977, Comparison of continental slope and shelf environments in the Upper Cambrian and lowest Ordovician of Nevada, *in* Cook, H.E., and Enos, P., eds., Deep-water carbonate environments: Society of Economic Paleontologists and Mineralogists, Special Publication 25, p. 51–81.

Dypvik, H., and Jansa, L.F., 2003, Sedimentary signatures and processes during marine bolide impacts: A review: Sedimentary Geology, v. 161, p. 309–337, doi: 10.1016/S0037-0738(03)00135-0.

Edwards, L.E., and Powars, D.S., 2003, Impact damage to dinocysts from the late Eocene Chesapeake Bay event: Palaios, v. 18, p. 275–285.

Epstein, A.G., Epstein, J.B., and Harris, L.E., 1977, Conodont color alteration—an index to organic metamorphism: U.S. Geological Survey Professional Paper 995, 27 p.

French, B.M., 1998, Traces of catastrophe: A handbook of shock-metamorphic effects in terrestrial meteorite impact structures: Lunar and Planetary Institute Contribution No. 954, 120 p.

French, B.M., and Short, N.M., editors, 1968, Shock metamorphism of natural materials: Baltimore, Mono Book Corporation, 644 p.

French, B.M., Underwood, J.R., Jr., and Fisk, E.P., 1974, Shock-metamorphic features in two meteorite impact structures, southeastern Libya: Geological Society of America Bulletin, v. 85, p. 1425–1428.

Grieve, R.A.F., Langenhorst, F., and Stöffler, D., 1996, Shock metamorphism in nature and experiment: II. Significance in geoscience: Meteoritics & Planetary Science, v. 31, p. 6–35.

Jansa, L.F., 1993, Cometary impacts into ocean: their recognition and the threshold constraint for biological extinctions: Palaeogeography, Palaeoclimatology, Palaeoecology, v. 104, p. 271–286, doi: 10.1016/0031-0182(93)90137-8.

Johnson, G.L., 1998, Nevada shaded relief map Nevada Bureau of Mines and Geology Open-File Report OF 98-3, scale 1:1,000,000, 1 sheet.

Kieffer, S.W., and Simonds, C.H., 1980, The role of volatiles and lithology in the impact cratering process: Reviews of Geophysics and Space Physics, v. 18, no. 1, p. 143–181.

Kieffer, S.W., Phakey, P.P., and Christie, J.M., 1976, Shock processes in porous quartzite: Transmission Electron Microscope observations and theory: Contributions to Mineralogy and Petrology, v. 59, p. 41–93.

Koeberl, C., Poag, C.W., Reimold, W.U., and Brandt, D., 1996, Impact origin of the Chesapeake Bay structure and the source of the North American tektites: Science, v. 271, p. 1263–1266.

Koeberl, C., Reimold, W.U., Brandt, D., Dallmeyer, R.D., and Powell, R.A., 1997, Target rocks and breccia from the Ames impact structure, Oklahoma, *in* Johnson, K.S., and Campbell, J.A., eds., Ames structure in northwest Oklahoma and similar features: Origin and petroleum production (1995 symposium): Oklahoma Geological Survey Circular 100, p. 169–198.

Koeberl, C., Huber, H., Morgan, M., and Warme, J.E., 2003, Search for an extraterrestrial component in the Late Devonian Alamo impact breccia (Nevada): Results of iridium measurements, *in* Koeberl, C., and Martínez-Ruiz, F.C., eds., Impact markers in the stratigraphic record: Berlin, Springer-Verlag, p. 315–332.

Königshof, P., 1992, Der Farbänderungsindex von Conodonten (CAI) in paläozoischen Gesteinen (Mitteldevon bis Unterkarbon) des Rheinischen Schiefergebirges: Eine Ergänzung zur Vitrinitreflexion: Courier Forschungsinstitut Senckenberg, v. 146, 118 p.

Kuehner, H.-C., 1997, The Late Devonian Alamo impact breccia, southeastern Nevada [Ph.D. thesis]: Golden, Colorado School of Mines, 327 p.

Leroux, H., Warme, J.E., and Doukhan, J.C., 1995, Shocked quartz in the Alamo breccia, southern Nevada: Evidence for a Devonian impact event: Geology, v. 23, p. 1003–1006, doi: 10.1130/0091-7613(1995)0232.3.CO;2.

Melosh, H.J., 1989, Impact cratering, a geological process: New York, Oxford University Press, 245 p.

Morrow, J.R., and Sandberg, C.A., 2001, Distribution and characteristics of multi-sourced shock-metamorphosed quartz grains, Late Devonian Alamo Impact, Nevada: Lunar and Planetary Institute Contribution No. 1080, abstract 1233, 2 p.

Morrow, J.R., and Sandberg, C.A., 2002, Distribution and characteristics of proximal to distal ejecta from Late Devonian Alamo Impact, Nevada: Geological Society of America Abstracts with Programs, v. 34, no. 6, p. 543–544.

Morrow, J.R., and Sandberg, C.A., 2003, Late Devonian Alamo Event, Nevada, USA; Multiple evidence of an off-platform marine impact: Lunar and Planetary Institute Contribution No. 1167, abstract 4055, 2 p.

Morrow, J.R., Sandberg, C.A., Warme, J.E., and Kuehner, H.-C., 1998, Regional and possible global effects of sub-critical Late Devonian Alamo Impact Event, southern Nevada, USA: Journal of the British Interplanetary Society, v. 51, p. 451–460.

Morrow, J.R., Sandberg, C.A., and Poole, F.G., 2001, New evidence for deeper water site of Late Devonian Alamo Impact, Nevada: Lunar and Planetary Institute Contribution No. 1080, abstract 1018, 2 p.

Morrow, J.R., Sandberg, C.A., and Poole, F.G., 2002, Deep-water deposits of Late Devonian Alamo Impact Breccia, northern Reveille and southern Hot Creek Ranges, Nevada: Geological Society of America Abstracts with Programs, v. 34, no. 4, p. A-52.

Norris, R.D., and Firth, J.V., 2002, Mass wasting of Atlantic continental margins following the Chicxulub impact event, *in* Koeberl, C., and MacLeod, K.G., eds., Catastrophic events and mass extinctions: Impacts and beyond: Geological Society of America Special Paper 356, p. 79–95.

Oberbeck, V.R., Marshall, J.R., and Aggarwal, H., 1993, Impacts, tillites, and the breakup of Gondwanaland: Journal of Geology, v. 101, p. 1–19.

Ormö, J., and Lindström, M., 2000, When a cosmic impact strikes the sea bed: Geological Magazine, v. 137, no. 1, p. 67–80, doi: 10.1017/S0016756800003538.

Poole, F.G., Stewart, J.H., Palmer, A.R., Sandberg, C.A., Madrid, R.J., Ross, R.J., Jr., Hintze, L.F., Miller, M.M., and Wrucke, C.T., 1992, Latest Precambrian to latest Devonian time; development of a continental margin, *in* Burchfiel, B.C., Lipman, P.W., and Zoback, M.L., eds., The Cordilleran Orogen: Conterminous U.S.: Boulder, Colorado, Geological Society of America, The geology of North America, Chapter 2, v. G-3, p. 9–56.

Quinlivan, W.D., and Rogers, C.L., 1974, Geologic map of the Tybo quadrangle, Nye County, Nevada: U.S. Geological Survey Miscellaneous Investigations Map I-821, scale 1:48,000, 1 sheet.

Rejebian, V.A., Harris, A.G., and Huebner, J.S., 1987, Conodont color and textural alteration: An index to regional metamorphism, contact metamorphism, and hydrothermal alteration: Geological Society of America Bulletin, v. 99, p. 471–479.

Sandberg, C.A., 1963, Spirorbal limestone in the Souris River(?) Formation of Late Devonian age at Cottonwood Canyon, Bighorn Mountains, Wyoming, *in* Short papers in geologic and hydrologic sciences: U.S. Geological Survey Professional Paper 475-C, Art. 63, p. C14–C16.

Sandberg, C.A., 1967, Measured sections of Devonian rocks in northern Wyoming: Wyoming Geological Survey Bulletin 52, 93 p.

Sandberg, C.A., and McMannis, W.J., 1964, Occurrence and paleogeographic significance of the Maywood Formation of Late Devonian age in the Gallatin Range, southwestern Montana, *in* Geological Survey Research 1964: U.S. Geological Survey Professional Paper 501-C, p. C50–C54.

Sandberg, C.A., and Morrow, J.R., 1998a, Role of conodonts in deciphering and dating Late Devonian Alamo Impact Megabreccia, southeastern Nevada, USA, *in* Bagnoli, G., ed., Seventh International Conodont Symposium held in Europe (ECOS VII): Bologna-Modena, Italy, Abstracts volume, p. 93–94.

Sandberg, C.A., and Morrow, J.R., 1998b, Tiny teeth forecast ancient comet showers: U.S. Geological Survey News Release, June 12, 1998, for ECOS VII meeting, http://www.usgs.gov/public/press/public_affairs/press_releases/pr553m.html.

Sandberg, C.A., and Warme, J.E., 1993, Conodont dating, biofacies, and catastrophic origin of Late Devonian (early Frasnian) Alamo Breccia, southern Nevada: Geological Society of America Abstracts with Programs, v. 25, no. 3, p. 77.

Sandberg, C.A., and Ziegler, W., 1996, Devonian conodont biochronology in geologic time calibration: Senckenbergiana lethaea, v. 76, p. 259–265.

Sandberg, C.A., Morrow, J.R., and Warme, J.E., 1997, Late Devonian Alamo Impact Event, global Kellwasser Events, and major eustatic events, eastern Great Basin, Nevada and Utah: Brigham Young University Geology Studies, v. 42, pt. 1, p. 129–160.

Sandberg, C.A., Morrow, J.R., and Ziegler, W., 2002, Late Devonian sea-level changes, catastrophic events, and mass extinctions, *in* Koeberl, C., and MacLeod, K.G., eds., Catastrophic events and mass extinctions: Impacts and beyond: Geological Society of America Special Paper 356, p. 473–487.

Sandberg, C.A., Morrow, J.R., Poole, F.G., and Ziegler, W., 2003, Middle Devonian to Early Carboniferous event stratigraphy of Devils Gate and Northern Antelope Range sections, Nevada, U.S.A.: Courier Forschungsinstitut Senckenberg 242, Frankfurt, a.M., p. 187–207.

Self-Trail, J.M., 2003, Shock-wave-induced fracturing of calcareous nannofossils from the Chesapeake Bay impact crater: Geology, v. 31, no. 8, p. 697–700, doi: 10.1130/G19678.1.

Shoemaker, E.M., Wolfe, R.F., and Shoemaker, C.S., 1990, Asteroid and comet flux in the neighborhood of Earth, *in* Sharpton, V.L., and Ward, P.D., eds., Global catastrophes in Earth history; An interdisciplinary conference on impacts, volcanism, and mass mortality: Geological Society of America Special Paper 247, p. 155–170.

Stewart, J.H., 1980, Geology of Nevada; a discussion to accompany the Geologic Map of Nevada: Nevada Bureau of Mines and Geology Special Publication 4, 136 p.

Stewart, J.H., and Carlson, J.E., 1978, Geologic map of Nevada: U.S. Geological Survey, scale 1:500,000, 2 sheets.

Stewart, K.G., 2003, Forcefully injected clastic dikes and sills associated with the K/T impact tsunami: Geological Society of America Abstracts with Programs, v. 35, no. 6, p. 602.

Stöffler, D., and Langenhorst, F., 1994, Shock metamorphism of quartz in nature and experiment: I. Basic observation and theory: Meteoritics, v. 29, p. 155–181.

Sturkell, E.F.F., and Ormö, J., 1997, Impact-related clastic injections in the marine Ordovician Lockne impact structure, Central Sweden: Sedimentology, v. 44, p. 793–804, doi: 10.1046/j.1365-3091.1997.d01-54.x.

von Dalwigk, I., and Ormö, J., 2001, Formation of resurge gullies at impacts at sea: The Lockne crater, Sweden: Meteoritics & Planetary Science, v. 36, p. 359–369.

Warme, J.E., 1991, The Alamo breccia: Catastrophic Devonian platform deposit in southeastern Nevada, *in* Event markers in Earth history: Program Abstracts, International Union of Geological Sciences, Joint Meeting of IGCP Projects 216, 293, and 303, August 28–30, 1991, Calgary, p. 75.

Warme, J.E., 2004, The many faces of the Alamo impact breccia: Geotimes, January 2004, p. 28–31.

Warme, J.E., and Kuehner, H.-C., 1998, Anatomy of an anomaly; The Devonian catastrophic Alamo impact breccia of southern Nevada: International Geology Review, v. 40, p. 189–216.

Warme, J.E., and Sandberg, C.A., 1995, The catastrophic Alamo breccia of southern Nevada; Record of a Late Devonian extraterrestrial impact: Courier Forschungsinstitut Senckenberg, v. 188, p. 31–57.

Warme, J.E., and Sandberg, C.A., 1996, Alamo megabreccia; Record of a Late Devonian impact in southern Nevada: GSA Today, v. 6, no. 1, p. 1–7.

Warme, J.E., Chamberlain, A.K., and Ackman, B.W., 1991, The Alamo Event: Devonian cataclysmic breccia in southeastern Nevada: Geological Society of America Abstracts with Programs, v. 23, no. 2, p. 108.

Warme, J.E., Morgan, M., and Kuehner, H.-C., 2002, Impact-generated carbonate accretionary lapilli in the Late Devonian Alamo Breccia, *in* Koeberl, C., and MacLeod, K.G., eds., Catastrophic events and mass extinctions: Impacts and beyond: Geological Society of America Special Paper 356, p. 489–504.

Wernicke, B.P., 1992, Cenozoic extensional tectonics of the U.S. Cordillera, *in* Burchfiel, B.C., Lipman, P.W., and Zoback, M.L., eds., The Cordilleran Orogen: Conterminous U.S.: Boulder, Colorado, Geological Society of America, The geology of North America, Chapter 2, v. G-3, p. 553–581.

Manuscript Accepted by the Society 9 August 2004

Printed in the USA

Geological Society of America
Special Paper 384
2005

Are there signs of a large Paleocene impact, preserved around Disko Bay, West Greenland? Nuussuaq spherule beds origin by impact instead of volcanic eruption?

A.P. Jones*
Research School of Geophysical Sciences, University and Birkbeck College London, Gower Street, London WC1E 6BT, UK

A.T. Kearsley
Department of Mineralogy, Natural History Museum, Cromwell Road, South Kensington, London SW7 5BD, UK

C.R.L. Friend
45, Stanway Road, Risinghurst, Headington, Oxford OX3 8HU, UK

E. Robin
Laboratoire C.E.A./C.N.R.S. des Sciences du Climat et de l'Environnement, 91198 Gifsur-Yvette, France

A. Beard
Research School of Geophysical Sciences, University and Birkbeck College London, Gower Street, London WC1E 6BT, UK

A. Tamura
Research School of Geophysical Sciences, University and Birkbeck College London, Gower Street, London WC1E 6BT, and
Graduate School of Natural Science and Technology, Kanazawa University, Kakuma, Kanazawa 920-1192, Japan

S. Trickett
Research School of Geophysical Sciences, University and Birkbeck College London, Gower Street, London WC1E 6BT, UK

P. Claeys
Department of Geology, Vrije Universiteit Brussel, Pleinlaan 2, B-1050 Brussels, Belgium

ABSTRACT

On the Nuussuaq peninsula, Western Greenland sedimentary deposits of glass spherules also contain high Ir, Co, Ni, and Cu anomalies. The iron-rich silicate glass spherules (to ~3 wt% NiO, ~35 wt% FeO) are highly circular in cross section. They show surface dissolution, smectite replacement and calcite infilling of vesicles, though many glasses are optically unaltered. They are strikingly heterogeneous, with schlieren outlining counter-flowing convection cells. Their pronounced Fe-Ni correlation is unlike volcanic suites, but is explained by mixing between basaltic melt and an

*adrian.jones@ucl.ac.uk

Jones, A.P., Kearsley, A.T., Friend, C.R.L., Robin, E., Beard, A., Tamura, A., Trickett, S., and Claeys, P., 2005, Are there signs of a large Paleocene impact, preserved around Disko Bay, West Greenland? Nuussuaq spherule beds origin by impact instead of volcanic eruption?, *in* Kenkmann, T., Hörz, F., and Deutsch, A., eds., Large meteorite impacts III: Geological Society of America Special Paper 384, p. 281–298. For permission to copy, contact editing@geosociety.org.

enriched iron-nickel source. Distinctive nickel-spinel (~7–10 wt% NiO) contains very nickel-rich cores. Occasional glass spherules show compositional gradients toward resorbed silicates, (plagioclase, clinopyroxene); isotropic plagioclase has anomalous texture comparable to impact-melted lunar breccias. Their anomalously high copper and sulfur (to ~1%) have lead to an explanation as products of fire-fountaining of exotic or picritic Disko lavas; they would be perhaps the only non-impact occurrence of Ni spinel. Since their discovery, better criteria for recognition of spherules ejected from large impacts have been established, and greater variations in meteorite chemistry as potential projectiles have been described. New mineralogical and petrographic textural data for the Nuussuaq spherules suggest they should be reinterpreted as impact ejecta; the highly oxidized Ni-spinel is a very characteristic signature of meteorite impact ejecta. Delicate preservation features rule out substantial sedimentary reworking, and spherule bed thicknesses imply a large source crater. Stratigraphically, the spherule beds are poorly constrained, but nannofossils and magnetostratigraphy place them close to onset of the West Greenland flood lavas (ca. 61–62 Ma). They share many characteristics with massive native iron localities in dykes and lavas up to >100 km away on Disko (Qeqertarsuaq) Island, but their precise relationship remains to be established.

Keywords: spherule, ferrosilicate glass, Ni spinel, impact, ejecta, Disko iron, Paleocene, Nuussuaq, nickel.

INTRODUCTION

The island of Qeqertarsuaq (Disko) and the adjacent peninsula Nuussuaq (Fig. 1) in Western Greenland hosts a rare terrestrial occurrence of native iron and iron alloys together with iron-nickel sulphides. First it was thought to be meteoritic iron impacted into basalt (Nordenskjöld, 1872), but others have argued for a terrestrial origin due to the interaction of basalt with reducing carbonaceous sediments (e.g., Steenstrup, 1882; Pauly, 1969; Bird et al., 1981). More recently, a wider range of origins have been proposed. A volcanic contamination model has been advocated by Pedersen (1975a, 1975b, 1978, 1979, 1981, 1985a, 1985b) and Klöck et al. (1986) although alternatives include formation by reduction of iron sulphides, or even as material from Earth's mantle or core (Bird and Weathers, 1977; Bird et al., 1981). In addition to this occurrence of iron, associated with Paleogene basalts, a number of glass-metal spherule beds occur in Paleogene sediments that outcrop stratigraphically just below the basalt lavas on the Nuussuaq peninsula (e.g., Pedersen, 1978; Robin et al., 1996; Pedersen et al., 1993).

It is clear from all the mapping carried out that there are Fe-enriched lavas as layers within contaminated units, e.g., the Asuk Member (Pedersen, 1985a) or the Kuganguaq Member (Pedersen, 1985b), as well as localized intrusive centers, e.g., the Hannekammen Complex (Ulff-Møller, 1990) and dike-like features that cut the highest stratigraphic units (e.g., Pedersen et al., 1993, 2002, 2003). From these different occurrences it would appear that there must have been several different events at different times contributing to the iron occurrences in the Qeqertarsuaq area. Thus, materials from Disko show intimate relationships between metals and silicates, which appear to be the products of diverse emplacement processes and complex siderophile element fractionation.

The glass-metal spherule beds occur over a 3 m thick interval, within a sedimentary sequence, ~10 m beneath the main lavas (e.g., Robin et al., 1996). This contribution focuses on the Nuussuaq spherule beds from Agatdal (70°30′N 53°30′W, Fig. 1), and presents new evidence to suggest they are probably meteoritic impact ejecta from an unknown crater. While there are several native iron localities in the vicinity, their stratigraphic relations suggest that they may not be directly related to the spherules, because many of the occurrences are in cross-cutting bodies emplaced in the Maligât Formation and are therefore too young. It is thus unlikely that conventional volcanic processes relate the spherules and all of the iron occurrences, and it is emphasized here that the spherules are likely to be something different again. However, since there may be implications for the origin of Disko iron, we also provide a brief review of that classic locality, and further work is in progress.

DISKO IRON

Large lumps of native nickeliferous iron metal were discovered in 1870 in the south of Disko Island by Nordenskjöld, who first described the locality near Uivfaq as meteoritic iron, which had impacted into basaltic lavas (Nordenskjöld, 1872). Metal and metal sulphides occur as disseminated microscopic particles, in all sizes up to massive boulders. Subsequently, virtually all published papers have reinterpreted the origin of the iron as terrestrial, produced by various chemical reactions between basaltic magmas and carbon-bearing sedimentary rocks (e.g., Pedersen, 1978; Klöck et al., 1986). One of the largest surviving blocks

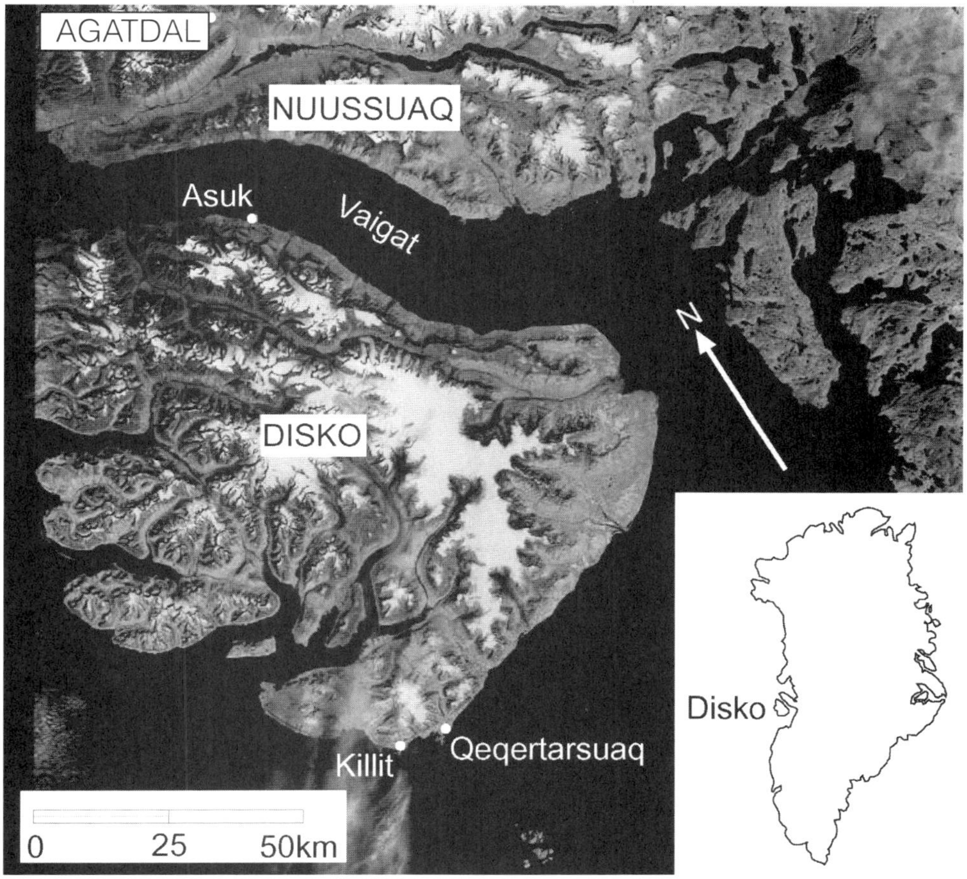

Figure 1. Location map (LANDSAT image) of the Nuussuaq Peninsula indicating the position of the spherule locality at Agatdal (Agatdalen). For detailed maps and sections, refer to Robin et al. (1996).

(~23 tons) remains intact outside of the Natural History Museum in Stockholm. A drill core through this block, carefully preserved in evacuated glass tubes, shows prominent coarsely crystalline bladed iron and cohenite crystals (~0.5–1.0 cm), with small dimpled pits, minor Fe-Ni sulphides and schreibersite, slight porosity (~5%), and average composition ~91%–94% Fe, 2%–3% Ni, 3%–5% C (Löfquist and Benedicks, 1941). Unfortunately, many other large Uivfaq iron blocks, which have been stored indoors in other museums, have suffered from advanced corrosion over the past ~100 yr, thought to be catalyzed by chlorine derived from seawater. Subsequent studies have discovered several other native iron localities (e.g., Pedersen, 1975a; Pedersen et al., 1993) with some described in detail (e.g., Ulff-Møller, 1990).

Papers up to the mid-1970s are summarized by Fundal (1975), who observed that north-trending native iron-bearing dykes cut the basal Maligât Formation lava succession and in places also cut the underlying basement gneiss, e.g., at Killit west of Qeqertarsuaq (Fig. 1). In accord with previous work, Fundal also summarily dismissed a meteoritic origin, but was surprisingly unconvinced by the widely accepted volcanic-xenolith assimilation theory, concluding that "a satisfactory explanation is still missing." He also noted that in south and central Disko, the Tertiary lava succession often directly overlies altered basement gneiss, generally with a boulder-rich landform conglomerate, passing up into massive basal picritic hyaloclastite breccias.

The flood basalts are ~5 km thick and divided into two main formations. The lower, the Vaigat Formation, comprises picritic flows with extraordinary, steeply dipping mega-scale, cross-bedded sets of basal glassy picrites and breccias (~0.5 km thick), that extend along the coastline for several tens of kilometers (Pedersen et al., 1998). The upper unit, the Maligât Formation has step-weathered plagioclase-rich flows giving classic flood lava topography several kilometers thick.

Large amounts of data on the lava sequence have provided support to the volcanic theory for assimilation, mainly through the discovery of more widespread contaminated, iron-rich, lava horizons, and "volcanic centers" at further localities across central western and northern Disko, including graphite-bearing andesitic tuffs (Pedersen, 1985a; Pedersen et al., 1989). Goodrich and Bird (1985) identified two types of metallic iron and produced a novel model involving immiscible high-C liquids and exothermic reactions to explain the iron-carbon phase relations. Also in the mid-1980s, Klöck et al. (1986) made the significant, though subsequently unappreciated, discovery that the detailed minor element chemistry of the native iron particles varies systematically as a function of their size. In particular, they showed that the siderophile elements Au, Ir, Co, and Ni increase by orders of magnitude from small microscopic metallic droplets toward meteoritic levels in the large meter-scale lumps. Klöck et al. (1986) concluded, on the basis of siderophile elements and element ratios and in particular on the abundance of copper, that the metals could not have either a meteoritic or mantle origin. Instead, they fitted their data to a special cumulate extraction model of selective partitioning between immiscible metallic sulphides and igneous silicate liquids. They involved the elements C, S, and P to lower the liquidus temperatures to ~1200 °C in order to have these two liquids coexisting. Their model, however, does not explain two very important points. First, how do large volumes of C get into magmas that have come through underlying Precambrian basement gneisses? Second, how are meter-scale lumps (up to ~23 tons) of coarsely crystalline high-density iron transported in relatively narrow dykes by basaltic melt? We have replotted the siderophile data of Klöck et al. (1986) for Ni, Co, Ir, and Au in iron-rich segregations from Disko Island (Fig. 2). The comparison with a range of iron meteorites (average iron, Libyan iron, octahedrite) and chondrite meteorite ("C1") is particularly interesting, and there is no clear similarity or trend between the Disko iron and average crust for these siderophile elements. Recent studies of carbon isotopes extracted from iron carbides and native iron-bearing samples from Disko (Smith et al., 2000) claim that the carbon is derived from the crust, yet the ^{13}C values for Disko ($\delta^{13}C < -25‰$) are not matched by the favored local crustal carbon reservoirs ($\delta^{13}C \sim -23‰$). Clearly the enigmatic iron at Disko, in our view, has still not been satisfactorily explained. In particular, perhaps insufficient comparison has been made with the more extensive native iron occurrences associated with intrusive rocks of the Siberian Traps, which have been well-studied in the Russian literature (Oleynikov et al., 1985), that contain a similar range of mineralogies and chemistries but include examples with additional C-phases of silicon carbide and pseudomorphs after diamond (cliftonite). The Siberian native iron occurrences retain no clear meteorite signature, but have been interpreted as impact-related breccias by Treiman et al. (2002).

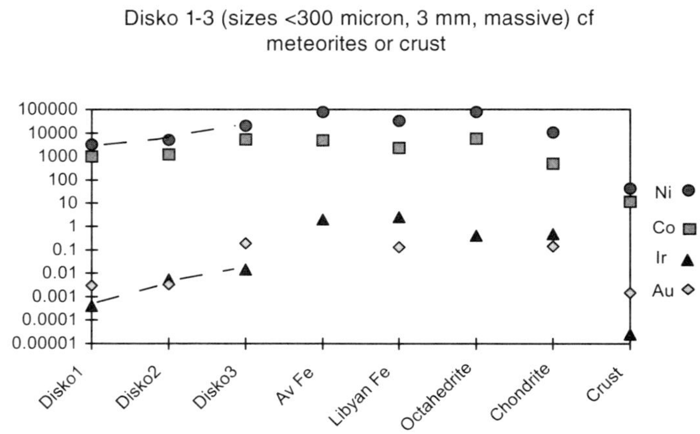

Figure 2. Siderophile trace element concentrations in metallic iron from Disko show orders of magnitude increasing contents of Ni (circles), Co (squares), Ir (triangles), and Au (diamonds) as a function of particle size, after Klöck and Palme (1988). Note similarity between largest size metallic iron particles from Disko and a range of meteorites, in contrast to very low abundances in average crust.

NUUSSUAQ SPHERULE BEDS

The spherule beds in sediments immediately underlying the volcanic succession on the Nuussuaq (formerly Nugssuaq) peninsula were first described by Pedersen (1978) at the type locality in Turritellakloft in Agatdal (Fig. 1). This was followed by detailed descriptions of the glassy spherules by Magde (1991) and Margolis et al. (1991). The significance of the spherule beds as possible ejecta from a meteorite impact crater was immediately recognized by Margolis et al. (1991), but not all members of the research group were in agreement, as incorporated basaltic materials show strong similarities with the nearby Disko native iron-bearing volcanic rocks. A decisive interpretation of the spherule beds as volcanic in origin was later published by Robin et al. (1996). In this model it was suggested that hyaloclastites and spherules were contemporaneous and formed by one or more subaerial fire-fountaining events of Vaigat Formation magmas to produce quenched glass containing both native metals and highly oxidized nickel-rich spinels. On the basis of our recent findings, presented in detail below, a simple volcanic origin now seems highly unlikely. The possibility that the Nuussuaq spherules are ejecta from a meteorite impact was first suggested by Margolis et al. (1991) due to their sphericity, the coincidence of Ni-spinel and associated high Ir contents.

Spherule Bed Locality: Agatdalen

Our material was derived from the same samples as previously collected and described in detail by Robin et al. (1996). In summary, four neighboring outcrop sections can be compared using the previously published stratigraphic logs, but the precise correlation between the sections remains uncertain. The rocks comprise glass spherule sandstone, with channels and additional sedimentary features including ripples, lithoclasts, and graded bedding, indicative of a shallow subaqueous environment. The lithoclasts include carbonate rocks, gneisses and basalts. At each section there are between 1 and 4 spherule layers, ranging in thickness from a few millimeters up to more than 10 cm. There are two main types of spherule layer, (*a*) in sandstone composed mostly of relatively close-packed glass spherules and shards derived from broken spherules, and (*b*) as minor components in sandstones with detrital felsic minerals and glass shards, and relatively dispersed spherules. Locally much of the cement is fine-grained carbonate (calcite), smectite, or palagonite. The spherules preserve copious quantities of unaltered glass and so must preserve their original chemistry, independent of any of the diagenetic and supergene alteration processes for which there is much evidence.

Spherule Bed Stratigraphy

On the Nuussuaq peninsula, Paleogene hyaloclastic volcanics overlie marine and nonmarine Paleogene sediments that are dominantly carbonaceous shales and sandstone units (e.g., Henderson et al., 1976; Clarke and Pedersen, 1976; Dam, 2002). These sediments have been described as being partly older and partly contemporaneous to the volcanics (see Pedersen, 1985a; Piasecki et al., 1992, and references therein). The volcanics are divided into two main formations (Fig. 3) but only the lowest, the Vaigat Formation, has relevance here (see Pedersen, 1985a; Larsen and Pedersen, 2000, and references therein). The Vaigat Formation is divided into three members, from lowest to highest, the Anaanaa, Naujánguit, and Ordlingassoq, which represent three different magmatic cycles. Additionally, there are also several less extensive members that are of only local importance, the lowest one of which is the Asuk Member (Pedersen, 1985a). Nuussuaq is located at the transition between marine and nonmarine sediments. The marine sediments belong to the Kangilia Formation and the unconformably overlying Agatdal Formation (e.g., Dam et al., 1998; Dam, 2002).

In Agatdal the described glass and metal spherules come from several layers within the shallow marine Abraham Member of the Agatdal Formation (Fig. 3). The layers occur some 10–15 m stratigraphically below the lowest basaltic hyaloclasites (e.g., Rosenkrantz, 1970; Robin et al., 1996). On the north coast of Nuussuaq, the Naujánguit Member is intercalated with Kangilia Formation mudstones (Piasecki et al., 1992), and the inference is that the stratigraphically higher Agatdal Formation must also be chronologically equivalent to some of the volcanics to the west. Thus it is postulated that a direct correlation may be made between the Abraham Member on Nuussuaq and the Asuk Member on Disko to the west (e.g., Pedersen, 1978, 1985a; Piasecki et al., 1992). In the Agatdal section, however, the first volcanic rocks occur only above the Agatdal Formation (e.g., Piasecki et al., 1992; Robin et al., 1996; Dam, 2002), and no equivalent volcanic rocks can be demonstrated.

The precise absolute age of the sedimentary units has proven difficult to determine. Direct $^{40}Ar/^{39}Ar$ dating on the spherules in the sediments has provided an age of 60.8 ± 1.4 Ma (Robin et al., 1992). More recently, $^{40}Ar/^{39}Ar$ ages from several stratigraphic levels of the overlying basalts have been obtained by Storey et al. (1998) where the lowest dated sample from the Naujánguit Member in the Vaigat Formation gave an age of 60.4 ± 0.5 Ma. It is not clear if this age was obtained above or below the position of the Asuk Member, although Riisager and Abrahamsen (1999) indicate it was taken from the Nuuk Killeq Member. Dates obtained from the upper part of the Formation were 60.3 ± 1.0 and 60.7 ± 0.5 Ma (Fig. 3). All of these dates are unresolvable within analytical error and the units can only be controlled stratigraphically. The oldest lavas, the Anaanaa, and lower parts of the Naujánguit members are all normally magnetized and considered to belong to magnetochron C27n while the upper parts are reversely magnetized and belong to C26r (Riisager and Abrahamsen, 1999;

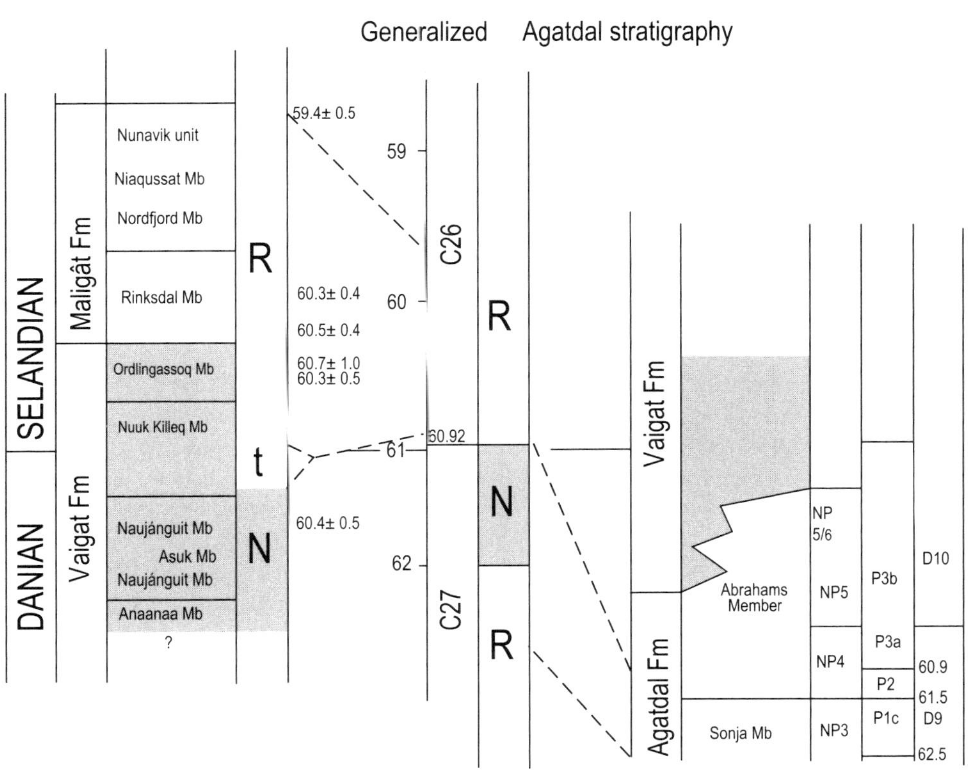

Figure 3. Stratigraphic correlations between the spherule bed localities in Agatdalen, with generalized stratigraphy for the Nuussuaq peninsula, see references in text. In Agatdalen, the glass spherules occur in several different beds: in the lower to middle part of the Agatdal Formation, in the shallow marine Abraham Member, some 10–15 m stratigraphically below the lowest basalt (Robin et al., 1996). See Robin et al. (1996) and Piasecki et al. (1992) for further details.

Riisager et al., 2003). The accepted date for the end of C27n is 60.92 Ma (Cande and Kent, 1995), and this transition is observed in the measured basaltic sections (Riisager et al., 2003).

Microfossil correlations with magnetostratigraphy (e.g., Ogg and Bardot, 2001) can also be utilized (e.g., Piasecki et al., 1992). Within the lower part of the Vaigat Formation a basalt conglomerate with a limestone matrix containing marine nannofossils was interpreted to correlate with Danian NP3 zone (Jürgensen and Mikkelsen, 1974). The Sonja Member, which underlies the Abraham Member has also been dated using planktonic foraminifera to early Danian P1b-c (Robin et al., 1996). More recent palynostratigraphy of the basalts using dinoflagellates and calcareous nanoplankton (Piasecki et al., 1992) suggests that the lowest parts of the Vaigat Formation equate to the upper parts of NP4 and that the higher parts still, including the Abraham Member, correlate with NP5-6. The unit overlying the Vaigat Formation, the Maligât Formation, is clearly younger, correlating with NP7-8 (Piasecki et al., 1992). More recently, a spherule bed has been discovered in the Asuk member that is contended to prove the correlation between the Abraham Member and the Asuk Formation (Piasecki et al., 1992; A.K. Pedersen, 2003, personal commun.).

PETROGRAPHIC FEATURES

Detailed descriptions have already been published, including maps and vertical sections, by Robin et al. (1996) to which the reader is referred, and the same specimen numbers apply. The regional setting of the sediments and lavas has been illustrated by Pedersen et al. (1993). Therefore, only a general description will be given here. Attention will be drawn to specific features of interest in the interpretation of the origins of the spherules and glass in particular.

Glass Spherules

Glass spherules show a high proportion of perfectly round spheres throughout each section, in contrast to volcanic spherules (quenched magma as Pele's hair and tears) where irregularity, flattening, and strings are predominant (e.g., Garcia et al., 1998; F. Stoppa, 2003, personal commun.). Uncommon flattened or oblate Nuussuaq spherules contain smaller internal flattened spherules with identical geometry, including smaller flattened carbonate and metal spherules (Fig. 4). We agree with Robin et al. (1996) that the glass and metallic spherules must have once been immiscible liquids to preserve these spherical relationships. Optically, the glass spherules are often completely fresh, unaltered isotropic glass, with irregular brown shades and internal swirls reflecting incomplete mixing and variable iron content (Fig. 5). We interpret some of these bands as schlieren, which preserve counter-convection in some spherules. Many glass spherules are partly altered to palagonite clay-like minerals, and some spherules are completely palagonatized (Fig. 5). Other spherules are partly filled by carbonate (calcite), with several distinctive textures (Figs. 4 and 6).

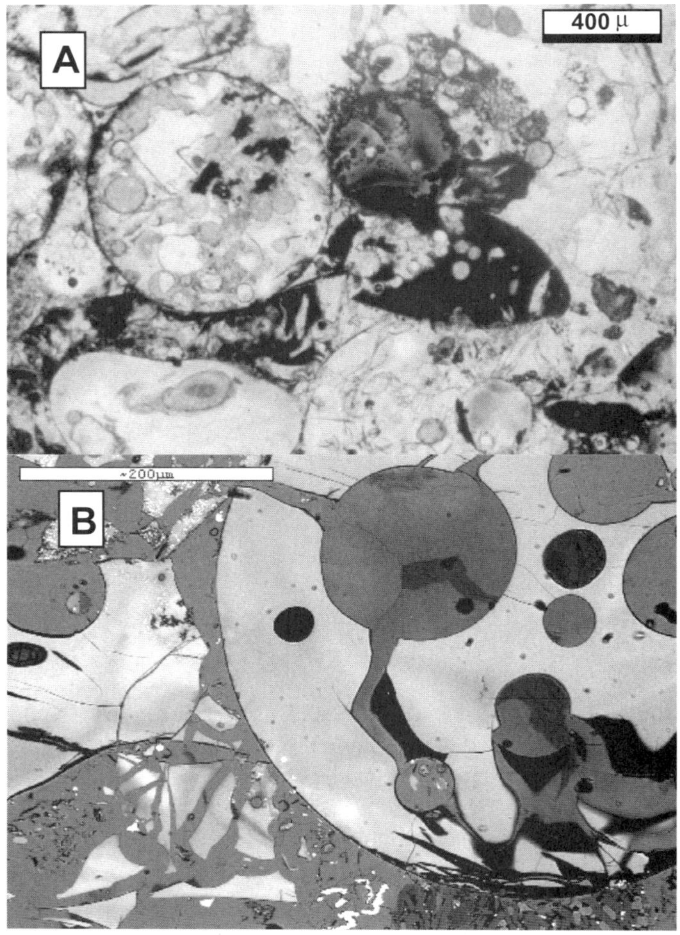

Figure 4. (A) Photomicrograph (PPL) general texture of sample KS475 with various spherules (scale bar 400 µm); large, pale, left-hand spherule has multiple small internal calcite spheres (filled vesicles?), dark material is mostly Fe-Ni-rich ferrosilicate glass forming oblate spherule and glass fragments in matrix which also contains basaltic fragments. (B) Scanning electron microscope image detail of part of a ferrosilicate spherule (scale bar 200 µm), with multiple internal calcite-filled vesicles (?), metal-rich spheres and partly fractured exterior; the delicate jig-saw-like remnants of a shattered glass sphere (lower left) in carbonate matrix (pale gray) are clearly derived in situ, suggesting minimal disturbance by sedimentary processes.

Carbonate Spherules

Carbonate spherules occur in large, but highly variable numbers within the fresh glass spherules. Many of these carbonate spherules are much smaller in diameter than the thickness of the petrographic thin sections (35 µm), and they appear to be completely isolated within the silicate glass, which is unfractured. The carbonate spherules (Fig. 7) may be formed of either (*a*) single crystals of calcite (perfectly round), or (*b*) radiating aggregates of elongated feathery calcite crystals. Some of these carbonate spherules satisfy criteria for immiscible carbonate melts and may have been trapped as fine droplets within the silicate liquid, as reported

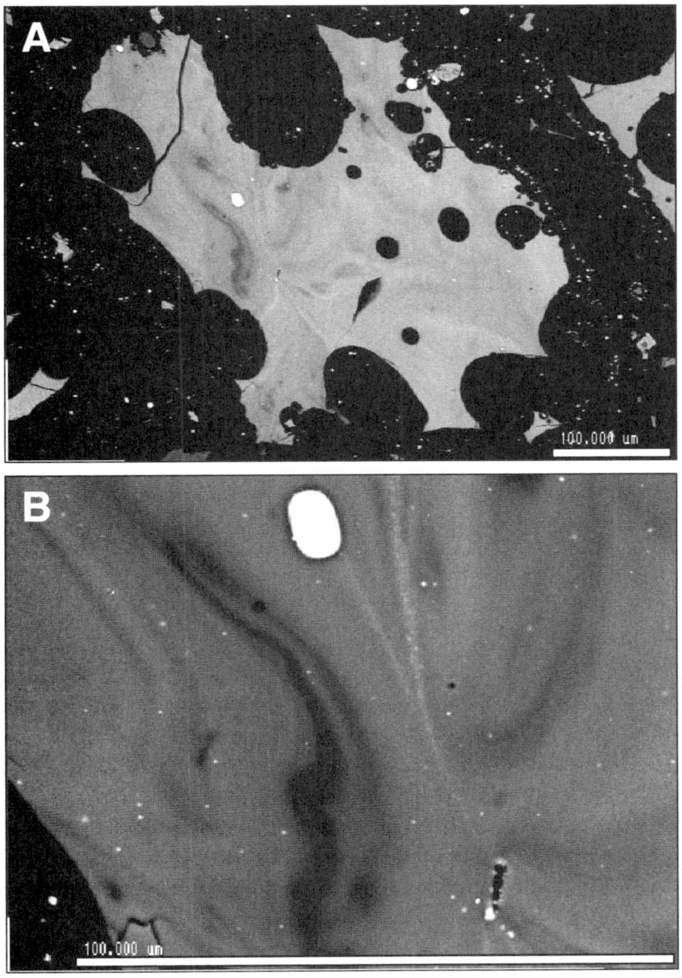

Figure 5. Scanning electron microscope images (backscatter) of spherules from sample KS475. (A) Heterogeneous ferrosilicate glass spherule (light gray) with calcite-filled and palagonatized (dark) vesicles or other spherules (scale bar 100 μm). Note the concentration of secondary sulphides (as bright irregular spots) around the former margins of the spherule. These sulphides are very different in composition to those contained within the fresh glass. (B) Detail showing Ni-rich sulphide droplet from A; note flow-banded glass (brighter colors here more Fe, Ni-rich) containing many tiny spherules of Ni-Fe-Cu-S (bright dots which are circular when magnified). Scale bar 100 μm.

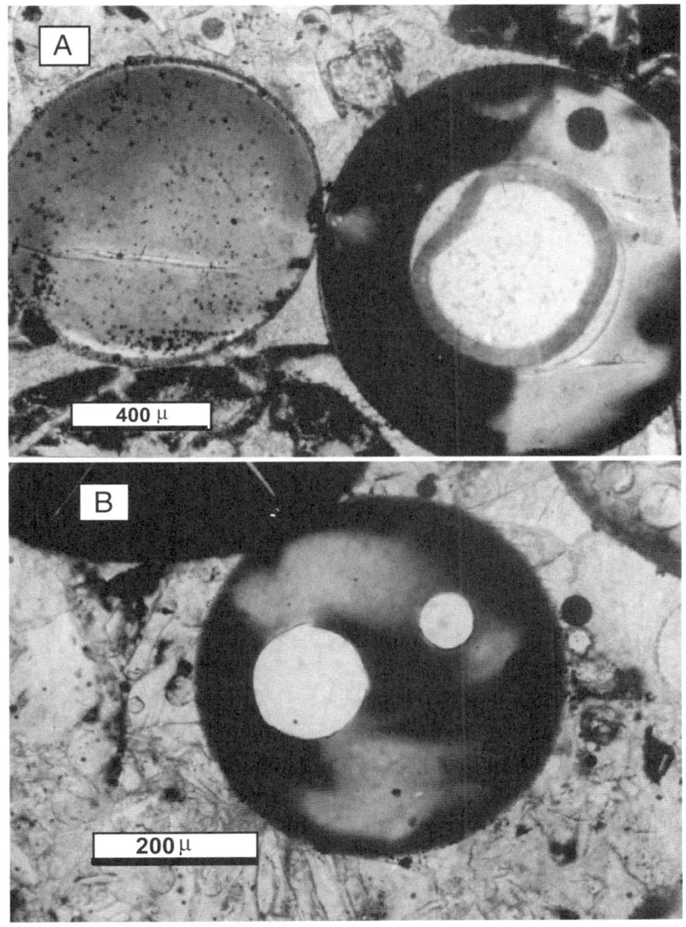

Figure 6. Photomicrographs in PPL. (A) Two contrasting ferrosilicate spherules. Left-hand image is homogeneous with microcrystals of Ni-spinel (scale bar 400 μm). Right-hand image is heterogeneous with central calcite sphere, strongly zoned peripheral ferrosilicate glass (dark is Fe-Ni-rich) and isolated metal spheres (Fe, Ni, S; black). (B) Heterogeneous ferrosilicate glass spherule with internal calcite spheres (white) and metal spheres (Fe, Ni, S) of many sizes (part of large metal sphere protrudes into top left of view). Scale bar 200 μm.

from impact suevites in craters within crustal carbonate-containing targets such as Ries (Graup, 1999), Chicxulub (Jones et al., 2000), and Haughton (Osinski and Spray, 2001). Crystalline carbonate, both sparry and microcrystalline varieties, also forms widespread cement to the glass spherules and shards, and gives the impression of closely postdating deposition of the spherules and shards.

Metal-Rich Microspheres

Microspheres of remarkably perfect circularity occur dispersed throughout the silicate spherules, and additionally can also be found in between spherules, broken glass shards and even through the matrix and cement (e.g., Figs. 4 and 6). These occur in varying sizes from ~100 μm down to around 1 μm (close to the limit of electron-microscope visibility). In reflected light many of these are sulphides, some are native metals and others are metal oxides occasionally with very fine scale exsolution textures.

Anomalous Plagioclase

Feldspar with a "tartan" texture is included within a few spheres in sample A338, and is optically isotropic or has anomalously low birefringence, similar to shocked plagioclase maskelynite. In some crystals, a fine-scale feathery texture extends from the edge of the crystal into the surrounding silicate glass (Fig. 8;

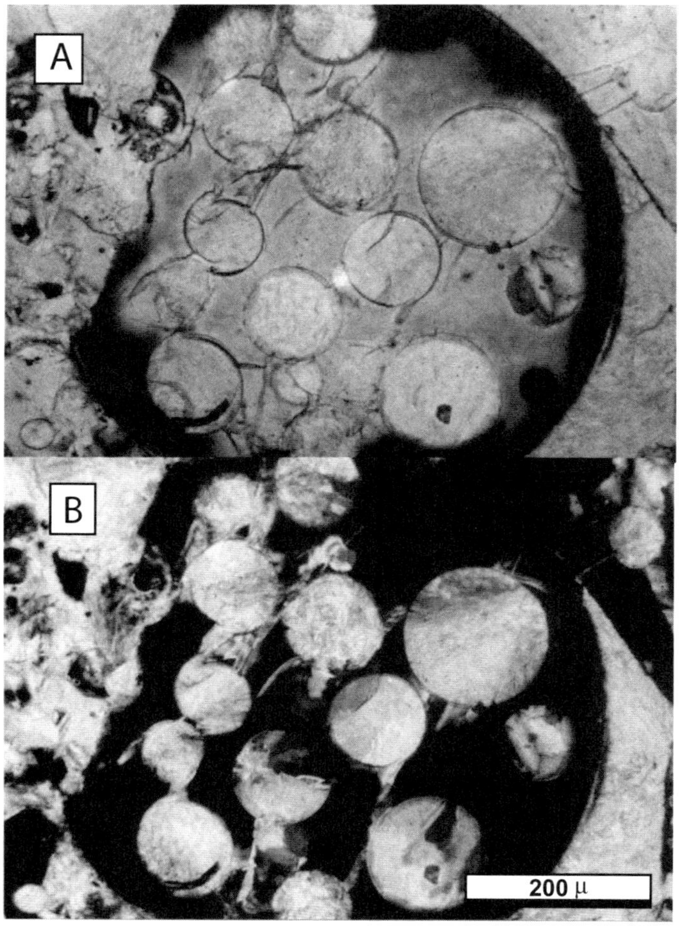

Figure 7. Photomicrographs sample KS475. (A) PPL multiple internal calcite filled spheres and metal rich spheres (black) in ferrosilicate spherule with Fe-Ni-rich margin (dark rim). (B) XPL shows isotropic nature of glass and some calcite spheres; those with feathery textures could represent immiscible droplets, or infilled vesicles. Scale bar 200 μm.

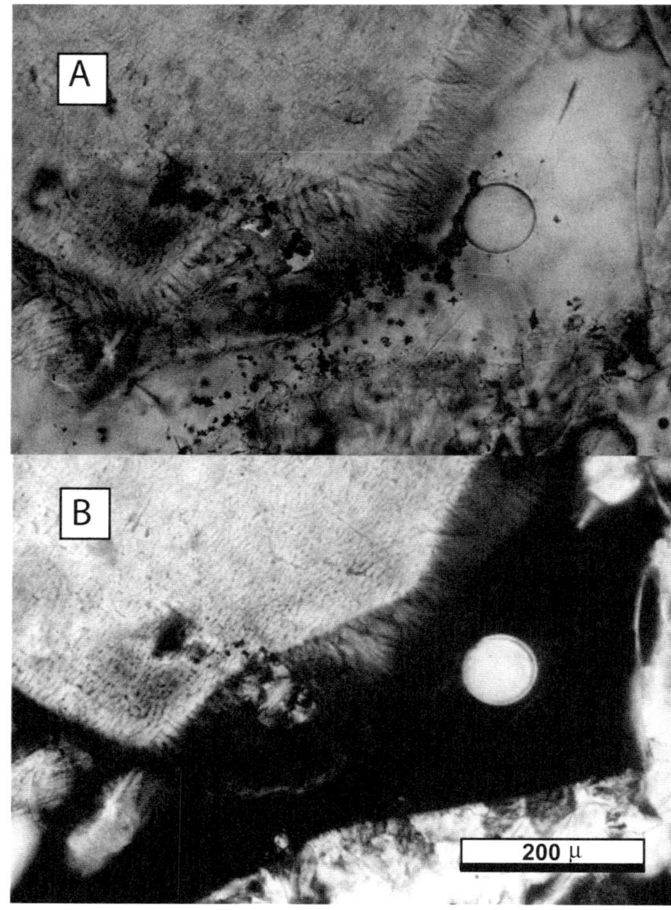

Figure 8. Photomicrographs in PPL (A) and XPL (B) show details at feathery contact between relic anomalous plagioclase and host ferrosilicate glass of spherule. Plagioclase texture (top left) contains tartan-like pattern of interconnected partial melt; surrounding glass is enriched in plagioclase component and contains small Ni-spinels and perfect calcite spheres. Plagioclase is thought to have been shocked, and in other grains is isotropic. Scale bar 200 μm.

bytownite An_{73-71}). Substantial enrichment of plagioclase components increases in the silicate glass (Ca/alkalies or "An" from <60 to >68, Al_2O_3 from 14 to >20%) toward melted plagioclase crystals in A258; this is the reverse of normal plagioclase growth from silicate melts, which show depletion in plagioclase components adjacent to phenocrysts. The identification of reverse zonation in volcanic-derived plagioclase crystals (as opposed to melt), might signify re-equilibration with host melt, as may be associated with magma mixing. The "reverse" zonation of silicate glass we observe in the spherules is very different from volcanic textures and implies very fast time scales (relative to volcanic processes) to inhibit diffusion within the melt. Conversely, zoned glasses around crystals imply time scales relatively slow compared with typical impact shock processes. However, comparable relic plagioclase-glass relationships are common in shock-melted breccias from the Moon (similarly Anorthite-rich; James et al., 2003), and impact glasses once considered to be homogeneous have been recognized to preserve diverse compositional signals diagnostic of protolith targets in terrestrial impact breccias (Osinski, 2003).

Partial Melt Textures

Melt textures are seen in some quartz grains, with irregular curvilinear fractures filled with SiO_2 (glass? Fig. 9A). Fine scale partial melt textures also form a delicate latticework across some of the anomalous plagioclase (Fig. 9B).

Ni-Spinel

This distinctive mineral occurs as clouds and trails of quench-textured euhedral crystals in many of the ferrosilicate glass spherules. At high magnification in optical thin sections, it is clear that many of these spinels are twinned with a three-

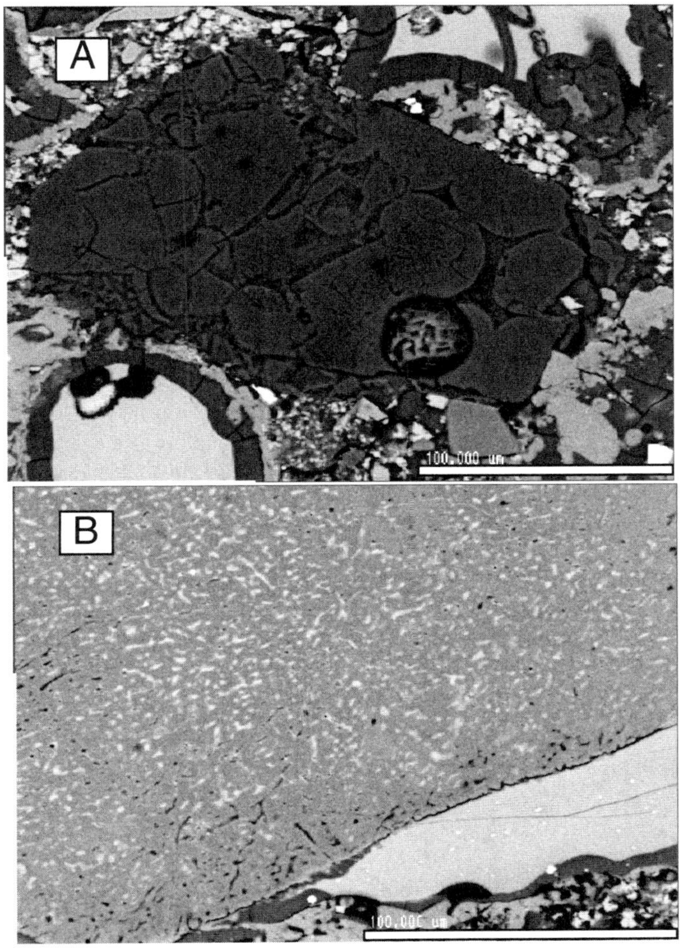

Figure 9. Unusual melt textures in scanning electron microscope images (backscatter). (A) Quartz grain with irregular melt-filled curvilinear fractures. Scale bar 100 µm. (B) detail of regular fine-scale tartan-like network of Fe-rich silicate melt in relic plagioclase within ferrosilicate spherule, (scale bar 100 µm).

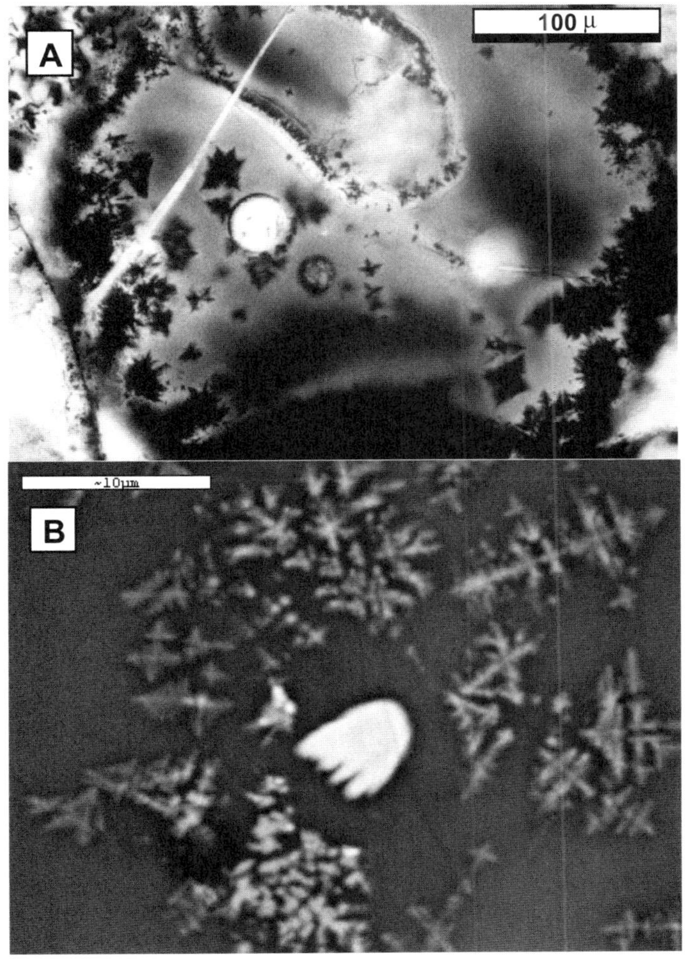

Figure 10. (A) Photomicrograph (PPL, scale bar 100 µm) of clouds and trails of quench textured euhedral Ni-spinels (7–10 wt% NiO) in ferrosilicate glass spherule (shown arrowed in Fig. 4A). Small bright sphere and linear fracture are calcite-filled. (B) Detail (scale bar 10 µm) of Ni-spinels with quench texture ("Christmas trees") in scanning electron microscope. The core region contains nearly pure Ni (see Table 3, analysis 9) grain (bright).

dimensional "Christmas tree" appearance, and many radiate from a central nucleus, or core (Fig. 10). Nickeliferous quench textured spinel is not known in igneous rocks, whose ultimate mantle origin is instead usually characterized by Cr-spinel (Dressler and Reimold, 2001). Ni-spinel is often found in cosmic spherules, meteorites, meteorite fusion crusts, and various cosmic debris heated and oxidized in Earth's atmosphere (see review in Robin et al., 1992). Ni-spinel is famously associated with Ir and other impact markers in sediments at the K-T boundary (e.g., Kyte and Smit, 1986). The only other locality where Ni-spinel (Nichromite) has been reported is from the very Ni-rich Bon Accord "palaeometeorite" in the Barberton basin (De Waal, 1977).

GLASS AND NI-SPINEL CHEMISTRY

New electron microprobe analyses were conducted on all representative materials in the spherule samples, and compared with data previously reported for the same samples by Robin et al. (1996). A few additional sections were prepared, and we were able to reproduce the earlier findings. In this work, we concentrated primarily on the silicate glasses and Ni-spinels, and on their mutual relationships especially in, for example, local regions of high compositional gradients. These items provide the data critical to our main reinterpretation of the spherules as impact-derived and non-volcanic.

Glass Spherules—Major Element Chemistry

On a microscopic scale, the silicate spherules are strikingly heterogeneous, both optically and chemically (Table 1). Such heterogeneity is often characteristic of terrestrial impact melts

TABLE 1. COMPOSITION OF NUUSSUAQ FERROSILICATE GLASS SPHERULES FROM SAMPLE K475 (1–10) COMPARED TO BULK ROCK COMPOSITIONS (11–13)

	1	2	3	4	5	6	7	8	9	10	11 Vaigat*	12 Asuk†	13 Howardite#
SiO_2	51.53	54.25	50.12	52.67	45.82	51.79	48.93	49.40	40.59	47.62	51.58	52.60	43.4–50.4
TiO_2	1.17	1.21	1.07	1.29	1.10	1.06	1.19	1.17	1.04	1.08	1.65	1.13	0.52–0.83
Al_2O_3	13.59	13.20	13.73	12.60	12.15	14.08	14.10	12.65	11.24	12.82	16.58	14.80	9.6–14
Cr_2O_3	0.21	0.15	0.20	0.22	0.18	0.10	0.23	0.19	0.15	0.18	0.05	n.a.	0.26–0.56
Fe_2O_3	16.19	14.31	18.41	16.12	23.80	17.55	20.51	19.45	33.22	19.53	10.33	10.01	17.8–32.1
MnO	0.11	0.14	0.11	0.10	0.19	0.17	0.17	0.17	0.11	0.15	0.15	0.18	~0.4–0.8
MgO	5.46	5.90	5.05	5.99	5.87	4.66	4.75	5.18	3.54	5.67	7.78	7.48	5.4–13.3
NiO	0.25	0.34	0.49	0.59	0.99	1.05	1.25	1.35	1.96	2.90	0.02	2.90	~108
CuO	0.23	0.47	0.48	0.38	0.41	0.61	0.71	0.63	0.50	1.42	0.02	n.a.	n.a.
CaO	8.63	7.78	8.11	7.70	8.24	7.91	7.90	7.98	6.09	8.10	10.05	9.25	5–12
Na_2O	1.95	2.05	2.04	1.89	1.53	1.81	1.46	1.61	0.67	1.87	1.96	2.05	0.07–0.28
K_2O	0.67	0.87	0.68	0.74	0.48	0.78	0.67	0.64	0.49	0.62	0.68	0.37	0.06–1.26
SO_3	0.23	n.d.	n.d.	n.d.	0.14	n.d.	n.d.	n.d.	n.d.	n.d.	n.a.	n.d.	n.a.
Total	100.22	100.67	100.49	100.29	100.90	101.57	101.87	100.42	99.60	101.96	100.85	100.77	

Note: Analyses 1–10 arranged in increasing Ni content. n.d.—not detected, n.a.—not available.
*Average (n = 8) of glass inclusions in magmatic olivine from the Vaigat Formation (data from Larsen and Pedersen, 2000).
†Average (n= 4) of contaminated Asuk Member lavas (data from Lightfoot et al., 1997).
#Brown heterogeneous iron-rich glass spherules (~1 mm, nonvesicular), in brecciated meteorite (howardite NWA 1664) with chemistry and certain optical properties similar to Nuussuaq glasses, data after Kurat et al. (2003) and Lorenz et al. (2001).

and glasses (e.g., Dressler and Reimold, 2001, and references therein). In fact, the glass spherules are notably iron-rich, even the palest colored glass containing >14 wt% Fe_2O_3. We therefore refer to it as a ferrosilicate glass. The oxidation state of the iron has not been measured directly, but is inferred to be largely ferric, due to the reddish brown color, and also the consistent analytical totals from electron microprobe when cast in this way. This is consistent with the stability of abundant Ni-spinel in many of the spherules, which has a high iron oxidation state ($Fe^{3+}/Fe^{3+} + Fe^{2+} > 80$ mol %) involving formation in an oxygen-rich environment (Robin et al., 1992; Siret and Robin, 2003). We have therefore assigned total iron as being ferric in Table 1, but emphasize this has not been measured directly, and the presence of metals and minor sulphides in the spherules is consistent with evidence for fine-scale heterogeneity which may extend to the oxidations state of iron. The most striking feature is the uniform positive correlation between Fe and Ni (NiO from 0.25 to 1.96 wt%), which is quite contrary to normal igneous behavior. In order to compare with volcanic rocks, total iron is recast as FeO and plotted against NiO in Figure 11. In particular, we are interested to compare the spherule glass chemistry with that of the Asuk Member lavas from Disko Island, which are considered parental or equivalent to the spherules (Pedersen, 1978; Piasecki et al., 1992; A.K. Pedersen, 2003, personal commun.). In terrestrial igneous rocks, nickel correlates with magnesium, due to its strong compatible element behavior partitioning into crystals rather than liquid and its common association with abundance of olivine, as illustrated by the Vaigat lavas (e.g., Lightfoot et al., 1997; Larsen and Pedersen, 2000; Larsen et al., 2003). In stark contrast, magnesium in the ferrosilicate glass spherules does not correlate either with iron or with nickel and shows relatively minor variation among all the glass analyses (3.54–5.99 wt% MgO; Table 1; Fig. 12). Similarly, minor titanium is also remarkably constant (1.04–1.29 wt% TiO_2). Silica (40.6–54.3 wt% SiO_2) varies from ultrabasic to basic in terms of igneous rock compositions. The other major elements (Al_2O_3, CaO, Na_2O, K_2O) all show rather small variations. It is also clear that the spherule glasses are anomalously enriched in copper (average ~0.5 wt% Cu), in common, for example, with the metal samples from Disko analyzed by Klöck et al. (1986) and Ulff-Møller (1990). Comparing these compositions with the copious analytical data for the adjacent volcanic rocks (Lightfoot et al., 1997; Larsen and Pedersen, 2000), it is clear that the spherules are generally more silicic and have exceptionally high nickel, copper, and iron contents (Table 1). Rather than use pillow glasses for comparison, approximations of the parental magma to the volcanics, preserved as quenched glass inclusions inside olivine have been used. These are viewed as parental liquids that have become more siliceous than the pillow glasses due to precipitation of olivine onto the inclusion wall (Larsen and Pedersen, 2000). The North Atlantic Tertiary igneous province is also regionally high in copper (Saunders et al., 1997; Andersen et al., 2002), but even viewed in this context, we agree with Klöck et al. (1986) that the metallic spherules contain anomalously high copper. This problem of resolving the high copper in the glasses is exacerbated when one compares the average sediments from the Nuussuaq area that only carry <75 ppm Cu (Table 1).

The simplest interpretation of the major element data is that the silicate glass spherules represent mixtures between basalt and

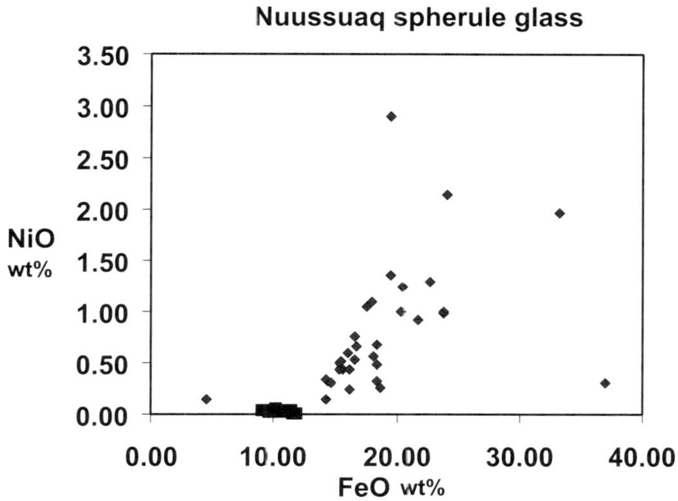

Figure 11. NiO wt% vs. FeO wt% (total iron) for Nuussuaq ferrosilicate glass spherules (diamonds). The glass spherules are likely very rich in ferric iron, and many are heterogeneous. Note strong positive correlation which is the opposite of compatible behavior of Ni in igneous rocks. Data cannot be explained by volcanic compositions from West Greenland (squares, Larsen and Pedersen, 2000), but may represent mixing between basaltic lava and a Fe-Ni metallic source (meteoritic iron).

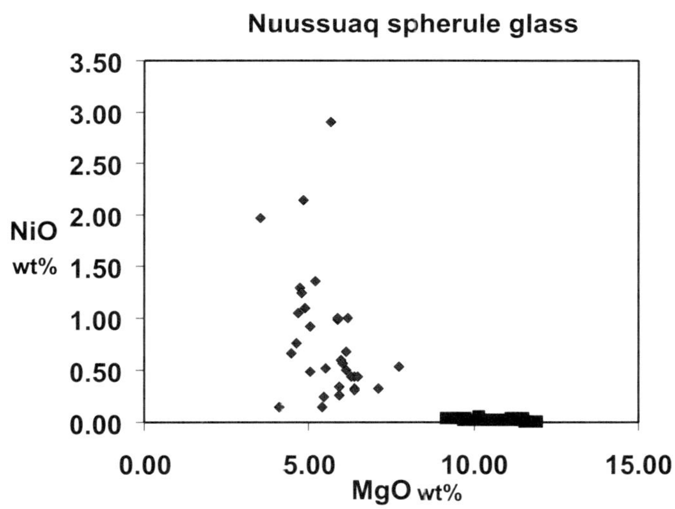

Figure 12. NiO wt% vs. MgO wt% for Nuussuaq ferrosilicate glass spherules. Note strong negative correlation, which is unlike behavior in igneous rocks. Symbols and comments as for Figure 11.

a siderophile element-enriched iron end member (e.g., Pedersen 1978; Klöck et al., 1986). However, there are chemical problems attached to this model. The basalts, including the contaminated versions, have approximately the expected Ni concentrations and the adjacent sediments also have expected abundances (e.g., Lightfoot et al., 1997; Larsen and Pedersen, 2000). The process for the concentration of Ni in the spherules must produce an average Fe/Ni ratio of the glass spherules of 12.5, which gives an end member "iron" source with 92.4% Fe, 7.4% Ni. This is, in fact, rather typical of the metallic iron component commonly found in meteorites, and we now consider the minor elements. Nearly identical, although unexplained, ferrosilicate glass spherules have been described in brecciated howardite meteorite (Kurat et al., 2003), which have similar major element chemistries and physical properties, although they lack vesicles. In particular, the detailed internal heterogeneities within individual spherules are comparable, where large variations in iron content correspond to color intensity (brown), and we follow the interpretation of Kurat et al. (2003) that the distinctive swirling patterns might be flow banding in a heterogeneous melt, or schlieren, outlining counter-flowing convection cells, which were suddenly quenched. Presumably, the very tiny FeNiS droplets in the Nuussuaq glass spherules were immiscible liquids that did not amalgamate before quenching.

Glass Spherules—Minor Element Chemistry

With their initial descriptions of the same Nuussuaq spherule samples used here, Robin et al. (1996) reported important minor element contents, and elemental ratios, from bulk analyses of separated spherules. The principal elemental ratios of interest are presented in Table 2. First, we note the Ni/Cr ratios of the spherules (1.5–13.1) are an order of magnitude higher than seen in any Vaigat lavas (typically Ni/Cr ~0.4–0.5), including picrites (~0.3), but the glasses do overlap ratios for chondritic meteorites (~4.0) and approach values for iron meteorites (>20).

In particular, the spherules contain anomalously high iridium, with 0.5–6.8 ng/g; higher values are associated with magnetic particles, which include the metal-rich phases, especially iron. Much higher values for iridium are reported for larger masses of native iron from Disko, which correlate positively with other siderophile elements (Co, Au, Ni) and with particle size (~1.5–150 ng/g Ir; Klöck et al., 1986) and intermediate values occur in coexisting sulphides (~30–60 ng/g: Klöck and Palme, 1988). In addition to the Disko native iron localities, the Nuussuaq spherule beds also constitute a distinctive Ir-rich deposit.

Ni-Spinel

The quench euhedral crystal forms of Ni-spinel are reminiscent of Ni-spinel from the global K-T boundary layer (Bohor et al., 1986; Kyte and Bostwick, 1995). The comparison with impact-derived spinel is supported by the analyses, which show that the Nuussuaq Ni-spinel has a compositional range quite unlike normal igneous spinel. We present analyses of the Nuussuaq Ni-spinel in Table 3, where calculations of the ferric iron ratio indicate consistently oxidized components (Fe^3/Fe^3 + Fe^2 = 0.68–0.79), in common with impact Ni spinel in general. Nickel spinel from the K-T boundary ejecta layer commonly has

TABLE 2. POTENTIAL SOURCES OF SIDEROPHILE ELEMENT RATIOS IN SPHERULE GLASSES

	Spherules*	Chondrite	Iron	CH metal†	Basalt§	Picrite#	Shale/gneiss#
Fe/Ni	~12.5	~10–15	10–15	14–21	>500	~350	~360
Ni/Co	~15	21	10–22	17–25	n.a.	~4.0	2.4
Ir/Cu	~2.5e–6	~5e–3	~5e–3	2e–3	n.a.	n.a.	n.a.
Ni/Cr	2.8–10	4.04	>20	8.8–65	~0.4	~0.3	0.14

*Spherule data from Robin et al. (1996).
†Heterogeneous metal in chondritic meteorites (Campbell and Humayan, 2004; Campbell et al., 2002).
§Average values for Disko basalts after Larsen and Pedersen (2000).
#Average values for picrite, country rock shale and gneiss after Lightfoot et al. (1997) and Larsen et al. (2003).

TABLE 3. ELECTRON MICROPROBE ANALYSES OF NICKEL-SPINEL (1–6), Cr-SPINEL (7), Fe-Ni OXIDE (8), AND Ni-METAL (9)

	1	2	3	4	5	6	7	8		9
	ni-sp	ni-sp	ni-sp	ni-sp	ni-sp	ni-sp	Cr-sp	Fe-Ni-oxide		Ni-metal
SiO_2	4.67	1.83	6.2	4.8	2.5	0.7	0.3	n.a.	Si	0.14
TiO_2	0.42	0.58	0.3	0.3	0.3	0.1	1.29	0.02	Ti	0.05
Al_2O_3	7.53	7.35	6.3	5.4	5.3	0.2	22.5	0.1	Al	n.d.
V_2O_3	0.55	n.d.	0.1	0.1	0.1	0.1	n.d.	n.a.		
Cr_2O_3	9.8	9.65	0.7	0.8	0.9	0.2	46.4	0.21	Cr	0.06
FeO	19.8	16.3	16.1	15.8	15.3	18.8	7.45	75.72	Fe	1.77
Fe_2O_3*	42.8	49.0	55.1	56.7	59.2	66.2	1.44			
MnO	n.a.	n.a.	0.2	0.2	0.1	0.1	4.55	n.d.	Mn	n.d.
MgO	6.53	5.67	2.7	2.6	3.1	0.4	16.5	0.57	Mg	n.d.
NiO	9.59	10.64	9.1	9.4	10.1	7.7	0.08	16.5	Ni	93.71
CaO	0.55	0.49	1.0	1.2	0.9	0.5	n.d.	n.d.	Ca	0.29
Total	102.24	101.45	97.8	97.3	97.8	95.0	100.51	93.57		96.02
Si	1.2608	0.5604	1.8100	1.4143	0.7389	0.2261	0.0712			
Ti	0.0848	0.1200	0.0655	0.0661	0.0663	0.0242	0.2312			
Al	2.3968	2.4016	2.1684	1.8758	1.8470	0.0762	6.3448			
V	0.1184	0.0000	0.0233	0.0234	0.0235	0.0257	0.0000			
Cr	2.0920	2.1152	0.1616	0.1864	0.2103	0.0511	8.7784			
Fe^{2}	4.4832	3.7752	3.9419	3.9043	3.7929	5.0922	0.2592			
Fe^{3}	8.6904	10.2184	12.0979	12.5645	13.1605	16.0798	1.4912			
Mn	0.0000	0.0000	0.1081	0.2183	0.10949	0.0598	0.9216			
Mg	2.6280	2.3432	1.1750	1.1497	1.3659	0.19257	5.8848			
Ni	2.0824	2.3720	2.1367	2.2276	2.4012	2.0001	0.0144			
Ca	0.1584	0.1448	0.3114	0.3771	0.2838	0.1723	0.0000			
SUM	24.000	24.000	24.000	24.000	24.000	24.000	24.000			
Fe^{3}/FeT	0.66	0.73	0.75	0.76	0.77	0.76	0.15			
Ni-ferr	0.26	0.30	0.27	0.28	0.30	0.25	0.002			

Note: 1–2—new analyses; 3–6—Robin et al; 7—new; 8—Fe-Ni-oxide-rich grain includes CuO 0.45; 9—nearly pure Ni-metal at core of Ni-spinel (all elements, oxygen-free analysis, Cu and Co both undetected). All data collected by energy dispersive spectra; n.d.—not detected, n.a.—not analysed. Ni-ferr—mole proportion nickel-ferrite ($Fe(2 - x)Ni \times O_4$) in nickel spinel.
*Fe_2O_3 calculated assuming stoichiometry.

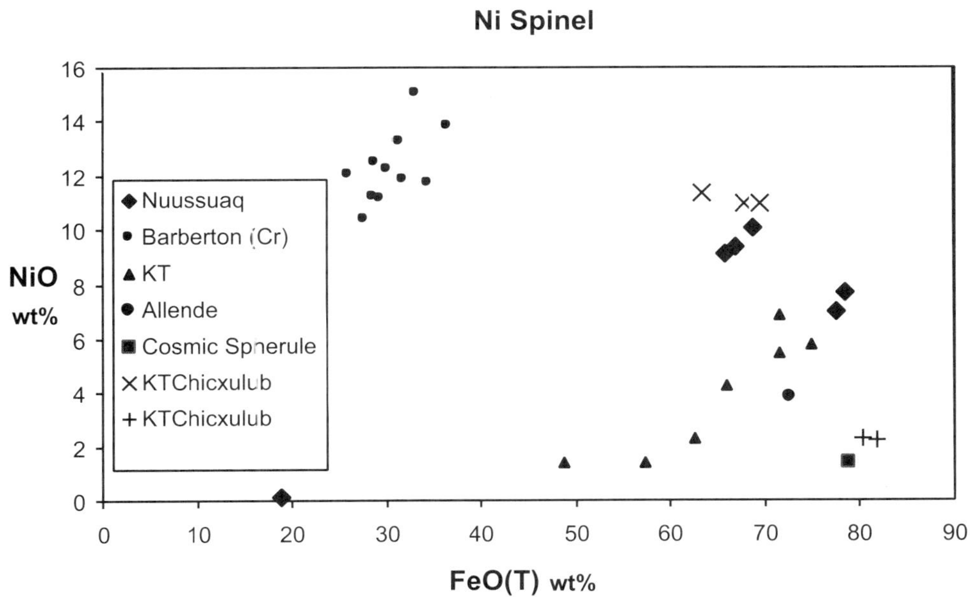

Figure 13. NiO wt% vs. FeO wt% (total iron) for Ni spinel from Nuussuaq spherule beds (Robin et al., 1996) compared with other Ni-spinel: impact spherule deposits are Barberton (small circles; Byerly and Lowe, 1994) and K-T boundary (triangles; Robin and Rocchia 1998) and K-T from Chicxulub crater (crosses; Jones et al., 2004). Also shown is Ni-spinel from cosmic spherule (square) and Allende (large circle; Kearsley unpublished data, see footnote 1). FeO is selected for ease of plotting probe data, and comparison, but the Ni spinels are strongly ferric-rich (typically $Fe^{3}/(Fe^{3}+Fe^{2}) > 0.7$). Note the single Nuussuaq spinel at 10.9% MgO, 18.9% FeO with near zero nickel, is a typical igneous Cr-spinel (see Table 3). Supplementary material available from the GSA Data Repository; see footnote 1.

ferric/ferrous iron ratios in the range of 3:1 or even 4:1 (Robin et al., 1992, 1993; Kearsley, unpublished data[1]). The Nuussuaq Ni-spinel contains modest to low amounts of aluminum (7.5–0.2 wt% Al_2O_3). Our data for Nuussuaq Ni-spinel confirms earlier measurements by Robin et al. (1996), and extends the range to much higher Cr (up to 9.8 wt% Cr_2O_3 new analyses compared with <1% reported in Robin et al., 1996) and higher MgO (total range from 0.4 to 6.53 wt% MgO). Despite the substantial range in varying proportions of the major elements, nickel contents are really remarkably uniform. Examination of Table 3 shows that the sum of trivalent to divalent cations deviates from ratio in standard spinel, suggesting that the nickel may not be present as stoichiometric $NiFe_2O_4$ (trevorite), but is more likely present as 25–30 mol% of non-stoichiometric nickel-ferrite ($Fe_{(2-x)}Ni_{(1+x)}O_4$). Since ferrite spinel has commercially important magnetic properties (nickel ferrite is also a carcinogen to humans) it is often synthesized (Park and Kim, 1999, and references therein) and phase equilibria studies show that from simple starting oxide components in air, the NiO-saturated spinel composition decreases from the ideal fraction Ni/(Ni + Fe) = 0.33 at low temperatures (~1000 °C) to 0.30 at its congruent melting point at ~1700 °C. nickel-ferrite. It is possible, therefore, that Ni-spinel in the Nuussuaq spherule beds formed toward the upper range of these temperatures (see also Siret and Robin, 2003). The range of compositions, and Ni contents in particular, are compared directly with Ni-spinel from established impact horizons (Barberton spherule beds, and K-T boundary) as well as with spinels from a meteorite (e.g., Allende) and from a cosmic spherule. These compositional fields appear distinct in terms of NiO versus FeO (Fig. 13), and also for NiO versus MgO (Fig. 14). However, we have presented only simplified data for the K-T spinel, which shows a larger range of compositions when individual grain zoning is taken into account. A database of Ni-spinels is available from the GSA Data Repository (see footnote 1). For the purposes of this paper, it is important to demonstrate that the Nuussuaq Ni-spinel is very similar to known impact spinel, and occupies a compositional field between data for the K-T boundary and Barberton Ni-spinel. The Barberton Ni-spinel is high in chromium, and under normal circumstances might be called chromite, but this is misleading given the high amounts of nickel (~11–15 wt% NiO). The overall strong anti-correlation between NiO and MgO is striking. One spinel plotted from Nuussuaq in Figures 13 and 14 also contains high chromium (43.4% Cr_2O_3), and plots with undetectable Ni at approximately ~11% MgO, ~19% FeO (Robin et al., 1996; Table 3, analysis 6). This is chromite, and is fairly representative of much igneous chrome-rich spinel in general; its lack of nickel reinforces exactly why Ni-spinel is both distinctive and special. Recalculations of the electron microprobe-derived Ni-spinel data from Nuussuaq, give consistently high ferric iron values ($Fe^{3+}/Fe^{3+}+Fe^{2+}$ = 0.66–0.77 mol%) as reported by Robin et al. (1996). Therefore, in common with other impact-derived Ni-spinel, the presence of Ni-spinel in the Nuussuaq spherule beds requires highly oxidized conditions to form and a source of nickel. This is the opposite to the reduction model required in the usual explanation of the Disko native iron (e.g., Klöck et al., 1986) and siderophile elements compositions of the spherules. The rare occurrence of nearly pure nickel and Fe-Ni-oxide (Table 3, analyses 8 and 9) at the cores of a few Ni-spinel grains is not fully understood, but is reminiscent of

[1]GSA Data Repository Item 2005029, Table DR1, Italian K-T boundary impact spherule spinel compositions, is available on request from Documents Secretary, GSA, P.O. Box 9140, Boulder, CO 80301-9140, USA, editing@geosociety.org, or at www.geosociety.org/pubs/ft2005.htm.

wildly varying metal alloy ratios between Fe-Ni-Co and near end-member native metals in dispersed meteorite-derived metals seen in shattered country rocks at Meteor and Ries impact craters (Kearsley et al., 2004).

DISCUSSION

In many respects, the Nuussuaq spherules resemble spherules associated with the K-T boundary layer, attributed to formation at the Chicxulub impact crater in the Yucatan peninsula. A feature of the K-T spherules is their wide variation in bulk chemistry and state of preservation, which reflects both primary and secondary processes. In addition, the spherules show a decrease in size away from the crater, and a decrease in total accumulated thickness. The Nuussuaq spherules are special for their striking major element heterogeneity, preserved in pristine, sometimes flow-textured glass, and the clear chemical signals that cannot be related to volcanic processes, although they are conventionally explained as being tuffs and fire-fountaining products related to exotic, contaminated volcanism in the Disko region. Geochemical arguments against an extraterrestrial source for the major siderophile elements have centered on non-meteoritic abundances and ratios controlled by, in particular, copper and chromium. Our reexamination of the chromium relative to nickel shows that the spherules have rather constant chromium, at unexceptional levels, and the variable ratio Ni/Cr is controlled by nickel; this ratio for Nuussuaq spherules (2.8–10) lies completely outside the ranges not only for basaltic and picritic rocks, but also for the potential country rock contaminants, sedimentary shale and metamorphic gneiss. Ordinarily in igneous rocks, such mixed siderophile/chalcophile element enrichments might reflect the often-cited pivotal role of sulfur, if sulfur were present, but the spherule analyses (Table 1) clearly reflect ferrosilicate glasses which are consistently sulfur-poor (S is mostly below EDS detection in the glasses). Minor immiscible sulphides were certainly also present, but their significance is very difficult to assess. For example, mass balance estimates are hindered by the small size of the sulphide droplets, and heterogeneous primary and secondary distribution. Our preliminary assessment is that the silicate contribution to the siderophile element budget in the spherules is massively predominant. If this is correct, then Ni-Cu-sulphide concentration models applied to, for example, distribution of chalcophile-siderophile elements within the upper mantle (e.g., Lorand and Alard, 2001) and in mantle-derived melts (e.g., Lightfoot et al., 1997) cannot be invoked to explain the formation of the Nuussuaq spherules, although they have been applied to the massive iron and sulphide localities in Disko Island (Pedersen, 1979; Ulff-Møller, 1985) and have been invoked for regional mantle plume models of the Greenland basalts (Andersen et al., 2002).

It is unlikely that the anomalous copper was derived from a meteoritic source. Conventional analyses present low Cu values in, for example, iron meteorites, except for one unusual iron IAB which has 0.62% Cu ("Oktibbeha County"; Choi et al., 1995), and recent microscopic dissection of the metal component in some chondritic meteorites has revealed unexpectedly large variations in copper (1–300 ppm Cu; Campbell and Humayun, 2004; Campbell et al., 2002). Copper is a highly volatile siderophile element and can occur in native form in chondritic meteorites, where it is considered to be a distinctive shock indicator (Rubin, 2003; Rubin et al., 2001). Generally, most meteoritic siderophile ratios normalized to copper are very different from the Nuussuaq spherule values.

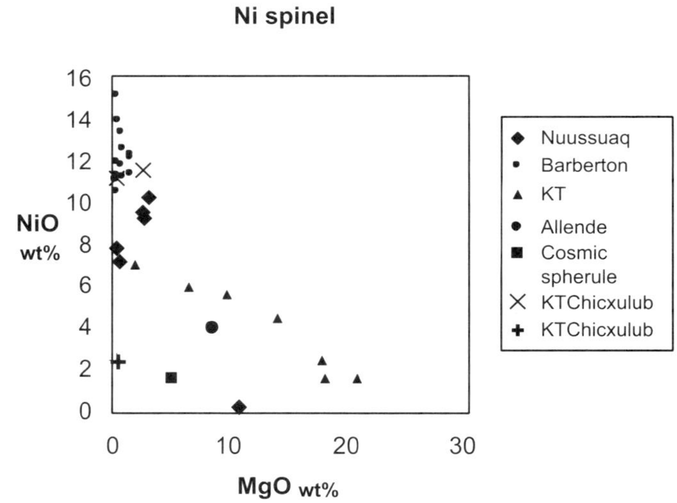

Figure 14. NiO wt% vs. MgO wt% for Ni-spinel from Nuussuaq spherule beds; symbols as in Figure 10. Ni-spinel from impacts, and from the K-T in particular, covers a wide range of compositions. Nuussuaq Ni-spinel is relatively Mg-rich, approaching similar levels to the Ni-spinel in the Precambrian Barberton spherule beds, which is also richer in Cr (34–49 wt% Cr_2O_3; Koeberl and Reimold, 1995).

The same fundamental processes considered to explain heterogeneities, element fractionation, and irregular grain distribution in meteorites are also applicable to impact ejecta. These include the initial prograde (energizing, non-thermodynamic) shock-related distribution between gas-solid-liquid and plasma, and the retrograde (thermodynamic) impact processes (see Melosh, 1989, for discussion). Much interpretation of meteoritic phenomena appears to have concentrated on the retrograde phase, using condensation models, and these can be applied to terrestrial ejecta models, suitably adjusted for interaction between the impact plume and Earth's atmosphere. In particular, the microscopic interaction of very high velocity particles with oxygen can produce a ram-jet effect which dramatically increases oxygen activity, even though oxygen molecules may be widely dispersed, as in Earth's upper atmosphere. This ram-jet effect is thought to be responsible for the formation of highly oxidized Ni-spinel, which characterizes fusion crusts of meteorites, and is also diagnostic of trans-atmospheric impact ejecta (Siret and Robin, 2003). The abundance of Ni-spinel in some of the brown ferrosilicate glass spherules is sometimes highest at the spherule rims, and along internal swirling concentric flow lines suggestive of differential flow-banding (on a microscopic scale) between heterogeneous melt(s).

Criteria for recognition of impact spherules in Archean terrestrial rocks are being established, which attempt to see through complicated secondary replacement histories by Simonson (2003), and some of these arguments, together with observations of impact melts and glasses (Dressler and Reimold, 2001), are pertinent to interpretation of the pristine Nuussuaq spherules. The foremost characteristic is "a predominance of highly spherical grains." This is certainly a feature of the Nuussuaq spherules, and Simonson lists additional petrographic criteria to help distinguish impact spherules from, for example, sedimentary ooids or volcanic accretionary lapilli. The debate over enigmatic spherule beds in classic localities has been involved and vigorous. For example, the coincidence between the spherule beds and komatiite volcanic rocks within the Barberton sequence provoked opposing views on their impact (Lowe and Byerly, 1986) versus volcanic origin (Buick, 1987; Koeberl and Reimold, 1995); if ever unusual volcanic processes were needed, surely komatiites represent an extreme (e.g., Jones, 2002), and detailed models were proposed to explain the presence of spinels with unusual compositions such as magnesioferrite or Ni-spinel (Kyte and Smit, 1986; Robin et al., 1992). A study of Cr-isotopes in the Barberton spherule beds confirmed an extraterrestrial signature (Shukolyukov et al., 2000; Kyte et al. 2003; Lowe et al., 2003) which had been argued previously on the grounds of petrographic textures (sphericity), high Ir and chondrite-like platinum group element anomalies (Lowe and Byerly, 1986; Lowe et al., 1989). Given the renewed confirmation of impact signatures in the Barberton spherule beds, it may be worth reinvestigating the Ni-rich Bon Accord deposit as the relic, albeit considerably metamorphosed (De Waal, 1979), of a meteoritic impactor. De Waal abandoned his earlier (1977) palaeometeorite idea (2002, personal commun.) since another comparable Ni-rich deposit was discovered at a similar stratigraphic level on the other side of the basin, and since meteorites should be rare, the chances of finding more than one was deemed unlikely.

The identification of spherule beds as impact-derived initially met with widespread skepticism (see review in Simonson, 2003), with typical objections including the failure to identify source craters (although, see Jones et al., 2002) and the failure to detect shocked minerals. Additional arguments are summarized in Koeberl and Reimold (1995). A recurrent theme for opposing arguments has been to call upon unusual or exotic volcanic processes, required both to explain the chemistry, and in many cases the anomalously energetic distribution process (eruption versus impact). At Nuussuaq, no unusual volcanic parental silicate magma is available (basalt, picrite, or andesite). Instead, authors have appealed to sustained interactions, at several different times, between basaltic magma and graphitic sedimentary rocks, re-equilibrated with a Fe-Ni-Cu sulphide melt (Pedersen, 1978, 1979; Ulff-Møller, 1990; Robin et al., 1996); these models have difficulty explaining much of the well-exposed field data (Bird and Weathers, 1977; Bird et al., 1981; Fundal, 1975). To be fair, all presentations prior to discovery of the Nuussuaq spherules by Pedersen (1978) were concerned only with the massive iron and the iron-rich, crust-contaminated volcanic deposits on Disko (Qeqertarsuaq) Island. Chemical reactions involving reduction of volcanic magma by carbon-bearing sediments to produce native iron cannot explain the metals within the Nuussuaq spherules, which are instead characterized by highly oxidized ferrosilicate glass containing ferric-Ni spinel. The unusual siderophile element similarities between Disko and the Nuussuaq impact-derived spherule beds (enriched in meteorite-like Fe, Co, Ni, Ir and terrestrial-like Cu) provide a chemical shared link with the "exotic" volcanic source in the Disko region (Robin et al., 1996), which is not yet fully understood.

CONCLUSIONS

The Nuussuaq spherules are abundant, nearly close-packed and surrounded by delicate spherule fragments of glass and altered glass. Their high degree of sphericity makes them unlike igneous or volcanic lapilli, and they are unlikely to have been substantially reworked. The heterogeneous ferrosilicate glass spherules have major element chemistry that cannot be explained by conventional igneous processes, including primarily a strong positive correlation between Fe and Ni. They contain extraordinary bulk chemical compositions that can be explained by rapid and incomplete mixing between a basaltic melt and an enriched iron-nickel source. Siderophile trace element ratios between Fe, Ni, Co, and Cr are strongly suggestive of a meteoritic, perhaps chondritic or even iron meteorite as the metal source for the iron end member. Anomalously high copper and sulfur are also present and have guided previous workers away from an impact explanation; we are unable to explain these additional anomalies. The Ni-spinel is comparable in morphology, size, and composition to Ni-spinel characteristic of confirmed impact deposits. It is characterized by high ferric/ferrous iron ratios and rather constant proportions (25–30 mol %) of nickel ferrite $NiFe_2O_4$ or magnesioferrite $MgFe_2O_4$. We have been unable to identify shocked quartz, although a few quartz grains have unusual melt textures. We have found anomalous isotropic and tartan-textured plagioclase which we interpret to be shocked, but whose shock history has been partly erased by subsequent high temperatures. We conclude that the Nuussuaq spherule beds represent the ejecta from an unknown impact crater, and substantial amounts of meteoritic metal are dissolved within the ferrosilicate glass spherules, together with minor immiscible sulphides, modified by unknown fractionation processes.

What may we conjecture about the impact event itself? The location of the conjectured source impact crater is unknown, and its size cannot be constrained, but clearly obtaining the most accurate dates for the glass spherules would be significant. However, the absence of shocked quartz, and abundance of basalt-derived fragments including relic melted and shocked calcic plagioclase suggests an impact into basaltic, perhaps a subaqueous target which would also be characterized by Ni-spinel (Robin et al., 1993). Some models for formation of Ni-spinel imply a ballistic transit through Earth's atmosphere, but the discovery of K-T

Ni-spinel within the Chicxulub crater itself (Jones et al., 2004) suggests origins by a single impact process, such as ejecta plume and vapor condensation models, may be too simple; Ni-spinel formation may also be related to impact melts and the availability of components, including iron, oxygen, and nickel. Byerly and Lowe (1994) used the average spherule diameter and Ir-fluence in Archean spherule beds correlated between Africa and Australia to estimate projectile size (24 km). Byerly and Lowe also considered that differences in atmospheric oxygen content between Archean and Mesozoic times may have contributed to differences in ferric iron ratios between Archean and, e.g., K-T impact nickel-spinels. The Nuussuaq spherule bed thickness and spherule size are rather comparable to the Archean spherule beds, but there is as yet no distal constraint or correlation with equivalent spherule beds outside of West Greenland. Therefore, the parent crater characteristics are less well constrained, and spherule and ejecta dimensions increase with proximity to the crater. We consider that the source could have been a medium-sized (20 km) proximal crater, or a large-sized (>50 km), more distal crater. We can probably rule out a very local much smaller impact crater, as the area is well exposed and has been very well mapped, and no additional impact materials have been reported. The proposed lateral equivalence of some of the lavas and the spherule beds (e.g., Piasecki et al., 1992) at least offers the possibility that there may be a relation to some of the massive iron localities in Nuussuaq and on Disko (Qeqertarsuaq) Island. This may have implications for the rest of the Disko iron occurrences, which could reverse the previous hundred years of research in favor of the original ideas of Nordenskjöld (1872). The Nuussuaq spherule beds contain possibly the best preserved and most spectacular terrestrial impact spherules yet described. Clearly it would be highly desirable to perform new isotopic studies to resolve some of the issues, using for example stable isotopes, as well as new techniques for Cr (Shukolyukov et al., 2000) and perhaps Cu (Russell et al., 2003).

ACKNOWLEDGMENTS

We are very grateful for the thoughtful and extensive comments provided by our reviewers. In particular, Frank Kyte and Bruce Simonson have helped us to better understand the potential implications of impact signatures in the Nuussuaq spherule beds, and Sager Pedersen has reminded us of the wealth and importance of geological information which has been made available through the Greenland Geological Survey. APJ is also grateful to Jan Nostrum for hospitality and access to the collection of Disko iron at the Swedish Museum of Natural History in Stockholm. Thomas Kinsman is thanked for his thorough guidance and editorship.

REFERENCES CITED

Andersen, J.C.O., Power, M.R., and Momma, P., 2002, Platinum-group elements in the Palaeogene North Atlantic Igneous Province, *in* Cabri, L.J., ed., The Geology, Geochemistry, Mineralogy and Mineral Beneficiation of Platinum-Group Elements: Canadian Institute of Mining, Metallurgy and Petroleum, p. 637–668.

Bird, J.M., and Weathers, M.S., 1977, Native iron occurrences of Disko Island, Greenland: Journal of Geology, v. 85, p. 359–371.

Bird, J.M., Goodrich, C.A., and Weather, J.S., 1981, Petrogenesis of Uivfaq iron, Disko Island, Greenland: Journal of Geophysical Research, v. 86, p. 11787–11805.

Bohor, B.F., Foord, E.E., and Ganapathy, R., 1986, Magnesioferrite from the Cretaceous-Tertiary boundary, Caravaca, Spain: Earth and Planetary Science Letters, v. 81, p. 57–66, doi: 10.1016/0012-821X(86)90100-7.

Buick, R., 1987, Comment on Lowe and Byerly's "Early Archean silicate spherules of probable impact origin, South Africa and Western Australia": Geology, v. 15, p. 180–181.

Byerly, G.R., and Lowe, D.R., 1994, Spinel from Archean impact spherules: Geochimica et Cosmochimica Acta, v. 58, p. 3469–3486, doi: 10.1016/0016-7037(94)90099-X.

Cande, S.C., and Kent, D.V., 1995, Revised calibration of the geomagnetic polarity timescale for the late Cretaceous and Cenozoic: Journal of Geophysical Research, v. 100, p. 6093–6095, doi: 10.1029/94JB03098.

Campbell, A.J. and Humayun, M., 2004, Formation of metal in the CH chondrites ALH 85085 and PCA 91467: Geochimica et Cosmochimica Acta, v. 68, p. 3409–3422.

Campbell, A.J., Humayun, M., and Weisberg, M.K., 2002, Siderophile element constraints on the formation of metal in the metal-rich chondrites Bencubbin, Gujba, and Weatherford: Geochimica et Cosmochimica Acta, v. 66, p. 631–644.

Choi, B., Ouyang, X., and Wasson, J.T., 1995, Classification and origin of IAB and IIICD iron meteorites: Geochimica et Cosmochimica Acta, v. 59, p. 593–612, doi: 10.1016/0016-7037(94)00384-X.

Clarke, D.B., and Pedersen, A.K., 1976, Tertiary volcanic province of West Greenland., *in* Escher, A., and Watt, W.S., eds., Geology of Greenland: Copenhagen, GGU, p. 364–385.

Dam, G., 2002, Sedimentology of magmatically and structurally controlled outburst valleys along rifted volcanic margins: examples from the Nuussuaq Basin: West Greenland: Sedimentology, v. 49, p. 505–532, doi: 10.1046/j.1365-3091.2002.00457.x.

Dam, G., Nør-Hansen, H., and Kennedy, W.J., 1998, The northernmost marine Cretaceous-Tertiary boundary section: Nuussuaq, West Greenland: Geology of Greenland Survey Bulletin, v. 180, p. 138–144.

De Waal, S.A., 1977, The nickel deposit at Bon Accord, Barberton, South Africa—a proposed palaeometeorite: Mineralization in metamorphic terranes: Geological Society of South Africa Special Publication 4, p. 87–98.

De Waal, S.A., 1979, The metamorphism of the Bon Accord nickel deposit by the Nelspruit granite: Transactions of the Geological Society of South Africa, v. 82, p. 335–342.

Dressler, B.O., and Reimold, W.U., 2001, Terrestrial impact melt rocks and glasses: Earth-Science Reviews, v. 56, p. 205–284, doi: 10.1016/S0012-8252(01)00064-2.

Fundal, E., 1975, The Uivfaq dike and related hybrid dikes from southern Disko, west Greenland: Meddelemser om Grønland, v. 195, p. 1–28.

Garcia, M.O., Rubin, K.H., Norman, M.D., Rhodes, M., Graham, D.W., Muenow, D., and Spencer, K., 1998, Petrology and geochronology of basalt breccia from the 1996 earthquake swarm of Loihi Seamount, Hawaii: Magmatic history of its 1996 eruption: Bulletin of Volcanology, v. 59, p. 577–592, doi: 10.1007/s004450050211.

Goodrich, C.A., and Bird, J.M., 1985, Formation of iron-carbon alloys in basaltic magma at Uivfaq, Disko Island: the role of carbon in mafic magmas: Journal of Geology, v. 93, p. 475–492.

Graup, G., 1999, Carbonate-silicate liquid immiscibility upon impact melting, Ries Crater, Germany: Meteoritics & Planetary Science, v. 34, p. 425–438.

Henderson, G., Rosenkrantz, A., and Schiener, E.J., 1976, Cretaceous-Tertiary sedimentary rocks of West Greenland, *in* Escher, A., and Watt, W.S., eds., Geology of Greenland: Copenhagen, GGU, p. 340–363.

James, O.B., Cohen, B.A., and Taylor, L.A., 2003, Lunar meteorite Dhofar 026: a shocked granulitic breccia, not an impact melt (abstract): Lunar and Planetary Science Conference XXXIV, abstract no. 1149.

Jones, A.P., 2002, Komatiites: new information on the type locality (Barberton), and some new ideas: Geology Today, v. 18, p. 23–25, doi: 10.1046/j.1365-2451.2002.00325.x.

Jones, A.P., Heuschkel, S., and Claeys, P., 2000, Impact melting: a review of experimental constraints for carbonate targets and applications to the Chicxulub crater, *in* Gilmour, I., and Koeberl, C., eds., Impacts and the Early Earth, Lecture Notes in Earth Sciences: Springer-Verlag, v. 91, p. 343–362.

Jones, A.P., Price, G.D., Price, N.J., DeCarli, P.S., and Clegg, R., 2002, Impact induced melting and the development of large igneous provinces: Earth and Planetary Science Letters, v. 202, p. 551–561, doi: 10.1016/S0012-821X(02)00824-5.

Jones, A.P., Vermeesch, P.M., Morgan, J., and Kearsley, A., 2004, Carbonate and silicate melts with Ni-spinel in Chicxulub suevites, from Yaxcopoil-1(abstract): European Geophysical Union 2004 Meeting, Nice, 04-A-07200.

Jürgensen, T., and Mikkelsen, N., 1974, Coccoliths from volcanic sediments (Danian) in Nûgssuaq, West Greenland: Bulletin of the Geological Society of Denmark, v. 23, p. 225–230.

Kearsley, A., Graham, G., McDonnell, T., Bland, P., Hough, R., and Helps, P., 2004, Early fracturing and impact residue emplacement: Can modelling help to predict their location in major craters?: Meteoritics & Planetary Science, v. 39, p. 247–265.

Klöck, H., and Palme, H.J., 1988, Partitioning of siderophile and chalcophile elements between sulphide, olivine, and glass in a naturally reduced basalt from Disko Island, Greenland: Proceedings 18th Lunar and Planetary Science Conference, p. 471–483.

Klöck, H., Palme, H.J., and Tobschall, H.J., 1986, Trace elements in natural metallic iron from Disko Island, Greenland: Contributions to Mineralogy and Petrology, v. 93, p. 273–282, doi: 10.1007/BF00389387.

Koeberl, C., and Reimold, W.U., 1995, Early Archaean spherule beds in the Barberton Mountain Land, South Africa: no evidence for impact origin: Precambrian Research, v. 74, p. 1–33, doi: 10.1016/0301-9268(94)00099-D.

Kurat, G., Varela, M.E., Zinner, E., Maruoka, T., and Brandstaetter, F., 2003, Major, minor and trace elements in some glasses from the NWA 1664 Howardite: Lunar and Planetary Science XXXIV, abstract no. 1733.

Kyte, F.T., and Bostwick, J.A., 1995, Magnesioferrite spinel in the Cretaceous-Tertiary boundary sediments of the Pacific basin: Remnants of hot, early ejecta from the Chicxulub impact?: Earth and Planetary Science Letters, v. 132, p. 113–127, doi: 10.1016/0012-821X(95)00051-D.

Kyte, F.T., and Smit, J., 1986, Regional variations in spinel compositions: An important key to the Cretaceous-Tertiary event: Geology, v. 14, p. 485–487.

Kyte, F.T., Shukolyukov, A., Lugmair, G.W., Lowes, D.R., and Byerly, G.R., 2003, Early Archean spherule beds: Chromium isotopes confirm origin through multiple impacts of projectiles of carbonaceous chondrite type: Geology, v. 31, p. 283–286, doi: 10.1130/0091-7613(2003)0312.0.CO;2.

Larsen, L.M., and Pedersen, A.K., 2000, Processes in high-Mg, high-T magmas: evidence from olivine, chromite and glass in Palaeogene picrites from West Greenland: Journal of Petrology, v. 41, p. 1071–1098, doi: 10.1093/petrology/41.7.1071.

Larsen, L.M., Pedersen, A.K., Sundvoll, B., and Frei, R., 2003, Alkali Picrites Formed by Melting of Old Metasomatized Lithospheric Mantle: Manîtdlat Member, Vaigat Formation—Palaeocene of West Greenland: Journal of Petrology, v. 44, p. 3–38, doi: 10.1093/petrology/44.1.3.

Lightfoot, P.C., Hawkesworth, C.J., Olshefsky, K., Green, T., Doherty, W., and Keays, R.R., 1997, Geochemistry of Tertiary tholeiites and picrites from Qeqertarssuaq (Disko Island) and Nuussuaq, West Greenland with implications for the mineral potential of comagmatic intrusions: Contributions to Mineralogy and Petrology, v. 128, p. 139–163, doi: 10.1007/s004100050300.

Löfquist, H., and Benedicks, C., 1941, Det stora Nordenskiöldska järnblocket från Ovifak: mikrostruktur och bildningssätt. Kungliga Svenska Vetenskapsakademiens Handlingar, Serie 3, Band 19, no. 3, 96 p. (in Swedish).

Lorand, J.-P., and Alard, O., 2001, Platinum-group element abundances in the upper mantle: new constraints from in situ whole-rock analyses of Massif Central xenoliths (France): Geochimica et Cosmochimica Acta, v. 65, p. 2789–2806, doi: 10.1016/S0016-7037(01)00627-5.

Lorenz, C., Nazarov, M., Kurat, G., Bradstaetter, F., and Ntaflos, Th., 2001, Clast population and chemical composition of the Dhofar 018 Howardite (abstract): Lunar and Planetary Science XXXII, abstract no. 1778.

Lowe, D.R., and Byerly, G.R., 1986, Early Archean silicate spherules of probable impact origin, South Africa and Western Australia: Geology, v. 14, p. 599–602.

Lowe, D.R., Byerly, G.R., Asaro, F., and Kyte, F.T., 1989, Geological and geochemical evidence for a record of 3,400 Ma–old terrestrial meteorite impacts: Science, v. 245, p. 959–962.

Lowe, D.R., Byerly, G.R., Kyte, F.T., Shukolykov, A., Asaro, F., and Krull, A., 2003, Spherule beds 3.47–3.24 billion years old in the Barberton Greenstone Belt, South Africa: A record of large meteorite impacts and their influence on early crustal and biological evolution: Astrobiology, v. 3, p. 7–48, doi: 10.1089/153110703321632408.

Magde, L.S., 1991, Unusual glass spherules from the west Greenland basalt province [undergraduate thesis]: San Diego, University of California.

Margolis, S.V., Swinburne, N.H.M., Asaro, F., Swisher, C.C., McWilliams, M., Hansen, H.J., and Pedersen, A.K. 1991, Possible impact ejecta in the Palaeocene flood basalt province of west Greenland: Eos (Transactions, American Geophysical Union), v. 72, p. 278.

Melosh, H.J., 1989, Impact Cratering: A Geologic Process: Oxford University Press, 245 p.

Nordenskjöld, A.E., 1872, Account of an expedition to Greenland in the year 1870: Geological Magazine, v. 9, p. 289.

Ogg, J.G., and Bardot, L., 2001, Aptian through Eocene magnetostratigraphic correlation of the Blake Nose Transect (Leg 171B), Florida continental margin, *in* Kroon, D., Norris, R.D., and Klaus, A., eds., Proceedings Ocean Drilling Program, Scientific Results, p. 171B (online).

Oleynikov, B.V., Okrugin, A.V., and Tomshin, M.D., 1985, Native metals formation in basic rocks of the Siberian Platform: Yakutian Branch of the Soviet Academy of Sciences, Yakutsk, 188 p. (in Russian).

Osinski, G.R., 2003, Impact glasses in fallout suevites from the Ries impact structure, Germany: An analytical SEM study: Meteoritics & Planetary Science, v. 38, p. 1641–1668.

Osinski, G.R., and Spray, J.G., 2001, Impact-generated carbonate melts: evidence from the Haughton structure, Canada: Earth and Planetary Science Letters, v. 194, p. 17–29, doi: 10.1016/S0012-821X(01)00558-1.

Park, B.H., and Kim, D.S., 1999, Thermodynamic properties of $NiCr_2O_4$-$NiFe_2O_4$ spinel solid solution: Bulletin of the Korean Chemical Society, v. 20, p. 939–942.

Pauly, H., 1969, White cast iron with cohenite, schreibersite and sulphides from Tertiary basalts on Disko: Meddelelser Dansk Forenschrift, v. 19, p. 8–26.

Pedersen, A.K., 1975a, New mapping in north-western Disko: Rapport Grønlands Geologiske Undersøegelse, v. 69, p. 25–32.

Pedersen, A.K., 1975b, New investigations of the native iron bearing volcanic rocks of Disko, central West Greenland: Rapport Grønlands Geologiske Undersøegelse, v. 75, p. 48–51.

Pedersen, A.K., 1978, Graphite andesite tuffs resulting from high-Mg tholeiite and sediment interaction; Nugssuaq, West Greenland: Bulletin of the Geological Society of Denmark, v. 27, p. 117–130.

Pedersen, A.K., 1979, Basaltic glass with high-temperature equilibrated immiscible sulphide bodies with native iron from Disko, central West Greenland: Contributions to Mineralogy and Petrology, v. 69, p. 397–407.

Pedersen, A.K., 1981, Armalcolite-bearing Fe-Ti oxide assemblages in graphite-equilibrated salic volcanic rocks with native iron from Disko, central west Greenland: Contributions to Mineralogy and Petrology, v. 77, p. 307–324, doi: 10.1007/BF00371560.

Pedersen, A.K., 1985a, Lithostratigraphy of the Tertiary Vaigat Formation on Disko, central West Greenland: Rapport Grønlands Geologiske Undersøgelse, v. 124, 30 p.

Pedersen, A.K., 1985b, Reaction between picritic magma and continental crust: early Tertiary silicic basalts and magnesian andesites from Disko, West Greenland: Bulletin Groenlands Geologiske Undersoegelse, v. 152, 126 p.

Pedersen, A.K., Larsen, L.M., and Ulff-Moller, F., 1989, Discovery of andesite tuffs with graphite from the Vaigat Formation of south central Nugssuaq: stratigraphical implications: Rapport Grønlands Geologiske Undersøgelse, v. 145, p. 42–45.

Pedersen, A. K., Larsen, L. M., and Dueholm, K. S., 1993, Geological section along the south coast of Nuussuaq, central West Greenland: Copenhagen, Geological Survey of Greenland, scale 1:20,000, colored geological sheet.

Pedersen, G.K., Larsen, L.M., Pedersen, A.K., and Hjortkjaer, B.F., 1998, The synvolcanic Naajaat lake, Palaeocene of West Greenland: Palaeogeography, Palaeoclimatology, Palaeoecology, v. 140, p. 271–278, doi: 10.1016/S0031-0182(98)00034-0.

Pedersen, A.K., Larsen, L.M., Riisager, P., and Dueholm, K.S., 2002, Rates of volcanic deposition, facies changes and movements in a dynamic basin: the Nuussuaq Basin, West Greenland, around the C27n–C26r transition, *in* Jolley, D.W., and Bell, B.R., eds., The North Atlantic Igneous Province: stratigraphy, tectonics, volcanic and magmatic processes: Geological Society [London] Special Publication 197, p. 157–181.

Pedersen, A.K., Larsen, L.M., Pedersen, G.K., and Dueholm, K.S., Compilers, 2003: Geological section along the south and south-west coast of Disko, central West Greenland: Copenhagen, Geological Survey of Denmark and Greenland, scale 1:20,000, colored geological sheet.

Piasecki, S., Larsen, L.M., Pedersen, A.K., and Pedersen, G.K., 1992, Palynostratigraphy of the Lower Tertiary volcanics and marine clastic sediments in the southern part of the West Greenland Basin: Rapport Grønlands Geologiske Undersøgelse, v. 154, p. 13–31.

Riisager, P., and Abrahamsen, N., 1999, Magnetostratigraphy of Palaeocene basalts from the Vaigat Formation of West Greenland: Geophysical Journal International, v. 137, p. 774–782, doi: 10.1046/j.1365-246x.1999.00830.x.

Riisager, J., Riisager, P., and Pedersen, A.K., 2003, The C27n-C26r geomagnetic polarity reversal recorded in the west Greenland flood basalt province: How complex is the transitional field?: Journal of Geophysical Research, v. 108, no. B3, 2155, doi:10.1029/2002JB002124.

Robin, E., and Rocchia, R., 1998, Ni-rich spinel at the Cretaceous-Tertiary boundary of El Kef, Tunisia: Bulletin de la Société Géologique Française, v. 169, no. 3, p. 365–372.

Robin, E., Bonté, Ph., Froget, L., Jéhanno, C., and Rocchia, R., 1992, Formation of spinels in cosmic objects during atmospheric entry: a clue to the Cretaceous-Tertiary event: Earth and Planetary Science Letters, v. 108, p. 181–190, doi: 10.1016/0012-821X(92)90021-M.

Robin, E., Froget, L., Jehanno, C., and Rocchia, R., 1993, Evidence for a K/T impact event in the Pacific Ocean: Nature, v. 363, p. 615–617, doi: 10.1038/363615a0.

Robin, E., Swinburne, N.H.M., Froget, L., Rocchia, R., and Gayraud, J., 1996, Characteristics and origin of the glass spherules from the Paleocene flood basalt province of western Greenland: Geochimica et Cosmochimica Acta, v. 60, p. 815–830, doi: 10.1016/0016-7037(95)00433-5.

Rosenkrantz, A., 1970, Marine Upper Cretaceous and lowermost Tertiary deposits in West Greenland: Meddelelser Dansk Geologiske Forenschrift, v. 19, p. 406–453.

Rubin, A.E., 2003, Chromite-plagioclase assemblages as a new shock indicator; implications for the shock and thermal histories of ordinary chondrites: Geochimica et Cosmochimica Acta, v. 67, p. 2695–2709, doi: 10.1016/S0016-7037(03)00107-8.

Rubin, A.E., Ulff-Møller, F., Wasson, J.T., and Carlson, W.D., 2001, The Portales Valley meteorite breccia: evidence for impact-induced melting and metamorphism of an ordinary chondrite: Geochimica et Cosmochimica Acta, v. 65, p. 323–342, doi: 10.1016/S0016-7037(00)00516-0.

Russell, S.S., Zhu, X., Guo, Y., Mullane, E., Gounelle, M., Mason, T.F.D., and Coles, B., 2003, Copper and zinc isotopic compositions of CR, CB, and CH-like meteorites: Geophysical Research Abstracts, v. 5, p. 11331.

Saunders, A.D., Fitton, J.G., Kerr, A.C., Norry, M.J., and Kent, R.W., 1997, The North Atlantic Igneous Province, *in* Mahoney, J.J., and Coffin, M.F., eds., Large Igneous Provinces: Continental, Oceanic and Planetary Volcanism: Washington D.C., American Geophysical Union Monograph v. 100, p. 45–94.

Shukolyukov, A., Kyte, F.T., Lugmair, G.W., Lowe, D.R., and Byerly, G.R., 2000, The oldest impact deposits on Earth—first confirmation of an extraterrestrial component, *in* Gilmour, I., and Koeberl, C., eds., Impacts and the Early Earth: Lecture Notes in Earth Science, Heidelberg, Springer-Verlag, v. 91, p. 99–115.

Simonson, B.M., 2003, Petrographic criteria for recognizing certain types of impact spherules in well-preserved Precambrian successions: Astrobiology, v. 3, p. 49–65, doi: 10.1089/153110703321632417.

Siret, D., and Robin, E., 2003, Spinel formation in an impact plume: a thermodynamic approach [abs.]: Lunar and Planetary Science Conference XXXIV, abstract no. 1865.

Smith, C., Franchi, I.A., Wright, I.P., Grady, M.M., and Pillinger, C.T., 2000, Disko island C-rich basalts: A terrestrial analogue for ureilite petrogenesis and their carbon isotope systematics: Lunar and Planetary Science Conference XXXI, abstract no. 1735.

Steenstrup, K.J.V., 1882, Om Forekomsten af Forstenigner i de kulførende Dannelser I Nord-Grønland: Meddelemser om Grønland, v. 4, p. 113–132.

Storey, M., Duncan, R.A., Pedersen, A.K., Larsen, L.M., and Larsen, H.C., 1998, $^{40}Ar/^{39}Ar$ geochronology of the West Greenland Tertiary volcanic province: Earth and Planetary Science Letters, v. 160, p. 569–589, doi: 10.1016/S0012-821X(98)00112-5.

Treiman, A.H., Lindstrom, D.J., Schwandt, C.S., Franchi, I.A., and Morgan, M.L., 2002, A "Mesosiderite" Rock from Northern Siberia, Russia: Not a Meteorite: Meteoritics & Planetary Science, v. 37, supplement, p. B13–B22.

Ulff-Møller, F., 1985, Solidification history of the Kitdlit lens: Immiscible metal and sulphide liquids from a basaltic dyke on Disko: West Greenland: Journal of Petrology, v. 26, p. 64–91.

Ulff-Møller, F., 1990, Formation of native iron in sediment-contaminated magma I: A case study of the Hannekammen Complex on Disko Island: West Greenland: Geochimica et Cosmochimica Acta, v. 54, p. 57–700.0, doi: 10.1016/0016-7037(90)90195-Q.

Manuscript Accepted by the Society 9 August 2004

Geological Society of America
Special Paper 384
2005

Distribution and origin of impact diamonds in the Ries crater, Germany

R.T. Schmitt
C. Lapke*
C.M. Lingemann
M. Siebenschock*
D. Stöffler
Institute of Mineralogy, Museum of Natural History, Humboldt-University of Berlin, Invalidenstrasse 43, D-10115 Berlin, Germany

ABSTRACT

The distribution and origin of impact diamonds in the ejecta blanket of the Ries crater, Germany, was investigated. Impact diamonds are present in the fallout suevite, whereas the cataclastic crystalline breccias, lithic impact breccia (Bunte Breccia) and clast-rich impact melt rock do not contain diamonds. No regional concentrations of impact diamonds in the fallout suevite could be detected. The average concentration of diamonds is ~0.1–0.2 ppm. The carriers of impact diamonds are specific suevite components, such as graphite-bearing crystalline rock fragments of shock stage III, and most likely small fragments thereof in melt fragments and suevite matrix. Impact diamonds occur as pseudo-hexagonal, transparent, and birefringent plates, which reach sizes up to 300 µm. Their color is commonly greenish, but can also be black, gray, yellow, or colorless. Most of the impact diamonds have a fibrous or spongy internal structure and extensional microfractures, which lead to a characteristic porosity. This is the result of a volume decrease due to the phase transformation of graphite to diamond. The Raman characteristics of these impact diamonds are discussed in detail. The strong morphologic similarity of impact diamonds to the precursor graphite from the crystalline target rocks indicate a solid state martensitic phase transformation which occurs at shock pressures of 45–55 GPa.

Key words: impact diamonds, Ries crater.

INTRODUCTION

Diamonds produced by shock metamorphism were discovered for the first time by Yerofeev and Lachinov (1888) in the ureilite Novo Urei. In 1891, Foot described impact diamonds in fragments of the Canyon Diablo iron meteorite associated with the Meteor (Barringer) Crater in Arizona, USA. In this case, polycrystalline diamonds were formed by shock transformation of carbonaceous matter that occurred within the projectile. Impact diamonds are generally different from kimberlite or lamproite diamonds due to their µm-sized, polycrystalline structure. Moreover, X-ray investigations revealed the occurrence of the hexagonal carbon phase lonsdaleite (Frondel and Marvin, 1967; Hannemann et al., 1967) in many of these impact diamond grains. Since these findings, impact diamonds were frequently reported from other iron meteorites and especially from ureilites (e.g., Anders, 1965; Frondel and Marvin, 1967; Vdovykin, 1970; Berkley et al., 1980; Bischoff and Stöffler, 1992).

*E-mail, Schmitt: ralf-thomas.schmitt@rz.hu-berlin.de. Present addresses: Lapke—Kleine Dorfstrasse 29, D-38524 Sassenburg, Germany; Siebenschock—Herrlache 14, D-79348 Freiamt, Germany

Schmitt, R.T., Lapke, C., Lingemann, C.M., Siebenschock, M., and Stöffler, D., 2005, Distribution and origin of impact diamonds in the Ries crater, Germany, *in* Kenkmann, T., Hörz, F., and Deutsch, A., eds., Large meteorite impacts III: Geological Society of America Special Paper 384, p. 299–314. For permission to copy, contact editing@geosociety.org.

Diamonds synthetically produced by shock compression were for the first time described by DeCarli and Jamieson (1961). They used mixtures of graphite and metal, and shock-loaded them at pressures above 15 GPa. Impact diamonds are produced if the shock temperature of the samples is above 1300 K and the post-shock temperature below 2000 K (DeCarli, 1995, 1998). A successful experimental shock transformation of graphite to diamond in natural graphite-bearing gneiss was reported by Kenkmann et al. (2002) above 45 GPa.

Impact diamonds in shocked terrestrial target rocks were first discovered in the Popigai crater, northern Siberia, Russia (Masaitis et al., 1972, 1995; Masaitis, 1998; Koeberl et al., 1995, 1997; Vishnevsky et al., 1997). In this crater, impact diamonds are found in shocked graphite-bearing gneiss clasts in suevite and impact melt rocks, in the suevite matrix, and in impact melt rocks. These diamond grains reach sizes up to 1 cm in diameter. They are yellow, gray, or black in color and rarely colorless. The diamond grains display the original shape of the precursor graphite, yet they are aggregates and consist of microcrystals with diameters of <1–5 μm. For such diamond grains, the term paramorphs was introduced (Shafranovsky, 1985). Impact diamonds are also found in alluvial deposits at Popigai due to weathering of the impactites and in this case, the grains are broken or have irregular shapes. Subsequently, impact diamonds were found in the Kara and Puchezh-Katunki craters in Russia (Vishnevsky et al., 1997), in the Ilyinets, Obolon, Ternovka, and Zapadnaja craters in Ukraine (Valter et al., 1992; Gurov et al., 1995, 1996; Masaitis et al., 1995; Vishnevsky et al., 1997), in Sudbury, Canada (Masaitis et al., 1997), in the Cretaceous-Tertiary boundary layer of Mexico, USA, and Canada (Hough et al., 1997a, 1997b), in the Lappajärvi crater in Finland (Masaitis et al., 1998; Hough et al., 1999; Moroz et al., 2003) and the Gardnos crater in Norway (Gilmour et al., 2003). The shape and size of the impact diamonds depend strongly on the shape and size of the precursor carbon phase (e.g., graphite, coal) in the respective target rocks, e.g., in the Sudbury crater impact, diamonds are generally anhedral. None of these craters contains impact diamonds suitable for commercial exploration.

GEOLOGY OF THE RIES CRATER

The Ries crater, Germany (diameter ~24 km; age 14.3–14.4 Ma; Fig. 1) occurs in a target of up to 0.75 km thick Cenozoic and Mesozoic sediments (limestone, shale, sandstone) overlying a complex Variscan crystalline basement (e.g., Graup, 1978; Stöffler and Ostertag, 1983, and references therein; Engelhardt, 1990; Hüttner and Schmidt-Kaler, 1999a, 1999b; Buchner et al., 2003; Laurenzi et al., 2003). The base of the ejecta deposits consists of a lithic, clastic matrix breccia (Bunte Breccia) with a thickness of up to 200 m and a radial extent of up to 42 km from the crater center (Stöffler and Ostertag, 1983; Engelhardt, 1990). This polymict breccia is dominated by material (shock stages 0–II) from the sedimentary cover mixed with locally derived material, especially in the outer parts of the ejecta blanket. Near to the crater rim and in the megablock zone between the crater rim and the inner ring (diameter ~12 km) crystalline breccias (shock stages 0–II) occur as irregular bodies within or on top of the lithic breccia (Stöffler and Ostertag, 1983; Engelhardt, 1990). These crystalline and polymict breccias are covered locally by patches of fallout suevite (Engelhardt, 1997, and references therein). In contrast to the Bunte Breccia, the suevite consists mainly of crystalline basement material of shock stages 0–IV (Engelhardt, 1997) and impact melt that is dispersed in form of melt fragments ("glass bombs") of up to 50 cm in size and of small, chilled melt particles within the matrix (Engelhardt, 1997; Osinski et al., 2003). The chemical composition of this impact melt reflects a mixture of crystalline target rocks, mainly granite and biotite-plagioclase-gneiss, and is rather homogeneous in composition across the entire Ries crater (Stähle, 1972; Engelhardt and Graup, 1984; Engelhardt, 1997). Melt fragments contain abundant mineral, crystalline, and sedimentary rock fragments of different shock stages. Inside the inner ring, a suevite lens (crater or fallback suevite), up to 400 m thick and deficient in melt fragments, was found

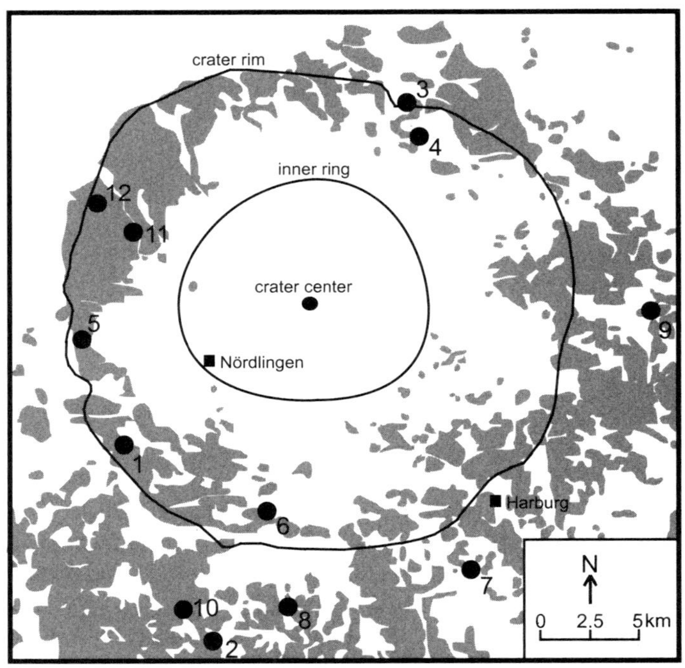

Figure 1. Simplified map of the Ries crater, Germany. The ejecta deposits (cataclastic crystalline breccias, lithic clastic impact breccia, fallout suevite, and clast-rich impact melt rock) are shown in gray. The crater rim (diameter ~24 km) and the inner ring (diameter ~12 km) are indicated by solid lines. Due to the erosion, ejecta deposits occur preferentially in the megablock zone between the inner ring and the crater rim, and in the southern and eastern surroundings of the crater. The diamond-bearing suevite localities are marked by filled circles: 1—Altenbürg, 2—Amerdingen, 3—Aumühle, 4—Hainsfarth, 5—Heerhof, 6—Hohenaltheim, 7—Mauren, 8—Oberringingen, 9—Otting, 10—Seelbronn, 11—Unterwilflingen, 12—Zipplingen. Map modified after Schmidt-Kaler (1978).

by drilling (Gudden, 1974; Bayerisches Geologisches Landesamt, 1977; Stöffler and Ostertag, 1983). The crater suevite is covered by up to 330 m thick, post-impact lake sediments. An impact melt rock layer does not occur in the central depression, although two small outcrops of red, clast-rich, recrystallized impact melt rock occur near to Polsingen and Amerbach at the eastern crater rim (Engelhardt et al., 1969; Stöffler and Ostertag, 1983; Hüttner and Schmidt-Kaler, 1999a, 1999b).

IMPACT DIAMONDS FROM THE RIES CRATER

The first impact diamonds in the Ries crater were discovered by Rost et al. (1977) in melt fragments from the suevite of Otting. Masaitis et al. (1995) described impact diamonds extracted from shocked gneiss fragments from the suevite deposits Otting and Aumühle, which occur in the form of paramorphs after graphite. Also, Abbott et al. (1996) found impact diamonds in a melt fragment from the Otting suevite. In the crater suevite of the scientific drill core Nördlingen 1973, Abbott et al. (1998) reported impact diamonds in suevite samples from depths of 384 and 494 m. Valter et al. (1998) and Thiele and Scherer (2000) extracted impact diamonds from suevite whole-rock samples from the Otting quarry. All of these findings show a similar appearance of the impact diamonds, closely resembling the impact diamonds occurring in the Popigai crater. The diamonds are flat platelets of variable color, 20–400 µm in size, which are interpreted as paramorphs after the precursor graphite. El Goresy et al. (2002, 2003) described a potentially new carbon phase of similar appearance from a shocked gneiss fragment of the suevite.

Different types of diamonds were discovered by Hough et al. (1995), who reported µm-sized skeletal aggregates of diamonds and diamond-silicon carbide intergrowths from a suevite whole-rock sample of the Otting quarry. These grains were interpreted as chemical vapor deposited from the ejecta plume. Still another type of diamond was detected by El Goresy et al. (2001), who found µm-sized grains of diamond together with graphite at graphite-garnet, graphite-sillimanite, and graphite-rutile interfaces in shocked garnet-cordierite-sillimanite and garnet-biotite-gneiss fragments of shock stage II from the Ries suevite.

The aim of our study was a systematic investigation of the distribution and abundance of impact diamonds in the various types of impactites and their components within the ejecta blanket of a complex impact structure. We chose the Ries crater because it exposes all types of ejecta deposits (e.g., lithic impact breccia, cataclastic crystalline breccia, fallout suevite, clast-rich impact melt rock). In contrast to previous studies at the Ries crater, this is the first systematic investigation that is based on a larger number of samples from widely different localities representative of the whole ejecta blanket. Therefore, we were able to assess the regional distribution of diamonds in the ejecta blanket and to provide quantitative estimates of the diamond concentration in different types of impactites. Another focus of our work was a systematic Raman investigation of a large number of impact diamonds from the Ries crater.

SAMPLES AND ANALYTICAL METHODS

Samples

For this study, 32 whole-rock samples of impactites (lithic impact breccia, cataclastic crystalline breccia, fallout suevite, clast-rich impact melt rock) from different localities in the ejecta blanket of the Ries crater were chosen (Table 1; Fig. 1). All major fallout suevite and lithic impact breccia deposits from the ejecta blanket were investigated during this study. The crater or fallback suevite was not analyzed because sufficient sample material was not available from the drill cores. Field samples of impactites were additionally split in the laboratory into their components and analyzed separately: 7 matrix samples of fallout suevite, 24 melt fragments from the suevite, and 70 crystalline fragments of different shock stages from the fallout suevite and the clast-rich impact melt rock (Table 1). During the preparation of melt fragments and matrix samples, all fragments with a diameter >0.5 cm were discarded. Each sample was mechanically processed, split by a hydraulic press with breaking jaws of stainless steel and subdivided into three aliquots for (1) thin section preparation, (2) chemical analysis, and (3) extraction of diamonds. Possible contamination by industrial diamonds and silicon carbide grains from saw blades and grinding devices were eliminated by avoiding any sawing and grinding during the dividing and processing of the aliquots used for the diamond extraction.

Petrographic and Geochemical Investigations

Thin sections were used for shock and rock classification, especially for the unshocked to moderately shocked crystalline fragments. The shock stage of the crystalline fragments was determined after Stöffler (1971). Shock stage 0 (shock pressure <10 GPa) is characterized by quartz and feldspar grains which display only fracturing. In shock stage I (shock pressure 10–35 GPa) quartz and feldspar show planar deformation features and reduced refractive index. Shock stage II (shock pressure 35–45 GPa) is indicated by diaplectic quartz glass with coesite and diaplectic feldspar glass, whereas rocks of shock stage III (shock pressure 45–60 GPa) display mainly vesiculated feldspar glass together with coesite-bearing diaplectic quartz glass. The whole-rock texture of shock stage III is generally porous to pumice-like. Major element chemistry of crystalline fragments was measured by X-ray fluorescence spectrometry to establish the chemical signature of heavily shocked fragments in comparison with a database of Ries crystalline rocks (Engelhardt and Graup, 1984). Based on the SiO_2 versus $Na_2O + K_2O$ diagram after Le Maitre (1984), precursors of granitic, granodioritic, and monzonitic to dioritic composition could be distinguished. Due to the large number of different crystalline rocks in the Ries crater (Graup, 1978; Engelhardt and Graup, 1984), a detailed rock classification based only on the composition of major elements was not possible in the present study.

Extraction of Diamonds

The extraction of impact diamonds was carried out by a cyclic acid dissolution technique. To accelerate the extraction procedure we did not use large, uncrushed whole-rock samples. Our starting material had a grain size of 90–560 µm, which was produced using a cutting mill (breaking jaws of stainless steel) and a planetary ball mill (stainless steel grinding jars), followed by sieving on a vibratory sieve shaker. Due to this procedure, most of the separated impact diamond grains were broken. The basic extraction procedure of impact diamonds was as follows: The silicates were dissolved by alternating cycles of HF–HCl and HCl mixtures at room temperature. Acids were cycled every 24 h by decanting. This procedure was repeated for 10–14 cycles. Afterwards most of the silicates were dissolved and the sample was treated with a HCl–H_3BO_3 mixture for 24 hours to completely remove HF, followed by a pure water cycle for 24 hours. The residues (~0.2–0.5 wt% of the initial sample mass) were filtered and inspected under the optical microscope. The residues contain the carbon polymorphs (graphite and impact diamonds) together with some partially dissolved minerals (e.g., zircon). Due to the filter paper used and the handling methods, only diamond grains with a diameter greater than 15–20 µm could be extracted and investigated with this method. For each sample, a minimum initial sample mass of 30 g was used for the extraction procedure. Depending on the size of the samples, up to 100 g of material was dissolved. The detection limit based on 30 g sample was ~0.02 ppm diamond assuming a homogeneous distribution of diamond grains with an average weight of 0.5 µg within the sample.

Investigation of Impact Diamonds

For the identification of impact diamonds, the grains in the acid dissolution residues were investigated by Micro Raman spectroscopy (DILOR LabRam; HeNe laser beam 632.8 nm; laser power on sample 0.65 mW·µm^{-2}; 50 × objective lens). This is the best method to distinguish impact diamonds from other, optically similar, phases. Confirmed impact diamond grains were analyzed under the optical microscope for size, shape, color, and birefringence. For an estimation of the diamond concentration

TABLE 1. COMPILATION OF SAMPLE MATERIALS FROM THE RIES CRATER USED IN THE PRESENT STUDY

Locality	Whole-rock samples	Matrix	Melt fragment	Crystalline fragment SIII	Crystalline fragment SII	Crystalline fragment SI/0
Lithic impact breccia (Bunte Breccia)						
Aumühle	2 / 0					
Otting	1 / 0					
Ronheim	1 / 0					
Cataclastic crystalline breccia						
Appetshofen	1 / 0					
Langenmühle	5 / 0					
Meyers Keller	1 / 0					
Unterwilflingen	3 / 0					
Fallout suevite						
Altenbürg	1 / 1	1 / 0	1 / 0	3 / 0		
Amerdingen	1 / 1					
Aumühle	4 / 0		2 / 2	17 / 0	8 / 0	1 / 0
Bollstadt			2 / 0			
Christgarten			1 / 0			
Hainsfarth			1 / 1			
Heerhof			2 / 1			
Hohenaltheim	1 / 1	1 / 1	1 / 1			
Mauren	1 / 1					
Oberringingen	1 / 1	1 / 1	3 / 3			
Otting	1 / 0	1 / 1	3 / 2	2 / 0	1 / 0	
Seelbronn	3 / 2	2 / 0	3 / 1	27 / 1		
Unterwilflingen	1 / 0	1 / 1	4 / 1			
Zipplingen	2 / 0		1 / 1	1 / 0		
Clast-rich impact melt rock						
Amerbach	1 / 0					
Polsingen	1 / 0			4 / 0	4 / 0	1 / 0
Total	32 / 7	7 / 4	24 / 13	52 / 1	16 / 0	2 / 0

Note: The samples are listed according to the type of impactite. The number before the slash gives the total number of samples investigated per locality and the number after the slash gives the number of diamond-bearing samples (see Table 2).

within the samples, each diamond grain was weighed with an ultra-microbalance (SARTORIUS SC 2). Chemical composition, shape, and surface of the impact diamonds were investigated with a scanning electron microscope (SEM; JEOL JSM 6300 with RÖNTEC EDX system).

RESULTS

Investigation of Impactite Whole-Rock Samples

The investigation of whole-rock samples from different impactites (cataclastic crystalline breccias, lithic impact breccia, fallout suevite, clast-rich impact melt rock) from the ejecta blanket of the Ries crater show clearly that impact diamonds were observed only in the fallout suevite (Fig. 2A; Tables 1 and 2). Impact diamonds were found in ~45% of the suevite whole-rock samples. In contrast to the Popigai crater (Masaitis, 1998), a preferred regional distribution of impact diamonds in the fallout suevite of the Ries crater could not be observed (Fig. 1). However, the concentration of impact diamonds in the fallout suevite displays great variability, even in single outcrop, ranging between <0.02 (proposed detection limit) and 0.26 ppm diamond (Fig. 2B; Table 2). If all of our analyzed suevite whole-rock samples were taken into consideration, the mean diamond concentration of the Ries fallout suevite could be estimated at ~0.1–0.2 ppm.

Distribution of Impact Diamonds in the Fallout Suevite

The distribution of impact diamonds in the fallout suevite was investigated by subdividing whole-rock samples into their specific components: matrix (clasts and melt fragments >0.5 cm diameter were discarded), crystalline, and melt fragments (Table 1). The results are shown in Figure 3A and Tables 1 and 2.

Impact diamonds were found in ~50% of the analyzed matrix samples and melt fragments. The large melt fragment Au-G from the suevite of Otting was subdivided for the diamond extraction into a core sample (Au-G-core; rich in vesicles) and a rim sample (Au-G-rim; poor in vesicles and rich in mineral and rock clasts). Both fractions reveal similar diamond concentration, which indicates that the diamonds are relatively homogenously distributed within this melt fragment. The melt fragment SIE988 from the dike suevite of Unterwilflingen was completely transformed into kaolinite and montmorillonite due to pervasive post-impact hydrothermal alteration. This sample contains impact diamond grains which have a similar appearance in comparison to impact diamonds from only slightly altered melt fragments. Therefore the impact diamonds are stable under the conditions of hydrothermal alteration.

In contrast to the matrix samples and melt fragments, only one diamond-bearing crystalline fragment of shock stage III could be detected, whereas ~50% of the fragments in this shock stage contain still recognizable graphite (Fig. 3A). Figure 3B indicates that the analyzed suite of crystalline fragments represents all major lithological units of the crystalline basement (Graup, 1978; Engelhardt and Graup, 1984; Engelhardt, 1997): (*a*) Granitic composition (22 samples) compared to 22% granites and 7% orthogneisses in the crystalline clast population of the suevite; (*b*) granodioritic composition (24 samples) compared to 21% biotite-plagioclase-gneiss, 10% mixed-, blastomylonite- and migmatic-gneisses, and 7% granodiorite in the crystalline

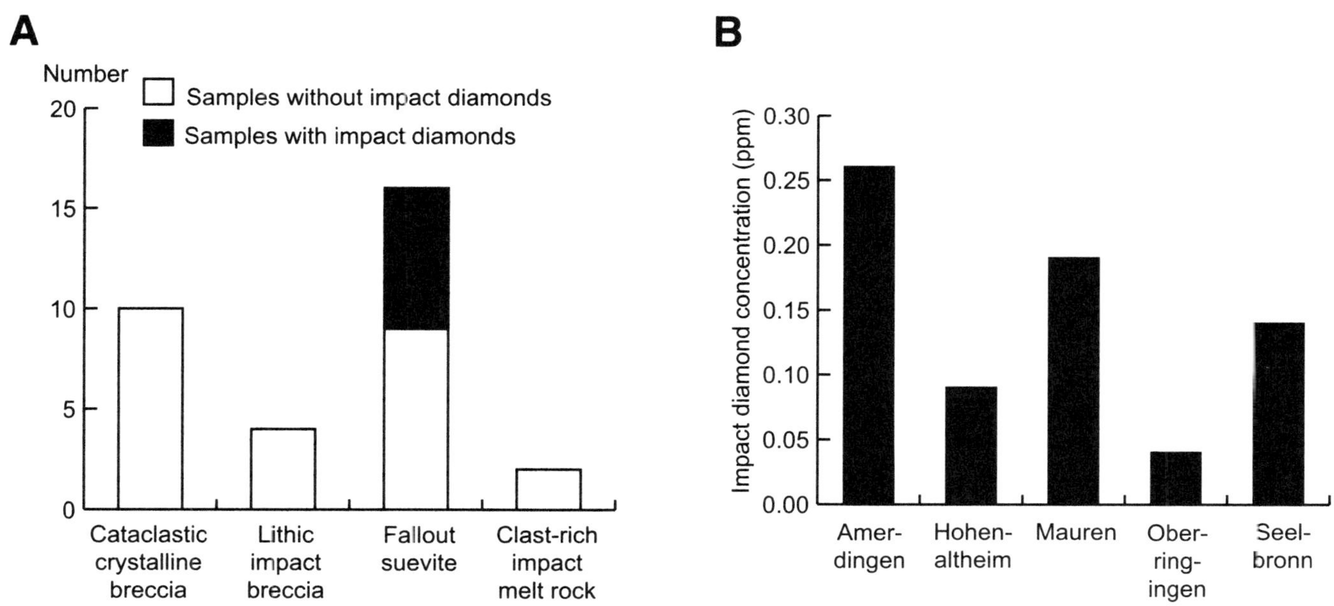

Figure 2. (A) Distribution of impact diamonds in the different lithological units (cataclastic crystalline breccias, lithic impact breccia, fallout suevite, clast-rich impact melt rock) of the Ries ejecta blanket; see also Tables 1 and 2. (B) Estimates of impact diamond concentrations in suevite whole-rock samples of selected localities (see Fig. 1 and Table 2).

TABLE 2. COMPILATION OF DIAMOND-BEARING SAMPLES

Locality	Lithology	Sample type	Sample number	Diamond concentration (ppm)	Number of diamond grains
Altenbürg	Suevite	Whole rock	SIE97121	n.d.	1
Amerdingen	Suevite	Whole rock	La9930 Am	0.26	3
Aumühle	Suevite	Melt fragment, core/rim	Au-G core/rim	0.04 / 0.02	4 / 3
Aumühle	Suevite	Melt fragment	SIE9781	0.21	15
Hainsfarth	Suevite	Melt fragment	Ha-G1	0.20	5
Heerhof	Suevite	Melt fragment	SIE97103	0.03	1
Hohenaltheim	Suevite	Whole rock	SIE97132	n.d.	1
Hohenaltheim	Suevite	Matrix	SIE97132M	0.03	1
Hohenaltheim	Suevite	Melt fragments	SIE97132G	0.10	4
Mauren	Suevite	Whole rock	La9934 Mau	0.19	6
Oberringingen	Suevite	Whole rock	La9928 Ob	0.04	1
Oberringingen	Suevite	Matrix	La9928M Ob	0.39	3
Oberringingen	Suevite	Melt fragment	La9927a Ob	1.48	24
Oberringingen	Suevite	Melt fragment	La9927b Ob	0.61	4
Oberringingen	Suevite	Melt fragment	La9927c Ob	0.17	7
Otting	Suevite	Matrix	SIE97120M	0.09	1
Otting	Suevite	Melt fragment	Ot-G	0.14	1
Otting	Suevite	Melt fragment	Ot-G3	0.02	2
Seelbronn	Suevite	Whole rock	SIE97123	n.d.	1
Seelbronn	Suevite	Whole rock	SIE985	0.14	1
Seelbronn	Suevite	Melt fragment	SIE9721	0.10	10
Seelbronn	Suevite	Crystalline fragment SIII	Se-2	0.62	31
Unterwilflingen	Suevite	Matrix (hydrothermally altered)	SIE983	0.42	9
Unterwilflingen	Suevite	Melt fragment (hydrothermally altered)	SIE988	0.37	5
Zipplingen	Suevite	Melt fragment	Zi-G1	0.33	22

Note: n.d.—no data

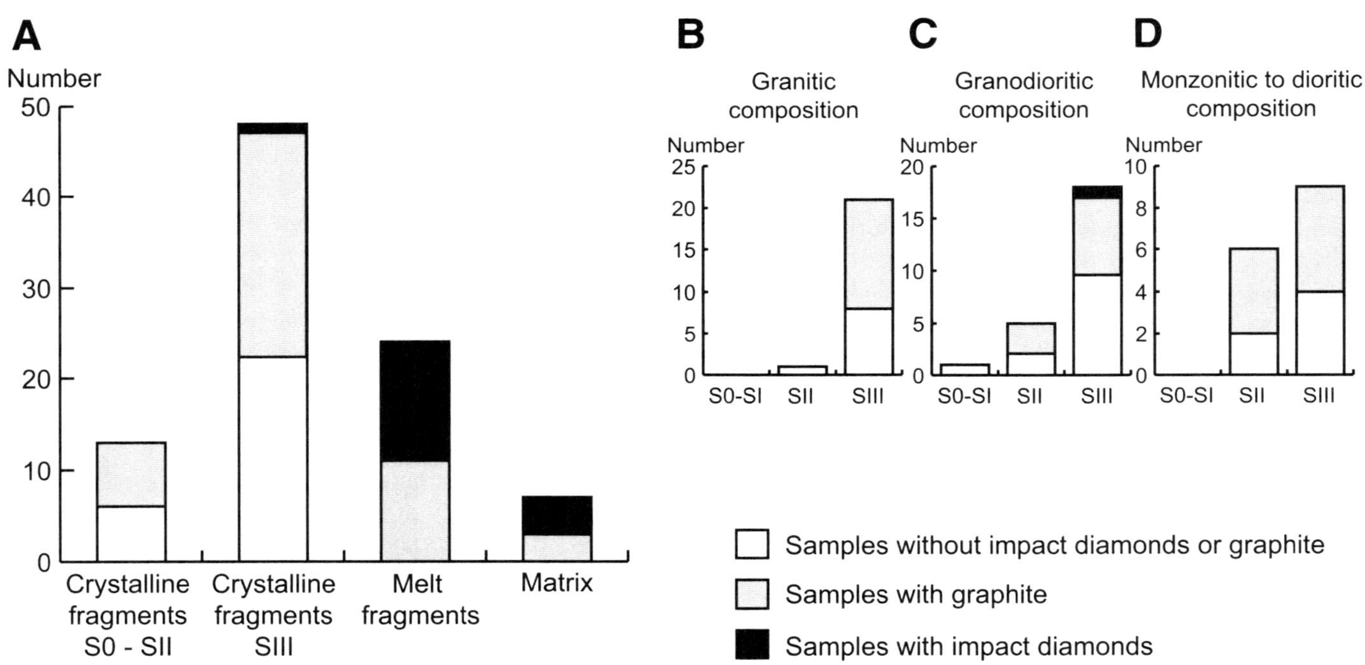

Figure 3. Distribution of impact diamonds and graphite in the different suevite components and shocked crystalline fragments of different chemical composition (see Tables 1 and 2). (A) Number of samples with and without impact diamonds or graphite from melt fragments, matrix samples and crystalline fragments of shock stage 0 to II and III. Note that the diamond-bearing melt fragments and matrix samples contain also graphite. (B–D) Samples with and without impact diamonds or graphite in shocked crystalline fragments of variable chemical composition. (B) Granitic composition includes granites and orthogneiss. (C) Granodioritic composition includes biotite-plagioclase-gneiss, mixed-, blastomylonite-, and migmatic-gneisses and granodiorite. (D) Monzonitic to dioritic composition includes amphibolite, sphene-fleck-rock, monzonite, diorite, gabbro, cordierite- and granulitic-gneisses, and dike rocks.

clast population of the suevite; and (*c*) monzonitic to dioritic composition (15 samples) compared to 14% amphibolite, 8% sphene-fleck-rock, 5% monzonite, diorite and gabbro, 3% cordierite- and granulitic-gneisses, and 3% dike rocks in the crystalline clast population of the suevite. Crystalline fragments that have high, primary contents of carbon (e.g., cordierite-gneisses) are rare within the suevite. They constitute ~2% of the crystalline clast population of the fallout suevite (Graup, 1978; Engelhardt, 1997), and these fragments are distinctly smaller in comparison to other crystalline clasts, e.g., granite. Therefore the probability of finding such highly shocked fragments containing impact diamonds is also rather low.

The diamond concentrations show great variability in the specific components of the fallout suevite similar to the suevite whole-rock samples (Fig. 4). The highest observed diamond concentration in this study was 1.48 ppm in a melt fragment from Oberringingen (sample La9927a Ob). Average diamond concentrations estimated on the basis of all analyzed samples are 0.3 ppm for the melt fragments and 0.2 ppm for the matrix samples.

Diamond-Bearing Crystalline Fragment Se-2

The diamond-bearing crystalline fragment Se-2 (Figs. 5A and 6A) from the suevite of Seelbronn was originally ~8 cm in diameter. This sample shows a pumice-like texture and the primary texture is largely destroyed. A relic gneiss foliation is indicated by a succession of white and dark schlieren. The high degree of shock metamorphism can be easily recognized macroscopically due to the pumice-like texture. This fragment belongs to shock stage III according to Stöffler (1971). Under the microscope, recrystallized feldspar melt, partially recrystallized diaplectic quartz glass, which contains tube-like coesite, and biotite, which displays birefringence and kink bands as well as thermal decomposition, have been observed (Fig. 6A). The shock features of the minerals indicate a shock pressure of 45–55 GPa (Stöffler, 1971, 1972; Stöffler and Langenhorst, 1994). Secondary mineralizations consist of quartz, calcite, and chlorite. The chemical analysis (Siebenschock et al., 1998) shows a granodioritic composition for this sample. Low TiO_2, P_2O_5, Cr, and V concentrations indicate that this fragment was originally most likely a biotite-plagioclase-gneiss when compared with chemical data of Engelhardt and Graup (1984). The only carbon phase observed in the extraction residue of this fragment is impact diamond. Therefore the phase transformation of graphite to diamond was complete.

Diamond-Bearing Melt Fragments of the Fallout Suevite

Diamond-bearing melt fragments (e.g., Figures 5C and 6B) show a fluidal texture, irregularly distributed vesicles, and numerous mineral and rock clasts of different shock stages, which are mainly derived from the crystalline basement. Mineral and rock clasts were discarded during the preparation, if possible. The matrix of all analyzed melt fragments is more or less recrystallized indicating that these samples come from the central, slowly cooled part of the suevite layer. In most of the analyzed melt fragments, schlieren-like zones of highly shocked mineral assemblages were observed, which consist of recrystallized or rare glassy feldspar and quartz melt, coesite and thermally decomposed biotite (Fig. 6B). In the melt fragment La9927b Ob, graphite could be observed within such a schlieren-like zone. The shock pressure estimated for the schlieren-like zones is similar to that of the crystalline fragment Se-2. Most likely these schlieren-like zones within the melt fragments are the carrier of impact diamonds. A test of this hypothesis would be the in situ detection of diamonds in such zones, but this was not successful during this study. In all extraction residues of melt fragments, the carbon phases were graphite and diamond.

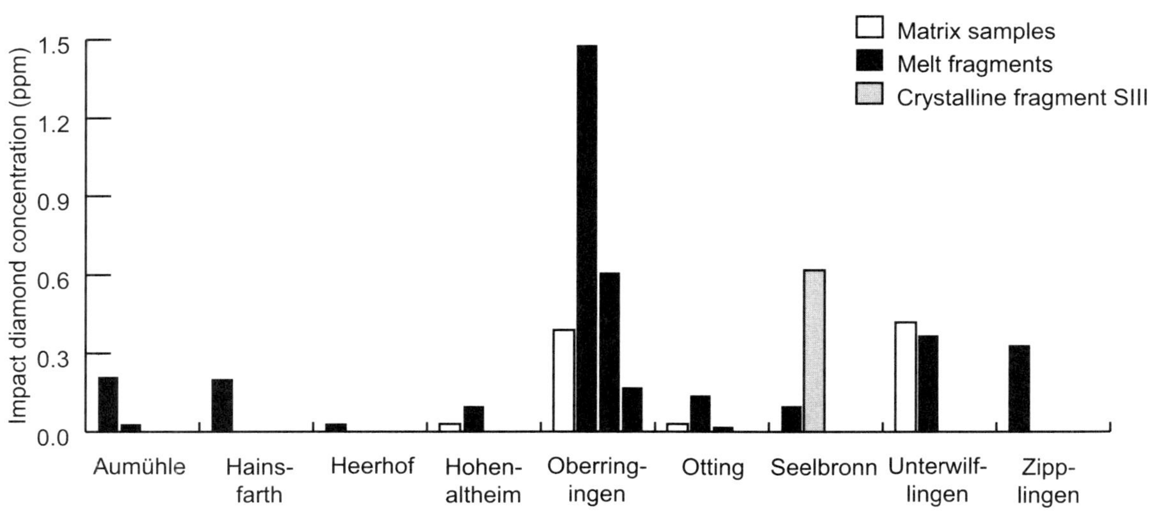

Figure 4. Estimates of impact diamond concentrations in crystalline fragments, melt fragments, and matrix samples of the fallout suevite from different localities (see Fig. 1 and Table 2).

A
m
fr
B
C
D
E
F
G
H
mf
m
m
I
J

Figure 5. Host rocks of impact diamonds and associated diamonds from the Ries crater. (A) Diamond-bearing crystalline fragment (fr; size 3.5 × 3 cm; sample Se-2) of shock stage III displaying relic metamorphic foliation embedded in the suevite matrix (m); Seelbronn. (B) Greenish diamond grain extracted from Se-2 (see A; Se-2/grain 11; diameter ~105 µm). (C) Melt fragment (size 27 × 11 cm) in the suevite displaying fluidal textures especially in the center of the fragment (sample Au-G; Aumühle quarry). For diamond extraction, this sample was subdivided into a core sample (Au-G-core; rich in vesicles) and a rim sample (Au-G-rim; poor in vesicles and rich in mineral and rock clasts). (D, E) Diamond grains extracted from the Au-G-rim sample (see C). (D) Olive-green diamond grain with some dark areas on the right hand side of the grain (Au-G-rim/grain 1; size 100 × 60 µm). (E) Light-olive-green diamond grain (Au-G-rim/grain 2; size 50 × 40 µm). (F) Large, olive-gray diamond grain from a melt fragment of the suevite of Mauren (La9934Mau/grain 1; size 140 × 60 µm). (G) Diamond grain with a greenish-brownish color in the rim region and a dark center from a melt fragment of the suevite of Zipplingen (Zi-G1/grain 2; size 120 × 75 µm). (H) Dike suevite sample from Unterwilflingen (9 × 7 cm). This sample consists of a greenish colored matrix (m; sample SIE983) and a heavily altered, white-colored melt fragment (mf; sample SIE988) displaying a relic fluidal texture. (I) Olive-brownish diamond grain extracted from the heavily altered melt fragment of the dike suevite (SIE988/grain 5; size 90 × 90 µm). (J) Light greenish diamond grain extracted from the matrix of the dike suevite (SIE 983/grain 3; size 85 × 35 µm). All diamond grains in this figure are shown in plane polarized light under the optical microscope.

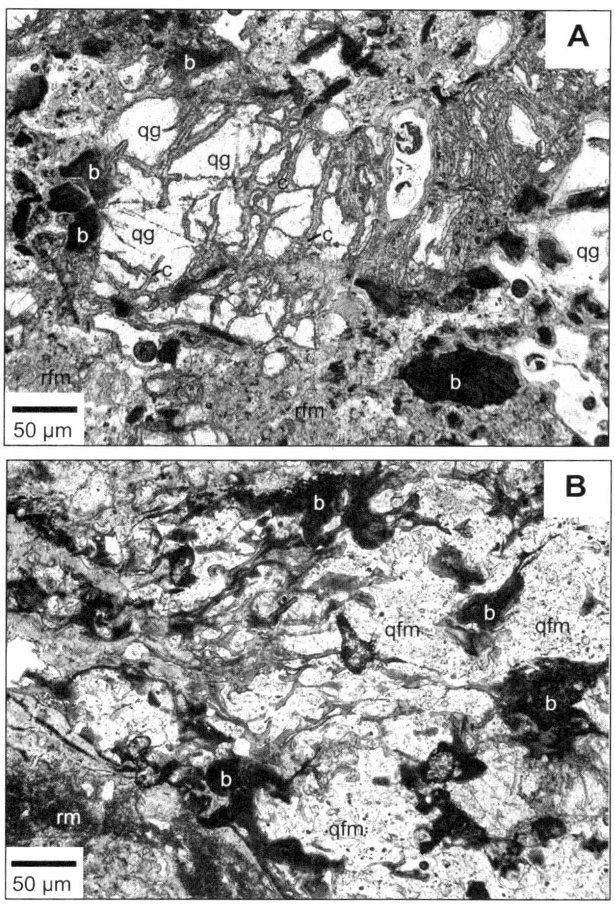

Figure 6. (A) Thin section of the highly shocked, diamond-bearing crystalline fragment Se-2 from the suevite of Seelbronn. The sample consists of diaplectic quartz glass (qg) with tube-like coesite aggregates (c), recrystallized feldspar melt (rfm) and biotite (b) which is (at the rims) partially thermally decomposed. This paragenesis indicates shock stage III after Stöffler (1971) and corresponds to a shock pressure of 45–55 GPa. (B) Schlieren-like zone of a highly shocked mineral assemblage in melt fragment La9927c Ob from the suevite of Oberringingen. In the recrystallized melt matrix (rm), thermally decomposed biotite (b) and recrystallized quartz and feldspar melt (qfm) can be observed. Both images in transmitted, plane polarized light.

Diamond-Bearing Fallout Suevite Matrix

The matrix of the fallout suevite is a clastic mixture of fine-grained mineral and various rock fragments of different shock stages together with fine-grained melt particles altered to montmorillonite and zeolites (Engelhardt and Graup, 1984; Osinski et al., 2003). Chemical investigations do not show great differences in composition between suevite whole-rock and matrix samples from any single locality. The diamond carrier in the suevite matrix is unknown. It is assumed that the diamond carriers could be small graphite-bearing crystalline fragments of shock stage III. Similar to the melt fragments, the residues of matrix samples show the carbon phases graphite and diamond.

Impact Diamonds

The impact diamonds show an identical appearance independent of their source (suevite whole rock, crystalline fragment, melt fragment, matrix) and locality (Figs. 5 and 7). They occur as thin, pseudo-hexagonal plates, which are composed of a parallel intergrowth of much smaller platelets. Therefore the plates show a distinct cleavage parallel to the platelets. The diameter of the diamond grains reaches up to 300 µm, but the thickness is much smaller, mostly <20 µm. Most of the grains display fractures which intersect with angles between 110° and 130° (Fig. 7C) or they are broken along such fracture systems. Some diamond grains display twinning striations in certain parts. The morphology of the impact diamond grains resembles unambiguously the precursor graphite. Due to their polycrystalline structure as shown by transmission electron microscopy (e.g., Siebenschock et al., 1999) the diamond grains are paramorphs after graphite. The term paramorphs instead of pseudomorphs is used to emphasize the internal polycrystalline structure of the diamond grains. The color is most commonly olive-green, grayish-green, or grayish-black. Yellow, pale green, brownish-black, or rarely, colorless grains were observed as well. The surface as seen in the optical microscope is uneven to rough and displays a vitreous to sub-metallic luster. Thin diamond grains are transparent and birefringent.

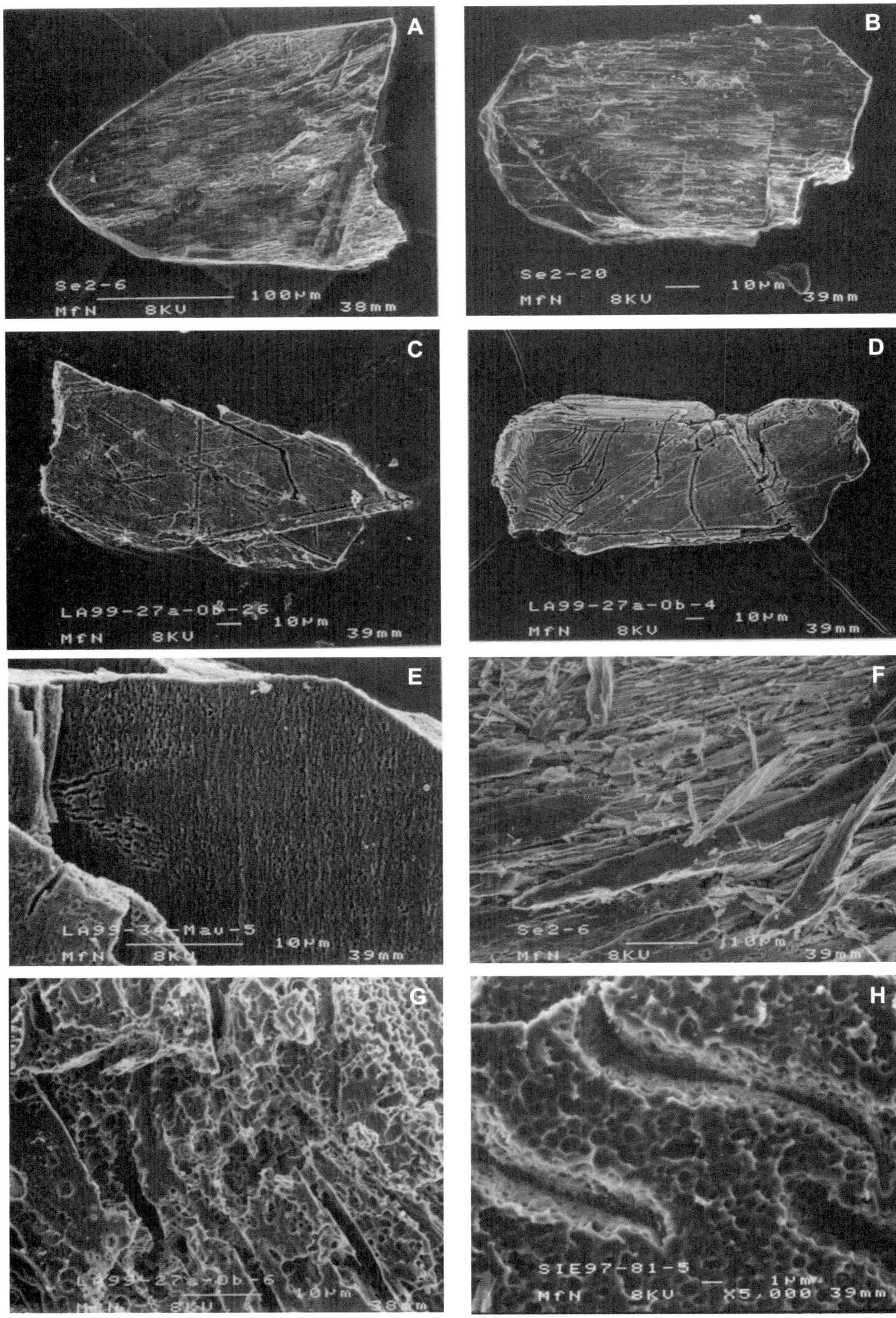

Figure 7. Scanning electron microscope images (secondary electrons) of impact diamonds from the Ries crater and their surfaces. (A, B) Impact diamonds from the crystalline fragment Se-2, suevite of Seelbronn. Both grains display a fibrous surface which is shown in detail in F. (C, D) Impact diamonds from melt fragment La9927a Ob, suevite of Oberringingen. Fracture systems could be observed in C whereas the surface of D displays in addition curved depressions. (E) Rough surface caused by small inhomogeneities; on the left-hand side occur some extensional microfractures (whole-rock sample La9934 Mau; Mauren suevite). (F) Fibrous surface and structure, enlargement of the upper right hand corner of the diamond grain shown in A. (G, H) Spongy surface and structure of diamond grains from melt fragments of the suevite (La9927a Ob from Oberringingen, and SIE9781 from Aumühle, respectively). The spongy diamond grain H is cut by two extensional microfractures.

At higher magnifications with the SEM, the surface of the impact diamond grains (Fig. 7) varies independent of their source even within one grain. The following types could be distinguished: (*a*) smooth surface; (*b*) isolated pores with diameters of ~1 μm occurring on more or less smooth surfaces; (*c*) linear or curved depressions crosscutting more or less smooth surfaces (Fig. 7D); (*d*) rough surface (Fig. 7E) caused by small inhomogeneities (diameter <1 μm) which define sometimes a lineation; (*e*) fibrous surface (Figs. 7A, 7B, and 7F) caused by elongated anhedral platelets with sizes up to 30 μm × 5 μm x < 0.5 μm; and (*f*) spongy surfaces (Figs. 7G and 7H) caused by closely arranged pores of ~1 μm diameter. Relatively frequent are the types (*e*) and (*f*), which display striped patterns and a streaky appearance, respectively, under the optical microscope. Smooth surfaces do not contain graphite based on Raman mapping and therefore cannot be interpreted as graphite coatings which were frequently observed by Masaitis (1998) on the surface of Popigai diamonds. In comparison, pure graphite grains from the crystalline basement rocks of the Ries crater have generally smooth surfaces and display sometimes striped surface patterns. Surface features similar to the types (*b–f*) of the impact diamonds could not be observed on graphite grains. The surface patterns (*b–d*) are a typical feature also of impact diamonds from other impact craters, e.g., Popigai, Puchezh-Katunki, and Lappajärvi (Vishnevsky et al., 1997; Masaitis, 1998; Masaitis et al., 1998). Such surface patterns are the result of a natural corrosion or etching process at the contact with impact melt, or of a heating process as proposed by Masaitis (1998). Cross sections of impact diamonds show that most of the grains have a fibrous or spongy internal structure similar to the surface patterns (*e*) and (*f*). This characteristic porosity is most likely the result of some specific volume decrease during the phase transformation of graphite to diamond, rather than the result of a natural corrosion or etching process. In addition to the above-described structures, the volume decrease causes relatively frequent extensional microfractures in the diamond grains (Figs. 7E and H). Compositional analysis and elemental mapping of the impact diamonds with the SEM-EDX reveal only carbon. Diamond-silicon carbide intergrowths could not be detected during this study.

Irrespective of their color, surface structure and source, ~75% of the diamond grains exhibit a strong photoluminescence (Fig. 8A), which makes the detection of Raman bands impossible. This is also a typical feature of impact diamonds from the Popigai (Phelps, 1997; Boudeulle et al., 1999) and Lappajärvi craters (Moroz et al., 2003). The remaining grains display a minor photoluminescence. A broad Raman band at 1305–1330 cm^{-1} and sometimes an additional broad Raman band at 1560–1575 cm^{-1} could be observed (Figs. 8B and 8C). The Raman band at 1305–1330 cm^{-1} is generally higher in intensity than the latter which reaches 0–90% of the intensity of the band at 1305–1330 cm^{-1}. In contrast, single crystal grains of natural (e.g., from Brasilia, sample MfN 1997–2465) as well as of synthetic diamond (e.g., diamonds from saw blade, diamond polishing paste with diamonds of 1–2 μm grain size) display a sharp Raman band at 1331–1333 cm^{-1}. This Raman band is shifted for impact diamond grains to lower frequencies (1305–1330 cm^{-1}) and reveals a FWHM (full width at half maximum) ranging from 70–170 cm^{-1} (Fig. 8D). There is no correlation between the wavenumber of this Raman band and the FWHM visible. A few impact diamonds and also highly shocked graphite grains (e.g., sample SIE9733; Lapke et al., 2000) from the Ries crater reveal a broad and strong Raman band at 1420–1440 cm^{-1} (Fig. 8E). Mapping of these grains with the SEM-EDX display only carbon, and exclude the presence of non-carbon phases. Therefore this Raman band is most likely caused by amorphous diamond-like carbon (Yoshikawa et al., 1995).

For comparison with the impact diamonds, graphite grains from crystalline breccias and crystalline fragments of the suevite of different shock stages were investigated. Graphites occur as black, non-transparent, more or less hexagonal crystals, which show a distinct cleavage. Some of the graphite grains display Veselowski twinning. The diameter of the graphite grains reaches up to 1 mm, whereas the thickness is much smaller, mostly <100 μm. Graphite grains have generally smooth surfaces and some irregular fractures. With increasing shock stage, an increasing number of parallel microfractures, oriented normal to the basal plane, were observed. As a consequence, highly shocked graphite in shock stage III displays striped surface patterns which are caused by parallel, densely packed microfractures. Graphite grains in shock stages II and III are frequently broken and occur, therefore, in form of blocky fragments. The Raman investigation (Lapke et al., 2000) of graphite of shock stage 0 (sample SIE9795) normal to the basal plane shows a sharp, strong Raman band at 1580 cm^{-1} and a weak, broad band at 1330–1345 cm^{-1} with an intensity of 5–20% in comparison to the 1580 cm^{-1} band. At shock stage I and II (samples SIE97116, SIE9777) the strong band decreases in intensity, moves to 1575–1578 cm^{-1}, and starts to broaden. The broad band at 1330–1340 cm^{-1} increases in intensity up to 30–40% of the band at 1575–1578 cm^{-1}. At shock stage III (sample SIE9733), the intensity of this band reaches 30–90% of the band at 1575–1578 cm^{-1}. Additionally, all of the graphite spectra of shock stage 0–III display a weak band at 1600–1620 cm^{-1}. None of the graphite grains show photoluminescence under the laser beam.

DISCUSSION

Formation of Impact Diamonds

Several mechanisms have been proposed for the graphite-diamond phase transition in the past including: (*a*) solid state martensitic phase transformation (Masaitis et al., 1990; Erskine and Nellis, 1991; DeCarli, 1998; Vishnevsky et al., 1997; Langenhorst et al., 1998; Masaitis, 1998); (*b*) thermally activated, diffusion controlled formation mechanism based on a solid-vapor/liquid-solid sequence of phase transformations (Kleiman et al., 1984; DeCarli, 1995, 1998; Vishnevsky et al., 1997; Kenkmann et al., 2002); (*c*) ultrafast annealing of a glassy carbon phase (Pujols and Boisard, 1970); (*d*) dissociation of twin boundaries,

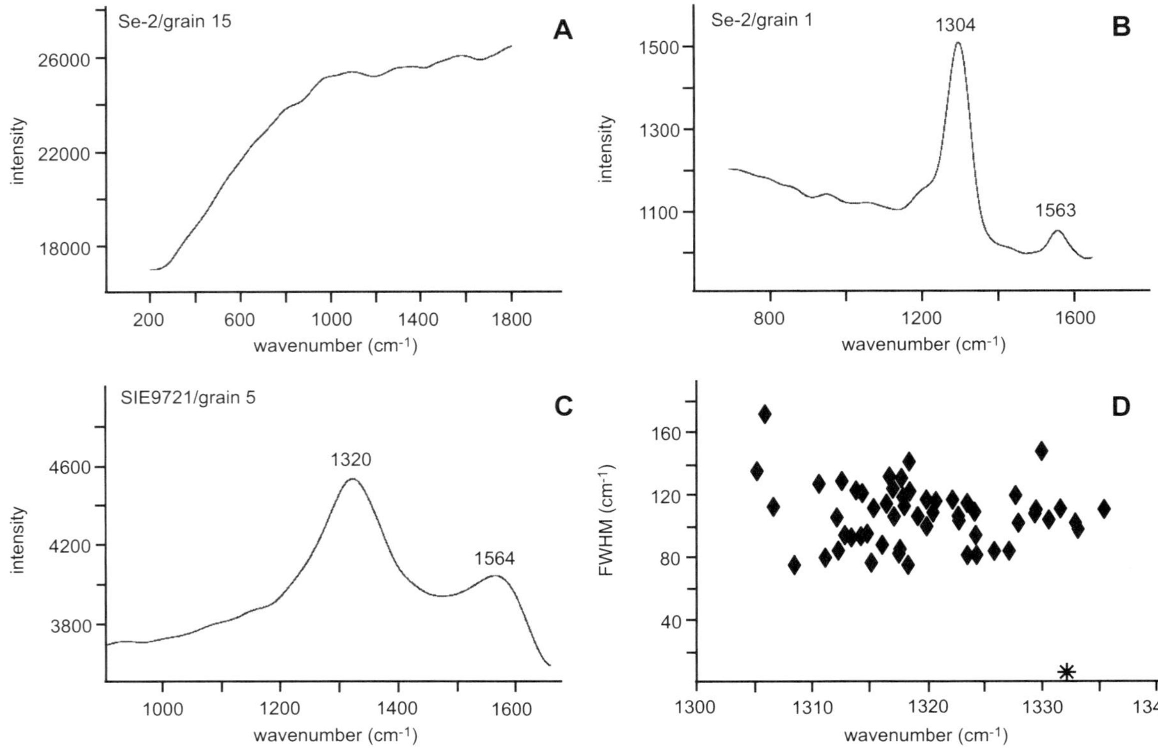

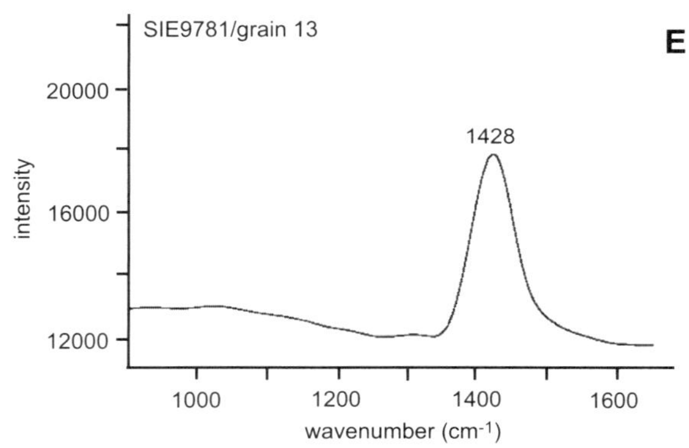

Figure 8. (A–C) Raman spectra of impact diamond grains from the Ries crater. (A) Diamond grain with strong photoluminescence. (B, C) Grains with minor photoluminescence reveal two broad Raman bands at 1304 and 1563 cm^{-1}, and 1320 and 1564 cm^{-1}, respectively. (D) Wavenumber position of the "diamond" Raman band versus FWHM (full width at half maximum) of this band for the impact diamonds of the Ries crater. The position of a typical sharp Raman band of single crystal diamond (1333 cm^{-1}) is marked by an asterisk. (E) Raman spectra of an impact diamond with strong photoluminescence and a broad Raman band at 1420–1440 cm^{-1} which is most likely produced by amorphous, diamond-like carbon.

which leads to partial dislocations and diamond-like stacking faults (Pujols and Boisard, 1970); and (*e*) vapor phase growth of diamond (Hough et al., 1995). Previous studies indicate that the completely different mechanisms (*a*) and (*b*) above should be primarily considered for the formation of impact diamonds. Erskine and Nellis (1991) propose that the mechanism (*a*) depends strongly on the shock temperature: a solid state martensitic phase transformation occurs at shock temperatures <2000 K, whereas at shock temperatures >3000 K, the transformation mechanism is thermally activated, diffusion controlled.

(1) Diamonds produced by the thermally activated, diffusion-controlled formation mechanism characteristically display aggregates of small diamond grains of single crystal structure in paragenesis with shocked graphite or other high-temperature phases of carbon, e.g., carbynes. For example, μm-sized grains of diamond displaying a sharp and unshifted Raman band were found together with graphite by El Goresy et al. (2001) at graphite-garnet, graphite-sillimanite and graphite-rutile interfaces in garnet-cordierite-sillimanite and garnet-biotite-gneiss fragments of shock stage II in the Ries crater. These rocks display an equilibrium shock pressure of 35–45 GPa and a post-shock temperature of 575–1175 K (Stöffler, 1971). Graphite grains which are not in contact with garnet, sillimanite, or rutile are shocked, but do not contain diamond. Therefore the phase transformation is controlled by the textural setting of the graphite grains in contact with minerals of high shock impedance leading to

distinct temperature spikes during the shock compression. These temperature spikes were neglected in the proposed transformation mechanism of El Goresy et al. (2001), but might reach or exceed the melting temperature of graphite (~4000 K at 20 GPa, ~3400 K at 35 GPa, and ~2800 K at 45 GPa, respectively; e.g., Zoltai and Stout, 1984). Therefore, in this case, the partial phase transformation of graphite to diamond can simply be explained by a thermally activated, diffusion-controlled process. In the shock experiments of Kenkmann et al. (2002), µm-sized single crystal grains of diamond as indicated by a sharp and unshifted Raman band were observed in contact with biotite melt, together with shocked graphite grains in contact with quartz and feldspar glass. In this case, the formation of diamond is thermally activated by the biotite melting which released water that produced hot spots during vaporization (Kenkmann et al., 2002). Close to these hot spots, the melting temperature of graphite was exceeded and the phase transformation of graphite to diamond based on a solid-vapor/liquid-solid sequence occurred at shock pressures of >45 GPa. Synthetically produced diamonds by shock loading of graphite-metal mixtures or porous graphite were formed also by this type of phase transformation at temperatures above 3000 K (Kleiman et al., 1984; DeCarli, 1998).

(2) The solid-state martensitic phase transformation of graphite to diamond is characterized by the occurrence of diamond paramorphs after graphite or other carbon phases (Shafranovsky, 1985; Masaitis et al., 1990). These paramorphs are well known as impact diamonds from many craters, all of similar appearance (Gurov et al., 1995, 1996; Masaitis et al., 1972, 1995, 1997, 1998; Masaitis, 1998; Koeberl et al., 1995, 1997; Vishnevsky et al., 1997; Valter et al., 1992). Depending on the shape of the precursor carbon phase (e.g., graphite, coal) the shape of the shock-produced diamond grains varies from pseudo-hexagonal plates to anhedral grains. These paramorphs consist internally of nm to at most µm-sized diamond grains, which possibly may also contain lonsdaleite (e.g., Vishnevsky et al., 1997; Langenhorst et al., 1998). Raman spectra of such diamond paramorphs display a characteristic broadening of the "diamond" Raman band caused by this microstructure (Masaitis et al., 1990; Yoshikawa et al., 1995; Vishnevsky et al., 1997; Boudeulle et al., 1999). Based on this microstructure, Erskine and Nellis (1991) proposed a two-stage phase transformation: the first step is a martensitic phase transformation of graphite to lonsdaleite followed by a second, possibly also martensitic phase transformation, of lonsdaleite to diamond. This type of phase transformation leads, above a certain threshold pressure, to the complete in situ transformation of the carbon precursor phase to diamond. If the post-shock temperature exceeds some threshold, a partial or complete back transformation of diamond to graphite occurs (Masaitis, 1998). In shock experiments with graphite crystals, this type of phase transformation was observed at shock pressures of 30 GPa within ~10 ns if the shock loading was applied normal to the basal planes of graphite (Erskine and Nellis, 1991). In this case the shock temperature (1800 K) was clearly lower than the melting temperature of graphite (~3550 K; e.g., Zoltai and Stout, 1984). In contrast DeCarli (1998) propose for this type of phase transformation a minimum pressure of 15 GPa, a shock temperature above 1300 K and a post-shock temperature lower than 2000 K.

This study shows unambiguously that the impact diamond grains from the Ries crater resemble the morphology of the precursor graphite. Raman spectra of these diamond grains show a broadening of the "diamond" Raman band, which is typical for microcrystalline paramorphs. Therefore, a solid state martensitic phase transformation of graphite to diamond is proposed for the formation of these impact diamonds, consistent with previous studies. The shock pressure necessary for the phase transformation is on the order of 45–55 GPa in the highly shocked Se-2 crystalline fragment, which displays a complete transformation of graphite to diamond. The impact diamonds in the melt fragments occur most probably in schlieren-like zones of highly shocked mineral assemblages which have suffered the same shock pressure as melt-rich zones of the diamond-bearing crystalline fragment Se-2. A similar source of impact diamonds is also proposed for the suevite matrix. The accompanying post-shock temperature for shock pressures of 45–55 GPa in the case of quartzo-feldspathic rocks is in the range of 1175–1775 K (Stöffler, 1971). In the Popigai crater, impact diamonds could be observed in an extended pressure range from 35 to 50 GPa and from 50 to 60 GPa as grains which were partially transformed back to graphite (Masaitis, 1998). In contrast to suevite, the shock pressures in the lithic impact breccia (Bunte Breccia) and the cataclastic, crystalline breccias of the Ries crater are too low to produce impact diamonds. In the clast-rich impact melt rock occurring at the east rim of the Ries crater, the post-shock temperature is probably too high and impact diamonds might be completely transformed back to graphite.

Silicon carbide grains with diameters up to 100 µm or diamond-silicon carbide intergrowths as described by Hough et al. (1995) could not be found in this study based on samples for which sawing and grinding procedures were avoided during the preparation. However, we found some green and blue silicon carbide grains of ~25 µm diameter in a sawed suevite whole-rock sample of Otting at the beginning of our study. These findings suggest contamination, and this sample was excluded therefore from our study. The occurrence of silicon carbide grains (Hough et al., 1995) is therefore doubtful and could neither be confirmed by our study nor by the work of other groups (Abbott et al., 1996, 1998; Valter et al., 1998; Thiele and Scherer, 2000; El Goresy et al., 2001, 2002, 2003).

Raman Characteristics of Impact Diamonds

Many impact diamond grains from the Ries crater display a strong photoluminescence, which makes the detection of Raman bands impossible. Also, impact diamonds from other craters, e.g., Popigai and Lappajärvi (Phelps, 1997; Boudeulle et al., 1999; Moroz et al., 2003) as well as many natural, endogenous diamonds display a strong photoluminescence, which is caused by impurities and lattice defects (e.g., Moroz et al., 2003).

Single crystals of natural, endogenous diamond (e.g., from Brasilia; sample MfN 1997-2465) and synthetical diamond (e.g., diamonds from saw blade, diamond polishing paste with diamonds of 1–2 μm grain size) display a sharp and intense Raman band at 1331–1333 cm^{-1}, which corresponds very generally to the main C-C bond vibration in diamond (Nasdala and Massone, 2000). In contrast, impact diamond grains (paramorphs) from the Ries crater that possess a minor photoluminescence show a broad Raman band, which is shifted to lower frequencies at 1305–1330 cm^{-1}, and an additional Raman band at 1560–1575 cm^{-1}, which is in comparison to the first one generally lower in intensity. Broadening and shifting of the "diamond" Raman band to lower frequencies is also observed in shock-produced synthetic diamonds (Yoshikawa et al., 1995), in diamonds of ureilites (Vishnevsky et al., 1997; Greshake et al., 2000; Mostefaoui et al., 2001) and the Canyon Diablo iron meteorite (Miyamoto, 1998), and in impact diamonds from the Kara crater (Vishnevsky et al., 1997). In contrast, impact diamonds from the Popigai crater reveal only a broadening of this Raman band (Masaitis et al., 1990; Vishnevsky et al., 1997; Boudeulle et al., 1999). Our standard Raman spectroscopy procedure generates unshifted and sharp Raman spectra of diamond even on isolated, single-crystal grains of 1–2 μm diameter (e.g., diamond polishing paste). Therefore, broadening and shifting of Raman bands in impact diamonds is not the result of a heating effect by the laser beam as proposed by Andreyev et al. (1992). Variations of the intensity and the exposure time of the laser beam for a single analysis point of an impact diamond grain do not influence either of these effects. Nevertheless, Raman mapping shows that the shifting and broadening varies distinctly even within one grain. One possibility for the broadening of the major Raman band is the presence of lonsdaleite within the grains (Vishnevsky et al., 1997). More likely, broadening and shifting could be explained by the small crystallite size (<1 μm) of the polycrystalline impact diamonds (Yoshikawa et al., 1995).

The additional Raman band at 1560–1575 cm^{-1} is also observed in impact diamonds from Popigai (Boudeulle et al., 1999) and shock-produced synthetic diamonds (Yoshikawa et al., 1995). In contrast, this Raman band does not occur in diamond spectra of ureilites (e.g., Greshake et al., 2000; Mostefaoui et al., 2001) or the Canyon Diablo iron meteorite (Miyamoto, 1998). This Raman band could be explained by the occurrence of graphite-like nanodomains in the impact diamond grains (Boudeulle et al., 1999).

Diamond Concentration in the Suevite of the Ries Crater

This study estimates a mean diamond concentration of the Ries fallout suevite of ~0.1–0.2 ppm, which is clearly lower than a previous estimate of ~5.5 ppm diamond by Hough et al. (1995). A simple calculation can be carried out based on the following assumptions: the crystalline target of the Ries crater contains 2 wt% graphite-bearing rocks. This assumption is based on the frequency of cordierite-gneisses in the fallout suevite (Graup, 1978; Engelhardt, 1997). The cordierite-gneisses have an average carbon concentration of ~5 wt% (estimate from an investigation of graphite-bearing rocks in the Variscan crystalline basement by Hofmann, 1989). The graphite content of the biotite-plagioclase-gneisses was neglected because their graphite content is much lower, presumably <0.1 wt%. Neglecting further a small amount of shock stage II material in the lithic impact breccia, the crystalline target material of shock stages II to IV occurs almost exclusively in the suevite. Within the fallout suevite, some 25% of the crystalline target material was shocked to shock stage III (Engelhardt, 1997) and may therefore contain impact diamonds. This leads to a potential diamond concentration of 0.25 ppm for the fallout suevite. This is in good agreement to the observed diamond concentration of ~0.1–0.2 ppm. Using the volume estimate of the fallout suevite (0.13–0.25 km^3; Stöffler, 1977) and the observed diamond concentration (0.1–0.2 ppm), a diamond content of 27–105 t could be assumed for the Ries fallout suevite. Within the fallback or crater suevite, the content of material shocked to shock stage III and the melt content is much lower (Engelhardt, 1997), and therefore the diamond concentration should be clearly lower than in the fallout suevite.

Impact Diamonds as Impact Criteria

The investigation of the highly altered dike suevite of Unterwilflingen reveals that impact diamonds have survived hydrothermal alteration without any change. In contrast, silicates typically used as shock indicators, e.g., quartz with planar deformation features, quartz or feldspar glass and coesite, were destroyed during the alteration process. Therefore, the impact origin of this dike material could simply be derived from the existence of impact diamonds if other impact criteria would be lacking. This makes impact diamonds a very diagnostic and reliable tool for the search for impact features in deeply eroded and altered or even metamorphosed impact structures.

CONCLUSIONS

Impact diamonds in the Ries crater are present in all types of suevite (fallout suevite, suevitic dikes, fallback or crater suevite) whereas the cataclastic crystalline breccias, lithic impact breccias, and clast-rich impact melt rocks do not contain diamonds. The mean diamond concentration in the fallout suevite is ~0.1–0.2 ppm. A preferred regional distribution of impact diamonds in the fallout suevite could not be detected. Impact diamonds are contained in three main components of the suevite: graphite-bearing crystalline rock fragments of shock stage III, melt fragments, and matrix. Within the latter two, the carriers of impact diamonds are most likely small graphite-bearing crystalline rock fragments of shock stage III.

Impact diamond grains show an invariant appearance independent of their source and locality. They occur as pseudo-hexagonal plates and reach sizes up to 300 μm. Their color is commonly greenish, but black, yellow, and colorless grains were also observed. Under the optical microscope, the diamond grains are transparent and birefringent. SEM shows that most of the diamond grains have a fibrous or spongy internal structure and extensional microfractures, which leads to a characteristic porosity. This is the result of a volume decrease due to the phase transformation of graphite to diamond. Internally, the diamond grains are polycrystalline (e.g., Siebenschock et al., 1999) and therefore paramorphs after the precursor graphite. Irrespective of their color, surface structure, and origin, ~75% of the impact diamond grains exhibit a strong photoluminescence, which makes the detection of Raman bands impossible. The remaining grains display a minor photoluminescence and show a broad Raman band at 1305–1330 cm^{-1} and sometimes an additional broad Raman band at 1560–1575 cm^{-1}, which is lower in intensity than the first one. This is the result of their micro-polycrystalline internal structure and the occurrence of graphite-like nanodomains. Some impact diamond grains with an additional Raman band at 1420–1440 cm^{-1} may contain amorphous, diamond-like carbon.

The impact diamonds were most likely formed by a complete in situ shock transformation of graphite in crystalline target rocks at pressures of 45–55 GPa. The accompanying post-shock temperature in the case of quartzo-feldspathic rocks is in the range of 1175–1775 K (Stöffler, 1971). In agreement with previous studies, a solid-state martensitic phase transformation of graphite to diamond is proposed due to the strong morphologic similarity of impact diamonds with the precursor graphite. Evidence for the formation of diamond by a chemical vapor deposition and the occurrence of silicon carbide as proposed by Hough et al. (1995) could not be confirmed.

ACKNOWLEDGMENTS

This work was supported by the Märker Zementwerke GmbH, Harburg, Germany (Forschungs- und Entwicklungsvertrag, 1997–1999) and the Deutsche Forschungsgemeinschaft (DFG) grant Sto 101/36. We appreciate technical assistance by P. Czaja, I. Felber, A. Kissner, H.-R. Knöfler, H.-J. Nier, and C. Schmitt. We thank I. Krueger for the translation of Russian publications. We are grateful to F. Hörz, V. L. Masaitis, and an anonymous reviewer for comments that significantly improved the manuscript.

REFERENCES CITED

Abbott, J.I., Hough, R.M., Gilmour, I., and Pillinger, C.T., 1996, Impact diamonds in glass from Otting quarry, Ries crater, Germany: Meteoritics & Planetary Science, v. 31, Supplement, p. A5.

Abbott, J.I., Hough, R.M., Gilmour, I., and Pillinger, C.T., 1998, Nördlingen drill core 1973–Diamonds and stable carbon isotopes from a selection of crater suevite samples: European Science Foundation workshop "Impacts and the early earth": Cambridge, Abstracts vol., p. 1.

Anders, E., 1965, Diamonds in meteorites: Scientific American, v. 213, p. 26–36.

Andreyev, V.D., Nachalnaya, T.A., and Gabrusenok, E.V., 1992, The peculiarities of Raman scattering of diamond micron powders: Superhard Materials, v. 15, p. 9–16.

Bayerisches Geologisches Landesamt, editor, 1977, Ergebnisse der Ries-Forschungsbohrung 1973: Struktur des Kraters und Entwicklung des Kratersees: Geologica Bavarica, v. 77, 470 p.

Berkley, J.L., Taylor, G.J., Keil, K., Harlow, G.E., and Prinz, M., 1980, The nature and origin of ureilites: Geochimica et Cosmochimica Acta, v. 44, p. 1579–1597, doi: 10.1016/0016-7037(80)90119-2.

Bischoff, A., and Stöffler, D., 1992, Shock metamorphism as a fundamental process in the evolution of planetary bodies: Information from meteorites: European Journal of Mineralogy, v. 4, p. 707–755.

Boudeulle, M., Koeberl, C., Langenhorst, F., and Masaitis, V., 1999, Diamond-graphite phase transitions in impact diamonds from the Popigai structure (Russia): European Union of Geosciences, v. 10, p. 636.

Buchner, E., Seyfried, H., and van den Bogaard, P., 2003, $^{40}Ar/^{39}Ar$ laser probe age determination confirms the Ries impact crater as the source of glass particles in Graupensand sediments (Grimmelfingen formation, North Alpine foreland basin): International Journal of Earth Sciences, v. 92, p. 1–6.

DeCarli, P.S., 1995, Shock wave synthesis of diamond and other phases: Material Research Society Symposium, Proceedings, v. 383, p. 21–31.

DeCarli, P.S., 1998, Direct synthesis of diamond in the laboratory and in impact craters: Meteoritics & Planetary Science, v. 33, Supplement, p. A39–A40.

DeCarli, P.S., and Jamieson, J.C., 1961, Formation of diamond by explosive shock: Science, v. 133, p. 1821–1822.

El Goresy, A., Gillet, P., Chen, M., Künstler, F., Graup, G., and Stähle, V., 2001, In situ discovery of shock-induced graphite-diamond phase transition in gneisses from the Ries crater, Germany: American Mineralogist, v. 86, p. 611–621.

El Goresy, A., Gillet, P., Mostefaoui, S., Chen, M., and Masaitis, V.L., 2002, A transparent, very hard, dense, and unusually disordered form of carbon in heavily shocked gneisses from Popigai, Russia: Petrographic settings and comparison with a similar phase in shocked gneisses from the Ries: Houston, Texas, USA, Lunar and Planetary Institute, Lunar and Planetary Science Conference, v. XXXIII, abstract no. 1031 (CD-ROM).

El Goresy, A., Dubrovinsky, L.S., Gillet, P., Mostefaoui, S., Graup, G., Drakopoulos, M., Simionovici, A.S., Swamy, Y., and Masaitis, V.L., 2003, A novel cubic, transparent and super-hard polymorph of carbon from the Ries and Popigai craters: Implications to understanding dynamic-induced natural high-pressure phase transitions in the carbon system: Houston, Texas, USA, Lunar and Planetary Institute, Lunar and Planetary Science Conference, v. XXXIV, abstract no. 1016 (CD-ROM).

Engelhardt, von W., 1990, Distribution, petrography and shock metamorphism of the ejecta of the Ries crater in Germany: A review: Tectonophysics, v. 171, p. 259–273, doi: 10.1016/0040-1951(90)90104-G.

Engelhardt, von W., 1997, Suevite breccia of the Ries impact crater, Germany: Petrography, chemistry and shock metamorphism of crystalline rock clasts: Meteoritics & Planetary Science, v. 32, p. 545–554.

Engelhardt, von W., and Graup, G., 1984, Suevite of the Ries crater, Germany: Source rocks and implications for cratering mechanics: Geologische Rundschau, v. 73, p. 447–481.

Engelhardt, von W., Stöffler, D., and Schneider, W., 1969, Petrologische Untersuchungen im Ries: Geologica Bavarica, v. 61, p. 229–296.

Erskine, D.J., and Nellis, W.J., 1991, Shock-induced martensitic phase transformation of oriented graphite to diamond: Nature, v. 349, p. 317–319, doi: 10.1038/349317a0.

Foot, A.E., 1891, A new locality for meteoritic iron with a preliminary notice of the discovery of diamonds in iron: American Journal of Science, v. XLII, no. 251, p. 413–417.

Frondel, C., and Marvin, U.B., 1967, Lonsdaleite, a hexagonal polymorph of diamond: Nature, v. 214, p. 587–589.

Gilmour, I., French, B.M., Franchi, I.A., Abbott, J.I., Hough, R.M., Newton, J., and Koeberl, C., 2003, Geochemistry of carbonaceous impactites from the Gardnos impact structure, Norway: Geochimica et Cosmochimica Acta, v. 67, p. 3889–3903, doi: 10.1016/S0016-7037(03)00213-8.

Graup, G., 1978, Das Kristallin im Nördlinger Ries: Stuttgart, Enke Verlag, 190 p.

Greshake, A., Kenkmann, T., and Schmitt, R.T., 2000, In situ Raman spectroscopy of diamond in the ureilite Hammadah al Hamra 126: Meteoritics & Planetary Science, v. 35, Supplement, p. A64–A65.

Gudden, H., 1974, Die Forschungsbohrung Nördlingen 1973: Durchführung und erste Befunde: Geologica Bavarica, v. 72, p. 11–31.

Gurov, E.P., Gurova, E.P., and Rakitskaya, R.B., 1995, Impact diamonds in the craters of the Ukrainian shield: Meteoritics, v. 30, p. 515.

Gurov, E.P., Gurova, E.P., and Rakitskaya, R.B., 1996, Impact diamonds in the Zapadnaya crater: Phase composition and some properties: Meteoritics & Planetary Science, v. 31, Supplement, p. A56.

Hannemann, R.E., Strong, H.M., and Bundy, F.P., 1967, Hexagonal diamonds in meteorites: Implications: Science, v. 155, p. 995–997.

Hofmann, K.-H., 1989, Das Auftreten von Graphit in metamorphen Gesteinen in Bayern: Geologische Blätter für Nordost-Bayern, v. 39, p. 59–98.

Hough, R.M., Gilmour, I., Pillinger, C.T., Arden, J.W., Gilkes, K.W.R., Yuan, J., and Milledge, H.J., 1995, Diamond and silicon carbide in impact melt rock from the Ries impact crater: Nature, v. 378, p. 41–44, doi: 10.1038/378041a0.

Hough, R.M., Gilmour, I., and Pillinger, C.T., 1997a, Diamonds in ejecta and fireball layers of the K/T boundary in the USA and Mexico: Houston, Texas,

USA, Lunar and Planetary Institute, Large Meteorite Impacts and Planetary Evolution Conference, Sudbury, Abstract no. 6093 (CD-ROM).

Hough, R.M., Gilmour, I., Pillinger, C.T., Langenhorst, F., and Montanari, A., 1997b, Diamonds from the iridium-rich K-T boundary layer at Arroyo el Mimbral, Tamaulipas, Mexico: Geology, v. 25, p. 1019–1022, doi: 10.1130/0091-7613(1997)0252.3.CO;2.

Hough, R.M., Vishnevsky, S., Abbott, J.I., Pal'chik, N., Raitala, J., and Gilmour, I., 1999, New data on the nature of impact diamonds from the Lappajärvi impact structure, Finland: Houston, Texas, USA, Lunar and Planetary Institute, Lunar and Planetary Science Conference, v. XXX, no. 1571 (CD-ROM).

Hüttner, R., and Schmidt-Kaler, H., 1999a, Die Geologische Karte des Rieses 1: 50000, Erläuterungen zu Erdgeschichte, Bau und Entstehung des Kraters sowie zu den Impaktgesteinen: Geologica Bavarica, v. 104, p. 7–76.

Hüttner, R., and Schmidt-Kaler, H., 1999b, Wanderungen in der Erdgeschichte Band 10: Meteoritenkrater Nördlinger Ries: München, Verlag Dr. Friedrich Pfeil, 144 p.

Kenkmann, T., Hornemann, U., and Stöffler, D., 2002, Transformation of graphite to diamond in shock experiments: A Raman study: Houston, Texas, USA, Lunar and Planetary Institute, Lunar and Planetary Science Conference, v. XXXIII, no. 1052 (CD-ROM).

Kleiman, J., Heimann, R.B., Hawken, D., and Salansky, N.M., 1984, Shock compression and flash heating of graphite/metal mixtures at temperatures up to 3200 K and pressures up to 25 GPa: Journal of Applied Physics, v. 56, p. 1440–1454, doi: 10.1063/1.334116.

Koeberl, C., Masaitis, V.L., Langenhorst, F., Stöffler, D., Schrauder, M., Lengauer, C., Gilmour, I., and Hough, R.M., 1995, Diamonds from the Popigai impact structure, Russia: Houston, Texas, USA, Lunar and Planetary Institute, Lunar and Planetary Science Conference, v. XXVI, p. 777–778.

Koeberl, C., Masaitis, V.L., Shafranovsky, G.I., Gilmour, I., Langenhorst, F., and Schrauder, M., 1997, Diamonds from the Popigai impact structure, Russia: Geology, v. 25, p. 967–970, doi: 10.1130/0091-7613(1997)025<0967:DFTPIS>2.3.CO;2.

Langenhorst, F., Shafranovsky, G., and Masaitis, V.L., 1998, A comparative study of impact diamonds from the Popigai, Ries, Sudbury, and Lappajärvi craters: Meteoritics & Planetary Science, v. 33, Supplement, p. A90–A91.

Lapke, C., Schmitt, R.T., Kenkmann, T., and Stöffler, D., 2000, Raman microspectrometry of shocked graphite and impact diamonds from the Ries crater, Germany: Houston, Texas, USA, Lunar and Planetary Institute, Lunar and Planetary Science Conference, v. XXXI, no. 1040 (CD-ROM).

Laurenzi, M.A., Bigazzi, G., Balestrieri, M.L., and Bouska, V., 2003, $^{40}Ar/^{39}Ar$ laser probe dating of the Central European tektite-producing impact event: Meteoritics & Planetary Science, v. 38, p. 887–893.

Le Maitre, R.W., 1984, A proposal by the IUGS subcommission on the systematics of igneous rocks for a chemical classification of volcanic rocks based on the total alkali silicia (TAS) diagram: Australian Journal of Earth Sciences, v. 31, p. 243–255.

Masaitis, V.L., 1998, Popigai crater: Origin and distribution of diamond-bearing impactites: Meteoritics & Planetary Science, v. 33, p. 349–359.

Masaitis, V.L., Futergendler, S.I., and Gnevushev, M.A., 1972, Diamonds in impactites of the Popigai meteorite crater: Proceedings, All-Union Mineralogical Society, v. 1, p. 108–112.

Masaitis, V.L., Shafranovsky, G.I., Ezersky, V., and Reshetnyak, N., 1990, Impact diamonds from ureilites and impactites: Meteoritika, v. 49, p. 180–195.

Masaitis, V.L., Shafranovsky, G.I., and Fedorova, I.G., 1995, The apographitic impact diamonds from astroblemes Ries and Popigai: Proceedings of the Russian Mineralogical Society Pt CXXIV, no. 4, p. 12–19 (in Russian).

Masaitis, V.L., Shafranovsky, G.I., Grieve, R.A.F., Langenhorst, F., Peredery, W.V., Balmasov, E.L., Fedorova, I.G., and Therriault, A., 1997, Discovery of impact diamonds at the Sudbury structure: Houston, Texas, USA, Lunar and Planetary Institute, Large Meteorite Impacts and Planetary Evolution Conference, Sudbury, Abstract no. 6027 (CD-ROM).

Masaitis, V.L., Shafranovsky, G.I., Fedorova, I.G., Koivisto, M., and Korhonen, J.V., 1998, Lappajärvi astrobleme: The first find of impact diamonds on the Fennoscandian shield: Houston, Texas, USA, Lunar and Planetary Institute, Lunar and Planetary Science Conference, Nördlingen, v. XXIX, no. 1171 (CD-ROM).

Miyamoto, M., 1998, Micro Raman spectroscopy of diamonds in the Canyon Diablo iron meteorite: Implications for the shock origin: Antarctic Meteorite Research, v. 11, p. 171–177.

Moroz, T.N., Vishnevsky, S.A., Palchik, N.A., and Raitala, J., 2003, Photoluminescence spectra of the Lappajärvi impact diamonds: Meteoritics & Planetary Science, v. 38, Supplement, p. A90.

Mostefaoui, S., El Goresy, A., Hoppe, P., Gillet, P., and Ott, U., 2001, In-situ discovery of diamond in Bencubbin: Evidence from Raman spectroscopy and cathodoluminescence: Meteoritics & Planetary Science, v. 35, Supplement, p. A141–A142.

Nasdala, L., and Massone, H.-J., 2000, Microdiamonds from the Saxonian Erzgebirge, Germany: in situ micro-Raman characterisation: European Journal of Mineralogy, v. 12, p. 495–498.

Osinski, G.R., Grieve, R.A.F., and Spray, J.G., 2003, The nature of the groundmass of surficial suevites from the Ries impact structure, Germany: Houston, Texas, USA, Lunar and Planetary Institute, Large Meteorite Impacts Conference, Nördlingen, Abstract no. 4022 (CD-ROM).

Phelps, A.W., 1997, Raman spectroscopic analysis of microdiamonds from the Popigai crater: Houston, Texas, USA, Lunar and Planetary Institute, Lunar and Planetary Science Conference, v. XXVIII, no. 1772 (CD-ROM).

Pujols, H., and Boisard, F., 1970, Effects of an intense shock wave on graphite: Carbon, v. 8, p. 781–782, doi: 10.1016/0008-6223(70)90103-X.

Rost, R., Dolgov, Y.A., and Vishnevsky, S.A., 1977, Gases in inclusions of impact glasses in the Ries crater, West Germany, and finds of high-pressure carbon polymorphs: Doklady AN SSSR, v. 3, p. 695–698.

Shafranovsky, G.I., 1985, The crystallomorphology of diamond paramorphs after graphite: Proceedings of the Russian Mineralogical Society, v. 1, p. 30–34.

Schmidt-Kaler, H., 1978, Geologische Übersichtskarte des Ries-Meteoriten-Kraters 1:100.000: München, Bayerisches Geologisches Landesamt.

Siebenschock, M., Schmitt, R.T., and Stöffler, D., 1998, Investigation of a diamond bearing gneiss clast in the suevite from Seelbronn, Ries crater, Germany: Houston, Texas, USA, Lunar and Planetary Institute, Lunar and Planetary Science Conference, v. XXIX, no. 1022 (CD-ROM).

Siebenschock, M., Langenhorst, F., Schmitt, R.T., and Stöffler, D., 1999, Impact diamonds in the dike suevite of Unterwilflingen, Ries crater, Germany: Houston, Texas, USA, Lunar and Planetary Institute, Lunar and Planetary Science Conference, v. XXX, CD-ROM # 1172.

Stähle, V., 1972, Impact glasses from the suevite of the Nördlinger Ries: Earth and Planetary Science Letters, v. 17, p. 275–293, doi: 10.1016/0012-821X(72)90287-7.

Stöffler, D., 1971, Progressive metamorphism and classification of shocked and brecciated crystalline rocks at impact craters: Journal of Geophysical Research, v. 76, p. 5541–5551.

Stöffler, D., 1972, Deformation and transformation of rock-forming minerals by natural and experimental shock processes, I, Behavior of minerals under shock compression: Fortschritte der Mineralogie, v. 49, p. 50–113.

Stöffler, D., 1977, Research drilling Nördlingen 1973: polymict breccias, crater basement, and cratering model of the Ries impact structure: Geologica Bavarica, v. 75, p. 443–458.

Stöffler, D., and Langenhorst, F., 1994, Shock metamorphism of quartz in nature and experiment: I. Basic observation and theory: Meteoritics, v. 29, p. 155–181.

Stöffler, D., and Ostertag, R., 1983, The Ries impact crater: Fortschritte der Mineralogie, v. 61: Beiheft 2, p. 71–116.

Thiele, I., and Scherer, T., 2000, Diamanten im Nördlinger Ries: Berichte der Deutschen Mineralogischen Gesellschaft, Beiheft zum European Journal of Mineralogy, v. 12, p. 213.

Valter, A.A., Yerjomenko, G.K., Kvasnitsa, V.N., and Polkanov, Y.A., 1992, Impact-metamorphic minerals of carbon: Kiev/Ukraine, Naukova Dumka Press, 171 p. (in Russian).

Valter, A.A., Eremenko, G.K., Polkano, Y.A., Khrenov, A.Y., Siebenschock, M., Schmitt, R.T., and Stöffler, D., 1998, Impact diamonds from suevites of Ries crater (Germany): New Results: Mineralogitscheskij Zhurnal, v. 20, no. 3, p. 3–12.

Vdovykin, G.P., 1970, Ureilites: Space Science Reviews, v. 10, p. 483–510, doi: 10.1007/BF00172536.

Vishnevsky, S.A., Afanasiev, V.P., Argunov, K.P., and Pal'chik, N.A., 1997, Impact diamonds: Their features, origin and significance: Novosibirsk, Russian Academy of Science, Siberian Branch, United Institute of Geology, Geophysics and Mineralogy, Transactions, Issue 835, 55 p.

Yerofeev, M.V., and Lachinov, P.A., 1888, About the Novo Urei meteorite: Zhurnal Russkogo Physiko-Khimicheskogo Obstchestva, v. XX, p. 185–213.

Yoshikawa, M., Mori, Y., Obata, H., Maegawa, M., Katagiri, G., Ishida, H., and Ishitani, A., 1995, Raman scattering from nanometer-sized diamond: Applied Physics Letters, v. 67, p. 694–696, doi: 10.1063/1.115206.

Zoltai, T., and Stout, J.H., 1984, Mineralogy: Minneapolis: Burgess Publishing Company, 505 p.

Manuscript Accepted by the Society 9 August 2004

Geological Society of America
Special Paper 384
2005

Planar microstructures and Dauphiné twins in shocked quartz from the Charlevoix impact structure, Canada

Claudia A. Trepmann
John G. Spray
Planetary and Space Science Centre, Department of Geology, University of New Brunswick, 2 Bailey Drive, Fredericton, New Brunswick, E3B 5A3, Canada

ABSTRACT

Shocked quartz from the Charlevoix impact structure has been investigated by optical and scanning electron microscopy, combined with electron backscatter diffraction techniques. The apparent shock pressure recorded by specific sets of planar deformation features (PDFs) in quartz shows a systematic variation with distance (0–10 km) from the center of the structure from ~5–20 GPa. The occurrence of basal PDFs at distances of ~2–10 km from the center of the structure indicates a high deviatoric stress component of the shock wave–associated stress tensor. Grain size effects and a greater mineralogical heterogeneity are proposed to be the main cause for slightly lower shock pressures recorded by PDFs in finer-grained granitic gneisses in the southeastern part of the structure, compared to coarse-grained charnockitic gneisses to the northwest at similar distances from the center of the structure. The influence of the crystallographic orientation of quartz on the orientation distribution of planar microstructures appears to superimpose an influence of the orientation of the impact-related stress field. Based on the appearance of Dauphiné twins that are associated with PDFs and the occurrence of PDFs with orientations that correspond to positive and negative rhombohedra, quartz is suspected to have locally been in the β-modification state. Dauphiné twinning is proposed to be mainly due to a reversion to α-quartz during cooling. These findings imply that the uplifted, preheated target rocks have locally been shock-heated to the α-β transition temperature.

Keywords: Charlevoix structure, Dauphiné twinning, shocked quartz, α-β quartz transition, planar microstructures, planar deformation features, electron backscatter diffraction, orientation contrast imaging, shock pressure estimation.

INTRODUCTION

A characteristic feature of shock waves that are generated by the hypervelocity impact of a projectile into rock is the generation of a highly dynamic, quasi-hydrostatic pressure of several tens of GPa. The pressure attenuates with time and distance from the contact point as the hemispherical shock front propagates through the target rocks. This is due to the expanding shock front reducing the overall energy density, and due to energy loss to the target rocks through heating, deformation, and acceleration (e.g., Melosh, 1989). At attenuated shock pressures, the conditions are no longer quasi-hydrostatic and the deviatoric stress component of the shock wave-associated stress tensor becomes effective (Melosh, 1989; Leroux and Doukhan, 1996). The conditions

Trepmann, C.A., and Spray, J.G., 2005, Planar microstructures and Dauphiné twins in shocked quartz from the Charlevoix impact structure, Canada, *in* Kenkmann, T., Hörz, F., and Deutsch, A., eds., Large meteorite impacts III: Geological Society of America Special Paper 384, p. 315–328. For permission to copy, contact editing@geosociety.org.

during and after hypervelocity impact change abruptly, causing complex and dynamic effects. Planar microstructures in quartz from naturally shocked rocks can be used to estimate the shock pressure (e.g., Robertson, 1975; Robertson and Grieve, 1977; Stöffler and Langenhorst, 1994). They have also been proposed to yield information on the shock wave propagation direction (Walzebuck and Engelhardt, 1979), as well as on temperature conditions (Langenhorst and Deutsch, 1994; Grieve et al., 1996). The Charlevoix impact structure in Quebec, Canada, offers an excellent opportunity to study impact-related planar microstructures in crystalline target rocks that have experienced different shock intensities (Robertson, 1975). This study uses the occurrence of planar microstructures in shocked quartz from these rocks to obtain further constraints on impact-related stress and temperature conditions.

GEOLOGIC SETTING

The Charlevoix impact structure (47°32′N; 70°18′W) is located ~105 km northeast of Quebec City along the northern shore of the St. Lawrence River (Fig. 1). The topographic expression of the ~54 km diameter complex structure is well defined. The prominent central peak, the Mont des Eboulements, marks the center of the structure. However, the southeastern part of the crater, in the St. Lawrence River, has been obliterated by thrusting along the Logan Line (Robertson, 1968): a southeast-dipping thrust fault that separates Precambrian Shield units to the northwest from Appalachian units to the southeast (Neale et al., 1961; Kumarapeli, 1985). The St. Laurent Fault (Fig. 1) divides the structure into northwestern and southeastern segments. It is a steeply dipping normal fault, which was initiated in Late Proterozoic–early Paleozoic times during creation of the Iapetus Ocean (Rondot, 1989), and reactivated during Paleozoic and Mesozoic times (Lemieux et al., 2003; Tremblay and Lemieux, 2001). The impact structure is eroded, and virtually all crater-fill deposits have been removed. K-Ar data from impact melt rocks imply an impact age of ~357 ± 15 Ma (Rondot, 1971). This result may be too young, owing to Ar loss, because new Ar-Ar ages imply an older Ordovician age for the impact event (Whitehead et al., 2003). The youngest Ordovician deposit of the St. Lawrence Platform within the impact structure and showing shock effects is the Utica formation (ca. 450 Ma, Rondot, 1989), which gives a maximum age for the impact (Robertson, 1968).

The target lithologies are predominantly parauthochtonous crystalline rocks of the Grenville Province of the Canadian Shield (Fig. 1). The oldest (early Proterozoic) unit (unit 1) comprises migmatitic paragneisses (1379 ± 66 Ma; Frith and Doig, 1973) and granitic gneisses, which are exposed in the northeastern part of the structure. The predominant unit consists of charnockitic gneisses (unit 2, Mesoproterozoic; 1080–1513 Ma; Frith and Doig, 1973). A major anorthosite complex (unit 3) occurs in the outer western part of the structure (1079 ± 22 Ma; Ashwal and Wooden, 1983). Ordovician arkoses, siltstones, and limestones of the St. Lawrence Platform (unit 5) outcrop in the annular trough and along the northern shore of the St. Lawrence River.

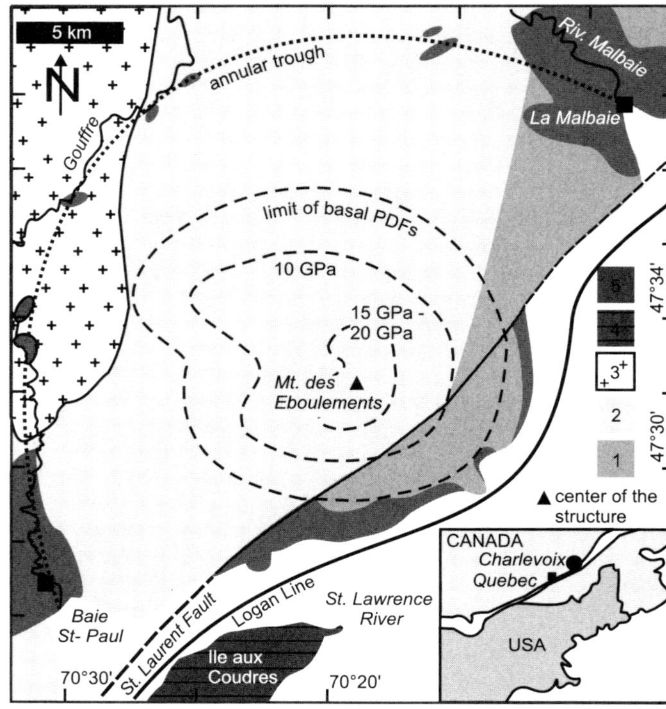

Figure 1. Simplified geological map of the Charlevoix impact structure with shock zoning based on planar deformation feature occurrence in quartz, modified after Robertson (1975). 1—Granitic gneisses and migmatites of the Grenville Province. 2—Charnockitic gneisses of the Grenville Province. 3—Anorthosite of the Grenville Province. 4—Cambrian sedimentary rocks of the Appalachians. 5—Ordovician sedimentary rocks of the St. Lawrence Platform.

These represent the sedimentary cover at the time of the impact (Robertson, 1968). Evidence for post-impact regional thermal overprinting in the Charlevoix area is lacking. Biotite K-Ar ages of ca. 900 Ma from the anorthosite complex in the outer western part of the structure (Wanless and London, 1961) indicate that the rocks of the Charlevoix area, that have not been affected by shock metamorphism, have remained below the closure temperature of biotite (~300 °C, e.g., Hodges, 1991) since that time.

METHODS

Quartz-rich target rocks from the Charlevoix structure were taken as oriented samples in the field. The samples were cut in two directions normal to the regional/Grenvillian foliation. The microstructures were examined in standard petrographic polished thin section (~30 μm) with a polarizing microscope. Orientations of planar deformation features (PDFs) in shocked quartz crystals were measured using a Leitz U-stage mounted on a Leitz polarizing microscope. A major problem in investigating the orientation distribution of PDFs is the restricted visibility of PDFs in the U-stage: only steeply dipping planes in the thin section can be observed; planes dipping with less than ~40° are not visible. To minimize this problem and to enable measurements of a

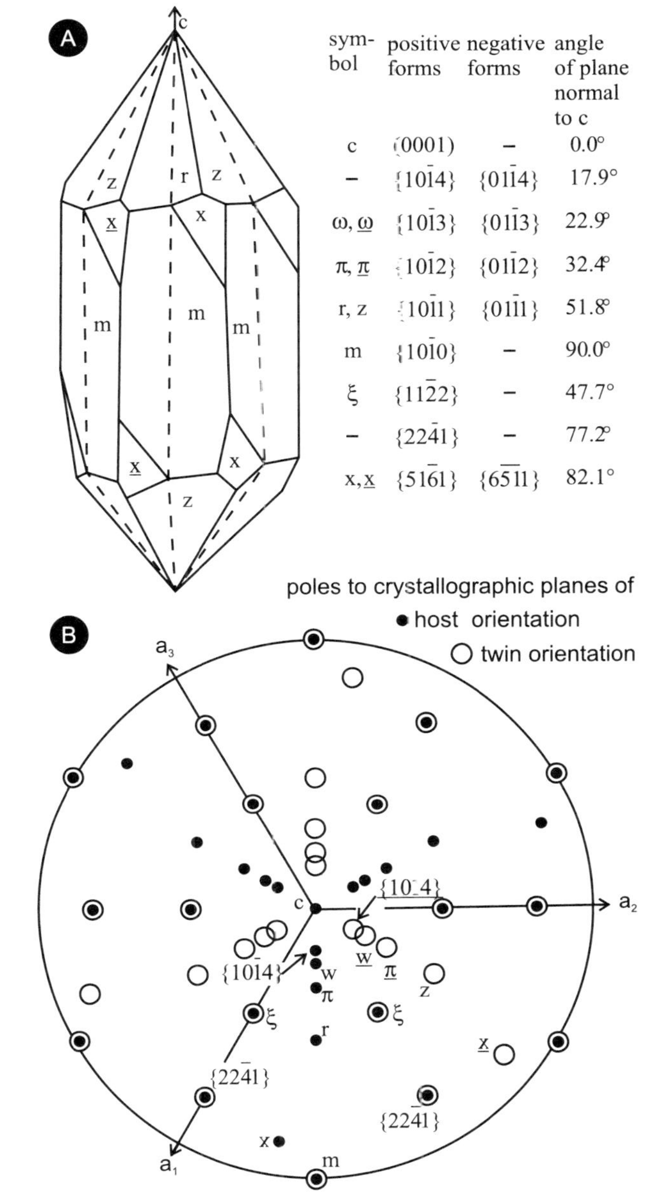

Figure 2. (A) Idealized scheme of Dauphiné twinned quartz crystal. (B) Stereographic projection (upper hemisphere, equal angle) showing host and twin orientations for the Dauphiné twin law in quartz. For simplification, negative forms are not displayed in this projection.

representative range of existing PDFs, two thin sections in an angle of 90° were investigated for each sample. The measurements were plotted in a lower hemisphere equal angle stereograph using the computer program StereoNett 2.0 (Duyster, 1996). To index the PDFs, these measurements were rotated to a vertical projection of the c-axes and then rotated around the c-axes to best fit a template of the crystallography of quartz (Engelhardt and Bertsch, 1969; Langenhorst, 2002). The error of U-stage measurements is generally ± 5°. An additional error for c-axis measurements may result from the undulatory extinction of the quartz crystals. However, the varying orientation of c-axes in undulatory crystals lies within the error of U-stage measurements. The PDFs are generally sharp and well defined, enabling measurements of good accuracy.

Dauphiné twin orientations in quartz were investigated by electron backscatter diffraction (EBSD) and orientation contrast (OC) imaging using a scanning electron microscope (SEM; LEO 1530) equipped with field emission gun and forescatter detector (at the Ruhr-University Bochum, Germany). The thin sections used for the EBSD and OC imaging techniques were chemically polished using a colloidal silicon suspension (SYTON®) to minimize surface damage and then coated with carbon to limit charging effects. The SEM was operated at an accelerating voltage of 25 kV and a working distance of 25 mm, with the thin section tilted at an angle of 70° with respect to the beam. EBSD provides the full crystallographic orientation of crystals in a thin section using a backscattered electron signal. This allows the distinction between positive and negative forms in quartz, and the detection of Dauphiné twins in which positive forms are parallel to negative forms of the host, and vice versa. This is not possible by optical microscopy, given the trigonal symmetry of quartz and the same orientation of c-axes for host and twin (Fig. 2). OC imaging reveals crystallographic misorientation between adjacent crystals by grayscale contrast. Although this contrast does not correlate with the magnitude of misorientation, OC images reveal the position and shape of Dauphiné twins. The combination of EBSD measurements of the quartz crystallographic orientation with U-stage measurements of the PDF orientation provides an additional and more accurate method to index PDFs.

SAMPLE DESCRIPTION

In this study, quartz-rich target rocks from the two units (1 and 2 in Fig. 1) of the Grenvillian Province have been investigated. Unit 1 is petrologically complex, consisting of migmatites (biotite-garnet-sillimanite gneisses, metaquartzites, amphibolites) and metagranites (biotite-muscovite gneisses, with pegmatitic, aplitic, and alaskitic intrusives). The quartz in the granitic gneisses typically occurs in aggregates of more or less isometric grains with a diameter of 0.1–1 mm (Fig. 3A; Table 1). However, grain sizes in almost mono-phase quartz-rich veins can be several millimeters in diameter (Figs. 3B and 3C). Unit 2 is more homogeneous and consists of charnockitic gneisses (quartz-two pyroxene-feldspar gneisses) with minor intrusions of anorthosite. The quartz in the charnockitic gneisses typically exhibits a ribbon structure (Fig. 3D). These ribbons have lengths up to a few millimeters and widths of 0.1–1 mm.

PLANAR MICROSTRUCTURES

The most striking feature of shocked quartz crystals from the Charlevoix target rocks are their planar microstructures (Figs. 3 and 4). There is a vast literature concerning planar microstructures in quartz, which comprise planar fractures, planar

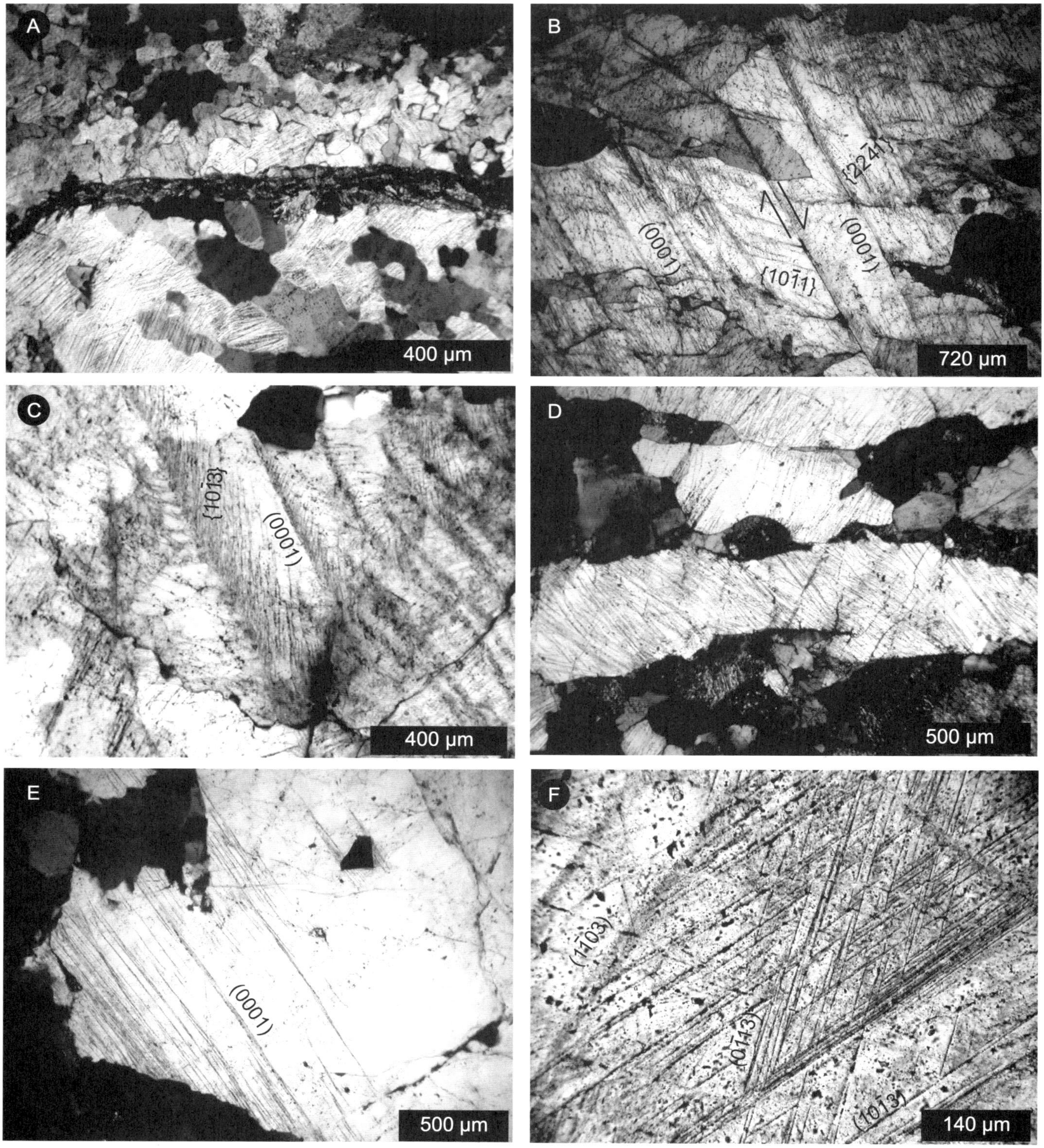

Figure 3. Photomicrographs (cross-polarized light) showing planar microstructures in quartz. (A) Shocked granitic gneiss (sample CT464, unit 1) with quartz of variable grain size; larger quartz crystals have markedly higher planar deformation feature (PDF) densities than smaller ones. (B) Microfaults with shear offset and multiple sets of PDFs (sample CT383, quartz vein, unit 1). (C) One set of PDFs in quartz that apparently has been initiated at a microfault (sample CT464, unit 1). (D) Shocked charnockitic gneiss (sample CT378, unit 2) in which quartz has a characteristic ribbon structure containing multiple sets of PDFs. (E) Concentration of basal PDFs in quartz at quartz grain boundaries and around an inclusion in a charnockitic gneiss (sample CT337, unit 2). (F) Multiple sets of PDFs in quartz parallel to all three crystallographic equivalent planes of {10$\bar{1}$3} in a charnockitic gneiss (sample CT384, unit 2).

TABLE 1. SHOCK PRESSURE ESTIMATED USING PLANAR MICROSTRUCTURES

Sample	Unit	Rock type	Average quartz grain size [mm] $d = \sqrt{lw}$*	Distance of sample from center of the structure [km]	Planar deformation features in quartz: % of crystals containing characteristic sets of PDFs					% PDFs parallel to negative forms from all rhomb. PDFs	% of crystals that contain PDFs parallel to negative forms	Number of crystals measured	Number of PDF sets per crystal	Max. of PDF sets per crystal	Planar features in K-feldspar	Estimated mean shock pressure [GPa] after the scheme of		
					no PDFs	(0001)	$\{10\bar{1}3\}$	$\{10\bar{1}1\}$ $\{11\bar{2}2\}$ $\{22\bar{4}1\}$	$\{10\bar{1}2\}$							Huffman et al. (1993)	Grieve and Robertson (1976)	Grieve et al. (1996)
CT292	2	charnockitic gneiss	~2.0	0.6	0	20.4	55.1	22.5	2.0	13.9	26.5	49	2.4	3	–	–	12	10
CH2	2	charnockitic gneiss	~2.0	1.0	0	8.3	75.0	8.3	8.3	20.5	33.3	24	2.1	3	–	–	12	11
CT384	2	charnockitic gneiss	~3.0	1.5	0	0	20.0	0.0	80.0	44.9	84.0	25	4.8	6	present	> 15	16	15
CT377	2	charnockitic gneiss	~3.0	3.9	0	3.2	53.2	25.8	17.7	16.7	25.4	63	3.6	6	–	–	14	12
CT378	2	charnockitic gneiss	~2.0	3.9	0	6.5	69.3	22.6	1.6	21.4	26.6	64	3.9	6	–	–	12	11
CT444	2	quartz vein	~7.04.2	5	90.0	5.0	0.0	0.0	0.0	0.0	35	1.2	3	–	–	–	7	5
CT337	2	charnockitic gneiss	~3.56.5	0	40.0	48.0	12.0	0.0	8.7	4.0	25	2.3	3	–	–	–	11	9
CT239	2	pegmatite	~7.0	8.7	0	53.8	38.5	7.7	0.0	11.7	7.7	13	2.1	3	–	–	10	8
CT414	1	quartz vein	~0.6	2.3	0	0.0	85.7	14.3	0.0	11.9	20.0	15	3.0	4	–	–	12	11
CT383	1	quartz vein	~5.0	3.4	0	29.8	13.4	47.8	9.0	5.7	3.0	67	2.1	3	–	–	13	10
CT336	1	mafic gneiss	no quartz	3.7	–	–	–	–	–	–	–	–	–		present	> 15	–	–
CT470	1	granitic gneiss	~0.5	4.0	85	15.0	0.0	0.0	0.0	0.0	0.0	50	1.0	1	–	–	6	5
CT464	1	granitic gneiss	~0.5, ~5.0	4.6	32	45.0	23.0	0.0	0.0	1.7	10.0	50	1.3	2	–	–	8	7
CT282	1	granitic gneiss	~0.5, ~5.0	5.4	90	10.0	0.0	0.0	0.0	0.0	0.0	60	1.0	1	–	–	6	5
CT379	1	granitic gneiss	~0.5	6.9	88	12.0	0.0	0.0	0.0	0.0	0.0	50	1.0	1	–	–	5	5

*The average grain size d has been estimated as the square root of the product of length (l) and width (w) of grains.

cleavages, microfaults, and PDFs (see, for example, the detailed review by Stöffler and Langenhorst, 1994 and Grieve et al., 1996). In the following, two different planar microstructures that occur in the Charlevoix target rocks are described.

Microfaults

Sets of planar features parallel to crystallographic planes occur in shocked quartz crystals, with a typical spacing of more than 50 μm (Figs. 3B and 3C). These features commonly show a distinct shear offset, therefore we refer to them as microfaults (Fig. 3B). In only a few quartz-rich, coarse-grained samples from the Charlevoix structure were quartz grains found to possess microfaults, but in these samples they are prevalent (Figs. 3B and 3C). Because of their rare occurrence, no systematic variation with respect to rock type or sample location is apparent. Microfaults appear to be mostly parallel to {11$\bar{2}$2}, {10$\bar{1}$1}, (0001), and, less abundantly, {51$\bar{6}$1}. Except for {11$\bar{2}$2}, these planes are common cleavage planes in quartz (Frondel, 1962).

Planar Deformation Features (PDFs)

PDFs are very common in shocked quartz from the Charlevoix target rocks. In the optical microscope, they appear as arrays of narrow planar features parallel to certain crystallographic planes decorated by fluid inclusions. In contrast to microfaults, generally no shear offset is apparent along PDFs, and they are more closely spaced, typically in the range of 50 μm to smaller than 5

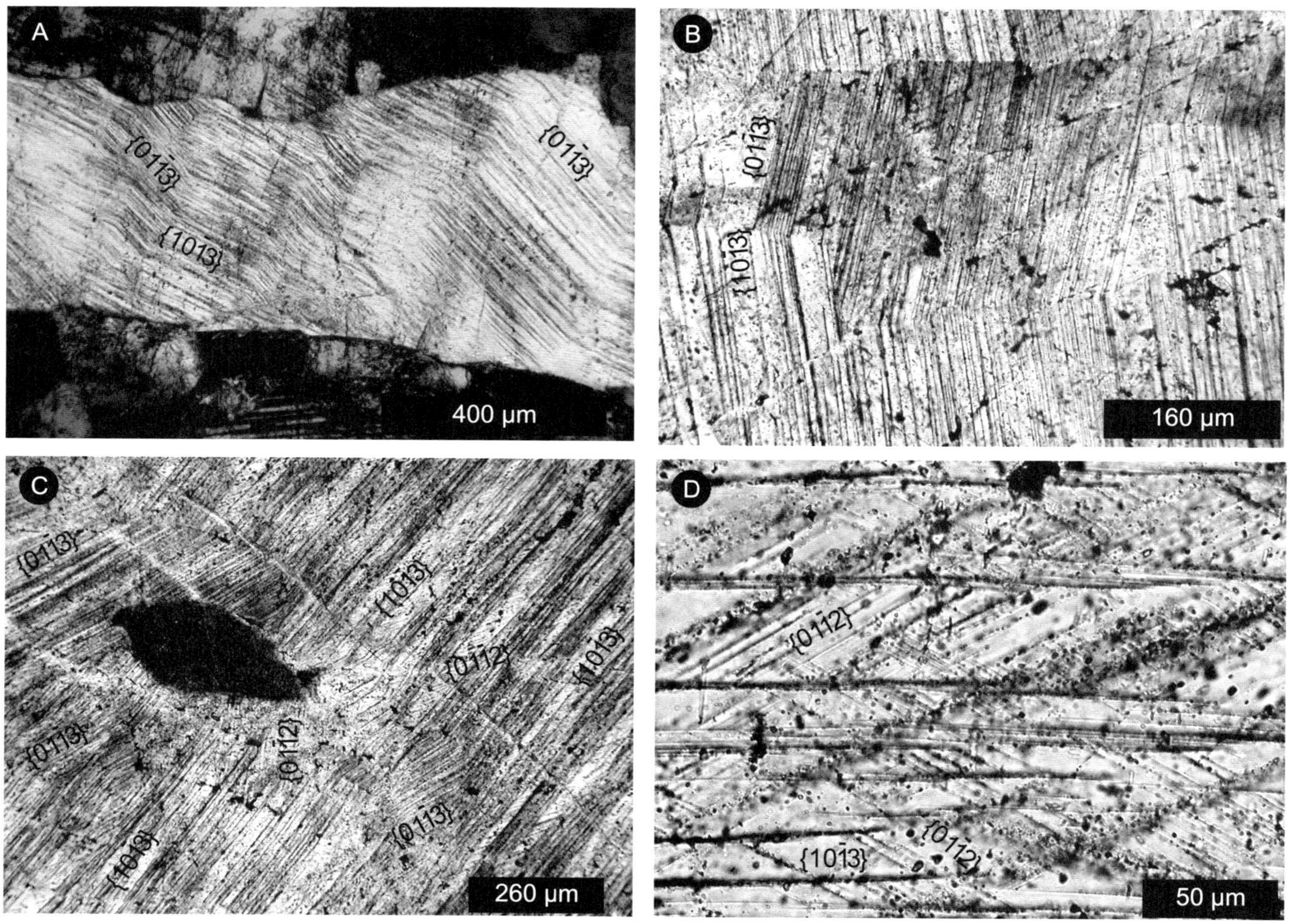

Figure 4. Photomicrographs showing planar deformation features (PDFs) in quartz with orientations corresponding to positive and negative rhombohedra. (A, B) Kinkband-like occurrence of different quartz-domains with PDFs parallel to planes of either positive {10$\bar{1}$3}or negative {01$\bar{1}$3} rhombohedra (cross-polarized light) in charnockitic gneisses (unit 1; A: sample CT 378, B: sample CT 377). (C) Coexisting PDFs parallel to the positive {10$\bar{1}$3}and the negative {01$\bar{1}$2} rhombohedra around an inclusion (charnockitic gneiss, sample CT384, unit 1). Note kinkband-like occurrence of domains that exhibit PDFs parallel to {01$\bar{1}$3} (plane-polarized light). (D) PDFs parallel to negative {01$\bar{1}$2} rhombohedra occur in the interspaces of PDFs parallel to positive {10$\bar{1}$3} rhombohedra (charnockitic gneiss, sample CT384, unit 1; plane-polarized light).

μm. Although shocked quartz crystals show a pronounced undulatory extinction and subgrains, PDFs are generally straight features (Figs. 3 and 4). In the optical microscope, they can only rarely appear to be bent, reflecting a continuous bending of the crystal lattice. PDFs are commonly concentrated around inclusions and along grain boundaries (Figs. 3E and 4C). In some cases, a certain PDF set appears to be restricted to the interspace of another set (Fig. 4D). PDFs can occur throughout the whole crystal, but they cross neither grain boundaries nor microfaults (Figs. 3A–3E). They can start from one side of a microfault to form a ladder-like structure (Figs. 3B and 3C). This observation could imply that these PDFs developed after the microfaults, as suggested by Stöffler and Langenhorst (1994) and Grieve et al. (1996).

At Charlevoix, the PDF density apparent in the optical microscope varies systematically with the radial distance from the central peak (Table 1). In samples near the center of the structure, the PDF density is markedly higher than at greater distances from the center. Quartz crystals in samples close to the center (<~0.5 km) contain very high PDF densities, exhibiting a characteristic browning, or "toasting," formed during post-shock annealing (Whitehead et al., 2002).

It is convenient to distinguish three different types of PDFs that occur in quartz from the Charlevoix structure by their crystallographic orientation. The first type is basal PDFs parallel to the (0001) plane (Fig. 3E). TEM investigations reveal that basal PDFs represent mechanical Brazil micro-twin lamellae (Goltrant et al., 1991, 1992a, 1992b). Brazil twins are formed due to high differential stresses of ~4 GPa (McLaren et al., 1967). The second type is rhombohedral PDFs parallel to planes of the $\{10\bar{1}n\}$ forms, with n = 1–4 (Figs. 3F and 4). These PDFs are characterized by a discontinuous amorphisation of quartz as the result of the high quasi-hydrostatic pressure related to the shock wave without the involvement of shear (Goltrant et al., 1992b). However, most rhombohedral PDFs in quartz from the Charlevoix target rocks are crystallized due to annealing (Goltrant et al., 1992b) and are now visible in the optical microscope as fluid inclusion trails (Figs. 3 and 4). The third type of PDF comprises less abundant PDFs parallel to $\{11\bar{2}2\}$, $\{22\bar{1}4\}$ and $\{51\bar{6}1\}$.

Amazingly, these different PDFs appear very similar in the optical microscope, despite their different nature. However, their occurrence with distance from the central peak reveals systematic differences. PDFs parallel to $\{10\bar{1}2\}$ are only present in the uplifted central part of the structure (~0–2 km). However, the most common rhombohedral PDFs are parallel to $\{10\bar{1}3\}$. They get less abundant with increasing distance from the center of the structure. Their occurrence diminishes at distances of ~8–9 km. Basal PDFs occur up to ~10 km (Fig. 1) but they are absent in samples from the center of the structure (at a radial distance <2 km).

PDFs with Orientations that Correspond to Positive and Negative Rhombohedra

In general, PDFs in shocked trigonal α-quartz do not occur with orientations that correspond to both positive and negative forms (Langenhorst and Deutsch, 1994; Grieve et al., 1996). Under the optical microscope, it is not possible to distinguish between PDFs parallel to positive and negative rhombohedra if only one form is present. However, if PDFs occur parallel to planes of both $\{10\bar{1}n\}$ and $\{01\bar{1}n\}$ forms (Fig. 2), an arbitrary differentiation is possible.

Observations by optical microscopy reveal crystals that exhibit PDFs with orientations that correspond to positive and negative rhombohedra (Fig. 4). The number of these crystals appears to increase with the recorded shock pressure (Table 1). In highly shocked samples, up to 84% of the measured crystals contain PDFs parallel to both positive and negative rhombohedra (Table 1). Distinct domains occur, where PDFs are parallel to either $\{10\bar{1}3\}$or $\{01\bar{1}3\}$, and single PDFs can be traced from one domain into the other (Figs. 4A, 4B, and 6C). The appearance of these domains with straight boundaries is commonly kinkband-like. Notably, a slight change in extinction is related to these domains. PDFs parallel to negative $\{01\bar{1}2\}$rhombohedra coexist with PDFs parallel to positive $\{10\bar{1}3\}$ rhombohedra (Figs. 4C and 4D). PDFs parallel to $\{10\bar{1}3\}$ are well-developed over almost the entire crystal, whereas PDFs parallel to $\{01\bar{1}2\}$appear to occur in the interspace of the PDFs parallel to $\{10\bar{1}3\}$ (Figs. 4C and 4D). An intersection of PDFs parallel to positive and negative forms of the same rhombohedra could not be observed with certainty.

Orientation Distribution of Planar Microstructures

A dependence of the development of PDFs on crystallographic orientation relative to the principle shock front has been reported from naturally shocked rocks (Walzebuck and Engelhardt, 1979) and experiments (e.g., Hörz, 1968; Gratz et al., 1988; Langenhorst and Deutsch, 1994). The orientation distribution of PDFs and microfaults at Charlevoix is displayed for three samples in geographic coordinates (Figs. 5A–5C), and for one sample in thin section coordinates (Fig. 5D). It was unfortunately not possible to take the latter sample oriented in the field.

Sample CT383 (Fig. 5A) is an almost mono-phase quartz vein, showing well-developed microfaults and multiple sets of PDFs (Fig. 3B; Table 1). Because their normals are parallel to the c-axes, basal PDFs reflect the crystallographic orientation of the quartz crystals. However, as crystals with a certain c-axis orientation (Fig. 5A) do not show basal PDFs, these orientations do not appear to favor the development of basal PDFs. This can also be observed for the samples CT377 and CT378 (Figs. 5B and 5C). Rhombohedral PDFs in Figure 5A show orientation maxima in the interspace between the orientation maxima of the basal PDFs. Microfaults are oriented in a broad girdle, however, lacking the orientation that is displayed by the PDFs (Fig. 5A). In the charnockitic samples, CT377 and CT378 (Table 1), the orientation of rhombohedral PDFs is concentrated in broad girdles, similar to the preferred orientation of quartz c-axes (Figs. 5B and 5C). In another sample of charnockitic gneiss, CT384 (Table 1), a strong crystallographic preferred orientation of the quartz crystals is apparent, with only two predominant orientations of the c-axes,

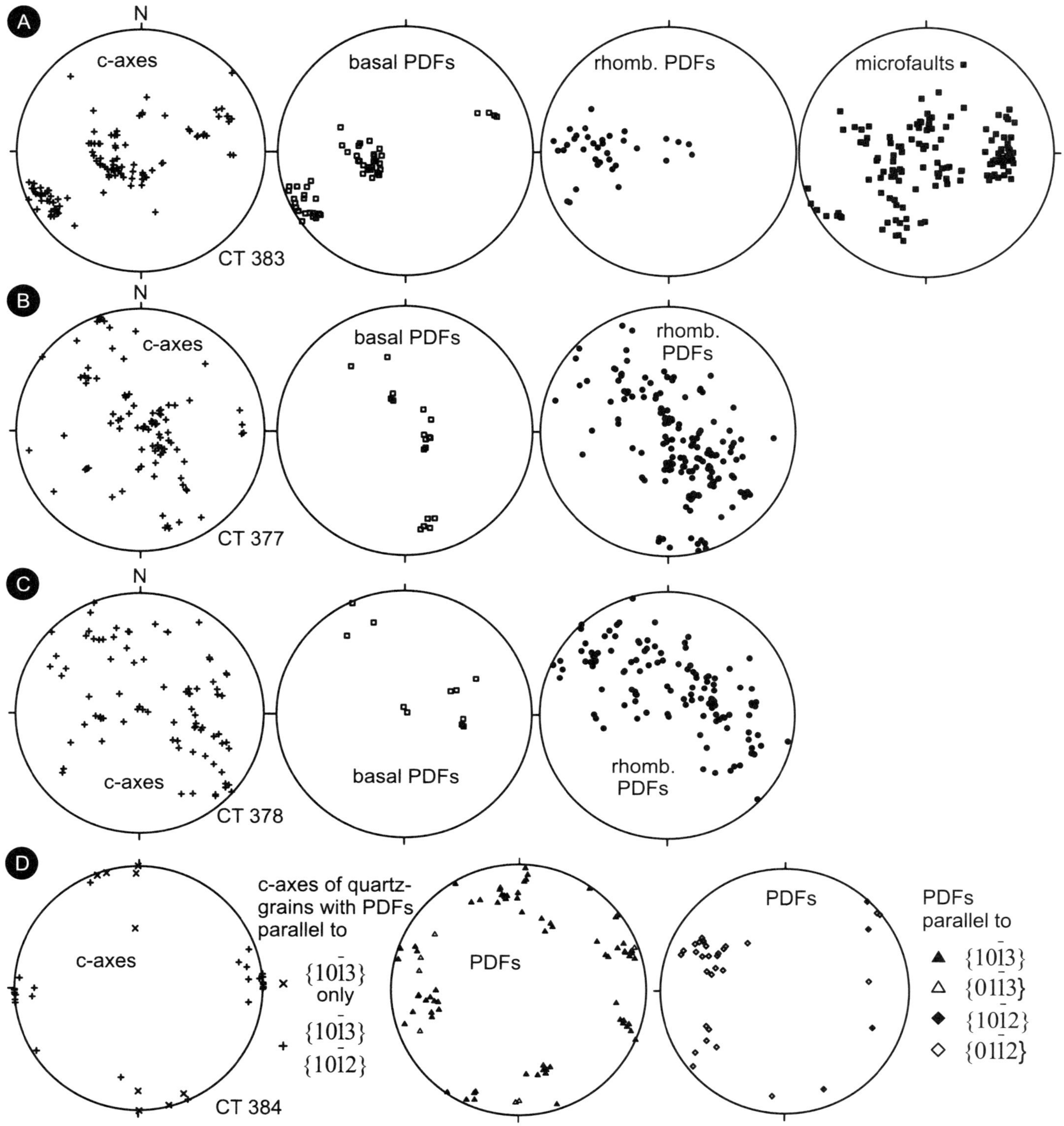

Figure 5. Stereographic projection (lower hemisphere, equal angle) of c-axes and poles to planar microstructures in quartz crystals from different samples at Charlevoix, as measured with a U-stage. (A) Orientation of c-axes, basal planar deformation features (PDFs), rhombohedral PDFs and microfaults in sample CT383 (quartz vein, unit 1). (B, C) Orientation of c-axes, basal PDFs and rhombohedral PDFs in samples CT377 and CT378 (charnockitic gneisses, unit 1), respectively. (D) Orientation of c-axes (sample CT384, charnockitic gneiss, unit 1) containing PDFs either exclusively parallel to {10$\bar{1}$3}, or parallel to both {10$\bar{1}$3} and {10$\bar{1}$2}, PDFs parallel to {10$\bar{1}$3}/{01$\bar{1}$3} and PDFs parallel to {10$\bar{1}$2}/{01$\bar{1}$2}.

oriented either N-S or E-W in the stereograph (Fig. 5D). Crystals with c-axes oriented N-S contain mostly PDFs parallel to $\{10\bar{1}3\}$, whereas crystals with c-axes parallel to E-W possess additional PDFs parallel to $\{10\bar{1}2\}$(Fig. 5D).

DAUPHINÉ TWIN ORIENTATIONS

The occurrence of Dauphiné twins is revealed by OC images in the SEM and by EBSD measurements (Fig. 6). The twin orientation is geometrically related to the host orientation by a rotation of 180° (or 60°) around the c-axis (Frondel, 1962). The twinning operation brings positive forms of the host to coincidence with negative forms of the twin, and vice versa (Figs. 2 and 6). The twins are visible by grayscale contrast in the OC images, while PDFs are represented by lines of pits (Fig. 6). The host orientation is assumed to be the orientation that exhibits PDFs parallel to positive forms. In kinkband-like features, where PDFs in the twin and host domain can occur parallel to both positive and negative forms, the differentiation has been made arbitrarily (Fig. 6C). Commonly, the twin boundaries are planar and aligned along PDFs parallel to $\{10\bar{1}3\}/\{01\bar{1}3\}$ (Fig. 6). This is notable, as Dauphiné twin boundaries are usually irregular (Frondel, 1962). The twins can be concentrated in the interspace between PDFs (Fig. 6A). PDFs can cross from the host orientation into the twin orientation, therefore changing the crystallographic orientation from a positive $\{10\bar{1}3\}$ to the corresponding negative $\{01\bar{1}3\}$ form (arrows in Fig. 6B). Different domains that have a Dauphiné twin relationship commonly show additionally small angles between the c-axes of generally ~2°–5°.

DISCUSSION

Shock Pressure Estimation

The abundance of different sets of PDFs in quartz crystals can be used to give an estimate of the shock pressure experienced by the rock (e.g., Robertson, 1975; Robertson and Grieve, 1977; Stöffler and Langenhorst, 1994). The experimental calibration was performed on small single crystals and for ~1 µs shock pulse duration and strain rates on the order of 10^6–10^9 s^{-1} (e.g., Huffman and Reimold, 1996; Grieve et al., 1996). In planetary impact, the shock duration is orders of magnitude longer and can be up to a few seconds, depending on the size of the projectile (e.g., Langenhorst, 2002). Strain rates are probably lower than in experiments (e.g., Huffman and Reimold, 1996), and a large rock volume is shocked at varying depths with gradients in pre-shock temperature and confining pressure (Grieve et al., 1996). The discrepancy between experiment and nature is apparent in the absence of basal PDFs and Brazil twins in experimentally shocked quartz crystals. Additionally, stress localization, as indicated by the varying PDF density on the grain scale, differences in mineralogy (Kieffer, 1971; Dressler et al., 1997), crystallographic orientation (Langenhorst, 1994; Langenhorst and Deutsch, 1994) and pre-shock temperature (Langenhorst, et al., 1992; Huffman et al., 1993; Huffman and Reimold, 1996), as well as grain size effects (Walzebuck and Engelhardt, 1979; Grieve and Robertson, 1976) limit the use of PDFs as a pressure gauge. Nevertheless, the occurrence of characteristic PDF sets can yield important information on shock pressure attenuation (e.g., Robertson, 1975). At Charlevoix, quartz in rocks from a radial distance of up to ~10 km from the center of the structure exhibits PDFs. Using the calibration scheme of Grieve and Robertson (1976), the shock pressure has been estimated for fifteen samples and compared to the calibration scheme of Grieve et al. (1996; Table 1, Fig. 7). Planar microstructures are also observed in K-feldspar at Charlevoix (Table 1; Robertson, 1975). However, the abundance of specific orientations does not appear to vary systematically with shock pressure (Robertson, 1975). The occurrence of planar microstructures in K-feldspar has been estimated to occur at >15 GPa (Huffman et al., 1993; Robertson, 1975).

The shock pressure estimation from the planar microstructures in this study is consistent with the shock isobars proposed by Robertson (1975) (Table 1; Figs. 1 and 7). Robertson (1975) recorded an asymmetry of the shock isobars with lower spacing of the shock zones to the southeast of the St. Laurent Fault. PDFs in quartz from most samples of unit 1, southeast of the St. Laurent Fault, are also found in this study to record slightly lower shock pressures compared to samples in the northwestern part at similar distances from the center of the structure (Table 1; Fig. 7). However, the difference is not very distinct, given the heterogeneity of the recorded shock pressures. Possible explanations of the asymmetry of the shock zones include:

(1) The recorded lower shock pressure is an original feature. As a large grain size appears to enhance PDF development (Walzebuck and Engelhardt, 1979; Grieve and Robertson, 1976), the generally finer-grained quartz aggregates in the granitic gneisses may record a lower shock pressure compared to the typically coarse-grained quartz ribbons from the charnockitic gneisses (Figs. 3A and 3D; Table 1). Additionally, the petrology of unit 1 is more complex, with layers of amphibolites, muscovite- and biotite-rich gneisses, which may have reacted as "soft" layers due to their higher compressibility relative to quartz-rich rocks, thus absorbing a larger part of the shock wave energy.

(2) Robertson (1975) proposed that the asymmetry of the shock zones could be the result of activity along the Logan Line during the post-impact Acadian orogeny. The absence of shock effects in the Ordovician rocks from Isle-aux-Coudres (Fig. 1) is evidence of major post-impact thrusting, which led to the truncation of the southeastern part of the crater (Robertson, 1968). However, the sense of movement along the Logan Line, as well as along the St. Laurent Fault, involves a down-faulting of the southeastern part of the impact structure (Fig. 1; e.g., Rondot, 1989; Carignan et al., 1997; Lemieux et al., 2003; Tremblay and Lemieux, 2001). This should result in higher shock pressures at a shorter distance to the center in comparison to the northwestern part of the impact structure. This is contrary to the observations.

Therefore, we suspect that the slightly lower recorded shock pressures in the southeastern part of the structure represent an orig-

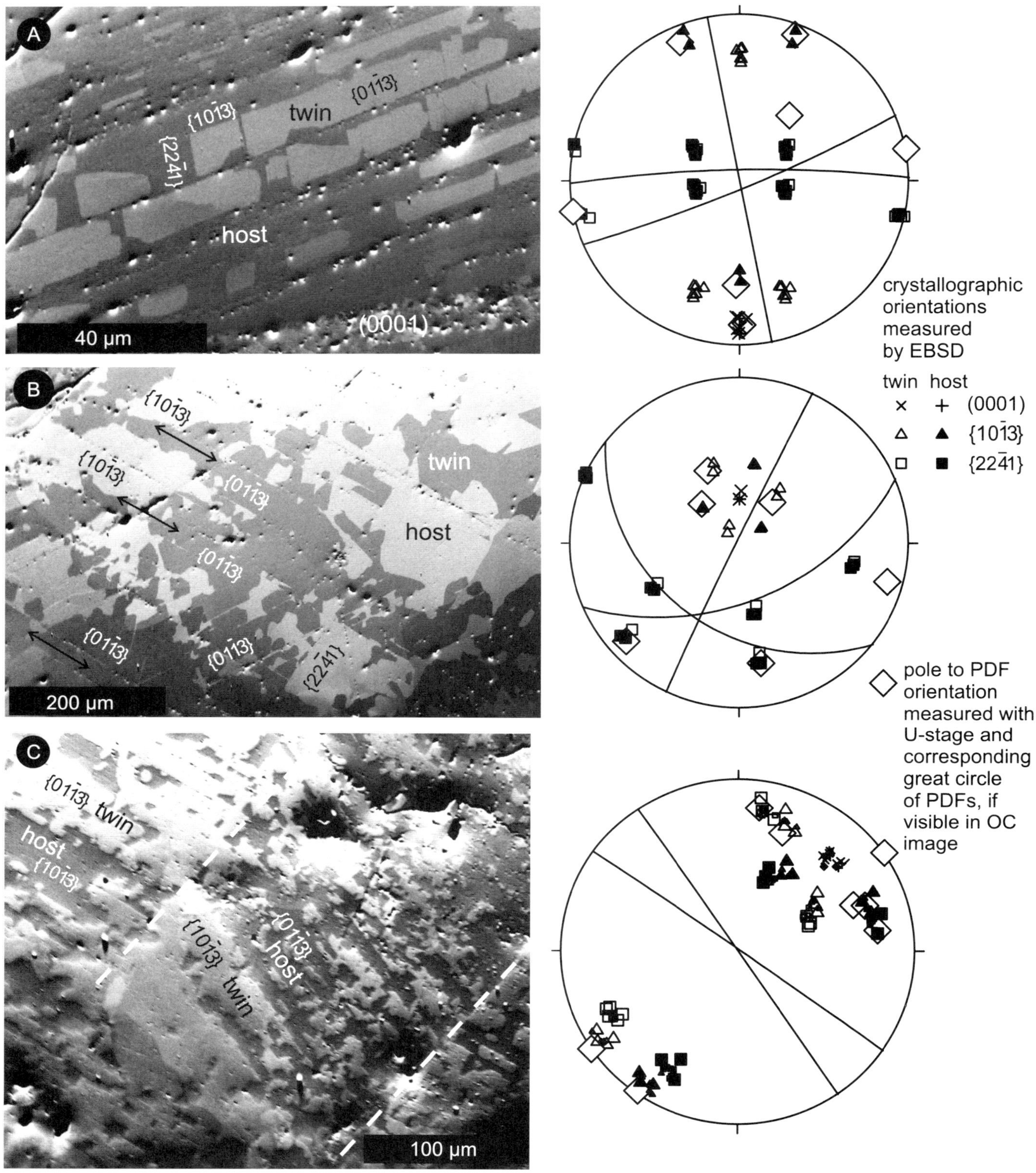

Figure 6. Orientation contrast (OC) images of shocked quartz on the left and corresponding stereographic projection (lower hemisphere, equal angle) of host and twin orientations, as measured by electron back scatter diffraction (EBSD), and poles to planar deformation features (PDFs) as measured with a U-stage, on the right. PDFs visible in the OC image are additionally displayed as great circles. Dauphiné twin orientations can be distinguished by different grayscale. Note that host orientation in A and B is assumed to be the orientation that exhibits PDFs parallel to positive rhombohedra. PDFs are indicated by lines of pits. (A) Dauphiné twin boundaries are planar and aligned parallel to PDFs. Twins occur in the interspace between rhombohedral PDFs. (B) Dauphiné twin orientation is patchy, but boundaries can be aligned parallel to PDFs. Note that PDFs can cross twin boundaries without changing orientation (black arrows). (C) OC-image of kinkband-like domain shown in Figure 4B. "Kinkband boundaries" are marked by dashed white lines. The differentiation between host and twin orientation has been made arbitrarily.

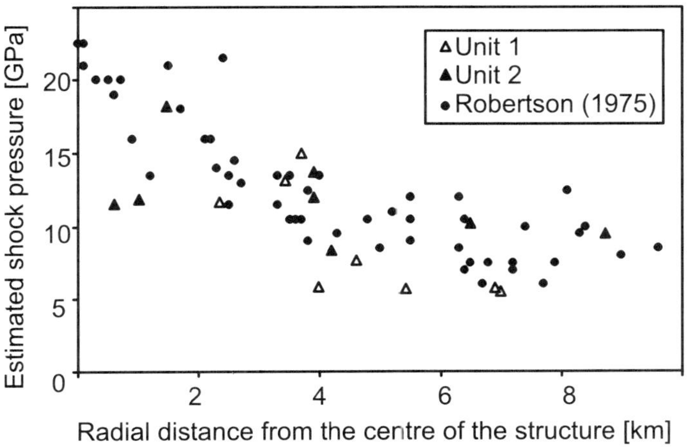

Figure 7. Variation in estimated shock pressure for samples of unit 1 and unit 2 with radial distance from the center of the structure in comparison with data from Robertson (1975).

inal feature that is due to the higher mineralogical heterogeneity and the generally smaller quartz grain size in the rocks of unit 1.

Deviatoric Stress

TEM investigations have shown that basal PDFs represent mechanical Brazil micro-twins (Goltrant et al., 1991, 1992a). Secondary Brazil twinning in naturally deformed rocks has been reported exclusively from impact structures. However, Brazil twins have not been produced by dynamic, high-pressure shock experiments, but they have been produced by static high-stress experiments (McLaren et al., 1967). Thus, the occurrence of basal PDFs implies high deviatoric stresses (Goltrant et al., 1992a; Leroux and Doukhan, 1996). The observation of the absence of basal PDFs in the highest recorded shock levels at the center of the structure implies that the deviatoric stress component of the shock wave–associated stress tensor increases relative to the hydrostatic pressure component with distance from the point of impact (Leroux and Doukhan, 1996).

Experiments yield a critical shear stress for Brazil twinning in the order of 2 GPa (McLaren et al., 1967), which implies that a differential stress of at least 4 GPa is required to twin a crystal in a favorable orientation to the stress field. Even higher differential stresses would be necessary for the twinning of crystals that are not in an optimal orientation to the stress field. At Charlevoix, basal PDFs occur in crystals at a wide range of orientations (Fig. 5) and in rocks at distances of ~2–10 km from the center of the structure (Fig. 1; Robertson, 1975). This implies that the shock-related differential stress probably exceeded 4 GPa in these rocks.

Orientation Distribution of Planar Microstructures

Since basal PDFs and microfaults are shear stress controlled, they should occur ideally at an angle of 45° to the shock compression direction, assuming that this is the direction of the highest compressive direction responsible for the shear deformation and assuming a more or less uniform orientation of the stress field. In contrast, rhombohedral PDFs are presumed to develop in planes parallel to the shock wave propagation direction (Goltrant et al., 1992b). Therefore, the orientation of rhombohedral PDFs should show optimally a girdle-maximum in stereographic projection, with the pole to this girdle representing the shock wave propagation direction. Thus, the orientation distribution of planar microstructures might be expected to yield information on the impact-related stress field. A preferred orientation of rhombohedral PDFs in a girdle has been reported from naturally shocked rocks of the Ries crater in Germany (Walzebuck and Engelhardt, 1979) and from experimentally shocked quartz (Gratz et al., 1988). In contrast, a preferred orientation of PDFs was not apparent in the results of experiments by Huffman et al. (1993). However, that investigation was undertaken in a thin section orientation perpendicular to the shock direction, so that a possible lack of PDF orientations at high angles to the shock direction was not detectable. Also, Dressler et al. (1998) reported a random orientation distribution of planar microstructures from a fine-grained chert from the Slate Island impact structure. However, these authors did not distinguish between shear stress controlled microstructures and those induced by quasi-hydrostatic pressure.

For the Charlevoix target rocks, the concentration of rhombohedral PDFs in girdle maxima from two charnockitic gneisses is similar to a preferred orientation of the quartz c-axes (Figs. 5B and 5C) and therefore reflects the texture of quartz rather than providing information on the stress field. Interestingly, the orientation pattern of c-axes, basal PDFs, rhombohedral PDFs, and microfaults in sample CT383 (Fig. 5A) show differences. However, a systematic variation, as may be expected from the different nature of the features, is not apparent. In sample CT384 (Fig. 5D), the observation that only crystals with a specific orientation developed PDFs parallel to $\{10\bar{1}2\}$ indicates a strong influence of the quartz anisotropy on PDF development, which has also been reported from experiments (Langenhorst and Deutsch, 1994). The influence of the crystallographic orientation of quartz on the orientation distribution of planar microstructures appears to superimpose an influence of the orientation of the impact-related stress field. Therefore, the use of the orientation distribution of PDFs as a stress field indicator is equivocal and not reliable for the Charlevoix target rocks.

Dauphiné Twinning and PDFs Parallel to Positive and Negative Forms

Secondary Dauphiné twinning in quartz can be due to the transformation from hexagonal β-quartz to trigonal α-quartz during cooling below the transition temperature, which is ~573 °C at atmospheric pressure (e.g., Frondel, 1962; Tullis, 1970; Markgraaff, 1986; Heaney and Veblen, 1991). Dauphiné twins are also known to form as shear stress–induced deformation twins in naturally deformed rocks over a wide range of geological situations, from diagenetic to amphibolite facies conditions (e.g., Tullis and Tullis, 1972; Lloyd, 2000). They have also been induced in shock experiments on preheated quartz crystals where they are found to be associated with PDFs and fractures (Gratz et al., 1992).

Rhombohedral PDFs have been observed to cross Dauphiné twin boundaries without changing direction (arrows in Fig. 6B), thereby changing their crystallographic orientation from a positive into a negative form. This indicates that twinning occurred after PDF development, as generally, PDFs do not occur with orientations that correspond to both positive and negative forms (Langenhorst and Deutsch, 1994; Grieve et al., 1996). Twin boundaries are commonly planar and aligned parallel to PDFs with the twins occurring in the interspaces between the PDFs (Fig. 6). These observations suggest that twinning is related, at least indirectly, to PDF development.

To explain the occurrence of Dauphiné twins and PDFs with orientations that correspond to positive and negative rhombohedra, the following possibility is discussed: Quartz might have locally been in the hexagonal β-modification, where planes of positive and negative forms are symmetrically equivalent (Frondel, 1962). This would require that the uplifted target rocks have been shock-heated above the α-β transition temperature, which is 573 °C at atmospheric pressures plus ~25 °C/kbar under confining pressure (e.g., Coe and Paterson, 1969). Structural observations indicate a maximum elevation of the rocks from the central uplift of ~6 km (Roy, 1979). At the time of impact, the Charlevoix area was a tectonically active region due to the ongoing Taconian orogeny. Assuming an average geothermal gradient of ~30 °C/km and erosion of ~1 km since the impact event (Robertson, 1975; Roy, 1979), pre-shock temperatures in the uplifted rocks would have been in the order of 200 °C. High shock and post-shock temperature can be expected from preheated rocks (Langenhorst and Deutsch, 1994; Langenhorst, 1994). At shock pressures of <20 GPa and pre-shock temperature of ~200 °C, generally a post-shock temperature below 573 °C has to be presumed (Langenhorst, 1994). However, fluid inclusions trapped along PDFs in quartz from charnockitic host gneisses of the Charlevoix structure indicate that temperatures reached >500 °C shortly after the impact event (Pagel and Poty, 1975). Thus, it may be possible that temperatures locally reached the transition temperature during shock. Transformation from the α- to the β-modification might have occurred either directly due to heating under shock pressure, or, more likely, during decompression of the shock-heated crystals. If so, Dauphiné twinning is probably due to reversion to α-quartz during cooling. Transition twinning is primarily a consequence of strain produced in the crystal due to crystallographic mismatch with the amount and distribution of twins being influenced by the rate of cooling, temperature gradient and the occurrence or absence of preexisting Dauphiné twin orientations (Frondel, 1962). Inclusions and irregularities can localize the twinning. This can explain the alignment of the twin boundaries parallel to PDFs (Fig. 6).

PDFs parallel to negative $\{01\bar{1}2\}$ rhombohedra are commonly found to coexist with PDFs parallel to positive $\{10\bar{1}3\}$ rhombohedra, and only rarely together with PDFs parallel to $\{01\bar{1}3\}$ (Figs. 4C and 4D). This relationship has also been reported from shock experiments (Langenhorst and Deutsch, 1994) and in naturally shocked quartz from the Ries crater in Germany (Engelhardt and Bertsch, 1969). The PDFs parallel to $\{10\bar{1}3\}$ are well-developed over almost the entire crystal, whereas PDFs parallel to $\{01\bar{1}2\}$ appear to occur in the interspace of the PDFs parallel to $\{10\bar{1}3\}$ (Figs. 4C and 4D), suggesting that the PDFs parallel to $\{01\bar{1}2\}$ developed later (Engelhardt and Bertsch, 1969). This might imply that a transformation from α-into the β-modification has occurred after, or related to, the generation of PDFs parallel to $\{10\bar{1}3\}$—eventually localized in the crystalline areas adjacent to the PDF, which may have acted as a source of heat (Langenhorst, 1994). Subsequently, PDFs parallel to $\{01\bar{1}2\}$may have developed in β-quartz, where the form $\{01\bar{1}2\}$is symmetrically equivalent to $\{10\bar{1}2\}$. The increase in the number of crystals that contain PDFs with orientations that correspond to positive and negative rhombohedra with increasing apparent shock pressure (Table 1) may be due to a more widespread occurrence of the β-quartz modification and therefore may reflect shock-temperature gradients. In kinkband-like features (Figs. 4A, 4B, and 6C), the change in direction of PDFs may be due to preexisting Dauphiné twins or other heterogeneities, as subgrains or microfractures.

Because it explains many of our observations, we suspect that quartz has locally been in the β-modification due to shock-heating and that Dauphiné twinning is mainly due to a reversion to α-quartz during cooling.

SUMMARY AND CONCLUSIONS

Detailed investigations of planar microstructures in target rocks of the Charlevoix impact structure provide constraints on the pressure-temperature conditions induced by hypervelocity impact. The shock pressure estimation from planar microstructures in this study is consistent with the shock isobars proposed by Robertson (1975). Grain size effects and greater mineralogical heterogeneity are proposed to be the main cause for slightly lower shock pressures being recorded in finer-grained granitic gneisses in the southeastern part of the structure, compared to coarse-grained charnockitic gneisses to the northwest at similar distances from the center of the structure. The influence of the crystallographic orientation of quartz appears to superimpose the effect of the orientation of the impact-related stress field on the orientation distribution of planar microstructures. The occurrence of basal PDFs at distances of ~2–10 km from the center of the structure indicates high deviatoric stresses. The occurrence of Dauphiné twins associated with PDFs and the alignment of their twin boundaries parallel to PDFs, indicate that twinning is related to PDF development. Based on the occurrence of PDFs with orientations that correspond to positive and negative rhombohedra and the appearance of Dauphiné twins, quartz is suspected to have locally existed in the β-modification state. The transformation from the α- into the β-modification might have occurred either directly due to shock-heating under compression, or, more likely, due to decompression of the shock-heated crystals. Dauphiné twinning is proposed to be primarily due to the reversion to α-quartz during cooling. These findings imply high shock temperatures of >573 °C and suggest that either shock heating in crystalline target rocks is more effective than predicted in theory or that the pre-shock temperature of the target rocks has been higher than assumed.

ACKNOWLEDGMENTS

This research was funded by the German Science Foundation (DFG-grant Tr 534/1-1 to CAT) and by National Sciences and Engineering Research Council of Canada research grants to JGS, which are gratefully acknowledged. Thanks are due to Bernhard Stöckhert and Rolf Neuser for facilitating EBSD analysis at the Ruhr-University Bochum and to Michael Dence for an introduction to the geology of Charlevoix. James Whitehead is thanked for discussions. Constructive reviews by Roman Skála and Ann Therriault are gratefully acknowledged. Planetary and Space Science Centre contribution 38.

REFERENCES CITED

Ashwal, L.D., and Wooden, J.L., 1983, Isotopic evidence from the eastern Canadian Shield for a geochemical discontinuity in the Proterozoic mantle: Nature, v. 306, p. 679–680.

Carignan, J., Gariépy, C., and Hillaire-Marcel, C., 1997, Hydrothermal fluids during Mesozoic reactivation of the St. Lawrence rift system, Canada: C, O, Sr and Pb isotopic characterization: Chemical Geology, v. 137, p. 1–21, doi: 10.1016/S0009-2541(96)00144-1.

Coe, R.S., and Paterson, M.S., 1969, The α-β inversion in quartz: A coherent phase transition under nonhydrostatic stress: Journal of Geophysical Research, v. 74, p. 4921–4948.

Dressler, B.O., Crabtree, D., and Schuraytz, B.C., 1997, Incipient melt formation and devitrification at the Wanapitei impact structure, Ontario, Canada: Meteoritics & Planetary Science, v. 32, p. 249–258.

Dressler, B.O., Sharpton, V.L., and Schuraytz, B.C., 1998, Shock metamorphism and shock barometry at a complex impact structure: Slate Islands, Canada: Contributions to Mineralogy and Petrology, v. 130, p. 275–287, doi: 10.1007/s004100050365.

Duyster, J., 1996, StereoNett 2.0.: University of Bochum, Shareware, http://homepage.ruhr-uni-bochum.de/Johannes.P.Duyster/stereo/stereoload.htm.

Engelhardt, W. von, and Bertsch, W., 1969, Shock induced planar deformation structures in quartz from the Ries crater, Germany: Contributions to Mineralogy and Petrology, v. 20, p. 204–234.

Frith, R.A., and Doig, R., 1973, Rb-Sr isotopic ages and petrologic studies of the rocks in the Lac Saint-Jean area, Quebec: Canadian Journal of Earth Sciences, v. 10, p. 881–899.

Frondel, C., 1962, The System of Mineralogy, 7th edition, Volume III, Silica Minerals: New York and London, John Wiley & Sons, inc., 334 p.

Goltrant, O., Cordier, P., and Doukhan, J.-C., 1991, Planar deformation features in shocked quartz; a transmission electron microscopy investigation: Earth and Planetary Science Letters, v. 106, p. 103–115, doi: 10.1016/0012-821X(91)90066-Q.

Goltrant, O., Doukhan, J.-C., Cordier, P., and Courtillot, V., 1992a, An investigation by transmission electron microscopy of planar deformation features in naturally shocked quartz: Terra Nova, v. 4, p. 405–412.

Goltrant, O., Leroux, H., Doukhan, J.-C., and Cordier, P., 1992b, Formation mechanisms of planar deformation features in naturally shocked quartz: Physics of the Earth and Planetary Interiors, v. 74, p. 219–240, doi: 10.1016/0031-9201(92)90012-K.

Gratz, A.J., Tyburczy, J., Christie, J., Ahrens, T., and Pongratz, P., 1988, Shock metamorphism of deformed quartz: Physics and Chemistry of Minerals, v. 16, p. 221–233, doi: 10.1007/BF00220689.

Gratz, A.J., Nellis, W.J., Christie, J.M., Brocious, W., Swegle, J., and Cordier, P., 1992, Shock metamorphism of quartz with initial temperatures –170 to +1000°C: Physics and Chemistry of Minerals, v. 19, p. 267–288, doi: 10.1007/BF00204005.

Grieve, R.A.F., and Robertson, P.B., 1976, Variations in shock deformation at the Slate Island impact structure, Lake Superior, Canada: Contributions to Mineralogy and Petrology, v. 58, p. 37–49.

Grieve, R.A.F., Langenhorst, F., and Stöffler, D., 1996, Shock metamorphism of quartz in nature and experiment: II. Significance in geoscience: Meteoritics & Planetary Science, v. 31, p. 6–35.

Heaney, P.J., and Veblen, D.R., 1991, Observation and kinetic analysis of a memory effect at the α-β quartz transition: American Mineralogist, v. 76, p. 1459–1466.

Hodges, K.V., 1991, Pressure-temperature-time paths: Annual Review of Earth and Planetary Science, v. 19, p. 207–236, doi: 10.1146/annurev.ea.19.050191.001231.

Hörz, F., 1968, Statistical measurements of deformation structures and refractive indices in experimentally shock-loaded quartz, *in* French, B.M., and Short, N.M., eds., Shock metamorphism of natural materials: Baltimore, Mono Book Corp., p. 243–254.

Huffman, A.R., and Reimold, W.U., 1996, Experimental constraints on shock-induced microstructures in naturally deformed silicates: Tectonophysics, v. 256, p. 165–217, doi: 10.1016/0040-1951(95)00162-X.

Huffman, A.R., Brown, J.M., Carter, N., and Reimold, W.U., 1993, The microstructural response of quartz and feldspar under shock loading at variable temperatures: Journal of Geophysical Research, v. 98, p. 22,171–22,197.

Kieffer, S.W., 1971, Shock metamorphism of the Coconino sandstone of Meteor crater, Arizona: Journal of Geophysical Research, v. 76, p. 5449–5473.

Kumarapeli, P.S., 1985, Vestiges of Iapetan rifting in the craton west of the northern Appalachians: Geoscience Canada, v. 12, p. 54–59.

Langenhorst, F., 1994, Shock experiments on pre-heated α- and β-quartz: II. X-ray and TEM investigations: Earth and Planetary Science Letters, v. 128, p. 683–698, doi: 10.1016/0012-821X(94)90179-1.

Langenhorst, F., 2002, Shock metamorphism of minerals from a microstructural point of view: Bulletin of the Czech Geological Survey, v. 77, p. 265–282.

Langenhorst, F., and Deutsch, A., 1994, Shock experiments on pre-heated α- and β-quartz: 1. Optical and density data: Earth and Planetary Science Letters, v. 125, p. 407–420, doi: 10.1016/0012-821X(94)90229-1.

Langenhorst, F., Deutsch, A., Stöffler, D., and Hornemann, U., 1992, Effect of temperature on shock metamorphism of single crystal quartz: Nature, v. 356, p. 507–509, doi: 10.1038/356507a0.

Lemieux, Y., Tremblay, A., and Lavoie, D., 2003, Structural analysis of supracrustal faults in the Charlevoix area, Quebec: relation to impact cratering and the St-Laurent fault system: Canadian Journal of Earth Sciences, v. 40, p. 221–235, doi: 10.1139/e02-046.

Leroux, H., and Doukhan, J.-C., 1996, A transmission electron microscope study of shocked quartz from the Manson impact structure, *in* Koeberl, C., and Anderson, R.R., eds., The Manson Impact Structure, Iowa: Anatomy of an Impact Crater: Boulder, Colorado, Geological Society of America Special Paper 302, p. 267–274.

Lloyd, G.E., 2000, Grain boundary contact effects during faulting of quartzite: an SEM/EBSD analysis: Journal of Structural Geology, v. 22, p. 1675–1693, doi: 10.1016/S0191-8141(00)00069-9.

Markgraaff, J., 1986, Elastic behavior of quartz during stress-induced Dauphiné twinning: Physics and Chemistry of Minerals, v. 13, p. 102–112, doi: 10.1007/BF00311900.

McLaren, A.C., Retchford, R.J.A., Griggs, D.T., and Christie, J.M., 1967, TEM study of Brazil twins and dislocations experimentally produced in natural quartz: Physica Status Solidi, v. 19, p. 631–644.

Melosh, H.J., 1989, Impact Cratering: A Geologic Process: New York, Oxford University Press, Oxford, 245 p.

Neale, E.R.W., Beland, J., Potter, R.R., and Poole, W.H., 1961, A preliminary tectonic map of the Canadian Appalachian region based on age of folding: Canadian Mining and Metallurgical Bulletin, v. 54, p. 687–694.

Pagel, M., and Poty, B., 1975, Fluid inclusion studies in rocks of the Charlevoix Structure (Quebec, Canada): Fortschritte der Mineralogie, Special Issue, I.M.A. Ninth general meeting, v. 52, p. 479–489.

Robertson, P.B., 1968, La Malbaie Structure, Québec—A Paleozoic meteorite impact site: Meteoritics, v. 4, p. 1–24.

Robertson, P.B., 1975, Zones of shock metamorphism at the Charlevoix impact structure, Québec: Geological Society of America Bulletin, v. 86, p. 1630–1638.

Robertson, P.B., and Grieve, R.A.F., 1977, Shock attenuation at terrestrial impact structures, *in* Roddy, D.J., Pepin, R.O., and Merrill, R.B., eds., Impact and Explosion Cratering: New York, Pergamon Press, p. 687–702.

Rondot, J., 1971, Impactite of the Charlevoix structure, Québec, Canada: Journal of Geophysical Research, v. 76, p. 5414–5423.

Rondot, J., 1989, Géologie de Charlevoix: Quebec, Canada, Ministère Énergie et Ressources, MB 89-21.

Roy, D.W., 1979, Origin and evolution of the Charlevoix cryptoexplosion structure [Ph.D. thesis]: Princeton, Princeton University, 190 p.

Stöffler, D., and Langenhorst, F., 1994, Shock metamorphism of quartz in nature and experiment: I. Basic observation and theory: Meteoritics, v. 29, p. 155–181.

Tremblay, A., and Lemieux, Y., 2001, Supracrustal faults of the St. Lawrence rift system between Cap-Tourmente and Baie-Saint-Paul, Québec: Current Research, part D: Geological Survey of Canada, Paper 01-D15, 19 p.

Tullis, J., 1970, Quartz: Preferred orientation in rocks produced by Dauphiné twinning: Science, v. 168, p. 1342–1344.

Tullis, J., and Tullis, T., 1972, Preferred orientation of quartz produced by mechanical Dauphiné twinning: Thermodynamics and axial experiments, *in* Heard, H.C., Borg, I.Y., Carter, N.C., and Raleigh, C.B., eds., Flow and Fracture of rocks: American Geophysical Union Geophysical Monograph, v. 16, p. 55–66.

Walzebuck, J.P., and Engelhardt, W. von, 1979, Shock deformation of quartz influenced by grain size and shock direction: observations on quartz-plagioclase rocks from the basement of the Ries Crater, Germany: Contributions to Mineralogy and Petrology, v. 70, p. 267–271.

Wanless, R.K., and London, J.A., 1961, Isotopic age measurements on coeval minerals and mineral pairs: Geological Survey of Canada, v. 61-17, 127 p.

Whitehead, J., Spray, J.G., and Grieve, R.A.F., 2002, Origin of "toasted" quartz in terrestrial impact structures: Geology, v. 30, p. 431–434, doi: 10.1130/0091-7613(2002)0302.0.CO;2.

Whitehead, J., Kelley, S., Sherlock, S.C., Grieve, R.A.F., Spray, J.G., and Trepmann, C.A., 2003, Structural and geochronologic constraints on the timing of the Charlevoix impact, Quebec, Canada: Nördlingen, 3rd International Conference on Large Meteorite Impacts, abstract no. 4084.

MANUSCRIPT ACCEPTED BY THE SOCIETY 9 AUGUST 2004

Geological Society of America
Special Paper 384
2005

Shock pressure distribution in the Vredefort impact structure, South Africa

Roger L. Gibson*
Wolf Uwe Reimold
Impact Cratering Research Group, School of Geosciences, University of the Witwatersrand, WITS, Johannesburg 2050, South Africa

ABSTRACT

A petrographic study of Archean gneisses exposed in the core of the 80-km-wide central uplift of the Vredefort impact structure has revealed widespread evidence of shock-related metamorphic effects in feldspar and ferromagnesian minerals in addition to the already documented quartz microdeformation features. The shock features are variably overprinted by annealing effects, which increase in intensity toward the center of the structure. The shock effects show a general increase in intensity toward the center of the dome but are most characteristically marked by extreme heterogeneity on a small (centimeter to millimeter) scale, indicating highly localized shock pressure heterogeneity. Closest to the center, the pre-impact Archean fabrics and textures have been locally partially destroyed by comprehensive melting and melt mobilization, which has given rise to distinctive granofelses and clast-laden melt breccias. Based on the features observed, background shock pressures in the currently exposed level of the core of the dome range from ~10 GPa at distances of ~20 km from the center to >30–35 GPa within 5 km of the center. Pressures responsible for localized melting and for the formation of the melt breccias in this central zone may have exceeded 45 GPa, although it is difficult to constrain this value more precisely owing to uncertainty about the pre-impact temperature of the rocks and the role of local syn-shock frictional heating in raising temperature. Apart from the localized heterogeneity in shock pressures, the background shock pressure gradient appears to increase toward the center of the dome, approaching ~4 GPa/km in the central parts.

Keywords: shock metamorphism, impact, Vredefort dome, impact heating.

INTRODUCTION

The 2.02 Ga Vredefort impact structure in South Africa is one of the three largest impact structures known on Earth (Grieve and Therriault, 2000; Gibson and Reimold, 2001). It is also the most deeply eroded impact structure presently known, with the level of erosion estimated at between 7 and 10 km (Gibson et al., 1998; Henkel and Reimold, 1998). Thus, it provides a rare opportunity to study the shock and post-shock response of a large volume of rock from deep crustal levels beneath a large impact crater, unencumbered by much of the structural complexity that typically characterizes the highly brecciated subsurface levels of large impact structures. The removal of the impact melt rocks and other crater fill breccias has exposed the roots of the central uplift, the Vredefort dome, in which shock metamorphic and other impact-related features abound (see review in Gibson and Reimold, 2001). However, an anomalous aspect of these features is that estimated shock pressures derived from planar deformation

*gibsonr@geosciences.wits.ac.za

Gibson, R.L., and Reimold, W.U., 2005, Shock pressure distribution in the Vredefort impact structure, South Africa *in* Kenkmann, T., Hörz, F., and Deutsch, A., eds., Large meteorite impacts III: Geological Society of America Special Paper 384, p. 329–349. For permission to copy, contact editing@geosociety.org.

feature (PDF) orientations in quartz in the rocks of the dome are inconsistent with expectations based on the size of the impact structure. For instance, Grieve et al. (1990) obtained relatively constant estimates of 10–15 GPa across much of the exposed central uplift using the widely applied statistical technique of Grieve and Robertson (1976). This pattern is contrary to the observations from other exposed central uplifts (e.g., Dressler, 1990; Dressler et al., 1998), in which peak shock pressures increase radially toward the center of the uplift, and is considerably lower than the >50 GPa maximum shock pressure estimates predicted by some hydrocode modeling results for a structure of this size (Ivanov and Deutsch, 1999). In view of both the recent hydrocode modeling results (Ivanov and Deutsch, 1999) and evidence from the central uplift of the Puchezh-Katunki structure (Masaitis and Pevzner, 1999), a new petrographic study of the rocks in the central parts of the Vredefort dome has been conducted to investigate the shock effects related to the Vredefort impact across a range of common minerals.

GEOLOGICAL SETTING

The Vredefort dome, located some 120 km southwest of Johannesburg, South Africa, is an ~80–90 km wide structure within Archean and Paleoproterozoic rocks of the central Kaapvaal craton that is partially obscured beneath Phanerozoic cover rocks (Fig. 1). Recent comprehensive reviews of the geology of the dome may be found in Reimold and Gibson (1996) and Gibson and Reimold (2001). A catastrophic origin for the dome was originally proposed based on the upturned to overturned nature of the rocks in the outer parts of the dome, and the ubiquitous presence of voluminous pseudotachylitic breccias (Shand, 1916). Later, the identification of bona fide shock deformation features (coesite and stishovite, e.g., Martini, 1991; shatter cones, e.g., Hargraves, 1961; PDFs in quartz, e.g., Lilly, 1981; Grieve et al., 1990; and planar features in zircon, e.g., Kamo et al., 1996; see full review in Gibson and Reimold, 2001), together with the recognition of chemical contamination by a meteoritic component in dikes of the Vredefort Granophyre (Koeberl et al., 1996), led to the confirmation of the dome as the central uplift of a large impact structure. Most estimates of the original crater diameter fall between 200 and 300 km; however, considerable uncertainty remains owing to the deep erosion of the structure, indicated by the absence of the crater form and the crater-fill impact breccias and melt sheet observed in other large structures such as Chicxulub and Sudbury (Grieve and Therriault, 2000).

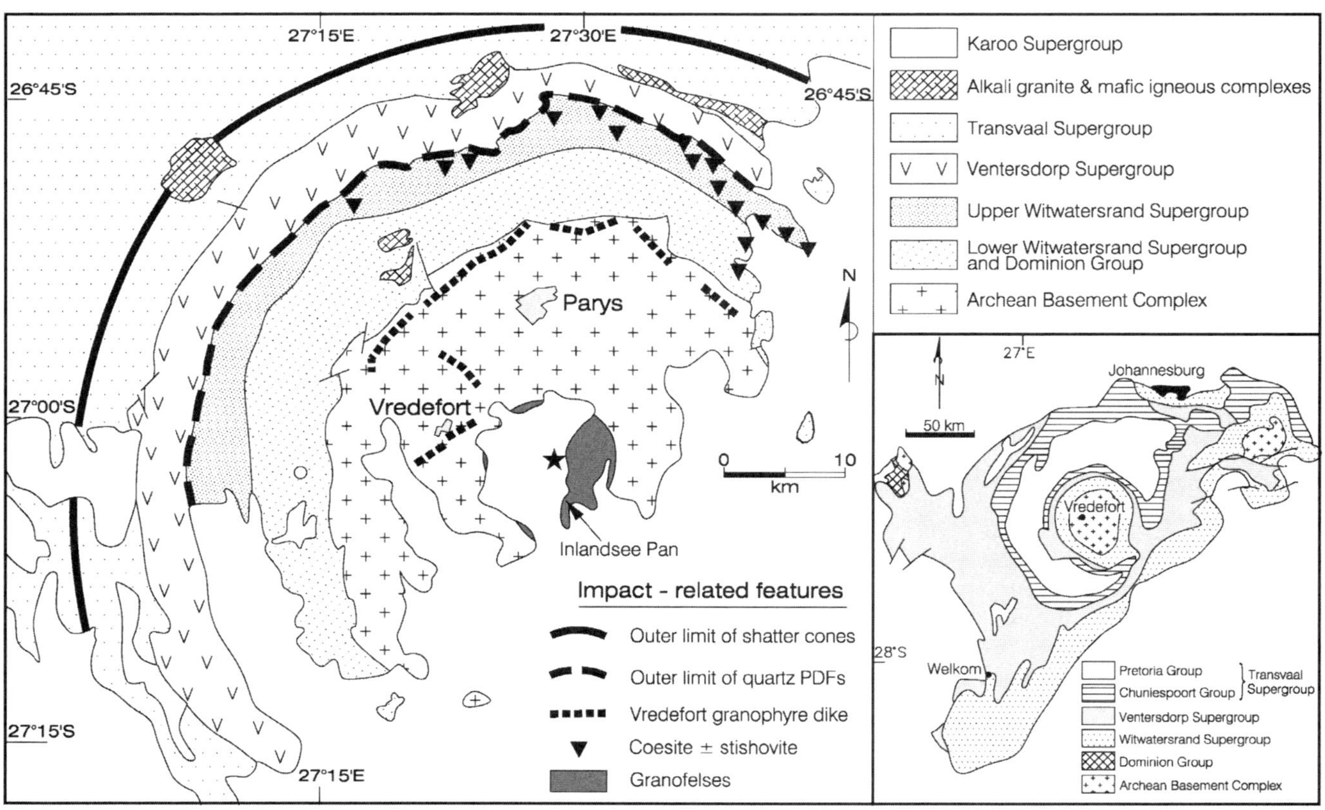

Figure 1. Simplified geological maps of the Vredefort impact structure (inset) and the Vredefort dome. The distributions of impact melt dikes, shatter cones, planar deformation features in quartz, high-pressure quartz polymorphs, and the impact-related granofelses are shown, as is the geographic center of the dome (star).

The estimated 7–10 km of erosion of the Vredefort impact structure is based on stratigraphic (McCarthy et al., 1990), thermobarometric (Gibson et al., 1998), and geophysical (Henkel and Reimold, 1998) considerations. One of the benefits of this deep section is that the thermal disturbance created by the impact event was sufficiently long-lived that texturally and chemically equilibrated metamorphic mineral parageneses were created following the impact event. The study of these parageneses has enabled quantification of the impact-related thermal overprint across the structure. This overprint has been shown to vary strongly from a background value of ~300 °C, in rocks at radial distances of >25 km from the center of the structure, to >700 °C within 10 km of the center (Gibson et al., 1998; Gibson and Reimold, 2000). Recently, Gibson (2002) attributed ultra-high temperature pelitic metamorphic parageneses and granitic anatectic melts from near the center of the dome to the attainment of post-impact temperatures in excess of 1000 °C. The magnitude and extent of this impact-induced thermal metamorphism goes a long way to explaining the anomalous character of shock-diagnostic features such as PDFs in quartz, as noted by, for example, Grieve et al. (1990). However, it also provides direct constraints on the shock pressures experienced by these rocks, because of the positive correlation between shock pressure and the magnitude of shock heating (Raikes and Ahrens, 1979; Stöffler, 1984) (Table 1). Estimates of the pre-impact temperature of rocks in the central parts of the dome prior to impact, based on structural reconstructions (Lana et al., 2003a, 2003b) and calculation of the Paleoproterozoic geotherm for the region (Gibson and Jones, 2002), suggest that shock heating of several hundred °C is needed to produce these post-shock temperatures. Comparison with Table 1 shows that shock pressures far in excess of 10–15 GPa are needed to achieve this.

Apart from the studies of Fricke et al. (1990) and Grieve et al. (1990), most of the work on shock effects in the Vredefort impact structure has concentrated on the supracrustal rocks exposed around the gneissic core of the Vredefort dome between 20 and 40 km from the center. Shatter cones were previously documented up to 40 km from the center (see review in Therriault et al., 1997), but Wieland et al. (2004) have reported a single shatter cone occurrence in quartzite in the rim syncline surrounding the dome at a radial distance of 65 km from the center that may require revision of the extent of development of this feature. PDFs in quartz are known up to a radial distance of ~30–35 km (Lilly, 1981), at which point the predominantly arenaceous sediments of the Witwatersrand Supergroup give way to quartz-absent basaltic lavas of the Ventersdorp Supergroup. Martini (1991), however, found coesite in a quartz vein cutting these lavas in a borehole drilled into the southern part of the dome, which would suggest that conditions for PDF development might also have extended farther than the present outcrop limits of arenaceous rocks. Other evidence of shock metamorphism comes from coesite ± stishovite at several localities between 21 and 28 km from the center of the dome (Martini, 1991) (Fig. 1). Apart from this, only non-diagnostic features such as kinkbands in biotite (Wilshire, 1971; Gibson et al., 1997) and fracture cleavage in garnet (Gibson et al., 1997) have been reported as being linked to the impact shock. Schreyer (1983) provided a brief description of annealed shocked plagioclase from near the center of the dome, although he believed that the shock event was somehow linked to a contemporaneous endogenic granulite-facies metamorphic event rather than impact.

SHOCK FEATURES IN THE ARCHEAN BASEMENT COMPLEX

Field Relationships

The Archean Basement Complex exposed in the 36–40 km wide core of the Vredefort dome (Fig. 1) comprises a heterogeneous sequence of coarse-grained (0.2–2 cm) trondhjemitic and tonalitic migmatitic gneisses with scattered mafic, ultramafic and sedimentary xenoliths, and small granitic to granodioritic bodies, that experienced mid-crustal upper amphibolite- to granulite-facies metamorphism (~700 °C, 0.45 GPa in the amphibolite-facies zone; 800 ± 50 °C, 0.5 GPa in the granulite-facies zone) at ca. 3.1 Ga (Stepto, 1990; Stevens et al., 1997; Hart et al., 1999; Lana, 2004; Lana et al., 2003a, 2003b, 2003c, 2004). The rocks are typically characterized by banding and aplitic and pegmatitic veining on a scale of a few centimeters to a few meters (Figs. 2A and 2B). The trondhjemitic and tonalitic gneisses are usually

TABLE 1. PRESSURES MARKING THE ONSET OF DEVELOPMENT OF ROCK AND MINERAL SHOCK DEFORMATION EFFECTS FOR QUARTZOFELDSPATHIC ROCKS

Approximate shock P (GPa)	Approximate post-shock T (°C)	Rock and mineral effects
<2	<100	Shatter cones, rock jointing, breccias
8–10	100	Quartz (0001) Brazil twins
>10	100	Planar deformation features in quartz, K-feldspar
12–15	150	Stishovite
15	150	Planar deformation features in plagioclase
20	170	Reduced refractive index and birefringence in quartz and feldspar
>30	275	Coesite, diaplectic quartz and feldspar glass
45	900	Feldspar melt
50–60	>1500	Whole-rock melt

Note: Data from Ostertag (1981); Stöffler and Langenhorst (1994); Huffman and Reimold (1996); French (1998).

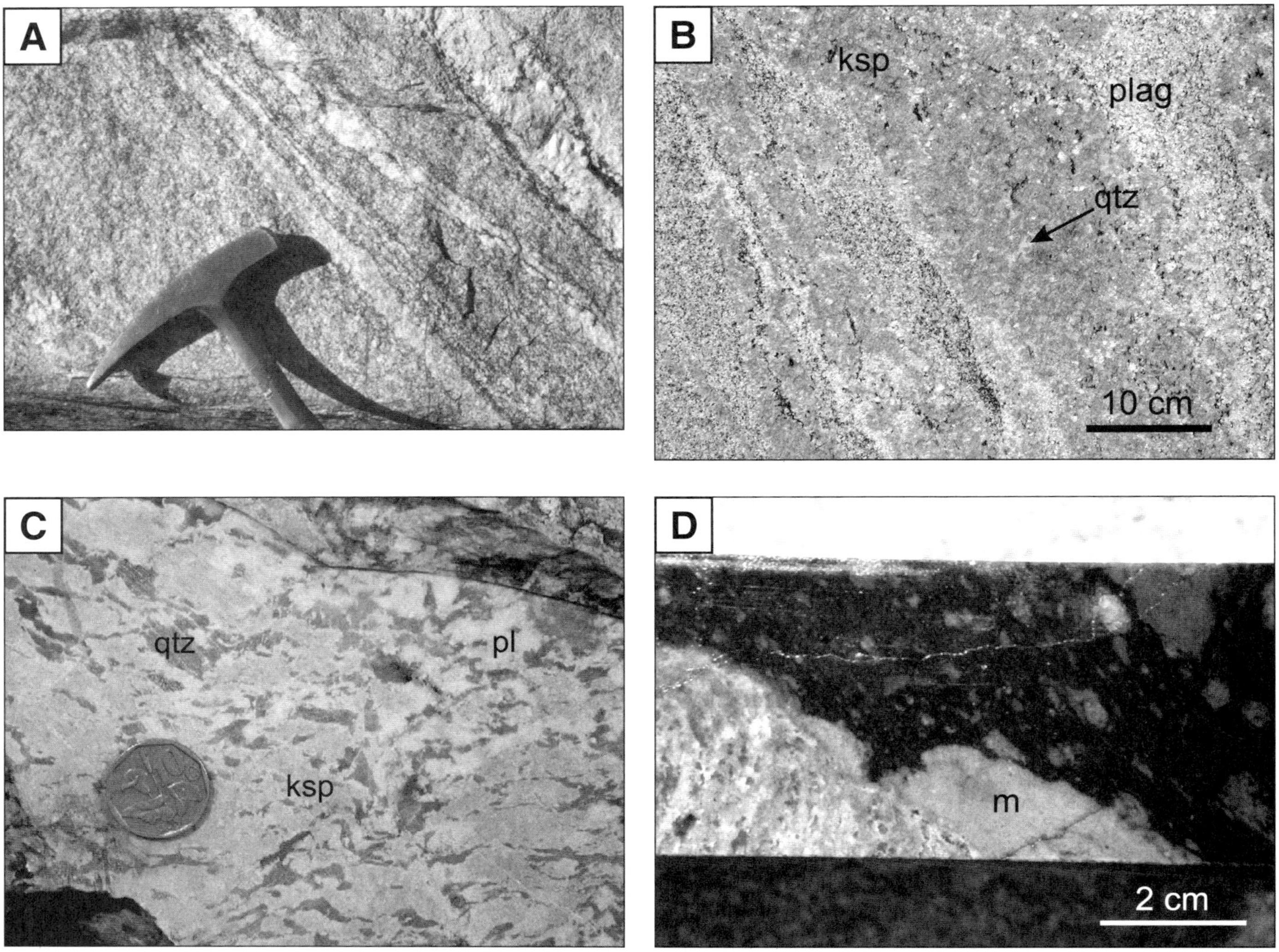

Figure 2. (A) Typical Archean Basement Complex migmatite from the outer parts of the core of the Vredefort dome, showing trondhjemitic (dark gray) and intrusive granitic (light gray) bands parallel to the strong S3 foliation. (B) Polished slab of migmatite showing very coarse-grained, poorly-foliated, late-D3 granite veins within a coarse-grained biotite trondhjemite matrix. Compare with C. (C) Typical granofels texture in granite from near the center of the dome (Inlandsee Pan, Fig. 1). Note the lack of clearly-defined feldspar crystal shapes compared with B, which reflects the near-complete recrystallization of the granite associated with the shock event. The irregular shape of the quartz aggregates suggests some plastic deformation of the rock associated with the shock event. Coin is 1 cm across. (D) Granulitic breccia in borehole core from the Inlandsee Pan area. The large clast of granite shows clear evidence of grain size reduction along its margin (m) and partial entrainment of this margin into the breccia matrix (dark gray). The smaller clasts are also recrystallized and elongated parallel to a flow fabric in the breccia matrix. qtz—quartz; ksp—K-feldspar; pl/plag—plagioclase.

well-foliated (S2 and/or S3; Lana et al., 2003a, 2004), while the granitic to granodioritic leucosomes and intrusive bodies are less well foliated, as befits their generally late-D3 intrusion age (Lana et al., 2003a, 2004) (Fig. 2B). The foliations are defined by compositional banding and aligned minerals and mineral aggregates. Within ~5 km of the geographic center of the dome, however, the coarse-grained foliated texture is variably replaced by a more massive, fine-grained (<0.2 mm) granoblastic texture, and the rocks locally display a vitreous luster, increased induration and conchoidal fracture (Fig. 2C). In general, this overprint is static in that the gneisses, for the most part, retain their gross pre-impact Archean fabric; however, closer to the center of the dome, some samples show more comprehensive destruction of oriented fabrics. Microscopically, these rocks are best classified as granofelses (Fig. 1). Closer examination shows that the feldspar component of the rock appears to be preferentially affected relative to quartz, and that some mafic minerals also show evidence of severe grain size reduction. Stepto (1990) also described isolated occurrences of "crush breccia," in which fragments of gneiss occur within a fine-grained quartzofeldspathic matrix. Gibson et al. (2002)

referred to these features (Fig. 2D) as granulitic breccias and interpreted them as annealed shock ± friction melts.

Conflicting estimates exist for the maximum depth of crust uplifted by the impact in the central parts of the dome. Hart et al. (1990) suggested that lower crust and even upper mantle were exposed around the Inlandsee Pan (Fig. 1), from which they concluded that the Pan marks the center of the dome and that the dome exposes a full, 36-km-deep, crustal section. Recent lithological and structural mapping and geochemical studies by Lana et al. (2003a, 2003b, 2004) have, however, found no evidence to support this hypothesis and suggest a maximum exposed pre-impact depth of ~25 km. Furthermore, the geometric center of the dome appears to lie ~4 km north of the Inlandsee Pan, in the vicinity of the farm Glenlivet 208 (Fig. 1). All estimates of radial distances reported here are based on this location.

Petrography

Recent mapping and geochemical studies of the Archean Basement Complex (Lana et al., 2003a, 2004) have indicated that the principal lithologies developed in the outer core of the dome and in the granofels zone, while highly heterogeneous on an outcrop scale, are essentially the same. Contrary to Hart et al. (1981, 1990) who proposed a major Archean terrane boundary between the inner and outer parts of the core, we conclude, thus, that the petrographic changes observed in the granofels zone reflect shock effects superimposed onto Archean gneisses that originally displayed similar pre-impact textures and mineralogy to the gneisses in the outer core. The only discernable difference is that the grade of Archean metamorphism in the inner parts of the core (granulite-facies) was slightly higher than in the outer parts (upper amphibolite-facies) (Stepto, 1990; Lana et al., 2003a, 2004). In the context of the geometry of the dome, this is consistent with rocks from the inner core representing deeper levels of the original Archean regional metamorphic terrane. Petrographically, this means that pre-impact pyroxene is absent from the gneisses in the outer 8 km of the core but that it increases proportionally at the expense of pre-impact biotite and amphibole toward the center of the dome.

The granitic, granodioritic and trondhjemitic-tonalitic gneisses are dominated by quartz + plagioclase + perthitic K-feldspar, with small amounts of biotite, amphibole and (in the central parts) clinopyroxene and/or orthopyroxene, as well as accessory ilmenite and/or magnetite, zircon and apatite. The mafic minerals and quartz in these felsic gneisses tend to occur as aggregates of 1–2 mm grains aligned parallel to the generally strong Archean foliations. Feldspars show a wide range of grain size, from 1 to 20 mm, although K-feldspar megacrysts in porphyritic granite and granodiorite bodies may reach 5 cm in length. Plagioclase ranges in composition from albite-oligoclase in the granitic bodies and leucosomes to oligoclase in the trondhjemitic gneisses and andesine in the tonalitic gneisses. The mafic gneisses are commonly homogeneous, but may also be banded on a millimeter scale, with plagioclase-rich bands alternating with plagioclase + clinopyroxene + hornblende + magnetite + quartz ± orthopyroxene ± biotite bands. Grain size is typically finer than in the felsic gneisses (1–2 mm), and the plagioclase is slightly more anorthitic (andesine to labradorite). The pelitic granulite assemblages have been described in detail elsewhere (Schreyer and Abraham, 1978; Stevens et al., 1997; Gibson, 2002). They comprise quartz + garnet + biotite ± plagioclase (oligoclase-andesine) ± K-feldspar ± orthopyroxene ± cordierite ± spinel, with grain size typically varying from ~1 mm to 10 mm. In general, these gneisses display coarse-grained granoblastic textures characterized by polygonal grain boundaries and triple junctions. Ironstone granulites intercalated with the pelitic granulites comprise clinopyroxene + quartz + magnetite ± orthopyroxene ± biotite ± olivine and have been described by Schreyer et al. (1978).

Analysis of more than 200 thin sections from outcrops across the core of the dome and several borehole cores from within a radius (r) of 5 km of the center of the dome has shown that the textures in the central gneisses differ significantly from those in the rest of the Basement Complex. Although the planar pre-impact Archean foliation typically remains visible in hand specimen, microscopically the gneisses are characterized by a strongly heterogranular texture, with relics of coarse pre-impact minerals interspersed with fine-grained mosaic aggregates that rim, cross-cut, or completely pseudomorph the older grains. The fine grain size of the aggregates (<200 μm and commonly <50 μm) is atypical for regionally metamorphosed granulites, which are usually considerably more coarse-grained because of the slow heating and cooling at high temperatures in regional metamorphic terranes, but is not unusual for an impact setting where intense heating can occur virtually instantaneously and where strong thermal gradients lead to relatively rapid cooling (Gibson et al., 2002). The aggregates show polygonal and/or subhedral grain boundaries, indicative of the attainment of textural equilibrium locally, but in many places, evidence of rapid cooling (for a granulite) is indicated by irregular, amoeboid, or interpenetrating grain boundaries. The transition between the textures in rocks in the outer and inner core of the dome is progressive and allows the mapping out of intermediate shock pressure zones.

Shock-Induced Deformation and Metamorphic Features

A wide range of shock-induced microdeformation and metamorphic features has been identified in the gneisses. There is a clear progression in intensity of development of these features toward the center of the dome; however, it is also apparent that the textures are heterogeneously developed on a local scale, with the intensity of features commonly increasing in close proximity to pseudotachylitic breccia veins. This is the subject of a separate study by our group, which seeks to determine how much of this heterogeneity can be related to increased shock intensity or annealing adjacent to the pseudotachylitic breccias.

Quartz

Planar features in quartz in the gneisses and quartzites of the dome have received considerable attention in the past (e.g., Lilly, 1981; Schreyer and Medenbach, 1981; Fricke et al., 1990; Grieve et al., 1990; Reimold, 1990; Hart et al., 1991; Leroux et al., 1994), and will not be described in any detail here. All these studies noted the unusually intense annealing of these features, which initially hampered absolute confirmation of their shock origin; however, Leroux et al. (1994) showed that features with (0001) orientations are bona fide Brazil twins of shock origin. One of the problematic aspects of these features, discussed previously, is that statistical techniques for estimating shock pressure using the crystallographic orientation of the PDFs (Grieve and Robertson, 1976) have failed to show either a progressive increase in shock pressure toward the center of the dome, or shock pressure values greater than ~10–15 GPa (Grieve et al., 1990). Schreyer and Medenbach (1981) and Fricke et al. (1990) used fluid inclusion thermometry to show that the PDFs in the Archean gneisses experienced a strongly variable thermal overprint of up to at least 800 °C after their formation. Grieve et al. (1990) proposed that the absence of PDF orientations that are diagnostic of high shock pressures might reflect preferential annealing of the delicate glass lamellae in these orientations during this thermal event. This was supported by Leroux et al. (1994), who suggested that the prevalence of (0001) PDFs reflects the greater robustness of the basal twins to annealing. Huffman and Reimold (1996), however, suggested that elevated pre-impact rock temperatures, consistent with the deep levels of burial of these rocks at the time of the impact event, may also have altered the shock deformation response of the quartz by increasing its plasticity, thereby inhibiting PDF development.

In the scores of thin sections examined during the course of our study, we encountered a small proportion of quartz grains with up to 3—but more commonly 1 or 2—sets of decorated PDFs defined by inclusion trails (Fig. 3A), but these are by no means common. In samples from $r < 5$ km from the center of the dome, these planar inclusion trails cut across polygonal quartz grain boundaries, or the boundaries themselves are pinned by these inclusion trails. We concur with Schreyer and Medenbach (1981) and Fricke et al. (1990) that these features reflect recrystallization of larger, shocked quartz grains. This indicates that lattice rotations have occurred following PDF formation and, thus, that the crystallographic orientations of the decorated PDFs in these rocks may be unconstrained with respect to the original c-axis orientation of the shocked quartz.

In addition to the decorated PDFs, quartz from pelitic granulites located at $r \sim 6$ km (i.e., farther from the center than the texture shown in Fig. 3A) contains multiple sets of planar features that comprise microcrystalline (20–100 µm) mosaics of anhedral to subhedral, polygonal to elongate quartz (Fig. 3B). The elongate quartz grains may be aligned parallel to the planar feature, or consistently oblique to it but parallel to one of the other sets in the grain. These planar features widen where two sets intersect and toward the margins of the host grain. In places, they are continuous with fluid inclusion trails that define decorated PDFs, or more irregular trails that might represent annealed fractures (Fig. 3C). Although most of these lamellae are planar, some are curved. However, nowhere was one of these features observed cutting a primary grain boundary (these relationships are somewhat qualitative as the recrystallization at grain boundaries is high). Where intersections are particularly intense, an irregular mass of fine-grained mosaic aggregate exists. The host grains commonly show mosaic extinction, with the boundaries of deformation bands aligned parallel to the planar features. The width of these zones (40–250 µm) and their spacing (100–500 µm) is too great for them to be normal PDFs (which typically are <2–3 µm wide and have a 2–10 µm spacing; e.g., Kieffer et al., 1976). However, there is no doubt that they are crystallographically controlled features, and, in places, they grade into decorated PDFs in the cores of the quartz grains. The absence of offsets along these features or their extension across grain boundaries rules out a microfault process. They may be similar to features identified by Huffman and Reimold (1996) in shocked, preheated samples of quartzite which developed fewer, wider PDFs, or annealed clustered PDFs such as those described by Langenhorst and Deutsch (1994). Quartz grains in the central granofelses are distinctly coarser (maximum 500–800 µm) than the aggregates in the crystallographically controlled features in Figure 3B. Grain size is also more homogeneous and the grains display increased polygonization or roundness relative to quartz in the Archean gneisses located farther from the center of the dome. The grains are largely strain-free and only rarely display traces of decorated PDFs (Fig. 3A). Polycrystalline quartz aggregates are commonly slightly disaggregated (glomerogranular texture; Figs. 3D and 3E), with K-feldspar ± plagioclase occurring in cusps and interstices between the quartz grains. An individual aggregate defines a similar shape and size to the lensoid pre-impact quartz aggregates in the gneisses (Fig. 3C), while the occurrence of PDFs continuing across quartz grain boundaries, or grain boundaries aligned along PDFs (e.g., Schreyer and Medenbach, 1981) (Fig. 3A), confirms that the quartz texture formed by recrystallization after the shock deformation.

In places in the granofelses, the boundaries between quartz and feldspar grains display thin (<100 µm) myrmekitic rims that are narrower and typically an order of magnitude finer-grained than the pre-impact myrmekites in the same rocks (Figs. 3D and 3E). Several small post-impact biotite- or hornblende-granite bodies and veins in the granofelses indicate that suprasolidus post-shock temperatures were sufficiently long-lived to allow some partial melting (Gibson, 2002). The myrmekites in the granofelses may, thus, indicate a similar eutectic melting phenomenon. This is supported by the development of myrmekite between feldspars and corroded quartz grains (e.g., Fig. 3D).

Plagioclase

At $r > 10$ km, plagioclase generally shows no diagnostic shock effects. Inferred impact-related deformation in these rocks is restricted to irregular fractures, the intensity of which increases

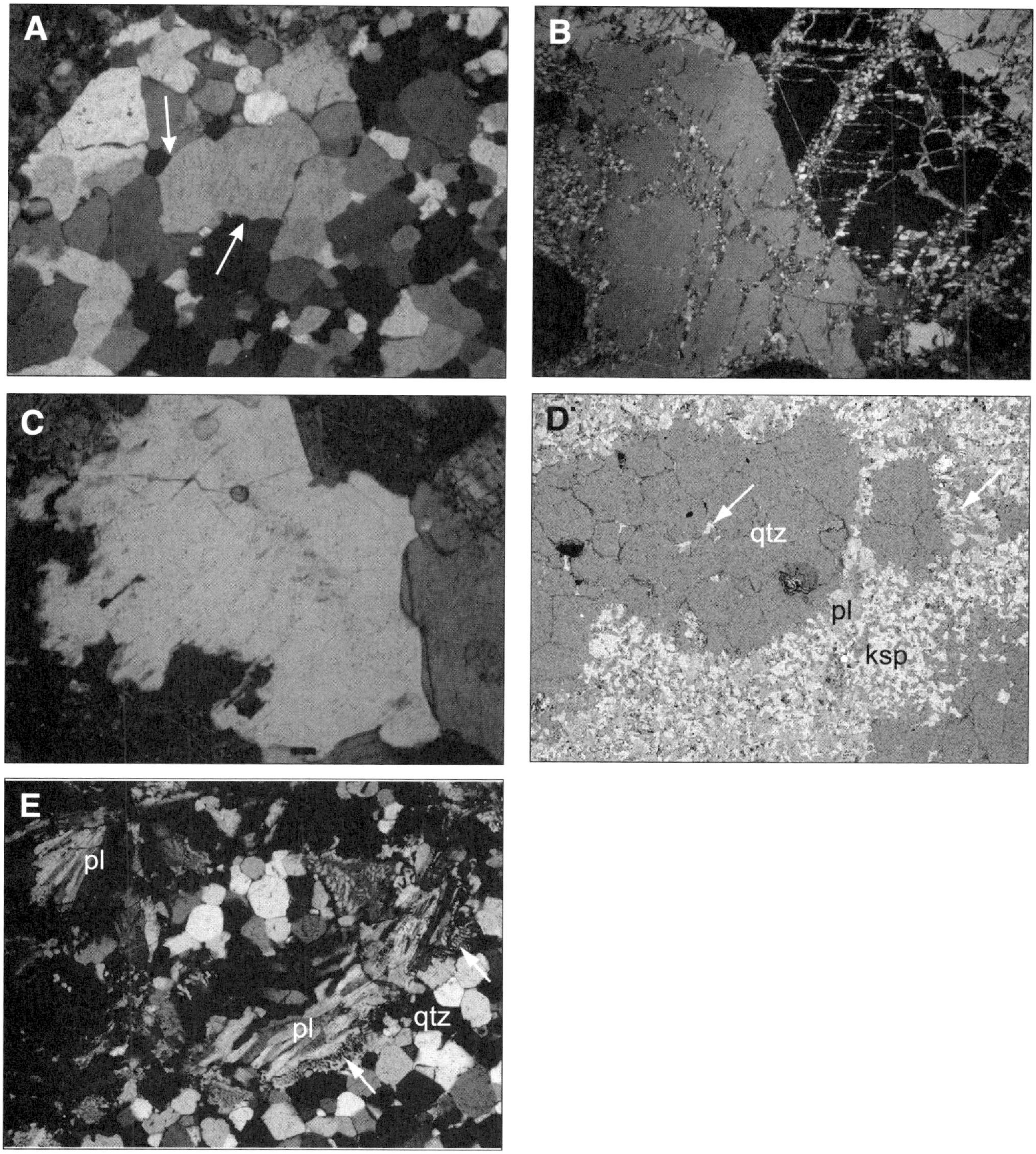

Figure 3. (A) Decorated planar deformation features (PDFs) in quartz defined by fluid inclusion trails (arrowed). Note that the main PDF set (NNE-SSW) cuts across polygonal grain boundaries, indicating that annealing followed PDF formation. A second set trending NNW-SSE is also visible. No meaningful crystallographic information can be obtained from such textures. The clouded appearance of the quartz is caused by an abundance of fluid inclusions in the plane of view. Granitic granofels IL29 from Inlandsee Pan (see Figure 9 for locality). Crossed polarizers. Length of photo 1.5 mm. (B) Mosaic recrystallization of quartz in crystallographically controlled planes. The irregular fractures are late-stage cracks with low-temperature alteration minerals. Pelitic granulite VH9, Jagers Vrede farm, SSW of Parys (Fig. 9). Crossed polarizers. Length of photo 7 mm. (C) Partial recrystallization of decorated PDFs in quartz in mafic granulite from same locality as Figure 3B. The lower left part of the grain displays strongly lobate boundaries consistent with replacement by a new quartz grain (dark gray). Crossed polarizers. Length of photo 1 mm. (D) Backscattered electron image of granofels shown in Figure 2C. The polygonal shape of the quartz grains (qtz) is outlined by grain boundary fractures and by interstitial K-feldspar (ksp) in cusps between the grains. Note the finer grain size of the intergrown plagioclase (pl) and K-feldspar in the matrix relative to the quartz. Myrmekitic reaction textures similar to Figure 2E are arrowed. Length of photo 2 mm. (E) Glomerogranular quartz aggregate, slightly disaggregated, with myrmekitic reaction against plagioclase. In contrast to the granoblastic texture in D, the plagioclase (pl) pseudomorph comprises a bladed aggregate of plagioclase crystals. Note the radiating bladed plagioclase aggregate in the top left corner. Crossed polarizers. Sample IL1 from near Inlandsee Pan. Length of photo 1.8 mm.

close to pseudotachylitic breccia veins. These fractures may cut across and displace grain boundaries and are commonly sites of preferential sericite or saussurite alteration. In places, they contain fine-grained (<10 μm) plagioclase subgrain mosaics. Adjacent to pseudotachylitic breccia-bearing fractures, grains may also show increasing strain (undulose or mosaic extinction, displaced twin lamellae), partial isotropization and recrystallization features (Fig. 4A). At its most intense, this recrystallization involves replacement of the parent grain by microcrystalline aggregates with the same composition as the parent. This is commonly associated with patchy or undulose extinction and deformation bands in the remainder of the parent grain. Immediately adjacent to the breccia, the feldspar may be replaced by plumose or spherulitic plagioclase aggregates. McIntyre (1968) and French (1998) described similar plumose features as the product of recrystallization of diaplectic glass. A similar origin has been proposed for microcrystalline plagioclase aggregates by Arndt et al. (1982) and Ostertag and Stöffler (1982).

Between ~7 and 10 km from the center, plagioclase grains commonly contain irregular to subplanar fractures that are filled by microcrystalline plagioclase mosaics which display an identical composition to the parent grain, and these features are no longer restricted to the vicinity of pseudotachylitic breccias. At $r < 7$ km, grains show a variety of features, including increased isotropization similar to that shown in Figure 4A, and partial replacement by irregular microcrystalline aggregates with, in places, the fine-grained aggregates selectively replacing alternate twin lamellae (Fig. 4B). Grains may show up to three sets of PDFs (Fig. 4C). Other features are irregular polysynthetic twin boundaries and planar to spindle-shaped features in alternate polysynthetic twin lamellae, which are either oblique to twin planes (Fig. 4D; see also Dressler, 1990, Fig. 2) or close to orthogonal to them, creating a ladder structure (Stöffler, 1966). These planar features have a spacing of ~10 μm and are <5 μm wide. The differential response evidenced both in the development of these features and in the recrystallization in alternate twin lamellae is reminiscent of the "asymmetric isotropization" described by, inter alia, Stöffler (1966) and Dressler (1990). Ostertag (1981) suggested that short fractures oblique to twin lamellae in plagioclase and microcline and reduced birefringence in alternate twins, are related to enhanced compressibility of the feldspar lattice along [100] and [010] versus [001]; however, the features observed in this study more closely resemble twin lamellae or deformation bands. Approaching the granofels zone (r ~5–7 km), the level of recrystallization in plagioclase increases, with both microcrystalline and plumose or sheaf-like aggregates found in close proximity, and a strongly heterogranular texture developed with rare remnants of the parent grains preserved (Figs. 4E and 4F). The static nature of this recrystallization is seen in the preservation of the original mineral textures, e.g., the subophitic texture in the mafic dike in Figure 4E.

Within the granofels zone ($r < 5$ km), coarse plagioclase grains are rare. Most pre-impact grains are pseudomorphed by fine-grained mosaics with a slightly coarser grain size (50–200 μm) than the aggregates farther from the center (Figs. 3D and 3E). Nonetheless, even where replacement is total, the aggregates commonly mimic the original outline of the pre-impact plagioclase grains, thereby preserving the gross Archean fabric. However, some evidence for bulk plastic deformation is suggested by the common flame-like lenticular texture observed in leucocratic granofelses (see Fig. 2C). In contrast to pre-impact plagioclase grains, which are typically anhedral and display complex albite and pericline twinning, grains within the aggregates typically display relatively simple Carlsbad and/or albite twinning and, where they are coarser, subhedral to polygonal shapes. Relict cores or parts of the precursor grains may be preserved locally. In these cases, the fine-grained grains show no difference in composition to the relict cores. In places, the pre-impact gneissic texture has been partially destroyed. Here, the fine-grained aggregates may comprise intergrown plagioclase and K-feldspar laths on a smaller scale than the grain size of the precursor assemblage (Figs. 3D and 3E). This suggests either that the precursor grains were perthitic or antiperthitic or that the aggregates (re)crystallized from melts of both plagioclase and alkali feldspar that were able to mix prior to quenching or crystallization. The presence of interstitial alkali feldspar ± plagioclase in the cusps between the coarse polygonal quartz grains in the glomerogranular aggregates (Fig. 3D) supports the formation of feldspar melts rather than diaplectic glass. Further evidence of plagioclase melting is provided by patches of highly oikocrystic K-feldspar containing, inter alia, plagioclase laths (see Fig. 5B), and the granulitic breccias (Fig. 2D). In places, the plagioclase laths in the breccia matrix are aligned parallel to the vein margins but show no internal strain, consistent with suprasolidus flow (Gibson et al., 2002). The clasts in these breccias are elongated in the same direction (Fig. 2D), which is also consistent with flow of a clast-crystal-melt mixture.

Ultramafic gneiss samples from borehole core from the Inlandsee Pan, which contain the most calcic plagioclase (An_{45-60}), show preservation of relict plagioclase cores, some displaying PDFs, surrounded by fine-grained aggregates of anhedral grains (Ashley et al., 1999). It is not clear if this reflects the greater resistance of more calcic plagioclase to shock transformation, or heterogeneity in the shock pressure. However, the preservation of biotite in the sample but not in adjacent rocks suggests that the latter may be the reason for the differences (see below).

Alkali Feldspar

Alkali feldspar shows similar textures to those found in plagioclase, such as subplanar fractures filled with fine-grained mosaics with identical composition to the parent grain; however, asymmetric and patchy partial replacement by microcrystalline aggregates is not as widespread as in the coexisting plagioclase. Most alkali feldspar is perthitic, with relatively coarse, aligned flames or blebs of plagioclase in the host grain. Diaplectic glass formation and subsequent recrystallization of such perthites may account for the textures shown in Figure 3E

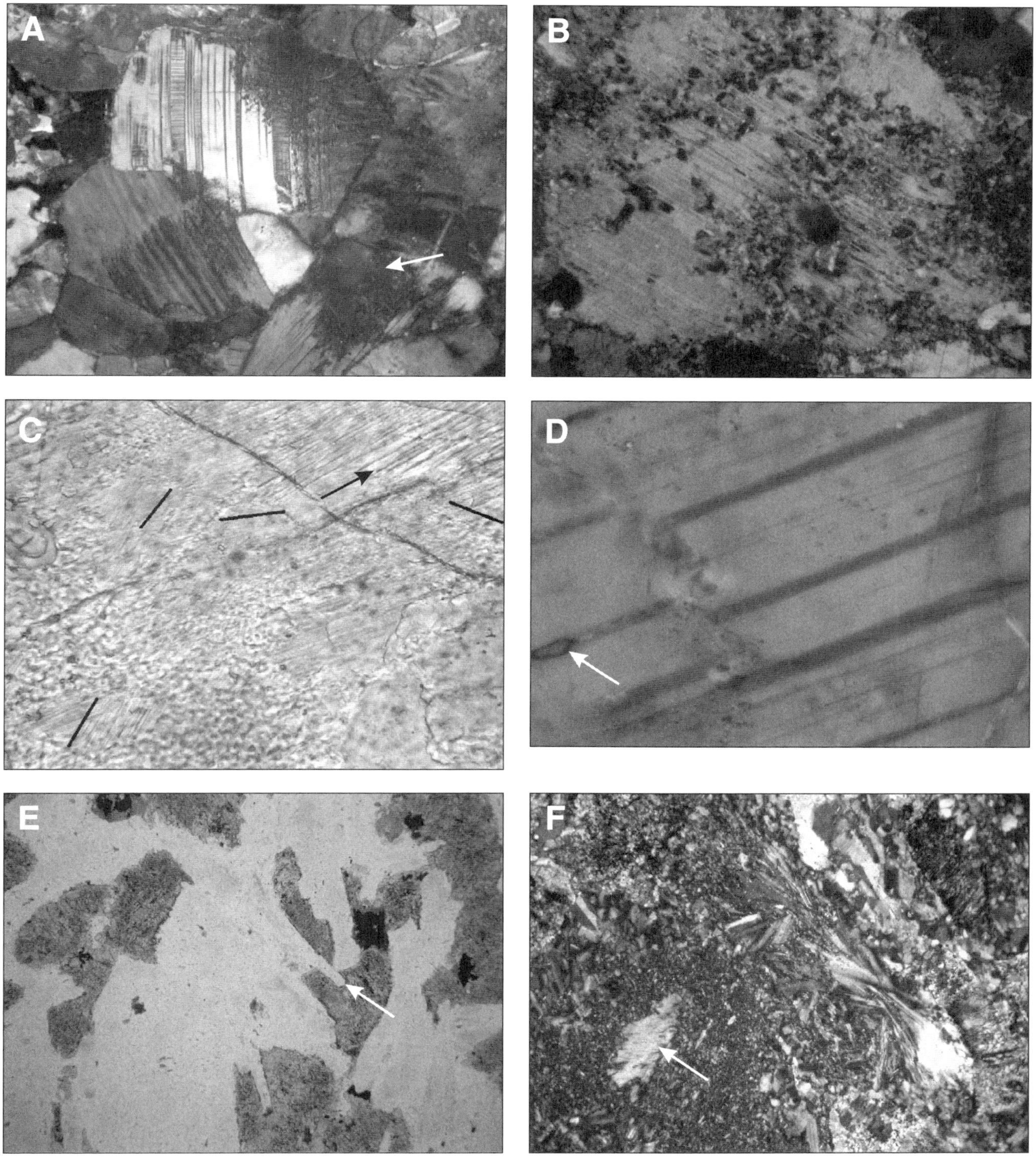

Figure 4. (A) Plagioclase grains in a mafic amphibolite (UP36) showing internal strain, isotropization (arrow) and asymmetric replacement by a fine-grained plagioclase mosaic that locally preferentially replaces alternate twin lamellae (left). The level of replacement here is unusual for this sample and is restricted to the vicinity of a pseudotachylitic breccia vein (out of view, top). Crossed polarizers. Length of photo 2 mm. (B) Microcrystalline aggregate replacing plagioclase. Note the typical patchy replacement and the localized selective replacement of alternate twin lamellae. Mafic granulite sample HM9, Jagers Vrede farm south-southwest of Parys. Crossed polarizers. Length of photo 0.7 mm. (C) Plagioclase grain showing at least three planar deformation feature sets. The lower part of the grain is recrystallized to a mosaic aggregate. The lamellae in the top right corner (arrowed) are polysynthetic twins. Pelitic granulite, Steynskraal farm, SSW of Parys. Plane-polarized light. Length of photo 0.2 mm. (D) Oblique twin lamellae or deformation bands, in alternate polysynthetic twins in plagioclase. Note that the twin lamellae are displaced along a partially recrystallized fracture, and are themselves partially recrystallized (arrow). Mafic granulite sample from same locality as VH9. Crossed polarizers. Length of photo 0.25 mm. (E) Subophitic-textured mafic dike showing strong impact-related recrystallization of plagioclase (see Fig. 4F) and pyroxene (see Fig. 7C). The preservation of the igneous texture (note the rectangular plagioclase shapes against the pyroxene, arrow) indicates that this dike intruded after the high-grade Archean metamorphism and that the impact-related metamorphism was static. Plane-polarized light. Sample VT590, SW core. Length of photo 2.5 mm. (F) Close-up of center of Figure 4E showing the strongly heterogranular recrystallization of plagioclase laths, with radiating and mosaic aggregates. Relict plagioclase core is arrowed. Crossed polarizers. Length of photo 1.8 mm.

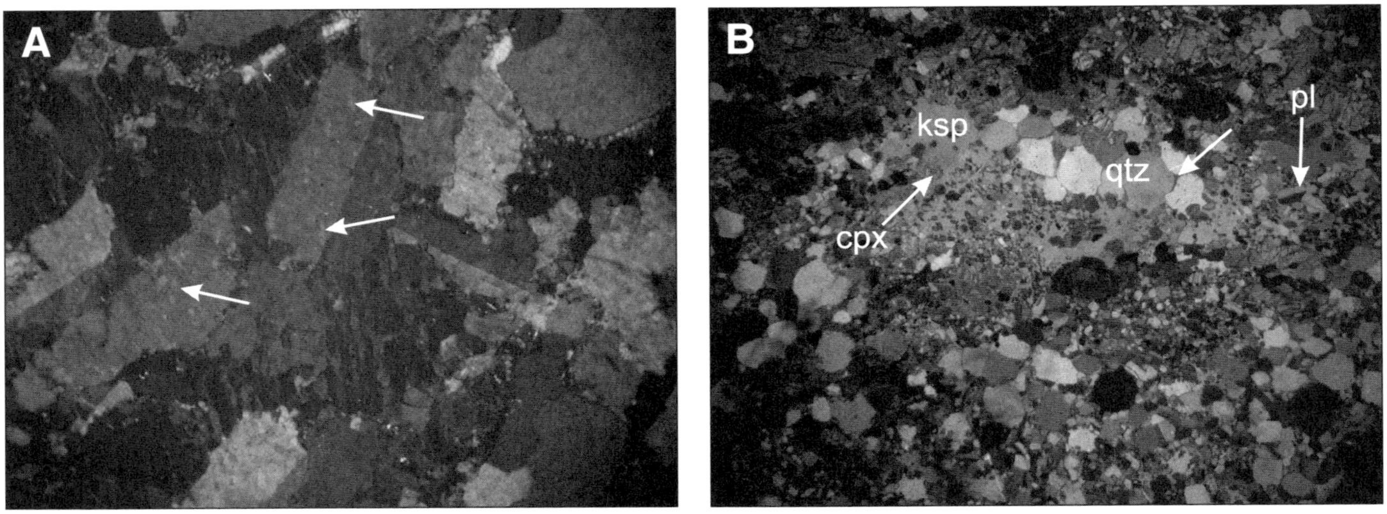

Figure 5. (A) Subhedral K-feldspar aggregate partially replacing anhedral perthitic Archean feldspar grain (in extinction, center and left). Note that the plagioclase blebs within the younger grains (arrows) are more rounded than in the Archean grain but are still slightly elongate in the same direction as the perthite flames in the latter. Myrmekitic albite-quartz rims are developed around most of the younger grains (top, top right). Note the simple Carlsbad twin and straight grain edges parallel to the twin direction. Sample IL24, Inlandsee Pan. Crossed polarizers. Length of photo 4 mm. (B) Poikilitic K-feldspar from granofels, showing inclusions of plagioclase (pl), clinopyroxene (cpx) and quartz (qtz). K-feldspar has also infiltrated along grain boundaries in the glomerogranular quartz aggregates (arrow), indicating a mobile melt stage. Crossed polarizers. Length of photo 4 mm. Inlandsee borehole core.

in the granofels zone, but mixing of plagioclase and K-feldspar melts seems more likely.

Pre-impact K-feldspar grains up to 1 cm in size are preserved in places within the granofels zone. These grains typically display anhedral to rounded shapes, are perthitic, and may display low birefringence, internal patchy or undulose extinction, and irregular fractures (Fig. 5A). In addition to the coarse, elongate but commonly irregular, perthite exsolution features, they may also contain extremely fine deformation bands or spindle twins that are up to 10 μm wide and have a similar spacing of ~10 μm. However, overgrowing these grains is K-feldspar that occurs as up to 2 mm long, unstrained, elongate subhedral to locally euhedral, laths characterized by Carlsbad twinning and well-developed faces parallel to the twin plane (Fig. 5A). Crystal faces are commonly well developed where these grains are in contact with the large anhedral perthitic K-feldspar plates but they can also be highly irregular where two subhedral laths meet. At these latter boundaries, grain margins are commonly cuspate-lobate or scalloped. The laths commonly display thin rims of albitic plagioclase and/or myrmekite (Fig. 5A). In contrast to myrmekites associated with the Archean granulite event, these myrmekites are extremely fine-grained (<100 μm wide). This confirms the evidence from the irregular grain boundaries that supports rapid crystallization. Locally, the laths may contain plagioclase blebs that, although distinctly more rounded than the elongate blebs in the pre-impact perthite, are nonetheless still slightly elongate parallel to the blebs in the latter. Overall, these textures suggest a recrystallization-replacement origin for the K-feldspar laths. The myrmekite textures are consistent with some separation of the K-feldspar and albitic components of the original perthitic K-feldspar. While this might, technically, be purely as a consequence of prolonged high-temperature annealing following the shock event, the fact that this is not associated with all K-feldspar grains in the sample is more consistent with involvement of localized shock processes—either recrystallization of diaplectic K-feldspar glass or crystallization from a shock melt derived from a perthitic feldspar—whereby sufficient time would have been available for significant (but not complete) segregation of the K-feldspar and plagioclase components. We favor recrystallization of diaplectic glass to explain these features, as the partial retention of the perthite texture seems unlikely if fusion had occurred. Some evidence for mobility of K-feldspar, probably as a melt, is, however, seen in the K-feldspar that has infiltrated glomerogranular quartz aggregates (Figs. 3D and 5B). The order-of-magnitude coarser grain size of the K-feldspar compared with the plagioclase mosaics in the same sample may reflect the greater ease of growth of K-feldspar or a lower threshold crystallization temperature for K-feldspar relative to the plagioclase.

The K-feldspar in the granulitic breccias in the granofels zone is typically poikilitic and contains euhedral to rounded inclusions of plagioclase, pyroxene, quartz, biotite, and magnetite, consistent with K-feldspar having formed during the latter stages of crystallization of a melt (Gibson et al., 2002). Similar textures are seen locally in the granofelses (Fig. 5B), which, nonetheless, appear to retain their pre-impact structural fabric. Gibson (2002) reported hypersolvus K-feldspar from highly aluminous pelitic granofelses from the Inlandsee Pan that was interpreted as indicative of the extremely high post-shock temperatures (>900 °C) experienced by these rocks.

Biotite

Biotite most commonly shows intense conjugate kinkband sets oriented at moderate to high angles to the (0001) cleavage (Fig. 6A). These kinkbands are typically spindle-shaped and are straight where they are present in low densities, but are curved where the density is higher. Although this feature is not necessarily diagnostic of shock, we concur with Wilshire (1971) that the kinking was related to the impact event based on the textural evidence for partial annealing of the kinkbands by the post-impact thermal event (see also Gibson et al., 1997). Single-crystal shock experiments by Hörz and Ahrens (1969) showed that kinking can occur at pressures as low as 0.9 GPa, and such a shock level is consistent with the presence of kinking in biotite and muscovite up to radial distances of 30 km from the center of the dome, where shatter cones and simple PDF geometries in quartz indicate shock pressures of <10 GPa. In places, the kinkbands are recrystallized to fine-grained unstrained biotite laths as a result of subsequent annealing (see also Gibson et al., 1997). In the Archean basement gneisses, however, annealing has more commonly taken the form of complete recrystallization of the biotite, producing anhedral, ragged, commonly poikilitic, strain-free grains with variable internal birefringence. Elsewhere, fine ilmenite or magnetite needles or aggregates occur along the kinkband boundaries and surrounding the biotite, or occur within and around the recrystallized biotite (Fig. 6A).

Biotite is scarce, but not completely absent, within the granofelses within 5 km of the center of the dome. This is partially due to the higher grade of the original Archean metamorphism, in which dehydration melting reactions led to the formation of pyroxene at the expense of biotite and amphibole. However, its scarcity also appears to be related to the presence within these rocks of fine-grained (20–50 µm) polymineralic aggregates comprising orthopyroxene + quartz + K-feldspar + ilmenite ± magnetite that define shapes reminiscent of pre-impact biotite grains elsewhere in the gneisses and even in the same thin section (Fig. 6B). In places, these aggregates contain skeletal biotite laths, some of which show evidence of partially annealed kinkbanding, suggesting that they are the relics of pre-impact biotite grains. However, in other aggregates, needle- to lath-shaped biotite grains show ragged, partially vermicular edges that suggest growth within the pseudomorph following the shock event. The presence of the biotite laths and the stoichiometry of the aggregates are consistent with them representing pseudomorphs after pre-impact biotite. The heterogeneous development of these textures within individual thin sections is also more consistent with a shock origin than a purely thermally controlled post-shock decomposition. Fel'dman (1994) suggested that substantial decomposition of biotite (to ilmenite + K-feldspar + orthopyroxene + glass) occurred at pressures as low as 20–25 GPa; however, Stöffler (1984) maintained that hydrous ferromagnesian minerals such as biotite are more likely to decompose incongruently than form glass under high shock pressures. Stöffler (1972) suggested that this would be most likely at shock pressures of >45–50 GPa. Lambert and Mackinnon (1984), on the other hand, noted congruent melting of biotite starting at 33 GPa. Recent shock experiments by Ogilvie et al. (2004a) confirm that biotite in pelitic granulites may start to melt congruently at ~30 GPa at room temperature, but at ~25 GPa in samples pre-heated to 400 °C. The biotite in the aggregates resembles the biotite needles that crystallized from the melt in the matrix of the granulitic breccias (Gibson et al., 2002). Based on the evidence, we interpret these pseudomorphs as crystallized biotite shock melts or annealed biotite shock melt glass.

Not all biotite within the granofels zone is replaced: kinked grains were found within harzburgitic gneiss described by Hart et al. (1990). In addition, in several samples, coarse, unkinked biotite grains are preserved. Given the constraints on shock pressure in the granofels zone and the apparent ease with which kinkbands form in biotite, we believe these are completely annealed shocked grains. No evidence was found of shock-induced changes in pleochroism or isotropization as seen, for instance, in the Manicouagan structure (Dressler, 1990), nor was haematization noted. This may be attributed to the high temperatures both before and after impact, which aided breakdown of biotite to more complex assemblages, including magnetite or ilmenite, and which promoted annealing.

Hornblende

Amphibole (typically hornblende) shows no distinctive shock-induced features in the outer parts of the dome (r > 12 km); however, between 12 and 7 km from the center, grains display subplanar fractures that largely follow one or both of the cleavages but may also occur as more random, somewhat irregular, features (Fig. 6C). Grain edges may be slightly displaced along these fractures. The fractures are dotted with <5 µm acicular to cubic magnetite crystals and subsidiary quartz or plagioclase. Similar magnetite is found along narrow (<5 µm) polysynthetic twin planes (Fig. 6C). Within 7 km of the center of the dome, these fractures become more planar and are filled with fine-grained polymineralic, slightly symplectitic to granoblastic aggregates of clinopyroxene + plagioclase + magnetite ± biotite, which also partially rim the hornblende grains (Fig. 6D). Within the granofels zone, the same granoblastic aggregate occurs as a pseudomorph after what is thus assumed to have been original hornblende (Fig. 6E). As with biotite, there is no indication of compositional zoning within these aggregates that might suggest reaction with adjacent minerals, and the aggregates appear stoichiometrically indistinguishable from average hornblende compositions for the Archean gneisses. Furthermore, the textures are also heterogeneously developed within single thin sections, consistent with a shock-induced origin.

Like biotite, hornblende appears not to form a shock glass, but rather melts incongruently at high shock pressures (Stöffler, 1972; Bischoff and Stöffler, 1992). Together with the evidence of breakdown along cleavage planes isolated from the rock matrix, we conclude from this that the texture is not the product of eutectic melting but that it represents selective mineral melting associated with shock. The selective development of these aggregates

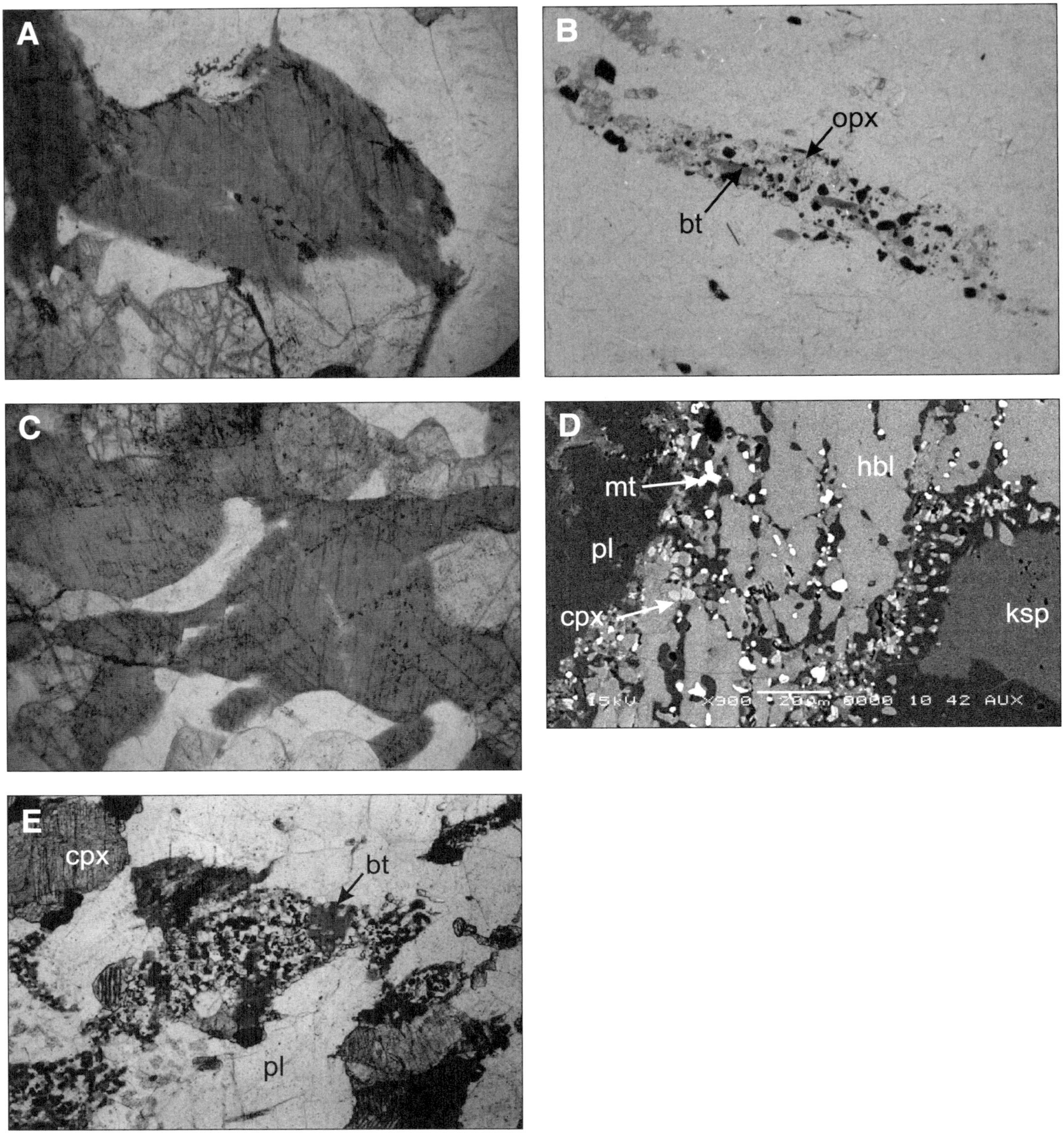

Figure 6. (A) Conjugate kinkbanding and partial oxidation of biotite, producing fine ilmenite needles at the margin and along kink bands. Mafic granulite sample VH11, Jagers Vrede farm, SW of Parys. Plane-polarized light. Length of photo 1.2 mm. (B) Polymineralic pseudomorph (orthopyroxene + ilmenite + quartz + K-feldspar) after biotite. Borehole core sample, Inlandsee Pan. Plane-polarized light. Length of photo 0.5 mm. (C) Magnetite aggregates defining fractures oriented mainly parallel to cleavage or deformation bands in hornblende. Sample VH13a, Jagers Vrede farm, SSW of Parys. Plane-polarized light, Length of photo 2.5 mm. (D) Backscattered electron image showing granoblastic aggregate of plagioclase (pl) + clinopyroxene (cpx) + magnetite (mt) replacing hornblende (hbl) along cleavage and along grain margin. Mafic granulite sample UP13, SW sector of the core. Length of photo 130 µm. (E) Pseudomorph after hornblende from granofels zone. Note the strong reduction in grain size relative to the Archean granulite assemblage, and the subplanar fractures in the Archean clinopyroxene (top, left). Borehole core sample, Inlandsee Pan. Plane-polarized light. Length of photo 2 mm.

along cleavage planes suggests that, at lower shock pressures, this melting occurred only along the hornblende cleavage.

Pyroxene

Pyroxene is the silicate mineral least affected by shock and post-shock metamorphic effects in the basement gneisses in the core of the Vredefort dome. Outside of the granofels zone, both orthopyroxene and clinopyroxene grains may display irregular fractures and undulose or patchy extinction. Orthopyroxene both in the granofels zone and the granulites that lie outside of this zone is generally fractured and partially altered to fine-grained amphibole along the fractures (Fig. 7A), but otherwise it shows no discernable internal shock effects. Clinopyroxene in the granofels zone contains equivocal planar and subplanar fractures (Fig. 6E) and fine-scale (<5 μm) polysynthetic twin lamellae. Some grains display fine-curved to spindle-shaped lamellae within alternate twins, which are oriented oblique to the twin planes (Fig. 7B). These features resemble the deformation band or micro-twin sets found in plagioclase (compare Fig. 4D). Pyroxene grains displaying the twin lamellae are typically riddled with extremely fine-grained opaque needles that are oriented roughly orthogonal to the lamellae (Fig. 7C). These needles are locally restricted to the twin lamellae, but in places they are longer than the width of the lamellae and cut across them. We believe this may reflect exsolution of Fe from the pyroxene lattice as a result of the post-shock thermal event.

Both clinopyroxene and orthopyroxene within the granofelses show similar grain size to that found in pre-impact assemblages elsewhere in the dome, but close to the center of the dome, some samples contain pyroxenes that are rimmed by fine-grained (<50 μm) polygonal-granular to symplectitic pyroxene aggregates that lack any evidence of internal twins or exsolution and that are indistinguishable compositionally from the larger grains (Fig. 7C). In places, individual grains appear to have been wholly replaced by aggregates, but adjacent pyroxene grains may be entirely unaffected. Unfortunately, while this may be a consequence of shock, the level of annealing of these rocks due to the extreme post-impact temperatures has obscured the initial cause of the grain-scale bimodality (mechanical fracturing, shock glass or melt, or post-shock metamorphic reactions). Where granular rims occur around larger grains, it is also not clear (given the high level of annealing) whether they may represent an annealed pseudomorph after hornblende or even biotite that might have rimmed the pyroxene in the pre-impact assemblage.

The bulk of the fine-grained granoblastic clinopyroxene and orthopyroxene in the granulitic breccias and adjacent granofelses (e.g., Fig. 5B) is interpreted as having grown from variably mobilized melt derived by shock metamorphism of biotite and/or hornblende (Gibson et al., 2002). The breccias typically contain larger grains of pyroxene that are interpreted as clasts of pre-impact pyroxene grains. However, the greater abundance of orthopyroxene as clasts relative to clinopyroxene may suggest that, at the highest shock pressures experienced in the dome, clinopyroxene also melted.

Garnet

Garnet is only found in the core of the dome in pelitic granulites between 5 and 8 km from the center (Stepto, 1990; Schreyer and Abraham, 1978; Stevens et al., 1997). Grains contain up to three sets of planar intragranular fractures that terminate at the grain boundary and do not extend into adjacent minerals (Fig. 7D). These features resemble the crystallographically controlled parting and cleavage observed by, for example, Dressler (1990) and Stähle (1975) in samples from the Manicouagan and Ries impact structures, respectively, and by Gibson et al. (1997) in metapelites from the collar of the Vredefort dome. Unlike the garnets in the collar rocks, however, the garnet in the pelitic granulites is commonly replaced along these fractures by fine-grained symplectites of cordierite + orthopyroxene ± plagioclase ± spinel ± biotite (Stevens et al., 1997; Gibson, 2002). Stähle (1975) attributed superficially similar symplectites in the Ries crater to crystallization of a garnet shock glass; however, this texture is also consistent with decompression-related breakdown of garnet under high-temperature conditions (Schreyer and Abraham, 1978; Gibson, 2002). Stevens et al. (1997) and Gibson (2002) have shown that the symplectites are continuous with complex coronas around the garnet and that the composition of the garnet and orthopyroxene in and around these symplectites is variable, indicating reaction between pre-impact phases rather than exclusive derivation from the garnet. This appears to be fundamentally different to the cleavage-controlled symplectites in hornblende described above. Both the degree of breakdown of the garnets and the grain size of the symplectites increases toward the center of the dome as a result of the higher initial post-shock temperatures and consequent longer duration of high-grade metamorphic conditions in the rocks closer to the center (P. Ogilvie, 2003, personal commun.).

Oxides

Although magnetite and ilmenite were not studied in detail, there is evidence of recrystallization to extremely fine-grained, commonly acicular, crystal aggregates and partial mobilization of oxides into the silicate matrix (Fig. 7A). New ilmenite and magnetite was derived from the breakdown of biotite, hornblende and possibly pyroxene, and magnetite needles by exsolution from clinopyroxene (Fig. 7C), possibly as a result of the post-shock heating.

Accessory Minerals

Several papers have referred to the occurrence of distinctive conjugate planar features in zircon from the central parts of the Vredefort dome (Kamo et al., 1996; Gibson et al., 1997; Reimold et al., 2002). The occurrence of similar planar to near-planar features in detrital grains from up to 28 km from the center suggests that these features may form at relatively low shock pressures (Reimold et al., 2002). To our knowledge, no definite evidence for melting along such features in zircon has been observed, whereas scanning electron microscopy strongly suggests that this deformation phenomenon represents crystallographically

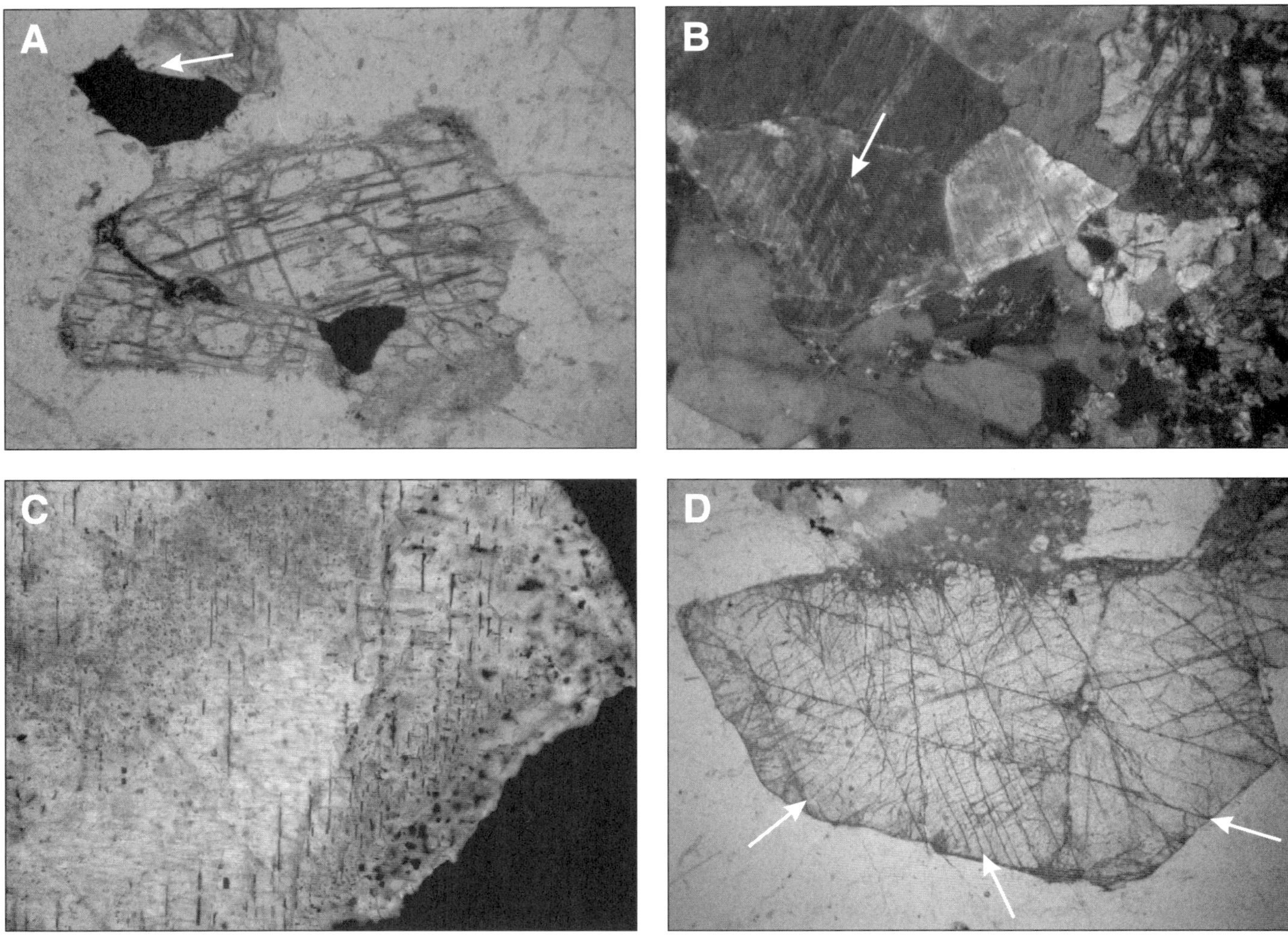

Figure 7. (A) Fractured orthopyroxene with syntaxial overgrowth of fine-grained amphibole in the cracks. Note the fine magnetite needles around the large magnetite grain (arrow). Mafic granulite from borehole core, Inlandsee Pan. Plane-polarized light. Length of photo 1 mm. (B) Clinopyroxene displaying polysynthetic twin lamellae, with oblique spindle-shaped lamellae in alternating twins. Borehole core mafic granulite sample, Inlandsee Pan. Length of photo 2.5 mm. (C) Close-up of clinopyroxene grain showing magnetite exsolution and a granoblastic-symplectitic clinopyroxene-quartz-magnetite rim against a large magnetite grain (black). Same sample (VT590) as Figure 4E. Plane-polarized light. Length of photo 0.4 mm. (D) Three sets of crystallographically controlled fractures in garnet (arrowed). Pelitic granulite sample VH9, Jagers Vrede farm, SSW of Parys. Plane-polarized light. Length of photo 3.5 mm.

controlled fracturing (Gucsik et al., 2003). In addition, several zircon grains from granulitic breccias and granofelses show a polycrystalline texture resembling the "strawberry texture" described by Bohor et al. (1993) and Kamo et al. (1996).

Schreyer (1983) described apatite with shock-induced lamellae parallel to $(10\bar{1}1)$ from near Parys, 10–12 km from the center of the dome; however, no evidence of such features was found in apatite in samples examined in the present study.

PSEUDOTACHYLITIC BRECCIAS

During the course of this study, numerous examples were found of local enhancement of shock effects in minerals adjacent to millimeter-scale pseudotachylitic breccias and the fracture networks within which discontinuous pods of breccia are found (Figs. 4A and 8). This corroborates Martini's (1991) suggestion, based on the spatial association between <1 mm thick pseudotachylitic breccias and coesite ± stishovite at 17 sites around the dome (see Fig. 1), that these thin breccias formed during the shock compression phase of the impact event. He proposed that the localized elevation of shock pressure needed to generate the high-pressure silica polymorphs was achieved by collapse of pre-impact joints and that brecciation and melting may have been aided by friction along the joint surfaces. However, he distinguished these thin veins from veins and dikes >1 mm thick that he attributed to frictional heating along faults associated with

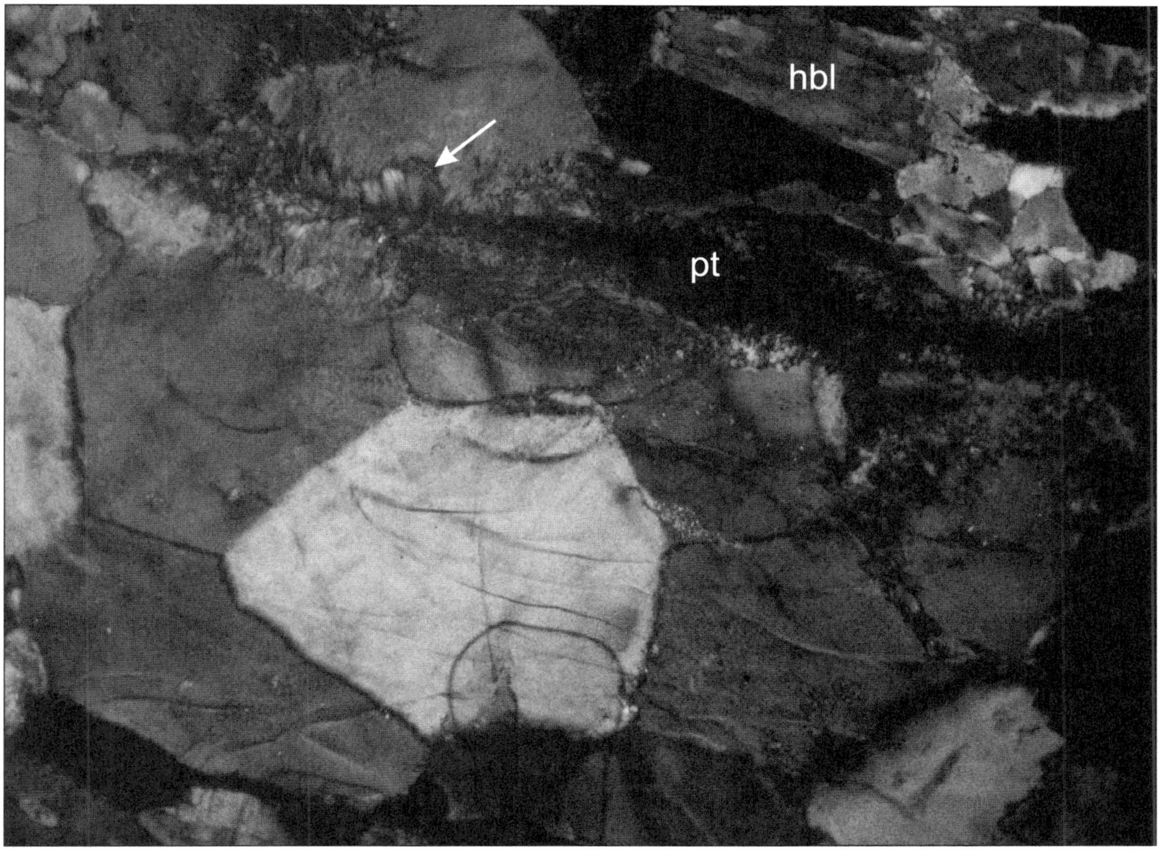

Figure 8. Pseudotachylitic breccia veinlet (pt) flanked by annealed damage zones in plagioclase attributed to shock. Note the plumose aggregates (arrow) immediately adjacent to the veinlet and the crescentic annealed fractures farther away. Crossed polarizers. Mafic amphibolite UP36, same sample as Figure 4A. Length of photo 3 mm. hbl—hornblende.

the development of the central uplift. A similar volumetric distinction between so-called "shock-type" and "endogenic-type" pseudotachylites was proposed by Spray (1998) as a general rule in large impact structures, but this is not supported by all impact workers (see review by Dressler and Reimold, 2004).

Despite the occurrence of pseudotachylitic breccia dikes up to tens of meters across and hundreds of meters long, no conclusive evidence of suitable large slip magnitude faults has been found in the Vredefort dome that would support a purely frictional origin for the voluminous breccias (Reimold and Colliston, 1994; Lana et al., 2003b). We have also found examples of partially annealed enhanced shock effects adjacent to pseudotachylitic breccia veins up to 10 cm wide, suggesting that melt volume cannot be used to discriminate between different origins for the breccias. We believe, instead, that the pseudotachylitic breccias in the dome formed largely through shock melting, with or without a frictional component, and that the variation in melt volumes in veins and dikes relates to the efficiency of ponding of the clast-laden melts in extensional sites after their formation.

A mechanism involving local enhancement of shock pressures to generate the pseudotachylitic breccias is supported by our and Martini's (1991) petrographic evidence. Both void collapse (Martini, 1991) and interference of secondary shock fronts caused by refraction or reflection off the myriad discontinuities in the strongly layered target rocks could do this. The increase in both the intensity of breccia development and overall volumes of breccia veins and dikes from the collar to the outer core of the dome suggests that melting was facilitated by the rising background shock pressure. In the inner core ($r < 10$ km), however, individual breccia volumes and overall intensity of veining appear to decrease. This phenomenon may reflect the attainment of the threshold shock pressure necessary to effect substantial shock modification of feldspar throughout the rock mass, probably assisted by the increased residual temperature of the rocks. We believe that this could have led to a fundamental change in the way the shock wave interacted with the rocks, partially subduing the shock pressure perturbations seen at lower background shock pressures, and thereby restricting the development of the pseudotachylitic breccias. Nonetheless, the fact that the background pressure in these rocks was closer to the pressure necessary to generate shock melts ensured that some melts did form. The proportion of in situ shock melt appears to increase in the granofels zone (Gibson et al., 2002).

ESTIMATION OF SHOCK PRESSURES

The relationship between peak shock pressure and the sequence of shock-metamorphic effects in rocks and minerals has been investigated experimentally and theoretically in numerous studies (see, e.g., reviews in Stöffler and Langenhorst, 1994; Grieve et al., 1996; French, 1998) and is reasonably well understood, although problems remain with extrapolation of experimental data to account, for instance, for the more prolonged nature of natural shock events compared to the short-duration experimental shock conditions, and for the more complex nature of rock types (e.g., De Carli et al., 2002). Shock-metamorphic effects observed in rocks and minerals in impact sites are the product of both the transient passage of the compressional shock wave and the subsequent release to ambient pressure by the rarefaction wave. Owing to the nonisentropic nature of shock compression, residual heat remains in the shocked material after this release. The extent of the effects caused by this post-shock heating, and the degree of preservation of the shock pressure effects in a particular mineral, depend both on the magnitude of the post-shock temperature increase (which is directly proportional to the shock pressure) and the rate of cooling, which is a function of the location of the shocked sample within the impact structure.

Despite the high level of annealing in the Vredefort samples, the mineral textures in the rocks in the core of the dome are consistent with the well-established hierarchy of shock-induced changes known from observation of experiments and naturally shocked rocks:

1. Quartz is dominated by post-shock recrystallization features which have largely obliterated shock textures, thereby reducing the usefulness of quartz for estimation of shock pressures in the core of the dome (e.g., Grieve et al., 1990). A feature of particular interest, however, is the glomerogranular texture in the granofelses at $r < 5$ km. The polygonal to rounded shape of the individual quartz grains and their slight disaggregation could imply recrystallization of quartz "ballen-structure" (Carstens, 1975). This feature is typically believed to form at shock pressures above 30 GPa (Bischoff and Stöffler, 1984) and, although Carstens (1975) favored melting of quartz to produce disaggregated ballen-structure, it is more likely that this feature forms in diaplectic glass.

2. Shock effects in plagioclase grade inwards from intergranular fractures in the outer parts of the core of the dome at $r > 10$ km, to intragranular, largely crystallographically controlled, fractures, followed by isotropization, asymmetric twin or deformation band effects and PDF formation at r ~5–10 km. At $r < 5$ km, grains are mostly replaced by fine-grained aggregates. This replacement is most likely recrystallization of maskelynite to microcrystalline pseudomorphs. Locally, in this central zone, plagioclase must have melted in the granulitic breccias and then crystallized while undergoing magmatic flow; however, in situ melting is also suggested by the euhedral to subhedral plagioclase inclusions in the K-feldspar oikocrysts (Fig. 5B).

3. K-feldspar shows similar intergranular and intragranular fractures to plagioclase in the outer parts of the core of the dome but, unlike plagioclase, does not show any conclusive evidence of the intermediate shock pressure effects. Although reduced birefringence was not specifically tested, grains in the central parts of the core commonly show low birefringence. The replacement of perthites in the granofelses by comparatively coarse K-feldspar grains is inferred to have proceeded via an intermediate glass stage. In contrast to this, individual granofels samples show evidence for complete K-feldspar melting and crystallization of oikocrysts.

4. Biotite exhibits a transition from increasing intensities of kinkbanding to partial replacement by ilmenite and/or magnetite, to pseudomorphous replacement by orthopyroxene + K-feldspar + quartz + ilmenite/magnetite, most likely after shock melt. Biotite survives in some samples even in the center of the dome, either because shock pressure effects were heterogeneous or through complete recrystallization of shocked grains.

5. Hornblende displays irregular fractures locally associated with magnetite ± quartz/plagioclase in the outer core of the dome, which gradually give way to cleavage-parallel fractures, with or without more complex breakdown assemblages. At $r < 5$ km, hornblende may be comprehensively pseudomorphed by clinopyroxene + plagioclase + magnetite ± biotite, probably after a shock-melt glass, although samples containing these pseudomorphs typically also contain intact hornblende grains.

6. Clinopyroxene displays fractures and ubiquitous polysynthetic twins, with oblique polysynthetic twins or deformation band sets developed in the central parts of the core, whereas orthopyroxene shows only fracturing. Fine-grained granoblastic aggregates of clinopyroxene are found in the most highly shocked rocks, but their origin is unclear.

7. Garnet has only a restricted radial distribution in the dome. It shows both irregular and crystallographically controlled fractures and partings. These are progressively exploited by cordierite-orthopyroxene symplectites toward the center. Gibson (2002) speculated that unusual aluminous granofelses close to the center of the dome may be crystallized shock melt derived from garnetiferous pelitic granulites; however, recent shock experiments on garnet-bearing pelite (Ogilvie et al., 2004a) suggest that, even at 60 GPa, garnet does not melt. A combination of shock fusion of other minerals and subsequent resorption of the garnet into the melt at high temperatures thus appears to be necessary for this to happen.

Figure 9 provides a summary of the distribution of the main shock pressure effects in the Vredefort dome, and inferred pressures based on the observed textures. The key features are the onset of PDF development in plagioclase, which, according to Huffman et al. (1993) and Huffman and Reimold (1996), is largely independent of pre-shock temperature and occurs at ~15 GPa; and the inferred development of diaplectic feldspar and quartz glass above ~30–35 GPa (Ostertag, 1981; Stöffler, 1984) that we infer to have been annealed to generate the distinctive granofelses. Maximum shock pressures in the granofelses do not

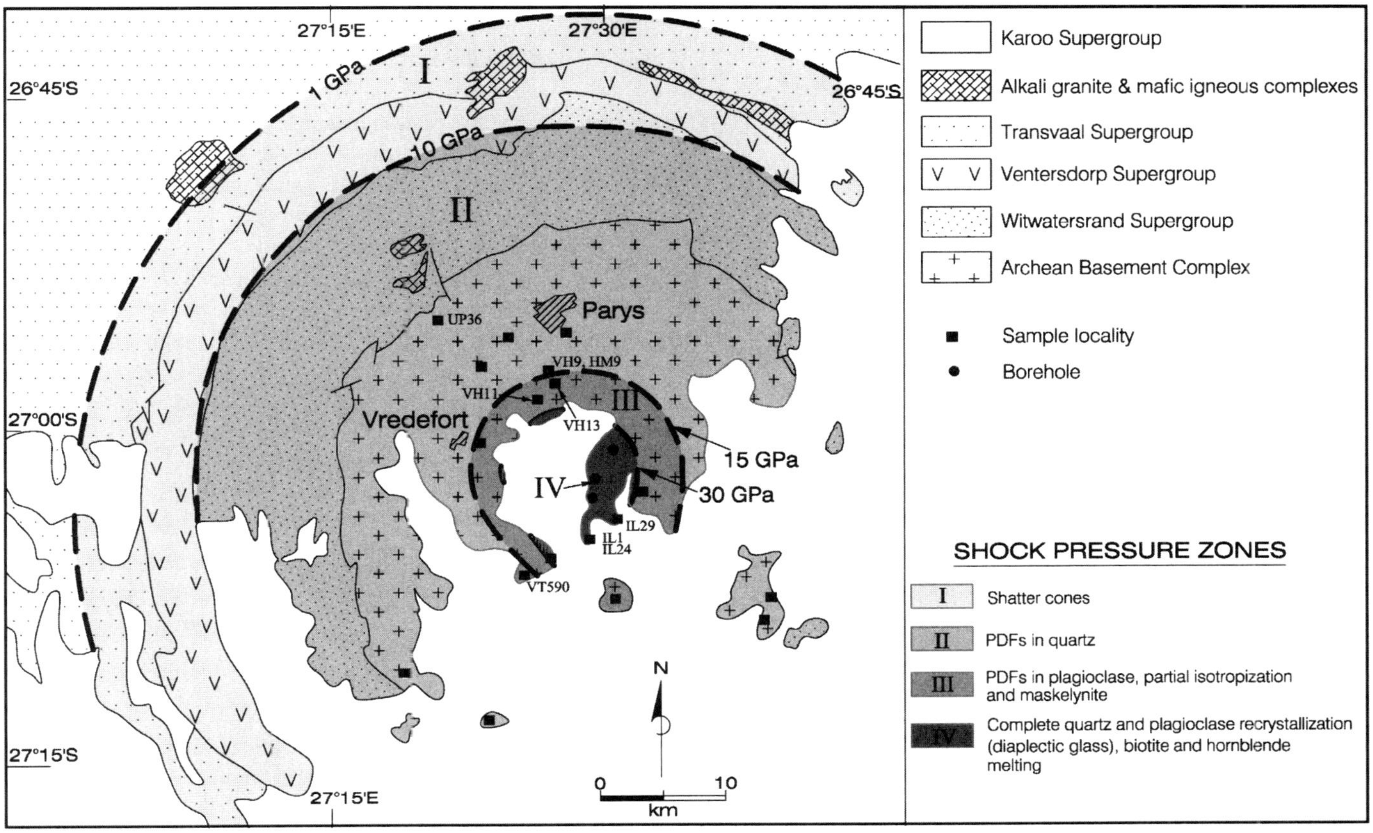

Figure 9. Shock isobar pattern for the Vredefort dome based on observed mineral and rock textures. Samples specifically referred to in the text are labeled.

appear to have been high enough (45–50 GPa) to generate substantial feldspar shock melts, although the local existence of melt is indicated by the plagioclase-dominated granulitic breccias and some K-feldspar textures. A maximum background shock pressure range of 35–45 GPa for the central granofelses is compatible with the textures displayed by biotite and hornblende (congruent or incongruent melting; Lambert and Mackinnon, 1984; Ogilvie et al., 2004a), and pyroxene and garnet (no significant effects other than twinning and/or fracturing below 45 GPa).

While most of the shock features have been destroyed or at least substantially modified by the post-shock thermal effects, Gibson et al. (1998) and Gibson (2002) used the post-shock assemblages to establish that the Vredefort dome is marked by a strongly variable thermal overprint related to the impact event. This helps to further refine the shock pressure estimates. These authors found that maximum post-shock temperatures increased from a regional background value of ~300 °C at $r > 25$ km to values of ~500 °C at r ~20 km, and ≥800–1000 °C at $r < 5$ km, from the center. Additional constraints on both the pre-impact geotherm (~15 °C/km; Gibson and Jones, 2002) and the maximum pre-impact burial depth of rocks presently exposed at the surface (~25 km; Lana et al., 2003b) provide a rough estimate of the contribution to these values by the pre-impact temperatures of the rocks exposed in different parts of the dome and, hence, constrain the amount of shock heating required to generate the post-impact temperatures. These constraints suggest that the amount of shock heating must have ranged from 50 to 100 °C at $r > 25$ km to 150–200 °C at r ~20 km, and at least 400–600 °C, and possibly as high as 900 °C (Gibson, 2002), in the central parts of the dome. These values are consistent with the numerical modeling results obtained by Ivanov and Deutsch (1999) for an impact structure of comparable size to Vredefort (i.e., of the order of 250 km diameter). Comparison of these values with Table 1 suggests, thus, that peak shock pressures in the center of the dome must have exceeded ~35 GPa and may locally have had to have been as high as at least 45 GPa.

It is important to note that this is a qualitative approach, as the noted and inferred shock-metamorphic effects are both heterogeneously distributed on a small scale and highly annealed. Our observations suggest that shock effects increase markedly in the vicinity of pseudotachylitic breccia veins (e.g., Fig. 8). Fortunately, in most samples, these effects are restricted to zones <2 mm wide and can be distinguished from the background effects in a single thin section; however, even then, shock effects can differ strongly between adjacent grains. It is for this reason that the planar features and suspected patches of maskelynite in plagioclase in sample UP36B from the outer parts of the core (Figs. 4A and 8) in the northwest sector of the dome have not been used to estimate background shock pressure. If they had, then they would indicate $P > 30$ GPa at $r = 19$ km. Also, even in the center of the dome, relics of pre-impact minerals survive. What is clear, however, is that Fricke et al. (1990) and Grieve et al. (1990) were correct in suspecting that the 10–15 GPa estimates obtained from statistical analysis of PDFs in quartz for much of the expanse of the core of the dome do not represent a true picture of the shock pressure pattern.

DISCUSSION

Previous attempts to constrain the shock pressure distribution in the Vredefort impact structure have focused mainly on quartz. As outlined above, Grieve et al. (1990) used the Grieve and Robertson (1976) method to estimate shock pressures in the core of the Vredefort dome but were puzzled by the absence of any clear increase in shock pressure estimates toward the center of the dome (in fact, they found that, using this method, pressure estimates decrease toward the center of the dome). They also found that the PDFs die out in abundance in the same direction, and their crystallographic orientations differ from those in other large impact structures. At the time, Grieve et al. (1990) were only able to speculate on the possible causes of these patterns. Without any other constraints, they estimated that uplift of hot, preheated, deep crustal levels had been a major contributing factor in the annealing of most of the PDFs, explaining their absence. Based on their experimental results with preheated samples, however, Huffman et al. (1993) and Huffman and Reimold (1996) suggested, alternatively, that fewer PDFs may have originally formed in the hot target rocks at deeper crustal levels.

Such debates are important if we are to better understand large and deeply eroded impact structures. The more holistic approach adopted here, while unable to improve on the precision of the pressure estimates, at least appears to have improved the accuracy of the estimates and is in line with the predictions of numerical modeling (Ivanov and Deutsch, 1999), both in terms of the shock pressure values and in the persistence of shock pressure effects to extreme levels beneath such large structures.

Reasons for Shock Pressure Heterogeneity

Exhumed impact structures provide an opportunity to examine the extent and spatial distribution of shock metamorphic effects in the basement underlying the impact crater. Modeling of impact cratering events suggests that the gross distribution of shock isobars in a target subjected to an impact should be broadly hemispherical about the point of impact (e.g., Melosh, 1989). In large impact structures, however, the attainment of maximum shock pressures is followed by substantial lateral and vertical displacements at deep levels within the crater basement during crater modification. This may not only greatly attenuate the shock isobar pattern due to the inward and upward movement of rocks during formation of the central uplift but may also severely disrupt it along faults that may involve individual displacements exceeding several kilometers. In some impact structures, this disruption seems to have produced a mega-breccia beneath the crater (e.g., Ivanov and Deutsch, 1999). In the Manicouagan Structure, Dressler (1990) reported a broadly concentric shock isobar pattern for the crater basement; however, he noted that anomalously low shock pressures exist in the center of the structure. He attributed this to either post-shock impact-related faulting that disturbed the initial isobar pattern, or enhanced exhumation in the center of the structure caused by a combination of impact-induced uplift and subsequent erosion. In the latter explanation, uplift and erosion are supposed to have exposed rocks in the center of the impact structure that were originally more deeply buried and, thus, were subjected to lower shock pressures, than rocks exposed farther from the center. Detailed examination of the distribution of shock features in rocks from the deep borehole core through the central uplift of the Puchezh-Katunki structure led Masaitis and Pevzner (1999) to conclude that local inconsistencies were the product of differential movements along faults related to the rise and collapse of the central uplift.

The large field and petrographic database available for the Vredefort impact structure confirms that, at least at deep levels in a complex impact structure, shock effects appear to be largely restricted to the central uplift and that a broad radial increase in shock pressure does, indeed, exist toward the center. The Vredefort section is sufficiently deep that any heterogeneity in the impactor itself is unlikely to have played a part in causing any shock pressure heterogeneity. Lana et al. (2003b) failed to detect any large-scale impact-related faults in the core of the dome and proposed that the differential rotations necessary to effect doming may have occurred in a more distributed way, rather than through the formation of large mega-blocks. This may have been supported by the elevated rock temperatures, caused, at least in part, by the pre-impact geotherm. Consequently, any anomalies in the shock isobar pattern should have an alternative explanation. Our work has shown that shock pressure heterogeneity exists on a hand specimen–scale and even mineral grain–scale (see also Gibson et al., 2002). Further refinement of shock pressure estimates may, therefore, be impossible, and qualitative estimates may be more relevant.

Attenuation Rate

When the data presented here are combined with the evidence from the outer parts of the dome, they suggest an exponential increase in shock pressure from ~2 to 4 GPa in the outer parts of the dome (shatter cones; e.g., Baratoux and Melosh, 2003), through 5–10 GPa in the inner collar of the dome (Lilly, 1981; Grieve et al., 1990; Martini, 1991 [allowing for local pressure excursions to 30 GPa to generate coesite]; Therriault et al., 1997), to >30 GPa in the center of the dome. The absence of abundant PDFs in quartz indicative of shock pressures in the range from 10 to 25 GPa, or of reduced refractive index in feldspar in the same pressure range, is attributed to annealing due to the high post-impact temperatures in the rocks (caused by elevated pre-impact temperatures in addition to the shock heating) and the longer cooling time of these rocks after the impact (caused by their deep levels of burial); although the pre-impact temperature of >200–400 °C may also have affected the shock behavior of these minerals (Huffman et al., 1993; Huffman and Reimold, 1996). The pattern of shock isobars shown in Figure 9 implies that shock pressure attenuation across the dome increases from <1 GPa/km at r > 10 km to ~4 GPa/km in the inner 10 km. This jump in shock pressure attenuation at r ~ 10 km is indirectly corroborated by the cordierite-orthopyroxene symplectites after garnet in the pelitic granulites (Schreyer and Abraham, 1978; Stevens et al., 1997; Gibson, 2002). These rocks, which occur in a narrow band between 5 and 8 km from the center of the dome, show an order-of-magnitude increase in symplectite grain size, and a significant increase in the extent of garnet replacement, toward the center (Ogilvie et al., 2004b). This signifies a significant temperature gradient across this 3-km-long traverse, which cannot be adequately explained by the low pre-impact geotherm. Consequently, the temperature increase must have its origin in differential shock heating and, thus, in a strong rate of increase in background shock pressure.

It is important to recognize that the current disposition of the shock isobars and, thus, the measured radial attenuation rate is at least partially influenced by the differential rotations accompanying formation of the central uplift. Lana et al. (2003b) showed that the amount of rotation of pre-impact structures in the Vredefort dome is relatively minor up to r ~ 14–17 km (using the revised center location), beyond which uplift-related rotation is in the order of 90°. This suggests that any shock isobars that formed with an initial subvertical orientation, as portrayed in numerical simulations by, for example, Ivanov and Deutsch (1999), should now have a shallow inward dip in the outer parts of the core and in the collar of the dome. Thus, the observed radial attenuation rate in these parts of the dome is likely to be significantly smaller than the actual original attenuation rate. Given the apparently minor rotations in the central core of the dome (Lana et al., 2003b), the calculated attenuation rate there may reflect more closely the actual figure during the shock event. It must be noted, however, that the jump in the radial attenuation rate does not coincide exactly with the zone of maximum differential rotation identified by Lana et al. (2003b), but lies closer to the center of the dome. This suggests that the variation may have another origin.

The ~4 GPa/km attenuation rate deduced for the inner core of the dome is similar to the 4.5 GPa/km estimate obtained by Dressler et al. (1998) from the more shallowly eroded, 8-km-wide, central uplift of the Slate Islands Structure. Given that the maximum shock pressures in the Slate Islands Structure appear to have been too low (~20–25 GPa) to generate substantial shock melting in the core of the structure, it seems unlikely that the exponential increase observed in the center of the Vredefort dome requires enhanced post-shock attenuation of (partially) molten crust to be invoked. The marked change in the shock pressure attenuation rate at r ~10 km does, however, coincide with the increase in shock effects in feldspar. This suggests that such jumps in the shock pressure profile across an impact structure most likely reflect the consumption of proportionally more shock energy where the threshold pressure for substantial change in the shock behavior of a major matrix mineral is reached. This obviously has implications for layered targets containing minerals with very different shock responses, to such a degree that it might well affect shock attenuation profiles calculated numerically.

CONCLUSIONS

Petrographic analysis of quartz, feldspars, and ferromagnesian minerals in the Archean basement gneisses in the central parts of the Vredefort dome has revealed that:

1. Shock pressures were significantly higher than could be previously estimated using statistical techniques applied to PDF orientations in quartz.

2. The high post-shock temperatures and the comparatively slow cooling of the rocks have largely obscured many of the original shock features, including the crucial issue of whether feldspars converted to diaplectic glass or melt in many samples; however, compelling microtextural evidence exists for the attainment of shock pressures conducive to, at least, diaplectic feldspar glass formation.

3. The extensive shock-induced modification of feldspars suggests that maximum shock pressures in the central parts of the dome exceeded 30–35 GPa and, at least locally, may have exceeded 45 GPa.

4. Based on a reconstruction of the central uplift (Lana et al., 2003b) and using a calculated pre-impact geotherm of 15 °C/km (Gibson and Jones, 2002), post-impact metamorphic assemblages indicate that shock heating must have accounted for an average temperature increase of at least 500 °C, and possibly as much as 900 °C, in the central parts of the dome. This is compatible with the shock pressure estimates derived from the observed and inferred shock metamorphic effects.

5. Shock pressure distribution was highly heterogeneous on a local (mm- to dm-scale), presumably as a result of enhancement caused by reflection and refraction of the shock wave in the heterogeneous target rocks.

6. Shock pressure attenuation increases from <1 GPa/km to ~4 GPa/km toward the center of the central uplift, probably at least partly as a result of the increase in shock modification of feldspars at higher shock pressures.

ACKNOWLEDGMENTS

Funding for this work was provided by the University of the Witwatersrand Research Committee. A. Mathebula prepared the thin sections and L. Whitfield assisted with the drawings. R. Grieve and V. Masaitis are thanked for their constructive reviews. University of the Witwatersrand Impact Cratering Research Group Contribution No. 71.

REFERENCES CITED

Arndt, J., Hummel, W., and Gonzalez-Cabeza, I., 1982, Diaplectic labradorite glass from the Manicouagan impact crater, I: Physical properties, crystallization, structural and genetic implications: Physics and Chemistry of Minerals, v. 8, p. 230–239, doi: 10.1007/BF00309482.

Ashley, A.J., Gibson, R.L., Koeberl, C., Reimold, W.U., and Greshake, A., 1999, A new type of melt rock and first evidence of shock deformation in plagioclase from the Vredefort impact structure, South Africa: Meteoritics & Planetary Science, v. 34 (supplement), p. A9–A10.

Baratoux, D., and Melosh, H.J., 2003, The formation of shatter cones by shock wave interference during impacting: Earth and Planetary Science Letters, v. 6817, p. 1–12.

Bischoff, A., and Stöffler, D., 1984, Chemical and structural changes induced by thermal annealing of shocked feldspar inclusions in impact melt rocks from Lappajärvi Crater, Finland, Proceedings Lunar and Planetary Science Conference, 14th: Journal of Geophysical Research, v. 89, p. B645–B656.

Bischoff, A., and Stöffler, D., 1992, Shock metamorphism as a fundamental process in the evolution of planetary bodies: Information from meteorites: European Journal of Mineralogy, v. 4, p. 707–755.

Bohor, B.F., Betterton, W.J., and Krogh, T.E., 1993, Impact-shocked zircons: Discovery of shock-induced textures reflecting increasing degrees of shock metamorphism: Earth and Planetary Science Letters, v. 119, p. 419–424, doi: 10.1016/0012-821X(93)90149-4.

Carstens, H., 1975, Thermal history of impact melt rocks in the Fennoscandian Shield: Contributions to Mineralogy and Petrology, v. 50, p. 145–155.

De Carli, P.S., Bowden, E., Jones, A.P., and Price, G.D., 2002, Laboratory impact experiments versus natural impacts, *in* Koeberl, C., and MacLeod, K.G., eds., Catastrophic Events and Mass Extinctions: Impacts and Beyond, Boulder, Colorado: Geological Society of America Special Paper 356, p. 595–605.

Dressler, B.O., 1990, Shock metamorphic features and their zoning and orientation in the Precambrian rocks of the Manicouagan structure, Quebec, Canada: Tectonophysics, v. 171, p. 229–245, doi: 10.1016/0040-1951(90)90101-D.

Dressler, B.O., and Reimold, W.U., 2004, Order or chaos? Origin and mode of emplacement of breccias in floors of large impact structures: Earth-Science Reviews, v. 67, p. 1–54.

Dressler, B.O., Sharpton, V.L., and Schuraytz, B.C., 1998, Shock metamorphism and shock barometry at a complex impact structure, Slate Islands, Canada: Contributions to Mineralogy and Petrology, v. 130, p. 275–287, doi: 10.1007/S004100050365.

Fel'dman, V.I., 1994, The conditions of shock metamorphism, *in* Dressler, B.O., Grieve, R.A.F., and Sharpton, V.L., eds., Large Meteorite Impacts and Planetary Evolution: Geological Society of America Special Paper 293, p. 121–132.

French, B.M., 1998, Traces of catastrophe: Houston, Lunar and Planetary Institute, 120 p.

Fricke, A., Medenbach, O., and Schreyer, W., 1990, Fluid inclusions, planar elements and pseudotachylites in the basement rocks of the Vredefort structure, South Africa: Tectonophysics, v. 171, p. 169–183, doi: 10.1016/0040-1951(90)90097-R.

Gibson, R.L., 2002, Impact-induced melting of Archaean granulites in the Vredefort Dome, South Africa, I: Anatexis of metapelitic granulites: Journal of Metamorphic Geology, v. 20, p. 57–70, doi: 10.1046/J.0263-4929.2001.00358.X.

Gibson, R.L., and Jones, M.Q.W., 2002, Late Archaean to Palaeoproterozoic geotherms in the Kaapvaal craton, South Africa: Constraints on the thermal evolution of the Witwatersrand Basin: Basin Research, v. 14, p. 169–181, doi: 10.1046/J.1365-2117.2002.00168.X.

Gibson, R.L., and Reimold, W.U., 2000, Deeply exhumed impact structures: A case study of the Vredefort Structure, South Africa, *in* Gilmour, I., and Koeberl, C., eds., Lecture Notes in Earth Sciences 91: Impacts and the Early Earth: Berlin, Springer-Verlag, p. 249–277.

Gibson, R.L., and Reimold, W.U., 2001, The Vredefort Impact Structure, South Africa: The scientific evidence and a two-day excursion guide: Memoir of the Council for Geoscience, v. 92, 111 p.

Gibson, R.L., Reimold, W.U., and Wallmach, T., 1997, Origin of pseudotachylite in the lower Witwatersrand Supergroup, Vredefort Dome (South Africa): Constraints from metamorphic studies: Tectonophysics, v. 283, p. 241–262, doi: 10.1016/S0040-1951(97)00072-3.

Gibson, R.L., Reimold, W.U., and Stevens, G., 1998, Thermal-metamorphic signature of an impact event in the Vredefort Dome, South Africa: Geology, v. 26, p. 787–790, doi: 10.1130/0091-7613(1998)0262.3.CO;2.

Gibson, R.L., Reimold, W.U., Ashley, A.J., and Koeberl, C., 2002, Metamorphism on the Moon: A terrestrial analogue in the Vredefort Dome, South Africa?: Geology, v. 30, p. 475–478, doi: 10.1130/0091-7613(2002)0302.0.CO;2.

Grieve, R.A.F., and Robertson, P.B., 1976, Variations in shock metamorphism at the Slate Islands impact structure, Lake Superior, Canada: Contributions to Mineralogy and Petrology, v. 58, p. 37–49.

Grieve, R.A.F., and Therriault, A.M., 2000, Vredefort, Sudbury, Chicxulub: Three of a kind?: Annual Review of Earth and Planetary Sciences, v. 28, p. 305–338, doi: 10.1146/ANNUREV.EARTH.28.1.305.

Grieve, R.A.F., Coderre, J.M., Robertson, P.B., and Alexopoulos, J., 1990, Microscopic planar deformation features in quartz of the Vredefort structure: Anomalous but still suggestive of an impact origin: Tectonophysics, v. 171, p. 185–200, doi: 10.1016/0040-1951(90)90098-S.

Grieve, R.A.F., Langenhorst, F., and Stöffler, D., 1996, Shock metamorphism of quartz in nature and experiment: II. Significance in geoscience: Meteoritics & Planetary Science, v. 31, p. 6–35.

Gucsik, A., Koeberl, C., Brandstätter, F., Libowitsky, E., and Reimold, W.U., 2003, Cathodoluminescence, scanning electron microscopy, and Raman spectroscopy of experimentally shock metamorphosed zircon crystals and naturally shocked zircon from the Ries impact crater, *in* Dypvik, H., Burchell, M., and Claeys, P, eds., Cratering in Marine Environments and on Ice Impact Studies Series: Berlin-Heidelberg, Springer-Verlag, v. 5, p. 281–322.

Hargraves, R.B., 1961, Shatter cones in the rocks of the Vredefort Ring: Transactions of the Geological Society of South Africa, v. 64, p. 147–161.

Hart, R.J., Welke, H.J., and Nicolaysen, L.O., 1981, Geochronology of the deep profile through the Archaean Basement of Vredefort, with implications for early crustal evolution: Journal of Geophysical Research, v. 86, p. 10663–10680.

Hart, R.J., Andreoli, M.A.G., Smith, C.B., Otter, M.L., and Durrheim, R., 1990, Ultramafic rocks in the centre of the Vredefort Structure: Possible exposure of the upper mantle: Chemical Geology, v. 83, p. 233–248, doi: 10.1016/0009-2541(90)90282-C.

Hart, R.J., Andreoli, M.A.G., Reimold, W.U., and Tredoux, M., 1991, Aspects of the dynamic and thermal metamorphic history of the Vredefort cryptoexplosion structure: Implications for its origin: Tectonophysics, v. 192, p. 313–331, doi: 10.1016/0040-1951(91)90106-3.

Hart, R.J., Moser, D., and Andreoli, M.A.G., 1999, Archaean age for the granulite facies metamorphism near the center of the Vredefort structure, South Africa: Geology, v. 27, p. 1091–1094, doi: 10.1130/0091-7613(1999)0272.3.CO;2.

Henkel, H., and Reimold, W.U., 1998, Integrated geophysical modelling of a giant, complex impact structure: Anatomy of the Vredefort Structure, South Africa: Tectonophysics, v. 287, p. 1–20, doi: 10.1016/S0040-1951(98)80058-9.

Hörz, F., and Ahrens, T.J., 1969, Deformation of experimentally shocked biotite: American Journal of Science, v. 267, p. 1213–1229.

Huffman, A.R., and Reimold, W.U., 1996, Experimental constraints on shock-induced microstructures in naturally deformed silicates: Tectonophysics, v. 256, p. 165–217, doi: 10.1016/0040-1951(95)00162-X.

Huffman, A.R., Brown, J.M., Carter, N.L., and Reimold, W.U., 1993, The microstructural response of quartz and feldspar under shock loading at variable temperatures: Journal of Geophysical Research, v. 98, p. 22171–22197.

Ivanov, B.A., and Deutsch, A., 1999, Sudbury impact event: Cratering mechanics and thermal history, *in* Dressler, B.O. and Sharpton, V.L., eds., Large Meteorite Impacts and Planetary Evolution II: Geological Society of America Special Paper 339, p. 389–397.

Kamo, S.L., Reimold, W.U., Krogh, T.E., and Colliston, W.P., 1996, A 2.023 Ga age for the Vredefort impact event and a first report of shock metamorphosed zircons in pseudotachylitic breccias and granophyre: Earth and Planetary Science Letters, v. 144, p. 369–388, doi: 10.1016/S0012-821X(96)00180-X.

Kieffer, S.W., Phakey, P.P., and Christie, J.M., 1976, Shock processes in porous quartzite: Transmission electron microscope observations and theory: Contributions to Mineralogy and Petrology, v. 59, p. 41–93.

Koeberl, C., Reimold, W.U., and Shirey, S.B., 1996, Re-Os isotope and geochemical study of the Vredefort Granophyre: Clues to the origin of the Vredefort Structure, South Africa: Geology, v. 24, p. 913–916, doi: 10.1130/0091-7613(1996)0242.3.CO;2.

Lambert, P., and Mackinnon, I.D.R., 1984, Micas in experimentally shocked gneiss: Journal of Geophysical Research, v. 89 (supplement), p. B685–B699.

Lana, C., 2004, Structural and petrogenetic studies related to the geological evolution of the Archaean basement complex of the Vredefort dome, South Africa, [unpublished Ph.D. thesis]: Johannesburg, University of the Witwatersrand, 328 p.

Lana, C., Gibson, R.L., Kisters, A.F.M., and Reimold, W.U., 2003a, Archean crustal structure of the Kaapvaal craton, South Africa—evidence from the Vredefort dome: Earth and Planetary Science Letters, v. 206, p. 133–144, doi: 10.1016/S0012-821X(02)01086-5.

Lana, C., Gibson, R.L., and Reimold, W.U., 2003b, Impact tectonics in the core of the Vredefort dome, South Africa: Implications for central uplift formation in very large impact structures: Meteoritics & Planetary Science, v. 38, p. 1093–1107.

Lana, C., Gibson, R.L., Reimold, W.U., and Minnitt, R.C.A., 2003c, Geology and geochemistry of a granite-greenstone association in the southeastern Vredefort dome, South Africa: South African Journal of Geology, v. 107, p. 291–314.

Lana, C., Reimold, W.U., Gibson, R.L., Koeberl, C., and Siegesmund, S., 2004, Nature of the Archaean mid-crust in the central Kaapvaal craton, South Africa—evidence from the Vredefort dome: Geochimica et Cosmochimica Acta, v. 68, p. 623–642, doi: 10.1016/S0016-7037(03)00447-2.

Langenhorst, F., and Deutsch, A., 1994, Shock experiments on pre-heated α- and β-quartz: I. Optical and density data: Earth and Planetary Science Letters, v. 125, p. 407–420, doi: 10.1016/0012-821X(94)90229-1.

Leroux, H., Reimold, W.U., and Doukhan, J.-C., 1994, A TEM investigation of shock metamorphism in quartz from the Vredefort Dome, South Africa: Tectonophysics, v. 230, p. 223–239, doi: 10.1016/0040-1951(94)90137-6.

Lilly, P.A., 1981, Shock metamorphism in the Vredefort collar: Evidence for internal shock sources: Journal of Geophysical Research, v. 86, p. 10689–10700.

Martini, J.E.J., 1991, The nature, distribution and genesis of the coesite and stishovite associated with the pseudotachylite of the Vredefort Dome, South Africa: Earth and Planetary Science Letters, v. 103, p. 285–300, doi: 10.1016/0012-821X(91)90167-G.

Masaitis, V.L., and Pevzner, L.A., eds., 1999, Deep drilling in the Puchezh-Katunki Impact Structure: Saint Petersburg, VSEGEI Press, 392 p.

McCarthy, T.S., Stanistreet, I.G., and Robb, L.J., 1990, Geological studies related to the origin of the Witwatersrand Basin and its mineralization—an introduction and a strategy for research and exploration: South African Journal of Geology, v. 93, p. 1–4.

McIntyre, D.B., 1968, Impact metamorphism at Clearwater Lake, Quebec, *in* French, B.M., and Short, N.M., eds., Shock Metamorphism of Natural Materials: Baltimore, Mono Book Corporation, p. 363–366.

Melosh, H.J., 1989, Impact Cratering—A Geologic Process: New York, Oxford University Press, 245 p.

Ogilvie, P., Gibson, R.L., Reimold, W.U., and Deutsch, A., 2004a, Experimental investigation of shock effects in a pelitic granulite, *in* Lunar and Planetary Science XXXV [abs.]: Houston, Lunar and Planetary Institute, (CD-ROM).

Ogilvie, P., Gibson, R.L., and Stevens, G., 2004b, Rates and scales of chemical and textural equilibration during impact-induced thermal metamorphism in the Vredefort dome, South Africa [abs.]: Johannesburg, Geoscience Africa, p. 500–501.

Ostertag, R., 1981, Annealing experiments on experimentally shocked feldspar single crystals: Meteoritics, v. 16, p. 373.

Ostertag, R., and Stöffler, D., 1982, Thermal annealing of experimentally shocked feldspar crystals: Journal of Geophysical Research, v. 87, p. A457–A463.

Raikes, S.A., and Ahrens, T.S., 1979, Post-shock temperatures in minerals: Royal Astronomical Society Geophysical Journal, v. 58, p. 717–747.

Reimold, W.U., 1990, The controversial microdeformations in quartz from the Vredefort Structure, South Africa: South African Journal of Geology, v. 93, p. 645–663.

Reimold, W.U., and Colliston, W.P., 1994, Pseudotachylites of the Vredefort Dome and the surrounding Witwatersrand Basin, South Africa, *in* Dressler, B.O., Grieve, R.A.F., and Sharpton, V.L., eds., Large Meteorite Impacts and Planetary Evolution: Geological Society of America Special Paper 293, p. 177–196.

Reimold, W.U., and Gibson, R.L., 1996, Geology and evolution of the Vredefort Impact Structure, South Africa: Journal of African Earth Sciences, v. 23, p. 125–162, doi: 10.1016/S0899-5362(96)00059-0.

Reimold, W.U., Leroux, H., and Gibson, R.L., 2002, Shocked and thermally metamorphosed zircon from the Vredefort impact structure, South Africa: A transmission electron microscopic study: European Journal of Mineralogy, v. 14, p. 859–868, doi: 10.1127/0935-1221/2002/0014-0859.

Schreyer, W., 1983, Metamorphism and fluid inclusions in the basement of the Vredefort Dome, South Africa: Guidelines to the origin of the structure: Journal of Petrology, v. 24, p. 26–47.

Schreyer, W., and Abraham, K., 1978, Symplectitic cordierite-orthopyroxene-garnet assemblages as products of contact metamorphism of pre-existing basement granulites in the Vredefort structure, South Africa: Contributions to Mineralogy and Petrology, v. 68, p. 53–62.

Schreyer, W., and Medenbach, O., 1981, CO_2-rich fluid inclusions along planar elements in quartz in basement rocks of the Vredefort dome, South Africa: Contributions to Mineralogy and Petrology, v. 77, p. 93–100.

Schreyer, W., Stepto, D., Abraham, K., and Mueller, W.F., 1978, Clinoeulite (magnesian clinoferrosilite) in a eulysite of a metamorphosed iron formation in the Vredefort structure, South Africa: Contributions to Mineralogy and Petrology, v. 65, p. 351–361.

Shand, S.J., 1916, The pseudotachylyte of Parijs (Orange Free State), and its relation to "trap-schotten gneiss" and "flinty crush-rock": Quarterly Journal of the Geological Society London, v. 72, p. 198–221.

Spray, J.G., 1998, Localized shock- and friction-induced melting in response to hypervelocity impact, *in* Grady, M.M., Hutchinson, R., McCall, G.J.H., Rothery, D.A., eds., Meteorites: Flux with Time and Impact Effects: Geological Society of London Special Publication 140, p. 195–204.

Stähle, V., 1975, Natural shock behaviour of almandite in metamorphic rocks from the Ries crater, Germany: Earth and Planetary Science Letters, v. 25, p. 71–81, doi: 10.1016/0012-821X(75)90212-5.

Stepto, D., 1990, The geology and gravity field in the central core of the Vredefort structure: Tectonophysics, v. 171, p. 75–103, doi: 10.1016/0040-1951(90)90091-L.

Stevens, G., Gibson, R.L., and Droop, G.T.R., 1997, Mid-crustal granulite facies metamorphism in the central Kaapvaal Craton: The Bushveld Complex connection: Precambrian Research, v. 82, p. 113–132, doi: 10.1016/S0301-9268(96)00043-5.

Stöffler, D., 1966, Zones of impact metamorphism in the crystalline rocks at the Nördlinger Ries Crater: Contributions to Mineralogy and Petrology, v. 12, p. 15–24.

Stöffler, D., 1972, Deformation and transformation of rock-forming minerals by nature and experimental shock processes: I. Behaviour of minerals under shock compression: Fortschritte der Mineralogie, v. 49, p. 50–113.

Stöffler, D., 1984, Glasses formed by hypervelocity impact: Journal of Non-Crystalline Solids, v. 67, no. 1-3, p. 465–502, doi: 10.1016/0022-3093(84)90171-6.

Stöffler, D., and Langenhorst, F., 1994, Shock metamorphism of quartz in nature and experiment: I. Basic observation and theory: Meteoritics, v. 29, p. 155–181.

Therriault, A.M., Grieve, R.A.F., and Reimold, W.U., 1997, The Vredefort Structure: Original size and significance for geological evolution of the Witwatersrand Basin: Meteoritics & Planetary Science, v. 32, p. 71–77.

Wieland, F., Gibson, R.L., and Reimold, W.U., 2004, Field studies in the central uplift of the Vredefort impact structure, *in* Lunar and Planetary Science XXXV, Abstract: Houston, Lunar and Planetary Institute (CD-ROM).

Wilshire, H.G., 1971, Pseudotachylite from the Vredefort Ring, South Africa: Journal of Geology, v. 79, p. 195–206.

Manuscript Accepted by the Society 9 August 2004

Printed in the USA

Geological Society of America
Special Paper 384
2005

Evaporative differentiation of impact-produced melts: Laser-simulation experiments and comparison with impact glasses from the Logoisk crater

M.V. Gerasimov*
Space Research Institute, Russian Academy of Sciences, Profsoyuznaya Street, 84/32 Moscow, 117997, Russia

O.I. Yakovlev
Vernadsky Institute of Geochemistry and Analytical Chemistry, Russian Academy of Sciences, Kosygin Street, 19, Moscow, 117975, Russia

Yu.P. Dikov
Institute of Ore Deposits, Petrography, Mineralogy and Geochemistry, Russian Academy of Sciences, Staromonetny pereulok, 35, Moscow, 109017, Russia

F. Wlotzka
Max-Plank-Institut für Chemie, Abteilung Kosmochemie, P.O. Box 3060, D-55020, Mainz, Germany

ABSTRACT

Evaporative modification of silicates during impacts by large meteorites is the result of high-temperature processing of colliding silicate materials. There are still unresolved questions about the influence of vaporization on the chemical compositions of impact glasses. Here we present the results of experimental modeling of impact-related high-temperature pulse heating of silicates with compositions representative of Earth's crust. Trends involving differential volatilization of the main rock-forming elements were determined experimentally and compared with those predicted by thermodynamic equilibrium calculations. Our results indicate that impact-induced vaporization is significant for melts at temperatures of 3000–5000 K. As vaporization begins, melts lose Si, Fe, and alkalis and are enriched in Mg, Ca, Al, and Ti. With increased mass loss, melts start to lose Mg while Si and even Na can still be present. Still greater mass loss produces melts enriched in Ca, Al, and Ti. We then attempted to use the experimentally determined trends to look for signs of evaporative differentiation in impact glasses from the Logoisk structure (Byelorussia). Fresh impact glasses from suevites in the Logoisk crater have compositions that are in good agreement with the differentiation trends obtained both experimentally and via thermodynamic calculations in the range of 3000–5000 K. The most useful methods for recognizing evaporative differentiation in the natural glasses are the conservation of the initial Ca/Al ratio and the anti-correlation of Si and Al.

Keywords: impact-induced vaporization, impact glasses, differential volatilization of elements, fractionation trends.

*mgerasim@mx.iki.rssi.ru

Gerasimov, M.V., Yakovlev, O.I., Dikov, Yu.P., and Wlotzka, F., 2005, Evaporative differentiation of impact-produced melts: Laser-simulation experiments and comparison with impact glasses from the Logoisk crater, *in* Kenkmann, T., Hörz, F., and Deutsch, A., eds., Large meteorite impacts III: Geological Society of America Special Paper 384, p. 351–366. For permission to copy, contact editing@geosociety.org.

INTRODUCTION

Large meteorites characteristically impact the surfaces of terrestrial planets at velocities of over 10 km/s. Impacts at such high velocities produce strong thermal effects in colliding silicate materials that result in melting and partial vaporization of an amount of target material which is several times the mass of the meteorite (e.g., O'Keefe and Ahrens, 1977; Melosh, 1989). High-temperature processing during an impact provides good opportunities for chemical modification of the silicate materials involved. Despite the obvious likelihood of evaporative modification of silicates during hypervelocity impacts, questions about fractionation trends and the degree to which glasses in impact structures have been modified by evaporation are still open. The impact-induced loss of highly volatile species such as water, CO_2, and other gaseous components is universally accepted (e.g., Lange and Ahrens, 1982; Ivanov et al., 1996; Gerasimov, 2002). Attempts to find clear evidence of evaporative differentiation of rock-forming elements in impactites have not been so fruitful. Large impact structures usually have heterogeneous target rocks (e.g., Masaitis et al., 1980; Grieve, 1987, Valter and Ryabenko, 1977), and projectiles several kilometers or more across are also likely to be heterogeneous in composition. Therefore it is very difficult to distinguish compositional differences among impact melts due to evaporative effects from the effects of mixing heterogeneous materials. Since the heating effect decreases with distance from an impact center (Melosh, 1989; Basilevsky et al., 1983; Grieve et al., 1977), large volumes of peripheral low-temperature melt similar to volcanic magmas are produced that do not experience significant evaporative processing. The dilution of small amounts of high-temperature impact-produced melt by larger volumes of melt with low thermal effects obscures the possible effect of evaporative differentiation in the most heated portions of the melt. The bulk of this central high-temperature melt, which involves most of the projectile material, is disrupted into millimeter- to micrometer-sized droplets (Melosh and Vickery, 1991), ejected from the crater, and redeposited after traveling for very large distances (Artemieva and Shuvalov, 2002; Basilevsky et al., 1983). The unusual chemical compositions of such glasses and in some cases an uncertain stratigraphic position hinders their correlation with specific target rocks and craters (e.g., Simonson et al., 2000a, 2000b; Simonson and Glass, 2004; Schulte et al., 2003).

The possibility of chemical differentiation of impact-produced melts has been proven in small-scale experiments using two-stage light-gas-gun facilities. Hörz et al. (1983) investigated the composition of impact-produced melts after an impact of a Na-Ca glass projectile into a gold target at velocities of 5–6.4 km/s. Analysis of impact-produced melts indicated moderate changes in Si and Ca content but noticeably higher changes in alkalis. Yakovlev et al. (1992, 1995, 2000) investigated the composition of impact-produced condensates and in some cases melts after impacts of a copper projectile into ultramafic, mafic, and acidic targets at velocities of ~6 km/s. A strong Si enrichment (~60 rel%) was detected in condensates produced from ultramafic and mafic targets (Yakovlev et al., 1995). Comparable impacts in granite and obsidian targets had the opposite effect (Yakovlev et al., 2000). Melts which were produced in experiments with obsidian showed a strong mobility of alkalis, but only moderate changes in the content of other rock-forming elements. The velocities of projectiles in light-gas-gun facilities are usually <7 km/s, which is too low for efficient modeling of impact-induced vaporization effects.

Some progress was made in modeling impact-related vaporization of silicates using a laser pulse (LP) vaporization facility (Gerasimov et al., 1999). Investigation of glass spherules produced in LP impact simulation experiments with augite (Gerasimov et al., 2000) revealed a consistent sequence of mobility of all of the main rock-forming elements during progressive vaporization.

Nevertheless, there are no systematic investigations of the impact-induced evaporative differentiation of silicate materials. The aim of the present work was to investigate the differentiation trends of the main rock-forming elements during high-temperature flash vaporization of silicates similar to planetary crust material. Another goal of the work was to use experimental data obtained to examine possible effects of vaporization among glasses of natural, large impact structures.

EXPERIMENTAL TECHNIQUE

Experimental Procedure

Simulation of impact vaporization of samples was performed using the method and the laser pulse installation described by Gerasimov et al. (1999). The scheme of the experiment is presented in Figure 1. A sample was placed in a sealed aluminum wall cylindrical cell with an internal volume of ~500 cm^3. A powerful beam from an unmodulated Nd-glass (wavelength 1.06 μm)

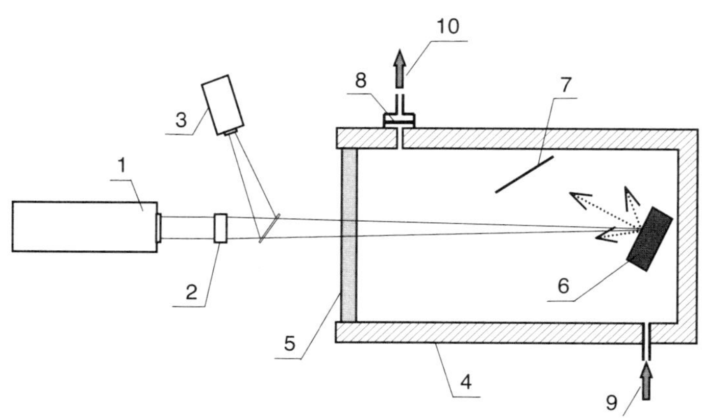

Figure 1. General schematic of the laser pulse experimental setup: 1—laser; 2—focusing lens; 3—calorimeter; 4—sample chamber; 5—quartz window; 6—sample; 7—foil; 8—filter; 9—inflow of helium; 10—outflow of helium.

pulse laser was focused on the sample surface through an optical quartz window. The luminous energy output of the laser pulse was in the range from 400 to 600 J. Duration of the laser pulse was ~10^{-3} s. By adjusting the diameter of the beam spot from 2 to 5 mm on the sample's surface, it was possible to vary the power of a laser pulse between 2×10^6 and 2×10^7 W/cm^2. Based on considerations developed by Gerasimov et al. (1999), the temperature of vaporization was estimated as 4000–5000 K. About 10–20 mg of a sample was vaporized during one pulse, producing a crater 3–6 mm in diameter with a residue of melt inside. Experiments were performed in an atmosphere of helium kept at pressure of 1 atm to exclude any vapor-atmosphere interaction chemistry in the expanding vapor plume The sample surface was inclined at angle of ~30° to the direction of the optical laser axes so that the expansion of the vapor plume could proceed outside the laser beam, thereby avoiding additional heating. A Ni-foil (4 × 3 cm) was mounted at a distance of 5–6 cm from the sample in the path of the spreading evaporated cloud to collect condensing vapor products.

Dispersed droplets of melt solidified in the vapor cloud and were found on the foil as micrometer-sized spherules adhering to the surface of precipitated condensate. Figure 2 shows a view of the condensed film with an abundance of 1–20 µm-sized glass droplets. Large pieces of melt which did not solidify completely during their flight were smeared across the foil surface and had irregular shapes. A portion of the melt still adhered to the walls of craters. Chemical analyses of glass spherules and residual crater melts were performed using field emission scanning electron microscope (LEO-1530) with energy dispersive X-ray spectroscopy at 15–20 kV acceleration voltage and 100 nA sample current. Chemically analyzed mineral standards from the United States National Museum (Washington, D.C.) (Jarosewich et al., 1980) were used as reference samples. We used an energy dispersive Si-Li detector together with a commercial data reduction program from Link-Oxford.

Most of the glass droplets had dimensions in the range of 1–20 µm. Analyses were done by pointing the focused electron beam at the center of a selected droplet without removing it from the foil. This may have resulted in some contribution from the surrounding condensate, especially for droplets smaller than ~2 µm. Quantitative analysis of individual unpolished microparticles using electron beam instruments results in an increase in the error of up to ~20 rel% (Armstrong and Buseck, 1975). Therefore only compositional changes greater than 20% should be considered significant with measurements of this kind.

Analyses of the crater melt and starting material were done in polished cross sections using standard methods and had high reliability. Cross sections were made after LP experiments through the centers of the crater and perpendicular to sample surfaces. Before cutting, the samples were covered by epoxy to keep pieces of melt in place on the crater walls. Crater melts were analyzed by random probing using the focused electron beam. Analyses of starting samples were done in the same polished cross sections by focused and area (20 × 20 µm) probing in the region 50–500 µm beneath the lower limit of the zone modified by the laser pulse. These limits were easily defined on scanning-

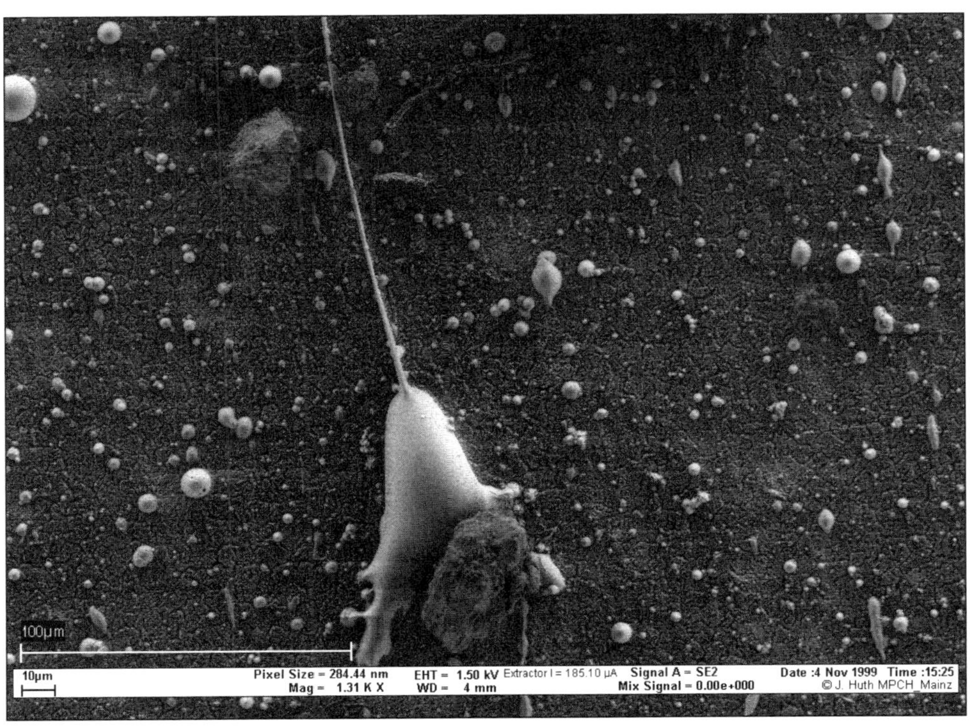

Figure 2. A typical scanning electron microscope picture of the condensed film, which was precipitated on the foil after a laser pulse experiment. Melt droplets and splashes overlay the surface of the condensate. (Used with permission.)

electron-microscope images by the difference in color between melt and unmelted sample and by a system of cracks in the melt formed by thermal contraction during cooling.

Analysis of the composition of the precipitated films of condensed materials was performed using X-ray photo-electron spectroscopy technique. Methodological details of the measurements can be found in Gerasimov et al. (2002).

Samples

Two principal criteria were used to select samples for this set of experiments: they had to be (1) highly homogeneous and (2) representative of Earth's crustal materials in terms of chemical composition. Homogeneity was important to avoid problems of mixing of different target materials in the interpretation of any compositional modifications detected in the melts formed.

For the LP experiments, we used monominerallic blocks of augite and obsidian that were ~1 cm thick and had a flat surface ~2 cm wide. Samples were selected from among pieces of larger blocks after hammer crushing. Fresh surfaces without any mechanical and chemical treatment were then selected for irradiation by laser pulses. The laser pulses in this series of experiments were adjusted to consistently produce craters with diameters of 3–4 mm. Examinations of polished cross sections of the samples demonstrated rather high homogeneity of both starting augite and obsidian. Chemical compositions of starting augite and of starting obsidian are presented in Table 1.

RESULTS AND DISCUSSION

We performed five LP experiments with the augite samples and two LP experiments with the obsidian. We analyzed 55 droplets from the augite experiments and 83 droplets from the obsidian experiments. Droplets in all of the experiments show rather high departures from their original chemical compositions, greatly exceeding 20 rel% for all of the main rock-forming elements. Crater melt compositions for augite and obsidian also deviated noticeably from their starting compositions. No systematic change in the crater melt composition as a function of depth through its thickness was detected.

Figures 3 and 4 show the relative abundances of the elements in glass spherules from experiments with augite and obsidian, respectively. Comparison of Figures 3 and 4 demonstrate, close similarity in the behavior of elements for both samples. Al, Ca, and Ti behaved normally for refractory elements and their concentration increased up to 3–4 times in droplets due to volatilization of other elements. Na and K behaved normally for volatile elements and escaped almost totally from most spherules. Si and Fe also behaved as volatile elements and were depleted several fold in some glass spherules as compared to the starting concentrations. Mg in both cases showed bimodal behavior producing enrichments of up to twofold in some spherules and depletions from several fold up to almost total loss in others.

The degree of enrichment of refractory elements (Al, Ca, and Ti) is indicative of the amount of mass loss before solidification of glass spherules since the increased concentration of refractories can only be due to the preferential volatilization of other elements. The maximum increase observed in the concentration of refractory elements was fourfold, which corresponds to the loss of more than 75% of a spherule's mass. Some fraction of the refractory elements was also volatilized during the vaporization process as indicated by the presence of reduced quantities of these elements in the condensate. The presence of refractory elements in the condensate means that the calculated mass loss based on the degree of enrichment of the refractories is an underestimation. The chemical compositions of the condensates are shown in Table 1.

Differences in the compositions of glass spherules reflect

TABLE 1. CHEMICAL COMPOSITION OF STARTING AUGITE AND OBSIDIAN AND OF THEIR CONDENSATES OBTAINED IN LASER PULSE EXPERIMENT.

Oxides	Augite				Obsidian			
	Starting sample*	$\pm\sigma$	Condensate	$\pm\sigma$	Starting sample†	$\pm\sigma$	Condensate	$\pm\sigma$
SiO_2	49.29	0.10	54.69	0.90	57.90	0.19	62.16	0.86
TiO_2	1.13	0.03	0.74	0.08	1.32	0.09	1.13	0.55
Al_2O_3	9.98	0.09	7.57	0.48	15.02	0.15	12.56	0.64
FeO	8.22	0.10	8.11	0.36	9.31	0.18	5.27	0.66
MgO	13.09	0.08	12.06	0.69	5.11	0.10	6.58	1.39
CaO	15.46	0.11	12.33	1.28	7.37	0.10	4.23	0.85
MnO	0.07	0.02	<0.1					
Na_2O	2.75	0.14	4.51	0.28	2.99	0.09	7.25	0.83
K_2O					0.53	0.04	1.08	0.36

* Average of 18 analyses.
† Average of 10 analyses.

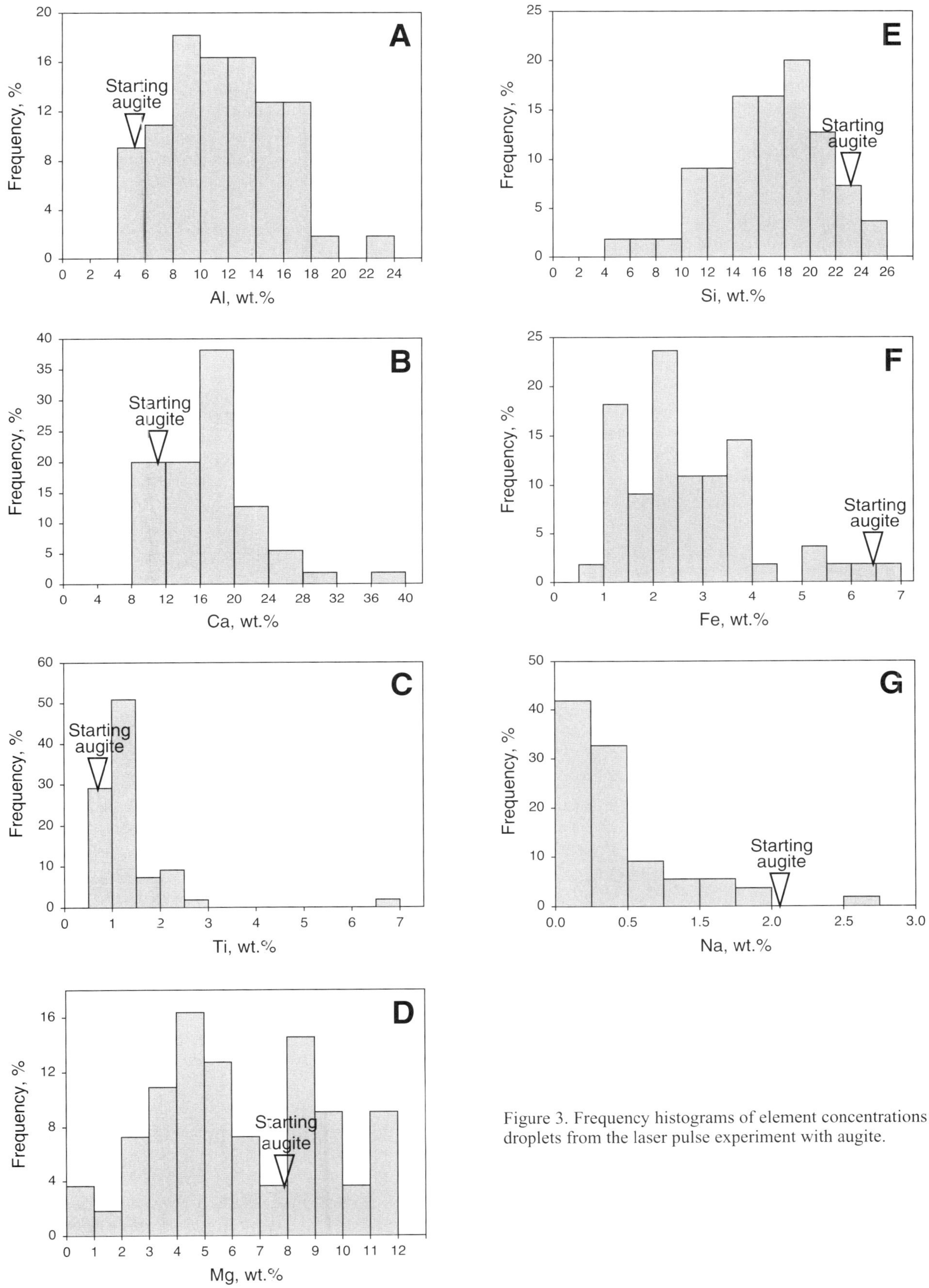

Figure 3. Frequency histograms of element concentrations in glass droplets from the laser pulse experiment with augite.

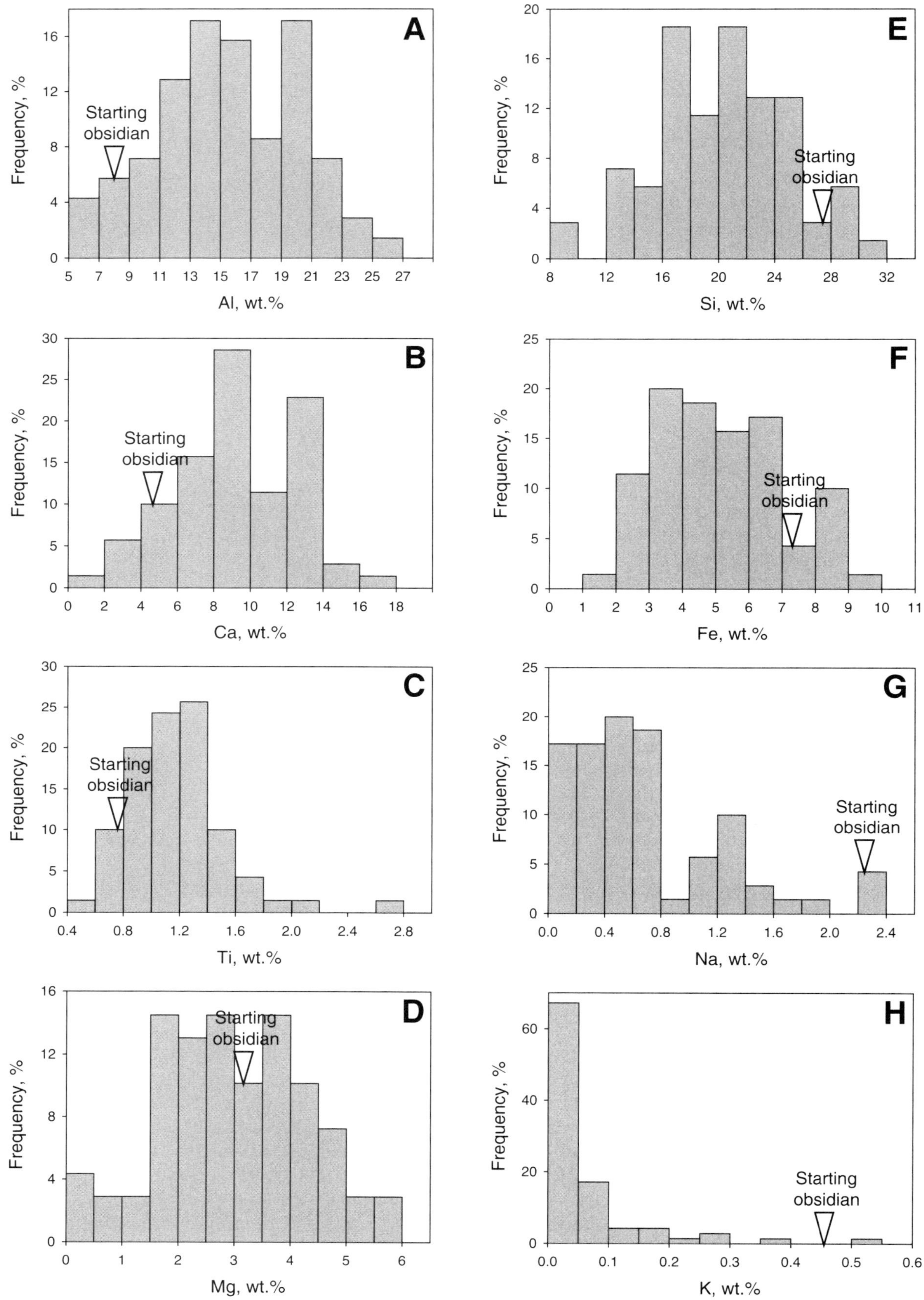

Figure 4. Frequency histograms of element concentrations in glass droplets, from the laser pulse experiment with obsidian.

their different volatilization histories. Some spherules were formed from volatilized crater melts and were ejected into the central parts of the expanding vapor cloud. Other spherules were formed from relatively colder melts at the periphery of the crater and were rapidly ejected outside of the plume. Spherules in the first case are exposed to higher temperatures for longer periods of time than spherules in the second case. This results in different degrees of volatilization for spherules along different flight-paths through the vapor cloud. Statistically, the experimental spherules should have a continuous range of chemical compositions reflecting progressive volatilization as a function of high-temperature processing. In the case of an actual impact, individual glasses should show a continuous range of compositions reflecting their degree of volatilization as a function of both their dimensions and exposure to high temperatures.

Thermodynamic evaluation of the behavior of elements during the high-temperature volatilization was done using the MAGMA computer code, which was developed by Fegley and Cameron (1987) to model vaporization of the planet Mercury during its formation. The code considers volatilization of elements in atomic and/or oxide form but the concentrations of these unbound (or free) metal oxides in a complex silicate melt is decreased by formation of complex oxide and silicate pseudocomponents. Such pseudocomponents reduce the activity of atomic and/or oxide forms of elements and control the volatilization process. These pseudocomponents are assumed to mix ideally in the melt. The equilibrium distribution of the metal oxides and pseudocomponents is calculated in the code by a standard method of Gibbs' free energy minimization. Since there are no thermodynamic data on related activities of oxides and pseudocomponents in the high-temperature range of interest here, our calculations in the range of 2000–5000 K are only approximations. Nevertheless, these calculations can help us to evaluate the expected results in light of those obtained experimentally.

Refractory Elements

Since the concentrations of refractory elements in high-temperature melts are increasing due to the loss of other elements, their relative proportions should remain unchanged because of conservation of their absolute quantity in a specific volume. Figures 5A and 5B show the correlation between Ca and Al, and Figures 5C and 5D between Ti and Al for glass spherules and crater melts in comparison with those of starting samples of augite and obsidian. Thermodynamic evaluations of volatilization trends are illustrated for 2000, 3000, 4000, and 5000 K. The calculated compositions progressively deviate from the starting value with increasing degrees of mass loss. The calculations were arbitrarily terminated once 80% of the mass of the melt was lost. For both augite and obsidian, calculations indicate near-perfect conservation of Ca and Al until these elements become the main components of the melt. Figure 5B shows a linear trend in the compositional evolution of both crater melt and spherules from obsidian. The correlation between Ca and Al contents in crater melt for augite (Fig. 5A) is also linear, but most of the spherules are noticeably depleted in Ca. The Ca loss is more obvious in Figures 5E and 5F where the same data points are replotted using the Ca/Al ratios of both melts and starting materials. In the absence of volatilization of Ca and Al, conservation of Ca/Al ratio can be expected for silicate melts during high-temperature processing. Compositions should therefore plot linearly with increasing Al and Ca but with about the same Ca/Al ratio as that of the starting rocks. For both augite and obsidian, the Ca/Al ratios of the spherules consistently fall below the conservation line (Figs. 5E and 5F). This indicates differential loss of Ca that may be due to the volatility of Ca-containing molecular clusters during vaporization of silicate melts at very high temperatures (Gerasimov et al., 1999, Yakovlev et al., 1995). Besides the volatility of Ca in atomic and oxide forms, there is a possible volatility of Ca in the form of a "wollastonite" cluster (Ca:Si = 1:1). The MAGMA code calculations support this interpretation in that the "$CaSiO_3$" pseudocomponent binds more than 50% of available Ca at appropriate temperatures. No thermodynamic data exist on the volatility of molecular clusters and the MAGMA code does not account for the possible volatility of such clusters.

The behavior of Ti is similar to the behavior of Ca. It also has a positive and linear correlation with Al which is nearly perfect for crater melts whereas spherules show some loss of Ti. Thermodynamic calculations indicate that the loss of Ti should be greater at higher temperatures and, given the degree of the mass loss, is believed to be responsible for some of the enrichment of Al and Ca. Aluminum appears to be the most refractory element in high-temperature silicate melts, so it is the best one to use as a reference to assess the volatility of other elements.

Behavior of Silica

Figures 5G and 5H show an anti-correlation between Si and Al in the high-temperature melts. The loss of Si is high for melts from both augite and obsidian. The Si concentration decreases from 23 wt% to ~5 wt% in some spherules from augite and likewise from 27 to ~7 wt% in the case of obsidian. The melt compositions fall within a rather narrow range, especially in the case of obsidian. Thermodynamic calculations also give a rather narrow compositional range during volatilization. Crater melts again are in good agreement with calculations whereas spherules have slightly higher Si concentrations than those predicted theoretically for certain Al contents. To investigate possible contamination from the background condensate to the analyses of the glass spherules, we divided the population of the spherules from obsidian into three size categories (<2 μm, 2–5 μm, and >5 μm) and plotted the results separately (e.g., Fig. 5H). If there were a significant contribution from the background condensate, one would expect a systematic difference in composition between the smaller and larger spherules, but this does not appear to be the case. In Figure 5H, for example, spherules <2 μm across are widely distributed and show the same range of composition as spherules that are 2–5 μm across. Larger spherules mostly plot

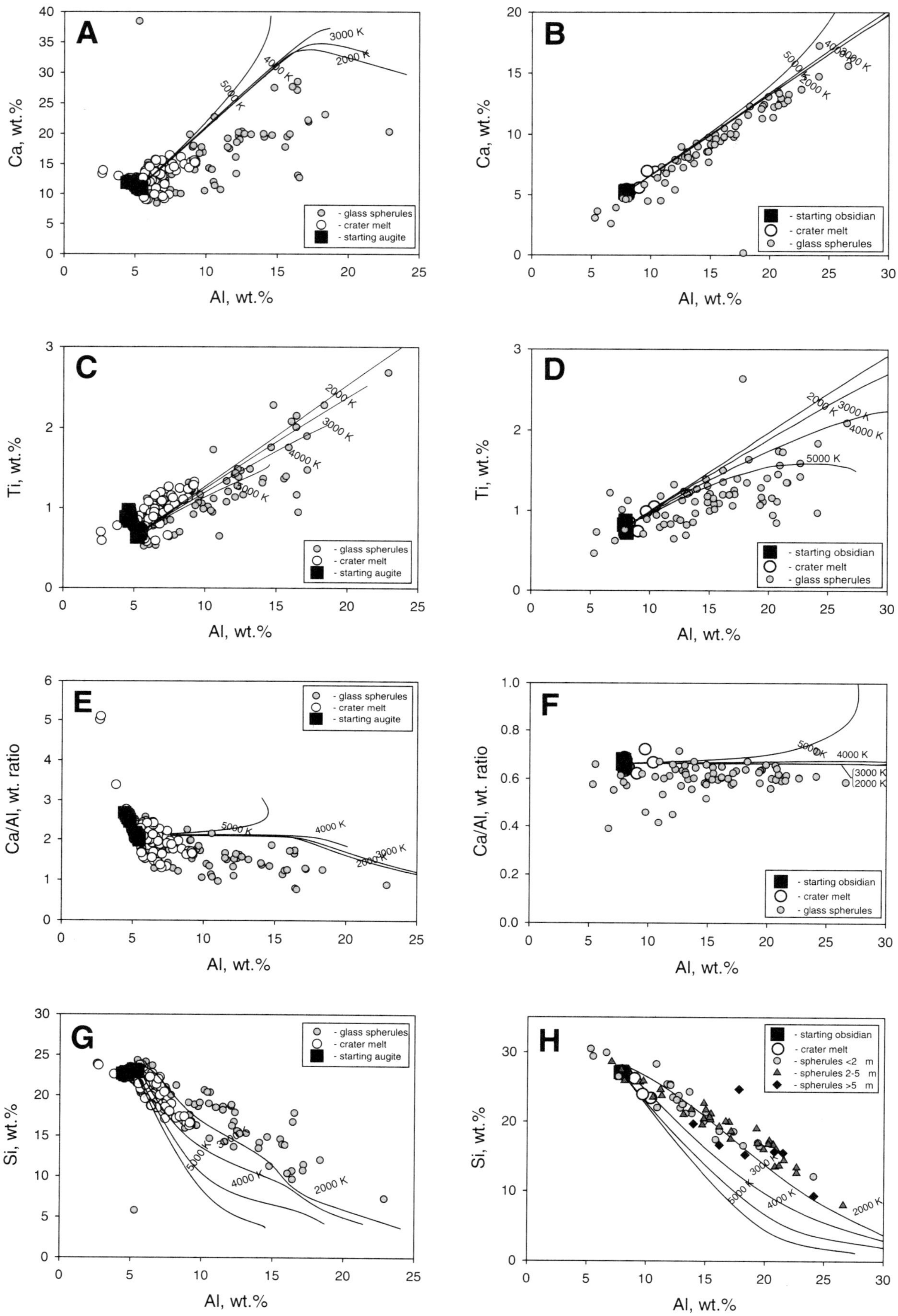

A
B
C
D
E
F
G
H
Ca, wt.%
Ti, wt.%
Ca/Al, wt. ratio
Si, wt.%
Al, wt.%
2000 K
3000 K
4000 K
5000 K
- glass spherules
- crater melt
- starting augite
- starting obsidian
- spherules <2 m
- spherules 2-5 m
- spherules >5 m

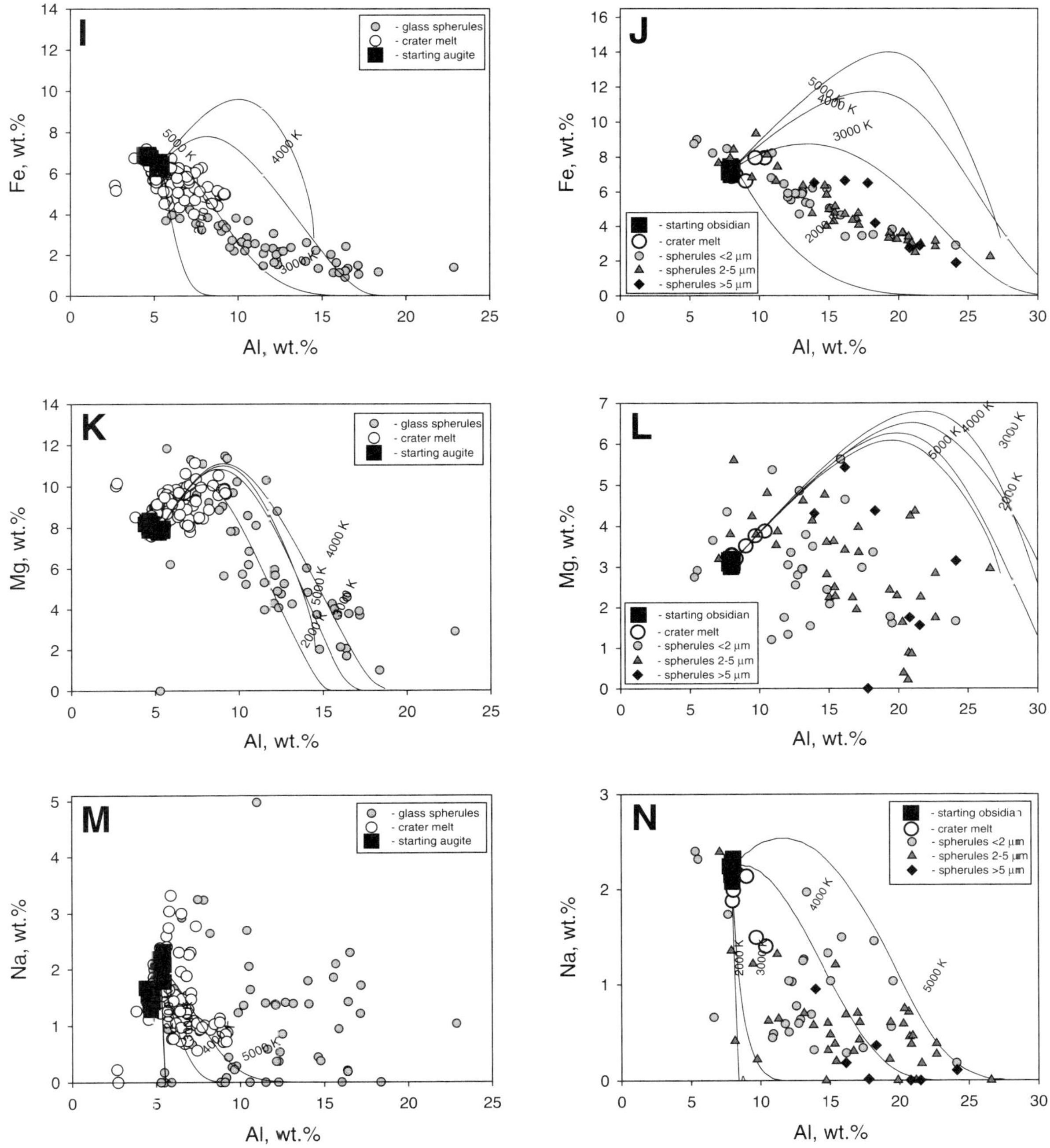

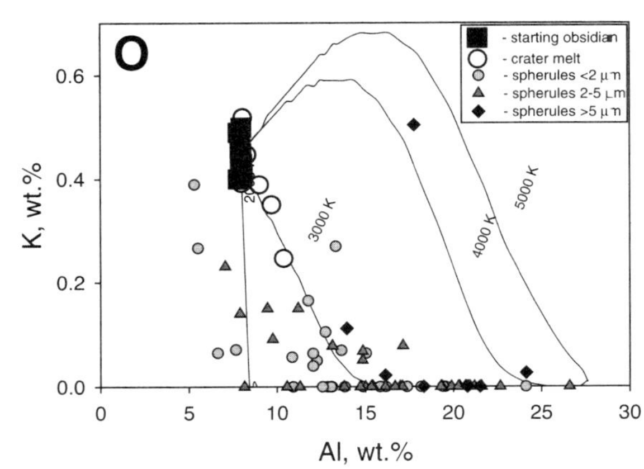

Figure 5 (*on this and previous page*). Chemical compositions of starting samples, crater melt, and glass spherules from laser pulse experiments. Figures A, C, E, G, I, K, and M show the Ca, Ti, Ca/Al, Si, Fe, Mg, and Na versus Al graphs respectively for experiments with augite. Figures B, D, F, H, J, L, N, and O show the Ca, Ti, Ca/Al, Si, Fe, Mg, Na, and K versus Al graphs, respectively, for experiments with obsidian. All 18 analyses for starting augite and all 10 analyses for starting obsidian are plotted in the graphs. Solid lines show the isothermal behavior of the average starting compositions at the temperatures indicated using the MAGMA code.

along the most volatilized compositional range of the smaller spherules, but there are fewer large spherules to plot, so this may be a statistical artifact. In addition, several small spherules have Si concentrations higher than the starting obsidian; this could be a result of background input, but other small spherules plot along the most depleted edge. Moreover, some crater melts for both augite and obsidian show an increase in Si content over the starting material, and there is no possible background input in these analysis. This effect deserves further investigation because other experimental impacts using a two-stage light-gas-gun using granite and obsidian targets also indicated depletion of Si in the condensates and, hence, enrichment of Si in the melts (Yakovlev et al., 2000). In summary, we believe the Si to Al dependency in our measurements must be taken as a real trend because we do not see any significant deviations attributable to background contamination in our analyses.

Behavior of Iron

Figures 5I and 5J are plots of the Fe versus Al contents in augite and obsidian respectively and the high-temperature melts produced from them. Melts from both starting materials show progressive depletions of Fe along very similar trends that are significantly different from the behavior predicted by thermodynamic calculations. This strongly suggests the mechanism of volatilization of Fe from high-temperature silicate melts is different from the one utilized in the MAGMA code. We suspect the vaporization of Fe proceeds via reduction into nano-sized metallic droplets which are then dispersed from the melt mechanically by the flow of the vapor (Yakovlev et al., 2002, Gerasimov et al., 2001). Such a process must result in progressive depletion of Fe from the melt as its high-temperature exposure increases. No significant differences in compositional trends were observed between groups of spherules with different sizes (see Fig. 5J).

Behavior of Magnesium

Figures 5K and 5L are plots of the Mg versus Al contents in augite and obsidian respectively and the high-temperature melts produced from them. In both cases, the compositions of the crater melts coincide closely to those predicted by the thermodynamic calculations. The compositions of the spherules also coincide rather well with those calculated in the case of augite, but in the case of obsidian, the spherules show great compositional diversity and diverge widely from the calculated path. Alone amongst the elements analyzed, Mg shows a bimodal distribution in which significant numbers of melted droplets deviated in both directions from the starting material, i.e., numerous droplets show enrichments in Mg while many others show depletions. (see Figs. 3 and 4). The calculations performed can help explain this unusual behavior, as follows: during volatilization, there is generally an increase of Mg in the beginning, but this gives way to rapid Mg depletion at higher mass losses. Therefore the droplets enriched in Mg represent the population that involved low degrees of volatilization, whereas droplets with low Mg concentrations form the population that involved high degrees of volatilization. The reason why Mg behavior correlates well with calculated trends for augite but not for obsidian is not clear yet. The explanation may be related to the formation and volatilization of different Mg-containing molecular clusters in the two situations, and this is not accounted for in the code. Figure 5L also shows that Mg shows a greater compositional discrepancy between spherules of different sizes than other elements. Most notably, larger spherules tend to have compositions closer to the calculated volatilization trend than smaller spherules do.

Behavior of Sodium and Potassium

Figures 5M and 5N are plots of the Na versus Al contents in augite and obsidian, respectively, and the high-temperature melts produced from them. Thermodynamic calculations for these elements define a triangle-shaped field for high-temperature melt compositions in which Na rapidly decreases to zero as Al increases. The compositions of the experimental melts coincide with such a behavior for Na, and the K in melts derived from obsidian follows a similar trend (Fig. 5N). However, it is remarkable that the concentrations of highly volatile Na and K do not decrease immediately with mass loss at increasing temperatures. Some melts even show an increase in their Na and K contents, which presumably reflects a more intense loss of Si in the beginning of the vaporization process and the incorporation of Na and K into pseudocomponents, primarily $NaAlSiO_4$, $NaAlSi_3O_8$, $NaAlO_2$ and $KAlSiO_4$, $KAlO_2$, $KAlSi_2O_6$. Moreover, a number of the crater melt and glass spherules plotted in Figure 5M have concentrations of Na high enough to remain above the 5000 K isotherm, or even above the starting Na concentration. Most of these analyses of crater melt were done within 1–3 μm of the surface of the melt. It is evident from Figures 5N and 5O that Na and K concentrations are generally higher in small spherules than in large ones. The enrichments in Na near the surface of crater melt volumes and in both Na and K in smaller spherules both indicate there is efficient back flow of vapor components onto melt surfaces during progressive cooling of the expanding vapor cloud.

GLASSES FROM NATURAL IMPACT STRUCTURES

The experimental data and thermodynamic evaluation of evaporative trends presented here clearly demonstrate that evaporative processing of a homogeneous target can form glasses with a wide range of chemical compositions. These compositions form a continuous spectrum and mainly reflect differences in temperature history and degree of mass loss. Natural impact glasses are produced from planetary surface and meteoritic materials via impact-induced melting followed by rapid cooling. Impact-induced melting occurs when shock-wave loading exceeds the melting threshold of a given material, and the post-shock temperature of the melt becomes progressively higher as the amplitude of the shock wave increases (Zel'dovich and

Raizer, 1967). The mass loss via partial vaporization is most efficient when the post-shock temperature in an impact-induced melt is so high that the vapor pressure exceeds the ambient pressure, leading to boiling of the melt (Melosh, 1989; Basilevsky et al., 1983). Different parts of the melt formed during an impact event can experience very different temperature histories, so even impact glasses formed from more or less homogeneous materials should show variations in final composition across a continuous spectrum. Compositional changes during impact-induced high-temperature volatilization are expected to follow trends analogous to those revealed in the experiments described here.

All of the compositional changes observed in high-temperature glasses produced in these experiments were entirely a result of the selective volatilization of elements. Both experimental results and thermodynamic calculations demonstrate the general evaporative behavior of silicate melts at high temperatures. The main problem in using the experimentally determined trends to interpret natural impact glasses lies in distinguishing the effects of vaporization from potentially similar effects due to the mixing of heterogeneous target materials and/or post-impact alteration of the glasses, e.g., via hydrothermal alteration as the melt cools, diagenetic processes in a sedimentary succession, and/or subsequent metamorphism.

Experiments have clearly shown that dispersed melt spherules show a much higher degree of evaporative modification than larger bodies or sheets of melt sheets that remain in the crater on a sample (see Figs. 5A–5O). Thus, one can expect the effects of evaporative differentiation to be more readily recognized among distant, highly dispersed spherules compared to the melt retained within a crater. On the other hand, relating distal spherules to original compositions is hampered by their being found at great distances from the site of the impact and possibly differing substantially in composition from the target materials, both of which create uncertainties in correlations to parent impact structures.

In many cases, we have to deal with impact melts which were deposited in a crater as rather "cold" and homogenized mixtures of large, melted volumes. An approach we want to propose here is to apply a combination of petrographic similarities and experimentally derived trends to identify objects that have a similar genesis among melts related to one crater. To avoid changes due to post-impact alteration, only fresh glasses can be used for this approach. Once such glasses are divided into petrographically identical groups, experimentally derived trends can be used to determine whether or not evaporative effects are sufficient to explain their genesis. As a case study, we will now demonstrate how such an approach can be applied to impact glasses from the Logoisk impact structure (Glazovskaya et al., 1991).

Impact Glasses from Logoisk Impact Structure (Byelorussia)

Logoisk impact structure is located in Byelorussia ~8–10 km northwest of the city Logoisk. A detailed description of the structure can be found in the monograph by Glazovskaya et al. (1991). The target material of the crater has a two-layer structure consisting of a granite-gneiss basement overlain by sedimentary rocks, mainly sandstones. The mean chemical compositions of the granite-gneiss and sandstones are presented in Table 2. The Logoisk impact is believed to have taken place in the mid-Devonian, and the crater is currently covered by younger sediments and has flat relief. Based on geophysical investigations and the drilling of 28 holes, there is a depression with a diameter of 8–9 km in the basement, but distortion of the Devonian sediments suggest a larger original diameter of 17–22 km. Impact glasses were collected from the drill cores and were divided by Glazovskaya et al. (1991) into two groups: "monomineralic" and "polymineralic." Monomineralic glasses were formed from target rocks with only one type of mineral. In contrast, polymineralic glasses were formed from mixtures of minerals, so they reflect the bulk composition of the target material better. We only used samples of fresh, polymineralic glasses, mostly millimeter-sized fragments, in our search for the effects of vaporization. From 168 analyses of polymineralic glasses made by Glazovskaya et al. (1991), we selected 58 analyses of fresh glasses from suevites in holes drilled no more than ~3 km from the center of the crater. Analyses of these glasses are presented in Table 3. The glasses we selected were divided into five groups based on their petrographic character, as follows: brown glasses (26 analyses), colorless glasses (7), green and gray glasses (11), tawny glasses (8), and spherules (6). Our basic assumptions were that all these glasses were derived from the granite-gneiss target rocks and that the diversity of their compositions mainly reflected evaporative differentiation. We believe this differentiation happened during the short period the high-temperature melts (they "fragmented" later) were in flight in the plume and before they were mixed with unmelted rock and mineral clasts to form suevites. Our aim was to determine whether or not these glasses followed the trends of

TABLE 2. CHEMICAL COMPOSITION OF TARGET ROCKS OF LOGOISK CRATER

	Granite-gneisses		Sandstones	
	Composition*	±σ	Composition†	±σ
SiO_2	68.38	1.98	74.08	5.82
TiO_2	0.64	0.12	0.61	0.69
Al_2O_3	14.40	0.62	11.36	2.33
FeO	4.72	1.04	2.73	1.08
MgO	1.63	0.44	1.30	0.98
CaO	2.00	0.55	1.71	0.85
MnO	0.073	0.33	0.46	0.28
Na_2O	2.72	0.31	0.76	0.50
K_2O	4.65	1.08	6.89	2.36
P_2O_5	0.12	0.04	0.06	0.02

Note: Reprinted from Glazovskaya et al. (1991), Table 5.
*Average of analyses of 15 samples.
†Average of analyses of 9 samples.

TABLE 3. CHEMICAL COMPOSITION OF SELECTED FRESH GLASSES FROM SUEVITES OF LOGOISK CRATER

Description of glasses	Sampling information*	Chemical composition of glasses (wt%)									
		SiO_2	TiO_2	Al_2O_3	FeO	MnO	MgO	CaO	Na_2O	K_2O	Cr_2O_3
brown	111/583	65.55	0.95	17.38	5.09	0.04	1.62	2.30	1.92	5.14	0.00
brown	111/583	65.93	0.83	17.25	4.69	0.29	1.73	2.03	1.94	5.29	0.01
brown	111/583	65.73	0.80	18.02	4.21	0.38	1.02	1.86	2.26	5.71	0.01
brown	117/341	67.27	0.94	16.16	5.10	0.21	1.59	1.56	1.95	5.22	0.00
brown	117/341	66.01	0.90	16.01	5.09	0.44	1.52	1.88	2.12	6.01	0.02
brown	117/341	66.55	0.91	16.20	4.76	0.18	1.71	1.74	2.23	5.58	0.15
brown	117/341	66.62	1.04	16.45	5.00	0.21	1.51	1.61	1.89	5.61	0.05
brown	117/341	66.70	0.81	16.27	4.91	0.48	1.33	1.67	1.83	5.91	0.00
brown	117/341	68.37	0.81	16.03	4.47	0.35	1.37	1.67	1.46	5.46	0.00
brown	23П/380	66.82	0.69	15.86	4.42	0.22	0.82	1.61	3.62	5.91	0.03
brown	23П/380	66.59	0.71	16.27	4.09	0.14	1.00	1.89	2.67	4.66	0.00
brown	23П/380	66.62	0.84	16.26	3.79	0.19	1.17	1.81	3.41	5.73	0.17
brown	23П/380	68.30	0.79	15.65	4.13	0.15	0.83	1.66	3.34	5.16	0.00
brown	23П/380	67.22	0.80	15.94	4.49	0.00	1.05	2.01	3.42	5.08	0.00
brown	23П/380	64.43	0.72	17.31	4.66	0.20	1.45	2.17	3.74	5.18	0.19
brown	23П/380	67.32	0.91	15.45	4.22	0.06	0.78	1.86	3.80	5.60	0.02
brown	23П/380	66.67	0.94	16.11	4.33	0.48	0.67	2.08	3.33	5.39	0.00
brown	23П/380	68.11	0.82	15.59	4.51	0.35	0.82	1.56	2.70	5.40	0.13
brown	23П/380	67.96	0.68	15.64	5.19	0.10	0.39	1.89	2.74	5.39	0.00
brown	109/377	68.42	1.01	15.85	5.28	0.00	0.84	2.07	1.56	4.67	0.31
brown	109/377	66.51	0.48	17.90	3.73	0.24	0.96	3.07	1.92	5.15	0.02
brown	109/377	66.85	0.58	15.44	5.42	0.47	1.66	2.59	2.06	4.93	0.00
brown	109/377	67.51	0.79	14.72	5.88	0.00	1.76	2.43	1.73	5.01	0.19
brown	109/377	68.61	0.78	14.54	5.71	0.33	2.04	2.46	1.43	3.97	0.13
brown	109/377	67.70	0.95	15.80	5.72	0.31	2.07	2.81	1.22	3.64	0.08
brown	109/469.5	65.30	0.90	17.50	5.00	0.20	1.40	3.00	4.90	1.50	0.20
spherules	119/455	65.72	0.68	17.25	4.87	0.44	1.71	2.18	2.13	5.03	0.00
spherules	119/455	67.36	0.74	16.18	4.63	0.10	1.27	1.92	2.02	5.73	0.03
spherules	119/455	66.60	0.93	16.54	4.93	0.02	1.81	1.96	2.04	5.10	0.08
spherules	119/455	66.84	0.76	16.25	4.66	0.10	1.60	2.03	2.13	5.65	0.00
spherules	119/455	64.64	0.81	17.86	4.92	0.11	1.78	2.42	2.17	5.26	0.03
spherules	119/455	67.78	0.71	16.62	4.52	0.55	1.59	1.73	1.53	4.97	0.00
colorless	119/445	65.97	0.78	17.20	4.53	0.52	1.70	1.86	1.74	5.51	0.19
colorless	119/445	66.49	0.79	17.47	4.83	0.25	1.45	2.09	1.28	5.34	0.09
colorless	119/445	65.98	0.97	16.91	4.52	0.30	1.65	2.24	1.51	5.93	0.00
colorless	119/445	66.42	0.96	16.84	4.44	0.28	1.72	1.89	1.76	5.69	0.00
colorless	119/445	66.14	1.15	17.02	4.90	0.13	1.99	1.92	1.38	5.37	0.00
colorless	119/556	64.30	1.00	17.90	4.50	0.03	1.60	2.20	2.10	5.90	0.20
colorless	119/556	65.20	0.90	18.30	5.50	0.30	1.50	2.30	2.30	3.70	0.00
green&gray	119/445	65.81	0.71	18.41	4.26	0.28	1.46	2.44	1.38	5.22	0.02
green&gray	119/445	66.79	0.74	17.34	4.24	0.22	1.55	1.91	1.43	5.79	0.00
green&gray	78/402	67.59	0.96	15.72	4.67	0.27	0.27	2.30	2.86	5.36	0.00
green&gray	78/402	67.58	0.82	14.70	4.86	0.54	0.44	2.07	3.32	5.54	0.13
green&gray	78/402	66.14	1.02	16.56	4.92	0.00	0.42	1.94	3.35	5.65	0.00
green&gray	78/402	67.95	0.91	15.40	4.54	0.48	0.48	1.99	2.72	5.53	0.00
green&gray	78/402	68.66	0.87	15.71	4.63	0.00	0.68	1.96	2.40	5.09	0.00
green&gray	78/402	67.55	0.94	16.33	4.70	0.00	0.66	1.89	2.29	5.49	0.15
green&gray	78/402	69.76	0.73	16.20	4.20	0.26	0.61	1.76	1.42	5.03	0.00
green&gray	78/402	65.77	0.92	16.68	5.03	0.28	1.74	2.04	2.09	5.38	0.05
green&gray	78/402	66.09	0.55	16.64	5.10	0.52	1.42	2.27	2.16	5.26	0.00
tawny	109/383	67.61	0.85	16.25	4.48	0.18	1.61	1.68	2.07	5.19	0.07
tawny	109/383	65.36	0.88	17.60	5.18	0.08	1.64	2.11	1.85	5.28	0.03
tawny	109/383	66.45	0.88	17.07	5.03	0.00	1.20	1.75	1.94	5.46	0.21
tawny	109/383	67.39	0.72	16.24	5.14	0.10	1.36	1.59	2.07	5.35	0.05
tawny	109/383	64.91	0.72	18.12	5.32	0.08	1.61	2.13	2.14	4.93	0.04
tawny	109/383	62.59	0.80	18.52	6.33	0.36	2.26	2.81	1.70	4.62	0.00
tawny	109/383	66.06	1.03	16.96	4.49	0.42	1.61	1.84	2.07	5.49	0.04
tawny	109/383	65.47	0.77	17.41	4.70	0.33	1.52	1.99	2.20	5.62	0.00

Note: Reprinted from Glazovskaya et al. (1991), Table 8. The accuracy of microprobe analyses was ~2% of elements quantity. Totals were adjusted to 100%.

*The number of the drill hole/depth of the sample in the core from the surface.

evaporative differentiation documented in the LP experiments and derived from thermodynamic calculations.

Figure 6A shows a plot of Si versus Al compositions of both target rocks and selected glasses. The solid lines show thermodynamic predictions for isothermal volatilization trends for the granite-gneiss target. The simultaneous increase in Si and Al contents in the beginning of volatilization along the calculated 2000 K isotherm is caused by selective loss of alkalis. The trend of volatilization along the 3000 K isotherm is a particularly good match for the compositions of the glasses selected. The dashed lines define the compositional limits of the experimentally produced glasses within the Si-Al range of the target granite-gneisses compositions. The compositional trends of the natural glasses are in good agreement with those determined experimentally.

Figure 6B is a plot of Ca/Al ratios versus Al within the selected glasses and target rocks. Again, all compositions plot well inside the range defined by the diverse granite-gneiss compositions. Glasses within the groups selected also show good agreement with the conservation line of the Ca/Al ratio. Brown glasses have two distinct populations, one of which is below the conservation line whereas the other is along the uppermost limit of the target granite-gneiss. All of the brown glasses with high Ca/Al ratio come from the same drill core (no. 109), so they may all come from a starting material different in composition from the other brown glasses. It is also worth pointing out that on average, the natural glasses plot slightly below the mean Ca/Al value of the target material, which is something we also observed in our experimental glasses.

Ti is a refractory element and consistently increases in abundance with increasing Al content in the glasses selected (Fig. 6C). The measured values are in agreement with calculated trends.

Mg has a wide variety of concentrations in the glasses selected, especially the brown and green and gray glass groups (Fig. 6D). The natural glasses are generally depleted in Mg, some to a high degree, even though the calculations predict no loss of Mg at any temperatures for the degree of mass loss implied by of the observed range of Al content. Nevertheless, Mg in the natural glasses behaved very much like the Mg in the LP experimental glasses from obsidian (see Fig. 5L).

The Fe versus Al dependency of the glasses selected is shown in Figure 6E. The loss of Fe is not as pronounced in the natural glasses as it was in the LP experimental glasses. Nevertheless, the compositional changes in all of the Logoisk glasses selected are in good agreement with thermodynamically predicted volatilization trends for the target granite-gneiss. Bear in mind that the effect of evaporative differentiation on the Logoisk glasses cannot be as high as it was in the LP experimental glasses because the increase of Al content was much lower. The Al content of the natural glasses increased no more than ~3 wt% (implying ~20% mass loss), whereas it increased up to ~20% (implying ~70% mass loss) in the experimental glasses. In fact, the Fe dependency on Al of the natural glasses (Fig. 6E) compares very favorably with that of the experimental glasses from obsidian within the range where the content of Al increases 3% (Fig. 5J).

K versus Al compositional trends in the glasses selected (Fig. 6F) are in close agreement with thermodynamic predictions for temperatures of 3000–5000 K. The same agreement with thermodynamic predictions for temperatures in the range 3000–5000 K was found for Na versus Al compositions for the selected glasses (Fig. 6G). Na versus Al compositional trends of selected glasses coincide with those of glasses in the LP experiments, whereas K shows greater depletions in the experimental glasses than is observed in the Logoisk glasses.

SUMMARY

Compositions of selected groups of Logoisk impact glasses form trends that agree with results of both LP experiments and thermodynamic evaluation of high-temperature evaporative trends. Figures 6A, 6F, and 6G show that only temperatures higher than 3000 K are consistent with the observed trends. The MAGMA code gives the vapor pressures of the granite-gneiss target as 0.052, 0.35, 31, and 630 bar at temperatures 2000, 3000, 4000, and 5000 K, respectively. This means that the vapor pressure exceeds 1 atm at temperatures slightly greater than 3000 K, at which point vaporization can become quite efficient. In other words, vapor pressures at temperatures below 3000 K are too low to permit significant loss of mass from the melt in a dense atmosphere. The consistency of compositional trends of Logoisk impact glasses with evaporative trends requiring temperatures in excess of 3000 K is strong evidence that selective evaporation of elements is responsible for the compositional variation observed in these impact glasses.

The effect of volatilization on the Logoisk impact glasses involves element concentrations changes of only several wt%, which is rather small compared to the full compositional range of all impact glasses. The Logoisk glasses came from within the crater, so they probably do not represent highly processed material. The effects of volatilization on the Logoisk glasses are more comparable to the effects of volatilization on the crater melt than the spherules in the experiment with obsidian. Forming such a diverse suite of glass compositions by simply mixing heterogeneous target materials seems unlikely. The compositional limits of the starting granite-gneiss materials are defined by error bars for Ti (Fig. 6C), Mg (Fig. 6D), and Na (Fig. 6G), and the compositions observed in the selected glasses exceed these limits in ways that are consistent with volatilization trends. If the mixing of granite-gneiss with sandstone were responsible for the decreased Mg content in some glasses (Fig. 6D), it would also affect Fe (Fig. 6E) and K (Fig. 6F) to the same degree, and this is not observed.

The LP experiments clearly demonstrate that rapid and significant differentiation of silicate melts is possible during high-temperature flash heating via the selective volatilization of elements. The efficiency of selective volatilization depends mainly on the individual volatility of elements present in the system. There are three main groups of elements: highly volatile alkalis (Na and K), moderately volatile (Si, Fe and Mg), and

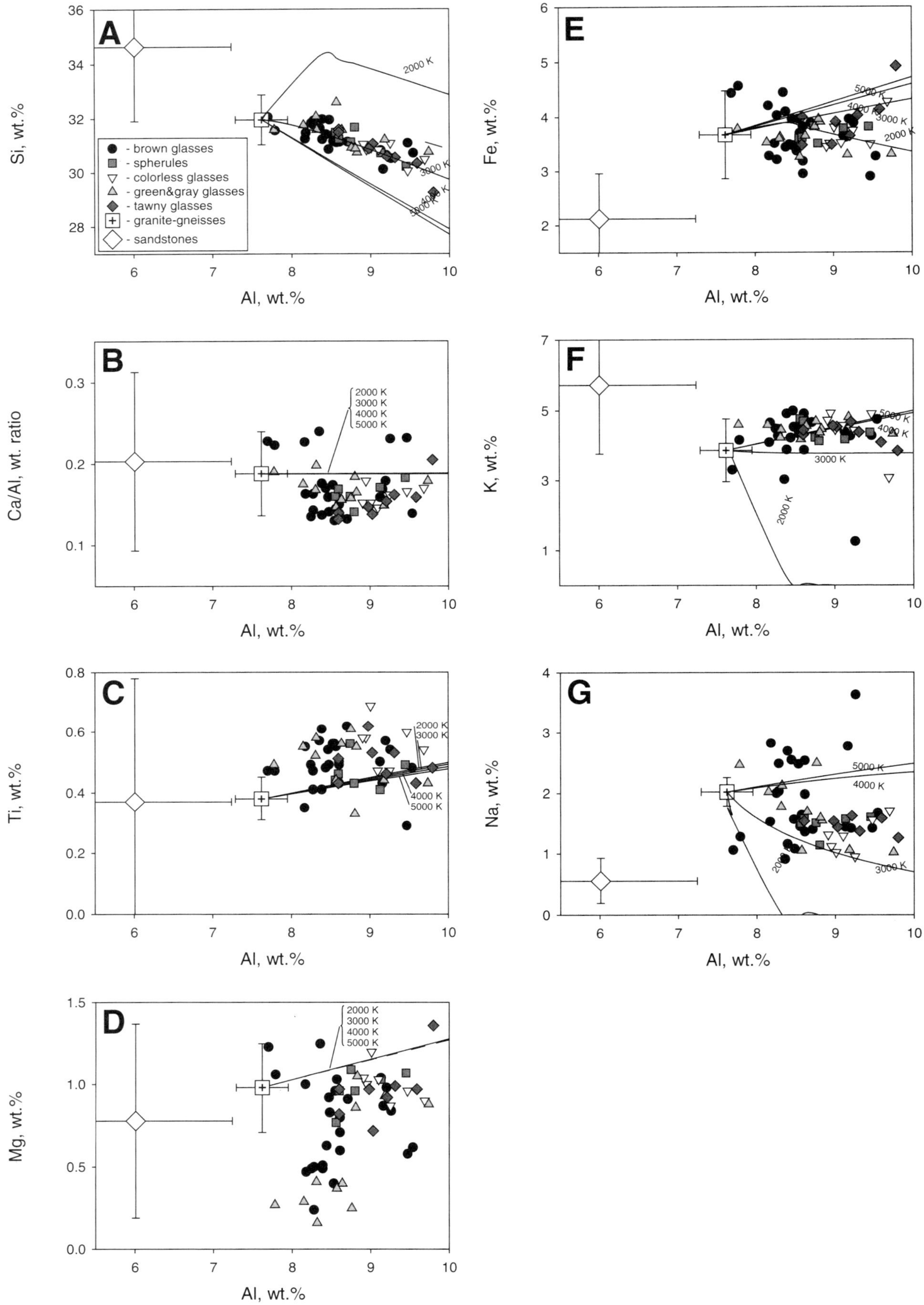

Figure 6. Chemical compositions of selected groups of impact glasses from suevites and of target granite-gneisses and sandstones from the Logoisk impact structure. Figures A, B, C, D, E, F, and G show the Si, Ca/Al, Ti, Mg, Fe, K, and Na vs. Al graphs, respectively. Solid lines show the isothermal behavior of the average granite-gneisses composition at indicated temperatures calculated using the MAGMA code.

refractory (Ti, Ca, and Al). Alkali elements escape from high-temperature silicate melts most efficiently, but the rates of volatilization of alkali elements relative to those of refractory elements decrease significantly as temperature increases. At temperatures over ~4000 K, alkalies are not completely lost even when half the mass of the melt is vaporized, while at temperatures below ~2000 K alkalis are completely vaporized before any significant loss of other elements takes place. Moderately volatile elements such as Si, Fe, and Mg can also volatilize noticeably during an impact. Si and Fe consistently show progressive depletion as mass loss increases at high temperatures, but the behavior of Mg is more complicated. Silicate melts with compositions similar to those used in the experiments are enriched in Mg when vaporization first begins due to the preferential loss of Si, Fe, and alkalis. In later stages, Mg escapes from the melt more efficiently as its composition becomes richer in Ca and Al and poorer in Si, to the point where Mg volatilizes and departs from silicate melts more rapidly than Si. All refractory elements consistently become more concentrated as melts lose mass by the preferential vaporization of elements with higher volatility. After the volatile elements are lost, competition between the remaining refractory elements results in the preferential loss of Ti. Some preferential depletion of Ca was also detected in the experiments, which could reflect the loss of Ca bound in molecular cluster complexes. Highly depleted residua melts were not investigated in this study.

High-temperature melts from homogeneous targets, like the ones produced in these experiments, have a wide range of possible compositions. In such cases, the variability of Mg, Na, and K versus Al in melt compositions (see Figs. 5K–5O) is entirely a result of temperature variations during the vaporization process. Every melt fragment experiences its own unique history of temperature change, and the local composition of the melt evolves through time as its temperature changes according to the successive sequence of isotherms. The final extent of the depletion in each particular melt fragment is the integral result of its exposure to a range of high temperatures, so the final melt can wind up anywhere between the maximum and minimum temperature evolutionary trends across the whole range of compositions. Hence, the formation of melts with a wide range of Mg, Na, and K versus Al compositions is a regular and predictable consequence of the high-temperature evaporative differentiation of silicates.

Experiments show that high-temperature volatilization of silicate melts can increase variability in the concentrations of refractory elements relative to starting materials to a much higher degree than would be inferred from thermodynamic calculations (see Figures 5A–5D). The increase in variability is common to both glass spherules and crater melts. It cannot be explained by a simple model invoking diverse temperature histories and the mixing of melts whose compositions differ as a result of different degrees of mass loss because there is no intrinsic mechanism for the differential loss of refractory elements. The effect may be a result of volatility of molecular cluster complexes not currently incorporated into the calculations and/or interactions between the melt and the dense vapor cloud and deserves further investigation.

Chemical analysis of the crater melts formed in the LP experiments show that the depletion of volatile elements is not depth-dependent. If elements were being volatilized preferentially close to the surface of the melt, this would produce elemental gradients in the glass that were not observed. An increase was detected in the volatile content (e.g., Na) within a micron of the surface layer, but this may be due to the back precipitation of vapor during the cooling of the melt. The loss of volatile elements from the melt is a volumetric process whose main mechanism is the release of volatile-rich components into vapor bubbles forming inside the boiling melt. These escape by coalescing into larger bubbles until they disrupt the main body of the melt. The boiling of the melt also produces small droplets of melt which achieve much higher degrees of depletion during their flight in the expanding vapor cloud. A bubble growing inside a melt is a closed system with high vapor pressure inducing a strong flow of volatile components back into the melt. This interaction of the vapor and the melt helps keep the vapor-melt system well equilibrated, which helps to explain why the experimental crater melt compositions are in such good agreement with the thermodynamic calculations. Melt droplets in the expanding vapor cloud experience a rapid pressure decrease, which induces higher rates of volatilization from the surface of the melt and, as a result, higher degree of depletion. Vapor-melt equilibration in melt droplets is a question of kinetics dependent on pressure and temperature variations and, if it is ever achieved, it terminates at some points as the vapor cloud rapidly expands.

The fact that the compositions of selected groups of Logoisk impact glasses show good agreement with experimentally determined trends of evaporative differentiation in high-temperature silicate melts offers strong evidence that the Logoisk glasses experienced impact-induced evaporative modification. Nevertheless, confirming the genesis of these glasses via evaporative differentiation requires more detailed analyses, including studies of their trace elements content, thermodynamic modeling, etc. Evaporative differentiation is only one of many effects influencing the compositions of impact-produced glasses. Other important effects such as the heterogeneity of the target and projectile materials, the dynamics of their mixing, and post-impact alteration via diagenesis and/or metamorphism must be also taken into account. Our aim here was to demonstrate an approach that can be used to detect evaporative effects among glasses from large impact structures on Earth, Moon, and other planets. We found the consistency of the Ca/Al ratio (the "conservation line") and the anti-correlation of Si versus Al to be the most telling signs of evaporative differentiation. It also proved worthwhile to perform the experiments using materials with compositions similar to those of likely targets and projectiles and to involve thermodynamic calculations. Our results may not be applicable to large volumes of impact melt as they are unlikely to boil off to any large degree. However, they should be important in the interpretation of impact-produced glasses found in distal ejecta since such melts are much more likely to have undergone extensive evaporative differentiation during their long ballistic flight. We hope that the proposed method will be helpful in such investigations.

ACKNOWLEDGMENTS

We thank Christian Koeberl and Dominik Hezel for reading an earlier version of the paper and for their valuable suggestions that led to additions and improvements to the manuscript. Special thanks to Bruce Simonson for his valuable editorial work to improve the manuscript. This research was supported by the Russian Fund of Basic Research grant 04-05-64338.

REFERENCES CITED

Armstrong, J.T., and Buseck, P.R., 1975, Quantitative chemical analysis of individual microparticles using the electron microprobe: Theoretical: Analytical Chemistry, v. 47, p. 2178–2192.

Artemieva, N.N., and Shuvalov, V.V., 2002, Extraterrestrial material deposition after impacts into continental and oceanic sites, *in* Buffetaut, E., and Koeberl, C., eds., Impact Studies: Berlin, Heidelberg, Springer-Verlag, p. 249–263.

Basilevsky, A.T., Ivanov, B.A., Florensky, K.P., Yakovlev, O.I., Feldman, V.I., and Granovsky, L.V., 1983, Impact craters on the Moon and planets: Moscow, Nauka, 200 p. (in Russian).

Fegley, B., Jr, and Cameron, A.G.W., 1987, A vaporization model for iron/silicate fractionation in the Mercury protoplanet: Earth and Planetary Science Letters, v. 82, p. 207–222, doi: 10.1016/0012-821X(87)90196-8.

Gerasimov, M.V., 2002, Toxins produced by meteorite impacts and their possible role in a biotic mass extinction, *in* Koeberl, C., and MacLeod, K.G., eds., Catastrophic Events and Mass Extinctions: Impacts and Beyond: Geological Society of America Special Paper 356, p. 705–716.

Gerasimov, M.V., Ivanov, B.A., Yakovlev, O.I., and Dikov, Yu.P., 1999, Physics and chemistry of impacts, *in* Ehrenfreund, P., Krafft, K., Kochan, H., and Pirronello, V., eds., Laboratory Astrophysics and Space Research: Astrophysics and Space Science Library: Dordrecht, Kluwer Academic Publishers, v. 236, p. 279–330.

Gerasimov, M.V., Dikov, Yu.P., Yakovlev, O.I., and Wlotzka, F., 2000, The formation of differentiated melted spherules during an impact: Meteorite Impacts in Precambrian Shields: 4th ESF Workshop, Geological Survey of Finland and University of Helsinki, p. 54.

Gerasimov, M.V., Dikov, Yu.P., Yakovlev, O.I., and Wlotzka, F., 2001, Reduction of Iron During an Impact: Impact markers in the stratigraphic record: 6th ESF-Impact Workshop, CSIC-UGR, Granada, Spain, Abstracts, p. 31–32.

Gerasimov, M.V., Dikov, Yu.P., Yakovlev, O.I., and Wlotzka, F., 2002, Experimental investigation of the role of water in the impact vaporization chemistry: Deep-Sea Research. Part II, Topical Studies in Oceanography, v. 49, no. 6, p. 995–1009, doi: 10.1016/S0967-0645(01)00138-2.

Glazovskaya, L.I., Gromov, E.I., Parfenova, O.V., and Il'kevich, G.I., 1991, Logoisk astrobleme: Nauka press, Moscow, 135 p. (in Russian).

Grieve, R.A.F., 1987, Terrestrial Impact Structures: Annual Review of Earth and Planetary Sciences, v. 15, p. 245–270, doi: 10.1146/ANNUREV.EA.15.050187.001333.

Grieve, R.A.F., Dence, M.R., and Robertson, P.B., 1977, Cratering process: As interpreted from the occurrence of impact melts, *in* Roddy, D.J., Pepin, R.O., and Merrill, R.B., eds., Impact and Explosion Cratering: Planetary and Terrestrial Implications: New York, Pergamon Press, p. 791–814.

Hörz, F., Fechtig, H., Janike, J., and Schneider, E., 1983, Morphology and chemistry of projectile in small experimental impact craters, Proceedings of the 14th Lunar and Planetary Science Conference: Journal of Geophysical Research, v. 88, Supplement, p. B353–B363.

Ivanov, B.A., Badukov, D.D., Yakovlev, O.I., Gerasimov, M.V., Dikov, Yu.P., Pope, K.O., and Ocampo, A.C., 1996, Degassing of sedimentary rocks due to Chicxulub impact: Hydrocode and physical simulations, *in* Ryder, G., Fastovsky, D., and Gartner, S., eds., The Cretaceous-Tertiary event and other catastrophes in Earth history: Geological Society of America Special Paper 307, p. 125–139.

Jarosewich, E., Nelen, J.A., and Norberg, J.A., 1980, Reference samples for electron microprobe analysis: Geostandards Newsletter, v. 4, p. 43–47.

Lange, M.A., and Ahrens, T.J., 1982, The evolution of an impact-generated atmosphere: Icarus, v. 51, p. 96–120, doi: 10.1016/0019-1035(82)90031-8.

Masaitis, V.L., Danilin, A.N., Mashchak, M.S., Raykhlin, A.I., Selivanovskaya, T.V., and Shadenkov, Ye.M., 1980, The Geology of Astroblems: Leningrad, Nedra, 231 p. (in Russian).

Melosh, H.J., 1989, Impact Cratering, A Geologic Process: New York, Oxford University Press, 245 p.

Melosh, H.J., and Vickery, A.M., 1991, Melt Droplet formation in energetic impact events: Nature, v. 350, p. 494–497, doi: 10.1038/350494A0.

O'Keefe, J.D., and Ahrens, T.J., 1977, Impact-induced energy partitioning, melting, and vaporization on terrestrial planets, *in* Proceedings of the Eighth Lunar Science Conference, p. 3357–3374.

Schulte, P., Stinnesbeck, W., Stüben, D., Kramar, U., Berner, Z., Keller, G., and Adatte, T., 2003, Fe-rich and K-rich mafic spherules from slumped and channelized Chicxulub ejecta deposits in the northern La Sierrita area, NE Mexico: International Journal of Earth Sciences, v. 92, no. 1, p. 114–142.

Simonson, B.M., Koeberl, C., McDonald, I., and Reimold, W.U., 2000a, Geochemical evidence for an impact origin for a Late Archean spherule layer, Transvaal Supergroup, South Africa: Geology, v. 28, p. 1103–1106, doi: 10.1130/0091-7613(2000)0282.3.CO;2.

Simonson, B.M., Hassler, S.W., and Beukes, N.J., 2000b, Late Archean impact spherule layer in South Africa that may correlate with a Western Australian layer, *in* Dressler, B.O., Sharpton, V.L., eds., Large Meteorite Impacts and Planetary Evolution II: Geological Society of America Special Paper 339, p. 249–261.

Simonson, B.M., and Glass, B.P., 2004, Spherule Layers—Records of Ancient Impacts: Annual Review of Earth and Planetary Sciences, v. 32, p. 329–361, doi: 10.1146/ANNUREV.EARTH.32.101802.120458.

Valter, A.A., and Ryabenko, V.A., 1977, Explosive Craters on Ukrainian Shield: Kiev, Naukova Dumka, 154 p. (in Russian).

Yakovlev, O.I., Dikov, Yu.P., and Gerasimov, M.V., 1992, Shock vaporization of granites: Experimental results: Lunar and Planetary Science XXIII, Lunar and Planetary Institute, Houston, Texas, Abstracts, p. 1557–1558.

Yakovlev, O.I., Dikov, Yu.P., and Gerasimov, M.V., 1995, Experimental study of shock and pulse evaporation of ultrabasite: Geokhimiya, no. 8, p. 1235–1247.

Yakovlev, O.I., Dikov, Yu.P., and Gerasimov, M.V., 2000, Differentiation caused by impact-induced vaporization during the Earth's accretion: Geochemistry International, v. 38, no. 10, p. 937–954.

Yakovlev, O.I., Dikov, Yu.P., Gerasimov, M.V., Wlotzka, F., and Huth, J., 2002, The behavior of Pt in silicate melts during impact-simulated high temperature heating: Lunar and Planetary Science XXXIII: Houston, Texas, Lunar and Planetary Institute, Abstract no. 1271 (CD-ROM).

Zel'dovich, Ya.B., and Raizer, Yu.P., 1967, Physics of Shock Waves and High Temperature Hydrodynamic Phenomena: New York, Academic Press, 916 p.

Manuscript Accepted by the Society 9 August 2004

Geological Society of America
Special Paper 384
2005

Geochemistry of target rocks, impact-melt particles, and metallic spherules from Meteor Crater, Arizona: Empirical evidence on the impact process

David W. Mittlefehldt*
Friedrich Hörz
SR, National Aeronautics and Space Administration Johnson Space Center, Houston, Texas 77058, USA

Thomas H. See
C23, Lockheed Martin SO, 2400 Nasa Road 1, Houston, Texas 77058, USA

Edward R.D. Scott
Hawai'i Institute of Geophysics and Planetology, University of Hawai'i, Honolulu, Hawai'i 96822, USA

Stanley A. Mertzman
Department of Geosciences, Franklin & Marshall College, Lancaster, Pennsylvania 17604, USA

ABSTRACT

We have done lithophile- and siderophile-element analyses of a large suite of target rocks, ballistically dispersed impact-melt particles and ballistically dispersed metallic spherules from Meteor Crater, Arizona. The Moenkopi Formation (topmost unit) has a unique lithophile-element signature that confirms it as a major component of the impact-melt particles. The Kaibab Formation is very heterogeneous, containing dolomite-rich and quartz-rich layers. The lithophile-element compositions of the impact-melt particles can be entirely explained as mixtures of Moenkopi and Kaibab depleted in CO_2. The Toroweap and Coconino Formations (lowest units) are not required components, but small contributions from them cannot be excluded. We conclude the impact-melt particles were formed entirely in the upper portion of the section, above the lower two units.

The impact-melt particles average ~14 wt% projectile material. Most siderophile-element ratios of the impact-melt particles are unchanged from those of the projectile. Many samples are depleted in Au; the most extreme depletions are in impact-melt particles with the highest Kaibab component. Kaibab rocks are highest in Br, and we suggest loss of volatile Au halides may have caused the fractionation.

Ballistically dispersed metallic spherules are enriched in Co, Ni, Ir, and Au compared to Canyon Diablo metal. Element/Ni ratios deviate slightly from projectile ratios, and are inversely correlated with susceptibility to oxidation relative to Ni. We attribute this to partial oxidation of molten metal spherules during flight. Spherule compositions suggest some selective melting of graphite-troilite-schreibersite inclusions of the projectile, consistent with enhanced shock melting of these lower density phases.

Keywords: Meteor Crater, Arizona, impact-melt particles, geochemistry, target rocks, siderophile elements, projectile-target mixing

*Corresponding author: david.w.mittlefehldt@nasa.gov

Mittlefehldt, D.W., Hörz, F., See, T.H., Scott, E.R.D, and Mertzman, S.A., 2005, Geochemistry of target rocks, impact-melt particles, and metallic spherules from Meteor Crater, Arizona: Empirical evidence on the impact process, *in* Kenkmann, T., Hörz, F., and Deutsch, A., eds., Large meteorite impacts III: Geological Society of America Special Paper 384, p. 367–390. For permission to copy, contact editing@geosociety.org.

"Special features, such as the presence of layering or volatiles in the target are not shown here. The effects of such circumstances are largely unknown at present."
—H.J. Melosh, *Impact Cratering: A Geologic Process,* p. 48

INTRODUCTION

The study of impact processes has progressed along a five-pronged front: (*i*) field and petrologic study of terrestrial impact craters and their products; (*ii*) laboratory-scale impact experiments; (*iii*) large-scale explosion experiments; (*iv*) theoretical modeling of the impact process; and (*v*) astronomical and spacecraft study of craters on planets, moons and asteroids (see Melosh, 1989). Geochemical studies have contributed to the identification of meteoritic signatures in impact melts (e.g., King et al., 2002), attempted to identify the type of impactor (Morgan et al., 1975; Palme, 1982; Palme et al., 1978), or established the process of mixing of target rocks to form unusual impact-melt rocks (e.g., Dence, 1971; Grieve, 1982; Simonds et al., 1978). Chemical studies aimed at a more detailed understanding of the cratering process have been limited for several reasons. Older structures have been eroded and/or buried, making detailed reconstruction of the pre-impact chemical stratigraphy difficult. Impact-melt sheets in larger structures may have suffered hydrothermal alteration (see Hagerty and Newsom, 2003), which affected the composition of the melt sheet. Finally, older and larger structures do not have preserved meteoritic material available for study (Grieve, 1991), precluding unambiguous identification of the impactor. As a consequence, the depth of the melt zone is poorly defined, and little is known regarding fractionation of impactor and target rocks during melting.

A few younger, smaller impact craters with preserved projectile material have been studied. Attrep et al. (1991) did geochemical studies of a few samples from Henbury and Wolfe Creek Craters. These authors showed that siderophile-element ratios in "impactites" are fractionated relative to those of the preserved projectile material. We did an extensive study of samples from Wabar Crater that confirmed that siderophile-element fractionations occurred during formation of impact-glass particles (Mittlefehldt et al., 1992). We suggested that because lithophile elements were not fractionated in the glasses, the siderophile-element fractionation must have occurred before mixing of projectile and target materials. We also inferred that projectile material was preferentially mixed with the upper stratigraphic horizons of the target.

We have continued our study of natural impacts through field, petrologic, and geochemical studies of Meteor Crater, Arizona, the largest crater with preserved projectile material (Grieve, 1991). Meteor Crater is almost ideal for this study. The stratigraphy is variegated, allowing a more detailed look at the projectile-target interaction; it is young and fresh; abundant impact-melt materials already reside in collections; and remnants of the projectile have been well characterized. The results of our detailed study of the mineralogy and major-element chemistry of the target rocks and the petrology of the impact glasses have been reported (Hörz et al., 2002; See et al., 2002). Here we present our completed study of lithophile and siderophile elements for target rocks, impact-melt particles, and metallic spherules.

GEOLOGIC SETTING AND BACKGROUND

Target Rocks

Meteor Crater is one of the best-studied terrestrial impact structures. The synopsis given here is taken from Roddy (1978), Roddy et al. (1975) and Shoemaker (1963; 1987). See et al. (2002) presented details of the major-element chemistry and mineralogy of a stratigraphic sequence systematically collected from the crater walls.

Meteor Crater was formed in a sequence of flat-lying Mesozoic and Paleozoic sediments. The uppermost unit, the Triassic Moenkopi Formation, consists of calcareous sandstone and siltstone, estimated to have averaged ~8.5 m in thickness in the target area (Roddy, 1978). The Moenkopi is more carbonate-rich at the top and more quartz-rich at the bottom (See et al., 2002). Below the Moenkopi is the Permian Kaibab Formation, an ~80 m thick carbonate sequence consisting largely of sandy dolomite, but including some sandstone. The composition of this unit is highly variable, reflecting wide ranges in quartz and dolomite content; calcite is a minor component. The upper half of the unit is richer in carbonate than is the lower half (See et al., 2002). Below the Kaibab is the ~1.5 m thick Toroweap Formation consisting dominantly of dolomitic sandstone. The crater bottoms in the ~220 m thick Coconino Formation of very pure sandstone (Kieffer, 1971). Below the Coconino is the Permian Supai Formation of sandstone and siltstone. There is no evidence that the transient crater excavated this formation.

Canyon Diablo Iron Meteorite

The projectile was the Canyon Diablo IAB iron meteorite, estimated to have been ~33 m in diameter for an assumed impact velocity of 15 km/sec (Roddy, 1978). The IAB irons, especially Canyon Diablo, have been extensively studied. Wasson and Ouyang (1990) have done a detailed compositional study of Canyon Diablo, while Choi et al. (1995) have presented a thorough study of the composition of the metal phase of most IAB irons. Choi et al. (1995) present an average composition of Canyon Diablo metal which we use for our discussions (Appendix 1). The composition of the Canyon Diablo iron, with 70 mg/g Ni, is typical of that of the majority of IAB irons (Choi et al., 1995).

Buchwald (1975) gave a detailed description of the petrology of Canyon Diablo. It is a coarse octahedrite containing ~8.5% by volume of graphite-troilite inclusions. These inclusions have varying ratios of troilite to graphite, but Buchwald (1975) estimated that on average, they have roughly equal amounts of troilite and graphite by volume. The inclusions are typically surrounded by schreibersite and cohenite (Benedix et al., 2000; Buchwald, 1975). Silicate inclusions are present in many members of the IAB group and are mostly chondritic in composition (e.g., Benedix et al., 2000; Bunch et al., 1970; Mittlefehldt et al., 1998). Buchwald (1975) reported that rare silicates are present in some graphite-troilite nodules in Canyon Diablo.

Meteor Crater and Impact Products

Meteor Crater is a bowl-shaped impact crater ~1 km in diameter. Structural controls in the target rocks give it a squarish shape in map view (Roddy, 1978). The crater was formed ~50 k.y. ago (Nishiizumi et al., 1991; Phillips et al., 1991). Shocked and frothy Coconino sandstone occurs in the crater floor below a layer of alluvium, but no continuous impact-melt sheet is present (Roddy et al., 1975). Shocked and brecciated target rock occurs on the crater floor, in the walls and in the overturned flap (Roddy et al., 1975). Impact melt composed of mixtures of target rock and projectile are confined to particles up to a few cm in size ballistically dispersed over the surrounding plains (Nininger, 1956). Metallic spherules a few mm in diameter are also found on the surrounding plains (Nininger, 1956). These are melt-droplets of the Canyon Diablo iron and are not mixed with target rock (Blau et al., 1973). Some samples of Canyon Diablo show petrographic evidence for shock deformation superimposed on the normal structure of the iron. These are solid fragments of the impactor spalled from the backside of the projectile as the shock wave reflected off the free surface (Buchwald, 1975).

SAMPLES AND ANALYTICAL METHODS

Samples

We have analyzed three types of materials for this study: (*i*) representative samples of target rocks, mostly from within the crater or the overturned flap, (*ii*) ballistically dispersed impact-melt particles, and (*iii*) ballistically dispersed metallic spherules. A few of the target-rock samples analyzed early on are representative hand samples of the Kaibab and Coconino Formations, but are from undocumented locations. A set of documented samples from the crater was collected to obtain a representative suite of target rocks; these are briefly described in Appendix 2. A final set of samples was collected to obtain a continuous stratigraphic sequence from within the crater for geochemical and mineralogical characterization. Major-element analyses by X-ray fluorescence (XRF), and mineralogy determined by X-ray diffraction (XRD) of these samples are given in See et al. (2002). The impact-melt particle and metallic spherule samples were obtained from the collection of the Center for Meteorite Studies, Arizona State University. H.H. Nininger collected these samples from the plains surrounding the crater, but their precise collection locations are not known. The impact-melt particles were selected by one of us (FH) to represent the range of materials in the collection. Brief macroscopic descriptions of these are given in Appendix 3. The petrology of the impact-melt particles has been discussed in detail in Hörz et al. (2002).

Analytical Methods

The target-rock and impact-melt particle samples were analyzed by instrumental neutron activation analysis (INAA) at Johnson Space Center (JSC) in a series of irradiations. The target rocks labeled Ma, Ka, Ta, Ca, etc., are splits of the pooled subsection samples of See et al. (2002). For the other target rocks, several grams of each sample were ground and homogenized, and splits taken for analysis. Some of the impact-melt particles contained variable amounts of adhering alteration material, or fine-grained soil particles. These contaminants were removed by scraping with a dental pick and ultrasonication in distilled water. The impact-melt particles were coarsely crushed and the cleanest material was handpicked. Many of the impact-melt particles are finely vesicular, and alteration material was found even in interior vesicles. (It is possible that the alteration phases in interior vesicles represent material mobilized during quenching and cooling of the particles, and not later, secondary alteration phases.) It proved impossible to eliminate all alteration material. The occurrence of alteration material in tiny interior vesicles indicates that it likely formed from elements leached out of the glass. Samples of alteration material from three impact-melt particles were analyzed in order to evaluate the potential effects of this material on analyses. For large impact-melt particles, clean material was crushed and homogenized and splits taken for INAA. For small particles, the entire clean sample was used for INAA.

Samples, standards, and international standard rocks used as controls were sealed in ultra-pure silica glass tubes, and irradiated at the University of Missouri Research Reactor Facility. The samples were counted several times after irradiation to obtain data for nuclides of differing half-lives. Irradiation times, neutron fluxes, and counting schedules were slightly different for the different irradiations. Data reduction was done using standard JSC procedures (Mittlefehldt and Lindstrom, 1991, 1993). The impact-melt particles and some target rocks have high Mg and low Na contents. A correction to the data was applied to account for Na produced by (n,p) reactions on Mg. The ultra-pure silica contains a tiny amount of La that can significantly affect the data for very low La samples. A blank correction was made when needed.

The metallic spherules were analyzed by INAA at the University of California at Los Angeles (UCLA) in the mid 1970s. The samples consisted of metal cores with oxide coatings. Following Kelly et al. (1974), the samples were abraded in a mixer-mill with a sandpaper lining until the oxide coatings appeared to have been removed. The spherule samples, six samples of the North Chile hexahedrite (IIA iron) Tocopilla serving as standards, and the IVA iron meteorite Rembang serving as a control, were irradiated at the Ames Laboratory Research Reactor at a flux of 2×10^{13} n cm^{-2} sec^{-1} for 10 h, and counted several times to obtain data for elements of differing half-lives. Subsequently, four of the samples were reirradiated at the UCLA reactor for radiochemical determination of Ga and Ge following the procedures of Wasson and Kimberlin (1967).

Splits of the larger impact-melt particles were powdered and fused to glass on a Mo metal strip in an Ar atmosphere (Brown, 1977) for major element analysis by electron microprobe (EMPA). We experienced two problems using this procedure. First, the impact-melt particles are not in internal equilibrium, especially with respect to oxidation state—they contain Fe^{3+}, Fe^{2+}, and Fe^0 (Hörz et al., 2002). This caused some oxidation of Mo which dissolved in the glass at up to a few wt%. Second, some of the Fe remained in the metallic state, and was not part of the homogeneous glass. Thus, there is poor agreement between FeO determined by INAA and EMPA. However, major-element data are only used to determine which of the target rocks (all Fe-poor) contributed to the impact-melts. Most of the Fe in the impact-melt particles is derived from the impactor (See et al., 2002; this study). To first order, FeO can be ignored in evaluating target-rock contributions to the impact-melt particles. The glass beads were analyzed with the JSC SX100 electron microprobe using a 15 kV potential, 15 nA sample current, and the electron beam rastered over a 10 × 10 μm area.

See et al. (2002) presented major-element analyses by XRF of many of the target rocks studied here. Additional target-rock samples plus a composite of 15 impact-melt particles were analyzed as part of this study. The analytical methods follow standard procedures used at Franklin & Marshall College, described in detail by Boyd and Mertzman (1987) and Mertzman (2000). An exception to these procedures is that the samples were first ground and homogenized in an agate mortar and pestle at JSC, and splits sent to Franklin & Marshall College for analysis. Final grinding and sieving were done at Franklin & Marshall. Major-element contents were determined by XRF analysis, with all Fe determined as Fe_2O_3. The ferrous iron content was determined by titration, and loss on ignition was determined by heating an aliquot at 950 °C for one hour.

RESULTS

Table 1 presents the major-element analyses of the target rocks plus the composite impact-melt particle (H11; see Appendix 3). The XRF analysis determines major elements on a volatile-free basis, and all Fe as Fe_2O_3. In Table 1, the data have been recast to a volatile-bearing basis, and with Fe split between FeO and Fe_2O_3 as determined by titration. The low sum quoted for the composite impact-melt particle is an artifact of the recalculation—the XRF sum was 99.31. The INAA data on target rocks, including those studied by See et al. (2002) are given in Table 2. Our INAA data on impact-melt particles are given in Table 3, and major-element analyses of select impact-melt particles determined by EMPA on fused beads are presented in Table 4. These latter analyses are normalized to 100%. The analyses included variable amounts of MoO_2 (<0.01–4.56 wt%) derived from the strip heater, and the Fe in the glass likely is a mixture of FeO and Fe_2O_3. These analyses are used to evaluate which target rocks dominate in the impact-melt particles, and devolatilized, Fe-free (projectile-free) compositions are used for this. Table 5 contains our INAA data on the ballistically dispersed metallic spherules. We have also done INAA on a few miscellaneous samples from the impact-melt particle suite—a lithic clast, samples of secondary alteration products and an oxide bead. These data are given in Appendix 4 for completeness.

DISCUSSION

Comparing geochemical data on the target rocks with those on impact-melt particles is not straightforward. The high-temperature formation of the impact-melt particles caused devolatilization of carbonates and hydrous phases of the target components, and mixing with projectile material diluted the target components. To overcome these problems, we use normalized compositions. The impact-melt particle compositions are normalized to an Fe-Ni-Co-free basis. The volatile content of these particles is not known, but they should be nearly volatile-free. We use the loss on ignition (LOI) determination on the composite of 15 impact-melt particles (H11, Table 1) to correct for volatile content. The target-rock compositions are normalized to a volatile-free basis using the LOI determinations, and to an iron-free basis to make them directly comparable to the impact-melt particles. These corrected impact-melt and target-rock compositions are referred to as adjusted compositions. The impact-melt particles studied here are bulk particles, including glass, quenched mineral phases, undigested clasts, and remnant metal and sulfide beads, and thus are not directly comparable to the pure glass compositions reported by Hörz et al. (2002).

Geochemical Stratigraphy of Meteor Crater

The upper portion of the target stratigraphy shows substantial major-element compositional heterogeneity (Hörz et al., 2002; See et al., 2002). The Moenkopi Formation shows generally decreasing CaO and LOI, and increasing SiO_2 with depth, indicating increasing sand and decreasing carbonate. The Kaibab exhibits substantial fluctuation in SiO_2 and MgO+CaO throughout the section reflecting variations in quartz and dolomite. On a finer scale, we found that the quartz content of the Kaibab varies from ~100% to ~15% within the section (Hörz et al., 2002; See et al., 2002). These variations are non-systematic—the Kaibab sample with the lowest quartz content is adjacent to a sample with one of the highest (see Hörz et al., 2002; Figure 15; samples K106.2 and K104.7). The Toroweap and Coconino are distinguishable from other target rocks by their high SiO_2 contents.

Variations in adjusted major-element content of the target rocks with depth in the crater are shown in Figure. 1. The Moenkopi is distinct in having much higher Ti and Al contents, while the Kaibab is distinct in having a much higher Mg content, and generally lower but variable Si content. The high Al content of the Moenkopi indicates a higher terrigenous component (clays). This is also seen in Sc, rare-earth elements, and Ta contents

(Fig. 2). The two lowest Moenkopi samples are also rich in Hf, but the upper portion has Hf contents like those of the Kaibab. The Kaibab is distinct in showing a general increase in U content with depth, and the lower portion of the section has the highest U content in the depth profile (Fig. 2).

The Toroweap and Coconino have lower adjusted contents of all major and trace elements, excluding Si (Figs. 1 and 2). Hence, these units would only act as diluents in lithophile-element mixing relations should they be a component of the impact-melt particles. (The Toroweap is a minuscule volume fraction of the section, and could not significantly affect the composition of impact-melt particles.)

Lithophile Elements and Target Rock Mixing

Kargel et al. (1996) suggested that impact-melt particles from Meteor Crater are composed of mixtures of projectile and rock from the Kaibab Formation, with no clear signature for either Moenkopi or Coconino rocks. Our preliminary synopsis of target rock and impact-melt particle compositions led us to suggest that both Moenkopi and Kaibab strata are important terrestrial components of the particles (Mittlefehldt et al., 2000). This was supported by our modeling of the major-element compositions of glasses contained in the impact-melt particles. We identified three distinct melt compositional types and concluded that mixtures of Moenkopi and Kaibab in variable proportions could explain the glasses, with Moenkopi composing as much as half of the target component in some (Hörz et al., 2002). We could not exclude Toroweap or Coconino as minor components of some glasses, however. We suggested that the melt zone was a relatively small fraction of the transient crater volume, and that melting occurred at depths of <30 m for many, and possibly all, of the glasses (Hörz et al., 2002).

Mixing relations for major elements Mg-Al, Ca-Al and Ti-Si between target rocks and bulk impact-melt particles are shown in Figure 3. All impact-melt particles fall between the fields of Moenkopi and Kaibab samples, and none have compositions that would require addition of Toroweap or Coconino into the mix.

TABLE 1. MAJOR-ELEMENT COMPOSITIONS OF TARGET ROCKS AND A COMPOSITE OF SEVERAL IMPACT-MELT PARTICLES DETERMINED BY XRF AND IRON TITRATION.

	Moenkopi		Kaibab					Toroweap
	MC-01	MC-07	K-1	K-2	MC-06	MC-10	MC-12	MC-08
SiO_2	83.94	57.76	14.52	2.41	45.36	35.31	44.12	95.74
TiO_2	0.44	0.49	0.04	0.01	0.12	0.10	0.09	0.08
Al_2O_3	6.95	6.14	0.78	0.13	2.09	2.28	1.48	1.62
Fe_2O_3	0.86	1.58	0.11	0.28	0.19	0.16	0.70	0.00
FeO	0.34	0.14	0.07	0.20	0.07	0.16	0.06	0.11
MnO	0.03	0.06	0.02	0.03	0.04	0.03	0.02	0.00
MgO	0.39	1.44	18.12	20.77	10.60	12.18	10.70	0.03
CaO	2.01	16.05	26.82	30.83	15.60	19.25	16.65	0.54
Na_2O	0.02	0.04	0.07	0.07	0.07	0.06	0.07	0.00
K_2O	1.24	1.17	0.15	0.03	0.44	0.45	0.26	0.46
P_2O_5	0.13	0.13	0.03	0.03	0.17	0.45	0.46	0.03
LOI	3.72	15.23	39.19	44.43	24.88	28.93	25.11	1.31
sum	100.07	100.23	99.92	99.22	99.63	99.36	99.72	99.92

	Coconino							impact-melt
	C-3†	MC-02	MC-04	MC-05	MC-05	MC-09	MC-11a	H11‡
SiO_2	71.83	97.81	95.48	96.30	96.12	92.15	99.12	48.27
TiO_2	0.02	0.04	0.09	0.06	0.07	0.14	0.02	0.20
Al_2O_3	0.96	1.14	2.63	2.17	2.26	4.11	0.72	2.90
Fe_2O_3	0.02	0.05	0.20	0.29	0.08	0.13	0.14	7.05
FeO	0.05	0.06	0.11	0.05	0.06	0.06	0.06	16.48
MnO	0.00	0.00	0.00	0.00	0.00	0.00	0.00	0.08
MgO	0.78	0.00	0.04	0.05	0.06	0.22	0.00	8.24
CaO	14.55	0.10	0.13	0.10	0.11	0.45	0.05	12.00
Na_2O	0.00	0.00	0.00	0.00	0.01	0.00	0.00	0.09
K_2O	0.13	0.18	0.52	0.59	0.60	0.60	0.07	0.32
P_2O_5	0.13	0.02	0.03	0.03	0.03	0.04	0.02	0.44
LOI	12.01	0.53	0.96	0.75	0.75	1.77	0.37	1.40
sum	100.48	99.93	100.19	100.39	100.15	99.67	100.57	97.47

†This sample from an undocumented location is anomalous in CaO content, likely reflecting secondary calcite cement.
‡H11 is a composite of 15 impact-melt particles. See Appendix 3.

TABLE 2. COMPOSITIONS OF METEOR CRATER TARGET ROCKS DETERMINED BY INSTRUMENTAL NEUTRON ACTIVATION ANALYSIS

		Moenkopi Formation							Kaibab Formation																
		Ma	Mb	Mc	Md	Me	MC-01	MC-07	Ka	Kb	Kc	Kd	Ke	Kf	Kg	Kh	Ki	Kj	Kk	Kl	MC-06	MC-10	MC-12	K-1	K-2
mass†	mg	105.33	81.18	70.64	72.05	84.83	62.99	73.91	109.88	98.15	94.31	102.02	80.19	87.40	81.43	93.46	91.30	92.20	101.52	87.05	66.05	79.30	63.85	72.87	66.84
Na	µg/g	284	358	363	355	372	322	349	229	364	300	272	295	263	270	259	236	222	249	213	478	269	246	389	222
K_2O	wt%	1.16	1.46	1.69	1.53	1.44	1.22	1.24	0.48	0.46	0.72	0.53	0.56	0.58	0.260	0.61	0.269	0.50	0.58	0.70	0.45	0.49	*0.234*	0.222	0.043
CaO	wt%	7.2	9.3	8.4	10.6	13.1	2.0	15.0	13.5	18.0	11.7	17.4	17.2	17.4	24.6	15.7	24.6	18.0	18.6	15.4	15.2	19.2	16.0	26	29
Sc	µg/g	3.98	6.54	8.02	7.29	6.51	3.84	4.87	0.96	0.91	1.17	1.07	1.01	1.23	0.604	1.63	0.89	0.772	1.41	2.38	0.722	1.08	0.96	0.625	0.159
Cr	µg/g	24.0	29.3	32.4	24.1	27.3	15.7	37.2	30.6	23.9	32.9	32.3	27.3	28.6	8.2	15.6	8.0	22.6	14.4	23.8	28.0	25.4	34.8	12.2	5.6
Fe	mg/g	6.49	15.46	19.10	20.25	17.26	8.21	13.20	2.13	2.18	1.53	3.11	2.09	2.68	2.14	2.67	3.45	3.49	4.14	11.70	1.68	2.04	4.80	1.29	3.72
Co	µg/g	7.48	5.13	5.81	10.5	8.8	2.47	4.32	3.08	2.86	1.65	1.99	1.55	1.60	0.83	0.96	1.01	1.08	2.09	15.0	1.70	1.82	2.86	1.49	2.14
Ni	µg/g					14	16		7	8	8	11	13	10	7	7	6	5	9	37	8	10	20	10	10
Zn	µg/g	33	44	58	82	59	22.2	36	19.6	21.6	18.5	19.1	16.1	18.2	24.0	24.5	27.9	22.1	26.8	64	27.8	13.0	132		
As	µg/g	7.3	3.30	4.14	6.2	4.21	1.55	3.12	1.48	1.27	1.44	6.38	2.40	1.61	2.14	4.71	1.65	3.13	9.8	22.0	9.4	*1.00*	11.4	2.16	1.91
Se	µg/g		0.4						1.2	1.4	1.1	0.88	1.3	0.9	0.10	0.22			0.14		1.0	1.49	1.3		
Br	µg/g	0.42	0.46	0.45	0.43	0.53	*0.18*	0.40	1.10	1.26	1.00	2.0	1.46	1.02	2.2	1.28	1.10	1.07	0.81	0.79	1.36	1.07	1.00	3.0	1.07
Rb	µg/g	33	48	56	46	46	38	40	13.3	12.0	19.0	14.4	14.8	16.0	7.2	16.7	8.2	14.0	16.8	20.9	12.9	13.9	6.6	*6.1*	
Sr	µg/g	74	80	82	73	93	72	88	310	89	75	108	101	149	87	107	97	109	102	101	67	97	70	110	87
Zr	µg/g	460	360	230	150	210	370	470	140	100	200	140	160	150	58	160	61	150	90	200	140	100	100	55	
Sb	ng/g	374	451	530	440	440	343	389	203	160	199	285	176	177	131	197	104	166	218	500	287	140	560	118	*40*
Cs	µg/g	1.82	5.22	5.96	3.37	4.34	1.97	3.59	0.64	0.65	0.89	0.89	0.66	0.89	0.319	0.75	0.434	0.71	1.20	1.36	0.61	0.84	0.419	0.330	*0.052*
Ba	µg/g	207	440	308	304	269	191	207	198	88	229	180	137	151	64	193	78	114	105	140	142	97	63	65	*16*
La	µg/g	9.8	16.4	17.8	14.4	16.7	11.2	19.5	4.82	3.60	4.84	4.19	4.04	6.57	2.26	4.18	2.68	5.07	5.00	7.78	4.02	4.45	8.8	2.13	0.736
Ce	µg/g	21.0	33.1	36.1	29.5	33.0	25.0	38.0	8.4	6.5	8.7	7.4	7.3	8.5	4.06	7.4	4.56	8.0	8.4	14.1	7.6	7.9	7.2	4.20	*1.33*
Nd	µg/g	11	14	17	13	16	13	21	5.1	3.0	4.9	3.7	3.7	5.1	2.2	4.2	2.9	5.2	4.7	8.4	4.1	4.6	7	*1.8*	
Sm	µg/g	2.65	3.65	3.84	3.45	3.74	2.79	4.32	1.24	0.74	0.96	0.87	0.83	1.11	0.463	0.89	0.541	0.93	0.89	1.65	0.69	0.87	1.40	0.426	0.155
Eu	µg/g	0.65	0.84	0.88	0.81	0.86	0.69	0.92	0.289	0.159	0.211	0.191	0.181	0.233	0.091	0.203	0.110	0.191	0.182	0.365	0.148	0.197	0.286	0.089	*0.029*
Tb	µg/g	0.415	0.56	0.56	0.52	0.55	0.411	0.66	0.173	0.099	0.129	0.126	0.111	0.160	0.060	0.126	0.074	0.124	0.116	0.240	0.086	0.120	0.198	0.057	*0.026*
Yb	µg/g	1.82	2.17	1.91	1.68	1.98	1.51	2.65	0.55	0.432	0.61	0.57	0.55	0.73	0.271	0.57	0.345	0.55	0.52	0.94	0.446	0.52	0.87	0.256	*0.083*
Lu	µg/g	0.292	0.336	0.294	0.253	0.303	0.235	0.420	0.090	0.071	0.103	0.096	0.090	0.118	0.040	0.090	0.055	0.088	0.081	0.146	0.077	0.085	0.130	0.043	0.027
Hf	µg/g	13.0	10.2	6.2	4.64	6.1	11.1	16.0	4.18	2.87	5.6	4.11	4.46	4.12	1.53	4.14	1.63	3.88	2.55	5.39	4.68	3.11	2.99	1.71	0.36
Ta	ng/g	580	540	500	390	500	570	690	197	144	211	167	151	163	81	137	86	138	144	254	177	167	108	68	9
W	ng/g	770	800	820	650	710	630	850	270	200	270	210	200	200	200	230	160	240	280	460	400	180	310	210	*80*
Au	ng/g			*0.6*					1.1	*0.4*	*0.6*	2.8	*0.4*	*0.5*	1.2	*0.5*	1.0	2.0	2.3		0.7	*0.6*	20.2	1.4	0.7
Th	µg/g	3.69	4.87	5.26	4.25	4.54	3.44	5.97	1.09	0.91	1.27	1.03	1.00	1.19	0.461	0.94	0.58	1.02	1.12	1.80	0.90	0.96	0.92	0.55	*0.090*
U	µg/g	1.73	1.55	1.60	1.83	1.56	1.16	2.25	3.06	2.40	2.85	2.41	2.76	2.08	2.20	1.40	1.70	1.41	1.20	1.53	3.37	2.04	4.7	1.34	1.86

(continued)

TABLE 2. COMPOSITIONS OF METEOR CRATER TARGET ROCKS DETERMINED BY INSTRUMENTAL NEUTRON ACTIVATION ANALYSIS *(continued)*

		Toroweap Formation				Coconino Formation														
		Ta	Tb	MC-08	MC-13	Ca	Cb	Cc	Cd	MC-02	MC-03	MC-04	MC-05	MC-09	MC-11a	MC-11b	C-1	C-2	C-3‡	C-4
mass†	mg	105.98	97.53	77.33	65.78	117.32	108.48	133.77	117.47	71.98	64.39	76.93	76.07	65.33	65.16	72.15	116.90	127.84	80.82	85.3
Na	µg/g	74.1	119	49.6	154	132	90.2	55.9	60.5	39.1	62.5	31.2	37.5	126	*22.9*	*16.4*	128	73	88	27.3
K_2O	wt%	0.284	0.43	0.192	0.66	0.216	0.238	0.158	0.184	0.137	0.236	0.115	0.146	0.46	*0.085*	*0.038*	0.35	0.28	0.135	0.106
CaO	wt%	0.15	1.53	*0.30*	3.4	0.07	0.09	0.06	*0.13*					0.26					13.7	0.05
Sc	µg/g	0.514	1.18	0.393	0.775	0.714	1.05	0.490	0.479	0.496	1.03	0.479	0.353	1.27	0.421	0.222	1.25	0.639	0.369	0.342
Cr	µg/g	17.6	18.5	12.6	36.4	9.4	18.2	11.0	12.7	4.3	7.6	4.0	7.0	14.3	*2.6*	*1.7*	11.9	7.6	9.2	*3.0*
Fe	mg/g	0.892	1.017	0.378	0.696	0.530	0.570	0.363	0.438	0.338	0.527	0.357	0.284	0.830	1.524	0.469	0.581	0.427	0.277	0.257
Co	µg/g	0.83	0.70	0.187	0.427	1.25	0.426	0.239	0.203	0.166	0.265	0.104	0.118	0.61	0.328	0.311	4.31	0.389	0.64	0.157
Ni	µg/g	4							2										9	
Zn	µg/g	7.3	9.9	2.5	5.5	7.2	5.7	2.9	3.6	3.4	4.5	2.2	2.2	10.4	12.3	5.1				
As	µg/g	5.84	2.55	*0.59*	1.65	0.69	1.12	0.42	1.17	0.57	1.51	0.78	0.79	2.01	5.7	2.11	1.42	0.77	1.06	0.49
Se	µg/g		0.14												0.95	0.44			0.15	
Br	µg/g	0.40	0.46	0.36	0.62	0.32	0.38	0.29	0.35	0.23	0.20	0.22	0.27	0.38	0.28	0.28	0.44	0.31	1.91	0.34
Rb	µg/g	8.0	11.8	5.2	19.3	6.2	7.0	4.5	5.3	4.0	7.0	3.9	3.6	12.9	2.5	*1.3*	10.1	7.7	3.4	3.4
Sr	µg/g	137	188	*8*	600	29	38	28	38	25	33	16	36	48	11	*8*	43	23	90	23
Zr	µg/g	73	130	150	150	110	120	43	70	40	120	48	32	120	21	21	100	16	28	17
Sb	ng/g	186	170	155	196	72	99	76	102	67	95	64	74	1000	104	63	119	82	68	64
Cs	µg/g	0.417	0.57	0.276	1.02	0.270	0.399	0.268	0.283	0.295	0.468	0.263	0.245	0.69	0.233	0.138	0.51	0.38	0.217	0.197
Ba	µg/g	49	312	154	186	374	151	215	163	88	154	272	101	510	8	7	520	184	240	*17*
La	µg/g	3.62	4.22	1.04	13.8	3.63	5.15	2.94	2.94	3.31	5.46	2.46	2.60	5.36	2.74	2.06	6.24	3.27	2.40	2.14
Ce	µg/g	6.8	8.2	1.97	28.1	6.70	9.4	4.98	5.09	6.1	10.7	3.98	4.47	10.2	5.37	4.29	12.6	6.1	4.06	3.46
Nd	µg/g	2.8	4.5	*1.2*	17	3.5	4.6	2.9	3.1	3.0	4.5	*1.7*	3.2	5.1	2.8	*2.0*	7	3.3	3.6	*1.3*
Sm	µg/g	0.619	0.79	0.233	2.82	0.71	0.96	0.560	0.550	0.607	0.94	0.424	0.445	1.10	0.506	0.354	1.17	0.627	0.530	0.456
Eu	µg/g	0.136	0.189	0.058	0.614	0.149	0.203	0.117	0.103	0.124	0.192	0.089	0.081	0.246	0.112	0.075	0.246	0.139	0.103	0.102
Tb	µg/g	0.082	0.115	0.075	0.257	0.088	0.145	0.066	0.059	0.076	0.119	0.061	0.047	0.137	0.064	0.045	0.133	0.072	*0.064*	0.069
Yb	µg/g	0.305	0.403	0.55	0.61	0.412	0.66	0.232	0.243	0.277	0.53	0.246	0.169	0.55	*0.186*	*0.159*	0.64	0.246	*0.217*	0.258
Lu	µg/g	0.047	0.066	0.087	0.095	0.065	0.102	0.0363	0.039	0.040	0.082	0.034	0.0251	0.089	0.028	0.022	0.106	0.038	0.037	0.038
Hf	µg/g	1.99	3.41	4.45	4.70	2.91	3.39	1.18	1.90	1.31	3.61	1.34	0.87	3.77	0.70	0.58	4.18	0.76	1.43	0.64
Ta	ng/g	81	101	62	222	79	154	49	59	53	149	49	36	163	42	*23*	179	57	*44*	29
W	ng/g	220	240	140	510	200	280	110	130	100	220	110	60	300	*80*	*60*	300	*80*	*140*	60
Au	ng/g	1.6	1.6	*0.7*	*1.2*	1.9	1.8	0.36	0.9	1.8		*0.5*	*0.3*	*1.4*			1.8	2.1	2.3	0.44
Th	µg/g	0.86	0.87	0.56	2.25	0.83	1.21	0.63	0.67	0.73	1.22	0.63	0.50	1.32	0.60	0.42	1.47	0.76	0.60	0.47
U	µg/g	1.40	1.50	0.89	1.73	0.50	0.91	0.44	0.75	0.38	0.69	0.46	0.48	0.90	*0.24*	*0.15*	0.81	0.38	1.14	0.36

Note: In this table, Ni is given in units of µg/g. Ir in all target rocks was below the 2σ detection limit of ~1 ng/g. Typical 1σ relative uncertainties are: ≤2%—Na, Sc, Fe, Co, La, Sm; 1–3%—Cr, As, Cs, Ce, Eu, Yb, Hf, Th, U; 2–5%—K_2O, CaO, Zn, Sb, Ba, Tb, Lu, Ta; 4–8%—Br, Rb; 5–10%—Se; 5–20%—Sr, Au; 10–20%—Zr, Nd W; 15–50%—Ni. Values in italics have uncertainties 1.33–3× the upper end of the typical uncertainty range. Underlined values have uncertainties 3–6× the upper end of the typical uncertainty range. In all cases, the proper number of significant digits is used.

†Mass refers to weight in mg of analyzed material; samples were splits of homogenized powders of much larger mass.

‡This sample from an undocumented location is anomalous in CaO content, likely reflecting secondary calcite cement. The CaO content of most Coconino samples was below the instrumental neutron activation analysis detection limit.

TABLE 3. COMPOSITIONS OF METEOR CRATER IMPACT-MELT PARTICLES DETERMINED BY INSTRUMENTAL NEUTRON ACTIVATION ANALYSIS

samples‡		IIG-1a	IIG-1b	IIG-2a	IIG-2b	IIG-3a	IIG-3b	IIG-3c	IIG-3d	IIG-4	SIG-1	SIG-2	SIG-3	SIG-4	M1	M3a	M3b	M4a	M5	M6	M7	M8a	M8b
mass†	mg	83.16	107.47	78.63	98.38	100.52	87.19	20.05	31.29	25.5	9.43	79.08	0.84	27.05	50.34	52.81	50.47	50.93	38.54	48.57	48.94	49.70	63.17
Na	µg/g	433	421	608	496	507	415	464	558	130	240	125	209	179	214	157	150	151	171	284	157	206	384
K_2O	wt%	0.35	0.32	0.38	0.31	0.34	0.32	0.289	0.381	0.134	0.218	0.099	*0.17*	0.353	0.202	0.216	0.201	0.134	0.140	0.26	*0.097*	0.40	0.50
CaO	wt%	*12.1*	*12*	*7.3*	*8*	*10*	9.7	10.4	10.1	16.0	*14*	*13*	13	13.4	14.8	10.0	10.0	14.3	12.8	10.9	13.6	9.1	8.4
Sc	µg/g	2.00	1.98	2.38	2.17	2.19	1.96	2.30	2.47	2.36	1.99	1.89	1.86	1.37	1.50	1.59	1.55	1.74	1.90	2.22	1.69	2.36	2.90
Cr	µg/g	84	84	80	81	83	85	80	80	86	80	91	*91*	84	82	82	84	78	81	79	86	73	72
Fe	mg/g	180	184	178	190	180	184	183	179	210	183	245	202	206	196.2	210.6	215.7	171.5	201.7	163.3	220.5	149.3	135.5
Co	µg/g	910	880	824	910	826	870	828	772	940	870	1030	930	1010	912	960	1000	782	776	722	1020	687	589
Ni	mg/g	14.0	12.6	12.5	14.0	11.5	12.4	11.5	10.4	15.3	14.2	17.7	15.9	16.3	13.7	13.6	14.2	11.9	12.3	9.6	15.9	9.5	8.1
Zn	µg/g														32			28	34	35	33		62
As	µg/g	7.2	6.3	7.1	6.7	6.1	6.0	7.7	6.4	2.58	2.9	13.4	3.6	7.3	5.1	14.3	14.1	2.49	7.6	11.6	2.4	6.6	6.6
Se	µg/g	1.0	0.7		0.8	0.8	0.7	0.7														0.6	
Br	µg/g	0.37	0.31	0.39	*0.32*	0.35	0.35	0.70	0.43	0.45	1.0	0.26	6.8	0.48	0.76	0.74	0.81	0.30	1.0	1.9	*0.21*	0.59	0.65
Rb	µg/g			14	17	16	11	14															20
Sr	µg/g																						
Zr	µg/g	130	150	180	110	160	150	180	240	220	230	180		160	150	150	120	140	210	250	190	170	160
Sb	ng/g	260	230	300	260	230	220	280	260	160	*170*	170		210	170	220	220	110	260	270	120	320	380
Cs	µg/g	0.82	0.81	1.05	0.98	0.91	0.81	0.68	1.05	0.27	0.7			0.74	0.42	0.44	0.41	0.26	0.30	0.62	0.26	1.18	1.53
Ba	µg/g	430	470	540	550	470	500	530	610	300	*190*	270		1960	530	820	810	240	690	410	180	320	300
La	µg/g	13.2	13.9	16.6	17.4	11.6	10.0	19.5	16.6	18.0	9.3	11.7	7.0	5.81	4.90	8.6	8.3	9.0	10.2	13.3	22.2	11.0	15.0
Ce	µg/g	25.7	28.3	34.2	37.7	20.6	17.8	33.6	29.4	42.2	17.1	23.3	21	9.2	9.5	13.3	12.9	14.6	14.2	18.7	23.5	16.9	23.0
Nd	µg/g	*15*	16	21	15	14	14	15	13	16		11			6			11	12	13	24	17	16
Sm	µg/g	2.39	2.50	3.01	3.11	2.06	1.83	3.39	2.98	3.49	1.90	2.33	1.56	1.11	1.06	1.60	1.54	1.80	1.98	2.34	3.75	1.95	2.66
Eu	µg/g	0.57	0.58	0.71	0.73	0.48	0.43	0.78	0.70	0.80	*0.41*	0.57	0.32	*0.23*	0.252	0.38	0.37	0.43	0.47	0.57	0.92	0.46	0.62
Tb	µg/g	0.36	0.41	0.49	0.42	0.32	0.26	0.45	0.39	0.43	*0.35*	0.41		*0.18*	0.15	0.26	0.21	0.26	0.28	0.33	0.53	0.25	0.36
Yb	µg/g	1.04	1.10	1.32	1.27	1.00	0.91	1.44	1.29	1.43	0.97	1.05	0.8	0.68	0.64	0.84	0.75	0.88	1.00	1.16	1.42	0.83	1.06
Lu	µg/g	0.161	0.165	0.197	0.186	0.144	0.138	0.22	0.196	0.201	*0.14*	0.149	0.12	*0.12*	0.091	0.126	0.116	0.125	0.133	0.172	0.194	0.134	0.153
Hf	µg/g	3.86	4.03	4.17	3.72	3.80	3.87	4.5	4.3	5.0	4.6	4.4	3.4	4.7	4.5	4.5	4.2	4.6	5.2	5.9	4.4	4.43	4.9
Ta	ng/g	250	250	250	280	270	240	230	250	360	400	260		230	240	240	220	270	270	270	220	260	280
W	ng/g	400	300	400		350	360	300	200			270		320	280		350			270		290	
Ir	ng/g	450	410	360	410	370	390	410	360	285	360	460	*350*	580	460	530	560	382	448	388	374	359	302
Au	ng/g	160	142	124	150	101	113	87	117	*11*	62	24	37	314	110	251	256	28.7	101	71	50	124	115
Th	µg/g	2.1	2.1	2.4	2.4	2.0	1.81	2.4	2.3	3.3	1.6	2.1	*2.5*	*1.4*	1.43	1.66	1.7	1.7	1.78	2.0	1.6	2.06	2.59
U	µg/g	1.20	1.03	0.86	0.70	0.95	1.69	0.83	0.91	*0.23*	0.87	0.43		1.69	1.10	2.33	2.29	0.39	0.74	0.66	0.71	0.93	1.11

(continued)

TABLE 3. COMPOSITIONS OF METEOR CRATER IMPACT-MELT PARTICLES DETERMINED BY INSTRUMENTAL NEUTRON ACTIVATION ANALYSIS *(continued)*

		M10	M11	M12	M13a	M13b	M14	M15	M16	M17	M18a	M18b	M19	M20	M21	M22	M23	M24	M25	M26	M27	H11	H11a
mass†	mg	50.33	50.56	44.06	49.01	44.40	47.86	37.00	57.45	29.48	48.89	28.67	25.59	5.10	32.17	35.20	30.47	33.75	32.93	9.54	22.52	78.25	67.24
Na	µg/g	232	411	425	356	204	114	263	166	264	753	841	384	1870	234	153	200	176	125	167	570	299	208
K_2O	wt%	0.32	0.45	1.23	0.26	0.160	0.167	0.52	0.31	0.256	0.49	0.28	0.41	*0.49*	0.169	0.154	0.158	0.222	0.079	0.127	0.30	0.32	0.33
CaO	wt%	12.9	16.0	14.6	8.4	11.1	9.0	18.6	11.1	12.0	18.0	18.4	13.3	2.8	17.7	12.2	18.2	*10.8*	13.4	15.2	16.1	11.2	12.9
Sc	µg/g	1.67	2.17	4.20	2.87	1.84	1.46	1.64	1.83	1.54	1.63	1.75	1.32	3.81	2.19	3.50	2.07	1.82	1.66	1.74	1.86	2.26	1.84
Cr	µg/g	80	58	34.4	87	87	86	58	76	74	51.5	54	58	69	66	86	66	86	89	74	67	78	75
Fe	mg/g	191.3	71.0	42.9	187.4	186.9	181.4	79.5	141.7	151.4	55.2	58.0	97.7	67.0	100.3	235.0	101.5	192.8	218.0	177.0	129.0	173.1	159.9
Co	µg/g	891	311	169	700	785	792	372	584	694	244	248	452	285	417	757	435	903	1020	733	518	763	744
Ni	mg/g	11.3	4.71	2.48	9.0	9.1	11.0	5.8	7.1	9.2	3.87	3.88	6.1	*1.48*	7.2	13.2	7.4	13.1	15.9	12.0	8.5	10.0	9.4
Zn	µg/g		*25*	*16*				15	33		13	18		77	25	21					21	29	101
As	µg/g	5.3	5.4	*0.28*	28.9	11.7	14.1	1.52	8.0	19.4	2.95	3.36	7.0	13.4	6.5	30.2	5.70	12.4	4.4	6.2	5.4	7.6	5.2
Se	µg/g	1.2					0.9						0.7										0.6
Br	µg/g	0.8	2.4	*0.10*	4.4	3.4	1.7	1.23	3.2	2.2	1.24	2.0	1.40	5.0	0.60	0.53	0.40	0.65	0.36	0.73	0.32	1.1	0.57
Rb	µg/g	16	15	39				18	11		17	10	16								8		
Sr	µg/g										580	640		<u>350</u>									
Zr	µg/g	160	230	320	190	190	110	200	260	200	240	220	190	500	160	310	180	190	160	210	230	150	160
Sb	ng/g	190	310	150	440	220	220	150	370	160	160	90	160	*130*	140	260	100	310	*80*	*130*	170	260	200
Cs	µg/g	0.63	1.01	3.23	0.53	0.27	0.52	0.92	0.52	0.44	0.85	0.50	0.67	0.56	0.44	0.24	0.29	0.49	0.21	0.25	0.52	0.78	0.65
Ba	µg/g	930	470	273	600	430	270	580	420	4900	10100	13400	6400	2110	200	390	220	470	160	320	690	540	670
La	µg/g	8.0	7.87	14.6	23.1	10.6	5.23	7.83	13.2	4.69	6.75	5.44	4.83	9.0	9.1	24.6	6.05	10.6	6.62	8.3	7.83	12.2	7.86
Ce	µg/g	13.9	14.6	28.1	27.8	14.0	9.8	12.3	18.3	9.3	11.6	11.1	9.0	25.9	19.5	47.3	18.8	19.1	10.8	13.0	13.7	22.4	14.3
Nd	µg/g	11	7	13	24	13		7	14	8	6	7			11	29	6	10	5		6	13	
Sm	µg/g	1.48	1.50	2.98	4.63	1.97	1.06	1.47	2.47	1.04	1.33	1.23	0.98	1.92	2.08	6.23	1.46	1.98	1.43	1.72	1.64	2.33	1.54
Eu	µg/g	0.34	0.341	0.65	1.11	0.47	0.246	0.322	0.59	0.239	0.304	0.288	0.226	0.43	0.50	1.51	0.350	0.47	0.36	0.41	0.37	0.54	0.354
Tb	µg/g	0.20	0.215	0.419	0.65	0.29	0.15	0.202	0.36	0.15	0.203	0.185	0.14	0.32	0.30	0.93	0.22	0.27	0.21	0.24	0.24	0.34	0.22
Yb	µg/g	0.83	0.86	1.62	1.85	1.02	0.65	0.88	1.15	0.65	0.82	0.76	0.66	1.18	0.95	2.30	0.85	0.94	0.72	1.03	0.90	1.095	0.94
Lu	µg/g	0.135	0.138	0.248	0.258	0.145	0.096	0.146	0.167	0.107	0.122	0.114	0.103	0.17	0.146	0.334	0.126	0.139	0.102	0.146	0.139	0.163	0.141
Hf	µg/g	5.2	5.9	8.3	5.9	4.9	4.4	5.3	6.6	4.98	6.1	6.2	5.3	13.4	5.2	6.6	5.6	4.7	3.92	4.8	5.5	5.1	5.1
Ta	ng/g	220	270	520	260	200	220	240	260	210	270	290	220	570	310	340	290	240	200	*210*	240	240	230
W	ng/g	600	400	570	410	220	200	290	200	400	380	260	320	400	190	440	190						500
Ir	ng/g	479	169	<u>2.4</u>	421	405	437	133	321	433	93	101	255	91	160	400	183	518	485	403	259	380	350
Au	ng/g	164	85	<u>2.3</u>	157	139	202	21.7	112	180	19.8	15.1	122		*7.0*	24.5	*4.0*	243	57	40	25.9	124	93
Th	µg/g	1.6	1.93	4.05	3.3	1.74	1.35	1.66	1.8	1.44	1.70	1.76	1.35	3.9	2.12	5.3	2.03	1.81	1.36	1.62	1.75	2.31	1.8
U	µg/g	1.6	1.57	1.91	0.55	0.53	1.13	2.08	0.33	1.52	1.98	1.31	1.85	0.88	0.32	0.66	0.32	0.78	0.26	0.77	0.89	1.28	1.65

Note: In this table, Ni is given in units of mg/g. Typical 1σ relative uncertainties are: ≤2%—Na, Sc, Fe, Co, Ni, La, Sm; 1–3%—Cr; 1–4%—As, Ce, Eu, Hf, Ir; 1–5%—Yb, Au; 2–6%—K_2O, CaO, Lu; 2–8%—Ba, Th; 2–12%—Tb, U; 2–16%—Cs; 4–5%—Sr; 4–12%—Sb; 5–20%—Zn, Br, Ta; 10–35%—Zr; 10–50%—Rb, Nd, W; 25–45%—Se. Values in italics have uncertainties 1.33–3× the upper end of the typical uncertainty range. Underlined values have uncertainties 3–6× the upper end of the typical uncertainty range. In all cases, the proper numbers of significant digits are used.

†Mass refers to weight in mg of analyzed material and represents most or all of the mass of sample prepared. H11 is an exception; this is a split of homogenized powder prepared by combining 15 impact-melt particles.

‡Samples IIG-1a, IIG-1b, etc., and M3a, M3b are different splits of individual impact-melt particles run as homogeneity tests. Sample M8a is hand-picked fresh interior glass, while M8b is fine dust produced during crushing of an individual impact-melt particle. Samples M13a and M13b are exterior and interior sample pairs, respectively, from an individual impact-melt particle. Samples M18a and M18b are interior and exterior sample pairs, respectively, from an individual impact-melt particle. H11a is an individual impact-melt particle from the same group as used for sample H11.

TABLE 4. MAJOR-ELEMENT COMPOSITIONS OF IMPACT-MELT PARTICLES DETERMINED BY ELECTRON MICROPROBE ANALYSIS OF FUSED BEADS

	IIG-1a	IIG-2a	IIG-3a	IIG-4	M1	M3	M4	M7	M8	M10
SiO_2	47.9	51.8	50.9	41.7	43.0	47.7	43.4	45.3	54.4	47.6
TiO_2	0.19	0.20	0.18	0.22	0.17	0.19	0.18	0.19	0.20	0.23
Al_2O_3	3.10	3.47	3.30	3.41	2.54	3.04	2.74	2.92	3.88	3.23
Cr_2O_3	0.01	0.02	0.01	0.01	0.01	0.02	0.02	0.02	0.01	0.01
FeO	25.3	27.4	25.5	22.7	26.3	28.4	24.7	27.7	20.9	25.2
MnO	0.12	0.17	0.10	0.12	0.03	0.07	0.05	0.05	0.06	0.06
MgO	8.67	5.93	7.83	12.0	10.7	9.44	10.9	9.06	8.6	8.70
CaO	13.6	9.38	11.0	19.2	16.6	10.4	17.4	13.8	11.1	14.0
Na_2O	0.07	0.08	0.07	0.02	0.03	0.03	0.03	0.05	0.03	0.05
K_2O	0.39	0.42	0.37	0.19	0.19	0.27	0.15	0.21	0.42	0.42
P_2O_5	0.74	1.14	0.63	0.47	0.35	0.43	0.44	0.77	0.37	0.46

	M11	M13	M14	M15	M16	M17	M18	M19	SIG-4
SiO_2	55.6	50.0	51.9	52.0	51.9	52.1	54.5	57.9	44.6
TiO_2	0.24	0.22	0.16	0.20	0.26	0.18	0.21	0.16	0.18
Al_2O_3	3.40	3.19	2.70	2.95	3.24	2.79	3.25	2.68	2.86
Cr_2O_3	0.01	0.02	0.02	0.01	0.05	0.01	0.01	0.01	0.02
FeO	9.05	26.0	24.9	11.3	20.3	20.7	7.93	12.7	30.3
MnO	0.07	0.09	0.05	0.08	0.11	0.03	0.04	0.03	0.07
MgO	12.4	8.78	9.09	13.2	10.6	10.2	13.8	11.1	8.47
CaO	18.5	11.0	10.7	19.4	12.8	13.3	19.5	14.7	12.5
Na_2O	0.06	0.06	0.03	0.04	0.03	0.04	0.12	0.06	0.03
K_2O	0.43	0.29	0.21	0.32	0.38	0.26	0.39	0.37	0.38
P_2O_5	0.25	0.38	0.30	0.48	0.30	0.33	0.33	0.31	0.58

TABLE 5. COMPOSITIONS OF METALLIC SPHERULES AND THE REMBANG IVA IRON DETERMINED BY INSTRUMENTAL NEUTRON ACTIVATION ANALYSIS*

Spherule #:		1	2	3	4	5	6	7	8	9	10	11	12	14
mass	(mg):	9.03	12.58	8.53	9.35	11.24	3.32	12.83	8.60	9.40	5.08	3.96	4.36	2.12
Fe	mg/g:	818	716	815	714	800	570	768	740	772	834	759	637	614
Co	mg/g:	7.6	9.3	8.9	5.6	9.1	7.5	10.4	11.1	10.0	7.3	10.7	3.1	3.2
Ni	mg/g:	128	152	152	86.6	148	124	169	181	166	115	171	50.2	38.9
Ga	μg/g:	92.3[†]	88.2[†]	120	83.5[†]	110	170	81.8[†]	–	50	–	–	80	90
Ge	μg/g:	514[†]	772[†]	–	485[†]	–	–	795[†]	–	–	–	–	–	–
As	μg/g:	44.1	49.5	48.0	24.0	48.2	66.8	51.9	63.8	52.1	25.8	64.9	19.0	27.5
Ir	μg/g:	4.48	5.53	5.4	3.14	4.9	9.15	6.07	6.62	5.83	4.18	6.75	2.61	1.96
Au	μg/g:	3.17	3.85	3.86	2.11	3.74	1.99	4.3	4.57	4.26	2.72	4.69	1.28	0.63
Σ(Fe, Co, Ni) :		954	877	976	806	957	702	947	932	948	956	941	690	656

Spherule # :		15	16	17	18	19	20	21	23	Rembang[‡]			
										this work	WR01	WR01	S96
mass	(mg):	5.21	5.24	0.72	3.25	9.39	3.80	2.21	2.38	16.7			
Fe	mg/g:	820	748	676	783	726	745	653	788	943	–	–	–
Co	mg/g:	8.0	11.3	3.3	9.7	12.6	9.6	2.8	7.7	4.0	4.03	4.03	4
Ni	mg/g:	130	182	41.1	155	216	164	31.3	137	89.6	85.4	88.4	–
Ga	μg/g:	–	–	–	–	–	–	–	–	–	–	–	–
Ge	μg/g:	–	–	–	–	–	–	–	–	–	–	–	–
As	μg/g:	41.7	94.9	20.7	47.9	84.8	44.4	13.7	43.0	9.4	8.85	8.93	9
Ir	μg/g:	4.73	6.77	2.24	5.69	7.57	5.76	1.59	5.09	1.19	1.16	1.16	1.14
Au	μg/g:	3.51	4.57	0.75	4.11	5.59	4.09	0.55	3.58	1.19	1.78	1.75	1.91
Σ(Fe, Co, Ni):		958	941	720	948	955	919	687	933	1037			

*Typical instrumental neutron activation analysis 1σ relative uncertainties are: 1%—Fe, Co; 2%—Au; 1–3%—Ni; 2–4%—Ir; 5–10%—As; 30–50%—Ga.

[†]These analyses done by radiochemical neutron activation analysis, typical 1σ relative uncertainties are 4%.

[‡]WR01 – Wasson and Richardson (2001); S96—Scott et al. (1996).

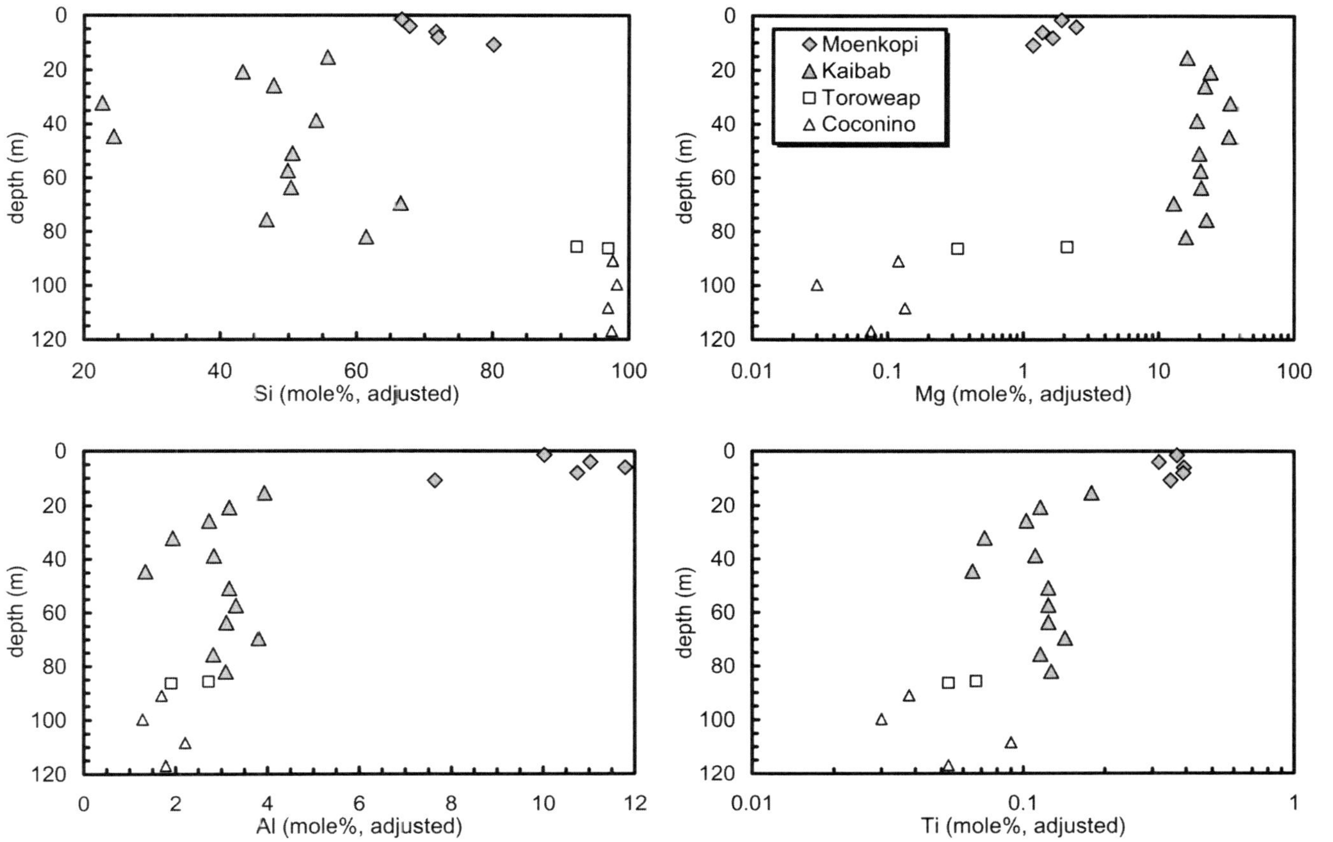

Figure 1. Variation in major-element compositions of Meteor Crater target rocks with depth demonstrating the geochemical characteristics of the different formations. On all figures, lithophile element data are adjusted to a volatile- and Fe-free basis to correct for differences in CO_2 and H_2O in the target rocks, and differences in projectile components in impact-melt particles. See text for a discussion of this. Note that Mg and Ti are plotted on log scales. Data from See et al. (2002).

However, the scatter in the data would *permit* a small Toroweap or Coconino component. The expansion of the impact-melt field in the Mg-Al plot demonstrates that the data distribution could obscure at most ~10% Coconino contribution. Because there are no distinguishing elements in the Toroweap or Coconino, they are difficult to fingerprint in impact melts.

Lithophile-trace-element contents of impact-melt particles and target rocks are in accord with the above inference. In Ca-Sc and Ca-Ta plots, the impact-melt particles plot between the fields occupied by Kaibab and Moenkopi rocks and away from the trace-element-poor Toroweap and Coconino (Fig. 4). The Toroweap and Coconino rocks have the lowest lithophile-trace-element contents (Table 2, Fig. 2), and these elements are generally correlated in the target rocks. Thus, on diagrams such as Ta versus Sc (Fig. 4), Toroweap and Coconino rocks could be interpreted as one end member of a mixing trend. However, their stratigraphic location argues against this. The Ta-Sc plot clearly shows that the topmost unit, the Moenkopi, is a significant component of the impact-melts, and that all of them lie between the fields for Moenkopi and Kaibab rocks. Because the Kaibab is a thick unit between the Moenkopi and Toroweap/Coconino, it is more plausible that the Kaibab is the most significant trace-element-poor component in the impact-melts.

The impact-melt particles exhibit substantial heterogeneity in lithophile element contents (Figs. 3 and 4). Some particles fall within the field of Kaibab rocks in major element composition (Fig. 3) and could contain only this target component. None fall within the field of Moenkopi rocks for major elements (Fig. 3), although some do so for trace elements (Fig. 4). (We have trace element data for a larger suite of samples than we do for major elements.) The Moenkopi and Kaibab rocks each show substantial ranges in composition, thus impact-melt particles that plot within the fields for either of the target rocks are not necessarily composed of only that target component. Note that all of the impact-melt particles (this work) and pure glasses (Hörz et al., 2002) that we have analyzed contain substantial Al and Si; none of them are pure carbonate melts.

The lithophile-element contents of some impact-melt particles are outside the ranges for Moenkopi and Kaibab rocks. For example, the adjusted La content of a few are higher than

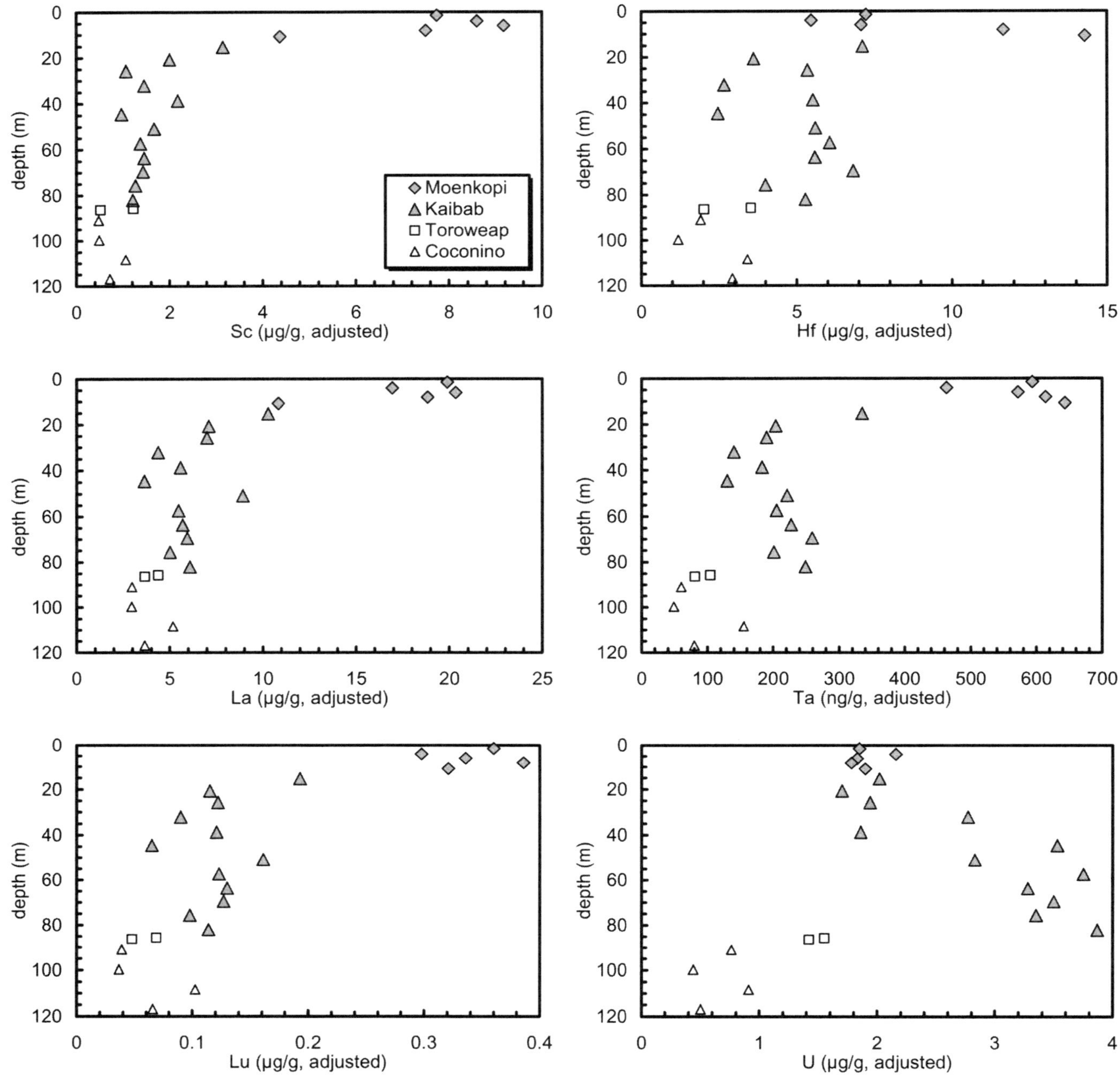

Figure 2. Variation in trace-element compositions of Meteor Crater target rocks with depth demonstrating the geochemical characteristics of the different formations.

observed in our suite of Moenkopi rocks, while the U and K contents of many are lower than those of Kaibab or Moenkopi rocks (Fig. 5). The high La contents determined for a few impact-melt particles likely indicates that our sampling of the Moenkopi formation at ~2.5 m scale was inadequate to fully define its compositional variations.

The low K and U contents for many impact-melt particles suggest possible fractionation during or after formation. Potassium could have been partially volatilized during the high temperature phase. Uranium is normally considered a refractory element, but some U halides are quite volatile (Lide, 2001). The Kaibab formation typically contains 1–2 μg/g Br (Table 2) and likely much higher Cl contents (seawater Cl/Br ~290). Hence, loss of species such as UCl_4 (boiling point 1064°K, Lide, 2001) may have occurred. Alternatively, the low K and U contents might indicate loss during low-temperature alteration of glasses in the particles.

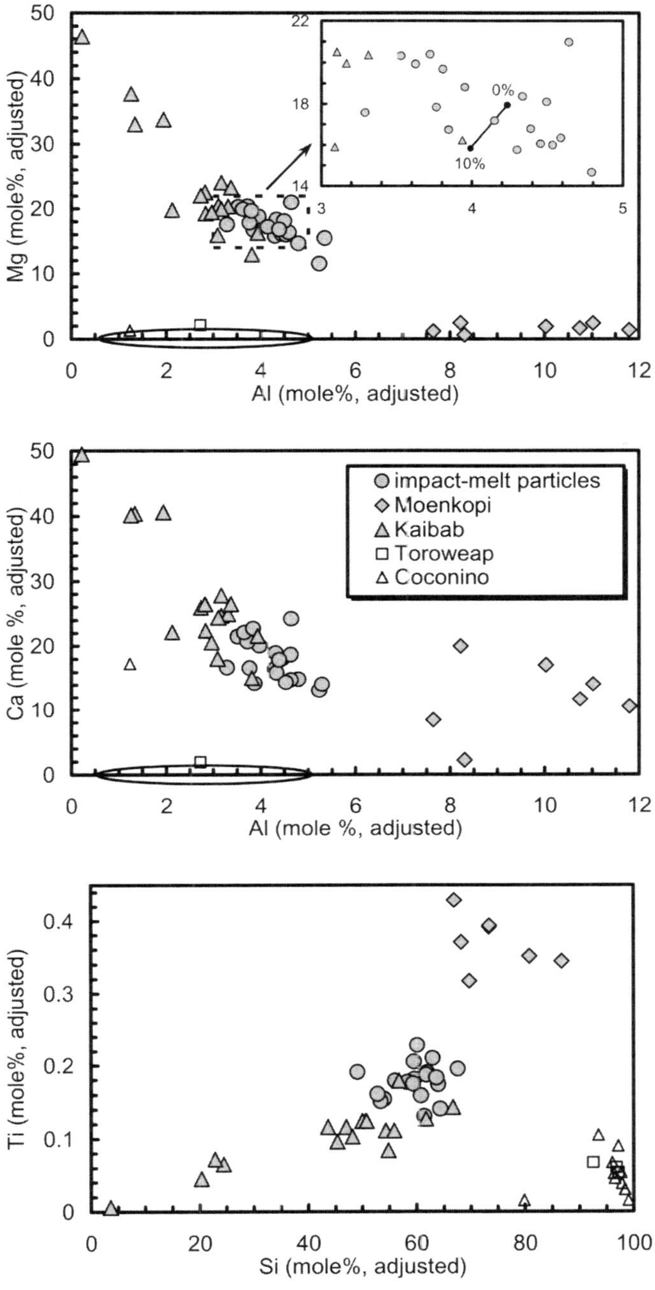

Figure 3. Major-element variation diagrams comparing Meteor Crater target rocks with impact-melt particles demonstrating that the Moenkopi and Kaibab Formations are the major target-rock components. The inset in the Mg vs. Al diagram shows that mixing of at most 10% of Coconino Formation could be hidden in the data scatter, but no Coconino or Toroweap components are required by the data. The ovals enclose Toroweap and Coconino samples with very low Mg and Ca contents. Some target-rock data from See et al. (2002).

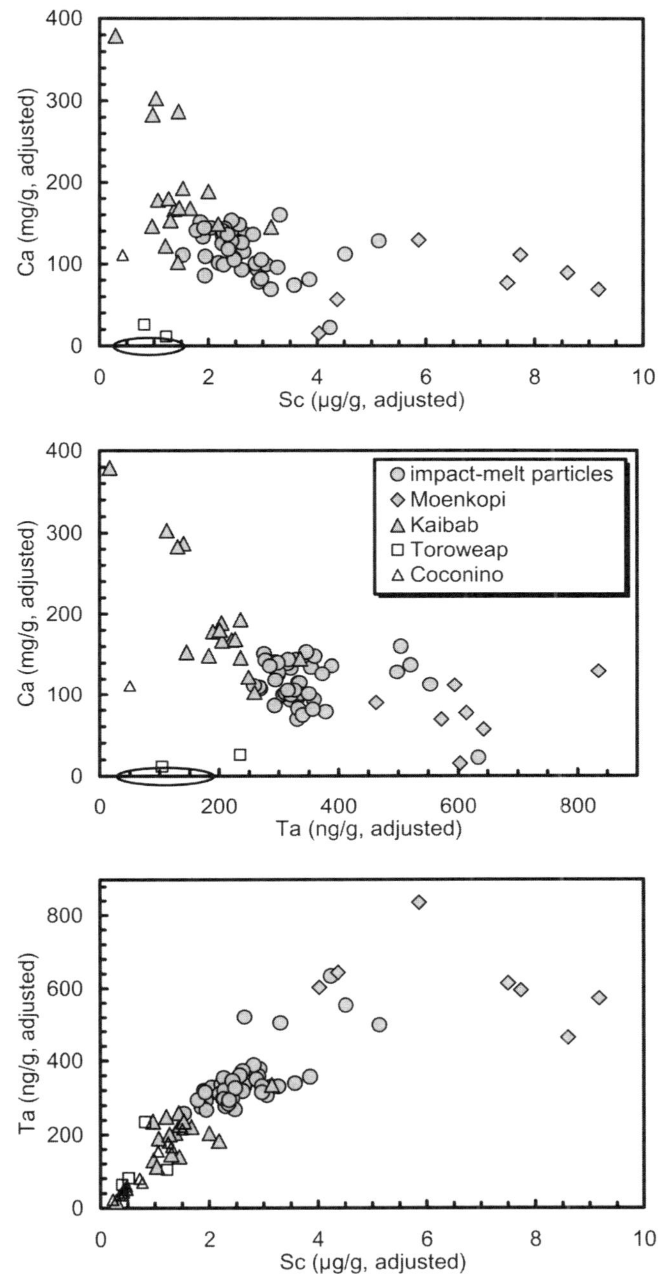

Figure 4. Ca vs. Sc, Ca vs. Ta, and Ta vs. Sc diagrams comparing Meteor Crater target rocks with impact-melt particles demonstrating that the Moenkopi and Kaibab Formations are the major target-rock components. The ovals enclose Toroweap and Coconino samples with very low Ca contents.

We documented late oxidation of Fe in glasses (Hörz et al., 2002), and interior vesicles in some impact-melt particles contain secondary phases indicating leaching from the particles. Uranium is subject to oxidation and mobilization during alteration, and alkali elements are typically easily mobilized. Regardless, the K and U contents of the impact-melt particles do not indicate that Toroweap or Coconino are significant components. Those impact-melt particles with K and U contents like those of Toroweap and Coconino rocks nevertheless contain high Ca contents demonstrating a dominant Kaibab component.

Based on the somewhat variable major-element compositions of glasses in impact-melt particles, we concluded that incomplete mixing rocks from the upper 30 m of the target formed many, if not most of them (Hörz et al., 2002). We have not attempted to do mixing calculations for bulk impact-melt particles presented here because the compositional ranges in the target rocks (Fig. 2) preclude unique solutions. We also concluded that the higher SiO_2 contents of some of the glasses *may* indicate a small component of Coconino is present, but this was not *required* by the data (Hörz et al., 2002). The major-element contents of the impact-melt particles would allow for as much as ~10% Coconino (Fig. 3). However, because all major- and trace-element data (excluding potentially mobile elements) for the impact-melt particles lie between the fields for Moenkopi and Kaibab, we believe it to be more plausible that only these two rock units were mixed in the melt zone to form the particles, and thus the melt zone extends no deeper than the base of the Kaibab, ~85 m.

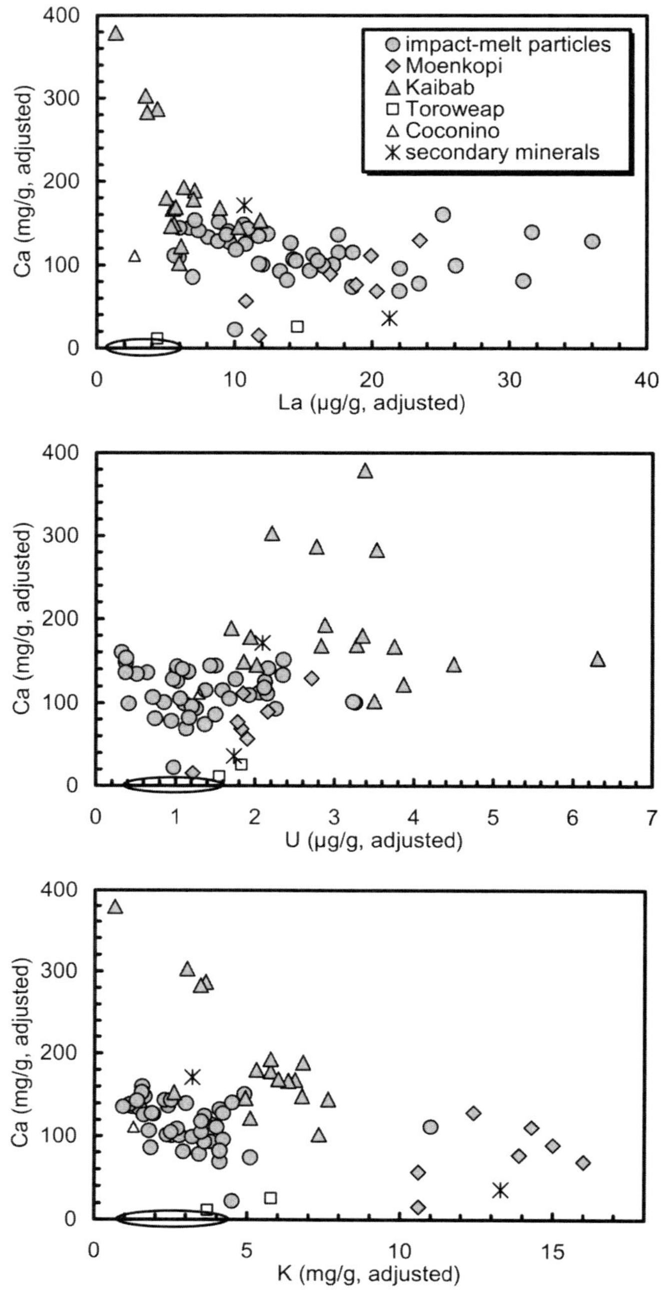

Figure 5. Ca vs. trace-element diagrams comparing Meteor Crater target rocks with impact-melt particles. The ovals enclose Toroweap and Coconino samples with very low Ca contents. Secondary mineral data are from Appendix 4.

Siderophile Elements and Projectile-Target Mixing

The impact-melt particles contain a substantial projectile component. Their Co, Ni, and Ir contents normalized to the mean Canyon Diablo metal composition (Appendix 1) reach 22%–27% that of the projectile (Fig. 6). However, CO_2 was volatilized from the carbonate-rich target rocks, and the actual maximum projectile component is less. There is a reasonably good correlation between target-rock CaO+MgO and LOI (See et al., 2002). We have used this to estimate the mass of target rock lost during impact melting for particles for which we have major-element data (Table 4, plus H11, Table 1). We estimate that between 7% and 15% of the mass was lost, and that the maximum projectile component based on Co, Ni, and Ir was between 20% and 22%. Using an average mass correction of 11%, we find that the impact-melt particles average 14% projectile component. This is higher than we found for ballistically dispersed impact-melt particles from Wabar Crater (Fig. 6), where average projectile components were ~7% and ~9% for large and small glass beads, with a maximum of ~17% for one small bead (Mittlefehldt et al., 1992). As was the case for lithophile elements, the impact-melt particles are quite heterogeneous in siderophile element contents (Fig. 6).

Few other craters have both impact melts and impactor preserved, and very few analyses of impact melts have been done. Attrep et al. (1991) found projectile components of ~4% and ~10% (based on Co and Ni) for two impact melts from Wolfe Creek Crater (Fig. 6), and <1% for three impact melts from Henbury Crater. The types of impact melts analyzed were not specified. The very low projectile component in the Henbury Crater samples most likely indicates that they represent massive-melt objects. We previously showed that large melt specimens (>100 g) from Wabar contain a much smaller projectile component than

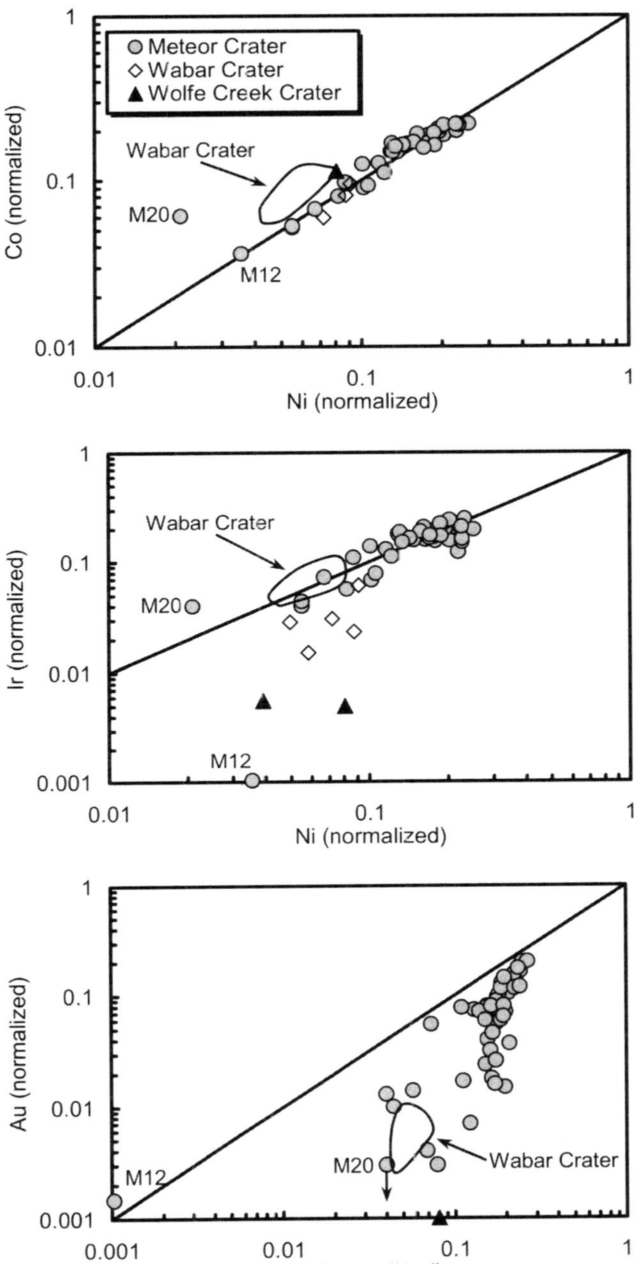

Figure 6. Element-element diagrams for siderophile elements in Meteor Crater impact-melt particles, with comparative data from Wabar Crater (Mittlefehldt et al., 1992) and Wolfe Creek Crater (Attrep et al., 1991). The data are normalized to projectile compositions. The Meteor Crater samples have Co/Ni and Ir/Ni ratios equivalent to the Canyon Diablo iron, but show strong fractionations in Au/Ir. Lines indicate Canyon Diablo element ratios. Arrow on M20 signifies Au is plotted at the 2σ upper limit. Fields enclose the majority of Wabar Crater samples; the few samples plotted have Co/Ni and/or Ir/Ni ratios distinct from the majority. See text.

do ballistically dispersed, mm-sized melt beads (Mittlefehldt et al., 1992). The Henbury samples will not be discussed further.

The Meteor Crater impact-melt particles have relatively unfractionated Co/Ni and Ir/Ni ratios, but Au is strongly fractionated as exemplified by Au/Ir ratios (Fig. 6). The Au fractionation is independent of the amount of projectile component in the particles. Samples with projectile-normalized Au/Ni ratios as low as 0.03 have normalized Co contents ~20% that of the Canyon Diablo iron, and two samples with normalized Au/Ni of 0 82 and 0.90 have normalized Co contents of 7% and 10% that of the projectile (Fig. 7). This shows that the Au fractionation is independent of the target-projectile mixing process, i.e., fractional condensation from a vapor cloud, or selective vaporization of siderophile elements from impact melts, is not the cause (see Gibbons et al., 1976; Kelly et al., 1974 for discussions of proposed fractionation processes). Hydrocode simulations of the Meteor Crater impact also show that for plausible impact velocities, projectile vaporization seems unimportant (Schnabel et al., 1999).

Two samples have anomalous siderophile-element characteristics. Particle M20 has high Co/Ni and Ir/Ni ratios due to depletion in Ni—its Ir/Co ratio is within the range of other impact-melt particles. Particle M20 contains clear glass and two types of altered glass intimately mixed (Hörz et al., 2002). The clear glass contains ~7% projectile component (total Fe as FeO, 8.43 wt%) and has an Fe/Ni ratio like that of the projectile, while the altered glasses are more iron-rich (total Fe as FeO, 11.1–14.2 wt%) with Fe/Ni ratios much greater than that of the Canyon Diablo iron (Hörz et al., 2002). The bulk sample of M20 studied here is like the clear glass in Fe content, but like the two types of altered glass in Na, K, and Ca. Possibly, the anomalous characteristics of M20 may simply reflect heterogeneous distribution of siderophile elements among different glass types. Particle M12 has a very low Ir/Ni ratio due to an anomalous Ir content. This particle has a Au/Ir ratio like that of Canyon Diablo, but this is anomalous compared to all other impact-melt particles with low Ir and Au contents (Fig. 6). Particle M12 contains only clear glass and the lowest projectile component found (Hörz et al., 2002). This is also observed in the bulk impact-melt particle, with ~3.5% projectile based on Co and Ni, but only ~0.1% based on Ir and Au (Fig. 6).

The siderophile-element characteristics of the Meteor Crater impact-melt particles show some similarities and some differences with those from Wabar and Wolfe Creek Craters (Fig. 6). Most Wabar samples have Co/Ni ratios significantly elevated above that of the impactor, while only Meteor Crater particle M20 is so endowed. The single Wolfe Creek Crater sample plots close to the impactor line. Most Wabar impact-melt particles have Ir/Ni ratios like that of the impactor. A few have anomalously low Ir/Ni ratios, but none are as anomalous as M12. Both Wolfe Creek Crater samples show anomalous Ir/Ni ratios, but again, not to the extent of M12. All samples from Wabar Crater and the single sample from Wolfe Creek Crater have Au/Ir ratios less than that of the projectile. Although only three craters have been sampled, with only a single Au determination for one, the data suggest Au fractionation may be a common process occurring during formation of ballistically dispersed impact melts at terrestrial craters.

Attrep et al. (1991) and Mittlefehldt et al. (1992) noted that because siderophile-element ratios are fractionated at some terrestrial craters, attempts to identify impactor types from terrestrial and lunar impact melts should be treated cautiously. We still believe this to be true for terrestrial craters, but the observed fractionations may be uniquely terrestrial. There is a general tendency for those Meteor Crater impact-melt particles with the highest molar (Ca+Mg)/Si (highest Kaibab content) to have lower normalized Au/Ir ratios (Fig. 8). Kaibab samples have higher Br contents (Fig. 8), and Se, a proxy for S, was detected almost exclusively in Kaibab rocks (Table 2). The Au fractionation may be tied to the volatile content of the target, and thus loss of volatile Au complexes may have caused the fractionation. If true, then siderophile-element fractionation may not occur during impact melting on the volatile-poor Moon.

The case is not clear-cut; substantial scatter in the Meteor Crater data (Fig. 8) obscures compositional correlations. This may result from the heterogeneous nature of the Kaibab (Figs. 1 and 2; Hörz et al., 2002; See et al., 2002), the heterogeneous melting process, or signify that the trend is not real. Wolfe Creek Crater was formed in Precambrian quartzite, but we have no detailed compositional information on the target rocks. Wabar Crater was formed in sandstone. The low Ca contents of Wabar impact glasses and target materials indicate that there was little carbonate in the target (Hörz et al., 1989; Mittlefehldt et al., 1992). We do not have Br or Se data for the Wabar target-rock samples, and thus, cannot compare possible halogen or sulfate contents with Meteor Crater. Similar data sets from additional craters with preserved projectiles are needed to further evaluate the cause of Au fractionation.

We deduced that more projectile material was mixed into melts formed in the upper portions of the melt zone at Wabar crater (Mittlefehldt et al., 1992). The Meteor Crater data give some support to this. The Kaibab rocks have the highest molar (Ca+Mg)/Si and (Ca+Mg)/Al ratios among the target rocks (Fig. 9). These ratios should then increase in impact-melt particles formed as the melt zone penetrates deeper into the section. Impact-melt particles with (Ca+Mg)/Si and (Ca+Mg)/Al ratios less than or at the low end of the range for the Kaibab rocks have projectile-normalized Ir contents, on average, higher than those with the highest (Ca+Mg)/Si and (Ca+Mg)/Al ratios (Fig. 9). Those samples with higher (Ca+Mg)/Si and (Ca+Mg)/Al ratios had higher carbonate contents and have lost the most mass through CO_2 devolatilization. Thus, their normalized Ir contents have been enhanced more by this process than those with low (Ca+Mg)/Si and (Ca+Mg)/Al. The data are thus consistent with a generally decreasing projectile/target mixing ratio as the projectile penetrates. Samples with the highest normalized Ir contents have intermediate (Ca+Mg)/Si (~0.5–0.6) and (Ca+Mg)/Al (~7–8) ratios (Fig. 9), and may indicate that the peak projectile/target mixing ratio was reached at some intermediate depth.

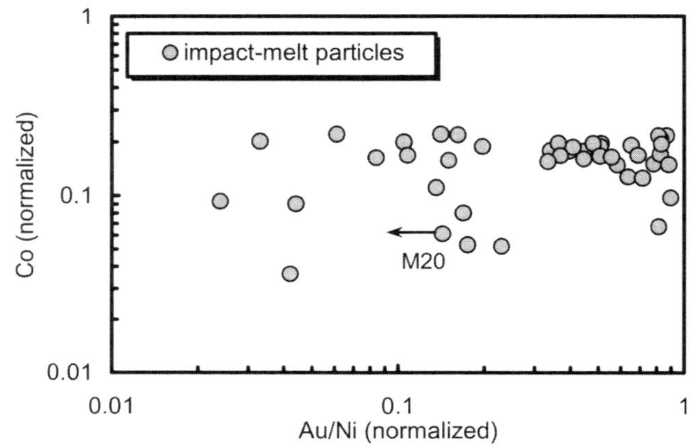

Figure 7. Co vs. Au/Ni for impact-melt particles. Particles covering nearly the entire rage in Au/Ni ratios can have similar Co contents, and samples with similar Au/Ni can have substantially different Co contents. Thus the Au/Ni fractionation is not related to the mixing ratio of projectile and target materials. Arrow on M20 signifies Au/Ni is plotted using the Au 2σ upper limit.

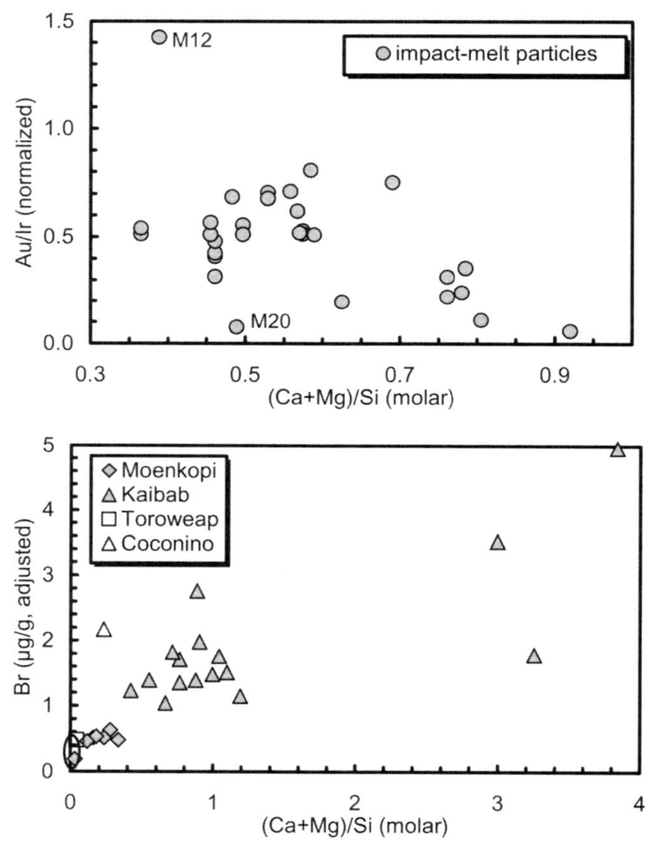

Figure 8. Au/Ir vs. (Ca+Mg)/Si of impact-melt particles shows that those particles with a high dolomite component tend to have lower Au/Ir ratios. Br vs. (Ca+Mg)/Si of target rocks show that the halide Br is roughly correlated with the dolomite content of the target rock. These suggest that volatilization of Au-halides may be responsible for the observed fractionation. Gold was below the detection limit in M20; its Au/Ir ratio was calculated using the Au 2σ upper limit. The oval encloses Toroweap and Coconino samples with very low Br contents.

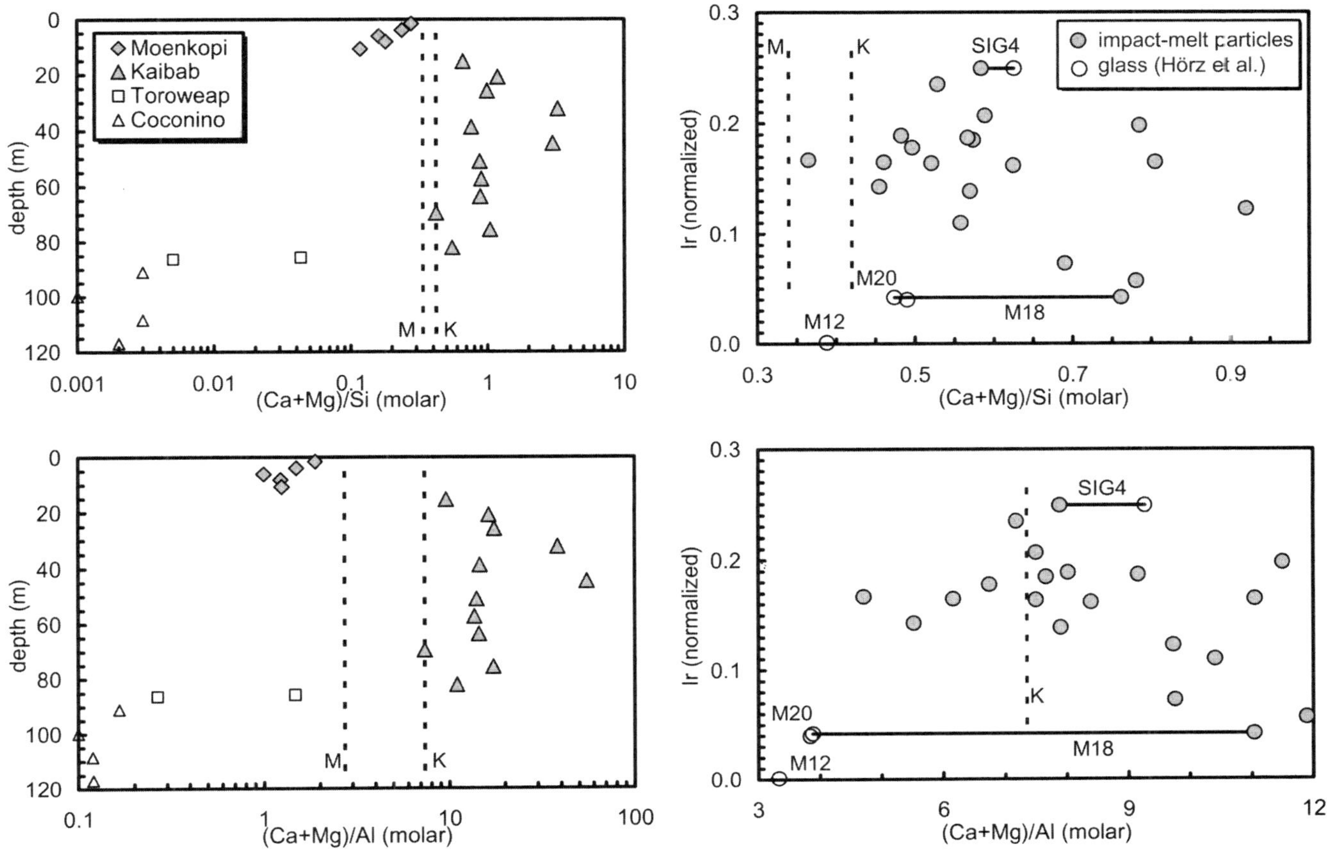

Figure 9. Variations in (Ca+Mg)/Si and (Ca+Mg)/Al with depth in target rocks (left) and with Ir content in impact-melt particles (right). Dashed lines labeled K mark the lower limits of (Ca+Mg)/Si and (Ca+Mg)/Al observed for Kaibab samples, while the line labeled M marks the upper limit of (Ca+Mg)/Si in Moenkopi samples—all Moenkopi samples have (Ca+Mg)/Al <3. Tie lines in right-side diagrams connect particles plotted using both bulk particle major element data (Table 4) and clear glass data (Hörz et al., 2002). See text.

Two impact-melt particles, M12 and M20, have lithophile-element contents indicating a higher Moenkopi component, yet have very low normalized Ir contents (Fig. 9), contrary to the general trend just discussed. The data for these are from clear glass analyses of Hörz et al. (2002); the samples were too small for bulk major element analysis. These authors inferred that the M12 and M20 glasses contained a higher Moenkopi component than most clear glasses, and that their low FeO contents indicated a low projectile component. Our data for K, Ca, and Fe on bulk M12 match well the clear glass data (Hörz et al., 2002, Table 3) suggesting that most of the bead was composed of such glass. As discussed above, M12 has anomalous siderophile-element contents compared to all other impact-melt particles. The bulk data for M20 do not match the glass data. Bulk particle M20 has the lowest CaO content (Table 3), yet the clear glass has among the highest (Hörz et al., 2002, Table 3). Altered glass in M20 has much lower CaO contents (Hörz et al., 2002), similar to what we found for the bulk particle. However, the altered glass has much higher Fe contents than the clear glass, and our bulk particle datum matches the clear glass. M20 also has somewhat anomalous siderophile-element contents (Fig. 6). Thus, M12 and M20 should be treated cautiously when attempting to infer projectile-target mixing details.

We have major-element compositions for bulk particles (Table 4) and clear glasses (Hörz et al., 2002) from M18 and SIG-4. Bulk SIG-4 has (Ca+Mg)/Al and (Ca+Mg)/Si very similar to those of the clear glass (Fig. 9). Bulk M18 is very different from its clear glass in major-element contents (Fig. 9). It contains higher CaO and MgO, and lower Al_2O_3 than does the clear glass, suggesting incorporation of dolomite clasts, an additional Kaibab component not present in the clear glass. Thus, it is difficult to determine whether M18 violates the projectile-target mixing generalization discussed above.

Hörz et al. (2002) showed that there are two basic types of glasses in the impact-melt particles: those with low Fe and Ni and Ni/Fe like that of the projectile, and those with high Fe, low Ni, and Ni/Fe much lower than that of the projectile. The bulk particles have a continuum of Fe contents covering the total range observed for glasses, but all have Ni/Fe like that of the impactor. Thus bulk impact-melt particles have variable contents of finely disseminated, unfractionated (except for Au) projectile material. Some glasses on the other hand, contain a fractionated projectile component that must be compensated by other phases, probably metal or sulfide blebs, in the bulk particles. Electron microprobe analyses of metal blebs in impact-melt particles show that they can be highly enriched in Ni and Co relative to Canyon Diablo metal (Brett, 1967; Kelly et al., 1974; our unpublished data), and can balance the low Ni contents of pure glasses with low Ni/Fe observed by Hörz et al. (2002).

Metallic Spherules and Melting of the Projectile

Blau et al. (1973) studied a suite of metallic spherules from Meteor Crater. They concluded that the dendritic texture demonstrated that the spherules solidified by quenching of molten droplets, and that the textures of some are similar to those observed for Fe-Ni alloys containing C. They also noted that the spherules are enriched in Ni, S, and P compared to Canyon Diablo metal and suggested that this was due to slight selective melting of troilite-schreibersite-rich regions of the meteorite. Finally, Blau et al. (1973) argued that high Co contents of the spherules compared to Canyon Diablo metal was the result of oxidation of the spherules, enhancing the Ni and Co contents through loss of Fe oxide. This process also led to some oxidative loss of S and P (Blau et al., 1973).

In contrast, Kelly et al. (1974) noted that the S content of the spherules—9–28 mg/g (Blau et al., 1973)—is not enhanced relative to an estimated bulk meteorite S content of 22 mg/g, and that Ni/Co and Ni/Cu ratios are like those of the bulk meteorite. Kelly et al. (1974) concluded that the spherules were formed by melting bulk Canyon Diablo material, not selective melting of troilite-schreibersite-rich regions. Kelly et al. (1974) explained the high Ni, Co, and Cu contents relative to bulk meteorite by oxidation and removal of Fe.

Buchwald (1975, p. 392) estimated the bulk meteorite S content is ~9–10 mg/g, about half that estimated by Kelly et al. (1974) using older modal data. Thus the S content of the spherules may be enhanced by up to a factor of ~3 compared to the bulk meteorite. Buchwald (1975, p. 392) also estimates a bulk meteorite P content of 2.5 mg/g. The spherules contain 7–14 mg/g (Blau et al., 1973), suggesting they may be enriched up to ~6 times in P. This seems to support the contention of Blau et al. (1973) that some selective melting of troilite-schreibersite-rich regions occurred. However, Blau et al. (1973) did their analyses by electron microprobe using an "integrated line scan technique" on heterogeneous samples. True bulk spherule S and P contents need to be determined.

Figure 10 shows our metallic spherule data and those of Xue et al. (1995). Most spherules form a trend away from Canyon Diablo toward higher Ni but lower Fe, consistent with simple loss of Fe (arrow, upper panel). A few spherules plot below the Fe-loss trend. Either these samples contained some iron oxides–hydrous oxides that were not removed prior to analysis, or they are richer in C, P, and/or S than bulk Canyon Diablo. Note that the oxide-shell samples analyzed by Xue et al. (1995) plot at a lower normalized Fe abundance than "ideal" limonite (FeO•OH•2.1H_2O) and with higher Fe/Ni than the metal spherules they were taken from (tie lines). This is consistent with enrichment in Fe relative to the metallic core coupled with dilution by H, O, C, P, and/or S. Because the oxide shells have lower bulk Fe than "ideal" limonite, some dilution by C, P, and/or S is indicated.

Cobalt-Ni, Ir-Ni, and Au-Ni are well correlated in the spherules (Fig. 10). Kelly et al. (1974) argued that the Co/Ni ratios of the spherules were identical to the bulk metal of Canyon Diablo. We find subtle variations in siderophile-element/Ni ratios in the spherules. The Co-Ni trend has a slightly lower slope than the Canyon Diablo ratio, while Au/Ni has a slightly higher slope. Regression lines pass through the mean Canyon Diablo composition (Fig. 10). This is consistent with the oxidation model of Kelly et al. (1974). Cobalt is more readily oxidized than Ni. Those spherules with the highest Ni, having suffered the most oxidation and loss of Fe, have also experienced some oxidative loss of Co and have low Co/Ni ratios. Gold is less readily oxidized than Ni and shows the opposite—high Au/Ni for the most Ni-rich spherules.

Four spherules have Ni contents lower than bulk Canyon Diablo (numbers 12, 14, 17, 21; Table 5). They also have among the lowest sums of Fe, Co and Ni (Table 5), indicating that they contain larger amounts of diluents than do most other spherules. Finally, they have high Co/Ni and low Au/Ni ratios (Fig. 10). All of these characteristics are consistent with their containing a substantial amount of the Fe-rich, Ni-Au–poor oxides produced during spherule formation. One would expect the oxide to form on the outer surface of the spherule. Thus, should these four spherules be enriched in oxides, the Co-Ni-Ir-Au–rich metallic core of the spherule somehow was lost during formation.

All of the spherules have As/Ni ratios in excess of that of Canyon Diablo (Fig. 10), contrary to expectations. Arsenic is more easily oxidized than Ni and should be depleted relative to it, but Ni-rich spherules have As/Ni ratios of 1.89–2.79 times Canyon Diablo. The As-Ni data also scatter considerably compared to the very linear Co-Ni array (Fig. 10). Our As-Ni data closely mimic the P-Ni data of Blau et al. (1973) (Fig. 10), suggesting that the anomalous enrichment of As may be a result of slight selective melting of schreibersite-troilite–rich regions of the impactor. Arsenic and P are incompatible elements during crystallization of molten Fe-Ni, and will be correlated in irons (e.g., Scott, 1972). We did not find data on the As content of schreibersite, but As is in the same chemical group as P, suggesting that schreibersite exsolved from the metal may be enriched in As. The high As contents of the spherules may thus support slight selective melting of troilite-schreibersite-rich regions of Canyon Diablo as the explanation for enrichments in P and S.

Canyon Diablo contains graphite-troilite inclusions that are typically rimmed by schreibersite and cohenite (Buchwald, 1975). These inclusions have lower mean density than the surrounding metal. This will cause a passing shock front to reverberate (e.g., Kieffer, 1971), resulting in localized stress and temperature concentrations along metal-inclusion margins (e.g., Melosh,

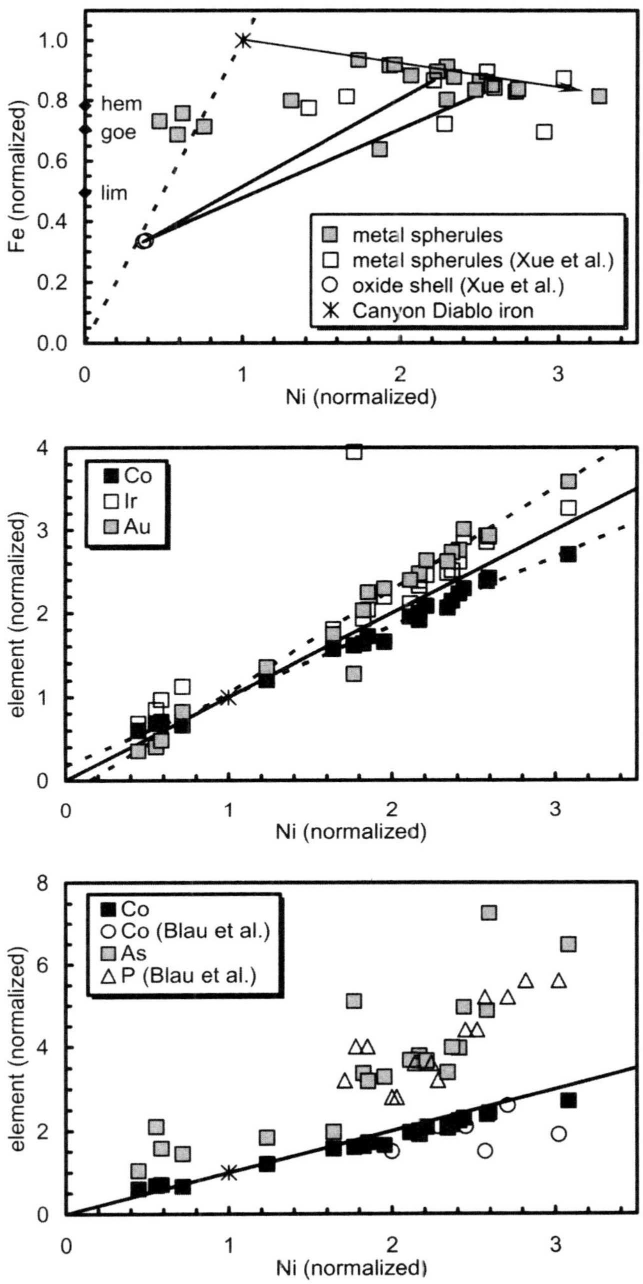

Figure 10. Siderophile-element data for metallic spherules compared to the Canyon Diablo iron. Most spherules have Fe and Ni contents that follow a trend of Fe loss (arrow, upper diagram). Some samples with low Fe are consistent with dilution by oxides. Equivalent normalized Fe contents for hematite, goethite and limonite are shown for comparison. Co-Ni, Ir-Ni and Au-Ni are highly correlated, with most samples enriched in these elements compared to the Canyon Diablo iron. Dashed lines are regression lines through the Co-Ni and Au-Ni data. The slopes of these lines are consistent with expectations from a simple oxidation model (see text). The metallic spherules show enrichments in As that are inconsistent with a simple oxidation model, but closely mimic the P-Ni trend. This suggests that some enhanced melting of graphite-troilite-schreibersite-rich regions of Canyon Diablo may have occurred (see text). Some data are taken from Blau et al. (1973) and Xue et al. (1995).

1989; Stöffler et al., 1991), leading to enhanced localized melting at metal-inclusion interfaces. Hydrocode simulations of the Meteor Crater impact show that this can occur along the trailing edge of the impactor at moderate shock pressures, i.e., moderate depths below the rear surface (Schnabel et al., 1999).

We conclude that both enhanced melting of graphite-troilite-schreibersite inclusions and oxidation and removal of Fe (and other elements) played roles in producing the chemical characteristics of the metallic spherules. Kelly et al. (1974) and we ground off the oxide coating on the spherules before analysis under the assumption these represented low-temperature rusting of the spherules. In retrospect, this most likely biased the data. The oxide coatings likely represent original high-temperature oxides produced as the spherules were airborne immediately after the impact. By grinding these off, we artificially increased the concentrations of the less oxidized elements that are concentrated in the metallic cores. True bulk analyses of spherules, including the oxide coatings, are needed to further evaluate their origin.

Geochemical Constraints on the Impact Process

We have shown that impact-melt particles contain material from the upper part of the target stratigraphy, that all are consistent with mixtures of Moenkopi and Kaibab rocks, and that those particles that seem to have been derived from lower portions of the melt zone contain less projectile material. These observations allow us to constrain the mechanism of impact melting.

One mechanism for producing dispersed impact-melts is through jetting, a process that occurs during the earliest phase of cratering when the free surface of a hypervelocity projectile forms an oblique angle with the target and results in target-projectile mixing (see Kieffer, 1977; Melosh, 1989). At Meteor Crater, this geometry occurred when a roughly spherical Canyon Diablo meteoroid contacted the target surface (Fig. 11), and continued until the projectile penetrated about halfway into the target (Melosh, 1989). Roddy (1978) calculated the kinetic energy of formation of Meteor Crater, and the projectile size for various assumed impact velocities. The lowest velocity assumed results in a maximum projectile radius of ~16 m, which sets the approximate maximum depth for jetting. We previously favored a shallow zone of melting for formation of some of the impact-melt glasses, <30 m (Hörz et al., 2002), compatible with jetting as the formation mechanism. However, none of the impact-melt particles (this study) or the clear glasses (Hörz et al., 2002) contains only Moenkopi and projectile components. Roddy (1978) estimated the Moenkopi averaged ~8.5 m thick in the target zone. We thus expect that some impact melts should be free of Kaibab material if jetting was a major formation mechanism. Because of this, we do not believe jetting was the dominant mechanism for production of the impact-melts. The bulk particles we have studied are mostly large, roughly cm-sized (Appendix 3). If jetting predominantly produced finer impact-melt spray, our studies could have missed those particles composed only of Moenkopi plus projectile. Thus we cannot rule out jetting as an important mechanism.

Another mechanism for formation of ballistically dispersed impact-melt particles is along the contact between the deforming projectile and the transient crater floor (Fig. 11). In this case, melting may extend to depths a few times the projectile radius (e.g., Melosh, 1989). Impact-melts produced would flow up the transient crater wall and could mix with solid target material during flow. Thus, melts formed after the deformed projectile was entirely below the original target surface would have the projectile component diluted by inclusion of clasts of target material. This is compatible with our data on bulk particles. Those particles with the lowest normalized Ir content (lowest projectile component) have the strongest Kaibab signature, suggesting formation deeper in the transient crater.

Because the melt compositions are variable, the melt-mixing process must have been chaotic, and clear signatures were not preserved in the impact-melt particles. We believe our data are compatible with both jetting and melting at the transient crater-deforming projectile interface as being important in the formation of the ballistically dispersed impact-melt particles at Meteor Crater.

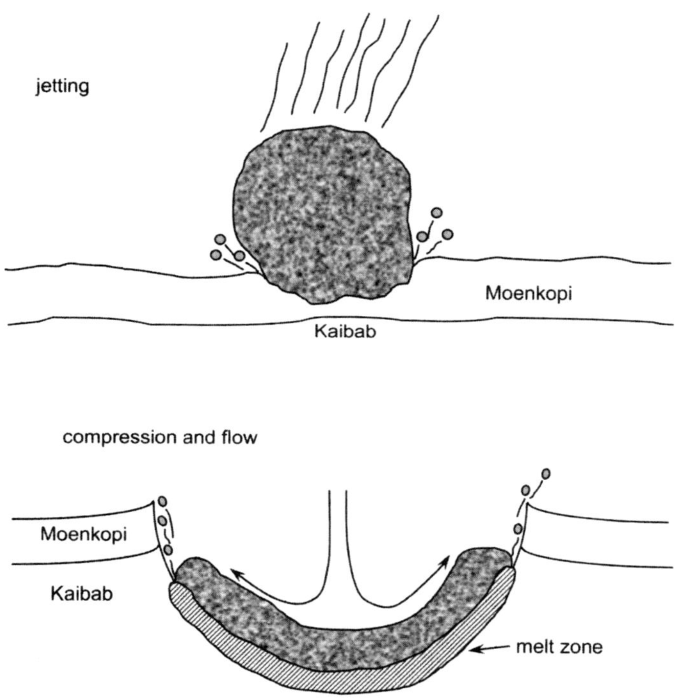

Figure 11. Schematic diagrams indicating two possible mechanisms for formation of impact-melt particles at Meteor Crater. Jetting (upper) occurs while the contact between the projectile and target remains at an oblique angle, and will continue until the projectile deforms substantially, roughly at a depth equal to its radius (Melosh, 1989). The thickness of the Moenkopi is estimated to have been ≥0.5 the projectile radius (Roddy, 1978). Thus jetting ought to have produced some impact-melt particles composed dominantly of Moenkopi plus meteorite. We have not found any of these, but the compositions of some melts indicate formation at shallow levels (<30 m) of the target (Hörz et al., 2002). Impact-melts will also form at the interface between the deforming projectile and the transient crater floor (lower), and the melt zone will extend a few projectile radii below the surface (Melosh, 1989). Flow of the melt up the transient crater wall will promote mixing. The data suggest that many of the particles were formed by this latter mechanism.

CONCLUSIONS

Ballistically dispersed impact-melt particles from Meteor Crater are heterogeneous in composition and are dominantly composed of three components: Canyon Diablo meteorite, Moenkopi Formation and Kaibab Formation. The compositions of the impact-melt particles do not require that rocks from the lower portions of the target section—the Toroweap and Coconino Formations—be mixed in. The data do not rule out up to 10% contribution from these rocks, but it is more plausible that these sandstones are not components.

The bulk impact-melt particles contain on average ~14% Canyon Diablo iron, with a maximum of 20%–22%. This is higher than we previously found for ballistically dispersed impact-melts from the much smaller Wabar Crater (Mittlefehldt et al., 1992). Siderophile-element ratios are essentially unchanged from those of the Canyon Diablo iron, except for Au. Many samples have Au/Ni and Au/Ir ratios much lower than those of the projectile. The lowest ratios are found for impact-melt particles with the highest Kaibab component. The Br contents of Kaibab rocks are higher than those for other target rocks, and we suggest that loss of volatile Au halides may have caused the fractionation.

Metallic spherules are generally enriched in Co, Ni, Ir, and Au compared to Canyon Diablo metal. Element/Ni ratios show subtle deviations from Canyon Diablo ratios that are inversely correlated with ease of oxidation of the element relative to Ni. We attribute this to partial oxidation of molten metal spherules during flight. Because we ground off oxide coatings from the spherules prior to analysis, our data are biased in favor of the metallic cores. The spherule compositions are consistent with slight enhanced melting of schreibersite-troilite–rich regions of the projectile, and this can be explained by enhanced shock melting of lower density graphite-troilite-schreibersite inclusions present in Canyon Diablo.

ACKNOWLEDGMENTS

This paper is dedicated to the memory of David J. Roddy, field-guide, colleague, and friend.

We thank The Barringer Crater Company for permission to do field work in Meteor Crater, and the late D.J. Roddy for guiding our field collection. The senior author thanks J.H. Jones and A.H. Treiman for accompanying him in the field. We thank C.B. Moore and the Center for Meteorite Studies, Arizona State University, for providing the impact-melt particles and metallic spherules studied here. Reviews by H. Huber and an anonymous reviewer, and the editorial handling by A. Deutsch led to substantial improvements. The NASA Cosmochemistry Program supported DWM, and the Planetary Geology and Geophysics Program supported FH in these studies.

APPENDIX 1. CANYON DIABLO METEORITE

TABLE A1. AVERAGE COMPOSITION OF THE METAL PHASE OF CANYON DIABLO IRON METEORITE AND KEY ELEMENT RATIOS*

Cr	µg/g	25
Fe	mg/g	922
Co	µg/g	4660
Ni	mg/g	70.1
Ga	µg/g	81.8
Ge	µg/g	327
As	µg/g	13.1
Sb	ng/g	290
W	ng/g	1110
Ir	ng/g	2320
Au	ng/g	1560
Fe/Ni		13.2
Co/Ni		6.65×10^{-2}
Ir/Ni		3.31×10^{-5}
Au/Ni		2.23×10^{-5}
Au/Ir		0.672

*Average composition is taken from Choi et al. (1995) except for Fe. Iron is determined by difference from 1000 mg/g by subtracting Co and Ni from above, and C (1.6 mg/g) and P (2.0 mg/g) estimated from Buchwald (1975).

APPENDIX 2. TARGET ROCKS

Samples K-1, K-2, and C-1 through C-4 (Table 2) are representative samples of Kaibab and Coconino collected by Hörz from undocumented locations. Samples labeled Ma, Ka, Ta, Ca, etc., are from a continuous stratigraphic section collected by Hörz and See for this study (See et al., 2002). The samples described below were collected by Mittlefehldt from the crater and its vicinity.

Moenkopi Formation

MC-01: Light pink, massive sandstone; north crater wall, ~2–2.5 m above Kaibab contact.

MC-07: Reddish siltstone/clay-rich facies; block in talus, southwest crater wall.

Kaibab Formation

MC-06: Shattered gray dolomite; southwest crater wall, ~15 m above formation base.

MC-10: Gray dolomite; overturned flap, west crater rim.

MC-12: Gray dolomite; boulder float in stream, Oak Creek Canyon.

Toroweap Formation

MC-08: Gray sandstone; overturned flap, southwest crater rim.

MC-13: Knobby gray sandstone; overturned flap; southwest crater rim.

Coconino Formation

MC-02: White, frothy, and powdery sandstone fallback deposit; shaft dump, crater floor.

MC-03: Shattered gray sandstone; southwest crater wall, ~70 m below Kaibab.

MC-04: Clast-free gray matrix of brecciated sandstone; ~10 m above MC-03.

MC-05: Gray sandstone; southwest crater wall, ~5 m below Kaibab.

MC-09: Highly shocked sandstone ejecta debris; south crater rim.

MC-11: Gray sandstone; stream cut, Oak Creek Canyon.

APPENDIX 3. IMPACT-MELT PARTICLES

H11: Composite of 15 individual irregularly shaped cm-sized impact-melt particles. The particles are vesicular, with dark gray interiors and thin brownish rims. The particles were individually ground and the powders mixed and homogenized to produce a bulk sample of about 4 grams.

H11a: Individual cm-sized irregularly shaped impact-melt particle from the same group used to prepare H11.

IIG-1: Irregularly shaped dark brown vesicular impact-melt particle ~1.8 × 1.3 × 1.3 cm in size. Vesicles are up to ~4 mm in size. Numerous small clasts are present. Some secondary minerals are present on the surface and in interior vesicles. IIG-1a and IIG-1b were hand-picked from the freshest material.

IIG-2: Irregularly shaped dark brown impact-melt particle ~1.8 × 1.4 × 1.4 cm in size containing very few vesicles, ~1–2 mm in size. Numerous small clasts are present. Some secondary minerals are present on the surface and in interior vesicles. IIG-2a and IIG-2b were hand-picked from the freshest material.

IIG-3: Irregularly shaped dark brown impact-melt particle ~1.9 × 1.1 × 1.1 cm in size containing few vesicles ≤1 mm in size. Little clastic material is present. Some secondary minerals occur on the surface and in interior vesicles. Samples IIG-3a through IIG-3d were hand-picked from the freshest material.

IIG-4: Smooth, highly irregularly shaped impact-melt particle. The particle is composed of black to red-black glass and yellow-brown, melted clastic debris. The interior is cored with melted clastic material. Nearly pure impact-glass (IIG-4) and clast (IIG-4c) samples were hand-picked.

M1: Large (7 mm) accretionary sphere of black glass with many accretionary promontories and splashes, some flattened and transitional to the dark matrix glass. The bead is vesicular, almost a hollow shell. Some interior surfaces and small vesicles contain fine-grained secondary minerals.

M2: Large (5 × 3 mm), dense, black, shiny spheroid with sugary surface texture. It contains no prominent accretions, but has one hemispherical depression. The bead is very magnetic, and is an oxide bead.

M3: Large (9 mm) rounded brownish bead. The interior is highly porous and weathered with a distinctly orange, oxidized skin. The exterior surface is dull and pitted with sugary appearance. Quartz clasts are present. Samples M3a and M3b are two different samples of hand-picked material.

M4: Large (8 mm) ellipsoid of dark glass with grayish hue. It is a mixture of dark glass and a lighter grayish glass. Many accretionary particles are present. The interior is vesicular, nearly hollow, and contains secondary minerals. A patch of fine-grained, white sugary material may be a glassy clast. Sample M4a is hand-picked, glassy material; M4b is hand-picked material enriched in secondary minerals.

M5: Stubby ellipsoid (4 mm) of dark colored, fresh-appearing glass. The surface is dull and sugary with many small pits. The interior is very porous, essentially frothy. Accretionary features are abundant. Interior vesicles contain some secondary minerals.

M6: Large (12 mm), broken, ropy, and cylindrical lapillus of dull glass with pitted surface with many accretionary particles on the surface. Few clasts are present. The interior displays a dark, very porous glass matrix and tan inclusions of melt or finely crystalline material.

M7: Round sphere (3 mm) of dark glass with numerous accretionary beads. Slightly lighter colored glass makes distinct promontories. The interior contains numerous microscopic pores, some clasts, and a small amount of fine-grained secondary minerals.

M8: A highly porous, slag-like, very friable weathered brown glass containing numerous clasts and pockets of different colored melts. The rim is modestly oxidized. Sample M8a is hand-picked coarse fragments; M8b is the fine debris produced during crushing.

M10: A dull, dark colored impactite. It is coarsely porous and somewhat weathered with an oxidized rim, but only a small amount of fine-grained secondary minerals.

M11: A large (12–15 mm) flattened ellipsoid of porous, tan melt. The surface is unusually sculptured with polygons forming a cauliflower-like texture. The sample is highly altered.

M12: An unusual particle of tektite-like, dense black glass; almost volcanic in appearance. The sample is very fresh, and clast-poor; possibly clast-free.

M13: A very large (14 mm) aerodynamically shaped impactite of rusty-brown colored glass. The surface is dull and sugary in appearance with no good accretionary features. The interior is frothy and has a red colored weathering rind surrounding a dark gray/black core. Sample M13a is reddish rind glass; M13b is black interior glass.

M14: A large (8 mm long) elongated, ropy lapillus fragment. The surface has numerous weathered knobs of dull, secondary material and some clasts. The interior is frothy, containing dark and honey-yellow glass with little clastic material.

M15: A large, ropy lapillus fragment (~15 mm long) with many accretionary features of dark glass. Some secondary minerals are present on the surface. The interior shows yellowish weathering rind glass with a fresh black glass core. Some clastic material is present. The sample was very tough to break.

M16: A distinctly rusty-colored, highly porous aerodynamically shaped melt object. Most of the particle is altered, with little fresh, dark gray to black glass in the interior. It contains some clastic quartz.

M17: A large (11 mm) ropy, elongate lapillus of dense dark melt with many accretionary knobs on the surface. The interior includes fresh black glass and reddish, altered glass.

M18: A large ovoid lapillus (16 mm) with a grayish-white surface of secondary minerals and bleached glass. Numerous honey-colored pockets are visible. The interior displays intimate mixing of the different colored glasses. The particle contains a distinct, bleached rind. Sample M18a is hand-picked interior glass containing some bleached glass; M18b is enriched in bleached glass; M18c is exterior material enriched in secondary minerals.

M19: A flat, pancake shaped object displaying prominent ridges showing material flow and aerodynamic shaping. The interior is porous, somewhat weathered black glass with distinctly rusty-colored oxidation rim. Secondary minerals are present on the exterior.

M20: Small irregular fragment of a ballistic bead of dense mostly honey-colored glass.

M21: A very fresh appearing bead of yellow glass with accretionary promontories of black and yellow glasses—a mixed melt. No secondary minerals are present.

M22: A hollow hemisphere—a fragment of a hollow sphere—with a smooth outer surface. The interior glass is dark and contains numerous microvesicles.

M23: An elongate bead of yellow glass with prominent accretionary knobs. Some black glass is present in the interior.

M24: A large (9 × 2 mm) dumbbell of relatively dense, dark glass, somewhat sugary, and with a pitted surface. A very thin weathering rind is present.

M25: A fresh, small (3–4 mm) bead of dark glass with a distinctly shiny surface with an uneven distribution of accretionary features. The interior contains some vesicles, and a few clasts.

M26: A cylindrical object of unusual dark wine-red color. The particle is fresh with shiny surfaces and no prominent accretionary features. The interior contains some black glass.

M27: A sphere of very fresh dense dark glass with a smooth, shiny surface, fresh, dark, frothy glass in the interior, and with a thin alteration rind.

SIG-1: An oblong hollow spheroid ~5 × 4 × 4 mm in size with few protuberances. One end contains many <1 mm size vesicles. The interior cavity is filled with secondary minerals. Several fresh, black glass fragments were hand-picked for analysis.

SIG-2: A hollow sphere ~4 mm in size with a few protuberances and many <0.5 mm size vesicles. This sample contained relatively little secondary material. One piece of fresh glass was hand-picked for analysis.

SIG-3: Flattened (on one side) spheroid ~5 mm in size with a large protuberance on the flattened side. There are a few <1 mm size vesicles. This sample was partially hollow—a clast occupies part of the center. Fragments of impact glass (SIG-3) and a sample of secondary minerals (SIG-3w) were selected for analysis.

SIG-4: Slightly vesicular, smooth, brown, hollow glass ball. A red-brown weathering rind surrounds fresh black glass. The interior cavity is small, and partially filled with white secondary minerals.

APPENDIX 4. INAA DATA ON MISCELLANEOUS SAMPLES

TABLE A2. COMPOSITIONS OF MISCELLANEOUS SAMPLES FROM METEOR CRATER DETERMINED BY INSTRUMENTAL NEUTRON ACTIVATION ANALYSIS

		IIG-4c clast‡		SIG-3w sec min‡		M4b sec min‡		M18c sec min‡		M2 oxide‡	
mass†	mg	16.73	±	3.99	±	1.05	±	20.91	±	54.58	±
Na	μg/g	324	4	1800	20	169	4	1010	10	12.6	0.8
K_2O	wt%	0.91	0.02	1.25	0.03			0.28	0.01	0.014	0.007
CaO	wt%	18	1	3.9	0.5	50	2	17.4	0.6		
Sc	μg/g	4.05	0.05	4.25	0.05	0.12	0.01	2.11	0.02	0.11	0.03
Cr	μg/g	46	1	34	2			56.7	0.9	15	1
Fe	mg/g	54.6	0.6	13.2	0.4	2.1	0.2	56.5	0.6	712	7
Co	μg/g	213	2	18.0	0.2	15.6	0.3	254	3	2730	30
Ni	mg/g	3.27	0.08	0.11	0.02	0.2	0.1	3.53	0.07	14.1	0.3
Zn	μg/g							23	3		
As	μg/g	3.15	0.08	3.0	0.2	2.6	0.3	3.3	0.1	0.9	0.2
Br	μg/g	0.48	0.05	1.3	0.2	28	1	2.4	0.1		
Rb	μg/g	28	4	42	5			11	2		
Sr	μg/g			120	60			350	20		
Zr	μg/g	260	40	450	60			240	30		
Sb	ng/g	220	20	480	40			210	10		
Cs	μg/g	2.39	0.06	2.76	0.08			0.85	0.03		
Ba	μg/g	690	20	310	30	1000	200	3440	80		
La	μg/g	14.2	0.2	16.6	0.2	4.4	0.1	7.77	0.09	0.35	0.01
Ce	μg/g	28.0	0.5	29.6	0.8			16.2	0.3		
Nd	μg/g	12	3					9	3		
Sm	μg/g	3.04	0.04	2.72	0.04	0.51	0.03	1.62	0.02	0.053	0.005
Eu	μg/g	0.65	0.02	0.54	0.02			0.377	0.009		
Tb	μg/g	0.42	0.03	0.38	0.04			0.24	0.01		
Yb	μg/g	1.60	0.04	1.49	0.08			0.94	0.03		
Lu	μg/g	0.245	0.008	0.22	0.02	0.10	0.05	0.145	0.008		
Hf	μg/g	7.5	0.2	12.2	0.3			7.0	0.1		
Ta	ng/g	480	40	470	40			300	20		
W	ng/g	380	80	1100	300					700	300
Ir	ng/g	38	3					92	2		
Au	ng/g	4.7	0.7			24	9	10.9	0.9	33	2
Th	μg/g	4.0	0.1	4.3	0.2			2.10	0.06		
U	μg/g	1.56	0.06	1.4	0.1			1.52	0.06		

Note: In this table, Ni is given in units of mg/g. Individual uncertainties are given in the ± columns.

†Mass refers to mass in mg of analyzed material and represents most or all of the mass of sample prepared.

‡Clast—lithic clast separated from impact melt particle IIG-4; sec min—secondary minerals found in impact melt particles SIG-3, M4 and M18; oxide-rounded oxide bead.

REFERENCES CITED

Attrep, M., Jr., Orth, C.J., Quintana, L.R., Shoemaker, C.S., Shoemaker, E.M., and Taylor, S.R., 1991, Chemical fractionation of siderophile elements in impactites from Australian meteorite craters [abs.]: Lunar and Planetary Science, v. 22, p. 39–40.

Benedix, G.K., McCoy, T.J., Keil, K., and Love, S.G., 2000, A petrologic study of the IAB iron meteorites: Constraints on the formation of the IAB-Winonaite parent body: Meteoritics & Planetary Science, v. 35, p. 1127–1141.

Blau, P.J., Axon, H.J., and Goldstein, J.I., 1973, Investigation of the Canyon Diablo metallic spheroids and their relationship to the breakup of the Canyon Diablo meteorite: Journal of Geophysical Research, v. 78, p. 363–374.

Boyd, F.R., and Mertzman, S.A., 1987, Composition and structure of the Kaapvaal lithosphere, southern Africa, *in* Mysen, B.O., ed., Magmatic Processes: Physicochemical Principles: University Park, Pennsylvania, Geochemical Society, p. 13–24.

Brett, R., 1967, Metallic spherules in impactite and tektite glasses: American Mineralogist, v. 52, p. 721–733.

Brown, R.W., 1977, A sample fusion technique for whole rock analysis with the electron microprobe: Geochimica et Cosmochimica Acta, v. 41, p. 435–438.

Buchwald, V.F., 1975, Handbook of Iron Meteorites: Berkeley, University of California Press, 1418 p.

Bunch, T.E., Keil, K., and Olsen, E., 1970, Mineralogy and petrology of silicate inclusions in iron meteorites: Contributions to Mineralogy and Petrology, v. 25, p. 297–340.

Choi, B.-G., Ouyang, X., and Wasson, J.T., 1995, Classification and origin of IAB and IIICD iron meteorites: Geochimica et Cosmochimica Acta, v. 59, p. 593–612, doi: 10.1016/0016-7037(94)00384-X.

Dence, M.R., 1971, Impact melts: Journal of Geophysical Research, v. 76, p. 5552–5556.

Gibbons, R.V., Hörz, F., Thompson, T.D., and Brownlee, D.E., 1976, Metal spherules in Wabar, Monturaqui, and Henbury impactites: Proceedings, Lunar Science Conference, 7th, New York: New York, Pergamon Press, p. 863–880.

Grieve, R.A.F., 1982, The record of impact on Earth: Implications for a major Cretaceous/Tertiary impact event, *in* Silver, L.T., and Schultz, P.H., eds., Geological implications of impacts of large asteroids and comets on the earth: Geological Society of America Special Paper 190, p. 25–37.

Grieve, R.A.F., 1991, Terrestrial impact: The record in the rocks: Meteoritics, v. 26, p. 175–194.

Hagerty, J.J., and Newsom, H.E., 2003, Hydrothermal alteration at the Lonar Lake impact structure, India: Implications for impact cratering on Mars: Meteoritics & Planetary Science, v. 38, p. 365–381.

Hörz, F., See, T.H., Murali, A.V., and Blanchard, D.P., 1989, Heterogeneous dissemination of projectile materials in the impact melts from Wabar crater, Saudi Arabia: Proceedings, 19th Lunar and Planetary Science Conference: Houston, Lunar and Planetary Institute, p. 697–709.

Hörz, F., Mittlefehldt, D.W., See, T.H., and Galindo, C., 2002, Petrographic studies of the impact melts from Meteor Crater, Arizona, USA: Meteoritics & Planetary Science, v. 37, p. 501–531.

Kargel, J.S., Coffin, P., Kraft, M., Lewis, J.S., Moore, C., Roddy, D., Shoemaker, E.M., and Wittke, J.H., 1996, Systematic collection and analysis of meteoritic materials from Meteor Crater, Arizona [abs.]: Lunar and Planetary Science, v. 27, p. 645–646.

Kelly, W.R., Holdsworth, E., and Moore, C.B., 1974, The chemical composition of metallic spheroids and metallic particles within impactite from Barringer Meteorite Crater, Arizona: Geochimica et Cosmochimica Acta, v. 38, p. 533–543, doi: 10.1016/0016-7037(74)90039-8.

Kieffer, S.W., 1971, Shock metamorphism of the Coconino Sandstone at Meteor Crater, Arizona: Journal of Geophysical Research, v. 76, p. 5449–5473.

Kieffer, S.W., 1977, Impact conditions for formation of melt by jetting in silicates, *in* Roddy, D.J., Pepin, R.O., and Merrill, R.B., eds., Impact and Explosion Cratering: New York, New York, Pergamon Press, p. 751–769.

King, D.T., Jr., Neathery, T.L., Petruny, L.W., Koeberl, C., and Hames, W.E., 2002, Shallow-marine impact origin of the Wetumpka structure (Alabama, USA): Earth and Planetary Science Letters, v. 202, p. 541–549, doi: 10.1016/S0012-821X(02)00803-8.

Lide, D.R., ed., 2001, CRC Handbook of Chemistry and Physics, 82nd Edition: Boca Raton, Florida, CRC Press.

Melosh, H.J., 1989, Impact Cratering: A Geologic Process: New York, Oxford University Press, 245 p.

Mertzman, S.A., 2000, K-Ar results from the southern Oregon-northern California Cascade Range: Oregon Geology, v. 62, p. 99–122.

Mittlefehldt, D.W., and Lindstrom, M.M., 1991, Generation of abnormal trace element abundances in Antarctic eucrites by weathering processes: Geochimica et Cosmochimica Acta, v. 55, p. 77–87, doi: 10.1016/0016-7037(91)90401-P.

Mittlefehldt, D.W., and Lindstrom, M.M., 1993, Geochemistry and petrology of a suite of ten Yamato HED meteorites: Proceedings, National Institute of Polar Research Symposium on Antarctic Meteorites 17th, no. 6, p. 268–292.

Mittlefehldt, D.W., See, T.H., and Hörz, F., 1992, Dissemination and fractionation of projectile materials in the impact melts from Wabar Crater, Saudi Arabia: Meteoritics, v. 27, p. 361–370.

Mittlefehldt, D.W., McCoy, T.J., Goodrich, C.A., and Kracher, A., 1998, Nonchondritic meteorites from asteroidal bodies, *in* Papike, J.J., ed., Planetary Materials: Reviews in Mineralogy, v. 36, p. 4-1–4-195.

Mittlefehldt, D.W., Hörz, F., and See, T.H., 2000, Trace element chemistry of Meteor Crater impact-melt particles and target rocks—Empirical evidence on the cratering process [abs.]: Lunar and Planetary Science v. 31, no. 1798.

Morgan, J.W., Higuchi, H., Ganapathy, R., and Anders, E., 1975, Meteoritic material in four terrestrial meteorite craters: Proceedings, Lunar Science Conference, 6th, New York: New York, Pergamon Press, p. 1609–1623.

Nininger, H.H., 1956, Arizona's Meteorite Crater: Denver, Colorado, World Press, 232 p.

Nishiizumi, K., Kohl, C.P., Shoemaker, E.M., Arnold, J.R., Klein, J., Fink, D., and Middleton, R., 1991, In situ ^{10}Be-^{26}Al exposure ages at Meteor Crater, Arizona: Geochimica et Cosmochimica Acta, v. 55, p. 2699–2703, doi: 10.1016/0016-7037(91)90388-L.

Palme, H., 1982, Identification of projectiles of large terrestrial impact craters and some implications for the interpretation of Ir-rich Cretaceous/Tertiary boundary layers, *in* Silver, L.T., and Schultz, P.H., eds., Geological implications of impacts of large asteroids and comets on the earth: Geological Society of America Special Paper 190, p. 223–233.

Palme, H., Janssens, M.-J., Takahashi, H., Anders, E., and Hertogen, J., 1978, Meteoritic material at five large impact craters: Geochimica et Cosmochimica Acta, v. 42, p. 313–323, doi: 10.1016/0016-7037(78)90184-9.

Phillips, F.M., Zreda, M.G., Smith, S.S., Elmore, D., Kubik, P.W., Dorn, R.I., and Roddy, D.J., 1991, Age and geomorphic history of Meteor Crater, Arizona, from cosmogenic ^{36}Cl and ^{14}C in rock varnish: Geochimica et Cosmochimica Acta, v. 55, p. 2695–2698, doi: 10.1016/0016-7037(91)90387-K.

Roddy, D.J., 1978, Pre-impact geologic conditions, physical properties, energy calculations, meteorite and initial crater dimensions and orientations of joints, faults and walls at Meteor Crater, Arizona: Proceedings, Lunar and Planetary Science Conference, 9th New York: New York, Pergamon Press, p. 3891–3930.

Roddy, D.J., Boyce, J.M., Colton, G.W., and Dial, A.L., Jr., 1975, Meteor Crater, Arizona, rim drilling with thickness, structural uplift, diameter, depth, volume, and mass-balance calculations: Proceedings, Lunar Science Conference, 6th, New York: New York, Pergamon Press, p. 2621–2644.

Schnabel, C., Pierazzo, E., Xue, S., Herzog, G.F., Masarik, J., Cresswell, R.G., di Tada, M.L., Liu, K., and Fifield, L.K., 1999, Shock melting of the Canyon Diablo impactor: Constraints from Nickel-59 contents and numerical modeling: Science, v. 285, p. 85–88, doi: 10.1126/SCIENCE.285.5424.85.

Scott, E.R.D., 1972, Chemical fractionation in iron meteorites and its interpretation: Geochimica et Cosmochimica Acta, v. 36, p. 1205–1236, doi: 10.1016/0016-7037(72)90046-4.

Scott, E.R.D., Haack, H., and McCoy, T.J., 1996, Core crystallization and silicate-metal mixing in the parent body of the IVA iron and stony-iron meteorites: Geochimica et Cosmochimica Acta, v. 60, p. 1615–1631, doi: 10.1016/0016-7037(96)00031-2.

See, T.H., Hörz, F., Mittlefehldt, D.W., Varley, L., Mertzman, S., and Roddy, D., 2002, Major element analyses of the target rocks at Meteor Crater, Arizona: NASA/TM-2002-210787, 23 p.

Shoemaker, E.M., 1963, Impact mechanics at Meteor Crater, Arizona, *in* Middlehurst, B.M., and Kuiper, G. P., eds., The Moon, Meteorites, and Comets: Chicago, Illinois, University of Chicago Press, p. 301–336.

Shoemaker, E.M., 1987, Meteor Crater, Arizona, *in* Beus, S.S., ed., Rocky Mountain Section of the Geological Society of America Centennial Field Guide: Boulder, Colorado, Geological Society of America, v. 2, p. 399–404.

Simonds, C.H., Floran, R.J., McGee, P.E., Phinney, W.C., and Warner, J.L., 1978, Petrogenesis of melt rocks, Manicouagan, impact structure, Quebec: Journal of Geophysical Research, v. 83, p. 2773–2788.

Stöffler, D., Keil, K., and Scott, E.R.D., 1991, Shock metamorphism of ordinary chondrites: Geochimica et Cosmochimica Acta, v. 55, p. 3845–3867, doi: 10.1016/0016-7037(91)90078-J.

Wasson, J.T., and Kimberlin, J., 1967, The chemical classification of iron meteorites, II: Irons and pallasites with germanium concentrations between 8 and 100 ppm: Geochimica et Cosmochimica Acta, v. 31, p. 2065–2093, doi: 10.1016/0016-7037(67)90143-3.

Wasson, J.T., and Ouyang, X., 1990, Compositional range in the Canyon Diablo meteoroid: Geochimica et Cosmochimica Acta, v. 54, p. 3175–3183, doi: 10.1016/0016-7037(90)90132-5.

Wasson, J.T., and Richardson, J.W., 2001, Fractionation trends among IVA meteorites: Contrasts with IIIAB trends: Geochimica et Cosmochimica Acta, v. 65, p. 951–970, doi: 10.1016/S0016-7037(00)00597-4.

Xue, S., Herzog, G.F., Hall, G.S., Klein, J., Middleton, R., and Juenemann, D., 1995, Stable nickel isotopes and cosmogenic beryllium-10 and aluminum-26 in metallic spheroids from Meteor Crater, Arizona: Meteoritics, v. 30, p. 303–310.

Manuscript Accepted by the Society 9 August 2004

Printed in the USA

Geological Society of America
Special Paper 384
2005

Shock metamorphism of siliceous volcanic rocks of the El'gygytgyn impact crater (Chukotka, Russia)

Eugene P. Gurov
Institute of Geological Sciences, National Academy of Sciences of the Ukraine, 55b Oles Gontchar Street, Kiev 01054, Ukraine

Christian Koeberl*
Department of Geological Sciences, University of Vienna, Althanstrasse 14, A-1090 Vienna, Austria

Wolf Uwe Reimold
Impact Cratering Research Group, School of Geosciences, University of the Witwatersrand, Private Bag 3, P.O. Wits 2050, Johannesburg, South Africa

Franz Brandstätter
Natural History Museum, P.O. Box 417, A-1014 Vienna, Austria

Kassa Amare
Department of Geological Sciences, University of Vienna, Althanstrasse 14, A-1090 Vienna, Austria

ABSTRACT

The 18-km-diameter El'gygytgyn crater is located on the Chukotka peninsula, northeastern Russia. It represents the only currently known impact structure formed in siliceous volcanics, including tuffs. The impact melt rocks and target rocks provide an excellent opportunity to study shock metamorphism of volcanic rocks. The shock-induced changes observed in porphyritic volcanic rocks from El'gygytgyn can be applied to a general classification of shock metamorphism of siliceous volcanic rocks.

Strongly shocked volcanic rocks with phenocrysts converted to diaplectic quartz glass and partially melted feldspars as well as cryptocrystalline matrices are widespread in the El'gygytgyn crater. In particular, the following different stages of shock metamorphism are observed: (*i*) weakly to moderately shocked lavas and tuffs with phenocrysts and clasts of quartz and feldspars; (*ii*) moderately shocked volcanic rocks and tuffs with diaplectic glasses of quartz and feldspars; (*iii*) strongly shocked lavas and tuffs with phenocrysts of diaplectic quartz glass and fused glasses of feldspars in melted matrixes; and (*iv*) impact melt rocks and impact glasses. In addition, thin glassy coatings of voids in impact melt rocks have been observed.

While the shock-induced changes of clasts of framework silicates in these volcanic rocks do not differ from respective changes in other crystalline rocks, the fine-grained matrix of porphyritic rocks is converted into fused glass at the same shock pressures as feldspar minerals. No remnants of fine-grained quartz are preserved in matrix converted into fused glass by shock.

Keywords: El'gygytgyn crater, shocked volcanic rocks, shocked rhyolite, Chukotka.

*Corresponding author: christian.koeberl@univie.ac.at.

Gurov, E.P., Koeberl, C., Reimold, W.U., Brandstätter, F., and Amare, K., 2005, Shock metamorphism of siliceous volcanic rocks of the El'gygytgyn impact crater (Chukotka, Russia), *in* Kenkmann, T., Hörz, F., and Deutsch, A., eds., Large meteorite impacts III: Geological Society of America Special Paper 384, p. 391–412.

INTRODUCTION

The El'gygytgyn impact structure is located in the far northeastern part of Russia (centered at 67°30′ N and 172°05′ E) in the Late Mesozoic Ochotsk–Chukotsky Volcanic Belt of Northeast Asia (Fig. 1). The structure was discovered and described as a gigantic volcanic crater in 1933 (Obruchev, 1957). The first suggestion that this structure could be of impact origin was made by Nekrasov and Raudonis (1963), but they searched in vain for coesite in thin sections of volcanic rocks from the crater rim and, consequently, concluded that the "El'gygytgyn basin" had a tectonic and volcanic origin. Zotkin and Tsvetkov (1970) included this structure in a list of probable terrestrial impact craters. Dietz and McHone (1976) studied satellite imagery of the structure and concluded that El'gygytgyn was probably the largest Quaternary impact crater preserved on Earth. Later, Dietz (1977) suggested that El'gygytgyn could be the source structure of the Australasian tektites. Gurov and workers visited the El'gygytgyn structure in 1977 and confirmed its impact origin after finding shock metamorphosed rocks and impact melt rock (Gurov et al., 1978, 1979a, 1979b). Investigations of the El'gygytgyn crater by this group continued into the 1980s and 1990s (Gurov and Gurova, 1991). Further work was done by Feldman et al. (1981), who gave a short description of the geology of the crater and its target. The main types of impact melt rocks and highly shocked volcanic rocks (named slags and pumices) were described. An average chemical composition of the crater target rocks was derived from the study of the compositions of volcanic rock pebbles from eleven locations on the beach of Lake El'gygytgyn. A preliminary geophysical investigation of the crater was carried out by Dabizha and Feldman (1982). The geological structure of the crater rim was described by Gurov and Gurova (1983) and Gurov and Yamnichenko (1995).

First age determinations for the El'gygytgyn impact crater were obtained by fission track dating (4.52 ± 0.11 Ma; Storzer and Wagner, 1979) and K-Ar dating (3.50 ± 0.50 Ma; Gurov et al., 1979b). Later fission track work indicated an age for the crater of 3.45 ± 0.15 Ma (Komarov et al., 1983). More recently, the ^{40}Ar-^{39}Ar dating method yielded an age for impact glasses of 3.58 ± 0.04 Ma (Layer, 2000), in good agreement with some of the earlier results. Thus, the suggestion of Dietz that El'gygytgyn might be the source of the Australasian tektites (of 0.8 Ma age) was not confirmed.

Although the impact origin of the El'gygytgyn structure had been recognized and confirmed more than 20 years ago, an endogenic origin for this structure was once again proposed during the 1990s by Beliy (1982, 1998).

In recent years, the post-impact geological and climatic history of the area around El'gygytgyn crater and, especially, its sedimentary record, have been extensively investigated in a joint research program of the Alfred-Wegener-Institute, Bremerhaven (Germany), the University of Massachusetts, Amherst (USA),

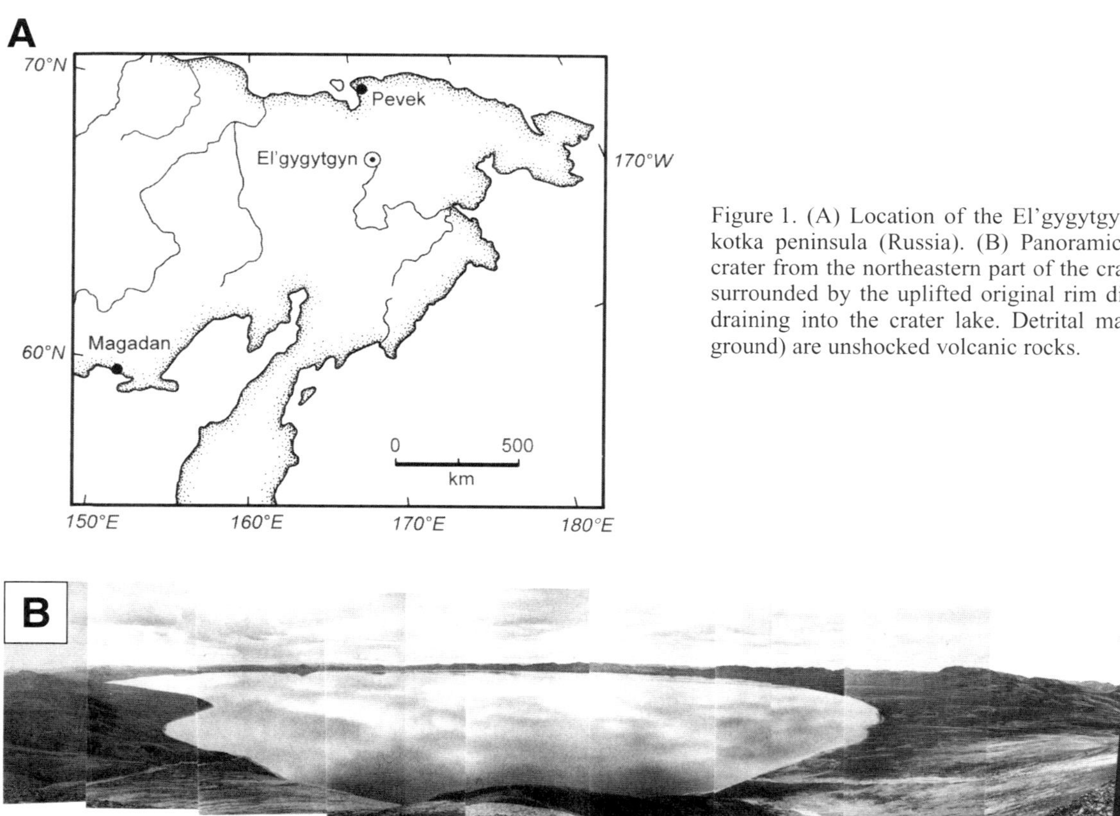

Figure 1. (A) Location of the El'gygytgyn impact structure at Chukotka peninsula (Russia). (B) Panoramic view of the El'gygytgyn crater from the northeastern part of the crater rim. The crater basin is surrounded by the uplifted original rim dissected by several streams draining into the crater lake. Detrital material (left and right foreground) are unshocked volcanic rocks.

and the North-East Interdisciplinary Scientific Research Institute, Magadan (Russia) (cf. Brigham-Grette, 2002). A 12.7-m-long core of lake sediments was studied for paleoclimatic information of this continental Arctic region (Brigham-Grette et al., 1998; Nowaczyk et al., 2002). The dynamics and morphology of ice covering the Lake El'gygytgyn were studied with radar satellite data by Nolan et al. (2003), who also speculated on the location of a possible central uplift buried by lake sediments.

Recently, petrographic and geochemical data on selected impact melt bombs, and related rocks from El'gygytgyn, have been reported by Gurov and Koeberl (2004).

GEOLOGICAL BACKGROUND

The El'gygytgyn impact structure occurs in the Late Mesozoic Ochotsk-Chukotsky Volcanic Belt, specifically in the central part and at the southeastern slope of the Academician Obruchev Ridge of Central Chukotka. The crater forms a flat-floored circular basin with a rim-to-rim diameter of ~18 km (Fig. 2). The crater is one of the best-preserved impact structures on Earth

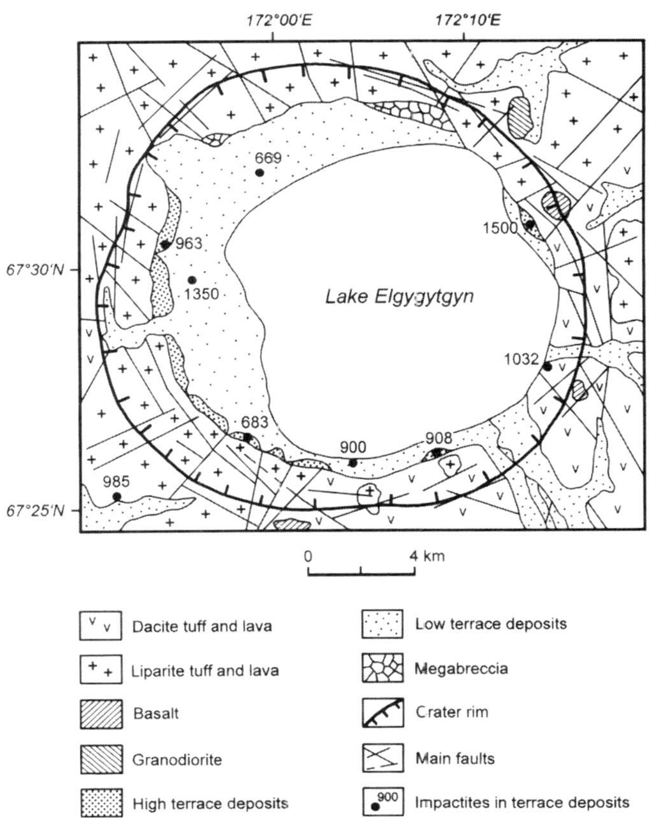

Figure 2. Schematic geological map of the El'gygytgyn impact crater, after Gurov and Gurova (1991). The locations of some more important samples on the terraces, from which impact rocks were studied, are indicated by solid circles and sample numbers.

with excellent morphological expression (Grieve et al., 1988; Dietz and McHone, 1976). The crater floor, ~14 km in diameter, is largely covered by the nearly circular Lake El'gygytgyn, which is 12 km in diameter and up to 170 m deep in its central part; the lake is somewhat offset from the center of the crater rim. A complex system of lacustrine terraces surrounds the lake. The highest terrace is elevated ~80 m above the lake level, and the most modern terrace is 1–3 m high.

A central peak is not exposed on the recent surface of the crater floor, nor is it evident in bathymetric data of the lake bottom. However, from gravity measurements Dabizha and Feldman (1982) suggested the presence of a ~2-km-wide central peak underneath post-crater sediments, and centered relative to the crater outline. Nolan et al. (2003) suggested that the central uplift is centered within the outline of the lake, which, however, would offset the central uplift relative to the crater center. In contrast, recent seismic work cited by Melles et al. (2003) seems to confirm that the central uplift is centered relative to the crater rim, not the lake. The geological structure of the crater is mostly known from the work of Gurov et al. (1978), Gurov and Gurova (1983), and Gurov and Yamnichenko (1995). The crater is surrounded by an uplifted rim that has an asymmetrical cross section, with steep inner walls and gentle outer slopes. The average height of the rim, ~180 m above the lake level and 140 m above the surrounding area, was calculated from 24 radial morphological profiles. A low (10–14 m high), concentric outer feature at a distance of ~1.75 crater radii from the crater center was discovered during morphological investigations of the structure by Gurov and Yamnichenko (1995).

The crater was formed in a sequence of volcanic rocks forming a monoclinic structure that dips to the east by angles of 6°–10°. The target rocks are weakly disturbed in the vicinity of the crater by a complex system of faults that extend to a distance of 2.7 crater radii (~24 km) from the center of the structure (Gurov and Gurova, 1983). The volcanic rocks of the crater area were described as the Pykarvaam and Milguveem series of the Late Cretaceous (Beliy, 1969, 1982; Feldman et al., 1981). ^{40}Ar-^{39}Ar dating of some volcanic rocks of the crater area gave an age of 86 Ma (Layer, 2000).

The stratigraphy of the volcanic country rocks was compiled from numerous sections of volcanic rocks exposed along the inner walls and outer slopes of the crater rim. The average stratigraphy for the largest part of the target area (SW, W, NW, N, and NE parts of the crater area) includes (from the top of the section): ignimbrites—250 m; tuffs and rhyolitic lava—200 m; tuffs and andesitic lava—70 m; and ash tuffs and welded tuffs of rhyolitic and dacitic compositions—100 m. Thus, rhyolitic ignimbrites, lava, and tuffs amount to 89% and andesitic lava and tuffs to 11% of the target composition.

Altogether, a package of 600 m of volcanic rocks is exposed in the crater walls and their outer slopes. The relics of a flat-lying and 110-m-thick basalt sheet occur in the NE part of the crater rim, covering an area of 0.7 km^2. While shocked basalt is only very rarely observed among the clasts of target rocks in melt breccias, we cannot exclude that this basalt sheet—at pre-impact time—covered at least some part of the target area.

Ignimbrites and rhyolitic and dacitic tuffs are the most abundant rock types of the crater basement. Mineral clasts and phenocrysts of those rocks are quartz, orthoclase/alkali feldspar ($Or_{80}Ab_{20}$ to $Or_{60}Ab_{40}$), and plagioclase, mainly of oligoclase and oligoclase-andesine ($Ab_{80}An_{20}$ to $Ab_{70}An_{30}$) composition.

Refractive indices of unshocked orthoclase are: n_γ = 1.527–1.529, n_α = 1.520–1.523, birefringence 0.007, and of plagioclase: n_γ = 1.548–1.553, n_α = 1.541–1.547, birefringence 0.007. Mafic minerals are biotite and, rarely, amphibole and clinopyroxene.

Fine-grained matrices of lavas and tuffs are composed mostly of quartz and feldspars. Textures of lava matrices range from fluidal glassy to fine-grained granular and spherulitic. Andesites and andesite tuffs contain phenocrysts and clasts of andesine ($Ab_{55}An_{45}$ to $Ab_{60}An_{40}$), clinopyroxene and amphibole.

Chemical compositions of the main types of volcanic rocks of the crater basement are presented in Table 1. The average composition given in Table 1 was calculated on the basis of relative proportions of rock types that likely contributed to the target volume, as detailed above.

A somewhat different target rock sequence is indicated by exposures at the eastern and southeastern parts of the crater rim, where dacites as well as dacitic and andesitic tuffs are dominant.

Rocks of the crater rim do not display any characteristic shock metamorphic effects. Only weak cataclasis of these volcanic rocks is displayed in the NE parts of the crater wall. However, megabreccia is widespread in some areas of the inner crater wall, especially in the northern and northwestern sectors (Gurov and Gurova, 1983; Gurov and Yamnichenko, 1995).

Shock metamorphosed rocks and impact melt rocks occur in the El'gygytgyn crater as redeposited material in lacustrine terraces inside the crater and, locally, in terraces of little streams on the outer slope of the crater rim. Deep drilling has not yet been carried out in the El'gygytgyn crater; lithic impact breccia, suevite, and impact melt rock are expected from comparison with other impact structures to occur under the lake sediments in the central part of the crater.

The source of the impact rocks in the terrace deposits were ejecta that have been completely eroded. The original ejecta blanket was composed of lithic impact breccia, suevite, and impact melt breccia. The material is thought to have been transported to the areas of final deposition in the terraces due to slumping off the rim, thus only covering short distances. Fragments of unshocked and shocked rocks are mixed together with fragments of impact melt rocks and impact melt breccias in terrace deposits. Fragments of shocked rocks are generally of irregular form and have generally not been rounded by disintegration of ejecta over short transport distances. Rounded cobbles and pebbles of impact rocks occur only in the recent terraces of the crater. The size of fragments of shock metamorphosed rocks and impactites ranges from 2 to 15 cm. Remnants of dark impact melt glass on the surface of some fragments of shocked rocks are evidence that they originally were part of suevite and/or impact melt breccia.

TABLE 1. COMPOSITION OF VOLCANIC TARGET ROCKS FROM THE EL'GYGYTGYN IMPACT CRATER

	1 (8)	2 (11)	3 (5)	4 (5)	5 (3)	6
SiO_2	69.94	72.05	72.98	63.00	73.37	70.72
TiO_2	0.32	0.23	0.38	0.56	0.15	0.29
Al_2O_3	14.52	13.09	12.49	16.57	12.30	13.90
Fe_2O_3	1.90	1.33	1.28	3.40	1.26	1.76
FeO	1.00	0.76	0.75	1.37	0.49	0.86
MnO	0.08	0.05	0.05	1.12	0.07	0.06
MgO	0.73	0.50	0.48	1.59	0.61	0.72
CaO	2.25	1.37	0.95	3.73	1.66	2.01
Na_2O	2.95	1.97	2.44	3.29	2.30	2.57
K_2O	3.95	4.93	5.79	3.08	5.50	4.48
Li_2O (ppm)	84	91	83	65	82	86
Rb_2O (ppm)	168	176	240	134	238	186
Cs_2O (ppm)	7	10	3	5	2	7
P_2O_5	0.07	0.09	0.04	0.14	0.24	0.10
CO_2	0.65	0.38	0.98	0.91	0.29	0.56
H_2O^-	0.38	0.52	0.31	0.56	0.29	0.39
LOI	1.29	2.24	1.08	1.54	1.47	1.58
Total	100.30	99.80	100.33	101.06	100.32	100.28
Fe_2O_3/FeO	1.90	1.75	1.71	2.48	2.57	2.05

Note: Major elements in wt%, rare alkalies in ppm. Number in parentheses indicates number of samples. 1—rhyolitic ignimbrite; 2—rhyolitic tuff; 3—rhyolite; 4—andesite, andesitic tuff; 5—rhyolitic ash tuff (lower horizon); 6—regional composition, based on the respective thickness of the various rock types in the target region.

Fragments of impact melt breccia in terrace deposits are up to 1 m in size. Aerodynamically shaped glass bombs occur together with shock metamorphosed rocks in the lacustrine terraces inside the crater and also in terraces of some streams around it. All types of impactite are generally fresh and most of them do not display significant post-impact hydrothermal alteration and weathering.

SAMPLES AND METHODS OF INVESTIGATIONS

The samples (E series) for this study of shock metamorphism of volcanic rocks were collected by E. Gurov and E. Gurova during expeditions to the El'gygytgyn crater between 1977 and 1980. Additional samples (the "G" series) were studied by thin section only. Detailed sample descriptions are given in the Appendix.

The shock metamorphism of siliceous volcanic rocks was studied by investigation of the progressive shock-induced changes of quartz and feldspar phenocrysts and clasts of porphyritic volcanic rocks and tuffs. It is obviously difficult to investigate the shock-induced changes of the fine-grained matrices in the weakly to moderately shocked volcanic rocks, and at higher shock degrees the entire matrices are converted to glass. Accordingly, shock metamorphism of fine-grained matrix material of porphyritic rocks is evident only in strongly shocked volcanics and tuffs. Refractive indices of minerals and glasses were determined with the immersion method. Two main refractive indices were measured for shocked and unshocked feldspars: the highest n_γ and the lowest n_α, whereas n_β was not determined. Refractive indices of quartz and feldspar minerals were used for the estimation of shock pressures of the rocks, using calibrations by Stöffler (1974), Stöffler and Langenhorst (1994), and Langenhorst and Deutsch (1994). Additional indications of shock pressure were obtained by X-ray diffractometry (following methods of Hörz and Quaide, 1973). Determination of shock pressures included 3–4 measurements on separate quartz grains per sample. Measurements were also made of orientations of planar fractures (PFs) and planar deformation features (PDFs) in quartz by universal stage analysis. Additional PDF orientation measurements (on sample E963-12) were performed in Vienna in 2003.

The presence of high-pressure polymorphs of silica was determined by X-ray diffractometry on insoluble fractions of shocked quartz after their partial dissolution in hydrofluoric acid. Coesite in diaplectic quartz glass of moderately to strongly shocked rocks was also identified by optical microscopy in several thin sections.

Graphite is rare in volcanic rocks and, thus, also in the impact rocks of this crater. Isolation of shocked graphite from impact melt rocks was made by high-temperature alkaline dissolution of crushed samples, using the method described by Kashkarov and Polkanov (1970). Crystal structural analysis of carbon phases was then carried out by X-ray diffractometry.

The chemical compositions of El'gygytgyn rocks are given in Tables 1–5. The composition of basement rocks and impact melt rocks was studied by wet chemical methods at the Institute of Geological Sciences of the Ukraine (Tables 1, 3, and 4). The composition of some feldspar diaplectic and melt glasses (Table 2) was studied by quantitative wavelength-dispersive microprobe analysis at the Natural History Museum in Vienna, using an ARL-SEMQ instrument (acceleration voltage 15 kV, beam current 20 nA). Data reduction was done with standard ZAF procedures. Major (and some minor) element compositions of 14 selected samples of shocked rocks, impact breccias, and impact glasses were determined by X-ray fluorescence (XRF) spectrometry at the University of the Witwatersrand, and trace element composition of the same samples was analyzed at the University of Vienna by instrumental neutron activation analysis (INAA), following procedures described by Reimold et al. (1994) and Koeberl (1993), respectively (Table 5).

TABLE 2. COMPOSITION OF SHOCKED FELDSPARS FROM VOLCANIC ROCKS OF THE EL'GYGYTGYN IMPACT CRATER (ELECTRON MICROPROBE DATA)

	963-3/1-1	963-3/1-2	963-14/1-2	963-14/1-3	963-14/2-2
SiO_2	64.78	65.58	65.49	58.17	55.38
Al_2O_3	19.13	17.86	18.51	25.16	30.45
FeO	0.07	0.26	0.14	0.37	0.28
MgO	<0.01	0.02	0.05	0.05	0.02
CaO	0.06	0.17	0.69	8.60	9.19
Na_2O	2.00	2.10	4.77	5.63	5.90
K_2O	11.98	11.93	10.23	0.56	0.51
Total	98.03	97.92	99.88	98.54	101.73
Or	70.60	70.10	59.50	2.80	2.80
Ab	16.80	17.30	40.40	47.20	49.80
An	0.30	1.10	3.60	42.60	45.60
Q	7.10	7.80	–	5.60	–
Cor	2.80	1.10	–	–	4.50

Note: E963-3/1-1—orthoclase melt glass from moderately shocked rhyolitic tuff; E963-3/1-2—orthoclase melt glass, same sample; E963-14/1-2—orthoclase diaplectic glass from shocked dacite; E963-14/1-3—andesine diaplectic glass from the same sample; E963-14/2-2—andesine diaplectic glass from the same sample.

TABLE 3. COMPOSITION OF PARTIALLY MELTED VOLCANIC ROCKS AND TUFFS FROM THE EL'GYGYTGYN IMPACT CRATER

	E908-6B	E908-8B	E908-14B	E669-32B	E908-11B	E908-16B	E908-18B
SiO_2	70.28	70.34	73.40	68.88	71.56	70.60	70.50
TiO_2	0.18	0.12	0.18	0.30	0.26	0.30	0.33
Al_2O_3	13.34	14.40	12.82	15.29	15.52	15.70	15.41
Fe_2O_3	1.61	1.21	0.31	0.81	0.61	0.81	1.11
FeO	1.43	1.07	1.07	1.79	1.79	1.79	1.07
MnO	0.03	0.04	0.03	0.07	0.04	0.07	0.03
MgO	1.10	0.59	0.14	0.90	0.36	0.54	0.96
CaO	0.54	0.48	0.41	1.81	1.09	1.84	0.70
Na_2O	2.90	2.60	2.56	4.00	1.90	1.10	1.52
K_2O	5.14	5.14	4.44	4.22	3.72	3.81	3.55
P_2O_5	0.11	0.11	0.14	0.15	0.16	0.14	0.08
CO_2	–	–	–	–	–	–	1.13
H_2O^-	0.32	0.42	0.32	0.17	0.19	0.30	0.84
LOI	3.33	3.91	4.57	1.73	2.86	3.45	2.30
Total	100.31	100.43	100.39	100.12	100.06	100.45	99.53
Fe_2O_3/FeO	1.12	1.13	0.29	0.45	0.34	0.45	1.04

Note: Brief sample information: E908-6B—partially melted ignimbrite with phenocrysts of diaplectic quartz glass and melt glasses of feldspars in vesicular glassy matrix; E908-8B—as 908-6B; E908-14B—partially melted porphyritic rock with phenocrysts of diaplectic quartz glass with coesite; phenocrysts of glassy vesicular feldspars and vesicular matrix; E669-32B—partially melted porphyritic rock with phenocrysts of diaplectic quartz glass, glassy feldspars and strongly shocked pyroxene in vesicular glassy matrix; E908-11B—partially melted rhyolitic tuff; E908-16B—partially melted rhyolitic tuff with relics of layering, clasts of diaplectic quartz glass and melted feldspars in vesicular, glassy groundmass. E908-18B—as 908-16B.

A classification scheme for shock metamorphosed siliceous volcanic rocks was initially proposed by Gurov and Gurova (1979). The classification scheme for granitic rocks by Stöffler (1971) was used for basic comparison, but the peculiarities of these porphyritic rocks and tuffs from El'gygytgyn had to be taken into account. It should be noted, though, that in some cases samples show characteristics of a variety of shock pressures within one specimen. This is common in impactites. Assignment of shock stages was done based on the majority of criteria belonging to one stage.

SHOCK METAMORPHISM OF SILICEOUS VOLCANIC ROCKS

Shock metamorphism of siliceous volcanic rocks is still poorly understood. A first description of strongly shocked tuffs from the Sedan nuclear crater was published by Short (1968). Shocked tuffs ejected from the Sedan crater are vesicular, glassy rocks of low density. Aerodynamic shapes of tuff particles were obviously formed during their ejection and flight. Clasts of quartz and feldspars in shocked tuffs preserve their initial shapes, but are transformed into glass or have low birefringence (Short, 1968). In addition, shocked dacites were described from ejecta of the Acraman impact crater (Gostin et al., 1986; Williams, 1986); however, their role in relation to the impact rocks in this crater is still to be investigated. A rather detailed investigation of shock metamorphosed Deccan Trap basalts of the Lonar impact crater was made by Fredriksson et al. (1973), who described progressive changes of main minerals by shock. Fracturing, formation of PDFs, and

TABLE 4. COMPOSITION OF IMPACT MELT ROCKS AND IMPACT GLASSES FROM THE EL'GYGYTGYN IMPACT CRATER

	1 (25)	2 (18)	3 (3)
SiO_2	69.70	69.21	63.57
TiO_2	0.36	0.34	0.50
Al_2O_3	15.12	14.99	16.03
Fe_2O_3	0.92	1.23	1.74
FeO	2.33	1.85	2.73
MnO	0.07	0.07	0.08
MgO	1.20	1.01	1.46
CaO	2.74	2.32	3.97
Na_2O	2.90	3.15	3.16
K_2O	3.76	3.70	3.19
Li_2O	116	85	–
Rb_2O	139	147	–
Cs_2O	11	10	–
P_2O_5	0.70	0.11	0.12
CO_2	0.21	0.21	1.31
H_2O^-	0.11	0.18	0.52
LOI	0.26	1.37	1.07
Total	100.65	99.98	99.45
Fe_2O_3/FeO	0.39	0.66	0.64

Note: Major elements in wt%; Li, Rb, and Cs in ppm. Numbers in parentheses indicate number of samples measured. 1—Glass bombs from the whole crater area. 2—Impact melt rocks from the SW, W, NW, N, NE, and E parts of the crater. 3—Impact melt rocks from the S part of the crater.

TABLE 5. MAJOR AND TRACE ELEMENT COMPOSITIONS OF SELECTED SHOCKED ROCKS, IMPACT BRECCIAS, AND IMPACT GLASSES FROM THE EL'GYGYTGYN IMPACT STRUCTURE, RUSSIA

Sample #:	E-699-33b	E-908-70	E-908-73	E-908-74	E-963-3	E-963-4	E-963-5	E-669-820	E-900-12	E-908-55	E-908-21b	E-963-12	E-963-14	E-985-101
Shock Stage:	II isotropic volc. rock	II altered volc. rock	II Porphyritic volc. rock	II altered volc. rock	II glass (fr. tuff?)	II glass (fr. tuff?)	II glass (fr. tuff?)	IV Impact melt breccia	IV impact melt rock	IV impact glass	IV impact glass	IV polymict lithic l.br.	IV polymict lithic l.br.	IV impact glass bomb
SiO_2	69.16	63.05	60.48	63.16	71.38	61.70	65.27	66.94	65.93	70.22	73.07	63.90	69.34	69.37
TiO_2	0.26	0.60	0.53	0.63	0.20	0.47	0.74	0.47	0.54	0.37	0.25	0.49	0.37	0.39
Al_2O_3	15.20	17.20	17.83	17.24	15.34	20.20	18.73	15.40	16.21	15.79	14.30	16.42	15.13	14.69
Fe_2O_3	2.93	4.68	4.97	4.61	2.73	4.22	3.85	3.60	4.12	2.86	1.92	4.19	3.64	3.31
MnO	0.05	0.08	0.10	0.09	0.05	0.09	0.10	0.09	0.08	0.08	0.04	0.09	0.09	0.07
MgO	0.24	1.50	1.60	1.44	0.18	0.78	0.72	1.24	1.27	0.90	0.47	1.27	0.69	0.97
CaO	0.50	4.37	4.20	2.37	0.91	2.08	2.52	2.75	3.47	2.73	1.46	3.56	2.44	2.63
Na_2O	2.84	2.93	4.03	2.75	1.77	5.84	2.71	3.47	3.42	3.13	2.62	3.72	4.01	3.24
K_2O	3.37	3.02	3.25	3.26	4.42	3.11	2.37	3.81	3.41	4.36	4.12	3.40	2.16	4.10
P_2O_5	0.08	0.15	0.17	0.15	0.03	0.12	0.06	0.11	0.13	0.09	0.06	0.14	0.05	0.07
LOI	5.44	2.87	2.92	3.08	2.09	1.59	1.93	1.33	0.75	-0.09	2.09	2.51	1.09	0.20
Total	100.07	100.45	100.08	98.78	99.10	100.20	99.00	99.21	99.33	100.44	100.40	99.69	99.01	99.04
Sc	8.39	11.4	10.7	10.7	12.0	11.9	13.3	8.47	10.1	5.54	6.69	9.77	6.46	9.53
V	22	85	74	82	17	34	70	54	67	39	24	24	36	42
Cr	3.07	15.9	130	8.47	3.99	145	8.06	86.0	17.2	12.8	61.5	11.2	136	139
Co	1.84	6.92	7.94	6.16	0.71	4.18	6.59	5.84	6.48	3.61	4.75	4.68	4.27	4.92
Ni	<6	7	58	9	7	68	7	40	16	9	4	8	76	66
Cu	12	11	53	25	18	15	49	8	9	<6	17	14	12	9
Zn	66.4	95.3	95.3	80.5	124	116	109	78.6	87.1	37.5	66.3	80.4	215	49.8
As	5.56	5.42	6.97	2.74	6.24	6.27	9.03	9.14	15.7	2.65	2.70	16.8	3.53	5.72
Se	0.53	0.31	0.13	0.40	0.34	0.32	0.49	0.32	0.30	0.37	0.12	0.44	0.33	0.52
Br	0.5	0.6	0.2	0.1	0.4	0.6	0.4	0.4	0.4	0.5	0.5	0.2	0.2	0.5
Rb	112	116	110	107	186	98.8	137	127	122	13.9	133	143	91.0	146
Sr	138	503	403	384	120	412	317	259	358	284	167	172	774	215
Y	32	27	25	26	49	33	35	28	28	25	21	42	24	38
Zr	192	215	225	217	305	290	226	183	210	191	155	205	229	217
Nb	15	13	13	14	16	17	15	14	14	14	12	15	13	17
Sb	0.65	0.44	0.84	1.24	1.91	0.52	0.93	1.17	0.83	0.75	1.03	0.54	1.0	1.23
Cs	3.89	8.50	9.02	23.5	9.99	15.1	15.1	7.42	5.35	6.75	8.39	2.45	26.8	8.01
Ba	1003	1109	1155	1690	897	1549	858	904	933	891	665	976	1696	987
La	35.0	28.3	27.4	25.1	46.0	45.8	26.9	30.6	31.0	32.9	32.0	40.1	21.7	36.1
Ce	66.5	57.5	52.3	54.1	93.5	92.9	56.6	58.0	60.8	62.6	54.4	80.4	40.3	70.1
Nd	37.5	28.6	27.1	16.1	51.9	38.6	29.3	26.2	27.0	30.1	24.7	40.3	24.1	39.6
Sm	6.42	4.48	4.40	3.76	8.84	6.79	5.63	4.26	5.59	4.28	4.04	6.93	3.85	6.30
Eu	0.81	1.14	1.15	1.02	0.89	1.41	1.07	0.82	1.03	0.80	1.07	0.82	0.99	0.92
Gd	6.44	5.28	4.93	4.37	8.85	7.20	6.46	4.73	6.14	4.81	4.57	7.51	4.78	6.62
Tb	1.03	0.83	0.70	0.74	1.30	1.18	1.05	0.75	0.96	0.79	0.73	1.18	0.68	0.90
Tm	0.61	0.45	0.45	0.47	0.85	0.68	0.52	0.57	0.45	0.55	0.42	0.76	0.38	0.52
Yb	3.97	2.09	2.52	3.12	4.49	3.92	3.77	3.38	3.08	2.85	2.44	4.29	2.0	3.09
Lu	0.56	0.32	0.35	0.36	0.75	0.57	0.50	0.55	0.41	0.43	0.38	0.6	0.31	0.54
Hf	6.18	5.36	5.37	5.18	8.77	7.12	5.63	4.81	6.34	4.86	4.66	5.46	5.01	6.05
Ta	0.66	0.49	0.51	0.47	0.78	0.60	0.58	0.52	0.57	0.61	0.47	0.79	0.48	0.68
Ir (ppb)	<1	<1.5	<1.5	<1.5	<0.5	<1	<1	<1	<0.6	<0.8	<0.9	<0.8	<0.6	<1
Au (ppb)	0.2	0.4	0.8	0.3	0.5	0.3	0.3	0.4	0.4	0.3	0.5	0.3	0.3	0.3
Th	15.1	10.2	8.73	9.69	16.3	12.4	9.0	12.0	11.6	14.0	11.9	15.7	7.46	13.3
U	3.81	3.02	2.44	3.04	4.35	3.04	2.32	3.21	2.88	3.54	2.44	4.58	2.18	3.78
K/U	10614	12000	15984	12868	12193	12276	12259	14243	14208	14780	20262	8908	11890	13016
Th/U	3.96	3.38	3.58	3.19	3.75	4.08	3.88	3.74	4.03	3.95	4.88	3.43	3.42	3.52
La/Th	2.32	2.77	3.14	2.59	2.82	3.69	2.99	2.55	2.67	2.35	2.69	2.55	2.91	2.71
Zr/Hf	31.1	40.1	41.9	41.9	34.8	40.7	40.1	38.0	33.1	39.3	33.3	37.5	45.7	35.9
Hf/Ta	9.36	10.9	10.5	11.0	11.2	11.9	9.71	9.25	11.1	7.97	9.91	6.91	10.4	8.90
La_N/Yb_N	5.96	9.15	7.35	5.44	6.92	7.90	4.82	6.12	6.80	7.80	8.86	6.32	7.33	7.89
Eu/Eu*	0.38	0.72	0.75	0.77	0.31	0.62	0.54	0.56	0.54	0.54	0.76	0.35	0.71	0.44

Note: Major elements in wt%; trace elements in ppm (except as noted). All Fe is Fe_2O_3. N—chondrite-normalized; fr—fragmental.

partial and complete isotropization were found in plagioclase. Fused plagioclase glasses and opaque, fine-grained masses after pyroxene occur in highly shocked rocks of the Lonar crater.

Shock metamorphosed volcanic rocks from El'gygytgyn crater yielded the following information.

Shock Stage 0

Volcanic rocks and tuffs of shock stage 0 (0–10 GPa according to the classification of Stöffler [1971]) do not exhibit any characteristic changes of their framework silicates that could be regarded as characteristic effects of shock deformation, at the scale of the optical microscope. The only microdeformation effect occasionally observed is weak kink banding of biotite phenocrysts, which, by itself, is not a shock-diagnostic deformation effect. Phenocrysts and clasts of quartz and feldspar sometimes are weakly fractured, but never contain PDFs and PFs. Shock pressures of 4 ± 2 GPa were determined for quartz of two large clasts of an ignimbrite from impact melt breccia in the northern part of the crater with the X-ray diffraction method proposed by Hörz and Quaide (1973). Quartz in both samples is weakly fractured, and its refractive indices are not lowered.

The volcanic rocks from the crater rim and from megabreccias at the base of the northern slope do not display any traces of shock deformation at all.

Shock Stage I

Weakly to moderately shocked volcanic rocks with phenocrysts and clasts of shock metamorphosed quartz and feldspar belong to stage I, the upper limit of which is placed at shock pressures necessary for the transition of quartz into diaplectic glass (30–35 GPa).

Weakly to moderately shocked rocks preserve some similarities to their unshocked counterparts in hand samples. Shatter cones up to 2 cm long were observed on three samples of shocked rhyolitic tuff from the terraces in the southern and northeastern parts of the crater (El-1352, El-1500-105, etc.). The most noticeable sign of shock metamorphism of moderately shocked volcanic rocks is the conversion of quartz phenocrysts into whitish, brittle masses with a dense network of planar fractures (PFs), similar to narrow cleavage of feldspar. Sets of PFs and planar deformation features (PDFs) in quartz are visible in thin section (Fig. 3). We measured the orientations of PFs and PDFs in quartz grains from several samples by U-stage (Fig. 4). The PFs are parallel open fissures with spacings on the order of 5 µm. Their orientations are mostly parallel to $\{10\bar{1}1\}$, $\{22\bar{4}1\}$, $\{11\bar{2}2\}$ and (0001) (Fig. 4A).

PDFs in quartz grains are always non-decorated. They are oriented parallel to $\{10\bar{1}3\}$, rarely $\{10\bar{1}2\}$, $\{10\bar{1}1\}$ and $\{51\bar{6}1\}$ (Figs. 4B–4D). High-resolution scanning electron microscope images of PFs show that they are open fissures that separate neighboring blocks of a crystal with weak displacement and rotation. Refractive indices measured in some quartz grains are

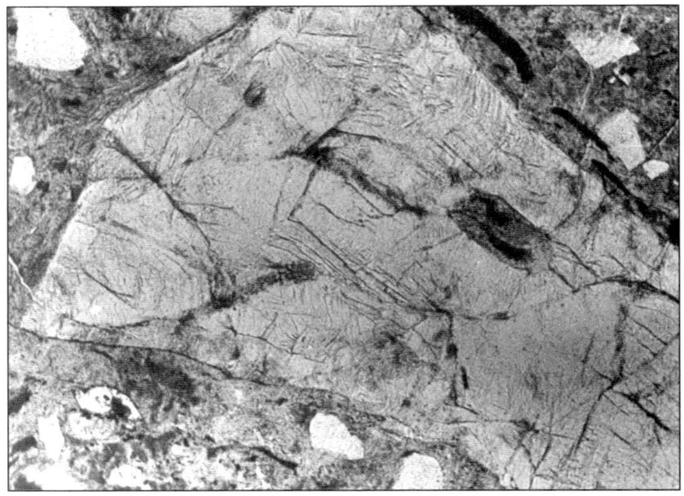

Figure 3. Weakly shocked quartz clast in rhyolite. Prominent irregular fractures, planar to subplanar fractures, and (very faint) multiple planar deformation features occur in quartz (sample E1032-7, 2.0 mm wide, crossed polarizers).

slightly lowered (n_e = 1.548–1.550 and n_o = 1.540–1.542) in comparison to normal values of unshocked quartz, and correspond to shock pressures of 25–28 GPa (Langenhorst and Deutsch, 1994). Shock pressures in excess of 20 GPa were determined by measurements of X-ray asterism of various quartz grains from such samples following the method of Hörz and Quaide (1973).

Orthoclase and plagioclase have lowered refractive indices and birefringence. Some feldspar grains in lavas and tuffs are weakly cataclastic (Fig. 5). Biotite has kink bands and PDFs. The fine-grained matrix of these samples does not exhibit any noticeable signs of shock metamorphism.

More noticeable changes occur in volcanic rocks and tuffs shocked to 30–35 GPa. Systems of open cracks up to 2 cm long and 1 mm wide occur in some tuff clasts of 4–8 cm size (Fig. 6). Quartz in rocks of this shock degree forms semi-transparent, opalescent grains with low refractive indices (1.493–1.475) and birefringence of 0.002–0.003, corresponding to shock pressures between 30 and 35 GPa (Langenhorst and Deutsch, 1994). Numerous PDFs occur in all quartz grains (Fig. 7). The $\{10\bar{1}2\}$ system is dominant (Fig. 3).

High-pressure silica phases were also diagnosed in quartz from stage I. Stishovite was identified in three samples of shocked rhyolite and rhyolitic tuff (samples E987-9, E699-a, and E699-11b) (Gurov et al., 1979a) that had been shocked at pressures between 30 and 35 GPa. Grains of shocked quartz with stishovite are white, very fine-grained, brittle masses with refractive indices of 1.485 ± 0.002. Crystallites with high birefringence, up to 2 µm long and 0.5 µm wide in diaplectic quartz are interpreted as probable stishovite. X-ray patterns of such quartz contain weak main lines of quartz (d = 3.35 Å), stishovite (d = 2.95 Å) and coesite (d = 3.10 Å). X-ray patterns of insoluble residues after dissolution of these quartz fractions in hydrofluoric acid correspond to

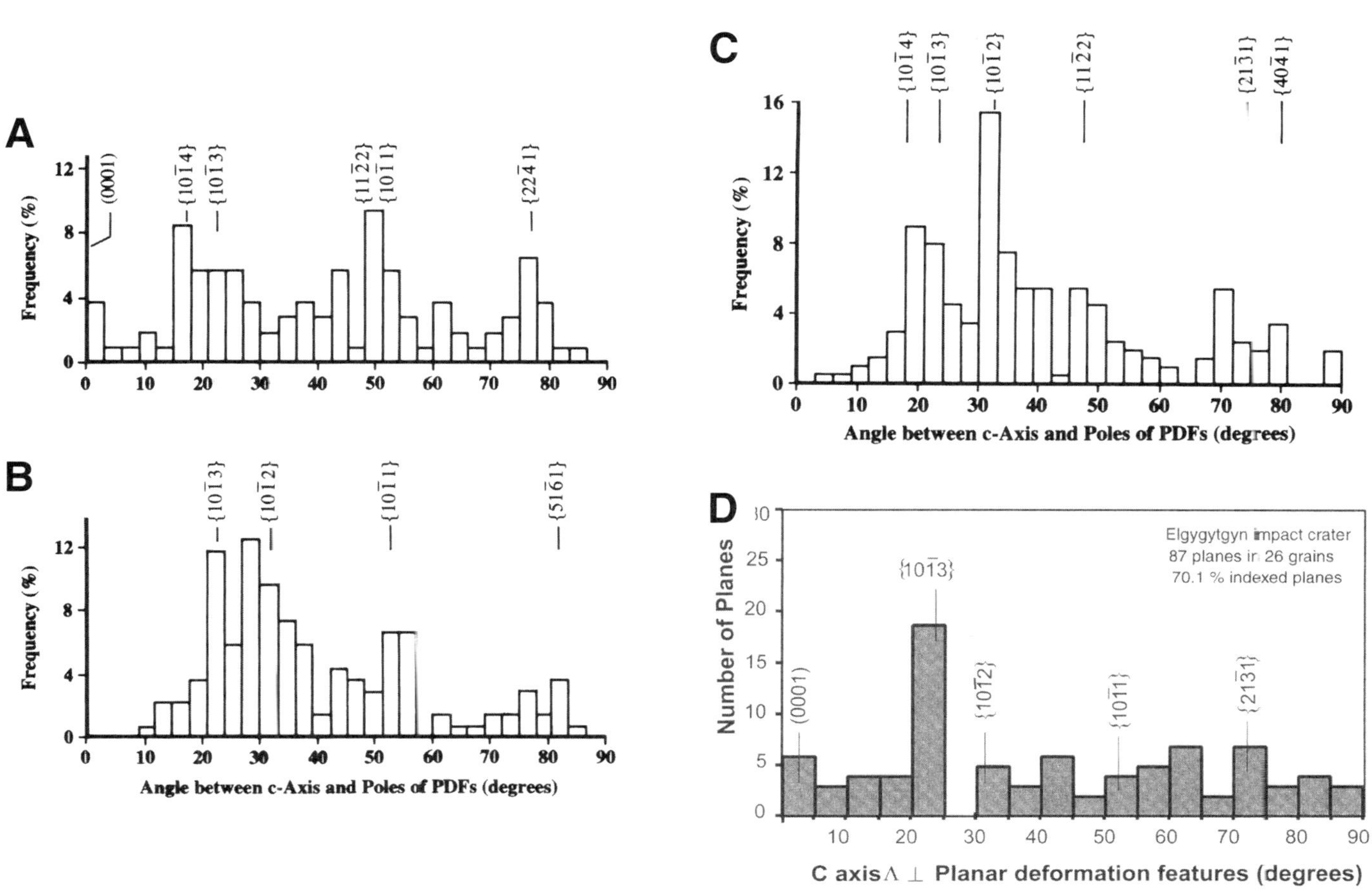

Figure 4. Frequency histograms of angles between the c-axis and poles of PFs (in percent) (A, 52 measurements in 20 grains) and planar deformation features (PDFs) (B, 63 measurements in 21 grains) in weakly shocked quartz. For rhyolite sample E1032-7. (C) Histogram for shocked quartz in moderately shocked rhyolite sample E987-9 (102 measurements in 18 grains). (D) PDFs in quartz in polymict lithic impact breccia E963-12, 87 planes in 25 grains. Of 100 quartz grains, 51 were unshocked, 5 had one, 7 had two, 10 had three, 3 had four, 4 had five, and 2 had six sets of PDFs, respectively (most of the rest were diaplectic quartz). Histograms A–C were measured in Kiev (3° binning) and histogram D was measured in Vienna (5° binning).

Figure 5. Weakly shocked oligoclase clast with polysynthetic twins in rhyolite. Microfracturing and displacement of twins are visible (sample E1032-7, crossed polarizers, 2.5 mm wide).

Figure 6. Moderately shock metamorphosed rhyolite tuff with open, irregularly shaped cracks in fine-grained groundmass (sample E987-8, parallel polarizers, 9.2 mm wide).

those of pure stishovite. Coesite fractions were also isolated by partial dissolution of shocked quartz, but only low abundances of this mineral were found in quartz from volcanic rocks of stage I.

Orthoclase is extensively isotropized in these moderately shocked rocks. Its refractive indices range from 1.502 to 1.510 and birefringence is 0.002–0.003; complete isotropization is characteristic for the most strongly shocked rocks of this stage. An isotropic feldspar clast in tuff, with a composition of $Or_{55}Ab_{43}An_2$, has a refractive index of 1.503 ± 0.002.

Polysynthetic twins are preserved in plagioclase phenocrysts and clasts, which have reduced refractive indices (n_γ = 1.522, n_α = 1.517) and birefringence values of 0.005. PDFs are only rarely observed in plagioclase, then being short and only occurring in specific twin lamellae.

Biotite contains kink bands and PDFs. Partial replacement of biotite by opaque material along fractures and PDFs is observed. X-ray patterns still correspond to that of normal biotite, but they also contain some additional peaks corresponding to magnetite.

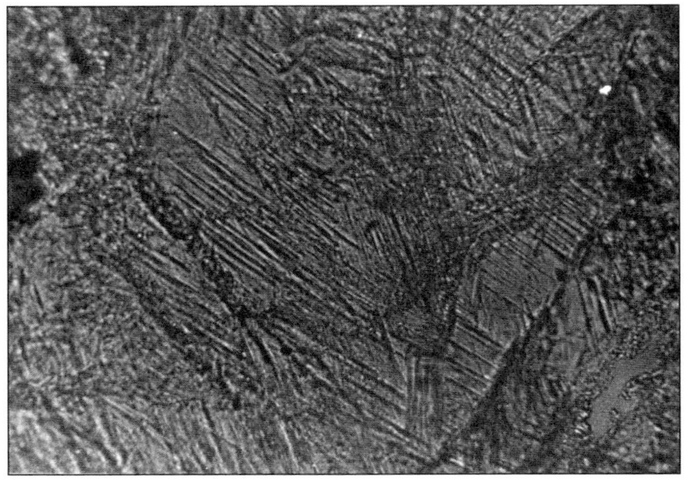

Figure 7. Multiple planar deformation features in shocked quartz from rhyolite tuff (sample E699-A, 0.9 mm wide, crossed polarizers).

Shock Stage II

Siliceous volcanic rocks that could be assigned to shock stage II (35–45 GPa) include some rhyolites, ignimbrites and tuffs, with phenocrysts and clasts of diaplectic quartz glass. The lower shock pressure limit of stage II was determined from the complete transformation of quartz into diaplectic glass, and the upper limit from the onset of formation of feldspar melt glasses.

Quartz in volcanic rocks and tuffs of stage II is transformed into isotropic diaplectic glass with refractive indices that range from 1.462 to 1.467. In some cases, remnant areas with low birefringence and relics of PDFs are visible in diaplectic glass. Segregations of coesite with high refractive index occur in some of these diaplectic quartz glass grains. Refractive indices of coesite measured are n_γ = 1.594, n_α = 1.590, and birefringence is 0.004. X-ray diffraction patterns of pure coesite were obtained in some instances. Stishovite was never identified in diaplectic quartz glass from the volcanic rocks of this stage, in accordance with post-shock temperatures that are too high for its preservation. Thus, the upper limit for formation and preservation of stishovite in tuffs and volcanic rocks is ~35 GPa, which is somewhat lower in comparison with other crystalline rocks (Grieve et al., 1996).

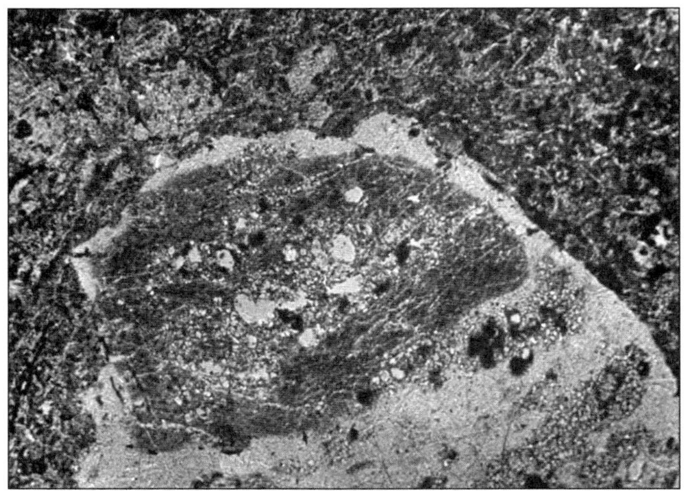

Figure 8. Clast of orthoclase feldspar converted into diaplectic glass in strongly shocked rhyolite tuff. Rare vesicles are visible in the central part of the grain (sample E908-11B, 1.6 mm wide, parallel polarizers).

Orthoclase is completely converted into isotropic diaplectic glass, whereby phenocrysts and clasts preserve the initial shapes of the mineral grains (Fig. 8). Roundish vesicles often occur in such phenocrysts. Refractive indices are 1.504–1.507, still higher than refractive indices of fused glass of orthoclase composition.

Phenocrysts and clasts of plagioclase partly preserve the crystalline state; polysynthetic twins are still visible. Oligoclase from ignimbrites has refractive indices of: n_γ = 1.522–1.527, n_α = 1.517–1.522, and birefringence of 0.003–0.005. Part of samples of shocked rocks of stage II contain diaplectic glass of plagioclase, with refractive indices ranging from 1.513 ± 0.001 to 1.517 ± 0.002.

Two types of shocked volcanic rocks are placed in the second shock stage: (1) porphyric rocks with phenocrysts of diaplectic glass of quartz and orthoclase, and phenocrysts of oligoclase with lowered refractive indices and birefringence; and (2) porphyric rocks with phenocrysts of diaplectic glass of quartz, orthoclase, and oligoclase. We suggest that the first rock type was shocked in the range 35–40 GPa, whereas the coexistence of diaplectic glasses of quartz, orthoclase, and oligoclase in some rocks indicates their transformation at 40–45 GPa. These observations are in good agreement with data of Ahrens et al. (1969), who stated, "Above the elastic limit, extending to ~300 kb and ~400 kb, transition regions of anomalously high compression are observed for microcline and oligoclase." Complete transformation of orthoclase into diaplectic glass at 32 GPa was determined by Kleeman (1971).

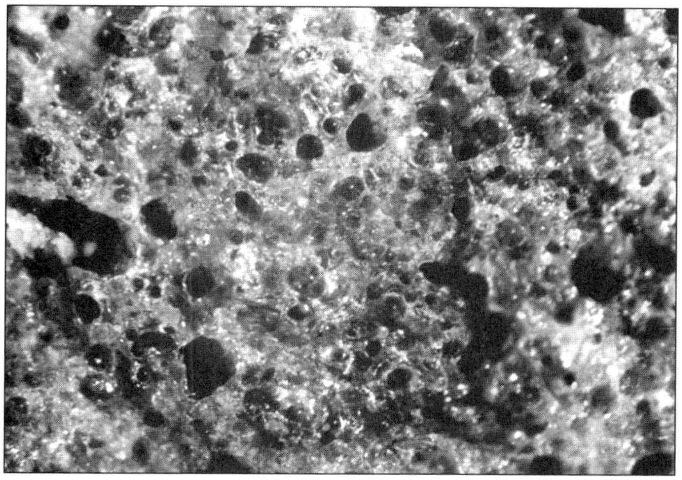

Figure 9. Surface of vesicular glassy matrix of strongly shocked and partially melted rhyolitic tuff (sample E908-7B, reflected light, 3 mm wide).

Biotite phenocrysts are replaced by fine-grained opaque masses of ore minerals, but still preserve their initial grain shapes. The fine-grained matrix of porphyritic rocks and the cement of tuffs are completely transformed into isotropic glass; however, glass without any signs of flow.

According to Stöffler and Hornemann (1972), shock melting of feldspars takes place at pressures above 45 GPa. Thus, the upper boundary of stage II for volcanic rocks is at ~45 GPa.

The major and trace element composition of seven rocks belonging to this shock stage is given in Table 5. Major element compositions of the samples are much more similar than the trace element contents. For example, silica contents vary from ~60 to 71 wt%, and most other major elements vary by not more than a factor of two to three (excluding MgO and CaO, which show more variation). On the other hand, contents of the siderophile elements (e.g., Cr, Co, Ni) vary by more than an order of magnitude. This is not likely due to significant contents of an extraterrestrial component, as Ir contents are very low in El'gygytgyn impact glasses (see Gurov and Koeberl, 2004). Lithophile elements (including the rare earth elements) show much less variation in contents that are fairly typical for volcanic rocks.

Shock Stage III

For siliceous volcanic rocks and tuffs, this shock stage is defined by the appearance of melted (fused) feldspars and fine-grained matrix, whereas phenocrysts and clasts of diaplectic quartz glass remain unmelted. Volcanic rocks and tuffs of the third stage are highly vesicular (Fig. 9), inhomogeneous glassy rocks with relics of phenocrysts of diaplectic quartz glass and melted glasses of feldspars. Some lithic clasts have aerodynamic shapes. This indicates that these rock fragments were partially melted and viscous during their transportation through the atmosphere. Aerodynamically shaped bodies of strongly shocked tuffs ejected from the Sedan nuclear crater were described by Short (1968). Tuff clasts sometimes preserve their layered structure, but often it is difficult to determine the original nature of rocks of shock stage III. The rocks have a low density, and some clasts float in water.

Phenocrysts and clasts of quartz are converted into transparent diaplectic glass of the same refractive index as in rocks of stage II. Segregations and veinlets with high refractive indices abundantly occur in such diaplectic glass (Fig. 10A). Detailed

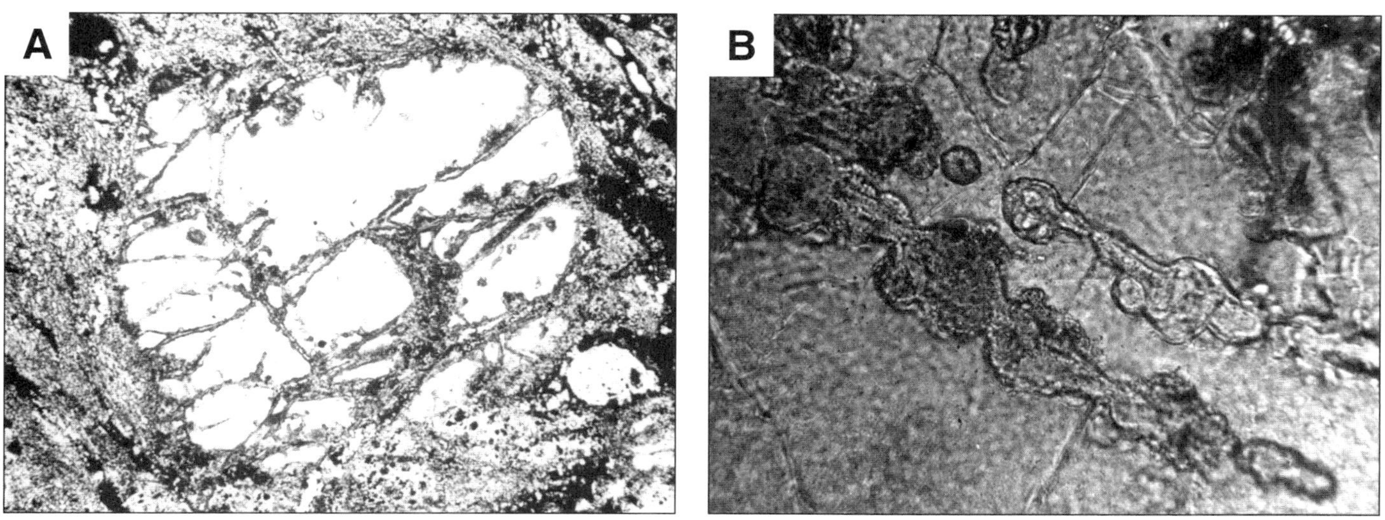

Figure 10. (A) Transparent diaplectic quartz glass grain with very thin veinlets of coesite and secondary quartz in strongly shocked rhyolitic tuff. Matrix is melted and converted into inhomogeneous fluidal glass (sample E908-7B, 3.0 mm wide, parallel polarizers). (B) Segregations of coesite and secondary quartz in diaplectic quartz glass. Coesite forms grains with high relief in central parts of veinlets, while their peripheral zones are fine-grained aggregates of secondary quartz (sample E908-7B, 0.65 mm wide, parallel polarizers).

study of such material reveals that it represents coesite and secondary quartz. Kidney-shaped segregations of coesite form the central parts of the veinlets, while their marginal zones are formed by secondary quartz (Fig. 10B). Thus, coesite is present in rocks of stages I–III of shock metamorphism of siliceous volcanic rocks. The occurrence of coesite in shocked crystalline rocks of stages I–III was described before by Grieve et al. (1996).

Orthoclase melt glass forms lensoid grains of transparent, highly vesicular, and colorless glass, with a refractive index of 1.488–1.492, which is lower in comparison with the refractive index of diaplectic orthoclase. The composition of this type of melt glass corresponds to a high content of normative orthoclase, with a content of ~17% of albite and a low content of normative anorthite (Table 2). The presence of normative quartz up to 7.8% and corundum up to 2.8% within the glass is a probable consequence of the loss of alkali elements by evaporation. In comparison, composition of orthoclase diaplectic glass from dacitic andesite (Table 2) is characterized by a higher content of normative albite of 40.4% and does not contain any normative quartz and corundum. Plagioclase diaplectic glasses from dacitic andesite correspond to andesine with low content of normative orthoclase: $Or_3Ab_{47}An_{43}$ and $Or_3\,Ab_{50}\,An_{46}$ (Table 2).

Plagioclase phenocrysts and clasts are highly vesicular, colorless melt glasses. Their forms vary and include irregular and lens-like shapes (Fig. 11), but in some cases are the initial shapes of phenocrysts preserved. Plagioclase and potassium feldspar melt glasses are very similar in thin section, and their distinction is only possible by measurement of refractive indices. Refractive indices of plagioclase glasses range from 1.505 to 1.517.

Biotite is converted into fine-grained masses of opaque melt glass that is partly mixed with the matrix glass. Remnants of opaque phases may still occur. Phenocrysts of strongly shocked pyroxene occur in some rocks of shock stage III. These phenocrysts are recognized as dark, fine-grained matter with a refractive index of ~1.685. X-ray patterns correspond to augite, but contain additional peaks of magnetite. The density of shocked pyroxene is 3.24 g/cm^3. Similar observations were reported for pyroxene from the Lonar Lake crater, shocked in the range of 40–60 GPa (Fredriksson et al., 1973).

The matrix of stage III rocks is a frothy, highly vesicular glass of light-gray color with dark spots and schlieren. The glass has a fluidal texture, with schlieren surrounding the few remnant phenocrysts or clasts of diaplectic quartz glass. Refractive indices of matrix glass vary from 1.485 to 1.500–1.505, and up to 1.530–1.540 in dark-colored schlieren and spots. Some micro-inclusions in glass have refractive indices of ~1.590. X-ray diffraction patterns of glassy matrix show only weak lines corresponding to d = 0.310 nm, the value for the main diffraction line of coesite.

In summary, the main shock effect in siliceous volcanic rocks of shock stage III is a selective melting of both feldspars and fine-grained matrices of quartz-feldspar–rich compositions. The post-impact temperatures are still lower than the temperature of quartz melting (1513 °C). High viscosity of the melt and rapid quenching during transportation of clasts in the atmosphere are responsible for the immiscibility of the feldspar melts and matrix melt. At the same time, softening of these highly shocked rocks contributed to formation of aerodynamically shaped bodies during their flight through the atmosphere.

The chemical compositions of highly shocked volcanic rocks and tuffs of shock stage III are mostly similar to the compositions of corresponding unshocked samples (Table 3); however, there are differences to the partially melted rocks. For example, the Fe_2O_3/FeO ratio of the partially melted rocks varies from 0.45 to 1.04, in comparison with ratios of 1.7–2.05 of corresponding unshocked rocks.

The upper boundary of shock pressures of the rocks of stage III is 55–60 GPa, corresponding to the shock pressure of melting of quartz and siliceous rocks according to Stöffler (1971).

Shock Stage IV

For volcanic rocks and tuffs, this shock stage is represented by impact melt rocks and impact melt glasses, produced by complete shock melting of volcanic rocks and tuffs. At the current level of erosion, such rocks were found together with other shock metamorphosed rocks in terraces near the Lake El'gygytgyn.

Two main types of impact melt rocks occur in the El'gygytgyn crater. Massive impact melt breccia represents the first type. Debris of impact melt breccia occurs within terrace deposits inside the crater. The second type of impact melt rock is represented by aerodynamically shaped glass bombs. Rocks of this type were formed by solidification of melt particles during their transportation through the atmosphere. The occurrence of such glassy bombs is not restricted to the crater itself, but they were also found in terrace deposits of some streams on the outer slopes of the crater rim. Rare rounded glass pebbles occur in the terrace deposits of the Enmivaam River at distances of tens of kilometers from the crater (see Gurov and Koeberl, 2004, for details).

Impact melt rocks and impact melt breccias occur as irregular clasts and lumps with maximum sizes of up to 1 m. The color of these rocks is dark gray or black. The rocks are composed of vesicular glass and clasts of highly shocked volcanic rocks, glasses, and minerals. Clasts have sharp contacts with the melt, and their abundance varies from 5% to 10% in impact melt rocks and up to 40%–50% in impact melt breccias. The matrix of impact melt rocks is vesicular glass that is transparent in thin section. Glasses are light gray, light brown, or colorless. Their fluidal structures are locally enhanced by brown schlieren and trails. Fresh colorless glass has refractive indices between 1.505 and 1.512. Some cloudy areas of devitrified glass are light brown in thin section. Their X-ray diffraction patterns have characteristic peaks of plagioclase, with the strongest line at d = 4.04 Å. Prismatic crystallites of pyroxene, up to 0.1 mm long, sometimes occur in the melt (Fig. 12). Refractive indices of pyroxene are: $n_\gamma = 1.674 \pm 0.003$, $n_\alpha = 1.658 \pm 0.003$. The main peaks of X-ray diffraction patterns are at 2.51, 2.13, 1.61 Å, corresponding to the pattern of diopside.

Clasts in impact melt breccias are predominantly particles of vesicular, black, gray, and greenish-gray glass, rarely less

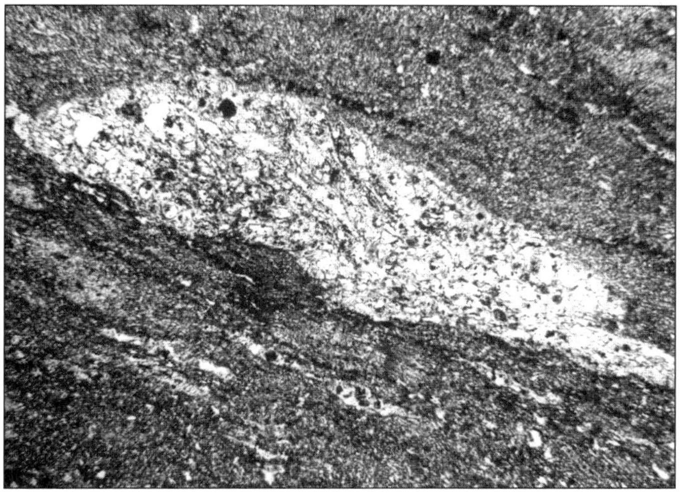

Figure 11. Lensoid grain of partially melted, vesicular feldspar glass in melted matrix (sample E908-22, 2.5 mm wide, parallel polarizers).

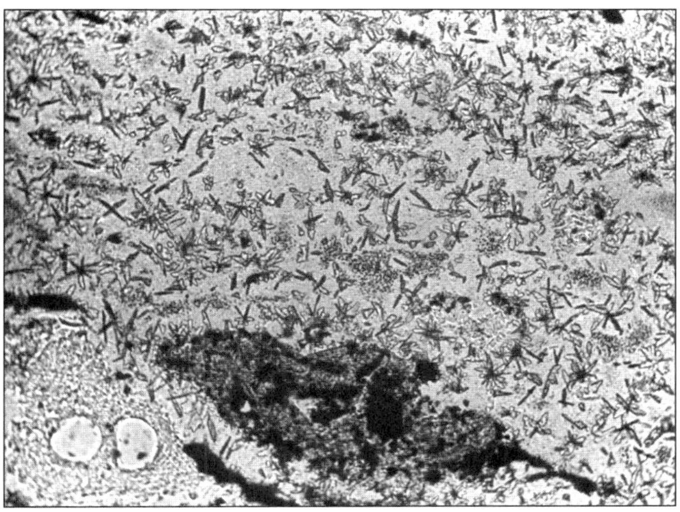

Figure 12. Impact melt rock with glassy matrix and prismatic microlites of pyroxene. Some areas of dark, basically opaque glass are extensively devitrified. Grains (in the lower left) are vesicular lechatelierite and the large, dark clast (lower center) represents a shock metamorphosed mafic particle (sample E985-60, 2.2 mm wide, parallel polarizers).

shocked fragments of volcanic rocks and tuffs. Inclusions of unshocked and weakly shocked rocks do sometimes occur. The size of such clasts ranges from millimeters to several centimeters; they generally have sharp contacts with the glass, which is further evidence for rapid cooling of the melt. Mineral clasts include—very rarely—unshocked quartz fragments, shocked quartz with PDFs, lechatelierite, and—rarely—quartz with ballen structure. Some very rare fragments of pyroxene and biotite, unshocked or weakly shocked, have also been recorded.

Rare grains of graphite and diamond-bearing graphite were extracted from these impact melt rocks. They are black, tabular, or irregular shaped grains that are 0.1–0.3 mm in size. X-ray diffraction patterns of some shocked graphite grains show that they have complex phase compositions; besides graphite, X-ray reflections of lonsdaleite (at 2.18 Å and 2.06 Å) were observed, but pure lonsdaleite has not been identified in the impact melt rocks of El'gygytgyn crater.

The chemical compositions of impact melt rocks (Tables 4 and 5) are similar to the compositions of target volcanic rocks, especially to the composition of rhyolitic ignimbrites of the upper level of the target stratigraphy (Table 1). However, impact melt rocks are more reduced than their unshocked counterpart. The Fe_2O_3/FeO ratio is 0.64–0.66 in impact melt rocks, whereas it is 1.8–2.05 for unshocked volcanic rocks. The trace element composition of seven samples analyzed (Table 5) shows similar variations as discussed for rocks of shock stage II—lithophile elements show little to moderate variation, whereas siderophile elements show significant differences between the various samples. The chondrite-normalized rare earth element distribution patterns (Fig. 13) of the various rocks show similar overall patterns with a variation in the extent of the negative Eu anomalies.

A more basic composition of impact melt rocks was determined from the lacustrine deposits in the southern part of the crater, where the silica content of melt rocks is as low as 63.92 wt% (Table 4). This possibly reflects a more basic composition of target rocks in the southern and southeastern parts of the El'gygytgyn crater.

Glasses in the El'gygytgyn crater occur in the form of aerodynamically shaped bombs (Fig. 14). The bombs have the forms of drops, ropes, cakes, and cylinders, and they rarely have irregular forms. The surface of bombs is rough and lusterless. Rounding of glass bodies from post-impact transportation by water is rare. Bomb diameters range from 1 to 15 cm and most of their masses vary from several grams to 500 g, but bodies of up to 2 kg have also been found. Deep open cracks are a main characteristic of the bomb surfaces. These cracks were probably formed by contraction of the melt during cooling and solidification while in transit through the atmosphere (see Koeberl and Gurov, 2002; Gurov and Koeberl, 2004).

The density of glass ranges from 2.40 ± 0.05 to 2.50 ± 0.05 g/cm^3. The color of glass is black, rarely dark gray in hand samples, and colorless to pale yellow and brown in thin sections. Bands of brown glass enhance its fluidal structure (Fig. 15). Refractive indices vary mostly from 1.505 to 1.515, but reach up to 1.540 in brown glass. Glass is fresh and devitrification is rare. Mineral inclusions in glass are shocked quartz, as well as diaplectic quartz glass with coesite and lechatelierite. The bombs are different in their morphology and appearance from those of partly devitrified glasses of massive impact melt rocks.

Analyses of glass bombs from eight areas of the El'gygytgyn impact crater show only a limited range of compositions (Table 4) (see also Gurov and Koeberl, 2004). The SiO_2 content varies from 68.60 to 71.48 wt%. The content of Na_2O ranges from 2.66

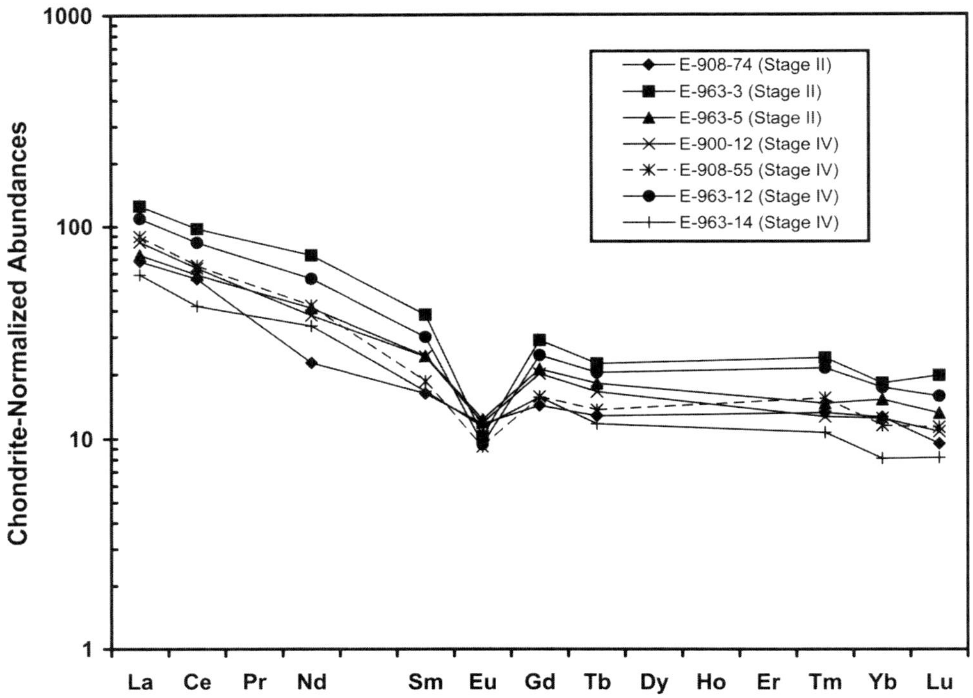

Figure 13. Chondrite-normalized rate earth element distribution patterns of selected impactites from El'gygytgyn, representing shock stages II and IV. Normalization factors from Taylor and McLennan (1985).

to 3.16 wt% and that of K_2O from 3.30 to 4.37 wt%. The compositions of the glass bombs are similar to the calculated average composition of the target (Table 1), but the closest agreement is with the composition of rhyolitic ignimbrite of the uppermost part of the volcanic stratigraphy of the crater area. We assume that rhyolitic ignimbrites were the main source of melt for bomb formation during the earliest stage of impact melting of the target. A difference between the compositions of impact melt glass and volcanic country rocks are higher contents of MgO, CaO, and FeO and lower contents of alkali elements in the glass. One of the main characteristics of the glass composition is a low ratio of Fe_2O_3/FeO of 0.36 due to reduction in high-temperature melt.

Figure 14. Aerodynamically shaped glass bomb with black, shiny glass on the fresh surface. Diameter of sample is 5 cm.

Shock Stage V

A shock stage V for crystalline rocks was also proposed by Stöffler (1971), involving silicate condensates formed as a consequence of vaporization processes in the innermost zone of the target. The glassy cover on the inner surface of voids in the impact melt rocks of the El'gygytgyn crater could be such silicate condensate. Some samples of vesicular impact melt rocks and impact melt breccias contain gaseous voids of irregular form and up to 1–1.5 cm in size. Their surface is covered with a colorless transparent glassy crust ~0.2 mm thick. Clusters of white glass up to 1 mm in size occur on the surface of the biggest voids (Fig. 16). The glass is semitransparent, and its refractive indices range from 1.440 to 1.465. The composition of the glass cluster (wet chemistry data, wt%) is: SiO_2—87.0; Al_2O_3—2.90; FeO_{total}—0.30; Na_2O—0.40; K_2O—0.70; LOI—4.85; H_2O^-—3.93; TiO_2, MnO, MgO, CaO, P_2O_5—traces; in total—100.06%. There is no direct proof for a condensate origin of this glass, though formation of that glass by impact melting or by hydrothermal activity is likely problematic. Similar glassy condensates covering the surface of shocked rocks were described from the Rainer underground nuclear explosion (Rawson, 1968). An origin as condensate was

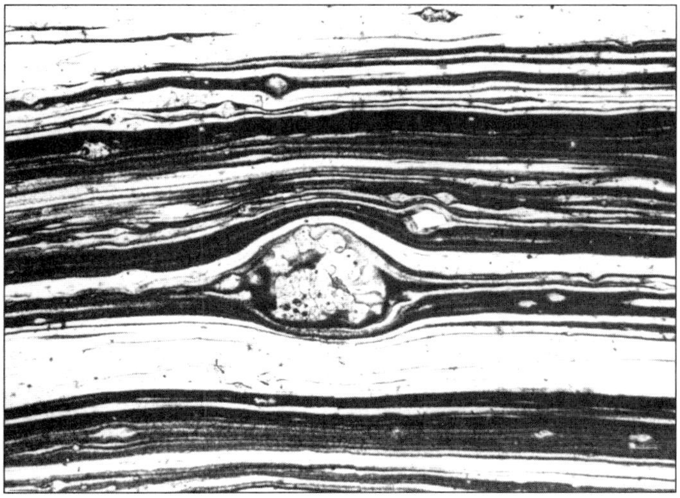

Figure 15. Fluidal structure of impact glass. The variably colored lamina of felsic and mafic glass have been affected by the rotation of the clast composed of lechatelierite (sample E690, 3.5 mm wide, parallel polarizers).

Figure 16. Cluster of possible condensates (silicate minerals) within a void in impact melt rock (E669-H, 2.0 mm wide, reflected light).

also assumed for glassy clusters on some Apollo 11 lunar rocks (McKay et al., 1970).

SUMMARY AND CONCLUSIONS

The 18-km-diameter El'gygytgyn impact crater is located in the central mountainous region of the Chukotka Peninsula, northeastern Russia. The crater is a flat-floored, circular basin, surrounded by an uplifted rim of up to 200 m elevation. The deepest part of the basin is occupied by Lake El'gygytgyn of 12 km diameter.

The crater was formed in a series of Late Cretaceous volcanic rocks, including lavas and tuffs of rhyolitic, dacitic and andesitic composition. The entire sequence of volcanic rocks exposed on the inner walls and outer slopes of the crater rim is more than 600 m thick. The uppermost horizon of the volcanic series comprises rhyolitic ignimbrites of 250 m thickness. Such a target lithology is unique among known terrestrial impact structures and provides a good possibility for the study of shock metamorphic effects in siliceous volcanic rocks.

The volcanic rocks of the crater rim and its slopes do not exhibit shock deformation, and no remnants of ejecta have been preserved around the crater. Shock metamorphosed rock, impact melt rocks and glasses were redeposited at the current erosion level in lacustrine terraces inside the crater and in terraces of some streams on the outer slopes of the rim. Unshocked debris from the rim dominates these lacustrine deposits, whereas impactites occur in subordinate quantity. Most fragments of impactites are not rounded and preserve their initial form and size according to short transportation of impact material by slumping down the slopes of the crater rim. All types of impact rocks are rather fresh and do not contain signs of strong hydrothermal alteration and weathering. Impact rocks found on the terraces are derived from ejecta and flows of impact melt on the crater rim and its slopes. Massive bombs of up to several tens of cm in size have been found as well. These rocks have flown as individual objects through the air and they exhibit signs of aerodynamic ablation.

The stages of shock metamorphism of volcanic rocks and tuffs observed range from weakly shocked rocks to partially and completely melted rocks. The classification scheme of shock metamorphosed crystalline rocks of Stöffler (1971) was used to classify the shocked siliceous volcanic rocks of the El'gygytgyn crater (Gurov and Gurova, 1979). Five stages of shock metamorphism and shock melting of siliceous volcanic rocks of rhyolitic to dacitic composition are recognized:

Stage 0: Shock pressures up to 10 GPa. Unshocked and weakly shocked volcanic rocks without any sign of shock metamorphism.

Stage I: Shock pressures of 10 to 30–35 GPa. Tuff and lava contain weakly to moderately shocked phenocrysts and clasts of quartz and feldspars. Stishovite and coesite occur in quartz. Shock metamorphism of the fine-grained matrix is not detectable.

Stage II: Shock pressures up to 45 GPa. Phenocrysts and clasts of quartz and feldspars in lava and tuff are converted into diaplectic glasses. Coesite is abundant in diaplectic quartz glass, but stishovite was not determined in diaplectic quartz glass. Groundmasses are isotropic.

Stage III: Shock pressures up to 55–60 GPa. Phenocrysts of diaplectic quartz glass are the last phase that remains unmelted. Coesite is still abundant in diaplectic quartz glass. Feldspar phenocrysts and clasts are melted and have irregular or lensoid, vesicular forms, but mainly preserve sharp contacts with matrix. The matrix is composed of heterogeneous vesicular glass. The most highly shocked rocks of this stage are transitional to the rocks of stage IV.

Stage IV: Shock pressures up to 80 GPa. Impact melt rocks and glasses formed by the complete melting of volcanic rocks.

In addition, glassy coatings on inner surfaces of some voids in impact melt rocks were found; these could be the possible product of condensation of the siliceous vapor.

APPENDIX: SAMPLE DESCRIPTIONS

The Appendix lists the petrographic characteristics of rocks from all different shock stages. A number of characteristic petrographic features are illustrated in Figures 17 and 18.

Stage 0, Unshocked Rocks

G19 (Dark basalt-textured country rock fragment). Porphyroblasts of olivine and aggregates of plagioclase laths occur in a fine-grained matrix of plagioclase+pyroxene+mesostasis. Olivine is partially altered to iddingsite, and the mesostasis is also partially altered. Minor fracturing is the only evidence of the very weak deformation of this sample.

E649 (Ignimbrite). Porphyritic rock with phenocrysts of quartz, feldspar, biotite, and rare amphibole in fine-grained matrix. Matrix is fluidal, microcrystalline glass, with very fine-grained granular and spherulitic structures. Clasts of recrystallized tuff are abundant.

E649 (Rhyolite). Phenocrysts of quartz, potassium feldspar, plagioclase, biotite, and, rarely, amphibole in brown fluidal devitrified glass.

E655 (Rhyolitic tuff). Crystal clasts of quartz, plagioclase and potassium feldspar in fine-grained groundmass composed of particles of devitrified glass.

3-207a (Porphyritic volcanic rock). The sample contains medium-grained euhedral feldspar porphyroblasts, minor clinopyroxene, and some euhedral oxides in an aphanitic groundmass. The matrix and some of the porphyroblasts are strongly altered. Some long and narrow vesicles are filled with quartz and phyllosilicates. Besides minor fracturing, no shock deformation effects were found.

E1544 (Rhyolitic ash tuff). Groundmass is extensively devitrified glass; some glass particles preserve their initial shapes. Rare mineral clasts are quartz and altered feldspars. Glass particles show subparallel orientation.

E665 (Dark-brown andesitic tuff). Clasts of dark-brown microporphyritic rocks and clasts of andesine and pyroxene in a finest-grained clastic matrix.

Stage I, Weakly to Moderately Shocked Rocks

E1032-7 (Moderately shocked rhyolite). Phenocrysts of quartz with PFs and PDFs, $n_o = 1.548$, $n_e = 1.540$, birefringence 0.008; potassium feldspar with decreased refractive indices and birefringence, $n_\gamma = 1.516$, $n_\alpha = 1.512$; phenocrysts are mostly isotropic; oligoclase ($Ab_{28}An_{72}$) with preserved polysynthetic twins, $n_\gamma = 1.533$, $n_\alpha = 1.527$; biotite with kink bands; matrix is devitrified microcrystalline fluidal glass. Shock pressure ~28 GPa.

E1032-8 (Moderately shocked rhyolite, similar to 1032-7). Phenocrysts of quartz with PFs and PDFs, refractive indices: $n_o = 1.550$, $n_e = 1.541$, birefringence 0.009; potassium feldspar with lowered refractive indices; plagioclase (oligoclase $Ab_{73}An_{27}$) preserves polysynthetic twins; biotite forms dark-brown phenocrysts with kink bands; matrix with microgranular texture with fluidal structures. Shock pressure ~26 GPa.

E699a (Moderately to strongly shocked rhyolite). The rock is dissected by quartz veinlets of up to 1 mm thickness. Phenocrysts and veinlets of quartz are fractured and contain sets of PDFs, n = 1.483 ± 0.003, birefringence ~0.002; potassium feldspar phenocrysts are converted to diaplectic glass; polysynthetic twins preserved in plagioclase phenocrysts. Shock pressure ~32 GPa.

E699-11b (Moderately to strongly shocked rhyolite). Phenocrysts of quartz with sets of PDFs, n = 1.485 ± 0.002, and birefringence of ~0.002; stishovite and coesite were found in quartz; K-feldspar forms phenocrysts of transparent glass with rare bubbles, n = 1.505; oligoclase forms phenocrysts with $n_\gamma = 1.528$, $n_\alpha = 1.524$, birefringence 0.004. Shock pressure ~32 GPa.

E963-7 (Strongly shocked rhyolite tuff). Phenocrysts of isotropized quartz and feldspar in isotropic groundmass. Quartz is almost isotropic, refractive index is 1.464 ± 0.002, birefringence ~0.003, but relics of PDFs are locally visible; potassium feldspar is completely isotropized; plagioclase forms irregular clasts with refractive indices $n_\gamma = 1.522$, $n_\alpha = 1.517$ and birefringence ~0.005. The groundmass is gray glass. Shock pressure ~35 GPa.

Stage II, Strongly Shocked Rocks

E699-33B (Completely isotropic volcanic rock). The original groundmass was either extremely fine-grained or even aphanitic. Rare, angular, sometimes rectangular porphyroblasts (presumably feldspar) are likewise converted to diaplectic glass. Some patches of dark oxides occur, in part intermingled with a glass phase that contains numerous tiny fluid inclusions. A handful of crystalline feldspar blasts has strongly reduced birefringence. No PDFs observed in remnant crystalline patches of clasts.

E908-73 (Porphyritic volcanic rock). The sample is similar to 3-207a, but stronger secondary alteration and also stronger shock deformation. Feldspar porphyroblasts are partially converted to maskelynite; clinopyroxene shows local twinning (that was not seen in the unshocked sample of this rock type). Extensive fracturing to local brecciation of large crystals. Twinning and intense cleavage in clinopyroxene occurs on (010) crystal faces. Some feldspar is shock melted and recrystallized.

E908-74 (Altered and strongly shocked volcanic rock). The sample contains large feldspar porphyroblasts in a glassy matrix. Feldspar blasts are completely transformed to maskelynite. Several dark blobs are probably relics of primary opaque minerals (magnetite or ilmenite). Blasts and groundmass are partially converted to secondary phyllosilicates.

E963-4 (Strongly shocked volcanic rock). The sample contains numerous euhedral to subhedral plagioclase crystals that are mostly (>90%) converted to maskelynite. The groundmass is dark-gray and presumably mafic, and completely glassy. The fabric with strong alignment of porphyroblasts in thin bands is suggestive of an origin of this rock as a tuff. Some relics of possibly sphene are noted. In places, fused glass occurs with flow structures. Some glass is greenish (after clinopyroxene?). At the scale of a whole thin section, the sample has a distinct lamination.

E963-5 Similar to E963-4, but the layering and/or lamination is less distinct, as there are several lensoid zones intercalated with lamellar zones. The mineralogy and shock degree are the same as for the other sample. Many feldspar (maskelynite) grains are quite angular, similar to crystal fragments in a tuff. Still crystalline feldspar blasts show intense internal deformation (largely unresolvable with the optical microscope) involving some dense arrays of PDFs.

E988-2 (Strongly shocked rhyolitic tuff). Clasts of quartz and feldspar are embedded in an isotropic groundmass. Quartz is converted into diaplectic glass with n = 1.462. Coesite occurs very rarely in diaplectic glass. Clasts of potassium feldspar are diaplectic glass with rare voids; its refractive index is n = 1.505 ± 0.002. Plagioclase clasts have preserved polysynthetic twins; their refractive indices are $n_\gamma = 1.527$, $n_\alpha = 1.522$, birefringence is 0.005; some areas in plagioclase grains are isotropized; the groundmass is isotropic.

E987-11 (Strongly shocked rhyolitic tuff). The sample has open cracks; the initial structure is partly preserved. Clasts of diaplectic quartz glass (n = 1.463) contain coesite. Potassium feldspar is converted into diaplectic glass (n = 1.505 ± 0.001), with vesicles in central parts of clasts; plagioclase (oligoclase $Ab_{75}An_{25}$) has refractive indices of $n_\gamma = 1.528$, $n_\alpha = 1.525$, a birefringence of 0.003, and polysynthetic twins are preserved.

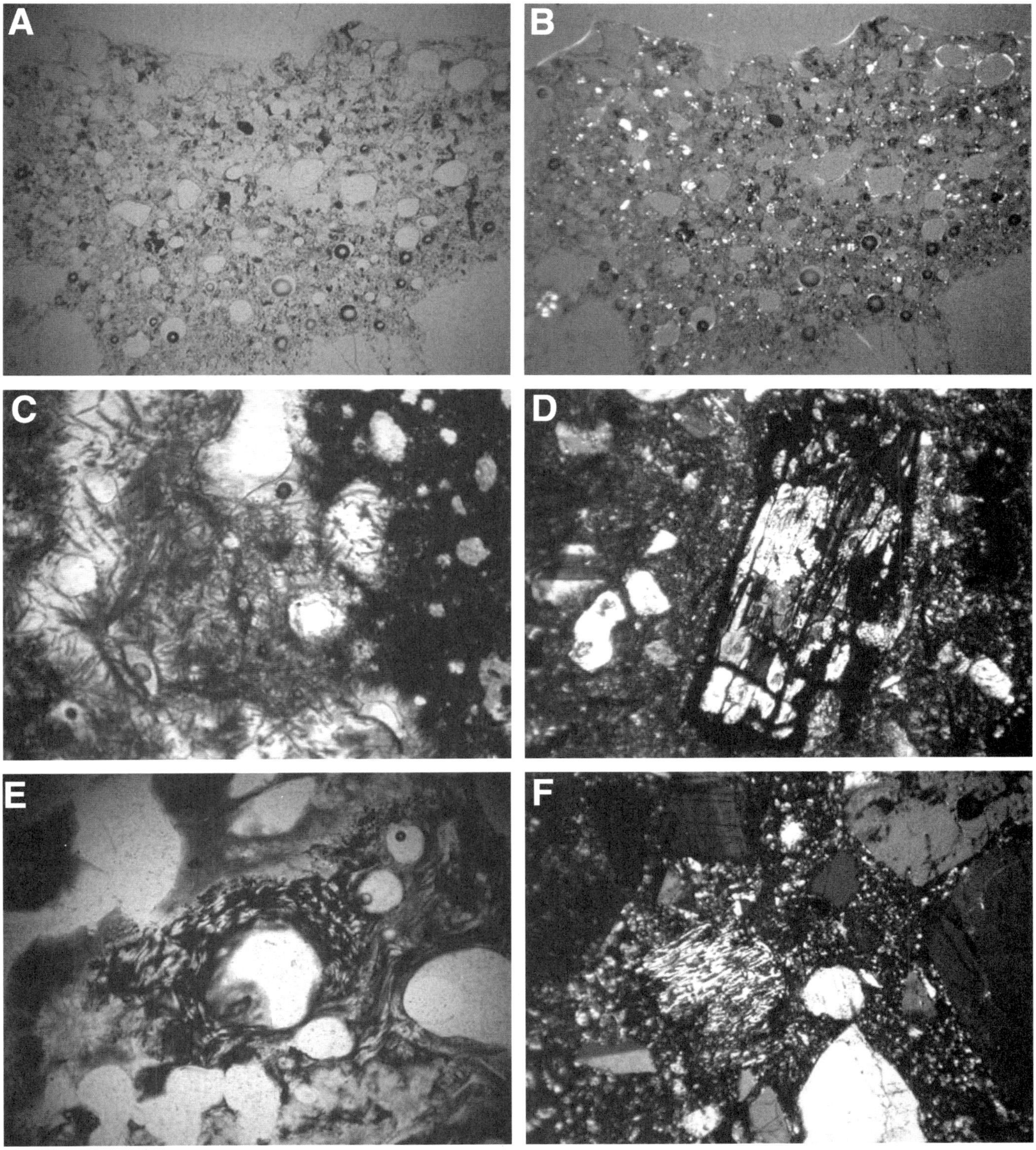

Figure 17. (A, B) Impact glass G32: Whole thin section image of vesiculated impact glass with melted felsic clasts. Note the small component of crystalline remnants of biotite, quartz and feldspar, all of which are unshocked. (A) plane polarized light; (B) cross-polarized light. Width of field of view: 0.8 mm. (C) Impact melt breccia E669-820: Two types of glassy groundmass (light with incipient crystallization and dark, aphanitic, with magnetite and/or ilmenite micro-crystals). Note the abundant remnants of still crystalline biotite. Rare clasts are quartz and feldspar that, with rare exceptions, have been converted to diaplectic glass. Plane-polarized light, width of image 1.1 mm. (D) Impact glass G10a: Microclast-rich impact glass with a large inclusion of a mafic mineral (presumably originally amphibole or clinopyroxene). The clast is largely converted to secondary minerals including carbonate and quartz. Note the dark, oxidic reaction rim. The matrix is completely glassy; however, the microclasts are partially crystalline and unshocked. Cross-polarized light, width of image 2.5 mm. (E) Impact glass G12: Strongly vesiculated, fluidal-textured (note the obvious flow structure emphasized by parallel oriented microliths) glass matrix. Schlieren of different colors are noted. Some light-colored areas represent melted and finest-grained annealed felsic clasts. Some of these are strongly distended, indicative of their plastic behavior. Plane-polarized light, width of image 1.1 mm. (F) Clast-rich impact melt breccia G23: Most clasts are unshocked or, at best, weakly shocked quartz and feldspar. The matrix is cryptocrystalline melt. A few clasts were melted and then recrystallized and are now recognizable by their finest-grained annealing texture. Cross-polarized light, width of image 1.1 mm.

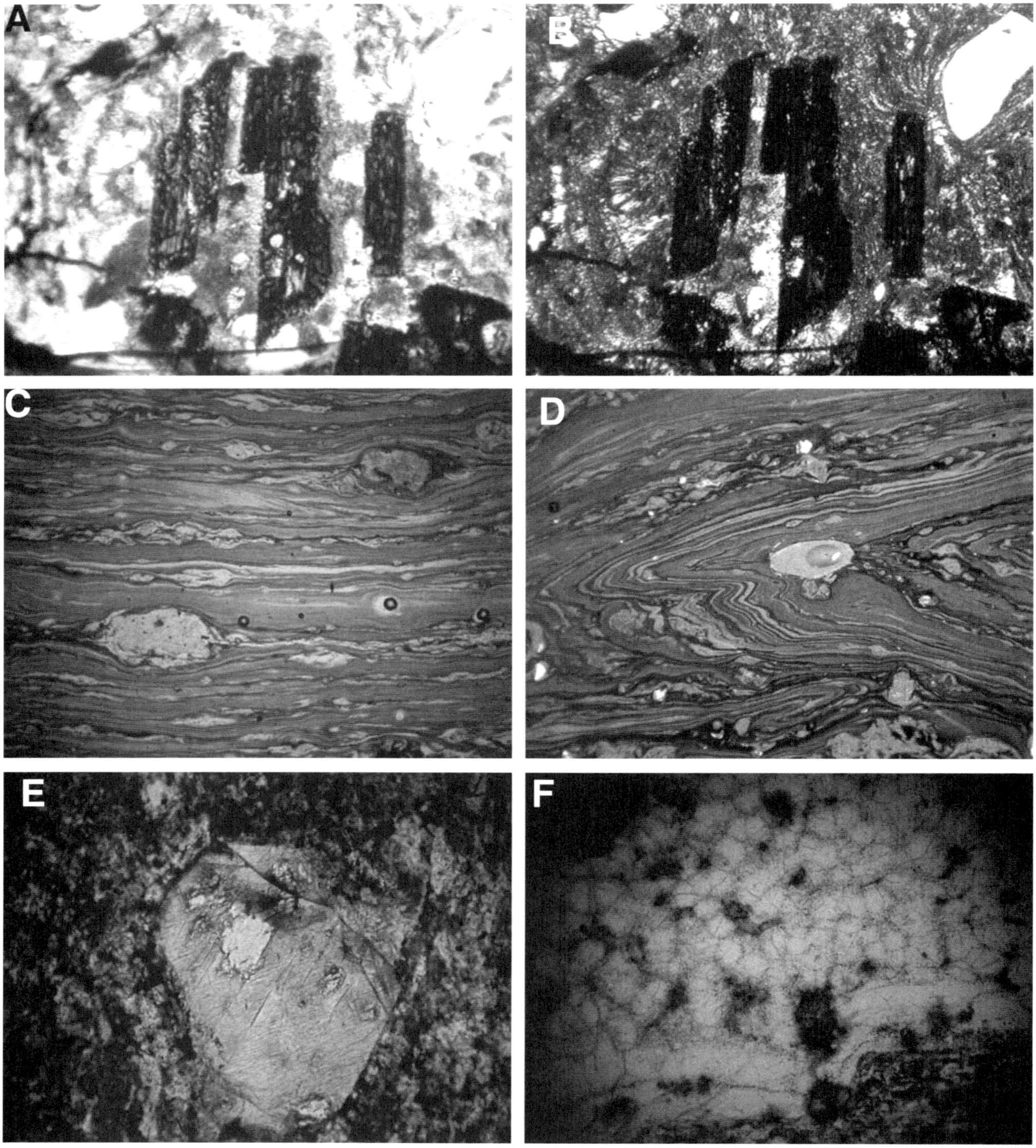

Figure 18. (A, B) Impact melt breccia G17: Remnants of a biotite crystal that has been partially melted—with crystallographic control. Note the slightly darker, biotite-derived melt in between the remnant laths. These laths are largely oxidized and only partially retained primary crystallinity. The matrix glass is finest-grained crystallized and crystallites show locally flow texture. Other clasts visible are quartz and apparently unshocked. (A) Plane-polarized light. (B) Cross-polarized light. Width of images: 1.0 mm. (C) Impact glass E985-101: Finely laminated and variegated impact glass with well-defined flow structures and micro-fold structures. The feldspar clasts are all melted themselves and strongly extended. Plane-polarized light, width of image 1.0 mm. (D) Variegated impact glass G25/1: Another example of variegated impact glass with intricate flow folding. Plane-polarized light, width: 1.1 mm. (E) Polymict lithic impact breccia E963-12. Quartz clast with planar fractures (trending northeast-southwest) and multiple sets of planar deformation features in quartz clast. Plane-polarized light, width of image 1.0 mm. (F) Impact melt breccia E900-12. Ballen quartz texture in impact melt; devitrified impact melt recognizable in lower right of image. Plane-polarized light, width of image 1.0 mm.

E908-64 (Strongly shocked ignimbrite). Relics of porphyritic structure are preserved. Phenocrysts of quartz and feldspars are converted into diaplectic glasses; most plagioclase phenocrysts are converted into maskelynite, but some grains still preserve a low birefringence.

Stage III, Partially Melted Rocks

E908-6 (Strongly shocked, partially melted rhyolite). In hand specimen, it is a fine-grained, vesicular, dark-gray rock that retains primary porphyritic structure. White vesicular phenocrysts of feldspar are visible in a dark matrix. In thin section, phenocrysts of diaplectic quartz glass with coesite and melt glasses of feldspar occur in a melted vesicular matrix. Feldspars have formed vesicular, colorless glass that preserves the initial shapes of the phenocrysts. The matrix is composed of vesicular gray glass. Dark opaque areas in glass are remnants of biotite.

E908-11 (Strongly shocked and partially melted ash tuff). In hand specimen, the rock is highly porous with rare white mineral clasts. In thin section, the groundmass is composed of vesicular fluidal glass; rare clasts are diaplectic quartz glass with coesite, and strongly vesiculated feldspar glass.

E908-16 (Strongly shocked and partially melted rhyolitic tuff). The hand specimen is vesicular, layered rock with gray and light-gray layers up to 3 mm wide. Relics of feldspar clasts form lens-like, white grains. In thin section, clasts of diaplectic quartz glass with coesite are observed; feldspars are converted to highly vesicular, transparent (fused) glass of lensoidal shapes; the groundmass is a gray, vesicular, fluidal glass.

E699-15B (Strongly shocked and partially melted volcanic rock). The hand specimen is a highly vesicular, porous rock of low density with rare light crystal clasts. In thin section, grains of diaplectic quartz glass exist that contain coesite veinlets; feldspar grains are converted into very porous masses of colorless glass. The matrix is a gray, vesicular, fluidal glass with bands of opaque glass.

Stage IV, Impact Glass and Melt Breccias

G4 (Pumice-like impact glass). The sample contains diaplectic mineral clasts and relatively few crystalline clasts. Clasts are generally very small. Groundmass glass shows some small birefringent needles as indication of beginning crystallization. Glass is slightly altered. The color of the glass is generally gray, but it contains several centimeter-long schlieren of colorless glass that could represent original silica-rich areas. The fabric of this sample is that of a laminated tuff. There are also a few darker bands in the glass that could originally have been some mesocratic material. The rare crystalline clasts all have very low birefringence. Some resemble toasted quartz, and some unshocked relics of plagioclase are also present. Dark-brown elongated patches are probably relics after a mafic mineral such as biotite.

G10a (Fluidal-textured impact glass). The specimen contains several large (0.5 cm) clasts that are themselves fluidized melt, but which still form aggregates of relic crystals held together by light-colored schlieren of feldspathic and/or silica-rich glass. Relic clasts are quartz and plagioclase and represent ~25–35 vol% of the sample. All these clasts are unshocked. Some of these feldspar relics are entirely replaced by carbonate. Vesicles in groundmass are filled with nontronite and/or calcite. Mafic minerals are largely oxidized (Fig. 17D). Some large quartzitic aggregates resemble the completely annealed quartz clasts in the Vredefort Granophyre. Amphibole clasts are complexly altered and replaced by carbonate and chlorite as well as oxides. Biotite is oxidized and dusted with opaques. A shocked biotite crystal shows growth of opaque needles, obviously from a melt phase. Roundish vesicles in this phase are partially filled with secondary minerals. Thus, a dichotomy of shock deformation is observed—unshocked/weakly shocked clasts and melt phases.

G11 (Impact glass). Completely isotropic, well-laminated sample, which may originally have been a tuff. Very thin laminae occur, and some of them contain reddish alteration and/or oxidation products. Some laminae are very vesicular, presumably because they contained a hydrous mineral. No remnants of crystalline material are observed. In some narrow bands, devitrification and/or alteration has begun, and tiny needles of a birefringent mineral have formed.

G12 (Fluidal impact glass). Light-beige sample with several marginal pockets of suevitic breccia. In transparent parts, this glass is similar to G13 and E908-55, but it also has several brownish, more mafic schlieren, and a few crystalline relics are embedded in the brownish, oxide-phyllosilicate matrix (Fig. 17E). The sample contains a dark-gray clast (zircon or perhaps sphene?) of a strongly tempered, unidentified mineral that is surrounded by an isotropic halo of high refractive index glass. The breccia at the sample margin consists of crystalline fragments in a partially melted granitoid clast. This clast includes several quartz grains that contain multiple sets of PDFs, and a brownish glass patch in the shape of the precursor crystal (presumably biotite) set in a fine-grained, phyllosilicate rich groundmass that also carries tiny mineral clasts, including K-feldspar with PDFs.

G13 (Clear impact glass). Similar to E908-55, but has far less inclusions and only a few vesicles filled with secondary minerals.

G14 (Vesicular impact glass). Clasts are recognizable in plane-polarized light but are completely isotropic in cross-polarized mode: they are entirely converted to diaplectic glass. A few tiny relics of crystalline biotite and some oxide relics are observed. There are also schlieren of black oxides, presumably relics of amphibole and opaque minerals. Locally, incipient devitrification of glass is observed, also along the edges of some vesicles where a volatile phase must have reacted with the glass host. A second section marked G14 represents clear glass with only very rare fluid inclusions, identical to G32 below.

G25/1 (Variegated impact glass). Clear, light-gray and dark-gray, and brownish glasses alternate in millimeter-wide schlieren (Fig. 18D). These glasses are strongly devitrified (crystallized), with this effect ranging from holocrystalline to only rare microliths in patches of glass. Often, microliths combine to form plumose aggregates. Some local patches of melt of irregular to roundish shapes are related to individual melt phases from clastic components including lithic fragments, as derived from rare mineral remnants in some of such patches. One patch comprises a remnant of crystalline feldspar, a part of which has been melted (the glass contains many tiny fluid inclusions).

G32 (Impact glass). This clear transparent sample contains a few "droplets" of dark-brown, oxidic material that sometimes occurs in narrow trails (Figs. 17A and 17B). Fluid inclusions are also present. At the edge of the thin section, a more heterogeneous patch (probably a melted lithic fragment) with vesicles, remnants of biotite, and some oxides. There are also a few felsic patches in the sample that represent partially melted, partially diaplectic felsic minerals.

E669-820 (Impact melt breccia). The sample is composed of several types of glass and/or melt fragments, including a clear felsic melt and a dark and highly vesiculated variety that is also altered to a small degree. In contrast to the impact glasses described above, this sample has experienced a significant degree of devitrification resulting in growths of tiny needle-shaped crystals (pyroxene?). A single inclusion in the form of a dark-brown "blob" probably represents a melted biotite grain. There are several felsic diaplectic glass inclusions (Fig. 17C).

E689 (Impact melt glass). This is a drop-like bomb with striated surface, 7.7 × 14.0 × 14.5 cm. The glass appears black on fresh surfaces. Colorless fluidal glass has rare bubbles and very rare inclusions of lechatelierite.

E696 (Black glass). Colorless, fluidal glass with rare inclusions of lechatelierite up to 1 mm in size. Round voids in glass are up to 0.5 mm.

E908-21B (Highly vesicular impact glass). The sample comprises mostly clear glass, but locally small patches or schlieren occur of yellowish-brownish, even greenish glass. Flow structures are restricted to small areas. A few vesicles are filled with finest-grained and unidentified secondary minerals—mostly phyllosilicate. The sample is nearly hyaline; the only crystalline particle observed is a partially diaplectic feldspar grain. Otherwise, no diaplectic glass clasts exist. This sample is very fresh, with crystallization and/or alteration basically confined to the rare greenish schlieren.

E908-55 (Impact glass). This is a clear glass with rare inclusions (diaplectic glass, lechatelierite, some oxide particles, fluid inclusions, a single patch of sericite flakes, a few annealed microclasts), similar to G32 and G14. It contains relatively few vesicles.

E985-101 (Glass bomb sample with prominent flow-banding). Vesicles are well aligned in individual bands and intermingled with dark (oxide-rich) schlieren, suggesting that the narrow bands could have been mesocratic (mafic mineral enriched) bands in a finely laminated tuff (Fig. 18C). At one edge of the section, a remnant of a microclast rich zone (presumably the suevitic material in which the bomb was embedded). Some glass schlieren form intricate fold structures. A few crystalline microclasts, mostly quartz, occur. There is no evidence of devitrification or alteration. A few vesicles have crystalline fills of unidentified minerals.

E699-37 (Dark-gray impact melt rock). Clasts of gray, vesicular glass and strongly shocked volcanic rocks up to 7 cm in diameter are visible in hand specimen. Mineral inclusions are lechatelierite and diaplectic quartz glass. Impact melt is gray, semi-transparent devitrified glass with prismatic microliths of pyroxene. The inner surface of lens-like voids in impact melt glass is coated with white and colorless glass of probable condensation origin (see text for discussion).

E669-1 (Dark-gray impact melt rock). This impact melt rock comprises vesicular, semi-transparent devitrified glass, with abundant clasts of strongly shocked volcanic rocks and vesicular glasses. Mineral clasts are lechatelierite and diaplectic quartz glass with coesite.

E900-12 (Devitrified, light-gray impact melt breccia). Melt of variable gray tones with well-developed schlieren structure of <1 mm to >5 mm spacing. These melts are basically completely crystallized to aggregates of tiny laths of what is probably pyroxene. Several clasts of ballen quartz are noted, in places melted and/or annealed (Fig. 18F). Also, several clasts of lithic, probably granitoid-derived material are in part completely melted or may still show some small, angular quartz and feldspar relics in melt. These fragments are not diaplectic and do not contain PDFs. But most of them have somewhat reduced birefringence. One large lithic clast has a quartz fragment with mosaicism, as well as several smaller felsic and mafic remnants with some melt and patches of oxidized material.

G1-2 (Impact melt breccia). The sample is macroscopically similar to G25/2, but shows more schlieren. Schlieren are characterized by very different colors including clear, light and dark brown, and greenish. Some parts of the glass groundmass are devitrified (microliths). The clast content varies, but is generally high (~20 vol%). The specimen includes unshocked, euhedral feldspar crystals, totally melted felsic and mafic (dark-brown) glass fragments, a completely oxidized mafic phase with small feldspar laths (microporphyritic texture) in dark-brown, oxidized groundmass that also contains some angular, unshocked quartz and feldspar fragments; largely annealed diaplectic quartz, one shock fractured (not the typical straight PFs), large fragment of porphyritic volcanic with melted matrix and unshocked porphyroblasts of plagioclase and clinopyroxene; a lot of diaplectic quartz glass, some clasts of which have local veins of annealed melt. A large fragment of granitoid with melted and oxidized remnants of biotite and felsic—presumably feldspathic—melt; has a large angular feldspar crystal with three sets of likely (narrow-spaced) shock-induced cleavage. Locally, the groundmass to this sample has flow structures.

G17 (Pinkish impact melt breccia). The sample is shown in Figures 18A and 18B. The coloration is the result of alteration of dark glass schlieren. The glass displays a prominent flow-banding. Mafic minerals in clasts are oxidized, but some entirely fresh biotite occurs as well. Alkali feldspar is locally "boiled," with some of the resulting melt zones having been strongly altered. Ghost clasts still preserve some original textures, indicating that both granitoid and a tuffaceous phase were precursor rocks. Few diaplectic quartz and feldspar glass fragments and very rare felsic mineral clasts with strongly reduced birefringence (probably feldspar) occur. No PDFs were observed. One finds a predominance of unshocked mineral and rare lithic clasts. Even in very fresh and apparently undeformed biotite crystals, there may be narrow zones parallel to (010) that are melted and/or oxidized. Some feldspar crystals are strongly zoned, as observed in some of the samples of unshocked porphyritic volcanics.

G18 (Altered impact melt breccia). The specimen resembles Onaping breccia samples from the Sudbury impact structure. The matrix is cryptocrystalline and/or holocrystalline. In the matrix, distinctive areas can be recognized in cross-polarized light that are ghost clasts—entirely melted and finest-grained recrystallized material. In places, the matrix is fluidal-textured. Clasts are derived from granitoid precursors. In some ghost clast areas, in plane-polarized light, ballen-texture is recognizable. Feldspar is strongly altered, in clasts, to sericite and carbonate, plus some other, yet unidentified brownish phyllosilicate. Remnants of mafic minerals are likely after biotite and amphibole (and are partially oxidized). Some fragments are partially melted and exhibit melt trails of finest-grained devitrified glass into matrix. Some vesicles are noted in groundmass and are filled with chlorite and nontronite. Quartz and K-feldspar crystals in clasts or as clasts are mostly unshocked or barely shocked and only display fracturing. Some K-feldspar crystals, however, appear "boiled" with local melt pockets. Many of the larger, altered feldspar blades were also melted. In summary, most of the clasts are unshocked or hardly shocked, but there is a significant proportion of clasts with melting effects. PDFs and diaplectic glass are absent.

G21 (Impact melt breccia). The reddish specimen is composed of several melt phases. A glassy matrix contains angular glass clasts. The matrix is very fragment-rich, containing both angular mineral as well as roundish glass clasts. The glass clasts in part contain microliths. The entire assemblage is isotropic. Thus, the sample is an impact melt breccia with impact glass and diaplectic mineral clasts. Microliths in clasts are generally plagioclase laths, sometimes showing H-shapes (Carstens, 1975). Ovoid inclusions are filled with polycrystalline material, often in radial growths, and likely represent silica-filled vesicles. However, the section is too thin to verify this optically. A few relic crystalline clasts are noted, including some K-feldspar with parallel internal features that could represent PDFs. Some glass clasts are altered to chlorite.

G23 (Glassy matrix breccia). The sample has a clast population of 40–50 vol%. Clasts range from unshocked quartz and feldspar fragments to completely melted felsic mineral clasts and ghost clasts after lithic clast precursors (Fig. 17F). The groundmass is cryptocrystalline. Mafic minerals (mostly biotite) are strongly oxidized or still relatively fresh but then appear bleached ("boiled"). Some alkali feldspar clasts are partially melted. Kink bands in biotite occur occasionally. No PDFs—no diaplectic glass is recorded.

G25/2 (Impact melt breccia). Partially crystallized (microliths) impact melt with numerous, partially melted and partially crystalline clasts derived from a feldspar porphyric volcanic. A few large quartz clasts are weakly shocked (irregular fracturing). Melt phases in clasts are often annealed. Some of the remaining crystalline porphyroblasts have reduced birefringence, but do not exhibit diaplectic glass or PDFs. The melt groundmass to this sample is slightly altered in brownish patches and some microliths are also altered. This groundmass is strongly vesicular. Note that another section of G25 (G25/1) has been classified above as impact melt. It is not clear what the relationship between these two portions of G25 is.

E963-12 (Polymict lithic impact breccia). Very fine-grained matrix is full of fine-grained clasts (clast size covers the entire range from submicroscopic to several 100 μm); in addition, at least 50 vol% of this sample are larger clasts. Most of them are medium-grained quartz and feldspar clasts of all kind of shapes from well rounded to angular. Clasts are generally shocked, with shock degrees ranging from weak (fracturing only) to moderate (several systems of PDFs) to strong (beginning and pronounced formation of diaplectic glass). The numerous quartz fragments with PDFs have mostly three or more systems. More than 6 systems per grain have been noted repeatedly. PDFs are set very densely (Fig. 18E). This effect also results in a strong reduction of birefringence, with many grains appearing nearly isotropic. Several small-scale melt bombs of dark brown or gray color occur as well and have elongated shapes. Feldspar and maskelynite are considerably altered. A single large grain of sphene occurs and has a globular domain texture, a strongly reduced birefringence, and several sets of PDFs. A few feldspar crystals exhibit local melting and annealing.

ACKNOWLEDGMENTS

We are grateful to E. Gurova for optical measurements of impact rocks, and to R. Rakitskaya for X-ray investigations. We thank D. Jalufka (University of Vienna) for some of the drawings. We also appreciate the helpful and constructive reviews by V. Stähle and two anonymous reviewers, as well as editorial comments by T. Kenkmann. This work was supported by the Austrian Science Foundation (to CK), by the International Exchange Programs of the Austrian Academy of Science and the University of Vienna (to EG and CK), and by the Austrian Academic Exchange Service (to KA). This is Impact Cratering Research Group Contribution No. 82.

REFERENCES CITED

Ahrens, T.J., Peterson, C.F., and Rosenberg, J.T., 1969, Shock compression of feldspars: Journal of Geophysical Research, v. 74, p. 2727–2746.

Beliy, V.F., 1969, Volcanic formations and stratigraphy of the northern part of the Ochotsk–Chukotsky Belt: Moscow: Nauka, 138 p. (in Russian).

Beliy, V.F., 1982, Basin of the El'gygytgyn Lake—meteorite crater or geological structure of the recent stage of the Central Chukotka development?: Tichookeanskaya geologia, v. 1982, no. 5, p. 85–92 (in Russian).

Beliy, V.F., 1998, Impactogenesis and volcanism of the El'gygytgyn depression: Journal of Petrology, v. 6, p. 86–99 (in Russian).

Brigham-Grette, J., 2002, El'gygytgyn Lake Workshop Report: Science Results and Plans for Deep Drilling; International Continental Drilling Program (ICDP): Department of Geosciences, University of Massachusetts, Amherst, 72 p.

Brigham-Grette, J., Glushkova, O., Minyuk, P., Melles, M., Overdun, P.P., Zielke, A., Nowacyzk, N., Nolan, M., Stone, D., and Layer, P., 1998, Preliminary lake coring results from El'gygytgyn crater, eastern Siberia [abs.]: Eos (Transactions, American Geophysical Union), v. 79, no. 46, suppl., p. F477.

Carstens, H., 1975, Thermal history of impact melt rocks in Fennoscandian Shield: Contributions to Mineralogy and Petrology, v. 50, p. 145–155.

Dabizha, A.I., and Feldman, V.I., 1982, Geophysical characteristics of some astroblemes of the USSR: Meteoritika, v. 40, p. 91–101 (in Russian).

Dietz, R.S., 1977, El'gygytgyn Crater, Siberia: Probable source of Australasian tektite field (and bediasites from Popigai): Meteoritics, v. 12, p. 145–157.

Dietz, R.S., and McHone, J.F., 1976, El'gygytgyn: Probable world largest meteorite crater: Geology, v. 4, p. 391–392.

Feldman, V.I., Granovsky, L.B., Kapustkina, I.G., Karoteeva, N.N., Sazonova, L.V., and Dabizha, A.I., 1981, Meteorite crater El'gygytgyn, *in* Marakhushev, A.A., ed., Impactites: Moscow State University Press, p. 70–92 (in Russian).

Fredriksson, K., Dube, A., Milton, D.J., and Balasandram, M., 1973, Lonar Lake, India: An impact crater in basalt: Science, v. 180, p. 862–864.

Gostin, V.A., Haines, P.V., Jenkins, R.J.F., Compston, W., and Williams, I.S., 1986, Impact ejecta horizon within Late Precambrian shales, Adelaide Geosyncline, South Australia: Science, v. 233, p. 198–200.

Grieve, R.A.F., Garvin, J.B., and McHone, J.F., 1988, Astronaut's Guide to Terrestrial Impact Craters: Lunar and Planetary Institute Technical Report Number 88-03, Houston, 89 p.

Grieve, R.A.F., Langenhorst, F., and Stöffler, D., 1996, Shock metamorphism of quartz in nature and experiment: II. Significance in geoscience: Meteoritics and Planetary Science, v. 31, p. 6–35.

Gurov, E.P., and Gurova, E.P., 1979, Stages of shock metamorphism of volcanic rocks of acid composition on example of the El'gygytgyn crater (Chukotka): Doklady Academii Nauk USSR, v. 249, p. 1197–1201 (in Russian).

Gurov, E.P., and Gurova, E.P., 1983, Regularities of fault spreading around meteorite craters (on example of the El'gygytgyn crater): Doklady Academii Nauk USSR, v. 275, p. 958–961 (in Russian).

Gurov, E.P., and Gurova, E.P., 1991, Geological structure and rock composition of impact structures: Kiev, Naukova Dumka Press, 160 p. (in Russian).

Gurov, E.P., and Koeberl, C., 2004, Shocked rocks and impact glasses from the El'gygytgyn impact structure (Russia): Meteoritics and Planetary Science, v. 39, p. 1495–1508.

Gurov, E.P., and Yamnichenko, A.Y., 1995, Morphology of rim of complex terrestrial craters [abs.]: Lunar and Planetary Science, v. 26, p. 533–534.

Gurov, E.P., Valter, A.A., Gurova, E.P., and Serebrennikov, A.I., 1978, Meteorite impact crater El'gygytgyn in Chukotka: Doklady Academii Nauk USSR, v. 240, p. 1407–1410 (in Russian).

Gurov, E.P., Gurova, E.P., and Rakitskaya, R.B., 1979a, Stishovite and coesite in shock metamorphosed rocks of the El'gygytgyn crater in Chukotka: Doklady Academii Nauk USSR, v. 248, p. 213–216 (in Russian).

Gurov, E.P., Valter, A.A., Gurova, E.P., and Kotlovskaya, F.I., 1979b, El'gygytgyn impact crater, Chukotka: Shock metamorphism of volcanic rocks [abs.]: Lunar and Planetary Science, v. 10, p. 479–481.

Hörz, F., and Quaide, W.L., 1973, Debye–Scherrer investigations of experimentally shocked silicates: The Moon, v. 6, p. 45–82.

Kashkarov, I.F., and Polkanov, Yu.A., 1970, Discovery of some rare minerals by chemical dressing of heavy concentrates: Razvedka y ochrana nedr, v. 12, p. 49–50 (in Russian).

Kleeman, J.D., 1971, Formation of diaplectic glass by experimental shock loading of orthoclase: Journal of Geophysical Research, v. 76, p. 5499–5503.

Koeberl, C., 1993, Instrumental neutron activation analysis of geochemical and cosmochemical samples: A fast and proven method for small sample analysis: Journal of Radioanalytical and Nuclear Chemistry, v. 168, p. 47–60.

Koeberl, C., and Gurov, E.P., 2002, Geochemistry of glassy impactite "bombs" from the El'gygytgyn impact structure (Russia): A progress report [abs.]: Lunar and Planetary Science, v. 33, abs. no. 1353 (CD-ROM).

Komarov, A.N., Koltsova, T.V., Gurova, E.P., and Gurov, E.P., 1983, Determination of the El'Gygytgyn crater impactites age by the fission track method: Doklady Academii Nauk UkrSSR, v. 10, p. 11–13 (in Russian).

Langenhorst, F., and Deutsch, A., 1994, Shock experiments on pre-heated α- and β-quartz: I. Optical and density data: Earth and Planetary Science Letters, v. 125, p. 407–420, doi: 10.1016/0012-821X(94)90229-1.

Layer, P.W., 2000, Argon-40/Argon-39 age of the El'gygytgyn event, Chukotka, Russia: Meteoritics and Planetary Science, v. 35, p. 591–599.

McKay, D.S., Greenwood, W.R., and Morrison, D.A., 1970, Morphology and related chemistry of small particles from Tranquility Base: Science, v. 167, p. 654–656.

Melles, M., Minyuk, P., Brigham-Grette, J., and Niessen, F., 2003, Successful completion of pre-site survey for deep-drilling at El'gygytgyn crater lake: Eos (Transactions, American Geophysical Union), v. 84 no. 46, Fall Meeting supplement, p. F896.

Nekrasov, I.A., and Raudonis, P.A., 1963, Meteorite craters: Priroda, no. 1, p. 102–104 (in Russian).

Nolan, M., Liston, G., Prokein, P., Brigham-Grette, J., Sharpton, V.L., and Huntzinger, R., 2003, Analysis of lake ice dynamics and morphology on Lake El'gygytgyn, NE Siberia, using synthetic aperture radar (SAR) and Landsat: Journal of Geophysical Research, v. 108, doi: 10.1029/2001JD000934, p. 3-1–3-12.

Nowaczyk, N.R., Minyuk, P., Melles, M., Brigham-Grette, J., Glushkova, O., Nolan, M., Lozhkin, A.V., Stetsenko, T.V., Andersen, P.M., and Forman, S.L., 2002, Magnetostratigraphic results from impact crater Lake El'gygytgyn, northeastern Siberia: A 300 kyr long high-resolution terrestrial palaeoclimatic record from the Arctic: Geophysical Journal International, v. 150, p. 109–126, doi: 10.1046/J.1365-246X.2002.01625.X.

Obruchev, S.V., 1957, Across the tundra and mountains of Chukotka: Moscow, State Press of Geography, 198 p. (in Russian).

Rawson, D.E., 1968, Characteristics of glasses resulted from nuclear explosions in silicate rocks, *in* French B.M. and Short, N.M., eds., Shock

Metamorphism of Natural Materials: Baltimore, Mono Book Corporation, p. 626–627.

Reimold, W.U., Koeberl, C., and Bishop, J., 1994, Roter Kamm impact crater, Namibia: Geochemistry of basement rocks and breccias: Geochimica et Cosmochimica Acta, v. 58, p. 2689–2710, doi: 10.1016/0016-7037(94)90138-4.

Short, N.M., 1968, Nuclear-explosion-induced microdeformations of rocks: An aid to the recognition of meteorite impact structures, *in* French, B.M., and Short, N.M., eds., Shock Metamorphism of Natural Materials: Baltimore, Mono Book Corporation, p. 185–210.

Stöffler, D., 1971, Progressive metamorphism and classification of shocked and brecciated crystalline rocks: Journal of Geophysical Research, v. 76, p. 5541–5551.

Stöffler, D., 1974, Deformation and transformation of rock-forming minerals by natural and experimental shock processes II, Physical properties of shocked minerals: Fortschritte der Mineralogie, v. 51, p. 256–289.

Stöffler, D., and Hornemann, U., 1972, Quartz and feldspar glasses produced by natural and experimental shock: Meteoritics, v. 7, p. 371–394.

Stöffler, D., and Langenhorst, F., 1994, Shock metamorphism of quartz in nature and experiment: I. Basic observations and theory: Meteoritics, v. 29, p. 155–181.

Storzer, D., and Wagner, G.A., 1979, Fission track dating of Elgygytgyn, Popigai and Zhamanshin craters: No source for Australasian or North American tektites: Meteoritics, v. 14, p. 541–542.

Taylor, S.R., and McLennan, S.M., 1985, The Continental Crust: Its Composition and Evolution: Blackwell, Oxford, 312 p.

Williams, G.E., 1986, The Acraman impact structure: Source of ejecta in late Precambrian shales, South Australia: Science, v. 233, p. 200–203.

Zotkin, I.T., and Tsvetkov, V.I., 1970, Search of meteorite craters on the Earth: Astronomitcheskiy Vestnik, v. 4, p. 55–65 (in Russian).

Manuscript Accepted by the Society 9 August 2004

Printed in the USA

Geological Society of America
Special Paper 384
2005

Experimentally shock-loaded anhydrite: Unit-cell dimensions, microstrain and domain size from X-ray powder diffraction

Roman Skála
Bayerisches Geoinstitut, Universität Bayreuth, D-95440 Bayreuth, Germany, and *Czech Geological Survey, Klárov 3/131, CZ-118 21 Praha 1, Czech Republic*

Friedrich P. Hörz
National Aeronautics and Space Adminstration, Johnson Space Center, SR, Houston, Texas 77058, USA

Falko Langenhorst
Bayerisches Geoinstitut, Universität Bayreuth, D-95440 Bayreuth, Germany and *Institut für Geowissenschaften, Friedrich-Schiller-Universität Jena, Burgweg 11, D-07749 Jena, Germany*

ABSTRACT

The Cretaceous-Tertiary boundary is characterized by mass extinctions triggered by a large body impact into predominantly limestone-, dolomite-, and anhydrite-bearing sediments of the Yucatan peninsula, Mexico. Decomposition of these volatile-rich minerals and associated deterioration of the atmosphere and hydrosphere rank among the most prominent kill mechanisms during this global catastrophe. As a consequence, we conducted optical and scanning electron microscopy and X-ray diffraction studies of anhydrite ($CaSO_4$) that was experimentally shock-loaded to pressures from 4 to 64 GPa to determine the shock damage and potential loss of volatiles as a function of shock stress. We did not find any decomposition products in any of the recovered samples. It appears that anhydrite is stable over a wide pressure range (up to 64 GPa). Peak widths of the X-ray diffraction powder patterns increase with peak shock pressure up to ~50 GPa, yet the peaks become narrower again above this pressure, implying some recrystallization.

Keywords: anhydrite, shock metamorphism, shock experiments, K-T extinction, X-ray diffraction.

INTRODUCTION

The Cretaceous-Tertiary (K-T) boundary is associated with one of the most dramatic mass extinctions in Earth's history. A number of kill mechanisms have been suggested to contribute to the widespread extinctions of Cretaceous biota 65 Ma, including severe global deterioration of the atmosphere and hydrosphere from the shock-induced release of CO_2 and SO_x from carbonate and sulfate-bearing target rocks, respectively (e.g., Emiliani et al., 1981; Crutzen, 1987; Brett, 1992; Sigurdsson et al., 1992; Chen et al., 1994; Pope et al., 1997; Yang and Ahrens, 1998; Pierazzo et al., 1998; Gupta et al., 2001). Recent calculations by Pierazzo (2001) reveal that the global warming caused by CO_2 release was considerably less important than the cooling due to SO_x gases potentially released during the K-T Chicxulub impact event. However, the response of sulfates to shock waves is currently not

Skála, R. Hörz, F.P., and Langenhorst, F., 2005, Experimentally shock-loaded anhydrite: Unit-cell dimensions, microstrain and domain size from X-ray powder diffraction, *in* Kenkmann, T., Hörz, F., and Deutsch, A., eds., Large meteorite impacts III: Geological Society of America Special Paper 384, p. 413–425. For permission to copy, contact editing@geosociety.org

well understood. Earlier studies on anhydrite ($CaSO_4$) devolatilization by Chen et al. (1994) and Yang and Ahrens (1998) have been modified by Gupta et al. (2001), who found that the pressures for incipient and complete vaporization along the Hugoniot for the solid anhydrite are ~32 and ~120 GPa, respectively. Schmitt and Hornemann (1998) reported on unit-cell dimensions of anhydrite-quartz mixtures experimentally shock-loaded to 60 GPa; they observed no change in unit cell dimensions for the shocked anhydrite. No clear correlation of unit-cell parameters and shock pressure has also been noted in experimentally shocked anhydrite by Skála and Hörz (2001). However, Langenhorst et al. (2003) found a correlation between the length of parameter ***a*** and shock pressure. Both Skála and Hörz (2001) and Langenhorst et al. (2003) reported significant broadening of anhydrite reflections in diffractograms with increasing pressure up to ~50 GPa. As a consequence, we conducted additional shock experiments on anhydrite over a pressure range of 4–64 GPa. We report here on the detailed optical, scanning electron microscopy, and X-ray diffraction analyses of these samples.

EXPERIMENTAL AND ANALYTICAL METHODS

The shock-recovery experiments employed a 20-mm-caliber powder propellant gun at NASA Johnson Space Center, Houston, as detailed by, e.g., Hörz (1970). Double-polished disks (diameter 7.3 mm, thickness 1.0 or 0.7 mm) of polycrystalline, massive anhydrite from Bees, Canton Vaud/Waadt, Switzerland, were encased in metal target holders made of aluminum (Al2024-T4), stainless steel (SS304), or a high-density W-Ni-Cu alloy ("Tungsten HD17"; previously traded as "Fansteel 77"). These assemblies were impacted by Lexan projectiles carrying metal flyer plates, also made from the above metals or pure W (for the experimental design see Figure 1 in Skála et al., 2002). Peak shock pressures achieved by multiple shock reverberations were obtained with a numerical solution of the graphical impedance match method of Duvall (1962) using the equations of state (EOS) and the actual impact velocity measured via IR laser occultation methods. The EOS data for Al2024, SS304, and W were taken from Marsh (1980), those for Tungsten HD17 from Jones et al. (1966) and Isbell et al. (1968); the anhydrite EOS data have been measured by Simakov et al. (1974). Details of the experiments are given in Table 1. Regarding shock and post-shock temperatures, we refer to Ivanov et al. (2004), who recently published on anhydrite subjected to single shocks.

Carbon-coated polished thin sections of the recovered materials were prepared and imaged using back-scattered electrons in a scanning electron microscope (SEM; LEO Gemini equipped with a Schottky type gun; Bayerisches Geoinstitut [BGI], Bayreuth). The accelerating voltage was 20 kV and the working distance ~15 mm. The chemical composition was

TABLE 1. EXPERIMENTAL CONDITIONS FOR THE SHOCK RECOVERY EXPERIMENTS ON ANHYDRITE

Shot no.	Flyer	Container	v_i (km/s)	P_1 (GPa)	NR	P_f (GPa)	t (μs)
3370	Lexan	Al2024	1.208	4.00	6	4.31	1.1667
3283	Al2024	Al2024	1.212	9.34	5	10.39	0.0163
3324	Al2024	SS304	1.310	10.24	3	16.31	0.0716
3284	Al2024	SS304	1.533	12.38	4	19.65	0.0813
3325	SS304	SS304	1.179	12.81	4	24.13	0.1030
3285	HD17	SS304	1.124	14.28	3	27.73	0.0130
3327	SS304	HD17	1.245	13.70	4	33.44	0.0992
3286	W	SS304	1.370	18.96	3	37.03	0.0257
3287	HD17	HD17	1.240	16.19	3	40.20	0.0486
3329	HD17	HD17	1.154	14.77	6	44.79	0.0786
3288	W	HD17	1.365	18.86	3	47.63	0.0373
3328	HD17	HD17	1.239	16.18	7	49.35	0.0160
3289	W	HD17	1.415	19.77	3	49.78	0.0275
3332	HD17	HD17	1.346	17.99	6	53.88	0.0581
3334	HD17	HD17	1.403	18.97	5	55.33	0.0319
3335	W	HD17	1.409	19.67	5	58.84	0.0374
3336	W	HD17	1.448	20.38	5	60.82	0.0433
3339	W	HD17	1.437	20.17	6	61.74	0.0043
3337*	W	HD17	1.480	20.97	5	62.48	0.0470
3342	W	HD17	1.517	21.65	5	64.39	0.0301

Note: Impact velocity (v_i); initial shock pressure in the sample upon arrival of the first shock wave (P_1); number of reverberations (NR); final peak shock pressure in the sample (P_f) and its duration (t). Lexan—polycarbonate; Al2024—aluminum alloy with 3.8–4.9 wt.% Cu, 1.2–1.8 wt.% Mg, and 0.3–0.9 wt.% Mn; SS304—stainless steel containing 18–20 wt.% Cr, 8–10.5 wt.% Ni, max. 2 wt.% Mn, and max. 1 wt.% Si; and HD17 = 90 wt.% W, 6 wt.% Ni, and 4 wt.% Cu.

*Velocity counters for the shot no. 3337 failed, and the targeted velocity is given.

measured qualitatively using an Oxford Link energy dispersive Si(Li) detector with an ultrathin window to enable the analysis of light elements. Analysis of the energy-dispersive X-ray spectra employed the Oxford Inca software. Uncoated, loose grains were observed with secondary electrons utilizing an inlens detector in the same microscope at an accelerating voltage of 1 kV and ~2–3 mm working distance.

Three powder X-ray diffraction patterns were acquired for each sample using the STOE STADI P and Siemens D-5000 diffractometers at the BGI, Bayreuth, Germany. The STOE diffractometer is equipped with a linear, position-sensitive detector, a Co X-ray tube, and a primary Ge (111) monochromator. One set of diffraction data recorded with the STOE equipment was used for the refinement of unit-cell dimensions and the other for the detailed analysis of line broadening. The angular range for the unit cell refinement was 24.5–120 °2θ and for the analysis of the peak broadening was 24.5–68 °2θ. The data acquired with the Siemens diffractometer in reflecting Bragg-Brentano geometry were used for domain size–microstrain analysis applying the fundamental approach method (see below). This diffractometer was equipped with a Cu X-ray tube, the samples were placed onto low-background silicon wafers, and the diffractograms were recorded in the range from 21 to 64 °2θ with 0.02 °2θ step width. The National Institute of Standards and Technology (NIST) Standard Reference Material (SRM) standards were used to calibrate the data. Their silicon SRM 640b (a = 5.43094 Å) was utilized as an internal standard to calibrate peak positions for unit-cell refinements and their LaB_6 SRM 660 was employed for the peak broadening study.

RESULTS

Appearance and Microscopic Observations

The unshocked anhydrite is a massive and compact, nonporous rock. Individual grains are clear and transparent under binoculars. In thin section, the starting material is largely composed of subhedral to anhedral anhydrite grains, typically ~0.3–0.5 mm across, and a mosaic texture. Anhydrite grains frequently display cleavage, and twinning has been observed in a few grains. Next to anhydrite, the rock contains rare, irregular patches of polycrystalline and very fine-grained dolomite, up to several mm across, and rare quartz grains.

Shock-loaded samples recovered from their metal containers are pervasively fractured and tend to disintegrate easily. Under the binocular microscope, individual grains are no longer clear and transparent; instead, they became white and have a matte surface. At moderate to high magnifications, ubiquitous fracturing, mostly along cleavage planes, is apparent. At peak pressures >40 GPa, tiny veinlets of metal injected into the samples from the molten metal containers are visible along the original margins of the sample disks. Observations with a petrographic microscope (Fig. 1) corroborate the binocular observations and reveal pervasive fragmentation of the samples and a distinctly higher density of cleavage planes than the starting material. Frequently, undulatory extinction is observed in the shocked samples. Also, twinning is more frequent and of different appearance compared to the unshocked rock. Samples shocked to pressures above 50 GPa commonly contain grains composed of a fibrous felt-like aggregates of lowered birefringence, commonly first-order gray and white-gray (Fig. 1F).

Back-scattered electron (BSE) as well as secondary electron (SE) images also reveal ubiquitous fragmentation. In BSE images (Fig. 2), shock-loaded samples generally display a considerably higher number of parallel cleavage planes than the starting material. The number of fissures or the size of individual grains, however, cannot easily be correlated with peak shock pressure. Instead, it appears that the distribution of shock effects is not completely homogeneous throughout the volume of the sample, and some regions display more severe deformation than others. BSE imaging also confirmed the presence of metallic veins in highly shocked materials due to the melting of container material during the experiments. Melts were injected into anhydrite and reacted locally with it producing calcium tungstate and nickel sulfides. Also, some fractionation of the metals was noticed, giving rise to droplets of pure tungsten and copper.

The SE images (Fig. 3) display narrow zones of intense comminution at all pressures. Grains in these zones are usually irregular and of variable size, frequently as small as a few microns. Often, shock-loaded samples display a "scaly" fabric consisting of thin and flat grains (Fig. 3C). This kind of fabric can be attributed to the extremely good cleavage of anhydrite. In addition, both SE and BSE images of some highly shocked materials can be interpreted to show shear features, suggesting that local zones or spots experienced temperatures significantly higher than the average sample.

Phase Identification

The X-ray diffraction patterns acquired without an internal standard were used for qualitative phase analysis. Phase identification employed the Bede Search/Match software and an International Center for Diffraction Data (ICDD) PDF2 database (1998 release). All X-ray powder diffraction patterns yielded anhydrite as the dominant, if not only, phase present. No more than two or three additional peaks were observed in some patterns. They were consistent with low contents of gypsum, dolomite, and quartz and represent the most intense peaks of these accessories. This was confirmed with compositional data acquired with an SEM. The concentrations of these accessories were evaluated adopting a Reference-Intensity-Ratio (RIR) approach, taking into account peak heights of the phases present in the mixture, and RIR values published in the ICDD PDF database. Results of this semiquantitative phase analysis are presented in Table 2. Dolomite may, however, be underestimated. The most intense dolomite peak lies close to the 102 anhydrite reflection and leads to substantial overlap. Consequently, low dolomite contents cannot be reliably identified in such highly shocked samples.

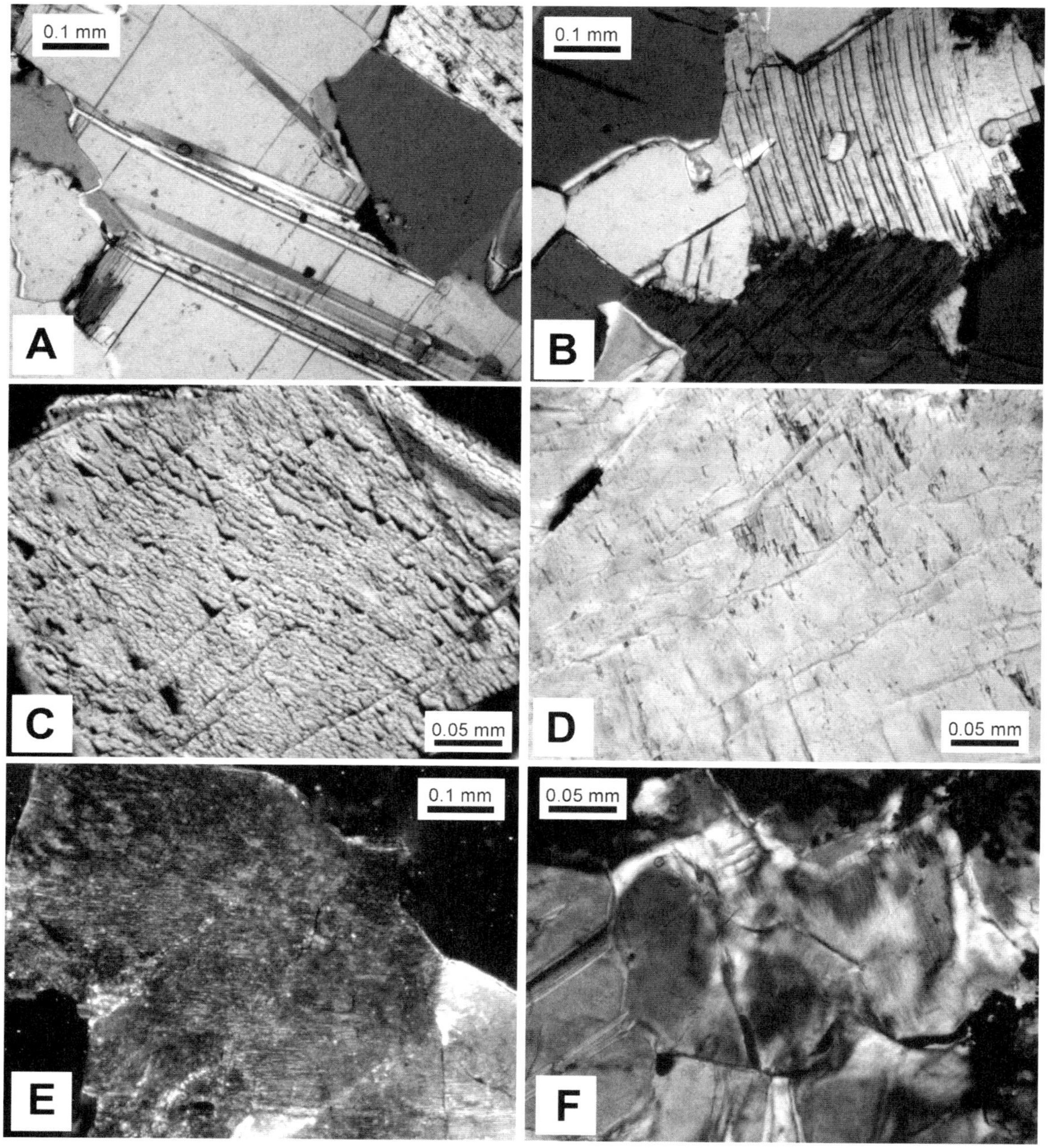

Figure 1. Optical micrographs. (A–B) Unshocked anhydrite composed of subhedral to anhedral grains with a mosaic texture and common cleavage. Twinning was observed in a few grains with twin domains occurring as relatively wide rectangular lamellae with straight boundaries, typically encompassing the entire grain. (C) Sample shocked to 4 GPa: pronounced fracturing along two perpendicular cleavage systems, resulting in jagged grain boundaries. (D) 10 GPa experiment reveals straight fractures parallel to cleavage directions as well as curved fractures cutting the cleavage planes at varying angles. (E) 40 GPa experiment: typical undulatory extinction and twinning, in the form of narrow, needle-shaped, lenticular parallel domains. (F) 60 GPa experiment: felt-like aggregates (center and top right) displaying decreased birefringence, commonly first-order gray and white-gray coexisting with grains of no obvious damage (left).

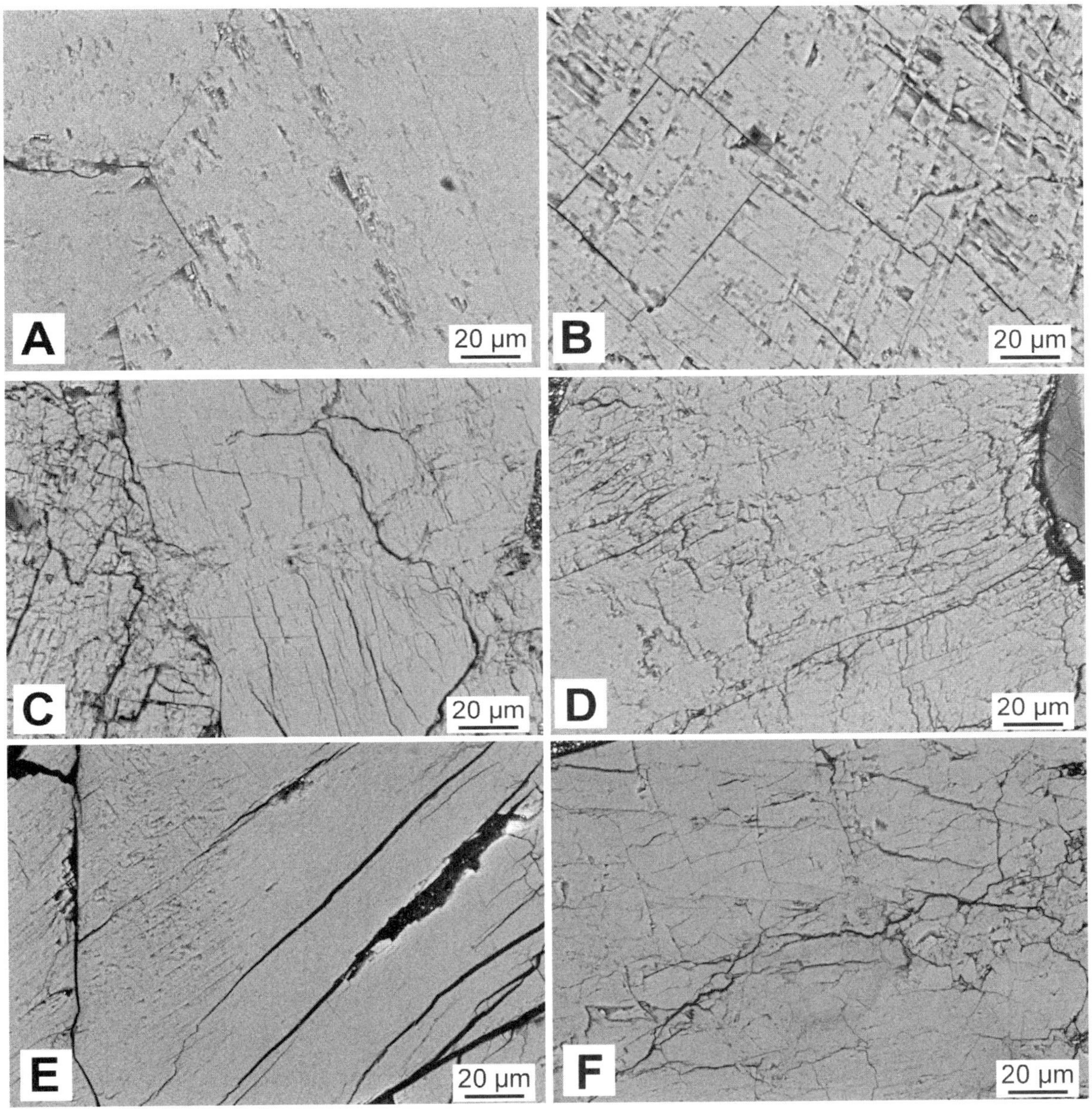

Figure 2. Back-scattered electron images. (A) Unshocked reference sample containing some cleavage planes. (B) 4 GPa experiment: substantially increased number of parallel cleavage planes compared to A. (C) 33 GPa and (D) 40 GPa: irregular fractures cross-cutting the anhydrite crystals and resulting in a mosaic of grains occasionally less than a few microns in diameter. (E) 49 GPa: parallel and irregular fractures; the grain size increases to the right of the image. (F) 60 GPa experiment: variable grain sizes and irregular fracturing and the orientation of fractures indicates some shear.

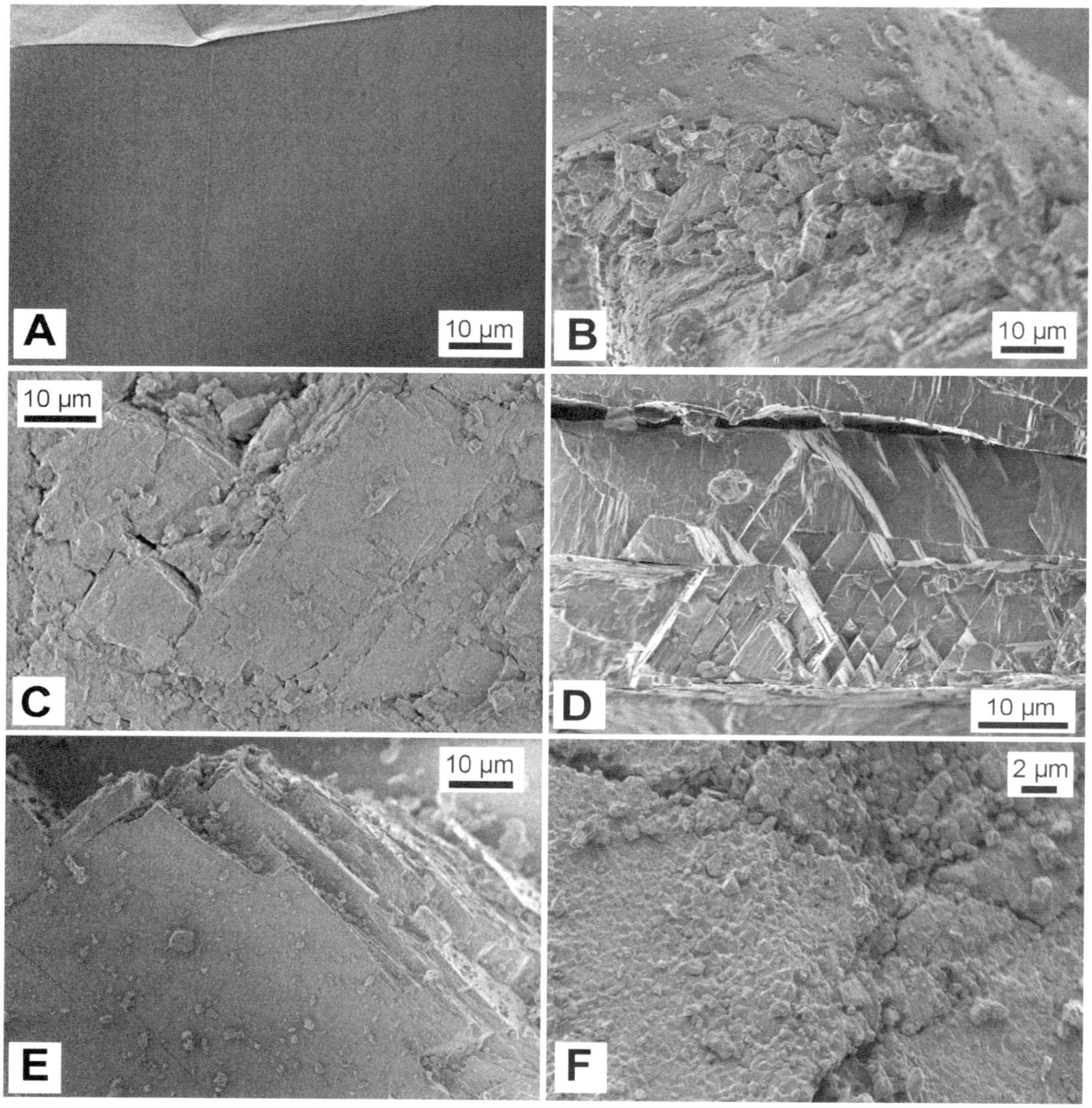

Figure 3. Secondary electron images. (A) Starting material usually consists of relatively large undeformed grains. (B) 33 GPa experiment: narrow zones of intense comminution which are typical for samples recovered from lower pressures. Grains in these zones only occasionally display cleavage planes in more than one direction and consequently they are usually irregular; their size is variable, frequently as small as a few microns. (C) 40 GPa experiment: "scaly" fabric consisting of rather thin and flat grains. (D) 49 GPa experiment: narrow zone with multiple cross-cut cleavages; grains in this zone are completely bound by cleavage planes. (E) 60 GPa experiment: local increase in grain size. (F) 64 GPa experiment: region composed of extremely fine-grained material consisting of irregular grains which may be indicative of solid-state recrystallization or crystallization from a melt.

TABLE 2. SEMI-QUANTITATIVE PHASE ANALYSIS OF THE SAMPLES STUDIED

Shot no.	P_f (GPa)	Dominant phase	Minor phase(s)	Content(s) of minor phase (wt.%)
STD*		anhydrite	dolomite	0.2
3370	4	anhydrite	–	–
3283	10	anhydrite	gypsum, calcite	0.7, 0.2
3324	16	anhydrite	–	–
3284	20	anhydrite	quartz	1.4
3325	24	anhydrite	–	–
3285	28	anhydrite	quartz	0.1
3327	33	anhydrite	gypsum, dolomite	2.8, 0.1
3286	37	anhydrite	dolomite	0.3
3287	40	anhydrite	–	–
3329	45	anhydrite	gypsum	2.8
3288	48	anhydrite	–	–
3328	49	anhydrite	–	–
3289	50	anhydrite	†	†
3332	54	anhydrite	gypsum	3.1
3334	55	anhydrite	–	–
3335	59	anhydrite	gypsum	2.8
3336	61	anhydrite	gypsum	1.5
3339	62	anhydrite	gypsum	3.0
3337	62	anhydrite	gypsum	2.5
3342	64	anhydrite	gypsum	3.3

Note: For the analysis the data were acquired without an internal standard using STOE and Siemens diffractometers. Dolomite may be present in more samples than indicated in the table, however, its most intense reflection might be obscured by broadened profiles of highly shocked anhydrite.
*STD—starting material.
†Sample no. 3289 is contaminated with ~wt.% Si standard.

Unit-Cell Dimensions

Individual peak positions for the refinement of unit-cell parameters were obtained with the program XFit (Coelho and Cheary, 1997). Peak profiles were fitted with a symmetrical pseudo-Voigt profile-shape function. Alternative fits with split Pearson VII profile-shape function did not reveal any significant peak asymmetry and substantiated the use of the above symmetrical pseudo-Voigt function. The maxima corresponding to the silicon internal standard were compared to the theoretical values, and a correction function was established from the observed differences; the latter was subsequently applied to the measured peak positions of the anhydrite samples.

The crystal structure of anhydrite by Hawthorne and Ferguson (1975) was used to generate theoretical powder data with the program FullProf (Rodríguez-Carvajal, 2003) to assign diffraction indices to individual peaks. Although the anhydrite patterns are not complex at first sight, the refinement of the unit-cell dimensions is complicated by the fact that two of the cell dimensions are very close (***a*** ~ 7.000 Å and ***b*** ~ 6.995 Å in the *Amma* setting). Consequently, several peaks represent pairs of reflections with indices *hkl* and *khl*; such pairs were excluded from the present analysis, because they would bias the results. The unit-cell dimensions were refined using the program UnitCell (Holland and Redfern, 1997) using the positions of the ten most prominent, non-overlapping diffraction lines (with *hkl* indices in the *Amma* setting: 111, 102, 220, 122, 031, 013, 302, 322, 142, and 224), and corrected with the silicon internal standard as detailed above. Table 3 summarizes the results, and Figure 4 shows the correlation of the parameters as a function of shock pressure. Parameters ***a*** and, to a lesser extent, ***b***, progressively increase with increasing peak shock pressure up to ~55–60 GPa; at still higher pressures they become smaller again, yet they do not reach the low values of the starting material. The cell volume shows corresponding variations. Parameter ***c*** increases with increasing pressure as well, but does not seem to decrease even at the highest pressures. The variation in cell parameters is small, however, within 0.1 rel% for unit cell edges and <0.15 rel% for the cell volume. Also, most changes are within experimental errors given by one sigma in Table 3.

Peak Broadening, Domain Size and Microstrain

Line broadening is prominent already at 10 GPa (Fig. 5) and increases with pressure, albeit not linearly, up to ~50 GPa. Above this pressure, the broadening trend reverses and approaches values typical for pressures around ~20 GPa. Simultaneously, the peak height decreases to ~50% of that for the 10 GPa experiment, but peak heights remain constant at >50 GPa. Figure 6 summarizes the data on broadening for selected samples. Some of the reflections (322, 122, and 302) of samples shocked to above ~30 GPa are broadened by more than 200%, others (200 and 020, 400 and 040, and 102) display significantly less broadening.

Domain size and microstrain are known to be manifested in peak shape of X-ray diffraction powder patterns (e.g., Klug and Alexander, 1974). These parameters were evaluated by several independent techniques, such as (1) modeling of peak shapes as a convolution of individual diffractometer contributions to the diffraction pattern following Cheary and Coelho (1998a, 1998b, 1998c), (2) whole powder pattern fitting (Rodríguez-Carvajal, 2003), (3) individual peak fitting followed by extracting the Gaussian and Lorentzian components and solving for the size and strain parameters (Delhez et al., 1993), and (4) multiple-line Voigt method (Balzar, 1999).

The fundamental parameter method developed by Cheary and Coelho (1998a, 1998b, 1998c) is based on modeling the observed peak shapes by the convolution of instrumental parameters characterizing any specific diffractometer unit, combined with parameters of the sample, including crystallite size and microstrain. For this approach, we used the data from the Siemens diffractometer as determined by the program Xfit (Coelho and Cheary, 1997). This program was operated to refine (besides the peak positions and intensities) the values of domain size and strain. Numerically, domain size varies between 320 Å and 2860 Å and microstrain ranges from 0.09 × 10^{-3} to 3.5 × 10^{-3}. Domain size decreases with increasing pressure up to ~50 GPa, and from this point on, it increase a bit again. The microstrain reveals the opposite trend.

TABLE 3. UNIT CELL DIMENSIONS OF SHOCK-LOADED ANHYDRITE COMPARED TO THOSE OF STARTING MATERIAL

Shot no.	P_f (GPa)	a (Å)	σ_a (Å)	b (Å)	σ_b (Å)	c (Å)	σ_c (Å)	V (Å^3)	σ_V (Å^3)
STD*		7.0018	0.0006	6.9918	0.0004	6.2400	0.0003	305.48	0.02
3370	4	7.0025	0.0008	6.9919	0.0005	6.2396	0.0004	305.50	0.03
3283	10	7.0039	0.0011	6.9924	0.0008	6.2397	0.0006	305.58	0.05
3324	16	7.0035	0.0015	6.9929	0.0010	6.2402	0.0008	305.61	0.06
3284	20	7.0037	0.0013	6.9937	0.0009	6.2406	0.0007	305.68	0.05
3325	24	7.0030	0.0012	6.9927	0.0008	6.2408	0.0007	305.61	0.05
3285	28	7.0050	0.0020	6.9938	0.0014	6.2402	0.0011	305.72	0.08
3327	33	7.0037	0.0019	6.9936	0.0013	6.2408	0.0010	305.68	0.08
3286	37	7.0058	0.0032	6.9933	0.0022	6.2406	0.0017	305.75	0.13
3287	40	7.0042	0.0023	6.9929	0.0016	6.2407	0.0012	305.67	0.09
3329	45	7.0057	0.0036	6.9938	0.0025	6.2400	0.0019	305.74	0.15
3288	48	7.0061	0.0021	6.9939	0.0014	6.2404	0.0011	305.78	0.09
3328	49	7.0049	0.0028	6.9937	0.0019	6.2397	0.0015	305.68	0.12
3289	50	7.0054	0.0032	6.9935	0.0022	6.2410	0.0017	305.76	0.13
3332	54	7.0046	0.0021	6.9926	0.0015	6.2424	0.0011	305.76	0.09
3334	55	7.0056	0.0023	6.9939	0.0016	6.2410	0.0012	305.79	0.10
3335	59	7.0049	0.0031	6.9949	0.0021	6.2409	0.0017	305.80	0.13
3336	61	7.0048	0.0033	6.9941	0.0023	6.2402	0.0018	305.72	0.14
3339	62	7.0049	0.0028	6.9923	0.0019	6.2407	0.0015	305.67	0.12
3337	62	7.0023	0.0014	6.9926	0.0010	6.2421	0.0008	305.64	0.06
3342	64	7.0026	0.0019	6.9917	0.0013	6.2405	0.0010	305.54	0.08

Note: This analysis uses data acquired with an internal silicon standard and employing the STOE diffractometer. Peak positions were obtained with Xfit program (Coelho and Cheary, 1997). The unit-cell dimensions were refined with the program UnitCell (Holland and Redfern, 1997). The sigmas shown represent one sigma values.
*STD—starting material.

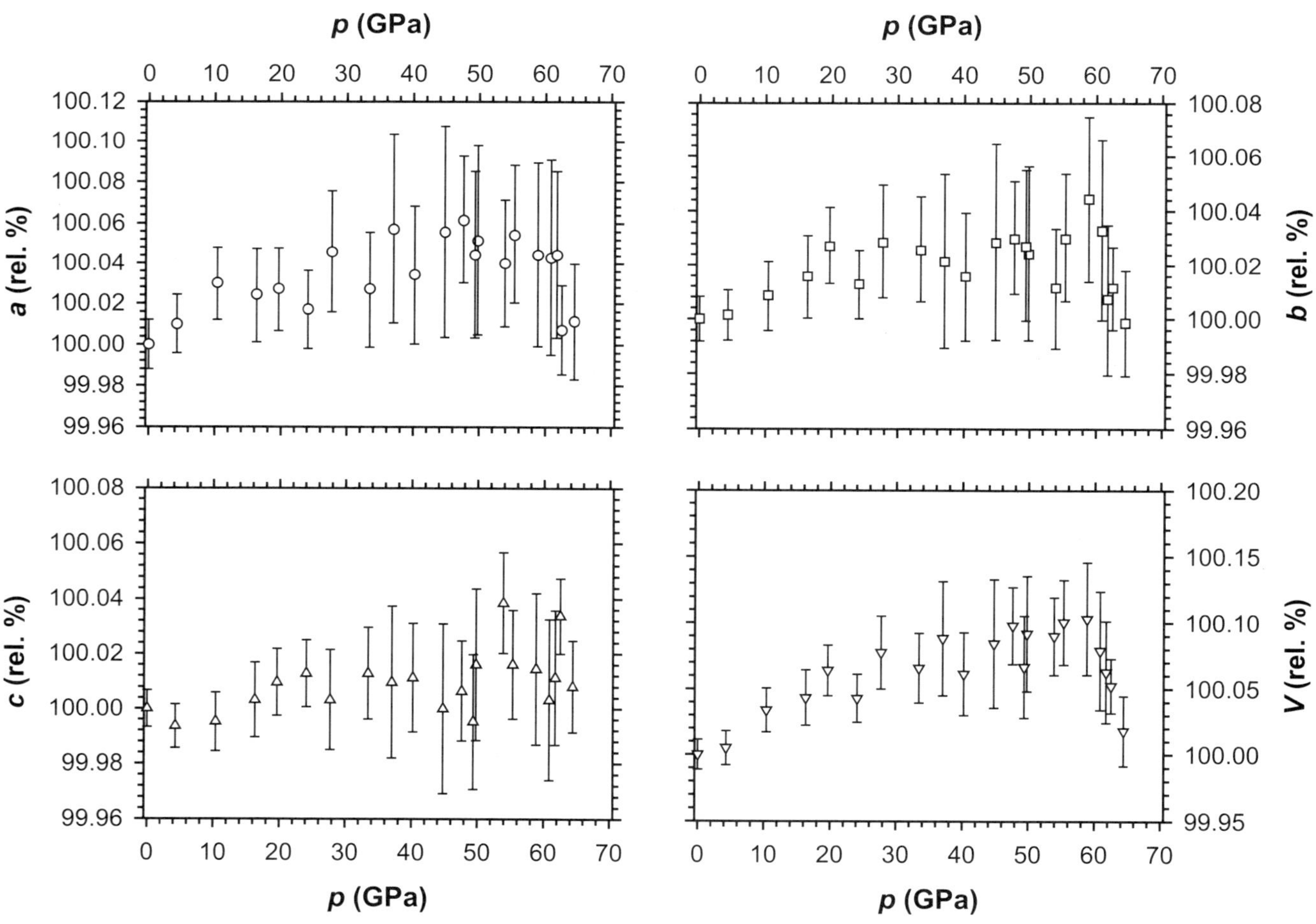

Figure 4. Normalized values of the unit-cell dimensions of experimentally shocked anhydrite. Error bars are 1σ.

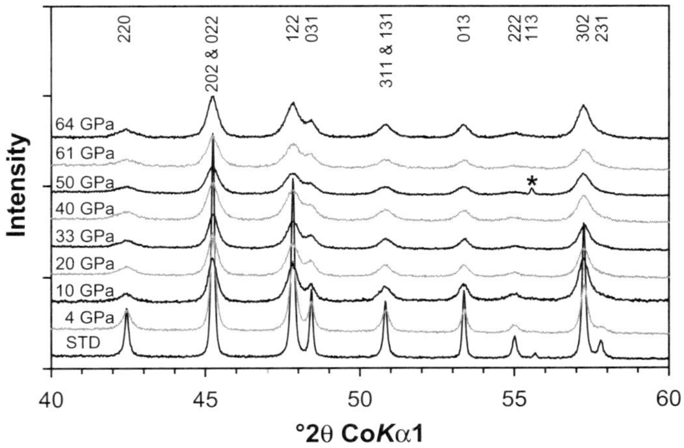

Figure 5. Partial X-ray powder patterns of anhydrite samples shocked to 4 GPa, 10 GPa, 20 GPa, 33 GPa, 40 GPa, 50 GPa, 61 GPa, and 64 GPa compared with unshocked reference material (STD). The shock-induced peak broadening increases with increasing pressure up to 50 GPa, and decreases again at higher pressures. Asterisk indicates an admixture of Si in the 50 GPa shot; *hkl* indices are shown for anhydrite peaks.

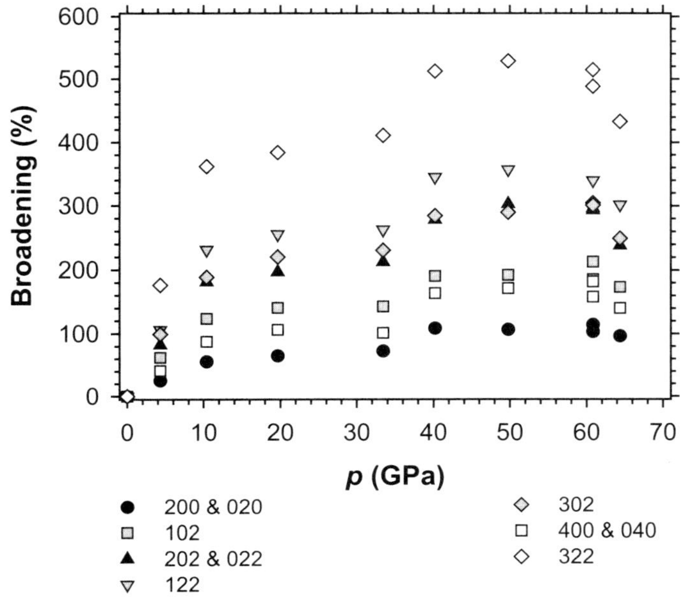

Figure 6. Peak broadening of selected reflections and reflection groups of shocked anhydrite. Peak broadening is defined as $100 \times (W_e - W_s)/W_s$, where W_e is the full width at half maximum of the sample and W_s is the half-width of the reference specimen. Legend shows *hkl* indices. Duplicate measurements for the 62 GPa shot illustrate good reproducibility of the data.

The whole powder pattern fitting technique was applied to data acquired with the STOE diffractometer. It employed the Thompson-Cox-Hastings profile shape function as defined in the FullProf program (Rodríguez-Carvajal, 2003) and the instrument resolution function based on the refined peak parameters of the LaB_6 standard. This instrument resolution function was constructed with the program WinPlotr (Roisnel and Rodríguez-Carvajal, 2003). The resulting numerical values for both grain size and strain parameters were about half of those inferred from the fundamental parameter approach, yet the general trends were the same.

The procedure for deriving the size of coherently scattering domains and for the strain from individual peak shapes included the fitting of individual peaks with pseudo-Voigt profile shape function as implemented in Xfit program (Coelho and Cheary, 1997), followed by a separation of the Gaussian and Lorentzian contributions (de Keijser et al., 1983) and a correction for instrumental broadening using the powder data of LaB_6 and formulae given in Delhez et al. (1993). For this approach, we used the powder patterns collected with the STOE diffractometer. The values of crystallite size and microstrain differed for individual reflections, implying non-isotropic behavior, either related to the direction of shock and/or to the intrinsic crystal structure of anhydrite. However, the mean values agree well with those obtained by the method of Cheary and Coelho (1998a, 1998b, 1998c).

The multiple-line Voigt method, as detailed in, e.g., Balzar (1999), used the same data corrected for instrumental broadening as described in the previous paragraph. The analysis was performed with the program BREADTH (Balzar, 1997). Based on Balzar (1999), this approach is considered more reliable than the others. Therefore, we consider the resulting values based on this method to be most representative for the domain size and strain, and we list them in Table 4. This approach also provides the distribution functions of domain size and strain as illustrated in Figure 7 for selected experiments. Normalized domain size and microstrain obtained from all the approaches are summarized in Figure 8.

CONCLUSIONS

The results of this X-ray diffraction study combined with optical and SEM observations show that anhydrite is a rather stable mineral under the conditions simulated by our experiments, including peak shock pressure of 65 GPa. This is contrary to Gupta et al. (2001), who found incipient vaporization of anhydrite at ~32 GPa. While we observed severe mechanical deformation, including fracturing, comminution and shearing, and marked increase in microstrain, combined with decreasing domain size with progressive shock pressure, we observed neither a phase change, including high-pressure polymorphs, nor any decomposition products. Our observations draw strong support from Ivanov et al. (2004), who calculated the shock and post-shock temperatures of anhydrite. These calculations show that the post-shock temperatures for pressures up to 64 GPa, obtained via multiple shock reverberation methods, are well

TABLE 4. DOMAIN SIZE AND MICROSTRAIN DETERMINED WITH MULTIPLE-LINE VOIGT METHOD

Shot no.	P_f (GPa)	$<D>_s$* (Å)	σ	$<D>_v$† (Å)	σ	ε_s§ (× 10^{-3})	σ	ε_v# (× 10^{-3})	σ
STD‡		1800	65	3600	130	0.79	0.01	0.73	0.01
3370	4.31	930	31	1421	85	1.22	0.02	1.07	0.02
3283	10.39	513	21	780	40	1.74	0.05	1.49	0.05
3324	16.31	478	12	689	18	1.63	0.04	1.40	0.04
3284	19.65	449	11	830	79	1.97	0.04	1.65	0.04
3325	24.13	291	6	581	11	2.26	0.07	1.99	0.07
3285	27.73	334	6	544	20	2.53	0.06	2.20	0.05
3327	33.44	350	7	460	23	2.42	0.05	2.18	0.05
3286	37.03	322	11	422	14	2.57	0.09	2.30	0.09
3287	40.20	371	9	515	18	2.49	0.06	2.24	0.06
3329	44.79	287	5	550	75	2.34	0.07	1.98	0.06
3288	47.63	315	10	456	21	2.46	0.11	2.21	0.10
3328	49.35	314	10	457	19	2.49	0.10	2.18	0.10
3289	49.78	319	8	521	24	2.47	0.08	2.15	0.08
3332	53.88	279	6	549	69	2.22	0.09	1.93	0.09
3334	55.33	326	9	436	12	2.27	0.08	2.02	0.08
3335	58.84	275	8	438	20	2.53	0.12	2.16	0.11
3336	60.82	316	15	451	25	2.39	0.18	2.20	0.16
3339	61.74	302	8	457	15	2.06	0.10	1.80	0.09
3337	62.48	341	11	496	20	2.30	0.09	2.00	0.09
3342	64.39	453	16	529	17	2.16	0.06	2.03	0.06

Note: This analysis uses the data acquired without an internal standard using the STOE diffractometer. Peak parameters were refined with program Xfit (Coelho and Cheary, 1997). Domain sizes and microstrains were evaluated with the BREADTH program of Balzar (1997). The sigmas shown represent one sigma values.
*Surface-weighted domain size.
†Volume-weighted domain size.
§Root mean square strain averaged over the distance $<D>_s/2$.
#Root mean square strain averaged over the distance $<D>_v/2$.
‡STD—starting material.

below the decomposition temperature of anhydrite, even if the partial SO_2 pressure is reduced.

Slight changes in unit-cell dimensions with pressure are consistent with the observations on anhydrite + quartz mixtures shocked to 60 GPa (Schmitt and Hornemann, 1998), and with our preliminary anhydrite results (Skála and Hörz, 2001), as well as with Langenhorst et al. (2003), who found a correlation between the length of parameter ***a*** and shock pressure (maximum difference *ca* 0.2 rel%). The data presented here, however, show the magnitude of change to be only one-half of that reported by Langenhorst et al. (2003). The quantitative difference in unit-cell dimensions can possibly be attributed to differences in the shock-loading experiments; the flyer plate thickness and thus shock pulse duration vary by a factor of 5 between these experiments. Material examined by Langenhorst et al. (2003) was prepared using large massive containers, whereas the experiments carried out at the NASA Johnson Space Center employ smaller containers and flyer plates, which may result in more shearing of samples.

The peak shifts and broadening observed in X-ray patterns may be attributed to some residual stress in the shocked samples due to lattice misfit along crystal structure defects. Our data on the behavior of peak widths, domain size, and microstrain generally agree with those presented in Langenhorst et al. (2003) but changes are less pronounced in our samples. The gradual narrowing of the peaks above ~50 GPa may support the idea of Langenhorst et al. (2003) that either a solid-state recrystallization or melting with subsequent crystallization of anhydrite occurred due to the high post-shock temperatures at pressures exceeding 40–50 GPa. Melting of volatile-rich targets in a closed system of shock recovery experiments has been observed previously for calcite by Ivanov et al. (2002) and Langenhorst et al. (2002) and is supported by thermodynamic calculations of Ivanov and Deutsch (2002). Recently, the melting of shocked anhydrite in the pressure range of 80–90 GPa was predicted by Ivanov et al. (2004), but an additional shear component could lead to melting at lower pressures.

The other plausible explanation for decreased line broadening at the highest pressures is the annealing of lattice defects due to enhanced post-shock temperatures. It has to be noted, however, that the majority of the material remains heavily damaged, because the peak shape did not completely revert to that of the unshocked material, even at the highest shock pressures. Instead, their shape characteristics approach those of the peaks recorded for samples shocked to ~20 GPa. Further transmission electron microscopy is required to completely understand the reasons for

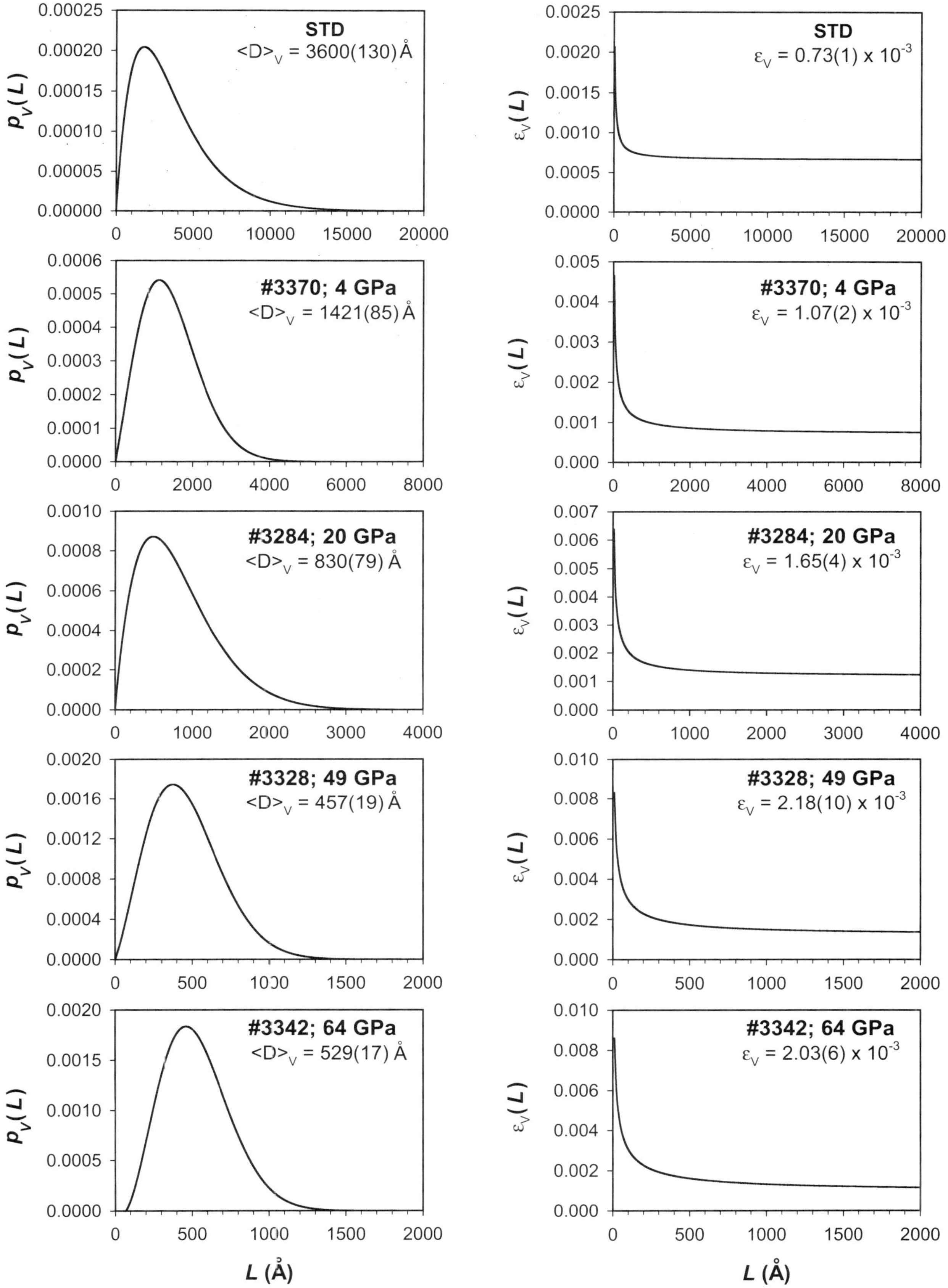

Figure 7. Volume-weighted column-length distribution function $p_v(L)$ (left column) and mean square strain (right column) as a function of averaging distance (in Å) for selected shock-loaded anhydrite samples. STD—starting material.

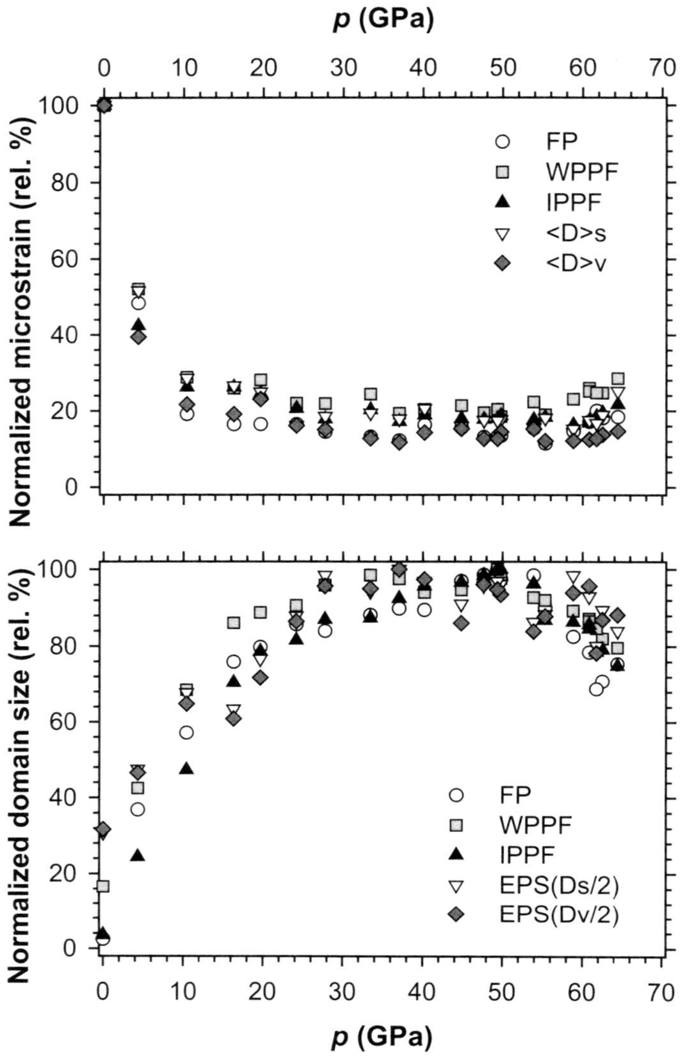

Figure 8. Normalized domain size and microstrain obtained by a variety of methods. FP—fundamental parameters convolution approach; WPPF—whole powder pattern fitting with pre-defined instrument-resolution function; IPPF—individual peak fitting with pseudo-Voigt function; <D>s—surface-weighted domain size yielded with multiple-line Voigt method; <D>v—volume-weighted domain size; EPS(Ds/2)—root mean square strain averaged over the distance <D>$_s$/2; and EPS(Dv/2)—root mean square strain averaged over the distance <D>$_v$/2. The trends observed are internally consistent, all suggesting that deformation increases with increasing pressure to attain a maximum at ~40–50 GPa, and that part of the lattice defects must be removed at still higher stresses.

enlarged lattice parameters of shocked anhydrite and the narrowing of X-ray peaks at high shock pressures.

ACKNOWLEDGMENTS

RS is grateful for an EU Marie Curie Programme fellowship at the Bayerisches Geoinstitut, Universität Bayreuth, Bayreuth, Germany. Participation of RS at the Third International Conference on Large Meteorite Impacts held in Nördlingen in 2003 was supported by the Grant Agency of the Czech Republic Project No. 205/02/1101. We also acknowledge the financial support provided by the Deutsche Forschungsgemeinschaft through grants DE 401/15-5, HO 1446/3-5, and LA 830/4-5. We are indebted to Alex Deutsch, Isabelle Martinez, and Milan Rieder for their constructive reviews.

REFERENCES CITED

Balzar, D., 1997, BREADTH—a PC program for analyzing diffraction-line broadening: http://www.boulder.nist.gov/div853/balzar/.

Balzar, D., 1999, Voigt function model in diffraction-line broadening analysis, *in* Snyder, R.L., Fiala, J., and Bunge, H.J., eds., Defects and microstructure analysis by diffraction: International Union of Crystallography Monograph on Crystallography 10, Oxford, Oxford University Press, p. 94–126.

Brett, R., 1992, The Cretaceous-Tertiary extinction: A lethal mechanism involving anhydrite target rocks: Geochimica et Cosmochimica Acta, v. 56, p. 3603–3606, doi: 10.1016/0016-7037(92)90406-9.

Chen, G., Tyburczy, J.A., and Ahrens, T.J., 1994, Shock-induced devolatilization of calcium sulfate and implication for K-T extinctions: Earth and Planetary Science Letters, v. 128, p. 615–628, doi: 10.1016/0012-821X(94)90174-0.

Cheary, R.W., and Coelho, A.A., 1998a, Axial divergence in a conventional X-ray powder diffractometer. I. Theoretical foundations: Journal of Applied Crystallography, v. 31, p. 851–861, doi: 10.1107/S0021889898006876.

Cheary, R.W., and Coelho, A.A., 1998b, Axial divergence in a conventional X-ray powder diffractometer. II. Realization and evaluation in a fundamental-parameter profile fitting procedure: Journal of Applied Crystallography, v. 31, p. 862–868, doi: 10.1107/S0021889898006888.

Cheary, R.W., and Coelho, A.A., 1998c, An experimental investigation of the effects of axial divergence on diffraction line profiles: Powder Diffraction, v. 13, p. 100–106.

Coelho, A.A., and Cheary, R.W., 1997, X-ray Line Profile Fitting Program, XFIT: ftp://ftp.minerals.csiro.au/pub/xtallography/koalariet.

Crutzen, P.J., 1987, Mass extinctions—acid-rain at the K/T boundary: Nature, v. 330, p. 108–109, doi: 10.1038/330108A0.

de Keijser, T.H., Mittemeijer, E.J., and Rozendaal, H.C.F., 1983, The determination of crystallite-size and lattice-strain parameters in conjunction with the profile-refinement method for the determination of crystal structures: Journal of Applied Crystallography, v. 16, p. 309–316, doi: 10.1107/S0021889883010493.

Delhez, R., de Keijser, T.H., Langford, J.I., Louër, D., Mittemeijer, E.J., and Sonneveld, E.J., 1993, Crystal imperfection broadening and peak shape in the Rietveld method, *in* Young, R.A., ed., The Rietveld method: International Union of Crystallography Monograph on Crystallography 5: Oxford, Oxford University Press, p. 132–166.

Duvall, G.E., 1962, Concepts of shock wave propagation: Bulletin of the Seismological Society of America, v. 4, p. 869–893.

Emiliani, C., Kraus, E.B., and Shoemaker, E.M., 1981, Sudden death at the end of Mesozoic: Earth and Planetary Science Letters, v. 55, p. 317–334, doi: 10.1016/0012-821X(81)90161-8.

Gupta, S.C., Ahrens, T.J., and Yang, W., 2001, Shock-induced vaporization of anhydrite and global cooling from the K/T impact: Earth and Planetary Science Letters, v. 188, p. 399–412, doi: 10.1016/S0012-821X(01)00327-2.

Hawthorne, F.C., and Ferguson, R.B., 1975, Anhydrous sulfates II. Refinement of the crystal structure of anhydrite: Canadian Mineralogist, v. 13, p. 289–292.

Holland, T.J.B., and Redfern, S.A.T., 1997, Unit cell refinement from powder diffraction data: The use of regression diagnostics: Mineralogical Magazine, v. 61, p. 65–77.

Hörz, F., 1970, A Small Ballistic Range for Impact Metamorphism Studies: NASA Technical Note, v. D-5787, p. 18.

Isbell, W.H., Shipman, F.H., and Jones, A.H., 1968, Hugoniot equation of state measurements for eleven materials to five megabars: General Motors Corporation, Material Science Laboratory, Report MSL-68-13.

International Center for Diffraction Data (ICDD) 1998, PDF2 database, 1998: International Center for Diffraction Data, Newton Square, Pennsylvania, USA.

Ivanov, B.A., and Deutsch, A., 2002, The phase diagram of $CaCO_3$ in relation to shock compression and decomposition: Physics of the Earth and Planetary Interiors, v. 129, p. 131–143, doi: 10.1016/S0031-9201(01)00268-0.

Ivanov, B.A., Langenhorst, F., Deutsch, A., and Hornemann, U., 2002, How strong was impact-induced CO_2 degassing in the K/T event? Numerical modeling of laboratory experiments, *in* Koeberl, C., and MacLeod, K.G., eds., Catastrophic Events and Mass Extinctions: Impacts and Beyond: Boulder, Colorado, Geological Society of America Special Paper 356, p. 587–594.

Ivanov, B.A., Langenhorst, F., Deutsch, A., and Hornemann, U., 2004, Anhydrite EOS and phase diagram in relation to shock decomposition: Lunar and Planetary Institute Conference Abstracts, v. 35, Abstract no. 1489 (CD-ROM).

Jones, A.H., Isbell, W.H., and Maiden, C.J., 1966, Measurements of the very-high-pressure properties of materials using a light-gas gun: Journal of Applied Physics, v. 37, p. 3493–3499.

Klug, H.A., and Alexander, L.E., 1974, X-ray diffraction procedures for polycrystalline and amorphous materials, 2nd edition: New York, John Wiley & Sons, 966 p.

Langenhorst, F., Boustie, M., Deutsch, A., Hornemann, U., Matignon, Ch., Migault, A., and Romain, J.P., 2002, Experimental techniques for the simulation of shock metamorphism: A case study on calcite, *in* Davison, L., Horie, Y., and Sekine, T., eds., High-Pressure Shock Compression of Solids V—Shock Chemistry and Meteoritic Applications: Berlin–Heidelberg–New York, Springer p. 1–27.

Langenhorst, F., Deutsch, A., Hornemann, U., Ivanov, B.A., and Lounejeva, E., 2003, On the shock behaviour of anhydrite: experimental results and natural observations: Lunar and Planetary Institute Conference Abstracts, v. 34, Abstract no. 1638 (CD-ROM).

Marsh, S.P., 1980, LASL Shock Hugoniot Data: Berkeley, California, University of California Press, 658 p.

Pierazzo, E., 2001, Climate forcing from the stratospheric injection of impact-produced sulfur: Lunar and Planetary Institute Conference Abstracts, v. 32, Abstract no. 1196 (CD-ROM).

Pierazzo, E., Kring, D.A., and Melosh, H.J., 1998, Hydrocode simulation of the Chicxulub impact event and the production of climatically active gases: Journal of Geophysical Research, v. 103, p. 28607–28626, doi: 10.1029/98JE02496.

Pope, K.O., Baines, K.H., Ocampo, A.C., and Ivanov, B.A., 1997, Energy, volatile production, and climatic effects of the Chicxulub Cretaceous/Tertiary impact: Journal of Geophysical Research, v. 102, p. 21645–21664, doi: 10.1029/97JE01743.

Rodríguez-Carvajal, J., 2003, FullProf2.k—a computer program: ftp://ftp.cea.fr/pub/llb/divers/fullprof.2k/Windows/.

Roisnel, T., and Rodríguez-Carvajal, J., 2003, WinPlotr—a computer program: http://www.llb-cea.fr/fullweb/winplotr/winplotr.htm.

Schmitt, R.T., and Hornemann, U., 1998, Shock experiments with a natural anhydrite-quartz-sample: Lunar and Planetary Institute Conference Abstracts, v. 29, Abstract no. 1019 (CD-ROM).

Sigurdsson, H., D'Hondt, S., and Carey, S., 1992, The impact of the Cretaceous Tertiary bolide on evaporite terrane and generation of major sulfuric acid aerosol: Earth and Planetary Science Letters, v. 109, p. 543–560, doi: 10.1016/0012-821X(92)90113-A.

Simakov, G.V., Pavlovskiy, M.N., Kalashnikov, N.G., and Trunin, R.F., 1974, Udarnaya szhimaemost' dvenadtsati mineralov [Shock compressibility of twelve minerals.]: Izvestiya AN SSSR, Fizika Zemli, v. 1974 (8), p. 11–17 (in Russian).

Skála, R., and Hörz, F., 2001, X-ray diffraction study of experimentally shocked anhydrite—A preliminary report [abs.]: Meteoritics & Planetary Science, v. 36, Supplement, p. A192.

Skála, R., Ederová, J., Matějka, P., and Hörz, F., 2002, Mineralogical investigations of experimentally shocked dolomite: Implications for the outgassing of carbonates, *in* Koeberl, C., and MacLeod, K.G., eds., Catastrophic Events and Mass Extinctions: Impacts and Beyond: Boulder, Colorado, Geological Society of America Special Paper 356, p. 571–585.

Yang, W., and Ahrens, T.J., 1998, Shock vaporization of anhydrite and global effects of the K/T bolide: Earth and Planetary Science Letters, v. 156, p. 125–140, doi: 10.1016/S0012-821X(98)00006-5.

Manuscript Accepted by the Society 9 August 2004

Printed in the USA

Geological Society of America
Special Paper 384
2005

Giant impact-induced blow-off of primordial atmosphere

Sidao Ni
Thomas J. Ahrens
Seismological Laboratory 252-21, California Institute of Technology, Pasadena, California 91125, USA

ABSTRACT

The surface motion from a large impact upon an attenuation-free fluid sphere was studied and numerically simulated. An analytic solution for the free-surface velocity shows that close to the source, the acoustic wave due to the free-surface interaction (a "quasi-surface wave") is not separable from the direct wave. At >90°, the quasi-surface wave separates and has a larger amplitude than the direct body wave. Near the antipode the quasi-surface wave amplitude is much larger than the direct body wave and is comparable to the direct wave amplitude immediately near the source at 0°. The resulting solution covers both the wave interference range as defined in the asymptotic theory of near-surface wave propagation developed by Russian physicist V.S. Buldyrev reported in 1968, as well as in the geometric ray range. The geometric range theory was described in several papers in terms of multi-geometry reflection by R. Burridge, H. Jeffreys, and E.R. Lapwood in England in 1957 through 1963. For a large surface excitation (e.g., giant ~10^{27} J impact) a portion of the atmosphere above a plane tangent to Earth at the impact point is launched to velocities greater than the escape velocity. The resulting antipodal free-surface velocity achieved is ~1.9 km/s, which is sufficient to launch a comparable fraction of the atmosphere to escape.

Keywords: impact, blow-off, atmosphere.

INTRODUCTION

It has long been recognized that the terrestrial planets are depleted relative to carbonaceous meteorites in volatile molecules (H_2O, CO_2, CO, SO_2, NH_3 and CH_4) as well as noble gases (Ringwood, 1979).

Experiments conducted on serpentine and other water-bearing minerals (Ahrens et al., 1989) suggest that for planetesimals impacting Earth at >1 km/s, partial devolatilization takes place. As a result, during accretion of Earth and other terrestrial planets, primitive atmospheres whose major component is water (plus CO_2, and other gases) form. Because of the infrared opacity of a water-bearing atmosphere, the impact energy of accretion is unable to be radiated efficiently to space and a molten rock (magma) ocean forms after Earth accretes ~0.2 of its final mass, as described by Abe and Matsui (1985).

During the period of planet formation, both planets and planetesimal impactors grow via accretion. Larger impactors colliding with the nearly grown Earth are hypothesized to have been Mars-sized objects. Such late-stage accretionary collisions result in large amplitude body wave and even larger amplitude surface waves. These large amplitude waves resulted in free-surface motion of the magma ocean–covered Earth. As a result, the motion of Earth's free surface drives a wave into the proto-atmosphere. Because of the exponential density distribution of the atmosphere, the particle velocity in the atmosphere is accelerated to higher values with increasing altitude (Zel'dovich and Raizer, 1967). Upon sufficient energetic impact, particle velocities at high altitude that exceed the Earth's escape velocity are achieved. Therefore, a portion of the upper atmosphere can be ejected upon accretion of the primordial Earth and other planets. Numerical studies of compressional body wave driven motion (Chen and

Ni, S., and Ahrens, T.J., 2005, Giant impact-induced blow-off of primordial atmosphere, *in* Kenkmann, T., Hörz, F., and Deutsch, A., eds., Large meteorite impacts III: Geological Society of America Special Paper 384, p. 427–432. For permission to copy, contact editing@geosociety.org.

Ahrens, 1997; Genda and Abe, 2003) show that a lunar-sized impactor (with an energy of $\approx 10^{31}$ J) could launch a portion of the entire (present) Earth atmosphere to greater than escape velocity. Therefore, Earth and its sister planets could have lost a large portion of their initial, primordial atmospheres. Further, Ahrens (1993) proposed that the impact-induced mechanism that erodes an atmosphere most efficiently is related to the planet free-surface-atmosphere interactions. There appears to be at least two regimes for atmospheric blow-off relating to planetary impact.

Regime A. This regime corresponds to the flow in the vicinity of the impact, where the bolide radius a, is $a < h$. Here, h is the atmospheric scale height. Initially, Melosh and Vickery (1989) discussed this regime in connection with the loss of Mars's primordial atmosphere. Later analysis and simulations suggested that atmosphere is indeed eroded via a Regime A impact. However, the assumption of self-similar gas flow used in Melosh and Vickery's model may overestimate the mass of atmosphere blown-off in this regime. Newman et al. (1999) showed that impact-induced gas flows in the atmosphere are strongly affected by gravity.

Regime B. In Regime B, $a > h$. To relate the motion of the atmosphere to the energy of the impactor, we calculate the acoustic wave field from a subsurface explosion in a cavity of radius a (Fig. 1). This approximates the initial energy distribution from a giant impact. In this case, impact-induced flows of the atmosphere induce atmospheric blow-off in the vicinity of the impact, as in Regime A. Also, in Regime B, a wave is transmitted radially through Earth. Upon interaction of this wave with the solid planet-atmosphere interface, a wave is launched into the atmosphere which may induce further global atmospheric blow-off.

Jeffreys and Lapwood (1957), and Burridge (1966) used asymptotic properties of Legendre functions to study reflections of waves (which are approximated by geometrical rays) against the free-surface of a sphere. They derived asymptotic results for seismic phases that involve several reflections, e.g., PP, pP and PPP. For a large number of reflections (e.g., P^n, where n is very large), only a limiting number of reflections M defined by Buldyrev (1968). If $n > M$, ray theory is no longer applicable near the free-surface because of the existence of a boundary layer where wave interference predominates. Therefore, it is desirable to obtain a solution which reduces to the above solution for single-pass waves and reverberations, so that the free-surface motion can be studied in detail for a realistic excitation.

METHOD

The transfer of energy into a magma-covered planet from an impactor is complex. We approximate the impact by the detonation of a shallow buried charge in the acoustic approximation (Fig. 1). Thus, effectively we only address impact on the magma ocean–covered Earth. The relation of near-surface explosive coupling and stress wave attenuation in the simulation of impact was previously studied in porous media (Oberbeck, 1971). Hughes et al. (1977) verified that the shock from a buried explosion source is a good approximation to that of a large impact for the case of

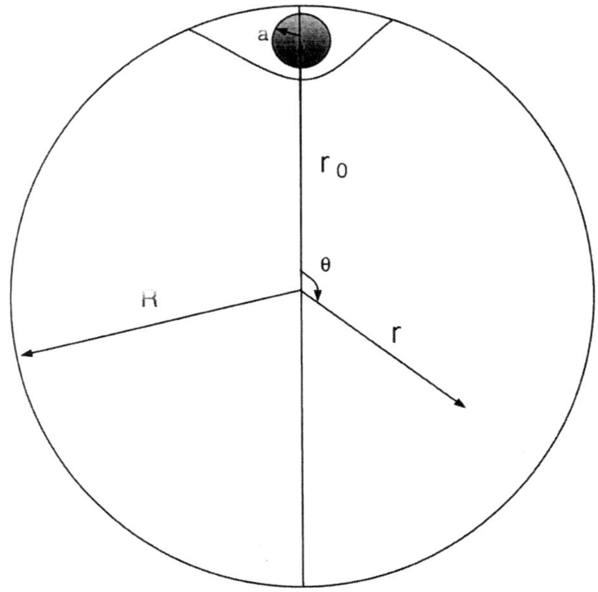

Figure 1. Simplified geometry employed in this paper. A fluid sphere (radius a at high pressure) is produced upon detonating an explosive, gas-filled spherical cavity, where $P = P_0$, is the source.

P-wave generation from an impact on the Moon. At present, we consider only a molten planet; the additional complexity of adding a shear modulus, for a solid uniform Earth. The effect of radial Earth structure is eventually of interest. We also note that the gravity-restoring force, neglected here, only accounts for ≈25% of the restoring force for the gravest poloidal mode of the free-oscillation of Earth. We argue that initially the fluid case can be applied to the very large amplitude P-wave case, at least qualitatively. Thus, in an acoustic fluid sphere, a large amplitude surface-wave–like motion may be induced by surface interference. Numerical experiments simulating impact-induced shock wave sources demonstrate that, for similar impact, solid planets tend to have surface motion amplitudes 0.1–0.3 times those on fluid planets (Hughes et al., 1977). Moreover, the conclusion that quasi-surface wave amplitude is larger than the direct-wave amplitude at $\Delta > 90°$ is still expected to be valid for the case of a fluid planet. Our idealized model planet consists of a homogeneous inviscid fluid sphere, without gravity. The initial fluid isentropic bulk modulus

$$K_s = \rho \left(\frac{dP}{d_\rho} \right)_s$$

has a value equal to P_0. The pressure at $t = 0$ is P_0 in the spherical source zone. The pressure is zero outside the source zone. The radius of the source sphere is a, and the center of the source sphere is located at $(r = r_0, \theta = 0)$ (spherical coordinate) (Fig. 1). From fluid dynamics, the linearized equations for conservation of mass and momentum are:

$$\frac{\partial \rho}{\partial t} + \rho_0 \nabla \cdot \vec{u} = 0 \quad (1)$$

$$\rho_0 \frac{\partial \vec{u}}{\partial t} + \nabla P = 0 \quad (2)$$

and the isentropic equation of state is $c^2 = (\partial P/\partial \rho)_s$, where $(\partial P/\partial \rho)_s$ is the isentropic partial derivative of pressure (P) with respect to density (ρ). ρ_0 is the unperturbed density, $\vec{u}$ is the particle velocity vector, and c is the isentropic acoustic sound speed. Equations (1) and (2) yield:

$$\partial^2 P / \partial^2 t = c^2 \nabla^2 P \quad (3)$$

Initial and boundary conditions are:

$\vec{u}(r,\theta,t)|_{t=0} = 0$; all particles are at rest;

$P(r,\theta,t)|_{r=R} = 0$; free surface;

$P(r,\theta,t)|_{t=0} = P_0$; if (r,θ) in the source zone defined by spherical region of radius a, (c.f. Fig. 1);

= 0; otherwise.

The solution of wave Equation (3) is:

$$P(r,\theta,t) = 2\pi \sum_{l,n}^{\infty} A_{\rm ln} j_{\rm l}\left(\frac{k_{\rm ln} r}{R}\right) P_{\rm l}(\cos\theta) \cos\left(\frac{k_{\rm ln} ct}{R}\right) \quad (4)$$

The radial particle velocity is:

$$u_r(r,\theta,t) = 2\pi \sum_{l,n}^{\infty} A_{\rm ln} j_{\rm l}'\left(\frac{k_{\rm ln} r}{R}\right) P_l(\cos\theta) \sin\left(\frac{k_{\rm ln} ct}{R}\right) \quad (5)$$

where $j_l(x)$ is the l th spherical Bessel function (Watson, 1922) which is defined as

$j_l(x) = (\pi/2x) J_{l+1/2}(x)$ and

$J_v(x)$ is the Bessel function.

$j'_l(x)$ is the derivative of $j_l(x)$.

$P_l(x)$ is the ordinary Legendre polynomial.

k_{ln} is n th zero of $j_l(x)$, and is usually called the wave number.

$$A_{\rm ln} = (2l+1) \frac{\int_0^R \int_0^\pi j_{\rm l}(k_{\rm ln} r/R) P_l(\cos\theta) P(r,\theta,t=0) \sin(\theta)(r/R)^2 \, dr d\theta}{k_{\rm ln} j_{\rm l+1}^2(k_{\rm ln})}$$

where A_{ln} is the excitation coefficient.

For the special case of a spherical source with uniform pressure, A_{ln} has the form (Ni and Ahrens, 2005):

$$A_{\rm ln} = P_0 \left(\frac{a}{R}\right)^2 j_{\rm l}\left(\frac{k_{ln} a}{R}\right) j_{\rm l}\left(\frac{k_{\rm ln} r_0}{R}\right) / \left(k_{\rm ln} j_{\rm l+1}^2(k_{\rm ln})\right) \quad (6)$$

Utilizing the recurrence relation for spherical Bessel functions, we obtained a solution that is simplified for radial velocity at the free surface:

$$u_r(r=R,\theta,t) = 2\pi \left(\frac{a}{R}\right)^2 \frac{P_0}{\rho c} \sum_{l,n}^{\infty} (2l+1) j_{\rm l}\left(\frac{k_{\rm ln} a}{R}\right) \frac{j_l\left(\frac{k_{\rm ln} r_0}{R}\right)}{k_{\rm ln} j_{\rm l+1}(k_{\rm ln})} P_l(\cos\theta) \sin\left(\frac{k_{\rm ln} tc}{R}\right) \quad (7)$$

Equations (6) and (7) are the main analytical results of this paper and u_r is plotted in Figure 2. Here, P_0 is the pressure in the source region (in Pa), c is the sound velocity (in km/s) and ρ is the density of the planet (in kg/m^3), and R is the radius of the planet (in meters). The given solution of u_r is in km/s. A more detailed mathematical treatment of this problem is given by Ni and Ahrens (2005) who have examined the excitation of body and surface waves for different values of r_0/R and a/R.

The method of expressing $u_r(r,\theta,t)$ and $P(r,\theta,t)$ as sums of some basis functions is called the normal mode summation method (Kanamori, 1993). Sato (1961) and Sato and Usami (1967) define the methodology for studying normal modes of homogeneous and radially heterogeneous spheres. These authors do not give explicit values of A_{ln}. Bateman (1930) obtained the solution of waves excited by sources distributed on the free-surface of a sphere. Numerical evaluation of u_r is possible if the appropriate algorithm for calculating w_{ln} and $j_l(x)$ is used. In Equation 5, u_r is given as an infinite summation. Its convergence can be demonstrated because the asymptotic behavior of $j_l(x)$ is known. Numerical tests show that, for given l, when $w_{ln} \geq M/\varepsilon$, ($M \geq 5$), truncation at n terms results in an error of less than ε. When $l \geq M/\varepsilon$ ($M \geq 5$), the error due to truncation of l is also $< \varepsilon$.

RESULTS AND DISCUSSION

We calculated a nominal case where $r_0 = 0.96R$ and $a = 0.02R$. (Ni and Ahrens [2005] present results for variously placed sources). Scaling to Earth, the source region has a radius of 127.42 km and contains a mass of rock of 4.78×10^{19} kg (assuming an average Earth density of $\rho = 5.51 \times 10^3$ kg/m^3, bulk sound velocity of $c = 10$ km/s; the pressure in the source region is 5.51×10^{11} *Pa* which is derived from $P_0 = K_s = c^2\rho$, where K_s is the mean isentropic bulk modulus). Here $u_r(r = R,\theta,t)$ is plotted in Figure 2. It is shown for θ from 0° to 180°, at 10° intervals. The peak amplitude (km/s) at each angle is given along the right-hand side of the figure. The curve plotted in the figure shows arrivals predicted by geometrical rays. The first arrivals are direct waves. For large angles (>90°, for $r_0 = 0.96R$), there are later arriving waves corresponding to propagation along the free surface. We note that the particle velocity at the antipode (180°) reaches its maximum at time 2001 s. Thus, a quasi–surface wave in a compressible fluid sphere is excited by a surface impact. The free-surface velocity amplitude associated with this wave is larger than

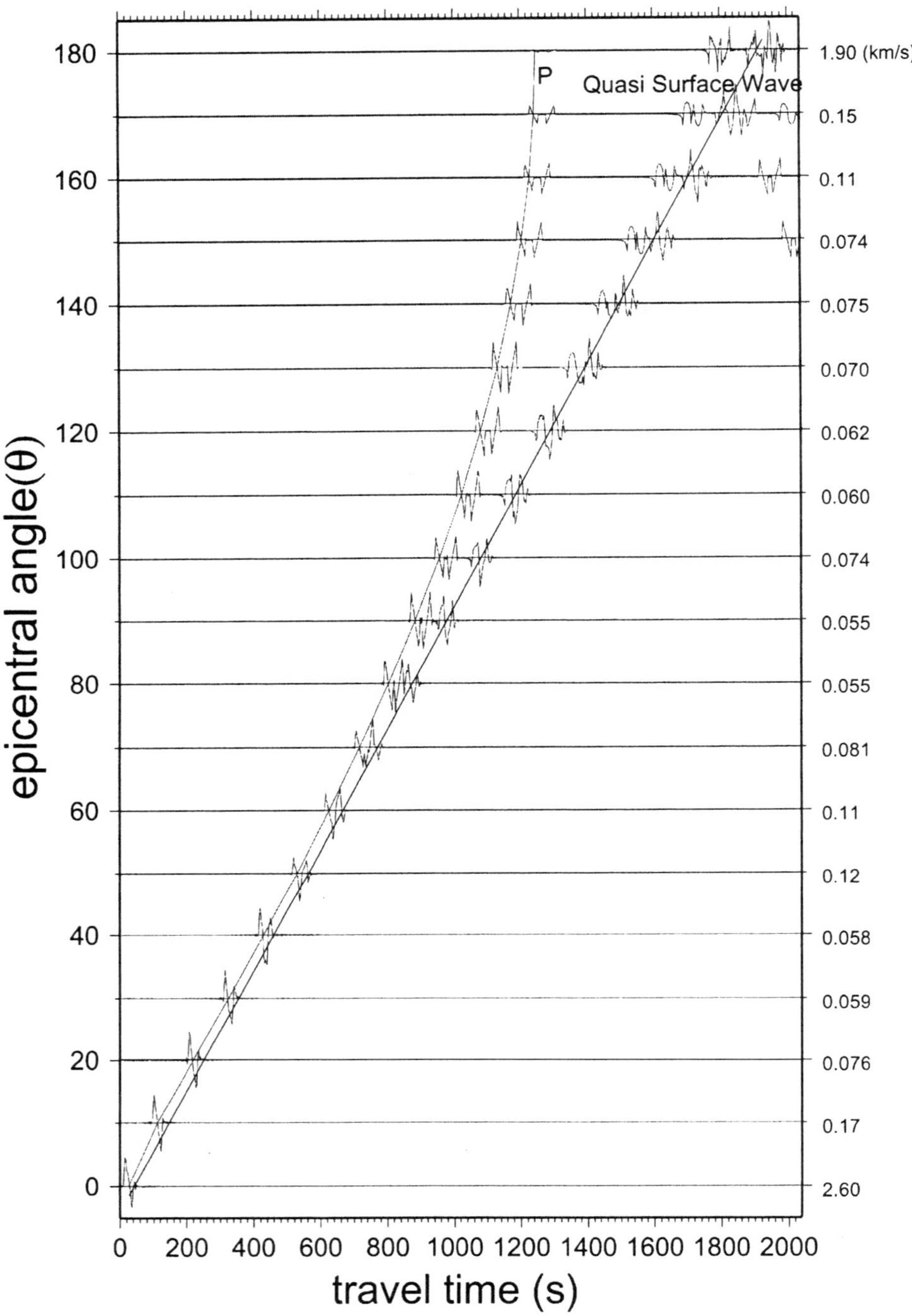

Figure 2. Particle free-surface velocity versus time for different epicentral angles. Source region contains 2.39×10^{27} ergs at $t = 0$. Peak free surface velocity at each station is given along right-hand margin of plot. Solid curve and line mark the geometrical ray arrival of *P* wave and quasi–surface wave, respectively.

the direct waves for angles greater than 90° on an Earth-sized planet. The energy density within the spherical source of Figure 1 may be estimated from the Grüneisen formula:

$$\Delta E = K_s/(\rho\gamma) \tag{8}$$

where γ is the Grüneisen parameter within the spherical source region. Assuming that $K_s/\rho = 100$ (km/s)^2 and $\gamma = 2$, the impact yields an energy density 5.0×10^6 J/kg. The 127.42 km radius source region thus contains 2.39×10^{27} J. For this energy, the antipodal surface point achieves a peak velocity of 1.9 km/s. Notably, we expect that the energy radiated seismically is ~10^{-2} times the energy of the impact. Hence we expect that an equivalent seismic source of 2.39×10^{27} J might be equivalent to an impact with some 10^2 times more energy than expected for planetary impactors in the 10–40 km/s velocity range (for the values of Fig. 2). According to the calculation (Figure 4) of Chen and Ahrens (1997), this antipodal velocity will cause escape of ~10^{-5} of the ambient Earth atmosphere. Recently, Genda and Abe (2003) have examined atmospheric loss as a function of impact energy.

In the present study, quasi–surface waves are excited by multiple reflections, with a propagation velocity of C.

Buldyrev (1968) called these "surface waves of interference nature." Watson (1922) found that the $j_l(x)$'s first zero is at $x \approx \nu + 1.8557\nu^{1/3}$ where $\nu = l + 1/2$, and the first maximum of the function is at $\approx \nu + 0.8086\nu^{1/3}$. Upon asymptotic expansion at $x \approx l$, it appears that $j_l(x)$ has a width of $l^{-2/3}$ when $x \approx l$. From Equation (5) and the recurrence relations for $j_l(x)$, we find that the quasi–surface wave has a skin depth of $k^{-2/3}$. This coincides with Buldyrev's result.

Elasticity Scaling of Present Results

In an expansion of the present paper, Ni and Ahrens (2005) examined the ground motion that is induced by varying the radius of the source center from r_0/R = 0.90–0.96, also the effect of varying the size of the source from a/R from 0.001 to 0.035.

We briefly discuss the scaling of the elastic results and do not consider the scaling of large velocity impacts. The present result given in mathematical form by Equations (6) and (7) and shown for the case of $r_0 = 0.96R$ and $a = 0.02R$ and $(K_s/\rho)^{1/2}$ = 100 km/s) in Figure 2 are scalable by varying five parameters, $r_0, a, K_s/\rho$, ρ, and γ. In this problem, scaling of distance and time stems from Earth's radius R and mean sound velocity c, and mass scaling stems from mean density ρ. We note that r_0 describes the placement of the equivalent explosive source. The effects of varying r_0 and a are not studied in this work. For a huge impact, we assumed $r_0 = 0.96R$. The radius from the center of Earth to the source region, in part, controls the volume of the source region, and might be considered approximately equivalent to a characteristic dimension of the terrestrial crater from a giant high velocity impact. Since $P_0 = K_s$, the mean isentropic bulk modulus and the density control both the wave propagation velocity and, via Equation 8, the energy density of the source. Finally the value of Grüneisen parameter γ specifies the internal energy which is associated with the gas pressure P_0.

Comparison of Shock and Elastic Wave Propagation, Limits to Present Approach

As the present wave propagation model is elastic, it is useful to make some limited comparisons as to the expected effect of an impact-induced shock propagation versus the present elastic assumption.

As discussed above, the source region assumed is a spherical 127 km region with an energy density of $\Delta E = 5 \times 10^6$ J/kg. If we assume that the peak particle velocity achieved at the source u_s corresponds to $\Delta E/2$, we have

$$u_s^2/2 = \Delta E/2 \tag{9}$$

or u_s = 2.2 km/sec. We note the free-surface velocity in Figure 2, immediately above the source, is 2.6 km/sec, corresponding to a particle velocity of 1.3 km/sec. Thus, the elastic solution at 0° is 1.3 km/sec, compared to a shock particle of 2.2 km/sec, a deficiency of ≈ 40%. We note that above peak amplitude attenuation over distances greater than a factor of ≈ 10^1 occur when the radius of the source is small compared to propagation distance, as for example, at 45–90°.

$$\frac{u}{u_s} = \left(\frac{r}{a}\right)^{-n} \tag{10}$$

where r is the chord distance to the source and u and u_s are the free field particle velocity and particle velocity at the source. Here n = 1.1 to 1.2 between the elastic wave of $n = 1$, but less than the $n = 2$ value expected of strong shock propagation (Melosh, 1989).

Although a fluid material may withstand strong compression, a fluid can withstand only very weak tensional loading. Hence, upon application to the present results (Genda and Abe, 2003) to pumping of the atmosphere, we restrict tensional unloading of the free surface velocity, u_{fs} to be:

$$\frac{\partial u_{fs}}{\partial t} \geq -\frac{GM}{R_0^2} \tag{11}$$

where G is the gravitational constant and M planetary mass. Therefore, we use the results summarized in Figure 2 to describe only the peak particle velocity upon compression. Decompression is assumed to occur ballistically as specified by Equation 11.

CONCLUDING REMARKS

Previously, Vickery and Melosh (1990) calculated that energetic impactors will remove a substantial portion (10^{-5}) (see Newman et al., 1999) of the atmosphere mass at the impact point, via a Regime A flow. This energy level (3×10^{27} J) is similar to that considered in the present paper, and corresponds to an asteroidal impactor some 100 km in diameter. (This is ~10^3 more energetic than that resulting in such terrestrial impact structures as Chicxulub, Vredefort, or Sudbury, and Imbrium on the Moon).

Chen and Ahrens (1997) demonstrated that a giant impact-induced solid Earth free surface velocity of ~8 km/s produces an upward particle velocity in the atmosphere in excess of the 11.2 km/sec escape velocity of Earth, and hence, virtually the entire atmosphere escapes. We finally note that for an impact with an energy of ~10^{30} J (a 10^3 km radius projectile), although some atmosphere is ejected, the impact brings more mass to the Earth than is ejected, and hence, the impact remains an accretional event (O'Keefe and Ahrens, 1977).

ACKNOWLEDGMENTS

We warmly thank Hiroo Kanamori and Donald V. Helmberger for technical suggestions that made this project work. This paper has benefited from a helpful review of Jay Melosh and an anonymous reviewer. Research supported by NASA/Goddard Grant NAG5-10198. This is contribution #8584, Division of Geological and Planetary Science, California Institute of Technology, Pasadena, California 91125.

REFERENCES CITED

Abe, Y., and Matsui, T., 1985, The formation of an impact-generated H_2O atmosphere and its implications for the early thermal history of the Earth: Journal of Geophysical Research, v. 90, p. C545–C559.

Ahrens, T.J., O'Keefe, J.D., and Lange, M.A., 1989, Formation of atmospheres during accretion of the terrestrial planets, *in* Atreya, S.K., et al., eds., Origin and Evolution of Planetary and Satellite Atmospheres: University of Arizona Press, p. 328–385.

Ahrens, T.J., 1993, Impact erosion of terrestrial planetary atmospheres: Annual Reviews of Earth and Planetary Sciences, v. 21, p. 525–555, doi: 10.1146/ANNUREV.EA.21.050193.002521.

Bateman, H., 1930, A solution of the wave-equation: Annals of Mathematics, v. 31, p. 158–162.

Buldyrev, V.S., 1968, Wave propagation near the curved surface of an inhomogeneous body, Spectral theory and problems in diffraction, *in* Birman, M.S., ed., Topics in Mathematical Physics: New York, Consultants Bureau, p. 47–65.

Burridge, R., 1963, Reflection of a pulse in a solid sphere: Proceedings of the Royal Society, Series A, p. 367–379.

Burridge, R., 1966, The Legendre function of the second kind with complex argument in the theory of wave propagation: Journal of Mathematical Physics, v. 45, p. 322–330.

Chen, G.Q., and Ahrens, T.J., 1997, Erosion of terrestrial planet atmosphere by surface motion after a large impact: Physics of the Earth and Planetary Interiors, v. 100, no. 1-4, p. 21–26.

Genda, H., and Abe, Y., 2003, Survival of a protoatmosphere through the stage of giant impacts: The mechanical aspects: Icarus, v. 164, p. 149–162, doi: 10.1016/S0019-1035(03)00101-5.

Hughes, H.G., App, F.N., and McGetchen, T.R., 1977, Global effects of basin-forming impacts: Physics of the Earth and Planetary Interiors, v. 15, p. 251–263, doi: 10.1016/0031-9201(77)90034-6.

Jeffreys, H., and Lapwood, E.R., 1957, The reflection of pulse within a sphere: Proceedings of the Royal Society. v. A241, p. 455–479.

Kanamori, H., 1993, Excitation of Jovian normal modes by an impact source: Geophysical Research Letters, v. 120, p. 2921–2924.

Melosh, H.J., 1989, Impact Cratering, A Geologic Process: New York, Oxford University Press, 245 p.

Melosh, H.J., and Vickery, A.M., 1989, Impact erosion of the primordial Martian atmosphere: Nature, v. 338, p. 487–489, doi: 10.1038/338487A0.

Newman, W.I., Symbalisty, E.M.D., Ahrens, T.J., and Jones, E.M., 1999, Impact erosion of planetary atmospheres: Some surprising results: Icarus, v. 138, p. 224–240, doi: 10.1006/ICAR.1999.6076.

Ni, S., and Ahrens, T.J., 2005, Surface motion of a fluid planet induced by impacts: Geophysical Journal International (in press).

Oberbeck, V.R., 1971, Laboratory simulation of impact cratering with high explosives: Journal of Geophysical Research, v. 76, p. 5732–5749.

O'Keefe, J.D., and Ahrens, T.J., 1977, Meteorite impact ejecta: Dependence on mass and energy lost on planetary escape velocity: Science, v. 198, p. 1249–1251.

Ringwood, A.E., 1979, Origin of the Earth and Moon: Berlin, New York, Springer-Verlag, 295 p.

Sato, Y., 1961, Normal model interpretation of the sound propagation in whispering galleries: Nature, v. 189, p. 475–476.

Sato, Y., and Usami, T., 1967, Theoretical seismograms of spheroidal type on surface of a homogeneous gravitating spherical Earth: Geophysical Journal of the Royal Astronomical Society, v. 13, p. 370–384.

Vickery, A.M., and Melosh, H.J., 1990, Atmospheric erosion and impactor retention in large impacts, with applications to mass extinctions, *in* Sharpton, V.L., and Ward, P.D., eds., Global Catastrophes in Earth History: Geological Society of America Special Paper 247, p. 289–300.

Watson, G.N., 1922, A Treatise on the Theory of Bessel Functions: Cambridge, UK, Cambridge University Press, p. 195–261.

Zel'dovich, Y.B., and Raizer, Y.P., 1967, Physics of Shock Waves and High-Temperature Hydrodynamic Phenomena: New York, Academic Press, 452 p.

Manuscript Accepted by the Society 9 August 2004

Printed in the USA

Geological Society of America
Special Paper 384
2005

A review of Martian impact crater ejecta structures and their implications for target properties

Nadine G. Barlow
Department of Physics and Astronomy, Northern Arizona University, Flagstaff, Arizona 86011-6010, USA

ABSTRACT

Most fresh Martian impact craters are surrounded by layered ("fluidized") ejecta which were emplaced as flow deposits. Observational data, laboratory experiments, and numerical modeling strongly suggest that particle size, particle density, atmospheric density and pressure, and the presence of subsurface volatiles all contribute to the features observed in these craters and their ejecta blankets. In this contribution, we review the evidence that both subsurface volatiles and the thin Martian atmosphere contribute to the morphologic, morphometric, and thermophysical characteristics of Martian impact craters and their ejecta deposits.

Keywords: Mars craters, subsurface volatiles, ejecta blanket.

BACKGROUND

Impact cratering produces different morphologic and morphometric features on Mars than on the Moon. These differences result primarily from the presence of a thin atmosphere and the probable existence of subsurface volatile reservoirs (ice and possibly liquid). Variations in particle size and density of surface layers and the concentrations of subsurface volatiles likely combine to produce the variety of ejecta morphologies observed on Mars. Approximately 25% of all Martian impact craters retain an observable ejecta blanket. The distribution of these relatively fresh craters is global, occurring on terrain units which have formed throughout Martian history. Analysis of the morphologic, morphometric, and thermophysical properties of Martian impact crater ejecta blankets provides important insights to the characteristics of the Martian surface and near-surface throughout time.

Visible and infrared imagery from the Viking Orbiter cameras, *Mars Global Surveyor* (MGS) Mars Orbiter Camera (MOC), and *Mars Odyssey* Thermal Emission Imaging System (THEMIS) have revealed a variety of ejecta structures surrounding fresh Martian impact craters. The most common ejecta morphologies are the layered ("fluidized") structures, which surround 9486 (89%) of the 10,651 cataloged craters ≥5 km diameter that display ejecta across the entire surface of Mars. The layered morphologies are subdivided into single layer ejecta (SLE), double layer ejecta (DLE), and multiple layer ejecta (MLE) classes (Fig. 1), based on the number of ejecta layers surrounding the crater (1, 2, or 3+, respectively; Barlow et al, 2000). Analysis of data in the *Catalog of Large Martian Impact Craters* (Barlow, 2003a) indicates that 8119 of the 9486 layered ejecta morphology craters are SLE (86%), 854 (9%) are DLE, and 513 (5%) are MLE.

The layered ejecta morphologies constitute the continuous ejecta blanket, which typically lies within 1.5–2 crater radii from the crater rim. *MGS* Mars Orbiter Laser Altimeter (MOLA) data indicate that most if not all layered ejecta blankets terminate in a distal rampart, which is topographically elevated above the surrounding terrain and the adjacent ejecta blanket (Garvin et al., 2000, 2003). Few secondary craters are observed within the layered ejecta deposits (Barlow, 2003b). But beyond the rampart, MOC and THEMIS imagery often reveals strings of secondary craters extending many crater radii further. McEwen et al. (2003) have recently identified a 10-km-diameter crater on young volca-

Barlow, N.G., 2005, A review of Martian impact crater ejecta structures and their implications for target properties, *in* Kenkmann, T., Hörz, F., and Deutsch, A., eds., Large meteorite impacts III: Geological Society of America Special Paper 384, p. 433–442. For permission to copy, contact editing@geosociety.org.

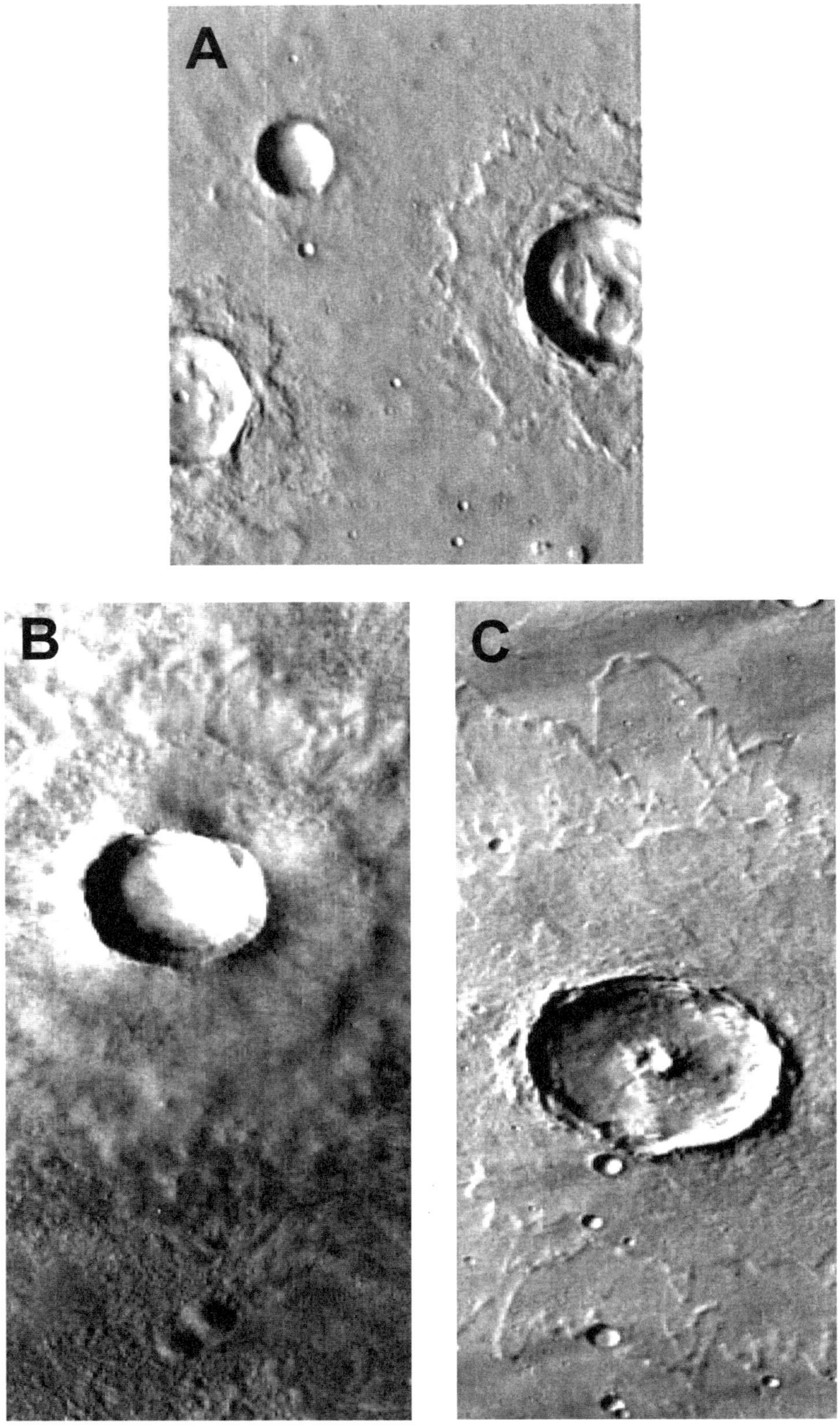

Figure 1. Example images of the three layered ejecta morphologies. (A) Single layer ejecta crater. Crater on right is 12 km in diameter and located near 24°N 101°E (THEMIS image I02493005). (B) Double layer ejecta crater. Crater is 17 km in diameter and located near 49°N 230.5°E (THEMIS image I03350005). (C) Multiple layer ejecta crater. Crater is 25 km in diameter and located 6°N 304°E (THEMIS image I03218002).

nic plains in the Cerberus region of Mars with secondaries that extend more than 800 km from the crater rim. Such extensive secondary crater fields have implications for crater formation models as well as providing important constraints on the cohesiveness of the target material (Head et al., 2002).

MORPHOLOGIC CHARACTERISTICS OF MARTIAN LAYERED EJECTA

Layered ejecta morphologies exhibit strong dependencies on crater diameter and geographic location (Fig. 2). The SLE morphology is typically found around craters between ~5 and 20 km in diameter within the equatorial region of Mars (±30° latitude), but cover the diameter range <1–65 km at higher latitudes (Mouginis-Mark, 1979; Kuzmin et al., 1988; Costard, 1989; Barlow and Bradley, 1990). Craters displaying the DLE morphology are typically 5–25 km in diameter and are located predominantly in the northern plains (Costard, 1989; Barlow and Perez, 2003), although recent analysis using MOC and THEMIS imagery find some concentrations in high southern latitudes and at diameters as small as 2.5 km (Barlow, 2004a). The MLE morphology is seen primarily around craters in the 20–45 km diameter range within the equatorial region of Mars, particularly along the highlands-lowlands dichotomy boundary (Barlow and Bradley, 1990; Barlow and Perez, 2003). Morphology does not appear to correlate with elevation or terrain age, and shows only a very weak trend with terrain type (Mouginis-Mark, 1979; Costard, 1989; Barlow and Bradley, 1990).

The layered ejecta morphologies are sufficiently different from the radial ejecta patterns surrounding lunar and mercurian craters that mechanisms other than simple ballistic emplacement have been proposed for their formation. The morphologic features associated with the layered ejecta structures suggest that the ejected material is emplaced by some type of fluidization mechanism. One formation model proposes that impact into and vaporization of subsurface volatiles produces a vapor cloud which

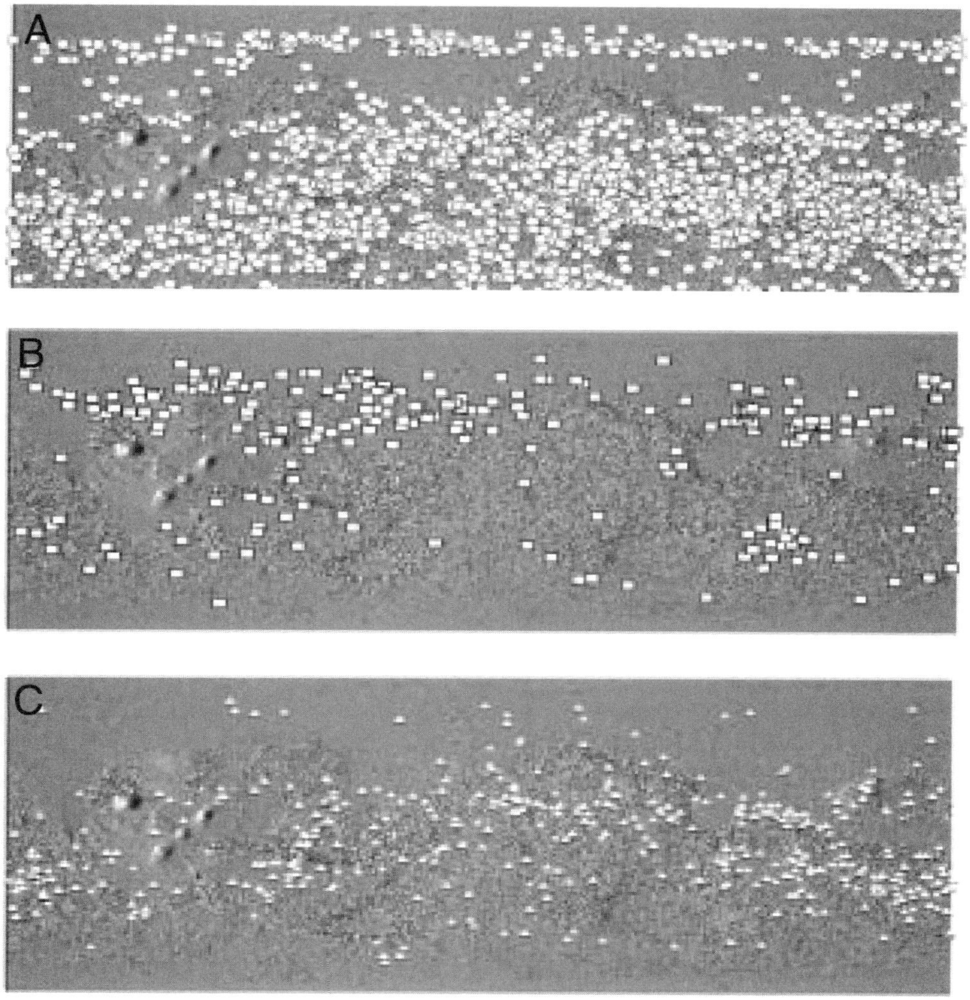

Figure 2. Geographic distribution of layered ejecta morphology craters. (A) Distribution of single layer ejecta craters. (B) Distribution of double layer ejecta (DLE) craters. (C) Distribution of multiple layer ejecta (MLE) craters. Region shown in each image is the ±65° latitude zone. 0° longitude is the center of each map. Base map is a Mars Orbiter Laser Altimeter shaded relief map. Data from *The Catalog of Large Martian Impact Craters*.

emplaces the entrained ejecta as a flow deposit around the crater (Carr et al., 1977; Wohletz and Sheridan, 1983; Mouginis-Mark, 1987). The other major formation model argues that the thin Martian atmosphere serves as the medium in which the ejecta are entrained (Schultz and Gault, 1979; Schultz, 1992a, 1992b; Barnouin-Jha and Schultz, 1998; Barnouin-Jha et al., 1999a, 1999b).

The next two sections discuss these two formation models and their implications for target characteristics.

Atmospheric Model

A number of laboratory experiments and numerical simulations have studied how ejecta are affected when an atmosphere is present (Schultz and Gault, 1979; Schultz, 1992a, 1992b; Barnouin-Jha and Schultz, 1998; Barnouin-Jha et al., 1999a, 1999b). These studies show that a vortex ring is produced by turbulence in the atmosphere. This vortex ring entrains, transports, and deposits fine-grained material from the ejecta curtain into a layered pattern. Larger material is ballistically emplaced prior to the passage of winds associated with the vortex ring, but can be remobilized as the vortex winds pass. These larger clasts are piled up into the peripheral rampart which characterizes the terminus of the ejecta deposit.

The atmospheric model of layered ejecta formation predicts the following:

- Ejecta emplacement style changes as a function of atmospheric density. Continuous ejecta deposits terminating in a distal rampart and superimposed on ballistically emplaced ("radial") deposits occur for atmospheric pressures between 0.06 and 0.3 bar. Ejecta flow lobes occur when atmospheric pressure is between 0.3 and 0.7 bar, and radial patterns are seen for atmospheric pressures >0.8 bar (Schultz, 1992a). Although these pressures are greater than the current atmospheric pressure on Mars (~0.006 bar), compression of the atmosphere during the impact process might allow the necessary pressures to be achieved (Schultz, 1992a).
- Variations in particle size, particle density, and target layering can further affect the resulting ejecta morphology (Schultz and Gault, 1979; Schultz, 1992a).
- The maximum ejecta extent scaled to crater radius (R) should increase as $R^{0.5}$ (Schultz, 1992a).
- DLE and MLE structures can be explained by flow separation. Flow separation occurs behind the leading ejecta front, causing ejecta entrained in the flow to run over and beyond the terminus of the inner ejecta deposits. Thus, outer flow layers should be thinner and composed of finer-grained material than the inner layers. These thinner outer layers would be more easily destroyed than the inner layers (Schultz, 1992a).
- Instabilities developing in the vortex ring will result in waves, which produce the sinuous lobes observed at the edges of layered ejecta deposits. The number of ejecta sinuosity lobes should increase as $R^{0.75}$ (Barnouin-Jha and Schultz, 1998). Incorporation of subsurface volatiles is not expected to be a major contributor to the number of ejecta lobes produced because such volatiles are dispersed early in the vapor cloud, before the vortex ring forms.

Subsurface Volatile Model

The distribution of subsurface volatiles, particularly H_2O, has been studied using hydrologic and geothermal considerations (Fanale, 1976; Clifford, 1993). Clifford considered the possible ranges of geothermal heat flow, thermal conductivities, and melting temperatures to derive his estimates of cryosphere (ice-rich subsurface region) thickness as a function of latitude (see Table 4 in Clifford, 1993). His nominal estimates range between ~2.3 km depth at the equator to 6.5 km at the poles. Below these depths, Clifford's analysis suggests that groundwater reservoirs could be present. Kuzmin, et al. (1988), Barlow and Bradley (1990), and Barlow et al. (2001), using standard depth-diameter relationships, have estimated how deeply craters with different ejecta morphologies are excavating at different latitudes. The transient diameter (D_t) of the crater is estimated from the current crater rim diameter (D_r) using the relationship developed by Croft (1985):

$$D_t = D_{sc}^{0.15 \pm 0.04} D_r^{0.85 \pm 0.04} \quad (1)$$

D_{sc} is the simple to complex transition diameter, which is ~7 km for Martian impact craters (Garvin et al., 2000). For small (simple) craters, the transient diameter is approximately equal to the current rim diameter. Excavation depths (d_e) are then computed assuming $d_e \sim D_t/10$ (Melosh, 1989). The results indicate that SLE and DLE craters are excavating within the region predicted to be ice-rich. Barlow and Bradley (1990) found that MLE craters are excavating below the cryosphere in the equatorial regions and proposed that liquid water may play a role in the production of the MLE structure.

These studies, together with analysis of flow-type features seen on layered ejecta blankets, has led to the suggestion that H_2O ice and possibly liquid are responsible for the formation of these layered morphologies (Wohletz and Sheridan, 1983; Mouginis-Mark, 1987; Costard, 1989; Barlow and Bradley, 1990). This mechanism proposes that the vaporization of ice and/or water produces a vapor plume which then interacts with the ejecta curtain. Essentially the mechanism is similar to that proposed for the atmospheric model but uses vapor produced from subsurface volatiles rather than the Martian atmosphere. Recent shock wave experiments in H_2O ice show that the pressures necessary to produce melting and vaporization are lower than previously estimated and well within the range expected for impact events on Mars (Stewart and Ahrens, 2003).

Other volatiles such as carbon dioxide (Hoffman, 2000) and methane clathrates (Max and Clifford, 2001) have been proposed to exist in the Martian substrate. Methane clathrates would not be expected to exist in large quantities unless the Martian substrate contains an extensive biosphere, for which no evidence currently exists. Stewart and Nimmo (2002) used computer models to test the idea that the small gullies seen on steep slopes (Malin and Edgett, 2000) might be produced by release of subsurface carbon dioxide (Musselwhite et al., 2001). Their results suggest that H_2O is likely to remain stable in the substrate longer than CO_2. In addition, *Mars Odyssey's* Gamma Ray Spectrometer (GRS) has detected a larger presence of H_2O than CO_2 in the upper meter of the Martian substrate except over the winter polar caps (Boynton et al., 2002; Feldman et al., 2002, 2003). All these lines of evidence suggest that H_2O is the dominant volatile within the Martian substrate.

Additional support for the role of subsurface volatiles in the formation of the layered ejecta morphologies includes the following:

- Layered ejecta morphologies are often found in association with other geologic features proposed to form in ice-rich and water-rich environments. These features include polygonal terrain, "terrain softened" topography, mottled terrain, possible thermokarst features, and proximity to valley network and outflow channels (Lucchitta, 1981; Rossbacher and Judson, 1981; Squyres and Carr, 1986; Costard and Kargel, 1995; Carr, 1996; Seibert and Kargel, 2001; Barlow and Perez, 2003). Central pits, proposed to result from degassing of subsurface volatiles during or shortly after crater formation (Wood et al., 1978), are more abundant within Martian impact craters than they are for lunar and mercurian craters, and the greater heights of central peaks in Martian craters have been attributed to the presence of subsurface volatiles (Cordell et al., 1974; Garvin et al., 2003).
- The diameter of the smallest craters around which layered ejecta morphologies are present decreases at higher latitudes, which is consistent with the proposed shallower depths of subsurface volatiles in the polar regions (Kuzmin et al., 1988; Boyce et al., 1998; Barlow et al., 2001).
- DLE craters are concentrated in the topographically low regions of Arcadia, Acidalia, and Utopia, the same areas where high near-surface H_2O ice content (25%–50%) is reported from the *Mars Odyssey* GRS instrument suite. MLE craters display a general correlation with the distribution of 8%–10% H_2O regions (possibly indicating hydrated minerals) determined from GRS analysis (Feldman et al., 2003; Barlow and Perez, 2003). This might indicate some connection between the surface materials sampled by GRS (<1 m depth) and the deeper reservoirs revealed by impact craters (typically >100 m depth).
- DLE morphology craters are concentrated in topographic lows in the northern plains and in the outwash regions of large outflow channels (Costard, 1994; Barlow and Perez, 2003). These areas are believed to be collection areas for ice-rich sediments deposited by fluvial or lacustrine activity (Costard, 1994; Head et al., 1998). A commonly cited formation mechanism for the DLE morphology involves impact into target material which contains stratigraphic layers with different concentrations of volatiles (Soderblom et al., 1973; Costard, 1994; Costard and Kargel, 1995).
- Craters displaying the MLE morphology are often associated with valley network and outflow channels, indicative of liquid water which has been released from depth (Dohm et al., 2001; Barlow and Perez, 2003). A new analysis using MOC and THEMIS imagery is finding that MLE craters often contain central pits, another putative indicator of volatile-rich targets (Barlow and Dohm, 2004).
- Laboratory experiments and computer models of impacts into subsurface volatile reservoirs accurately reproduce many of the morphologic and morphometric characteristics of the layered ejecta morphologies (Gault and Greeley, 1978; Greeley et al., 1980; Wohletz and Sheridan, 1983; Stewart et al., 2001).

Discussion of the Two Formation Hypotheses

The atmospheric model of layered ejecta formation has a considerable amount of laboratory and numerical testing to support it (Barnouin-Jha et al., 1999a, 1999b). Similar experimental and theoretical support for the subsurface volatile model has only recently started to be conducted (Stewart et al., 2001; Stewart and Ahrens, 2003). Atmospheric interaction with the ejecta curtain definitely explains the layered ejecta morphologies associated with impact craters on Venus (Schultz, 1992b). However, layered ejecta morphologies are seen around impact craters on the icy moon Ganymede, whose atmospheric density is too low to play the sole role in layered ejecta formation (Horner and Greeley, 1982; Neal and Barlow, 2004). Analysis of terrestrial impacts show that water (both vapor and liquid) affects the resulting ejecta blankets in ways ranging from the amount of melt production (Kieffer and Simonds, 1980) and emplacement of the continuous ejecta deposits (Hörz et al., 1983) to hydrothermal alteration of ejecta deposits (Newsom et al., 1986). Possible evidence of layered ejecta morphologies has been reported for Meteor Crater, Arizona (Grant and Schultz, 1993) and Puchezh-Katunk, Russia (Masaitis et al., 1993). Most likely, both the atmosphere and subsurface volatiles contribute to the variety of ejecta patterns observed on Mars. In fact, recent hydrocode models suggest that the SLE morphology results from impact into subsurface ice reservoirs, but that the MLE morphology may result from the combination of impact into subsurface ice and ejecta interaction with the Martian atmosphere (Stewart et al., 2001). Comparisons between the model predictions and actual observations will be described further in the Discussion section.

MORPHOMETRIC CHARACTERISTICS OF MARTIAN IMPACT CRATERS

Martian impact craters show many of the same characteristics as impact craters on other solar system bodies (Melosh, 1989). Craters ≤7 km in diameter display a bowl-shaped appearance which is typical of simple craters (Garvin and Frawley, 1998;

Garvin et al., 2000). Complex craters on Mars, like their counterparts on other bodies, have shallower depths relative to their diameters than simple craters and display features such as central peaks, central pits, summit pits, and/or wall terraces. Craters larger than ~140 km in diameter often display a peak-ring structure, while the multi-ring basin morphology occurs for craters larger than ~600 km in diameter (Pike, 1988; Strom et al., 1992). MOLA analysis of fresh Martian impacts reveals that the transition between simple and complex crater morphologies occurs near a diameter of 7 km for craters in the equatorial region, decreasing to 6 km in the ice-rich terrain of higher latitudes (Garvin and Frawley, 1998; Garvin et al., 2000). These simple-to-complex transition diameters are lower than expected, considering the gravity on Mars, and have commonly been attributed to the presence of subsurface ice on the planet (Garvin et al., 2000). Latitude and terrain variations in the crater depth-to-diameter ratio and other morphometric values also likely reflect changes in the subsurface properties, such as particle size, target strength, and/or volatile concentrations (Garvin et al., 2003).

As noted previously, the ejecta blankets surrounding fresh Martian impact craters typically display a flow-emplaced layered structure, unlike the radial pattern seen on the Moon and Mercury. Morphologic features visible in most Martian layered ejecta blankets include flow features, surface striations, hummocky material, and occasionally small channels, indicating that the material was highly fluidized during emplacement (Mouginis-Mark, 1987). Analysis of MOLA data indicates that layered ejecta morphologies almost always terminate in a distal ridge, or rampart (Garvin et al., 2003). The distance from the crater rim to the most distant rampart of the continuous ejecta blanket (the "runout distance") depends on latitude, terrain, and type of morphology. It is generally believed to provide information about the viscosity of the ejecta flow, although few detailed studies have been conducted to date (Woronow, 1981; Wohletz and Sheridan, 1983; Stewart et al., 2001). This runout distance can be normalized to the crater radius using a parameter called the ejecta mobility (EM) ratio (Mouginis-Mark, 1979):

$$\text{EM} = \frac{\text{(average extent of continuous ejecta layer)}}{\text{(crater radius)}} \quad (2)$$

A number of studies using image resolutions of ~100 m/pixel have shown that EM varies with latitude, ranging from an average value of 1.43 near the equator to over 2.0 at between 40°N and 50°N (Mouginis-Mark, 1979; Woronow, 1981; Costard, 1989; Barlow and Pollak, 2002). EM shows little variation among different terrain units. The three types of layered ejecta morphologies show differences in EM values (Table 1).

Another characteristic of the layered ejecta morphologies is the sinuosity of their ramparts, quantified by a parameter called lobateness. Lobateness (Γ) is defined in terms of the ejecta area (A) and perimeter (P) (Kargel, 1989; Barlow, 1994):

$$\Gamma = P/(4\pi A)^{1/2} \quad (3)$$

TABLE 1. EJECTA MOBILITY AND LOBATENESS VALUES

	Ave. EM	Max. EM	Min. EM	Ave. Γ	Max. Γ	Min. Γ
SLE	1.53	6.60	0.20	1.10	1.00	3.57
DLE						
Inner	1.49	3.30	0.40	1.04	1.38	1.00
Outer	3.24	10.60	1.20	1.14	2.27	1.01
MLE	2.17	4.70	0.30	1.18	1.74	1.02

Note: EM—ejecta mobility; Γ—lobateness; SLE—single layer ejecta; DLE—double layer ejecta; MLE—multiple layer ejecta

Perimeter (P) is measured as the linear distance along the outer edge of the ejecta deposit. An ejecta blanket with $\Gamma = 1$ indicates that the ejecta is circular; $\Gamma > 1$ implies increasingly larger sinuosities. At image resolutions of ~100 m/pixel, no statistically significant variation in Γ with latitude is observed once ejecta morphology is eliminated as a variable. As shown in Table 1, the outer layers of the DLE and MLE morphologies tend to show higher sinuosity than the SLE layer or the inner layer of the DLE morphology. Only a slight variation in lobateness is observed for craters on different terrain units.

THERMOPHYSICAL PROPERTIES

THEMIS infrared data allow study of the surface temperature variations throughout the course of a Martian day. Areas which appear infrared bright are warmer than areas which are infrared dark. Such temperature measurements can be translated into information about the thermal inertia of the materials comprising the upper few centimeters of the surface. Thermal inertia (I) depends on the thermal conductivity (K), bulk density (ρ), and specific heat (c) of the material:

$$I = (K\rho c)^{1/2} \quad (4)$$

Thermal conductivities and specific heats do not vary much for most geologic materials (at least those which do not contain water and/or ice). Thus, variations in I are primarily due to differences in ρ, which translates to differences in particle size and/or shape. Smaller clasts and grains have large surface areas compared to their volumes and can therefore lose heat more rapidly than larger objects, resulting in lower thermal inertia values. Low thermal inertia surfaces typically imply fine-grained unconsolidated materials within the upper few centimeters, while high thermal inertia regions contain larger concentrations of boulders or exposed bedrock (Jakosky et al., 2000).

THEMIS provides thermal imaging of the Martian surface during both day and night. Regions covered by fine-grained materials such as dust tend to be infrared bright (warm) during the day and infrared dark (cooler) at night. Thus THEMIS imagery provides a mechanism for rapidly determining whether a sur-

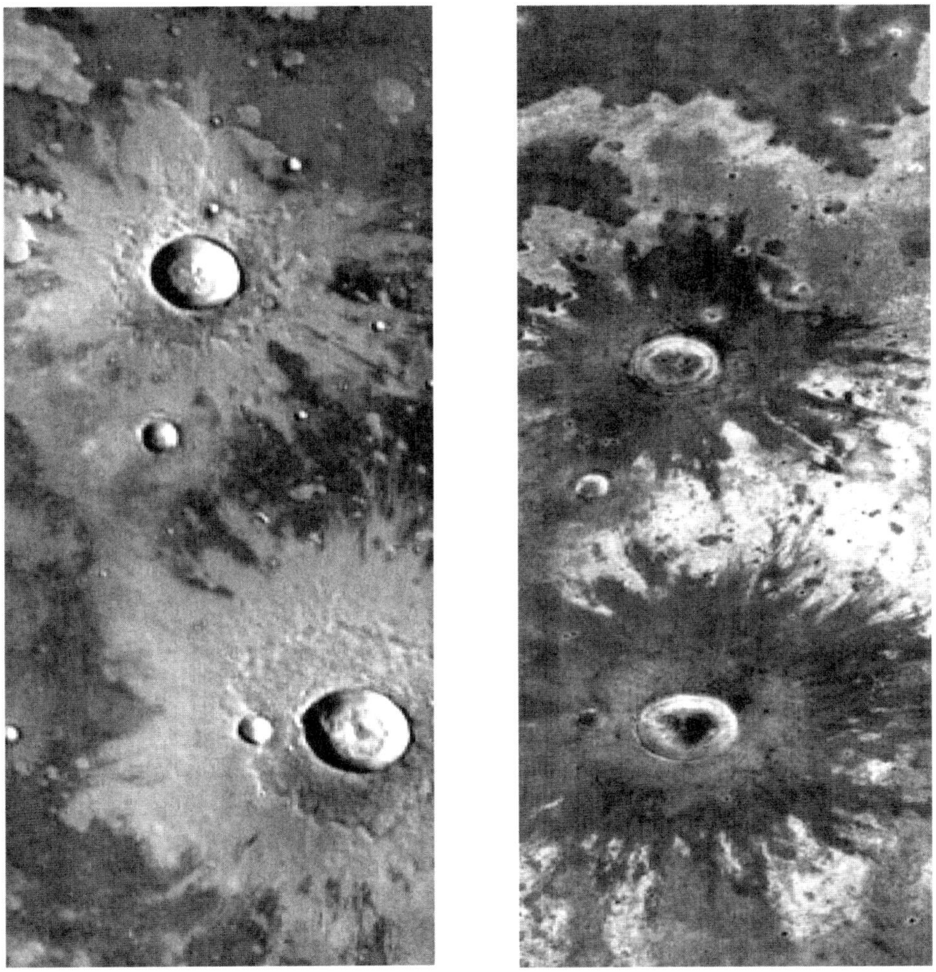

Figure 3. THEMIS day (left) and night (right) images of craters near 22.5°N 324°E. The daytime-bright and nighttime-dark ejecta blankets indicate that the thermal signature of the ejecta blankets are dominated by smaller-sized particles than the surrounding terrain. Crater interior deposits appear dark at night, also indicating small sized clasts or grains. Surrounding areas appear dark in daytime infrared and bright in nighttime infrared, suggesting a rocky surface. THEMIS images I02685009 (left) and I01980010 (right).

face is primarily dust-rich, dust-free, boulder-rich, or some combination of those conditions. A newly formed impact crater will tend to show high inertia values in its interior, along its rim, and throughout its ejecta blanket due to the abundance of rocks and steep slopes (Fig. 3). As the crater ages, crater modifications by erosion and deposition result in lower inertia values (Christensen et al., 2003). Such modifications include the eolian deposition of dust, sand-blasting of surfaces, degradation by subsequent impacts, possible erosion by fluvial and/or glacial processes, and burial by ejecta deposits from other impact cratering events.

THEMIS infrared data not only allow us to identify young craters on the Martian surface but also provide clear views of high thermal inertia rays extending beyond the crater rim. Combined with the high resolution MOC and THEMIS visible camera data, scientists are now able to trace many secondary craters back to their originating crater. The results of high-resolution image analysis show that craters surrounded by any of the layered ejecta morphologies show very low abundances of secondary craters within the region of the layered ejecta (Barlow, 2003b). However, secondaries are apparent beyond the layered ejecta deposits and extend to distances of 100s to 1000s of kilometers from the crater rim even for relatively small impacts (McEwen et al., 2003). The number of secondary craters produced on Mars may be higher than the number produced by similar energy impacts on the Moon (McEwen et al., 2003), which in turn will affect the reported ages of surface units based on crater density analyses utilizing small crater statistics (Barlow, 2003b; McEwen, 2004).

DISCUSSION

Analysis of the morphologic, morphometric, and thermophysical properties of Martian impact craters provides information which allows us to test the two formation models of ejecta emplacement (subsurface volatiles versus atmospheric interactions). These comparisons include the following:

- The distributions of the SLE, DLE, and MLE morphologies have been compared to the proposed distributions of subsurface ice and liquid (Mouginis-Mark, 1979; Kuzmin et al., 1988; Costard, 1989; Barlow and Bradley, 1990; Barlow and Perez, 2003) and surface particle size distributions

(Schultz, 1992a). The size ranges of SLE craters as a function of latitude are consistent with the proposed distribution of subsurface ice based on hydrologic and geothermal considerations (Clifford, 1993). The regional concentrations of DLE craters are consistent with layered targets containing varying volatile concentrations in these areas (Costard, 1989; Barlow and Perez, 2003) or with fine-grained material which would enhance the atmospheric interaction process (Schultz, 1992a). MLE craters might be excavating into liquid water reservoirs (Barlow and Bradley, 1990) or may represent interactions of the ejecta curtain with the atmosphere (Schultz, 1992a; Stewart et al., 2001).

- The atmospheric model predicts that EM should vary with crater radius in the form $R^{0.5}$, so that EM should increase with larger crater size (Schultz, 1992a). EM is found to be higher for the larger MLE craters than for the generally smaller SLE craters, but over the size range of a specific ejecta morphology, no clear trend is discerned for higher EM with larger size (Barlow and Pollak, 2002). EM increases at higher latitudes, with a larger increase seen in the north than in the south (Costard, 1989; Barlow and Pollak, 2002). Craters on heavily cratered highlands and ridged plains regions tend to have higher EM values than craters on lightly cratered plains (Barlow and Pollak, 2002). These results are consistent with geographic variations in the distribution of subsurface volatiles (Costard, 1989; Barlow and Pollak, 2002) or with possible variations in particle size and atmospheric density due to elevation (Schultz, 1992a).
- The atmospheric model defines lobateness as the number of sinuous lobes seen along the edge of the ejecta (Barnouin-Jha and Schultz, 1998) rather than the definition used above as the degree from which the entire ejecta deviates from a circular planform (Kargel, 1989; Barlow, 1994). Instabilities in the vortex ring give rise to a larger number of flow lobes as transient crater size increase. Barnouin-Jha and Schultz (1998) found that the dispersion in the number of sinuous lobes for Martian craters was quite large and likely resulted from additional factors such as local particle size, atmospheric pressure changes, and entrained volatiles. Using the other definition of lobateness (Γ) finds that SLE and the inner layer of DLE have similar Γ values, while Γ for the outer DLE layer and for MLE are much higher (Barlow, 1994). Numerical and experimental studies of impacts into ice and their implications for lobateness are still in their infancy. However, hydrocode simulations suggest that the EM and Γ values for equatorial craters with the SLE morphology can be replicated by impact into target materials containing ~20% ice (Stewart et al., 2001). The same calculations showed that the MLE characteristics can be produced using similar ice concentrations but with the inclusion of an interaction between the ejecta curtain and the Martian atmosphere. No simulations have yet been conducted for the DLE morphology or for craters at higher latitudes.
- The atmospheric model predicts that much of the ejecta blanket of even very fresh craters should contain substantial amounts of fine-grained material (Schultz, 1992a). In this case, thermal inertia values of ejecta blankets could not be used to determine the freshness of the crater since low infrared night thermal inertia values could not be interpreted as simply due to later dust cover. However, Barlow (2004b) finds that there are strong correlations between thermal inertia values indicating blocky ejecta blankets with other morphologic and morphometric indicators of crater freshness.

All of these characteristics suggest that the layered ejecta morphologies may not be produced purely by impact into subsurface volatiles or by interaction of the ejecta curtain with the Martian atmosphere, but perhaps by a combination of these two processes. This is not surprising since the mechanism of layered ejecta emplacement for both models involves interaction of the ejecta curtain with either a vapor plume (subsurface volatiles model) or the atmosphere (atmospheric model). As orbiting spacecraft continue to provide more detailed information about the Martian surface, we can begin to compare the predicted features of each model with actual observations. Some features, such as the presence of central pits, are not easily explained by the atmospheric model, while other features, such as the long runout distances of MLE deposits, are most easily explained by interaction with the atmosphere. Comparisons with layered ejecta morphologies on Venus (Schultz, 1992b) and Ganymede (Neal and Barlow, 2004) may provide the end-member cases which can help constrain the relative roles of subsurface volatiles and atmospheres in the production of the crater features seen on Mars. In addition to new data from upcoming Mars missions, additional insights into the role of atmospheres and subsurface volatiles can be obtained by detailed geophysical and geochemical analysis of terrestrial impact craters, particularly those which formed in ice-rich and/or water-rich environments (Grieve, 1988; Tsikalas et al., 1998; Masaitis, 1999). This multi-pronged approach to crater analysis will provide the knowledge necessary to unravel the role of surface and near-surface properties in the production of features associated with impact craters.

ACKNOWLEDGMENTS

The author thanks Peter Schultz, James Whitehead, and Fred Hörz for their very constructive reviews, which have greatly improved this paper. The Martian crater analysis has been funded through NASA Mars Data Analysis Program Grants NAG5-8265 and NAG5-12510.

REFERENCES CITED

Barlow, N.G., 1994, Sinuosity of Martian rampart ejecta deposits: Journal of Geophysical Research, v. 99, p. 10927–10935, doi: 10.1029/94JE00636.

Barlow, N.G., 2003a, Revision of the "Catalog of Large Martian Impact Craters": 6th International Conference on Mars: Houston, Texas, Lunar and Planetary Institute, Abstract no. 3073.

Barlow, N.G., 2003b. Comparison of secondary crater production by Martian and lunar impact craters: Bulletin of the American Astronomical Society, v. 35, p. 924–925.

Barlow, N.G., 2004a, Martian impact crater ejecta and interior morphologies: New insights from MGS and Odyssey data analysis: Abstracts for the 7th Mars Crater Consortium Meeting: Flagstaff, Arizona, Abstract no. 0703.

Barlow, N.G., 2004b, Martian subsurface volatile concentrations as a function of time: Clues from layered ejecta craters: Geophysical Research Letters, v. 31, L05703, doi: 10.1029/2003GL019075.

Barlow, N.G., and Bradley, T.L., 1990, Martian impact craters: Correlations of ejecta and interior morphologies with diameter, latitude, and terrain: Icarus, v. 87, p. 156–179, doi: 10.1016/0019-1035(90)90026-6.

Barlow, N.G., and Dohm, J.M., 2004, Impact craters in Arabia Terra, Mars: Lunar and Planetary Science XXXV: Houston, Texas, Lunar and Planetary Institute, Abstract no. 1122.

Barlow, N.G., and Perez, C.B., 2003, Martian impact crater ejecta morphologies as indicators of the distribution of subsurface volatiles: Journal of Geophysical Research, v. 108, 5085, doi: 10.1029/2002JE002036.

Barlow, N.G., and Pollak, A., 2002, Comparisons of ejecta mobility ratios in the northern and southern hemispheres of Mars: Lunar and Planetary Science XXXIII: Houston, Texas, Lunar and Planetary Institute, Abstract no. 1322.

Barlow, N.G., Boyce, J.M., Costard, F.M., Craddock, R.A., Garvin, J.B., Sakimoto, S.E.H., Kuzmin, K.O., Roddy, D.J., and Soderblom, L.A., 2000, Standardizing the nomenclature of Martian impact crater ejecta morphologies: Journal of Geophysical Research, v. 105, p. 26733–26738, doi: 10.1029/2000JE001258.

Barlow, N.G., Koroshetz, J., and Dohm, J.M., 2001, Variations in the onset diameter for Martian layered ejecta morphologies and their implications for subsurface volatile reservoirs: Geophysical Research Letters, v. 28, p. 3095–3098, doi: 10.1029/2000GL012804.

Barnouin-Jha, O.S., and Schultz, P.H., 1998, Lobateness of impact ejecta deposits from atmospheric interactions: Journal of Geophysical Research, v. 103, p. 25739–25756, doi: 10.1029/98JE02025.

Barnouin-Jha, O.S., Schultz, P.H., and Lever, J.H., 1999a, Investigating the interactions between an atmosphere and an ejecta curtain. 1. Wind tunnel tests: Journal of Geophysical Research, v. 104, p. 27105–27115, doi: 10.1029/1999JE001026.

Barnouin-Jha, O.S., Schultz, P.H., and Lever, J.H., 1999b, Investigating the interactions between an atmosphere and an ejecta curtain. 2. Numerical experiments: Journal of Geophysical Research, v. 104, p. 27117–27131, doi: 10.1029/1999JE001027.

Boyce, J.M., Roddy, D.J., and Soderblom, L.A., 1998, Distribution of onset diameters of rampart ejecta craters on Mars: Lunar and Planetary Science XXVIII: Houston, Texas, Lunar and Planetary Institute, Abstract no. 1404.

Boynton, W.V., Feldman, W.C., Squyres, S.W., Prettyman, T., Brückner, J., Evans, L.G., Reedy, R.C., Starr, R., Arnold, J.R., Drake, D.M., Englert, P.A.J., Metzger, A.E., Mitrofanov, I., Trombka, J.I., d'Uston, C., Wänke, H., Gasnault, O., Hamara, D.K., Janes, D.M., Marcialis, R.L., Maurice, S., Mikheeva, I., Taylor, G.J., Tokar, R., and Shinohara, C., 2002, Distribution of hydrogen in the near-surface of Mars: Evidence for subsurface ice deposits: Science, v. 297, p. 81–85, doi: 10.1126/SCIENCE.1073722.

Carr, M.H., 1996, Water on Mars: New York, Oxford University Press, 229 p.

Carr, M.H., Crumpler, L.S., Cutts, J.A., Greeley, R., Guest, J.E., and Masursky, H., 1977, Martian impact craters and emplacement of ejecta by surface flow: Journal of Geophysical Research, v. 82, p. 4055–4065.

Christensen, P.R., Bandfield, J.L., Bell, J.F., Gorelick, N., Hamilton, V.E., Ivanov, An., Jakosky, B.M., Kieffer, H.H., Lane, M.D., Malin, M.C., McConnochie, T, McEwen, A.S., McSween, H.Y., Mehall, G.L., Moersch, J.E., Nealson, K.H., Rice, J.W., Richardson, M.I., Ruff, S.W., Smith, M.D., Titus, T.N., and Wyatt, M.B., 2003, Morphology and composition of the surface of Mars: *Mars Odyssey* THEMIS Results: Science, v. 300, p. 2056–2061.

Clifford, S.M., 1993, A model for the hydrologic and climatic behavior of water on Mars: Journal of Geophysical Research, v. 98, p. 10973–11016.

Cordell, B.M., Lingenfelter, R.E., and Shubert, G., 1974, Martian cratering and central peak statistics: Mariner 9 results: Icarus, v. 21, p. 448–456, doi: 10.1016/0019-1035(74)90147-X.

Costard, F.M., 1989, The spatial distribution of volatiles in the Martian hydrolithosphere: Earth, Moon, and Planets, v. 45, p. 265–290.

Costard, F., 1994, Unusual concentrations of rampart craters at the mouths of outflow channels, Mars: Lunar and Planetary Science XXV, p. 287–288.

Costard, F.M., and Kargel, J.S., 1995, Outwash plains and thermokarst on Mars: Icarus, v. 114, p. 93–112, doi: 10.1006/ICAR.1995.1046.

Croft, S.K., 1985, The scaling of complex craters: Proceedings of the 15th Lunar and Planetary Science Conference, Part 2: Journal of Geophysical Research, v. 90, p. C828–C842.

Dohm, J.M., Ferris, J.C., Baker, V.R., Anderson, R.C., Hare, T.M., Strom, R.G., Barlow, N.G., Tanaka, K.L., Klemaszewski, J.E., and Scott, D.H., 2001, Ancient drainage basin of the Tharsis region, Mars: Potential source for outflow channel systems and putative oceans or paleolakes: Journal of Geophysical Research, v. 106, p. 32943–32958, doi: 10.1029/2000JE001468.

Fanale, F.P., 1976, Martian volatiles: Their degassing history and geochemical fate: Icarus, v. 28, p. 179–202, doi: 10.1016/0019-1035(76)90032-4.

Feldman, W.C., Boynton, W.V., Tokar, R.L., Prettyman, T.H., Gasnault, O., Squyres, S.W., Elphic, R.C., Lawrence, D.J., Lawson, S.L., Maurice, S., McKinney, G.W., Moore, K.R., and Reedy, R.C., 2002, Global distribution of neutrons from Mars: Results from *Mars Odyssey*: Science, v. 297, p. 75–78, doi: 10.1126/SCIENCE.1073541.

Feldman, W.C., Prettyman, T.H., Boynton, W.V., Squyres, S.W., Bish, D.L., Elphic, R.C., Funsten, H.O., Lawrence, D.J., Maurice, S., Moore, K.R., Tokar, R.L., and Vaniman, D.T., 2003, The global distribution of near-surface hydrogen on Mars: 6th International Conference on Mars: Houston, Texas, Lunar and Planetary Institute, Abstract no. 3218.

Garvin, J.B., and Frawley, J.J., 1998, Geometric properties of Martian impact craters: Preliminary results from the Mars Orbiter Laser Altimeter: Geophysical Research Letters, v. 25, p. 4405–4408, doi: 10.1029/1998GL900177.

Garvin, J.B., Sakimoto, S.E.H., Frawley, J.J., and Schnetzler, C., 2000, North polar region craterforms on Mars: Geometric characteristics from the Mars Orbiter Laser Altimeter: Icarus, v. 144, p. 329–352, doi: 10.1006/ICAR.1999.6298.

Garvin, J.B., Sakimoto, S.E.H., and Frawley, J.J., 2003, Craters on Mars: Global geometric properties from gridded MOLA topography: 6th International Conference on Mars: Houston, Texas, Lunar and Planetary Institute, Abstract no. 3277.

Gault, D.E., and Greeley, R., 1978, Exploratory experiments of impact craters formed in viscous-liquid targets: Analogs for Martian rampart craters?: Icarus, v. 34, p. 486–495, doi: 10.1016/0019-1035(78)90040-4.

Grant, J.A., and Schultz, P.H., 1993, Erosion of ejecta at Meteor Crater, Arizona: Journal of Geophysical Research, v. 98, p. 15033–15047.

Greeley, R., and Fink, J., Gault, D.E., Snyder, D.B., Guest, J.E., and Schultz, P.H., 1980, Impact cratering in viscous targets: Laboratory experiments: Proceedings of the 11th Lunar and Planetary Science Conference: New York, Pergamon Press, p. 2075–2097.

Grieve, R.A.F., 1988, The Haughton impact structure: Summary and synthesis of the results of the HISS project: Meteoritics, v. 23, p. 249–254.

Head, J.W., Kreslavsky, M., Hiesinger, H., Ivanov, M., Pratt, S., Siebert, N., Smith, D.E., and Zuber, M.T., 1998, Oceans in the past history of Mars: Tests for their presence using Mars Orbiter Laser Altimeter (MOLA) data: Geophysical Research Letters, v. 25, p. 4401–4404, doi: 10.1029/1998GL900116.

Head, J.N., Melosh, H.J., and Ivanov, B.A., 2002, Martian meteorite launch: High-speed ejecta from small craters: Science, v. 298, p. 1752–1756, doi: 10.1126/SCIENCE.1077483.

Hoffman, N., 2000, White Mars: A new model for Mars' surface and atmosphere based on CO_2: Icarus, v. 146, p. 326–342, doi: 10.1006/ICAR.2000.6398.

Horner, V.M., and Greeley, R., 1982, Pedestal craters on Ganymede: Icarus, v. 51, p. 549–562, doi: 10.1016/0019-1035(82)90145-2.

Hörz, F., Ostertag, R., and Rainey, D.A., 1983, Bunte Breccia of the Ries: Continuous deposits of large impact craters: Reviews of Geophysics and Space Physics, v. 21, p. 1667–1725.

Jakosky, B.M., Mellon, M.T., Kieffer, H.H., Christensen, P.R., Stacy Varnes, E., and Lee, S.W., 2000, The thermal inertia of Mars from the Mars Global Surveyor Thermal Emission Spectrometer: Journal of Geophysical Research, v. 105, p. 9643–9652, doi: 10.1029/1999JE001088.

Kargel, J.S., 1989, First and second-order equatorial symmetry of Martian rampart crater ejecta morphologies, *in* Fourth International Conference on Mars: Tucson, University of Arizona, p. 132–133.

Kieffer, S.W., and Simonds, C.H., 1980, The role of volatiles and lithology in the impact cratering process: Reviews of Geophysics and Space Physics, v. 18, p. 143–181.

Kuzmin, R.O., Bobina, N.N., Zabalueva, E.V., and Shashkina, V.P., 1988, Structural inhomogeneities of the Martian cryosphere: Solar System Research, v. 22, p. 121–133.

Lucchitta, B.K., 1981, Mars and Earth: Comparison of cold-climate features: Icarus, v. 45, p. 264–303, doi: 10.1016/0019-1035(81)90035-X.

Malin, M.C., and Edgett, K.S., 2000, Evidence for recent groundwater seepage and surface runoff on Mars: Science, v. 288, p. 2330–2335, doi: 10.1126/SCIENCE.288.5475.2330.

Masaitis, V.L., 1999, Impact structures of northeastern Eurasia: The territories of Russia and adjacent countries: Meteoritics & Planetary Science, v. 34, p. 691–711.

Masaitis, V.L., Naumov, M.V., and Ivanov, B.A., 1993, Fluidized ejecta of Puchezh-Katunki impact crater: Possible implications to rampart craters of Mars: Vernadsky Institute-Brown University Microsymposium, v. 18, p. 45–56.

Max, M.D., and Clifford, S.M., 2001, Initiation of Martian outflow channels: Related to the dissociation of gas hydrate?: Geophysical Research Letters, v. 28, p. 1787–1790, doi: 10.1029/2000GL011606.

McEwen, A.S., 2004, New Age Mars: Lunar and Planetary Science XXXV: Houston, Texas, Lunar and Planetary Institute, Abstract no. 1756.

McEwen, A., Turtle, E., Burr, D., Milazzo, M., Lanagan, P., Christensen, P., and Boyce, J., and the THEMIS science team, 2003, Discovery of a large rayed crater on Mars: Implications for recent volcanic and fluvial activity and the origin of Martian meteorites: Lunar and Planetary Science XXXIV: Houston, Texas, Lunar and Planetary Institute, Abstract no. 2040.

Melosh, H.J., 1989, Impact Cratering: A Geologic Process: New York, Oxford University Press, 245 p.

Mouginis-Mark, P., 1979, Martian fluidized ejecta morphology: Variations with crater size, latitude, altitude, and target material: Journal of Geophysical Research, v. 84, p. 8011–8022.

Mouginis-Mark, P.J., 1987, Water or ice in the Martian regolith? Clues from rampart craters seen at very high resolution: Icarus, v. 71, p. 268–286, doi: 10.1016/0019-1035(87)90152-7.

Musselwhite, D.S., Swindle, T.D., and Lunine, J.I., 2001, Liquid CO_2 breakout and the formation of recent small gullies on Mars: Geophysical Research Letters, v. 28, p. 1283–1285, doi: 10.1029/2000GL012496.

Neal, J.E., and Barlow, N.G., 2004, Layered ejecta craters on Ganymede: Comparisons with Martian analogs: Lunar and Planetary Science XXXV: Houston, Texas, Lunar and Planetary Institute, Abstract no. 1121.

Newsom, H.E., Graup, G., Sewards, T., and Keil, K., 1986, Fluidization and hydrothermal alteration of the suevite deposit at the Ries Crater, West Germany, and implications for Mars: Proceedings of the 17th Lunar and Planetary Science Conference, Part 1: Journal of Geophysical Research, v. 91, p. E239–E251.

Pike, R.J., 1988, Geomorphology of impact craters on Mercury, *in* Vilas, F., Chapman, C.R., and Matthews, M.S., eds., Mercury: Tucson, University of Arizona Press, p. 165–273.

Rossbacher, L.A., and Judson, S., 1981, Ground ice on Mars: Inventory, distribution, and resulting landforms: Icarus, v. 45, p. 39–59, doi: 10.1016/0019-1035(81)90005-1.

Schultz, P.H., 1992a, Atmospheric effects on ejecta emplacement: Journal of Geophysical Research, v. 97, p. 11623–11662.

Schultz, P.H., 1992b, Atmospheric effects on ejecta emplacement and crater formation on Venus from Magellan: Journal of Geophysical Research, v. 97, p. 16183–16248.

Schultz, P.H., and Gault, D.E., 1979, Atmospheric effects on Martian ejecta emplacement: Journal of Geophysical Research, v. 84, p. 7669–7687.

Seibert, N.M., and Kargel, J.S., 2001, Small-scale Martian polygonal terrain: Implications for liquid surface water: Geophysical Research Letters, v. 28, p. 899–902, doi: 10.1029/2000GL012093.

Soderblom, L.A., Kriedler, T.J., and Masursky, H., 1973, Latitudinal distribution of a debris mantle on the Martian surface: Journal of Geophysical Research, v. 78, p. 4117–4122.

Squyres, S.W., and Carr, M.H., 1986, Geomorphic evidence for the distribution of ground ice on Mars: Science, v. 231, p. 249–252.

Stewart, S.T. and Ahrens, T.J., 2003, Shock Hugoniot of H_2O ice: Geophysical Research Letters, v. 30, no. 6, 1332, doi: 10.1029/2002GL016789.

Stewart, S.T., and Nimmo, F., 2002, Surface runoff features on Mars: Testing the carbon dioxide formation hypothesis: Journal of Geophysical Research, v. 107, 5069, doi: 10.1029/2000JE001465.

Stewart, S.T., O'Keefe, J.D., and Ahrens, T.J., 2001, The relationship between rampart crater morphologies and the amount of subsurface ice: Lunar and Planetary Science XXXII: Houston, Texas, Lunar and Planetary Institute, Abstract no. 2092.

Strom, R.G., Croft, S.K., and Barlow, N.G., 1992, The Martian impact cratering record, *in* Kieffer, H.H., Jakosky, B.M., Snyder, C.W., and Matthews, M.S., eds., Mars: Tucson, University of Arizona Press, p. 383–423.

Tsikalas, F., Gudlaugsson, S.T., and Faleide, J.I., 1998, The anatomy of a buried complex impact structure: The Mjølnir structure, Barents Sea: Journal of Geophysical Research, v. 103, p. 30469–30483, doi: 10.1029/97JB03389.

Wohletz, K.H., and Sheridan, M.F., 1983, Martian rampart crater ejecta: Experiments and analysis of melt-water interaction: Icarus, v. 56, p. 15–37, doi: 10.1016/0019-1035(83)90125-2.

Wood, C.A., Head, J.W., and Cintala, M.J., 1978, Interior morphology of fresh Martian craters: The effects of target characteristics: Proceedings of the 9th Lunar and Planetary Science Conference, p. 3691–3709.

Woronow, A., 1981, Preflow stresses in Martian rampart ejecta blankets: A means of estimating the water content: Icarus, v. 45, p. 320–330, doi: 10.1016/0019-1035(81)90037-3.

Manuscript Accepted by the Society 9 August 2004

Geological Society of America
Special Paper 384
2005

Starting conditions for hydrothermal systems underneath Martian craters: Hydrocode modeling

E. Pierazzo*
Planetary Science Institute, 1700 E. Ft. Lowell Road, Suite 106, Tucson, Arizona 85719, USA

N.A. Artemieva*
B.A. Ivanov*
Institute for Dynamics of Geospheres, Russian Academy of Sciences, Leninsky pr., 38-6, 117334, Moscow, Russia

ABSTRACT

Mars is the first place to look for any sign of present or past extraterrestrial life. Its surface shows many features indicative of the presence of surface and subsurface water, while impact cratering and volcanism have provided temporary and local surface heat sources throughout Mars' geologic history. In particular, impact-generated hydrothermal systems could have been some of the most favorable sites for the origin of life on Mars.

We present preliminary results of hydrocode simulations of impacts on Mars aimed at constraining the initial conditions for modeling the onset and evolution of a hydrothermal system on the red planet. Simulations of the early stages of impact cratering can identify the role of target composition and inhomogeneities in characterizing shock melting due to asteroidal and cometary impacts on the Martian surface. We find that cometary impacts produce significantly larger melt volumes than asteroidal impacts, thus affecting the thickness of the resulting melt sheet and its cooling history. The description of mixed targets is the main challenge of this study: unfortunately at this time, the lack of specific experimental data does not allow us to identify what is the best modeling approach (mixed cells versus mixed material equation of state) for mixed material targets. Modeling crater collapse is a necessary step to determine the final thermal state of the target underneath. We find that the combination of shock and plastic heating and the structural uplift of initially deeper strata create a water-bearing zone at depths where water is in the liquid stability field. In the central uplift, the high temperatures cause water to evaporate (steam-driven circulation). The simulations show that for a mid-sized crater (rim diameter around 30 km) the hydrothermal circulation is probably restricted to a "column" contained well within the final crater.

Keywords: impact craters, hydrothermal systems, Mars, numerical simulations, melt production.

*E-mails: Pierazzo—betty@psi.edu; Artemieva—nata_art@mtu-net.ru; Ivanov—ivanov@idg.chph.ras.ru.

Pierazzo, E., Artemieva, N.A., and Ivanov, B.A., 2005, Starting conditions for hydrothermal systems underneath Martian craters: Hydrocode modeling, *in* Kenkmann, T., Hörz, F., and Deutsch, A., eds., Large meteorite impacts III: Geological Society of America Special Paper 384, p. 443–457. For permission to copy, contact editing@geosociety.org.

INTRODUCTION

Mars is the most Earth-like of the Solar System's planets and the first place to look for any sign of present or past extraterrestrial life. With the scientific debate that spurred from the announcement that the Mars meteorite ALH84001 appeared to exhibit signatures of past life (McKay et al., 1996), Mars has become a renewed international focus for space exploration. One of the most important points of Mars research consists in understanding the evolution of water, one of the key ingredients for the origin of life. The surface of Mars shows many features that have been used as indicators of the presence of surface and subsurface water throughout Martian history, the most prominent being the spectacular outflow channels and rampart craters, (e.g., Baker et al., 1992; Squyres et al., 1992; Carr, 1996). This hypothesis has received further support from the most recent missions to Mars (such as the *Mars Odyssey*, *Mars Exploration Rover*, and *Mars Express* missions).

Impact craters are widely used ubiquitous indicators for the presence of subsurface water or ice on Mars. Impact cratering affects the surface of a planetary body not only by creating a visible crater, but also by modifying, locally and temporarily, the thermal state of the region around the crater, as well as by ejecting and distributing material in various physical states (melted, shocked, pulverized, etc.) over a wide region around the excavated crater. These processes and their resulting surface expressions are largely controlled by the physical (and chemical) properties of the target. Therefore, the study of crater and ejecta characteristics ultimately provides clues to these properties. On Mars, the morphology of crater ejecta blankets (rampart craters), and crater degradation features (terrain softening) have been shown to provide a good qualitative indicator for the presence or absence of subsurface liquid water or ice in the impacted target region (e.g., Mouginis-Mark, 1987; Squyres et al., 1992; Barlow and Perez, 2003).

Regardless of the state of water in the Martian subsurface, it has long been proposed (e.g., Newsom, 1980; Berkley and Drake, 1981; Allen et al., 1982; Newsom et al., 1986) that every crater of relatively large dimensions (>30 km in diameter) should have created favorable conditions for local hydrothermal activity (Daubar and Kring, 2001; Newsom et al. 2001; Thorsos et al., 2001, 2002; Rathbun and Squyres, 2002). Hydrothermal systems are also associated with volcanism, which has been the main process in resurfacing the Martian surface over the history of the planet (Hartmann and Neukum, 2001). Its effectiveness, however, tends to be limited to specific regions. Craters, on the other hand, cover the entire Martian surface, reaching areas where the aquifer may be particularly close to the surface (such as at high latitudes). The great depth of impact craters, up to several kilometers relative to the surrounding terrain, can allow the breaching of local aquifers, thus providing a source of water for lakes and hydrothermal systems. The duration of the hydrothermal activity is very much dependent on the magnitude of the impact event. Newsom et al. (2001) suggest that impact melt and uplifted basement heat sources in craters larger than ~50 km in diameter should be sufficient to drive substantial hydrothermal activity and keep crater lakes from freezing for thousands of years, even under cold climatic conditions. For craters of 200–300 km in diameter the cooling period of the crater region could be as long as 1–10 Myr (Ivanov and Deutsch, 1999). Large craters, thus, may represent giant Petri dishes for promoting biogeochemical processes and culturing preexisting life on Mars (Newsom et al., 2001). A thorough understanding and quantification of the processes involved in impact cratering is therefore crucial for answering questions on the physical, geological, and biogeochemical state of Mars. Numerical simulations, and to a somewhat lesser degree (due mainly to scaling problems) laboratory experiments, are the best tools for such an undertaking.

Specific hydrocode simulations of impacts on a mixed ice-rock target on Mars, needed to constrain the initial conditions for modeling the onset and evolution of hydrothermal systems, are not yet available. We present initial results of two-dimensional (2D) and three-dimensional (3D) simulations of impacts on Mars aimed at constraining the initial conditions for the development of an impact-related hydrothermal system on the red planet. The simulations of the early stages of impact cratering allow us to characterize the pressure-temperature distribution in the target and to evaluate the amount of shock melting caused by various impacts on the Martian surface. The late stage of crater collapse is necessary to determine the final thermal state of the target, including crater uplift, distribution of heated target material (including the melt pool) and hot ejecta around the crater. Simulations of various size craters can provide important and detailed information, such as (i) estimates of melt volumes; (ii) estimate of the central crater uplift; (iii) temperature distribution around the crater through the crater formation process; and (iv) distribution of "hot" ejecta (i.e., molten material that may influence the thermal state of the subsurface) around the crater.

MODELING APPROACH

The cratering process is a rapid, highly dynamic, continuous sequence of events, and it is strongly influenced by material characteristics (Melosh, 1989). Material models are thus a crucial component of hydrocode simulations. They are composed of two parts: the equation of state and the constitutive equation (e.g., see overview in Pierazzo and Collins, 2003). The equation of state describes the material's thermodynamic state over a wide range of pressures, temperatures, and specific volumes (or densities) and it is critical in the early stages of the impact. The constitutive equation is fundamental for modeling the late stages of impact cratering, when material strength determines the final shape and characteristics of the crater. Limited experimental data and the difficulty of modeling complex material mixtures have made it difficult to simulate complex targets and projectiles. It is common to model impactors as spherical objects and targets as rectangular slabs of uniform composition. Furthermore, limitations in computer hardware have resulted in a historical separation, in modeling efforts, between the early and late stages of impact cratering. Early stages of impact cratering focus on evaluating shock states in the target, including shock melting and vaporization, and require high spatial resolution near the impact point. The late stages of impact cratering investigate the process of crater collapse, modeling the formation of the final impact structure, and require long integration time and a large spatial mesh. Continuous improvement in material modeling, as well as advances in computer hardware, is approaching the capability of combining the two modeling stages. So far, however, the most common approach to accurate modeling of the impact process is by modeling separately early and late stages. This is also the approach used in this study.

We performed high-resolution 3D hydrocode simulations of the early stages of impact cratering events using the hydrocode SOVA (Shuvalov, 1999) coupled to tabular versions of the ANEOS equations of state (Thompson and Lauson, 1972) for the materials of interest. The objective of this study is to investigate target shock melting and its dependence on target composition. The 2D hydrocode SALEB (Amsden et al., 1980; Ivanov, 1994; Ivanov and Kostuchenko, 1997; Ivanov, 2003b), coupled to ANEOS-based tabular materials' equations of state, is used to model the late stage of impact cratering and the formation of the final crater. The final temperature field around an impact crater depends both on the shock compression-decompression cycle and friction heating due to plastic deformation. Because of the temperature increase with depth due to the geothermal heat flow gradient, material uplifted from below during the formation of the central peak or peak ring in complex craters is at higher temperatures than the surrounding material, thus providing an additional source of heat. The entire crater-forming event, from impact to the final crater must be followed to obtain a complete understanding of the thermal field underneath an impact crater.

Codes Used

SOVA (Shuvalov, 1999) is a two-step Eulerian code developed at the Institute for Dynamics of Geospheres (Russia) that can model multidimensional, multi-material, large deformation, strong shock wave physics. It is based on the same principles utilized in the well-known hydrocode CTH (McGlaun et al., 1990), developed at the Sandia National Laboratories (Albuquerque, New Mexico). In particular, care has been used in SOVA to develop a more appropriate formalism for the simultaneous conservation of energy and momentum. We use the code in 3D geometry; in this case it is common practice to use bilateral symmetry, which will allow us to model only the positive half-space originating from the impact plane (plane perpendicular to the target identified by the impact direction), thus significantly reducing the rather large mesh size (and accompanying computational requirements) needed for the simulation. 3D benchmark tests have shown that SOVA produces shock melting and vaporization patterns and volumes comparable to the well-known CTH (Pierazzo et al., 2001; Artemieva and Ivanov, 2001). Tabular equations of state were constructed using a revised version of the ANEOS code (Thompson and Lauson, 1972; Melosh, 2000). The availability of reliable, wide range equations of state relating thermodynamic parameters of materials, such as pressure, density, and temperature, is vitally important for numerical simulations. By using different physical treatments in different domains of validity, ANEOS provides a thermodynamically consistent equation of state whose validity extends over a wide range of temperatures and pressures. ANEOS also offers a limited treatment of materials' phase changes. As a result, this code can model the thermodynamic evolution of a material well beyond the initial shock stage, accounting for melting and vaporization and providing consistent estimates of materials' energies and entropies. A new ANEOS equation of state for basalt has been developed for this study (see Appendix), and we used the available water ice equation of state (Turtle and Pierazzo, 2001).

SALE (Simplified Arbitrary Lagrangian Eulerian; Amsden et al., 1980) was developed originally as an explicit finite difference code for calculating 2D fluid flows at all speeds; the spatial mesh may be treated as Lagrangian (moving with the fluid) or Eulerian (held fixed). The code uses axial symmetry, and the impact space corresponds to a rotation of the half plane by 360°. Significant modifications of the original code have been implemented, and some major modules were added to enable its use for crater modeling. The upgraded version used in this work, SALEB, is used in Eulerian mode and includes a strength routine that employs different failure mechanisms to treat solid materials, including gradual shear damage accumulation, thermal softening, and acoustic fluidization (Ivanov, 1994; Ivanov and Kostuchenko, 1997). Mechanical properties of target rocks include dry friction of damaged rocks (Byerlee, 1978) with thermal softening, i.e., decrease of strength and friction to zero at the melting point (e.g., Ohnaka, 1995; Wünnemann and Ivanov, 2003). A detailed description of the strength model used in SALEB is provided in Collins et al. (2004). SALEB has already been used to model the formation and thermal evolution of several terrestrial impact structures of various sizes, from Kärdla (Estonia, D ~ 4 km; Versh et al., 2003), to Puchezh-Katunki (Russia, D ~ 40 km; Ivanov, 2002), Chicxulub (Mexico, D ~ 180 km, Ivanov, 2003a), and Sudbury (Canada, D ~ 250 km; Ivanov and Deutsch, 1999). The constitutive model parameters used for basalt, representative of a general brittle rock behavior, are based on the work of Lundborg (1968) and Stesky et al. (1974). The acoustic fluidization (AF) model used in SALEB was initially proposed by Melosh (1979), and is described in detail in Melosh and Ivanov (1999). The AF model parameters are scaled to different crater sizes and planetary gravity with a technique proposed by Ivanov and Artemieva (2002). The extensive testing of the model by Wünnemann and Ivanov (2003) show that with a proper parameter fit, AF reproduces depth-diameter relations for complex impact craters. However, further model improvements are needed to reproduce fine details of crater morphology. For the ~30 km diameter craters modeled in this study, the decay time of AF oscillations is assumed to be 40 s, and the effective viscosity of the fluidized rock media is assumed to be $6 \times 10^4\ m^2s$ (for kinematic viscosity; multiply by rock density to get a dynamic viscosity value).

Early Stage Modeling

Because the objective of this study is to characterize impact-related conditions for the onset of hydrothermal systems, we have begun our study by investigating the importance of target composition, specifically the presence of water ice, in shock propagation and target melting and vaporization. Mixed materials in the target may affect the shock distribution and consequently the final melting of the target in two ways. First, inhomogeneities in the target influence the propagation of the shock wave. If the target is a mixture of materials that have very different impedance, as is the case for rocks and water ice, the shock wave propagation may be affected by the interaction of the original shock wave with shocks reflected at material boundaries. Macroscopically, this results in increased shock levels near material boundaries, and it is normally used in laboratory shock reverberation experiments to increase the maximum shock pressure generated in the sample up to peak shock pressures typical of planetary impact velocities. Microscopically, this effect may be responsible for the generation of melt veins in moderately shocked rocks, as observed for meteorites (e.g., see Stöffler et al., 1991). In shock events, it is only the first shock that heats the material (as illustrated, for example, in model calculations by Ivanov et al., 2002), while secondary reflected shocks increase the final shock pressure, with a minor contribution to the temperature. As a result, single- and multiple-shocked materials experiencing the same peak shock pressure will reach substantially different thermodynamic states, with the multiple-shocked material having lower shock temperature and entropy. In this case, shock melting cannot be estimated on the basis of peak shock pressure alone. The presence of water may also affect a material's phase diagram. For example, the presence of water in basalt strongly decreases solidus and liquidus temperatures and alters crystalline phase assemblages (e.g., Best and Christiansen, 2000). On the other hand, how water affects shock melting of rocks is still not well characterized. It is generally believed that the presence of volatiles lowers the shock pressure for melting, thus implying that wet basalt should melt at lower shock pressures than its dry equivalent. However, experimental studies to evaluate this effect are not available. In this study, we can address the first case, i.e., the influence of target structure (inhomogeneities) on the shock wave propagation and maximum compression. Accurate estimates of shock melting for a "wet target" cannot be determined without the support of new experimental data. However, we can provide estimates of target volumes shocked to specific isobars that are reasonable thresholds for shock melting.

Inhomogeneous targets may be modeled in different ways, reflecting natural distributions of inclusions and their size with respect to the projectile (or final crater) size. We carried out a series of 3D simulations on different target configurations, starting with (A) a "dry basalt target." Macroscale inhomogeneities can be modeled through (B) a "layered basalt-water ice target" (where each layer consists of several cells in the simulation mesh); we used an overall percentage of 20% for water ice in the target (this can be considered as an upper limit, assuming a saturated crust; for example, Mellon et al., 1997, assumed a porosity of 15% for the Martian regolith crust, and results of the *Odyssey* mission (Mitrofanov et al., 2003) indicate that at high latitudes, the uppermost 1 m of the Martian crust may contain up to 35% of ice by weight). This is equivalent to impacts in layered targets, discussed in Stöffler et al. (2002). Microscale inclusions in macroscale cratering (where modeling is restricted by computer capacity) can be modeled in two different ways. (C) A "mixed cell target" where each computational cell of the target can be modeled to consist of two (or more) materials. In this case, each cell is composed of a mixture of two materials, such as 80% basalt and 20% water ice. The hydrocode internal procedure allows for mechanical (i.e., pressure) equilibrium in each cell, but not thermodynamic (temperature) equilibrium. (D) A wet basalt (i.e., a new equation of state representing the "wet basalt" material) can be constructed by combining the equation of state of two (or more) initial materials. We built a "wet basalt target" with each cell made of a single material, containing 20% water ice, in complete mechanical and thermodynamic equilibrium (see Appendix B). At this initial stage of the investigation, it is not significant to have realistic estimates of the amount of water, because the main concern here is to understand the role of target inhomogeneities in the impact event.

The impactor is modeled either as a basalt sphere (rocky asteroid) impacting with a velocity of 8 km/s (typical impact velocity for asteroids on Mars; Steel, 1998) or an ice sphere (icy comet) impacting with a velocity of 15.5 km/s (median impact velocity for short-period comets on Mars; data from Olsson-Steel, 1987), with a diameter of 2.4 km. Our simulations model impact angles of 45° as well as 90° from the surface in the Martian gravity field of 3.72 m/s^2. Using the pi-scaling law (Schmidt and Housen, 1987), and the observation that transient crater diameter is roughly 50%–65% that of the final crater (e.g., Melosh, 1989), we aimed at final crater rims around 30 km (roughly 25–35 km) in diameter (code used for estimates developed by H.J. Melosh: see www.lpl.arizona.edu/tekton/crater.html). A very thin CO_2 atmosphere was also included in the simulations to model the present-day Martian atmosphere, although this does not influence the thermal characterization of the target. A spatial resolution of 20–25 cells per projectile radius is maintained around the impact point, followed by regions of progressively lower resolution. For accurate estimates of melt volumes we distributed up to two million massless tracers around the impact point. Volume estimates are determined by adding up the volume of tracers experiencing shock pressures above a given threshold (as described in Pierazzo et al., 1997). Each simulation was extended long enough

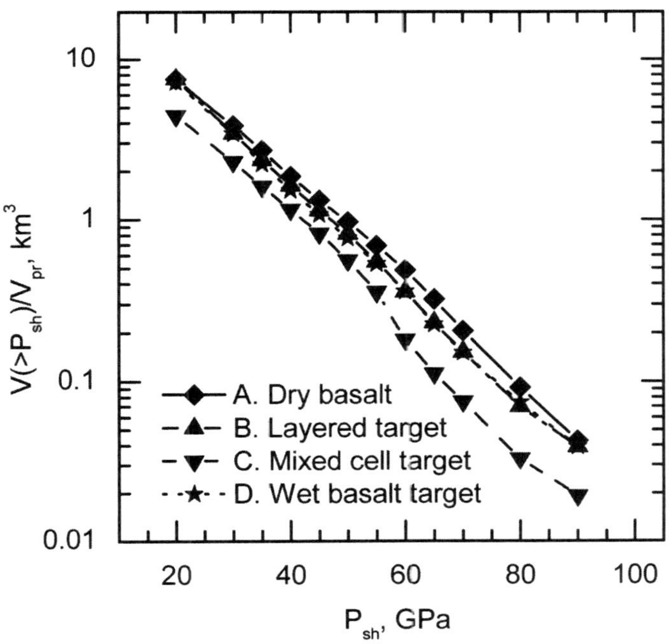

Figure 1. Volume of target subject to shock pressures above a given value normalized to projectile volume for: A, dry basalt target; B, layered target, with alternate layers of basalt and water ice in the proportion 80–20; C, mixed cell target, where each cell is composed of a mixture of two materials: 80% basalt and 20% water ice; D, uniform "wet basalt" target, whose tabular equation of state was constructed by combining the basalt and water ice equations of state in a 80–20 percentage (each cell is made of a single material).

to allow the shock wave propagating in the target to decay well below typical melting thresholds of rocks (20 GPa). Figure 1 summarizes the volume of target subjected to shock pressures above a given threshold normalized to the volume of the projectile for the various simulations.

Typical threshold pressures for shock melting of dry rocks are around 50–60 GPa (Stöffler, 1984), while for hydrous rocks localized shock melting can occur from ~30 GPa (e.g., Lambert and Lange, 1984; the absence of experimental data on the shock behavior of "wet basalts" does not allow us to use more precise values). We thus consider 30–50 GPa as the typical range for shock melting of rocks.

Late Stage Modeling

We model the vertical impact of a projectile 2 km in diameter, simulating an asteroid (basalt) impacting at 8 km/s, or a comet (ice) impacting at 15.5 km/s, on the Martian surface (basalt), both producing a final crater around 30 km in diameter. Tabular ANEOS equations of state are coupled to SALEB.

A large range of heat conductivities has been assumed for the upper ~1 km of the Martian crust (Fanale, 1976; Clifford and Fanale, 1985; Clifford 1993). Although the heat flux is not well constrained, general comparisons among terrestrial planets combined with the composition of the Martian meteorites allow for rough estimates of Mars' thermal gradient. Babeyko and Zharkov (2000) model a thermal gradient ranging from 6 to 20 K/km, with an "average" estimate of 13 K/km below a depth of roughly 10–11 km, where closure of fracture by lithostatic pressure causes fractured rocks to behave almost like crystalline rocks. In this study, we use a surface temperature of 220 K, and a constant temperature gradient of 13 K/km. The target initially is balanced in a gravity field with acceleration of 3.72 m/s^2.

The simulations model the well-known sequence of crater formation: initial penetration of the projectile, transient cavity growth, and crater gravity collapse. The code is permitted to run until the velocity of material around the crater decayed to the level of acoustic noise, i.e., around 1–3 m/s. A model resolution of 10 cells per projectile radius, corresponding to a 100 m cell size, is maintained throughout the simulation. This relatively low resolution (compared to the early stage 3D simulations) may result in underestimating the volume of highly shocked rocks (e.g., impact melt; see discussion in Pierazzo et al., 1997), but these simulations are mainly designed to investigate the large-scale thermal field under a fresh impact crater. Here the important effect is the structural uplift of lower, warmer strata to the surface, and the extent of the impact melt zone is not too critical for our estimates.

Simulations that employ dry rock targets may be reasonable when small amounts of water and/or ice are present in fractures. On the other hand, the Martian subsurface may include a large amount of either liquid or frozen water. Results of the *Odyssey* mission indicate that at high latitudes the uppermost 1 m of the Martian crust may contain up to 35 ± 15% of ice by weight (e.g., Boynton et al., 2002; Mitrofanov et al., 2003); hence the mechanical properties of a rock-ice mixture target should be investigated in more detail. This requires overcoming two main problems: (1) the lack of a good model for strength and friction of ice-rock mixtures, and (2) the poor knowledge of the real porosity of the upper kilometer of the Martian crust.

The mechanical behavior of ice-rock mixtures has mostly been studied for low strain rates (see, for example, Mangold et al., 2002, and references herein), when ice creep plays the main role in the strain-stress history. Triaxial tests of sand/ice mixtures (Fig. A3) have been conducted at relatively high temperatures, between −10 and −4 °C, where ice does not behave as a very brittle substance (e.g., Parameswaran and Jones, 1981). Most triaxial experiments with icy mixtures involve homogeneous soil-ice or sand-ice mixtures, and the behavior may be quite different from the case of ice in rock fractures. Nevertheless, we are planning on using available data to develop a model to describe the mechanical behavior of ice-rock mixtures.

Knowing the real structure of the Martian crust on a scale relevant to the excavation of impact craters tens to hundreds of kilometers in diameter requires seismic and other geophysical studies. In the absence of real data, Clifford (1993) has used gravity scaling of the lunar megaregolith porosity (measured by seismic sounding) to estimate porosity in the upper 10–20 km of the Martian crust. It should be kept in mind that the lunar megaregolith is believed to have formed mainly during the late heavy bombardment period, which ended ca. 3 Ga. In contrast, wide areas of the Martian surface were geologically active well after 3 Ga (e.g., Hartmann and Neukum, 2001). Hence a lunar-like, gradual decrease in porosity with depth (with the possibility of the pores being water-saturated) may not be representative of the crust as a whole, as some areas may be complicated by non-impact depositions of lava flows, aeolian and maybe even lacustrine deposits of unknown porosity.

In this initial study, we carry out simulations of crater collapse in a homogeneous dry basalt target, i.e., we do not attempt, at this stage, to change rock friction to include the effects of water and/or ice. This approximation does not provide a realistic description of the temperature field underneath a crater created in a "wet" target (when water ice amounts are significant). However, the results for a dry target are still reasonable for the case of a target containing small amounts of water. Even in the case of a dry target, however, water may flow into the heated target from nearby localized water-bearing layers triggering the onset of a hydrothermal circulation cell.

RESULTS AND DISCUSSION

Importance of Target Characteristics on Shock Compression

Figure 1 and Table 1 show the estimated volume of target rocks experiencing shock pressures above a given value for the four distinct targets modeled: (A) dry basalt; (B) layered ice-basalt target (layers with vertical extension larger than one cell size); (C) mixed cells target (where material mixing occurs at a sub-cell scale); and (D) wet basalt target (complete homogenization based on mass-weighted, average properties of the two materials, basalt and water). In the 30–50 GPa range typifying shock melting of rocks we find small differences (10%–20%) in volume estimates of impact melt between the simulations with the dry and wet basalt (i.e., mixed equation of state) as well as those with the layered target. In the basalt-ice mixed cells case, however, the volume of material shocked above a given pressure is always significantly (30%–40%) less than in the other three cases. This may be indicative of different thermodynamic effects involved when two materials are mixed together in individual target cells compared to uniform material cells. In general, the mixed cell case (C) results in overall lower values of shock compression, but manifests regions of "heat concentration" (i.e., cells with higher temperatures) for materials with lower impedance. Figure 2 shows the distribution of shock pressure in the plane of impact relative to the pre-impact target. In the layered target case, the shock-regions appear strikingly "ragged." This is due to the reflection of the shock wave at material interfaces, which does not occur in any of the other cases simulated. As a result, a given shock level may have been reached by multiple shock fronts, and the shocked material thus may have a much lower overall shock temperature than singly shocked material. Although it is remarkable that the "ragged" shock regions in the layered target ultimately result in overall volumes comparable to the uniform target cases, these volumes may not be representative of real melt regions. The target region subjected to shocks above 30–50 GPa appears to be less extended in the mixed cell case. The best modeling approach for mixed materials targets is not clear to us at this stage (both approaches, mixed cells and mixed material equation of state, have been used in modeling studies). The mixed-cells approach lacks thermodynamic equilibrium in individual computational cells, but allows us to see a separation between water and basalt in the vapor plume, as shown in Figure 3. The uniform target case (wet basalt, case D) achieves both mechanical and thermodynamic equilibria within each computational cell, but at any instant the vapor plume consists of wet basalt of constant (i.e., 20%) water content, which is unrealistic.

TABLE 1. NORMALIZED TARGET VOLUMES (km^3) ABOVE GIVEN SHOCK PRESSURES

P_{sh} (GPa)	V/V_{pr}, Comet			V/V_{pr}, Asteroid			
	Dry basalt target (A)	Mixed cells target (C)	Wet basalt target (D)	Dry basalt target (A)	Layered target (B)	Mixed cells target (C)	Wet basalt target (D)
20	9.88	7.16	9.58	7.52	7.55	4.46	7.21
30	5.76	5.10	6.09	3.86	3.44	2.29	3.42
40	4.44	3.92	4.61	1.86	1.62	1.15	1.52
50	3.51	3.07	3.56	0.97	0.82	0.97	0.77
60	3.76	2.38	2.75	0.48	0.36	0.18	0.36
70	2.06	1.73	1.98	0.20	0.15	0.07	0.15
80	1.43	1.19	1.34	0.09	0.07	0.03	0.07
90	0.97	0.81	0.89	0.04	0.02	0.02	0.04

Note: Comet: V_{proj} = 7.238 km^3; v_{imp} = 15.5 km/s; ρ_{ice} = 1.1 g/cm^3. Asteroid: V_{proj} = 7.238 km^3; v_{imp} = 8 km/s; ρ_{bas} = 2.86 g/cm^3. Impact angle: 45°. Final crater around 30 km in diameter.

Importance of Projectile Characteristics on Shock Compression

The effects of impactor characteristics on melt production are illustrated in Figure 4, showing the volume of a mixed cell target (case C) subject to pressure above a given value for asteroidal and cometary impacts and different impact angles. The results indicate that the difference in shocked volume between vertical and 45° impacts is minimal, yet impact velocity will have a significant effect on the total volume of possible melt. Because of their typically higher impact velocities, cometary impacts produce melt volumes that may be two to three times larger than asteroid impacts at constant crater size. Larger melt pools have longer cooling times and may contribute to a longer duration of the hydrothermal system.

Numerical studies (Stöffler et al., 2002) indicate that at least 30% of the shock melt (melt produced directly upon decompression from shock pressures) is ejected in impact events that produce craters around 30 km in diameter. On the other hand, the final impact melt sheet (as usually identified in terrestrial impact structures) includes unmelted material either by entrainment of material as the melt sheet flows along the transient cavity walls, or by incorporation of initially launched debris that falls back into the crater. For example, Simonds and Kieffer (1993) estimate that ~50% of the volume of the Manicouagan impact sheet constitutes digested material. It is thus reasonable to assume that the impact melt sheet corresponds to a volume roughly equivalent to the estimated shock melt (Table 1). This assumption allows us to make first-order estimates of the melt sheet thickness for the various cases modeled (Table 1), all related to a final rim crater diameter of roughly 30 km. According to Ugalde et al. (this volume), the melt sheet extends roughly over one-half of the total area of a mid-size crater. We thus find that in asteroidal impacts the melt sheet thickness may extend between ~40 and 160 m for 50 and 30 GPa melt thresholds, respectively. The higher melt volumes we find for cometary impacts mean that they could produce melt sheets 125–250 m thick, up to four times thicker than asteroidal impacts. The typical cometary impact velocities on Mars (we

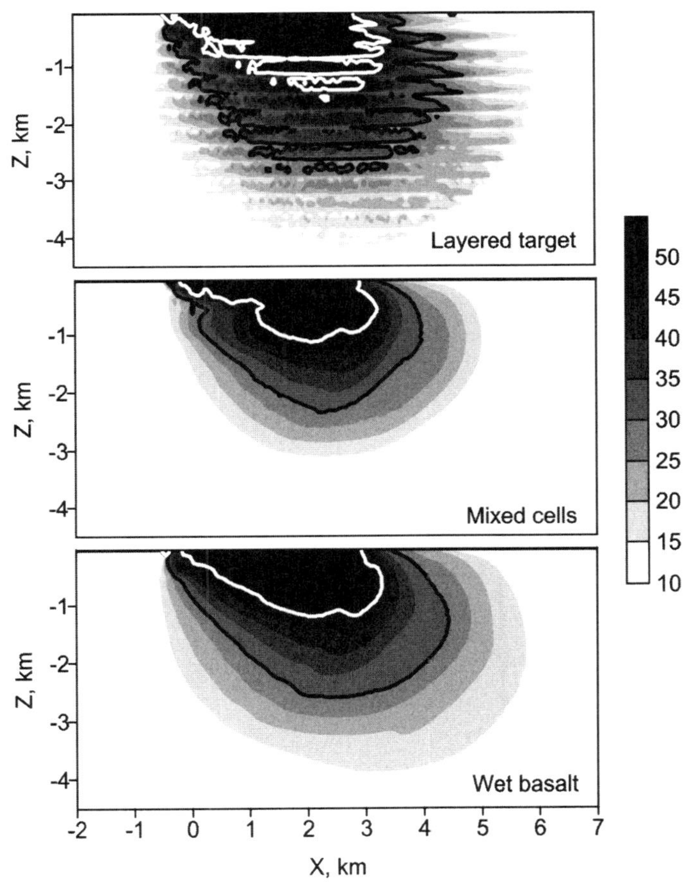

Figure 2. Shock pressure (grayscale, in GPa) distribution in the (pre-impact) target material for an asteroid impact (8 km/s, 45°) on a "Layered target" (case B), "Mixed cell target" (case C), and uniform "Wet basalt" target (case D of Fig. 1). Thick black and white lines represent the 30 and 50 GPa shock pressures, respectively. The same resolution (20 cells per projectile radius) is used in the three cases.

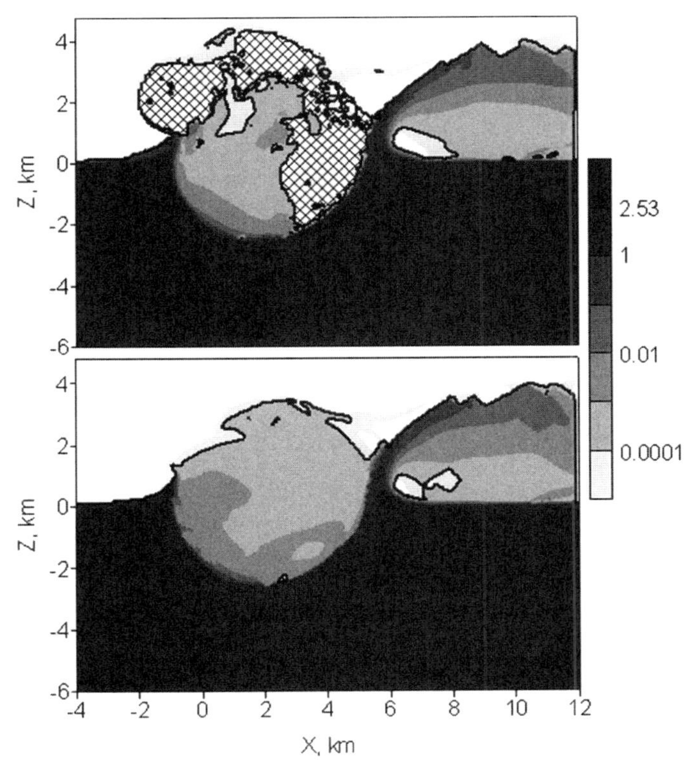

Figure 3. Asteroid impact (8 km/s, 45°) on a mixed cell target (upper) and uniform "wet basalt" (lower) target 1.2 seconds after impact. Grayscale represents basalt density. The plume region dominated by basalt vapor is shown in light gray, while the cross-hatched region represents the component of the plume dominated by water vapor.

used 15.5 km/s in the simulations) are similar to typical asteroidal impact velocities on Earth. Our melt thickness estimates for cometary impacts on Mars are in agreement with geological estimates of melt deposits in terrestrial craters of similar size: <300 m for the Boltysh crater (rim diameter around 25 km; Hölker and Deutsch, 1996), and 60–80 m melt cliffs identified at the Mistastin crater (rim diameter around 28 km; Robertson and Grieve, 1975).

Thermal Field under the Crater

For the late stage modeling, we use the pure water-ice phase diagram (Fig. 5) to identify the possible stable phase of water associated with the thermal state underneath the crater, providing that the pore water has the same temperature as the surrounding material. As described above, this estimate assumes inflow of water from a nearby water-rich region. Figures 6 and 7 show the resulting thermal fields underneath the transient and final crater created by asteroid (Fig. 6) and comet (Fig. 7) impacts in a dry target. Both modeled craters suggest the same general picture: the combination of shock and plastic heating and the structural uplift of initially deeper strata create a region under the crater where the stable phase for water is liquid (Fig. 5). In the simulations, heating comes mainly from the shock wave propagation combined with the structural uplift of rocks underneath the crater; deviatoric stresses (mostly friction from disrupted rocks) contribute a maximum of ~100 K to the heating, mainly concentrated near the surface, under the crater center. The main difference between the two cases, asteroid versus cometary impact, is that there is a larger volume of hot rocks in the central peak of the comet-produced crater, which is a natural result of higher initial shock pressures near the impact point due to the higher comet impact velocity (e.g., see Table 1). In the central uplift, the high temperatures would cause water to evaporate (steam-driven circulation). The liquid and vapor water stability zone identify the hydrothermal circulation cell. Figures 6 and 7 show that for a midsized crater (rim diameter ~30 km) in an arid region of Mars, any hydrothermal circulation is probably restricted to a "column" contained well within the final crater. These initial models provide crater shapes that are in reasonable agreement with MOLA data for fresh midsize craters; however, we are planning more systematic tests to optimize model parameters for acoustic fluidization on Mars.

There is no clear indication of the melt pool in Figures 6 and 7. This apparent contrast with the early stage simulations is due mainly to resolution problems. Because realistic durations of impact simulations require a balance between the need of high resolution and the necessity of long integration times, crater collapse studies are usually carried out at much lower resolution than early stage impact simulations. Resolution studies (e.g., Pierazzo et al., 1997) have shown that this may result in lower estimates of melt production. In this study, we used a cell size of 100 m × 100 m for the late stage modeling, a resolution too coarse to clearly identify a melt pool of few tens to a hundred meters deep.

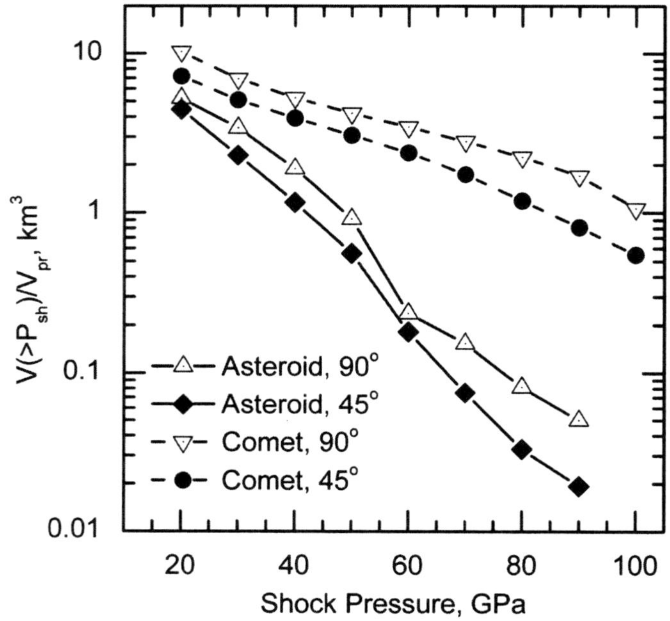

Figure 4. Volume of target subject to shock pressures above a given value normalized to projectile volume for asteroidal and cometary impacts, and for different impact angles (45°, 90°) on a mixed cell target (case C).

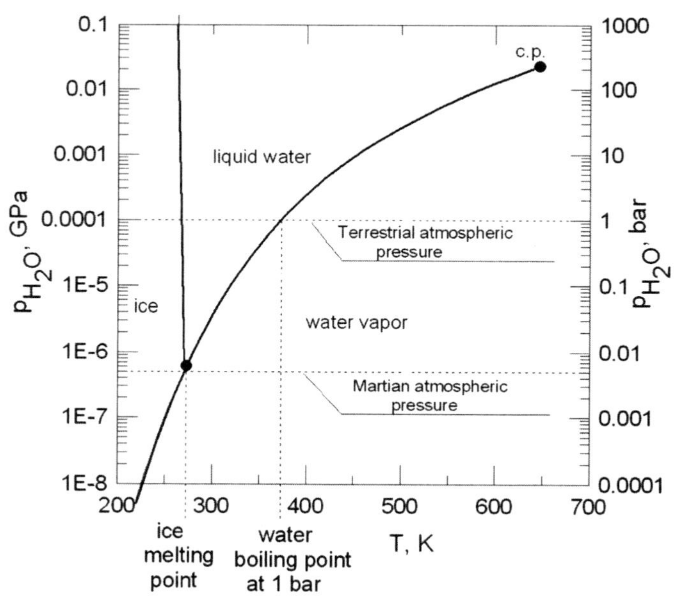

Figure 5. Phase diagram for pure water. At low temperatures, pure water ice is stable. At high temperatures and pressures, ice melts to liquid water. Below the triple point ice is in equilibrium with liquid and vapor. The boiling curve separates liquid water and water vapor up to the critical point (c.p.). At temperatures above the critical temperature there is separation among solid, liquid, or vapor phase; compression/decompression of water changes its state from fluid to vapor without a separation into different phases.

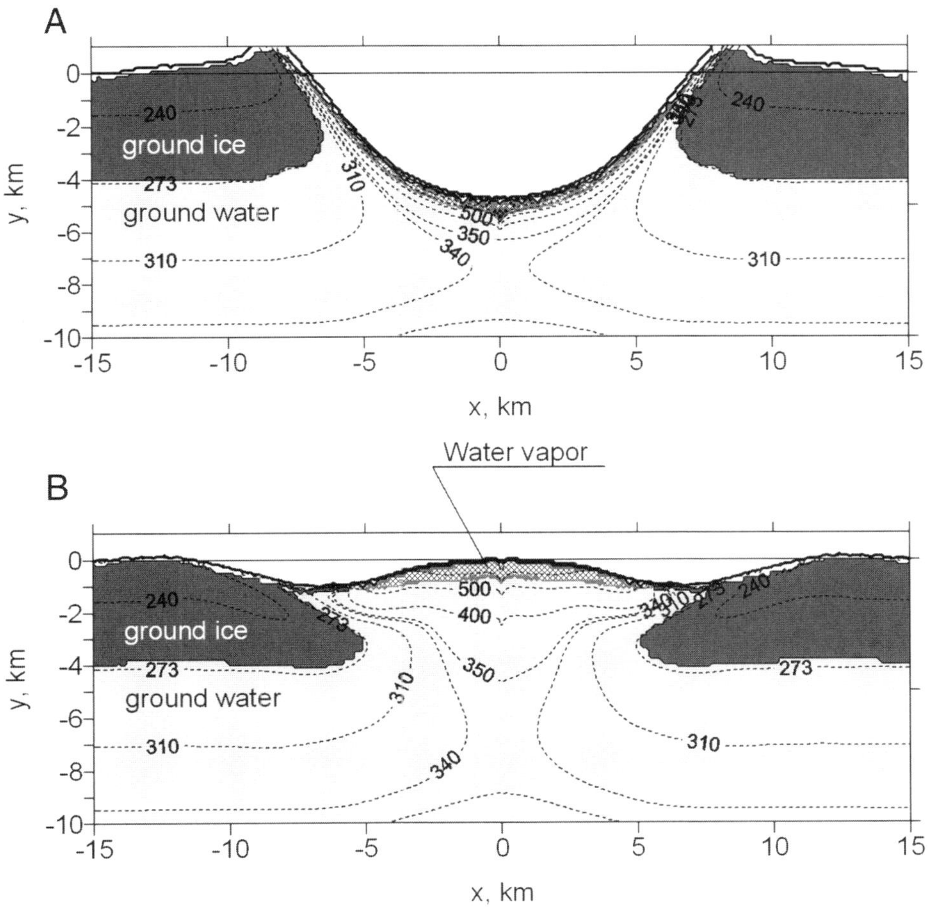

Figure 6. Isotherms and phase state of pure water (relative to the local temperatures and pressures) below a Martian crater with final rim diameter of ~30 km formed by an asteroid impact (A) at the transient cavity stage (i.e., right before crater collapse), and (B) after crater collapse (forming the final complex crater). Ice is shown in dark gray and liquid water in light gray; the cross-hatched area represents the region of stability for water vapor. At a depth of ~4 km, the melting line (for the assumed thermal gradient) is slightly above the 273 K isotherm, due to the decrease of the melting temperature at elevated lithostatic pressures. The lithostatic pressure also increases the boiling point; thus liquid water may exist at a temperature up to the critical point (~650 K) below the central peak.

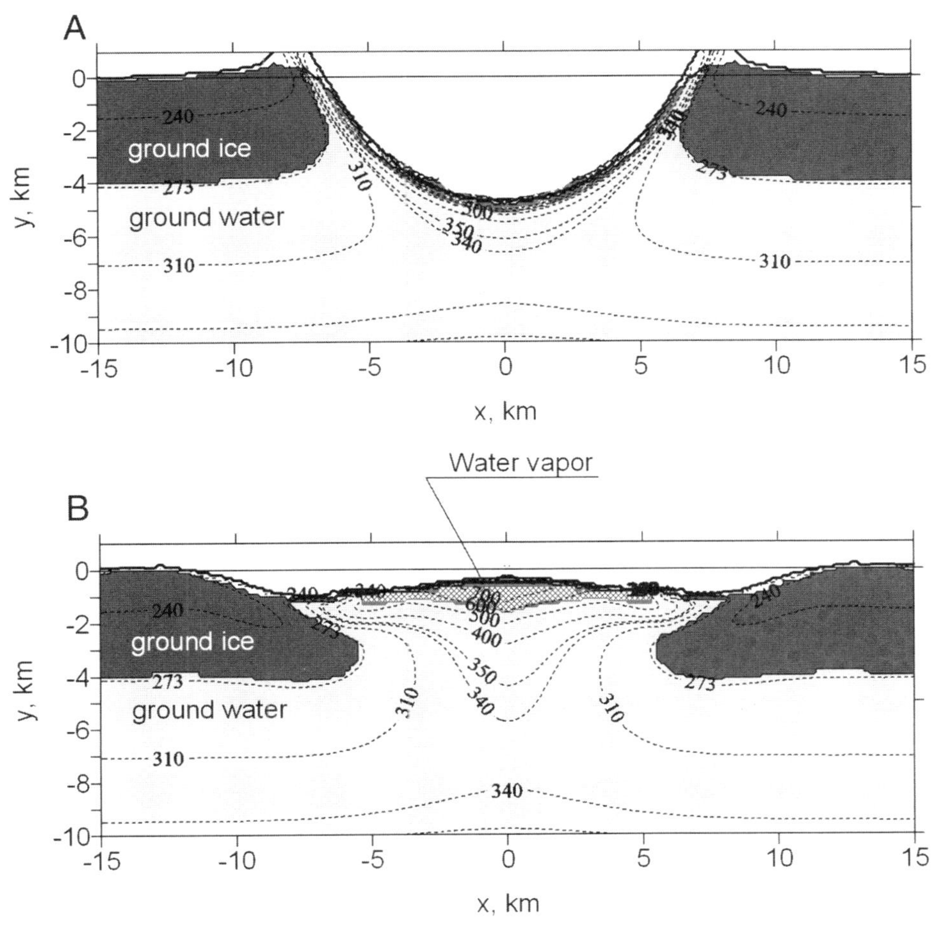

Figure 7. Isotherms and phase state of pure water (relative to the local temperatures and pressures) below a Martian crater with rim diameter of ~30 km formed by a comet impact (A) at the transient cavity stage (i.e., right before crater collapse), and (B) after crater collapse (forming the final complex crater). Same grayscale scheme as in Figure 6.

CONCLUSIONS

Impact-related hydrothermal systems generated under large Martian craters are considered sites favorable for the origin and/or development of early life (Newsom et al., 2001; Cabrol et al., 2001). While sophisticated computer codes are available to model the evolution of a hydrothermal system (e.g., HYDROTHERM; Hayba and Ingebritsen, 1994), the characterization of the initial conditions for modeling the onset of a hydrothermal system is still based on general scaling relationships (Daubar and Kring, 2001; Thorsos et al., 2001; 2002). We have begun a study aimed at using detailed hydrocode simulations of impacts to characterize the initial conditions for modeling the onset and evolution of hydrothermal systems on Mars.

High-resolution 3D SOVA simulations indicate that the amount of melt generated in impacts is sensitive to the impact velocity. In particular, cometary impacts produce 2–3 times more melting than asteroidal impacts for roughly the same final crater. The particular distribution of ice in the target (mixed cells versus layering) does not appear to significantly affect the overall shock decay. However, the description of the mixed material target, mixed cells of individual materials (basalt-ice) versus mixed material equation of state, may affect the overall results and deserves a more detailed investigation, both theoretical and experimental.

Two-dimensional SALEB simulations were carried out to evaluate the thermal state of a dry target underneath a midsize (roughly 30 km) Martian crater. The simulations show that for typical Martian conditions, the hydrothermal circulation is probably restricted to a "column" contained well within the final crater. Cometary and asteroidal impacts seem to produce very similar conditions, with cometary impacts producing larger volumes of "hot rocks" in the central uplift. These results are still preliminary. Needed improvements include the development of an adequate strength model for a mixture of rock and ice and the optimization of the parameter describing acoustic fluidization for craters on Mars.

APPENDIX A. A MATERIAL MODEL FOR BASALT

Basalt is by far the most abundant igneous rock on Earth's surface, and it is also abundant on the Moon, Mars, and Venus. It is thus important to develop a good material model (equation of state and strength model) for basalt.

Basalt Equation of State

The major difficulty in developing the equation of state for basalt lies in the fact that it is a complex mixture of minerals, each with its own thermodynamic properties. Furthermore, planetary basalts have a wide range of composition, as illustrated in Table A1. In developing the basalt ANEOS equation of state, we assume the MORB chemical composition of Table A1. In this case, we introduce eight main elements with the following elemental mass fraction: *O: 59.6%; Si: 16.8%; Al: 6.1%; Ca: 5.75%; Fe: 5.25%; Mg: 3.75%; Na: 1.73%; Ti: 0.53%.*

The input parameters for the basalt ANEOS equation of state (besides a list of elements used and their mass fraction) are provided in Table A2. The current version of ANEOS (Thompson and Lauson, 1972) cannot deal simultaneously with liquid-solid and solid-solid transitions. Therefore, it is necessary to choose which transition to model in the development of an ANEOS equation of state. Shock experiments performed on basalt (Ahrens and Gregson, 1964; Jones et al., 1968; Van Thiel, 1977; Nakazawa et al., 1997) indicate the presence of a solid-solid phase transition. Nakazawa et al. (1997) identify a "low-pressure phase" and a "high-pressure phase" for basalt, with a "mixed-phase" corresponding to the region of 5–23 GPa (0.4 and 1.4 km/s for material velocity). This prompted us to model a solid-solid phase transition for basalt, while giving up a detailed model of the melt curve. The phase transition causes an abrupt change in the Hugoniot's slope producing a modeled curve in good agreement with the available experimental data, as shown in Figure A1. The solid-solid phase transition is described in ANEOS (Table A2) by the use of five parameters, named D_1 through D_5. D_1 corresponds to the density below which the low-pressure phase is stable, while D_2 corresponds to the density above which the high-pressure phase is stable. In Table A2, $D_2 > D_1$, which is indicative of a first-order solid-solid transition (i.e., presence of a mixed phase region between low- and high-pressure phase regions). D_3 is the value of the cold pressure above which the first odd behavior is recorded in the Hugoniot. According to Nakazawa

TABLE A1. CHEMICAL COMPOSITION (BY WEIGHT) OF VARIOUS BASALTS

Oxide	Primitive Basalt (1)	MORB (2)	Komatiite (3)	Lunar Mare (4)	Mars (Shergotty) (5)	Vacaville Basalt (6)	NTS Basalt (7)
SiO_2	44.2	50.5	45.2	43.6	50.1	50.3	51.88
TiO_3	3.7	1.6	0.2	2.6	0.9	2.4	1.32
Al_2O_3	12.1	15.3	3.7	7.9	6.7	13.8	17.56
FeO	10.9	10.5	11.0	21.7	18.7	9.1	9.37
MnO	0.2	-	0.2	0.3	0.5	0.24	N.D.
MgO	13.1	7.5	32.2	14.9	9.4	4.4	5.54
CaO	10.1	11.5	5.3	8.3	10.0	7.7	6.74
Na_2O	3.6	2.6	0.4	0.2	1.3	3.3	4.26
K_2O	1.3	0.2	0.2	0.1	0.2	1.6	2.34
P_2O_5	0.8	0.1	0.02	-	0.7	1.2	N.D.
Cr_2O_3	0.1	N.D.	N.D.	0.9	0.2	N.D.	N.D.

Note: N.D.—no data. Sources: (1) Primitive basanite, Ross Island, Antarctica; BVSP81, p. 413; (2) Average basaltic glass from Atlantic, Pacific, and Indian Ocean spreading centers, BVSP81, p. 139; (3) Olivine spinifex-textured sample, Barberton, BVSP81, p. 16; (4) *Apollo 12* olivine basalt, no.6 on p. 239 of BVSP81; (5) Shergotty, BVSP81, p. 221. (6) From Van Thiel (1977), relative to Vacaville basalt. (7) Van Thiel compilation relative to the Buckboard hole 3 and area 18, Nevada Test Site (NTS).

et al. (1997), $D_3 \geq 5$ GPa. D_4 and D_5 are related to the first and second derivatives of the cold pressure with respect to density. (When $D_2 = D_3 = D_4 = D_5 = 0$, no solid-solid transition occurs, even if $D_1 \neq 0$.)

The introduction of the molecular cluster treatment in ANEOS (parameters R_{bond}, E_{bond}, and *IntDOF* in Table A2) allows for biatomic molecules in the gas phase (see Melosh, 2000), a major improvement over the old atomic gas approach in ANEOS. Combined with the introduction in ANEOS of the Lennard-Jones cold potential for the description of the pressure at the absolute zero temperature (last two parameters in ANEOS, Table A2, *flag* and P_c *exp*), it allows for a more realistic description (lower critical point) of the liquid-vapor phase change, and of the vapor phase in general (Fig. A2).

For any rock, a single melting and vaporization temperature is a considerable simplification, since the phase diagram for a mixture will be, in general, quite complicated. Nevertheless, the average melting and vaporization temperatures of the major constituents of the mixture determine the specific internal energy and entropy of the mixture in the phase change region. For basalt, we use $T_m \sim 1400$ K (Table A2) as typical incipient melting temperature from estimates for its main components, namely plagioclases and pyroxenes (Ahrens and O'Keefe, 1972). Other useful parameters reported by Ahrens and O'Keefe (1972) are: Grüneisen parameter, $\Gamma_0 = 1.39$, and Debye temperature, $\theta_0 = 1055$ K, while Nakazawa et al. (1997) reported a value of $K_0 = 25$ GPa for the average bulk modulus of Kinosaki (Hyogo, Japan) basalt, based on longitudinal (4.1–4.64 km/s) and shear (2.6–2.92 km/s) wave velocities.

Incipient and complete melting and vaporization entropies are normally used to estimate the shock pressure for melting and vaporization when experimental data are not available. Given the complexity of basalt it is unreasonable to use the hard spheres formalism to deter-

TABLE A2. ANEOS EQUATION OF STATE PARAMETERS FOR BASALT

Parameter	Value	Description
Nel	8	Number of elements in the compound
EOS	4	Type of equation of state (solid/gas with electronic terms and detailed treatment of liquid/vapor region)
ρ_0	2.86	Reference density (g/cm^3) (e.g., full density of Vacaville basalt)
T_0	0	Reference temperature (K) (0 = reference T: 298K)
P_0	0	Reference pressure (GPa) (0 = reference P)
C	5.0	Coefficient of shock versus material velocity linear relation
Γ_0	1.39	Reference Grüneisen coefficient
θ_0	1056	Reference Debye temperature (K)
S	2.6	Slope of shock versus material velocity linear relation
$3\gamma_0$	2	$3 \times$ limit of Γ for large compressions
E_{sep}	1.1×10^7	Zero-temperature separation energy (J/g)
T_m	1393	Melting temperature (K)
C_{53}	0	Parameter for low-density modification of critical point (0=not used)
C_{54}	0	Same as above
H_0	0	Parameter for thermal conduction (0=not included)
C_{41}	0	Same as above
ρ_{min}	0	Lowest allowed solid density (g/cm^3) (0 = automatic fix to $0.8\rho_0$)
D_1	3.1	ρ_1: start of solid-solid phase transition (g/cm^3)
D_2	4.0	ρ_2: end of solid-solid phase transition (g/cm^3)
D_3	5.0	Pressure of phase transition at zero temperature (GPa)
D_4	1.4×10^{12}	$dp/d\mu$ at the high pressure phase transition
D_5	0.8×10^{12}	$d^2p/d\mu^2$ at the high pressure phase transition
H_{fus}	0	Heat of fusion for melting (0 = not included)
ρ_{liq}/ρ_s	0	Liquid/solid density at melting point
U_p	0	Upper limit for cold compression (expanded states) (0 = default: 1)
L_0	0	Lower limit for cold compression (expanded states) (0 = default: 0)
α	0	Parameter for liquid EOS correction to match p_{vap} data and boiling point (does not apply if melt transition is not modeled)
β	0	Same as above
γ	0	Same as above
C_{60}	0	Interpolation parameter in Γ model
C_{61}	0	Same as above
C_{62}	0.1	Interpolation parameter in free energy expression (to move critical point)
Flag	1	Ionization model flag (1 = Thomas Fermi)
E_{sh}	0	Energy shift for reactive chemistry model (0 = not included)
S_{sh}	0	Entropy shift for reactive chemistry model (0 = not included)
R_{bond}	0	Bond length of molecular cluster (cm)
E_{bind}	7.0	Binding energy of molecules (eV)
IntDOF	0	Number of internal degrees of freedom in molecule (0, 1, or 2)
Flag	1	Lennard-Jones cold potential for p_{cold} (0 = standard ANEOS cold pressure equation of state)
$p_c$exp	1.25	Exponent of the density ratio in the Lennard-Jones potential

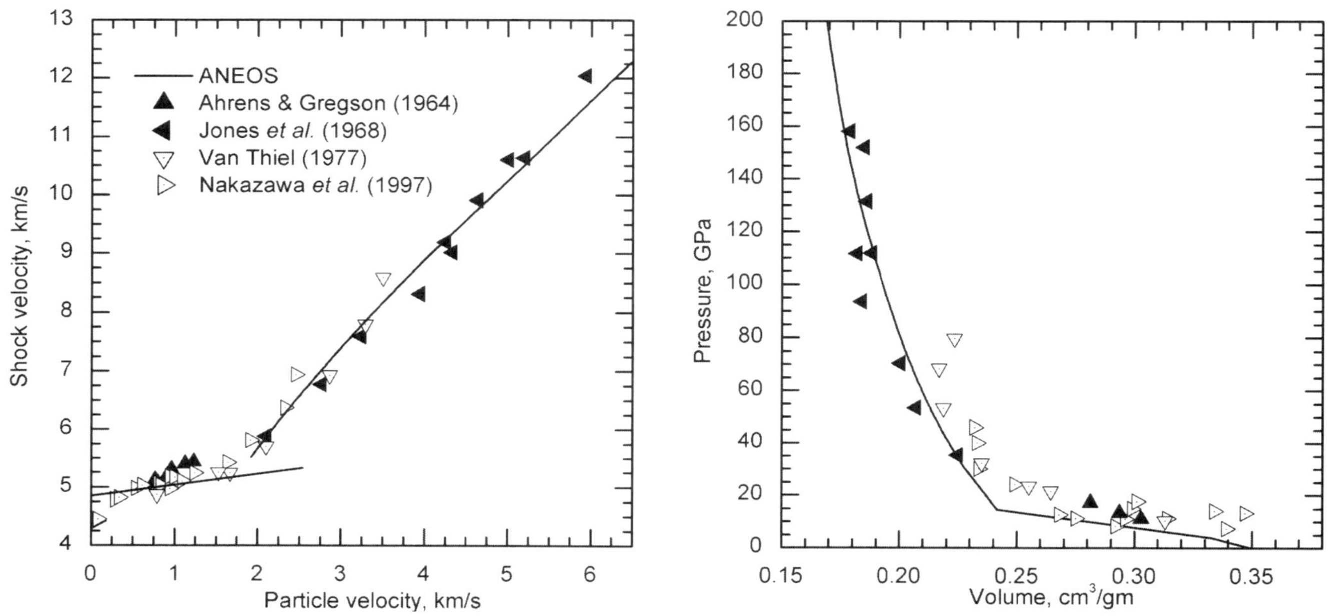

Figure A1. Shock data and ANEOS Hugoniot for basalt in shock versus material velocity and pressure versus volume space.

mine theoretically melting and vaporization entropies. As a first-order estimate, we assume that the incipient melting or vaporization entropy of basalt corresponds to the lowest incipient melting or vaporization of its main minerals (plagioclase, feldspar, augite), and the complete melting or vaporization entropy corresponds to the highest complete melting or vaporization of its main minerals (using estimates from Ahrens and O'Keefe, 1972). The shock pressures associated to the estimated entropies are shown in Table A3 for fully dense basalt (ρ = 2.86 g/cm^3) as well as a typical low porosity basalt (ρ = 2.65 g/cm^3). It must be emphasized that the entropy values represent rough upper limit estimates that do not take into account chemical interaction among the various mineral species in the basalt rock. The resulting shock pressure values thus can also be considered upper limit estimates.

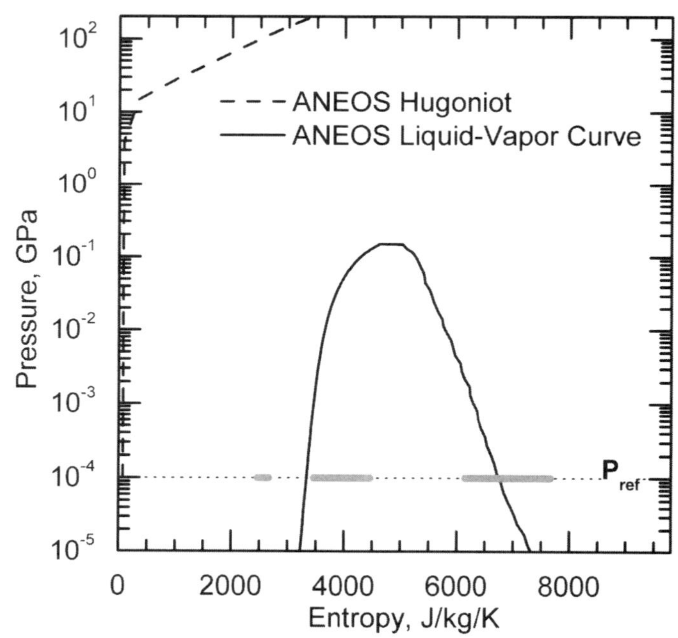

Figure A2. Vaporization diagram modeled by ANEOS. Incipient and complete melting and vaporization entropy ranges for basalt-minerals (e.g., anorthite, albite, augite; data from Ahrens and O'Keefe, 1972) are shown as thick horizontal gray lines at reference pressure (1 atm).

Basalt Strength Model

The generic term basalt encompasses a variety of rocks with strength properties mainly depending on the rock's initial porosity. As a first step to characterize the strength properties of the Martian crust we have collected a set of published data on triaxial compressive failure tests (Griggs et al., 1960; Clark, 1966; Volarovich, 1978; Angenheister, 1982; Morrow and Lockner, 2001). These data are used to determine the deviatoric stress invariants used in the hydrocode, i.e., pressure, $p = (\sigma_1 + \sigma_2 + \sigma_3)/3$, and the second stress invariant of the stress tensor (also called shear stress intensity), $T_s = \{[(\sigma_1 - \sigma_2)^2 + (\sigma_1 - \sigma_3)^2 + (\sigma_3 - \sigma_2)^2]/6\}^{1/2}$, where σ_1 refers to the main compressive stress and $\sigma_2 = \sigma_3$ refer to the confining stresses.

The strength model used in SALEB is described in detail in Collins et al. (2004). In brief, we use Lundborg (1968) analytical fit to available data for the yield surface of an intact material, Y_i:

$$Y_i = Y_0 + \frac{\mu_i p}{1 + \mu_i p/(Y_M - Y_0)}$$

where Y_0 is the shear strength at zero pressure, μ_i is the coefficient of internal friction and Y_M is known as the von Mises plastic limit of the material. For totally fragmented rock material, we use a Coulomb dry-friction law, where the yield strength is $Y_d = \min[\mu_d\, p;\ Y_d\,(p)]$ and μ_d is the coefficient of friction for damaged material. In SALEB deviatoric stresses computations are reduced to fit the condition $T_s \le Y_i$ for undamaged material, and $T_s \le Y_d$ for damaged material. Figure A3

TABLE A3. BASALT INCIPIENT AND COMPLETE MELTING AND VAPORIZATION PARAMETERS

Parameter	Entropy* (J/kg/K)	P_{sh}** [for dense material] (GPa)	P_{sh}** [for ρ_0 = 2.65 g/cm³] (GPa)
Incipient melting	2467	96.0	79.4
Complete Melting	2682	115.5	95.2
Incipient vaporization	3462	205.5	172.0
Complete vaporization	6144	>1138	N.D.

Note: N.D.—no data.
* Data from Ahrens and O'Keefe (1972).
** Estimated from ANEOS.

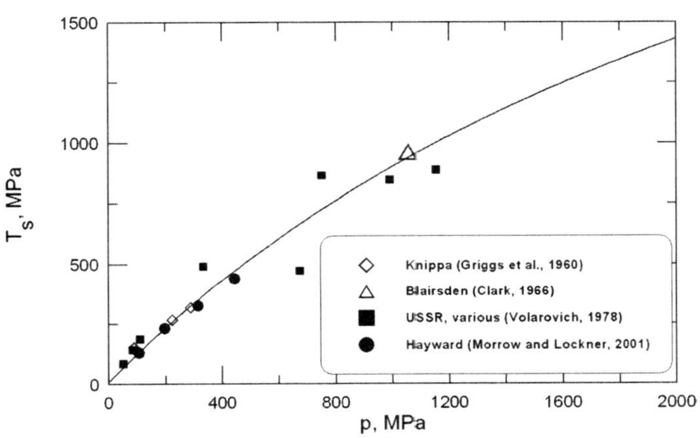

Figure A3. Experimental data (triaxial tests results) relative to the strength for various basalt samples expressed in the form of the invariant variables used in SALEB. Solid line represents the Lundborg analytical fit. (See text.)

illustrates the fit of Lundborg's approximation to triaxial compressive strength data for initially intact basalt with parameters: Y_d = 10 MPa; μ_i = 1.2; Y_M = 3500 MPa; and μ_d = 0.6 (the latter parameter is based on experiments by Byerlee, 1978).

In large scale impact events the target (the upper kilometers of a planetary crust) is obviously not as strong as described by Lundborg's intact material equation (with the given coefficients). This can be approximated by assuming an initial partial damage for the rock in computational cells. The coefficient of friction for damaged material, μ_d, is thus much more important than the exact behavior of the yield surface for the intact material. The similarity of this coefficients for most igneous rocks, as discussed by Byerlee (1978), makes specific measurements of friction for basalt not too crucial: in large solid bodies such as Earth, Moon, and Mars, most igneous rocks a few kilometers below the surface should have similar behaviors when totally damaged.

APPENDIX B. A MIXED MATERIAL EQUATION OF STATE

Let α be the content of ice (by mass) in a mixture of density, ρ, and temperature, T. In thermal equilibrium the temperature is assumed to be the same for each components of the mixture. The density of the components, ρ_b and ρ_i, (as well as all other thermodynamic values such as entropy, sound speed, etc.), may be found from the individual (basalt and ice) equations of state under the condition that the mixture components have the same pressure (i.e., mechanical equilibrium): $P_b(\rho_b, T) = P_i(\rho_i, T)$ with $\rho = \rho_b(1 - \alpha) + \rho_i\alpha$, or $\rho_b = (\rho - \rho_i\alpha)/(1 - \alpha)$. These conditions set up a system of nonlinear equations that can be solved numerically. For a perfect gas at constant temperature, the relationship between pressure and density is linear and the equations may be solved analytically (standard problem of gas mixtures). Figure B1 illustrates the construction of a mixed material equation of state for different mixture densities and temperatures. Figure B1A corresponds to a mixture of water and basalt vapors (pressure is linearly proportional to density). Figure B1B corresponds to a mixture density of 2.526 g/cm³ and temperature of 1000 K. The mixture's pressure is ~2 GPa, water density is 1.34 g/cm³ and basalt density –2.82 g/cm³. Dry basalt with the same density and temperature is in tension with a negative pressure value of –1 GPa, while pure water would be in a strongly compressed state with pressure of 74 GPa. Figure B1C

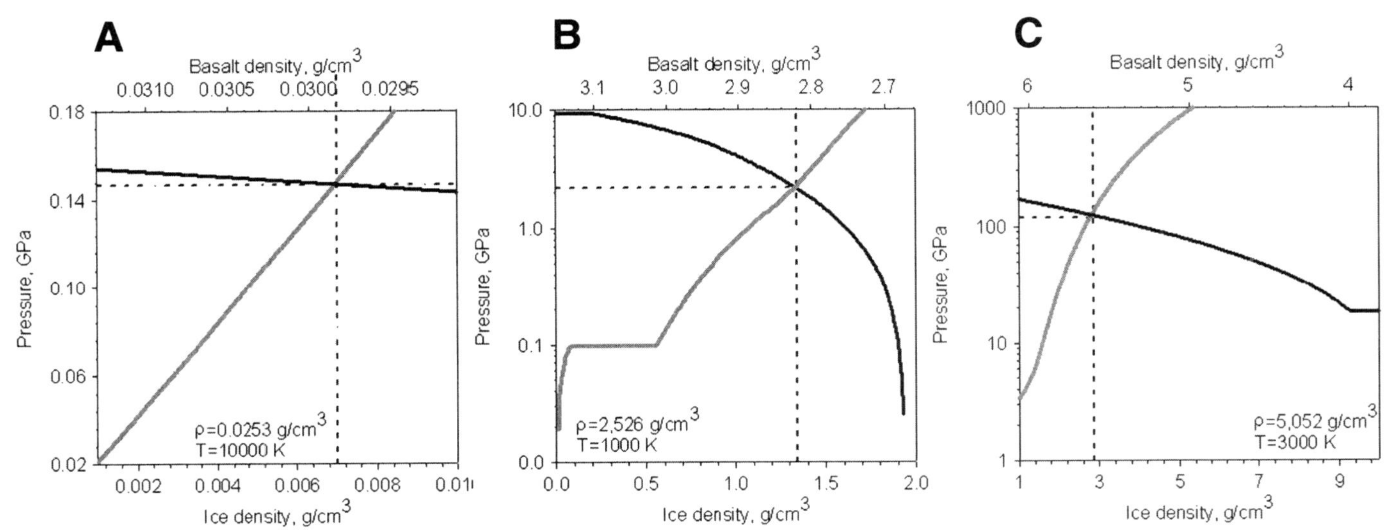

Figure B1. Pressure versus density for ice (bottom x-axis) and basalt (top x-axis). The intersection of the two curves gives the pressure value for the wet basalt equation of state and provides the density values for the components.

shows a strongly compressed mixture (5.052 g/cm^3) where density is 5.6 g/cm^3 for basalt, and 2.8 g/cm^3 for water. The mixture's pressure is 120 GPa—just between that of pure dry basalt (70 GPa) and pure water (830 GPa) at the same density. We were not able to construct an equation of state for basalt-water mixture at low (gaseous) densities and at temperatures below the vaporization temperature for basalt (~2000 K). In this case, the ANEOS basalt produces zero pressure, while water vapor has nonzero pressure. Physically, this means that it is impossible to uniformly mix the materials in these states: water vapor separates from basalt, and the mixture becomes solid or molten basalt particles suspended in water vapor. Then, the hydrocode itself should be modified to model the motion of condensed basalt particles in an ambient water vapor atmosphere.

ACKNOWLEDGMENTS

This work was supported by the NASA's Mars Fundamental Research Program under grant NAG5-12688 to the Planetary Science Institute. We are very grateful to two anonymous reviewers for the constructive criticism provided on an earlier draft of this paper, and to Fred Hörz for his constructive reviews and continuous assistance during manuscript revision. This manuscript is PSI contribution No. 373.

REFERENCES CITED

Ahrens, T.J., and Gregson, V.G., 1964, Shock compression of crystal rocks: Data for quartz, calcite, and plagioclase rocks: Journal of Geophysical Research, v. 69, p. 4839–4874.

Ahrens, T.J., and O'Keefe, J.D., 1972, Shock melting and vaporization of lunar rocks and minerals: The Moon, v. 4, p. 214–249.

Allen, C.C., Gooding, J.L., and Keil, K., 1982, Hydrothermally altered impact melt rock and breccia: Contributions to the soil of Mars: Journal of Geophysical Research, v. 87, p. 10083–10101.

Amsden, A.A., Ruppel, H.M., and Hirt, C.V., 1980, SALE: A simplified ALE computer program for fluid flow at all speeds: Report, LA-8095, Los Alamos National Laboratories, Los Alamos, New Mexico, 101 p.

Angenheister, G., ed., 1982, Landolt-Börnstein: Numerical Data and Functional Relationships in Science and Technology, Group V, Geophysical and Space Research, v. 1: Physical Properties of Rocks, sub-volume b: Berlin-Heidelberg-New York, Springer-Verlag Press, 604 p.

Artemieva, N.A., and Ivanov, B.A., 2001, Numerical simulations of oblique impact melt and transient cavity size: Lunar and Planetary Science Conference 32, Houston, Texas, Abstract no. 1431 (CD-ROM).

Babeyko, A.Yu., and Zharkov, V.N., 2000, Martian crust: a modeling approach: Physics of the Earth and Planetary Interiors, v. 117, p. 421–435, doi: 10.1016/S0031-9201(99)00111-9.

Baker, V.R., Carr, M.H., Gulick, V.C., Williams, C.R., and Marley, M.S., 1992, Channels and valley networks, *in* Kieffer, H.H., Jakosky, B.M., Snyder, C.W., and Matthews, M.S., eds., Mars: Tucson, Arizona, University of Arizona Press, p. 493–522.

Barlow, N.G., and Perez, C.B., 2003, Martian impact crater ejecta morphologies as indicators of the distribution of subsurface volatiles: Journal of Geophysical Research (Planets), v. 108, 5085, p. 4-1, doi 10.1029/2002JE002036.

Berkley, J.L., and Drake, M.J., 1981, Weathering on Mars: Antarctic analog studies: Icarus, v. 45, p. 231–249 doi: 10.1016/0019-1035(81)90016-6.

Best, M.G., and Christiansen, E.H., 2000, Igneous Petrology: Blackwell Science Inc., 455 p.

Boynton, W.V., Feldman, W.C., Squyres, S.W., Prettyman, T.H., Brückner, J., Evans, L.G., Reedy, R.C., Starr, R., Arnold, J.R., Drake, D.M., Englert, P.A.J., Metzger, A.E., Mitrofanov, I., Trombka, J.I., de Uston, C., Waenke, H., Gasnault, O., Hamara, D.K., Janes, D.M., Marcialis, R.L., Maurice, S., Mikheeva, I., Taylor, G.J., Tokar, R., and Shinohara, C., 2002, Distribution of hydrogen in the near surface of Mars: Evidence for subsurface ice deposits: Science, v. 297, p. 81–85.

Byerlee, J., 1978, Friction of rocks: Pure and Applied Geophysics, v. 116, p. 615–626.

Cabrol, N.A., Wynn-Williams, D.D., Crawford, D.A., and Grin, E.A., 2001, Recent aqueous environments in Martian impact craters: An astrobiological perspective: Icarus, v. 154, p. 98–112, doi: 10.1006/ICAR.2001.6661.

Carr, M.H., 1996, Water on Mars: Oxford University Press, New York, New York, 229 p.

Clark, S.P., editor, 1966, Handbook of Physical Constants: Boulder, Colorado, Geological Society of America Memoir 97, 587 p.

Clifford, S.M., 1993, A model for the hydrologic and climatic behavior of water on Mars: Journal of Geophysical Research, v. 98, p. 10973–11016.

Clifford, S., and Fanale, P., 1985, The thermal conductivity of the Martian crust (Abstract): Lunar and Planetary Science Conference 16, Houston, Texas, p. 144–145.

Collins, G.S., Melosh, H.J., and Ivanov, B.A., 2004, Damage and deformation in numerical impact simulations: Meteoritics and Planetary Sciences, v. 39, p. 217–231.

Daubar, I.J., and Kring, D.A., 2001, Impact-induced hydrothermal systems: Heat sources, and lifetimes: Lunar and Planetary Science Conference 32, Houston, Texas, Abstract no. 1727 (CD-ROM).

Fanale, P.P., 1976, Martian volatiles: their degassing history and geochemical fate: Icarus, v. 28, p. 179–202, doi: 10.1016/0019-1035(76)90032-4.

Griggs, D.T., Turner, F.J., and Heard, H.C., 1960, Deformation of rocks at 500° to 800°, *in* Griggs, D., and Handin, J., eds., Rock Deformation: Boulder, Colorado, Geological Society of America Memoir 79, p. 39–104.

Hartmann, W.K., and Neukum, G., 2001, Cratering chronology and evolution of Mars, *in* Kallenbach, R., Griss, J., and Hartmann, W.K., eds., Chronology and Evolution of Mars: Dordrecht-Boston-London, Kluwer Academic Press, Space Science Series of ISSI, v. 12, p. 165–194.

Hayba, D.O., and Ingebritsen, S.E., 1994, The computer model HYDROTHERM, a three-dimensional finite-difference model to simulate groundwater flow and heat transport in the temperature range of 0 to 1,200 degrees Celsius: U.S. Geological Survey Water-Resources Investigative Report 94-4045, p. 85.

Hölker, T., and Deutsch, A., 1996, The Boltysh impact structure, Ukraine: Geochemistry of the melt sheet (Abstract): Lunar and Planetary Science Conference 27, Houston, Texas, p. 555–556.

Ivanov, B.A., 1994, Geomechanical models of impact cratering: Puchezh-Katunki structure, *in* Dressler, B.O., Grieve, R.A.F., and Sharpton, V.L., eds., Large Meteorite Impacts and Planetary Evolution: Boulder, Colorado, Geological Society of America Special Paper 293, p. 81–91.

Ivanov, B.A., 2002, Deep drilling results and numerical modeling: Puchezh-Katunki impact crater, Russia: Lunar and Planetary Science Conference 32, Houston, Texas, Abstract no. 1286 (CD-ROM).

Ivanov, B.A., 2003a, Large impact crater modeling: Chicxulub: Third International Conference on Large Meteorite Impacts, Nördlingen, Germany, Abstract no. 4067 (CD-ROM).

Ivanov, B.A., 2003b, Complex crater formation: Verification of numerical models, *in* Herrick, R., and Pierazzo, E., eds., Results of the Workshop on "Impact Cratering: Bridging the Gap between Modeling and Observations," February 2003: Lunar and Planetary Institute Contribution 1162, Abstract no. 1155.

Ivanov, B.A., and Artemieva, N.A., 2002, Numerical modeling of the formation of large impact craters, *in* Koeberl, C., and MacLeod, K.G., eds., Catastrophic Events and Mass Extinctions: Impacts and Beyond: Boulder, Colorado, Geological Society of America Special Paper 356, p. 619–630.

Ivanov, B.A., and Deutsch, A., 1999, Sudbury impact event; cratering mechanics and thermal history, *in* Dressler, B., and Grieve, R.A.F., eds., Large Meteorite Impacts and Planetary Evolution II: Boulder, Colorado, Geological Society of America Special Paper 339, p. 389–397.

Ivanov, B.A., and Kostuchenko, V.N., 1997, Block oscillation model for impact crater collapse: Lunar and Planetary Science Conference 27, Houston, Texas, Abstract no. 1655 (CD-ROM).

Ivanov, B.A., Langenhorst, F., Deutsch, A., and Hornemann, U., 2002, How strong was impact-induced CO_2 degassing in the Cretaceous-Tertiary event? Numerical modeling of laboratory experiments, *in* Koeberl, C., and MacLeod, K.G., eds., Catastrophic Events and Mass Extinctions: Impacts and Beyond: Boulder, Colorado, Geological Society of America Special Paper 356, p. 587–594.

Jones, A.H., Isbell, W.M., Shipman, F.H., Perkins, R.D., Green, S.J., and Maiden, C.J., 1968, Interim Report 48090, Contract NAS Feb. 27: General Motors Tech. Center (Warren, Michigan).

Lambert, P., and Lange, M.A., 1984, Glasses produced by shock melting and devolatilization of hydrous silicates: Journal of Non-Crystalline Solids, v. 67, p. 521–542.

Lundborg, N., 1968, Strength of rock-like materials: International Journal of Rock Mechanics and Mining Sciences, v. 5, p. 427–454, doi: 10.1016/0148-9062(68)90046-6.

Mangold, N., Allemand, P., Duval, P., Geraud, Y., and Thomas, P., 2002, Experimental and theoretical deformation of ice-rock mixtures: Implications on rheology and ice content of Martian permafrost: Planetary and Space Science, v. 50, p. 385–401, doi: 10.1016/S0032-0633(02)00005-3.

McGlaun, J.M., Thompson, S.L., and Elrick, M.G., 1990, CTH: A three-dimensional shock wave physics code: International Journal of Impact Engineering, v. 10. p. 351–360.

McKay, D.S., Gibson, E.K., Thomas-Keprta, K.L., Vali, H., Romanek, C.S., Clemett, S.J., Chiller, X.D.F., Maechling, C.R., and Zare, R.N., 1996, Search for past life on Mars: Possible relic biogenic activity in Martian meteorite ALH84001: Science, v. 273, p. 924–930.

Mellon, M.T., Jakosky, B.M., and Postawko, S.E., 1997, The persistence of equatorial ground ice on Mars: Journal of Geophysical Research, v. 102, no. E8, p. 19357–19369, doi: 10.1029/97JE01346.

Melosh, H.J., 1979, Acoustic fluidization: A new geologic process?: Journal of Geophysical Research, v. 84, p. 7513–7520.

Melosh, H.J., 1989, Impact Cratering: A Geologic Process: New York, New York, Oxford University Press, 245 p.

Melosh, H.J., 2000, A new and improved equation of state for impact computations: Lunar Planetary Science Conference 31, Houston, Texas, Abstract no. 1903 (CD-ROM).

Melosh, H.J., and Ivanov, B.A., 1999, Impact crater collapse: Annual Review of Earth and Planetary Sciences, v. 27, p. 385–415, doi: 10.1146/ANNUREV.EARTH.27.1.385.

Mitrofanov, I.G., Zuber, M.T., Litvak, M.L., Boynton, W.V., Smith, D.E., Drake, D., Hamara, D., Kozyrev, A.S., Sanin, A.B., Shinohara, C., Saunders, R.S., and Tretyakov, V., 2003, CO_2 snow depth and subsurface water-ice abundance in the Northern Hemisphere of Mars: Science, v. 300, p. 2081–2084, doi: 10.1126/SCIENCE.1084350.

Morrow, C.A., and Lockner, D.A., 2001, Hayward Fault Rocks: Porosity, Density and Strength Measurements: Menlo Park, California, U.S. Geological Survey Open-File Report 01-421, 28 p.

Mouginis-Mark, P.J., 1987, Water or ice in the Martian regolith?—Clues from rampart craters seen at very high resolution: Icarus, v. 71, p. 268–286, doi: 10.1016/0019-1035(87)90152-7.

Nakazawa, S., Watanabe, S., Kato, M., Iijima, Y., Kobayashi, T., and Sekine, T., 1997, Hugoniot equation of state of basalt: Planetary and Space Science, v. 45, p. 1489–1492, doi: 10.1016/S0032-0633(97)00070-6.

Newsom, H.E., 1980, Hydrothermal alteration of impact melt sheets with implications for Mars: Icarus, v. 44, p. 207–216, doi: 10.1016/0019-1035(80)90066-4.

Newsom, H.E., Graup, G., Seward, T., and Keil, K., 1986, Fluidization and hydrothermal alteration of the suevite deposit at the Ries crater, West Germany, and implications for Mars, *in* Proceedings of the 17th Lunar and Planetary Science Conference: Journal of Geophysical Research, v. 91, p. E239–E251.

Newsom, H.E., Hagerty, J.J., and Thorsos, I.E., 2001, Location and sampling of aqueous and hydrothermal deposits in Martian impact craters: Astrobiology, v. 1, p. 71–87, doi: 10.1089/153110701750137459.

Ohnaka, M., 1995, A shear failure strength law of rock in the brittle-plastic transition regime: Geophysical Research Letters, v. 22, p. 25–28, doi: 10.1029/94GL02791.

Olsson-Steel, D., 1987, Collisions in the Solar System—IV: Cometary impacts upon the planets: Monthly Notices of the Royal Astronomical Society, v. 227, p. 501–524.

Parameswaran, V.R., and Jones, S.J., 1981, Triaxial testing of frozen sand: Journal of Glaciology, v. 27, p. 147–155.

Pierazzo, E., and Collins, G.S., 2003, A brief introduction to hydrocode modeling of impact cratering, *in* Henning, D., Burchell, M., and Claeys, P., eds., Impact Studies: Cratering in Marine Environments and on Ice: Heidelberg, Germany, Springer Press, p. 323–340.

Pierazzo, E., Vickery, A.M., and Melosh, H.J., 1997, A reevaluation of impact melt production: Icarus, v. 127, p. 408–423, doi: 10.1006/ICAR.1997.5713.

Pierazzo, E., Artemieva, N.A., and Spitale, J.N., 2001, The Ries impact event: A tale of two hydrocodes (abs.): Granada, Spain, ESF-IMPACT 5: Catastrophic Events and Mass Extinctions: Impacts and Beyond.

Rathbun, J.A., and Squyres, S.W., 2002, Hydrothermal systems associated with Martian Impact Craters: Icarus, v. 157, p. 362–372, doi: 10.1006/ICAR.2002.6838.

Robertson, P.B., and Grieve, R.A.F., 1975, Impact structures in Canada: their recognition and characteristics: The Journal of the Royal Astronomical Society of Canada, v. 69, p. 1–21.

Schmidt, R.M., and Housen, K.R., 1987, Some recent advances in the scaling of impact and explosion cratering: International Journal of Impact Engineering, v. 5, p. 543–560.

Shuvalov, V.V., 1999, Multi-dimensional hydrodynamic code SOVA for interfacial flows: Application to the thermal layer effect: Shock Waves, v. 9, p. 381–390, doi: 10.1007/S001930050168.

Simonds, C.H., and Kieffer, S.W., 1993, Impact and volcanism: A momentum scaling law for erosion: Journal of Geophysical Research, v. 83, p. 2773–2788.

Squyres, S.W., Clifford, S.M., Kuz'min, R.O., Zimbelman, J.R., and Costard, F.M., 1992, Ice in the Martian regolith, *in* Kieffer, H.H., Jakosky, B.M., Snyder, C.W., and Matthews, M.S., eds., Mars: Tucson, Arizona, University of Arizona Press, p. 523–554.

Steel, D., 1998, Distributions and moments of asteroid and comet impact speeds upon the Earth and Mars: Planetary and Space Science, v. 46, p. 473–478, doi: 10.1016/S0032-0633(97)00232-8.

Stesky, R.M., Brace, W.F., Riley, D.K., and Bobin, P.Y., 1974, Friction in faulted rock at high temperature and pressure: Tectonophysics, v. 23, p. 177–203 doi: 10.1016/0040-1951(74)90119-X.

Stöffler, D., 1984, Glasses formed by hypervelocity impact: Journal of Non-Crystalline Solids, v. 67, p. 465–502 doi: 10.1016/0022-3093(84)90171-6.

Stöffler, D., Keil, K., and Scott, E.R.D., 1991, Shock metamorphism of ordinary chondrites: Geochimica et Cosmochimica Acta, v. 55, p. 3845–3867 doi: 10.1016/0016-7037(91)90078-J.

Stöffler, D., Artemieva, N.A., and Pierazzo, E., 2002, Modeling the Ries-Steinheim impact event and the formation of the Moldavite strewn field: Meteoritics & Planetary Science, v. 37, p. 1893–1907.

Thompson, S.L., and Lauson, H.S., 1972, Improvements in the Chart-D radiation hydrodynamic code III: Revised analytical equation of state: Sandia National Laboratories Report SC-RR-710714, Albuquerque, New Mexico, 119 p.

Thorsos, I.E., Newsom, H.E., and Davies, A.G., 2001, Availability of heat to drive hydrothermal systems on large Martian impact craters: Lunar and Planetary Science Conference 32, Houston, Texas, Abstract # 2011 (CD-ROM).

Thorsos, I.E., Newsom, H.E., and Davies, A.G., 2002, Impact-induced hydrothermal systems and mineral deposits on Mars: Lunar and Planetary Science Conference 33, Houston, Texas, Abstract # 1912 (CD-ROM).

Turtle, E.P., and Pierazzo, E., 2001, Thickness of an Europan Ice Shell from Impact Crater Simulations: Science, v. 294, p. 1326–1328, doi: 10.1126/SCIENCE.1062492.

Van Thiel, M., 1977, Compendium of shock wave data: Livermore, California, Lawrence Livermore National Laboratories Report UCRL-50108.

Versh, E., Kirsimäe, K., Jõeleht, A., and Plado, J., 2003, Impact induced hydrothermal system at Kärdla impact crater: Development and biological consequences: Third International Conference on Large Meteorite Impacts, Nördlingen, Germany, Abstract no. 4120 (CD-ROM).

Volarovich, M.P., editor, 1978, Databook for Physical Properties of Minerals and Rocks at High Thermodynamic Parameters (Spravochnik po Fizicheskim Svoistvam Mineralov i Gornyh Porod pri Vysokih Termodynaicheskih Parametrah): Moscow, Russia, Nedra Press, 237 p. (in Russian).

Wünnemann, K., and Ivanov, B.A., 2003, Numerical modelling of the impact crater depth-diameter dependence in an acoustically fluidized target: Planetary and Space Science, v. 51, p. 831–845, doi: 10.1016/J.PSS.2003.08.001.

Manuscript Accepted by the Society 9 August 2004

Printed in the USA

Index

D

G

H

N

O

P

Q

W

Y

Z